P9-CDX-179

HOLT MIDDLE SCHOOL Math

Course 2

Jennie M. Bennett
David J. Chard
Audrey Jackson
Jim Milgram
Janet K. Scheer
Bert K. Waits

HOLT, RINEHART AND WINSTON
A Harcourt Education Company
Austin • Orlando • Chicago • New York • Toronto • London • San Diego

Staff Credits

Editorial

Lila Nissen, *Editorial Vice President*
Robin Blakely, *Associate Director*
Joseph Achacoso, *Assistant Managing Editor*
Threasa Boyar, *Editor*

Student Edition
Jennifer Gribble, *Editor*
Kerry Milam, *Associate Editor*
Chris Rankin, *Associate Editor*

Teacher's Edition
Kelli Flanagan, *Senior Editor*

Ancillaries
Mary Fraser, *Executive Editor*
Higinio Dominguez, *Associate Editor*

Technology Resources
John Kerwin, *Executive Editor*
Robyn Setzen, *Senior Editor*
Patricia Platt, *Senior Technology Editor*
Manda Reid, *Technology Editor*

Copyediting
Denise Nowotny, *Copyediting Supervisor*
Patrick Ricci, *Copyeditor*

Support
Jill Lawson, *Senior Administrative Assistant*
Benny Carmona, III, *Editorial Coordinator*

Design

Book Design
Marc Cooper, *Design Director*
Tim Hovde, *Senior Designer*
Lisa Woods, *Designer*
Teresa Carrera-Paprota, *Designer*
Bruce Albrecht, *Design Associate*
Ruth Limon, *Design Associate*
Holly Whittaker, *Senior Traffic Coordinator*

Teacher's Edition
José Garza, *Designer*
Charlie Taliaferro, *Design Associate*

Cover Design
Pronk & Associates

Image Acquisition
Curtis Riker, *Director*
Tim Taylor, *Photo Research Supervisor*
Terry Janecek, *Photo Researcher*
Elaine Tate, *Art Buyer Supervisor*
Joyce Gonzalez, *Art Buyer*
Sam Dudgeon, *Senior Staff Photographer*
Victoria Smith, *Staff Photographer*
Lauren Eischen, *Photo Specialist*

New Media Design
Ed Blake, *Design Director*

Media Design
Dick Metzger, *Design Director*
Chris Smith, *Senior Designer*

Graphic Services
Kristen Darby, *Director*
Eric Rupprath, *Ancillary Designer*
Linda Wilbourn, *Image Designer*

Prepress and Manufacturing

Mimi Stockdell, *Senior Production Manager*
Susan Mussey, *Production Supervisor*
Rose Degollado, *Senior Production Coordinator*
Sara Downs, *Production Coordinator*
Jevara Jackson, *Senior Manufacturing Coordinator*
Ivania Lee, *Inventory Analyst*
Wilonda Ieans, *Manufacturing Coordinator*

Printed in the United States of America

ISBN 0-03-071009-X

4 5 6 7 8 9 048 10 09 08 07 06 05 04

Authors

Jennie M. Bennett, Ed.D., is the Instructional Mathematics Supervisor for the Houston Independent School District and president of the Benjamin Banneker Association.

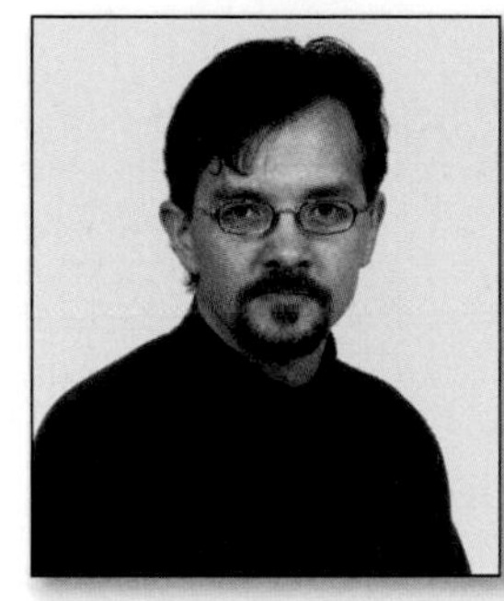

David J. Chard, Ph.D., is an Assistant Professor and Director of Graduate Studies in Special Education at the University of Oregon. He is the President of the Division for Research at the Council for Exceptional Children, is a member of the International Academy for Research on Learning Disabilities, and is the Principal Investigator on two major research projects for the U.S. Department of Education.

Audrey Jackson is a Principal in St. Louis, Missouri, and has been a curriculum leader and staff developer for many years.

Jim Milgram, Ph.D., is a Professor of Mathematics at Stanford University. He is a member of the Achieve Mathematics Advisory Panel and leads the Accountability Works Analysis of State Assessments funded by The Fordham and Smith-Richardson Foundations. Most recently, he has been named lead advisor to the Department of Education on the implementation of the Math-Science Initiative, a key component of the No Child Left Behind legislation.

Janet K. Scheer, Ph.D., Executive Director of Create A Vision™, is a motivational speaker and provides customized K-12 math staff development. She has taught internationally and domestically at all grade levels.

Bert K. Waits, Ph.D., is a Professor Emeritus of Mathematics at The Ohio State University and co-founder of T^3 (Teachers Teaching with Technology), a national professional development program.

Consulting Authors

Paul A. Kennedy is a Professor in the Mathematics Department at Colorado State University and has recently directed two National Science Foundation projects focusing on inquiry-based learning.

Mary Lynn Raith is the Mathematics Curriculum Specialist for Pittsburgh Public Schools and co-directs the National Science Foundation project PRIME, Pittsburgh Reform in Mathematics Education.

Reviewers

Thomas J. Altonjy
Assistant Principal
Robert R. Lazar Middle School
Montville, NJ

Jane Bash, M.A.
Math Education
Eisenhower Middle School
San Antonio, TX

Charlie Bialowas
District Math Coordinator
Anaheim Union High School District
Anaheim, CA

Lynn Bodet
Math Teacher
Eisenhower Middle School
San Antonio, TX

Louis D'Angelo, Jr.
Math Teacher
Archmere Academy
Claymont, DE

Troy Deckebach
Math Teacher
Tredyffrin-Easttown Middle School
Berwyn, PA

Mary Gorman
Math Teacher
Sarasota, FL

Brian Griffith
Supervisor of Mathematics, K–12
Mechanicsburg Area School District
Mechanicsburg, PA

Ruth Harbin-Miles
District Math Coordinator
Instructional Resource Center
Olathe, KS

Kim Hayden
Math Teacher
Milford Jr. High School
Milford, OH

Susan Howe
Math Teacher
Lime Kiln Middle School
Fulton, MD

Paula Jenniges
Austin, TX

Ronald J. Labrocca
District Mathematics Coordinator
Manhasset Public Schools
Manhasset, NY

Victor R. Lopez
Math Teacher
Washington School
Union City, NJ

George Maguschak
Math Teacher/Building Chairperson
Wilkes-Barre Area
Wilkes-Barre, PA

Dianne McIntire
Math Teacher
Garfield School
Kearny, NJ

Kenneth McIntire
Math Teacher
Lincoln School
Kearny, NJ

Francisco Pacheco
Math Teacher
IS 125
Bronx, NY

Vivian Perry
Edwards, IL

Vicki Perryman Petty
Math Teacher
Central Middle School
Murfreesboro, TN

Jennifer Sawyer
Math Teacher
Shawboro, NC

Russell Sayler
Math Teacher
Longfellow Middle School
Wauwatosa, WI

Raymond Scacalossi
Math Chairperson
Hauppauge Schools
Hauppauge, NY

Richard Seavey
Math Teacher–Retired
Metcalf Jr. High
Eagan, MN

Sherry Shaffer
Math Teacher
Honeoye Central School
Honeoye Falls, NY

Gail M. Sigmund
Math Teacher
Charles A. Mooney Preparatory School
Cleveland, OH

Jonathan Simmons
Math Teacher
Manor Middle School
Killeen, TX

Jeffrey L. Slagel
Math Department Chair
South Eastern Middle School
Fawn Grove, PA

Karen Smith, Ph.D.
Math Teacher
East Middle School
Braintree, MA

Bonnie Thompson
Math Teacher
Tower Heights Middle School
Dayton, OH

Mary Thoreen
Mathematics Subject Area Leader
Wilson Middle School
Tampa, FL

Paul Turney
Math Teacher
Ladue School District
St. Louis, MO

Welcome to Holt Middle School Math North Carolina Edition!

Have you ever wondered if anyone really uses the math that you learn in school? As you work through this book, you will find out that they certainly do. Not only that, but people are using math almost right in your backyard.

We have found the places where math is being used in your home state. In each chapter, you will "visit" one or two North Carolina locations, or you will learn about a specific aspect of North Carolina's culture, history, or industry to see how math is being used near you. It will be like taking a math tour of North Carolina without leaving your classroom!

Chapter 1 Analyze data about the instruments in the North Carolina Symphony, one of the first state symphonies in the U.S.

Chapter 2 Use scientific notation to describe astronomical distances at the Morehead Planetarium and Science Center.

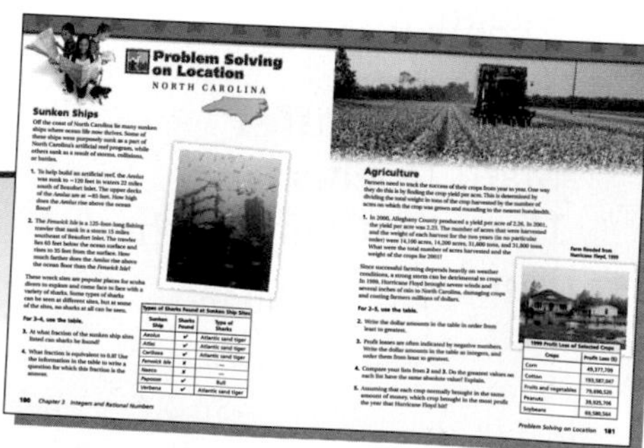

Chapter 3 Integers are used to describe the depths of sunken ships off North Carolina's coast.

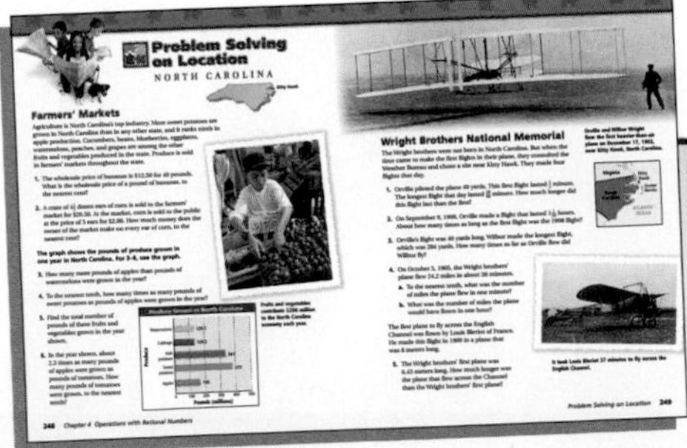

Chapter 4 Relive the first flight at the Wright Brothers National Memorial. You'll calculate the times and distances of early attempts at air travel.

Chapter 5 Relationships between rate, time, and distance will help you plan a road trip on North Carolina's highways.

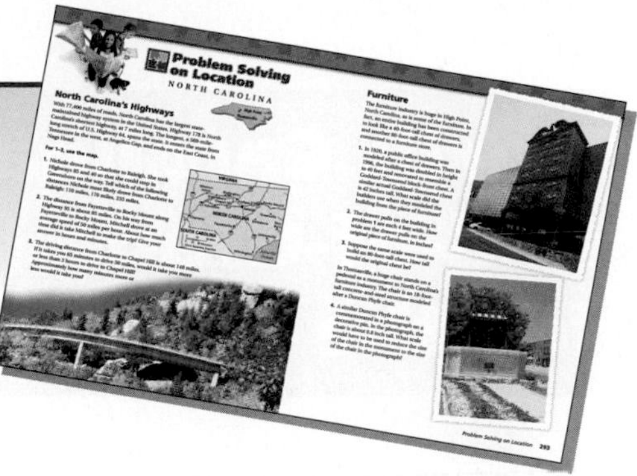

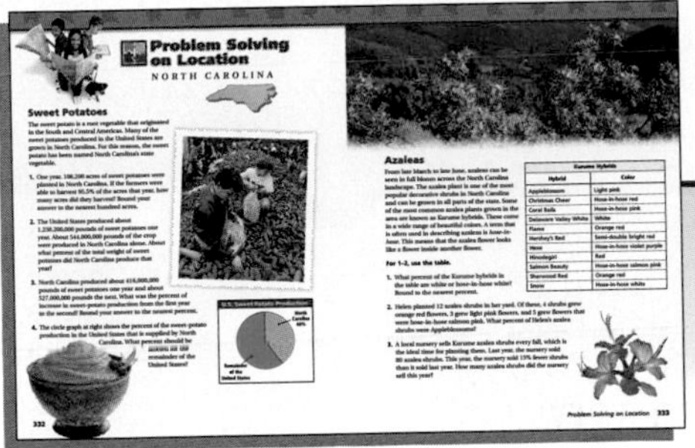

Chapter 6 Azaleas can be grown all over North Carolina. Use percents to describe the different varieties and colors of azaleas.

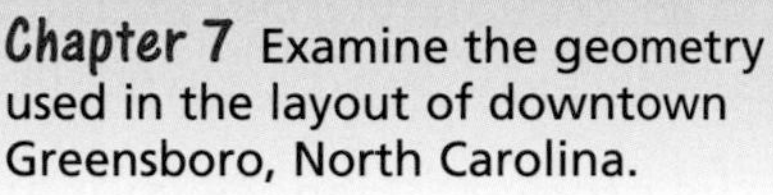

Chapter 7 Examine the geometry used in the layout of downtown Greensboro, North Carolina.

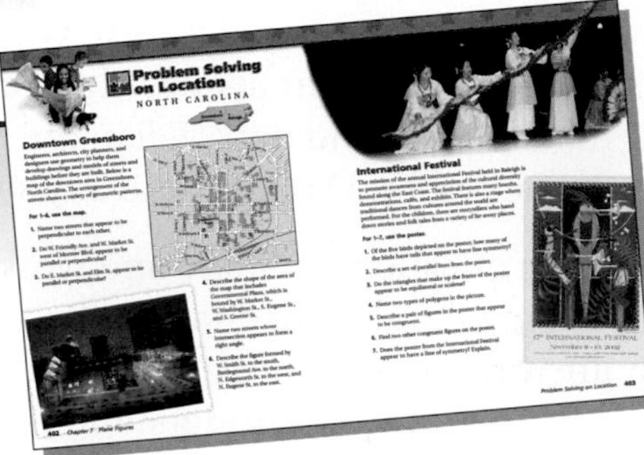

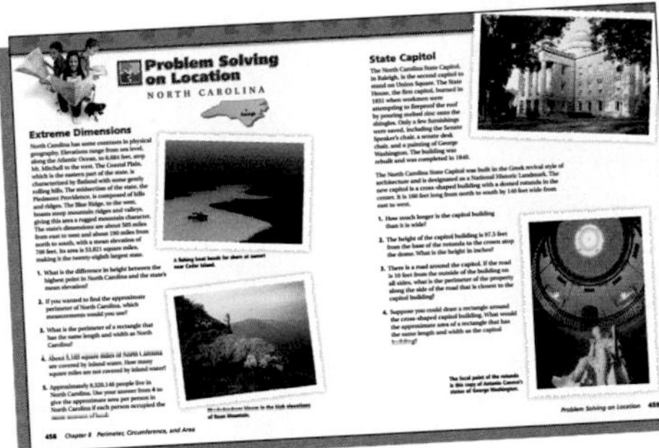

Chapter 8 Take some measurements of the North Carolina State Capitol, in Raleigh.

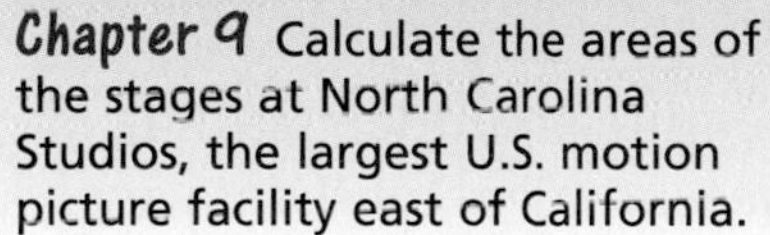

Chapter 9 Calculate the areas of the stages at North Carolina Studios, the largest U.S. motion picture facility east of California.

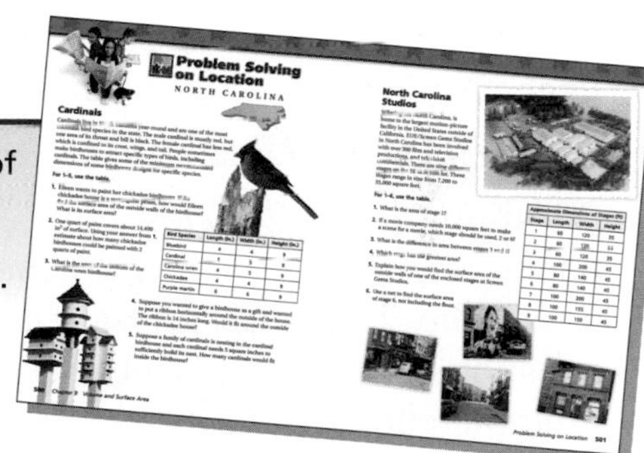

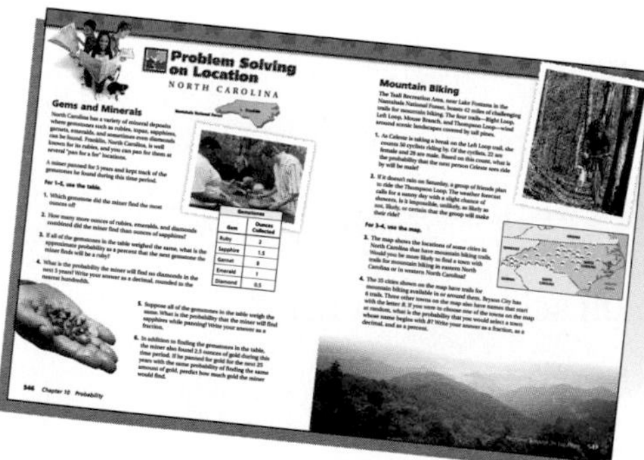

Chapter 10 What is the probability that you'll find a precious gemstone in North Carolina? Learn about North Carolina's gem and mineral deposits.

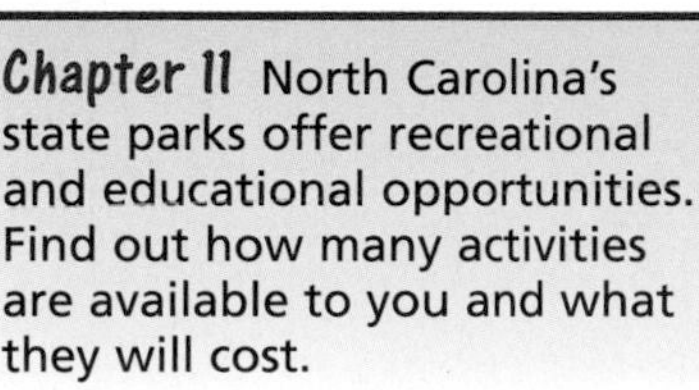

Chapter 11 North Carolina's state parks offer recreational and educational opportunities. Find out how many activities are available to you and what they will cost.

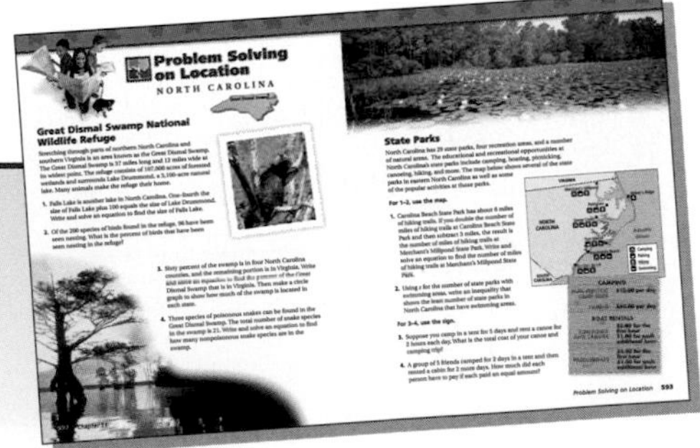

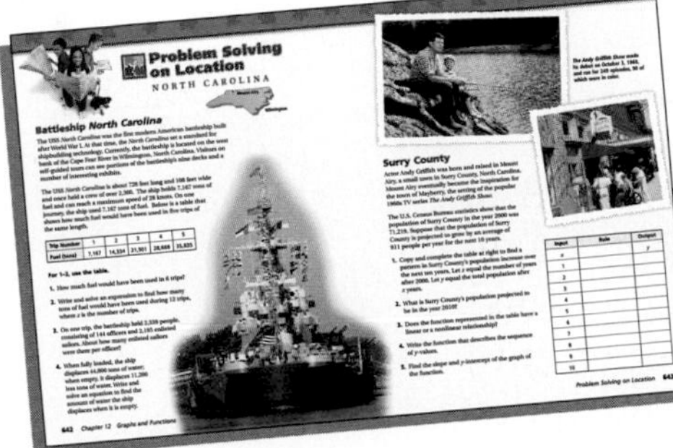

Chapter 12 Learn facts and statistics about the *USS North Carolina.*

North Carolina End-of-Grade (EOG) Test for the Middle Grades

The North Carolina End-of-Grade (EOG) Test is a multiple-choice test. To answer questions on the EOG test, you will fill in an answer sheet. It is very important to fill in your answer sheet correctly. When shading in circles, make your marks heavy and dark. Fill in the circles completely, but do not shade outside the circles. Do not make any stray marks on your answer sheet.

Read each question carefully and work the problem. Choose your answer from among the answer choices given, and fill in the corresponding circle on your answer sheet. If your answer is not one of the choices, read the question again. Be sure that you understand the problem. Check your work for possible errors.

Some questions on the EOG test are calculator inactive, and others are calculator active. You may not use a calculator on the calculator inactive questions, but you may on the calculator active questions.

Sample Question

Questions on the EOG test may require an understanding of number and operations, algebra, geometry, measurement, and data analysis and probability. Drawings, grids, or charts may be included for certain types of questions. Try the following practice question to prepare for taking a multiple-choice test. Choose the best answer from the choices given.

In a group of 30 students, 27 are middle school students and the others are high school students. If one person is selected at random from this group, what is the probability that the person selected will be a high school student?

A $\frac{1}{30}$

B $\frac{1}{10}$

C $\frac{3}{10}$

D $\frac{9}{10}$

Think About the Solution

First consider the total number of people in the group (30). If 27 out of 30 are middle school students, how many of them are high school students? (3) If one person is selected at random, there is a probability of 3 out of 30. This can be written as a ratio (3:30), a fraction $\left(\frac{3}{30}\right)$, a decimal (0.1), or a percent (10%). None of these solutions is listed as one of the choices, so you must look for a solution that is equivalent. The fraction $\frac{3}{30}$ can be simplified to $\frac{1}{10}$. Since $\frac{1}{10}$ is given as one of your answer choices, B is the correct response.

Indicate your response by filling in the circle that contains B.

Test-Taking Tip

Sometimes you can find the best solution to a test question by understanding what is wrong with some of the choices. Read the sample question again. Why are A, C, and D incorrect?

Response A is $\frac{1}{30}$.

You might think A is correct because there are 30 people and you are selecting one. However, this answer indicates that only 1 of the 30 people is a high school student. Since that is not what the problem states, A cannot be correct.

Response C is $\frac{3}{10}$.

This answer indicates that 3 out of 10 people are high school students. The numerator is correct, since there are three high school students in the group. However, the denominator must show the relationship 3 out of 30. C is not correct.

Response D is $\frac{9}{10}$.

If you chose D, read the problem again. The problem asks you to find the probability that a high school student will be selected. Answer D would be the best choice if you wanted to find the probability that a *middle school* student will be selected, but D does not match the question that was asked.

Practice, Practice, Practice

On the following pages, you will find a practice standardized test. This test has been designed to resemble the EOG test and the types of questions it contains. Use this test to practice answering these questions, as well as to review some of the math that you will be tested on. The more comfortable and familiar you can become with the EOG test, the better your chances of success!

EOG Test Practice

Directions: Read each question and choose the correct response. You may *not* use a calculator on this part of the test.

1. Noah is going camping and wants to waterproof all the outer surfaces of his pup tent, including the floor. A drawing of his tent is shown below.

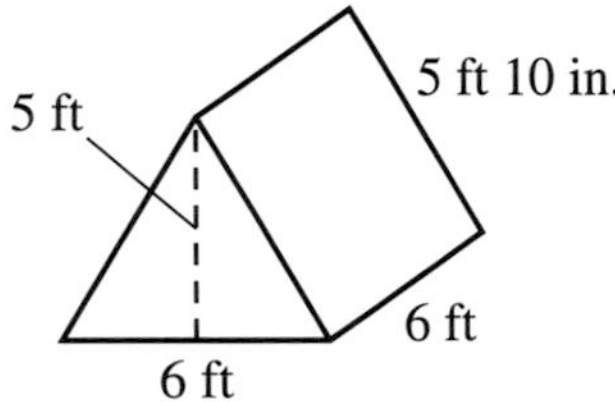

In order to determine how much water repellant to buy, Noah will need to know the surface area of his tent. What is the surface area of Noah's tent?

A 86 square feet

B 100 square feet

C 136 square feet

D 166 square feet

2. Jo wants an average of 90 on her 8 science tests. What is the *least* number of total points that she will need for all 8 tests to achieve her goal?

A 90

B 360

C 720

D 900

3. If a 32-ounce pitcher is $\frac{5}{8}$ full, how much liquid does it contain?

A 4 ounces

B 16 ounces

C 20 ounces

D 24 ounces

4. On a map, one inch represents 600 miles. Find the actual distance between two cities if they are $2\frac{1}{2}$ inches apart on the map.

A 1,500 miles

B 1,200 miles

C 300 miles

D 240 miles

5. The mean amount of rainfall on three consecutive days was 2 inches. On the first two days, there were 5 inches of rain and 1 inch of rain.

   How much rain fell on the third day?

   A 0 inches
   B 1 inch
   C 4 inches
   D 6 inches

6. Lou worked $1\frac{3}{4}$ hours on Friday and $3\frac{1}{2}$ hours on Saturday. Ani worked $2\frac{1}{4}$ hours on Friday and $1\frac{1}{2}$ hours on Saturday.

   How much longer did Lou work than Ani?

   A 0.5 hour
   B 0.75 hour
   C 1.0 hour
   D 1.5 hours

7. $8\frac{4}{5} \div 11$

   A $\frac{8}{11}$
   B $\frac{4}{5}$
   C $1\frac{1}{5}$
   D $1\frac{1}{4}$

8. At a local restaurant, the cost of a large orange juice is $2.25. This is $0.85 more than the cost of a small orange juice.

   How much does a small orange juice cost?

   A $1.40
   B $1.50
   C $1.85
   D $2.25

9. What is the volume of the figure below?

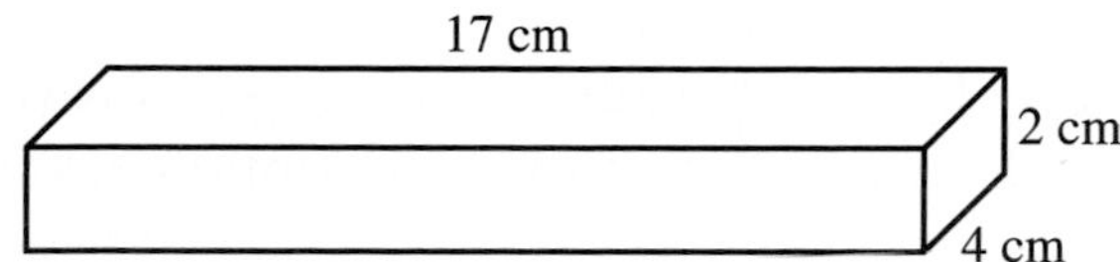

   A 23 cubic centimeters
   B 72 cubic centimeters
   C 136 cubic centimeters
   D 216 cubic centimeters

Go On

EOG Test Practice

10. The Statue of Liberty in New York City has a 25-window observation platform in the crown, from which visitors can view the city. The platform can hold about 40 viewers at one time. Suppose that the statue is open for 9 hours each day, and that each visitor spends 45 minutes on the platform.

    About how many people can view the city from the platform in one day?

    A 200   C 12,000

    B 480   D 16,200

11. 10% of what number is 6?

    A 0.6   C 60

    B 1.67   D 600

12. The admission prices at a movie theater are shown in the table below.

REGULAR ADMISSION PRICES

Adults	$8.50
Children	$5.00

**Matinee tickets are half price.

Suppose a family of two adults and three children want to see a movie. How much money will they save by going to a matinee instead of a regular show?

A $13.50   C $23.50

B $16.00   D $32.00

13. A subscription to a magazine cost $10.00 two years ago. This year, it costs $1\frac{1}{4}$ times as much. How much does the subscription cost this year?

    A $2.50   C $10.25

    B $8.00   D $12.50

14. What is the value of $4^3 - 3^2$?

    A 55   C 64

    B 61   D 73

15. The principal needs 685 copies of an $8\frac{1}{2}$-page list of all the courses offered at the high school. A package of copy paper contains 250 sheets. How many packages of paper will the principal need?

A 24

B 25

C 20,147

D 20,148

16. For which value of x will the expression $^{-}3 \cdot 5 \cdot x$ be positive?

A $x = 1$

B $x = 0$

C $x = \frac{1}{2}$

D $x = {}^{-}6$

17. On the map below, Springhill and Hampton are 6 centimeters apart.

What is the actual distance in kilometers between Springhill and Hampton?

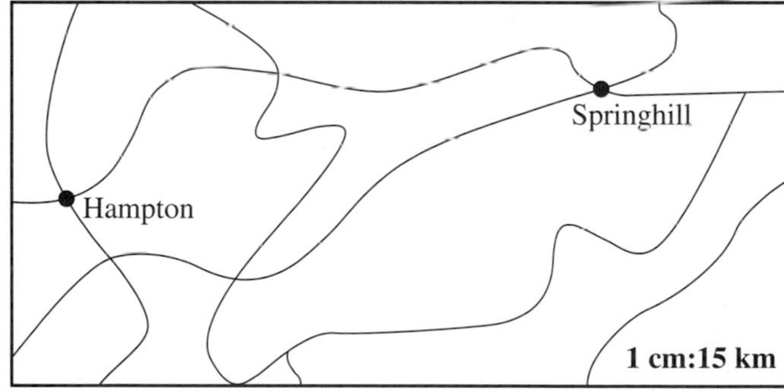

A 6 kilometers

B 15 kilometers

C 60 kilometers

D 90 kilometers

18. A tour group left North Carolina and traveled 120 miles by bus before stopping for a break. This distance was 60% of the total distance to their next stop. What was the total distance to their next stop?

A 72 miles

B 180 miles

C 192 miles

D 200 miles

Go On

19. For his science project, Ben planted seeds at two different temperatures and recorded the number of days that it took each seed to sprout. His results are given in the table.

Seed	1	2	3	4	5	6	7	8
Temperature (°C)	25	25	25	25	30	30	30	30
Time to Sprout (days)	8	8	5	6	5	4	7	6

What is the mean number of days it took the seeds planted at 25°C to sprout?

A 5.5 C 6.125

B 6.33 D 6.75

20. Your muscles make up about 40% of your overall mass. What is the muscle mass of a person whose total body mass is 60 kilograms?

A 20 kilograms C 33 kilograms

B 24 kilograms D 100 kilograms

21. Devon can type 40 words in one minute. At this rate, how long will it take her to type 900 words?

A 20.5 minutes

B 21 minutes

C 22 minutes

D 22.5 minutes

22. A video game is regularly priced at $40. When Benny bought the game, it was on sale for 30% off the regular price. What was the cost of the discounted video game?

A $12 C $28

B $24 D $36

23. At a health club, the swimming pool contains 86,000 liters of water. The maintenance staff uses 20 liters of chlorine to purify the water.

How many liters of water can be purified by one liter of chlorine?

A 43 C 43,000

B 4,300 D 430,000

24. How many degrees difference is there between 42°C and $^{-}7$°C?

A 35°C
B 48°C
C 49°C
D 50°C

25. The Bassett family paid $120 in cooling costs for the month of July. During August, they lowered the thermostat setting by 1°F. Their cooling costs for August totaled $132. What is the percent increase in cost from July to August?

A 10%
B 12%
C 90%
D 110%

26. Alison is borrowing $700 for one year. The loan company is charging her 10% yearly simple interest. How much interest will Alison have to pay?

A $10
B $70
C $710
D $7,000

27. Mr. Whitman made the scale drawing of his porch shown below.

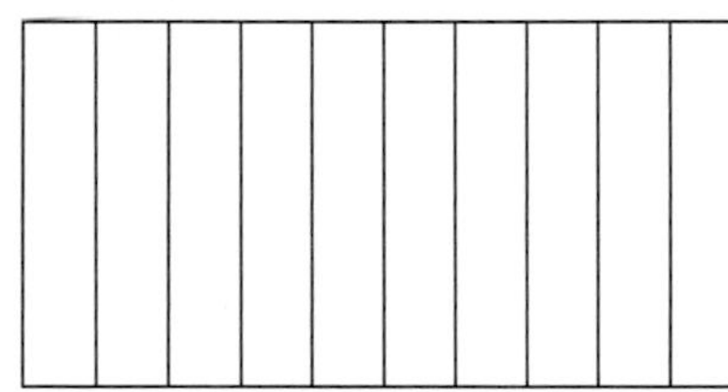

What is the actual width of the porch?

A 30 feet
B 36 feet
C 40 feet
D 44 feet

28. In 1903, Wilbur Wright flew the *Kitty Hawk* about 826 feet at an average speed of 14 feet per second. How long did the flight last?

A 4 seconds
B 14 seconds
C 59 seconds
D 77 seconds

Go On

EOG Test Practice

29. Grain is stored in cylindrical storage tanks called silos. What is the volume, in cubic meters, of the silo shown below? Use 3.14 as an approximation for π.

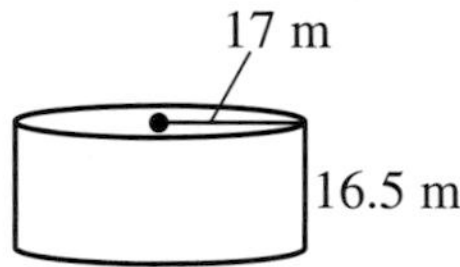

A 14,973 cubic meters

B 4,769 cubic meters

C 14,306 cubic meters

D 1,762 cubic meters

30. The road map below shows the course that Manny bicycles every day. If the entire course takes Manny 36 minutes to complete, how long does he take to bicycle one mile? Assume that Manny rides his bike at a constant speed.

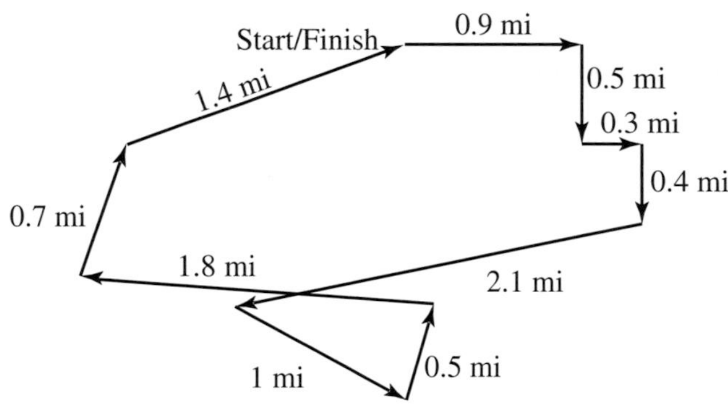

A 0.26 minute

B 3.75 minutes

C 9.6 minutes

D 38 minutes

Directions: Read each question and choose the correct response. You may use a calculator on this part of the test.

31. Susan has 3 times as many quarters as dimes. She has 220 dimes and quarters altogether.

Which equation could be used to represent this situation?

A $d + 3d = 220$

B $3d - d = 220$

C $3d = 220$

D $3d = d$

32. To display his math class's test scores, Mr. Fuentes made a line plot.

MATH TEST SCORES

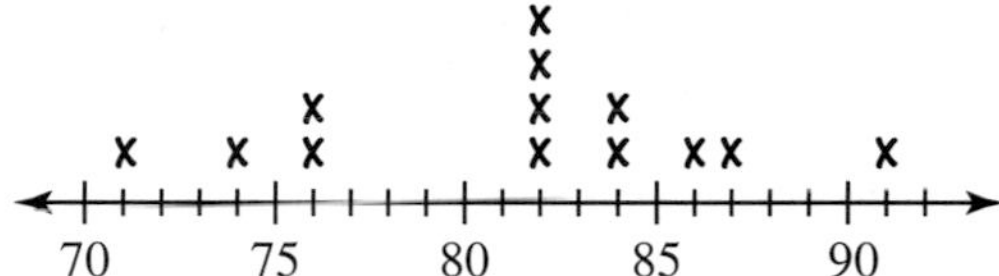

What is the mode of the test scores?

A 71

B 76

C 82

D 91

33. The Ames Middle School cafeteria staff surveyed 100 students about their favorite lunch foods. The survey results are shown in the table below.

FAVORITE FOODS

Food	Pizza	Tacos	Hamburgers	Chicken
Number of Students	42	29	18	11

There are 1,000 students at Ames Middle School. How many of these students would probably select pizza as their favorite food?

A 42

B 420

C 840

D 942

Go On

34. A quilt has 15 stripes. The first stripe is white, the second stripe is blue, and the third stripe is pink. If this pattern continues, how many times does the pattern appear in the quilt?

A 3
B 4
C 5
D 6

35. A package of 4 light bulbs costs $3.76. What is the unit cost?

A $0.94
B $1.88
C $3.76
D $37.60

36. If two polygons are congruent, which of the following statements *must* be true?

A The polygons share a common side.
B The polygons may be different sizes.
C Corresponding angles are congruent.
D The polygons together tessellate the plane.

37. These figures are similar. What is the value of x?

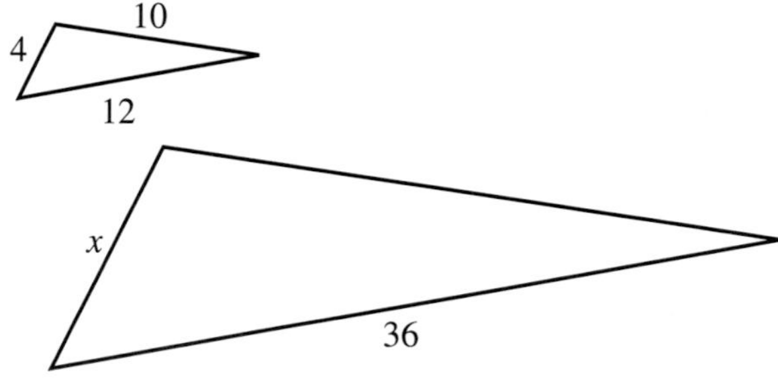

A 24
B 18
C 15
D 12

38. If the pattern below continues, what are the next two terms in this sequence?

48,000; 24,000; 12,000; 6,000

A 0; $^{-}$6,000
B 3,000; 0
C $^{-}$6,000; $^{-}$12,000
D 3,000; 1,500

39. Scientists estimate that 75% of all animals are arthropods. A percent of all arthropods are beetles. Which of the following describes the percent of all animals that are beetles?

A less than 75%

B more than 75%

C equal to 75%

D cannot be determined

40. Kaylie surveyed 9 of her friends about how many people lived in their households. Her results are shown in the table below.

Number of People Living in a Household								
3	4	3	6	2	7	10	4	4

What is the median of Kaylie's data set?

A 2

B 4

C 5

D 8

41. What is the next figure in this pattern?

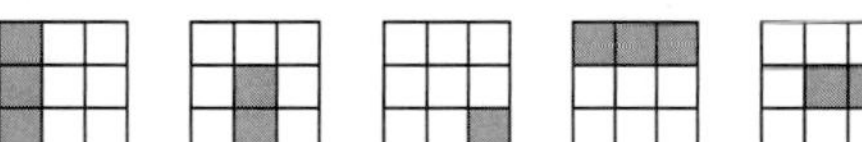

A

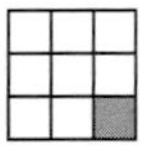

B

C 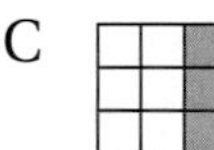

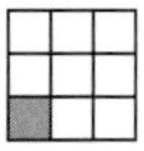

D 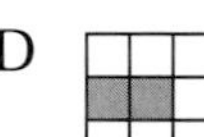

42. Look at the function table.

x	y
$^{-}2$	7
$^{-}1$	9
0	11
1	13
2	15
3	

What is the value of y when $x = 3$?

A 15

B 16

C 17

D 18

Go On

43. The histogram below shows the heights of students in the seventh grade at one middle school.

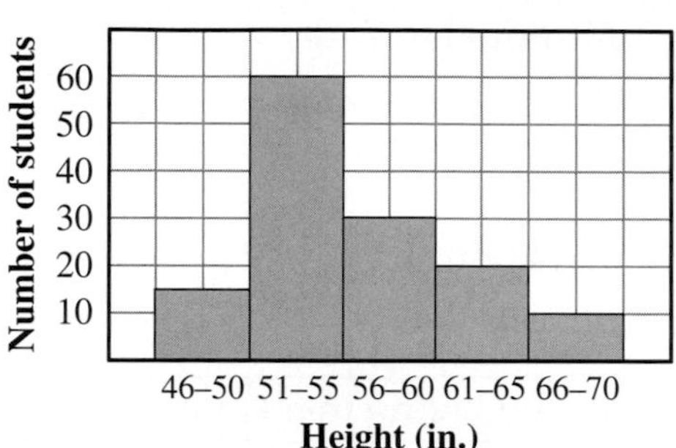

What is the *greatest* possible height of a seventh-grader at this school?

A 55 inches

B 60 inches

C 66 inches

D 70 inches

44. Wanda babysits in her neighborhood to earn money. She charges $2.00 per job plus an hourly fee. The graph below shows Wanda's earnings.

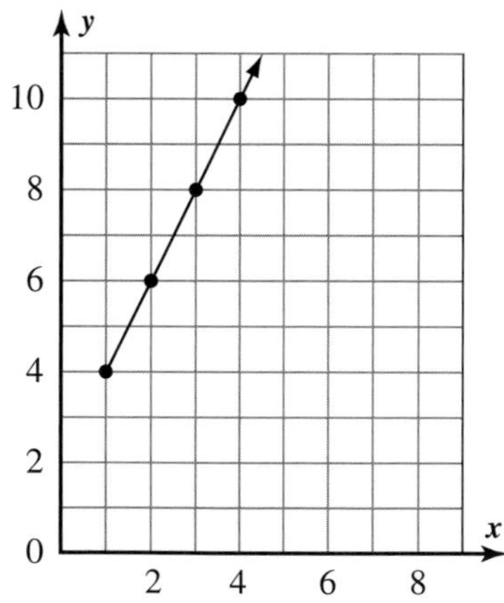

How much money does Wanda charge per hour?

A $2.00

B $4.00

C $8.00

D $9.00

45. Which proportion can be used to find the number of pints in 5 quarts?

A $\frac{2}{1} = \frac{p}{5}$

B $\frac{2}{p} = \frac{5}{1}$

C $\frac{2}{1} = \frac{5}{p}$

D $\frac{1}{2} = \frac{p}{5}$

46. Each net below can be folded along the dotted lines to form a rectangular prism. Which prism has the *greatest* surface area?

A

6 cm
3 cm
4 cm
3 cm
4 cm
3 cm

C

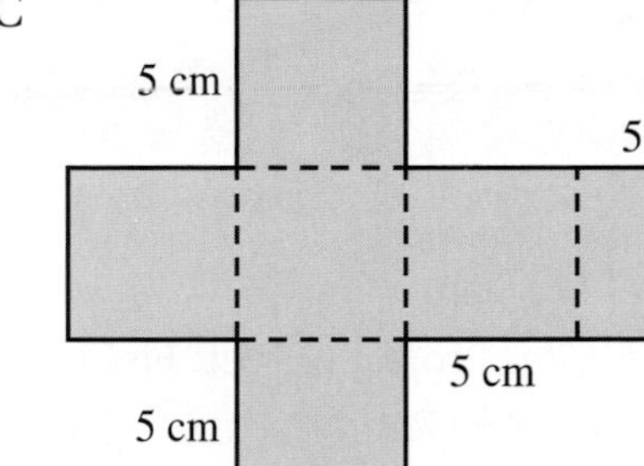

B

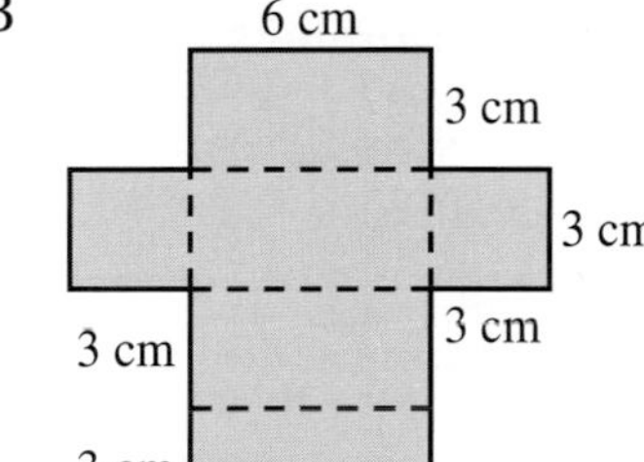

D

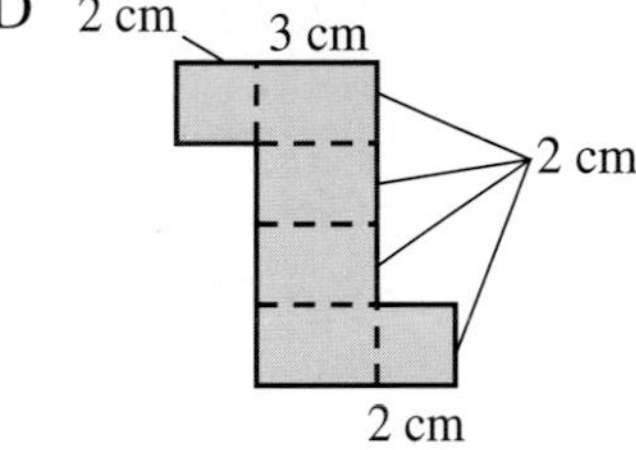

47. The graph below shows the high and low temperatures over five days.

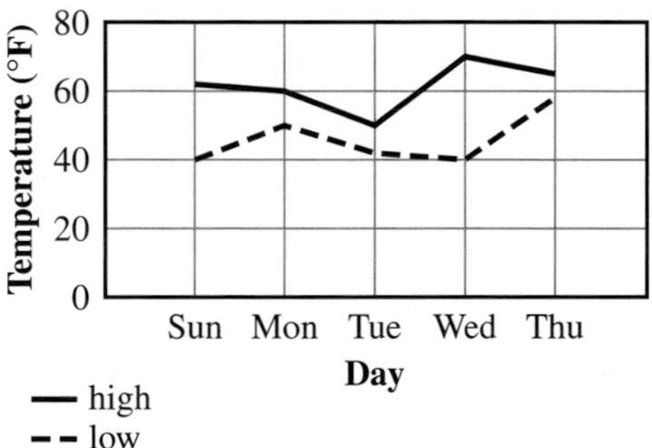

On which day is the difference between the high and low temperatures the *greatest*?

A Sunday

B Monday

C Tuesday

D Wednesday

Stop

Problem Solving Handbook

Data Toolbox

Assessment

Student Help

internet connect

Homework Help Online

6, 12, 16, 22, 26, 30, 37, 42, 46

KEYWORD: MS4 HWHelp

Algebra *Indicates algebra included in lesson development*

Number Theory & Algebraic Reasoning

Assessment

Interdisciplinary LINKS

Student Help

internet connect

Homework Help Online

KEYWORD: MS4 HWHelp

Integers and Rational Numbers

Interdisciplinary LINKS

Student Help

internet connect go.hrw.com

Homework Help Online

132, 136, 142, 148, 152, 158, 164, 168, 172, 176

KEYWORD: MS4 HWHelp

Assessment

go.hrw.com EOG Test Preparation Online . . . . . KEYWORD: MS4 TestPrep

Algebra *Indicates algebra included in lesson development*

Operations with Rational Numbers

Assessment

Interdisciplinary LINKS

Student Help

internet connect

Homework Help Online

194, 198, 204, 208, 212, 216, 224, 228, 232, 238, 242, 246

KEYWORD: MS4 HWHelp

Proportional Reasoning

Interdisciplinary LINKS

Assessment

EOG Test Preparation Online . . . . KEYWORD: MS4 TestPrep

Student Help

internet connect

Homework Help Online

262, 266, 270, 274, 282, 286, 290

KEYWORD: MS4 HWHelp

Algebra *Indicates algebra included in lesson development*

Percents

Plane Figures

Interdisciplinary LINKS

Student Help

internet connect

Homework Help Online

346, 350, 356, 364, 370, 376, 380, 384, 390, 394, 398

KEYWORD: MS4 HWHelp

Assessment

Algebra *Indicates algebra included in lesson development*

Perimeter, Circumference & Area

Interdisciplinary LINKS

Student Help

internet connect

Homework Help Online
416, 422, 426, 432, 436, 440, 446, 452
KEYWORD: MS4 HWHelp

Assessment

EOG Test Preparation Online . . . KEYWORD: MS4 TestPrep

Volume and Surface Area

Student Help

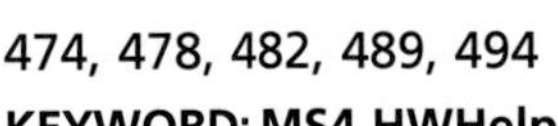

Homework Help Online
474, 478, 482, 489, 494
KEYWORD: MS4 HWHelp

Assessment

Algebra *Indicates algebra included in lesson development*

Probability

Homework Help Online

KEYWORD: MS4 HWHelp

Multistep Equations and Inequalities

Assessment

EOG Test Preparation Online .... KEYWORD: MS4 TestPrep

Interdisciplinary LINKS

Student Help

internet connect

Homework Help Online

562, 566, 570, 576, 580, 584, 588

KEYWORD: MS4 HWHelp

Algebra *Indicates algebra included in lesson development*

Problem Solving Handbook

The Problem Solving Plan

In order to be a good problem solver, you first need a good problem-solving plan. A plan or strategy will help you to understand the problem, to work through a solution, and to check that your answer makes sense. The plan used in this book is detailed below.

UNDERSTAND the Problem

■ **What are you asked to find?**	Restate the problem in your own words.
■ **What information is given?**	Identify the important facts in the problem.
■ **What information do you need?**	Determine which facts are needed to solve the problem.
■ **Is all the information given?**	Determine whether all the facts are given.

Make a PLAN

■ **Have you ever solved a similar problem?**	Think about other problems like this that you successfully solved.
■ **What strategy or strategies can you use?**	Determine a strategy that you can use and how you will use it.

SOLVE

■ **Follow your plan.**	Show the steps in your solution. Write your answer as a complete sentence.

LOOK BACK

■ **Have you answered the question?**	Be sure that you answered the question that is being asked.
■ **Is your answer reasonable?**	Your answer should make sense in the context of the problem.
■ **Is there another strategy you could use?**	Solving the problem using another strategy is a good way to check your work.
■ **Did you learn anything while solving this problem that could help you solve similar problems in the future?**	Try to remember the problems you have solved and the strategies you used to solve them.

Graphs and Functions

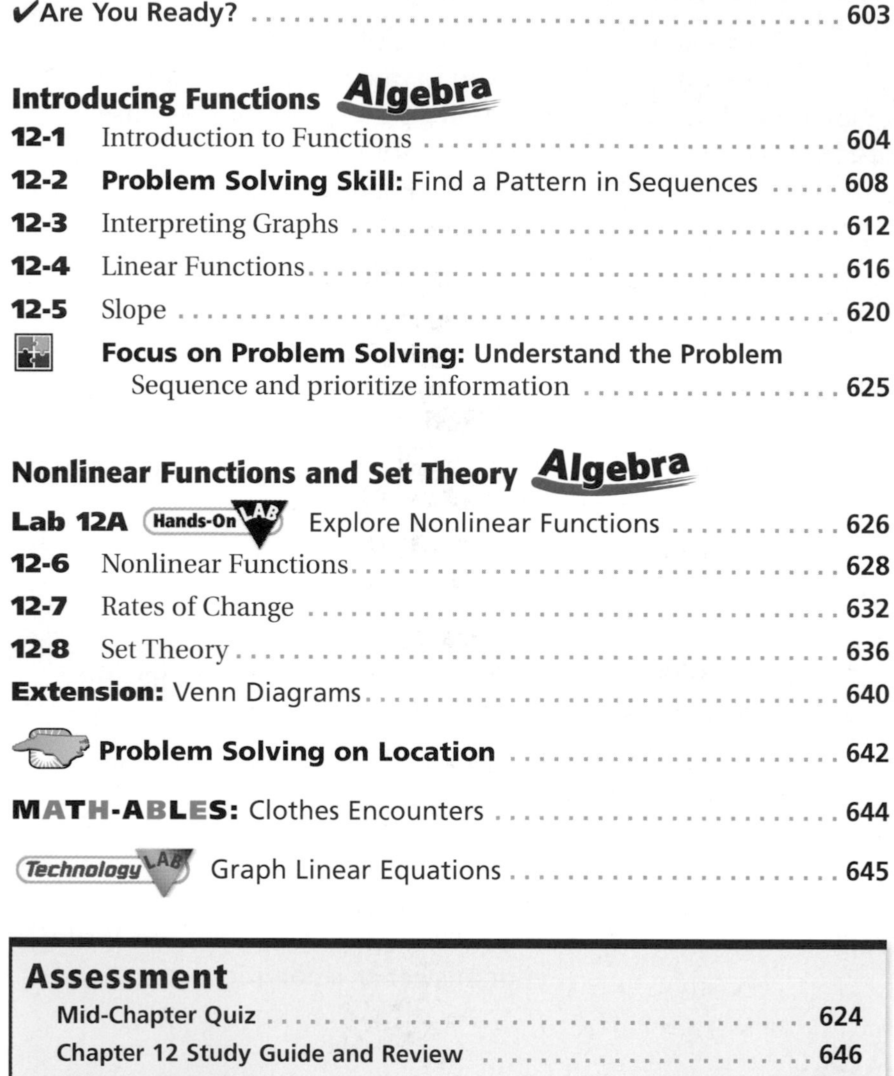

Introducing Functions *Algebra*

Nonlinear Functions and Set Theory *Algebra*

Assessment

Interdisciplinary LINKS

Student Help

internet connect

Homework Help Online

606, 610, 614, 618, 622, 630, 634, 638

KEYWORD: MS4 HWHelp

Student Handbook

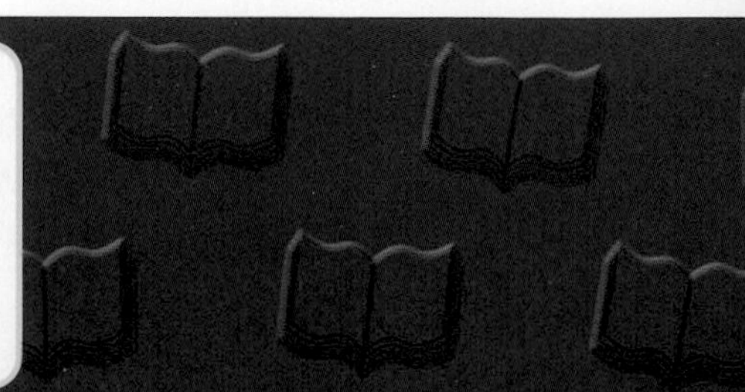

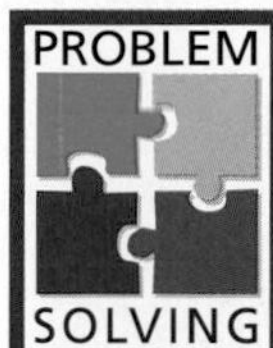

Using the Problem Solving Plan

During summer vacation, Ricardo will go to space camp and then to visit his relatives. He will be gone for 5 weeks and 4 days and will spend 11 more days with his relatives than at space camp. How long will Ricardo stay at each place?

UNDERSTAND the Problem

List the important information.

- Ricardo will be gone for 5 weeks and 4 days.
- He will spend 11 more days with his relatives than at space camp.

The answer will be how long Ricardo stays at each place.

Make a PLAN

You can **draw a diagram** to show how long he will stay at each place. Use boxes for the length of each stay. The length of each box will represent the length of each stay.

SOLVE

Think: There are 7 days in a week, so 5 weeks and 4 days is a total of 39 days. Your diagram might look like this:

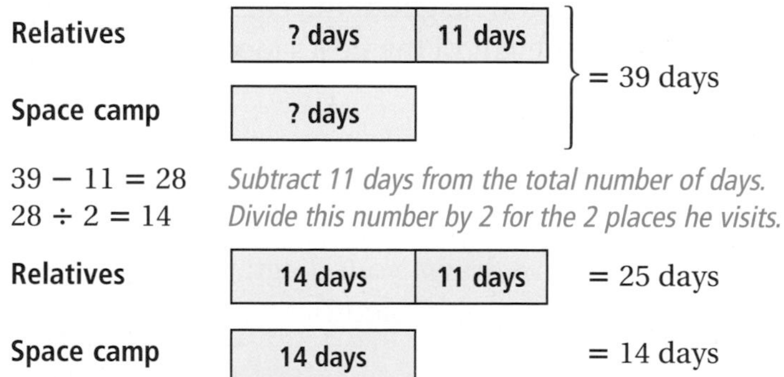

So Ricardo will stay with his relatives for 25 days and at space camp for 14 days.

LOOK BACK

Twenty-five days is 11 days longer than 14 days. The total length of the two stays is 25 + 14 = 39 days, or 5 weeks and 4 days. This solution fits the information given in the problem.

Draw a Diagram

When problems involve objects, distances, or places, you can **draw a diagram** to make the problem easier to understand. You can use the diagram to look for relationships among the given data and to solve the problem.

Problem Solving Strategies

Draw a Diagram	Make a Table
Make a Model	Solve a Simpler Problem
Guess and Test	Use Logical Reasoning
Work Backward	Use a Venn Diagram
Find a Pattern	Make an Organized List

A bald eagle has built a nest 18 feet below the top of a 105-foot-tall oak tree. The eagle sits on a limb 72 feet above the ground. What is the vertical distance between the eagle and its nest?

Understand the Problem

Identify the important information.

- The height of the tree is 105 feet.
- The eagle's nest is 18 feet from the top of the tree.
- The eagle is perched 72 feet above the ground.

The answer will be the vertical distance between the eagle and its nest.

Make a Plan

Use the information in the problem to **draw a diagram** showing the height of the tree and the locations of the eagle and its nest.

Solve

To find the height of the nest's location, subtract the distance of the nest from the top of the tree from the height of the tree.

105 feet − 18 feet = 87 feet

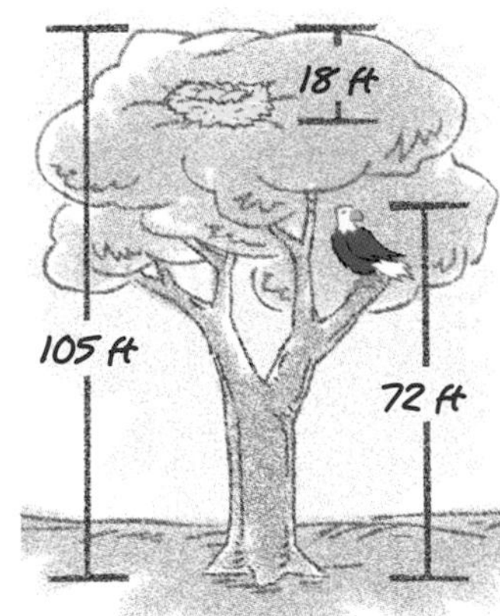

To find the vertical distance from the eagle to its nest, subtract the height of the eagle's location from the height of the nest's location.

87 feet − 72 feet = 15 feet

The vertical distance between the eagle and its nest is 15 feet.

Look Back

Be sure that you have drawn your diagram correctly. Does it match the information given in the problem?

PRACTICE

1. A truck driver travels 17 miles south to drop off his first delivery. Then he drives 19 miles west to drop off a second delivery, and then he drives 17 miles north to drop off another delivery. Finally, he drives 5 miles east for his last delivery. How far is he from his starting point?

2. A table that is standing lengthwise against a wall is 10 feet long and 4 feet wide. Sarah puts balloons 1 foot apart along the three exposed sides, with one balloon at each corner. How many balloons does she use?

Make a Model

When problems involve objects, you can **make a model** using those objects or similar objects. This can help you understand the problem and find the solution.

Problem Solving Strategies

Draw a Diagram	Make a Table
Make a Model	Solve a Simpler Problem
Guess and Test	Use Logical Reasoning
Work Backward	Use a Venn Diagram
Find a Pattern	Make an Organized List

A company packages 6 minipuzzles in a decorated 4 in. cube. They are shipped to the toy store in cartons shaped like rectangular prisms. Twenty cubes fit in each carton. If the height of each carton is 8 in., what are possible dimensions of the carton?

Understand the Problem

Identify the important information.

- Each cube is 4 inches on a side.
- Twenty cubes fit in one carton.
- The height of the carton is 8 inches.

The answer is the dimensions of the carton.

Make a Plan

You can use 20 cubes to **make a model** of cubes packed in a carton. Record possible values for length and width, given a height of 8 in.

Solve

Begin with a carton that is 8 in., or 2 cubes, high. Use all 20 cubes to make a rectangular prism.

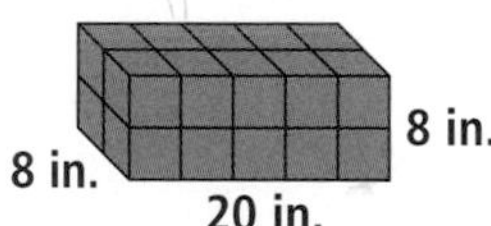

Possible dimensions of the carton are 20 in. × 8 in. × 8 in.

Look Back

The volume of each carton should equal the volume of the 20 cubes.

Volume of cartons: 8 in. × 20 in. × 8 in. = 1,280 in^3

Volume of 1 cube: 4 in. × 4 in. × 4 in. = 64 in^3

Volume of 20 cubes: 20 × 64 = 1,280 in^3

1,280 in^3 = 1,280 in^3 ✔

PRACTICE

1. Give two sets of possible dimensions of a rectangular prism made up of twenty 1-inch cubes.
2. John uses exactly eight 1-inch cubes to form a rectangular prism. Find the length, width, and height of the prism.

Guess and Test

If you do not know how to solve a problem, you can always make a **guess.** Then **test** your guess using the information in the problem. Use what you find out to make a second guess. Continue to **guess and test** until you find the correct answer.

Problem Solving Strategies

Draw a Diagram	Make a Table
Make a Model	Solve a Simpler Problem
Guess and Test	Use Logical Reasoning
Work Backward	Use a Venn Diagram
Find a Pattern	Make an Organized List

Shannon used equal numbers of quarters and nickels to buy an embossing template that cost $1.50. How many of each coin did she use?

Understand the Problem Identify the important information.

- Shannon used equal numbers of quarters and nickels.
- The coins she used total $1.50.

The answer will be the number of quarters and the number of nickels Shannon used.

Make a Plan Start with an educated **guess** in which the numbers of quarters and nickels are the same. Then **test** to see whether the coins total $1.50.

Solve Make a first guess of 4 quarters and 4 nickels, and find the total value of the coins.

Guess: 4 quarters and 4 nickels
Test: (4 × $0.25) + (4 × $0.05) = $1.00 + $0.20 = $1.20

$1.20 is too low. Increase the number of coins.

Guess: 6 quarters and 6 nickels
Test: (6 × $0.25) + (6 × $0.05) = $1.50 + $0.30 = $1.80

$1.80 is too high. The number of each coin must be between 4 and 6. So Shannon must have used 5 quarters and 5 nickels.

Look Back Test the answer to see whether the coins add up to $1.50.
(5 × $0.25) + (5 × $0.05) = $1.25 + $0.25 = $1.50 ✔

PRACTICE

1. The sum of Richard's age and his older brother's age is 63. The difference between their ages is 13. How old are Richard and his brother?
2. In the final game of the basketball season, Trinka scored a total of 25 points on 2-point shots and 3-point shots. She made 5 more 2-point shots than 3-point shots. How many of each did she make?

Work Backward

Some problems give you a sequence of information and ask you to find something that happened at the beginning. To solve a problem like this, you may want to start at the end of the problem and **work backward.**

Problem Solving Strategies

Draw a Diagram	Make a Table
Make a Model	Solve a Simpler Problem
Guess and Test	Use Logical Reasoning
Work Backward	Use a Venn Diagram
Find a Pattern	Make an Organized List

Tony is selling dried fruit snacks to help raise money for a new school computer. Half of the fruit snacks in the bag are apricots. Of the rest of the fruit snacks, half of them are bananas, and the other 8 are cranberries. How many fruit snacks are in the bag?

Understand the Problem

Identify the important information.

- Half of the fruit snacks are apricots.
- Half of the remaining fruit snacks are bananas.
- The final 8 fruit snacks are cranberries.

The answer will be the total number of fruit snacks in the bag.

Make a Plan

Start with the 8 cranberries, and **work backward** through the information in the problem to the total number of fruit snacks in the bag.

Solve

There are 8 cranberries. 8

The other half of the remaining fruit snacks are bananas, so there must be 8 bananas. $8 + 8 = 16$

The other half of the fruit snacks are apricots, so there must be 16 apricots. $16 + 16 = 32$

There are 32 fruit snacks in the bag.

Look Back

Using the starting amount of 32 fruit snacks, work from the beginning of the problem following the steps.

Start: 32
Half of 32: $32 \div 2 = 16$
Half of 16: $16 \div 2 = 8$
Minus 8: $8 - 8 = 0$ ✓

PRACTICE

1. In a trivia competition, each finalist must answer 4 questions correctly. Each question is worth twice as much as the question before it. The fourth question is worth \$1,000. How much is the first question worth?

2. The Ramirez family has 5 children. Sara is 5 years younger than her brother Kenny. Felix is half as old as his sister Sara. Kaitlen, who is 10, is 3 years older than Felix. Kenny and Celia are twins. How old is Celia?

Find a Pattern

In some problems, there is a relationship between different pieces of information. Examine this relationship and try to **find a pattern.** You can then use this pattern to find more information and the solution to the problem.

Problem Solving Strategies

Draw a Diagram	Make a Table
Make a Model	Solve a Simpler Problem
Guess and Test	Use Logical Reasoning
Work Backward	Use a Venn Diagram
Find a Pattern	Make an Organized List

John made a design using hexagons and triangles. The side lengths of each hexagon and triangle are 1 inch. What is the perimeter of the next figure in his design?

Understand the Problem

Identify the important information.

- The first 5 figures in the design are given.
- The side lengths of each hexagon and triangle are 1 inch.

The answer will be the perimeter of the sixth figure in the design.

Make a Plan

Try to **find a pattern** in the perimeters of the first 5 figures. Use the pattern to find the perimeter of the sixth figure.

Solve

Find the perimeter of the first 5 figures.

Figure	Perimeter (in.)	Pattern
1	6	
2	7	$6 + 1 = 7$
3	11	$7 + 4 = 11$
4	12	$11 + 1 = 12$
5	16	$12 + 4 = 16$

The pattern appears to be add 1, add 4, add 1, add 4, and so on. So the perimeter of the sixth figure will be $16 + 1$, or 17.

Look Back

Use another strategy. **Draw a diagram** of the sixth figure. Then find the perimeter.

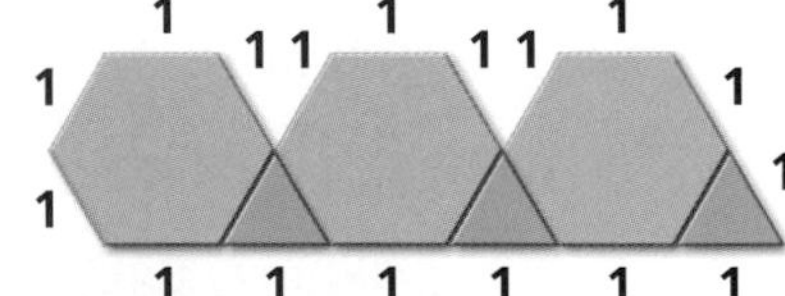

PRACTICE

Describe the pattern, and then find the next number.

1. 1, 5, 9, 13, 17, . . .

2. 1, 4, 16, 64, 256, . . .

Make a Table

When you are given a lot of information in a problem, it may be helpful to organize that information. One way to organize information is to **make a table.**

Problem Solving Strategies

Draw a Diagram	**Make a Table**
Make a Model	Solve a Simpler Problem
Guess and Test	Use Logical Reasoning
Work Backward	Use a Venn Diagram
Find a Pattern	Make an Organized List

On November 1, Wendy watered the Gribbles' yard and the Milams' yard. If she waters the Gribbles' yard every 4 days and the Milams' yard every 5 days, when is the next date that Wendy will water both yards?

Understand the Problem

Identify the important information.

- Wendy waters the Gribbles' yard every 4 days and the Milams' yard every 5 days. She watered both yards on November 1.

The answer will be the next date that she waters both yards again.

Make a Plan

Make a table using X's to show the days that Wendy waters each yard. Make one row for the Gribbles and one row for the Milams.

Solve

Start with an X in both rows for November 1. For the Gribbles, add an X on every fourth day after November 1. For the Milams, add an X every fifth day after November 1.

Date	1	2	3	4	5	6	7	8	9	10	11	12	13	14	15	16	17	18	19	20	21
Gribble	X				X				X				X				X				X
Milam	X					X					X					X					X

November 21 is the next date that Wendy will water both yards.

Look Back

The sum of 1 and five 4's should equal the sum of 1 and four 5's.
$1 + 4 + 4 + 4 + 4 + 4 = 21$ $\quad 1 + 5 + 5 + 5 + 5 = 21$ ✓

PRACTICE

1. Jess, Kathy, and Linda work on the math club's newspaper. One is the editor, one is the reporter, and one is the writer. Linda does not participate in sports. Jess and the editor play tennis together. Linda and the reporter are cousins. Find each person's job.

2. A toll booth accepts any combination of coins that total exactly \$0.75, but it does not accept pennies or half dollars. In how many different ways can a driver pay the toll?

Solve a Simpler Problem

Sometimes a problem may contain large numbers or require many steps to solve. It may appear complicated to solve. Try to **solve a simpler problem** that is similar to the original problem.

Problem Solving Strategies

Draw a Diagram	Make a Table
Make a Model	**Solve a Simpler Problem**
Guess and Test	Use Logical Reasoning
Work Backward	Use a Venn Diagram
Find a Pattern	Make an Organized List

Lawrence is making touch pools for a project about sea creatures. The pools are squares that will be arranged side by side. The side of each pool is a 1-meter-long piece of wood. How many meters of wood does Lawrence need to complete 20 square sections of pool?

Understand the Problem

Identify the important information.

- Each square side is a 1-meter-long piece of wood.
- There are 20 square sections set side by side.

The answer will be the total meters of wood needed.

Make a Plan

You could sketch all 20 pools and then count the number of meters of wood. However, it would be easier to first **solve a simpler problem**. Start with 1 square pool, and then move on to 2 and then 3. Then look for a way to solve the problem for 20 pools.

Solve

The first pool requires 4 sides to complete. After that, only 3 sides are needed for each pool.

1 square:

2 squares:

3 squares:

Notice that 1 pool requires 4 meters of wood, and the 19 other pools require 3 meters of wood each. So $4 + (19 \times 3) = 61$. The pools require 61 meters of wood.

Number of Squares	Number of Meters
1	$4(1) = 4$
2	$4 + (1 \times 3) = 7$
3	$4 + (2 \times 3) = 10$
4	$4 + (3 \times 3) = 13$

Look Back

If the pattern is correct, Lawrence would need 16 meters of wood for 5 pools. Complete the next row of the table to check this answer.

PRACTICE

1. The numbers 11; 444; and 8,888 all contain repeated single digits. How many numbers between 10 and 1,000,000 contain repeated single digits?
2. How many diagonals are there in a dodecagon (a 12-sided polygon)?

Use Logical Reasoning

Sometimes a problem may provide clues and facts that you must use to find a solution. You can use **logical reasoning** to solve this kind of problem.

Problem Solving Strategies

Draw a Diagram	Make a Table
Make a Model	Solve a Simpler Problem
Guess and Test	**Use Logical Reasoning**
Work Backward	Use a Venn Diagram
Find a Pattern	Make an Organized List

Jennie, Rachel, and Mia play the oboe, the violin, and the drums. Mia does not like the drums, and she is the sister of the oboe player. Rachel has soccer practice with the person who plays the drums. Which instrument does each person play?

Understand the Problem

Identify the important information.

- There are three people, and each person plays a different instrument.

Make a Plan

Start with clues given in the problem, and **use logical reasoning** to determine which instrument each person plays.

Solve

Make a table. Make a column for each instrument and a row for each person. Work with the clues one at a time. Write "Yes" in a box if the clue reveals that a person plays an instrument. Write "No" in a box if the clue reveals that a person does not play an instrument.

a. Mia does not like the drums, so she does not play the drums.

b. Mia is the sister of the person who plays the oboe, so she does not play the oboe.

c. Rachel has soccer practice with the person who plays the drums, so she does not play the drums.

	Oboe	Violin	Drums
Jennie			
Rachel			No
Mia	No		No

Jennie must play the drums, and Mia must play the violin. So Rachel must play the oboe.

Look Back

Compare your answer to the clues in the problem. Make sure none of your conclusions conflict with the clues.

PRACTICE

1. Kent, Jason, and Newman have a dog, a fish, and a hamster, though not in that order. Kent's pet does not have fur. The owner of the hamster has class with Jason. Match the owners with their pets.

2. Seth, Vess, and Benica are in the sixth, seventh, and eighth grades, though not in that order. Seth is not in seventh grade. The sixth-grader has band with Benica and the same lunchtime as Seth. Match the students with their grades.

Use a Venn Diagram

You can use a **Venn diagram** to display relationships among sets in a problem. Use ovals, circles, or other shapes to represent individual sets.

Problem Solving Strategies

Draw a Diagram	Make a Table
Make a Model	Solve a Simpler Problem
Guess and Test	Use Logical Reasoning
Work Backward	**Use a Venn Diagram**
Find a Pattern	Make an Organized List

At Landry Middle School, 127 students take French, 145 take Spanish, and 31 take both. How many students take only French? How many students take only Spanish?

Understand the Problem

Identify the important information.

- There are 127 students who take French, 145 who take Spanish, and 31 who take both.

Make a Plan

Use a Venn diagram to show the sets of students who take French and Spanish.

Solve

Draw and label two overlapping circles. Write "31" in the area where the circles overlap. This represents the number of students who take French and Spanish.

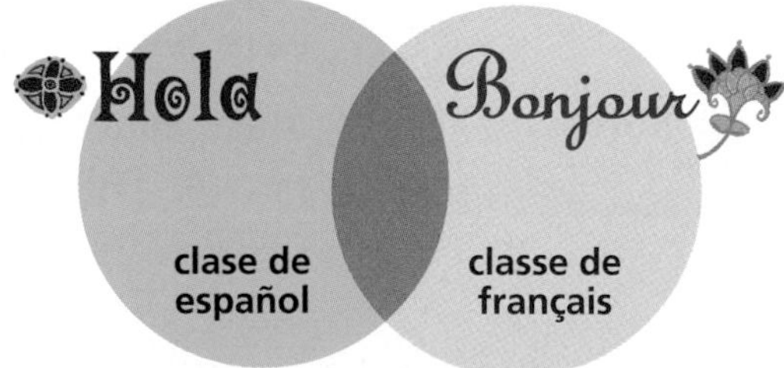

To find the number of students who take only French, subtract the number of students who take both French and Spanish from those who take French. To find the number of students who take only Spanish, subtract the number of students who take both French and Spanish from those who take Spanish.

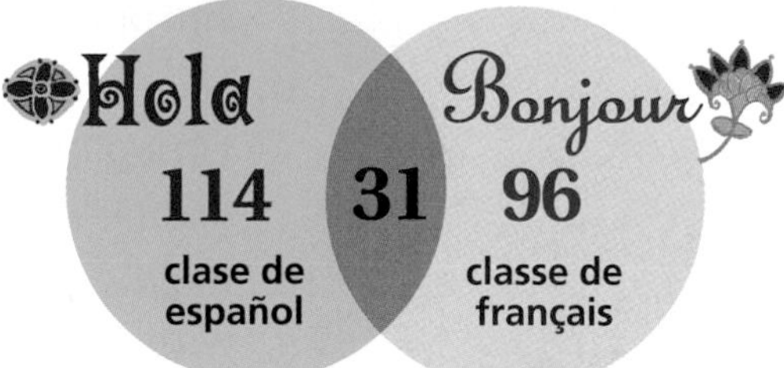

So 96 students take only French, and 114 students take only Spanish.

Look Back

Check your Venn diagram carefully against the information in the problem. Make sure your diagram agrees with the facts given.

PRACTICE

Responding to a survey, there were 60 people who said they like pasta, 45 who like chicken, and 70 who like hot dogs. There were 15 people who said they like both pasta and chicken, 22 who like both hot dogs and chicken, and 17 who like both hot dogs and pasta. Only 8 people said they like all 3.

1. How many people like only pasta?

2. How many people like only hot dogs?

Make an Organized List

In some problems, you will need to find out exactly how many different ways an event can happen. When solving this kind of problem, it is often helpful to **make an organized list**. This will help you count all the possible outcomes.

Problem Solving Strategies

Draw a Diagram	Make a Table
Make a Model	Solve a Simpler Problem
Guess and Test	Use Logical Reasoning
Work Backward	Use a Venn Diagram
Find a Pattern	**Make an Organized List**

A spinner has 4 different colors: red, blue, yellow, and white. If you spin the spinner 2 times, how many different color combinations could you get?

Understand the Problem

Identify the important information.

- You spin the spinner 2 times.
- The spinner is divided into 4 different colors.

The answer will be the total number of different color combinations the spinner can land on.

Make a Plan

Make an organized list to determine all the possible different color outcomes. List all the different combinations for each color.

Solve

First consider the color red. List all the different outcomes for the color red. Then consider blue, adding all the different outcomes, then yellow, and finally white.

Red	Blue	Yellow	White
RR	BB	YY	WW
RB	BY	YW	
RY	BW		
RW			

So there are 10 possible different color combinations.

Look Back

Make sure that all the possible combinations of color are listed and that each set of colors is different.

PRACTICE

1. The Pizza Planet has 5 different choices of pizza toppings: ham, pineapple, pepperoni, olive, and mushroom. You want to order a pizza with 2 different toppings. How many different combinations of toppings can you order?
2. How many ways can you make change for a fifty-cent piece by using a combination of dimes, nickels, and pennies?

Chapter 1

Data Toolbox

TABLE OF CONTENTS

Bird	Average Number of Sightings					
	Nov	Dec	Jan	Feb	Mar	Apr
Mourning dove	7.0	8.0	9.0	8.0	6.0	5.0
Red-bellied woodpecker	1.25	1.3	1.3	1.3	1.3	1.5
Carolina chickadee	2.8	2.8	2.8	2.7	2.5	2.4

Career Field Biologist

Field biologists spend time outdoors studying populations of fish, birds, and other living things. The information they collect is often used to determine whether populations are growing or declining.

Sometimes, amateur naturalists help scientists collect information. For example, people with bird feeders can report to Project FeederWatch the number and kinds of birds that they see at their feeders throughout the winter.

internet connect

Chapter Opener Online
go.hrw.com
KEYWORD: MS4 Ch1

ARE YOU READY?

Choose the best term from the list to complete each sentence.

1. A part of a line consisting of two endpoints and all points between those endpoints is called a(n) __?__.
2. A(n) __?__ is the space between the marked values on the __?__ of a graph.
3. The number of times an item occurs is called its __?__.

circle
frequency
interval
line segment
scale

Complete these exercises to review skills you will need for this chapter.

✔ Compare and Order Whole Numbers

Order the numbers from least to greatest.

4. 45, 23, 65, 15, 42, 18
5. 103, 105, 102, 118, 87, 104
6. 56, 65, 24, 19, 76, 33, 82
7. 8, 3, 6, 2, 5, 9, 3, 4, 2

✔ Add Whole Numbers

Find each sum.

8. 6 + 8 + 10 + 9 + 7
9. 10 + 23 + 19 + 17
10. 16 + 32 + 18 + 20 + 15 + 12
11. 102 + 98 + 64 + 32 + 80

✔ Subtract Whole Numbers

Find each difference.

12. 133 − 35
13. 54 − 29
14. 200 − 88
15. 1,055 − 899

✔ Locate Points on a Number Line

Copy the number line. Then graph each number.

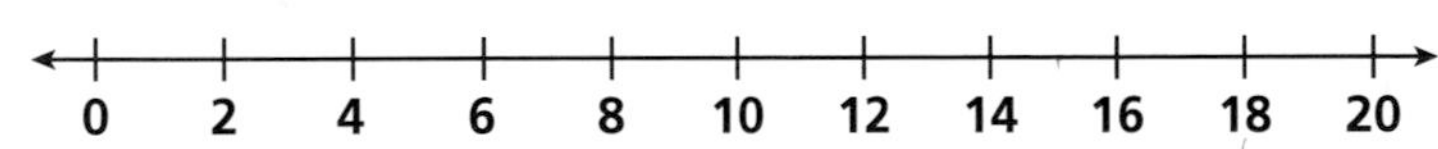

16. 15
17. 2
18. 18
19. 7

✔ Read a Table

Use the data in the table for Exercises 20 and 21.

20. Which animal is the fastest?
21. Which animal is faster, a rabbit or a zebra?

Top Speeds of Some Animals	
Animal	**Speed (mi/h)**
Elephant	25
Lion	50
Rabbit	35
Zebra	40

1-1 Populations and Samples

Learn to identify populations and random samples.

Vocabulary

population

sample

random sample

When you are gathering information about a group, the whole group is called the **population**. For your information to be absolutely accurate, you must collect data on every member of the population.

However, it is often not practical—or possible—to survey every member of a large group. For this reason, researchers often study a part of the group, called a **sample**.

When scientists track migratory animals, they cannot track each animal. Instead, they tag a sample of the entire population and track the sample. They use the sample to gather statistical data about the population.

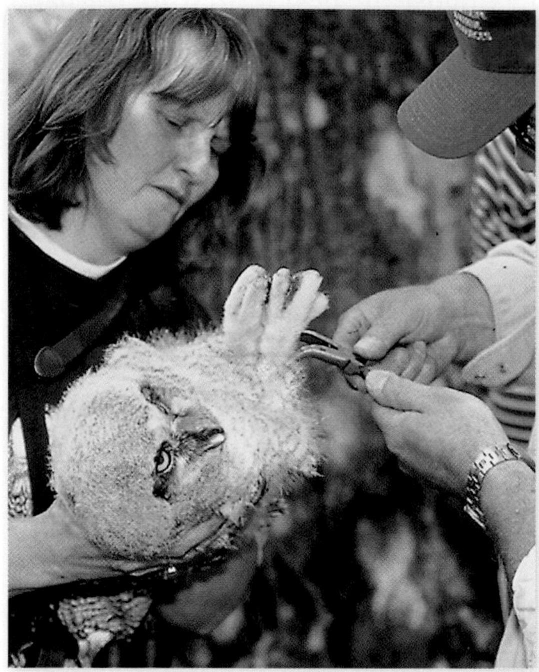

Biologists place a leg band on a young great horned owl.

EXAMPLE 1 Identifying Populations and Samples

Identify the population and sample in each situation.

A **A scientist studies lions on a wildlife preserve to learn about the parenting habits of lions.**

Population	Sample
All lions	The lions on the preserve

B **The school librarian surveys 100 students about the types of books they prefer.**

Population	Sample
All students in the school	The 100 students who are surveyed

Identify the population and sample in each situation.

C **A restaurant manager uses comment cards to find out about customer satisfaction.**

Population	Sample
All customers	The customers who fill out the comment card

For a sample to be useful, it must be representative of the population. If it is not, then the information gathered from the sample will be inaccurate. One type of representative sample, known as a **random sample**, gives *every* member of the population an equal chance of being chosen.

EXAMPLE 2 Identifying Random Samples

A newspaper reporter is gathering responses from Riverside Middle School band students about the style of their uniforms. Tell whether each sampling method is random. Explain your answer.

A **The reporter questions only the students he knows personally.**
This sampling method is not random. The students the reporter does not know have no chance of being chosen.

B **The reporter questions every tenth student on an alphabetized list of the students, starting with the first student on the list.**
While this method is representative of the population, it is not random. The students listed as 2 through 9, 11 through 19, etc., have no chance of being chosen.

C **The reporter writes each student's name on a card and puts all of the cards in a hat. He then questions the students whose names he draws.**
This sampling method is random. Each student has an equal chance of being chosen.

Think and Discuss

1. **Give an example** of a situation in which you would want to use a sample rather than poll the entire population.
2. **Explain** why it might be difficult to obtain a purely random sample of a large population.

1-1 Exercises

FOR EOG PRACTICE

see page 654

GUIDED PRACTICE

See Example 1 **Identify the population and sample in each situation.**

1. A scientist studies a pod of humpback whales to find out about migration patterns of humpback whales.
2. The decoration committee asks 25 students about their ideas for a theme for the seventh-grade party.

See Example 2 **The owner of a used car lot is conducting a customer survey. Tell whether each sampling method is random. Explain your answer.**

3. The owner surveys customers on the lot one Saturday morning.
4. The owner questions 100 customers who were randomly selected from a list of all customers.

INDEPENDENT PRACTICE

See Example 1 **Identify the population and sample in each situation.**

5. Scientists tag the ears of 50 moose to get information on the average life span of a moose in the wild.
6. A disc jockey asks the first ten listeners who call in if they like the last song that was played.
7. Researchers poll every fifteenth voter after a school board election.

See Example 2 **A librarian is gathering responses from community members. Tell whether each sampling method is random. Explain your answer.**

8. The librarian randomly selects members of the community by drawing their names from a hat.
9. The librarian sends out questionnaires to families with children.

PRACTICE AND PROBLEM SOLVING

Tell whether you would survey the population or use a sample. Explain.

10. You want to know the favorite painter of all the employees at the local art museum.
11. You want to know the types of calculators used by middle school students across the country.
12. You want to know the favorite magazine of all teens in your state.
13. You want to know where the students in your social studies class study for tests.

Give a reason why each sampling method may not be random.

14. A reporter calls 100 people from the phone book.

15. A reporter questions all of the customers at one of several entrances to a local supermarket.

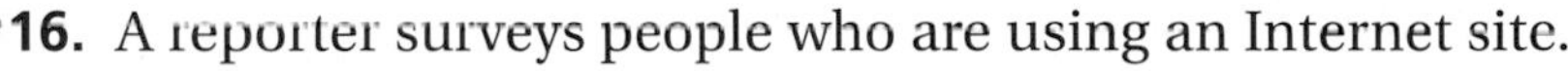

16. A reporter surveys people who are using an Internet site.

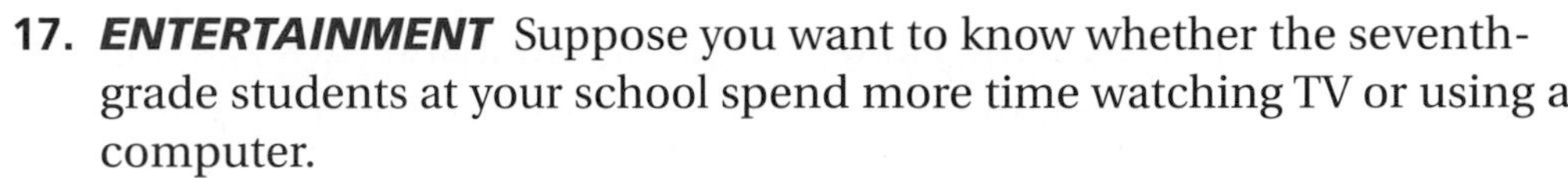

17. ***ENTERTAINMENT*** Suppose you want to know whether the seventh-grade students at your school spend more time watching TV or using a computer.
   a. What is the population for your survey?
   b. Describe how you might choose a random sample from the population.

18. ***SCHOOL*** The Mathematics Department at Truman Middle School is conducting a survey to determine in which class students use calculators most often, math or science. What is the population for this survey? Explain your answer.

19. ***BUSINESS*** A manufacturer selects every hundredth item to be tested. Tell whether this sampling method is random. Explain your answer.

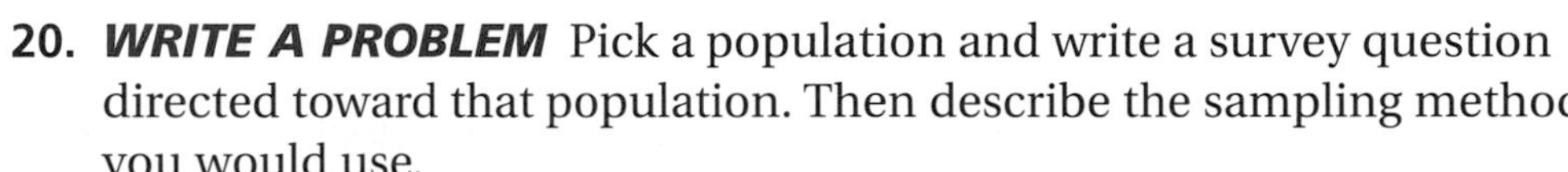

20. ***WRITE A PROBLEM*** Pick a population and write a survey question directed toward that population. Then describe the sampling method you would use.

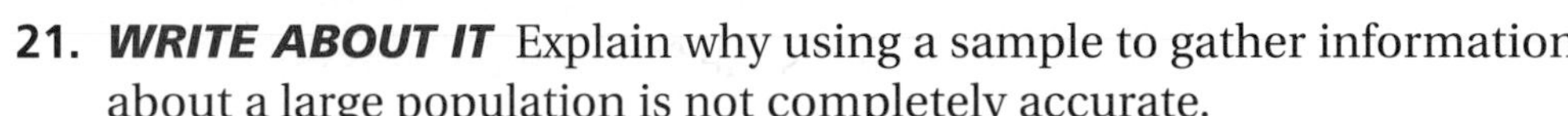

21. ***WRITE ABOUT IT*** Explain why using a sample to gather information about a large population is not completely accurate.

22. ***CHALLENGE*** Biased, or unfair, questions can affect the responses given in a survey. Explain why the following question is biased:
   "Do you prefer listening to noisy, ear-splitting rock music or quiet, relaxing classical music?"

Entertainment LINK

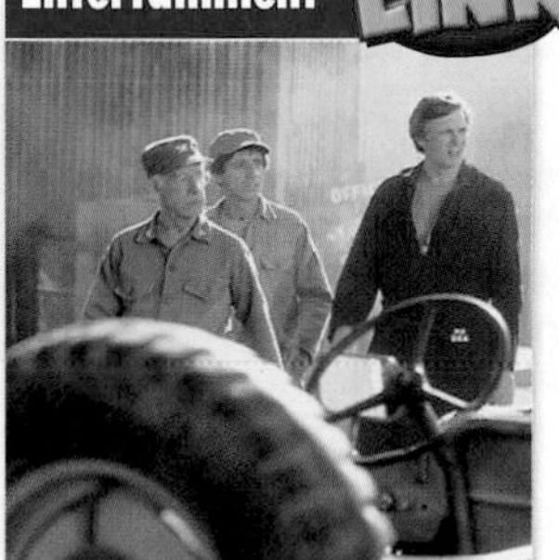

The television series *M*A*S*H* ran for 11 years and had 251 episodes.

go.hrw.com
KEYWORD: MS4 TV

Spiral Review

Simplify. (Previous course)

23. 15 + 27 + 5 + 3 + 11 + 16 + 7 + 4

24. 2 + 6 + 5 + 7 + 100 + 1 + 75

25. 2 + 9 + 8 + 12 + 6 + 8 + 5 + 6 + 7

26. 9 + 30 + 4 + 1 + 4 + 1 + 7 + 5 + 18 + 11

Divide. (Previous course)

27. 88 ÷ 8

28. 196 ÷ 7

29. 63 ÷ 9

30. 90 ÷ 10

31. **EOG PREP** A baseball stadium has 21,896 seats. To make room for a memorial to past great players, the club wants to remove 530 seats. How many seats will the stadium have after the change? (Previous course)

A 22,426 B 21,366 C 17,126 D 21,843

Technology

LAB 1A

Use with Lesson 1-1

Generate Random Numbers

You can use a spreadsheet to generate random numbers. By generating random numbers you can simulate real-world events, such as rolling a number cube.

Activity

1 Use a spreadsheet to generate five random decimal numbers between 0 and 1.

Type **=RAND()** into cell A1 and press **ENTER.** A random number will appear in cell A1.

	A	B	C
1	0.153794		
2			
3			

Click to highlight cell A1. Go to the **EDIT** menu and **COPY** the contents of A1. Then click cell A2 and drag your cursor down to cell A5. Cells A2 through A5 should be highlighted.

	A	B	C
1	0.808811		
2			
3			
4			
5			

Finally, go to the **EDIT** menu and use **PASTE** to fill cells A2 through A5 with the formula in cell A1. Notice that the random number in cell A1 changed when you filled the other cells.

	A	B	C
1	0.224473		
2	0.001325		
3	0.715995		
4	0.506313		
5	0.354271		

2 Use a spreadsheet to generate five random whole numbers between 1 and 10.

A2 = =10*RAND()

	A	B	C	D
1	1.97758			
2	2.624973			
3	5.451635			
4	6.931406			
5	5.059039			

RAND() generates numbers greater than or equal to 0 but less than 1. To generate random numbers between 0 and 10, you can multiply by 10.

Enter **=10*RAND()** into cell A1. Copy and paste the contents of A1 into cells A2 through A5.

Now you have decimal numbers greater than or equal to 0 but less than 10. However, you want whole numbers from 1 to 10.

Click on cell A1. Change the formula to **=INT(10*RAND())** and press **ENTER.** Copy and paste this formula into cells A2 through A5.

A2 =INT(10*RAND())

	A	B	C	D
1	8			
2	9			
3	4			
4	8			
5	0			

Now you have random whole numbers between 0 and 9, not 1 and 10.

To shift the range of the numbers up 1, change the formula in A1 to **=INT(10*RAND()) + 1** and press **ENTER.** Copy and paste this formula into cells A2 through A5.

A5 =INT(10*RAND())+1

	A	B	C	D
1	6			
2	9			
3	10			
4	1			
5	6			

The formula **=INT(10*RAND()) + 1** generates random whole numbers from 1 to 10.

3 Use a spreadsheet to simulate three rolls of a number cube.

A1 =INT(6*RAND())+1

	A	B	C	D
1	6			
2	2			
3	4			

A number cube has whole numbers between 1 and 6. Type **=INT(6*RAND()) + 1** into cell A1 and press **ENTER.** Copy and paste this formula into cells A2 and A3.

In this simulation, the number cube came up 6, 2, and 4.

Think and Discuss

1. Which part of the formula **=INT(10*RAND())** changes the decimal numbers to whole numbers?
2. Explain the effect of the number 10 in the formula **=INT(10*RAND()).**
3. How could you use a spreadsheet to choose the winner of a door prize fairly?

Try This

1. Generate five random decimal numbers between 1 and 10.
2. Generate ten random whole numbers between 0 and 19.
3. Generate ten random decimal numbers greater than or equal to 0 but less than 25. What spreadsheet formula did you use?
4. The makers of Berries and Bran cereal randomly insert collectible miniposters in their cereal boxes. The posters are numbered 1 through 8.
   a. Write a spreadsheet formula to generate random whole numbers from 1 to 8.
   b. Use your formula to simulate buying five boxes of cereal. List the numbers of the posters you simulated getting. Were they all different?

1-2 Mean, Median, Mode, and Range

Learn to find the mean, median, mode, and range of a data set.

Vocabulary

mean
median
mode
range
outlier

Navajo Code Talkers used the Navajo language as the basis of a code in World War II.

The most common letter in English is *e*. You can use this information to crack secret messages in code. Begin by listing the number of times each symbol of a code appears in a message. The symbol that appears the most often represents the *mode*, which likely corresponds to *e*.

The mode, along with the *mean* and the *median*, is a measure of *central tendency* used to represent the "middle" of a data set.

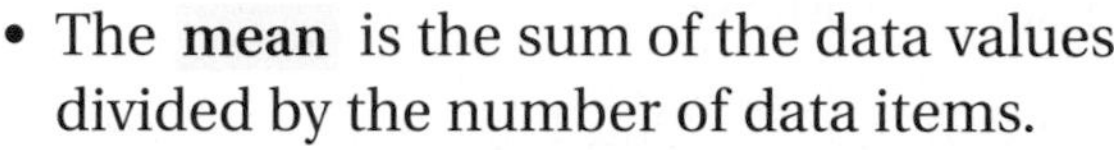

- The **mean** is the sum of the data values divided by the number of data items.
- The **median** is the middle value of an odd number of data items arranged in order. For an even number of data items, the median is the average of the two middle values.
- The **mode** is the value or values that occur most often. When all the data values occur the same number of times, there is no mode.

The **range** of a set of data is the difference between the greatest and least values. It is used to show the spread of the data in a data set.

EXAMPLE 1 Finding the Mean, Median, Mode, and Range of Data

Find the mean, median, mode, and range of the data set.

2, 1, 8, 0, 2, 4, 3, 4

mean:

$2 + 1 + 8 + 0 + 2 + 4 + 3 + 4 = 24$ *Add the values.*

$24 \div 8 = 3$ *Divide the sum by the number of items.*

The mean is 3.

median:

0, 1, 2, 2, 3, 4, 4, 8 *Arrange the values in order.*

$2 + 3 = 5$ $5 \div 2 = 2.5$ *Since there are two middle values, find the average of these two values.*

The median is 2.5.

mode:

The modes are 2 and 4. *The values 2 and 4 occur twice.*

range:

$8 - 0 = 8$ *Subtract the least value from the greatest value.*

The range is 8.

In the data set below, the value 12 is much less than the other values in the set. An extreme value such as this is called an **outlier**.

35, 38, 27, 12, 30, 41, 31, 35

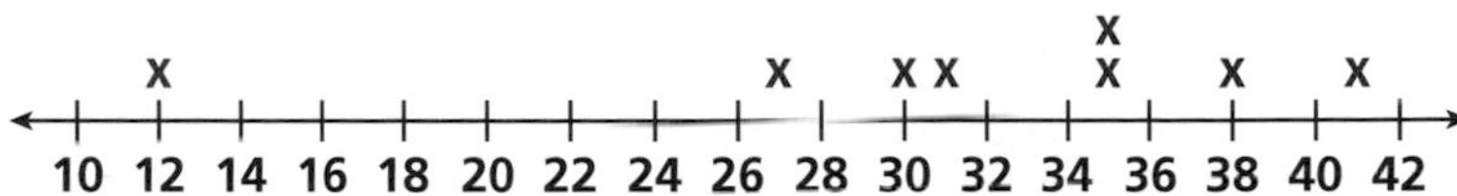

EXAMPLE 2 Exploring the Effects of Outliers on Measures of Central Tendency

The table shows the number of art pieces created by students in a glass-blowing workshop. Identify the outlier in the data set. Then determine how the outlier affects the mean, median, and mode of the data.

Name	Number of Pieces
Suzanne	5
Glen	7
Charissa	4
Eileen	6
Hermann	14
Tom	3

The outlier is 14. To determine its effect on the mean, median, and mode, calculate each value with and without the outlier.

Without the Outlier

mean:

$5 + 7 + 4 + 6 + 3 = 25$

$25 \div 5 = 5$

The mean is 5.

median:

3, 4, **5**, 6, 7

The median is 5.

mode:

There is no mode.

With the Outlier

mean:

$5 + 7 + 4 + 6 + 14 + 3 = 39$

$39 \div 6 = 6.5$

The mean is 6.5.

median:

3, 4, **5**, **6**, 7, 14

$5 + 6 = 11$ $\quad$ $11 \div 2 = 5.5$

The median is 5.5.

mode:

There is no mode.

Adding the outlier increased the mean by 1.5 and the median by 0.5. The mode did not change.

Think and Discuss

1. **Given** the mean, median, and mode for a set of data, which measure of central tendency must be a data value?
2. **Give an example** of a data set with an outlier. Describe how the outlier affects the mean, median, and mode of the data.

1-2 Exercises

FOR EOG PRACTICE

see page 654

internet connect
Homework Help Online
go.hrw.com Keyword: MS4 1-2

4.01, 4.02, 4.03, 4.04

GUIDED PRACTICE

See Example 1 **Find the mean, median, mode, and range of each data set.**

1. 5, 30, 35, 20, 5, 25, 20

2. 44, 68, 48, 61, 59, 48, 63, 49

See Example 2 **3.** The table shows the number of glasses of water consumed by several students in one day. Identify the outlier in the data set. Then determine how the outlier affects the mean, median, and mode of the data.

Water Consumption								
Name	Randy	Lori	Anita	Jana	Sonya	Victor	Mark	Jorge
Glasses	4	12	3	1	4	7	5	4

INDEPENDENT PRACTICE

See Example 1 **Find the mean, median, mode, and range of each data set.**

4.

Daily Low Temperatures							
Day	Sun	Mon	Tue	Wed	Thu	Fri	Sat
Temperature (°F)	68	65	68	61	68	67	65

5.

1995 NFL Touchdown Leaders							
Name	Faulk	Martin	Rhett	Sanders	Smith	Warren	Watters
TD's	11	14	11	11	25	15	11

See Example 2 **Identify the outlier in each data set. Then determine how the outlier affects the mean, median, and mode of the data.**

6. 13, 18, 20, 5, 15, 20, 13, 20

7. 45, 48, 63, 85, 151, 47, 88, 44, 68

PRACTICE AND PROBLEM SOLVING

8. The table shows the amount of snowfall in Colorado Springs during a ten-year period.

a. Find the mean, median, mode, and range of the data.

b. Would the mode be a good number to describe the amount of snowfall for these ten years? Explain.

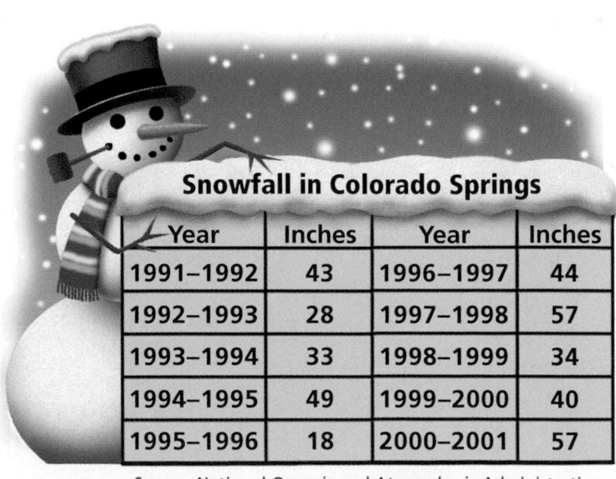

Snowfall in Colorado Springs			
Year	**Inches**	**Year**	**Inches**
1991–1992	43	1996–1997	44
1992–1993	28	1997–1998	57
1993–1994	33	1998–1999	34
1994–1995	49	1999–2000	40
1995–1996	18	2000–2001	57

Source: National Oceanic and Atmospheric Administration

9. ***ECONOMICS*** The table shows the monthly salaries for different grades of U.S. government employees. Find the range, mean, and median of the salaries.

Grade	Salary
GS-1	$1,461
GS-2	$1,591
GS-3	$1,793
GS-4	$2,013
GS-5	$2,252

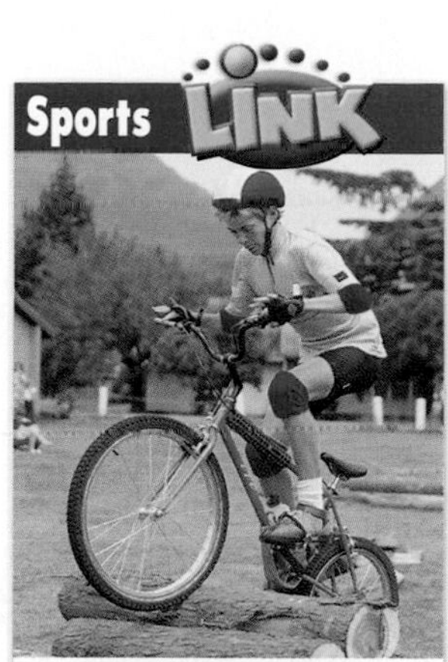

Mountain bikes account for over 50% of bicycle sales in the United States.

10. ***SPORTS*** The ages of the participants in a mountain bike race are 14, 23, 20, 24, 26, 17, 21, 31, 27, 25, 14, and 28.
    a. Find the mean, median, and mode of the ages.
    b. Which measure of central tendency best represents the ages of the participants? Explain.

11. Find the mean, median, and mode of the data displayed in the line plot. Then determine how the outlier affects the mean.

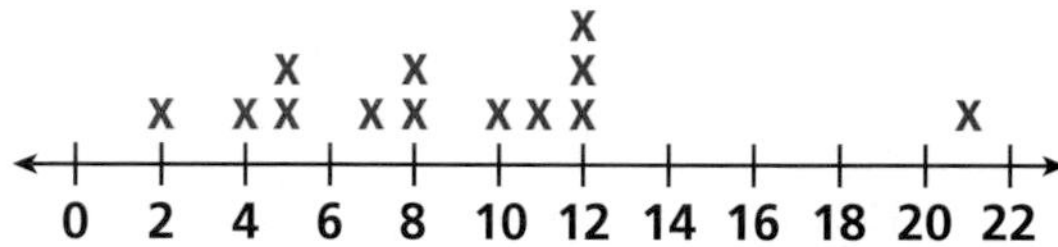

12. ***HEALTH*** Based on the data from three annual checkups, Jon's mean height is 62 in. At the first two checkups Jon's height was 58 in. and 61 in. What was his height at the third checkup?

13. ***WHAT'S THE QUESTION?*** The values in a data set are 10, 7, 9, 5, 13, 10, 7, 14, 8, and 11. What is the question about central tendency that gives the answer 9.5 for the data set?

14. ***WRITE ABOUT IT*** Which measure of central tendency is most often affected by including an outlier? Explain.

15. ***CHALLENGE*** Pick a measure of central tendency that describes each situation. Explain your choice.
    a. the number of siblings in a family
    b. the favorite color of students in your math class
    c. the number of days in a month

Spiral Review

Round each number to the nearest ten. (Previous course)

16. 52 **17.** 104 **18.** 75 **19.** 817

Round each number to the nearest hundred. (Previous course)

20. 234 **21.** 961 **22.** 1,720 **23.** 6,055

24. **EOG PREP** A VCR is on sale for $79 at Ezra's Electronics. The same VCR costs $113 at Smart Shoppe. Which is the *best* estimate of the difference in the two prices? (Previous course)

A $20 B $30 C $50 D $70

1-3 Frequency Tables and Stem-and-Leaf Plots

Learn to organize and interpret data in frequency tables and stem-and-leaf plots.

Vocabulary

frequency table

cumulative frequency

stem-and-leaf plot

IMAX® theaters, with their huge screens and powerful sound systems, make viewers feel as if they are in the middle of the action. In 2001, the top IMAX film of all time was *The Dream Is Alive,* with total box office receipts of over $149 million.

To see how common it is for an IMAX movie to attract such a large number of viewers, you could use a *frequency table.* A **frequency table** is a way to organize data into categories or groups. By including a **cumulative frequency** column in your table, you can keep a running total of the frequencies in each category.

EXAMPLE 1 Organizing Data in a Frequency Table

The list shows box office receipts in millions of dollars for 20 IMAX films. Make a cumulative frequency table of the data.

76, 51, 41, 38, 18, 17, 16, 15, 13, 13, 12, 12, 10, 10, 6, 5, 5, 4, 4, 2

Step 1: Look at the range to choose the scale and equal intervals for the data.

Step 2: Find the number of data values in each interval. Write these numbers in the "Frequency" column.

Step 3: Find the cumulative frequency for each row by adding all the frequency values that are above or in that row.

IMAX Films		
Receipts ($ million)	Frequency	Cumulative Frequency
0–19	16	16
20–39	1	17
40–59	2	19
60–79	1	20

A **stem-and-leaf plot** can be used to show how often data values occur and how they are distributed. Each *leaf* on the plot represents the right-hand digit in a data value, and each *stem* represents the remaining left-hand digits. The key shows the values of the data on the plot.

Stems	Leaves
2	4 7 9
3	0 6

*Key: **2** | **7** means 27*

EXAMPLE 2 **Organizing Data in a Stem-and-Leaf Plot**

The data shows the number of minutes students spent doing their homework. Make a stem-and-leaf plot of the data.

38, 48, 45, 32, 29, 32, 45, 36, 22, 21, 35, 45, 47, 26, 43, 48, 64

Step 1: Find the least data value and the greatest data value. Since the data values range from 21 to 64, use tens digits for the stems and ones digits for the leaves.

Step 2: List the stems from least to greatest on the plot.

Step 3: List the leaves for each stem from least to greatest.

Step 4: Add a key and a title.

The stems are the tens digits.

There are no data values in the 50's.

Minutes Doing Homework

Stems	Leaves
2	1 2 6 9
3	2 2 5 6 8
4	3 5 5 5 7 8 8
5	
6	4

Key: 3 | 2 means 32

The leaves are the ones digits.

The entries in the second row represent the data values 32, 32, 35, 36, and 38.

Think and Discuss

1. **Give an example** of when you might use a frequency table to organize data.
2. **Describe** how you would show the numbers 4 and 89 on a stem-and-leaf plot.
3. **Tell** which you would use to determine the number of data values in a set, a cumulative frequency table or a stem-and-leaf plot. Explain.

1-3 Exercises

FOR EOG PRACTICE
see page 655

internet connect
Homework Help Online
go.hrw.com Keyword: MS4 1-3

4.01

GUIDED PRACTICE

The table shows the ages of the first 18 American presidents when they took office.

President	Age	President	Age	President	Age
Washington	57	Jackson	61	Fillmore	50
Adams	61	Van Buren	54	Pierce	48
Jefferson	57	Harrison	68	Buchanan	65
Madison	57	Tyler	51	Lincoln	52
Monroe	58	Polk	49	Johnson	56
Adams	57	Taylor	64	Grant	46

See Example 1 1. Make a cumulative frequency table of the data.

See Example 2 2. Make a stem-and-leaf plot of the data.

INDEPENDENT PRACTICE

The table shows the states with the most drive-in movie screens in 2000.

See Example 1

3. Make a cumulative frequency table of the data.

See Example 2

4. Make a stem-and-leaf plot of the data.

State	Drive-ins	State	Drive-ins
Arizona	20	Missouri	21
California	59	New York	48
Florida	29	Ohio	52
Indiana	34	Pennsylvania	55
Kentucky	20	Washington	21
Michigan	24		

Source: USA Today

PRACTICE AND PROBLEM SOLVING

Use the stem-and-leaf plot for Exercises 5–8.

Stems	Leaves
0	4 6 6 9
1	2 5 8 8 8
2	0 3
3	1

Key: 1|2 means 12

5. What is the least number? What is the greatest number? What is the range?
6. Find the mean, median, and mode of the data set.
7. Which of the following is most likely the source of the data in the stem-and-leaf plot?
   a. shoe sizes of 12 middle-school students
   b. number of hours 12 adults exercised in one month
   c. number of boxes of cereal per household at one time
   d. monthly temperatures in degrees Fahrenheit in Chicago, Illinois
8. Make a cumulative frequency table of the data.

Life Science LINK

The map shows the number of endangered species in each country in South America. A species is classified as endangered when it faces a very high risk of extinction in the wild in the near future.

9. Which country has the fewest endangered species? Which has the most?

10. Make a cumulative frequency table of the data. How many countries have fewer than 20 endangered species?

11. Make a stem-and-leaf plot of the data. What is the mode of the data?

12. Find the mean and the median of the data set. Round to the nearest tenth, if necessary.

13. Identify the outlier in the data set. Then determine how the outlier affects the mean of the data.

14. ***WRITE ABOUT IT*** Explain how changing the size of the intervals you used in Exercise 10 affects your cumulative frequency table.

15. ***CHALLENGE*** In a recent year, the number of endangered species in the United States was 150. Show how to represent this number on a stem-and-leaf plot.

Source: United Nations Environment Programme

go.hrw.com
KEYWORD: MS4 Endangered
CNN Student News.

Spiral Review

Tell whether each number is divisible by 6. (Previous course)

16. 24
17. 91
18. 100
19. 426

20. **EOG PREP** Erin started watching a movie at 7:15 P.M. If the movie lasted 1 hour 35 minutes, what time did the movie end? (Previous course)

A 6:40 P.M. C 8:50 P.M.
B 8:45 P.M. D 9:00 P.M.

21. **EOG PREP** Fatima works 17 hr one week and 23 hr the next. If Fatima earns $6 an hr, how much does she earn during the two weeks? (Previous course)

A $50 C $138
B $102 D $240

Chapter 1

Mid-Chapter Quiz

LESSON **1-1** (pp. 4–7)

Identify the population and sample in each situation.

1. A scientist studies a group of green iguanas in Costa Rica to learn about the feeding behavior of green iguanas.
2. A local grocer asks 15 shoppers to name the vegetable they buy most often.

A pollster is gathering responses from students who use the track after school. Tell whether each sampling method is random. Explain your answer.

3. The pollster questions the first five students to run a mile on the track after school.
4. The pollster places the names of the students who use the track in a bag and questions the five students whose names he draws.

LESSON **1-2** (pp. 10–13)

Find the mean, median, mode, and range of each data set.

5. 64, 38, 29, 47, 52
6. 1, 3, 8, 7, 3, 5, 14, 7, 5, 6
7. 56,328; 29,133; 29,133; 48,716
8. 84, 84, 84, 86, 90, 84, 83

Use the table for problems 9–13.

9. Identify the outlier in the data set.
10. Find the mean of the data set with and without the outlier.
11. Find the median of the data set with and without the outlier.
12. Find the mode of the data set with and without the outlier.
13. Which measure of central tendency is most affected by the outlier? Explain.

Student	Hourly Wage
Michelle	\$5
Mary	\$3
Sean	\$3
Devin	\$6
Elliott	\$17
Anelise	\$8

LESSON **1-3** (pp. 14–17)

Use the stem-and-leaf plot for problems 14 and 15.

14. Find the mean, median, mode, and range of the data set. Round to the nearest tenth, if necessary.
15. Make a cumulative frequency table of the data.
16. Make a cumulative frequency table of the data 8, 10, 6, 7, 7, 8, 8, 8, 10, 6, 10, 7, 7, 8, 7.
17. Make a stem-and-leaf plot of the data 6, 8, 12, 23, 31, 21, 41, 7, 18.

Stems	Leaves
1	1
2	0 0 1 6
3	2 7
4	0 8

Key: 3|2 means 32

1.02a

Focus on Problem Solving

Solve

- **Choose an operation: addition or subtraction**

In order to decide whether to add or subtract to solve a problem, you need to determine what action is taking place in the problem. If you are combining or putting together numbers, you need to add. If you are taking away or finding how far apart two numbers are, you need to subtract.

Determine the action in each problem. Then determine which operation could be used to solve the problem. Use the table for problems 5 and 6.

1. Betty, Raymond, and Helen ran a three-person relay race. Their individual times were 48 seconds, 55 seconds, and 51 seconds. What was their total time?

2. The Scots pine and the sessile oak are trees native to Northern Ireland. The height of a mature Scots pine is 111 feet, and the height of a mature sessile oak is 90 feet. How much taller is the Scots pine than the sessile oak?

3. Mr. Hutchins has $35.00 to buy supplies for his social studies class. He wants to buy items that cost $19.75, $8.49, and $7.10. Does Mr. Hutchins have enough money to buy all of the supplies?

4. The running time for the 1998 movie *Antz* is 83 minutes. If Jordan has watched 25 minutes of the movie, how many minutes does he have left to watch?

Sizes of Marine Mammals

Mammal	Weight (kg)
Killer whale	3,600
Manatee	400
Sea lion	200
Walrus	750

5. The table gives the approximate weights of four marine mammals. How much more does the killer whale weigh than the sea lion?

6. Find the total weight of the manatee, the sea lion, and the walrus. Do these three mammals together weigh more or less than the killer whale?

1-4 Bar Graphs and Histograms

Learn to display and analyze data in bar graphs and histograms.

Vocabulary
- bar graph
- double-bar graph
- histogram

Hundreds of different languages are spoken around the world. The graph shows the numbers of native speakers of four languages.

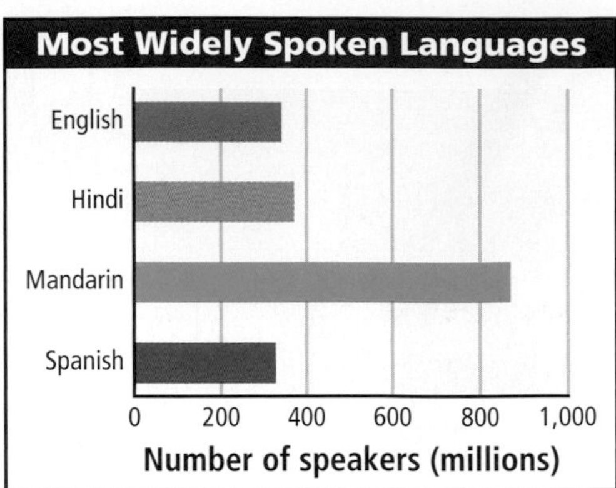

A **bar graph** can be used to display and compare data. The scale of a bar graph should include all the data values and be easily divided into equal intervals.

EXAMPLE 1 Interpreting a Bar Graph

Use the bar graph to answer each question.

A Which language has the most native speakers?

The bar for Mandarin is the longest, so Mandarin has the most native speakers.

B About how many more people speak Mandarin than speak Hindi?

About 500 million more people speak Mandarin than speak Hindi.

You can use a **double-bar graph** to compare two related sets of data.

EXAMPLE 2 Making a Double-Bar Graph

The table shows the life expectancies of people in three Central American countries. Make a double-bar graph of the data.

Country	Male	Female
El Salvador	67	74
Honduras	63	66
Nicaragua	65	70

Step 1: Choose a scale and interval for the vertical axis.

Step 2: Draw a pair of bars for each country's data. Use different colors to show males and females.

Step 3: Label the axes and give the graph a title.

Step 4: Make a key to show what each bar represents.

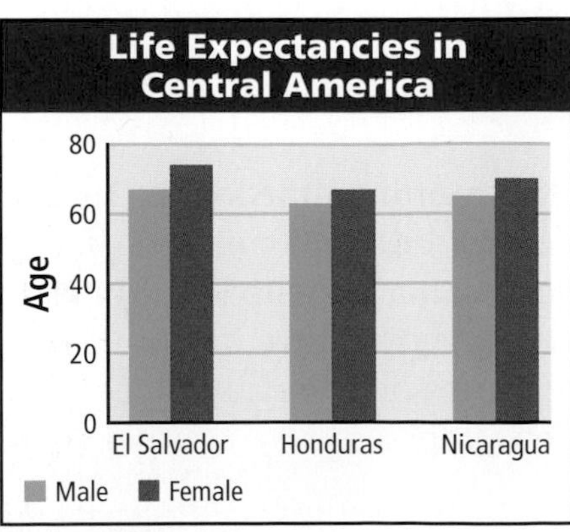

A **histogram** is a bar graph that shows the frequency of data within equal intervals. There is no space between the bars in a histogram.

EXAMPLE 3 Making a Histogram

The table below shows survey results about the number of CDs students own. Make a histogram of the data.

Number of CDs									
1	///	5	卌 /	9	卌 /	13	卌 ////	17	卌 ////
2	//	6	///	10	卌 卌	14	卌 卌 /	18	卌 //
3	卌	7	卌 ///	11	卌 卌 /	15	卌 卌 /	19	//
4	卌 /	8	卌 //	12	卌 卌	16	卌 卌 /	20	卌 /

Step 1: Make a frequency table of the data. Be sure to use equal intervals.

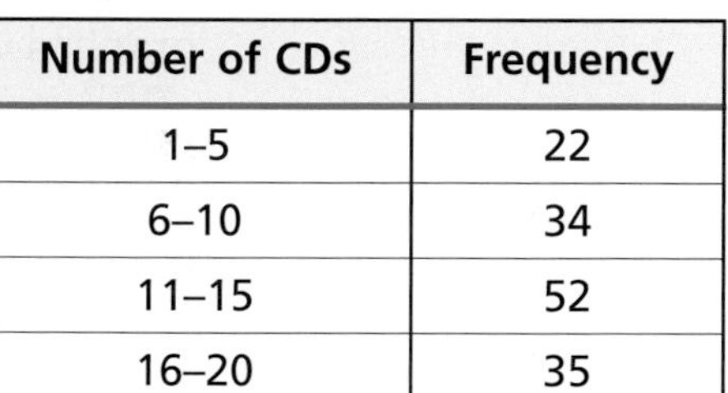

Number of CDs	Frequency
1–5	22
6–10	34
11–15	52
16–20	35

Step 2: Choose an appropriate scale and interval for the vertical axis. The greatest value on the scale should be at least as great as the greatest frequency.

Step 3: Draw a bar for each interval. The height of the bar is the frequency for that interval. Bars must touch but not overlap.

Step 4: Label the axes and give the graph a title.

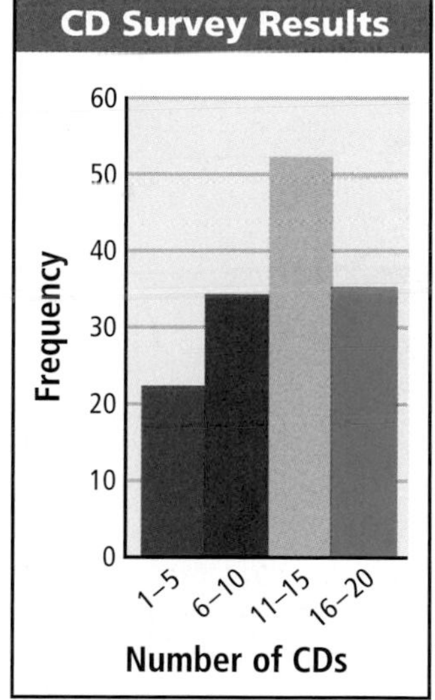

Think and Discuss

1. **Explain** why you might use a horizontal bar graph instead of a vertical bar graph to display data.
2. **Explain** why you might use a double-bar graph instead of two separate bar graphs to display data.
3. **Describe** the similarities and differences between a bar graph and a histogram.

1-4 Exercises

FOR EOG PRACTICE see page 656

internet connect — Homework Help Online go.hrw.com Keyword: MS4 1-4

4.01, 4.03, 4.05

GUIDED PRACTICE

See Example 1

The bar graph shows the average amount of fresh fruit consumed per person in the United States in 1997. Use the graph for Exercises 1 and 2.

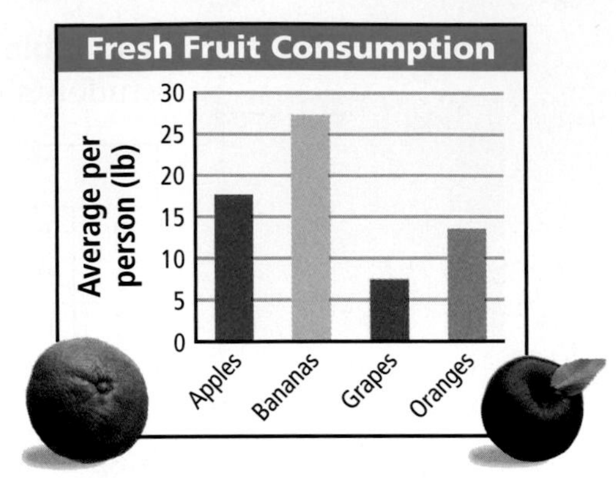

1. Which fruit was eaten the least?
2. About how many more pounds of bananas than pounds of oranges were eaten per person?

See Example 2

3. The table shows national average SAT scores for three years. Make a double-bar graph of the data.

Year	Verbal	Math
1980	502	492
1990	500	501
2000	505	514

See Example 3

4. The list below shows the ages of the musicians in a local orchestra. Make a histogram of the data.

   14, 35, 22, 18, 49, 38, 30, 27, 45, 19, 35, 46, 27, 21, 32, 30

INDEPENDENT PRACTICE

See Example 1

The bar graph shows the maximum precipitation in 24 hours for several states. Use the bar graph for Exercises 5 and 6.

Maximum 24-Hour Precipitation

Precipitation (in.): 0, 10, 20, 30, 40

Florida, Indiana, Oklahoma, Virginia

5. Which state received the most precipitation in 24 hours?
6. About how many more inches of precipitation did Oklahoma receive than Indiana?

See Example 2

7. The table shows the average annual income per capita for three Chinese cities. Make a double-bar graph of the data.

City	1994	2000
Beijing	$614	$1,256
Shanghai	$716	$1,424
Shenzhen	$1,324	$2,626

See Example 3

8. The list below shows the results of a typing test in words per minute. Make a histogram of the data.

   62, 55, 68, 47, 50, 41, 62, 39, 54, 70, 56, 47, 71, 55, 60, 42

Social Studies LINK

William Jennings Bryan

In 1896 and 1900, the same candidates ran for president of the United States. The candidates were William McKinley, a Republican, and William Jennings Bryan, a Democrat. The table shows the number of electoral votes each man received in these elections.

9. Use the data in the table to make a double-bar graph. Label the horizontal axis with the years.

Candidate	1896	1900
McKinley	271	292
Bryan	176	155

10. In which year was the difference between the number of electoral votes each candidate received the greatest?

11. In 1896, about how many more electoral votes did McKinley get than Bryan?

12. The frequency table shows the number of years the first 42 presidents spent in office. How many presidents spent fewer than six years in office?

13. Use the frequency table to make a histogram.

Years in Office	Frequency
0–2	7
3–5	22
6–8	12
9–11	0
12–14	1

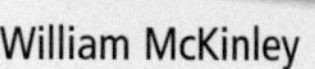

William McKinley

14. ***WRITE ABOUT IT*** What does your histogram show you about the number of years the presidents spent in office?

15. ***CHALLENGE*** Use the 1896 and 1900 election data to make a second double-bar graph. Label the horizontal axis with the candidates' names. Then write a question about the graph.

Spiral Review

Compare. Write < or >. (Previous course)

16. 37 ▒ 29 **17.** 165 ▒ 180 **18.** \$2.13 ▒ \$2.44

Order the numbers from least to greatest. (Previous course)

19. 402, 398, 417, 410 **20.** \$1.00, \$2.66, \$1.41, \$0.82 **21.** 8°F, 7°F, 14°F, 78°F, 41°F

22. **EOG PREP** Which rule *best* describes the number pattern 3, 6, 9, 12, 15? (Previous course)

A Begin with 3 and multiply by 2 repeatedly.
B Begin with 3 and add 3 repeatedly.
C Write the first five powers of 3.
D Begin with 3 and multiply by 3 repeatedly.

1-5 Reading and Interpreting Circle Graphs

Learn to read and interpret data presented in circle graphs.

Vocabulary

circle graph

sector

A **circle graph**, also called a pie chart, shows how a set of data is divided into parts. The entire circle contains 100% of the data. Each **sector**, or slice, of the circle represents one part of the entire data set.

The circle graph compares the number of species in each group of echinoderms. Echinoderms are marine animals that live on the ocean floor. The name *echinoderm* means "spiny-skinned."

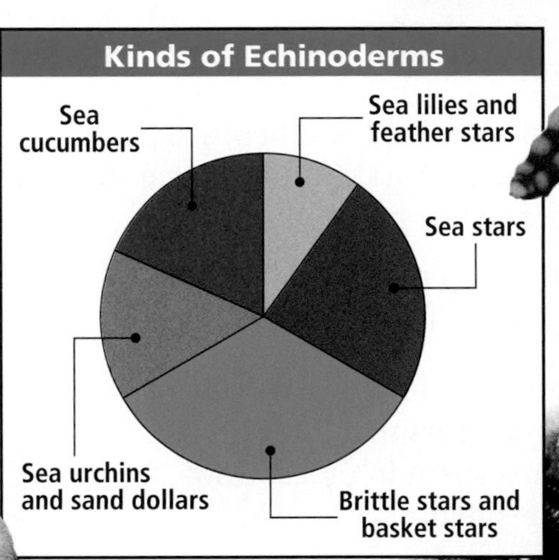

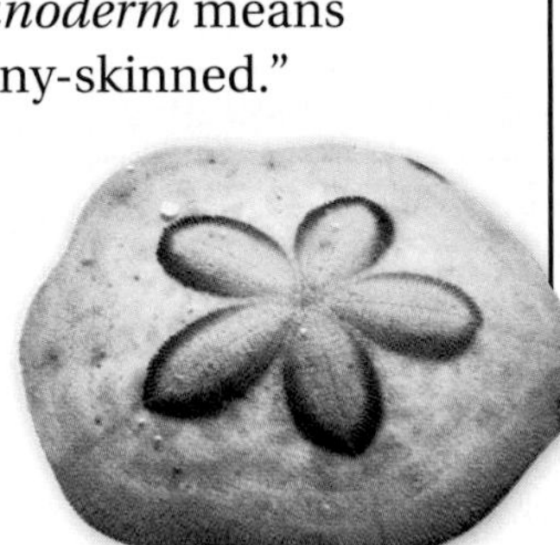

EXAMPLE 1 ***Life Science Application***

Use the circle graph to answer each question.

A **Which group of echinoderms includes the greatest number of species?**

The sector for brittle stars and basket stars is the largest, so this group includes the greatest number of species.

B **Approximately what percent of echinoderm species are sea stars?**

The sector for sea stars makes up about one-fourth of the circle. Since the circle shows 100% of the data, about one-fourth of 100%, or 25%, of echinoderm species are sea stars.

C **Which group is made up of fewer species—sea cucumbers or sea urchins and sand dollars?**

The sector for sea urchins and sand dollars is smaller than the sector for sea cucumbers. This means there are fewer kinds of sea urchins and sand dollars than kinds of sea cucumbers.

EXAMPLE 2 Interpreting Circle Graphs

Leon surveyed 30 people about pet ownership. The circle graph shows his results. Use the graph to answer each question.

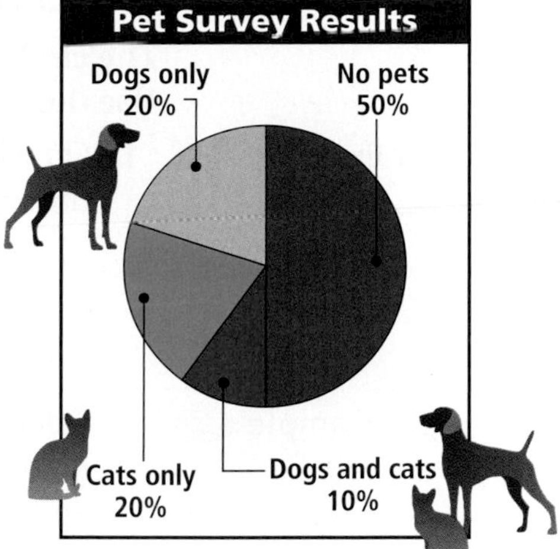

A How many people do not own pets?

The circle graph shows that 50%, or one-half, of the people surveyed do not own pets. One-half of 30 is 15, so 15 people do not own pets.

B Three people responded that they own both dogs and cats. How many people own cats only?

The circle graph shows that 10% of the people own both dogs and cats and that 20% own cats only. Since 10% is 3 people, 20% ($2 \cdot 10\%$) is $2 \cdot 3$, or 6 people. Six people own cats only.

EXAMPLE 3 Choosing an Appropriate Graph

Decide whether a bar graph or a circle graph would best display the information. Explain your answer.

A the percent of a nation's electricity supply generated by each of several fuel sources

A circle graph is the better choice because it makes it easy to see what part of the nation's electricity comes from each fuel source.

B the number of visitors to Arches National Park in each of the last five years

A bar graph is the better choice because it makes it easy to see how the number of visitors has changed over the years.

C the comparison between the time spent in math class and the total time spent in school each day

A circle graph is the better choice because the sector that represents the time spent in math class could be compared to the entire circle, which represents the total time spent in school.

Earth Science LINK

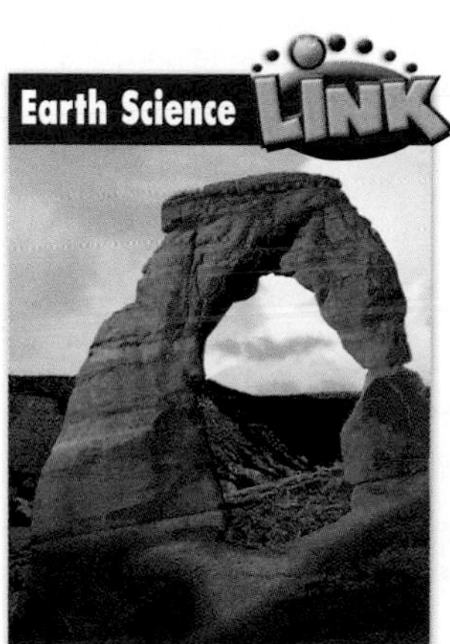

Arches National Park, located in southeastern Utah, covers 73,379 acres. The park is famous for its natural sandstone arches.

Think and Discuss

1. **Describe** two ways a circle graph can be used to compare data.
2. **Compare** the use of circle graphs with the use of bar graphs to display data.

FOR EOG PRACTICE
see page 657

internet connect
Homework Help Online
go.hrw.com Keyword: MS4 1-5

1.01, 4.01

GUIDED PRACTICE

The circle graph shows the estimated spending on advertising in 2000. Use the graph for Exercises 1–3.

See Example 1

1. On which type of advertising was the least amount of money spent?
2. Approximately what percent of spending was on radio and magazine advertising?

See Example 2

3. Television and magazine advertising made up about 50% of all advertising spending in 2000. If the total amount spent was $100,000, how much was spent on television and magazine advertising?

Source: *USA Today*

See Example 3

Decide whether a bar graph or a circle graph would best display the information. Explain your answer.

4. the lengths of the five longest rivers in the world
5. the percent of citizens who voted for each candidate in an election

INDEPENDENT PRACTICE

The circle graph shows the results of a survey of 100 teens who were asked about their favorite sports. Use the graph for Exercises 6–8.

See Example 1

6. Did more teens pick basketball or tennis as their favorite sport?
7. Approximately what percent of teens picked soccer as their favorite sport?

See Example 2

8. According to the survey, 5% of teens chose golf. What is the number of teens who chose golf?

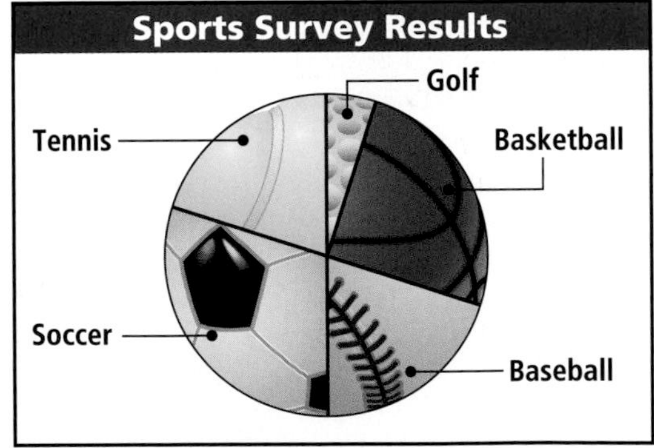

See Example 3

Decide whether a bar graph or a circle graph would best display the information. Explain your answer.

9. the number of calories eaten at breakfast compared with the total number of calories eaten in one day
10. the number of inches of rain that fell each month in Honolulu, Hawaii, during one year

PRACTICE AND PROBLEM SOLVING

The circle graph shows the percent of Earth's land area covered by each continent. Use the graph for Exercises 11–13.

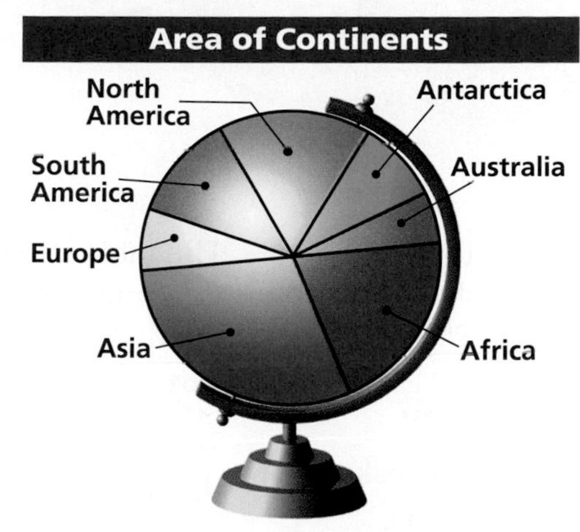

11. List the continents in order of size, from largest to smallest.

12. Approximately what percent of Earth's total land area is Asia?

13. Approximately what percent of Earth's total land area is North America and South America combined?

14. ***SOCIAL STUDIES*** The circle graph shows the diversity of the Hispanic population in Toledo, Ohio. Most of Toledo's Hispanic population are from which country?

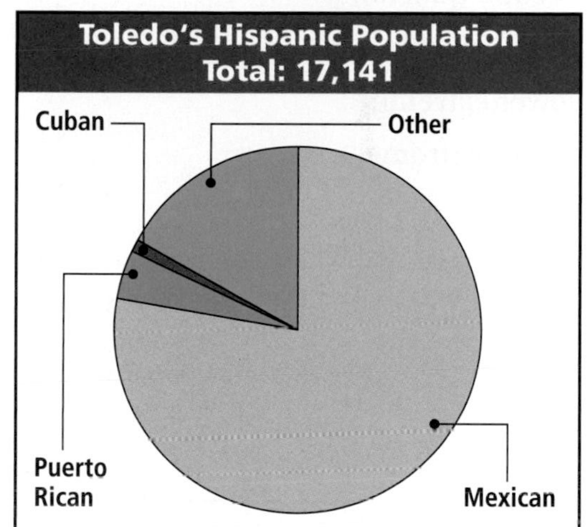

15. ***WHAT'S THE ERROR?*** A student wrote that more of Toledo's Hispanic population are from Cuba than from Puerto Rico. What did the student do wrong?

16. ***WRITE ABOUT IT*** What math skills do you use when interpreting information in a circle graph?

17. ***CHALLENGE*** Earth's total land area is approximately 57,900,000 square miles. Antarctica is almost 10% of the total area. What is the approximate land area of Antarctica in square miles?

Spiral Review

Find the mean of each data set. (Lesson 1-2)

18. 5, 9, 0, 6, 7, 5, 9, 3 **19.** 53, 73, 28, 79, 77 **20.** 626, 897, 786, 807

Find the median of each data set. (Lesson 1-2)

21. 2, 1, 8, 7, 6, 2, 9, 4, 3 **22.** 93, 48, 77, 84, 71, 11 **23.** 423, 298, 801, 944, 190

24. **EOG PREP** Which term most completely describes this figure? (Previous course)

A Square **B** Rectangle **C** Triangle **D** Trapezoid

1-6 Box-and-Whisker Plots

Learn to display and analyze data in box-and-whisker plots.

Vocabulary

box-and-whisker plot
first quartile
second quartile
third quartile
lower extreme
upper extreme

The table shows the stopping distances of different vehicles from 60 mi/h. To show the distribution of the data, you can use a **box-and-whisker plot**.

Vehicle	Stopping Distance (ft)
A	120
B	158
C	142
D	131
E	128
F	167
G	136

To make a box-and-whisker plot for a set of data, you need to divide the data into four parts using *quartiles*. The box-and-whisker plot below shows the distribution of the vehicles' stopping distances.

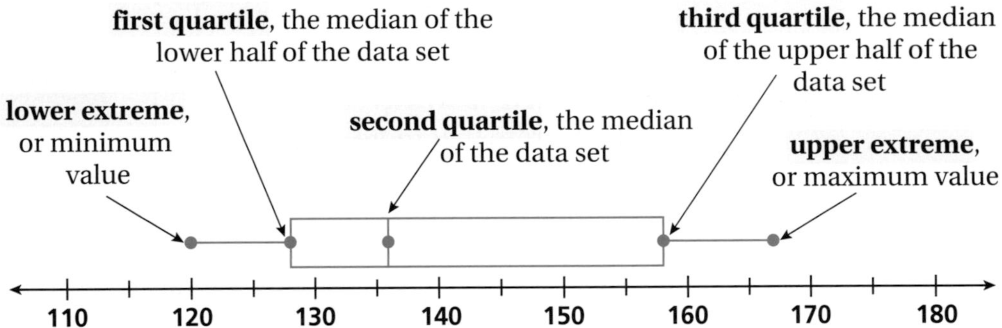

EXAMPLE 1 Reading a Box-and-Whisker Plot

Use the box-and-whisker plot of stopping distances to answer each question.

A **What is the median stopping distance?**

The median stopping distance is represented by the middle point on the box-and-whisker plot. This value is 136 ft.

B **What is the range of the stopping distances?**

The range is the difference between the lower extreme and the upper extreme: 167 − 120, or 47 ft.

C **About what fraction of the stopping distances are less than 158 ft?**

Three of the four parts of the box-and-whisker plot fall below 158 ft. This means about three-fourths of the stopping distances are less than 158 ft.

Remember!

The range of a set of data is the difference between the minimum value and the maximum value.

EXAMPLE 2 Making a Box-and-Whisker Plot

Use the data to make a box-and-whisker plot.

26 17 21 23 19 28 17 20 29

Step 1: Find the lower and upper extremes, the median, and the first and third quartiles.

17 17 19 20 21 23 26 28 29 — *Order the data from least to greatest.*

(17) 17 19 20 21 23 26 28 (29) — *Find the lower and upper extremes.*

(17 17 19 20) (21) (23 26 28 29) — *Find the median.*

(17 (17 19) 20) 21 (23 (26 28) 29) — *Find the first and third quartiles.*

$$\text{first quartile} = \frac{17 + 19}{2} = 18$$

$$\text{third quartile} = \frac{26 + 28}{2} = 27$$

Step 2: Draw a number line. Above the number line, plot points representing the lower and upper extremes, the median, and the first and third quartiles.

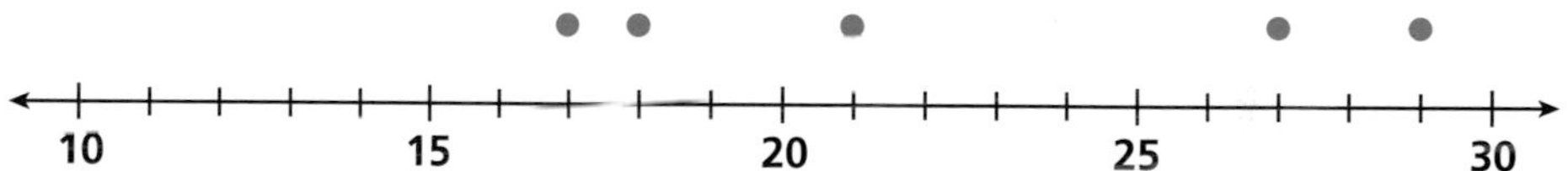

Step 3: Draw a box through the first and third quartiles. Inside the box, draw a vertical line through the median. Then draw lines from the first and third quartiles to the extremes. These lines are called *whiskers*.

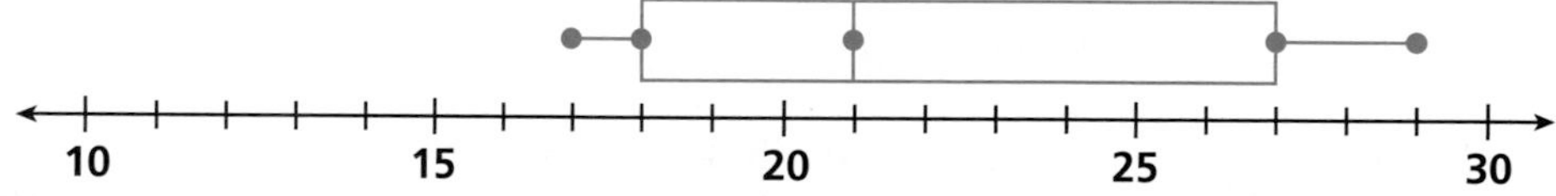

Remember!

The median is the middle number or the average of the two middle numbers in an ordered set of data.

Think and Discuss

1. **Describe** the distribution of the data in Example 2.
2. **Explain** why you cannot find the mean or the mode of a data set by looking at a box-and-whisker plot.
3. **Tell** whether a box-and-whisker plot shows how many data items are included in the data set. Explain.

1-6 Exercises

FOR EOG PRACTICE
see page 657

internet connect
Homework Help Online
go.hrw.com Keyword: MS4 1-6

4.01, 4.02, 4.03

GUIDED PRACTICE

See Example 1

Use the box-and-whisker plot of inches flown by paper airplanes for Exercises 1–3.

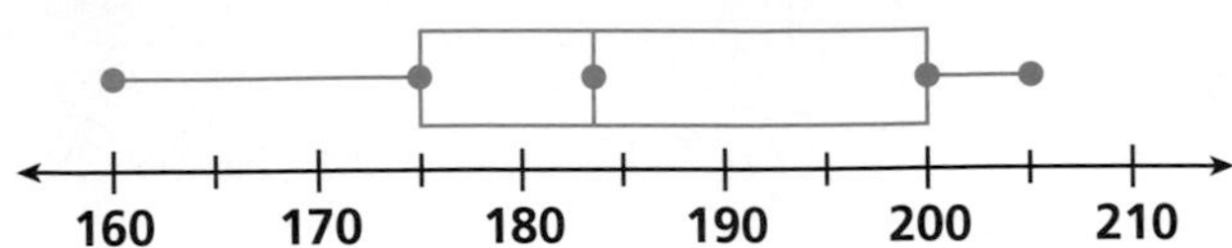

1. What is the first quartile?
2. What is the range of the distances flown?
3. About what fraction of the distances flown are greater than 200 inches?

See Example 2

4. Use the data below to make a box-and-whisker plot.

   46 35 46 38 37 33 49 42 35 40 37

INDEPENDENT PRACTICE

See Example 1

Use the box-and-whisker plot of apartment rental costs for Exercises 5–7.

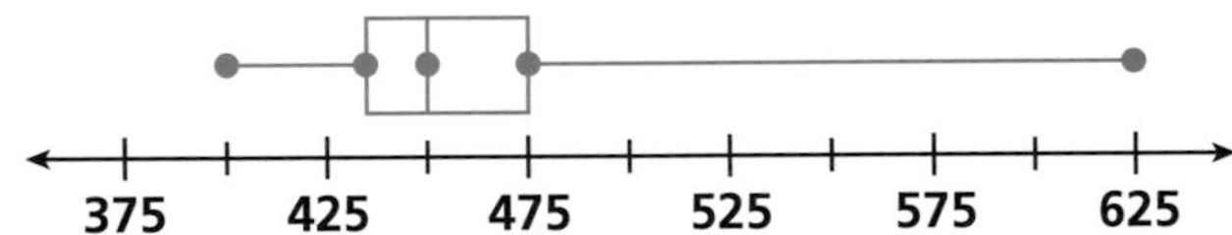

5. What is the third quartile?
6. What is the range of the rental costs?
7. About what fraction of the rental costs are between \$400 and \$450?

See Example 2

8. Use the data below to make a box-and-whisker plot.

   81 73 88 85 81 72 86 72 79 75 76

PRACTICE AND PROBLEM SOLVING

9. The data shows the number of points scored per game by a member of the basketball team.

   12 7 15 23 10 18 39 15 20 8 13

   a. Use the data to make a box-and-whisker plot.

   b. Make a second box-and-whisker plot of the data. This time do not include the outlier.

   c. How does the outlier affect the way the box-and-whisker plot looks?

10. ***SPORTS*** The table shows the countries that were the top 15 medal winners in the 2000 Olympics.

Country	Medals	Country	Medals	Country	Medals
USA	97	Russia	88	China	59
Australia	58	Germany	57	France	38
Italy	34	Cuba	29	Britain	28
Korea	28	Romania	26	Netherlands	25
Ukraine	23	Japan	18	Belarus	17

a. Make a box-and-whisker plot of the data.

b. Describe the distribution of the number of medals won.

11. ***MEASUREMENT*** The stem-and-leaf plot shows the heights in inches of a class of seventh graders.

a. What is the median height?

b. Make a box-and-whisker plot of the data.

c. Are more students taller or shorter than 5 feet? Explain.

Student Heights

Stems	Leaves
5	3 5 6 6 8 8 8 9 9
6	0 0 1 1 1 1 1 2 2 2 4

Key: 5|3 means 53

12. ***WHAT'S THE ERROR?*** A student made this box-and-whisker plot using the data 2, 9, 5, 14, 8, 13, 7, 5, 8. What did the student do wrong?

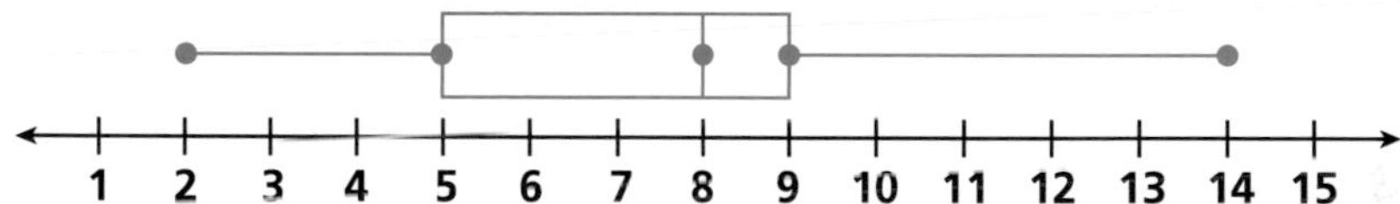

13. ***WRITE ABOUT IT*** Two box-and-whisker plots have the same median and equally long whiskers. If the box of one plot is longer, what can you say about the two data sets?

14. ***CHALLENGE*** The *interquartile range* is the difference between the third and first quartiles. Find the interquartile range for the data set 1, 2, 4, 2, 1, 0, 6, 8, 1, 6, 2.

Spiral Review

Find each sum or difference. (Previous course)

15. 499 + 231

16. 87 + 1645

17. 503 − 125

18. 900 − 281

19. **EOG PREP** Which is *not* a measure of central tendency for the data set 4, 6, 6, 7, 9, 10, 11, 11? (Previous course)

A 6
B 8
C 9
D 11

20. **EOG PREP** If Lorenzo earns $9 an hour, how many hours must he work to earn $120? (Previous course)

A 10 hours
B 12 hours
C 13 hours
D 14 hours

Use Technology to Create Graphs

Use with Lesson 1-7

There are several ways to display data, including bar graphs, line graphs, and circle graphs. A spreadsheet provides a quick way to create these graphs.

Activity

1 Use a spreadsheet to display the Kennedy Middle School Student Council budget shown in the table at right.

Student Council Budget	
Activity	**Amount ($)**
Assemblies	275
Dances	587
Spring Festival	412
Awards Banquet	384
Other	250

Open the spreadsheet program, and enter the data as shown below. Enter the activities in column A and the amount budgeted in column B. Include the column titles in row 1.

	A	B	C
1	Activity	Amount ($)	
2	Assemblies	275	
3	Dances	587	
4	Spring Festival	412	
5	Awards Banquet	384	
6	Other	250	
7			

Highlight the data by clicking on cell A1 and dragging the cursor to cell B6. Click the Chart Wizard icon. Then click **FINISH** to choose the first type of column graph.

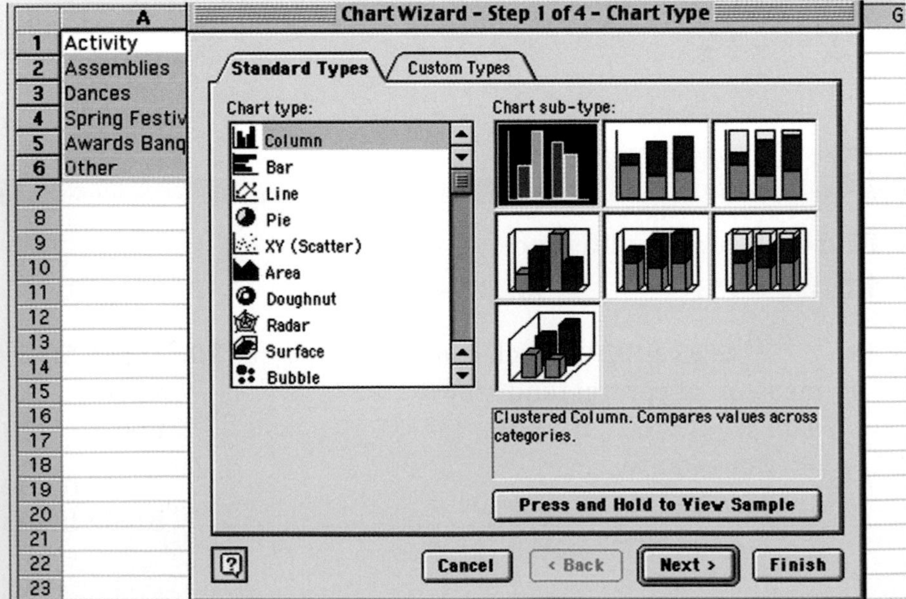

The bar graph of the data appears as shown. Resize or reposition the graph, if necessary.

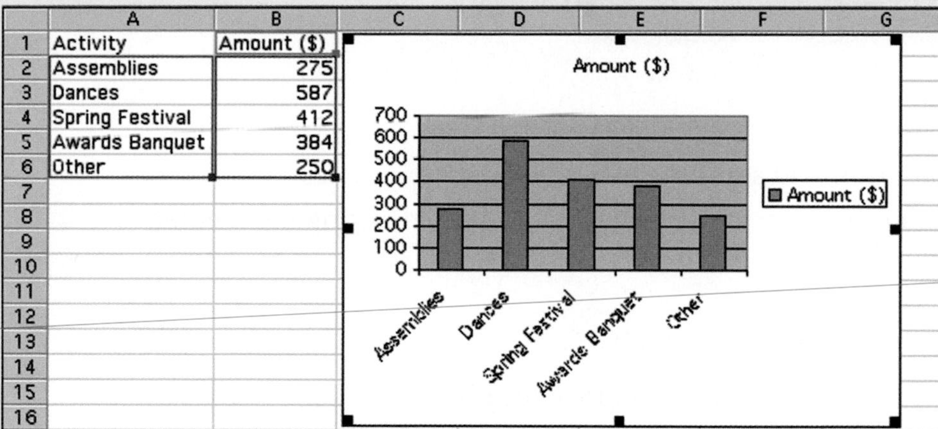

To see a line graph of the data, select the bar graph (as shown above). Click the Chart Wizard icon and choose the line graph. Then click **FINISH** to choose the first type of line graph.

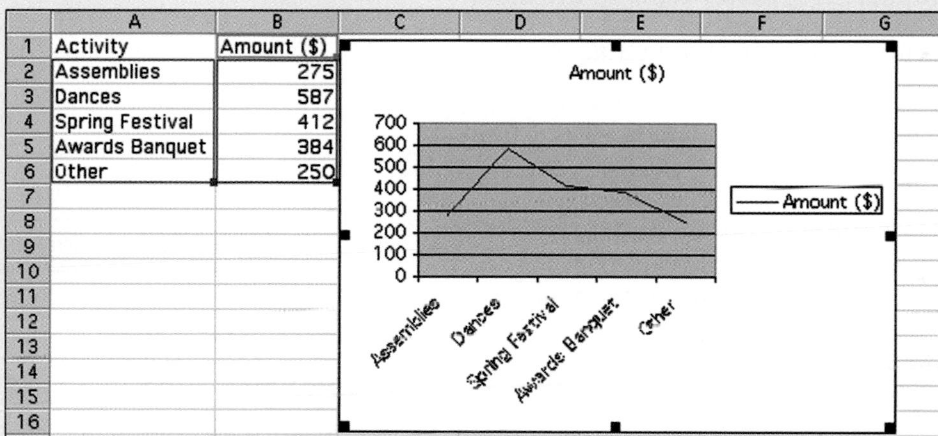

Think and Discuss

1. Which graph best displays the Student Council budget? Why?
2. How would you display the data in a circle graph?

Try This

1. The table shows the number of points scored by several members of a girls' basketball team. Use a spreadsheet to create a bar graph and a line graph of the data.

Player	Ana	Angel	Mary	Nia	Tina	Zoe
Points Scored	201	145	89	40	21	8

Chapter 1

Mid-Chapter Quiz

LESSON 1-1 (pp. 4–7)

Identify the population and sample.

1. A movie theater manager surveys the first 40 people exiting the theater.

LESSONS 1-2 AND 1-3 (pp. 10–17)

Use the data set 1, 3, 8, 7, 3, 5, 14, 7, 5, 6, 10, 8 for problems 2–5.

2. Find the mean, median, mode, and range of the data set. Round your answer to the nearest tenth, if necessary.
3. Which measure of central tendency best represents the data?
4. Make a cumulative frequency table of the data.
5. Make a stem-and-leaf plot of the data.

LESSON 1-4 (pp. 20–23)

6. The table shows the number of students in sixth and seventh grades who participated in school fairs. Make a double-bar graph of the data.
7. The list below shows the number of tracks on a group of CDs. Make a histogram of the data.

   13, 7, 10, 8, 15, 17, 22, 9, 11, 10, 16, 12, 9, 20

School Fair Participation

Fair	Sixth	Seventh
Book	55	76
Health	69	58
Science	74	98

LESSON 1-5 (pp. 24–27)

Use the circle graph for problems 8–10.

8. Did more students pick sausage or mushrooms as their favorite topping?
9. Approximately what percent of students picked cheese as their favorite topping?
10. Out of 200 students, 25% picked pepperoni as their favorite. How many students picked pepperoni?

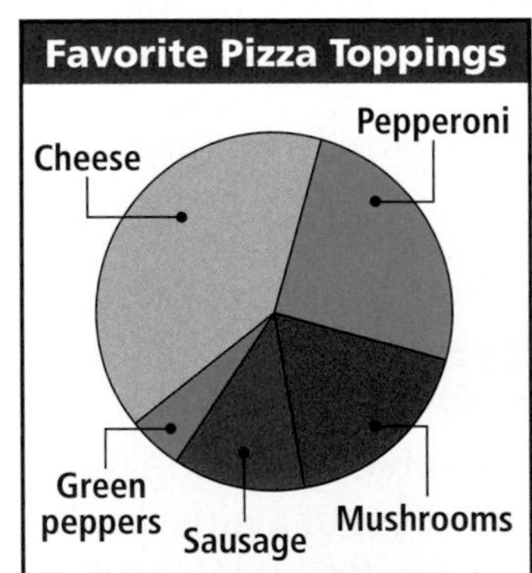

LESSON 1-6 (pp. 28–31)

11. Make a box-and-whisker plot of the data 14, 8, 13, 20, 15, 17, 1, 12, 18, 10.
12. What fraction of the data is greater than 13.5?

1-7 Line Graphs

Learn to display and analyze data in line graphs.

Vocabulary

line graph

double-line graph

You can use a *line graph* to show how data changes over a period of time. In a **line graph**, line segments are used to connect data points. The result is a visual record of change.

Line graphs can be used for a variety of reasons, including showing the growth of a cat over time.

EXAMPLE 1 Making a Line Graph

Make a line graph of the data in the table. Use the graph to determine during which 2-month period the kitten's weight increased the most.

Age (mo)	Weight (lb)
0	0.2
2	1.7
4	3.8
6	5.1
8	6.0
10	6.7
12	7.2

Step 1: Determine the scale and interval for each axis. Place units of time on the horizontal axis.

Helpful Hint

To plot each point, start at zero. Move *right* for the time and *up* for the weight.

Step 2: Plot a point for each pair of values. Connect the points using line segments.

Step 3: Label the axes and give the graph a title.

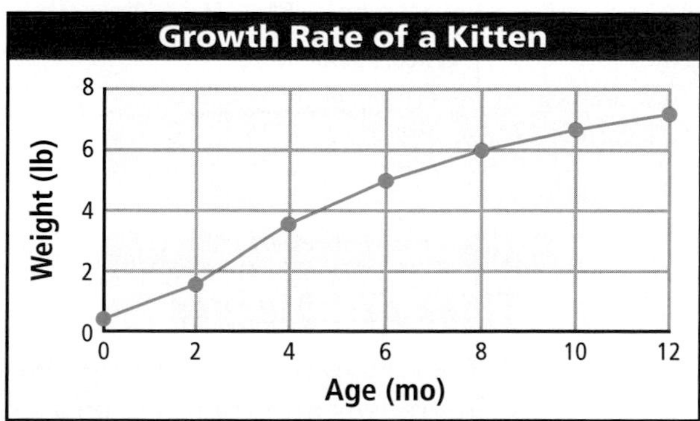

The graph shows the greatest increase between 2 and 4 months. This means the kitten's weight increased most between 2 and 4 months.

You can use a line graph to estimate values between data points.

EXAMPLE 2 Using a Line Graph to Estimate Data

Use the graph to estimate the population of Florida in 1990.

To estimate the population in 1990, find the point on the line between 1980 and 2000 that corresponds to 1990.

The graph shows about 12.5 million. In fact, the population was 12.9 million in 1990.

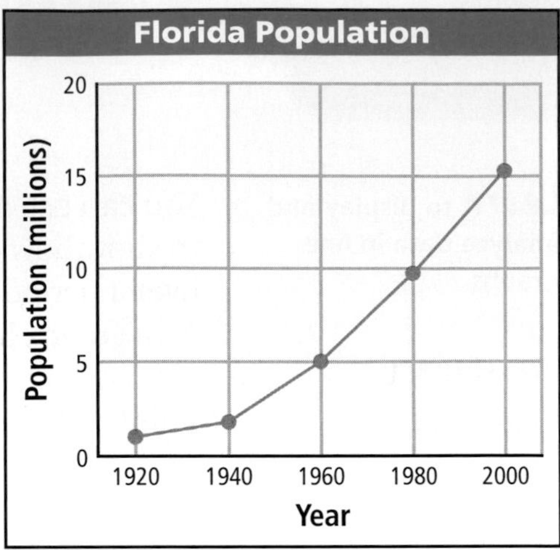

A **double-line graph** shows change over time for two sets of data.

EXAMPLE 3 Making a Double-Line Graph

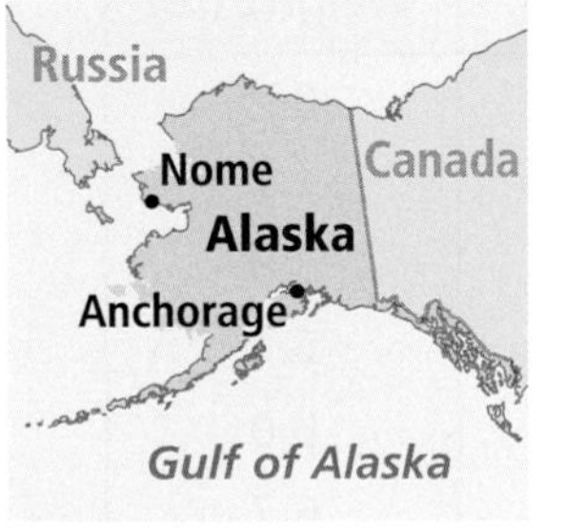

The table shows the normal daily temperatures in degrees Fahrenheit in two Alaskan cities. Make a double-line graph of the data.

Month	Nome	Anchorage
Jan	7	15
Feb	4	19
Mar	9	26
Apr	18	36
May	36	47
Jun	46	54

Average Temperatures

Temperature (°F): 0, 10, 20, 30, 40, 50, 60

Month: Jan, Feb, Mar, Apr, May, Jun

■ Nome ■ Anchorage

Plot a point for each temperature in Nome and connect the points. Then, using a different color, plot a point for each temperature in Anchorage and connect the points. Make a key to show what each line represents.

Think and Discuss

1. **Describe** how a line graph would look for a set of data that increases and then decreases over time.
2. **Give an example** of a situation that can be described by a double-line graph in which the two sets of data intersect at least once.

1-7 Exercises

FOR EOG PRACTICE

see page 658

internet connect
Homework Help Online
go.hrw.com Keyword: MS4 1-7

1.03, 4.01

GUIDED PRACTICE

The table at right shows average movie theater ticket prices in the United States. Use the table for Exercises 1 and 2.

Year	Ticket Price ($)
1965	1.01
1970	1.55
1975	2.05
1980	2.69
1985	3.55
1990	4.23
1995	4.35
2000	5.39

See Example 1

1. Make a line graph of the data in the table. Use the graph to determine during which 5-year period the average ticket price increased the least.

See Example 2

2. Use the graph to estimate the average ticket price in 1997.

See Example 3

3. The table below shows the amount of two varieties of cheese in pounds consumed per person in the United States. Make a double-line graph of the data.

	1992	1993	1994	1995	1996	1997
American	11.32	11.41	11.55	11.84	11.99	12.03
Italian	9.97	9.82	10.29	10.41	10.79	10.96

INDEPENDENT PRACTICE

The table at right shows the number of teams in the National Basketball Association (NBA). Use the table for Exercises 4–6.

Year	Teams
1965	9
1970	14
1975	18
1980	22
1985	23
1990	27
1995	27
2000	29

See Example 1

4. Make a line graph of the data in the table. Use the graph to determine during which 5-year period the number of NBA teams increased the most.

5. During which 5-year period did the number of teams increase the least?

See Example 2

6. Use the graph to estimate the number of NBA teams in 1988.

See Example 3

7. The table below shows the normal daily temperatures in degrees Fahrenheit in Peoria, Illinois, and Portland, Oregon. Make a double-line graph of the data.

	Jul	Aug	Sept	Oct	Nov	Dec
Peoria	76	73	66	54	41	27
Portland	68	69	63	55	46	40

PRACTICE AND PROBLEM SOLVING

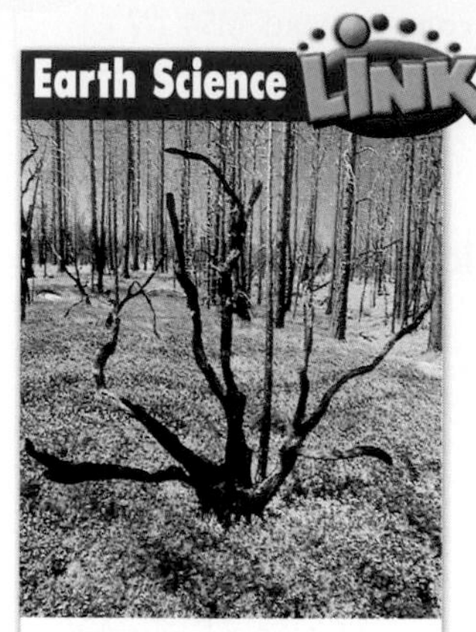

In 1988, forest fires burned millions of acres in Yosemite and Yellowstone National Parks.

8. ***EARTH SCIENCE*** The graph shows the number of acres burned by wildfires in the United States from 1995 to 2000.
   a. During which years did wildfires burn more than 6 million acres?
   b. Explain whether the graph would be useful in predicting future data values.

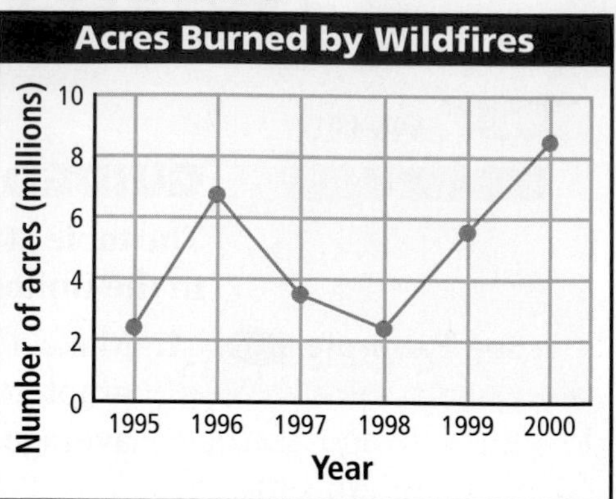

Source: National Interagency Fire Center

9. ***INDUSTRY*** The graph shows the number of new passenger cars imported into the United States from Germany and Japan.
   a. Approximately how many more cars were imported from Japan than from Germany in 1995?
   b. Describe how the number of cars imported from both countries changed from 1970 to 1995.

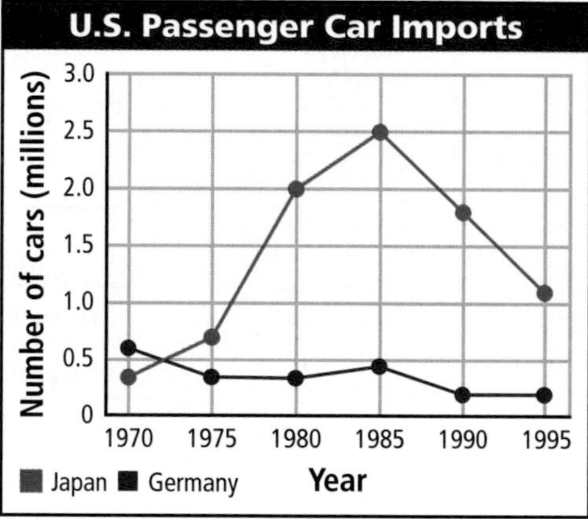

Source: World Almanac

10. ***WRITE A PROBLEM*** Write a question using the double-line graph in Exercise 9.

11. ***WRITE ABOUT IT*** Explain the benefit of drawing a double-line graph rather than two single-line graphs for related sets of data.

12. ***CHALLENGE*** The table shows the amount of vegetables in pounds a family sold at a farmers' market. Make a graph to display the data. Explain your choice of graph.

Vegetable	Day 1	Day 2
Cucumber	60	45
Onion	120	150
Potato	150	180
Tomato	240	225

Spiral Review

Write each number in standard form. (Previous course)

13. 100 + 60 + 8 **14.** 7,000 + 300 + 70 + 9 **15.** 40,000 + 4,000 + 60 + 2

Write each number in expanded form. (Previous course)

16. 306 **17.** 5,047 **18.** 27,684 **19.** 109,244

20. **EOG PREP** Each side of a square tile is 8 cm. A rectangle is formed by placing two of the tiles next to each other. What is the perimeter of the rectangle in centimeters? (Previous course)

A 32 cm B 48 cm C 64 cm D 128 cm

Explore Scatter Plots

Use with Lesson 1-8

4.01

Look at the graph at right. Do you think your shoe size has anything to do with your age? In this activity, you will explore data sets that may or may not have a relationship.

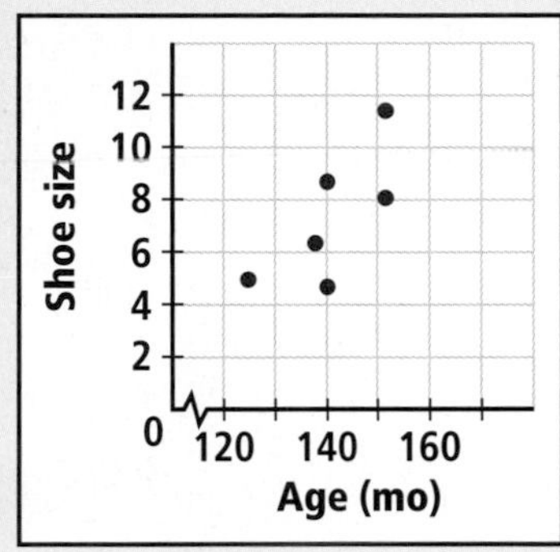

Activity

1 Follow the steps to create a scatter plot.

a. On a piece of graph paper, draw the horizontal and vertical axes for a graph.

b. Select two variables from the list below.

- shoe size
- height
- age in months
- length of forearm
- month of birth
- last two digits of phone number

c. Survey at least six people to find values for these two variables. Write the information you get from each person as an ordered pair. For example, (shoe size, last two digits of phone number) could be (7, 31).

d. Label the axes of your graph with the variables. Then plot the data you gathered as points on the graph.

Think and Discuss

1. Do the points on your graph form a pattern? Explain.
2. Do you have any point that does not fit your pattern? If so, how does this point compare with the other points?
3. Do the points appear to be almost in a straight line? If so, describe the line.

Try This

1. Graph the ordered pairs below. How is the pattern in this graph similar to or different from the pattern in the graph you made in the Activity?

(1, 2), (2, 2), (2, 3), (3, 4), (4, 3), (4, 5), (5, 5), (6, 4), (7, 6), (7, 8), (8, 7)

1-8 Scatter Plots

Learn to display and analyze data in scatter plots.

Vocabulary

scatter plot

positive correlation

negative correlation

no correlation

The supersaurus, one of the largest known dinosaurs, could weigh as much as 55 tons and grow as long as 100 feet from head to tail. The tyrannosaurus, a large meat-eating dinosaur, was about one-third the length of the supersaurus.

Two sets of data, such as the length and the weight of dinosaurs, may be related. To find out, you can make a *scatter plot* of the data values in each set. A **scatter plot** has two number lines, called *axes*—one for each set of data values. Each point on the scatter plot represents a pair of data values. These points may appear to be scattered or may cluster in the shape of a line or a curve.

EXAMPLE 1 Making a Scatter Plot

Use the data to make a scatter plot. Describe the relationship between the data sets.

Name	Length (ft)	Weight (tons)
Triceratops	30	6
Tyrannosaurus	39	7
Euhelopus	50	25
Brachiosaurus	82	50
Supersaurus	100	55

Step 1: Determine the scale and interval for each axis. Place units of length on the horizontal axis and units of weight on the vertical axis.

Step 2: Plot a point for each pair of values.

Step 3: Label the axes and title the graph.

The scatter plot shows that a dinosaur's weight tends to increase as its length increases.

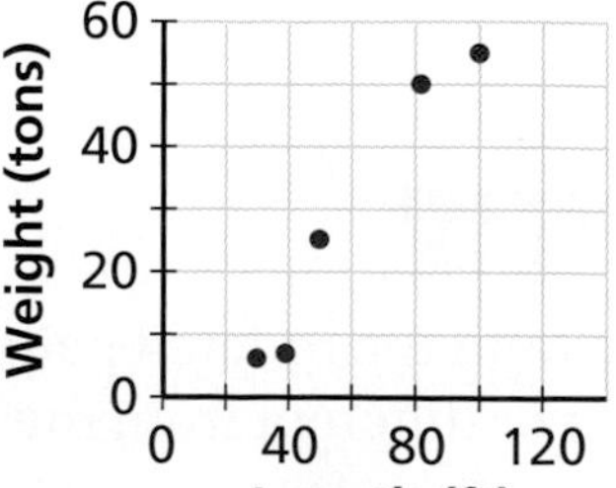

There are three ways to describe data displayed in a scatter plot.

Positive Correlation

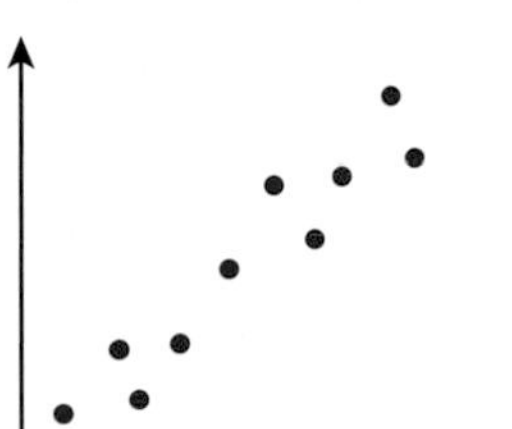

The values in both data sets increase at the same time.

Negative Correlation

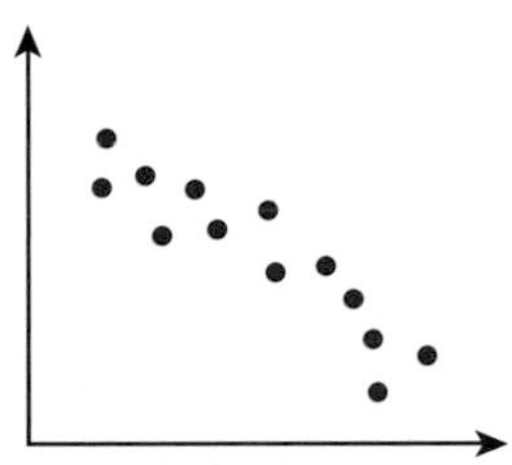

The values in one data set increase as the values in the other set decrease.

No Correlation

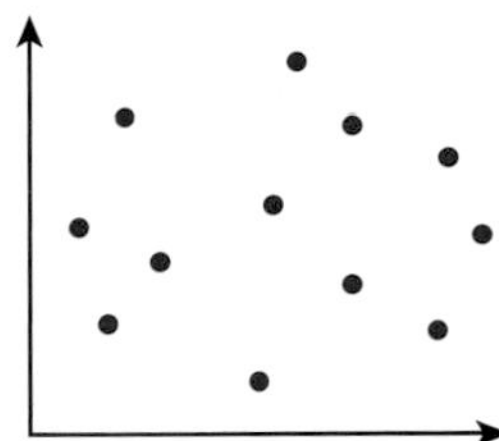

The values in both data sets show no pattern.

EXAMPLE 2 Determining Relationships Between Two Sets of Data

Write *positive correlation, negative correlation,* or *no correlation* to describe each relationship. Explain.

A

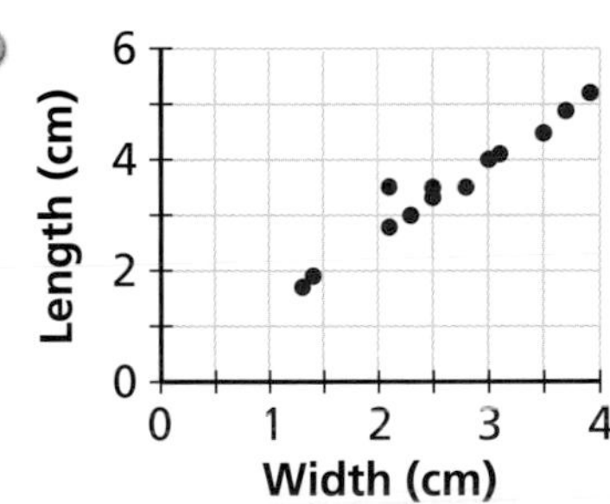

The graph shows that as width increases, length increases. So the graph shows a positive correlation between the data sets.

B

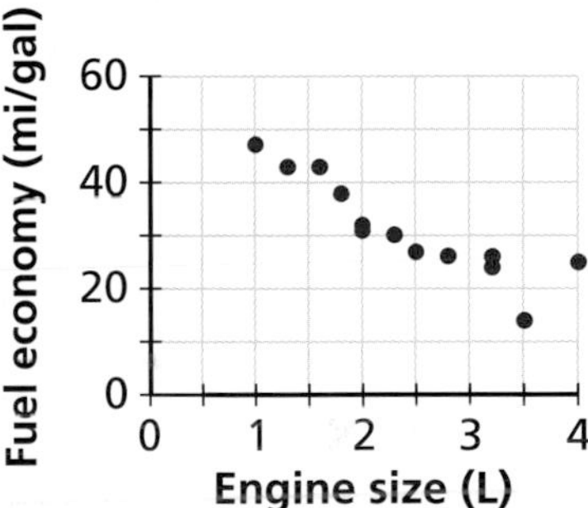

The graph shows that as engine size increases, fuel economy decreases. So the graph shows a negative correlation between the data sets.

C **the ages of people and the number of pets they own**

The number of pets a person owns is not related to the person's age. So there seems to be no correlation between the data sets.

Think and Discuss

1. **Describe** the type of correlation you would expect between the number of absences in a class and the grades in the class.
2. **Give an example** of a relationship between two sets of data that shows a negative correlation. Then give an example of a positive correlation.

1-8 Exercises

FOR EOG PRACTICE
see page 658

internet connect
Homework Help Online
go.hrw.com Keyword: MS4 1-8

4.01

GUIDED PRACTICE

See Example 1

1. The table shows the typical weights and heart rates of several mammals. Use the data to make a scatter plot. Describe the relationship between the data sets.

Mammal	Weight (kg)	Heart Rate (beats/min)
Ferret	0.6	360
Human	70	70
Llama	185	75
Red deer	110	80
Rhesus monkey	10	160

See Example 2

Write *positive correlation, negative correlation,* or *no correlation* to describe each relationship. Explain.

2. 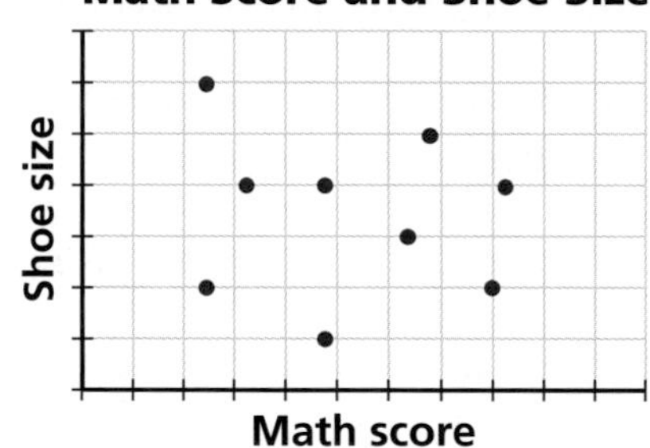

3. 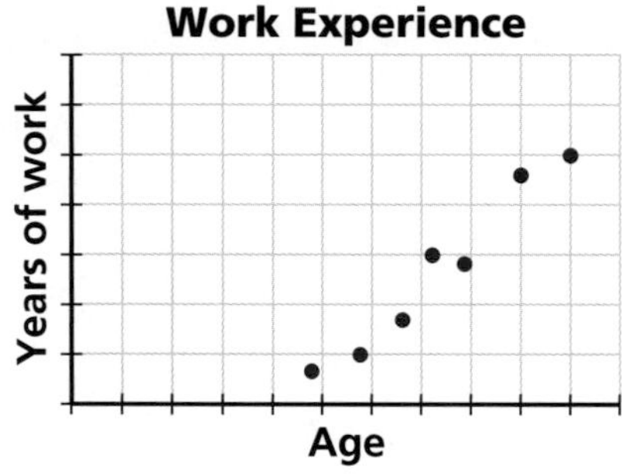

4. the time it takes to drive 100 miles and the driving speed

INDEPENDENT PRACTICE

See Example 1

5. The table shows solar energy cell capacity (in megawatts) over several years. Use the data to make a scatter plot. Describe the relationship between the data sets.

Year	Capacity	Year	Capacity
1990	13.8	1993	21.0
1991	14.9	1994	26.1
1992	15.6	1995	31.1

See Example 2

Write *positive correlation, negative correlation,* or *no correlation* to describe each relationship. Explain.

6. 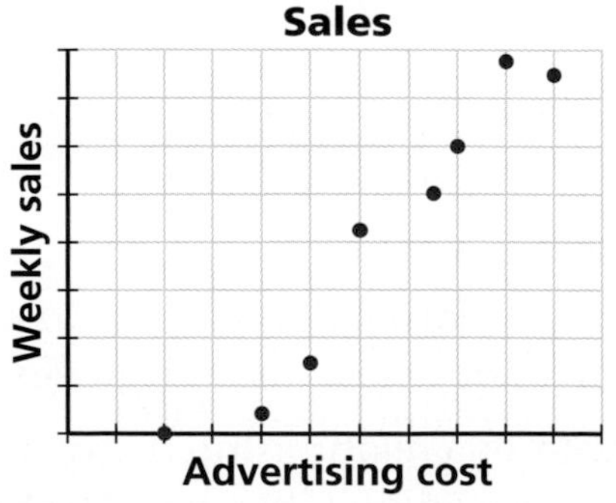

7. 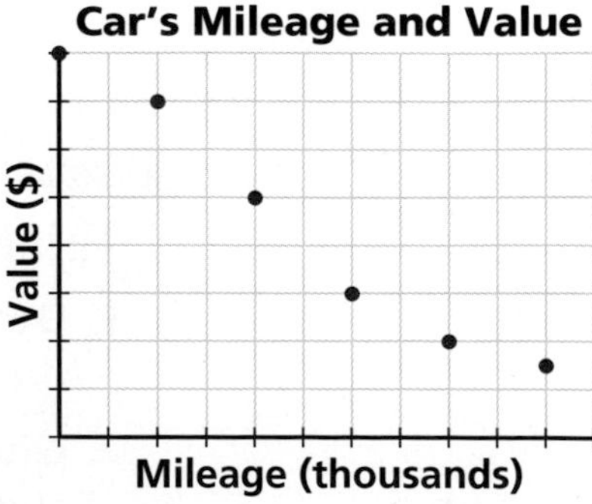

8. the number of students in a school district and the number of buses in the district

Science LINK

For Exercises 9–12, tell whether you would expect a positive correlation, a negative correlation, or no correlation. Explain your answers.

A scientist launching a weather balloon in Antarctica

9. the elevation of a city and its average daily temperature
10. the average temperature of a location and the amount of rainfall it receives each year
11. the latitude of a location and the amount of snow it receives each year
12. the number of hours of daylight and the amount of rainfall in a day

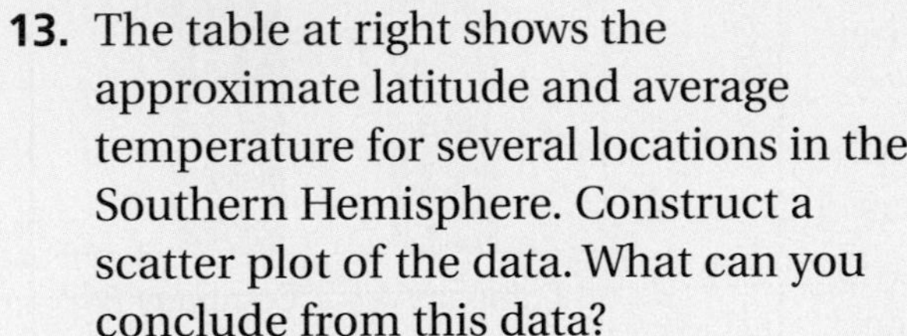

13. The table at right shows the approximate latitude and average temperature for several locations in the Southern Hemisphere. Construct a scatter plot of the data. What can you conclude from this data?

Location	Latitude	Temperature
Quito, Ecuador	0° S	55°F
Melbourne, Australia	38° S	43°F
Tucuman, Argentina	27° S	57°F
Tananarive, Madagascar	19° S	60°F
Halley Research Station, Antarctica	76° S	20°F

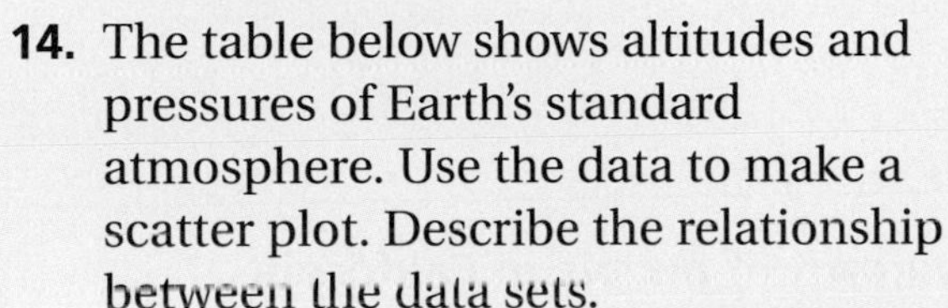

14. The table below shows altitudes and pressures of Earth's standard atmosphere. Use the data to make a scatter plot. Describe the relationship between the data sets.

Altitude (km)	Pressure (mb)
0	1013
5	540
10	265
15	121
20	55
25	25

15. ***CHALLENGE*** Suppose that a location's elevation is negatively correlated to its average temperature and positively correlated to the amount of snow it receives. What kind of correlation would you expect between temperature and the amount of snowfall? Explain.

Spiral Review

Find each quotient. (Previous course)

16. 45 ÷ 15
17. 24 ÷ 8
18. 63 ÷ 9
19. 36 ÷ 12

20. **EOG PREP** What is the value of 7 in 8.973? (Previous course)

A 7 ones
B 7 tenths
C 7 hundredths
D 7 thousandths

21. **EOG PREP** What is 12 ft 8 in. to the nearest foot? (Previous course)

A 13 ft
B 11 ft
C 12 ft
D 14 ft

1-9 Misleading Graphs

Learn to identify and analyze misleading graphs.

Advertisements and news articles often use data to support a point. Sometimes the data is presented in a way that influences how the data is interpreted. A data display that distorts information in order to persuade can be *misleading.*

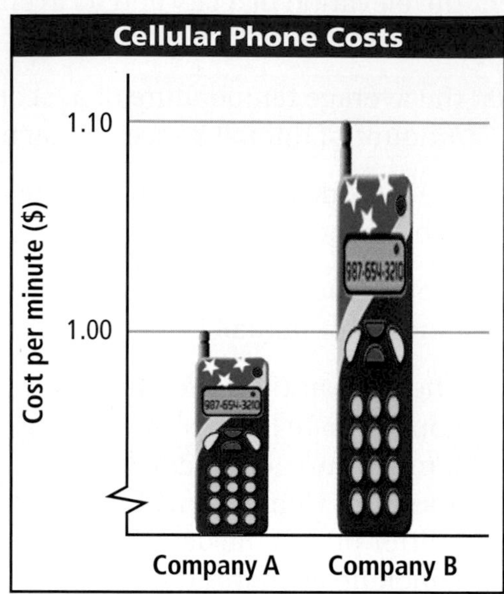

An axis in a graph can be "broken" to make the graph easier to read. However, a broken axis can also be misleading. In the graph at right, the cost per minute for service with Company B looks like it is twice as much as the cost for service with Company A. In fact, the difference is only $0.10 per minute.

EXAMPLE 1 *Social Studies Application*

Both bar graphs show the percent of people in California, Maryland, Michigan, and Washington who use seat belts. Which graph could be misleading? Why?

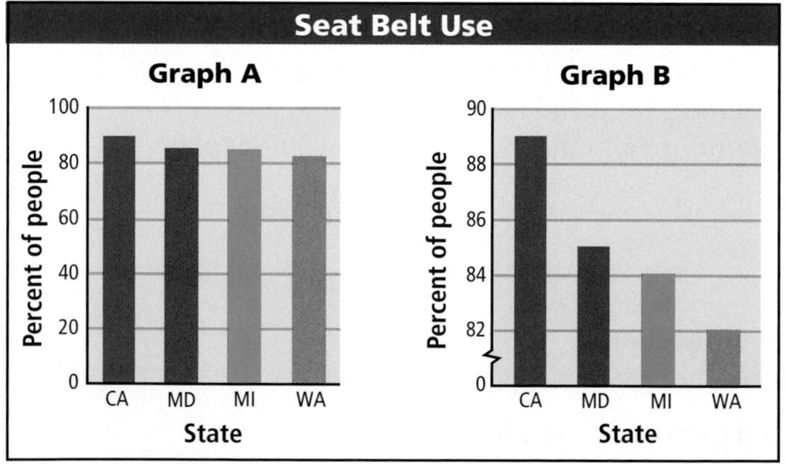

Graph B could be misleading. Because the vertical axis on graph B is broken, it appears that the percent of people in California who wear seat belts is twice as great as the percent in Michigan. In fact, it is only 5% greater. People might conclude from graph B that the percent of people in California who wear seat belts is much greater than the percents in the other states.

EXAMPLE 2 Analyzing Misleading Graphs

Explain why each graph could be misleading.

A

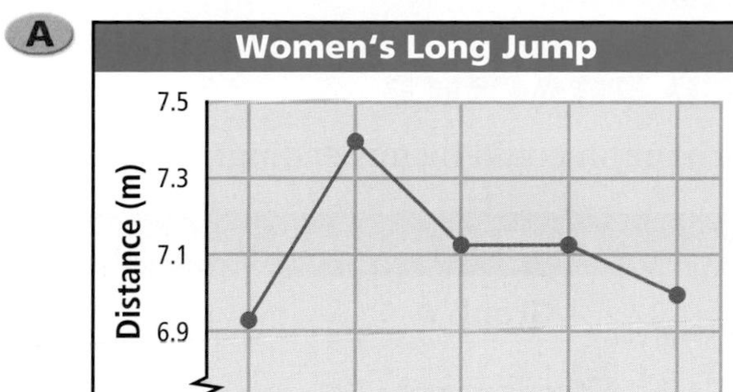

Because the scale on the vertical axis is broken, the distance jumped in 1988 appears to be over two times as far as in 1984. In fact, the distance jumped in 1988 is less than 0.5 meter greater than in the other years.

B

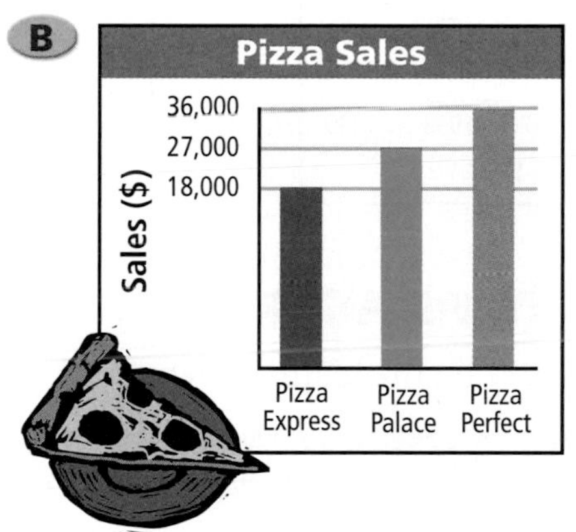

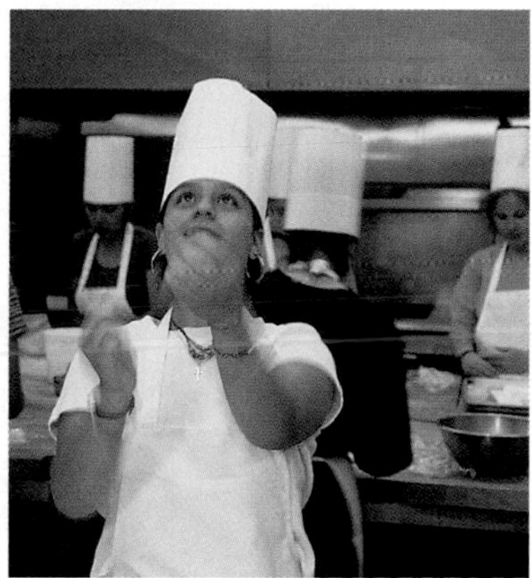

The scale of the graph is wrong. Equal distances on the graph should represent equal intervals of numbers, but in this graph, the first $18,000 in sales is larger than the next $18,000. Because of this, you can't tell from the bars that Pizza Perfect's sales were twice those of Pizza Express.

Think and Discuss

1. **Explain** how to use the scale of a graph to decide if the graph is misleading.
2. **Describe** what might indicate that a graph is misleading.
3. **Give an example** of a situation in which a misleading graph might be used to persuade readers.

1-9 Exercises

FOR EOG PRACTICE
see page 659

internet connect
Homework Help Online
go.hrw.com Keyword: MS4 1-9

GUIDED PRACTICE

See Example 1

1. Which graph could be misleading? Why?

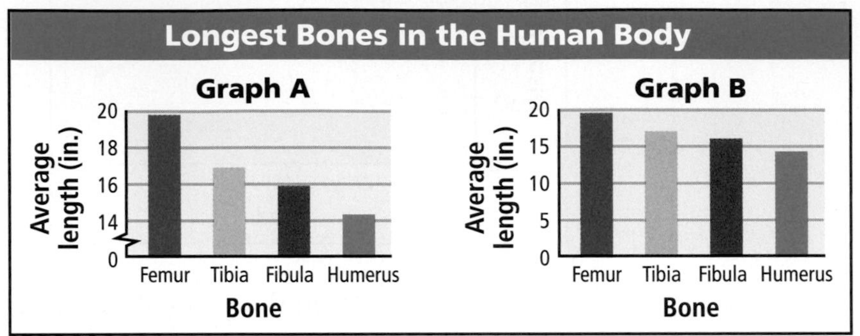

See Example 2

Explain why each graph could be misleading.

2.

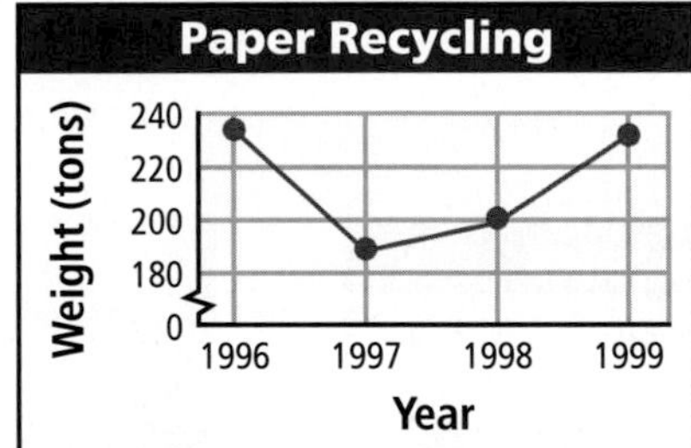

3.

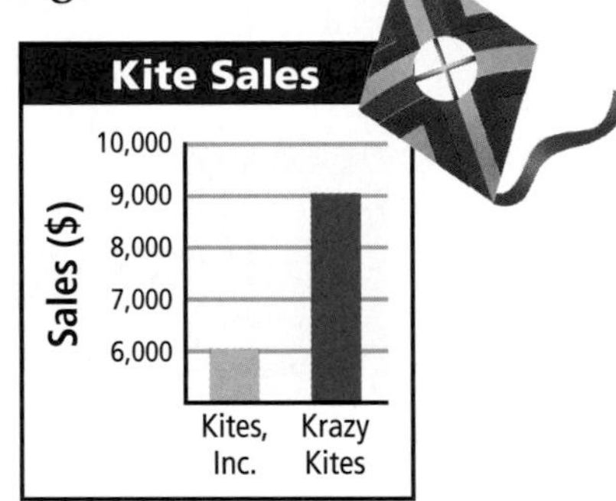

INDEPENDENT PRACTICE

See Example 1

4. Which graph could be misleading? Why?

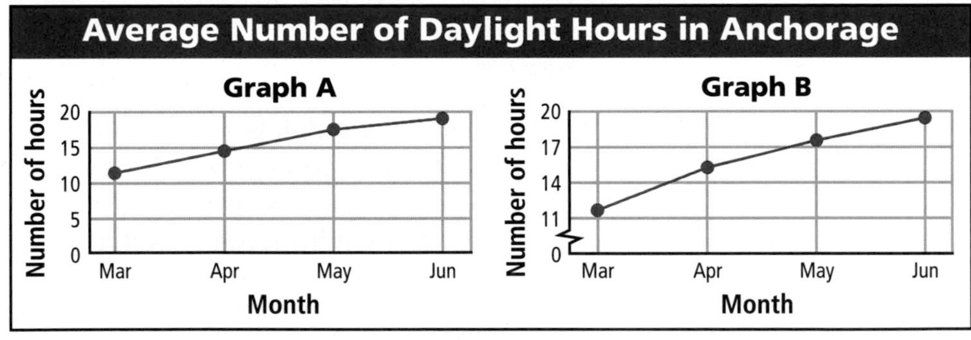

See Example 2

Explain why each graph could be misleading.

5. CD Sales

Sales ($): 3,000; 2,000; 1,000; 750; 500; 250; 0

CD Palace; Music World

6.

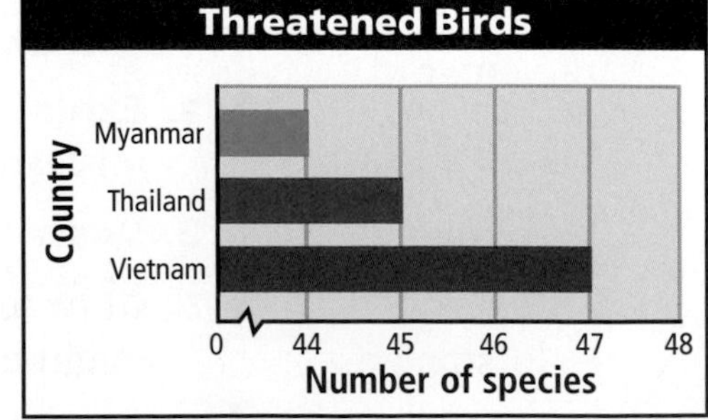

PRACTICE AND PROBLEM SOLVING

7. ***HOBBIES*** The Appalachian Trail is a marked footpath that runs about 2,160 miles from Maine to Georgia. The bar graph shows the number of miles of trail in three states.
   a. Redraw the graph so that it is not misleading.
   b. Compare the two graphs.

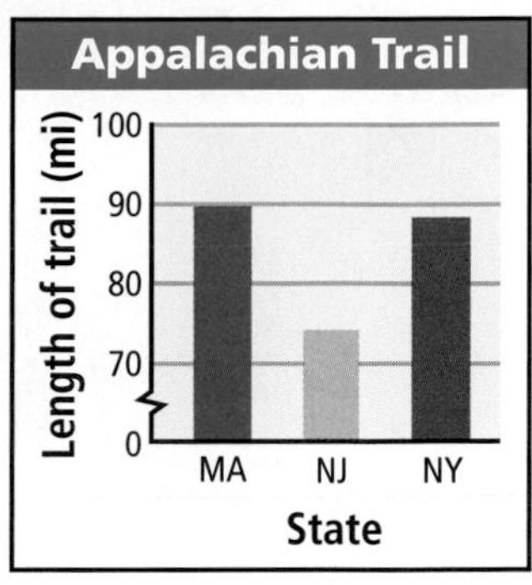

8. ***BUSINESS*** The weekly sales of a car dealer are shown. Explain why the graphs are misleading. Then tell how you can redraw them so that they are not misleading.

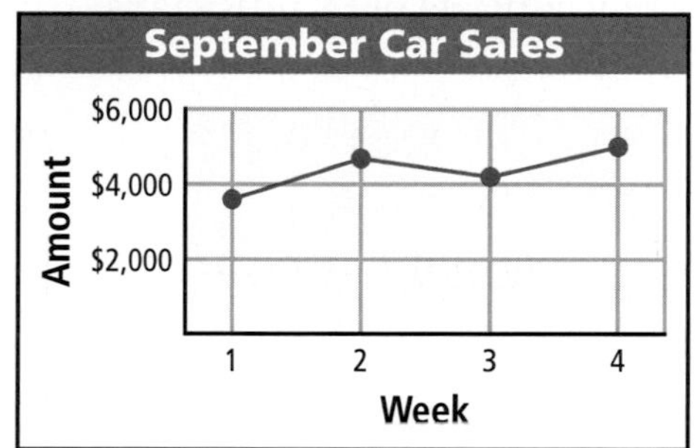

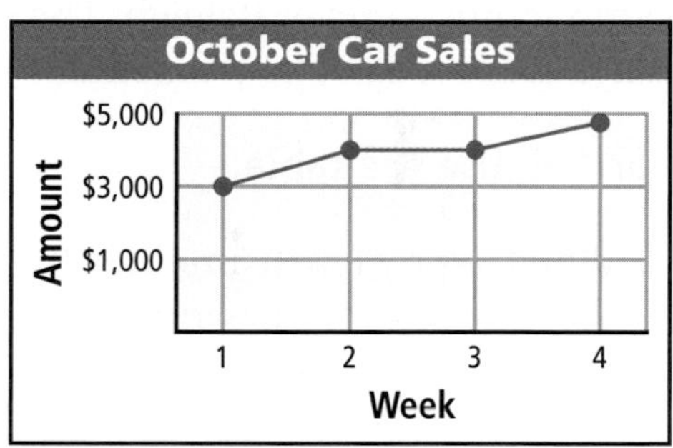

9. ***CHOOSE A STRATEGY*** Tanya had $1.19 in coins. None of the coins were dollars or 50-cent pieces. Josie asked Tanya for change for a dollar, but she did not have the correct change. Which coins did Tanya have?

10. ***WRITE ABOUT IT*** Explain why it is important to closely examine graphs in advertisements.

11. ***CHALLENGE*** A company asked 10 people about their favorite brand of toothpaste. Three people chose Sparkle, one person chose Smile, and six people chose Purely White. An advertisement for Sparkle states, "Three times as many people prefer Sparkle over Smile!" Explain why this statement is misleading.

Spiral Review

Simplify. (Previous course)

12. 15 • 7 • 4 **13.** 2 • 12 • 5 **14.** 3 • 14 • 3

Find the range of each data set. (Lesson 1-2)

15. 13, 6, 36, 3, 9 **16.** 20, 34, 25, 56, 46 **17.** 357, 395, 963, 873

18. **EOG PREP** Sergio has 50 stamps. This is twice as many as Tina has, plus 4. How many stamps does Tina have? (Previous course)

A 21 stamps B 23 stamps C 29 stamps D 104 stamps

NORTH CAROLINA

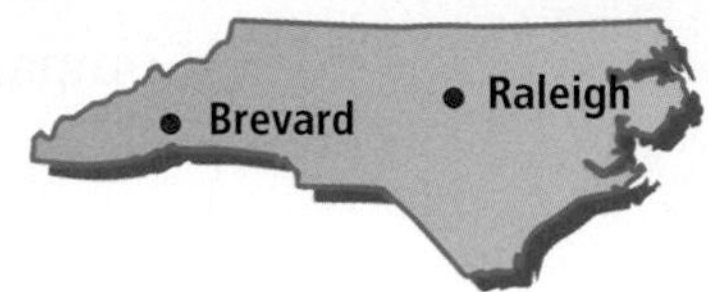

North Carolina Symphony

North Carolina was one of the first states to establish a state symphony. Founded in 1943, the North Carolina Symphony currently performs nearly 185 concerts each year; about 60 of those are performed in the symphony's home of Raleigh. The symphony consists of 68 musicians playing 18 types of instruments.

For 1–6, use the table.

1. Which type of instrument is there more of than any other type in the North Carolina Symphony?
2. Identify the outlier in the data set. How does this number relate to your answer in problem **1**?
3. Use the data to make a cumulative frequency table.
4. How many kinds of instruments are played by fewer than 11 musicians?
5. What number of musicians occurs most frequently? What measure of central tendency does this value represent?
6. Create a misleading bar graph using the data for the French horn, the viola, and the violin. How do the values appear to compare with each other, according to your graph? Are these comparisons correct? Explain.

Number of Musicians Who Play Each Type of Instrument

Instrument	Number of Musicians
Bass trombone	1
Bassoon	2
Cello	8
Clarinet	2
Double bass	5
English horn	1
Flute	3
French horn	5
Harp	1
Oboe	3
Percussion	2
Piccolo	1
Timpani	1
Trombone	3
Trumpet	2
Tuba	1
Viola	8
Violin	19

Brevard White Squirrels

Brevard, North Carolina, is home to a large population of white squirrels. As legend has it, a carnival truck overturned, releasing several white squirrels in Brevard and the other surrounding areas in Transylvania County. Soon after, the white squirrel population quickly grew throughout the county.

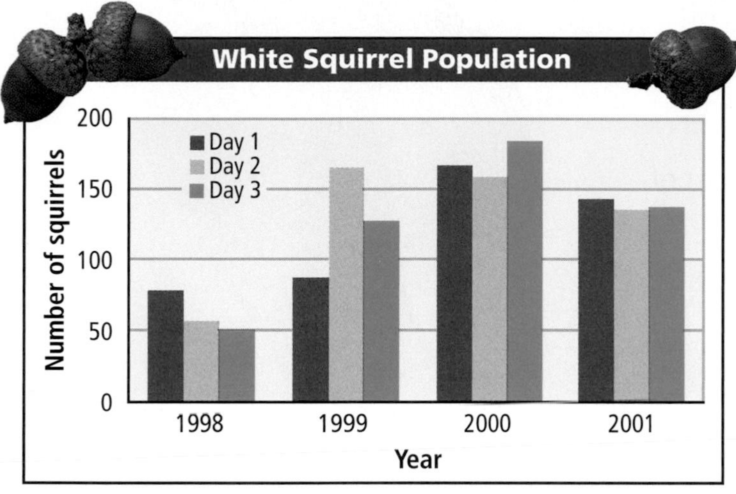

For 1–4, use the bar graph.

1. The bar graph shows the white squirrel count in Brevard during 3 days in October for 4 consecutive years.

a. The graph can be rearranged into 3 groups of 4 bars, where each group is one of the three days and the four bars in each group are the four years. Create a new bar graph with this arrangement.

b. In your rearranged bar graph, what trend can you see in the population of squirrels for each of the three days in October?

2. Find the approximate mean of the number of white squirrels for each of the four years.

3. What type of graph would be needed to show how many days the squirrel count was within certain intervals?

4. Make a graph to show the number of days over these four years that the white squirrel count was within certain intervals.

MATH-ABLES

Code Breaker

There are two main types of codes, or ciphers: *transposition* and *substitution*. In transposition, you switch the letters to make new words. Can you identify the following words?

REBMUN (Read the word backward.)

RGPAIHGN (Switch every other letter.)

THMA (Switch the first and last part of the word.)

Now try it with the sentence DOGS HAVE FOUR LEGS.

Substitution codes, such as *cryptograms*, are made by substituting different letters, numbers, or symbols for the originals. For example, suppose *RQH* represents *AND*. Each *A* in the puzzle would be written as *R*, each *N* as *Q*, and each *D* as *H*.

You can use math to solve a cryptogram. The following list tells you the probability for the occurrence of each letter of the English alphabet.

A	11	**B**	2	**C**	4	**D**	5	**E**	17	**F**	3
G	2	**H**	7	**I**	10	**J**	0	**K**	1	**L**	5
M	3	**N**	10	**O**	11	**P**	3	**Q**	0	**R**	8
S	9	**T**	13	**U**	4	**V**	1	**W**	3	**X**	0
Y	2	**Z**	0								

By knowing, for example, that *E* appears more often than any other letter, you can decode a message much more easily. Try decoding the following message by counting the number of times each letter appears.

AXMTV KDM VTAGTI YTVVKHTV KGT EJD. LGSIT NXJG XLD AXMT KDM IGN SI XJI XD NXJG EGSTDMV KDM EKYSCN.

Codes and secret messages are fun. Write your own code, and try it out on your friends and family.

Find Mean and Median

The scores on a math test that is worth 25 points were 23, 25, 10, 15, 22, 15, 9, 18, 17, 20, 15, 21, 5, 11, 24, 25, 20, 16, 20, 23, 17, 18, 23, 24, 10, 13, and 20.

internet connect
Lab Resources Online
go.hrw.com
KEYWORD: MS4 TechLab1

Activity

Use your graphing calculator to find the mean and median of the test scores.
To enter the values into list 1 (**L1**), press STAT 1.

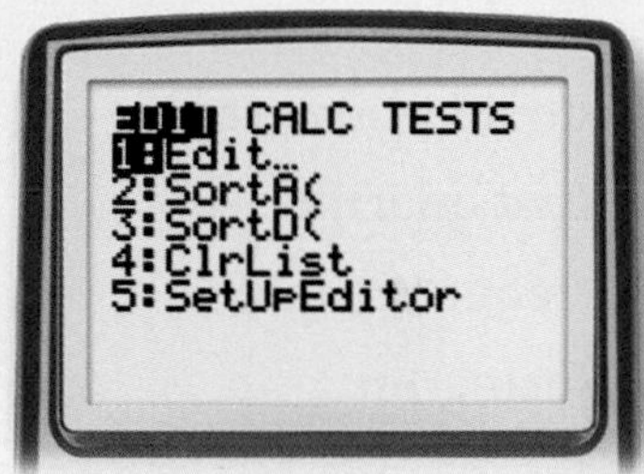

Type the data into **L1.** Press ENTER after each value.

To exit the list, press 2nd QUIT MODE. Then use the **LIST** menu to find the mean of the data.

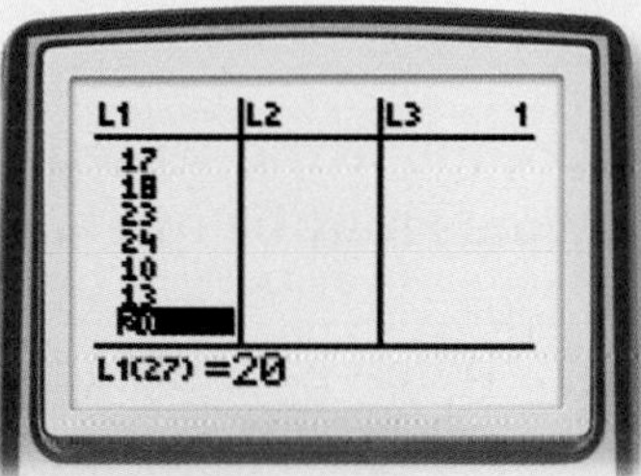

Press 2nd LIST STAT ▶ ▶ to highlight **MATH**. Press the down arrow twice to highlight **3: mean(.** Press ENTER.

Then press 2nd 1 ENTER to enter **L1** and see the mean.

The mean of the test scores, to the nearest hundredth, is 17.74.

To find the median, use the **LIST** menu again.

Press 2nd LIST STAT ▶ ▶ to highlight **MATH**. Press the down arrow three times to highlight **4: median(.** Press ENTER. Then press 2nd 1 ENTER. The median of the test scores is 18.

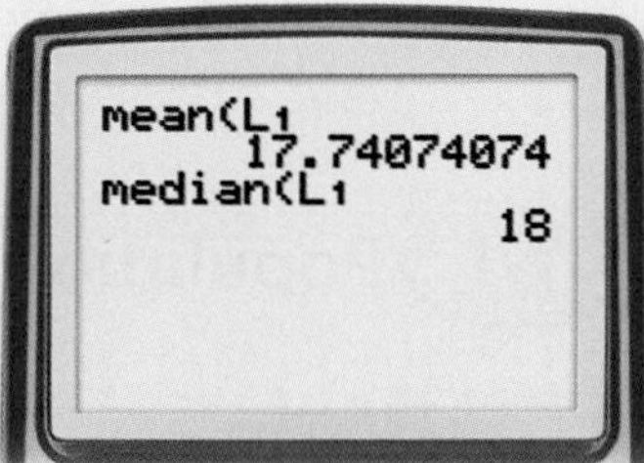

Think and Discuss

1. What do you think the **min** and **max** functions in the **LIST** menu will find?

Try This

Find the mean and median of each data set.

1. 1, 4, 5, 2, 1, 7, 9, 8, 5, 3, 5, 2, 8, 2, 7, 7

2. 55, 40, 70, 75, 60, 80, 70, 75, 70, 80

Chapter 1

Study Guide and Review

Vocabulary

Complete the sentences below with vocabulary words from the list above. Words may be used more than once.

1. When gathering information about a(n) ___?___, researchers often study part of the group, called a(n) ___?___.
2. The sum of the data values divided by the number of data items is called the ___?___ of the data.
3. A(n) ___?___ is a type of ___?___ that shows the frequency of data within equal intervals.
4. If the values in one data set increase as the values in the other set decrease, then the data sets show a(n) ___?___.

1-1 Populations and Samples (pp. 4–7)

EXAMPLE

■ **Identify the population and sample.**

A scientist tags 30 Canada geese to learn about the migration pattern of Canada geese.

Population	Sample
All Canada geese	The 30 Canada geese that were tagged

EXERCISES

Identify the population and sample.

5. The band director asks the first 5 students who enter the band hall after school how long they practice each afternoon.

1-2 Mean, Median, Mode, and Range (pp. 10–13)

EXAMPLE

- **Find the mean, median, mode, and range of the data set 3, 7, 10, 2, 3.**

mean: $3 + 7 + 10 + 2 + 3 = 25$ $\frac{25}{5} = 5$
median: 2, 3, 3, 7, 10
mode: 3 **range:** $10 - 2 = 8$

EXERCISES

Find the mean, median, mode, and range of each data set.

6. 2, 8, 0, 2, 3, 5, 7, 5, 4
7. 324, 233, 324, 399, 233, 299
8. 48, 39, 27, 52, 45, 47, 49, 37

1-3 Frequency Tables and Stem-and-Leaf Plots (pp. 14–17)

EXAMPLE

- **Make a stem-and-leaf plot of the data 27, 19, 35, 28, 25, 16, 39, 32, 29, 31, 16.**

Stems	Leaves
1	6 6 9
2	5 7 8 9
3	1 2 5 9

Key: 2|5 means 25

EXERCISES

Use the data set 35, 29, 14, 19, 32, 25, 27, 16, 8 for Exercises 9 and 10.

9. Make a cumulative frequency table of the data.
10. Make a stem-and-leaf plot of the data.

1-4 Bar Graphs and Histograms (pp. 20–23)

EXAMPLE

- **Make a bar graph of the chess club's results: W, L, W, W, L, W, L, L, W, W, W, L, W.**

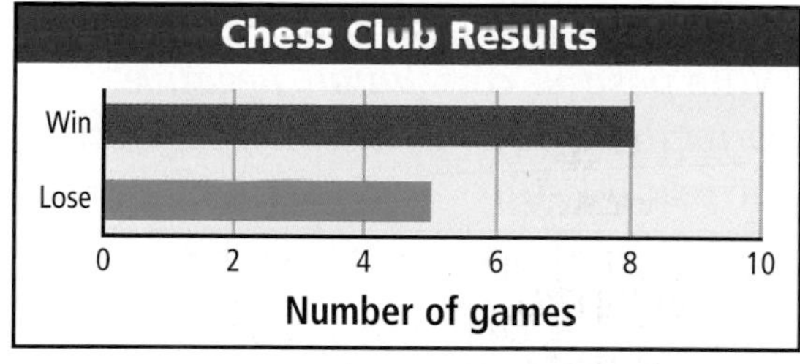

EXERCISES

11. Make a double-bar graph of the data.

Favorite Pet	Girls	Boys
Cat	42	31
Dog	36	52
Fish	3	10
Other	19	7

1-5 Reading and Interpreting Circle Graphs (pp. 24–27)

EXAMPLE

- **About what percent of people said yellow was their favorite color?**
about 25%

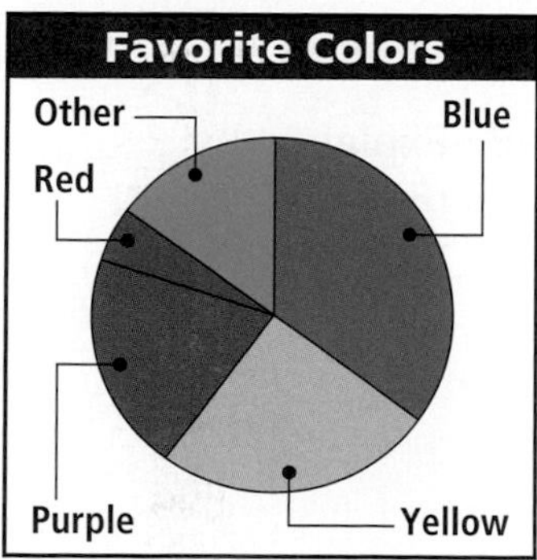

EXERCISES

Use the circle graph for Exercises 12 and 13.

12. Did more people choose purple or yellow as their favorite color?
13. According to the survey results, 35% of people chose blue as their favorite color. Out of the 100 people surveyed, how many people chose blue?

Study Guide and Review

1-6 Box-and-Whisker Plots (pp. 28–31)

EXAMPLE

■ **Use the data to make a box-and-whisker plot: 14, 10, 23, 16, 21, 26, 23, 17, 25.**

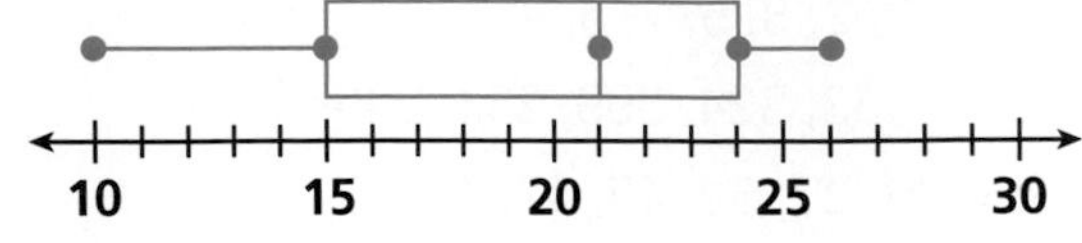

EXERCISES

Use the data to make a box-and-whisker plot.

14. 33, 38, 43, 30, 29, 40, 51, 27, 42, 23, 31

15. 11, 14, 18, 5, 16, 8, 19, 10, 17, 20, 34

1-7 Line Graphs (pp. 35–38)

EXAMPLE

■ **Make a line graph of the rainfall data: Apr, 5 in.; May, 3 in.; Jun, 4 in.; Jul, 1 in.**

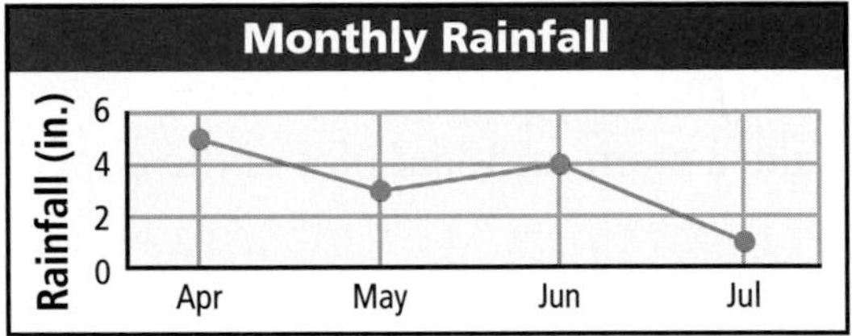

EXERCISES

16. Make a double-line graph of the data in the table.

U. S. Open Winning Scores					
	1995	1996	1997	1998	1999
Men	280	278	276	280	279
Women	278	272	274	290	272

1-8 Scatter Plots (pp. 40–43)

EXAMPLE

■ **Write *positive correlation, negative correlation,* or *no correlation* to describe the relationship.**

date of birth and eye color

There seems to be no correlation between the data sets.

EXERCISES

17. Use the data to make a scatter plot. Write *positive correlation, negative correlation,* or *no correlation* to describe the relationship.

Customers	47	56	35	75	25
Sales ($)	495	501	490	520	375

1-9 Misleading Graphs (pp. 44–47)

EXAMPLE

■ **Explain why the graph could be misleading.**

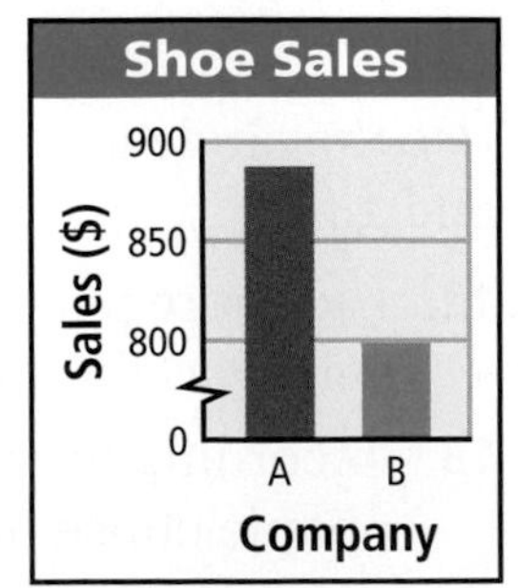

Because the vertical axis is broken, it appears that Company A's sales are more than twice Company B's sales.

EXERCISES

18. Explain why the graph could be misleading.

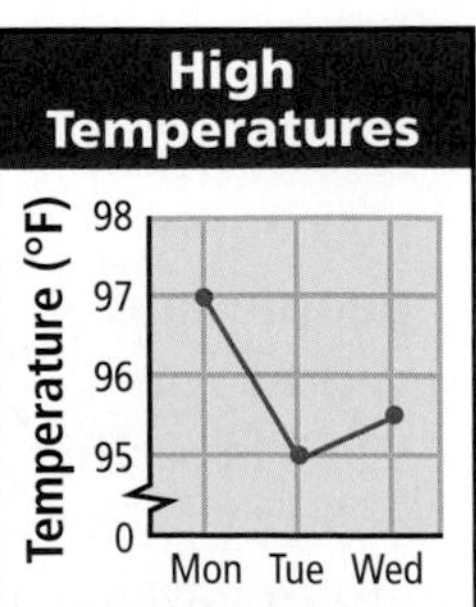

Chapter Test

1. A swim instructor asks 5 of his students which stroke they prefer. Identify the population and sample.

Use the data set 12, 18, 12, 22, 28, 23, 32, 10, 29, 36 for problems 2–9.

2. Find the mean, median, mode, and range of the data set.
3. How would the outlier 57 affect the mean of the data?
4. Make a cumulative frequency table of the data.
5. Make a stem-and-leaf plot of the data.
6. Make a histogram of the data.
7. Make a box-and-whisker plot of the data.
8. What is the third quartile?
9. What fraction of the data is less than 23?

Use the table for problems 10 and 11.

10. The table shows the weight in pounds of several mammals. Make a double-bar graph of the data.
11. Which mammal shows the greatest weight difference between the male and the female?

Mammal	Male	Female
Gorilla	450	200
Lion	400	300
Tiger	420	300

Use the circle graph for problems 12–14.

12. Which grade has the most students?
13. Approximately what percent of the students are seventh graders?
14. If the school population is 1,200 students, are more than 500 students in eighth grade? Explain.

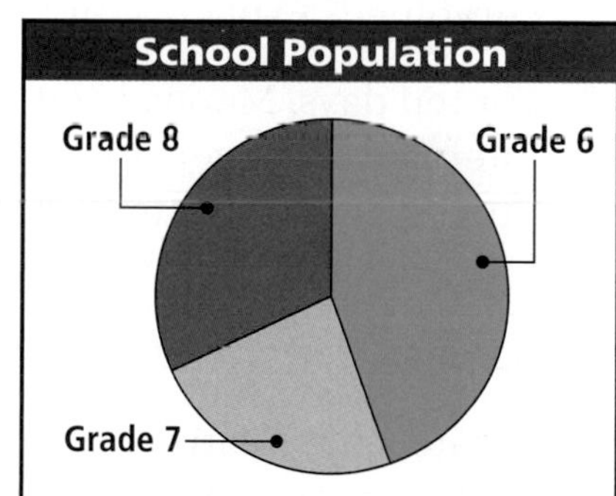

Use the table for problems 15–17.

15. The table shows passenger car fuel rates in miles per gallon for several years. Make a line graph of the data.
16. During which 2-year period did the fuel rate decrease?
17. Estimate the fuel rate in 1997.

Year	1992	1994	1996	1998
Rate	21.0	20.7	21.2	21.6

For problems 18 and 19, write *positive correlation, negative correlation,* or *no correlation* to describe each relationship.

18. size of hand and typing speed
19. height from which an object is dropped and time it takes to hit the ground
20. Explain why the graph at right could be misleading.

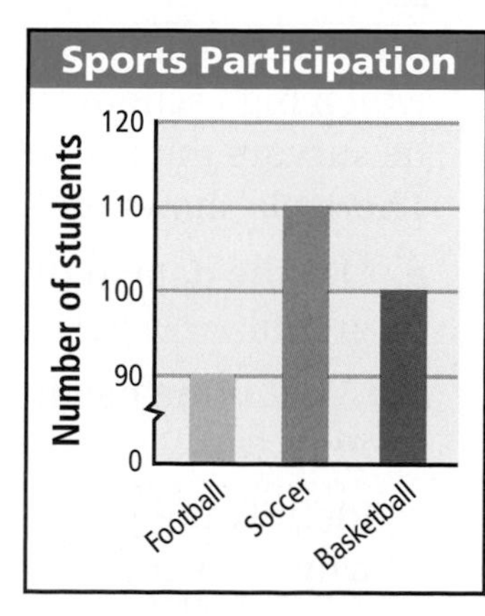

Chapter 1

Performance Assessment

Show What You Know

Create a portfolio of your work from this chapter. Complete this page and include it with your four best pieces of work from Chapter 1. Choose from your homework or lab assignments, mid-chapter quizzes, or any journal entries you have done. Put them together using any design you want. Make your portfolio represent what you consider your best work.

Short Response

1. Find the mean, median, mode, and range of the data set 48, 53, 46, 61, 63, 58, 52, 44, 51, and 63. Which measure of central tendency best represents the data? Explain your answer.

2. The circle graph shows the results of a survey about favorite kinds of pie. Do more people prefer apple pie than pecan and pumpkin pie combined? Explain your answer.

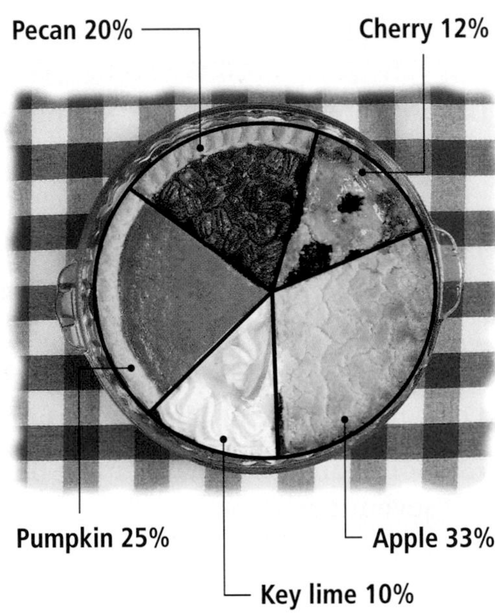

3. For ten days, Marcia went to the park and interviewed joggers for a report on physical fitness. The list shows the number of joggers she interviewed each day.

   12, 20, 5, 7, 8, 17, 13, 11, 3, 17

   Create a cumulative frequency table of the data. On how many days did Marcia interview fewer than 11 joggers?

Extended Problem Solving

4. Mr. Parker wants to identify the types of activities in which high school students participate after school, so he surveys the twelfth-graders in his science classes. The table shows the results of the survey.

Activity	Boys	Girls
Play sports	36	24
Talk to friends	6	30
Do homework	15	18
Work	5	4

   a. Use the data in the table to construct a double-bar graph.

   b. What is the mean number of girls per activity? Show your work.

   c. Identify the population and the sample in the survey. Explain why Mr. Parker's sampling method may not be random.

internet connect
EOG Practice Online
go.hrw.com Keyword: MS4 TestPrep

Getting Ready for EOG

Cumulative Assessment, Chapter 1

Use the numbers 8, 6, 7, 9, 8, 17, 8, 9 for items 1–3.

1. What is the mean of the numbers?

   A 8 C 8 and 9
   B 9 D 10

2. What is the mode of the numbers?

   A 8 C 8 and 9
   B 9 D 10

3. What is the median of the numbers?

   A 8 C 8 and 9
   B 9 D 10

Use the circle graph for items 4 and 5.

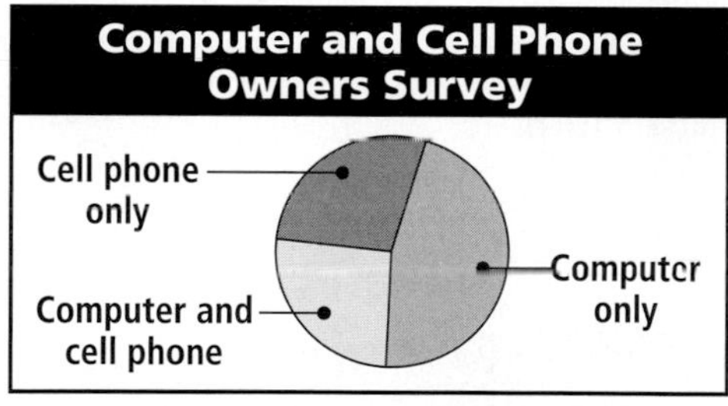

4. What percent of the people interviewed own a computer?

   A 46% C 72%
   B 52% D 100%

5. The circle graph shows the results of a survey of 200 people. How many of the people surveyed own a cell phone only?

   A 28 C 56
   B 45 D 108

6. What are the stems in a stem-and-leaf plot of the following data?
   18, 23, 15, 27, 16, 35, 38, 48, 19

   A 1, 2, 3, 4 C 1–9
   B 3, 5, 7, 8, 9 D 0–9

Use the line graph for items 7 and 8.

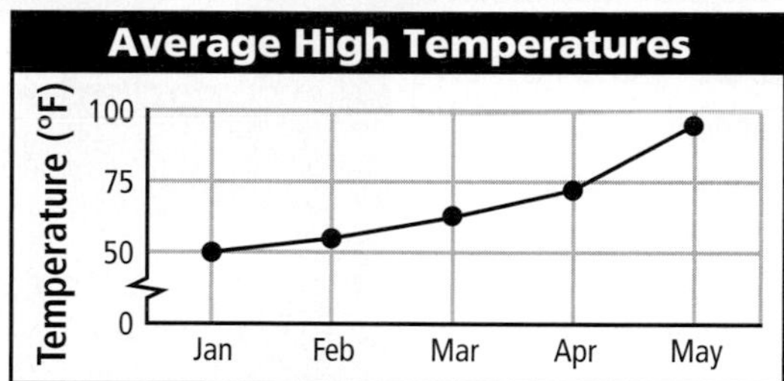

7. The line graph shows average high temperatures over a 5-month period. These temperatures increased the most between which two months?

   A Jan and Feb C Mar and Apr
   B Feb and Mar D Apr and May

TEST TAKING TIP!

Eliminate choices by using the definitions you have learned.

8. What is the range of the average high temperatures in the graph?

   A 45°F C 68°F
   B 62°F D 95°F

9. ***SHORT RESPONSE*** At the movie theater, Aimee and her brother bought 2 drinks for $2 each and a bucket of popcorn for $3. She gave the clerk a $10 bill. How much change should Aimee receive? Show your work.

10. ***SHORT RESPONSE*** A survey was conducted to determine which age group (0–19, 20–39, or 40–59) the most visitors at a local museum belonged to. The ages of the ten people surveyed were 12, 25, 17, 33, 31, 4, 15, 47, 51, and 7. Make a cumulative frequency table of the data. Then tell which age group the most visitors belonged to.

Chapter 2

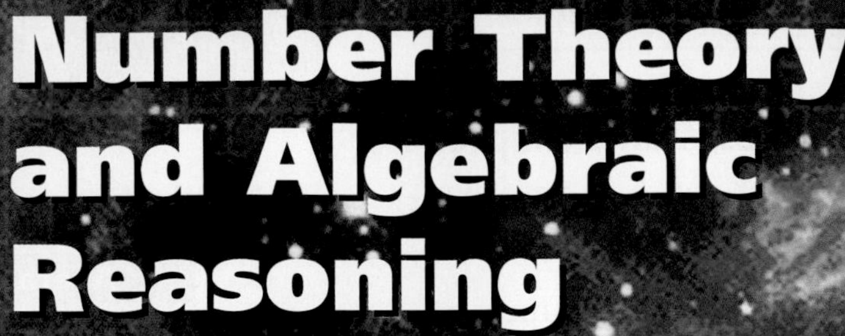

Number Theory and Algebraic Reasoning

TABLE OF CONTENTS

internet connect

Chapter Opener Online
go.hrw.com
KEYWORD: MS4 Ch2

Astronomical Distances	
Object	**Distance from the Sun (km)***
Mercury	5.80×10^7
Venus	1.082×10^8
Earth	1.495×10^8
Mars	2.279×10^8
Jupiter	7.780×10^8
Saturn	1.43×10^9
Uranus	2.90×10^9
Neptune	4.40×10^9
Pluto	5.80×10^9
Nearest star	3.973×10^{13}

*Distances of planets from the Sun are average distances.

Career *Cosmologist*

Dr. Stephen Hawking is a cosmologist. Cosmologists study the universe as a whole. They are interested in the origins, the structure, and the interaction of space and time.

The invention of the telescope has extended the vision of scientists far beyond nearby stars and planets. It has enabled them to view distant galaxies and structures that at one time were only theorized by astrophysicists such as Dr. Hawking. Astronomical distances are so great that we must use special mathematical notation to represent them conveniently.

ARE YOU READY?

Choose the best term from the list to complete each sentence.

1. The operation that gives the quotient of two numbers is __?__.
2. The __?__ of the digit 3 in 4,903,672 is thousands.
3. A number that is multiplied by another number is called a __?__.
4. The operation that gives the product of two numbers is __?__.
5. In the equation 15 ÷ 3 = 5, the __?__ is 5.

division
factor
multiplication
place value
product
quotient

Complete these exercises to review skills you will need for this chapter.

✔ Find Place Value

Give the place value of the digit 4 in each number.

6. 4,092 **7.** 608,241 **8.** 7,040,000 **9.** 4,556,890,100

10. 3,408,289 **11.** 34,506,123 **12.** 500,986,402 **13.** 3,540,277,009

✔ Use Repeated Multiplication

Find each product.

14. $2 \cdot 2 \cdot 2$ **15.** $9 \cdot 9 \cdot 9 \cdot 9$ **16.** $14 \cdot 14 \cdot 14$ **17.** $10 \cdot 10 \cdot 10 \cdot 10$

18. $3 \cdot 3 \cdot 5 \cdot 5$ **19.** $2 \cdot 2 \cdot 5 \cdot 7$ **20.** $3 \cdot 3 \cdot 11 \cdot 11$ **21.** $5 \cdot 10 \cdot 10 \cdot 10$

✔ Find Multiples

Find the first five multiples of each number.

22. 2 **23.** 9 **24.** 15 **25.** 1

26. 101 **27.** 54 **28.** 326 **29.** 1,024

✔ Find Factors

List all the factors of each number.

30. 8 **31.** 22 **32.** 36 **33.** 50

34. 108 **35.** 84 **36.** 256 **37.** 630

2-1 Exponents

Learn to represent numbers by using exponents.

Vocabulary

power

exponent

base

A DNA molecule makes a copy of itself by splitting in half. Each half becomes a molecule that is identical to the original. The molecules continue to split so that the two become four, the four become eight, and so on.

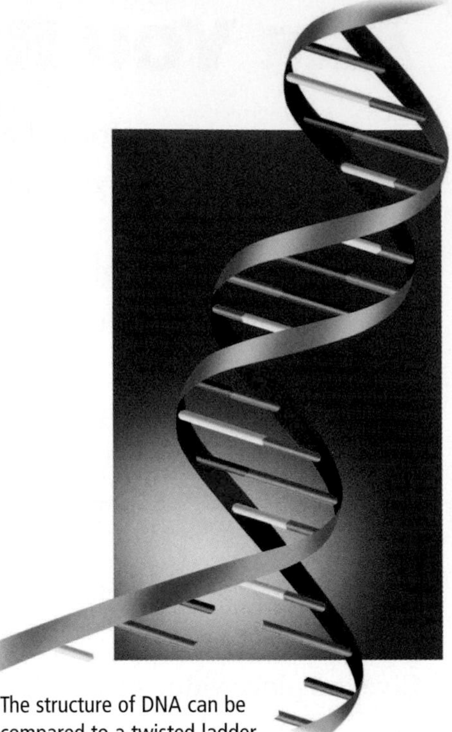

The structure of DNA can be compared to a twisted ladder.

Each time DNA copies itself, the number of molecules doubles. After four copies, the number of molecules is $2 \cdot 2 \cdot 2 \cdot 2 = 16$.

This multiplication can also be written as a **power**, using a *base* and an *exponent*. The **exponent** tells how many times to use the **base** as a factor.

Reading Math

Read 2^4 as "the fourth power of 2" or "2 to the fourth power."

EXAMPLE 1 Evaluating Powers

Find each value.

A 5^2

$5^2 = 5 \cdot 5$ *Use 5 as a factor 2 times.*

$= 25$

B 2^6

$2^6 = 2 \cdot 2 \cdot 2 \cdot 2 \cdot 2 \cdot 2$ *Use 2 as a factor 6 times.*

$= 64$

Recall that any number to the first power is equal to that number.

$6^1 = 6$ $\quad$ $13^1 = 13$ $\quad$ $25^1 = 25$

Any number to the zero power, except zero, is equal to 1.

$6^0 = 1$ $\quad$ $10^0 = 1$ $\quad$ $19^0 = 1$

Zero to the zero power is *undefined,* meaning that it does not exist.

To express a whole number as a power, write the number as the product of equal factors. Then write the product using the base and an exponent. For example, $10{,}000 = 10 \cdot 10 \cdot 10 \cdot 10 = 10^4$.

EXAMPLE 2 Expressing Whole Numbers as Powers

Write each number using an exponent and the given base.

A **49, base 7**

$49 = 7 \cdot 7$ *7 is used as a factor 2 times.*

$= 7^2$

B **81, base 3**

$81 = 3 \cdot 3 \cdot 3 \cdot 3$ *3 is used as a factor 4 times.*

$= 3^4$

EXAMPLE 3 *Earth Science Application*

Earth Science LINK

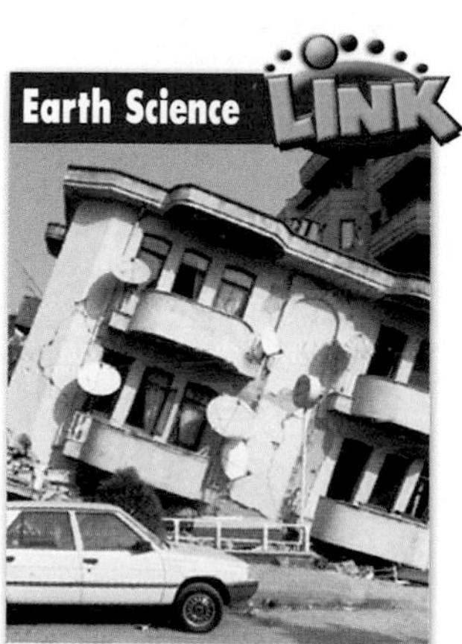

An earthquake measuring 7.2 on the Richter scale struck Duzce, Turkey, on November 12, 1999.

The Richter scale measures an earthquake's strength, or magnitude. Each category in the table is 10 times stronger than the next lower category. For example, a large earthquake is 10 times stronger than a moderate earthquake. How many times stronger is a great earthquake than a moderate one?

Earthquake Strength	
Category	**Magnitude**
Moderate	5
Large	6
Major	7
Great	8

An earthquake with a magnitude of 6 is 10 times stronger than one with a magnitude of 5.

An earthquake with a magnitude of 7 is 10 times stronger than one with a magnitude of 6.

An earthquake with a magnitude of 8 is 10 times stronger than one with a magnitude of 7.

$$10 \cdot 10 \cdot 10 = 10^3 = 1{,}000$$

A great earthquake is 1,000 times stronger than a moderate one.

Think and Discuss

1. **Describe** the relationship between 3^5 and 3^6.
2. **Tell** which power of 8 is equal to 2^6. Explain.
3. **Explain** why any number to the first power is equal to that number.

2-1 Exercises

FOR EOG PRACTICE

see page 660

internet connect

Homework Help Online
go.hrw.com Keyword: MS4 2-1

1.02

GUIDED PRACTICE

See Example 1 **Find each value.**

1. 2^5 **2.** 3^3 **3.** 6^2 **4.** 9^0

See Example 2 **Write each number using an exponent and the given base.**

5. 25, base 5 **6.** 16, base 4 **7.** 27, base 3 **8.** 100, base 10

See Example 3 **9.** On the Richter scale, a great earthquake is 10 times stronger than a major one, and a major one is 10 times stronger than a large one. How many times stronger is a great earthquake than a large one?

INDEPENDENT PRACTICE

See Example 1 **Find each value.**

10. 11^2 **11.** 3^5 **12.** 8^3 **13.** 1^0

14. 4^3 **15.** 3^4 **16.** 2^5 **17.** 5^1

18. 2^3 **19.** 5^3 **20.** 30^1 **21.** 10^4

See Example 2 **Write each number using an exponent and the given base.**

22. 81, base 9 **23.** 4, base 4 **24.** 64, base 4

25. 1, base 7 **26.** 32, base 2 **27.** 128, base 2

28. 1,600, base 40 **29.** 2,500, base 50 **30.** 100,000, base 10

See Example 3 **31.** In a game, a contestant had a starting score of one point. He tripled his score every turn for four turns. Write his score after four turns as a power. Then find his score.

PRACTICE AND PROBLEM SOLVING

Give two ways to represent each number using powers.

32. 81 **33.** 16 **34.** 64 **35.** 729

Find each value.

36. $2^2 + 2^5$ **37.** $3^2 + 4^2$ **38.** $8^2 + 9^2$

39. $3^4 + 3^3$ **40.** $6^2 + 6^0$ **41.** $125^1 + 125^0$

42. $1{,}254^0 \cdot 82$ **43.** $19^0 + 17^0 + 10^0$ **44.** $3^3 + 4^2 - 5^1$

45. To find the volume of a cube, find the third power of the length of an edge of the cube. What is the volume of a cube that is 6 inches long on an edge?

46. Domingo decided to save \$0.03 the first day and to triple the amount he saves each day. How much will he save on the seventh day?

47. ***SOCIAL STUDIES*** If the populations of the cities in the table double every 10 years, what will their populations be in 2029?

City	Population (1999)
Yuma, AZ	60,519
Phoenix, AZ	1,159,014

48. ***HOBBIES*** Malia is making a quilt with a pattern of rings. In the center ring, she uses four stars. In each of the next three rings, she uses three times as many stars as in the one before. How many stars does she use in the fourth ring? Write the answer as a power and find its value.

49. ***LIFE SCIENCE*** The cells of some kinds of bacteria divide every 30 minutes. If you begin with a single cell, how many cells will there be after 1 hour? 2 hours? 3 hours?

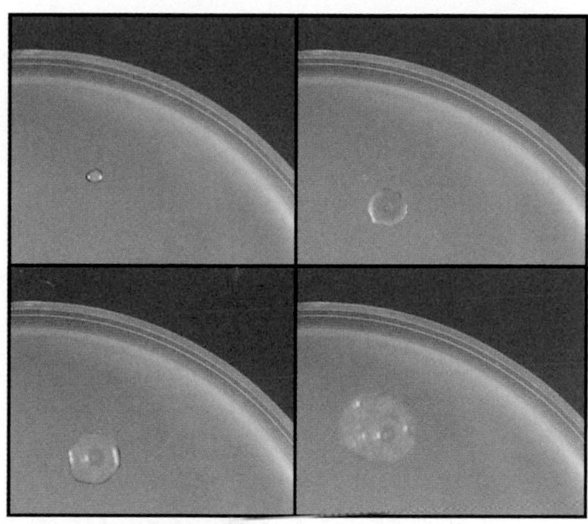
Bacteria divide by pinching in two. This process is called binary fission.

50. ***WHAT'S THE ERROR?*** A student wrote 64 as $8 \cdot 2$. What did the student do wrong?

51. ***WRITE ABOUT IT*** Is 2^5 greater than or less than 3^3? Explain your answer.

52. ***CHALLENGE*** What is the length of the edge of a cube if its volume is 1,000 cubic meters?

Spiral Review

53. Trent assigned each student in his class a number. Which of the following gives a random sample? (Lesson 1-1)

a. Trent puts the numbers on slips of paper in a bag and draws 10 numbers.

b. Trent puts the numbers on slips of paper and picks 10 even numbers.

Find the mean of each data set. (Lesson 1-2)

54. 54, 68, 29, 73

55. 2, 5, 3, 6, 8, 1, 7, 4, 2, 5

56. **EOG PREP** Given the range, mean, median, and mode of a set of data,which measure must be one of the values in the set? (Lesson 1-2)

A range
B mean
C median
D mode

57. **EOG PREP** The median of the lower half of a data set is called the __?__. (Lesson 1-6)

A first quartile
B second quartile
C third quartile
D lower extreme

2-2 Powers of Ten and Scientific Notation

Learn to express large numbers in scientific notation.

Vocabulary
standard form
scientific notation

The distance from Venus to the Sun is over 100,000,000 kilometers. You can write this number as a power of ten by using a base of ten and an exponent.

$10 \cdot 10 \cdot 10 \cdot 10 \cdot 10 \cdot 10 \cdot 10 \cdot 10 = 10^8$

Power of ten (points to 10^8)

The table shows several powers of ten.

Power of 10	Meaning	Value
10^1	10	10
10^2	$10 \cdot 10$	100
10^3	$10 \cdot 10 \cdot 10$	1,000
10^4	$10 \cdot 10 \cdot 10 \cdot 10$	10,000

Astronomers estimate that there are 100 billion billion, or 10^{20}, stars in the universe.

EXAMPLE 1 Multiplying by Powers of Ten

Multiply $21 \cdot 10^4$.

$21 \cdot 10^4 = 21 \cdot (10 \cdot 10 \cdot 10 \cdot 10)$ *Use 10 as a factor 4 times.*

$= 21 \cdot 10{,}000$ *Multiply.*

$= 210{,}000$

You can also find the product of a number and a power of ten simply by moving the decimal point of the number. For powers of ten with positive exponents, move the decimal point to the right.

EXAMPLE 2 Multiplying by Powers of Ten Mentally

Find each product.

A $137 \cdot 10^3$

3 places

$137 \cdot 10^3 = 137.000$ *Move the decimal point 3 places.*

$= 137{,}000$ *(You will need to add 3 zeros.)*

B $46.2 \cdot 10^5$

5 places

$46.2 \cdot 10^5 = 46.20000$ *Move the decimal point 5 places.*

$= 4{,}620{,}000$ *(You will need to add 4 zeros.)*

Numbers are usually written in **standard form**. For example, 17,900,000 is in standard form. **Scientific notation** is a kind of shorthand that can be used to write large numbers. Numbers expressed in scientific notation are written as the product of two factors. In scientific notation, 17,900,000 is written as

Writing Math

In scientific notation, it is customary to use a multiplication cross (×) instead of a dot.

$$1.79 \times 10^7$$

A number greater than or equal to 1 but less than 10 (1.79)

A power of 10 (10^7)

EXAMPLE 3 Writing Numbers in Scientific Notation

Write each number in scientific notation.

A **9,580,000**

$9{,}580{,}000 = 9{,}580{,}000.$ (6 places) — *Move the decimal point to get a number that is greater than or equal to 1 and less than 10.*

$= 9.58 \times 10^6$ — *The exponent is equal to the number of places the decimal point is moved.*

B **813,000,000**

$813{,}000{,}000 = 813{,}000{,}000.$ (8 places) — *Move the decimal point to get a number that is greater than or equal to 1 and less than 10.*

$= 8.13 \times 10^8$ — *The exponent is equal to the number of places the decimal point is moved.*

EXAMPLE 4 Writing Numbers in Standard Form

Pluto is about 3.7×10^9 miles from the Sun. Write this distance in standard form.

$3.7 \times 10^9 = 3.700000000$ — *Since the exponent is 9, move the decimal point 9 places to the right.*

$= 3{,}700{,}000{,}000$

Pluto is about 3,700,000,000 miles from the Sun.

Think and Discuss

1. **Tell** whether 15×10^9 is in scientific notation. Explain.
2. **Compare** $4 \cdot 10^3$ and $3 \cdot 10^4$. Explain how you know which is greater.

2-2 Exercises

FOR EOG PRACTICE

see page 660

internet connect
Homework Help Online
go.hrw.com Keyword: MS4 2-2

1.03

GUIDED PRACTICE

See Example 1 **Multiply.**

1. $15 \cdot 10^2$ **2.** $18 \cdot 10^3$ **3.** $11 \cdot 10^0$ **4.** $12 \cdot 10^4$

See Example 2 **Find each product.**

5. $208 \cdot 10^3$ **6.** $113 \cdot 10^7$ **7.** $47.2 \cdot 10^4$ **8.** $3.622 \cdot 10^2$

See Example 3 **Write each number in scientific notation.**

9. 3,600,000 **10.** 214,000 **11.** 8,000,000,000 **12.** 42,000

See Example 4 **13.** A drop of water contains about 2.0×10^{21} molecules. Write this number in standard form.

INDEPENDENT PRACTICE

See Example 1 **Multiply.**

14. $21 \cdot 10^2$ **15.** $8 \cdot 10^4$ **16.** $25 \cdot 10^5$ **17.** $40 \cdot 10^4$

18. $11 \cdot 10^2$ **19.** $19 \cdot 10^3$ **20.** $20 \cdot 10^0$ **21.** $14 \cdot 10^1$

See Example 2 **Find each product.**

22. $268 \cdot 10^3$ **23.** $105 \cdot 10^0$ **24.** $4.16 \cdot 10^4$ **25.** $2.164 \cdot 10^2$

26. $550 \cdot 10^7$ **27.** $1.020 \cdot 10^3$ **28.** $2{,}115 \cdot 10^5$ **29.** $70{,}030 \cdot 10^1$

See Example 3 **Write each number in scientific notation.**

30. 428,000 **31.** 1,610,000 **32.** 3,000,000,000 **33.** 60,100

34. 52.000 **35.** $29.8 \cdot 10^7$ **36.** 8,900,000 **37.** $500 \cdot 10^3$

See Example 4 **38.** Ancient Egyptians hammered gold into sheets so thin that it took 3.67×10^5 sheets to make a pile 2.5 centimeters high. Write this quantity in standard form.

39. In a vacuum, light travels at a speed of about nine hundred and eighty million feet per second. Write this speed in scientific notation.

PRACTICE AND PROBLEM SOLVING

Find the missing number or numbers.

40. $24{,}500 = 2.45 \times 10^{\square}$ **41.** $16{,}800 = \square \times 10^4$

42. $\square = 3.40 \times 10^2$ **43.** $280{,}000 = 2.8 \times 10^{\square}$

44. $5.4 \times 10^8 = \square$ **45.** $60{,}000{,}000 = \square \times 10^{\square}$

46. $\square = 5.92 \times 10^5$ **47.** $244{,}000{,}000 = \square \times 10^{\square}$

Earth Science 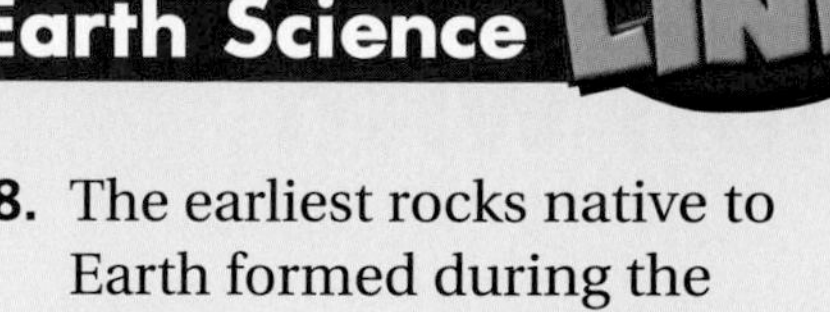

48. The earliest rocks native to Earth formed during the Archean eon. Calculate the length of this eon. Write your answer in scientific notation.

49. Dinosaurs lived during the Mesozoic era. After this era they were extinct. Calculate the length of the Mesozoic era. Write your answer in scientific notation.

50. Tropites were prehistoric marine animals. Because they lived for a relatively short time on Earth, their fossil remains can be used to date the rock formations in which they are found. Such fossils are known as *index fossils.* Tropites lived between 2.08×10^8 and 2.30×10^8 years ago. During what geologic time period did they live?

Geologic Time Scale

Eon	Era	Period
Phanerozoic (540 mya*–present)	**Cenozoic** (65 mya–present)	**Quaternary** (1.8 mya–present) **Holocene epoch** (11,000 yrs ago–present) **Pleistocene epoch** (1.8 mya–11,000 yrs ago) **Tertiary** (65 mya–1.8 mya) **Pliocene epoch** (5.3 mya–1.8 mya) **Miocene epoch** (23.8 mya–5.3 mya) **Oligocene epoch** (33.7 mya–23.8 mya) **Eocene epoch** (54.8 mya–33.7 mya) **Paleocene epoch** (65 mya–54.8 mya)
	Mesozoic (248 mya–65 mya)	**Cretaceous** (144 mya–65 mya) **Jurassic** (206 mya–144 mya) **Triassic** (248 mya–206 mya)
	Paleozoic (540 mya–248 mya)	**Permian** (290 mya–248 mya) **Pennsylvanian** (323 mya–290 mya) **Mississippian** (354 mya–323 mya) **Devonian** (417 mya–354 mya) **Silurian** (443 mya–417 mya) **Ordovician** (490 mya–443 mya) **Cambrian** (540 mya–490 mya)
Proterozoic (2,500 mya–540 mya)		
Archean (3,800 mya–2,500 mya)		
Hadean (4,600 mya–3,800 mya)		

*mya = million years ago

51. ***WRITE ABOUT IT*** Explain why scientific notation is especially useful in earth science.

52. ***CHALLENGE*** We live in the Holocene epoch. Write the age of this epoch in scientific notation.

Spiral Review

53. Make a stem-and-leaf plot of the data 48, 60, 57, 62, 43, 62, 45, 51. (Lesson 1-3)

54. The following data gives the number of students who attended the homecoming football game each year: 1995, 418 students; 1996, 330 students; 1997, 377 students; 1998, 403 students. Make a bar graph of the data. (Lesson 1-4)

55. **EOG PREP** What is the value of 4^6? (Lesson 2-1)

A 24
B 1,024
C 4,096
D 16,384

56. **EOG PREP** What is the value of $2^3 + 3^2$? (Lesson 2-1)

A 12
B 17
C 43
D 55

Scientific Notation with a Calculator

Use with Lesson 2-2

Scientists often have to work with very large numbers. For example, the Andromeda Galaxy contains over 200,000,000,000 stars. Scientific notation is a compact way of expressing large numbers such as this.

Lab Resources Online
go.hrw.com
KEYWORD: MS4 Lab2A

Activity

1 Show 200,000,000,000 in scientific notation.

Enter 200,000,000,000 on your graphing calculator. Then press ENTER.

2 E 11 on the calculator display means 2×10^{11}, which is 200,000,000,000 in scientific notation. Your calculator automatically puts very large numbers into scientific notation.

You can use the **EE** function (the second function for the *comma* key) to enter 2×10^{11} directly into the calculator.

Enter 2×10^{11} by pressing 2 2nd EE , 11 ENTER.

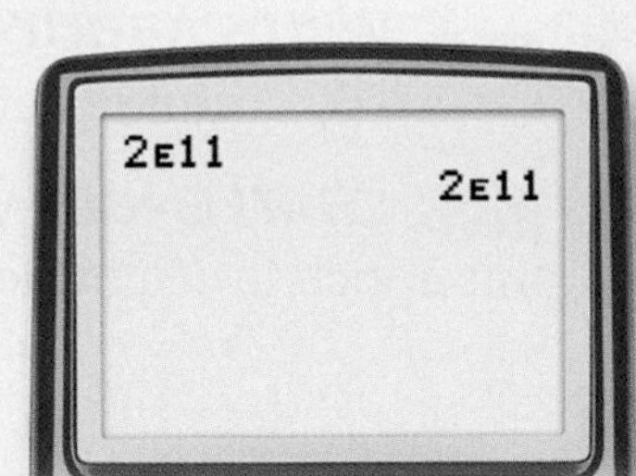

2 Evaluate 230,000,000 × 650,000.

Write 230,000,000 and 650,000 in scientific notation.

$230{,}000{,}000 = 2.3 \times 10^8$

$650{,}000 = 6.5 \times 10^5$

Press 2.3 2nd EE , 8 × 6.5 2nd EE , 5 ENTER.

In scientific notation, 230,000,000 × 650,000 is equal to 1.495×10^{14}.

3 Evaluate $(340{,}000{,}000{,}000 \div 1{,}235) \times 4{,}568$.

Write each number in scientific notation.

$340{,}000{,}000{,}000 = 3.4 \times 10^{11}$

$1{,}235 = 1.235 \times 10^{3}$

$4{,}568 = 4.568 \times 10^{3}$

Press (3.4 2nd EE , 11 ÷ 1.235 2nd EE , 3) ×

4.568 2nd EE , 3 ENTER .

$(340{,}000{,}000{,}000 \div 1{,}235) \times 4{,}568$ is equal to $1.257587045 \times 10^{12}$.

Because this calculator displays results to only ten decimal places, this answer is not exact. The exact answer is $1.25758704453441\,10 \times 10^{12}$.

Think and Discuss

1. What happens when you add 9,999,999,999 + 1 on a graphing calculator? Explain why this happens.

Try This

In space, light travels about 9,460,000,000,000 kilometers per year. This distance is known as a **light-year.** Use a calculator to find the distance in kilometers from Earth to each astronomical object. Write the results in scientific notation.

1. the star Altair, 16.3 light-years away
2. the Large Magellanic Cloud Galaxy, 180,000 light-years away
3. the Andromeda Galaxy, 2,900,000 light-years away

In 1999, astronomers at the Space Telescope Science Institute estimated that there are 125 billion galaxies in the observable universe.

4. The number 125 billion is equal to 125,000,000,000. Write this number in scientific notation.
5. Suppose the universe is made up of 125,000,000,000 galaxies, each containing 200,000,000,000 stars. Use this data to find the number of stars in the universe. Express your answer in scientific notation and in standard form.

2-3 Order of Operations

Learn to use the order of operations to simplify numerical expressions.

Vocabulary

numerical expression

order of operations

When you get ready for school, you put on your socks *before* you put on your shoes. In mathematics, as in life, some tasks must be done in a certain order.

A **numerical expression** is made up of numbers and operations. When simplifying a numerical expression, rules must be followed so that everyone gets the same answer. That is why mathematicians have agreed upon the **order of operations**.

ORDER OF OPERATIONS

1. Perform operations within grouping symbols.
2. Evaluate powers.
3. Multiply and divide in order from left to right.
4. Add and subtract in order from left to right.

EXAMPLE 1 Using the Order of Operations

Evaluate.

A $27 - 18 \div 6$

$27 - 18 \div 6$	*Divide.*
$27 - 3$	*Subtract.*
24	

B $36 - 18 \div 2 \cdot 3 + 8$

$36 - 18 \div 2 \cdot 3 + 8$	*Divide and multiply from left to right.*
$36 - 9 \cdot 3 + 8$	
$36 - 27 + 8$	*Subtract and add from left to right.*
$9 + 8$	
17	

C $5 + 6^2 \cdot 10$

$5 + 6^2 \cdot 10$	*Evaluate the power.*
$5 + 36 \cdot 10$	*Multiply.*
$5 + 360$	*Add.*
365	

EXAMPLE 2 Using the Order of Operations with Grouping Symbols

Evaluate.

A $36 - (2 \cdot 6) \div 3$

$36 - (2 \cdot 6) \div 3$ *Perform the operation inside the parentheses.*

$36 - 12 \div 3$

$36 - 4$

32

Helpful Hint

When an expression has a set of grouping symbols within a second set of grouping symbols, begin with the innermost set.

B $[(4 + 12 \div 4) - 2]^3$

$[(4 + 12 \div 4) - 2]^3$ *The parentheses are inside the brackets, so perform the operations inside the parentheses first.*

$[(4 + 3) - 2]^3$

$[7 - 2]^3$

5^3

125

EXAMPLE 3 *Career Application*

Maria works part-time in a law office, where she earns $20 per hour. The table shows the number of hours she worked last week. Evaluate the expression $(6 + 5 \cdot 3) \cdot 20$ to find out how much money Maria earned last week.

Day	Hours
Monday	6
Tuesday	5
Wednesday	5
Thursday	5

$(6 + 5 \cdot 3) \cdot 20$ *Perform the operations inside the parentheses.*

$(6 + 15) \cdot 20$

$21 \cdot 20$

420

Maria earned $420 last week.

Think and Discuss

1. **Apply** the order of operations to determine if the expressions $3 + 4^2$ and $(3 + 4)^2$ have the same value.
2. **Give** the correct order of operations for evaluating $(5 + 3 \cdot 20) \div 13 + 3^2$.
3. **Determine** where grouping symbols should be inserted in the expression $3 + 9 - 4 \cdot 2$ so that its value is 13.

2-3 Exercises

FOR EOG PRACTICE
see page 661

internet connect
Homework Help Online
go.hrw.com Keyword: MS4 2-3

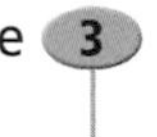
1.02a

GUIDED PRACTICE

See Example 1

Evaluate.

1. $43 + 16 \div 4$
2. $28 - 4 \cdot 3 \div 6 + 4$
3. $25 - 4^2 \div 8$

See Example 2

4. $26 - (7 \cdot 3) + 2$
5. $(3^2 + 11) \div 5$
6. $32 + 6(4 - 2^2) + 8$

See Example 3

7. Caleb earns \$10 per hour. He worked 4 hours on Monday, Wednesday, and Friday. He worked 8 hours on Tuesday and Thursday. Evaluate the expression $(3 \cdot 4 + 2 \cdot 8) \cdot 10$ to find out how much Caleb earned in all.

INDEPENDENT PRACTICE

See Example 1

Evaluate.

8. $3 + 7 \cdot 5 - 1$
9. $5 \cdot 9 - 3$
10. $3 - 2 + 6 \cdot 2^2$

See Example 2

11. $(3 \cdot 3 - 3)^2 \div 3 + 3$
12. $2^5 - (4 \cdot 5 + 3)$
13. $(3 \div 3) + 3 \cdot (3^3 - 3)$
14. $4^3 \div 8 - 2$
15. $(8 - 2)^2 \cdot (8 - 1)^2 \div 3$
16. $9{,}234 \div [3 \cdot 3(1 + 8^3)]$

See Example 3

17. Maki paid a \$14 basic fee plus \$25 a day to rent a car. Evaluate the expression $14 + 5 \cdot 25$ to find out how much it cost her to rent the car for 5 days.
18. Enrico spent \$20 per square yard for carpet and \$35 for a carpet pad. Evaluate the expression $35 + 20(12^2 \div 9)$ to find out how much Enrico spent to carpet a 12 ft by 12 ft room.

PRACTICE AND PROBLEM SOLVING

Compare. Write $<$, $>$, or $=$.

19. $8 \cdot 3 - 2$ ▢ $8 \cdot (3 - 2)$
20. $(6 + 10) \div 2$ ▢ $6 + 10 \div 2$
21. $12 \div 3 \cdot 4$ ▢ $12 \div (3 \cdot 4)$
22. $18 + 6 - 2$ ▢ $18 + (6 - 2)$
23. $[6(8 - 3) + 2]$ ▢ $6(8 - 3) + 2$
24. $(18 - 14) \div (2 + 2)$ ▢ $18 - 14 \div 2 + 2$

Insert grouping symbols to make each statement true.

25. $4 \cdot 8 - 3 = 20$
26. $5 + 9 - 3 \div 2 = 8$
27. $12 - 2^2 \div 5 = 20$
28. $4 \cdot 2 + 6 = 32$
29. $4 + 6 - 3 \div 7 = 1$
30. $9 \cdot 8 - 6 \div 3 = 6$
31. Bertha earned \$8.00 per hour for 4 hours babysitting and \$10.00 per hour for 5 hours painting a room. Evaluate the expression $8 \cdot 4 + 10 \cdot 5$ to find out how much Bertha earned in all.

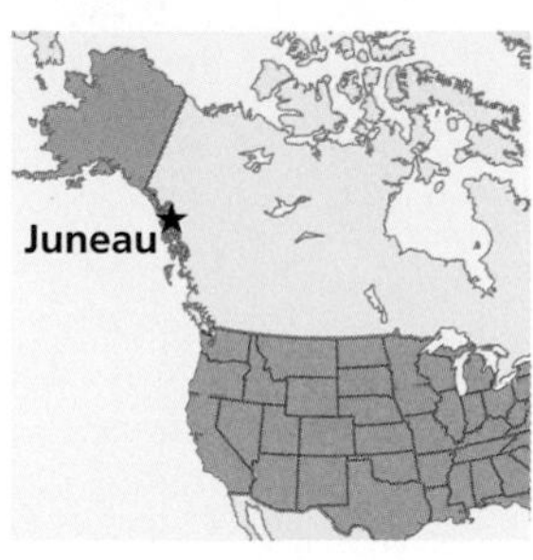

32. ***EARTH SCIENCE*** The graph shows the average temperatures for each of four months in Juneau, Alaska.

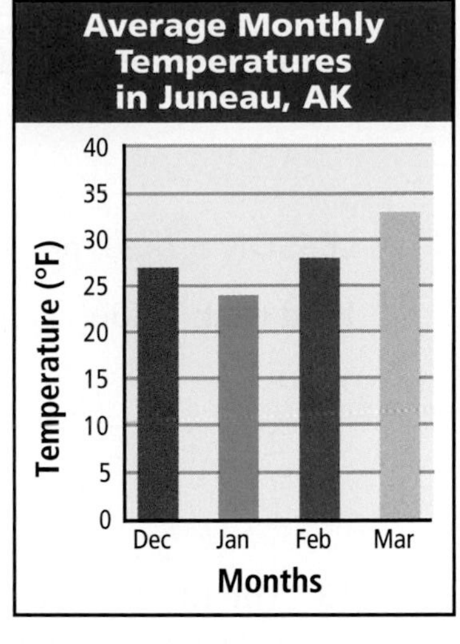

   a. Write and evaluate an expression to find the average temperature over the 4-month period.

   b. Write an expression that shows the difference between the range of average temperatures from December to February and the range of average temperatures from January to March.

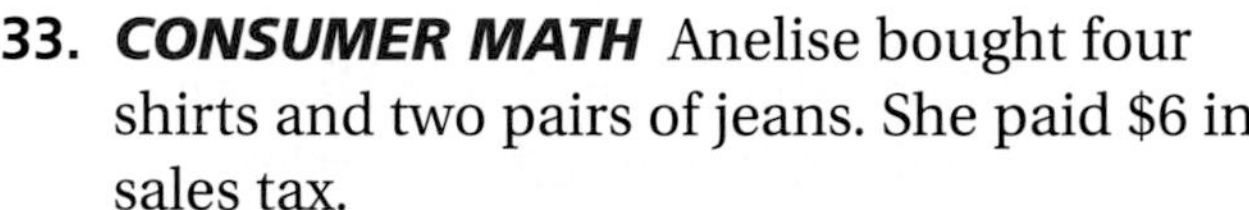

33. ***CONSUMER MATH*** Anelise bought four shirts and two pairs of jeans. She paid $6 in sales tax.

   a. Write an expression that shows how much she spent on shirts.

   b. Write an expression that shows how much she spent on jeans.

   c. Write and evaluate an expression to show how much she spent on clothes, including sales tax.

34. ***CHOOSE A STRATEGY*** There are four children in a family. The sum of the squares of the ages of the three youngest children equals the square of the age of the oldest child. How old are the children?

   A 1, 4, 8, 9 **B** 1, 3, 6, 12 **C** 4, 5, 8, 10 **D** 2, 3, 8, 16

35. ***WRITE ABOUT IT*** Describe in what order you would perform the operations to find the value of $[(2 + 4)^2 - 2 \cdot 3] \div 6$.

36. ***CHALLENGE*** Use the numbers 3, 5, 6, 2, 54, and 5 in that order to write an expression that has a value of 100.

Spiral Review

37. Hal recorded the following gas prices per gallon over a six-week period: $1.79, $1.88, $2.25, $1.90, $1.70, and $1.85. Construct a line graph to show the prices. (Lesson 1-7)

38. **EOG PREP** Which of the following is *not* equal to 64? (Lesson 2-1)

   A 64^0 C 2^6

   B 4^3 D 8^2

39. **EOG PREP** Which number is equivalent to 5.3×10^7? (Lesson 2-2)

   A 5,000,003 C 53,000,000

   B 5,300,000 D 530,000,000

Chapter 2

Mid-Chapter Quiz

Mid-Chapter Quiz

LESSON 2-1 (pp. 60–63)

Find each value.

1. 6^3 **2.** 12^2 **3.** 11^0 **4.** 3^4

5. 2^7 **6.** 10^6 **7.** 5^3 **8.** 7^2

9. 17^1 **10.** 9^2 **11.** 8^4 **12.** 4^5

13. The number of a certain bacteria doubles every hour. How many bacteria cells will there be after 8 hours if there is one cell to start? Write your answer as a power.

LESSON 2-2 (pp. 64–67)

Find each product.

14. $32 \cdot 10^3$ **15.** $147 \cdot 10^4$ **16.** $44.2 \cdot 10^2$

17. $23 \cdot 10^6$ **18.** $80.4 \cdot 10^3$ **19.** $140.02 \cdot 10^5$

Write each number in scientific notation.

20. 39,500,000 **21.** 5,400 **22.** $800 \cdot 10^4$

23. $21.6 \cdot 10^5$ **24.** 107,000 **25.** $5{,}010 \cdot 10^3$

LESSON 2-3 (pp. 70–73)

Evaluate.

26. $20 - 4 \cdot 3 + 2$ **27.** $16 + 34 - 2 \cdot 8$

28. $12 \div 4 \cdot 6 + 16 - 4$ **29.** $12 \cdot 2 + 2 \div 2 - 5$

30. $3 + 9 - 6 \cdot 4 \div 8$ **31.** $4 \cdot 7 - 14 \div 2$

32. $2 + (12 \div 4)^2 \div 3$ **33.** $[6 \cdot (5 + 7) \div 9]^2$

34. There are normally 365 days in a year. Every fourth year, called a *leap year,* has 366 days. Evaluate the expression $2(365 \cdot 3 + 366)$ to find out how many days there are in 8 consecutive years.

35. Audry spent \$8 per square foot for floor tile and \$20 for tile glue. Evaluate the expression $8(48 \cdot 30 \div 12^2) + 20$ to find out how much she spent to tile a 48-inch by 30-inch entryway.

36. Siras earns \$9 per hour plus a commission of \$12 for every sale he makes. Evaluate the expression $8 \cdot 9 + 14 \cdot 12$ to find out how much Siras earned during an 8-hour shift in which he made 14 sales.

1.02a

Focus on Problem Solving

Solve

• Choose an operation: multiplication or division

To solve a word problem, you must determine which mathematical operation you can use to find the answer. One way of doing this is to determine the action the problem is asking you to take. If you are putting equal parts together, then you need to multiply. If you are separating something into equal parts, then you need to divide.

Decide what action each problem is asking you to take, and tell whether you must multiply or divide. Then explain your decision.

1. Judy plays the flute in the band. She practices for 3 hours every week. Judy practices only half as long as Angie, who plays the clarinet. How long does Angie practice playing the clarinet each week?

2. Each year, members of the band and choir are invited to join the bell ensemble for the winter performance. There are 18 bells in the bell ensemble. This year, each student has 3 bells to play. How many students are in the bell ensemble this year?

3. For every percussion instrument in the band, there are 4 wind instruments. If there are 48 wind instruments in the band, how many percussion instruments are there?

4. A group of 4 people singing together in harmony is called a quartet. At a state competition for high school choir students, 7 quartets from different schools competed. How many students competed in the quartet competition?

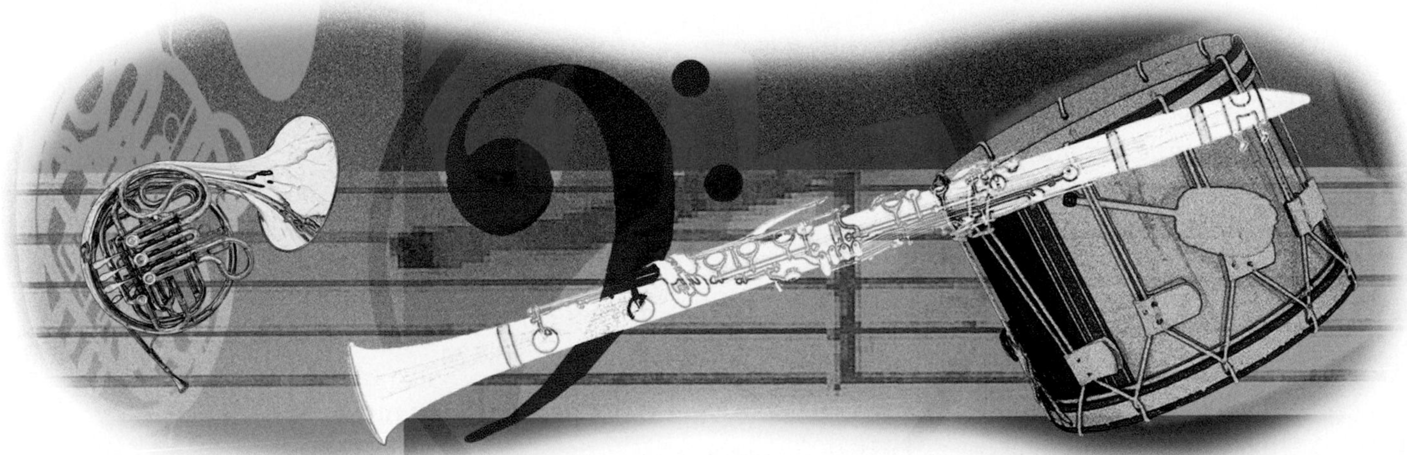

Hands-On LAB 2B

Divisibility

Use with Lesson 2-4

REMEMBER
- A number is divisible by another number if there is no remainder when you divide the numbers.
- Even numbers end in 0, 2, 4, 6, or 8.
- Odd numbers end in 1, 3, 5, 7, or 9.

Activity

The following numbers are divisible by 2: **12; 248; 3,006;** and **420.**

The following numbers are **not** divisible by 2: **81; 633; 5,977;** and **629.**

1. Write a rule for numbers that are divisible by 2. Write two different numbers that are divisible by 2.

The following numbers are divisible by 3: **387; 426; 8,004;** and **420.**

The following numbers are **not** divisible by 3: **782; 425; 1,451;** and **332.**

2. Write a rule for numbers that are divisible by 3. Write two different numbers that are divisible by 3.

The following numbers are divisible by 5: **2,000; 425; 860;** and **9,015.**

The following numbers are **not** divisible by 5: **3,046; 249; 551;** and **68.**

3. Write a rule for numbers that are divisible by 5. Write two different numbers that are divisible by 5.

The following numbers are divisible by 8: **5,248; 16,320;** and **14,864.**

The following numbers are **not** divisible by 8: **6,110; 14,596;** and **2,005.**

4. Write a rule for numbers that are divisible by 8. Write two different numbers that are divisible by 8.

The following numbers are divisible by 9: **18; 378;** and **6,057.**

The following numbers are **not** divisible by 9: **644; 5,817;** and **6,524.**

5. Write a rule for numbers that are divisible by 9. Write two different numbers that are divisible by 9.

The sieve of Eratosthenes is a method you can use to find prime numbers. Start by crossing out the number 1 in a table like the one at right. Next, circle the number 2, and then cross out all of the numbers that are divisible by 2. Next, circle the number 3, and then cross out all of the numbers that are divisible by 3. Continue in this pattern until all of the composite numbers are crossed out and all of the prime numbers are circled.

~~1~~	(2)	3	~~4~~	5	~~6~~	7	~~8~~	9	~~10~~
11	~~12~~	13	~~14~~	15	~~16~~	17	~~18~~	19	~~20~~
21	~~22~~	23	~~24~~	25	~~26~~	27	~~28~~	29	~~30~~
31	~~32~~	33	~~34~~	35	~~36~~	37	~~38~~	39	~~40~~
41	~~42~~	43	~~44~~	45	~~46~~	47	~~48~~	49	~~50~~
51	~~52~~	53	~~54~~	55	~~56~~	57	~~58~~	59	~~60~~
61	~~62~~	63	~~64~~	65	~~66~~	67	~~68~~	69	~~70~~
71	~~72~~	73	~~74~~	75	~~76~~	77	~~78~~	79	~~80~~
81	~~82~~	83	~~84~~	85	~~86~~	87	~~88~~	89	~~90~~
91	~~92~~	93	~~94~~	95	~~96~~	97	~~98~~	99	~~100~~

Use this method to find all of the prime numbers in the table.

Think and Discuss

1. Are all even numbers divisible by 2? Why or why not?
2. Are all odd numbers divisible by 3? Why or why not?

Try This

For each divisibility rule, name one number that would satisfy the rule and one that would not.

	A number is divisible by . . .	Divisible	Not Divisible
1.	2 if the last digit is even (0, 2, 4, 6, or 8).	☐	☐
2.	3 if the sum of the digits is divisible by 3.	☐	☐
3.	4 if the last two digits form a number divisible by 4.	☐	☐
4.	5 if the last digit is 0 or 5.	☐	☐
5.	6 if the number is divisible by 2 and 3.	☐	☐
6.	8 if the last three digits form a number divisible by 8.	☐	☐
7.	9 if the sum of the digits is divisible by 9.	☐	☐
8.	10 if the last digit is 0.	☐	☐

Use number tiles to find a number in which no digit is repeated for each description.

9. 4-digit number divisible by 5 and 10
10. 4-digit number divisible by 3 and 5
11. 4-digit number divisible by 3, 5, and 9
12. 4-digit number divisible by 4 and 9
13. 4-digit number divisible by 3 and 4
14. 3-digit number divisible by 2, 3, and 6

2-4 Prime Factorization

Learn to find the prime factorizations of composite numbers.

Vocabulary

prime number

composite number

prime factorization

In June 1999, Nayan Hajratwala discovered the first known *prime number* with more than one million digits. The new prime number, $2^{6,972,593} - 1$, has 2,098,960 digits.

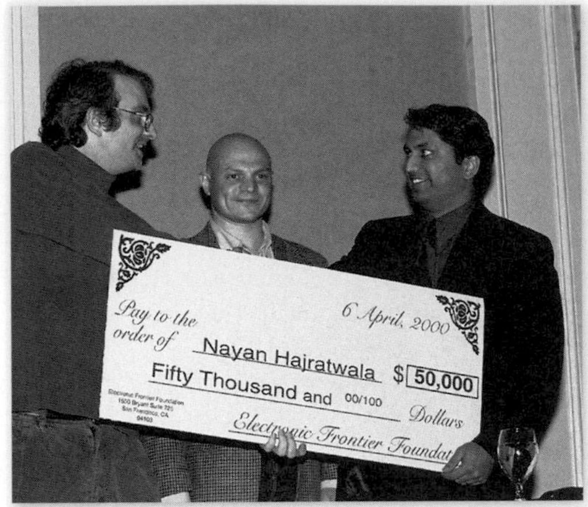

Nayan Hajratwala received a $50,000 award for discovering a new prime number.

A **prime number** is a whole number greater than 1 that has exactly two factors, 1 and itself. Three is a prime number because its only factors are 1 and 3.

A **composite number** is a whole number that has more than two factors. Six is a composite number because it has more than two factors—1, 2, 3, and 6. The number 1 has exactly one factor and is neither prime nor composite.

A composite number can be written as the product of its prime factors. This is called the **prime factorization** of the number.

You can use a factor tree to find the prime factors of a composite number.

EXAMPLE 1 **Using a Factor Tree to Find Prime Factorization**

Write the prime factorization of each number.

A 36

36

4 • 9 — *Write 36 as the product of two factors.*

2 • 2 • 3 • 3 — *Continue factoring until all factors are prime.*

The prime factorization of 36 is $2 \cdot 2 \cdot 3 \cdot 3$. Using exponents, you can write this as $2^2 \cdot 3^2$.

B 280

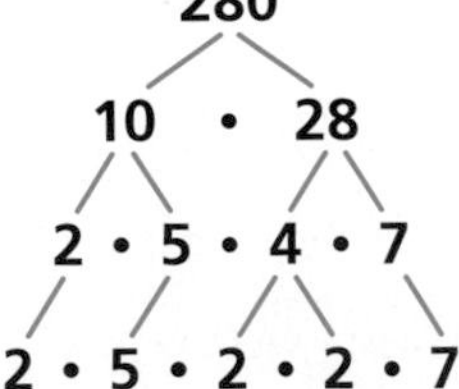

Write 280 as the product of two factors.

Continue factoring until all factors are prime.

The prime factorization of 280 is $2 \cdot 2 \cdot 2 \cdot 5 \cdot 7$, or $2^3 \cdot 5 \cdot 7$.

You can also use a step diagram to find the prime factorization of a number. At each step, divide by the smallest possible prime number. Continue dividing until the quotient is 1. The prime factors of the number are the prime numbers you divided by.

EXAMPLE 2 Using a Step Diagram to Find Prime Factorization

Write the prime factorization of each number.

A 252

$$\begin{array}{r|l} 2 & 252 \\ 2 & 126 \\ 3 & 63 \\ 3 & 21 \\ 7 & 7 \\ & 1 \end{array}$$

Divide 252 by 2. Write the quotient below 252. Keep dividing by a prime number.

Stop when the quotient is 1.

The prime factorization of 252 is $2 \cdot 2 \cdot 3 \cdot 3 \cdot 7$, or $2^2 \cdot 3^2 \cdot 7$.

B 495

$$\begin{array}{r|l} 3 & 495 \\ 3 & 165 \\ 5 & 55 \\ 11 & 11 \\ & 1 \end{array}$$

Divide 495 by 3. Keep dividing by a prime number.

Stop when the quotient is 1.

The prime factorization of 495 is $3 \cdot 3 \cdot 5 \cdot 11$, or $3^2 \cdot 5 \cdot 11$.

There is only one prime factorization for any given composite number. Example 2B began by dividing 495 by 3, the smallest prime factor of 495. Beginning with any prime factor of 495 gives the same result.

$$\begin{array}{r|l} 5 & 495 \\ 3 & 99 \\ 3 & 33 \\ 11 & 11 \\ & 1 \end{array} \qquad \begin{array}{r|l} 11 & 495 \\ 3 & 45 \\ 5 & 15 \\ 3 & 3 \\ & 1 \end{array}$$

The prime factorizations are $5 \cdot 3 \cdot 3 \cdot 11$ and $11 \cdot 3 \cdot 5 \cdot 3$, which are the same as $3 \cdot 3 \cdot 5 \cdot 11$.

Think and Discuss

1. **Explain** how to decide whether 47 is prime.
2. **Compare** prime numbers and composite numbers.
3. **Tell** how you know when you have found the prime factorization of a number.

2-4 Exercises

1.02a

FOR EOG PRACTICE

see page 662

internet connect

Homework Help Online
go.hrw.com Keyword: MS4 2-4

GUIDED PRACTICE

See Example 1

Write the prime factorization of each number.

1. 16
2. 54
3. 81
4. 105

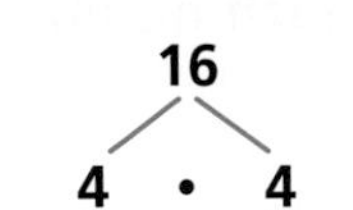

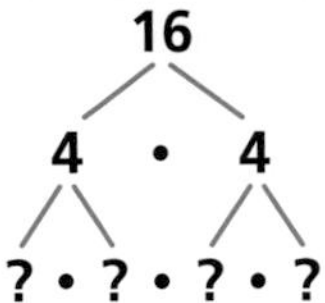

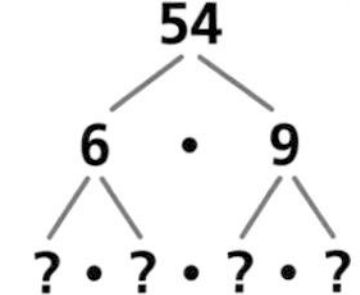

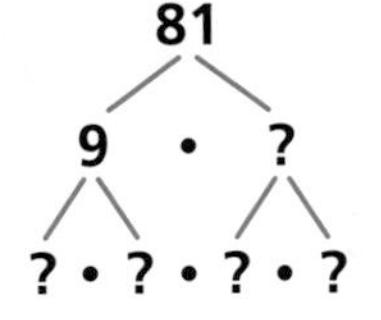

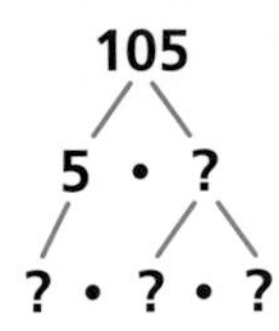

5. 18
6. 26
7. 45
8. 80

See Example 2

9. 50
10. 90
11. 100
12. 60
13. 63
14. 14
15. 1,000
16. 140

INDEPENDENT PRACTICE

See Example 1

Write the prime factorization of each number.

17. 68
18. 75
19. 120
20. 150
21. 135
22. 48
23. 154
24. 210
25. 800
26. 310
27. 625
28. 2,000

See Example 2

29. 315
30. 728
31. 189
32. 396
33. 242
34. 700
35. 187
36. 884
37. 1,225
38. 288
39. 360
40. 1,152

PRACTICE AND PROBLEM SOLVING

Complete each factor tree to find the prime factorization.

41. 144 → 9 • 16
42. 144 → 12 • 12
43. 200 → 20 • 10
44. 200 → 25 • 8

45. One way to factor 64 is 1 • 64. What other ways can 64 be written as the product of two factors?

46. How many prime factorizations of 64 are there?

Write the composite number for each prime factorization.

47. $2^2 \cdot 3^2 \cdot 5$
48. $2^6 \cdot 5$
49. $4^3 \cdot 7$
50. $3^2 \cdot 5^2 \cdot 7$
51. $2 \cdot 3 \cdot 7 \cdot 11$
52. $23 \cdot 29$
53. $3^2 \cdot 13$
54. $2^3 \cdot 5^3 \cdot 13$

55. If the prime factors of a number are all the prime numbers less than 10 and no factor is repeated, what is the number?

56. A number n is a prime factor of 28 and 63. What is the number?

57. A rectangular area on a farm has side lengths that are factors of 308. One of the side lengths is a prime number. Which of the areas in the diagram have the correct dimensions?

58. ***BUSINESS*** Eric is catering a party for 152 people. He wants to seat the same number of people at each table. He also wants more than 2 people but fewer than 10 people at a table. How many people can he seat at each table?

59. ***WRITE A PROBLEM*** Using the information in the table, write a problem using prime factorization that includes the number of calories per serving of the melons.

Fruit	Calories per Serving
Cantaloupe	66
Watermelon	15
Honeydew	42

60. ***WRITE ABOUT IT*** Describe how to use factor trees to find the prime factorization of a number.

61. ***CHALLENGE*** Find the smallest number that is divisible by 2, 3, 4, 5, 6, 7, 8, 9, and 10.

Spiral Review

Find the median of each data set. (Lesson 1-2)

62. 17, 25, 16, 94, 24

63. 7, 9, 6, 3, 5, 2, 7, 3

Find each value. (Lesson 2-1)

64. 8^2

65. 23^1

66. 18^0

67. 7^3

68. **EOG PREP** What is 2,430,000 in scientific notation? (Lesson 2-2)

A 243×10^4

B 24.3×10^5

C 2.43×10^5

D 2.43×10^6

69. **EOG PREP** Which of the following equals 16? (Lesson 2-3)

A $6 \cdot (2 + 4)$

B $6^2 + 4$

C $6 \cdot 2 + 4$

D $(6 \div 2) \cdot 4$

2-5 Greatest Common Factor

Learn to find the greatest common factor of two or more whole numbers.

Vocabulary

greatest common factor (GCF)

When getting ready for the Fall Festival, Sasha and David used the greatest common factor to make matching party favors.

The **greatest common factor (GCF)** of two or more whole numbers is the greatest whole number that divides evenly into each number.

One way to find the GCF of two or more numbers is to list all the factors of each number. The GCF is the greatest factor that appears in all the lists.

EXAMPLE 1 **Using a List to Find the GCF**

Find the greatest common factor (GCF).

24, 36, 48

24: 1, 2, 3, 4, 6, 8, (12), 24

36: 1, 2, 3, 4, 6, 9, (12), 18, 36

48: 1, 2, 3, 4, 6, 8, (12), 16, 24, 48

List all the factors of each number.

Circle the greatest factor that is in all the lists.

The GCF is 12.

EXAMPLE 2 **Using Prime Factorization to Find the GCF**

Find the greatest common factor (GCF).

A **60, 45**

$60 = 2 \cdot 2 \cdot (3) \cdot (5)$

$45 = (3) \cdot 3 \cdot (5)$

Write the prime factorization of each number and circle the common prime factors.

$3 \cdot 5 = 15$ *Multiply the common prime factors.*

The GCF is 15.

B **504, 132, 96, 60**

$504 = 2 \cdot 2 \cdot 2 \cdot (3) \cdot 3 \cdot 7$

$132 = 2 \cdot 2 \cdot (3) \cdot 11$

$96 = 2 \cdot 2 \cdot 2 \cdot 2 \cdot 2 \cdot (3)$

$60 = 2 \cdot 2 \cdot (3) \cdot 5$

Write the prime factorization of each number and circle the common prime factors.

$2 \cdot 2 \cdot 3 = 12$ *Multiply the common prime factors.*

The GCF is 12.

EXAMPLE 3 PROBLEM SOLVING APPLICATION

Sasha and David are making centerpieces for the Fall Festival. They have 50 small pumpkins and 30 ears of corn. What is the greatest number of matching centerpieces they can make using all of the pumpkins and corn?

1 Understand the Problem

Rewrite the question as a statement.

- Find the greatest number of centerpieces they can make.

List the **important information:**

- There are 50 pumpkins.
- There are 30 ears of corn.
- Each centerpiece must have the same number of pumpkins and the same number of ears of corn.

The **answer** will be the GCF of 50 and 30.

2 Make a Plan

You can write the prime factorizations of 50 and 30 to find the GCF.

3 Solve

$50 = \textcircled{2} \cdot \textcircled{5} \cdot 5$

$30 = \textcircled{2} \cdot 3 \cdot \textcircled{5}$ *Multiply the prime factors that are*

$2 \cdot 5 = 10$ *common to both 50 and 30.*

Sasha and David can make 10 centerpieces.

4 Look Back

If Sasha and David make 10 centerpieces, each one will have 5 pumpkins and 3 ears of corn, with nothing left over.

Think and Discuss

1. **Tell** what the letters GCF stand for and explain what the GCF of two numbers is.
2. **Discuss** whether the GCF of two numbers could be a prime number.
3. **Explain** whether every factor of the GCF of two numbers is also a factor of each number. Give an example.

2-5 Exercises

FOR EOG PRACTICE
see page 662

internet connect
Homework Help Online
go.hrw.com Keyword: MS4 2-5

1.02a

GUIDED PRACTICE

See Example 1 **Find the greatest common factor (GCF).**

1. 30, 42 **2.** 36, 45 **3.** 24, 36, 60, 84

See Example 2 **4.** 60, 231 **5.** 12, 28 **6.** 20, 40, 50, 120

See Example 3 **7.** The Math Club members are preparing identical welcome kits for the sixth-graders. They have 60 pencils and 48 memo pads. What is the greatest number of welcome kits they can prepare using all of the pencils and memo pads?

INDEPENDENT PRACTICE

See Example 1 **Find the greatest common factor (GCF).**

8. 60, 126 **9.** 12, 36 **10.** 75, 90

11. 22, 121 **12.** 28, 42 **13.** 38, 76

See Example 2 **14.** 28, 60 **15.** 54, 80 **16.** 30, 45, 60, 105

17. 26, 52 **18.** 11, 44, 77 **19.** 18, 27, 36, 48

See Example 3 **20.** Hetty is making identical gift baskets for the Senior Citizens Center. She has 39 small soap bars and 26 small bottles of lotion. What is the greatest number of baskets she can make using all of the soap bars and bottles of lotion?

PRACTICE AND PROBLEM SOLVING

Find the greatest common factor (GCF).

21. 5, 7 **22.** 12, 15 **23.** 4, 6

24. 9, 11 **25.** 22, 44, 66 **26.** 77, 121

27. 80, 120 **28.** 20, 28 **29.** 2, 3, 4, 5, 7

30. 4, 6, 10, 22 **31.** 14, 21, 35, 70 **32.** 6, 10, 11, 14

33. 6, 15, 33, 48 **34.** 18, 45, 63, 81 **35.** 13, 39, 52, 78

36. Which pair of numbers has a GCF that is a prime number, 48 and 90 or 105 and 56?

37. Find the prime factorization of 132.

38. Museum employees are preparing an exhibit of ancient coins. They have 49 copper coins and 35 silver coins to arrange on shelves. Each shelf will have the same number of copper coins and the same number of silver coins. How many shelves will the employees need for this exhibit?

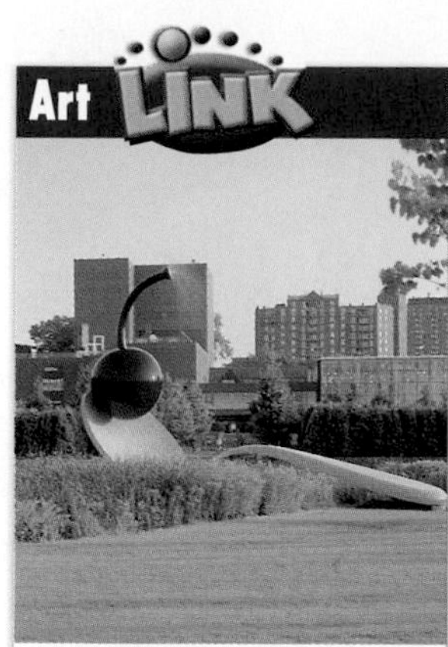

The sculpture *Spoonbridge and Cherry* by Claes Oldenburg and Coosje van Bruggen is 30 ft high and over 50 ft long.

39. ***ART*** A gallery is displaying a collection of 12 sculptures and 20 paintings by local artists. The exhibit is arranged into as many sections as possible, so that each section has the same number of sculptures and the same number of paintings. How many sections are in the exhibit?

40. ***SCHOOL*** Some of the students in the Math Club signed up to bring food and drinks to a party.

    a. If each club member gets the same amount of each item at the party, how many students are in the Math Club?

    b. How many cookies, pizza slices, cans of juice, and apples can each club member have at the party?

41. ***WHAT'S THE ERROR?*** A student used these factor trees to find the GCF of 50 and 70. The student decided that the GCF is 5. Explain the student's error and give the correct GCF.

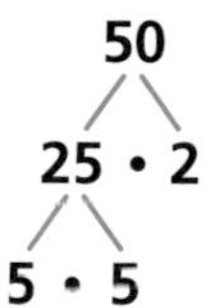

70
7 • 10
7 • 2 • 5

42. ***WRITE ABOUT IT*** The GCF of 1,274 and 1,365 is 91, or 7 • 13. Are 7, 13, and 91 factors of both 1,274 and 1,365? Explain.

43. ***CHALLENGE*** Find three *composite* numbers that have a GCF of 1.

Spiral Review

44. Find the range of the data set 19, 12, 8, 25, 33, 6, 8, and 8. (Lesson 1-2)

45. Make a stem-and-leaf plot of the data set 17, 19, 23, 39, 22, 27, 10, 25, 29, and 31. (Lesson 1-3)

Evaluate each expression. (Lesson 2-3)

46. $8 - 2^3 \div 4$

47. $12 + (9 - 4 \cdot 2)$

48. $24 \div (8 - 6)^2$

49. **EOG PREP** Which of the following is the prime factorization of 92? (Lesson 2-4)

    A $2^3 \cdot 3$

    B $2^2 \cdot 3^2$

    C $2^2 \cdot 23$

    D 2^5

50. **EOG PREP** Which of the following is the prime factorization of 6,600? (Lesson 2-4)

    A $2^3 \cdot 3 \cdot 5^2 \cdot 11$

    B $2^2 \cdot 3^2 \cdot 5^2$

    C $2 \cdot 3 \cdot 10^2 \cdot 11$

    D $6^{1,100}$

2-6 Least Common Multiple

Learn to find the least common multiple of two or more whole numbers.

Vocabulary

multiple

least common multiple (LCM)

The maintenance schedule on Ken's pickup truck shows that the tires should be rotated every 7,500 miles and that the oil filter should be replaced every 5,000 miles. What is the lowest mileage at which both services are due at the same time? To find the answer, you can use *least common multiples.*

A **multiple** of a number is the product of that number and a whole number. Some multiples of 7,500 and 5,000 are as follows:

7,500: 7,500, 15,000, 22,500, 30,000, 37,500, 45,000, . . .
5,000: 5,000, 10,000, 15,000, 20,000, 25,000, 30,000, . . .

A common multiple of two or more numbers is a number that is a multiple of each of the given numbers. So 15,000 and 30,000 are common multiples of 7,500 and 5,000.

The **least common multiple (LCM)** of two or more numbers is the common multiple with the least value. The LCM of 7,500 and 5,000 is 15,000. This is the lowest mileage at which both services are due at the same time.

EXAMPLE 1 Using a List to Find the LCM

Find the least common multiple (LCM).

A **3, 5**

3: 3, 6, 9, 12, (15), 18 — *List some multiples of each number.*
5: 5, 10, (15), 20, 25 — *Find the least value that is in both lists.*

The LCM is 15.

B **4, 6, 12**

4: 4, 8, (12), 16, 20, 24, 28 — *List some multiples of each number.*
6: 6, (12), 18, 24, 30 — *Find the least value that is in all the lists.*
12: (12), 24, 36, 48

The LCM is 12.

Sometimes, listing the multiples of numbers is not the easiest way to find the LCM. For example, the LCM of 78 and 110 is 4,290. You would have to list 55 multiples of 78 and 39 multiples of 110 to reach 4,290!

EXAMPLE 2 Using Prime Factorization to Find the LCM

Find the least common multiple (LCM).

A **78, 110**

$78 = \boxed{2} \cdot 3 \cdot 13$	*Write the prime factorization of each number.*
$110 = \boxed{2} \cdot 5 \cdot 11$	*Circle the common prime factors.*
②, 3, 13, 5, 11	*List the prime factors of the numbers, using the circled factors only once.*
$2 \cdot 3 \cdot 13 \cdot 5 \cdot 11$	*Multiply the factors in the list.*

The LCM is 4,290.

B **9, 27, 45**

$9 = \boxed{3} \cdot \boxed{3}$	*Write the prime factorization of each number.*
$27 = \boxed{3} \cdot \boxed{3} \cdot 3$	*Circle the common prime factors.*
$45 = \boxed{3} \cdot \boxed{3} \cdot 5$	
③, ③, 3, 5	*List the prime factors of the numbers, using the circled factors only once.*
$3 \cdot 3 \cdot 3 \cdot 5$	*Multiply the factors in the list.*

The LCM is 135.

EXAMPLE 3 *Recreation Application*

Charla and her little brother are running laps on a track. Charla runs one lap every 4 minutes, and her brother runs one lap every 6 minutes. They start together. In how many minutes will they be together at the starting line again?

Find the LCM of 4 and 6.

$4 = \boxed{2} \cdot 2$

$6 = \boxed{2} \cdot 3$

The LCM is ② $\cdot\ 2 \cdot 3 = 12$.

They will be together at the starting line in 12 minutes.

Think and Discuss

1. **Tell** what the letters LCM stand for and explain what the LCM of two numbers is.
2. **Describe** a way to remember the difference between GCF and LCM.
3. **List** four common multiples of 6 and 9 that are not the LCM.

2-6 Exercises

1.02a

FOR EOG PRACTICE

see page 663

internet connect

Homework Help Online
go.hrw.com Keyword: MS4 2-6

GUIDED PRACTICE

See Example  **Find the least common multiple (LCM).**

1. 4, 7 **2.** 14, 21, 28 **3.** 4, 8, 12, 16

See Example  **4.** 30, 48 **5.** 3, 9, 15 **6.** 10, 40, 50

See Example  **7.** Jerry and his dad are running around the same track. Jerry completes one lap every 8 minutes. His dad completes one full lap every 6 minutes. They start together. In how many minutes will they be together at the starting line again?

INDEPENDENT PRACTICE

See Example 1 **Find the least common multiple (LCM).**

8. 6, 9 **9.** 8, 12 **10.** 15, 20

11. 6, 14 **12.** 18, 27 **13.** 8, 10, 12

See Example 2 **14.** 6, 27 **15.** 16, 20 **16.** 12, 15, 22

17. 10, 15, 18, 20 **18.** 11, 22, 44 **19.** 8, 12, 18, 20

See Example  **20.** On her bicycle, Anna circles the block every 4 minutes. Her brother, on his scooter, circles the block every 10 minutes. They start out together. How many minutes will it be before they meet again at the starting point?

21. Rod helped his mom plant a vegetable garden. Rod planted a row every 30 minutes, and his mom planted a row every 20 minutes. If they started together, how long will it be before they both finish a row at the same time?

PRACTICE AND PROBLEM SOLVING

Find the least common multiple (LCM).

22. 3, 7 **23.** 2, 5 **24.** 4, 6

25. 9, 12 **26.** 22, 44, 66 **27.** 9, 27, 36

28. 80, 120 **29.** 10, 18 **30.** 3, 5, 7

31. 3, 6, 12 **32.** 5, 7, 9 **33.** 24, 36, 48

34. 2, 3, 4, 5 **35.** 4, 6, 10, 12 **36.** 14, 21, 35, 70

37. Jack mows the lawn every three weeks and washes the car every two weeks. If he does both today, how many days will pass before he does them both on the same day again?

38. Can two numbers have the same LCM as their GCF? Explain.

Social Studies LINK

Most calendars are based on cycles. Such calendars include the Mayan, the Chinese, and the standard western calendar.

39. The Mayan ceremonial calendar, or *tzolkin,* was 260 days long. It was composed of two independent cycles, a 13-day cycle and a 20-day cycle. At the beginning of the calendar, both cycles are at day 1. Will both cycles be at day 1 at the same time again before the 260 days are over? If so, when? Explain your answer.

40. The Chinese calendar has 12 months of 30 days each and 6-day weeks. The Chinese new year begins on the first day of a month and the first day of a week. Will the first day of a month and the first day of a week occur again at the same time before the 360-day year is over? If so, when? Explain your answer.

41. ***WRITE ABOUT IT*** The Julian Date calendar assigns each day a unique number. It begins on day 0 and adds 1 for each new day. So JD 2266296, or October 12, 1492, is 2,266,296 days from the beginning of the calendar. What are some advantages of using the Julian Date calendar? What are some advantages of using calendars that are based on cycles?

42. ***CHALLENGE*** The Mayan Long Count calendar used the naming system at right. Assuming the calendar began on JD 584285, express JD 2266296 in terms of the Mayan Long Count calendar. Start by finding the number of pictun that had passed up to that date.

Mayan Long Count Calendar
1 Pictun = 20 Baktun = 2,880,000 days
1 Baktun = 20 Katun = 144,000 days
1 Katun = 20 Tun = 7,200 days
1 Tun = 18 Winal = 360 days
1 Winal = 20 Kin = 20 days
1 Kin = 1 day

Spiral Review

Evaluate. (Lesson 2-1)

43. 6^0 **44.** 15^2 **45.** 10^3 **46.** 5^4

Write each number in scientific notation or in standard form. (Lesson 2-2)

47. 102.45 **48.** 62,100,000 **49.** 7.69×10^2 **50.** 8.00×10^5

51. **EOG PREP** What is the greatest common factor (GCF) of 30 and 105? (Lesson 2-5)

A 5 B 3 C 25 D 15

Chapter 2

Mid-Chapter Quiz

LESSONS 2-1 AND 2-3 (pp. 60–63 and 70–73)

Evaluate.

1. 12^2 **2.** $10^4 + 3 \cdot 9$ **3.** $81 \div 9^2$ **4.** $4 \div 2 + 7 \cdot 3$

5. $2^5 \div 8 + 6$ **6.** $6 \div 3 \cdot 15 + 8 - 1$ **7.** $(1 + 4^3) \div 5 \cdot 3^2$ **8.** $(1 + 4) \cdot (35 - 23)^2$

LESSON 2-2 (pp. 64–67)

Write each number in scientific notation.

9. 52,000 **10.** 109 **11.** 634,000,000 **12.** 82,000,000,000

13. $30.0 \cdot 10^2$ **14.** $5{,}210 \cdot 10^3$ **15.** $45.1 \cdot 10^5$ **16.** $70{,}800 \cdot 10^7$

LESSON 2-4 (pp. 78–81)

Complete each factor tree to find the prime factorization.

17. 24
6 • 4
? • ? • ? • ?

18. 140
14 • 10
? • ? • ? • ?

19. 45
3 • ?
3 • ? • ?

20.

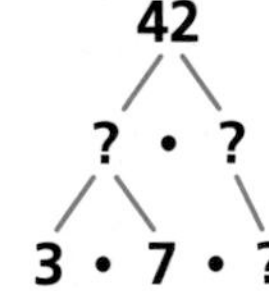

Write the prime factorization of each number.

21. 96 **22.** 125 **23.** 99 **24.** 105

LESSON 2-5 (pp. 82–85)

Find the greatest common factor (GCF).

25. 66, 96 **26.** 18, 27, 45 **27.** 16, 28, 44 **28.** 14, 28, 56

29. 85, 102 **30.** 76, 95 **31.** 52, 91, 104 **32.** 30, 75, 90

LESSON 2-6 (pp. 86–89)

Find the least common multiple (LCM).

33. 35, 40 **34.** 8, 25 **35.** 64, 72 **36.** 12, 20

37. 21, 33 **38.** 6, 30 **39.** 20, 42 **40.** 9, 13

41. Eddie goes jogging every other day, lifts weights every third day, and swims every fourth day. If Eddie begins all three activities on Monday, how many days will it be before he does all three activities on the same day again?

1.02d

Focus on Problem Solving

Look Back

- **Check that your answer is reasonable**

In some situations, such as when you are looking for an estimate or completing a multiple-choice question, check to see whether a solution or answer is reasonably accurate. One way to do this is by rounding the numbers to the nearest multiple of 10 or 100, depending on how large the numbers are. Sometimes it is useful to round one number up and another down.

Read each problem, and determine whether the given solution is too high, is too low, or appears to be correct. Explain your answer.

1. The cheerleading team is preparing to host a spaghetti dinner as a fund-raising project. They have set up and decorated 54 tables in the gymnasium. Each table can seat 8 people. How many people can be seated at the spaghetti dinner?

   Solution: 432 people

2. The cheerleaders need to raise $4,260 to attend a cheerleader camp. How much money must they charge each person if they are expecting 400 people at the spaghetti dinner?

   Solution: $4

3. To help out the fund-raising project, local restaurants have offered $25 gift certificates to give as door prizes. One gift certificate will be given for each door prize, and there will be six door prizes in all. What is the total value of all of the gift certificates given by the restaurants?

   Solution: $250

4. The total cost of hosting the spaghetti dinner will be about $270. If the cheerleaders make $3,280 in ticket sales, how much money will they have after paying for the spaghetti dinner?

   Solution: $3,000

5. Eighteen cheerleaders and two coaches plan to attend the camp. If each person will have an equal share of the $4,260 expense money, how much money will each person have?

   Solution: $562

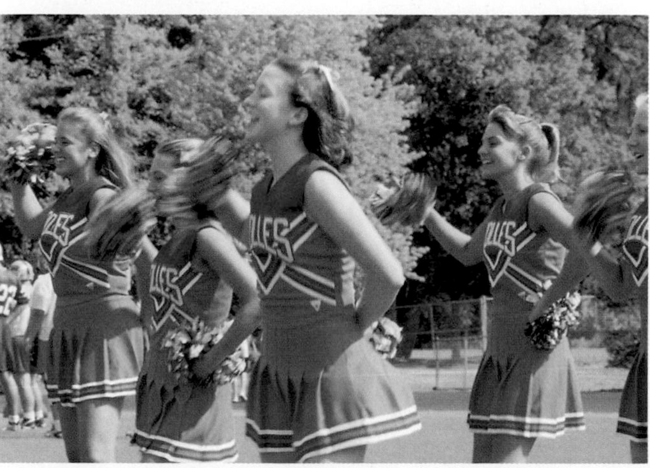

2-7 Variables and Algebraic Expressions

Learn to evaluate algebraic expressions.

Vocabulary

variable

constant

algebraic expression

evaluate

Ron Howard was born in 1954. You can find out what year Ron turned 16 by adding the year he was born to his age.

$$1954 + 16$$

In algebra, letters are often used to represent numbers. You can use a letter such as *a* to represent Ron Howard's age. When he turns *a* years old, the year will be

$$1954 + a.$$

The letter *a* has a value that can change, or *vary.* When a letter represents a number that can vary, it is called a **variable**. The year 1954 is a **constant** because the number cannot change.

Age	Year born + age = year at age	
16	1954 + 16	1970
18	1954 + 18	1972
21	1954 + 21	1975
36	1954 + 36	1990
a	1954 + *a*	■

An **algebraic expression** consists of one or more variables. It usually contains constants and operations. For example, 1954 + *a* is an algebraic expression for the year Ron Howard turns a certain age.

To **evaluate** an algebraic expression, substitute a number for the variable.

EXAMPLE 1 Evaluating Algebraic Expressions

Evaluate *n* + 7 for each value of *n*.

A $n = 3$

$n + 7$

$3 + 7$ *Substitute 3 for n.*

10 *Add.*

B $n = 5$

$n + 7$

$5 + 7$ *Substitute 5 for n.*

12 *Add.*

Multiplication and division of variables can be written in several ways, as shown in the table.

When evaluating expressions, use the order of operations.

Multiplication		Division	
$7t$	$7 \cdot t$	$\frac{q}{2}$	$q/2$
$7(t)$	$7 \times t$	$q \div 2$	
ab	$a \cdot b$	$\frac{s}{r}$	s/r
$a(b)$	$a \times b$	$s \div r$	

EXAMPLE 2 Evaluating Algebraic Expressions Involving Order of Operations

Evaluate each expression for the given value of the variable.

A $3x - 2$ for $x = 5$

$3(5) - 2$ *Substitute 5 for x.*

$15 - 2$ *Multiply.*

13 *Subtract.*

B $n \div 2 + n$ for $n = 4$

$4 \div 2 + 4$ *Substitute 4 for n.*

$2 + 4$ *Divide.*

6 *Add.*

C $6y^2 + 2y$ for $y = 2$

$6(2)^2 + 2(2)$ *Substitute 2 for y.*

$6(4) + 2(2)$ *Evaluate the power.*

$24 + 4$ *Multiply.*

28 *Add.*

EXAMPLE 3 Evaluating Algebraic Expressions with Two Variables

Evaluate $\frac{3}{n} + 2m$ for $n = 3$ and $m = 4$.

$\frac{3}{n} + 2m$

$\frac{3}{3} + 2(4)$ *Substitute 3 for n and 4 for m.*

$1 + 8$ *Divide and multiply from left to right.*

9 *Add.*

Think and Discuss

1. Write each expression another way. **a.** $12x$ **b.** $\frac{4}{y}$ **c.** $\frac{3xy}{2}$

2. Explain the difference between a variable and a constant.

2-7 Exercises

FOR EOG PRACTICE

see page 664

internet connect
Homework Help Online
go.hrw.com Keyword: MS4 2-7

5.03

GUIDED PRACTICE

See Example 1 **Evaluate $n + 9$ for each value of n.**

1. $n = 3$
2. $n = 2$
3. $n = 11$

See Example 2 **Evaluate each expression for the given value of the variable.**

4. $2x - 3$ for $x = 4$
5. $n \div 3 + n$ for $n = 6$
6. $5y^2 + 3y$ for $y = 2$

See Example 3 **Evaluate each expression for the given values of the variables.**

7. $\frac{8}{n} + 3m$ for $n = 2$ and $m = 5$
8. $5a - 3b + 5$ for $a = 4$ and $b = 3$

INDEPENDENT PRACTICE

See Example 1 **Evaluate $n + 5$ for each value of n.**

9. $n = 17$
10. $n = 9$
11. $n = 0$

See Example 2 **Evaluate each expression for the given value of the variable.**

12. $5y - 1$ for $y = 3$
13. $10b - 9$ for $b = 2$
14. $p \div 7 + p$ for $p = 14$
15. $n \div 5 + n$ for $n = 20$
16. $3x^2 + 2x$ for $x = 10$
17. $3c^2 - 5c$ for $c = 3$

See Example 3 **Evaluate each expression for the given values of the variables.**

18. $\frac{12}{n} + 7m$ for $n = 6$ and $m = 4$
19. $7p - 2t + 3$ for $p = 6$ and $t = 2$

PRACTICE AND PROBLEM SOLVING

Evaluate each expression for the given values of the variables.

20. $20x - 10$ for $x = 4$
21. $4d^2 - 3d$ for $d = 2$
22. $22p \div 11 + p$ for $p = 3$
23. $q + q^2 + q \div 2$ for $q = 4$
24. $\frac{16}{k} + 7h$ for $k = 8$ and $h = 2$
25. $f \div 3 + f$ for $f = 18$
26. $3t \div 3 + t$ for $t = 13$
27. $9 + 3p - 5t + 3$ for $p = 2$ and $t = 1$
28. Write $\frac{4ab}{3}$ another way.
29. You can factor $6a^2$ as $2 \cdot 3 \cdot a \cdot a$. Factor $10c^3d^2$ in the same manner.
30. The expression $60m$ gives the number of seconds in m minutes. Evaluate $60m$ for $m = 7$. How many seconds are there in 7 minutes?
31. ***MONEY*** Betsy has n quarters. You can use the expression $0.25n$ to find the total value of her coins. What is the value of 18 quarters?

32. ***PHYSICAL SCIENCE*** A color TV has a power rating of 200 watts. The expression $200t$ gives the power used by t color TV sets. Evaluate $200t$ for $t = 13$. How much power is used by 13 TV sets?

33. ***PHYSICAL SCIENCE*** The expression $1.8c + 32$ can be used to convert a temperature in degrees Celsius c to degrees Fahrenheit. What is the temperature in degrees Fahrenheit if the temperature is 30°C?

34. ***PHYSICAL SCIENCE*** The graph shows the changes of state for water.
   a. What is the boiling point of water in degrees Celsius?
   b. Use the expression $1.8c + 32$ to find the boiling point of water in degrees Fahrenheit.

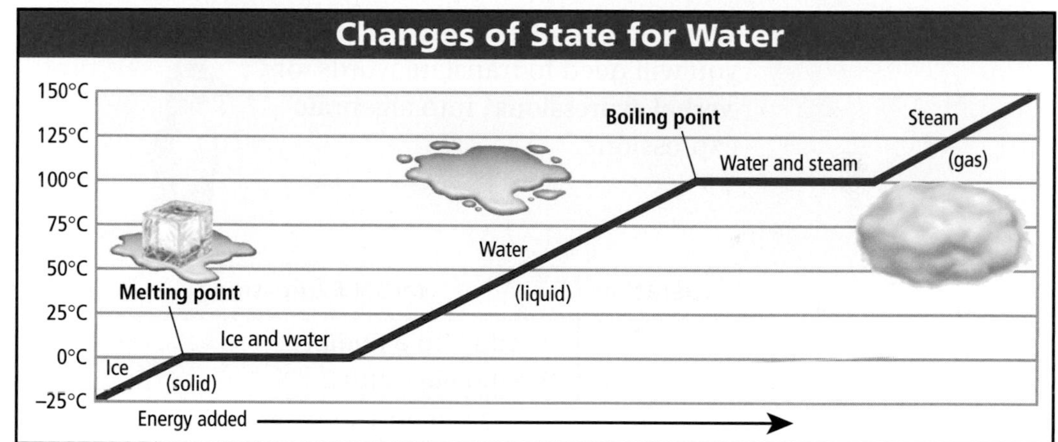

35. ***WHAT'S THE ERROR?*** A student was asked to identify the variable in the expression $72x + 8$. The student's answer was $72x$. What was the student's error?

36. ***WRITE ABOUT IT*** Explain why letters such as x, p, and n used in algebraic expressions are called variables. Use examples to illustrate your response.

37. ***CHALLENGE*** Evaluate the expression $\frac{x + y}{y - x}$ for $x = 6$ and $y = 8$.

Spiral Review

38. A box-and-whisker plot is made using the numbers 24, 9, 17, 35, 16, and 27. Which number is the upper extreme? (Lesson 1-6)

Write the prime factorization of each number. (Lesson 2-4)

39. 99 40. 24 41. 56 42. 80

43. **EOG PREP** Which expression does *not* simplify to 81? (Lesson 2-3)

A $9 \cdot (4 + 5)$ B $7 + 16 \cdot 4 + 10$ C $3 \cdot 25 + 2$ D $10^2 - 4 \cdot 5 + 1$

44. **EOG PREP** What is the least common multiple (LCM) of 9 and 12? (Lesson 2-6)

A 24 B 36 C 54 D 108

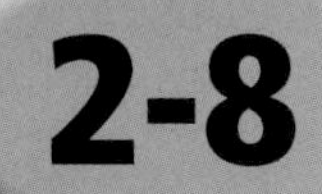

2-8 Translate Words into Math

Problem Solving Skill

Learn to translate words into numbers, variables, and operations.

Vocabulary

verbal expressions

Although they are closely related, a Great Dane weighs about 40 times as much as a Chihuahua. An expression for the weight of the Great Dane could be $40c$, where c is the weight of the Chihuahua.

When solving real-world problems, you will need to translate words, or **verbal expressions**, into algebraic expressions.

Operation	Verbal Expressions	Algebraic Expression
+	• add 3 to a number • a number plus 3 • the sum of a number and 3 • 3 more than a number • a number increased by 3	$n + 3$
−	• subtract 12 from a number • a number minus 12 • the difference of a number and 12 • 12 less than a number • a number decreased by 12 • take away 12 from a number • a number less 12	$x - 12$
×	• 2 times a number • 2 multiplied by a number • the product of 2 and a number	$2m$ or $2 \cdot m$
÷	• 6 divided into a number • a number divided by 6 • the quotient of a number and 6	$a \div 6$ or $\frac{a}{6}$

EXAMPLE 1 Translating Verbal Expressions into Algebraic Expressions

Write each phrase as an algebraic expression.

A **the product of 20 and t**

product means "multiply"

$20t$

B **24 less than a number**

less than means "subtract from"

$n - 24$

Write each phrase as an algebraic expression.

C **4 times the sum of a number and 2**

4 times the **sum** of a number and 2

$4 \cdot \quad n + 2$

$4(n + 2)$

D **the sum of 4 times a number and 2**

the **sum** of 4 times a number and 2

$4 \cdot n \quad + 2$

$4n + 2$

When solving real-world problems, you may need to determine the action to know which operation to use.

Action	Operation
Put parts together	Add
Put equal parts together	Multiply
Find how much more	Subtract
Separate into equal parts	Divide

EXAMPLE 2 Translating Real-World Problems into Algebraic Expressions

A **Jed reads p pages each day of a 200-page book. Write an algebraic expression for how many days it will take Jed to read the book.**

You need to *separate* the total number of pages *into equal parts.* This involves division.

$$\frac{\text{total number of pages}}{\text{pages read each day}} = \frac{200}{p}$$

B **To rent a certain car for a day costs \$84 plus \$0.29 for every mile the car is driven. Write an algebraic expression to show how much it costs to rent the car for a day.**

The cost includes \$0.29 per mile. Use m for the number of miles.

Multiply to *put equal parts together:* $0.29m$

In addition to the fee per mile, the cost includes a flat fee of \$84.

Add to *put parts together:* $84 + 0.29m$

Think and Discuss

1. Write three different verbal expressions that can be represented by $2 - y$. Then rewrite them so they can be represented by $y - 2$.

2. Explain how you would determine which operation to use to find the number of chairs in 6 rows of 100 chairs each.

2-8 Exercises

FOR EOG PRACTICE

see page 664

internet connect

Homework Help Online
go.hrw.com Keyword: MS4 2-8

5.02

GUIDED PRACTICE

See Example 1

Write each phrase as an algebraic expression.

1. the product of 7 and p
2. 3 less than a number
3. 3 times the sum of a number and 5

See Example 2

4. Carly spends \$5 for n notebooks. Write an algebraic expression to represent the cost of one notebook.
5. A company charges \$46 for cable TV installation and \$21 per month for basic cable service. Write an algebraic expression to represent the total cost of m months of basic cable service, including installation.

INDEPENDENT PRACTICE

See Example 1

Write each phrase as an algebraic expression.

6. the sum of 5 and a number
7. 2 less than a number
8. the quotient of a number and 8
9. 9 times a number
10. 10 less than the product of a number and 3

See Example 2

11. Video Express sells used tapes. Marta bought v tapes for \$45. Write an algebraic expression for the cost of each tape.
12. A 5-foot pine tree was planted and grew 2 feet each year. Write an algebraic expression for the height of the tree after t years.

PRACTICE AND PROBLEM SOLVING

Write each phrase as an algebraic expression.

13. m plus 6 times n
14. t less than 23 divided by u
15. 14 less than k times 6
16. 2 times the sum of y and 5
17. the quotient of 100 and the quantity 6 plus w
18. 35 multiplied by the quantity r less 45

Write a verbal expression for each algebraic expression.

19. $h + 3$
20. $90 \div y$
21. $s - 405$
22. $5(a - 8)$
23. $4p - 10$
24. $(r + 1) \div 14$

25. An ice machine can produce 17 pounds of ice in one hour. Write an algebraic expression to describe the following:
    a. the number of pounds of ice produced in n hours
    b. the number of pounds of ice produced in d days

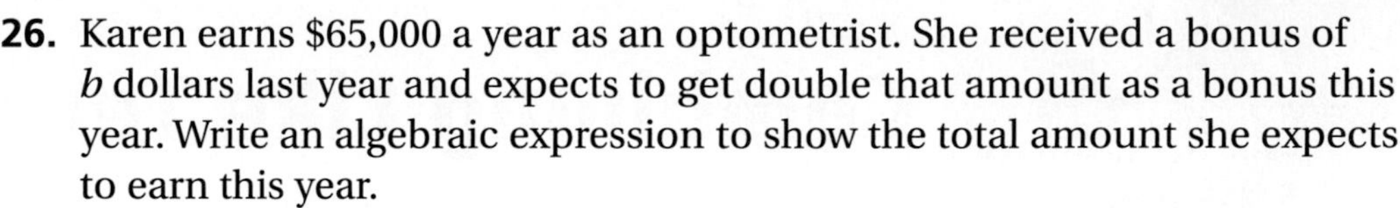

26. Karen earns \$65,000 a year as an optometrist. She received a bonus of b dollars last year and expects to get double that amount as a bonus this year. Write an algebraic expression to show the total amount she expects to earn this year.

Life Science LINK

Up to 25 follicle mite nymphs can hatch in a single hair follicle.

27. ***LIFE SCIENCE*** Follicle mites are tiny and harmless, and they live in our eyebrows and eyelashes. They are relatives of spiders and like spiders, they have eight legs. Write an algebraic expression for the number of legs in m mites.

28. ***NUTRITION*** The table shows the estimated number of grams of carbohydrates commonly found in various types of foods.

Food	Carbohydrates
1 c skim milk	12 g
1/2 c vegetable	5 g
1 piece of fruit	15 g
1 slice of bread	15 g
1 oz lean meat	0 g

   a. Write an algebraic expression for the number of grams of carbohydrates in y pieces of fruit and 1 cup of skim milk.

   b. How many grams of carbohydrates are in a sandwich made from t ounces of lean meat and 2 slices of bread?

29. ***WHAT'S THE QUESTION?*** Jimmy has twice as many baseball cards as Frank and four times as many football cards as Joe. The expression $2x + 4y$ can be used to show the total number of baseball and football cards Jimmy has. If the answer is y, then what is the question?

30. ***WRITE ABOUT IT*** If you are asked to compare two numbers, what two operations might you use? Why?

31. ***CHALLENGE*** In 1996, one U.S. dollar was equivalent, on average, to \$1.363 in Canadian money. Write an algebraic expression for the number of U.S. dollars you could get for n Canadian dollars.

Spiral Review

Find the mean of each data set. (Lesson 1-2)

32. 25, 18, 27, 30 **33.** 108, 77, 90, 97 **34.** 239, 247, 233, 263, 268

35. What is another way to write $8 \cdot 8 \cdot 8 \cdot 8$? (Lesson 2-1)

Find each product. (Lesson 2-2)

36. $612 \cdot 10^3$ **37.** $43.8 \cdot 10^6$ **38.** $590 \cdot 10^5$

39. **EOG PREP** What is the value of the expression $18 - 1 \cdot 9 \div 3$? (Lesson 2-3)

A 2 **B** 15 **C** 51 **D** 153

2-9 Combining Like Terms

Learn to combine like terms.

Vocabulary

term

coefficient

like terms

In the expression $7x + 5$, $7x$ and 5 are called *terms*. A **term** can be a number, a variable, or a product of numbers and variables. Terms in an expression are separated by + and −.

$$\underbrace{7x}_{\text{term}} + \underbrace{5}_{\text{term}} - \underbrace{3y^2}_{\text{term}} + \underbrace{y}_{\text{term}} - \underbrace{\frac{x}{3}}_{\text{term}}$$

In the term $7x$, 7 is called the *coefficient.* A **coefficient** is a number that is multiplied by a variable in an algebraic expression. A variable by itself, like y, has a coefficient of 1. So $y = 1y$.

Coefficient → 7x ← Variable

Term	$4a$	$\frac{2}{3}y$	$3k^5$	x^2	$\frac{x}{9}$	$4.7t$
Coefficient	4	$\frac{2}{3}$	3	1	$\frac{1}{9}$	4.7

Like terms are terms with the same variable raised to the same power. The coefficients do not have to be the same. Constants, like 5, $\frac{1}{2}$, and 3.2, are also like terms.

Like Terms	$3x$ and $2x$	w and $\frac{w}{7}$	5 and 1.8
Unlike Terms	$5x^2$ and $2x$ *The exponents are different.*	$6a$ and $6b$ *The variables are different.*	3.2 and n *Only one term contains a variable.*

EXAMPLE 1 Identifying Like Terms

Identify like terms in the list.

$$5a \quad \frac{t}{2} \quad 3y^2 \quad 7t \quad x^2 \quad 4z \quad k \quad 4.5y^2 \quad 2t \quad \frac{2}{3}a$$

Look for like variables with like powers.

$$5a \quad \frac{t}{2} \quad 3y^2 \quad 7t \quad x^2 \quad 4z \quad k \quad 4.5y^2 \quad 2t \quad \frac{2}{3}a$$

Like terms: $5a$ and $\frac{2}{3}a$ $\quad$ $\frac{t}{2}$, $7t$, and $2t$ $\quad$ $3y^2$ and $4.5y^2$

Helpful Hint

Use different shapes or colors to indicate sets of like terms.

Combining like terms is like grouping similar objects.

$$4x \quad + \quad 5x \quad = \quad 9x$$

To combine like terms that have variables, add or subtract the coefficients.

EXAMPLE 2 Combining Like Terms

Combine like terms.

A $7x + 2x$

$7x + 2x$ — *7x and 2x are like terms.*

$9x$ — *Add the coefficients.*

B $5x^3 + 3y + 7x^3 - 2y - 4x^2$

$5x^3 + 3y + 7x^3 - 2y - 4x^2$ — *Identify like terms.*

$(5x^3 + 7x^3) + (3y - 2y) - 4x^2$ — *Group like terms.*

$12x^3 + y - 4x^2$ — *Add or subtract the coefficients.*

C $3a + 4a^2 + 2b$

In this expression, there are no like terms to combine.

EXAMPLE 3 *Geometry Application*

Remember!

To find the perimeter of a figure, add the lengths of the sides.

Write an expression for the perimeter of the rectangle shown. Combine like terms in the expression.

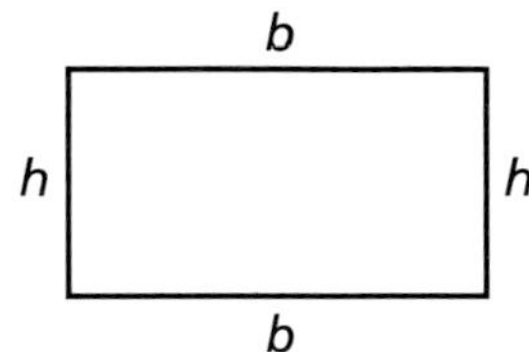

$b + h + b + h$ — *Write an expression using the side lengths.*

$(b + b) + (h + h)$ — *Identify and group like terms.*

$2b + 2h$ — *Add the coefficients.*

Think and Discuss

1. **Identify** the variable and the coefficient in each term.
   a. $11t$ b. $-3a$ c. $\frac{x}{2}$ d. $\frac{4}{5}n$
2. **Explain** whether $5x$, $5x^2$, and $5x^3$ are like terms.
3. **Explain** how you know which terms to combine in an expression.

2-9 Exercises

5.03, 5.04

FOR EOG PRACTICE

see page 664

internet connect

Homework Help Online
go.hrw.com Keyword: MS4 2-9

GUIDED PRACTICE

See Example 1 **Identify like terms in each list.**

1. $6b \quad 5x^2 \quad 4x^3 \quad \frac{b}{2} \quad x^2 \quad 2e$
2. $12a^2 \quad 4x^3 \quad b \quad 4a^2 \quad 3.5x^3 \quad \frac{5}{6}b$

See Example 2 **Combine like terms.**

3. $5x + 3x$
4. $6a^2 - a^2 + 16$
5. $4a^2 + 5a + 14b$

See Example 3

6. Write an expression for the perimeter of the rectangle. Combine like terms in the expression.

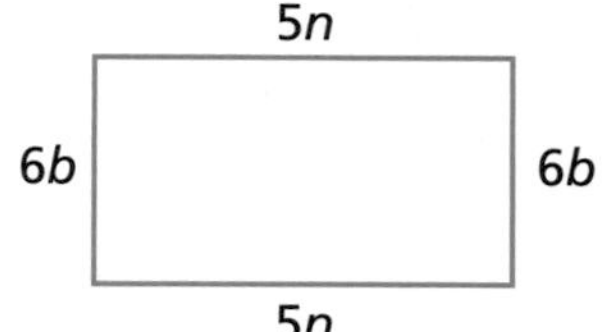

INDEPENDENT PRACTICE

See Example 1 **Identify like terms in each list.**

7. $2b \quad b^6 \quad b \quad x^4 \quad 3b^6 \quad 2x^2$
8. $6 \quad 2n \quad 3n^2 \quad 6m^2 \quad \frac{n}{4} \quad 7$
9. $10k^2 \quad m \quad 3^3 \quad \frac{p}{6} \quad 2m \quad 2$
10. $6^3 \quad y^3 \quad 3y^2 \quad 6^2 \quad y \quad 5y^3$

See Example 2 **Combine like terms.**

11. $3a + 2b + 5a$
12. $a + 2b + 2a + b + 2c$
13. $5b + 7b + 10$
14. $y + 4 + 2x + 3y$
15. $18 + 2d^3 + d + 3d$
16. $q^2 + 2q + 2q^2$

See Example 3

17. Write an expression for the perimeter of the given figure. Combine like terms in the expression.

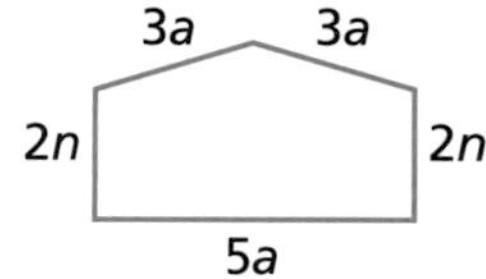

PRACTICE AND PROBLEM SOLVING

Combine like terms.

18. $4x + 5x$
19. $32y - 5y$
20. $4c^2 + 5c + 2c$
21. $5d^2 - 3d^2 + d$
22. $5f^2 + 2f + f^2$
23. $7x + 8x^2 - 3y$
24. $p + 9q + 9 + 14p$
25. $6b + 6b^2 + 4b^3$
26. $a^2 + 2b + 2a^2 + b + 2c$
27. Write an expression that has four terms and simplifies to $16x + 5$ when you combine like terms.
28. At a bake sale, Emily sold 4 pies, and Mark sold 3 pies and 2 cakes. Write an algebraic expression describing how many items they sold.

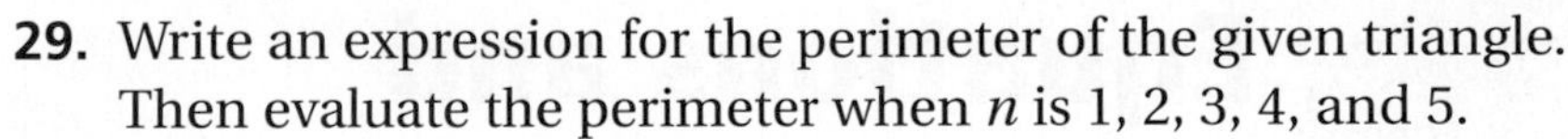

29. Write an expression for the perimeter of the given triangle. Then evaluate the perimeter when n is 1, 2, 3, 4, and 5.

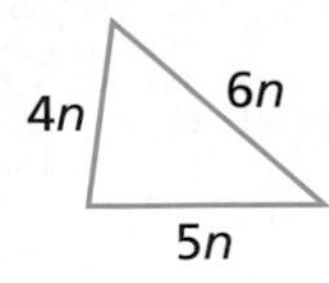

n	1	2	3	4	5
Perimeter					

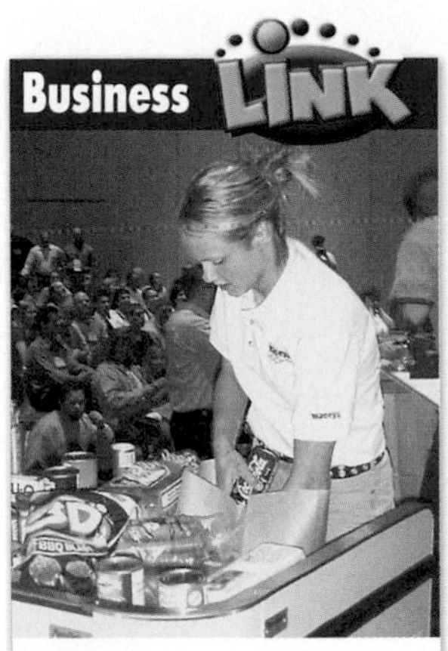

Twenty-three grocery baggers competed in the 2002 National Best Bagger Competition.

30. ***BUSINESS*** Ashley earns \$8 per hour working at a grocery store. Last week she worked h hours bagging groceries and twice as many hours stocking shelves. Write an expression for the amount of money Ashley earned last week and combine like terms.

31. ***BUSINESS*** Brad makes d dollars per hour as a cook at a deli. The table shows the number of hours he worked each week in June.

a. Write an expression for the amount of money Brad earned in June. Combine like terms in the expression.

b. Evaluate your expression from part **a** for $d = \$9.50$.

c. What does your answer to part **b** represent?

Hours Brad Worked in June

Week	Hours
1	21.5
2	23
3	15.5
4	19

32. The terms $23x$, $23x^2$, $6y^2$, $18x$, y^2 and one other term can be written in an expression which, when simplified, equals $5x + 7y^2$. Identify the missing term from the list and write the expression.

33. ***WHAT'S THE QUESTION?*** At one store, a pair of blue jeans costs \$29 and a shirt costs \$25. At another store, the same kind of blue jeans sells for \$26 and the same kind of shirt sells for \$20. The answer is $29j - 26j + 25s - 20s = 3j + 5s$. What is the question?

34. ***WRITE ABOUT IT*** Describe the steps for simplifying the expression $2x + 3 + 5x - 15$.

35. ***CHALLENGE*** A rectangle has a width of $x + 2$ and a length of $3x + 1$. Write an expression for the perimeter of the rectangle. Combine like terms.

Spiral Review

Write each number using an exponent and the given base. (Lesson 2-1)

36. 343, base 7 **37.** 243, base 3 **38.** 36, base 6 **39.** 125, base 5

Find the greatest common factor (GCF). (Lesson 2-5)

40. 45, 54 **41.** 81, 36 **42.** 84, 48 **43.** 132, 44

44. **EOG PREP** What is the value of $7t$ for $t = 9$? (Lesson 2-7)

A 16 B 54 C 61 D 63

2-10 Equations and Their Solutions

Learn to determine whether a number is a solution of an equation.

Vocabulary
equation
solution

Nicole has 82 CDs. This is 9 more than her friend Jessica has.

This situation can be written as an *equation.* An **equation** is a mathematical statement that two expressions are equal in value.

An equation is like a balanced scale.

Number of CDs Nicole has	is equal to	9 more than Jessica has.
82	=	$j + 9$
Left expression		Right expression

Just as the weights on both sides of a balanced scale are exactly the same, the expressions on both sides of an equation represent exactly the same value.

When an equation contains a variable, a value of the variable that makes the statement true is called a **solution** of the equation.

Reading Math
The symbol ≠ means "is not equal to."

$x + 3 = 10$ $\quad x = 7$ is a solution because $7 + 3 = 10$.

$12 = t + 9$ $\quad t = 4$ is not a solution because $12 \neq 4 + 9$.

EXAMPLE 1 Determining Whether a Number Is a Solution of an Equation

Determine whether each number is a solution of $18 = s - 7$.

A 11

$18 = s - 7$

$18 \stackrel{?}{=} \mathbf{11} - 7$ *Substitute 11 for s.*

$18 \stackrel{?}{=} 4$ ✗

11 **is not** a solution of $18 = s - 7$.

B 25

$18 = s - 7$

$18 \stackrel{?}{=} \mathbf{25} - 7$ *Substitute 25 for s.*

$18 \stackrel{?}{=} 18$ ✓

25 **is** a solution of $18 = s - 7$.

EXAMPLE 2 *Consumer Application*

Nicole has 82 CDs. This is 9 more than her friend Jessica has. The equation $82 = j + 9$ can be used to represent the number of CDs Jessica has. Does Jessica have 91 CDs, 85 CDs, or 73 CDs?

91 CDs

$82 = j + 9$

$82 \stackrel{?}{=} \mathbf{91} + 9$ *Substitute 91 for j.*

$82 \stackrel{?}{=} 100$ ✗

85 CDs

$82 = j + 9$

$82 \stackrel{?}{=} \mathbf{85} + 9$ *Substitute 85 for j.*

$82 \stackrel{?}{=} 94$ ✗

73 CDs

$82 = j + 9$

$82 \stackrel{?}{=} \mathbf{73} + 9$ *Substitute 73 for j.*

$82 \stackrel{?}{=} 82$ ✓

Jessica has 73 CDs.

EXAMPLE 3 Writing an Equation to Determine Whether a Number Is a Solution

Tyler wants to buy a new skateboard. He has \$57, which is \$38 less than he needs. Does the skateboard cost \$90 or \$95?

If s represents the price of the skateboard, then $s - 38 = 57$.

\$90

$s - 38 = 57$

$\mathbf{90} - 38 \stackrel{?}{=} 57$ *Substitute 90 for s.*

$52 \stackrel{?}{=} 57$ ✗

\$95

$s - 38 = 57$

$\mathbf{95} - 38 \stackrel{?}{=} 57$ *Substitute 95 for s.*

$57 \stackrel{?}{=} 57$ ✓

The skateboard costs \$95.

Think and Discuss

1. **Compare** equations with expressions.
2. **Give an example** of an equation whose solution is 5.

2-10 Exercises

FOR EOG PRACTICE

see page 665

internet connect
Homework Help Online
go.hrw.com Keyword: MS4 2-10

5.03

GUIDED PRACTICE

See Example 1 **Determine whether each number is a solution of $19 = x + 4$.**

1. 5 **2.** 14 **3.** 15 **4.** 23

See Example 2 **5.** Nancy has 94 baseball cards. This is 6 more than Claire has. The equation $94 = c + 6$ can be used to represent the number of baseball cards Claire has. Does Claire have 88 or 93 baseball cards?

See Example 3 **6.** Mavis wants to buy a book. She has \$25, which is \$9 less than she needs. Does the book cost \$34 or \$38?

INDEPENDENT PRACTICE

See Example 1 **Determine whether each number is a solution of $24 = 34 - n$.**

7. 58 **8.** 20 **9.** 14 **10.** 10

Determine whether each number is a solution of $p + 18 = 29$.

11. 18 **12.** 11 **13.** 9 **14.** 47

See Example 2  **15.** Nadia has 47 video games. This is 6 less than Benjamin has. The equation $47 = v - 6$ can be used to represent the number of video games Benjamin has. Does Benjamin have 39, 41, or 53 video games?

See Example 3  **16.** Curtis wants to buy a new snowboard. He has \$119, which is \$56 less than he needs. Does the snowboard cost \$165 or \$175?

PRACTICE AND PROBLEM SOLVING

Determine whether each given number is a solution of the given equation.

17. $j = 6$ for $15 - j = 21$

18. $x = 36$ for $48 = x + 12$

19. $m = 18$ for $16 = 34 - m$

20. $k = 23$ for $17 + k = 40$

21. $y = 8$ for $9y + 2 = 74$

22. $c = 12$ for $100 - 2c = 86$

23. $q = 13$ for $5q + 7 - q = 51$

24. $w = 15$ for $13w - 2 - 6w = 103$

25. $t = 12$ for $3(50 - t) - 10t = 104$

26. $r = 21$ for $4r - 8 + 9r - 1 = 264$

27. Monique has a collection of stamps from 6 different countries. She compared her collection with Jeremy's and found that Jeremy has stamps from 3 fewer countries than she does. Write an equation showing this, using j as the number of countries from which Jeremy has stamps.

Earth Science 

Source: Colorado Mall

28. The diagram shows approximate elevations for different climate zones in the Colorado Rockies. Use the diagram to write an equation that shows the vertical distance d from the summit of Mount Evans (14,264 ft) to the tree line, which marks the beginning of the alpine tundra zone.

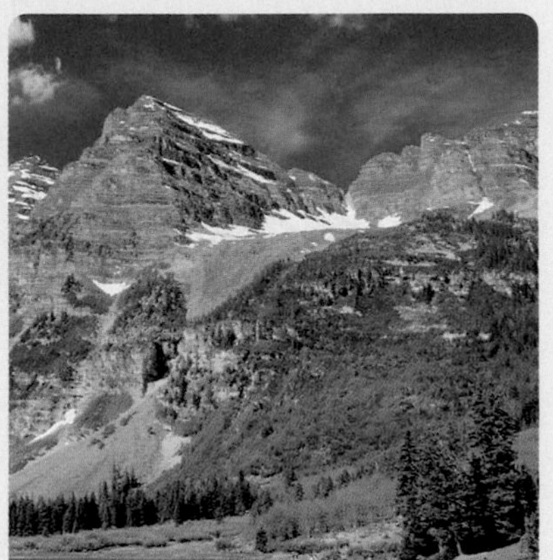

Maroon Bells in the Colorado Rockies

29. The top wind speed of an F5 tornado, the strongest known kind of tornado, is 246 mi/h faster than the top wind speed of an F1 tornado, the weakest kind of tornado. The top wind speed of an F1 tornado is 72 mi/h. Is the top wind speed of an F5 tornado 174 mi/h, 218 mi/h, or 318 mi/h?

30. In 2001, Tropical Storm Allison dropped an estimated 37 inches of rain on Houston in five days. This was only 9 inches less than Houston's average yearly rainfall. Is Houston's average yearly rainfall 28 inches, 46 inches or 49 inches?

31. ***WRITE A PROBLEM*** There has been an increase of about 1°F in the mean surface temperature of Earth from 1861 to 1998. In 1998, the mean surface temperature was about 60°F. Use this information to write a problem involving an equation with a variable.

32. ***CHALLENGE*** In the 1980s, about 9.3×10^4 acres of tropical forests were destroyed each year due to deforestation. About how many acres of tropical forests were destroyed during the 1980s?

Spiral Review

Evaluate. (Lesson 2-3)

33. $2(4 + 6) \div 5$ **34.** $8^2 - 9 \div (6 - 3)$ **35.** $14 \cdot 10 + 4 \div 2$ **36.** $36 \div 2 - 3 \cdot 6 + 5$

Write the prime factorization of each number. (Lesson 2-4)

37. 56 **38.** 72 **39.** 108 **40.** 522

41. **EOG PREP** Which of the following is *not* a common factor of 24 and 60? (Lesson 2-5)

A 3 B 4 C 5 D 6

Model Equations

Use with Lesson 2-11

KEY

[+] = 1

[+] = variable

REMEMBER

- In an equation, the expressions on both sides of the equal sign are equivalent.
- A variable can have any value that makes the equation true.
- To solve for a variable, you must get the variable alone on one side of the equal sign.

You can use algebra tiles to model and solve equations involving addition and subtraction.

Activity

Marcus has 16 CDs. If Jill gets 3 more CDs, she will have the same number of CDs as Marcus.

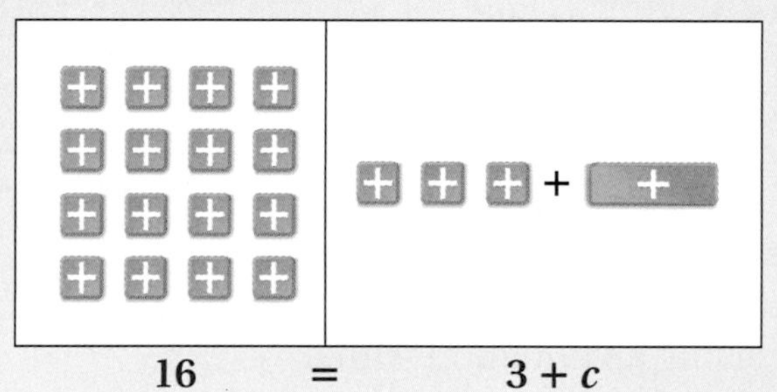

c = number of CDs that Jill has

1 Use algebra tiles to model each situation involving addition.

a. Jordan needs 2 more goals to tie the record of 10 goals in a single game.

b. Amy lives 7 miles from school. She lives 4 miles farther from school than Jack does.

c. There are 14 students in the school orchestra. Three students play the trombone, 1 plays the piano, 4 play the violin, and the rest play the drums.

At the county fair, five rides closed down when it started to rain. Only three rides continued to run in the rain.

n = total number of rides at the fair

2 Use algebra tiles to model each situation involving subtraction.

a. The French Club sold 15 handmade friendship bracelets on Tuesday for a fund-raiser and had 7 bracelets left over at the end of the day.

b. David caught 6 fish during a fishing trip, but he had to throw 2 of them back because they were too small.

c. Simone made 14 cornbread muffins. Her sister ate 2 of them, her brother ate 3 of them, and Simone ate some also. Then there were only 8 muffins left.

Think and Discuss

1. What would you need to do to solve the equation $16 = 3 + c$ with pencil and paper? Remember that to keep an equation equal, you must perform the same operations on both sides.
2. What would you need to do to solve the equation $n - 5 = 3$ with pencil and paper?
3. How could you check to see whether your solutions were correct?

Try This

Use algebra tiles to model each equation.

1. $4 + x = 6$ **2.** $3 + 5 = n$ **3.** $5 - r = 9$ **4.** $n - 9 = 12$

5. $p + 8 = 14$ **6.** $3 + y = 10$ **7.** $t - 8 = 3$ **8.** $11 - 5 = f$

9. $2 + h = 15$ **10.** $y + 7 = 9$ **11.** $13 - 7 = g$ **12.** $12 - c = 6$

Write an algebraic equation for each.

13.

14.

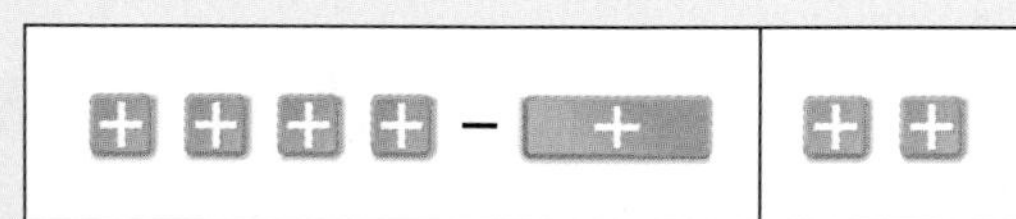

15.

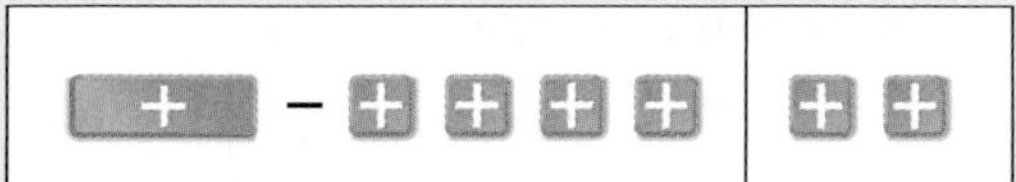

16.

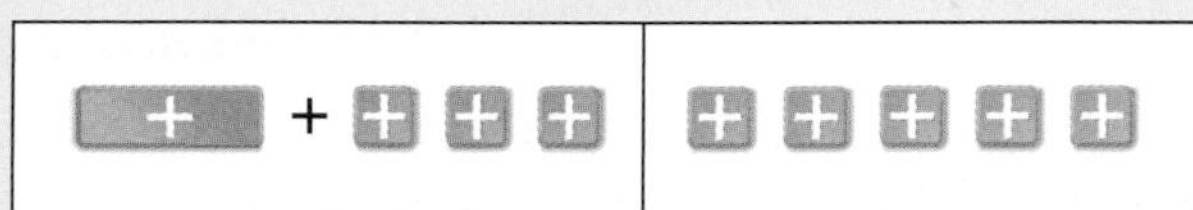

17.

18.

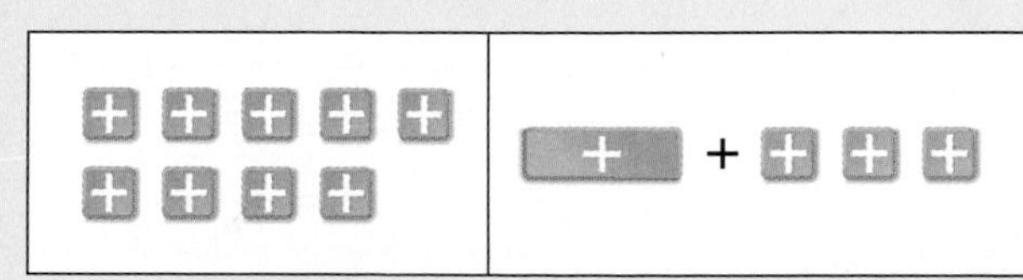

2-11 Solving Equations by Adding or Subtracting

Learn to solve one-step equations by using addition or subtraction.

Vocabulary

solve

isolate the variable

Addition Property of Equality

inverse operations

Subtraction Property of Equality

To **solve** an equation means to find a solution to the equation. To do this, **isolate the variable** —that is, get the variable alone on one side of the equal sign.

$x = 8 - 5$ $7 - 3 = y$	$x + 5 = 8$ $7 = 3 + y$
The variables are isolated.	The variables are *not* isolated.

Recall that an equation is like a balanced scale. If you increase or decrease the weights by the same amount on both sides, the scale will remain balanced.

ADDITION PROPERTY OF EQUALITY

Words	Numbers	Algebra
You can add the same amount to both sides of an equation, and the statement will still be true.	$2 + 3 = 5$ $\underline{+4} \quad \underline{+4}$ $2 + 7 = 9$	$x = y$ $\underline{+z} \quad \underline{+z}$ $x + z = y + z$

Use *inverse operations* when isolating a variable. Addition and subtraction are **inverse operations** , which means that they "undo" each other.

$$2 \boxed{+5} = 7 \longleftrightarrow 7 \boxed{-5} = 2$$

EXAMPLE 1 **Solving an Equation by Addition**

Solve the equation $x - 8 = 17$. Check your answer.

$x - 8 = 17$ — *Think: 8 is **subtracted** from x, so **add** 8 to both sides to isolate x.*

$\underline{+8} \quad \underline{+8}$

$x = 25$

Check

$x - 8 = 17$

$25 - 8 \stackrel{?}{=} 17$ — *Substitute 25 for x.*

$17 \stackrel{?}{=} 17$ ✔ — *25 is a solution.*

SUBTRACTION PROPERTY OF EQUALITY

Words	Numbers	Algebra
You can subtract the same amount from both sides of an equation, and the statement will still be true.	$\begin{array}{rcr} 4+7 & = & 11 \\ \underline{-3} & & \underline{-3} \\ 4+4 & = & 8 \end{array}$	$\begin{array}{rcr} x & = & y \\ \underline{-z} & & \underline{-z} \\ x-z & = & y-z \end{array}$

EXAMPLE 2 Solving an Equation by Subtraction

Solve the equation $a + 5 = 11$. Check your answer.

$$\begin{array}{rcl} a+5 & = & 11 \\ \underline{-5} & & \underline{-5} \\ a & = & 6 \end{array}$$

*Think: 5 is **added** to a, so **subtract** 5 from both sides to isolate a.*

Check

$a + 5 = 11$

$6 + 5 \stackrel{?}{=} 11$ *Substitute 6 for a.*

$11 \stackrel{?}{=} 11$ ✔ *6 is a solution.*

EXAMPLE 3 *Sports Application*

Michael Jordan's highest point total for a single game was 69. The entire team scored 117 points in that game. How many points did his teammates score?

Let p represent the points scored by the rest of the team.

Michael Jordan's points	+	*Rest of the team's points*	=	*Final team score*
69	+	p	=	117

$$\begin{array}{rcl} 69+p & = & 117 \\ \underline{-69} & & \underline{-69} \\ p & = & 48 \end{array}$$

Subtract 69 from both sides to isolate p.

The rest of the team scored 48 points.

Think and Discuss

1. Explain how to decide which operation to use in order to isolate the variable in an equation.

2. Describe what would happen if a number were added or subtracted on one side of an equation but not on the other side.

2-11 Exercises

FOR EOG PRACTICE
see page 665

internet connect
Homework Help Online
go.hrw.com Keyword: MS4 2-11

1.02, 5.03

GUIDED PRACTICE

See Example 1 **Solve each equation. Check your answer.**

1. $r - 77 = 99$ **2.** $102 = v - 66$ **3.** $x - 22 = 66$

See Example 2 **4.** $d + 83 = 92$ **5.** $45 = 36 + f$ **6.** $987 = 16 + m$

See Example 3 **7.** After a gain of 9 yards, your team has gained a total of 23 yards. How many yards had your team gained before the 9-yard gain?

INDEPENDENT PRACTICE

See Example 1 **Solve each equation. Check your answer.**

8. $n - 36 = 17$ **9.** $t - 28 = 54$

10. $b - 41 = 26$ **11.** $m - 51 = 23$

See Example 2 **12.** $x + 15 = 43$ **13.** $w + 19 = 62$

14. $110 = s + 65$ **15.** $x + 47 = 82$

16. $97 = t + 45$ **17.** $q + 13 = 112$

See Example 3 **18.** Hank is on a field trip. He has to travel 56 miles to reach his destination. He has traveled 18 miles so far. How much farther does he have to travel?

19. Sandy read 8 books in one month. If her book club requires her to read 6 books each month, how many more books did she read than what was required?

PRACTICE AND PROBLEM SOLVING

Solve each equation. Check your answer.

20. $p - 7 = 3$ **21.** $n + 17 = 98$

22. $356 = y - 219$ **23.** $105 = a + 60$

24. $651 + c = 800$ **25.** $f - 63 = 937$

26. $16 = h - 125$ **27.** $s + 841 = 1{,}000$

28. $63 + x = 902$ **29.** $z - 712 = 54$

30. After Renee deposited a check for $65, her new account balance was $315. Write and solve an equation to find the amount that was in her account before the deposit.

31. Adam collected 48 types of insects for his biology project. This was 34 more than he had collected in the first week. Write and solve an equation to find how many insects he found in the first week.

32. ***PHYSICAL SCIENCE*** An object weighs less when it is in water. This is because water exerts a *buoyant force* on the object. The weight of an object out of water is equal to the object's weight in water plus the buoyant force of the water. Suppose an object weighs 103 pounds out of water and 55 pounds in water. Write and solve an equation to find the buoyant force of the water.

33. ***MUSIC*** Jason wants to buy the trumpet advertised in the classified ads. He has saved \$156. Using the information from the ad, write and solve an equation to find how much more money he needs to buy the trumpet.

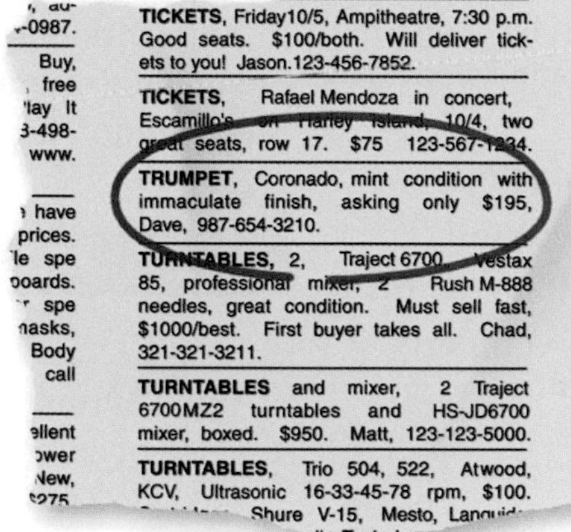

TICKETS, Friday10/5, Ampitheatre, 7:30 p.m. Good seats. \$100/both. Will deliver tickets to you! Jason.123-456-7852.

TICKETS, Rafael Mendoza in concert, Escamillo's on Harley Island, 10/4, two great seats, row 17. \$75 123-567-1234.

TRUMPET, Coronado, mint condition with immaculate finish, asking only \$195, Dave, 987-654-3210.

TURNTABLES, 2, Traject 6700, Vestax 85, professional mixer, 2 Rush M-888 needles, great condition. Must sell fast, \$1000/best. First buyer takes all. Chad, 321-321-3211.

TURNTABLES and mixer, 2 Traject 6700MZ2 turntables and HS-JD6700 mixer, boxed. \$950. Matt, 123-123-5000.

TURNTABLES, Trio 504, 522, Atwood, KCV, Ultrasonic 16-33-45-78 rpm, \$100. Shure V-15, Mesto,

34. ***WHAT'S THE ERROR?*** Describe and correct the error. $x = 50$ for $(8 + 4)2 + x = 26$

35. ***WRITE ABOUT IT*** Explain how you know whether to add or subtract to solve an equation.

36. ***CHALLENGE*** Kwan keeps a record of his football team's gains and losses on each play of the game. The record is shown in the table below. Find the missing information by writing and solving an equation.

Play	Play Gain/Loss	Overall Gain/Loss
1st down	Gain of 2 yards	Gain of 2 yards
2nd down	Loss of 5 yards	Loss of 3 yards
3rd down	Gain of 7 yards	Gain of 4 yards
4th down	■	Loss of 7 yards

Spiral Review

Find each value. (Lesson 2-1)

37. 8^6 **38.** 9^3 **39.** 4^5 **40.** 3^3

Find each product. (Lesson 2-2)

41. $147 \cdot 10^2$ **42.** $36.5 \cdot 10^4$

43. **EOG PREP** What is the value of the expression $60 - 30 \div 5 \cdot 3$? (Lesson 2-3)

A 2 C 42
B 18 D 58

44. **EOG PREP** What is the value of the expression $12 \cdot 3 - 6 \cdot 3$? (Lesson 2-3)

A 4 C 18
B 16 D 21

2-12 Solving Equations by Multiplying or Dividing

Learn to solve one-step equations by using multiplication or division.

Vocabulary

Multiplication Property of Equality

Division Property of Equality

Like addition and subtraction, multiplication and division are inverse operations. They "undo" each other.

$2 \cdot 5 = 10$

$10 \div 5 = 2$

MULTIPLICATION PROPERTY OF EQUALITY		
Words	**Numbers**	**Algebra**
You can multiply both sides of an equation by the same number, and the statement will still be true.	$3 \cdot 4 = 12$ $2 \cdot 3 \cdot 4 = 2 \cdot 12$ $6 \cdot 4 = 24$	$x = y$ $zx = zy$

If a variable is divided by a number, you can often use multiplication to isolate the variable. Multiply both sides of the equation by the number.

EXAMPLE 1 Solving an Equation by Multiplication

Solve the equation $\frac{x}{7} = 20$. Check your answer.

$\frac{x}{7} = 20$

$(7)\frac{x}{7} = 20(7)$ *Think: x is **divided** by 7, so **multiply** both sides by 7 to isolate x.*

$x = 140$

Check

$\frac{x}{7} = 20$

$\frac{140}{7} \stackrel{?}{=} 20$ *Substitute 140 for x.*

$20 \stackrel{?}{=} 20$ ✔ *140 is a solution.*

DIVISION PROPERTY OF EQUALITY		
Words	**Numbers**	**Algebra**
You can divide both sides of an equation by the same nonzero number, and the statement will still be true.	$5 \cdot 6 = 30$ $\frac{5 \cdot 6}{3} = \frac{30}{3}$ $5 \cdot \frac{6}{3} = 10$ $5 \cdot 2 = 10$	$x = y$ $\frac{x}{z} = \frac{y}{z}$ $z \neq 0$

Remember!

You cannot divide by 0.

If a variable is multiplied by a number, you can often use division to isolate the variable. You divide both sides of the equation by the number.

EXAMPLE 2 Solving an Equation by Division

Solve the equation $240 = 4z$. Check your answer.

$240 = 4z$ — *Think: z is **multiplied** by 4, so **divide** both sides by 4 to isolate z.*

$\frac{240}{4} = \frac{4z}{4}$

$60 = z$

Check

$240 = 4z$

$240 \stackrel{?}{=} 4(60)$ — *Substitute 60 for z.*

$240 \stackrel{?}{=} 240$ ✔ — *60 is a solution.*

EXAMPLE 3 *Health Application*

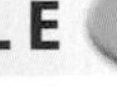

Health LINK

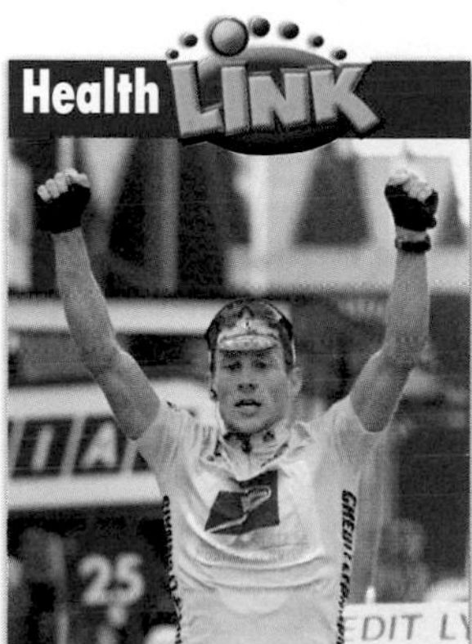

In 2002, Lance Armstrong completed the 2,051-mile Tour de France in 82 hours, 5 minutes, and 12 seconds.

go.hrw.com
KEYWORD: MS4 Lance

CNN Student News

If you count your heartbeats for 10 seconds and multiply that number by 6, you can find your heart rate in beats per minute. Lance Armstrong, who won the Tour de France four years in a row, from 1999 to 2002, has a resting heart rate of 30 beats per minute. How many times does his heart beat in 10 seconds?

Use the given information to write an equation, where b is the number of heartbeats in 10 seconds.

Beats in 10 s		*times 6*		*beats per minute*
b	$\cdot$	6	=	30

$6b = 30$ — *Think: b is **multiplied** by 6, so **divide** both sides by 6 to isolate b.*

$\frac{6b}{6} = \frac{30}{6}$

$b = 5$

Lance Armstrong's heart beats 5 times in 10 seconds.

Think and Discuss

1. **Explain** how to check your solution to an equation.
2. **Describe** how to solve $13x = 91$.
3. When you solve $5p = 35$, will p be greater than 35 or less than 35? **Explain** your answer.
4. When you solve $\frac{p}{5} = 35$, will p be greater than 35 or less than 35? **Explain** your answer.

2-12 Exercises

FOR EOG PRACTICE

see page 665

internet connect

Homework Help Online
go.hrw.com Keyword: MS4 2-12

1.02, 5.03

GUIDED PRACTICE

See Example 1

Solve the equation. Check your answer.

1. $\frac{s}{77} = 11$

2. $b \div 25 = 4$

3. $y \div 8 = 5$

See Example 2

4. $72 = 8x$

5. $3c = 96$

6. $x \cdot 18 = 18$

See Example 3

7. On Friday nights, a local bowling alley charges \$5 per person to bowl all night. If Carol and her friends paid a total of \$45 to bowl, how many people were in their group?

INDEPENDENT PRACTICE

See Example 1

Solve the equation. Check your answer.

8. $12 = s \div 4$

9. $\frac{k}{18} = 72$

10. $13 = \frac{z}{5}$

11. $\frac{c}{5} = 35$

12. $\frac{w}{11} = 22$

13. $17 = n \div 18$

See Example 2

14. $17x = 85$

15. $63 = 3p$

16. $6u = 222$

17. $97a = 194$

18. $9q = 108$

19. $495 = 11d$

See Example 3

20. It costs \$6 per ticket for groups of ten or more people to see a minor league baseball game. If Albert's group paid a total of \$162 for game tickets, how many people were in the group?

PRACTICE AND PROBLEM SOLVING

Solve the equation. Check your answer.

21. $9 = g \div 3$

22. $150 = 3j$

23. $7r = 84$

24. $5x = 35$

25. $b + 33 = 95$

26. $\frac{p}{15} = 6$

27. $504 = c - 212$

28. $8a = 288$

29. $21 = d \div 2$

30. $\frac{h}{20} = 83$

Translate each sentence into an equation. Then solve the equation.

31. A number d divided by 4 equals 3.

32. The product of a number b and 5 is 250.

33. Nine weeks from now Susan hopes to buy a bicycle that costs \$180. How much money must she save per week?

34. At a bake sale, the pies were cut into eight equally sized pieces before being sold. If there were 40 pieces in all, how many whole pies were there?

35. ***SCHOOL*** A school club is collecting toys for a children's charity. There are 18 students in the club. The goal is to collect 216 toys. Each member will collect the same number of toys. How many toys should each member collect?

36. ***TRAVEL*** Lissa drove from Los Angeles to New York City and averaged 45 miles per hour. Her driving time totaled 62 hours. Write and solve an equation to find the distance Lissa traveled.

37. ***BUSINESS*** A store rents space in a building at a cost of $19 per square foot. If the store is 700 square feet, how much is the rent?

38. ***WHAT'S THE ERROR?*** For the equation $16x = 102$, a student found the value of x to be 7. What was the student's error?

39. ***WRITE ABOUT IT*** How do you know whether to use multiplication or division to solve an equation?

40. ***CHALLENGE*** The graph shows the results of a survey about electronic equipment used by 8,690,000 college students. If you multiply the number of students who use portable CD players by 5 and then divide by 3, you get the total number of students represented by the survey. Write and solve an equation to find the number of students who use portable CD players.

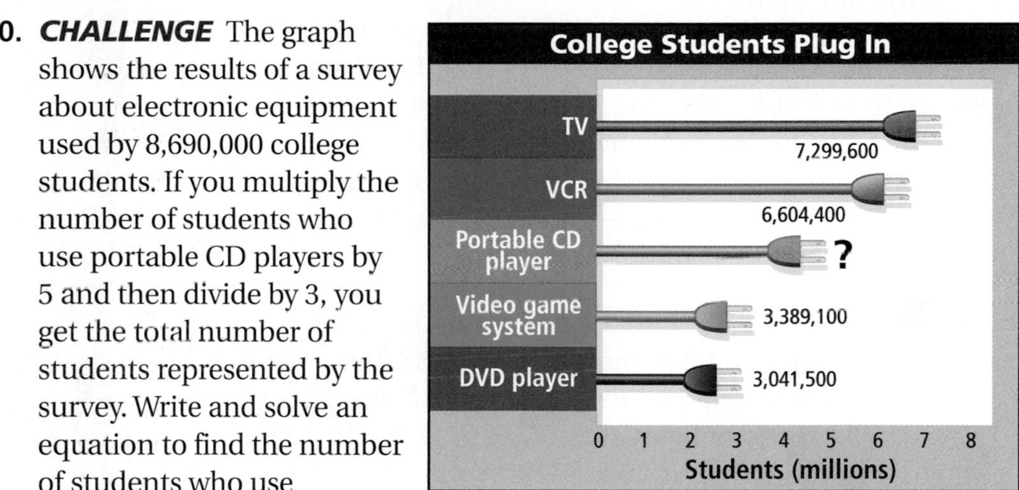

Spiral Review

Evaluate each expression for the given value of the variable. (Lesson 2-7)

41. $r^2 + 6$ for $r = 2$

42. $(11 - n)5$ for $n = 4$

Combine the like terms in each expression. (Lesson 2-9)

43. $8x^2 + 9 - x^2 - 3$

44. $n + 82m - n^4 + 5n - 20m$

45. **EOG PREP** Which expression is described by the phrase 7 times the sum of a number and 5? (Lesson 2-8)

A $7 \cdot 5$　　C $7 + n + 5$

B $7n + 5$　　D $7(n + 5)$

46. **EOG PREP** Which is a solution of $x - 28 = 7$? (Lesson 2-11)

A 21　　C 35

B 27　　D 43

Problem Solving on Location

NORTH CAROLINA

Morehead Planetarium and Science Center

The planetarium at the University of North Carolina in Chapel Hill offers a variety of exciting shows that bring you closer to the stars. One of the shows, *Sky Safari,* takes you on an imaginary ride through the solar system and to other parts of the universe.

1. A tour guide at the planetarium says that the planet Saturn is, at most, 1,658.5 million kilometers from Earth. A brochure explains that when Saturn and Earth are farthest apart, the distance between them is 1.6585×10^9 kilometers. Did the brochure give the same distance as the planetarium tour guide? Explain.

2. The distance from Earth to the nearest major galaxy, Andromeda, is about 2 million light-years. A light-year is the distance that light travels in one year. If you could travel at the speed of light, about how long would it take you to travel to the Andromeda galaxy and back? Write your answer in scientific notation.

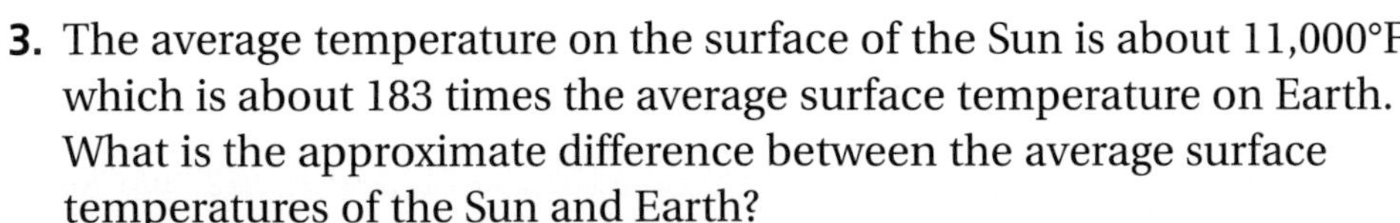

3. The average temperature on the surface of the Sun is about 11,000°F, which is about 183 times the average surface temperature on Earth. What is the approximate difference between the average surface temperatures of the Sun and Earth?

North Carolina State Fair

Every October, thousands of residents flock to the North Carolina State Fair in Raleigh. People can purchase discount tickets in advance or pay regular prices at the gate.

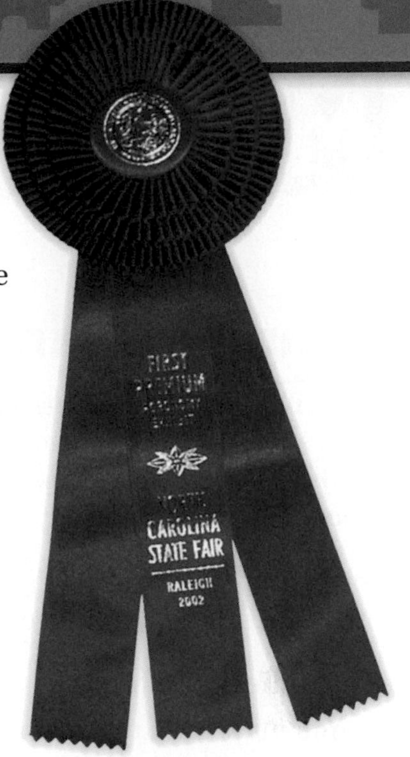

State Fair Ticket Prices		
	Regular ($)	Discount ($)
Adults (ages 13 and up)	6.00	5.00
Groups of 40 or more	6.00	4.75
Children (ages 6–12)	2.00	1.00
Children (ages 5 and under)	Free	Free
Senior citizens (ages 65 and up)	Free	Free

For 1–3, use the table.

1. When Sally's family decided to go to the state fair, Sally determined that the total cost for tickets would be $28. Before she bought the tickets, her 16-year-old brother decided to skip the fair and go on a campout instead. What was the new total cost for tickets for the rest of the family if they purchased them at the regular price?

2. A family of four paid a total of $10 at the regular price to get into the fair. In which age group does each family member belong if none of the family members are younger than age 6?

3. Elias and his friends bought their tickets to the fair in advance, so they paid the discount price. Elias is 13 years old, and his friends are 15, 12, and 14 years old. How much money did they save in all by buying their tickets in advance?

4. The *Carolinian,* the state's public transportation train, makes stops at the state fair. The northbound train arrives at the fairgrounds at 11:40 A.M. The southbound train departs at 4:23 P.M. If you took the train to the fair on both Saturday and Sunday for two weekends, about how many full hours would you get to spend at the fair?

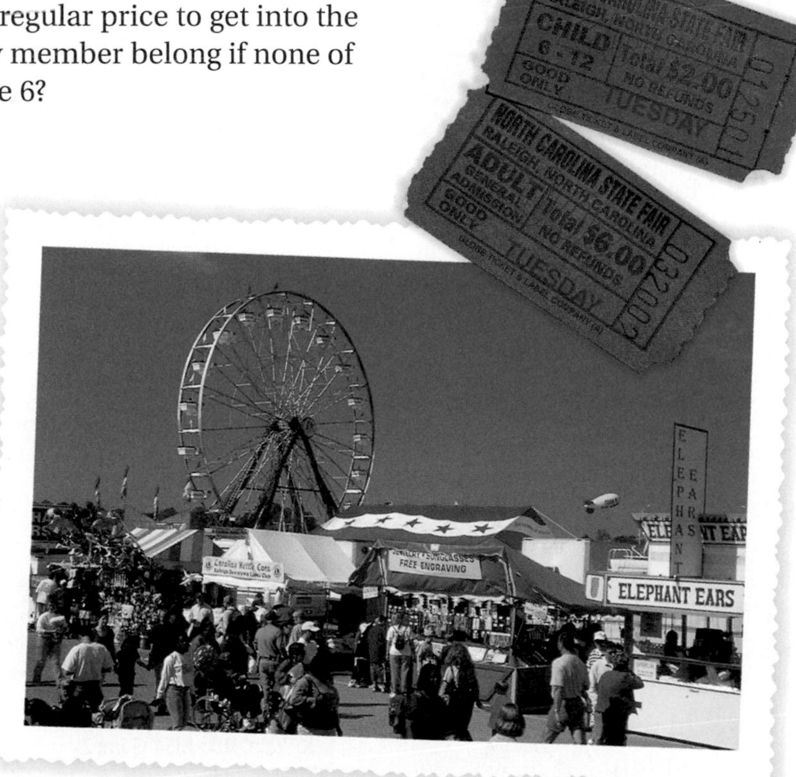

MATH-ABLES

Jumping Beans

You will need a grid that is 4 squares by 6 squares. Each square must be large enough to contain a bean. Mark off a 3-square by 3-square section of the grid. Place nine beans in the nine spaces, as shown below.

You must move all nine beans to the nine marked-off squares in the fewest number of moves.

Follow the rules below to move the beans.

1. You may move to any empty square in any direction.
2. You may jump over another bean in any direction to an empty square.
3. You may jump over other beans as many times as you like.

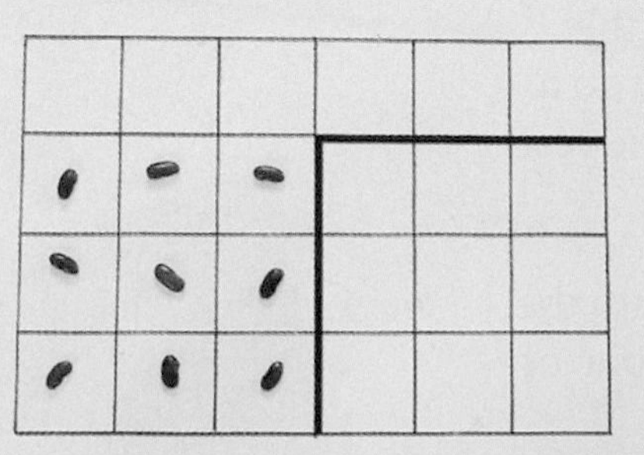

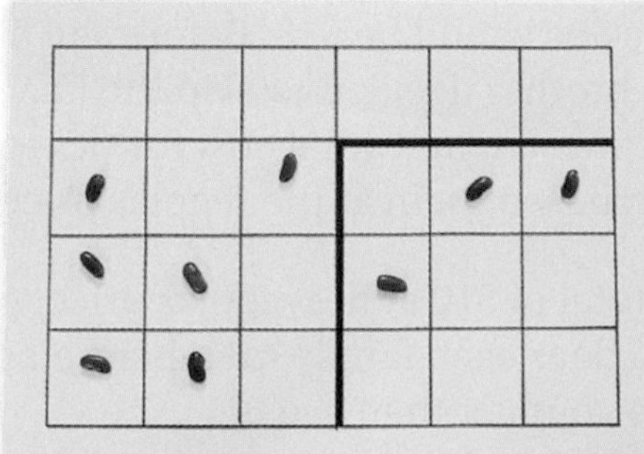

Moving all the beans in ten moves is not too difficult, but can you do it in nine moves?

Trading Spaces

The purpose of the game is to replace the red counters with the yellow counters, and the yellow counters with the red counters, in the fewest moves possible. The counters must be moved one at a time in an L-shape. No two counters may occupy the same square.

internet connect

For a complete copy of the rules and to print out a game board, go to ***go.hrw.com***
KEYWORD: MS4 Game2

Explore Order of Operations

REMEMBER

The order of operations

1. Perform operations within grouping symbols.
2. Evaluate powers.
3. Multiply and divide in order from left to right.
4. Add and subtract in order from left to right.

Many calculators have an x^2 key that allows you to find the square of a number. On calculators that do not have this key, or to use exponents other than 2, you can use the caret key, ^.

For example, to evaluate 3^5, press 3 ^ 5, and then press ENTER.

Activity

1 Evaluate $4 \cdot 2^3$ using paper and pencil. Check your answer with a calculator.

First evaluate the expression using paper and pencil: $4 \cdot 2^3 = 4 \cdot 8 = 32$.

Then evaluate $4 \cdot 2^3$ using your calculator.

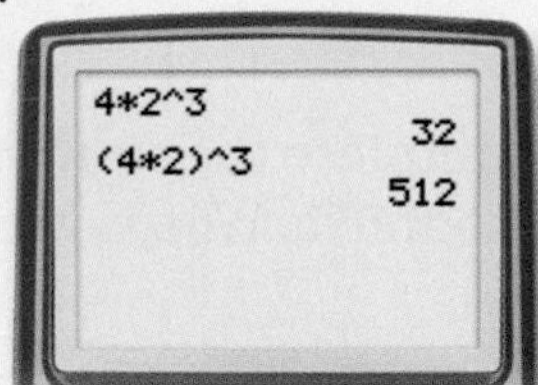

Notice that the calculator automatically evaluates the power first. If you want to multiply first, you must put that operation inside parentheses.

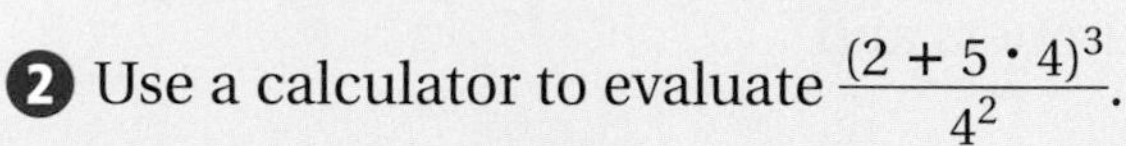

2 Use a calculator to evaluate $\frac{(2 + 5 \cdot 4)^3}{4^2}$.

Think and Discuss

1. Is $2 + 5 \cdot 4^3 + 4^2$ equivalent to $(2 + 5 \cdot 4^3) + 4^2$? Explain.

Try This

Evaluate each expression with pencil and paper. Check your answers with a calculator.

1. $3 \cdot 2^3 + 5$ **2.** $3 \cdot (2^3 + 5)$ **3.** $(3 \cdot 2)^2$ **4.** $3 \cdot 2^2$ **5.** $2^{(3 \cdot 2)}$

Use a calculator to evaluate each expression. Round your answers to the nearest hundredth.

6. $(2.1 + 5.6 \cdot 4^3) \div 6^4$ **7.** $[(2.1 + 5.6) \cdot 4^3] \div 6^4$

Chapter 2

Study Guide and Review

Vocabulary

Complete the sentences below with vocabulary words from the list above. Words may be used more than once.

1. The __?__ tells how many times to use the __?__ as a factor.
2. A(n) __?__ is a mathematical phrase made up of numbers and operations.
3. A(n) __?__ is a whole number with more than two factors.
4. A(n) __?__ consists of constants, variables, and operations.

2-1 Exponents (pp. 60–63)

EXAMPLE

- **Find the value of 4^3.**

  $4^3 = 4 \cdot 4 \cdot 4$
  $= 64$

EXERCISES

Find each value.

5. 9^2 **6.** 10^1 **7.** 2^7 **8.** 1^7 **9.** 11^2

2-2 Powers of Ten and Scientific Notation (pp. 64–67)

EXAMPLE

- **Find the product of $157 \cdot 10^4$.**

  $157 \cdot 10^4 = 1570000$
  $= 1,570,000$

EXERCISES

Find each product.

10. $144 \cdot 10^2$ **11.** $1.32 \cdot 10^3$ **12.** $22 \cdot 10^7$
13. $34 \cdot 10^1$ **14.** $56 \cdot 10^4$ **15.** $7.8 \cdot 10^2$

2-3 Order of Operations (pp. 70–73)

EXAMPLE

- **Evaluate $(18 + 6) \cdot 5$.**

  $(18 + 6) \cdot 5 = 24 \cdot 5 = 120$

EXERCISES

Evaluate.

16. $2 + (9 - 6) \div 3$ **17.** $12 \cdot 3^2 - 5$

2-4 Prime Factorization (pp. 78–81)

EXAMPLE

- **Write the prime factorization of 56.**

  $56 = 8 \cdot 7 = 2 \cdot 2 \cdot 2 \cdot 7$

EXERCISES

Write the prime factorization.

18. 88 **19.** 27 **20.** 162 **21.** 96

2-5 Greatest Common Factor (pp. 82–85)

EXAMPLE

- **Find the GCF of 32 and 12.**

  32: 1, 2, (4), 8, 16, 32
  12: 1, 2, 3, (4), 6, 12 The GCF is 4.

EXERCISES

Find the greatest common factor.

22. 120, 210 **23.** 81, 132
24. 36, 60, 96 **25.** 220, 440, 880

2-6 Least Common Multiple (pp. 86–89)

EXAMPLE

- **Find the LCM of 8 and 10.**

  8: 8, 16, 24, 32, (40)
  10: 10, 20, 30, (40) The LCM is 40.

EXERCISES

Find the least common multiple.

26. 5, 12 **27.** 4, 32 **28.** 3, 27
29. 15, 18 **30.** 6, 12 **31.** 5, 7, 9

2-7 Variables and Algebraic Expressions (pp. 92–95)

EXAMPLE

- **Evaluate $5a - 6b + 7$ for $a = 4$ and $b = 3$.**

  $5a - 6b + 7$
  $5(4) - 6(3) + 7 = 20 - 18 + 7 = 9$

EXERCISES

Evaluate for the given values.

32. $4x - 5$ for $x = 6$
33. $8y^3 + 3y$ for $y = 4$

Study Guide and Review

Study Guide and Review

2-8 Translate Words into Math (pp. 96–99)

EXAMPLE

■ **Write as an algebraic expression.**

5 times the sum of a number and 6

$5(n + 6)$

EXERCISES

Write as an algebraic expression.

34. 4 divided by the sum of a number and 12

35. 2 times the difference of a number and 11

2-9 Combining Like Terms (pp. 100–103)

EXAMPLE

■ **Combine like terms.**

$4x^3 + 5y + 8x^3 - 4y - 5x^2$

$\mathbf{4x^3} + 5y + \mathbf{8x^3} - 4y - 5x^2$

$\mathbf{12x^3} + y - 5x^2$

EXERCISES

Combine like terms.

36. $7b^2 + 8 + 3b^2$

37. $12a^2 + 4 + 3a^2 - 2$

38. $x^2 + x^3 + x^4 + 5x^2$

2-10 Equations and Their Solutions (pp. 104–107)

EXAMPLE

■ **Determine whether 22 is a solution.**

$24 \stackrel{?}{=} s - 13$

$24 \stackrel{?}{=} 22 - 13$

$24 \stackrel{?}{=} 9$ ✗ *22 is not a solution.*

EXERCISES

Determine whether each number is a solution of $36 = n - 12$.

39. 48 **40.** 54 **41.** 3

2-11 Solving Equations by Adding or Subtracting (pp. 110–113)

EXAMPLE

■ **Solve the equation. Then check.**

$b + 12 = 16$

$\underline{-12} \; \underline{-12}$

$b = 4$

$b + 12 \stackrel{?}{=} 16$

$4 + 12 \stackrel{?}{=} 16$

$16 \stackrel{?}{=} 16$ ✔

EXERCISES

Solve each equation. Then check.

42. $8 + b = 16$ **43.** $20 = n - 12$

44. $27 + c = 45$ **45.** $t - 68 = 44$

2-12 Solving Equations by Multiplying or Dividing (pp. 114–117)

EXAMPLE

■ **Solve the equation. Then check.**

$2r = 12$

$\frac{2r}{2} = \frac{12}{2}$

$r = 6$

$2r = 12$

$2(6) \stackrel{?}{=} 12$

$12 \stackrel{?}{=} 12$ ✔

EXERCISES

Solve each equation. Then check.

46. $n \div 12 = 6$ **47.** $3p = 27$

48. $\frac{d}{14} = 7$ **49.** $6x = 78$

Chapter Test

Find each value.

1. 6^2
2. 7^5
3. 8^6
4. 3^5

Find each product.

5. $148 \cdot 10^2$
6. $56.3 \cdot 10^3$
7. $6.89 \cdot 10^4$
8. $7.5 \cdot 10^4$

Evaluate.

9. $18 \cdot 3 \div 3^3$
10. $36 + 16 - 50$
11. $149 - (2^8 - 200)$
12. $(4 \div 2) \cdot 9 + 11$

Write the prime factorization of each number.

13. 30
14. 66
15. 78
16. 110

Find the greatest common factor (GCF).

17. 18, 27, 45
18. 16, 28, 44
19. 14, 28, 56
20. 24, 36, 64

Find the least common multiple (LCM).

21. 24, 36, 64
22. 24, 72, 144
23. 12, 15, 36
24. 9, 16, 25

Evaluate each algebraic expression for the given values of the variables.

25. $4a + 6b + 7$ for $a = 2$ and $b = 3$
26. $7y^2 + 7y$ for $y = 3$

Write each phrase as an algebraic expression.

27. a number increased by 12
28. the quotient of a number and 7
29. 5 less than the product of 7 and a number
30. the difference between three times a number and 4

Combine like terms.

31. $b + 2 + 5b$
32. $16 + 5b + 3b + 9$
33. $5a + 6t + 9 + 2a$

Determine whether each number is a solution of $30 = s + 6$.

34. 15
35. 24
36. 18

Solve each equation.

37. $x + 9 = 19$
38. $21 = y - 20$
39. $m - 54 = 72$
40. $136 = y + 114$
41. $16 = \frac{y}{3}$
42. $102 = 17y$
43. $\frac{r}{7} = 1{,}400$
44. $6x = 42$
45. A caterer charged $15 per person to prepare a meal for a banquet. If the total catering charge for the banquet was $1,530, how many people attended?

Chapter 2

Performance Assessment

Show What You Know

Create a portfolio of your work from this chapter. Complete this page and include it with your four best pieces of work from Chapter 2. Choose from your homework or lab assignments, mid-chapter quizzes, or any journal entries you have done. Put them together using any design you want. Make your portfolio represent what you consider your best work.

Short Response

1. You have enough pens to divide them into 12 equal groups or 10 equal groups. What is the least number of pens you could have? Explain how you found your answer.

2. What value of w makes the expressions $2w$ and $2 + w$ equal? What value of d makes the expressions $d - 2$ and $\frac{d}{2}$ equal? Show the steps that you used to find each answer.

3. Describe and correct the error: $2 + 3 \cdot 4 + 5 = 5 \cdot 9 = 45$.

Extended Problem Solving

4. The Lakemont Lions and the Hillcrest Hurricanes are buying new uniforms this year. Each team member receives a cap, a jersey, and a pair of pants.

a. Let l represent the number of Lakemont team members, and let h represent the number of Hillcrest team members. Write an expression that gives the total cost of the Lions' uniforms. Write a second expression for the total cost of the Hurricanes' uniforms.

b. Using the expressions from part **a,** write a new expression that represents the difference between the cost of the Hurricanes' uniforms and the cost of the Lions' uniforms.

c. Using $l = 15$ and $h = 18$, evaluate the expressions from part **a** and find the difference between the cost of the Hurricanes' uniforms and the cost of the Lions' uniforms. Then evaluate the expression from part **b.** Explain why the results are the same.

Performance Assessment

internet connect

EOG Practice Online

go.hrw.com Keyword: MS4 TestPrep

Getting Ready for EOG

Chapter 2

Cumulative Assessment, Chapters 1–2

1. What is the value of $9t$ for $t = 7$?

A 16 C 63

B 54 D 71

2. What is another way to write $5 \cdot 5 \cdot 5 \cdot 5 \cdot 5$?

A $5 \cdot 5$ C 55,555

B 5^5 D 5 cubed

3. What is the value of the expression $8 - 1 \cdot 4 \div 2$?

A 2 C 14

B 6 D 18

4. Which expression does **not** have a value of 38?

A $(8 + 2) \cdot 3 + 8$ C $45 - (15 + 8)$

B $8 + 2 \cdot 15$ D $(40 - 20) + 2 \cdot 9$

TEST TAKING TIP!

Eliminate choices by using the definition of *product*.

5. Which expression represents the product of 6 and a number n?

A $6n$ C $6 - n$

B $6 + n$ D $\frac{6}{n}$

6. Which expression is the prime factorization of 36?

A $4 \cdot 9$ C $2^3 \cdot 3$

B $2^2 \cdot 3^2$ D $2^3 \cdot 9$

Use the stem-and-leaf plot for Items 7–9.

Attendance at Weekly Baseball Game

Stems	Leaves
5	5 8 8 9
6	0 3 4 7
7	1 5 5 7 8
8	0 2 7 7 7 9

Key: 5 | 9 means 59

7. What is the range of the attendance at the baseball games?

A 30 C 75

B 34 D 87

8. What is the mode of the attendance at the baseball games?

A 34 C 75

B 58 D 87

9. What is the median of the attendance at the baseball games?

A 55 C 75

B 72 D 87

10. ***SHORT RESPONSE*** Paul is on the twentieth floor of a building. He climbs the stairs to the thirty-fifth floor. Represent this situation by writing an equation that includes a variable. Explain what the variable represents.

11. ***SHORT RESPONSE*** Use the following temperatures to make a box-and-whisker plot: 54°F, 70°F, 62°F, 55°F, 56°F, 66°F, and 62°F. Then give the median temperature.

Getting Ready for EOG

Chapter 3

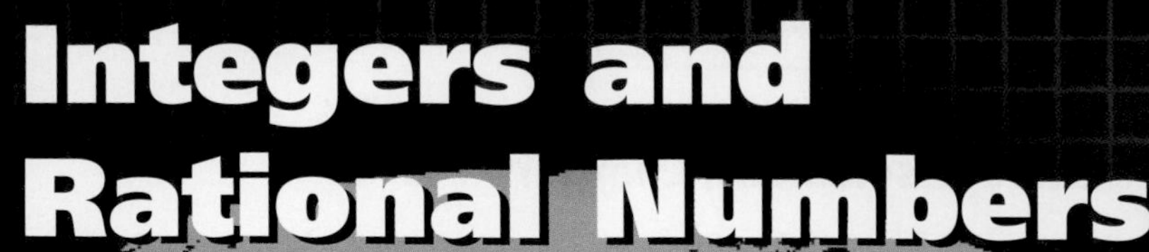

Integers and Rational Numbers

TABLE OF CONTENTS

internet connect

Chapter Opener Online
go.hrw.com
KEYWORD: MS4 Ch3

Speed of Sound Through Different Materials

Material	Speed (m/s)
Air at 20°C	344
Water at 20°C	1,500
Wood (oak) at 20°C	3,850
Glass at 20°C	4,540
Steel at 20°C	5,200

Career *Oceanographer*

Is Earth warming? Or is it cooling? The temperature of the oceans is a very important factor in answering these questions. Oceanographers have been studying the temperature of the Pacific Ocean by measuring the speed of sound waves in the water.

How does this work? The speed of sound is affected by the temperature of the material through which the sound travels. In air, for example, the speed of sound increases about 0.6 meter per second for every degree Celsius that the temperature rises. By measuring the speed of sound in water, scientists can tell the water's temperature.

ARE YOU READY?

Choose the best term from the list to complete each sentence.

1. To __?__ a number on a number line, mark and label the point that corresponds to that number.
2. The expression $5 > 3 > 1$ tells the __?__ of these three numbers on a number line.
3. A(n) __?__ is a mathematical statement showing two things are equal.
4. Each number in the set 0, 1, 2, 3, 4, 5, 6, 7, ... is a(n) __?__.
5. To __?__ an equation, find a value that makes it true.

whole number
expression
graph
solve
equation
order

Complete these exercises to review skills you will need for this chapter.

✔ Order of Operations

Simplify.

6. $7 + 9 - 5 \cdot 2$
7. $12 \cdot 3 - 4 \cdot 5$
8. $115 - 15 \cdot 3 + 9(8 - 2)$
9. $20 \cdot 5 \cdot 2(7 + 1) \div 4$

✔ Evaluate Expressions with Variables

Evaluate each expression for the given value of *n*.

10. $n + 3$ for $n = 2$
11. $3n - 9$ for $n = 10$
12. $\frac{6n}{5} + 2^2$ for $n = 5$
13. $501 + 9 - n^2$ for $n = 20$

✔ Use Inverse Operations to Solve Equations

Solve.

14. $n + 3 = 10$
15. $x - 4 = 16$
16. $9p = 63$
17. $\frac{t}{5} = 80$
18. $x - 3 = 14$
19. $\frac{q}{3} = 21$
20. $9 + r = 91$
21. $15p = 45$

✔ Words for Operations

Write an algebraic expression for each.

22. the sum of 3 and a number
23. the difference of 4 and a number
24. 6 minus the product of a number and 10

3-1 Integers

Learn to compare and order integers and to determine absolute value.

Vocabulary

opposite

integer

absolute value

The **opposite** of a number is the same distance from 0 on a number line as the original number, but on the other side of 0. Zero is its own opposite.

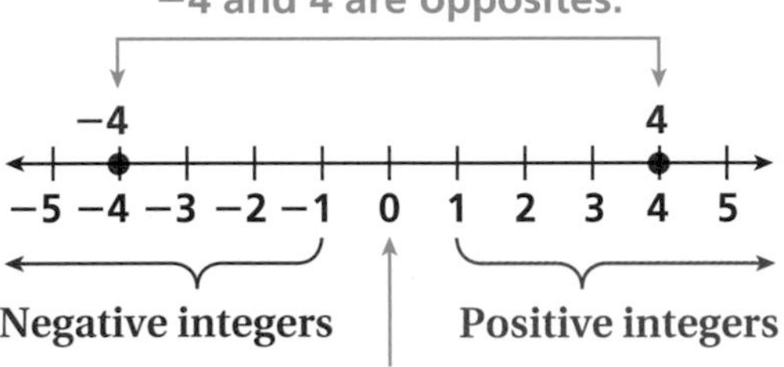

Dr. Sylvia Earle holds the world record for the deepest solo dive.

Remember!

The whole numbers are the counting numbers and zero: 0, 1, 2, 3, ...

The **integers** are the set of whole numbers and their opposites. By using integers, you can express elevations above, below, and at sea level. Sea level has an elevation of 0 feet. Sylvia Earle's record dive was to an elevation of −1,250 feet.

EXAMPLE 1 Graphing Integers and Their Opposites on a Number Line

Graph each integer and its opposite on a number line.

A 5

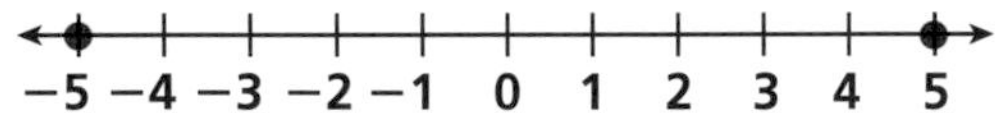

The opposite of 5 is −5.

B −3

−5 −4 −3 −2 −1 0 1 2 3 4 5

The opposite of −3 is 3.

Remember!

The symbol < means "is less than," and the symbol > means "is greater than."

Integers increase in value as you move to the right along a number line. They decrease in value as you move to the left.

$-6 < -2$ and $1 > -3$

EXAMPLE 2 Writing Integers in Order

Graph the integers on a number line, and then write them in order from least to greatest.

−2, 5, −4, 1, −1, 0

−5 −4 −3 −2 −1 0 1 2 3 4 5

−4, −2, −1, 0, 1, 5

A number's **absolute value** is its distance from 0 on a number line. Since distance can never be negative, absolute values are never negative. They are always positive or zero. The symbol $|\,|$ represents the absolute value of a number. This symbol is read as "the absolute value of." For example, $|-3|$ is the absolute value of −3.

EXAMPLE 3 Finding Absolute Value

Use a number line to find each absolute value.

A $|7|$

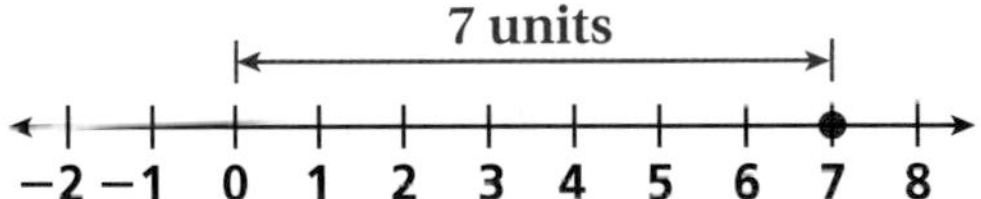

7 is 7 units from 0, so $|7| = 7$.

B $|-4|$

4 units

−6 −5 −4 −3 −2 −1 0 1 2 3 4

−4 is 4 units from 0, so $|-4| = 4$.

Think and Discuss

1. **Tell** whether two different integers can have the same absolute value. If yes, give an example. If no, explain why not.
2. **Name** the greatest negative integer and the least nonnegative integer. Then compare the absolute values of these integers.
3. **Give an example** in which a negative number has a greater absolute value than a positive number.

3-1 Exercises

FOR EOG PRACTICE

see page 666

internet connect

Homework Help Online
go.hrw.com Keyword: MS4 3-1

1.02

GUIDED PRACTICE

See Example 1 **Graph each integer and its opposite on a number line.**

1. 2 **2.** −9 **3.** −1 **4.** 6

See Example 2 **Graph the integers on a number line, and then write them in order from least to greatest.**

5. 6, −3, −1, −5, 4 **6.** 8, −2, 7, 1, −8 **7.** −6, −4, 3, 0, 1

See Example 3 **Use a number line to find each absolute value.**

8. $|-2|$ **9.** $|8|$ **10.** $|-7|$ **11.** $|-10|$

INDEPENDENT PRACTICE

See Example 1 **Graph each integer and its opposite on a number line.**

12. −4 **13.** 10 **14.** −12 **15.** 7

See Example 2 **Graph the integers on a number line, and then write them in order from least to greatest.**

16. −3, 2, −5, −6, 5 **17.** −7, −9, −2, 0, −5 **18.** 3, −6, 9, −1, −2

See Example 3 **Use a number line to find each absolute value.**

19. $|-16|$ **20.** $|12|$ **21.** $|-20|$ **22.** $|15|$

PRACTICE AND PROBLEM SOLVING

Compare. Write <, >, or =.

23. −25 ■ 25 **24.** 18 ■ −55 **25.** $|-21|$ ■ 21 **26.** −9 ■ −27

27. 34 ■ $|34|$ **28.** 64 ■ $|-75|$ **29.** 7 ■ −8 **30.** $|-3|$ ■ $|3|$

31. $|-2|$ ■ −10 **32.** 2 ■ −25 **33.** −100 ■ −82 **34.** $|-6|$ ■ $|-15|$

Find each absolute value.

35. $|-294|$ **36.** $|61|$ **37.** $|-45|$ **38.** $|-380|$

39. What is the opposite of $|32|$? **40.** What is the opposite of $|-29|$?

41. ***SOCIAL STUDIES*** Death Valley, California, is 282 feet below sea level. Write the elevation of Death Valley as an integer.

42. ***BUSINESS*** A company reported a net loss of $2,000,000 during its first year. In its second year it reported a profit of $5,000,000. Write each amount as an integer.

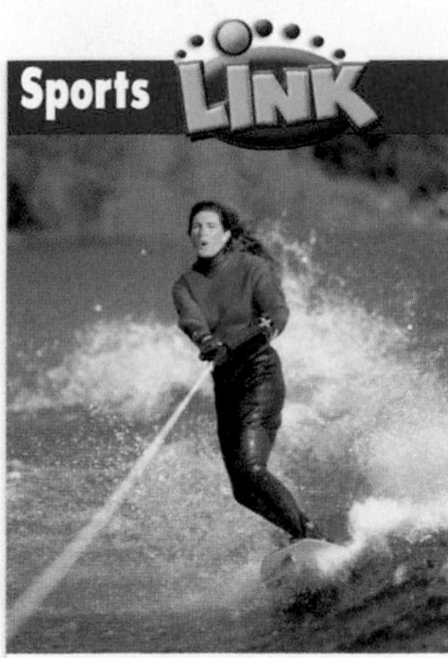

Wakeboarding is a combination of surfing and waterskiing. Tony Finn started the sport with his invention of the Skurfer. Herb O'Brien made improvements to the Skurfer, and the wakeboard was born.

43. ***SPORTS*** The graph shows how participation in several sports changed between 1999 and 2000 in the United States.

a. Which sport showed the greatest decrease in participation?

b. By about what percent did participation in racquetball increase or decrease?

c. By about what percent did participation in wall climbing increase or decrease?

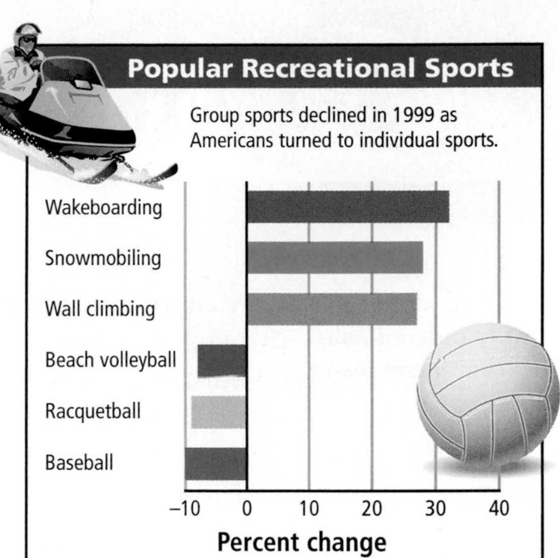

Source: *USA Today*, July 6, 2001

44. ***SOCIAL STUDIES*** Lines of latitude are imaginary lines that circle the globe in an east-west direction. They measure distances north and south of the equator. The equator represents 0° latitude.

a. What latitude is opposite of 30° north latitude?

b. How do these latitudes' distances from the equator compare?

45. ***BUSINESS*** In one year, a company reported a loss of $1,200,000 during its second quarter and a loss of $750,000 during its fourth quarter. During which quarter did the company report a greater loss?

46. ***WHAT'S THE ERROR?*** At 9 A.M. the outside temperature was −3°F. By noon, the temperature was −12°F. A newscaster said that it was getting warmer outside. Why is this incorrect?

47. ***WRITE ABOUT IT*** Explain how to compare two integers.

48. ***CHALLENGE*** What values can x have if $|x| = 11$?

Spiral Review

Find each value. (Lesson 2-1)

49. 8^2	**50.** 3^4	**51.** 15^0	**52.** 6^3
53. 2^6	**54.** 5^3	**55.** 4^4	**56.** 12^2

Evaluate each expression for $a = 2$ and $b = 9$. (Lesson 2-7)

57. $6a + 7b$	**58.** $b - a^2$	**59.** $44 \div (a + b)$	**60.** $10b + 9a \cdot b$
61. $b^3 - (6a)^2$	**62.** $b(a^2 - a) + 2b$	**63.** $a^3 \div 2a - a$	**64.** $7b + b^2 \div 3 - a^4$

65. **EOG PREP** Which is a solution of $b - 25 = 75$? (Lesson 2-11)

A $b = 3$	B $b = 50$	C $b = 100$	D $b = 150$

3-2 The Coordinate Plane

Learn to plot and identify ordered pairs on a coordinate plane.

Vocabulary

coordinate plane

x-axis

y-axis

origin

quadrant

ordered pair

A **coordinate plane** is a plane containing a horizontal number line, the ***x*-axis**, and a vertical number line, the ***y*-axis**. The intersection of these axes is called the **origin**.

The axes divide the coordinate plane into four regions called **quadrants**, which are numbered I, II, III, and IV.

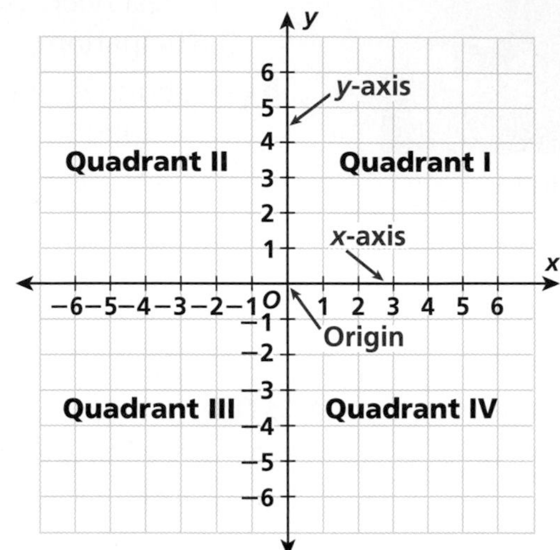

EXAMPLE 1 Identifying Quadrants on a Coordinate Plane

Identify the quadrant that contains each point.

A *P*

P lies in Quadrant II.

B *Q*

Q lies in Quadrant IV.

C *R*

R lies on the *x*-axis, between Quadrants II and III.

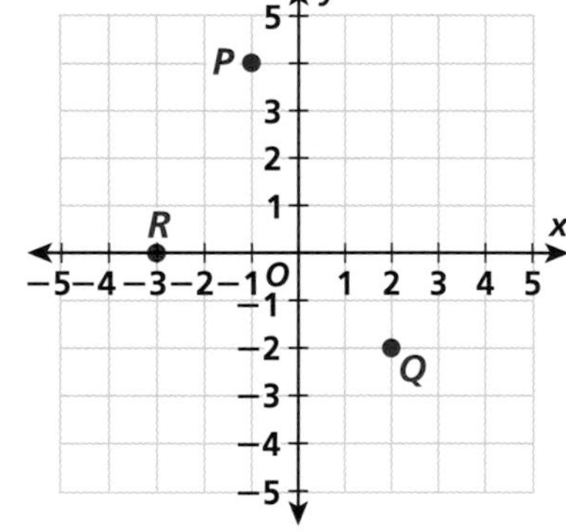

Remember!

A plane is a flat surface that extends indefinitely in all directions.

Points on a coordinate plane are identified by *ordered pairs.* An **ordered pair** consists of two numbers in a certain order. The origin is the point (0, 0).

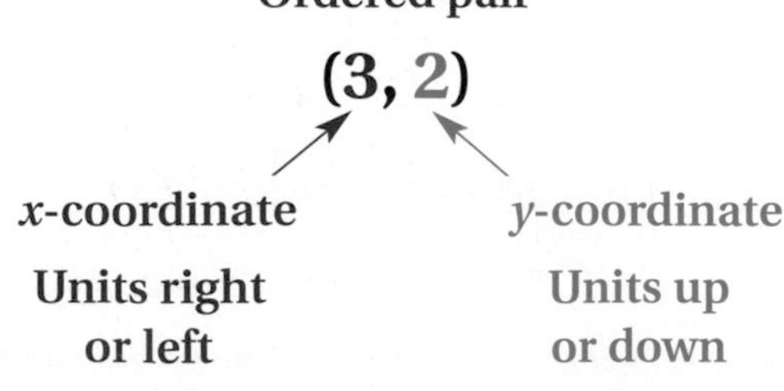

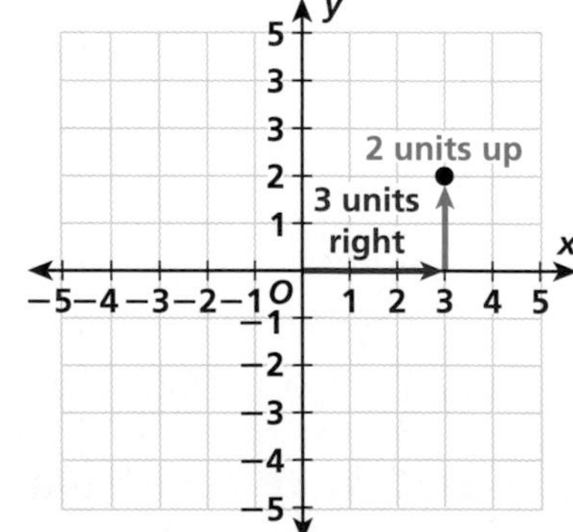

EXAMPLE 2 Plotting Points on a Coordinate Plane

Plot each point on a coordinate plane.

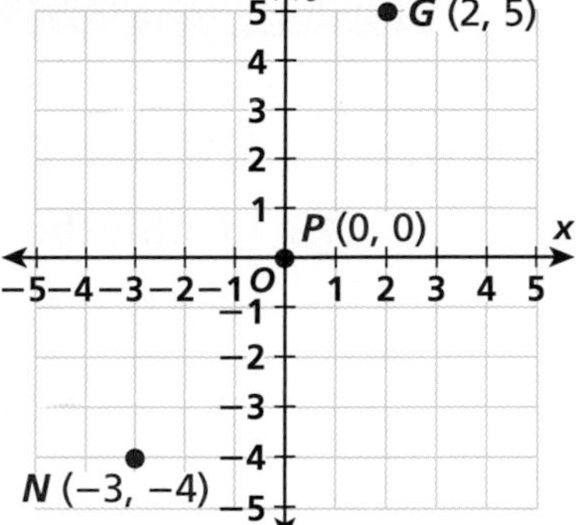

A $G(2, 5)$

Start at the origin. Move 2 units right and 5 units up.

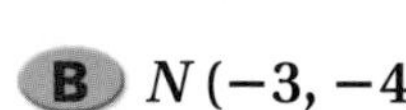

B $N(-3, -4)$

Start at the origin. Move 3 units left and 4 units down.

C $P(0, 0)$

Point P is on the origin.

EXAMPLE 3 Identifying Points on a Coordinate Plane

Give the coordinates of each point.

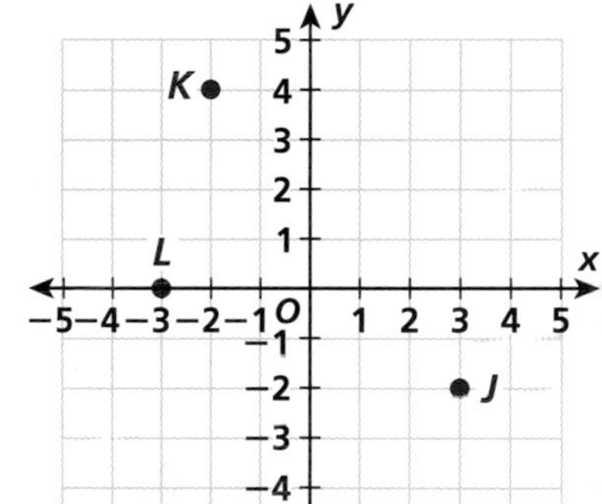

A J

Start at the origin. Point J is 3 units right and 2 units down.

$(3, -2)$

B K

Start at the origin. Point K is 2 units left and 4 units up.

$(-2, 4)$

C L

Start at the origin. Point L is 3 units left on the x-axis.

$(-3, 0)$

Think and Discuss

1. **Explain** whether point (4, 5) is the same as point (5, 4).
2. **Name** the x-coordinate of a point on the y-axis. Name the y-coordinate of a point on the x-axis.
3. **Suppose** the equator is the x-axis on a map of Earth and a line called the *prime meridian,* which passes through England, is the y-axis. Which of these directions—east, west, north, and south—are positive? Which are negative?

3-2 Exercises

FOR EOG PRACTICE
see page 666

internet connect
Homework Help Online
go.hrw.com Keyword: MS4 3-2

GUIDED PRACTICE

See Example 1

Identify the quadrant that contains each point.

1. *A* **2.** *B*

3. *C* **4.** *D*

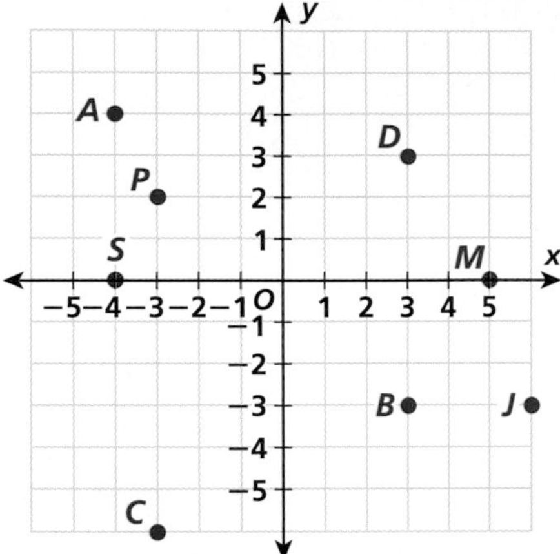

See Example 2

Plot each point on a coordinate plane.

5. (−1, 2) **6.** (2, −4)

7. (−3, −4) **8.** (5, 0)

See Example 3

Give the coordinates of each point.

9. *J* **10.** *P*

11. *S* **12.** *M*

INDEPENDENT PRACTICE

See Example 1

Identify the quadrant that contains each point.

13. *F* **14.** *J*

15. *K* **16.** *E*

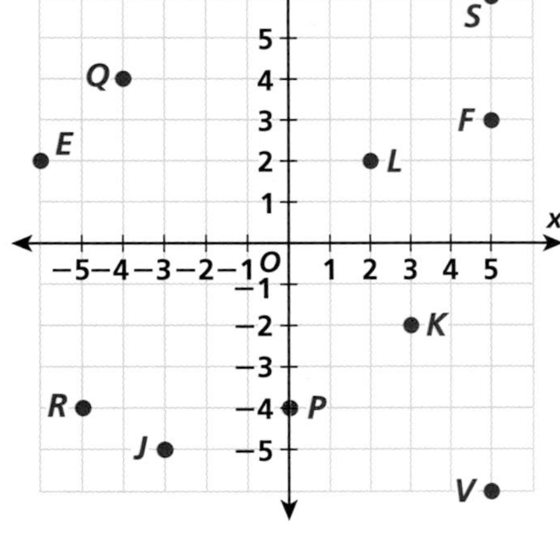

See Example 2

Plot each point on a coordinate plane.

17. (−1, 1) **18.** (2, −2)

19. (−5, −5) **20.** (0, −3)

See Example 3

Give the coordinates of each point.

21. *Q* **22.** *V*

23. *R* **24.** *P*

25. *S* **26.** *L*

PRACTICE AND PROBLEM SOLVING

For Exercises 27and 28, use graph paper to graph the ordered pairs. Use a different coordinate plane for each exercise.

27. (−8, 1); (4, 3); (−3, 6)

28. (−8, −2); (−1, −2); (−1, 3); (−8, 3)

29. Draw line segments to connect the points in Exercise 27 in the order listed. Name the figure and the quadrants in which it is located.

30. Draw line segments to connect the points in Exercise 28 in the order listed. Name the figure and the quadrants in which it is located.

Identify the quadrant of each point described below.

31. The x-coordinate and the y-coordinate are both negative.

32. The x-coordinate and the y-coordinate are both positive.

33. The x-coordinate is negative and the y-coordinate is positive.

34. The y-coordinate is negative and the x-coordinate is positive.

Weather LINK

When the wind speed of a tropical storm reaches 74 mi/h, it is called a hurricane or typhoon, depending on where in the world the storm is located.

go.hrw.com
KEYWORD: MS4 Hurricane

CNN Student News.

35. ***WEATHER*** The chart shows the coordinates of Hurricane Andrew in 1992. Use positive for north latitude and negative for west longitude.

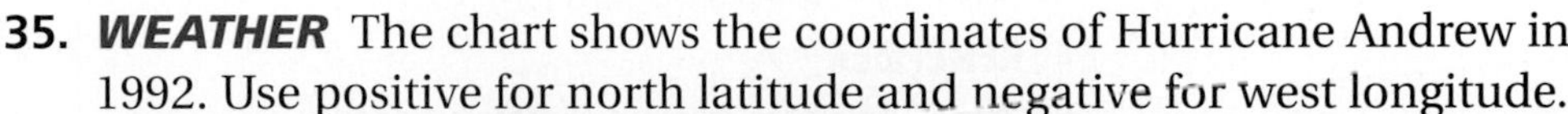

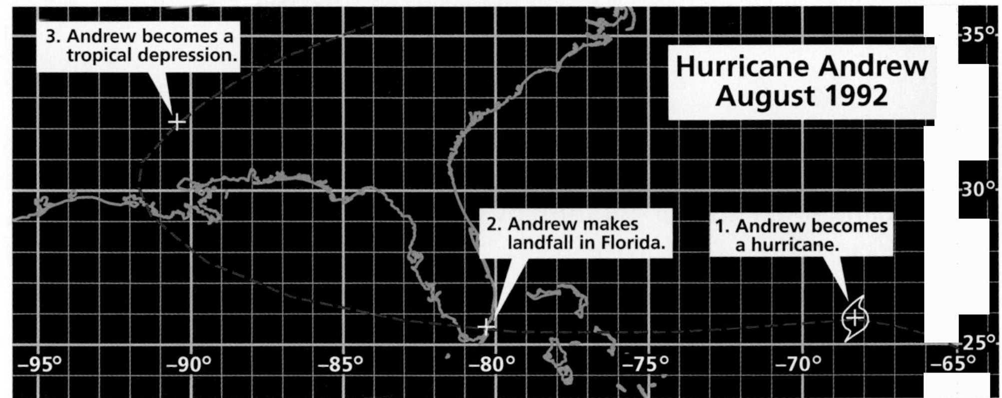

Source: National Hurricane Center

a. Estimate to the nearest integer the coordinates of the storm when it first became a hurricane.

b. Estimate to the nearest integer the coordinates of the storm when it made landfall in Florida.

c. Estimate to the nearest integer the coordinates of the storm when it weakened to a tropical depression.

36. ***WHAT'S THE ERROR?*** To plot (–12, 1), a student started at (0, 0) and moved 12 units right and 1 unit down. What did the student do wrong?

37. ***WRITE ABOUT IT*** Why is order important when graphing an ordered pair on a coordinate plane?

38. ***CHALLENGE*** Armand and Kayla started jogging from the same point. Armand jogged 4 miles south and 6 miles east. Kayla jogged west and 4 miles south. If they were 11 miles apart when they stopped, how far west did Kayla jog?

Spiral Review

Evaluate the expression $9y - 3$ for each given value of the variable. (Lesson 2-7)

39. $y = 2$ **40.** $y = 8$ **41.** $y = 10$ **42.** $y = 18$

43. **EOG PREP** What is the value of the expression $8(2 + 5) - 12$? (Lesson 2-3)

A 3 B 9 C 44 D 49

44. **EOG PREP** Which is the least common multiple of 12 and 20? (Lesson 2-6)

A 4 B 60 C 120 D 240

Model Integer Addition

Use with Lesson 3-3

KEY

= 1

= −1

\+ = 0

REMEMBER

Removing a zero from an expression does not change the expression's value.

You can model integer addition by using integer chips.

Activity

When you model adding numbers with the same sign, you can count the total number of chips to find the sum.

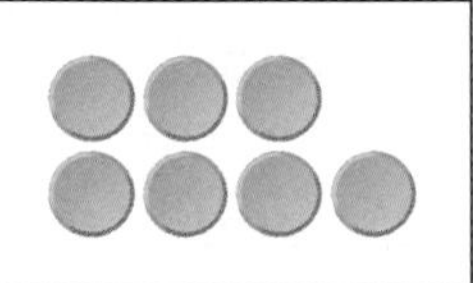

The total number of positive chips is 7.

$3 + 4 = 7$

The total number of negative chips is 7.

$-3 + (-4) = -7$

1 Model each expression.

a. $2 + 4$

b. $-2 + (-4)$

When you model adding numbers with different signs, you cannot count the chips to find their sum.

\+ = 2 and + = −2

but + = 0 *A red chip and a yellow chip make a neutral pair.*

When you model adding a positive and a negative number, you need to remove all of the neutral pairs that you can find—that is, all pairs of 1 red chip and 1 yellow chip. These pairs have a value of zero, so they do not affect the sum.

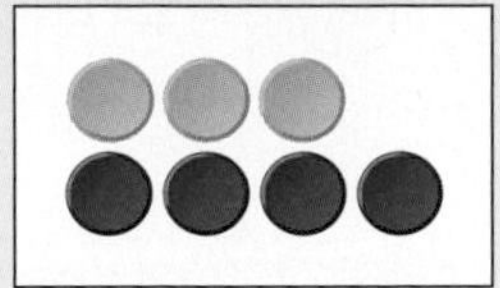

You cannot just count the colored chips to find their sum.

$3 + (-4) = \square$

Before you count the chips, you need to remove all of the neutral pairs.

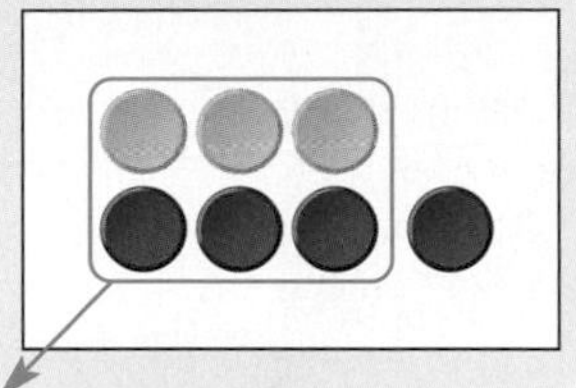

When you remove the neutral pairs, there is one red chip left. So the sum of the chips is −1.

$3 + (-4) = -1$

2 Model each expression.

a. $4 + (-6)$

b. $-5 + 2$

Think and Discuss

1. Will $8 + (-3)$ and $-3 + 8$ give the same answer? Why or why not?
2. If you have more red chips than yellow chips in a group, is the sum of the chips positive or negative?
3. If you have more yellow chips than red chips in a group, is the sum of the chips positive or negative?
4. Make a rule for the sign of the answer when negative and positive integers are added. Give examples.

Try This

Use integer chips to model and solve each addition problem.

1. $4 + (-7)$

2. $-5 + (-4)$

3. $-5 + 1$

Write the addition problems modeled below.

4.

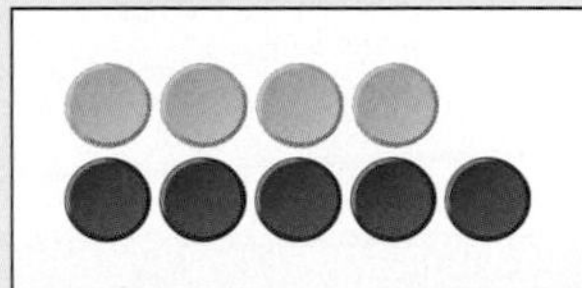

5.

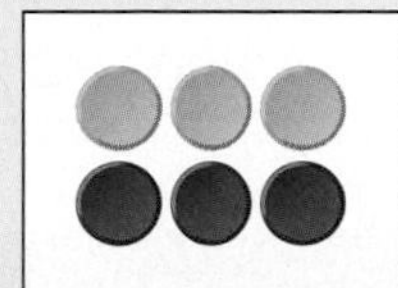

6.

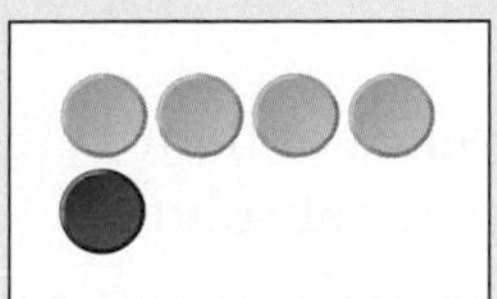

7.

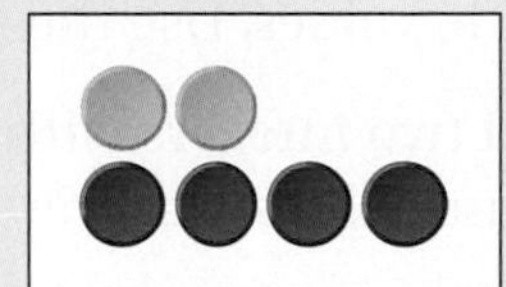

3-3 Adding Integers

Learn to add integers.

The Debate Club wanted to raise money for a trip to Washington, D.C. They began by estimating their income and expenses.

Income items are positive, and expenses are negative. By adding all your income and expenses, you can find your total earnings or losses.

Welcome to Washington

Club Ledger

Estimated Income and Expenses

Description	Amount
Car wash supplies	−\$25.00
Car wash earnings	\$300.00
Bake sale supplies	−\$50.00
Bake sale earnings	\$250.00
T-shirt decorating supplies	−\$65.00
T-shirt sale earnings	\$400.00

EXAMPLE 1 Modeling Integer Addition

Add.

A $-3 + (-6)$

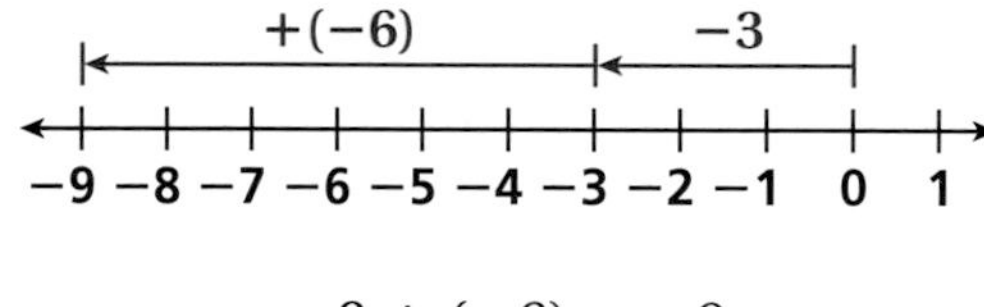

Start at 0. Move left 3 spaces. Then move left 6 more spaces.

$$-3 + (-6) = -9$$

B $4 + (-7)$

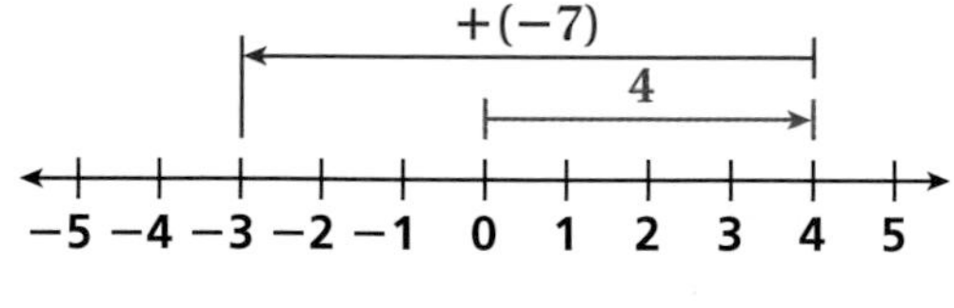

Start at 0. Move right 4 spaces. Then move left 7 spaces.

$$4 + (-7) = -3$$

Adding Integers

To add two integers with the same sign, find the sum of their absolute values. Use the sign of the two integers.

To add two integers with different signs, find the difference of their absolute values. Use the sign of the integer with the greater absolute value.

EXAMPLE 2 Adding Integers Using Absolute Values

Helpful Hint

To help you remember the rules for adding integers, think:
If the signs are the *same,* find the *sum.*
If the signs are *different,* find the *difference.*

Add.

A $8 + 6$

$8 + 6$	*The signs are the **same.***
	*Find the **sum** of the absolute values.*
	Think: 8 + 6 = 14.
14	*Use the sign of the two integers **(positive).***

B $-12 + (-4)$

$-12 + (-4)$	*The signs are the **same.***
	*Find the **sum** of the absolute values.*
	Think: 12 + 4 = 16.
-16	*Use the sign of the two integers **(negative).***

EXAMPLE 3 Evaluating Expressions with Integers

Evaluate *a* + *b* for the given values.

$a = 6, b = -10$

$a + b$

$6 + (-10)$	*Substitute for **a** and **b**.*
	*The signs are **different.***
	*Find the **difference** of the absolute values.*
	Think: 10 − 6 = 4.
-4	*Use the sign of the integer with the greater absolute value **(negative).***

EXAMPLE 4 *Banking Application*

The Debate Club's income from a car wash was $300, including tips. Supply expenses were $25. Use integer addition to find the club's total profit or loss.

$300 + (-25)$	*Use negative for the expenses.*
$300 - 25$	*Find the difference of the absolute values.*
275	*The answer is positive.*

The club earned $275.

Think and Discuss

1. **Explain** whether $-7 + 2$ is the same as $7 + (-2)$.
2. **Explain** whether $3 + (-5)$ is the same as $-5 + 3$.

3-3 Exercises

FOR EOG PRACTICE

see page 667

internet connect

Homework Help Online
go.hrw.com Keyword: MS4 3-3

1.02

GUIDED PRACTICE

See Example 1 **Add.**

1. $9 + 3$ **2.** $-4 + (-2)$ **3.** $7 + (-9)$ **4.** $-3 + 6$

See Example 2 **5.** $7 + 8$ **6.** $-1 + (-12)$ **7.** $-25 + 10$ **8.** $31 + (-20)$

See Example 3 **Evaluate $a + b$ for the given values.**

9. $a = 5, b = -17$ **10.** $a = 8, b = -8$ **11.** $a = -4, b = -16$

See Example 4 **12.** A football team gains 8 yards on one play and then loses 13 yards on the next. What is the team's total yardage?

INDEPENDENT PRACTICE

See Example 1 **Add.**

13. $-16 + 7$ **14.** $-5 + (-1)$ **15.** $4 + 9$ **16.** $-7 + 8$

17. $10 + (-3)$ **18.** $-20 + 2$ **19.** $-12 + (-5)$ **20.** $-95 + 6$

See Example 2 **21.** $-13 + (-6)$ **22.** $14 + 25$ **23.** $-22 + 6$ **24.** $35 + (-50)$

25. $-81 + (-7)$ **26.** $28 + (-3)$ **27.** $-70 + (15)$ **28.** $-18 + (-62)$

See Example 3 **Evaluate $c + d$ for the given values.**

29. $c = 6, d = -20$ **30.** $c = -8, d = -21$ **31.** $c = -45, d = 32$

See Example 4 **32.** The temperature dropped 17°F in 6 hours. If the final temperature was −3°F, what was the starting temperature?

PRACTICE AND PROBLEM SOLVING

Find each sum.

33. $-8 + (-5)$ **34.** $-12 + 16$ **35.** $-6 + (-9)$

36. $14 + (-7)$ **37.** $18 + 9$ **38.** $-41 + 15$

39. $-22 + (-18) + 22$ **40.** $27 + (-29) + 16$ **41.** $-30 + 71 + (-70)$

Compare. Write $<$, $>$, or $=$.

42. $-23 + 18$ ▢ -41 **43.** $59 + (-59)$ ▢ 0 **44.** $31 + (-20)$ ▢ 9

45. $-24 + (-24)$ ▢ 48 **46.** $25 + (-70)$ ▢ -95 **47.** $16 + (-40)$ ▢ -24

Evaluate each expression for $w = -12$, $x = 10$, and $y = -7$.

48. $w + 6$ **49.** $x + (-3)$ **50.** $w + y$

51. $x + y$ **52.** $w + x$ **53.** $w + x + y$

54. ***PERSONAL FINANCE*** Last week, Cody made deposits of $45, $18, and $27 into his checking account. He then wrote checks for $21 and $93. What is the overall change in Cody's account balance?

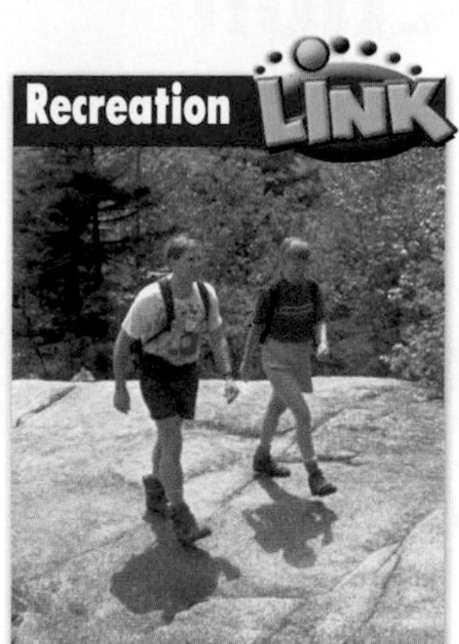

The Appalachian Trail extends about 2,160 miles from Maine to Georgia. It takes about 5 to 7 months to hike the entire trail.

55. ***RECREATION*** Hikers along the Appalachian Trail camped overnight at Horns Pond, at an of elevation 3,100 ft. Then they hiked along the ridge of the Bigelow Mountains to West Peak, which is one of Maine's highest peaks. Use the chart to determine the elevation of West Peak.

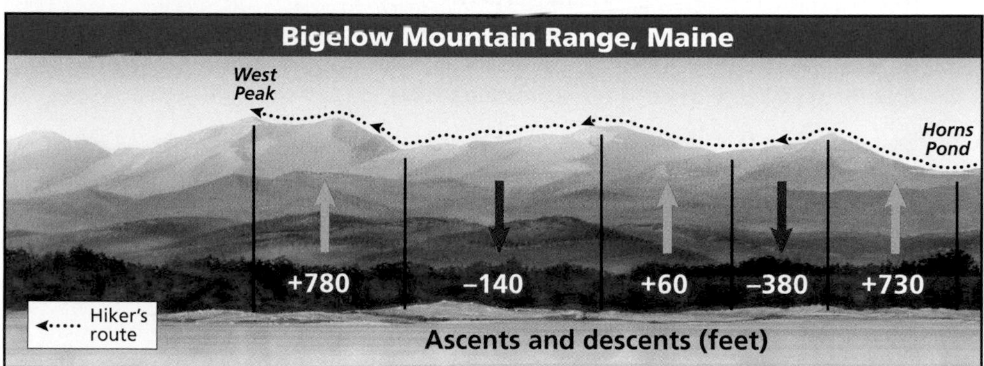

56. Hector and Luis play a game in which points can be gained or lost. In the game, each player starts with 0 points, and the player with the most points at the end wins. Hector gains 5 points, loses 3, loses 2, and then gains 3. Luis loses 5 points, gains 1, gains 5, and then loses 3. Who wins the game and by how much?

57. ***PERSONAL FINANCE*** Amanda borrowed $62 from her parents to buy a camera. She paid back $37 last week and $21 this week. How much does Amanda still owe?

58. ***WHAT'S THE QUESTION?*** The temperature was −8°F at 6 A.M. and rose 15°F by 9 A.M. The answer is 7°F. What is the question?

59. ***WRITE ABOUT IT*** Compare the method used to add integers with the same sign and the method used to add integers with different signs.

60. ***CHALLENGE*** A business had a loss of $225 million, a profit of $15 million, a profit of $125 million, a loss of $75 million, and a loss of $375 million. How much was its overall profit or loss?

Spiral Review

Write each number in scientific notation. (Lesson 2-2)

61. 839,000 **62.** 2,100,000 **63.** 4,023,000,000

64. **EOG PREP** Which is the greatest common factor of 40 and 24? (Lesson 2-5)

A 2
B 4
C 8
D 12

65. **EOG PREP** Which is a solution of the equation $6y = 96$? (Lesson 2-12)

A $y = 16$
B $y = 90$
C $y = 102$
D $y = 576$

Model Integer Subtraction

Use with Lesson 3-4

internet connect
Lab Resources Online
go.hrw.com
KEYWORD: MS4 Lab3B

KEY
(light chip) = 1
(dark chip) = –1
(light chip) + (dark chip) = 0

REMEMBER
Adding or removing a zero from an expression does not change the expression's value.

You can model integer subtraction by using integer chips.

Activity

These groups of chips show three different ways of modeling 2.

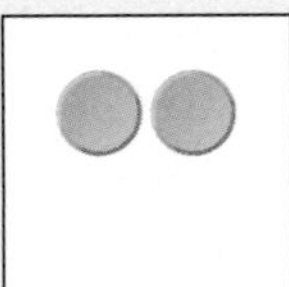

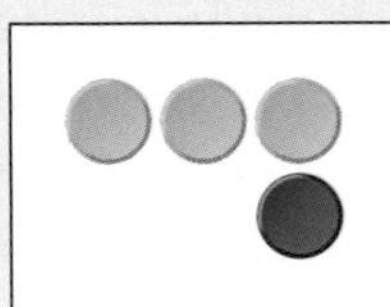

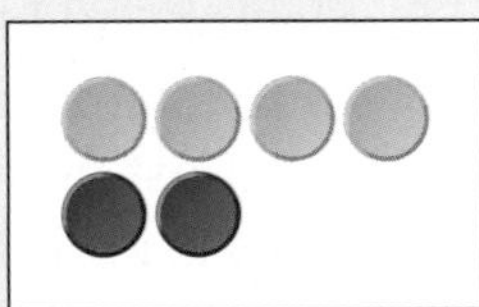

1. Show two other ways of modeling 2.

These groups of chips show two different ways of modeling −2.

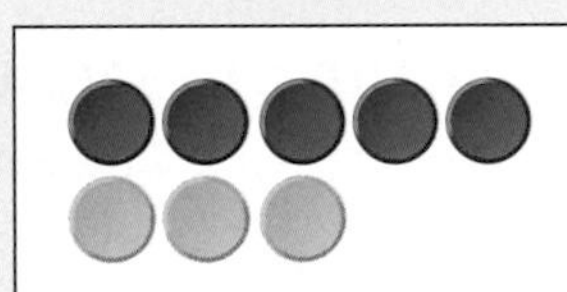

2. Show two other ways of modeling −2.

You can model some subtraction problems by taking away chips.

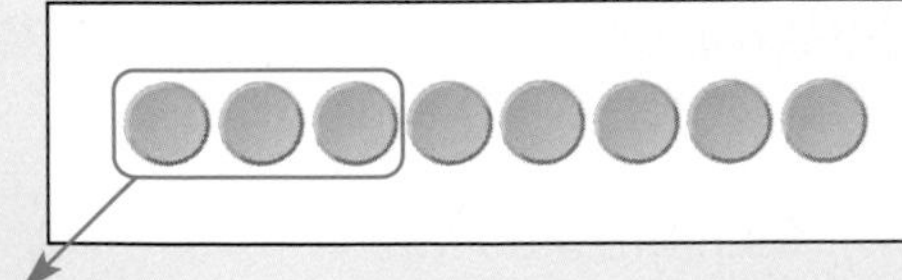

8 − 3 = 5

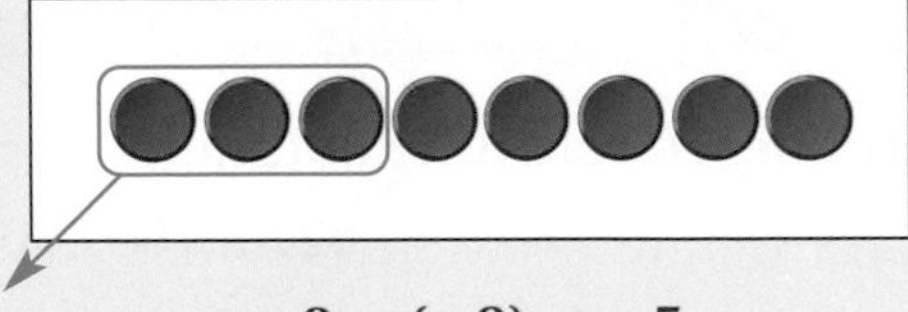

−8 − (−3) = −5

3. Model each expression.

a. 6 − 5 **b.** −6 − (−5)

To model some subtraction problems, such as $-6 - 3$, you will need to add neutral pairs before you can take chips away.

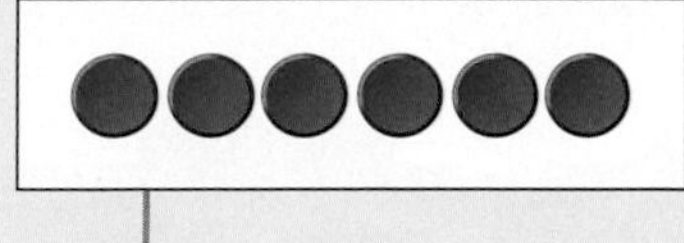

First place 6 red chips to represent −6.

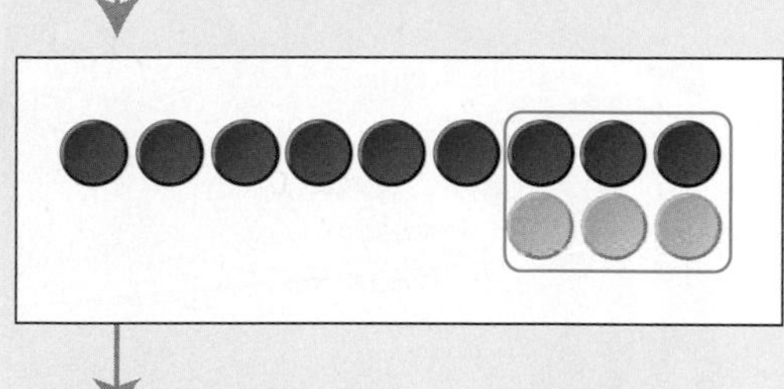

Since you cannot take away 3 yellow chips, add 3 yellow chips paired with 3 red chips.

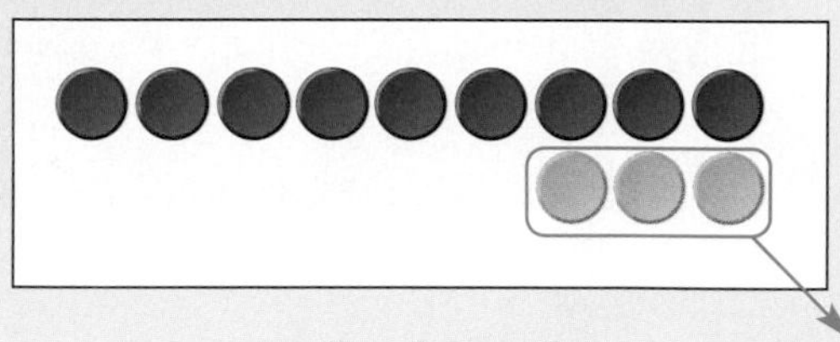

Now you can take away 3 yellow chips.

$$-6 - 3 = -9$$

4 Model each expression.

a. $-6 - 5$

b. $5 - (-6)$

c. $4 - 7$

d. $-2 - (-3)$

Think and Discuss

1. How could you model the expression $0 - 5$?

2. When you add neutral pairs to model subtraction using chips, does it matter how many neutral pairs you add?

3. Would $2 - 3$ have the same answer as $3 - 2$? Why or why not?

4. Make a rule for the sign of the answer when negative and positive integers are subtracted. Give examples.

Try This

Use integer chips to model and solve each subtraction problem.

1. $4 - 2$

2. $-4 - (-2)$

3. $-2 - (-3)$

4. $3 - 4$

5. $2 - 3$

6. $0 - 3$

7. $5 - 3$

8. $-3 - (-5)$

9. $6 - (-4)$

3-4 Subtracting Integers

Learn to subtract integers.

During its flight to and from Earth, the space shuttle may be exposed to temperatures as cold as −250°F and as hot as 3,000°F.

To find the difference in these temperatures, you need to know how to subtract integers with different signs.

You can model the difference between two integers using a number line. When you subtract a positive number, the difference is *less* than the original number, so you move to the *left*. To subtract a negative number, move to the *right*.

EXAMPLE 1 Modeling Integer Subtraction

Use a number line to find each difference.

A $3 - 8$

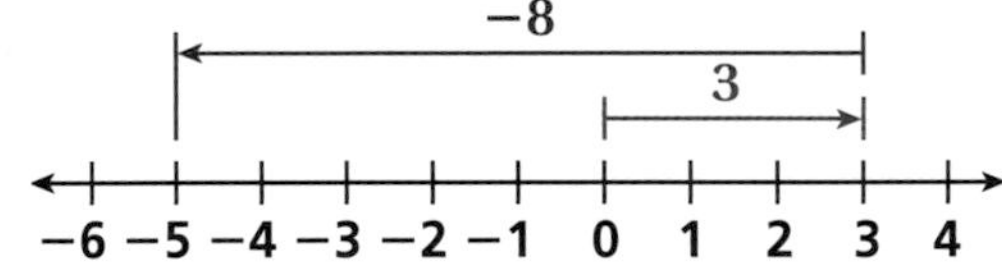

Start at 0.
Move right 3 spaces.
To subtract 8,
move to the left.

$3 - 8 = -5$

Helpful Hint

If the number being subtracted is less than the number it is subtracted from, the answer will be positive. If the number being subtracted is greater, the answer will be negative.

B $-4 - 2$

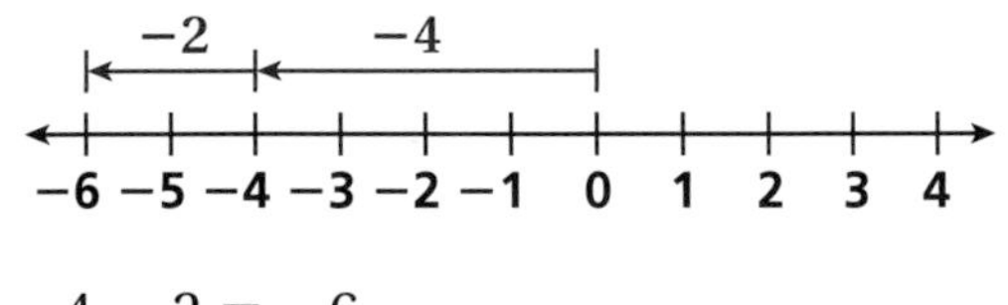

Start at 0.
Move left 4 spaces.
To subtract 2,
move to the left.

$-4 - 2 = -6$

C $2 - (-3)$

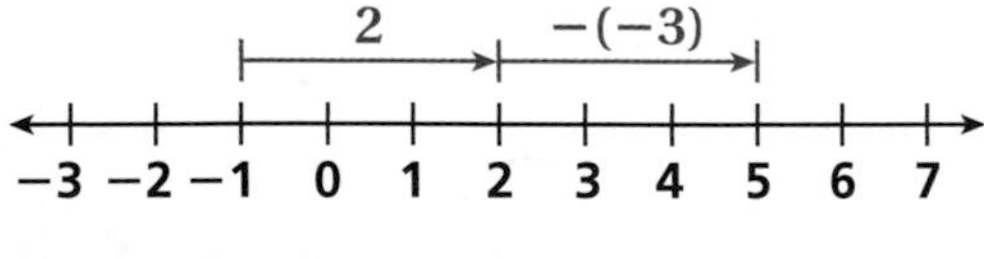

Start at 0.
Move right 2 spaces.
To subtract −3,
move to the right.

$2 - (-3) = 5$

Addition and subtraction are inverse operations—they "undo" each other. Instead of subtracting a number, you can *add its opposite.*

EXAMPLE 2 Subtracting Integers by Adding the Opposite

Subtract.

A $5 - 9$

$5 + (-9)$ *Add the opposite of 9.*

-4

B $-9 - (-2)$

$-9 + 2$ *Add the opposite of −2.*

-7

C $-4 - 3$

$-4 + (-3)$ *Add the opposite of 3.*

-7

EXAMPLE 3 Evaluating Expressions with Integers

Evaluate $a - b$ for each set of values.

A $a = -6, b = 7$

$a - b$

$-6 - 7 = -6 + (-7)$ *Substitute for a and b.*

$= -13$ *Add the opposite of 7.*

B $a = 14, b = -9$

$a - b$

$14 - (-9) = 14 + 9$ *Substitute for a and b.*

$= 23$ *Add the opposite of −9.*

EXAMPLE 4 *Temperature Application*

Find the difference between 3,000°F and −250°F, the temperatures the Space Shuttle must endure.

$3{,}000 - (-250)$

$3{,}000 + 250 = 3{,}250$ *Add the opposite of −250.*

The difference in temperatures the shuttle must endure is 3,250°F.

Think and Discuss

1. **Suppose** you subtract one negative integer from another. Will your answer be greater than or less than the number you started with?
2. **Tell** whether you can reverse the order of integers when subtracting and still get the same answer. Why or why not?

3-4 Exercises

FOR EOG PRACTICE
see page 667

1.02

GUIDED PRACTICE

See Example 1 **Use a number line to find each difference.**

1. $4 - 7$ **2.** $-6 - 5$ **3.** $2 - (-4)$ **4.** $-8 - (-2)$

See Example 2 **Subtract.**

5. $6 - 10$ **6.** $-3 - (-8)$ **7.** $-1 - 9$ **8.** $-12 - (-2)$

See Example 3 **Evaluate $a - b$ for each set of values.**

9. $a = 5, b = -2$ **10.** $a = -8, b = 6$ **11.** $a = 4, b = 18$

See Example 4 **12.** In 1980, in Great Falls, Montana, the temperature rose from −32°F to 15°F in seven minutes. How much did the temperature increase?
(*Source:* http://www.wrh.noaa.gov/greatfalls/topweather.html)

INDEPENDENT PRACTICE

See Example 1 **Use a number line to find each difference.**

13. $7 - 12$ **14.** $-5 - (-9)$ **15.** $2 - (-6)$ **16.** $7 - (-8)$

17. $9 - (-3)$ **18.** $-4 - 10$ **19.** $8 - (-8)$ **20.** $-3 - (-3)$

See Example 2 **Subtract.**

21. $-22 - (-5)$ **22.** $-4 - 21$ **23.** $27 - 19$ **24.** $-10 - (-7)$

25. $30 - (-20)$ **26.** $-15 - 15$ **27.** $12 - (-6)$ **28.** $-31 - 15$

See Example 3 **Evaluate $a - b$ for each set of values.**

29. $a = 9, b = -7$ **30.** $a = -11, b = 2$ **31.** $a = -2, b = 3$

32. $a = 8, b = 19$ **33.** $a = -10, b = 10$ **34.** $a = -4, b = -15$

See Example 4 **35.** In 1918, in Granville, North Dakota, the temperature rose from −33°F to 50°F in 12 hours. How much did the temperature increase?
(*Source:* http://www.infoplease.com/ipa/A0005317.html)

PRACTICE AND PROBLEM SOLVING

Simplify.

36. $2 - 8$ **37.** $-5 - 9$ **38.** $15 - 12 - 8$

39. $6 + (-5) - 3$ **40.** $1 - 8 + (-6)$ **41.** $4 - (-7) - 9$

42. $-11 - (-5) - (-6)$ **43.** $5 - (-8) - (-3)$ **44.** $10 - 12 + 2$

45. $(2 - 3) - (5 - 6)$ **46.** $(3 - 8) - (-2 + 9)$

47. $15 - 6 + 4 - 7 + 10$ **48.** $12 - (-5) - [3 - (-1)]$

Evaluate each expression for $m = -5$, $n = 8$, and $p = -14$.

49. $m - n + p$ **50.** $n - m - p$ **51.** $p - m - n$ **52.** $m + n - p$

Astronomy LINK

53. The temperature of Mercury, the planet closest to the Sun, can be as high as 873°F. The temperature of Pluto, the planet farthest from the Sun, is −393°F. What is the difference between these temperatures?

54. One side of Mercury always faces the Sun. The temperature on this side of Mercury can reach 873°F. The temperature on the other side can be as low as −361°F. What is the difference between the two temperatures?

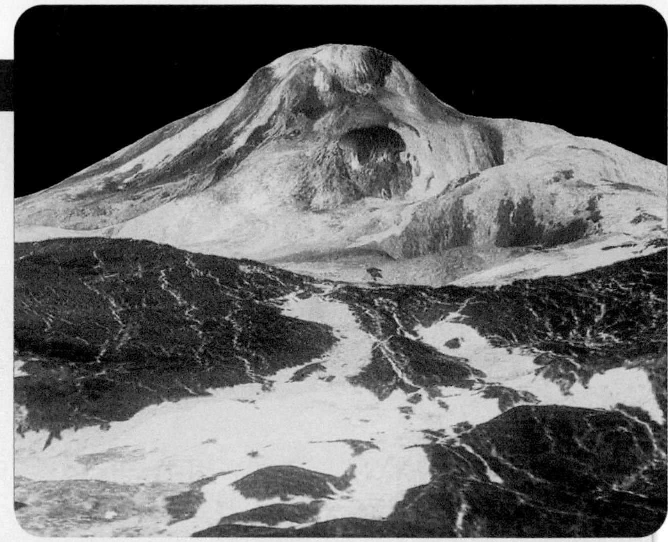

Maat Mons volcano on Venus
Source: NASA (computer-generated from the *Magellan* probe)

55. Earth's moon rotates relative to the Sun about once a month. The side facing the Sun at a given time can be as hot as 224°F. The side away from the Sun can be as cold as −307°F. What is the difference between these temperatures?

56. The highest recorded temperature on Earth is 136°F. The lowest is −129°F. What is the difference between these temperatures?

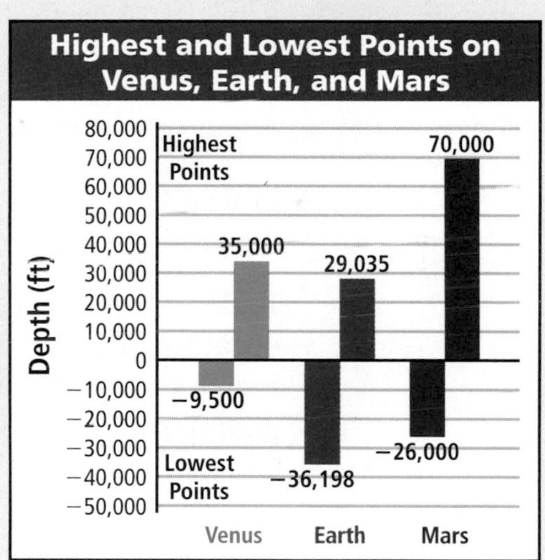

Use the graph for Exercises 57 and 58.

57. How much deeper is the deepest canyon on Mars than the deepest canyon on Venus?

58. ***CHALLENGE*** What is the difference between Earth's highest mountain and its deepest ocean canyon? What is the difference between Mars' highest mountain and its deepest canyon? Which difference is greater? How much greater is it?

Spiral Review

Simplify. (Lesson 2-3)

59. $6 + 4 \div 2$

60. $9 \cdot 1 - 4$

61. $5^2 - 3$

62. $8(10 + 2)$

63. $\left(\frac{14}{7} - 2\right)4$

64. $8 \div 1 + 3^3$

65. **EOG PREP** Which is the prime factorization of 36? (Lesson 2-4)

A $4 \cdot 9$
B $2^2 \cdot 3^2$
C $2^3 \cdot 3$
D $2 \cdot 3^2$

66. **EOG PREP** Which ordered pair is in Quadrant III? (Lesson 3-2)

A (3, 3)
B (1, −3)
C (−3, 1)
D (−1, −1)

3-5 Multiplying and Dividing Integers

Learn to multiply and divide integers.

You can think of multiplication as repeated addition.

$$3 \cdot 2 = 2 + 2 + 2 = 6 \text{ and } 3 \cdot (-2) = (-2) + (-2) + (-2) = -6$$

EXAMPLE 1 Multiplying Integers Using Repeated Addition

Find each product.

A $3 \cdot (-3)$

$3 \cdot (-3) = (-3) + (-3) + (-3)$ *Think: $3 \cdot (-3)$ means 3 groups of -3*

$= -9$

B $-4 \cdot 2$

$-4 \cdot 2 = (-4) + (-4)$ *Think: $-4 \cdot 2 = 2 \cdot (-4)$, or 2 groups of -4*

$= -8$

Remember!

The Commutative Property of Multiplication states that order does not matter when you multiply.

Example 1 suggests that when the signs of two numbers are different, the product is *negative.*

To decide what happens when both numbers are negative, look at the pattern at right. Notice that each product is 3 more than the preceding one. This pattern suggests that the product of two negative integers is *positive.*

$$-3 \cdot (2) = -6$$
$$-3 \cdot (1) = -3$$
$$-3 \cdot (0) = 0$$
$$-3 \cdot (-1) = 3$$
$$-3 \cdot (-2) = 6$$

EXAMPLE 2 Multiplying Integers

Multiply.

$-4 \cdot (-2)$

$-4 \cdot (-2) = 8$ *Both signs are negative, so the product is positive.*

Multiplication and division are inverse operations. They "undo" each other. You can use this fact to discover the rules for division of integers.

$$4 \cdot (-2) = -8 \qquad -4 \cdot (-2) = 8$$
$$-8 \div (-2) = 4 \qquad 8 \div (-2) = -4$$

Same signs → *Positive*; *Different signs* → *Negative*

The rule for division is like the rule for multiplication.

MULTIPLYING AND DIVIDING INTEGERS

If the signs are:		Your answer will be:
the same	⟶	positive
different	⟶	negative

EXAMPLE 3 Dividing Integers

Find each quotient.

A $72 \div (-9)$

$72 \div (-9)$ — *Think: 72 ÷ 9 = 8.*

-8 — *The signs are different, so the quotient is negative.*

B $-144 \div 12$

$-144 \div 12$ — *Think: 144 ÷ 12 = 12.*

-12 — *The signs are different, so the quotient is negative.*

C $-100 \div (-5)$

$-100 \div (-5)$ — *Think: 100 ÷ 5 = 20.*

20 — *The signs are the same, so the quotient is positive.*

EXAMPLE 4 Averaging Integers

Jonie recorded the temperature change every hour for five hours as a cold front approached. The table below shows her data. What was the average temperature change per hour?

Hour	1	2	3	4	5
Temperature Change (°F)	2	−2	−8	−10	−2

$2 + (-2) + (-8) + (-10) + (-2) = -20$ — *Find the sum of the changes in temperature.*

$\frac{-20}{5} = -4$ — *Divide to find the average.*

The average temperature change per hour was −4°F.

Weather LINK

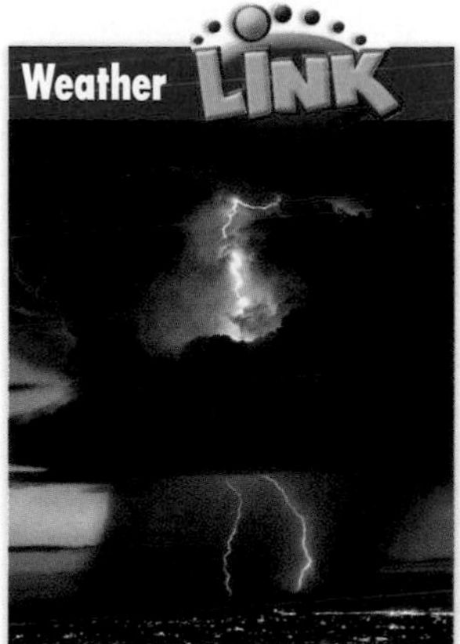

Each year there are about 16 million thunderstorms around the globe. Thunderstorms develop when moist air rises and encounters cooler air.

Think and Discuss

1. **List** at least four different multiplication examples that have 24 as their product. Use both positive and negative integers.
2. **Suppose** −3 is the answer to a division problem and −12 is the number being divided. What is the divisor? Explain your answer.

3-5 Exercises

FOR EOG PRACTICE

see page 667

internet connect

Homework Help Online
go.hrw.com Keyword: MS4 3-5

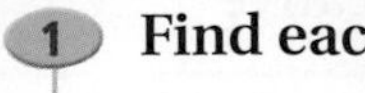

1.02

GUIDED PRACTICE

See Example 1 **Find each product.**

1. $5 \cdot (-3)$ **2.** $5 \cdot (-2)$ **3.** $-3 \cdot 5$

See Example 2 **Multiply.**

4. $-5 \cdot (-3)$ **5.** $-2 \cdot (-5)$ **6.** $-3 \cdot (-5)$

See Example 3 **Find each quotient.**

7. $32 \div (-4)$ **8.** $-18 \div 3$ **9.** $-20 \div (-5)$ **10.** $49 \div (-7)$

See Example 4 **11.** The table shows how the elevation changed as Denise walked along a path. What was the average change in elevation per minute?

Minute	1	2	3	4
Change in Elevation (ft)	−6	8	12	−2

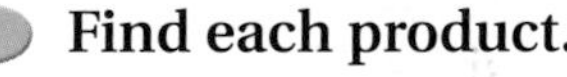

INDEPENDENT PRACTICE

See Example 1 **Find each product.**

12. $2 \cdot (-1)$ **13.** $-5 \cdot 2$ **14.** $-4 \cdot 2$

See Example 2 **Multiply.**

15. $-4 \cdot (-6)$ **16.** $-6 \cdot (-8)$ **17.** $-8 \cdot (-4)$

See Example 3 **Find each quotient.**

18. $48 \div (-6)$ **19.** $-35 \div (-5)$ **20.** $-16 \div 4$ **21.** $-64 \div 8$

See Example 4 **22.** The table shows temperature change over time. What was the average temperature change per hour?

Hour	1	2	3	4	5	6
Change in Temperature (°F)	−11	−18	−20	−15	12	16

PRACTICE AND PROBLEM SOLVING

Find each product or quotient.

23. $-4 \cdot 10$ **24.** $-3 \cdot (-9)$ **25.** $-45 \div 15$

26. $-3 \cdot 4 \cdot (-1)$ **27.** $-500 \div (-10)$ **28.** $5 \cdot (-4) \cdot (-2)$

29. $-4 \cdot (-6) \cdot (-5)$ **30.** $225 \div (-75)$ **31.** $-2 \cdot (-5) \cdot 9$

Simplify.

32. $(-3)^2$ **33.** $(-2)^4$ **34.** $(-5)^3$ **35.** $(-1)^5$

36. $8 \cdot (-7) + 9$ **37.** $-3 \cdot (4) - 12$ **38.** $25 - (-2) \cdot 4^2$ **39.** $8 + 6 \div (-2)$

Evaluate each expression for $a = -5$, $b = 6$, and $c = -12$.

40. $-2c + b$ **41.** $4a - b$ **42.** $ab + c$ **43.** $ac \div b$

44. ***EARTH SCIENCE*** The table shows the depths of major caves in the United States. Subtract the depth of the deepest cave from the depth of the shallowest cave to find their difference.

Depths of Major U.S. Caves

Cave	Depth (ft)
Carlsbad Caverns	–1,022
Caverns of Sonora	–150
Ellison's Cave	–1,000
Jewel Cave	–696
Kartchner Caverns	–137
Mammoth Cave	–379

Source: NSS U.S.A. Long Cave List, Caves over one mile long as of 10/18/2001

45. ***EARTH SCIENCE*** A scuba diver is swimming at a depth of -12 feet. Then she dives down to a coral reef that is at five times this depth. What is the depth of the coral reef?

46. ***PERSONAL FINANCE*** Does each person end up with more or less money than they started with? By how much?

a. Kevin spends \$24 a day for 3 days.

b. Devin earns \$15 a day for 5 days.

c. Evan spends \$20 a day for 3 days. Then he earns \$18 a day for 4 days.

47. ***BUSINESS*** The table shows the deposits and withdrawals made by the Purple Tomato Cafe to and from their bank account in one week. What was the average daily change in the account?

Day	1	2	3	4	5	6	7
Amount (\$)	−150	−280	160	190	250	−140	355

48. ***WHAT'S THE ERROR?*** A student writes, "The quotient of an integer divided by an integer of the opposite sign has the sign of the integer with the greater absolute value." What is the student's error?

49. ***WRITE ABOUT IT*** Explain how to find the product and the quotient of two integers.

50. ***CHALLENGE*** Use > or < to compare $-2 \cdot (-1) \cdot 4 \cdot 2 \cdot (-3)$ and $-1 + (-2) + 4 + (-25) + (-10)$.

Spiral Review

Combine like terms. (Lesson 2-9)

51. $6x + 2y - 4x$ **52.** $3x^2 + 5x + 3x - x^2$ **53.** $6a^2 + 2 + 2a^2 - 2$

54. $y - 3y^2 + y^2$ **55.** $x^2 + 2y + 6 - x^2$ **56.** $4a^2 - 6a^2 + 9a + 2a^2$

57. **EOG PREP** If $4 + b = 20$, what is the value of b? (Lesson 2-10)

A 5 B 80 C 24 D 16

58. **EOG PREP** If $q \cdot 7 = 21$, what is the value of q? (Lesson 2-12)

A 14 B 147 C 3 D 28

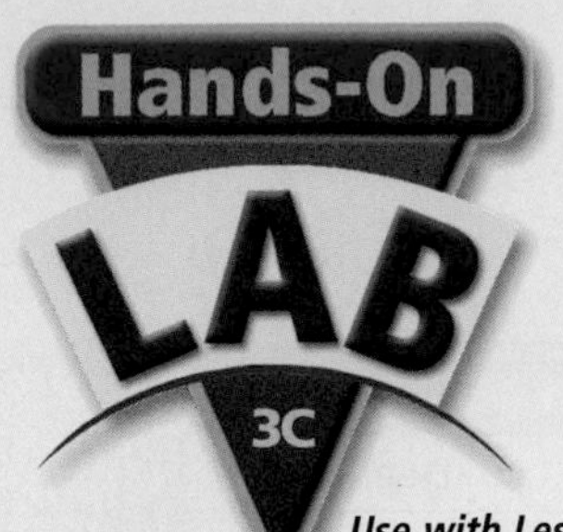

Model Integer Equations

Use with Lesson 3-6

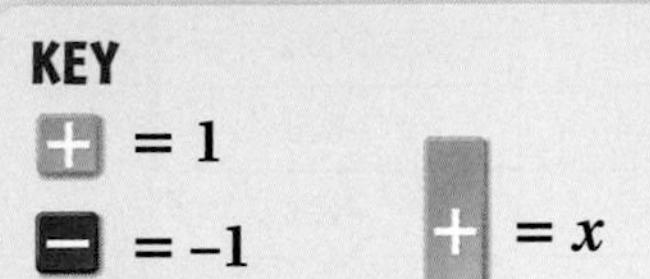

REMEMBER
Adding or removing a zero from an expression does not change the expression's value.

You can use algebra tiles to solve equations.

Activity

To solve the equation $x + 2 = 3$, you need to get x alone on one side of the equal sign. You can add or remove tiles as long as you add the same amount or remove the same amount on both sides.

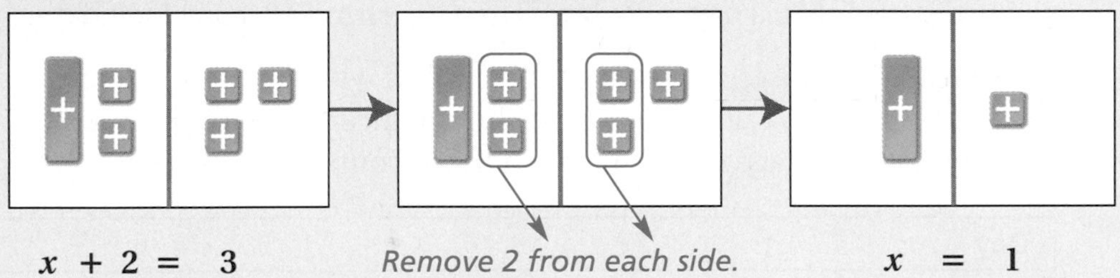

1 Use algebra tiles to model and solve each equation.

a. $x + 3 = 5$ **b.** $x + 4 = 9$ **c.** $x + 5 = 8$ **d.** $x + 6 = 6$

The equation $x + 6 = 4$ is more difficult to model because there are not enough tiles on the right side. You can use the fact that $1 + (-1) = 0$ to help solve the equation.

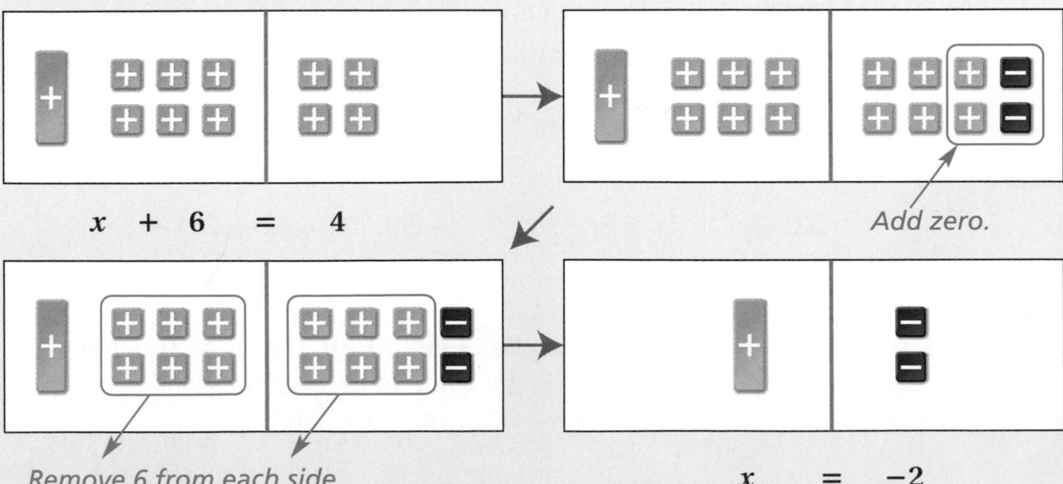

2 Use algebra tiles to model and solve each equation.

a. $x + 5 = 3$ **b.** $x + 4 = 2$ **c.** $x + 7 = -3$ **d.** $x + 6 = -2$

When modeling an equation that involves subtraction, such as $x - 6 = 2$, you must first rewrite the equation as an addition equation. For example, the equation $x - 6 = 2$ can be rewritten as $x + (-6) = 2$.

Modeling equations that involve addition of negative numbers is similar to modeling equations that involve addition of positive numbers.

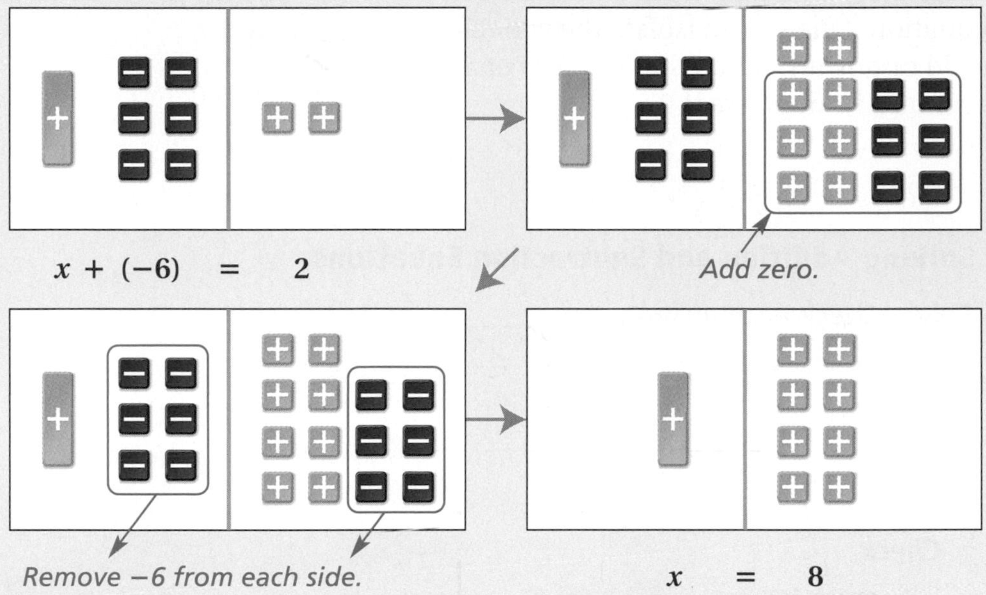

3 Use algebra tiles to model and solve each equation.

a. $x - 4 = 3$ **b.** $x - 2 = 8$ **c.** $x - 5 = -5$ **d.** $x - 7 = 0$

Think and Discuss

1. When you use neutral pairs to add zero to an equation, how do you know the number of yellow tiles and red tiles that you need?
2. When you remove tiles, what operation are you representing? When you add tiles, what operation are you representing?
3. How can you use the original model to check your solution?
4. To model $x - 6 = 2$, you must rewrite the equation as $x + (-6) = 2$. Why are you allowed to do this?

Try This

Use algebra tiles to model and solve each equation.

1. $x + 7 = 10$ **2.** $x - 5 = -8$ **3.** $x + (-5) = -4$ **4.** $x - 2 = 1$

5. $x + 4 = 8$ **6.** $x + 3 = -2$ **7.** $x + (-1) = 9$ **8.** $x - 7 = -6$

3-6 Solving Equations Containing Integers

Learn to solve one-step equations with integers.

When you are solving equations with integers, the goal is the same as with whole numbers—isolate the variable on one side of the equation. One way to isolate the variable is to add opposites. Recall that the sum of a number and its opposite is 0.

$3 + (-3) = 0$

$a + (-a) = 0$

EXAMPLE 1 Solving Addition and Subtraction Equations

Helpful Hint

$3 + (-3) = 0$
3 is the opposite of -3.

Solve. Check each answer.

A $-3 + y = -5$

$$\begin{aligned} -3 + y &= -5 \\ \underline{+3} \quad &\quad \underline{+3} \\ y &= -2 \end{aligned}$$

Add 3 to both sides to isolate the variable.

Check

$-3 + y = -5$

$-3 + (-2) \stackrel{?}{=} -5$ — *Substitute −2 for y in the original equation.*

$-5 \stackrel{?}{=} -5$ ✔ — *True. −2 is the solution to −3 + y = −5.*

B $n + 3 = -10$

$$\begin{aligned} n + 3 &= -10 \\ \underline{+\ (-3)} \ &\ \underline{+\ (-3)} \\ n &= -13 \end{aligned}$$

Add −3 to both sides to isolate the variable.

Check

$n + 3 = -10$

$-13 + 3 \stackrel{?}{=} -10$ — *Substitute −13 for n in the original equation.*

$-10 \stackrel{?}{=} -10$ ✔ — *True. −13 is the solution to n + 3 = −10.*

C $x - 8 = -32$

$$\begin{aligned} x - 8 &= -32 \\ \underline{+8} \quad &\quad \underline{+8} \\ x &= -24 \end{aligned}$$

Add 8 to both sides to isolate the variable.

Check

$x - 8 = -32$

$-24 - 8 \stackrel{?}{=} -32$ — *Substitute −24 for x in the original equation.*

$-32 \stackrel{?}{=} -32$ ✔ — *True. −24 is the solution to x − 8 = −32.*

EXAMPLE 2 Solving Multiplication and Division Equations

Solve. Check each answer.

A $\frac{a}{-3} = 9$

$$\frac{a}{-3} = 9$$

$$(-3)\left(\frac{a}{-3}\right) = (-3)9$$ *Multiply both sides by −3 to isolate the variable.*

$$a = -27$$

Check

$$\frac{a}{-3} = 9$$

$$\frac{-27}{-3} \stackrel{?}{=} 9$$ *Substitute −27 for a in the original equation.*

$$9 \stackrel{?}{=} 9 \checkmark$$ *True. −27 is the solution to $\frac{a}{-3} = 9$.*

B $-120 = 6x$

$$-120 = 6x$$

$$\frac{-120}{6} = \frac{6x}{6}$$ *Divide both sides by 6 to isolate the variable.*

$$-20 = x$$

Check

$$-120 = 6x$$

$$-120 \stackrel{?}{=} 6(-20)$$ *Substitute −20 for x in the original equation.*

$$-120 \stackrel{?}{=} -120 \checkmark$$ *True. −20 is the solution to −120 = 6x.*

EXAMPLE 3 *Business Application*

A shoe manufacturer made a profit of $800 million. This amount is $200 million more than last year's profit. What was last year's profit?

Let p represent last year's profit (in millions of dollars).

Profit this year = $p + 200$

Profit this year = $800 million

$$\begin{aligned} p + 200 &= 800 \\ \underline{-200} &\ \ \underline{-200} \\ p &= 600 \end{aligned}$$

Last year's profit was $600 million.

Think and Discuss

1. Tell what value of n makes $-n + 32$ equal to zero.

2. Explain why you would or would not multiply both sides of an equation by 0 to solve it.

3-6 Exercises

1.02, 5.03

FOR EOG PRACTICE

see page 667

internet connect

Homework Help Online
go.hrw.com Keyword: MS4 3-6

GUIDED PRACTICE

Solve. Check each answer.

1. $w - 6 = -2$ **2.** $x + 5 = -7$ **3.** $k = -18 + 11$

4. $\frac{n}{-4} = 2$ **5.** $-240 = 8y$ **6.** $-5a = 300$

7. Last year, a chain of electronics stores had a loss of \$45 million. This year the loss is \$12 million more than last year's loss. What is this year's loss?

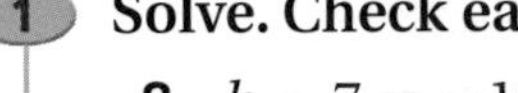

INDEPENDENT PRACTICE

See Example 1

Solve. Check each answer.

8. $b - 7 = -16$ **9.** $k + 6 = 3$ **10.** $s + 2 = -4$

11. $v + 14 = 10$ **12.** $c + 8 = -20$ **13.** $a - 25 = -5$

See Example 2

14. $9c = -99$ **15.** $\frac{t}{8} = -4$ **16.** $-16 = 2z$

17. $\frac{n}{-5} = -30$ **18.** $200 = -25p$ **19.** $\frac{l}{-12} = 12$

See Example 3

20. The temperature in Nome, Alaska, was −50°F. This was 18°F less than the temperature in Anchorage, Alaska, on the same day. What was the temperature in Anchorage?

PRACTICE AND PROBLEM SOLVING

Solve. Check each answer.

21. $9y = 900$ **22.** $d - 15 = 45$ **23.** $j + 56 = -7$

24. $\frac{s}{-20} = 7$ **25.** $-85 = -5c$ **26.** $v - 39 = -16$

27. $11y = -121$ **28.** $\frac{n}{36} = 9$ **29.** $w + 41 = 0$

30. $\frac{r}{238} = 8$ **31.** $-23 = x + 35$ **32.** $0 = -15m$

33. $4x = 2 + 14$ **34.** $c + c + c = 6$ **35.** $t - 3 = 4 + 2$

36. $4y + y = 10$ **37.** $23 + h - 4 = 39$ **38.** $k - 32 = 16 - 2 + 4$

39. The three angles of a triangle have equal measures. The sum of their measures is 180°. What is the measure of each angle?

40. ***SPORTS*** Herb has 42 days to prepare for a cross-country race. During his training, he will run a total of 126 miles. If Herb runs the same distance every day, how many miles will he run each day?

41. ***PERSONAL FINANCE*** Jared bought one share of stock for $225.
   a. He sold the stock for a profit of $55. What was the selling price of the stock?
   b. The price of the stock dropped $40 the day after Jared sold it. At what price would Jared have sold it if he had waited until then?
   c. What would Jared's profit have been if he had waited until the next day to sell his stock?

42. ***PHYSICAL SCIENCE*** On the Kelvin temperature scale, pure water boils at 373 K. The difference between the boiling point and the freezing point of water on this scale is 100 K. What is the freezing point of water?

The graph shows the most popular destinations for people who traveled over the 2001 Labor Day weekend. Use the graph shown for Exercises 43 and 44.

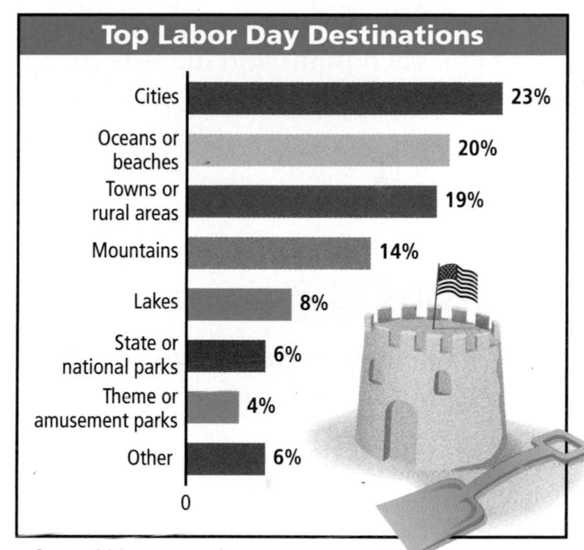

Source: AAA

43. ***RECREATION*** Which destination was 5 times more popular than theme or amusement parks?

44. ***RECREATION*** According to the graph, the mountains were as popular as state or national parks and what other destination combined?

45. ***CHOOSE A STRATEGY*** Matthew (M) earns $23 less a week than his sister Allie (A). Their combined salaries are $93. How much does each of them earn per week?

   A A: $35; M: $12 **B** A: $35; M: $58 **C** A: $58; M: $35

46. ***WRITE ABOUT IT*** Explain how to use inverse operations to isolate a variable in an equation.

47. ***CHALLENGE*** Write an equation that includes the variable p and the numbers 5, 3, and 31 so that the solution is $p = 16$.

Spiral Review

Order each set of integers from least to greatest. (Lesson 3-1)

48. −9, 12, −15, −1, 6 **49.** −3, −10, 7, 0, −8 **50.** 2, −3, −2, 1, −6

51. 4, −3, −1, 2, 0 **52.** −15, −21, 6, 9, −8 **53.** −2, 7, 10, −9, −1

54. **EOG PREP** Which ordered pair lies on the y-axis? (Lesson 3-2)

A (0, −6) B (4, 0) C (1, 8) D (−2, −9)

55. **EOG PREP** What is the value of the expression 3 + (−4)? (Lesson 3-3)

A 7 B −1 C 1 D −7

Chapter 3

Mid-Chapter Quiz

LESSON 3-1 (pp. 130–133)

Use a number line to find each absolute value.

1. $|-23|$ **2.** $|17|$ **3.** $|-20|$ **4.** $|14|$

LESSON 3-2 (pp. 134–137)

Plot each point and identify the quadrant in which it is contained.

5. $A\ (-4, 2)$ **6.** $B\ (1, 3)$

7. $C\ (3, 3)$ **8.** $D\ (-2, 4)$

LESSON 3-3 (pp. 140–143)

Evaluate $p + t$ for the given values.

9. $p = 5, t = -18$ **10.** $p = -4, t = -13$

11. $p = -37, t = 39$ **12.** $p = -25, t = 15$

LESSON 3-4 (pp. 146–149)

Subtract.

13. $-21 - (-7)$ **14.** $9 - (-11)$ **15.** $6 - 17$

16. $18 - 10 - 8$ **17.** $(5 - 9) - (-3 + 8)$ **18.** $14 - (-11) - (-5)$

LESSON 3-5 (pp. 150–153)

Find each product or quotient.

19. $-7 \cdot 3$ **20.** $48 \div 6$ **21.** $-5 \cdot (-9)$ **22.** $30 \div (-15)$

LESSON 3-6 (pp. 156–159)

Solve.

23. $3x = 30$ **24.** $k - 25 = 50$ **25.** $y + 16 = -8$ **26.** $\frac{90}{m} = -15$

27. This year, 72 students did projects for the science fair. This was 23 more students than last year. How many students did projects for the science fair last year?

1.02a

Focus on Problem Solving

Make a Plan

- **Choose a method of computation**

When you know the operation you must use and you know exactly which numbers to use, a calculator might be the easiest way to solve a problem. Sometimes, such as when the numbers are small or are multiples of 10, it may be quicker to use mental math.

Sometimes, you have to write the numbers to see how they relate in an equation. When you are working an equation, using a pencil and paper is the simplest method to use because you can see each step as you go.

For each problem, tell whether you would use a calculator, mental math, or pencil and paper to solve it. Explain your answer.

1. A scouting troop is collecting aluminum cans to raise money for charity. Their goal is to collect 3,000 cans in 6 months. If they set a goal to collect an equal number of cans each month, how many cans can they expect to collect each month?

2. The Grand Canyon is 29,000 meters wide at its widest point. The Empire State Building, located in New York City, is 381 meters tall. Laid end to end, how many Empire State Buildings would fit in the Grand Canyon at its widest point?

3. On a piano keyboard, all but one of the black keys are arranged in groups so that there are 7 groups with 2 black keys each and 7 groups with 3 black keys each. How many black keys are there on a piano?

4. Some wind chimes are made of rods. The rods are usually of different lengths, producing different sounds. The frequency (which determines the pitch) of the sound is measured in hertz (Hz). If one rod on a chime has a frequency of 55 Hz and another rod has a frequency that is twice that of the first rod's, what is the frequency of the second rod?

3-7 Fractions and Decimals

Learn to identify rational numbers and to place them on a number line.

Vocabulary

rational number

You can show -5 and 15 on a number line marked off by 5's.

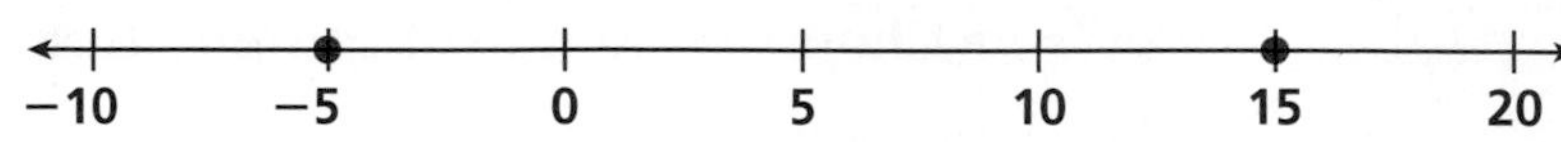

You can show -3 and 4 on a number line marked off by 1's.

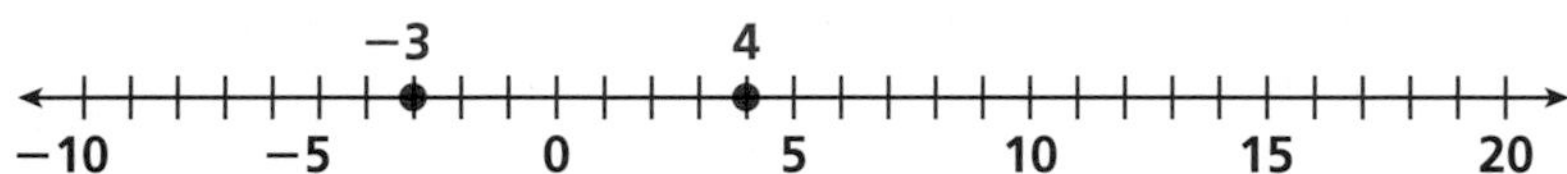

A number line can have as much detail as you want. The number line below shows that you can write numbers in many different ways.

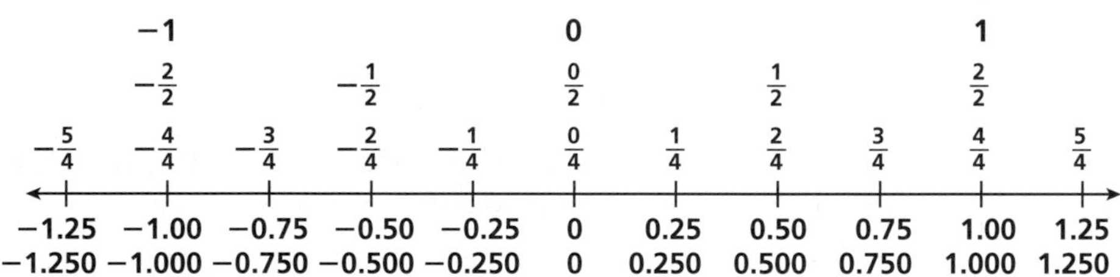

EXAMPLE 1 Graphing Numbers on a Number Line

Graph each number on a number line.

A $3\frac{1}{2}$

−5 −4 −3 −2 −1 0 1 2 3 4 5

$3\frac{1}{2}$ is between 3 and 4.

B -1.9

−5 −4 −3 −2 −1 0 1 2 3 4 5

-1.9 is between -1 and -2.

Remember!

The top number in a fraction is called the *numerator.* The bottom number is called the *denominator.* So in the fraction $\frac{1}{2}$, the numerator is 1 and the denominator is 2.

The numbers shown on the number lines in Example 1 are called *rational numbers.* **Rational numbers** are numbers that can be written as fractions, with integers for numerators and denominators. Integers and certain decimals are rational numbers because they can be written as fractions.

$$15 = \frac{15}{1} \qquad -5 = -\frac{5}{1} \qquad 0.75 = \frac{3}{4} \qquad -1.25 = -\frac{5}{4}$$

EXAMPLE 2 Writing Rational Numbers as Fractions

Show that each number is a rational number by writing it as a fraction.

A -0.500

$-0.500 = -\frac{1}{2}$

B 0.25

$0.25 = \frac{1}{4}$

C -0.75

$-0.75 = -\frac{3}{4}$

EXAMPLE 3 *Earth Science Application*

High tide in Astoria, Oregon, on July 1 was at 11:31 A.M. The graph shows how much earlier or later in minutes that high tide occurred in nearby towns.

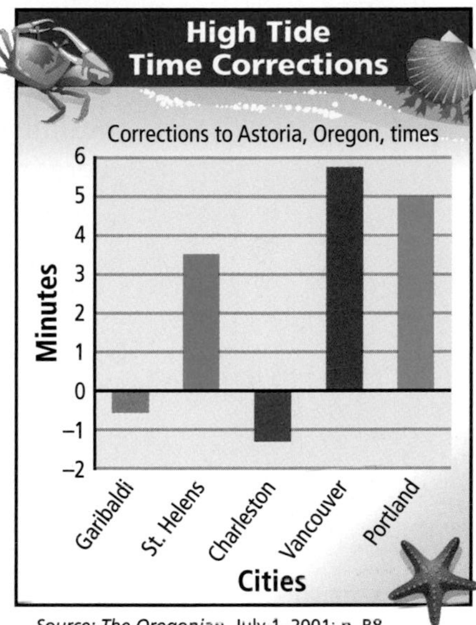

Source: *The Oregonian*, July 1, 2001; p. B8

A **Use a fraction to estimate how much later in minutes high tide occurred in St. Helens.**

The bar is about midway between 3 and 4.

$3\frac{1}{2}$ minutes later

B **Use a decimal to estimate how much earlier in minutes high tide occurred in Garibaldi.**

0.5 minutes earlier *The bar is about midway between 0 and −1.*

C **Use a fraction and a decimal to estimate the greatest value and the least value represented on the graph.**

The greatest value is about $5\frac{3}{4}$, or 5.75.

The least value is about $-1\frac{1}{4}$, or -1.25.

Think and Discuss

1. **Give** examples of three rational numbers that come between 1 and 2 on the number line.
2. **Tell** whether zero is a rational number. Explain your answer.
3. **Write** -1 and 1 as fractions with different denominators, and tell what these fractions have in common.

3-7 Exercises

FOR EOG PRACTICE
see page 668

internet connect
Homework Help Online
go.hrw.com Keyword: MS4 3-7

1.02

GUIDED PRACTICE

See Example 1 **Graph each number on a number line.**

1. $-2\frac{1}{4}$ 2. 3.25 3. −1.5 4. $\frac{1}{2}$

See Example 2 **Show that each number is a rational number by writing it as a fraction.**

5. −0.25 6. 0.750 7. −0.5

See Example 3

8. The graph shows the amount of rainfall above and below the average rainfall for 4 months.
   a. Use a fraction to estimate the amount of rainfall below the average for January.
   b. Use a decimal to estimate the amount of rainfall above the average for March.

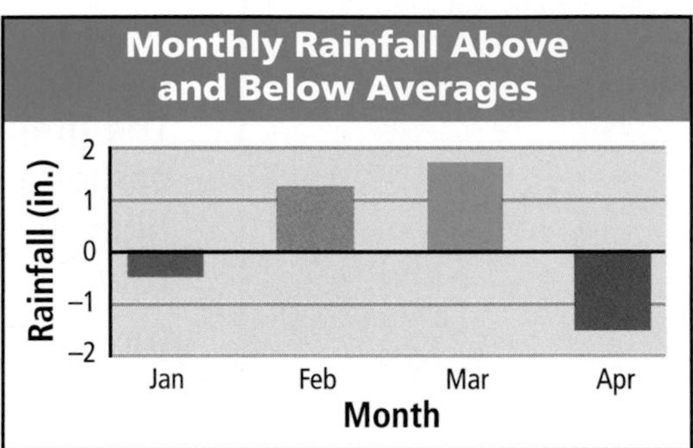

INDEPENDENT PRACTICE

See Example 1 **Graph each number on a number line.**

9. $\frac{2}{3}$ 10. 0.5 11. −1.1 12. $-4\frac{1}{2}$

13. $\frac{1}{5}$ 14. −2.25 15. $-\frac{3}{4}$ 16. 3.6

See Example 2 **Show that each number is a rational number by writing it as a fraction.**

17. 1.25 18. 0.250 19. −1

20. −0.750 21. 0 22. −1.250

See Example 3

23. The graph shows outdoor temperatures at different times during the day.
   a. Use a fraction to estimate the temperature at noon.
   b. Use a decimal to estimate the temperature at 11 A.M.

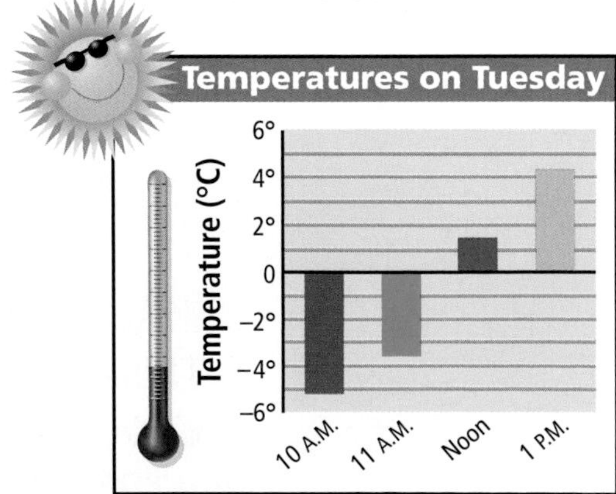

PRACTICE AND PROBLEM SOLVING

Graph each set of numbers on a number line.

24. 2.5, −3.75, $4\frac{1}{2}$, −1, 2 25. −4, 0, 3.25, $2\frac{3}{4}$, $-2\frac{1}{2}$

Show that each number is a rational number.

26. -1.00 **27.** -0.250 **28.** 6

29. 0.75 **30.** 1.250 **31.** 0.50

32. ***ENTERTAINMENT*** The circle graph shows the number of words in the titles of 20 Oscar-winning movies.

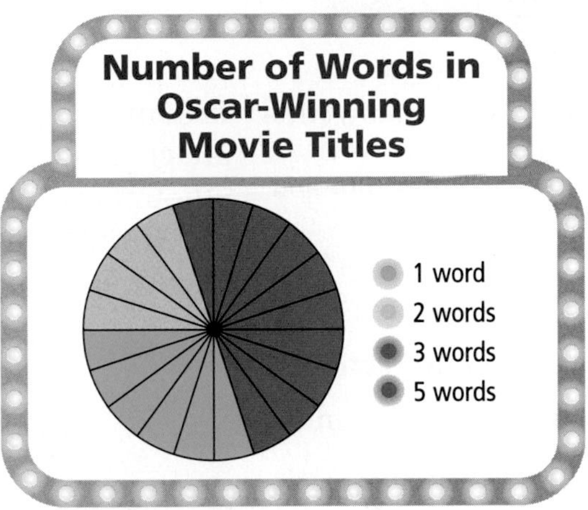

a. What fraction of the 20 movie titles are made up of three words?

b. Which two colors represent the movie titles that together make up $\frac{3}{4}$ of the graph?

33. ***WHAT'S THE QUESTION?*** The numbers 4 and -0.75 are rational numbers. The answer is $\frac{4}{1}$ and $-\frac{3}{4}$. What is the question?

34. ***WRITE ABOUT IT*** The Venn diagram at right shows how different sets of numbers are related. Some sets of numbers are parts of other sets of numbers, just as boxes are inside of other boxes in the diagram. Use the diagram to explain how integers, fractions, and rational numbers are related.

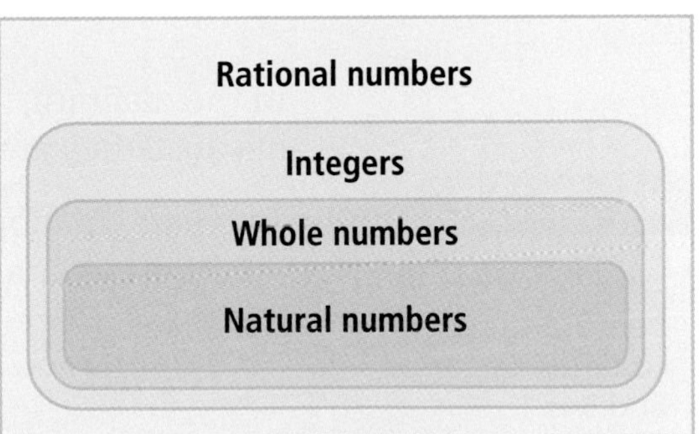

35. ***CHALLENGE*** If $0.25 = \frac{1}{4}$, $0.75 = \frac{3}{4}$, and $1.25 = \frac{5}{4}$, what is the fraction equivalent of 3.25?

Spiral Review

Find the mean of each set of numbers. (Lesson 1-2)

36. 5, 12, 16, 21, 21

37. 2.4, 3.6, 3.6, 4.1, 5.2, 8.1

Evaluate. (Lesson 2-3)

38. $5(3^2 - 1) + 6$

39. $12(18 \div 3 + 2) - 5^2$

40. **EOG PREP** The GCF of 18 and 45 is __?__. (Lesson 2-5)

A 1 C 18

B 9 D 4

41. **EOG PREP** In simplest form, $3x + 5x - 2x^2$ is __?__. (Lesson 2-9)

A $3x + 5x - 2x^2$ C $6x$

B $6x^2$ D $8x - 2x^2$

3-8 Equivalent Fractions and Mixed Numbers

Learn to identify, write, and convert between equivalent fractions and mixed numbers.

Vocabulary

equivalent fractions

improper fraction

mixed number

In some recipes the amounts of ingredients are given as fractions, and sometimes those amounts don't equal the fractions on a measuring cup. Knowing how fractions relate to each other can be very helpful.

Different fractions can name the same number.

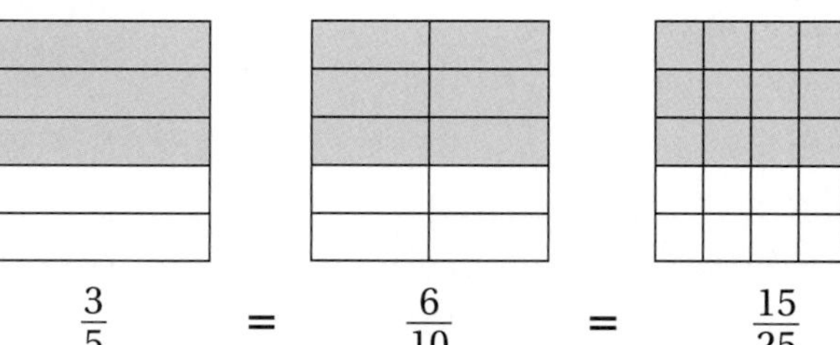

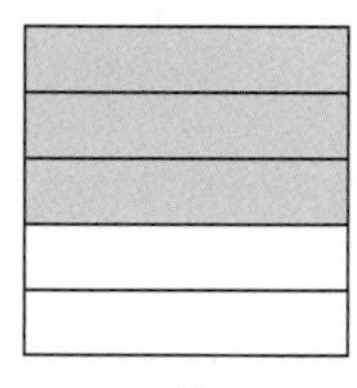

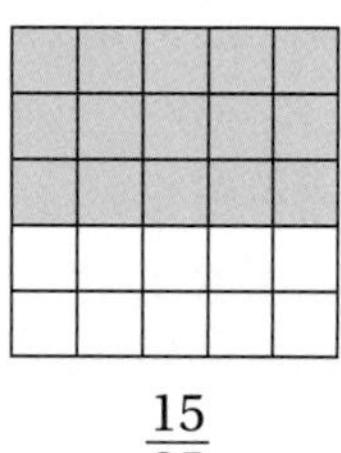

$$\frac{3}{5} = \frac{6}{10} = \frac{15}{25}$$

In the diagram, $\frac{3}{5} = \frac{6}{10} = \frac{15}{25}$. These are called **equivalent fractions** because they are different expressions for the same nonzero number.

Remember!

$\frac{3}{5}$ is in *simplest form* because the greatest common factor of 3 and 5 is 1.

To create fractions equivalent to a given fraction, multiply or divide the numerator and denominator by the same number.

$$\frac{3}{5} = \frac{3 \cdot 2}{5 \cdot 2} = \frac{6}{10} \qquad \frac{15}{25} = \frac{15 \div 5}{25 \div 5} = \frac{3}{5}$$

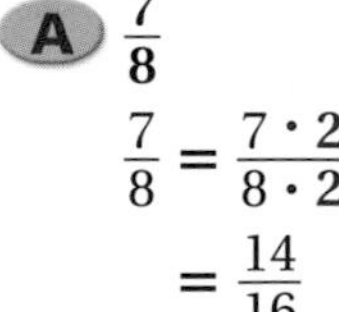

EXAMPLE 1 Finding Equivalent Fractions

Find a fraction equivalent to the given fraction.

A $\mathbf{\frac{7}{8}}$

$\frac{7}{8} = \frac{7 \cdot 2}{8 \cdot 2}$ *Multiply numerator and denominator by 2.*

$= \frac{14}{16}$

B $\mathbf{\frac{24}{36}}$

$\frac{24}{36} = \frac{24 \div 12}{36 \div 12}$ *Divide numerator and denominator by 12.*

$= \frac{2}{3}$

To determine if two fractions are equivalent, find a common denominator and compare the numerators.

EXAMPLE 2 Determining Whether Fractions Are Equivalent

Write the fractions with a common denominator. Then determine if they are equivalent.

A $\frac{6}{8}$ **and** $\frac{9}{12}$

Both fractions can be written with a denominator of 4.

$$\frac{6}{8} = \frac{6 \div 2}{8 \div 2} = \frac{3}{4}$$

$$\frac{9}{12} = \frac{9 \div 3}{12 \div 3} = \frac{3}{4}$$

The numerators are equal, so the fractions are equivalent.

B $\frac{18}{15}$ **and** $\frac{25}{20}$

Both fractions can be written with a denominator of **60**.

$$\frac{18}{15} = \frac{18 \cdot 4}{15 \cdot 4} = \frac{72}{60}$$

$$\frac{25}{20} = \frac{25 \cdot 3}{20 \cdot 3} = \frac{75}{60}$$

The numerators are *not* equal, so the fractions are *not* equivalent.

$\frac{8}{5}$ is an **improper fraction**. Its numerator is greater than its denominator.

$$\frac{8}{5} = 1\frac{3}{5}$$

$1\frac{3}{5}$ is a **mixed number**. It contains both a whole number and a fraction.

EXAMPLE 3 Converting Between Improper Fractions and Mixed Numbers

Remember!

Quotient → 5

4)21

−20

Remainder → 1

A **Write** $\frac{21}{4}$ **as a mixed number.**

First divide the numerator by the denominator.

$\frac{21}{4} = 5\frac{1}{4}$ *Use the quotient and remainder to write the mixed number.*

B **Write** $4\frac{2}{3}$ **as an improper fraction.**

First multiply the denominator and whole number, and then add the numerator.

$4\frac{2}{3} = \frac{3 \cdot 4 + 2}{3} = \frac{14}{3}$ *Use the result to write the improper fraction.*

Think and Discuss

1. **Explain** a process for finding common denominators.
2. **Describe** how to convert between improper fractions and mixed numbers.

3-8 Exercises

FOR EOG PRACTICE
see page 668

internet connect
Homework Help Online
go.hrw.com Keyword: MS4 3-8

1.01, 1.02

GUIDED PRACTICE

See Example 1 **Find a fraction equivalent to the given fraction.**

1. $\frac{1}{2}$ **2.** $\frac{3}{5}$ **3.** $\frac{10}{12}$ **4.** $\frac{15}{40}$

See Example 2 **Write the fractions with a common denominator. Then determine if they are equivalent.**

5. $\frac{3}{9}$ and $\frac{6}{8}$ **6.** $\frac{10}{12}$ and $\frac{20}{24}$ **7.** $\frac{8}{6}$ and $\frac{20}{15}$ **8.** $\frac{15}{8}$ and $\frac{19}{12}$

See Example 3 **Write each as a mixed number.**

9. $\frac{15}{4}$ **10.** $\frac{22}{5}$ **11.** $\frac{17}{13}$ **12.** $\frac{14}{3}$

Write each as an improper fraction.

13. $6\frac{1}{5}$ **14.** $1\frac{11}{12}$ **15.** $7\frac{3}{5}$ **16.** $2\frac{7}{16}$

INDEPENDENT PRACTICE

See Example 1 **Find a fraction equivalent to the given fraction.**

17. $\frac{1}{3}$ **18.** $\frac{5}{6}$ **19.** $\frac{18}{20}$ **20.** $\frac{25}{50}$

21. $\frac{3}{4}$ **22.** $\frac{2}{7}$ **23.** $\frac{9}{15}$ **24.** $\frac{42}{70}$

See Example 2 **Write the fractions with a common denominator. Then determine if they are equivalent.**

25. $\frac{5}{10}$ and $\frac{14}{28}$ **26.** $\frac{15}{20}$ and $\frac{20}{24}$ **27.** $\frac{125}{100}$ and $\frac{40}{32}$ **28.** $\frac{10}{5}$ and $\frac{18}{8}$

29. $\frac{2}{3}$ and $\frac{12}{18}$ **30.** $\frac{8}{12}$ and $\frac{24}{36}$ **31.** $\frac{54}{99}$ and $\frac{84}{132}$ **32.** $-\frac{25}{15}$ and $\frac{175}{75}$

See Example 3 **Write each as a mixed number.**

33. $\frac{19}{3}$ **34.** $\frac{13}{9}$ **35.** $\frac{81}{11}$ **36.** $\frac{71}{8}$

Write each as an improper fraction.

37. $25\frac{3}{5}$ **38.** $4\frac{7}{16}$ **39.** $9\frac{2}{3}$ **40.** $4\frac{16}{31}$

PRACTICE AND PROBLEM SOLVING

Write a fraction equivalent to the given number.

41. 5 **42.** 8 **43.** $6\frac{1}{2}$ **44.** $2\frac{2}{3}$

45. $\frac{8}{21}$ **46.** $9\frac{8}{11}$ **47.** $\frac{55}{10}$ **48.** $4\frac{26}{13}$

49. 101 **50.** $6\frac{15}{21}$ **51.** $\frac{475}{75}$ **52.** $11\frac{23}{50}$

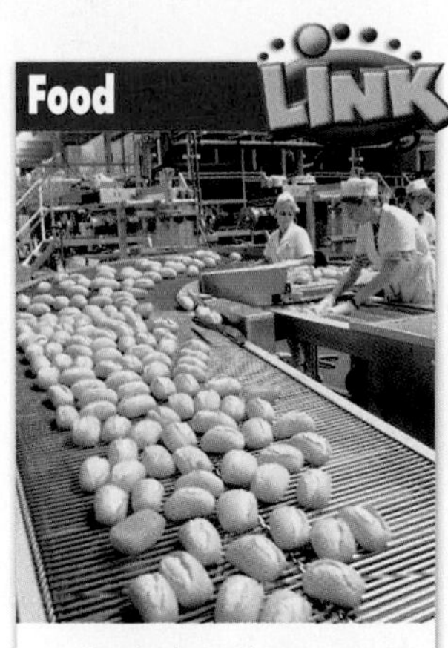

A single bread company can make as many as 1,217 loaves of bread each minute, and the average American eats about 53 pounds of bread each year.

Find the equivalent pair of fractions in each set.

53. $\frac{6}{15}, \frac{21}{35}, \frac{3}{5}$ 54. $\frac{7}{12}, \frac{12}{20}, \frac{6}{10}$ 55. $\frac{2}{3}, \frac{12}{15}, \frac{20}{30}, \frac{15}{24}$

There are 12 inches in 1 foot. Write a mixed number to represent each measurement in feet. (Example: 14 inches = $1\frac{2}{12}$ feet or $1\frac{1}{6}$ feet)

56. 18 inches 57. 25 inches 58. 100 inches

59. 362 inches 60. 42 inches 61. 965 inches

62. ***SOCIAL STUDIES*** A dollar bill is $15\frac{7}{10}$ centimeters long and $6\frac{13}{20}$ centimeters wide. Write each number as an improper fraction.

63. ***FOOD*** A bakery uses $37\frac{1}{2}$ cups of flour to make 25 loaves of bread each day. Write a fraction that shows how many $\frac{1}{4}$ cups of flour are used to make bread each day at the bakery.

64. ***PERSONAL FINANCE*** Every month, Adrian pays for his own long-distance calls made on the family phone. Last month, 15 of the 60 minutes of long-distance charges were Adrian's, and he paid $2.50 of the $12 long-distance bill. Did Adrian pay his fair share?

65. ***WRITE A PROBLEM*** Cal made a graph to show how he spends his time each day. Use the graph to write a problem involving fractions.

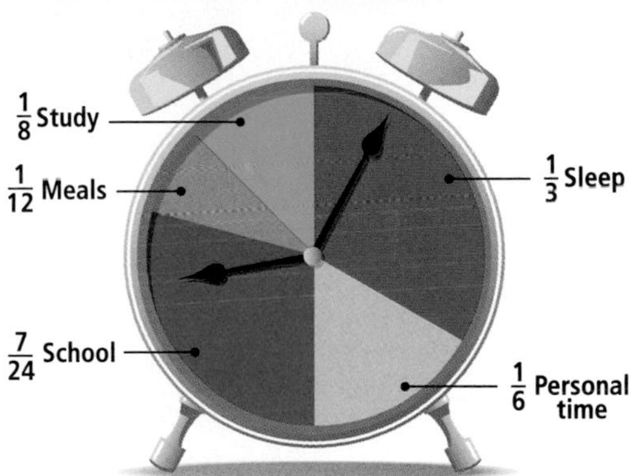

66. ***WRITE ABOUT IT*** Draw a diagram to show how you can use division to write $\frac{25}{3}$ as a mixed number. Explain your diagram.

67. ***CHALLENGE*** Kenichi spent $\frac{2}{5}$ of his $100 birthday check on clothes. How much did Kenichi's new clothes cost?

Spiral Review

Evaluate each number. (Lesson 2-1)

68. 5^3 69. 6^2 70. 2^8 71. 10^5

Write each number in scientific notation. (Lesson 2-2)

72. 1,230,000 73. 475,000 74. 968 75. 88

76. **EOG PREP** The prime factorization of 1,000 is __?__. (Lesson 2-4)

A $10 \cdot 10 \cdot 10$ C $2^3 \cdot 5^3$

B $2^2 \cdot 5^2$ D $8 \cdot 125$

77. **EOG PREP** The solution of $x - 4 = 12$ is __?__. (Lesson 2-10)

A 7 C 12

B 8 D 16

3-9 Equivalent Fractions and Decimals

Learn to write fractions as decimals, and vice versa, and to determine whether a decimal is terminating or repeating.

Vocabulary

terminating decimal

repeating decimal

In baseball, a player's batting average compares the number of hits with the number of times the player has been at bat. The statistics below are for the 2001 Major League Baseball season.

Sammy Sosa had 189 hits in the 2001 season.

Player	Hits	At Bats	$\frac{\text{Hits}}{\text{at Bats}}$	Batting Average (thousandths)
Mark Grace	142	476	$\frac{142}{476}$	$142 \div 476 \approx 0.298$
Cal Ripken, Jr.	114	477	$\frac{114}{477}$	$114 \div 477 \approx 0.239$
Ivan Rodriguez	136	442	$\frac{136}{442}$	$136 \div 442 \approx 0.308$
Sammy Sosa	189	577	$\frac{189}{577}$	$189 \div 577 \approx 0.328$

To convert a fraction to a decimal, divide the numerator by the denominator.

EXAMPLE 1 Writing Fractions as Decimals

Write each fraction as a decimal. Round to the nearest hundredth, if necessary.

A $\frac{3}{4}$

$$\begin{array}{r} 0.75 \\ 4\overline{)3.00} \\ \underline{-28} \\ 20 \\ \underline{-20} \\ 0 \end{array}$$

$\frac{3}{4} = 0.75$

B $\frac{6}{5}$

$$\begin{array}{r} 1.2 \\ 5\overline{)6.0} \\ \underline{-5} \\ 10 \\ \underline{-10} \\ 0 \end{array}$$

$\frac{6}{5} = 1.2$

C $\frac{1}{3}$

$$\begin{array}{r} 0.333 \\ 3\overline{)1.000} \\ \underline{-9} \\ 10 \\ \underline{-9} \\ 10 \\ \underline{-9} \\ 1 \end{array}$$

$\frac{1}{3} \approx 0.33$

Helpful Hint

You can use a calculator to check your division:

3 ÷ 4 = 0.75

6 ÷ 5 = 1.2

1 ÷ 3 = 0.333...

The decimals 0.75 and 1.2 in Example 1 are **terminating decimals** because the decimals come to an end. The decimal 0.333... is a **repeating decimal** because the decimal repeats a pattern forever. You can also write a repeating decimal with a bar over the repeating part.

$0.333\ldots = 0.\overline{3}$ $\quad 0.8333\ldots = 0.8\overline{3}$ $\quad 0.727272\ldots = 0.\overline{72}$

You can use place value to convert a terminating decimal to a fraction.

Place	Thousands	Hundreds	Tens	Ones	.	Tenths	Hundredths	Thousandths
Place Value	1,000	100	10	1		$0.1 = \frac{1}{10}$	$0.01 = \frac{1}{100}$	$0.001 = \frac{1}{1,000}$

EXAMPLE 2 Writing Decimals as Fractions

Write each decimal as a fraction in simplest form.

A **0.036**

$$0.036 = \frac{36}{1,000} = \frac{36 \div 4}{1,000 \div 4} = \frac{9}{250}$$

B **−0.8**

$$-0.8 = -\frac{8}{10} = -\frac{8 \div 2}{10 \div 2} = -\frac{4}{5}$$

C **1.88**

$$1.88 = \frac{188}{100} = \frac{188 \div 4}{100 \div 4} = \frac{47}{25}, \text{ or } 1\frac{22}{25}$$

EXAMPLE 3 *Sports Application*

Steve Young holds the record for the best career completion rate as a quarterback in the NFL. During his career, he completed 2,667 of the 4,149 passes he attempted. Find his completion rate. Write your answer as a decimal rounded to the nearest thousandth.

Fraction	What the Calculator Shows	Completion Rate
$\frac{2,667}{4,149}$	2667 ÷ 4,149 ENTER 0.642805495	0.643

His completion rate is 0.643.

Think and Discuss

1. **Tell** how to write a fraction as a decimal.
2. **Describe** the difference between a terminating decimal and a repeating decimal.
3. **Explain** how to use place value to convert a terminating decimal to a fraction.

3-9 Exercises

FOR EOG PRACTICE

see page 669

internet connect

Homework Help Online
go.hrw.com Keyword: MS4 3-9

1.02

GUIDED PRACTICE

See Example 1

Write each fraction as a decimal. Round to the nearest hundredth, if necessary.

1. $\frac{3}{5}$ **2.** $\frac{21}{8}$ **3.** $\frac{11}{6}$ **4.** $\frac{7}{9}$

See Example 2

Write each decimal as a fraction in simplest form.

5. 0.008 **6.** −0.6 **7.** −2.05 **8.** 3.75

See Example 3

9. After sweeping the Baltimore Orioles at home in 2001, the Seattle Mariners had a record of 103 wins out of 143 games played. Find the Mariners' winning rate. Write your answer as a decimal rounded to the nearest thousandth.

INDEPENDENT PRACTICE

See Example 1

Write each fraction as a decimal. Round to the nearest hundredth, if necessary.

10. $\frac{9}{10}$ **11.** $\frac{32}{5}$ **12.** $\frac{18}{25}$ **13.** $\frac{7}{8}$

14. $\frac{16}{11}$ **15.** $\frac{500}{500}$ **16.** $\frac{17}{3}$ **17.** $\frac{23}{12}$

See Example 2

Write each decimal as a fraction in simplest form.

18. 0.45 **19.** 0.01 **20.** −0.25 **21.** −0.08

22. 1.8 **23.** 15.25 **24.** 5.09 **25.** 8.375

See Example 3

26. On a test, Caleb answered 73 out of 86 questions correctly. What portion of his answers was correct? Write your answer as a decimal rounded to the nearest thousandth.

PRACTICE AND PROBLEM SOLVING

Give two numbers equivalent to each fraction or decimal.

27. $8\frac{3}{4}$ **28.** 0.66 **29.** 5.05 **30.** $\frac{8}{25}$

31. 15.35 **32.** $8\frac{3}{8}$ **33.** $4\frac{3}{1,000}$ **34.** $3\frac{1}{3}$

Determine if the numbers in each pair are equivalent.

35. $\frac{3}{4}$ and 0.75 **36.** $\frac{7}{20}$ and 0.45 **37.** $\frac{11}{21}$ and 0.55 **38.** 0.8 and $\frac{4}{5}$

39. 0.275 and $\frac{11}{40}$ **40.** $1\frac{21}{25}$ and 1.72 **41.** 0.74 and $\frac{16}{25}$ **42.** 0.35 and $\frac{7}{20}$

Economics 

Use the table for Exercises 43 and 44.

XYZ Stock Values (October 2001)				
Date	Open	High	Low	Close
Oct 15	16.9	17.9	16.4	17.89
Oct 16	17.89	18.05	17.5	17.8
Oct 17	18.01	18.04	17.15	17.95
Oct 18	17.84	18.55	17.81	18.20

Traders watch the stock prices change from the floor of a stock exchange and decide what stocks to buy or sell and when.

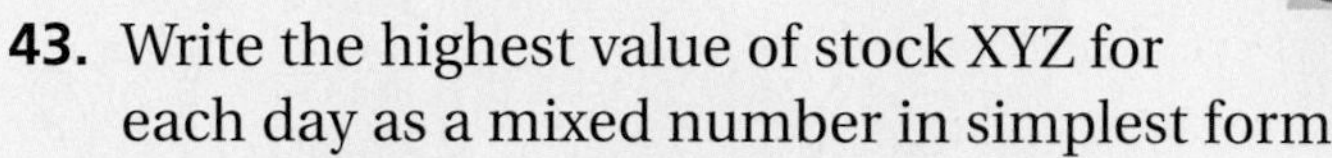

43. Write the highest value of stock XYZ for each day as a mixed number in simplest form.

44. On which of the four dates shown did the price of stock XYZ rise by $\frac{9}{25}$ of a dollar between the open and close of the day?

45. ***WRITE ABOUT IT*** Until recently, prices of stocks were expressed as mixed numbers, such as $24\frac{15}{32}$ dollars. The denominators of such fractions were multiples of 2, such as 2, 4, 6, 8, and so forth. Today, the prices are expressed as decimals to the nearest hundredth, such as 32.35 dollars.

a. What are some advantages of using decimals instead of fractions?

b. The old ticker tape machine punched stock prices onto a tape. Perhaps because fractions could not be shown using the machine, the stock values were punched as decimal equivalents of fractions. Write some decimal equivalents of fractions that the machine might print.

Before the days of computer technology, ticker-tape machines were used to punch the stock prices onto paper strands.

46. ***CHALLENGE*** Write $\frac{1}{9}$ and $\frac{2}{9}$ as decimals. Use the results to predict the decimal equivalent of $\frac{8}{9}$.

go.hrw.com
KEYWORD: MS4 Stock
CNN Student News

Spiral Review

Find the LCM of the numbers. (Lesson 2-6)

47. 16, 32 **48.** 8, 20 **49.** 9, 12 **50.** 3, 7, 10

Find the GCF of the numbers. (Lesson 2-5)

51. 18, 45 **52.** 25, 50 **53.** 16, 72 **54.** 15, 32

55. **EOG PREP** The value of $(-3) \cdot (-8) \div (-2)$ is __?__. (Lesson 3-5)

A -48 B -12 C 12 D 48

3-10 Comparing and Ordering Rational Numbers

Learn **to compare and order fractions and decimals.**

Which is greater, $\frac{7}{9}$ or $\frac{2}{9}$?

When two fractions have the same denominator, just compare the numerators.

$\frac{7}{9} > \frac{2}{9}$ because $7 > 2$.

$= \frac{7}{9}$

$= \frac{2}{9}$

I would like an extra-large pizza with $\frac{1}{2}$ pepperoni, $\frac{4}{5}$ sausage, $\frac{1}{3}$ anchovies on the pepperoni side, $\frac{3}{8}$ peanut butter fudge, $\frac{5}{11}$ pineapple, $\frac{2}{13}$ doggy treats…and extra cheese.

EXAMPLE 1 Comparing Fractions

Compare the fractions. Write < or >.

A $\frac{5}{11}$ ■ $\frac{3}{8}$

Both fractions can be written with a denominator of 88.

$\frac{5}{11} = \frac{5 \cdot 8}{11 \cdot 8} = \frac{40}{88}$

$\frac{3}{8} = \frac{3 \cdot 11}{8 \cdot 11} = \frac{33}{88}$

Write as fractions with common denominators.

$\frac{40}{88} > \frac{33}{88}$, and so $\frac{5}{11} > \frac{3}{8}$.

B $-\frac{3}{5}$ ■ $-\frac{5}{9}$

Both fractions can be written with a denominator of 45.

$-\frac{3}{5} = \frac{-3 \cdot 9}{5 \cdot 9} = \frac{-27}{45}$

$-\frac{5}{9} = \frac{-5 \cdot 5}{9 \cdot 5} = \frac{-25}{45}$

Write as fractions with common denominators. Put the negative signs in the numerators.

$\frac{-27}{45} < \frac{-25}{45}$, and so $-\frac{3}{5} < -\frac{5}{9}$.

To compare decimals, line up the decimal points and compare digits from left to right until you find the place where the digits are different.

EXAMPLE 2 **Comparing Decimals**

Compare the decimals. Write < or >.

0.841 ▬ 0.848

0.841 — *Line up the decimal points.*

The tenths and hundredths are the same.

Compare the thousandths: 1 < 8.

0.848

0.841 < 0.848

EXAMPLE 3 **Ordering Fractions and Decimals**

Order the numbers from least to greatest.

$\frac{3}{5}$**, 0.78, and 0.7**

Remember!

You can place zeros at the end of a decimal without changing its value.
2.045 = 2.045000

$\frac{3}{5} = 0.60$ $\quad$ $0.78 = 0.78$ $\quad$ $0.7 = 0.70$ — *Write as decimals with the same number of places.*

Graph the numbers on a number line.

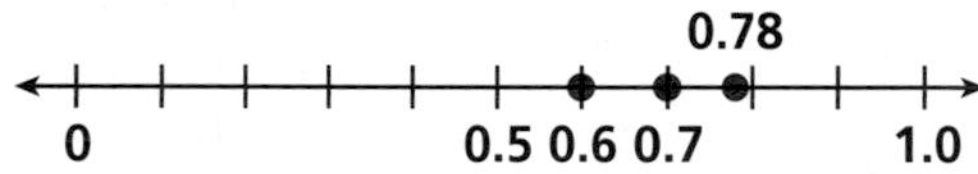

The values on a number line increase as we move from left to right.

$0.60 < 0.70 < 0.78$ — *Place the decimals in order.*

$\frac{3}{5}$, 0.7, 0.78

Think and Discuss

1. **Describe** how to compare two fractions with different denominators.
2. **Explain** why -0.61 is greater than -0.625 even though $2 > 1$.

3-10 Exercises

FOR EOG PRACTICE
see page 669

internet connect
Homework Help Online
go.hrw.com Keyword: MS4 3-10

1.02

GUIDED PRACTICE

See Example 1 **Compare the fractions. Write < or >.**

1. $\frac{3}{5}$ ■ $\frac{4}{5}$ **2.** $-\frac{5}{8}$ ■ $-\frac{7}{8}$ **3.** $-\frac{2}{3}$ ■ $-\frac{4}{7}$ **4.** $3\frac{4}{5}$ ■ $3\frac{2}{3}$

See Example 2 **Compare the decimals. Write < or >.**

5. 0.622 ■ 0.625 **6.** −0.405 ■ −0.45

7. −0.89 ■ −0.089 **8.** 3.822 ■ 3.819

See Example 3 **Order the numbers from least to greatest.**

9. 0.55, $\frac{3}{4}$, 0.505 **10.** 2.5, 2.05, $\frac{13}{5}$ **11.** $\frac{5}{8}$, −0.875, 0.877

INDEPENDENT PRACTICE

See Example 1 **Compare the fractions. Write < or >.**

12. $\frac{6}{11}$ ■ $\frac{7}{11}$ **13.** $-\frac{5}{9}$ ■ $-\frac{6}{9}$ **14.** $-\frac{5}{6}$ ■ $-\frac{8}{9}$ **15.** $10\frac{3}{4}$ ■ $10\frac{3}{5}$

16. $\frac{5}{7}$ ■ $\frac{2}{7}$ **17.** $-\frac{3}{4}$ ■ $\frac{1}{4}$ **18.** $\frac{7}{4}$ ■ $-\frac{1}{4}$ **19.** $-\frac{2}{3}$ ■ $\frac{4}{3}$

See Example 2 **Compare the decimals. Write < or >.**

20. 3.8 ■ 3.6 **21.** 0.088 ■ 0.109 **22.** −4.26 ■ 4.266

23. −1.902 ■ 0.920 **24.** −0.7 ■ −0.07 **25.** 3.08 ■ 3.808

See Example 3 **Order the numbers from least to greatest.**

26. 0.7, 0.755, $\frac{5}{8}$ **27.** 1.82, 1.6, $1\frac{4}{5}$ **28.** 2.25, 2.05, $\frac{21}{10}$

29. −3, −3.02, $1\frac{1}{2}$ **30.** 2.88, −2.98, $-2\frac{9}{10}$ **31.** $\frac{5}{6}$, $\frac{4}{5}$, 0.82

32. 2, $2\frac{2}{10}$, 2.02 **33.** −1.02, −1.20, $-1\frac{2}{5}$ **34.** $2\frac{5}{10}$, $2\frac{3}{5}$, 2.7

PRACTICE AND PROBLEM SOLVING

Choose the greater number.

35. $\frac{3}{4}$ or 0.7 **36.** 0.999 or 1.0 **37.** $\frac{7}{8}$ or $\frac{13}{20}$ **38.** −0.93 or 0.2

39. 0.32 or 0.088 **40.** $-\frac{1}{2}$ or −0.05 **41.** $-\frac{9}{10}$ or $-\frac{7}{8}$ **42.** 23.44 or 23

43. ***PHYSICAL SCIENCE*** Twenty-four karat gold is considered pure. If a gold coin is 22 karat, what is its purity as a fraction and as a decimal rounded to the nearest thousandth?

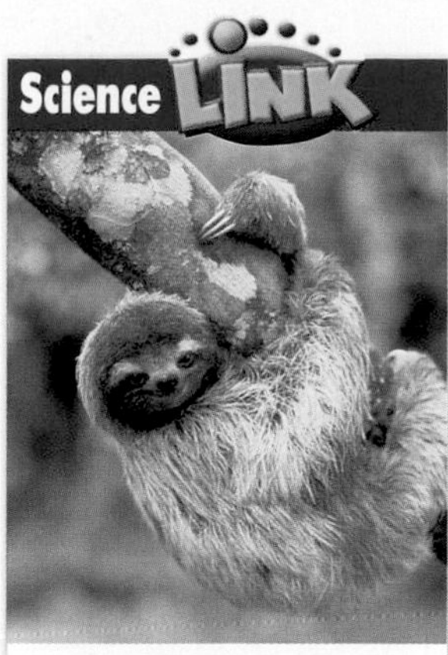

Sloths live in the trees of South and Central America. An algae that grows in their fur makes them look slightly green. This helps the sloths blend into the trees and stay out of sight from predators.

44. ***LIFE SCIENCE*** Sloths are tree-dwelling mammals that live in South and Central America. They sleep for up to 18 hours a day. Write the portion of the day a sloth spends sleeping as a fraction in simplest form and as a decimal.

45. ***EARTH SCIENCE*** Density is a measure of the amount of matter in a specific unit of space. The mean densities (measured in grams per cubic centimeter) of the planets of our solar system are given in order of the planets' distance from the Sun. Rearrange the planets from least to most dense.

Planet	Density	Planet	Density	Planet	Density
Mercury	5.43	Mars	3.93	Uranus	1.32
Venus	5.20	Jupiter	1.32	Neptune	1.64
Earth	5.52	Saturn	0.69	Pluto	2.05

46. ***ECOLOGY*** Of Beatrice's total household water use, $\frac{5}{9}$ is for bathing, toilet flushing, and laundry. How does her water use for these purposes compare with that shown in the graph?

Average Daily Household Use of Water

47. ***WHAT'S THE ERROR?*** A recipe for a large cake called for $4\frac{1}{2}$ cups of flour. The chef added 10 one-half cupfuls of flour to the mixture. What was the chef's error?

48. ***WRITE ABOUT IT*** Explain how to compare a mixed number with a decimal.

49. ***CHALLENGE*** Scientists estimate that Earth is approximately 4.6 billion years old. We are currently in what is called the Phanerozoic eon, which has made up about $\frac{7}{60}$ of the time that Earth has existed. The first eon, called the Hadean, made up approximately 0.175 of the time Earth has existed. Which eon represents the most time?

Spiral Review

Write each number in scientific notation. (Lesson 2-2)

50. 32,000 **51.** 108 **52.** 188,000 **53.** 10,000,000

Solve each equation. (Lesson 2-12)

54. $5x = 30$ **55.** $\frac{a}{2} = 14.2$ **56.** $1.1y = 44$ **57.** $\frac{c}{3.2} = 100$

58. **EOG PREP** The integers −2, −8, 4, and 2 written from least to greatest are __?__. (Lesson 3-1)

A −2, −8, 2, 4 B −8, −2, 2, 4 C 4, 2, −2, −8 D 4, 2, −8, −2

EXTENSION

Irrational Numbers

Learn to define and recognize irrational numbers.

Vocabulary

nonterminating decimal

irrational number

All rational numbers have decimal forms that either terminate or repeat. A terminating decimal, such as 0.5, is one that comes to an end. A **nonterminating decimal**, such as 0.33333…, goes on forever.

Repeating decimals are nonterminating decimals that have a digit or a sequence of digits that repeats without ever stopping. Examples are 0.333333… and 0.181818…. If a pattern is not too long, you can see it on your calculator screen.

Rational Numbers		
Rational Number	Decimal Form	Decimal Type
$\frac{1}{2}$	0.5	Terminating
$\frac{1}{3}$	0.33333… (written $0.\overline{3}$)	Nonterminating, repeating
$\frac{2}{11}$	0.181818… (written $0.\overline{18}$)	Nonterminating, repeating

Irrational Numbers		
Irrational Number	Decimal Form	Decimal Type
$\sqrt{2}$	1.4142135…	Nonterminating, nonrepeating
π	3.1415926…	Nonterminating, nonrepeating

Irrational numbers are numbers whose decimal forms are nonterminating, with *no* repeating pattern.

EXAMPLE 1 **Investigating Rational Numbers with a Calculator**

Use a calculator to determine if the decimal form of each rational number terminates or repeats. If it repeats, show the pattern by writing a bar over the digit or sequence of digits that repeats.

A $\frac{3}{16}$

3 ÷ 16 ENTER 0.1875

$\frac{3}{16}$ is a terminating decimal.

B $\frac{1}{27}$

1 ÷ 27 ENTER 0.037037037

$0.\overline{037}$

$\frac{1}{27}$ is a repeating decimal.

Calculators can reveal a repeating pattern only if the pattern occurs within the number of places the calculator shows. Some repeating decimals have very long patterns. For example, the decimal form of $\frac{1}{97}$ has a pattern that is 96 digits long!

Because you can write only an approximate value of irrational numbers in decimal form, symbols are sometimes used to represent the exact values.

Irrational Number Symbol (Exact)	Approximate Value
π	3.14159
$\sqrt{5}$	2.23607

EXAMPLE 2 Investigating Irrational Numbers with a Calculator

Use a calculator to evaluate each square root. Does there seem to be a repeating pattern?

A $\sqrt{6}$

There is no apparent pattern.

B $\sqrt{8}$

There is no apparent pattern.

Example 2 suggests, but does not prove, the following important fact: **If a whole number is not a perfect square, its square root is irrational.**

EXTENSION Exercises

1.03

Use a calculator to determine if the decimal form of each rational number terminates or repeats. If it repeats, show the pattern by writing a bar over the digit or sequence of digits that repeats.

1. $\frac{3}{16}$ **2.** $\frac{5}{16}$ **3.** $\frac{7}{16}$ **4.** $\frac{1}{54}$

5. $\frac{1}{9}$ **6.** $\frac{2}{9}$ **7.** $\frac{1}{99}$ **8.** $\frac{1}{82}$

9. $\frac{1}{7}$ **10.** $\frac{1}{37}$ **11.** $\frac{1}{22}$ **12.** $\frac{1}{39}$

Use a calculator to find the square root of each number. Tell whether each square root is rational or irrational.

13. $\sqrt{196}$ **14.** $\sqrt{169}$ **15.** $\sqrt{14}$ **16.** $\sqrt{144}$

17. $\sqrt{1{,}600}$ **18.** $\sqrt{18}$ **19.** $\sqrt{400}$ **20.** $\sqrt{444}$

21. $\sqrt{10}$ **22.** $\sqrt{100}$ **23.** $\sqrt{1{,}000}$ **24.** $\sqrt{10{,}000}$

Problem Solving on Location

NORTH CAROLINA

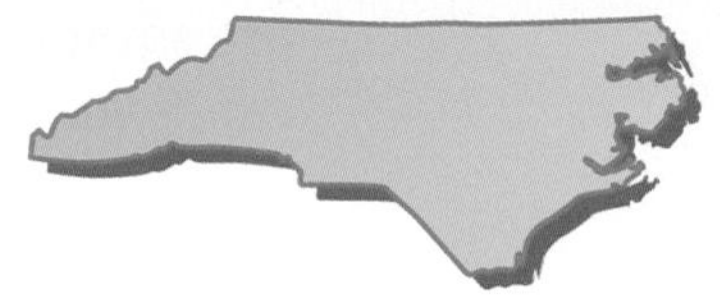

Sunken Ships

Off the coast of North Carolina lie many sunken ships where ocean life now thrives. Some of these ships were purposely sunk as a part of North Carolina's artificial reef program, while others sank as a result of storms, collisions, or battles.

1. To help build an artificial reef, the *Aeolus* was sunk to -120 feet in waters 22 miles south of Beaufort Inlet. The upper decks of the *Aeolus* are at -85 feet. How high does the *Aeolus* rise above the ocean floor?

2. The *Fenwick Isle* is a 125-foot-long fishing trawler that sank in a storm 15 miles southeast of Beaufort Inlet. The trawler lies 65 feet below the ocean surface and rises to 35 feet from the surface. How much farther does the *Aeolus* rise above the ocean floor than the *Fenwick Isle*?

These wreck sites are popular places for scuba divers to explore and come face to face with a variety of sharks. Some types of sharks can be seen at different sites, but at some of the sites, no sharks at all can be seen.

For 3–4, use the table.

3. At what fraction of the sunken ship sites listed can sharks be found?

4. What fraction is equivalent to 0.8? Use the information in the table to write a question for which this fraction is the answer.

Types of Sharks Found at Sunken Ship Sites

Sunken Ship	Sharks Found	Type of Sharks
Aeolus	✔	Atlantic sand tiger
Atlas	✔	Atlantic sand tiger
Caribsea	✔	Atlantic sand tiger
Fenwick Isle	✘	—
Naeco	✘	—
Papoose	✔	Bull
Verbena	✔	Atlantic sand tiger

Agriculture

Farmers need to track the success of their crops from year to year. One way they do this is by finding the crop yield per acre. This is determined by dividing the total weight in tons of the crop harvested by the number of acres on which the crop was grown and rounding to the nearest hundredth.

1. In 2000, Alleghany County produced a yield per acre of 2.26. In 2001, the yield per acre was 2.23. The number of acres that were harvested and the weight of each harvest for the two years (in no particular order) were 14,100 acres, 14,200 acres, 31,600 tons, and 31,800 tons. What were the total number of acres harvested and the weight of the crops for 2001?

Since successful farming depends heavily on weather conditions, a strong storm can be detrimental to crops. In 1999, Hurricane Floyd brought severe winds and several inches of rain to North Carolina, damaging crops and costing farmers millions of dollars.

For 2–5, use the table.

2. Write the dollar amounts in the table in order from least to greatest.

3. Profit losses are often indicated by negative numbers. Write the dollar amounts in the table as integers, and order them from least to greatest.

4. Compare your lists from **2** and **3**. Do the greatest values on each list have the same absolute value? Explain.

5. Assuming that each crop normally brought in the same amount of money, which crop brought in the most profit the year that Hurricane Floyd hit?

Farm flooded from Hurricane Floyd, 1999

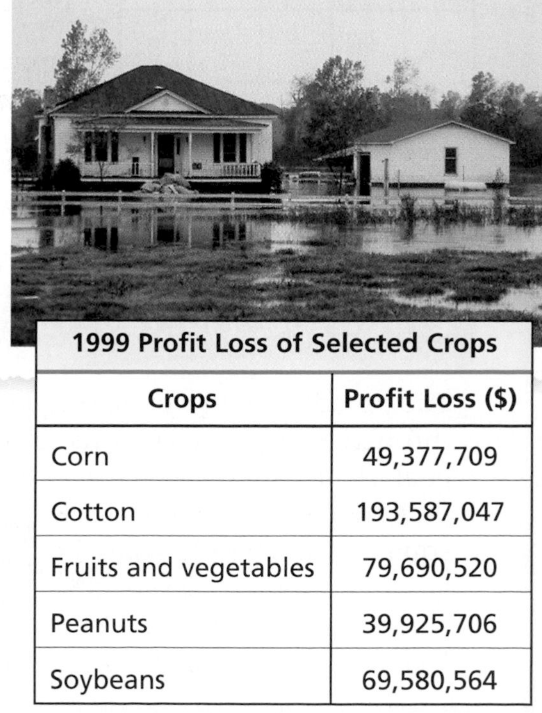

1999 Profit Loss of Selected Crops

Crops	Profit Loss ($)
Corn	49,377,709
Cotton	193,587,047
Fruits and vegetables	79,690,520
Peanuts	39,925,706
Soybeans	69,580,564

MATH-ABLES

Magic Squares

A magic square is a grid with numbers, such that the numbers in each row, column, and diagonal have the same "magic" sum. Test the square at right to see an example of this.

You can use a magic square to do some amazing calculations. Cover a block of four squares (2 × 2) with a piece of paper. There is a way you can find the sum of these squares without looking at them. Try to find it. (*Hint:* What number in the magic square can you subtract from the magic sum to give you the sum of the numbers in the block? Where is that number located?)

Here's the answer: To find the sum of any block of four numbers, take 65 (the magic sum) and subtract from it the number that is diagonally two squares away from a corner of the block.

18	10	22	14	1
12	4	16	8	25
6	23	15	2	19
5	17	9	21	13
24	11	3	20	7

$65 - 21 = 44$

18	10	22	14	1
12	4	16	8	25
6	23	15	2	19
5	17	9	21	13
24	11	3	20	7

$65 - 1 = 64$

The number you subtract must fall on an extension of a diagonal of the block. For each block that you choose, there will be only one direction you can go.

Try to create a 3 × 3 magic square with the numbers 1–9.

Modified Tic-Tac-Toe

The board has a row of nine squares numbered 1 through 9. Players take turns selecting squares. The goal of the game is for a player to select squares such that any three of a player's squares add up to 15. The game can also be played with a board numbered 1 through 16 and a sum goal of 34.

internet connect

For a complete copy of the rules and to print out a set of cards, go to ***go.hrw.com***
KEYWORD: MS4 Game3

Technology LAB

Graph Points Using a Calculator

internet connect
Lab Resources Online
go.hrw.com
KEYWORD: MS4 TechLab3

To graph on a calculator, you may need to use the **WINDOW** menu. This is where you choose the viewing boundaries and scale settings. If you do not choose these yourself, the calculator will use the standard viewing window, such as the one shown.

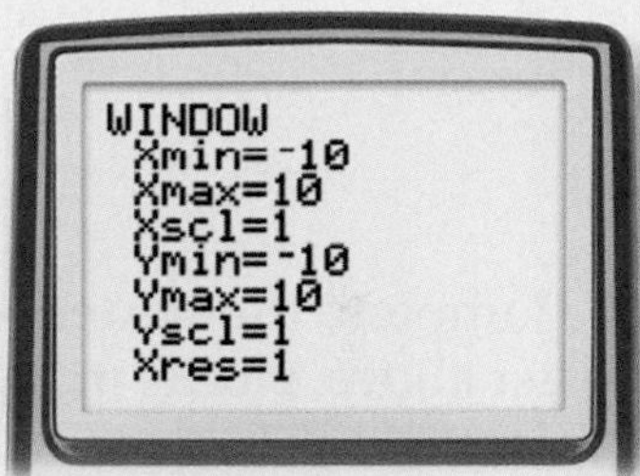

In the standard window, the x- and y-values go from -10 to 10. These are the boundary values. They are set by **Xmin, Xmax, Ymin,** and **Ymax**.

The scale values, **Xscl** and **Yscl**, give the distance between tick marks. In the standard window, tick marks are 1 unit apart.

Activity

1 Graph the points (1.5, 5), $\left(\frac{10}{3}, 3\right)$, $\left(-\frac{3}{2}, 6\right)$, and (5.75, −3.5) using the standard window.

To graph (1.5, 5), press 2nd PRGM (DRAW) ▶.

You will get the screen shown at right.

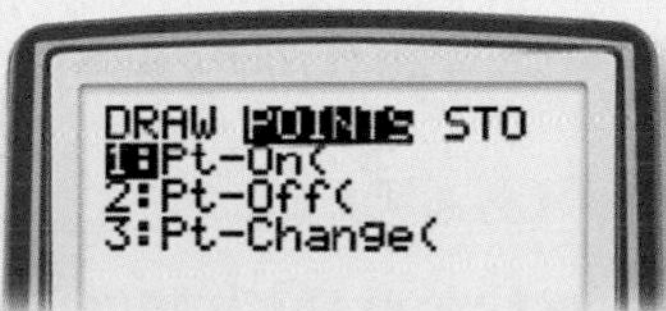

Then press ENTER 1.5 , 5 ENTER.

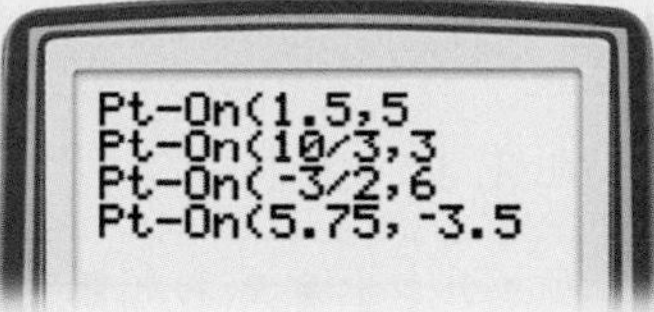

After you see the grid with a point at (1.5, 5), press 2nd MODE (QUIT).

Repeat the steps above to graph $\left(\frac{10}{3}, 3\right)$, $\left(-\frac{3}{2}, 6\right)$, and (5.75, −3.5).

Your graph should then look like the one shown at right.

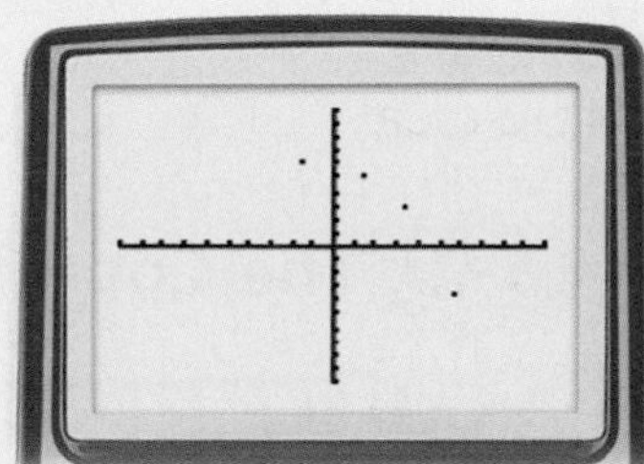

Think and Discuss

1. Explain when you might want to change the **WINDOW** dimensions.

Try This

1. Graph (−3, 9), (−2, 4), (13, 1), (2, 4), (3, 9), and (4, 16) using the standard window. How many points do you see? Why?

Chapter 3 Study Guide and Review

Vocabulary

absolute value	131	mixed number	167	rational number	162
coordinate plane	134	opposite	130	repeating decimal	171
equivalent fractions	166	ordered pair	134	terminating decimal	171
improper fraction	167	origin	134	*x*-axis	134
integer	130	quadrant	134	*y*-axis	134

Complete the sentences below with vocabulary words from the list above. Words may be used more than once.

1. A(n) __?__ can be written as the ratio of one __?__ to another and can be represented by a repeating or __?__.

2. When you graph a(n) __?__, the distance on the __?__ is always listed first.

3-1 Integers (pp. 130–133)

EXAMPLE

■ **Graph the integers on a number line, and then write them in order from least to greatest.**

3, 4, −2, 1, −3

−6 −4 −2 0 2 4 6

−3, −2, 1, 3, 4

EXERCISES

Graph the integers on a number line, and then write them in order from least to greatest.

3. −6, 4, 0, −2, 5

Use a number line to find each absolute value.

4. $|0|$ **5.** $|-17|$ **6.** $|6|$

3-2 The Coordinate Plane (pp. 134–137)

EXAMPLE

■ **Give the coordinates of each point and tell in which quadrant it lies.**

A (−3, 2); II
B (2, −3); IV
C (−2, −3); III
D (3, 2); I

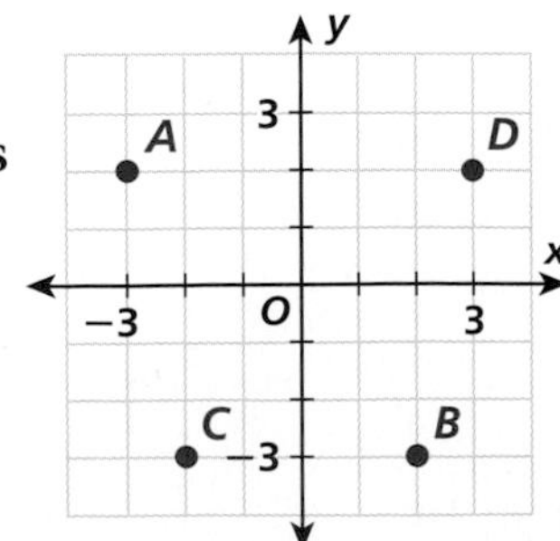

EXERCISES

Plot each point and tell in which quadrant it lies.

7. *A* (4, 2) **8.** *B* (−4, −2)

9. *C* (−2, 4) **10.** *D* (2, −4)

3-3 Adding Integers (pp. 140–143)

EXAMPLE

■ **Add.**

$$-7 + (-11)$$

$-7 + (-11)$ *The signs are the same.*

-18

EXERCISES

Add.

11. $-8 + 5$ **12.** $7 + (-6)$

13. $-16 + (-40)$ **14.** $-9 + 18$

15. $-2 + 16$ **16.** $12 + (-18)$

3-4 Subtracting Integers (pp. 146–149)

EXAMPLE

■ **Subtract.**

$-5 - (-3)$

$-5 + 3$ *Add the opposite of −3.*

-2

EXERCISES

Subtract.

17. $8 - 2$ **18.** $10 - 19$

19. $-6 - (-5)$ **20.** $-5 - 4$

3-5 Multiplying and Dividing Integers (pp. 150–153)

EXAMPLE

Find each product or quotient.

■ $12 \cdot (-3)$ *The signs are different, so*

-36 *the product is negative.*

■ $-16 \div (-4)$ *The signs are the same, so*

4 *the quotient is positive.*

EXERCISES

Find each product or quotient.

21. $5 \cdot (-10)$ **22.** $-27 \div (-9)$

23. $-2 \cdot (-8)$ **24.** $-40 \div 20$

25. $-3 \cdot 4$ **26.** $45 \div (-15)$

3-6 Solving Equations Containing Integers (pp. 156–159)

EXAMPLE

Solve.

■ $x - 12 = 4$

$\underline{+12} \quad \underline{+12}$ *Add 12 to each side.*

$x = 16$

■ $-10 = -2f$

$\frac{-10}{-2} = \frac{-2f}{-2}$ *Divide each side by −2.*

$5 = f$

EXERCISES

Solve.

27. $7y = 70$ **28.** $d - 8 = 6$

29. $j + 23 = -3$ **30.** $\frac{n}{36} = 2$

31. $-26 = -2c$ **32.** $28 = -7m$

3-7 Rational Numbers: Fractions and Decimals (pp. 162–165)

EXAMPLE

Show that each number is a rational number by writing it as a fraction.

■ 0.75

$0.75 = \frac{3}{4}$

■ −2.2

$-2.2 = -\frac{22}{10}$ or $-\frac{11}{5}$

EXERCISES

Graph each number on a number line.

33. 5.5 **34.** −4.25 **35.** $-3\frac{1}{4}$

Show that each number is a rational number by writing it as a fraction.

36. 12

37. 0.25

38. −3.7

3-8 Equivalent Fractions and Mixed Numbers (pp. 166–169)

EXAMPLE

■ Write $5\frac{2}{3}$ as an improper fraction.

$5\frac{2}{3} = \frac{3 \cdot 5 + 2}{3} = \frac{17}{3}$

■ Write $\frac{17}{4}$ as a mixed number.

$\frac{17}{4} = 4\frac{1}{4}$ *Divide the numerator by the denominator.*

EXERCISES

Write each as an improper fraction.

39. $4\frac{1}{5}$ **40.** $3\frac{1}{6}$ **41.** $10\frac{3}{4}$

Write each as a mixed number.

42. $\frac{10}{3}$ **43.** $\frac{5}{2}$ **44.** $\frac{17}{7}$

3-9 Equivalent Fractions and Decimals (pp. 170–173)

EXAMPLE

■ Write 0.75 as a fraction in simplest form.

$0.75 = \frac{75}{100} = \frac{75 \div 25}{100 \div 25} = \frac{3}{4}$

■ Write $\frac{-5}{4}$ as a decimal.

$\frac{-5}{4} = -5 \div 4 = -1.25$

EXERCISES

Write each decimal as a fraction in simplest form.

45. 0.25 **46.** −0.004 **47.** 0.05

Write each fraction as a decimal.

48. $\frac{7}{2}$ **49.** $-\frac{3}{5}$ **50.** $\frac{2}{3}$

3-10 Comparing and Ordering Rational Numbers (pp. 174–177)

EXAMPLE

■ Compare. Write < or >.

$-\frac{3}{4}$ ■ $-\frac{2}{3}$

$-\frac{3}{4} \cdot \frac{3}{3}$ ■ $-\frac{2}{3} \cdot \frac{4}{4}$

$-\frac{9}{12} < -\frac{8}{12}$

Write as fractions with common denominators.

EXERCISES

Compare. Write < or >.

51. $\frac{4}{5}$ ■ 0.81

52. 0.22 ■ $\frac{3}{20}$

53. $-\frac{3}{5}$ ■ −1.5

Study Guide and Review

Chapter Test

Graph the integers on a number line, and then write them in order from least to greatest.

1. $-4, 3, -2, 0, 1$ **2.** $7, -6, 5, -8, -3$

Compare. Write <, >, or =.

3. -5 ■ 5 **4.** $|-5|$ ■ 5 **5.** 8 ■ -7 **6.** -10 ■ -9

Plot each point and identify the quadrant in which it is contained.

7. $A\ (4, -3)$ **8.** $B\ (-5, 2)$ **9.** $C\ (7, 1)$ **10.** $D\ (-7, -2)$

Add or subtract.

11. $-7 + (-3)$ **12.** $-11 + 15$ **13.** $17 + (-27) + (-2)$ **14.** $102 + (-97) + 3$

15. $-6 - 3$ **16.** $10 - 15$ **17.** $12 - (-7)$ **18.** $17 - (-9) - 8$

Find each product or quotient.

19. $-3 \cdot 20$ **20.** $-36 \div 12$ **21.** $-400 \div (-10)$ **22.** $-5 \cdot (-2) \cdot 9$

Solve.

23. $w - 4 = -6$ **24.** $x + 5 = -5$ **25.** $-6a = 60$ **26.** $\frac{n}{-4} = 12$

Graph each number on a number line.

27. 4.1 **28.** $-3\frac{1}{3}$ **29.** -1.6 **30.** $1\frac{3}{8}$

Write the fractions with a common denominator. Then determine if they are equivalent.

31. $\frac{6}{12}$ and $\frac{13}{26}$ **32.** $\frac{17}{20}$ and $\frac{20}{24}$ **33.** $\frac{30}{24}$ and $\frac{35}{28}$ **34.** $\frac{5}{3}$ and $\frac{8}{5}$

Write each fraction as a decimal. Write each decimal as a fraction in simplest form.

35. $\frac{3}{50}$ **36.** $\frac{25}{10}$ **37.** 3.15 **38.** 0.004

Compare. Write < or >.

39. $\frac{2}{3}$ ■ 0.62 **40.** 1.5 ■ $1\frac{6}{20}$ **41.** $-\frac{9}{7}$ ■ -1 **42.** $\frac{11}{5}$ ■ $1\frac{2}{3}$

43. Three brothers stood in order of height from shortest to tallest for a family photo. Brad is $5\frac{1}{2}$ ft tall, Brandon is 5.29 ft tall, and Lee is 5.6 ft tall. Tell in what order the brothers stood in the photo.

Chapter 3

Performance Assessment

Show What You Know

Create a portfolio of your work from this chapter. Complete this page and include it with your four best pieces of work from Chapter 3. Choose from your homework or lab assignments, mid-chapter quiz, or any journal entries you have done. Put them together using any design you want. Make your portfolio represent what you consider your best work.

Short Response

1. Jana began the month with $102 in her checking account. During the month, she added $8 that she earned from babysitting, spent $10 for a CD, added $5 her aunt gave her, and spent $7 at the movies. Write an expression using integers to show the month's activity. Simplify your expression to find how much Jana had in her checking account at the end of the month.

2. After dark, the temperature dropped 5 degrees for every hour for 4 hours. Write an integer that represents the total drop in temperature 4 hours after dark. Show the steps that you used to find your answer.

3. Kyle needs a piece of 1 in. by 4 in. lumber that measures $2\frac{3}{4}$ feet in length and another piece of 1 in. by 2 in. lumber that measures $2\frac{5}{8}$ feet in length. Which piece of lumber is longer? Support your answer.

Extended Problem Solving

4. Gina and Alex are comparing recipes for smoothies. Gina prefers strawberry smoothies, and Alex prefers orange.
   a. Which recipe calls for more yogurt? Write an expression that justifies your answer.
   b. Write the amount of fruit called for by the strawberry-smoothie recipe as an improper fraction and as a decimal.
   c. Gina and Alex each make their favorite smoothie. Who makes more? Tell how you know.

Strawberry Smoothie

$2\frac{1}{2}$ cups strawberries
$1\frac{1}{2}$ cups frozen yogurt

Orange Smoothie

$\frac{1}{2}$ cup orange juice
1 cup mashed bananas
1 cup frozen yogurt

internet connect
EOG Practice Online
go.hrw.com Keyword: MS4 TestPrep

Getting Ready for EOG

Chapter 3

Cumulative Assessment, Chapters 1–3

1. Which is a list of equivalent numbers?

   A $0.8, \frac{2}{50}, \frac{8}{100}$ C $0.02, \frac{1}{50}, \frac{2}{100}$

   B $40, 0.4, \frac{1}{2}$ D $10, 0.01, \frac{1}{11}$

2. What is the sum of −6, 3, and −9?

   A 162 C −18

   B −12 D −162

3. What is the mean of 8, −12, and 4?

   A 0 C 6.67

   B 4 D 8

TEST TAKING TIP!

Eliminate choices: When finding the value of a power, you can eliminate any choices that are not divisible by the base of the power.

4. What is the value of 3^5?

   A 15 C 243

   B 125 D 320

5. What is 753,000 written in scientific notation?

   A 75.3×10^4 C 753×10^3

   B 7.53×10^3 D 7.53×10^5

6. What is the greatest common factor of 45 and 15?

   A 3 C 15

   B 5 D 45

7. What is the least common multiple of 12 and 20?

   A 240 C 20

   B 60 D 12

8. Which number is between 0 and −1?

   A −5.0 C 0.5

   B −0.3 D 3.0

9. ***SHORT RESPONSE*** The Spanish Club provided 48 sandwiches for a picnic. After the picnic, *s* sandwiches were left.

   a. Write an expression that shows how many sandwiches were eaten.

   b. Evaluate your expression for $s = 9$. What does your answer represent?

10. ***SHORT RESPONSE*** The circle graph shows how Amy spends her earnings each month. If Amy earned $100 in June, how much did she spend on entertainment and clothing combined? Show or explain how you got your answer.

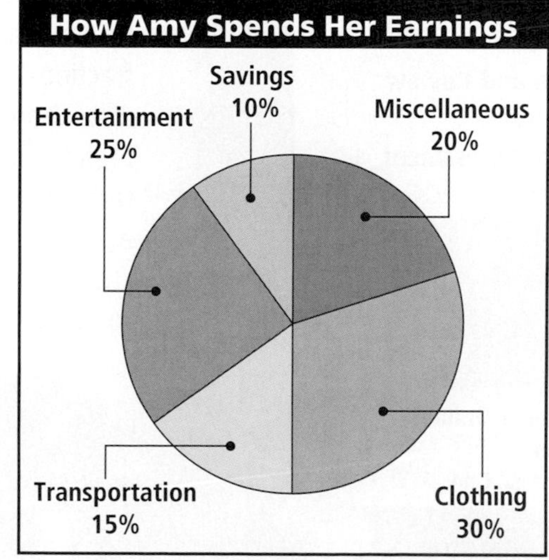

Chapter 4

Operations with Rational Numbers

TABLE OF CONTENTS

internet connect

Chapter Opener Online
go.hrw.com
KEYWORD: MS4 Ch4

Ingredients	10 Waffles	25 Waffles	50 Waffles
Flour	10 c	25 c	50 c
Salt	5 tsp	$12\frac{1}{2}$ tsp	25 tsp
Baking soda	$2\frac{1}{2}$ tsp	$6\frac{1}{4}$ tsp	$12\frac{1}{2}$ tsp
Buttermilk	$2\frac{1}{2}$ c	$6\frac{1}{4}$ c	$12\frac{1}{2}$ c
Butter (melted)	20 tsp	50 tsp	100 tsp

Career *Chef*

Tom Culbertson is a pastry chef. He develops and prepares all of the baked goods for his restaurant. In his work, Tom must often use fractions when measuring ingredients. He must also be able to multiply and divide fractions to increase or decrease the number of servings for a recipe. In addition to the breads and desserts that he creates, Tom is famous for his breakfast waffles. Tom often adds fresh fruits, such as blueberries, strawberries, or bananas, to his waffles.

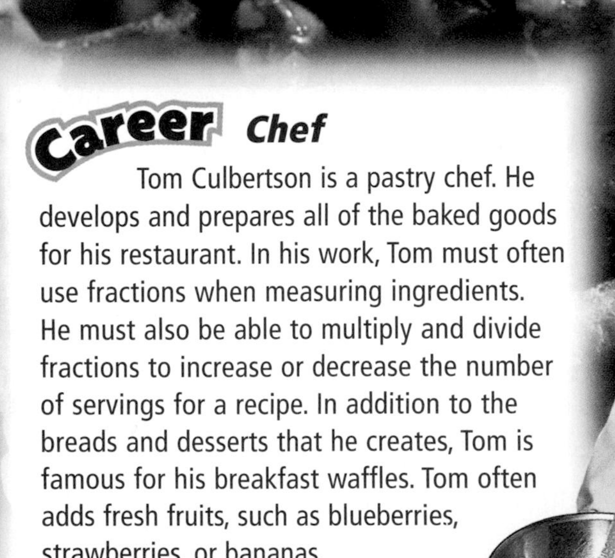

ARE YOU READY?

Choose the best term from the list to complete each sentence.

1. A(n) __?__ is a number that is written using the base-ten place value system.
2. An example of a(n) __?__ is $\frac{14}{5}$.
3. A(n) __?__ is a number that represents a part of a whole.

decimal

fraction

improper fraction

mixed number

simplest form

Complete these exercises to review the skills you will need for this chapter.

✔ Simplify Fractions

Write each fraction in simplest form.

4. $\frac{24}{40}$ **5.** $\frac{64}{84}$ **6.** $\frac{66}{78}$ **7.** $\frac{64}{192}$

8. $\frac{21}{35}$ **9.** $\frac{11}{99}$ **10.** $\frac{16}{36}$ **11.** $\frac{20}{30}$

✔ Write Mixed Numbers as Fractions

Write each mixed number as an improper fraction.

12. $7\frac{1}{2}$ **13.** $2\frac{5}{6}$ **14.** $1\frac{14}{15}$ **15.** $3\frac{2}{11}$

16. $3\frac{7}{8}$ **17.** $8\frac{4}{9}$ **18.** $4\frac{1}{7}$ **19.** $5\frac{9}{10}$

✔ Write Fractions as Mixed Numbers

Write each improper fraction as a mixed number.

20. $\frac{23}{6}$ **21.** $\frac{17}{3}$ **22.** $\frac{29}{7}$ **23.** $\frac{39}{4}$

24. $\frac{48}{5}$ **25.** $\frac{82}{9}$ **26.** $\frac{69}{4}$ **27.** $\frac{35}{8}$

✔ Add, Subtract, Multiply, or Divide Integers

Find each sum, difference, product, or quotient.

28. $-11 + (-24)$ **29.** $-11 - 7$ **30.** $-4 \cdot (-10)$ **31.** $-22 \div (-11)$

32. $23 + (-30)$ **33.** $-33 - 74$ **34.** $-62 \cdot (-34)$ **35.** $84 \div (-12)$

4-1 Estimate with Decimals

 Problem Solving Skill

Learn to estimate decimal sums, differences, products, and quotients.

Vocabulary

compatible numbers

Jessie earned $27.00 for baby-sitting. She wants to use the money to buy a ticket to a waterpark for $14.75 and a souvenir T-shirt for $13.20.

To find out if Jessie has enough money to buy both items, you can use estimation. To estimate the total cost of the ticket and the T-shirt, round each price to the nearest dollar, or integer. Then add the rounded values.

$14.75	*7 > 5, so round to $15.*	$15
$13.20	*2 < 5, so round to $13.*	+ $13
		$28

The estimated cost is $28, so Jessie does not have enough money to buy both items.

To estimate decimal sums and differences, round each decimal to the nearest integer and then add or subtract.

EXAMPLE 1 Estimating Sums and Differences of Decimals

Estimate by rounding to the nearest integer.

A **86.9 + 58.4**

86.9	→	87	*9 > 5, so round to 87.*
+ 58.4	→	+ 58	*4 < 5, so round to 58.*
		145	← *Estimate*

B **10.38 − 6.721**

10.38	→	10	*3 < 5, so round to 10.*
− 6.721	→	− 7	*7 > 5, so round to 7.*
		3	← *Estimate*

C **−26.3 + 15.195**

−26.3	→	−26	*3 < 5, so round to −26.*
+ 15.195	→	+ 15	*1 < 5, so round to 15.*
		−11	← *Estimate*

Remember!

To round to the nearest integer, look at the digit in the tenths place. If it is greater than or equal to 5, round to the next integer. If it is less than 5, keep the same integer.

You can use *compatible numbers* when estimating. **Compatible numbers** are numbers that replace the numbers in the problem and are easy to use.

Guidelines for Using Compatible Numbers	
When multiplying . . .	**When dividing . . .**
round numbers to the nearest nonzero integer or to numbers that are easy to multiply.	round numbers so that they divide without leaving a remainder.

EXAMPLE 2 Estimating Products and Quotients of Decimals

Helpful Hint

By rounding to multiples of 10, you can solve problems mentally.

Use compatible numbers to estimate.

A $32.66 \cdot 7.69$

$32.66 \longrightarrow 30$ *Round to the nearest multiple of 10.*

$\times 7.69 \longrightarrow \times 8$ *6 > 5, so round to 8.*

$240 \longleftarrow$ ***Estimate***

B $36.5 \div (-8.241)$

$36.5 \longrightarrow 36$ *37 is prime, so round to 36.*

$-8.241 \longrightarrow -9$ *−9 divides into 36 without a remainder.*

$36 \div (-9) = -4 \longleftarrow$ ***Estimate***

When you solve problems, using an estimate can help you decide whether your answer is reasonable.

EXAMPLE 3 *School Application*

On a math test, a student worked the problem $6.2\overline{)55.9}$ and got the answer 0.9. Use estimation to check whether the answer is reasonable.

$55.9 \longrightarrow 56$ *9 > 5, so round to 56.*

$6.2 \longrightarrow 7$ *7 divides into 56 without a remainder.*

$56 \div 7 = 8 \longleftarrow$ ***Estimate***

The estimate is almost ten times the student's answer, so 0.9 is not a reasonable answer.

Think and Discuss

1. **Explain** whether your estimate will be greater than or less than the actual answer when you round both numbers down in an addition or multiplication problem.
2. **Describe** a situation in which you would want your estimate to be greater than the actual amount.

4-1 Exercises

1.02c, 1.03

FOR EOG PRACTICE
see page 670

internet connect
Homework Help Online
go.hrw.com Keyword: MS4 4-1

GUIDED PRACTICE

See Example 1 **Estimate by rounding to the nearest integer.**

1. 37.2 + 25.83 **2.** 18.256 − 5.71 **3.** −9.916 + 12.4

See Example 2 **Use compatible numbers to estimate.**

4. 8.09 · 28.32 **5.** −3.45 · 73.6 **6.** 41.9 ÷ 6.391

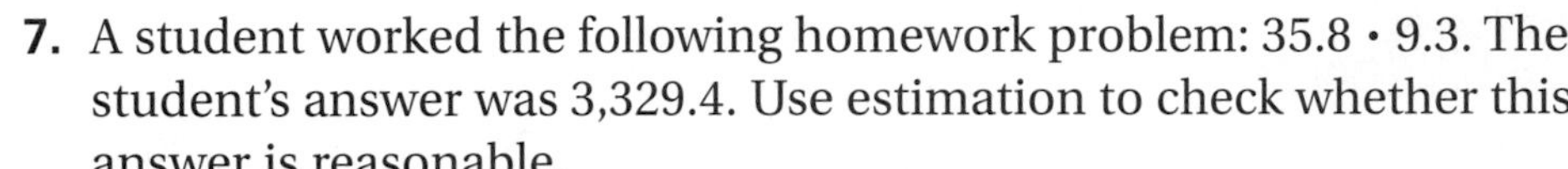

See Example 3 **7.** A student worked the following homework problem: 35.8 · 9.3. The student's answer was 3,329.4. Use estimation to check whether this answer is reasonable.

INDEPENDENT PRACTICE

See Example 1 **Estimate by rounding to the nearest integer.**

8. 5.982 + 37.1 **9.** 68.2 + 23.67 **10.** −36.8 + 14.217

11. 15.23 − 6.835 **12.** 6.88 + (−8.1) **13.** 80.38 − 24.592

See Example 2 **Use compatible numbers to estimate.**

14. 51.38 · 4.33 **15.** 46.72 ÷ 9.24 **16.** 32.91 · 6.28

17. −3.45 · 43.91 **18.** 2.81 · (−79.2) **19.** 28.22 ÷ 3.156

See Example 3 **20.** Amanda has a piece of ribbon that is 12.35 meters long. She wants to cut it into smaller pieces that are each 3.6 meters long. She thinks she will get about 3 smaller pieces of ribbon. Use estimation to check whether her assumption is reasonable.

PRACTICE AND PROBLEM SOLVING

Estimate.

21. 5.921 − 13.2 **22.** −7.98 − 8.1 **23.** −42.25 + (−17.091)

24. 98.6 + 43.921 **25.** 4.69 · (−18.33) **26.** 62.84 − 35.169

27. −48.28 + 11.901 **28.** 31.53 ÷ (−4.12) **29.** 35.9 − 24.71

30. 69.7 − 7.81 **31.** −6.56 · 14.2 **32.** 4.513 + 72.45

33. −8.9 · (−24.1) **34.** 6.92 · (−3.714) **35.** −78.3 ÷ (−6.25)

36. Jo needs 10 lb of ground beef for a party. She has packages that weigh 4.23 lb and 5.09 lb. Does she have enough?

37. ***CONSUMER MATH*** Ramón saves $8.35 each week. He wants to buy a video game that costs $61.95. For about how many weeks will Ramón have to save his money before he can buy the video game?

38. ***TRANSPORTATION*** Kayla stopped for gasoline at a station that was charging \$1.119 per gallon. If Kayla had \$5.25 in cash, approximately how many gallons of gas could she buy?

Social Studies LINK

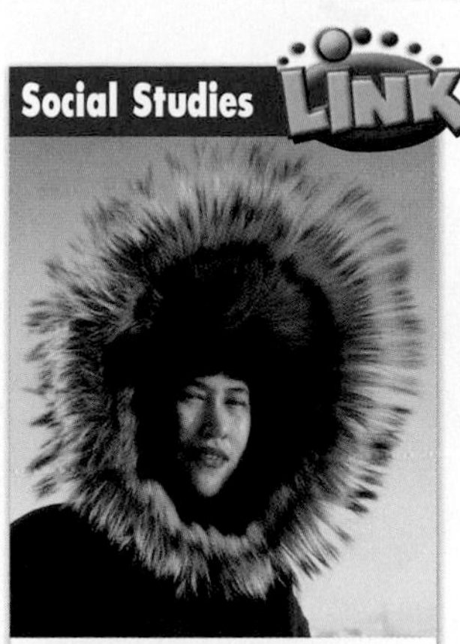

Inuktitut, the language of the Inuit people, is spoken by more than 60,000 people in Canada, Alaska, and Greenland. The Inuktitut word for "school" is *ilinniarvic.*

39. ***SOCIAL STUDIES*** The circle graph shows the languages spoken in Canada.

    a. Which is the most common language spoken in Canada?

    b. What is the approximate difference between the percent of people who speak English and the percent who speak French?

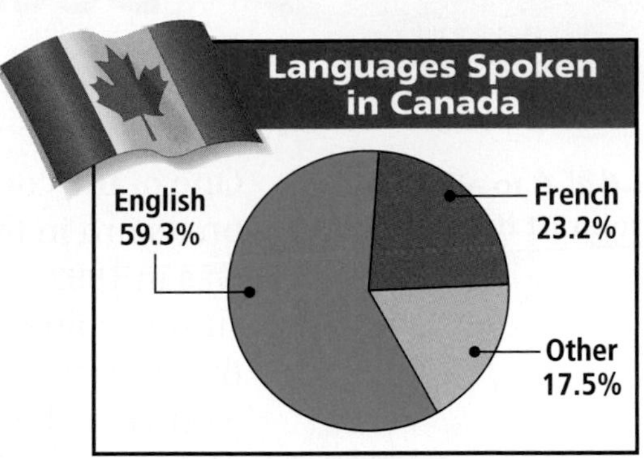

40. ***ASTRONOMY*** Jupiter is 5.20 astronomical units (AU) from the Sun. Neptune is almost 6 times as far from the Sun as Jupiter is. Estimate Neptune's distance from the Sun in astronomical units.

41. ***SPORTS*** Scott must earn a total of 27 points to advance to the final round in an ice-skating competition. He earns scores of 5.9, 5.8, 6.0, 5.8, and 6.0. Scott estimates that his total score will allow him to advance. Is his estimate reasonable? Explain.

42. ***WRITE A PROBLEM*** Write a problem that can be solved by estimating with decimals.

43. ***WRITE ABOUT IT*** Explain how an estimate helps you decide whether an answer is reasonable.

44. ***CHALLENGE*** Estimate.
$6.35 - 15.512 + 8.744 - 4.19 - 72.7 + 25.008$

Spiral Review

Order the integers from least to greatest. (Lesson 3-1)

45. −17, 9, −3, 6, −12 **46.** 14, 23, −18, −32, 0 **47.** 4, −5, 2, −1, −3

Simplify. (Lesson 3-3)

48. $-14 + 7$ **49.** $28 + (-18)$ **50.** $31 + (-50)$ **51.** $-102 + 67$

52. **EOG PREP** The low temperatures over a 6-day period in Fairbanks, Alaska, were −3°F, −10°F, −7°F, −9°F, −8°F, and −8°F. What are the mean, median, and mode of these temperatures? (Lesson 1-2)

A −7.5°F, −9°F, −8°F C 7.5°F, −8°F, −8°F

B −7.5°F, −8°F, −8°F D −7.5°F, −8°F, 8°F

53. **EOG PREP** Which is a solution to the equation $\frac{n}{-3} = 12$? (Lesson 3-6)

A −36 B −4 C 4 D 36

4-2 Adding and Subtracting Decimals

Learn to add and subtract decimals.

One of the coolest summers on record in the Midwest was in 1992. The average summertime temperature that year was 66.8°F. Normally, the average temperature is 4°F higher than it was in 1992.

To find the normal average summertime temperature in the Midwest, you can add 66.8°F and 4°F.

$$\begin{array}{r} 66.8 \\ +\ 4.0 \\ \hline 70.8 \end{array}$$

Use zero as a placeholder so that both numbers have the same number of digits after their decimal points.

Add each column just as you would add integers.

Line up the decimal points.

The normal average summertime temperature in the Midwest is 70.8°F.

EXAMPLE 1 Adding Decimals

Add. Estimate to check whether each answer is reasonable.

A 3.62 + 18.57

$$\begin{array}{r} 3.62 \\ +\ 18.57 \\ \hline 22.19 \end{array}$$

Line up the decimal points.

Add.

Estimate

4 + 19 = 23 *22.19 is a reasonable answer.*

B 9 + 3.245

$$\begin{array}{r} 9.000 \\ +\ 3.245 \\ \hline 12.245 \end{array}$$

Use zeros as placeholders.

Line up the decimal points.

Add.

Estimate

9 + 3 = 12 *12.245 is a reasonable answer.*

Remember!

When adding numbers with the same sign, find the sum of their absolute values. Then use the sign of the numbers.

Add. Estimate to check whether each answer is reasonable.

C $-5.78 + (-18.3)$

$-5.78 + (-18.3)$	*Think: 5.78 + 18.3.*
5.78	*Line up the decimal points.*
$+18.30$	*Use zero as a placeholder.*
24.08	*Add.*
$-5.78 + (-18.3) = -24.08$	*Use the sign of the two numbers.*

Estimate

$-6 + (-18) = -24$ — *−24.08 is a reasonable answer.*

EXAMPLE 2 Subtracting Decimals

Subtract.

A $12.49 - 7.25$

12.49	*Line up the decimal points.*
$-\ 7.25$	
5.24	*Subtract.*

B $14 - 7.32$

$\overset{13}{\cancel{14}}.\overset{9}{\cancel{0}}\overset{10}{\cancel{0}}$	*Use zeros as placeholders.*
$-\ 7.32$	*Line up the decimal points.*
6.68	*Subtract.*

EXAMPLE 3 *Transportation Application*

During one month in the United States, 492.23 million commuter trips were taken on buses, and 26.331 million commuter trips were taken on light rail. How many more trips were taken on buses than on light rail? Estimate to check whether your answer is reasonable.

492.230	*Use zero as a placeholder.*
$-\ 26.331$	*Line up the decimal points.*
465.899	*Subtract.*

Estimate

$490 - 30 = 460$ — *465.899 is a reasonable answer.*

465,899,000 more trips were taken on buses than on light rail.

Think and Discuss

1. Tell whether the addition is correct. If it is not, explain why not.

$$\begin{array}{r} 12.3 \\ +\ 4.68 \\ \hline 5.91 \end{array}$$

2. Describe how you can check an answer when adding and subtracting decimals.

4-2 Exercises

FOR EOG PRACTICE
see page 670

internet connect
Homework Help Online
go.hrw.com Keyword: MS4 4-2

1.02, 1.02c, 1.03

GUIDED PRACTICE

See Example 1 **Add. Estimate to check whether each answer is reasonable.**

1. 5.37 + 16.45 **2.** −5.62 + (−12.9) **3.** 7 + 5.826

See Example 2 **Subtract.**

4. 7.89 − 5.91 **5.** 18.31 − 8.66 **6.** 4.97 − 3.2

See Example 3 **7.** In 1990, international visitors to the United States spent $58.3 billion. In 1999, international visitors spent $95.5 billion. By how much did spending by international visitors increase from 1990 to 1999?

INDEPENDENT PRACTICE

See Example 1 **Add. Estimate to check whether each answer is reasonable.**

8. 7.82 + 31.23 **9.** 5.98 + 12.99 **10.** 4.917 + 12

11. 6 + 9.33 **12.** −3.29 + (−12.6) **13.** −9.82 + (−15.7)

See Example 2 **Subtract.**

14. 5.45 − 3.21 **15.** 12.87 − 3.86 **16.** 15.39 − 2.6

17. 5 − 0.53 **18.** 14 − 8.9 **19.** 41 − 9.85

See Example 3 **20.** Angela runs her first lap around the track in 4.35 minutes and her second lap in 3.9 minutes. What is her total time for the two laps?

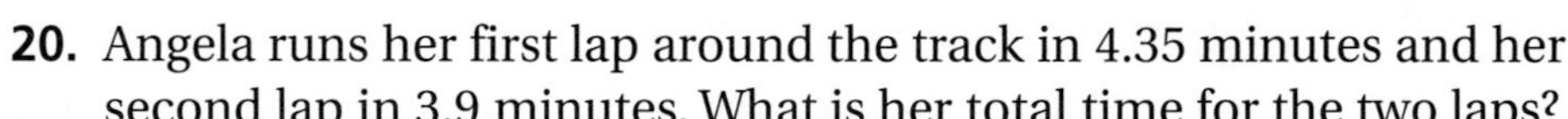

21. A jeweler has 122.83 grams of silver. He uses 45.7 grams of the silver to make a necklace and earrings. How much silver does he have left?

PRACTICE AND PROBLEM SOLVING

Add or subtract. Estimate to check whether each answer is reasonable.

22. −7.238 + 6.9 **23.** 4.16 − 9.043

24. 5.23 − (−9.1) **25.** −123 − 2.55

26. 32.6 − (−15.86) **27.** −32.7 + 62.82

28. 1.99 + 4.8 + 12.9 + 8.532 **29.** 219.7 + 43.92 + 7.482 + 390.8

30. 5.9 − 10 + 2.84 **31.** −8.3 + 5.38 − 0.537

32. ***PHYSICAL EDUCATION*** The students at Union Middle School are trying to run a total of 2,462 mi, which is the distance from Los Angeles to New York City. So far, the sixth grade has run 273.5 mi, the seventh grade has run 275.8 mi, and the eighth grade has run 270.2 mi. How many more miles do the students need to run to reach their goal?

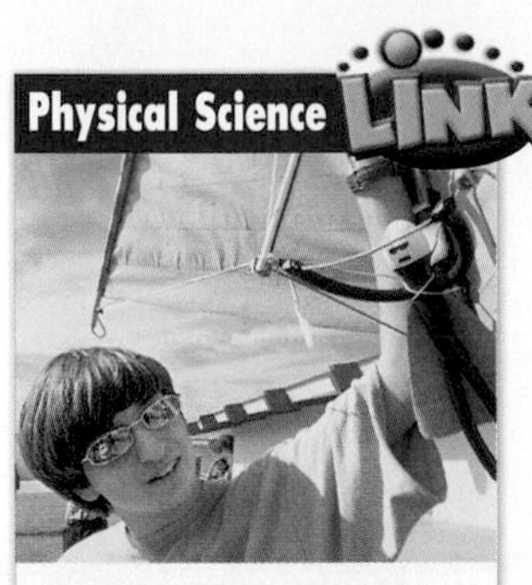

Egg-drop competitions challenge students to build devices that will protect eggs when they are dropped from as high as 100 ft.

33. ***PHYSICAL SCIENCE*** To float in water, an object must have a density of less than 1 gram per milliliter. The density of a fresh egg is about 1.2 grams per milliliter. If the density of a spoiled egg is about 0.3 grams per milliliter less than that of a fresh egg, what is the density of a spoiled egg? How can you use water to tell if an egg is spoiled?

34. ***WEATHER*** The graph shows the five coolest summers recorded in the Midwest. The average summertime temperature in the Midwest is 70.8°F.

   a. How much warmer was the average summertime temperature in 1950 than in 1915?

   b. In what year was the temperature 4.4°F cooler than the average summertime temperature in the Midwest?

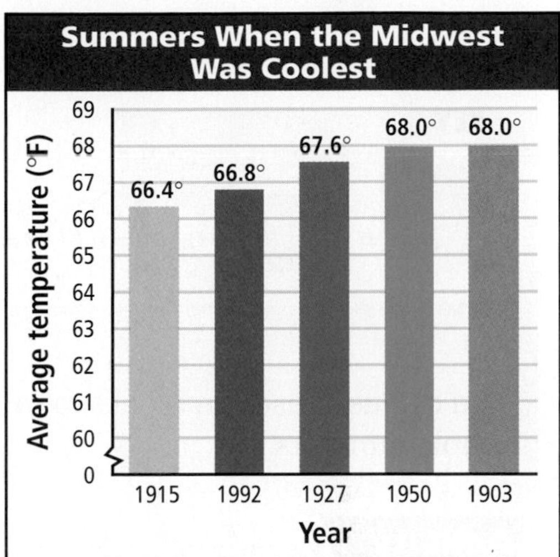

Source: Midwestern Regional Climate Center

35. ***CHOOSE A STRATEGY*** How much larger in area is Agua Fria than Pompeys Pillar?

   A 6.6 thousand acres

   B 71.049 thousand acres

   C 70.59 thousand acres

   D 20.1 thousand acres

National Monument	Area (thousand acres)
Agua Fria	71.1
Pompeys Pillar	0.051

36. ***WRITE ABOUT IT*** Explain how to find the sum or difference of two decimals.

37. ***CHALLENGE*** Find the missing number.

$$5.11 + 6.9 - 15.3 + \square = 20$$

Spiral Review

Evaluate. (Lesson 3-5)

38. $-5 \cdot 12$ **39.** $-63 \div (-9)$ **40.** $20 \cdot (-7)$ **41.** $18 \div (-3)$

Solve. (Lesson 3-6)

42. $x - 12 = -26$ **43.** $-7 + w = 12$ **44.** $-3y = 21$ **45.** $\frac{p}{6} = -11$

46. **EOG PREP** Which fraction shows $\frac{24}{36}$ in simplest form? (Lesson 3-8)

A $\frac{12}{18}$ B $\frac{4}{9}$ C $\frac{4}{6}$ D $\frac{2}{3}$

47. **EOG PREP** Which decimal is equivalent to $\frac{1}{8}$? (Lesson 3-9)

A 0.125 B 0.8 C 0.18 D 0.0125

Model Decimal Multiplication

Use with Lesson 4-3

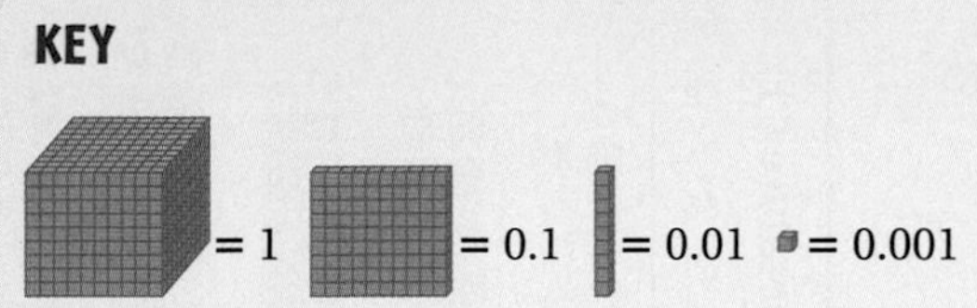

You can use base-ten blocks to model multiplying decimals by whole numbers.

Activity 1

Use base-ten blocks to find $3 \cdot 0.1$.

Multiplication is repeated addition, so $3 \cdot 0.1 = 0.1 + 0.1 + 0.1$.

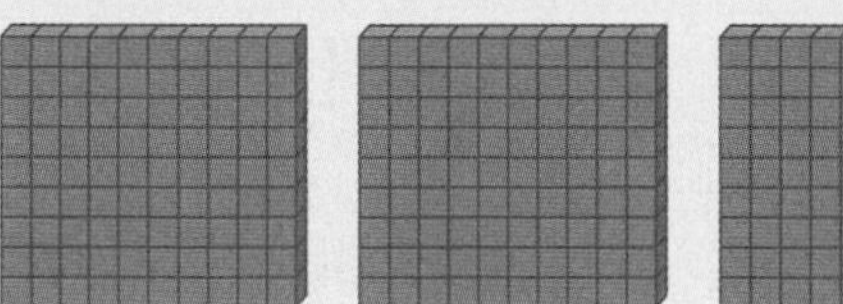

$3 \cdot 0.1 = 0.3$

1 Use base-ten blocks to model and evaluate each expression.

a. $2 \cdot 0.001$ **b.** $0.11 \cdot 3$ **c.** $5 \cdot 0.1$ **d.** $1.1 \cdot 4$

Use base-ten blocks to find $5 \cdot 0.03$.

$5 \cdot 0.03 = 0.03 + 0.03 + 0.03 + 0.03 + 0.03$

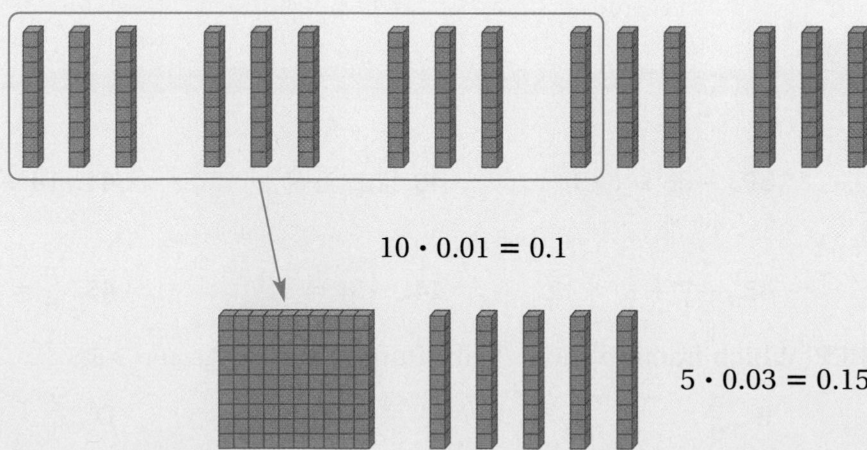

2 Use base-ten blocks to model and evaluate each expression.

a. $5 \cdot 0.2$ **b.** $8 \cdot 0.11$ **c.** $7 \cdot 0.15$ **d.** $6 \cdot 0.12$

Think and Discuss

1. Why can't you use base-ten blocks to model multiplying a decimal by a decimal?

Try This

Use base-ten blocks to model and evaluate each expression.

1. 4 • 0.5 **2.** 2 • 0.04 **3.** 3 • 0.16 **4.** 5 • 0.123

5. 6 • 0.2 **6.** 3 • 0.33 **7.** 0.25 • 5 **8.** 0.42 • 3

You can use decimal grids to model multiplying decimals by decimals.

Activity 2

Use a decimal grid to find 0.4 • 0.7.

Shade **0.4** horizontally.

Shade **0.7** vertically.

The area where the shaded regions overlap is the answer.

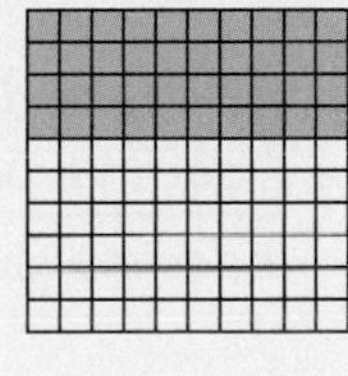 × 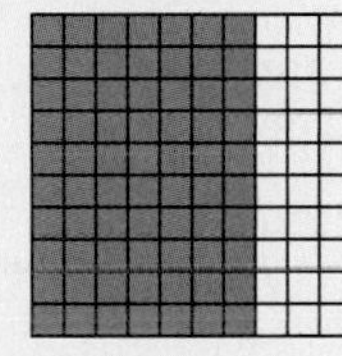 = 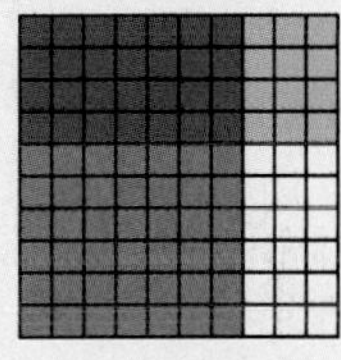

0.4 × 0.7 = 0.28

1 Use decimal grids to model and evaluate each expression.

a. 0.3 • 0.4 **b.** 0.1 • 0.1 **c.** 0.2 • 0.2 **d.** 0.1 • 0.2

Think and Discuss

1. Explain the steps you would take to model 0.5 • 0.5 with a decimal grid.

2. How could you use decimal grids to model multiplying a decimal by a whole number?

Try This

Use a decimal grid to model and evaluate each expression.

1. 0.6 • 0.6 **2.** 0.5 • 0.4 **3.** 0.3 • 0.8 **4.** 0.2 • 0.8

5. 3 • 0.3 **6.** 0.8 • 0.8 **7.** 2 • 0.5 **8.** 0.1 • 0.9

4-3 Multiplying Decimals

Learn to multiply decimals.

You can use decimal grids to model multiplication of decimals. Each large square represents 1. Each row and column represents 0.1. Each small square represents 0.01. The area where the shading overlaps shows the product of the two decimals.

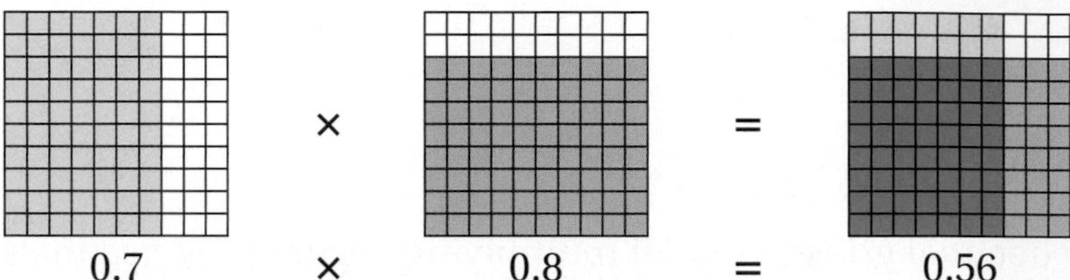

0.7 × 0.8 = 0.56

To multiply decimals, multiply as you would with integers. To place the decimal point in the product, count the number of decimal places in each factor. The product should have the same number of decimal places as the sum of the decimal places in the factors.

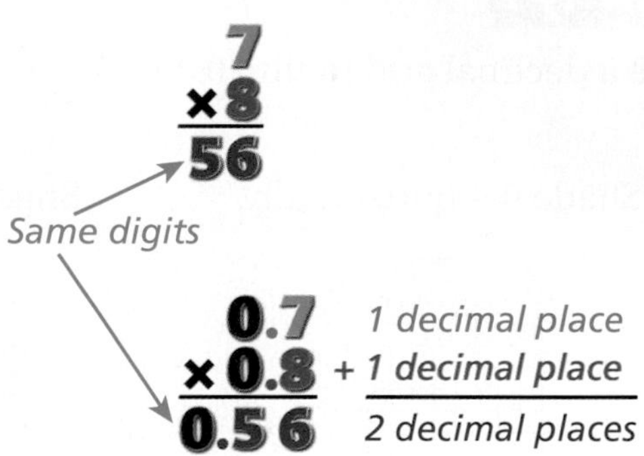

EXAMPLE 1 Multiplying Integers by Decimals

Multiply.

A $6 \cdot 0.1$

6	*0 decimal places*
× 0.1	*1 decimal place*
0.6	*0 + 1 = 1 decimal place*

B $-2 \cdot 0.04$

−2	*0 decimal places*
× 0.04	*2 decimal places*
−0.08	*0 + 2 = 2 decimal places. Use zero as a placeholder.*

C $1.25 \cdot 23$

1.25	*2 decimal places*
× 23	*0 decimal places*
3 75	
+ 25 00	
28.75	*2 + 0 = 2 decimal places*

EXAMPLE 2 Multiplying Decimals by Decimals

Multiply. Estimate to check whether each answer is reasonable.

A $1.2 \cdot 1.6$

$$\begin{array}{r} 1.2 \\ \times\ 1.6 \\ \hline 72 \\ 120 \\ \hline 1.92 \end{array}$$

1 decimal place
1 decimal place
1 + 1 = 2 decimal places

Estimate

$1 \cdot 2 = 2$ — *1.92 is a reasonable answer.*

B $-2.78 \cdot 0.8$

$$\begin{array}{r} -2.78 \\ \times\ 0.8 \\ \hline -2.224 \end{array}$$

2 decimal places
1 decimal place
2 + 1 = 3 decimal places

Estimate

$-3 \cdot 1 = -3$ — *−2.224 is a reasonable answer.*

EXAMPLE 3 *Earth Science Application*

On average, 0.36 kg of carbon dioxide is added to the atmosphere for each mile a single car is driven. How many kilograms of carbon dioxide are added for each mile the 132 million cars in the United States are driven?

$$\begin{array}{r} 132 \\ \times\ 0.36 \\ \hline 792 \\ 3960 \\ \hline 47.52 \end{array}$$

0 decimal places
2 decimal places
0 + 2 = 2 decimal places

Estimate

$130 \cdot 0.5 = 65$ — *47.52 is a reasonable answer.*

Approximately 47.52 million (47,520,000) kilograms of carbon dioxide are added to the atmosphere for each mile driven.

Think and Discuss

1. **Explain** whether the multiplication $2.1 \cdot 3.3 = 69.3$ is correct.
2. **Compare** multiplying integers with multiplying decimals.

4-3 Exercises

FOR EOG PRACTICE

see page 670

internet connect
Homework Help Online
go.hrw.com Keyword: MS4 4-3

1.02, 1.02c, 1.03

GUIDED PRACTICE

See Example 1
Multiply.

1. $-9 \cdot 0.4$
2. $3 \cdot 0.2$
3. $0.06 \cdot 3$

See Example 2
Multiply. Estimate to check whether each answer is reasonable.

4. $1.7 \cdot 1.2$
5. $2.6 \cdot 0.4$
6. $1.5 \cdot (-0.21)$

See Example 3

7. If Carla is able to drive her car 24.03 miles on one gallon of gas, how far could she drive on 13.93 gallons of gas?

INDEPENDENT PRACTICE

See Example 1
Multiply.

8. $8 \cdot 0.6$
9. $5 \cdot 0.07$
10. $-3 \cdot 2.7$
11. $6 \cdot 4.9$
12. $1.7 \cdot (-12)$
13. $43 \cdot 2.11$

See Example 2
Multiply. Estimate to check whether each answer is reasonable.

14. $2.4 \cdot 3.2$
15. $2.8 \cdot 1.6$
16. $5.3 \cdot 4.6$
17. $-5.14 \cdot 0.03$
18. $1.04 \cdot (-8.9)$
19. $4.31 \cdot (-9.5)$

See Example 3

20. Nicholas bicycled 15.8 kilometers each day for 18 days last month. How many kilometers did he bicycle last month?
21. While walking, Lara averaged 3.63 miles per hour. How far did she walk in 1.5 hours?

PRACTICE AND PROBLEM SOLVING

Multiply. Estimate to check whether each answer is reasonable.

22. $-9.6 \cdot 2.05$
23. $0.07 \cdot 0.03$
24. $-1.08 \cdot (-0.4)$
25. $1.46 \cdot (-0.06)$
26. $-325.9 \cdot (1.5)$
27. $14.7 \cdot 0.13$
28. $-7.02 \cdot (-0.05)$
29. $1.104 \cdot (-0.7)$
30. $0.3 \cdot 2.8 \cdot (-10.6)$
31. $1.3 \cdot (-4.2) \cdot (-3.94)$
32. $0.6 \cdot (-0.9) \cdot 0.05$
33. $-6.5 \cdot (-1.02) \cdot (-12.6)$
34. $-22.08 \cdot (-5.6) \cdot 9.9$
35. $-63.75 \cdot 13.46 \cdot 7.8$
36. ***FINANCE*** Wanda earns \$8.95 per hour plus commission. Last week, she worked 32.5 hours and earned \$28.75 in commission. How much did Wanda earn last week?

37. ***RECREATION*** The graph shows the results of a survey about river recreation activities.

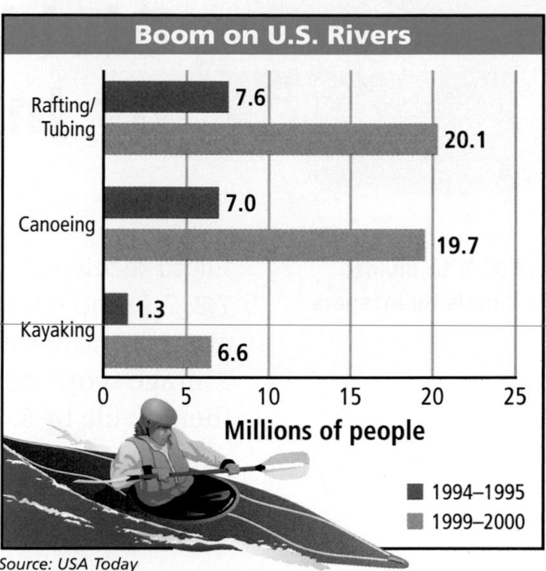

a. About how many people participated in these river recreation activities in 1999–2000?

b. In 1999–2000, almost 3 times as many people reported that they enjoyed canoeing than was reported in 1994–1995. If this trend continues, about how many people will report that they enjoy canoeing in 2004–2005?

38. ***WEATHER*** As a hurricane increases in intensity, the air pressure within its eye decreases. In a Category 5 hurricane, which is the most intense, the air pressure measures approximately 27.16 inches of mercury. In a Category 1 hurricane, which is the least intense, the air pressure is about 1.066 times that of a Category 5 hurricane. What is the air pressure within the eye of a Category 1 hurricane? Round your answer to the nearest hundredth.

39. ***WHAT'S THE QUESTION?*** In a collection, each rock sample has a mass of 4.35 kilograms. There are a dozen rocks in the collection. If the answer is 52.2 kilograms, what is the question?

40. ***WRITE ABOUT IT*** How do the products $4.3 \cdot 0.56$ and $0.43 \cdot 5.6$ compare? Explain.

41. ***CHALLENGE*** Evaluate $(0.2)^5$.

Spiral Review

Evaluate each expression for the given values of the variables. (Lesson 2-7)

42. $a^2 - 6a + 2$ for $a = 7$

43. $6m - 3p - 1$ for $m = 8$ and $p = 4$

Find each quotient. (Lesson 3-5)

44. $-49 \div (-7)$

45. $-75 \div 3$

46. $64 \div (-4)$

47. $-120 \div 5$

48. **EOG PREP** Which number is *less than* -1.8? (Lesson 3-10)

A $-1\frac{24}{25}$ B $-1\frac{8}{15}$ C 0 D 1.6

49. **EOG PREP** Amos has a gift certificate for \$10.00. He wants to buy a book for \$8.95 and a magazine for \$3.50. How much more money does he need? (Lesson 4-2)

A \$12.45 B \$6.50 C \$2.45 D \$1.05

4-4 Dividing Decimals by Integers

Learn to divide decimals by integers.

Elena received scores of 6.85, 6.95, 7.2, 7.1, and 6.9 on the balance beam at a gymnastics meet. To find her average score, add her scores and then divide by 5.

$6.85 + 6.95 + 7.2 + 7.1 + 6.9 = 35$

$35 \div 5 = 7$

Elena's average score was 7, or 7.0.

Notice that the sum of Elena's scores is an integer. But what if the sum is not an integer? You can find the average score by dividing a decimal by a whole number.

Remember!

Division can undo multiplication.
$0.2 \cdot 4 = 0.8$ and
$0.8 \div 4 = 0.2$

$0.8 \div 4$ — *0.8 divided into 4 equal groups.*

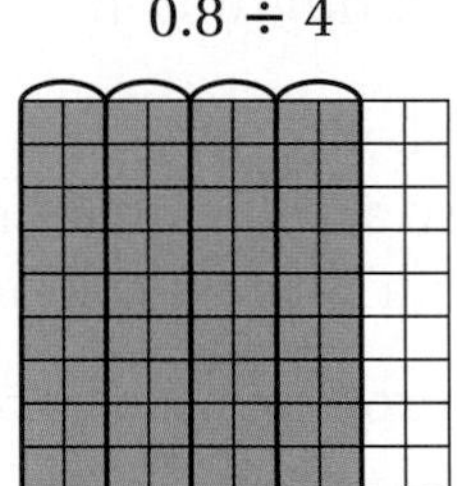

$0.8 \div 4 = 0.2$ — *The size of each group is the answer. Each group is 2 columns, or 0.2.*

EXAMPLE 1 Dividing Decimals by Integers

Divide. Estimate to check whether each answer is reasonable.

A $48.78 \div 6$

```
   8.13
6)48.78
 −48
   0 7
  −6
    18
   −18
     0
```

Place the decimal point for the answer directly above the decimal point under the division symbol.

Divide as with whole numbers.

Estimate

$48 \div 6 = 8$ — *8.13 is a reasonable answer.*

Divide. Estimate to check whether each answer is reasonable.

B 0.18 ÷ 2

$$\begin{array}{r} 0.09 \\ 2\overline{)0.18} \\ -18 \\ \hline 0 \end{array}$$

Place the decimal point for the answer directly above the decimal point under the division symbol.

Estimate

0.2 ÷ 2 = 0.1 *0.09 is a reasonable answer.*

C 71.06 ÷ (−34)

$$\begin{array}{r} 2.09 \\ 34\overline{)71.06} \\ -68 \\ \hline 3\ 06 \\ -3\ 06 \\ \hline 0 \end{array}$$

The signs are different. Think: 71.06 ÷ 34. Place the decimal point for the answer directly above the decimal point under the division symbol.

71.06 ÷ (−34) = −2.09

Estimate

68 ÷ (−34) = −2 *–2.09 is a reasonable answer.*

Remember!

When you divide two numbers with different signs, the answer is negative.

EXAMPLE 2 *Money Application*

For Mrs. Deece's birthday, her class bought her a pendant for $76.50 and a card for $2.25. If there are 25 students in the class, what is the average amount each student owes for the gift?

First find the total cost of the gift. Then divide by the number of students.

76.50 + 2.25 = 78.75 *The gift cost a total of $78.75.*

$$\begin{array}{r} 3.15 \\ 25\overline{)78.75} \\ -75 \\ \hline 3\ 7 \\ -2\ 5 \\ \hline 1\ 25 \\ -1\ 25 \\ \hline 0 \end{array}$$

Place the decimal point for the answer directly above the decimal point under the division symbol.

Each student owes an average of $3.15 for the gift.

Think and Discuss

1. Describe how to place the decimal point in the quotient when you divide a decimal by an integer.

2. Explain how to divide a positive decimal by a negative integer.

4-4 Exercises

FOR EOG PRACTICE
see page 670

1.02, 1.02c, 1.03

GUIDED PRACTICE

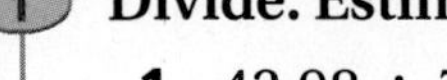

See Example 1 **Divide. Estimate to check whether each answer is reasonable.**

1. $42.98 \div 7$ **2.** $24.48 \div 8$ **3.** $64.89 \div (-21)$

4. $-94.72 \div 37$ **5.** $0.136 \div 8$ **6.** $1.404 \div 6$

See Example 2 **7.** Members of a reading group order books for \$89.10 and bookmarks for \$10.62. If there are 18 people in the reading group, how much does each person owe on average?

INDEPENDENT PRACTICE

See Example 1 **Divide. Estimate to check whether each answer is reasonable.**

8. $12.8 \div 4$ **9.** $80.1 \div (-9)$ **10.** $14.58 \div 3$

11. $-62.44 \div 7$ **12.** $7.2 \div 12$ **13.** $33.6 \div (-7)$

14. $0.108 \div 6$ **15.** $65.28 \div 32$ **16.** $-0.152 \div 8$

17. $21.47 \div 19$ **18.** $0.148 \div 4$ **19.** $79.82 \div (-26)$

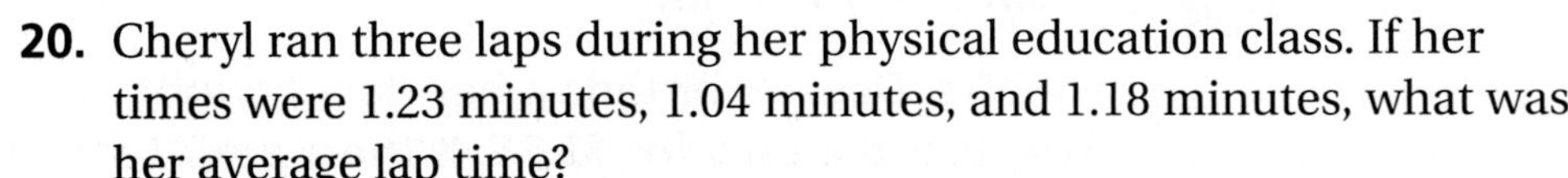

See Example 2 **20.** Cheryl ran three laps during her physical education class. If her times were 1.23 minutes, 1.04 minutes, and 1.18 minutes, what was her average lap time?

21. Randall spent \$61.25 on some CDs and a set of headphones. All of the CDs were on sale for the same price. The set of headphones cost \$12.50. If he bought 5 CDs, what was the sale price of each CD?

22. In qualifying for an auto race, one driver had lap speeds of 195.3 mi/h, 190.456 mi/h, 193.557 mi/h, and 192.575 mi/h. What was the driver's average speed for these four laps?

PRACTICE AND PROBLEM SOLVING

Divide. Estimate to check whether each answer is reasonable.

23. $-9.36 \div (-6)$ **24.** $48.1 \div (-13)$ **25.** $20.95 \div 5$

26. $0.84 \div 12$ **27.** $-39.2 \div 14$ **28.** $9.45 \div (-9)$

29. $47.75 \div (-25)$ **30.** $-94.86 \div (-31)$ **31.** $-0.399 \div 21$

Evaluate.

32. $6.95 \div 5 \cdot 3 - 1.6$ **33.** $0.29 + 18.6 \div 3$

34. $1 - 7.28 \div 4 + 0.9$ **35.** $(19.2 \div 16)^2$

36. $-6.8 \div 4 \cdot (-2.5)$ **37.** $-63.93 \cdot (-12.3) \div (-3)$

38. $-2.7 \div 9 \div 12$ **39.** $-99.25 \div (-5) \cdot 4.7$

40. ***BUSINESS*** A ticket broker bought two dozen concert tickets for \$455.76. To resell the tickets, he will include a service charge of \$3.80 for each ticket. What is the resale price of each ticket?

41. ***RECREATION*** The graph shows the number of visitors to the three most visited U.S. national parks in 2000. What was the average number of visitors to these three parks? Round your answer to the nearest hundredth.

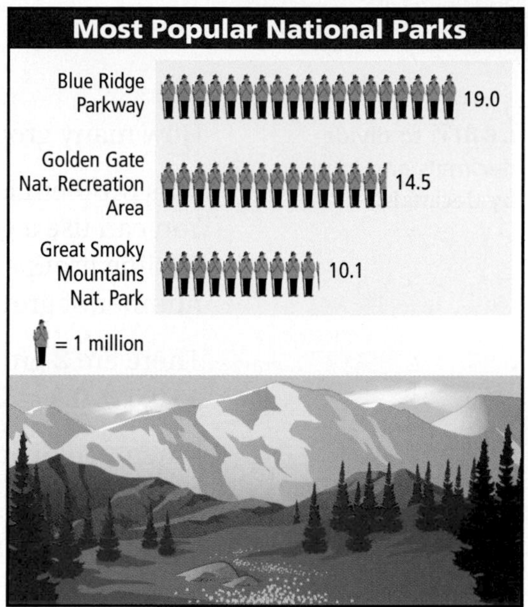

Source: USA Today

42. ***NUTRITION*** On average, each American consumed 261.1 lb of red meat and poultry in 2000. How many pounds of red meat and poultry did the average American eat during each month of 2000? Round your answer to the nearest tenth.

43. Jonathan wants a new mountain bike that costs \$216.99. If Jonathan can save enough money, his parents and grandparents have agreed to split the cost equally three ways with him. How much money must Jonathan save to get the mountain bike?

44. ***WRITE A PROBLEM*** Find some supermarket advertisements. Use the ads to write a problem that can be solved by dividing a decimal by a whole number.

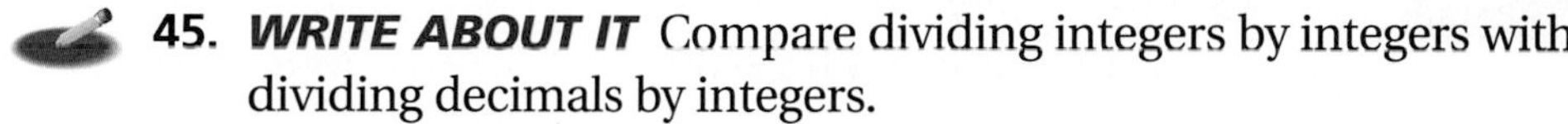

45. ***WRITE ABOUT IT*** Compare dividing integers by integers with dividing decimals by integers.

46. ***CHALLENGE*** Evaluate $0.0016 \div 4 + 0.0009 \div 3 - 0.008 \div 2$.

Spiral Review

Add. (Lesson 3-3)

47. $-15 + 10$ **48.** $8 + (-1)$ **49.** $97 + (-59)$ **50.** $-7 + (-25)$

Subtract. (Lesson 3-4)

51. $-19 - 4$ **52.** $-3 - (-61)$ **53.** $19 - (-12)$ **54.** $4 - 18$

55. **EOG PREP** Which of the following is equivalent to $(-5)^3$? (Lesson 3-5)

A -125 B -15 C 15 D 125

56. **EOG PREP** Which of the following is $0.8, \frac{1}{4}, \frac{1}{2}, 0.3$ in order from least to greatest? (Lesson 3-10)

A $0.8, \frac{1}{4}, \frac{1}{2}, 0.3$ B $\frac{1}{4}, 0.3, \frac{1}{2}, 0.8$ C $\frac{1}{2}, \frac{1}{4}, 0.3, 0.8$ D $\frac{1}{2}, 0.3, \frac{1}{4}, 0.8$

4-5 Dividing Decimals and Integers by Decimals

Learn to divide decimals and integers by decimals.

How many groups of 0.3 are in 0.6?

This problem is equivalent to $0.6 \div 0.3$. You can use a grid to model this division by circling groups of 0.3 and counting the number of groups.

There are 2 groups of 0.3 in 0.6, so $0.6 \div 0.3 = 2$.

When you divide two numbers, you can multiply *both numbers* by the same power of ten without changing the final answer.

Multiply both 0.6 and 0.3 by 10: $\mathbf{0.6 \cdot 10 = 6}$ and $\mathbf{0.3 \cdot 10 = 3}$

$$0.6 \div 0.3 = 2 \quad \text{and} \quad 6 \div 3 = 2$$

By multiplying both numbers by the same power of ten, you can make the divisor an integer. Dividing by an integer is much easier than dividing by a decimal.

EXAMPLE 1 Dividing Decimals by Decimals

Helpful Hint

Multiply both numbers by the least power of ten that will make the divisor an integer.

Divide.

A $\mathbf{4.32 \div 3.6}$

$4.32 \div 3.6 = 43.2 \div 36$ *Multiply both numbers by 10.*

$$\begin{array}{r} 1.2 \\ 36\overline{)43.2} \\ -36 \\ \hline 7\,2 \\ -7\,2 \\ \hline 0 \end{array}$$

Divide as with whole numbers.

B $\mathbf{12.95 \div (-1.25)}$

$12.95 \div (-1.25) = 1295 \div (-125)$ *Multiply both numbers by 100.*

$$\begin{array}{r} 10.36 \\ 125\overline{)1295.00} \\ -125 \\ \hline 45\,0 \\ -37\,5 \\ \hline 7\,50 \\ -7\,50 \\ \hline 0 \end{array}$$

Use zeros as placeholders.
Divide as with whole numbers.

$12.95 \div (-1.25) = -10.36$ *The signs are different.*

EXAMPLE 2 Dividing Integers by Decimals

Divide. Estimate to check whether each answer is reasonable.

A $9 \div 1.25$

$9.00 \div 1.25 = 900 \div 125$ — *Multiply both numbers by 100.*

$$\begin{array}{r} 7.2 \\ 125\overline{)900.0} \\ -875 \\ \hline 25\,0 \\ -25\,0 \\ \hline 0 \end{array}$$

Use zero as a placeholder.
Divide as with whole numbers.

Estimate $9 \div 1 = 9$ — *7.2 is a reasonable answer.*

B $-12 \div (-1.6)$

$-12.0 \div (-1.6) = -120 \div (-16)$ — *Multiply both numbers by 10.*

$$\begin{array}{r} 7.5 \\ 16\overline{)120.0} \\ -112 \\ \hline 8\,0 \\ -8\,0 \\ \hline 0 \end{array}$$

Divide as with whole numbers.

$-12 \div (-1.6) = 7.5$ — *The signs are the same.*

Estimate $-12 \div (-2) = 6$ — *7.5 is a reasonable answer.*

EXAMPLE 3 *Transportation Application*

If Sandy used 15.45 gallons of gas to drive her car 370.8 miles, what was her car's gas mileage?

> **Helpful Hint**
> To calculate gas mileage, divide the number of miles driven by the number of gallons of gas used.

$370.80 \div 15.45 = 37080 \div 1545$ — *Multiply both numbers by 100.*

$$\begin{array}{r} 24 \\ 1545\overline{)37080} \\ -3090 \\ \hline 6180 \\ -6180 \\ \hline 0 \end{array}$$

Divide as with whole numbers.

Sandy's car's gas mileage was 24 miles per gallon.

Think and Discuss

1. Explain whether $4.27 \div 0.7$ is the same as $427 \div 7$.

2. Explain how to divide an integer by a decimal.

4-5 Exercises

FOR EOG PRACTICE

see page 671

internet connect
Homework Help Online
go.hrw.com Keyword: MS4 4-5

1.02, 1.02c, 1.03

GUIDED PRACTICE

See Example 1

Divide.

1. 3.78 ÷ 4.2 **2.** 13.3 ÷ (−0.38) **3.** 14.49 ÷ 3.15

4. 1.06 ÷ 0.2 **5.** −9.76 ÷ 3.05 **6.** 263.16 ÷ (−21.5)

See Example 2

Divide. Estimate to check whether each answer is reasonable.

7. 3 ÷ 1.2 **8.** 84 ÷ 2.4 **9.** 36 ÷ (−2.25)

10. 24 ÷ (−1.2) **11.** −18 ÷ 3.75 **12.** 189 ÷ 8.4

See Example 3

13. Samuel used 14.35 gallons of gas to drive his car 401.8 miles. What was his car's gas mileage?

INDEPENDENT PRACTICE

See Example 1

Divide.

14. 81.27 ÷ 0.03 **15.** −0.408 ÷ 3.4 **16.** 38.5 ÷ (−5.5)

17. −1.12 ÷ 0.08 **18.** 27.82 ÷ 2.6 **19.** 14.7 ÷ 3.5

See Example 2

Divide. Estimate to check whether each answer is reasonable.

20. 35 ÷ (−2.5) **21.** 361 ÷ 7.6 **22.** 63 ÷ (−4.2)

23. 5 ÷ 1.25 **24.** 14 ÷ 2.5 **25.** −78 ÷ 1.6

See Example 3

26. Lonnie used 26.75 gallons of gas to drive his truck 508.25 miles. What was his truck's gas mileage?

27. Mitchell walked 8.5 laps in 20.4 minutes. If he walked each lap at the same pace, how long did it take him to walk one full lap?

PRACTICE AND PROBLEM SOLVING

Divide. Estimate to check whether each answer is reasonable.

28. −24 ÷ 0.32 **29.** 153 ÷ 6.8 **30.** −2.58 ÷ (−4.3)

31. 4.12 ÷ (−10.3) **32.** −17.85 ÷ 17 **33.** 64 ÷ 2.56

Evaluate.

34. 11.5 ÷ 4.6 − 5.8 **35.** 2 · 6.8 ÷ 3.4 + 1.9

36. 12 − 6.4 ÷ 2.56 − 1.2 **37.** $(11.7 \div 2.6 - 0.5)^2$

38. (1.8 + 2.34) ÷ 0.75 **39.** (−7.9 − 12.4) ÷ 3.5

40. 1.6 ÷ 3.2 · 1.6 **41.** 127 ÷ (−12.7) · (−25.32)

42. −24.63 · (−3.9) ÷ 0.03 **43.** 96.3 · 0.3 ÷ (−1.07)

Earth Science LINK

A glacier in Col Ferret, a pass in the Swiss Alps

44. Glaciers form when snow accumulates faster than it melts and thus becomes compacted into ice under the weight of more snow. Once the ice reaches a thickness of about 18 m, it begins to flow. If ice were to accumulate at a rate of 0.0072 m per year, how long would it take to start flowing?

45. An alpine glacier is estimated to be flowing at a rate of 4.75 m per day. At this rate, how long will it take for a marker placed on the glacier by a researcher to move 1,140 m?

46. If the Muir Glacier in Glacier Bay, Alaska, retreats at an average speed of 0.73 m per year, how long will it take to retreat a total of 7.9 m? Round your answer to the nearest year.

47. The table shows the thickness of a glacier as measured at five different points using radar. What is the average thickness of the glacier?

Location	Thickness (m)
A	180.23
B	160.5
C	210.19
D	260
E	200.22

48. The Harvard Glacier in Alaska is advancing at a rate of about 0.055 m per day. At this rate, how long will it take the glacier to advance 20 m? Round your answer to the nearest hundredth.

49. ***CHALLENGE*** Hinman Glacier, on Mount Hinman, in Washington State, had an area of 1.3 km^2 in 1958. The glacier has lost an average of 0.06875 km^2 of area each year. In what year was the total area 0.2 km^2?

go.hrw.com
KEYWORD: MS4 Ice
CNN Student News

Spiral Review

Find the prime factorization of each number. (Lesson 2-4)

50. 54 **51.** 88 **52.** 92 **53.** 225

Multiply. Estimate to check whether each answer is reasonable. (Lesson 4-3)

54. 1.8 · (−0.7) **55.** −3.2 · 1.04 **56.** 4.17 · 36 **57.** −0.09 · (−5.34)

58. EOG PREP Kay swims every 4 days, Nathan every 3 days, and Julia every 6 days. They all swam on May 1. On what day will they all swim again? (Lesson 2-6)

A May 6 C May 13
B May 9 D May 16

59. EOG PREP Jamison owes his sister $15. He pays her back $4 one day and $7 the next day. How much does he still owe his sister? (Lesson 3-3)

A $26 C $8
B $11 D $4

4-6 Solving Equations Containing Decimals

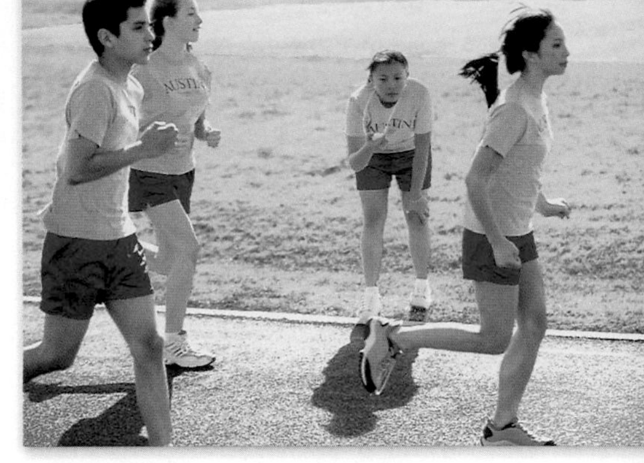

Learn to solve one-step equations that contain decimals.

Students in a physical education class were running 40-yard dashes as part of a fitness test. The slowest time in the class was 3.84 seconds slower than the fastest time of 7.2 seconds.

You can write an equation to represent this situation. The slowest time s minus 3.84 is equal to the fastest time of 7.2 seconds.

$$s - 3.84 = 7.2$$

EXAMPLE 1 Solving Equations by Adding or Subtracting

Remember!

You can solve an equation by performing the same operation on both sides of the equation to isolate the variable.

Solve.

A $\mathbf{s - 3.84 = 7.2}$

$$\begin{aligned} s - 3.84 &= 7.20 \\ \underline{+3.84} &\quad \underline{+\ 3.84} \\ s &= 11.04 \end{aligned}$$

Add to isolate s.

B $\mathbf{y + 20.51 = 26}$

$$\begin{aligned} y + 20.51 &= \overset{5\ 9\ 10}{2\not{6}.\not{0}\not{0}} \\ \underline{-\ 20.51} &\quad \underline{-\ 20.51} \\ y &= 5.49 \end{aligned}$$

Subtract to isolate y.

EXAMPLE 2 Solving Equations by Multiplying or Dividing

Solve.

A $\mathbf{\frac{w}{3.9} = 1.2}$

$$\frac{w}{3.9} = 1.2$$

$$\frac{w}{3.9} \cdot 3.9 = 1.2 \cdot 3.9$$

Multiply to isolate w.

$$w = 4.68$$

B $\mathbf{4 = 1.6c}$

$$4 = 1.6c$$

$$\frac{4}{1.6} = \frac{1.6c}{1.6}$$

Divide to isolate c.

$$\frac{4}{1.6} = c$$

Think: 4 ÷ 1.6 = 40 ÷ 16.

$$2.5 = c$$

EXAMPLE 3 PROBLEM SOLVING APPLICATION

Yancey wants to buy a new snowboard that costs $396.00. If she earns $8.25 per hour at work, how many hours must she work to earn enough money to buy the snowboard?

1 Understand the Problem

Rewrite the question as a statement.

- Find the number of hours Yancey must work to earn $396.00.

List the **important information**:

- Yancey earns $8.25 per hour.
- Yancey needs $396.00 to buy a snowboard.

2 Make a Plan

Yancey's pay is equal to her hourly pay times the number of hours she works. Since you know how much money she needs to earn, you can write an equation with h being the number of hours.

$$8.25h = 396$$

3 Solve

$$8.25h = 396$$

$$\frac{8.25h}{8.25} = \frac{396}{8.25} \qquad \textit{Divide to isolate } h.$$

$$h = 48$$

Yancey must work 48 hours.

4 Look Back

You can round 8.25 to 8 and 396 to 400 to estimate how many hours Yancey needs to work.

$400 \div 8 = 50$

So 48 hours is a reasonable answer.

Think and Discuss

1. **Describe** how to solve the equation $-1.25 + x = 1.25$. Then solve.
2. **Explain** how you can tell if 1.01 is a solution of $10s = -10.1$ without solving the equation.

4-6 Exercises

FOR EOG PRACTICE
see page 671

internet connect
Homework Help Online
go.hrw.com Keyword: MS4 4-6

1.02, 1.02d, 1.03, 5.03

GUIDED PRACTICE

See Example 1 **Solve.**

1. $w - 5.8 = 1.2$ **2.** $x + 9.15 = 17$

3. $k + 3.91 = 28$ **4.** $n - 1.35 = 19.9$

See Example 2

5. $\frac{b}{1.4} = 3.6$ **6.** $\frac{x}{0.8} = 7.2$

7. $3.1t = 27.9$ **8.** $7.5 = 5y$

See Example 3

9. Jeff bought a sandwich and a salad for lunch. His total bill was $7.10. The salad cost $2.85. How much did the sandwich cost?

INDEPENDENT PRACTICE

See Example 1 **Solve.**

10. $v + 0.84 = 6$ **11.** $c - 32.56 = 12$ **12.** $d - 14.25 = -23.9$

13. $3.52 + a = 8.6$ **14.** $w - 9.01 = 12.6$ **15.** $p + 30.34 = -22.87$

See Example 2

16. $3.2c = 8$ **17.** $72 = 4.5z$ **18.** $21.8x = -124.26$

19. $\frac{w}{2.8} = 4.2$ **20.** $\frac{m}{0.19} = 12$ **21.** $\frac{a}{21.23} = -3.5$

See Example 3

22. At the fair, a pack of 25 food tickets costs $31.25. What is the cost of each ticket?

23. To climb the rock wall at the fair, you must have 5 ride tickets. If each ticket costs $1.50, how much does it cost to climb the rock wall?

PRACTICE AND PROBLEM SOLVING

Solve.

24. $1.2y = -1.44$ **25.** $\frac{n}{8.2} = -0.6$

26. $w - 4.1 = -5$ **27.** $r + 0.48 = 1.2$

28. $x - 5.2 = -7.3$ **29.** $1.05 = -7m$

30. $a + 0.81 = -6.3$ **31.** $60k = 54$

32. $\frac{h}{-7.1} = 0.62$ **33.** $\frac{t}{-0.18} = -5.2$

34. $7.9 = d + 12.7$ **35.** $-1.8 + v = -3.8$

36. $-k = 287.658$ **37.** $-n = -12.254$

38. The Drama Club at Smith Valley Middle School is selling cookie dough in order to raise money for costumes. If each tub of cookie dough costs $4.75, how many tubs must members sell to make $570.00?

Social Studies LINK

From 1892 to 1924, more than 22 million immigrants came to Ellis Island, New York.

go.hrw.com
KEYWORD: MS4 Ellis

39. ***SOCIAL STUDIES*** The table shows the most common European ancestral origins of Americans (in millions), according to a Census 2000 supplementary survey. In addition, 19.6 million people stated that their ancestry was "American."

Ancestral Origins of Americans	
European Ancestry	**Number (millions)**
English	28.3
French	9.8
German	46.5
Irish	33.1
Italian	15.9
Polish	9.1
Scottish	5.4

a. How many people claimed ancestry from the countries listed, according to the survey?

b. If the data were placed in order from greatest to least, between which two nationalities would "American" ancestry be placed?

40. ***CONSUMER MATH*** Gregory bought a computer desk at a thrift store for \$38. The regular price of a similar desk at a furniture store is 4.5 times as much. What is the regular price of the desk at the furniture store?

41. ***SOCIAL STUDIES*** Pennies made before 1982 are made mostly of copper and have a density of 8.85 g/cm^3. Because of an increase in the cost of copper, the density of pennies made after 1982 is 1.71 g/cm^3 less. What is the density of pennies minted today?

42. ***WHAT'S THE ERROR?*** A student's solution to the equation $m + 0.63 = 5$ was $m = 5.63$. What is the error? What is the correct solution?

43. ***WRITE A PROBLEM*** Using the fact that 2.54 cm = 1 in., write a problem, an equation to represent the problem, and the solution.

44. ***CHALLENGE*** Solve the equation $-2.8 + (b - 1.7) = -0.6 \cdot 9.4$.

Spiral Review

Evaluate the expression $y^2 - 3$ for each given value of the variable. (Lesson 2-7)

45. $y = 3$ **46.** $y = 7$ **47.** $y = 12$ **48.** $y = 20$

Order the numbers from least to greatest. (Lesson 3-10)

49. $\frac{3}{5}$, 3.5, −3 **50.** −0.9, −6, $-\frac{1}{2}$ **51.** 1.4, −2, $-1\frac{1}{4}$

52.  **EOG PREP** Which of the following is *not* equal to −12? (Lesson 3-3)

A −7 + 5 B −8 + (−4) C −6 + (− 6) D −18 + 6

53. **EOG PREP** Mandy earned \$79.90 for working 8.5 hours. How much did she earn per hour? (Lesson 4-5)

A \$9.40 B \$9.04 C \$8.90 D \$8.09

Chapter 4

Mid-Chapter Quiz

LESSON 4-1 (pp. 192–195)

Estimate.

1. 163.2 · 5.4 **2.** 37.19 + 100.94 **3.** 376.82 − 139.28 **4.** 33.19 ÷ 8.18

5. −6.66 · 5.17 **6.** 67.78 + 85.76 **7.** −65.63 − 24.12 **8.** 29.05 ÷ 3.73

LESSON 4-2 (pp. 196–199)

Add or subtract.

9. 4.73 + 29.68 **10.** −6.89 − (−29.4) **11.** 23.58 − 8.36 **12.** −15 + (−9.44)

13. 66.84 + (−75.13) **14.** −6.48 + 2.3 **15.** −5.21 − 3.64 **16.** −81.6 − (−17.02)

LESSON 4-3 (pp. 202–205)

Multiply.

17. 3.4 · 9.6 **18.** −2.66 · 0.9 **19.** −7 · (−0.06) **20.** 6.94 · (−24)

21. −49.7 · (−9.6) **22.** 7.55 · (−31.21) **23.** −35.3 · (−8.6) **24.** 67.1 · 7.35

LESSON 4-4 (pp. 206–209)

Divide.

25. 10.8 ÷ (−4) **26.** 6.5 ÷ 2 **27.** −45.6 ÷ 12 **28.** −99.36 ÷ (−4)

29. 31.08 ÷ (−8) **30.** 12.5 ÷ 5 **31.** −52.6 ÷ 8 **32.** 48.01 ÷ 2

LESSON 4-5 (pp. 210–213)

Divide.

33. 10.4 ÷ (−0.8) **34.** 18 ÷ 2.4 **35.** −3.3 ÷ 0.11 **36.** −36 ÷ (−0.9)

37. 55 ÷ 12.5 **38.** 46.134 ÷ (−6.6) **39.** −130 ÷ 1.6 **40.** −126.45 ÷ (−4.5)

LESSON 4-6 (pp. 214–217)

Solve.

41. $3.4 + n = 8$ **42.** $x - 1.75 = -19$ **43.** $-3.5 = -5x$ **44.** $10.1 = \frac{s}{8}$

45. Pablo earns $5.50 per hour. His friend Raymond earns 1.2 times as much. How much does Raymond earn per hour?

Focus on Problem Solving

Look Back

- **Does your solution answer the question in the problem?**

Sometimes, before you solve a problem, you first need to use the given data to find additional information. Any time you find a solution for a problem, you should ask yourself if your solution answers the question being asked, or if it just gives you the information you need to find the final answer.

Read each problem, and determine whether the given solution answers the question in the problem. Explain your answer.

1. At one store, a new CD costs $15.99. At a second store, the same CD costs 0.75 as much. About how much does the second store charge?

   Solution: The second store charges about $12.00.

2. Bobbie is 1.4 feet shorter than her older sister is. If Bobbie's sister is 5.5 feet tall, how tall is Bobbie?

   Solution: Bobbie is 4.1 feet tall.

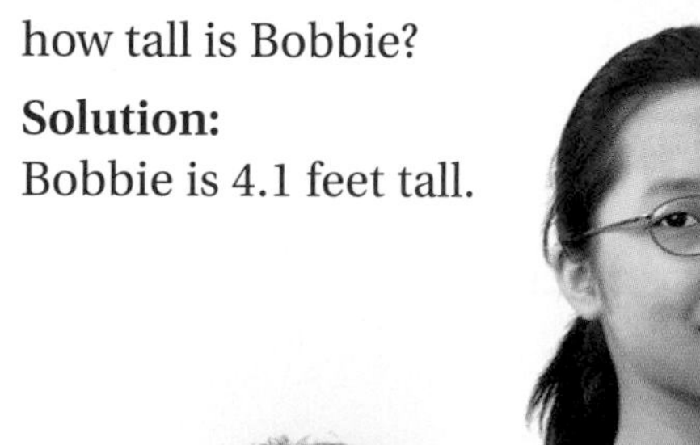

3. Juanita ran the 100-yard dash 1.12 seconds faster than Kellie. Kellie's time was 0.8 seconds faster than Rachel's. If Rachel's time was 15.3 seconds, what was Juanita's time?

   Solution: Kellie's time was 14.5 seconds.

4. The playscape at a local park is located in a triangular sandpit. Side A of the sandpit is 2 meters longer than side B. Side B is twice as long as side C. If the side C is 6 meters long, how long is side A?

   Solution: Side B is 12 meters long.

5. Both Tyrone and Albert walk to and from school every day. Albert has to walk 1.25 miles farther than Tyrone does each way. If Tyrone's house is 0.6 mi from school, how far do the two boys walk altogether?

   Solution: Albert lives 1.85 mi from school.

Model Fraction Multiplication and Division

Use with Lessons 4-7 and 4-8

KEY

REMEMBER

- The multiplication sign can mean "of."

internet connect
Lab Resources Online
go.hrw.com
KEYWORD: MS4 Lab4B

You can use pattern blocks to model multiplying and dividing fractions.

Activity

1 Determine the fraction that each shape represents. Draw a sketch to explain your answer.

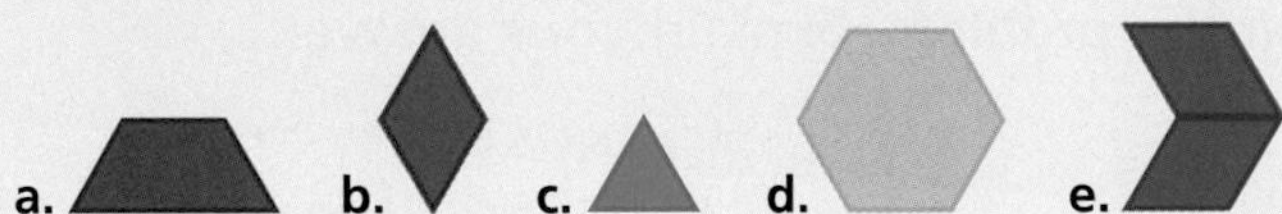

You can use pattern blocks to model multiplying a fraction by a whole number.

To evaluate $\frac{1}{12} \cdot 5$, arrange pattern blocks as shown.

$$\frac{1}{12} \cdot 5 = \frac{5}{12}$$

2 Use pattern blocks to evaluate each expression.

a. $\frac{1}{2} \cdot 2$ **b.** $\frac{1}{6} \cdot 2$ **c.** $\frac{1}{3} \cdot 2$ **d.** $\frac{2}{12} \cdot 3$

When you multiply two fractions that are each less than 1, the product will always be less than either fraction.

To evaluate $\frac{1}{8} \cdot \frac{2}{3}$, arrange pattern blocks as shown.

Show $\frac{2}{3}$.

Find which shape uses 8 blocks to exactly cover $\frac{2}{3}$.

$\frac{1}{8}$ of $\frac{2}{3}$ is $\frac{1}{12}$. $\frac{1}{8} \cdot \frac{2}{3} = \frac{1}{12}$

3 Use pattern blocks to evaluate each expression.

a. $\frac{1}{4} \cdot \frac{1}{3}$ **b.** $\frac{1}{2} \cdot \frac{2}{3}$ **c.** $\frac{1}{3} \cdot \frac{1}{2}$ **d.** $\frac{5}{6} \cdot \frac{1}{2}$

You can use pattern blocks to model dividing fractions.

To evaluate $\frac{1}{2} \div \frac{1}{4}$, arrange pattern blocks as shown.

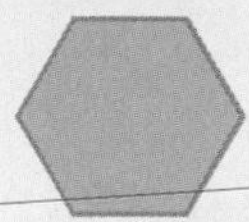

Show $\frac{1}{2}$.

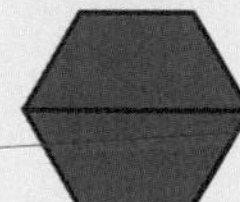

Find the number of $\frac{1}{4}$'s that cover $\frac{1}{2}$ exactly.

$\frac{1}{2} \div \frac{1}{4} = 2$

4 Use pattern blocks to evaluate each expression.

a. $\frac{1}{3} \div \frac{1}{6}$ **b.** $\frac{1}{2} \div \frac{1}{12}$ **c.** $1 \div \frac{1}{4}$ **d.** $\frac{1}{2} \div \frac{1}{6}$

Think and Discuss

1. Are $\frac{1}{4} \cdot \frac{1}{3}$ and $\frac{1}{3} \cdot \frac{1}{4}$ modeled the same way? Explain.

2. If you used [hexagon] to model one whole, what would be the value of [trapezoid]? of [triangle]? of [rhombus]?

3. Use pattern blocks to order the fractions from least to greatest.

$\frac{1}{3}$ $\frac{1}{6}$ $\frac{1}{4}$ $\frac{1}{2}$ $\frac{1}{12}$

What do you notice?

Try This

Model each expression with pattern blocks using

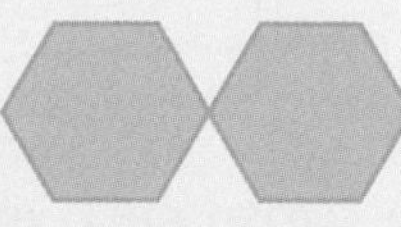

as one whole.

1. $\frac{3}{4} \cdot 1$ **2.** $\frac{1}{2} \cdot \frac{1}{6}$ **3.** $\frac{1}{2} \div \frac{1}{2}$ **4.** $\frac{2}{3} \cdot \frac{3}{6}$

5. $\frac{3}{4} \cdot \frac{1}{3}$ **6.** $2 \div \frac{1}{4}$ **7.** $\frac{1}{4} \cdot \frac{2}{3}$ **8.** $\frac{2}{3} \div \frac{1}{6}$

9. It took $\frac{1}{2}$ hour for your sister to drive to your aunt's house. If she was in traffic for $\frac{1}{3}$ of that time, how long was she in traffic?

4-7 Multiplying Fractions and Mixed Numbers

Learn to multiply fractions and mixed numbers.

The San Francisco–Oakland Bay Bridge, which opened in 1936, is a toll bridge used by drivers traveling between the two cities. In 1977, the toll for a car crossing the bridge was $\frac{3}{8}$ of the toll in 2001. To find the toll in 1977, you will need to multiply the toll in 2001 by a fraction.

EXAMPLE 1 *Transportation Application*

In 2001, the Bay Bridge toll for a car was \$2.00. In 1977, the toll was $\frac{3}{8}$ of the toll in 2001. What was the toll in 1977?

$2 \cdot \frac{3}{8} = \frac{3}{8} + \frac{3}{8}$

$= \frac{6}{8}$

$= \frac{3}{4}$ *Simplify.*

$= 0.75$ *Write the fraction as a decimal.*

$\frac{1}{8}\ \frac{1}{8}\ \frac{1}{8}$ + $\frac{1}{8}\ \frac{1}{8}\ \frac{1}{8}$

$\frac{1}{8}\ \frac{1}{8}\ \frac{1}{8}\ \frac{1}{8}\ \frac{1}{8}\ \frac{1}{8}$

$\frac{1}{4}\ \frac{1}{4}\ \frac{1}{4}$

The Bay Bridge toll for a car was \$0.75 in 1977.

To multiply fractions, multiply the numerators to find the product's numerator. Then multiply the denominators to find the product's denominator.

EXAMPLE 2 Multiplying Fractions

Remember!

You can write any integer as a fraction with a denominator of 1.

Multiply. Write each answer in simplest form.

A $-15 \cdot \frac{2}{3}$

$-15 \cdot \frac{2}{3} = -\frac{15}{1} \cdot \frac{2}{3}$ *Write –15 as a fraction.*

$= -\frac{\overset{5}{\cancel{15}} \cdot 2}{1 \cdot \cancel{3}_{1}}$ *Simplify.*

$= -\frac{10}{1} = -10$ *Multiply numerators. Multiply denominators.*

Helpful Hint

The product of two proper fractions is less than either fraction.

Multiply. Write each answer in simplest form.

B $\frac{1}{4} \cdot \frac{4}{5}$

$\frac{1}{4} \cdot \frac{4}{5} = \frac{1 \cdot \cancel{4}^{1}}{{}_{1}\cancel{4} \cdot 5}$ — *Simplify.*

$= \frac{1}{5}$ — *Multiply numerators. Multiply denominators.*

C $\frac{3}{4} \cdot \left(-\frac{1}{2}\right)$

$\frac{3}{4} \cdot \left(-\frac{1}{2}\right) = -\frac{3 \cdot 1}{4 \cdot 2}$ — *The signs are different, so the answer will be negative.*

$= -\frac{3}{8}$ — *Multiply numerators. Multiply denominators.*

EXAMPLE 3 Multiplying Mixed Numbers

Multiply. Write each answer in simplest form.

A $\frac{1}{3} \cdot 4\frac{1}{2}$

$\frac{1}{3} \cdot 4\frac{1}{2} = \frac{1}{3} \cdot \frac{9}{2}$ — *Write the mixed number as an improper fraction.*

$= \frac{1 \cdot \cancel{9}^{3}}{{}_{1}\cancel{3} \cdot 2}$ — *Simplify.*

$= \frac{3}{2}$ or $1\frac{1}{2}$ — *Multiply numerators. Multiply denominators.*

B $3\frac{3}{5} \cdot 1\frac{1}{12}$

$3\frac{3}{5} \cdot 1\frac{1}{12} = \frac{18}{5} \cdot \frac{13}{12}$ — *Write mixed numbers as improper fractions.*

$= \frac{{}^{3}\cancel{18} \cdot 13}{5 \cdot \cancel{12}_{2}}$ — *Simplify.*

$= \frac{39}{10}$ or $3\frac{9}{10}$ — *Multiply numerators. Multiply denominators.*

C $5\frac{1}{7} \cdot 2\frac{3}{4}$

$5\frac{1}{7} \cdot 2\frac{3}{4} = \frac{36}{7} \cdot \frac{11}{4}$ — *Write mixed numbers as improper fractions.*

$= \frac{{}^{9}\cancel{36} \cdot 11}{7 \cdot \cancel{4}_{1}}$ — *Simplify.*

$= \frac{99}{7}$ or $14\frac{1}{7}$ — *Multiply numerators. Multiply denominators.*

Think and Discuss

1. **Describe** how to multiply a mixed number and a fraction.
2. **Explain** why $\frac{1}{2} \cdot \frac{1}{3} \cdot \frac{1}{4} = \frac{1}{24}$ is or is not correct.
3. **Explain** why you may want to simplify before multiplying $\frac{2}{3} \cdot \frac{3}{4}$. What answer will you get if you don't simplify first?

4-7 Exercises

FOR EOG PRACTICE
see page 672

1.02, 1.02b

GUIDED PRACTICE

See Example 1

1. On average, people spend $\frac{1}{4}$ of the time they sleep in a dream state. If Maxwell slept 10 hours last night, how much time did he spend dreaming? Write your answer in simplest form.

See Example 2

Multiply. Write each answer in simplest form.

2. $-8 \cdot \frac{3}{4}$
3. $\frac{2}{3} \cdot \frac{3}{5}$
4. $\frac{1}{4} \cdot \left(-\frac{2}{3}\right)$

See Example 3

5. $4 \cdot 3\frac{1}{2}$
6. $\frac{4}{9} \cdot 5\frac{2}{5}$
7. $1\frac{1}{2} \cdot 1\frac{5}{9}$

INDEPENDENT PRACTICE

See Example 1

8. Sherry spent 4 hours exercising last week. If $\frac{5}{6}$ of the time was spent jogging, how much time did she spend jogging? Write your answer in simplest form.
9. A cookie recipe calls for $\frac{1}{3}$ tsp of salt for 1 batch. Doreen is making cookies for a school bake sale and wants to bake 5 batches. How much salt does she need? Write your answer in simplest form.

See Example 2

Multiply. Write each answer in simplest form.

10. $5 \cdot \frac{1}{8}$
11. $4 \cdot \frac{1}{8}$
12. $3 \cdot \frac{5}{8}$
13. $6 \cdot \frac{2}{3}$
14. $\frac{2}{5} \cdot \frac{5}{7}$
15. $\frac{3}{8} \cdot \frac{2}{3}$
16. $\frac{1}{2} \cdot \left(-\frac{4}{9}\right)$
17. $-\frac{5}{6} \cdot \frac{2}{3}$

See Example 3

18. $7\frac{1}{2} \cdot 2\frac{2}{5}$
19. $6 \cdot 7\frac{2}{5}$
20. $2\frac{4}{7} \cdot \frac{1}{6}$
21. $2\frac{5}{8} \cdot 6\frac{2}{3}$
22. $\frac{2}{3} \cdot 2\frac{1}{4}$
23. $1\frac{1}{2} \cdot 1\frac{5}{9}$
24. $7 \cdot 5\frac{1}{8}$
25. $3\frac{3}{4} \cdot 2\frac{1}{5}$

PRACTICE AND PROBLEM SOLVING

Multiply. Write each answer in simplest form.

26. $\frac{5}{8} \cdot \frac{4}{5}$
27. $4\frac{3}{7} \cdot \frac{5}{6}$
28. $-\frac{2}{3} \cdot 6$
29. $2 \cdot \frac{1}{6}$
30. $\frac{1}{8} \cdot 5$
31. $-\frac{3}{4} \cdot \frac{2}{9}$
32. $4\frac{2}{3} \cdot 2\frac{4}{7}$
33. $-\frac{4}{9} \cdot \left(-\frac{3}{16}\right)$
34. $3\frac{1}{2} \cdot 5$
35. $\frac{1}{2} \cdot \frac{2}{3} \cdot \frac{3}{5}$
36. $\frac{6}{7} \cdot 5$
37. $1\frac{1}{2} \cdot \frac{3}{5} \cdot \frac{7}{9}$
38. $-\frac{2}{3} \cdot 1\frac{1}{2} \cdot \frac{2}{3}$
39. $\frac{8}{9} \cdot \frac{3}{11} \cdot \frac{33}{40}$
40. $\frac{1}{6} \cdot 6 \cdot 8\frac{2}{3}$
41. $-\frac{8}{9} \cdot \left(-1\frac{1}{8}\right)$

Complete each multiplication sentence.

42. $\frac{1}{2} \cdot \frac{\blacksquare}{8} = \frac{3}{16}$
43. $\frac{\blacksquare}{7} \cdot \frac{2}{3} = \frac{10}{21}$
44. $\frac{2}{3} \cdot \frac{\blacksquare}{4} = \frac{1}{2}$

45. ***PHYSICAL SCIENCE*** The weight of an object on the moon is $\frac{1}{6}$ its weight on Earth. If a bowling ball weighs $12\frac{1}{2}$ pounds on Earth, how much would it weigh on the moon?

46. ***CONSUMER*** In a survey, 200 students were asked what most influenced them to buy their latest CD. The results are shown in the circle graph.

  a. How many students said radio most influenced them?

  b. How many more were influenced by radio than by a music video channel?

  c. How many said a friend or relative influenced them or they heard the CD in a store?

Influences for Buying CDs

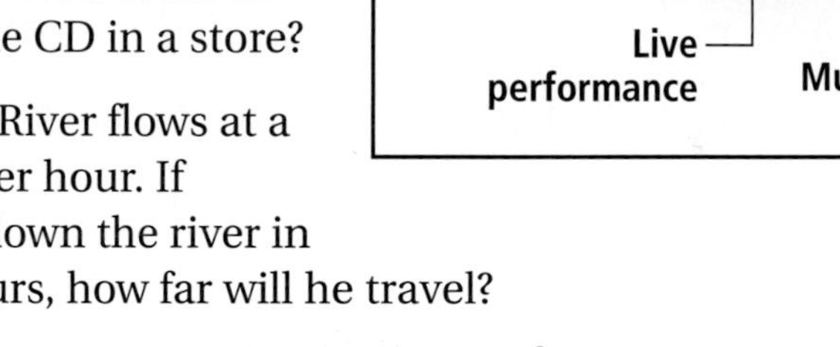

47. The Mississippi River flows at a rate of 2 miles per hour. If Eduardo floats down the river in a boat for $5\frac{2}{3}$ hours, how far will he travel?

48. ***MEASUREMENT*** A standard paper clip is $1\frac{1}{4}$ in. long. If you laid 75 paper clips end to end, how long would the line of paper clips be?

49. ***CHOOSE A STRATEGY*** What is the product of $\frac{1}{2} \cdot \frac{2}{3} \cdot \frac{3}{4} \cdot \frac{4}{5}$?

A $\frac{1}{5}$ **B** 5 **C** $\frac{1}{20}$ **D** $\frac{3}{5}$

50. ***WRITE ABOUT IT*** Explain why the product of two positive proper fractions is always less than either fraction.

51. ***CHALLENGE*** Write three multiplication problems to show that the product of two fractions can be less than, equal to, or greater than 1.

Spiral Review

Write each number in scientific notation. (Lesson 2-2)

52. 54,000 **53.** 3,430,000 **54.** 863

Find each product or quotient. (Lesson 3-5)

55. $3 \cdot (-9)$ **56.** $(-8) \div (-4)$ **57.** $6 \cdot (-2) \cdot (-5)$

58. **EOG PREP** Which is the solution of the equation $x + 3 = -4$? (Lesson 3-6)

A $x = 7$ C $x = -1$

B $x = 1$ D $x = -7$

59. **EOG PREP** Which is the product of 0.8 and 0.02? (Lesson 4-3)

A 0.0016 C 0.16

B 0.016 D 1.6

4-8 Dividing Fractions and Mixed Numbers

Learn to divide fractions and mixed numbers.

Vocabulary
reciprocal

When you divide 8 by 4, you find how many 4's there are in 8. Similarly, when you divide 2 by $\frac{1}{3}$, you find how many $\frac{1}{3}$'s there are in 2.

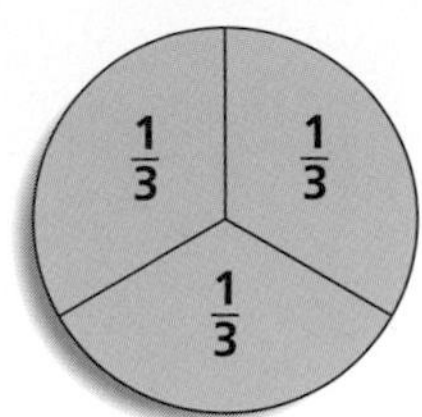

There are six $\frac{1}{3}$'s in 2.

Reciprocals can help you divide by fractions. Two numbers are **reciprocals** if their product is 1. The reciprocal of $\frac{1}{3}$ is 3 because

$\frac{1}{3} \cdot 3 = \frac{1}{3} \cdot \frac{3}{1} = \frac{3}{3} = 1.$

To divide by a fraction, find its reciprocal and then multiply.

$$2 \div \frac{1}{3} = 2 \cdot 3 = 6$$

EXAMPLE 1 Dividing Fractions

Divide. Write each answer in simplest form.

A $\frac{2}{3} \div \frac{1}{5}$

$\frac{2}{3} \div \frac{1}{5} = \frac{2}{3} \cdot \frac{5}{1}$ *Multiply by the reciprocal of $\frac{1}{5}$.*

$= \frac{2 \cdot 5}{3 \cdot 1}$

$= \frac{10}{3}$ or $3\frac{1}{3}$

B $\frac{3}{5} \div 6$

$\frac{3}{5} \div 6 = \frac{3}{5} \cdot \frac{1}{6}$ *Multiply by the reciprocal of 6.*

$= \frac{\overset{1}{\cancel{3}} \cdot 1}{5 \cdot \underset{2}{\cancel{6}}}$ *Simplify.*

$= \frac{1}{10}$

EXAMPLE 2 Dividing Mixed Numbers

Divide. Write each answer in simplest form.

A $4\frac{1}{3} \div 2\frac{1}{2}$

$4\frac{1}{3} \div 2\frac{1}{2} = \frac{13}{3} \div \frac{5}{2}$ *Write mixed numbers as improper fractions.*

$= \frac{13}{3} \cdot \frac{2}{5}$ *Multiply by the reciprocal of $\frac{5}{2}$.*

$= \frac{26}{15}$ or $1\frac{11}{15}$

Divide. Write each answer in simplest form.

B $\frac{5}{6} \div 7\frac{1}{7}$

$\frac{5}{6} \div 7\frac{1}{7} = \frac{5}{6} \div \frac{50}{7}$ *Write $7\frac{1}{7}$ as an improper fraction.*

$= \frac{5}{6} \cdot \frac{7}{50}$ *Multiply by the reciprocal of $\frac{50}{7}$.*

$= \frac{\overset{1}{\cancel{5}} \cdot 7}{6 \cdot \cancel{50}_{10}}$ *Simplify.*

$= \frac{7}{60}$

C $4\frac{4}{5} \div \frac{6}{7}$

$4\frac{4}{5} \div \frac{6}{7} = \frac{24}{5} \div \frac{6}{7}$ *Write $4\frac{4}{5}$ as an improper fraction.*

$= \frac{24}{5} \cdot \frac{7}{6}$ *Multiply by the reciprocal of $\frac{6}{7}$.*

$= \frac{\overset{4}{\cancel{24}} \cdot 7}{5 \cdot \cancel{6}_{1}}$ *Simplify.*

$= \frac{28}{5}$ or $5\frac{3}{5}$

EXAMPLE 3 *Social Studies Application*

Social Studies LINK

The German mark plummeted in value following World War I. By November 1923, a single loaf of bread cost 2,000,000,000 marks. People used the worthless paper money for many unusual purposes, such as building kites.

Use the bar graph to determine how many times longer a \$100 bill is expected to stay in circulation than a \$1 bill.

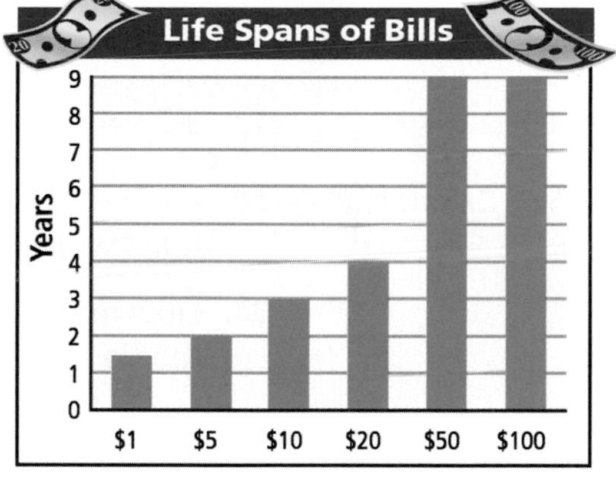

The life span of a \$1 bill is $1\frac{1}{2}$ years. The life span of a \$100 bill is 9 years.

Think: How many $1\frac{1}{2}$'s are there in 9?

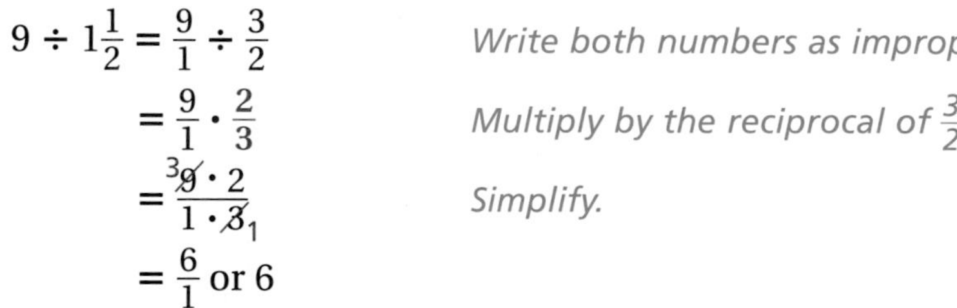

$9 \div 1\frac{1}{2} = \frac{9}{1} \div \frac{3}{2}$ *Write both numbers as improper fractions.*

$= \frac{9}{1} \cdot \frac{2}{3}$ *Multiply by the reciprocal of $\frac{3}{2}$.*

$= \frac{\overset{3}{\cancel{9}} \cdot 2}{1 \cdot \cancel{3}_{1}}$ *Simplify.*

$= \frac{6}{1}$ or 6

A \$100 bill is expected to stay in circulation 6 times longer than a \$1 bill.

Think and Discuss

1. **Explain** whether $\frac{1}{2} \div \frac{2}{3}$ is the same as $2 \cdot \frac{2}{3}$.
2. **Compare** the steps used in multiplying mixed numbers with those used in dividing mixed numbers.

4-8 Exercises

FOR EOG PRACTICE

see page 672

1.02

GUIDED PRACTICE

See Example 1 **Divide. Write each answer in simplest form.**

1. $6 \div \frac{1}{3}$
2. $\frac{3}{5} \div \frac{3}{4}$
3. $\frac{3}{4} \div 8$

See Example 2

4. $\frac{5}{6} \div 3\frac{1}{3}$
5. $5\frac{5}{8} \div 4\frac{1}{2}$
6. $10\frac{4}{5} \div 5\frac{2}{5}$

See Example 3

7. Kareem has $12\frac{1}{2}$ yards of material. A cape for a play takes $3\frac{5}{6}$ yards. How many capes can Kareem make with the material?

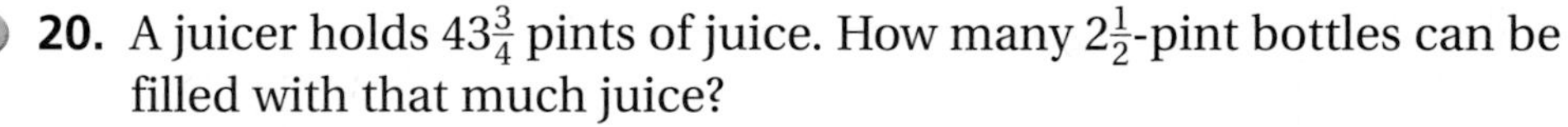

INDEPENDENT PRACTICE

See Example 1 **Divide. Write each answer in simplest form.**

8. $2 \div \frac{7}{8}$
9. $10 \div \frac{5}{9}$
10. $\frac{3}{4} \div \frac{6}{7}$
11. $\frac{8}{9} \div \frac{1}{4}$
12. $\frac{4}{9} \div 12$
13. $\frac{9}{10} \div 6$

See Example 2

14. $\frac{7}{11} \div 4\frac{1}{5}$
15. $\frac{3}{4} \div 2\frac{1}{10}$
16. $22\frac{1}{2} \div 4\frac{2}{7}$
17. $3\frac{5}{7} \div 9\frac{1}{7}$
18. $14\frac{2}{3} \div 1\frac{1}{6}$
19. $7\frac{7}{10} \div 2\frac{2}{5}$

See Example 3

20. A juicer holds $43\frac{3}{4}$ pints of juice. How many $2\frac{1}{2}$-pint bottles can be filled with that much juice?
21. How many $24\frac{1}{2}$ in. pieces of ribbon can be cut from a roll of ribbon that is 147 in. long?

PRACTICE AND PROBLEM SOLVING

Evaluate. Write each answer in simplest form.

22. $9 \div 1\frac{2}{3}$
23. $\frac{2}{3} \div \frac{8}{9}$
24. $-1\frac{7}{11} \div \left(-\frac{9}{11}\right)$
25. $6\frac{2}{3} \div \frac{7}{9}$
26. $\frac{1}{2} \div 4\frac{3}{4}$
27. $\frac{4}{21} \div 3\frac{1}{2}$
28. $4\frac{1}{2} \div 3\frac{1}{2}$
29. $1\frac{3}{5} \div 2\frac{1}{2}$
30. $\frac{7}{8} \div 2\frac{1}{10}$
31. $1\frac{3}{5} \div \left(2\frac{2}{9}\right)$
32. $\left(\frac{1}{2} + \frac{2}{3}\right) \div 1\frac{1}{2}$
33. $\left(2\frac{3}{4} + 3\frac{2}{3}\right) \div \frac{11}{18}$
34. $\frac{4}{5} \cdot \frac{3}{8} \div \frac{9}{10}$
35. $-\frac{12}{13} \cdot \frac{13}{18} \div 1\frac{1}{2}$
36. $\frac{3}{7} \div \frac{15}{28} \div \left(-\frac{4}{5}\right)$
37. Three friends are driving round-trip to an amusement park that is $102\frac{3}{4}$ mi from their town. If each friend drives the same distance, how far does each drive?
38. How many $\frac{1}{4}$ lb hamburger patties can be made from a $10\frac{1}{4}$ lb package and an $11\frac{1}{2}$ lb package of ground meat?

Industrial Arts LINK

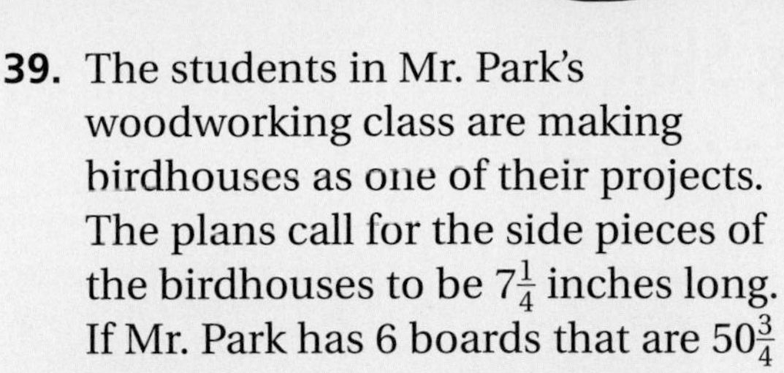

39. The students in Mr. Park's woodworking class are making birdhouses as one of their projects. The plans call for the side pieces of the birdhouses to be $7\frac{1}{4}$ inches long. If Mr. Park has 6 boards that are $50\frac{3}{4}$ inches long, how many side pieces can be cut?

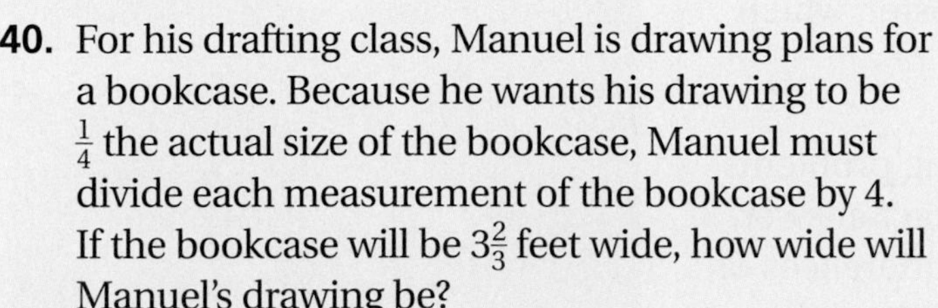

40. For his drafting class, Manuel is drawing plans for a bookcase. Because he wants his drawing to be $\frac{1}{4}$ the actual size of the bookcase, Manuel must divide each measurement of the bookcase by 4. If the bookcase will be $3\frac{2}{3}$ feet wide, how wide will Manuel's drawing be?

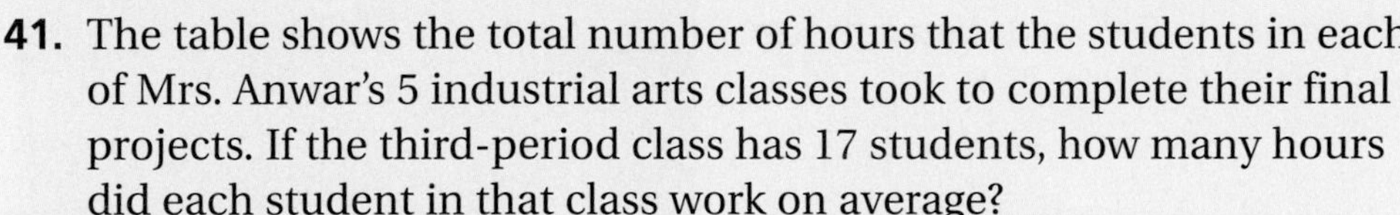

41. The table shows the total number of hours that the students in each of Mrs. Anwar's 5 industrial arts classes took to complete their final projects. If the third-period class has 17 students, how many hours did each student in that class work on average?

Period	Hours
1st	$200\frac{1}{2}$
2nd	$179\frac{2}{5}$
3rd	$199\frac{3}{4}$
5th	$190\frac{3}{4}$
6th	$180\frac{1}{4}$

42. Brandy is stamping circles from a strip of aluminum. If each circle is $1\frac{1}{4}$ inches tall, how many circles can she get from an $8\frac{3}{4}$-inch by $1\frac{1}{4}$-inch strip of aluminum?

43. ***CHALLENGE*** Alexandra is cutting wood stencils to spell her first name with capital letters. Her first step is to cut squares of wood that are $3\frac{1}{2}$ in. long on a side for each letter in her name. Will Alexandra be able to make all of the letters of her name from a single piece of wood that is $7\frac{1}{2}$ in. wide and 18 in. long? Explain your answer.

Spiral Review

Write the prime factorization of each number. (Lesson 2-4)

44. 102 **45.** 320 **46.** 150

47. **EOG PREP** What is the GCF of 18 and 12? (Lesson 2-5)

A 1 B 6 C 36 D 216

48. **EOG PREP** Which of the following is $\frac{1}{2}$, 0.4, -0.8, $-\frac{2}{3}$, and 0.04 ordered from least to greatest? (Lesson 3-10)

A $\frac{1}{2}, 0.4, 0.04, -\frac{2}{3}, -0.8$

B $-0.8, -\frac{2}{3}, 0.4, 0.04, \frac{1}{2}$

C $-\frac{2}{3}, -0.8, 0.04, 0.4, \frac{1}{2}$

D $-0.8, -\frac{2}{3}, 0.04, 0.4, \frac{1}{2}$

4-9 Estimate with Fractions

 Problem Solving Skill

Learn to estimate sums and differences of fractions and mixed numbers.

One of the largest lobsters ever caught was found off the coast of Nova Scotia, Canada, and weighed $44\frac{3}{8}$ lb. About how much heavier was this than an average lobster, which may weigh $3\frac{1}{4}$ lb?

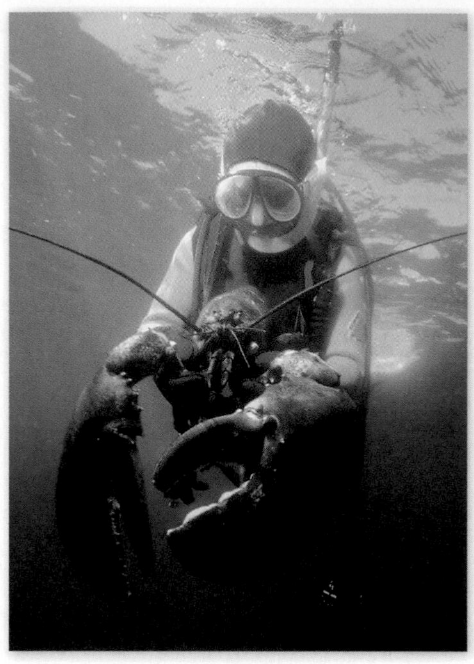

Sometimes, when solving problems, you may not need an exact answer. To estimate sums and differences of fractions and mixed numbers, round each fraction to 0, $\frac{1}{2}$, or 1. You can use a number line to help.

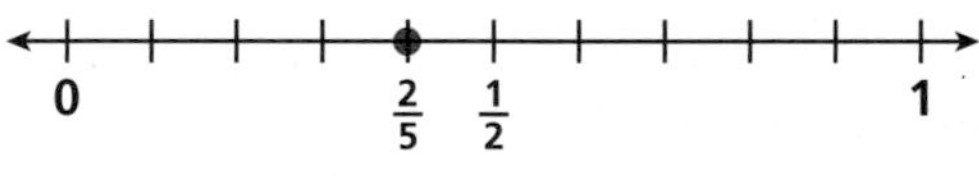

$\frac{2}{5}$ is closer to $\frac{1}{2}$ than to 0.

You can also round fractions by comparing numerators with denominators.

Guidelines for Rounding Fractions		
Round to **0** if the numerator is much smaller than the denominator. Examples: $\frac{1}{9}, \frac{3}{20}, \frac{2}{11}$	Round to $\frac{1}{2}$ if the numerator is about half the denominator. Examples: $\frac{2}{5}, \frac{5}{12}, \frac{7}{13}$	Round to **1** if the numerator is nearly equal to the denominator. Examples: $\frac{8}{9}, \frac{23}{25}, \frac{97}{100}$

EXAMPLE 1 *Measurement Application*

One of the largest lobsters ever caught weighed $44\frac{3}{8}$ lb. Estimate how much more this lobster weighed than an average $3\frac{1}{4}$ lb lobster.

$44\frac{3}{8} - 3\frac{1}{4}$

$44\frac{3}{8} \longrightarrow 44\frac{1}{2}$ $\quad 3\frac{1}{4} \longrightarrow 3\frac{1}{2}$ *Round each mixed number.*

$44\frac{1}{2} - 3\frac{1}{2} = 41$ *Subtract.*

The lobster weighed about 41 lb more than an average lobster.

Remember!

Round $\frac{1}{4}$ to $\frac{1}{2}$ and $\frac{3}{4}$ to 1.

EXAMPLE 2 Estimating Sums and Differences

Estimate each sum or difference.

A $\frac{4}{7} - \frac{13}{16}$

$\frac{4}{7} \longrightarrow \frac{1}{2}$ $\quad \frac{13}{16} \longrightarrow 1$ *Round each fraction.*

$\frac{1}{2} - 1 = -\frac{1}{2}$ *Subtract.*

B $3\frac{3}{8} + 3\frac{2}{9}$

$3\frac{3}{8} \longrightarrow 3\frac{1}{2}$ $\quad 3\frac{2}{9} \longrightarrow 3$ *Round each mixed number.*

$3\frac{1}{2} + 3 = 6\frac{1}{2}$ *Add.*

C $5\frac{7}{8} + \left(-\frac{2}{5}\right)$

$5\frac{7}{8} \longrightarrow 6$ $\quad -\frac{2}{5} \longrightarrow -\frac{1}{2}$ *Round each number.*

$6 + \left(-\frac{1}{2}\right) = 5\frac{1}{2}$ *Add.*

When exact answers are not needed for a problem, you can use estimation. When exact answers are needed, an estimate can help you decide if your answer is reasonable.

EXAMPLE 3 Deciding If an Estimate Is Enough

Decide if you need an exact answer or if an estimate is enough.

A **In 1988, the population of Phoenix, Arizona, was $1\frac{1}{5}$ million and the population of New York City was $7\frac{2}{5}$ million. About how many more people lived in New York City than in Phoenix?**

The question asks *about* how many more people lived in New York City than in Phoenix, so an estimate is enough. The problem does not ask for an exact answer.

B **A concert hall pays a musician $\frac{1}{5}$ of the total amount collected from ticket sales plus a flat fee of $1,000. If 2,532 people attend the concert and pay $15 each, how much money does the musician receive?**

Since the problem is asking for the exact amount of money the musician will receive, an estimate would not be appropriate.

Helpful Hint

Words like *about* and *around* signal that an exact answer is not necessary.

Think and Discuss

1. Demonstrate how to round $\frac{5}{12}$ and $5\frac{1}{5}$.

2. Give an example of a problem using fractions or mixed numbers in which an exact answer is not needed.

4-9 Exercises

1.02c, 1.03

FOR EOG PRACTICE

see page 672

internet connect
Homework Help Online
go.hrw.com Keyword: MS4 4-9

GUIDED PRACTICE

See Example 1

1. The length of a large SUV is $18\frac{9}{10}$ feet, and the length of a small SUV is $15\frac{1}{8}$ feet. Estimate how much longer the large SUV is than the small SUV.

See Example 2

Estimate each sum or difference.

2. $\frac{5}{6} + \frac{5}{12}$
3. $\frac{15}{16} - \frac{4}{5}$
4. $2\frac{1}{6} + 3\frac{6}{11}$

See Example 3

Decide if you need an exact answer or if an estimate is enough.

5. If cashews cost \$6.75 per pound, how much does $\frac{3}{4}$ pound of cashews cost?
6. Kevin has $3\frac{3}{4}$ pounds of pecans and $6\frac{2}{3}$ pounds of walnuts. About how many more pounds of walnuts than pecans does Kevin have?

INDEPENDENT PRACTICE

See Example 1

7. Sarah's bedroom is $14\frac{5}{6}$ feet long and $12\frac{1}{4}$ feet wide. Estimate the difference between the length and width of Sarah's bedroom.

See Example 2

Estimate each sum or difference.

8. $\frac{4}{9} + \frac{3}{5}$
9. $2\frac{5}{9} + 1\frac{7}{8}$
10. $8\frac{3}{4} - 6\frac{2}{5}$
11. $\frac{7}{8} - \frac{2}{5}$
12. $15\frac{1}{7} - 10\frac{8}{9}$
13. $8\frac{7}{15} + 2\frac{7}{8}$

See Example 3

Decide if you need an exact answer or if an estimate is enough.

14. Ellen uses about $\frac{3}{4}$ of a bag of bird seed a week to feed her birds. About how many bags of bird seed does she use in one month?
15. Jamil needs $7\frac{1}{2}$ cups of flour for a bread recipe. If he has $3\frac{3}{4}$ cups of flour, how much more does he need?

PRACTICE AND PROBLEM SOLVING

Estimate each sum or difference.

16. $\frac{7}{9} - \frac{3}{8}$
17. $\frac{3}{5} + \frac{6}{7}$
18. $8\frac{11}{20} - 4\frac{9}{11}$
19. $3\frac{7}{8} + \frac{2}{15}$
20. $23\frac{5}{11} - 16\frac{9}{10}$
21. $5\frac{3}{5} - 4\frac{1}{3}$
22. $\frac{3}{8} + 3\frac{5}{7} + 6\frac{7}{8}$
23. $8\frac{4}{5} + 6\frac{1}{12} + 3\frac{2}{5}$
24. $5\frac{6}{13} + 7\frac{1}{7} + 2\frac{7}{8}$
25. $10\frac{1}{3} + \left(-\frac{8}{15}\right)$
26. $7\frac{3}{8} + 8\frac{5}{11}$
27. $14\frac{2}{3} + 1\frac{7}{9} - 11\frac{14}{29}$

28. ***BUSINESS*** October 19, 1987, is known as Black Monday because the stock market fell 508 points. Xerox stock began the day at $\$70\frac{1}{8}$ and finished at $\$56\frac{1}{4}$. Approximately how far did Xerox's stock price fall during the day?

29. ***RECREATION*** While on a camping trip, Monica and Paul hiked $5\frac{3}{8}$ miles on Saturday and $4\frac{9}{10}$ miles on Sunday. Estimate the number of miles Monica and Paul hiked while camping.

30. ***LIFE SCIENCE*** The diagram shows the wingspans of different species of birds.

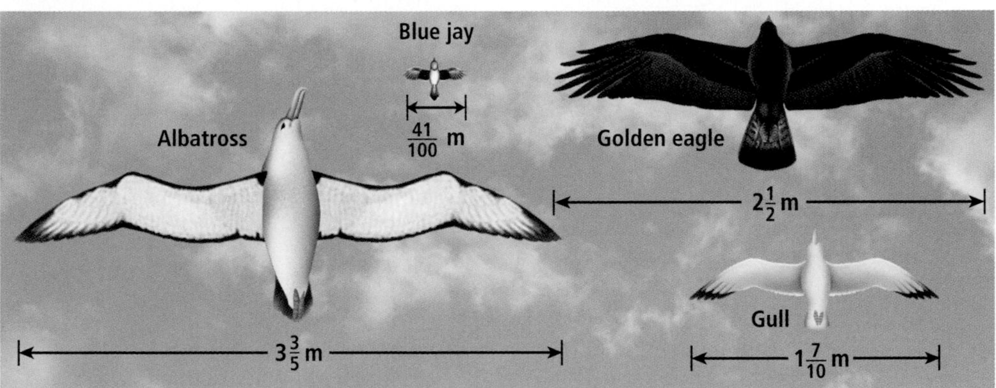

a. Approximately how much longer is the wingspan of an albatross than the wingspan of a gull?

b. Approximately how much longer is the wingspan of a golden eagle than the wingspan of a blue jay?

31. ***WRITE A PROBLEM*** Using mixed numbers, write a problem in which an estimate is enough to solve the problem.

32. ***WRITE ABOUT IT*** Is the estimate of $7\frac{1}{2}$ greater than or less than the actual difference of $12\frac{7}{9} - 5\frac{3}{8}$? Explain.

33. ***CHALLENGE*** Suppose you had bought 10 shares of Xerox stock on October 16, 1987, for $73 per share and sold them at the end of the day on October 19. Approximately how much money would you have lost?

Spiral Review

Combine like terms. (Lesson 2-9)

34. $3x + 5x - x$

35. $5b - 5a + 7a$

36. $5(p + 3) - 4p$

37. $-2y + 6z - 3y$

38. **EOG PREP** Which is the decimal form of $\frac{7}{8}$? (Lesson 3-9)

A 675
B $87.\overline{5}$
C 0.675
D 0.875

39. **EOG PREP** Which is the value of $3\frac{1}{8} \div \frac{5}{8}$? (Lesson 4-8)

A $\frac{64}{125}$
B 5
C $1\frac{61}{64}$
D $\frac{1}{5}$

Model Fraction Addition and Subtraction

Use with Lesson 4-10

internet connect
Lab Resources Online
go.hrw.com
KEYWORD: MS4 Lab4C

Fraction bars can be used to model and evaluate addition and subtraction of fractions.

Activity

You can use fraction bars to evaluate $\frac{3}{8} + \frac{2}{8}$.

Place the fraction bars side by side.

$\frac{1}{8}$	$\frac{1}{8}$	$\frac{1}{8}$	$\frac{1}{8}$	$\frac{1}{8}$

$\frac{3}{8} + \frac{2}{8} = \frac{5}{8}$

1 Use fraction bars to evaluate each expression.

a. $\frac{1}{3} + \frac{1}{3}$ **b.** $\frac{2}{4} + \frac{1}{4}$ **c.** $\frac{3}{12} + \frac{2}{12}$ **d.** $\frac{1}{5} + \frac{2}{5}$

You can use fraction bars to evaluate $\frac{1}{3} + \frac{1}{4}$.

Place the fraction bars side by side. Which kind of fraction bar placed side by side will exactly fill the space below? (*Hint:* What is the LCM of 3 and 4?)

$\frac{1}{3}$				$\frac{1}{4}$		
$\frac{1}{12}$	$\frac{1}{12}$	$\frac{1}{12}$	$\frac{1}{12}$	$\frac{1}{12}$	$\frac{1}{12}$	$\frac{1}{12}$

$\frac{1}{3} + \frac{1}{4} = \frac{7}{12}$

2 Use fraction bars to evaluate each expression.

a. $\frac{1}{2} + \frac{1}{3}$ **b.** $\frac{1}{2} + \frac{1}{4}$ **c.** $\frac{1}{3} + \frac{1}{6}$ **d.** $\frac{1}{4} + \frac{1}{6}$

You can use fraction bars to evaluate $\frac{1}{3} + \frac{5}{6}$.

Place the fraction bars side by side. Which kind of fraction bar placed side by side will exactly fill the space below? (*Hint:* What is the LCM of 3 and 6?)

$\frac{1}{3}$		$\frac{1}{6}$	$\frac{1}{6}$	$\frac{1}{6}$	$\frac{1}{6}$	$\frac{1}{6}$
$\frac{1}{6}$	$\frac{1}{6}$	$\frac{1}{6}$	$\frac{1}{6}$	$\frac{1}{6}$	$\frac{1}{6}$	$\frac{1}{6}$

$\frac{1}{3} + \frac{5}{6} = \frac{7}{6}$

When the sum is an improper fraction, you can use the 1 bar along with fraction bars to find the mixed-number equivalent.

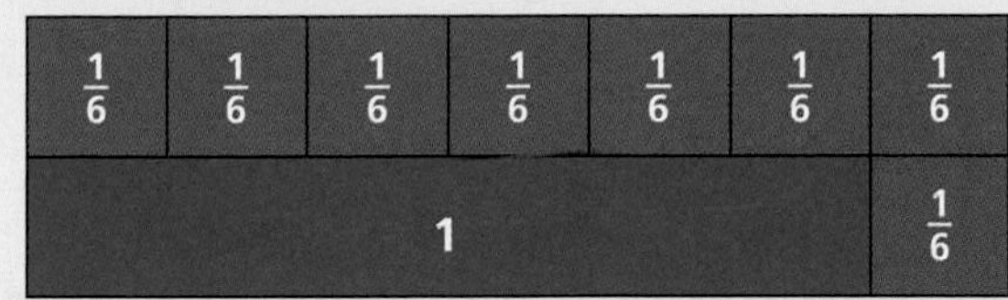

$\frac{7}{6} = 1\frac{1}{6}$

3 Use fraction bars to evaluate each expression.

a. $\frac{3}{4} + \frac{3}{4}$ **b.** $\frac{2}{3} + \frac{1}{2}$ **c.** $\frac{5}{6} + \frac{1}{4}$ **d.** $\frac{3}{8} + \frac{3}{4}$

You can use fraction bars to evaluate $\frac{2}{3} - \frac{1}{2}$.

Place a $\frac{1}{2}$ bar beneath bars that show $\frac{2}{3}$, and find which fraction fills in the remaining space.

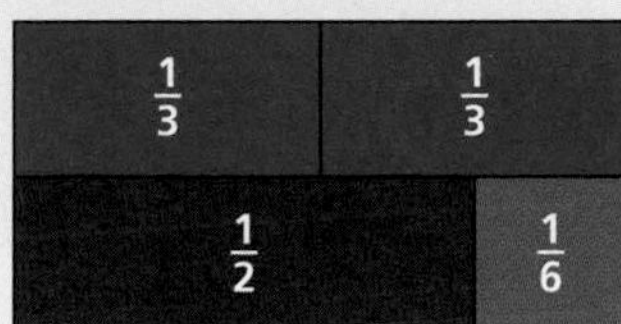

$\frac{2}{3} - \frac{1}{2} = \frac{1}{6}$

4 Use fraction bars to evaluate each expression.

a. $\frac{2}{3} - \frac{1}{3}$ **b.** $\frac{1}{4} - \frac{1}{6}$ **c.** $\frac{1}{2} - \frac{1}{3}$ **d.** $\frac{3}{4} - \frac{2}{3}$

Think and Discuss

1. Model and solve $\frac{3}{4} - \frac{1}{6}$. Explain your steps.
2. Two students solved $\frac{1}{4} + \frac{1}{3}$ in different ways. One got $\frac{7}{12}$ for the answer, and the other got $\frac{2}{7}$. Use models to show which student is correct.
3. Find three different ways to model $\frac{1}{2} + \frac{1}{4}$.

Try This

Use fraction bars to evaluate each expression.

1. $\frac{1}{2} + \frac{1}{2}$
2. $\frac{2}{3} + \frac{1}{6}$
3. $\frac{1}{4} + \frac{1}{6}$
4. $\frac{1}{3} + \frac{7}{12}$
5. $\frac{5}{12} - \frac{1}{3}$
6. $\frac{1}{2} - \frac{1}{4}$
7. $\frac{3}{4} - \frac{1}{6}$
8. $\frac{2}{3} - \frac{1}{4}$
9. You ate $\frac{1}{4}$ of a pizza for lunch and $\frac{5}{8}$ of the pizza for dinner. How much of the pizza did you eat in all?
10. It is $\frac{5}{6}$ mile from your home to the library. After walking $\frac{3}{4}$ mile, you stop to visit a friend on your way to the library. How much farther do you have left to walk to reach the library?

4-10 Adding and Subtracting Fractions

Learn to add and subtract fractions.

From January 1 to March 14 of any given year, Earth completes approximately $\frac{1}{5}$ of its orbit around the Sun, while Venus completes approximately $\frac{1}{3}$ of its orbit. The illustration shows what the positions of the planets would be on March 14 if they started at the same place on January 1 and their orbits were circular. To find out how much more of its orbit Venus completes than Earth, you need to subtract fractions.

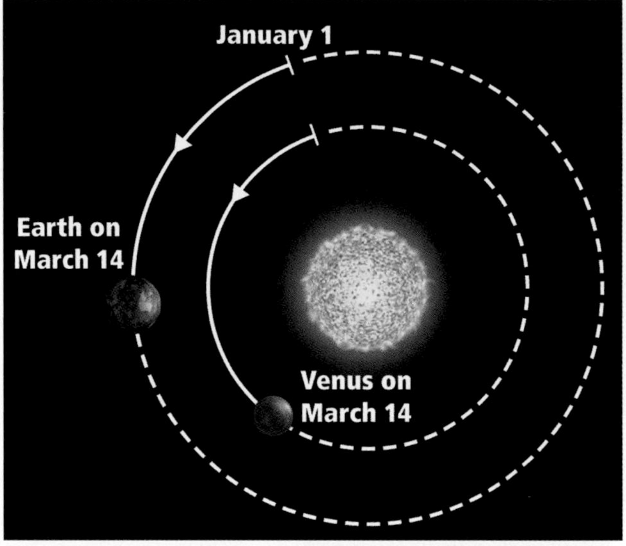

EXAMPLE 1 **Adding and Subtracting Fractions with Like Denominators**

Add or subtract. Write each answer in simplest form.

A $\frac{3}{10} + \frac{1}{10}$

$\frac{3}{10} + \frac{1}{10} = \frac{3 + 1}{10}$ *Add the numerators and keep the common denominator.*

$= \frac{4}{10} = \frac{2}{5}$ *Simplify.*

B $\frac{7}{9} - \frac{4}{9}$

$\frac{7}{9} - \frac{4}{9} = \frac{7 - 4}{9}$ *Subtract the numerators and keep the common denominator.*

$= \frac{3}{9} = \frac{1}{3}$ *Simplify.*

To add or subtract fractions with different denominators, you must rewrite the fractions with a common denominator.

Helpful Hint

The LCM of two denominators is the lowest common denominator (LCD) of the fractions.

Two Ways to Find a Common Denominator
• Find the LCM (least common multiple) of the denominators.
• Multiply the denominators.

EXAMPLE 2 Adding and Subtracting Fractions with Unlike Denominators

Add or subtract. Write each answer in simplest form.

A $\frac{3}{8} + \frac{5}{12}$

$\frac{3}{8} + \frac{5}{12} = \frac{3 \cdot 3}{8 \cdot 3} + \frac{5 \cdot 2}{12 \cdot 2}$ *The LCM of the denominators is 24.*

$= \frac{9}{24} + \frac{10}{24}$ *Write equivalent fractions using the common denominator.*

$= \frac{19}{24}$ *Add.*

B $\frac{1}{10} - \frac{5}{8}$

$\frac{1}{10} - \frac{5}{8} = \frac{1 \cdot 8}{10 \cdot 8} - \frac{5 \cdot 10}{8 \cdot 10}$ *Multiply the denominators.*

$= \frac{8}{80} - \frac{50}{80}$ *Write equivalent fractions using the common denominator.*

$= -\frac{42}{80} = -\frac{21}{40}$ *Subtract. Then simplify.*

C $-\frac{2}{3} + \frac{5}{8}$

$-\frac{2}{3} + \frac{5}{8} = -\frac{2 \cdot 8}{3 \cdot 8} + \frac{5 \cdot 3}{8 \cdot 3}$ *Multiply the denominators.*

$= -\frac{16}{24} + \frac{15}{24}$ *Write equivalent fractions using the common denominator.*

$= -\frac{1}{24}$ *Add.*

EXAMPLE 3 *Astronomy Application*

From January 1 to March 14, Earth completes about $\frac{1}{5}$ of its orbit, while Venus completes about $\frac{1}{3}$ of its orbit. How much more of its orbit does Venus complete than Earth?

$\frac{1}{3} - \frac{1}{5} = \frac{1 \cdot 5}{3 \cdot 5} - \frac{1 \cdot 3}{5 \cdot 3}$ *The LCM of the denominators is 15.*

$= \frac{5}{15} - \frac{3}{15}$ *Write equivalent fractions.*

$= \frac{2}{15}$ *Subtract.*

Venus completes $\frac{2}{15}$ more of its orbit than Earth does.

Think and Discuss

1. **Describe** the process for subtracting fractions with different denominators.
2. **Explain** whether $\frac{3}{4} + \frac{2}{3} = \frac{5}{7}$ is correct.

4-10 Exercises

FOR EOG PRACTICE
see page 673

internet connect
Homework Help Online
go.hrw.com Keyword: MS4 4-10

1.02

GUIDED PRACTICE

See Example 1 **Add or subtract. Write each answer in simplest form.**

1. $\frac{2}{3} - \frac{1}{3}$
2. $\frac{1}{12} + \frac{1}{12}$
3. $\frac{16}{21} - \frac{7}{21}$

See Example 2
4. $\frac{1}{6} + \frac{1}{3}$
5. $\frac{9}{10} - \frac{3}{4}$
6. $\frac{2}{3} + \frac{1}{8}$

See Example 3
7. Parker spends $\frac{1}{4}$ of his earnings on rent and $\frac{1}{6}$ on entertainment. How much more of his earnings does Parker spend on rent than on entertainment?

INDEPENDENT PRACTICE

See Example 1 **Add or subtract. Write each answer in simplest form.**

8. $\frac{2}{3} + \frac{1}{3}$
9. $\frac{3}{20} + \frac{7}{20}$
10. $\frac{5}{8} + \frac{7}{8}$
11. $\frac{7}{12} - \frac{5}{12}$
12. $\frac{5}{6} - \frac{1}{6}$
13. $\frac{8}{9} - \frac{5}{9}$

See Example 2
14. $\frac{1}{5} + \frac{2}{3}$
15. $\frac{1}{6} + \frac{1}{12}$
16. $\frac{5}{6} + \frac{3}{4}$
17. $\frac{21}{24} - \frac{1}{2}$
18. $\frac{3}{4} - \frac{11}{12}$
19. $\frac{1}{2} - \frac{2}{7}$

See Example 3
20. Seana picked $\frac{3}{4}$ quart of blackberries. She ate $\frac{1}{12}$ quart. How much was left?
21. Armando lives $\frac{2}{3}$ mi from his school. If he has walked $\frac{1}{2}$ mi already this morning, how much farther must he walk to get to his school?

PRACTICE AND PROBLEM SOLVING

Find each sum or difference. Write your answer in simplest form.

22. $\frac{4}{5} + \frac{6}{7}$
23. $\frac{5}{6} - \frac{1}{9}$
24. $\frac{1}{2} - \frac{3}{4}$
25. $\frac{5}{7} + \frac{1}{3}$
26. $\frac{1}{2} - \frac{7}{12}$
27. $\frac{3}{4} + \frac{2}{5}$
28. $\frac{7}{8} + \frac{2}{3} + \frac{5}{6}$
29. $\frac{3}{5} + \frac{1}{10} - \frac{3}{4}$
30. $\frac{3}{10} + \frac{5}{8} + \frac{1}{5}$
31. $-\frac{1}{2} + \frac{3}{8} + \frac{2}{7}$
32. $\frac{1}{3} + \frac{3}{7} - \frac{1}{9}$
33. $\frac{2}{9} - \frac{7}{18} + \frac{1}{6}$
34. $\frac{9}{35} - \frac{4}{7} - \frac{5}{14}$
35. $\frac{1}{3} - \frac{5}{7} + \frac{8}{21}$
36. $-\frac{2}{9} - \frac{1}{12} - \frac{7}{18}$
37. ***COOKING*** One fruit salad recipe calls for $\frac{1}{2}$ cup of sugar. Another recipe calls for 2 tablespoons of sugar. Since 1 tablespoon is $\frac{1}{16}$ cup, how much more sugar does the first recipe require?
38. It took Earl $\frac{1}{2}$ hour to do his science homework and $\frac{1}{3}$ hour to do his math homework. How long did Earl work on homework?

39. ***MUSIC*** In music written in $^4/_4$ time, a half note lasts for $\frac{1}{2}$ measure and an eighth note lasts for $\frac{1}{8}$ measure. In terms of a musical measure, what is the difference in the duration of the two notes?

40. ***STATISTICS*** The circle graph shows what the chances would be of having different numbers of boys and girls in a four-child family if gender were determined by chance rather than genetics.

Four-Child Family Probabilities

2 girls, 2 boys $\frac{3}{8}$

4 boys $\frac{1}{16}$

1 girl, 3 boys $\frac{1}{4}$

3 girls, 1 boy $\frac{1}{4}$

4 girls $\frac{1}{16}$

a. What is the sum of the chances of having 4 children of the same gender?

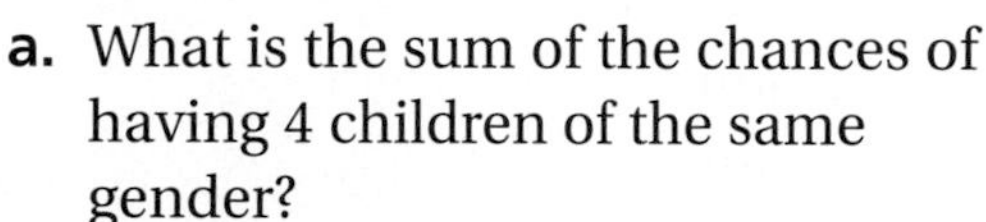

b. What is the sum of the chances of having only 1 child of a certain gender?

c. What is the difference in the chances of having 2 boys and 2 girls and of having 4 girls?

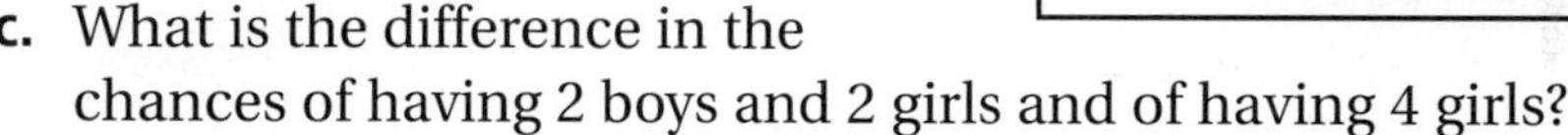

41. ***LIFE SCIENCE*** A shrew weighs $\frac{3}{16}$ lb. A hamster weighs $\frac{1}{4}$ lb.

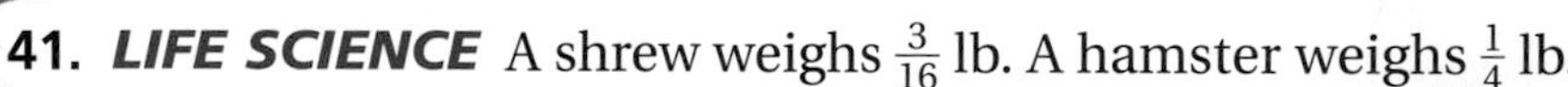

a. How many more pounds does a hamster weigh than a shrew?

b. There are 16 oz in 1 lb. How many more ounces does the hamster weigh than the shrew?

42. To make $\frac{3}{4}$ lb of mixed nuts, how many pounds of cashews would you add to $\frac{1}{8}$ lb of almonds and $\frac{1}{4}$ lb of peanuts?

43. ***WRITE A PROBLEM*** Use facts you find in a newspaper or magazine to write a problem that can be solved using addition or subtraction of fractions.

44. ***WRITE ABOUT IT*** Explain the steps you use to add or subtract fractions that have different denominators.

45. ***CHALLENGE*** The sum of two fractions is 1. If one fraction is $\frac{3}{8}$ greater than the other, what are the two fractions?

Spiral Review

Evaluate. (Lesson 2-3)

46. $12^2 \div (6 + 3) \cdot 2$

47. $4 + (8 - 2)^2 \div 9$

Solve each equation. Check your answers. (Lesson 2-11 and 2-12)

48. $x + 8 = 12$

49. $5y = 35$

50. **EOG PREP** Which of the following expressions is equal to 0.6? (Lesson 4-5)

A $36 \div 0.6$ B $3.6 \div 0.6$ C $3.6 \div 6$ D $0.36 \div 6$

4-11 Adding and Subtracting Mixed Numbers

Learn to add and subtract mixed numbers.

On average, adult males who live in the Netherlands are $1\frac{4}{5}$ inches taller than those who live in the United States. The average height of an adult male in the United States is $70\frac{4}{5}$ inches. To find the average height of an adult male in the Netherlands, you can add $70\frac{4}{5}$ and $1\frac{4}{5}$.

(Source: USA Today)

Bicyclists rest at a windmill in the Netherlands.

EXAMPLE 1 *Measurement Application*

What is the average height of an adult male in the Netherlands?

$70\frac{4}{5} + 1\frac{4}{5} = 71 + \frac{8}{5}$ *Add the integers and add the fractions.*

$= 71 + 1\frac{3}{5}$ *Rewrite the improper fraction as a mixed number.*

$= 72\frac{3}{5}$ *Add.*

The average height of an adult male in the Netherlands is $72\frac{3}{5}$ in.

Helpful Hint

A mixed number is the sum of an integer and a fraction: $3\frac{4}{5} = 3 + \frac{4}{5}$.

EXAMPLE 2 Adding Mixed Numbers

Add. Write each answer in simplest form.

A $3\frac{4}{5} + 4\frac{2}{5}$

$3\frac{4}{5} + 4\frac{2}{5} = 7 + \frac{6}{5}$ *Add the integers and add the fractions.*

$= 7 + 1\frac{1}{5}$ *Rewrite the improper fraction as a mixed number.*

$= 8\frac{1}{5}$ *Add.*

B $1\frac{2}{15} + 7\frac{1}{6}$

$1\frac{2}{15} + 7\frac{1}{6} = 1\frac{4}{30} + 7\frac{5}{30}$ *Find a common denominator.*

$= 8 + \frac{9}{30}$ *Add the integers and add the fractions.*

$= 8\frac{9}{30} = 8\frac{3}{10}$ *Add. Then simplify.*

Sometimes, when you subtract mixed numbers, the fraction portion of the first number is less than the fraction portion of the second number. In these cases, you must regroup before subtracting.

> **Remember!**
> Any fraction in which the numerator and denominator are the same is equal to 1.

REGROUPING MIXED NUMBERS	
Words	**Numbers**
Regroup.	$7\frac{1}{8} = 6 + 1 + \frac{1}{8}$
Rewrite 1 as a fraction with a common denominator.	$= 6 + \frac{8}{8} + \frac{1}{8}$
Add.	$= 6\frac{9}{8}$

EXAMPLE 3 Subtracting Mixed Numbers

Subtract. Write each answer in simplest form.

A $10\frac{7}{9} - 4\frac{2}{9}$

$10\frac{7}{9} - 4\frac{2}{9} = 6 + \frac{5}{9}$ *Subtract the integers and subtract the fractions.*

$= 6\frac{5}{9}$ *Add.*

B $12\frac{7}{8} - 5\frac{17}{24}$

$12\frac{7}{8} - 5\frac{17}{24} = 12\frac{21}{24} - 5\frac{17}{24}$ *Find a common denominator.*

$= 7 + \frac{4}{24}$ *Subtract the integers and subtract the fractions.*

$= 7\frac{4}{24}$ *Add.*

$= 7\frac{1}{6}$ *Simplify.*

C $72\frac{3}{5} - 63\frac{4}{5}$

$72\frac{3}{5} - 63\frac{4}{5} = 71\frac{8}{5} - 63\frac{4}{5}$ *Regroup.* $72\frac{3}{5} = 71 + \frac{5}{5} + \frac{3}{5}$

$= 8 + \frac{4}{5}$ *Subtract the integers and subtract the fractions.*

$= 8\frac{4}{5}$ *Add.*

Think and Discuss

1. **Describe** the process for subtracting mixed numbers.
2. **Explain** whether $2\frac{3}{5} + 1\frac{3}{5} = 3\frac{6}{5}$ is correct. Is there another way to write the answer?
3. **Demonstrate** how to regroup to simplify $6\frac{2}{5} - 4\frac{3}{5}$.

4-11 Exercises

1.02

FOR EOG PRACTICE
see page 673

internet connect
Homework Help Online
go.hrw.com Keyword: MS4 4-11

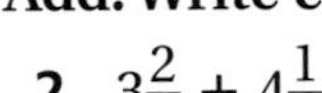

GUIDED PRACTICE

See Example 1

1. Chrystelle's mother is $1\frac{2}{3}$ ft taller than Chrystelle is. If Chrystelle is $3\frac{1}{2}$ ft tall, how tall is her mother?

See Example 2

Add. Write each answer in simplest form.

2. $3\frac{2}{5} + 4\frac{1}{5}$
3. $2\frac{7}{8} + 3\frac{3}{4}$
4. $1\frac{8}{9} + 4\frac{4}{9}$

See Example 3

Subtract. Write each answer in simplest form.

5. $6\frac{2}{3} - 5\frac{1}{3}$
6. $8\frac{1}{6} - 2\frac{5}{6}$
7. $3\frac{2}{3} - 2\frac{3}{4}$

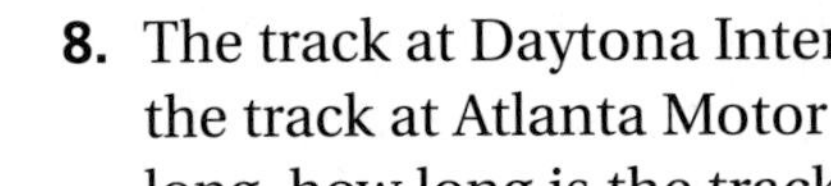

INDEPENDENT PRACTICE

See Example 1

8. The track at Daytona International Speedway is $\frac{24}{25}$ mi longer than the track at Atlanta Motor Speedway. If the track at Atlanta is $1\frac{27}{50}$ mi long, how long is the track at Daytona?

See Example 2

Add. Write each answer in simplest form.

9. $6\frac{1}{4} + 8\frac{3}{4}$
10. $3\frac{3}{5} + 7\frac{4}{5}$
11. $3\frac{5}{6} + 1\frac{5}{6}$
12. $2\frac{3}{10} + 4\frac{1}{2}$
13. $6\frac{1}{8} + 8\frac{9}{10}$
14. $6\frac{1}{6} + 5\frac{3}{10}$

See Example 3

Subtract. Write each answer in simplest form.

15. $2\frac{1}{14} - 1\frac{3}{14}$
16. $4\frac{5}{12} - 1\frac{7}{12}$
17. $8 - 2\frac{3}{4}$
18. $8\frac{3}{4} - 6\frac{2}{5}$
19. $3\frac{1}{3} - 2\frac{5}{8}$
20. $4\frac{2}{5} - 3\frac{1}{2}$

PRACTICE AND PROBLEM SOLVING

Add or subtract. Write each answer in simplest form.

21. $7\frac{1}{3} + 8\frac{1}{5}$
22. $14\frac{3}{5} - 8\frac{1}{2}$
23. $9\frac{1}{6} + 4\frac{6}{9}$
24. $3\frac{5}{8} + 2\frac{7}{12}$
25. $25\frac{1}{3} + 3\frac{5}{6}$
26. $1\frac{7}{9} - \frac{17}{18}$
27. $3\frac{1}{2} + 5\frac{1}{4}$
28. $6\frac{3}{4} + 2\frac{3}{4}$
29. $1\frac{7}{15} + 2\frac{7}{10}$
30. $4\frac{2}{3} + 1\frac{7}{8} + 3\frac{1}{2}$
31. $5\frac{1}{6} + 8\frac{2}{3} - 9\frac{1}{2}$
32. $12\frac{1}{2} - 3\frac{3}{4} - 6\frac{1}{3}$

Compare. Write <, >, or =.

33. $12\frac{1}{4} - 10\frac{3}{4}$ ■ $5\frac{1}{2} - 3\frac{7}{10}$
34. $4\frac{1}{2} + 3\frac{4}{5}$ ■ $4\frac{5}{7} + 3\frac{1}{2}$
35. $13\frac{3}{4} - 2\frac{3}{8}$ ■ $5\frac{5}{6} + 4\frac{2}{9}$
36. $4\frac{1}{3} - 2\frac{1}{4}$ ■ $3\frac{1}{4} - 1\frac{1}{6}$

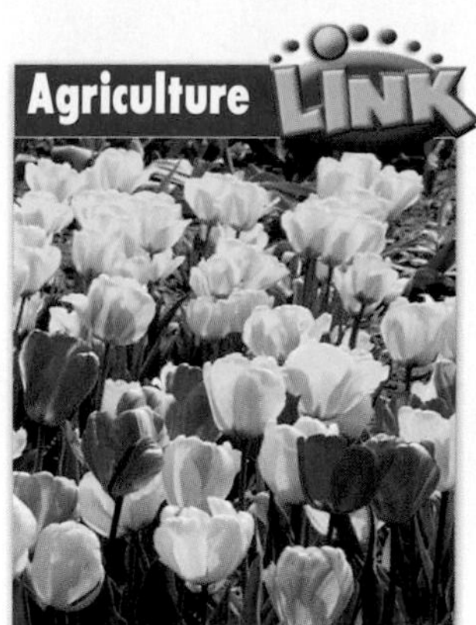

The Netherlands produces more than 3 billion tulips each year.

37. ***AGRICULTURE*** From January through September of 2001, the United States imported $\frac{49}{50}$ of its tulip bulbs from the Netherlands and $\frac{1}{100}$ of its tulip bulbs from New Zealand. What fraction more of tulip imports came from the Netherlands?

38. ***TRAVEL*** The table shows the distances in miles between four cities. To find the distance between two cities, locate the square where the row for one city and the column for the other city intersect.

   a. How much farther is it from Charleston to Dixon than from Atherton to Baily?

   b. If you drove from Charleston to Atherton and then from Atherton to Dixon, how far would you drive?

	Atherton	Baily	Charleston	Dixon
Atherton	X	$40\frac{2}{3}$	$100\frac{5}{6}$	$16\frac{1}{2}$
Baily	$40\frac{2}{3}$	X	$210\frac{3}{8}$	$30\frac{2}{3}$
Charleston	$100\frac{5}{6}$	$210\frac{3}{8}$	X	$98\frac{3}{4}$
Dixon	$16\frac{1}{2}$	$30\frac{2}{3}$	$98\frac{3}{4}$	X

39. ***RECREATION*** Kathy wants to hike to Candle Lake. The waterfall trail is $1\frac{2}{3}$ miles long, and the meadow trail is $1\frac{5}{6}$ miles long. Which route is shorter and by how much?

40. ***CHOOSE A STRATEGY*** Spiro needs to draw a 6-inch-long line. He does not have a ruler, but hc has sheets of notebook paper that are $8\frac{1}{2}$ inches wide and 11 inches long. Describe how Spiro can use the notebook paper to measure 6 inches.

41. ***WRITE ABOUT IT*** Explain why it is sometimes necessary to regroup a mixed number when subtracting.

42. ***CHALLENGE*** Todd had d pounds of nails. He sold $3\frac{1}{2}$ pounds on Monday and $5\frac{2}{3}$ pounds on Tuesday. Write an expression to show how many pounds he had left and then simplify it.

Spiral Review

Add. (Lesson 3-3)

43. $-3 + 8$ **44.** $-2 + (-7)$ **45.** $8 + (-12)$

Write each fraction as a decimal. (Lesson 3-9)

46. $\frac{3}{4}$ **47.** $\frac{5}{8}$ **48.** $\frac{3}{16}$ **49.** $\frac{2}{9}$

50. **EOG PREP** What is 154,000,000,000,000 written in scientific notation? (Lesson 2-2)

A 15.4×10^{13} B 1.54×10^{13} C 1.54×10^{14} D 0.154×10^{15}

51. **EOG PREP** Which number is a solution of $2.2b = 1.1$? (Lesson 4-6)

A 0.05 B 0.5 C 5 D 50

4-12 Solving Equations Containing Fractions

Learn to solve one-step equations that contain fractions.

Gold classified as 24 karat is pure gold, while gold classified as 18 karat is only $\frac{3}{4}$ pure. The remaining $\frac{1}{4}$ of 18-karat gold is made up of one or more different metals, such as silver, copper, or zinc. The color of gold varies, depending on the type and amount of each metal added to the pure gold.

Equations can help you determine the amounts of metals in different kinds of gold. The goal when solving equations that contain fractions is the same as when working with other kinds of numbers—*to isolate the variable* on one side of the equation.

EXAMPLE 1 Solving Equations by Adding or Subtracting

Solve. Write each answer in simplest form.

A $x - \frac{1}{5} = \frac{3}{5}$

$$x - \frac{1}{5} = \frac{3}{5}$$

$$x - \frac{1}{5} + \frac{1}{5} = \frac{3}{5} + \frac{1}{5} \quad \textit{Add to isolate x.}$$

$$x = \frac{4}{5} \quad \textit{Add.}$$

B $\frac{5}{12} + y = \frac{2}{3}$

$$\frac{5}{12} + y = \frac{2}{3}$$

$$\frac{5}{12} + y - \frac{5}{12} = \frac{2}{3} - \frac{5}{12} \quad \textit{Subtract to isolate y.}$$

$$y = \frac{8}{12} - \frac{5}{12} \quad \textit{Find a common denominator.}$$

$$y = \frac{3}{12} = \frac{1}{4} \quad \textit{Subtract. Then simplify.}$$

Helpful Hint

You can also isolate the variable y by adding the opposite of $\frac{5}{12}$, $-\frac{5}{12}$, to both sides.

C $\frac{7}{18} + u = -\frac{14}{27}$

$$\frac{7}{18} + u = -\frac{14}{27}$$

$$\frac{7}{18} + u - \frac{7}{18} = -\frac{14}{27} - \frac{7}{18} \quad \textit{Subtract to isolate u.}$$

$$u = -\frac{28}{54} - \frac{21}{54} \quad \textit{Find a common denominator.}$$

$$u = -\frac{49}{54} \quad \textit{Subtract.}$$

EXAMPLE 2 Solving Equations by Multiplying

Solve. Write each answer in simplest form.

A $\frac{2}{3}x = \frac{4}{5}$

$$\frac{2}{3}x = \frac{4}{5}$$

$$\frac{2}{3}x \cdot \frac{3}{2} = \frac{{}^{2}\cancel{4}}{5} \cdot \frac{3}{\cancel{2}_1}$$ *Multiply by the reciprocal of $\frac{2}{3}$. Then simplify.*

$$x = \frac{6}{5} \text{ or } 1\frac{1}{5}$$

Remember!

To undo multiplying by $\frac{2}{3}$, you can divide by $\frac{2}{3}$ or multiply by its reciprocal, $\frac{3}{2}$.

B $3y = \frac{6}{7}$

$$3y = \frac{6}{7}$$

$$3y \cdot \frac{1}{3} = \frac{{}^{2}\cancel{6}}{7} \cdot \frac{1}{\cancel{3}_1}$$ *Multiply by the reciprocal of 3. Then simplify.*

$$y = \frac{2}{7}$$

EXAMPLE 3 *Physical Science Application*

Pink gold is made up of gold, silver, and copper. The amount of pure gold in pink gold is $\frac{11}{20}$ more than the amount of copper. If pink gold is $\frac{3}{4}$ pure gold, how much of pink gold is copper?

Let c represent the amount of copper in pink gold.

$$c + \frac{11}{20} = \frac{3}{4}$$ *Write an equation.*

$$c + \frac{11}{20} - \frac{11}{20} = \frac{3}{4} - \frac{11}{20}$$ *Subtract to isolate c.*

$$c = \frac{15}{20} - \frac{11}{20}$$ *Find a common denominator.*

$$c = \frac{4}{20}$$ *Subtract.*

$$c = \frac{1}{5}$$ *Simplify.*

The amount of copper in pink gold is $\frac{1}{5}$.

Think and Discuss

1. **Show** the first step you would use to solve $m + 3\frac{5}{8} = 12\frac{1}{2}$.
2. **Describe** how to decide if $\frac{2}{3}$ is a solution of $\frac{7}{8}y = \frac{3}{5}$.
3. **Explain** why solving $\frac{2}{5}c = \frac{8}{9}$ by multiplying both sides by $\frac{5}{2}$ is the same as solving it by dividing both sides by $\frac{2}{5}$.

4-12 Exercises

1.02, 5.01, 5.03

FOR EOG PRACTICE

see page 673

internet connect
Homework Help Online
go.hrw.com Keyword: MS4 4-12

GUIDED PRACTICE

See Example 1 **Solve. Write each answer in simplest form.**

1. $a - \frac{1}{2} = \frac{1}{4}$
2. $m + \frac{1}{6} = \frac{5}{6}$
3. $p - \frac{2}{3} = \frac{5}{6}$

See Example 2

4. $\frac{1}{5}x = 8$
5. $\frac{2}{3}r = \frac{3}{5}$
6. $3w = \frac{3}{7}$

See Example 3

7. Kara has $\frac{3}{8}$ cup less oatmeal than she needs for a cookie recipe. If she has $\frac{3}{4}$ cup of oatmeal, how much oatmeal does she need?

INDEPENDENT PRACTICE

See Example 1 **Solve. Write each answer in simplest form.**

8. $n - \frac{1}{5} = \frac{3}{5}$
9. $t - \frac{3}{8} = \frac{1}{4}$
10. $s - \frac{7}{24} = \frac{1}{3}$
11. $x + \frac{2}{3} = 2\frac{7}{8}$
12. $h + \frac{7}{10} = \frac{7}{10}$
13. $y + \frac{5}{6} = \frac{19}{20}$

See Example 2

14. $\frac{1}{5}x = 4$
15. $\frac{1}{4}w = \frac{1}{8}$
16. $5y = \frac{3}{10}$
17. $6z = \frac{1}{2}$
18. $\frac{5}{8}x = \frac{2}{5}$
19. $\frac{5}{8}n = 1\frac{1}{5}$

See Example 3

20. Carbon-14 has a half-life of 5,730 years. After 17,190 years, $\frac{1}{8}$ of the carbon-14 in a sample will be left. If 5 grams of carbon-14 are left after 17,190 years, how much was in the original sample?

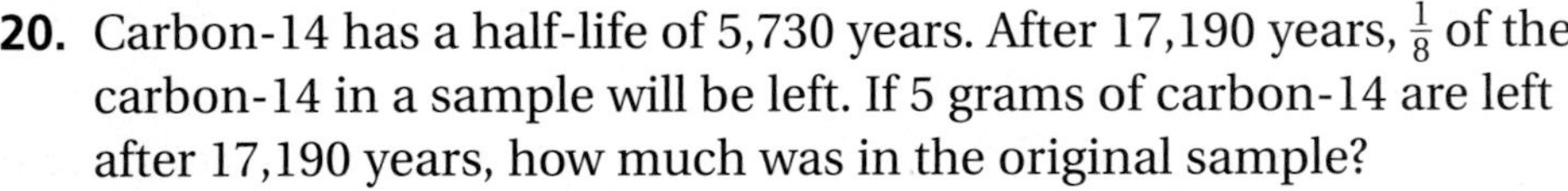

PRACTICE AND PROBLEM SOLVING

Solve. Write each answer in simplest form.

21. $\frac{4}{5}t = \frac{1}{5}$
22. $m - \frac{1}{2} = \frac{2}{3}$
23. $\frac{1}{8}w = \frac{3}{4}$
24. $\frac{8}{9} + t = \frac{17}{18}$
25. $\frac{5}{3}x = 1$
26. $j + \frac{5}{8} = \frac{11}{16}$
27. $\frac{4}{3}n = 3\frac{1}{5}$
28. $z + \frac{1}{6} = 3\frac{9}{15}$
29. $\frac{3}{4}y = \frac{3}{8}$
30. $-\frac{5}{26} + m = -\frac{7}{13}$
31. $-\frac{8}{77} + r = -\frac{1}{11}$
32. $y - \frac{3}{4} = -\frac{9}{20}$
33. $h - \frac{3}{8} = -\frac{11}{24}$
34. $-\frac{5}{36}t = -\frac{5}{16}$
35. $-\frac{8}{13}v = -\frac{6}{13}$
36. $4\frac{6}{7} + p = 5\frac{1}{4}$
37. $d - 5\frac{1}{8} = 9\frac{3}{10}$
38. $6\frac{8}{21}k = 13\frac{1}{3}$

39. ***FOOD*** Each person in Finland drinks an average of $24\frac{1}{4}$ lb of coffee per year. This is $13\frac{1}{16}$ lb more than the average person in Italy consumes. On average, how much coffee does an Italian drink each year?

40. ***WEATHER*** Yuma, Arizona, receives $102\frac{1}{100}$ fewer inches of rain each year than Quillayute, Washington, which receives $105\frac{9}{50}$ inches per year. (*Source:* National Weather Service). How much rain does Yuma get in one year?

41. ***LIFE SCIENCE*** Scientists have discovered $1\frac{1}{2}$ million species of animals. This is estimated to be $\frac{1}{10}$ the total number of species thought to exist. About how many species do scientists think exist?

42. ***HISTORY*** The circle graph shows the birthplaces of some of the first presidents of the United States.

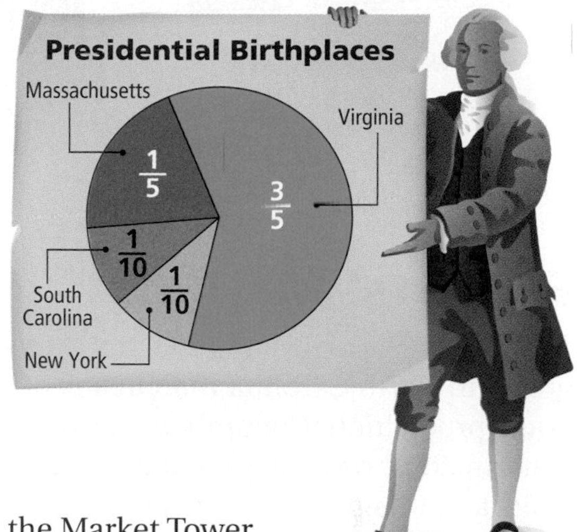

   a. If six of the presidents represented in the graph were born in Virginia, how many presidents are represented in the graph?

   b. Based on your answer to **a**, how many of the presidents were born in Massachusetts?

43. ***ARCHITECTURE*** In Indianapolis, the Market Tower has $\frac{2}{3}$ as many stories as the Bank One Tower. If the Market Tower has 32 stories, how many stories does the Bank One Tower have?

44. ***BUDGET*** Each week, Jennifer saves $\frac{1}{5}$ of her allowance and spends some of the rest on lunches. This week, she had $\frac{2}{15}$ of her allowance left after buying her lunch each day. What fraction of her allowance did she spend on lunches?

45. ***WHAT'S THE ERROR?*** A student solved $\frac{3}{5}x = \frac{2}{3}$ and got $x = \frac{2}{5}$. Find the error.

46. ***WRITE ABOUT IT*** Solve $3\frac{1}{3}z = 1\frac{1}{2}$. Explain why you need to write mixed numbers as improper fractions when multiplying and dividing.

47. ***CHALLENGE*** Solve $\frac{3}{5}w = 0.9$. Write your answer as a fraction and as a decimal.

Spiral Review

Write each improper fraction as a mixed number in simplest form. (Lesson 3-8)

48. $\frac{20}{3}$	**49.** $\frac{11}{5}$	**50.** $\frac{19}{4}$
51. $\frac{39}{9}$	**52.** $\frac{47}{4}$	**53.** $\frac{22}{14}$

54. **EOG PREP** Which of the following is equivalent to $14 + 12 \cdot 3 - 8^2$? (Lesson 2-3)

A 14   B −14   C 4,900   D −1,586

55. **EOG PREP** What is the sum of 3.9 and 0.35? (Lesson 4-2)

A 3.25   B 0.425   C 4.25   D 3.55

NORTH CAROLINA

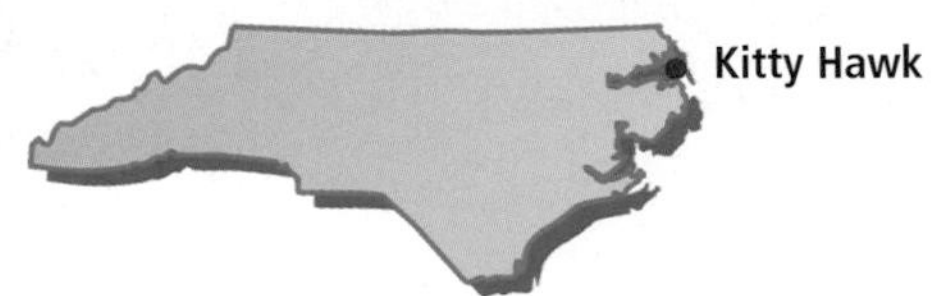

Farmers' Markets

Agriculture is North Carolina's top industry. More sweet potatoes are grown in North Carolina than in any other state, and it ranks ninth in apple production. Cucumbers, beans, blueberries, eggplants, watermelons, peaches, and grapes are among the other fruits and vegetables produced in the state. Produce is sold in farmers' markets throughout the state.

1. The wholesale price of bananas is \$12.50 for 40 pounds. What is the wholesale price of a pound of bananas, to the nearest cent?

2. A crate of $4\frac{1}{2}$ dozen ears of corn is sold to the farmers' market for \$20.50. At the market, corn is sold to the public at the price of 5 ears for \$2.00. How much money does the owner of the market make on every ear of corn, to the nearest cent?

The graph shows the pounds of produce grown in one year in North Carolina. For 3–6, use the graph.

3. How many more pounds of apples than pounds of watermelons were grown in the year?

4. To the nearest tenth, how many times as many pounds of sweet potatoes as pounds of apples were grown in the year?

5. Find the total number of pounds of these fruits and vegetables grown in the year shown.

6. In the year shown, about 2.3 times as many pounds of apples were grown as pounds of tomatoes. How many pounds of tomatoes were grown, to the nearest tenth?

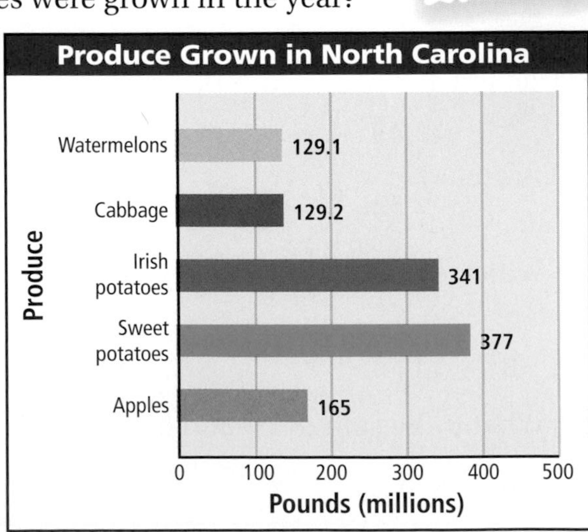

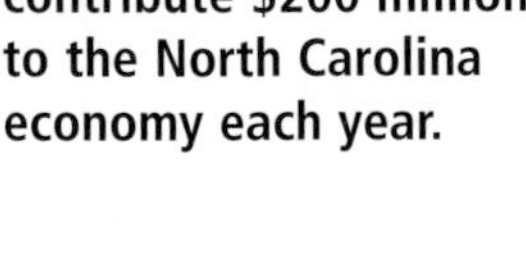
Fruits and vegetables contribute \$200 million to the North Carolina economy each year.

Wright Brothers National Memorial

Orville and Wilbur Wright flew the first heavier-than-air plane on December 17, 1903, near Kitty Hawk, North Carolina.

The Wright brothers were not born in North Carolina. But when the time came to make the first flights in their plane, they consulted the Weather Bureau and chose a site near Kitty Hawk. They made four flights that day.

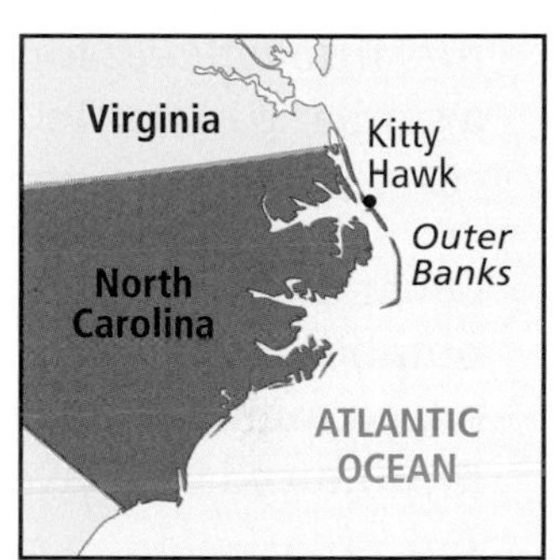

1. Orville piloted the plane 40 yards. This first flight lasted $\frac{1}{5}$ minute. The longest flight that day lasted $\frac{59}{60}$ minute. How much longer did this flight last than the first?
2. On September 9, 1908, Orville made a flight that lasted $1\frac{1}{30}$ hours. About how many times as long as the first flight was the 1908 flight?
3. Orville's flight was 40 yards long. Wilbur made the longest flight, which was 284 yards. How many times as far as Orville flew did Wilbur fly?
4. On October 5, 1905, the Wright brothers' plane flew 24.2 miles in about 38 minutes.
   - **a.** To the nearest tenth, what was the number of miles the plane flew in one minute?
   - **b.** What was the number of miles the plane would have flown in one hour?

The first plane to fly across the English Channel was flown by Louis Bleriot of France. He made this flight in 1909 in a plane that was 8 meters long.

It took Louis Bleriot 37 minutes to fly across the English Channel.

5. The Wright brothers' first plane was 6.43 meters long. How much longer was the plane that flew across the Channel than the Wright brothers' first plane?

MATH-ABLES

Number Patterns

The numbers one through ten form the pattern below. Each arrow indicates some kind of relationship between the two numbers. Four relates to itself. Can you figure out what the pattern is?

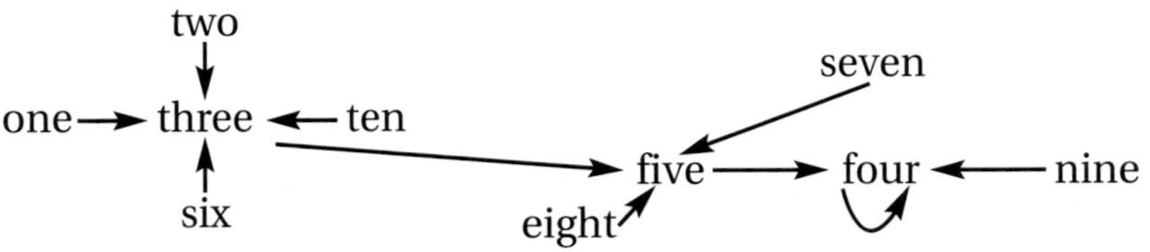

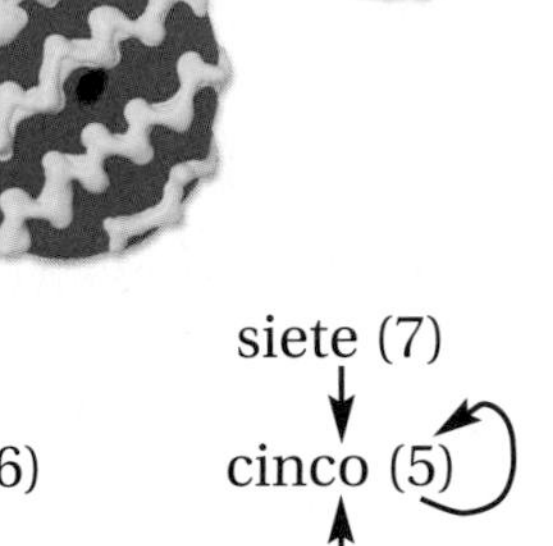

The Spanish numbers *uno* through *diez* form a similar pattern. In this case, *cinco* relates to itself.

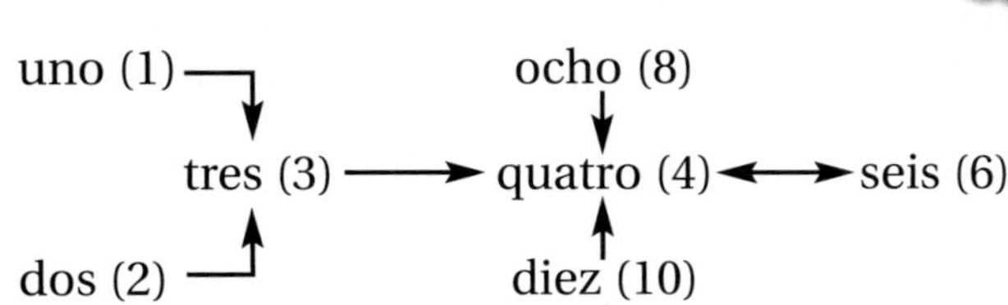

siete (7)
cinco (5)
nueve (9)

Other interesting number patterns involve cyclic numbers. Cyclic numbers sometimes occur when a fraction converts to a repeating nonterminating decimal. One of the most interesting cyclic numbers is produced by converting the fraction $\frac{1}{7}$ to a decimal.

$\frac{1}{7} = 0.142857142857142\ldots$

Multiplying 142857 by the numbers 1–6 produces the same digits in a different order.

$1 \cdot 142857 = 142857$ $\quad$ $3 \cdot 142857 = 428571$ $\quad$ $5 \cdot 142857 = 714285$

$2 \cdot 142857 = 285714$ $\quad$ $4 \cdot 142857 = 571428$ $\quad$ $6 \cdot 142857 = 857142$

Fraction Action

Roll four 1–6 number cubes and use the numbers to form two fractions. Add the fractions and try to get a sum as close to 1 as possible. To determine your score on each turn, find the difference between the sum of your fractions and 1. Keep a running total of your score as you play. The winner is the player with the lowest score at the end of the game.

internet connect

For a complete copy of the rules, go to ***go.hrw.com***
KEYWORD: MS4 Game4

go.hrw.com

Fraction Operations

You can use a calculator to complete fraction operations.

internet connect
Lab Resources Online
go.hrw.com
KEYWORD: MS4 TechLab4

Activity

Many calculators will not display fractions with the numerator above the denominator. On these calculators, enter the fraction as a division expression.

1 Evaluate $\frac{2}{3} + \frac{4}{7}$ using a calculator.

2 Evaluate $\frac{1}{2} \cdot \frac{5}{6}$ using a calculator.

Think and Discuss

1. A mixed number is the sum of an integer and a fraction. How would you enter the mixed number $7\frac{9}{11}$ on a calculator?

Try This

Evaluate each expression using a calculator.

1. $\frac{9}{10} + \frac{8}{15}$

2. $\frac{12}{13} \cdot \frac{1}{6}$

3. $1\frac{9}{11} + 8\frac{2}{3}$

4. $2\frac{1}{8} \cdot 4\frac{1}{6}$

Chapter 4

Study Guide and Review

Vocabulary

compatible numbers 192

reciprocal 226

Complete the sentences below with vocabulary words from the list above.

1. When estimating products or quotients, you can use ___?___ that are close to the original numbers.
2. The fractions $\frac{3}{8}$ and $\frac{8}{3}$ are ___?___ because they multiply to give 1.

4-1 Estimate with Decimals (pp. 192–195)

EXAMPLE

■ **Estimate.**

$$\begin{array}{r} 43.55 \longrightarrow \\ \times\ 8.65 \longrightarrow \end{array} \quad \begin{array}{r} 40 \\ \times\ 9 \\ \hline 360 \end{array}$$

EXERCISES

Estimate by rounding or using compatible numbers.

3. 54.4 + 55.99
4. 11.48 − 5.6
5. 24.77 · 3.45
6. 37.8 ÷ 9.3

4-2 Adding and Subtracting Decimals (pp. 196–199)

EXAMPLE

■ **Add.**

5.67 + 22.44

$$\begin{array}{r} 5.67 \\ +\ 22.44 \\ \hline 28.11 \end{array}$$

EXERCISES

Add or subtract.

7. 4.99 + 22.89
8. −6.7 + (−44.5)
9. 18.09 − 11.87
10. 47 + 5.902
11. 23 − 8.905
12. 4.68 + 31.2

4-3 Multiplying Decimals (pp. 202–205)

EXAMPLE

■ **Multiply.**

1.44 · 0.6

$$\begin{array}{r} 1.44 \\ \times\ 0.6 \\ \hline 0.864 \end{array}$$

EXERCISES

Multiply.

13. 7 · 0.5
14. −4.3 · 9
15. 4.55 · 8.9
16. 7.88 · 7.65
17. 63.4 · 1.22
18. −9.9 · 1.9

4-4 Dividing Decimals by Integers (pp. 206–209)

EXAMPLE

■ **Divide.**

$2.8 \div 7$

$$\begin{array}{r} 0.4 \\ 7\overline{)2.8} \\ -2\,8 \\ \hline 0 \end{array}$$

EXERCISES

Divide.

19. $16.1 \div 7$
20. $102.9 \div (-21)$
21. $0.48 \div 6$
22. $17.4 \div (-3)$
23. $8.25 \div (-5)$
24. $81.6 \div 24$

4-5 Dividing Decimals and Integers by Decimals (pp. 210–213)

EXAMPLE

■ **Divide.**

$0.96 \div 1.6$

$$\begin{array}{r} 0.6 \\ 16\overline{)9.6} \\ -9\,6 \\ \hline 0 \end{array}$$

EXERCISES

Divide.

25. $7.65 \div 1.7$
26. $9.483 \div (-8.7)$
27. $126.28 \div (-8.2)$
28. $2.5 \div (-0.005)$
29. $9 \div 4.5$
30. $13 \div 3.25$

4-6 Solving Equations Containing Decimals (pp. 214–217)

EXAMPLE

■ **Solve.**

$$\begin{aligned} n - 4.77 &= 8.60 \\ +4.77 &\quad +4.77 \\ n &= 13.37 \end{aligned}$$

EXERCISES

Solve.

31. $x + 40.44 = 30$
32. $\frac{s}{1.07} = 100$
33. $0.8n = 0.0056$
34. $k - 8 = 0.64$

4-7 Multiplying Fractions and Mixed Numbers (pp. 222–225)

EXAMPLE

■ **Multiply. Write the answer in simplest form.**

$$4\frac{1}{2} \cdot 5\frac{3}{4} = \frac{9 \cdot 23}{2 \cdot 4} = \frac{207}{8} \text{ or } 25\frac{7}{8}$$

EXERCISES

Multiply. Write each answer in simplest form.

35. $1\frac{2}{3} \cdot 4\frac{1}{2}$
36. $\frac{4}{5} \cdot 2\frac{3}{10}$
37. $4\frac{6}{7} \cdot 3\frac{5}{9}$
38. $3\frac{4}{7} \cdot 1\frac{3}{4}$

4-8 Dividing Fractions and Mixed Numbers (pp. 226–229)

EXAMPLE

■ **Divide.**

$$\frac{3}{4} \div \frac{2}{5} = \frac{3}{4} \cdot \frac{5}{2} = \frac{15}{8} \text{ or } 1\frac{7}{8}$$

EXERCISES

Divide. Write each answer in simplest form.

39. $\frac{1}{3} \div 6\frac{1}{4}$
40. $\frac{1}{2} \div 3\frac{3}{4}$
41. $\frac{11}{13} \div \frac{11}{13}$
42. $2\frac{7}{8} \div 1\frac{1}{2}$

4-9 Estimate with Fractions (pp. 230–233)

EXAMPLE

■ **Estimate the difference.**

$7\frac{3}{4} - 4\frac{1}{3}$

$7\frac{3}{4} \longrightarrow 8,\ 4\frac{1}{3} \longrightarrow 4\frac{1}{2}$

$8 - 4\frac{1}{2} = 3\frac{1}{2}$

EXERCISES

Estimate each sum or difference.

43. $11\frac{1}{7} + 12\frac{3}{4}$

44. $5\frac{2}{3} - 3\frac{1}{5}$

45. $5\frac{5}{7} - 13\frac{10}{17}$

46. $8\frac{1}{6} + 14\frac{6}{11}$

47. $9\frac{7}{8} + \left(-7\frac{1}{13}\right)$

48. $11\frac{8}{9} - 11\frac{1}{20}$

4-10 Adding and Subtracting Fractions (pp. 236–239)

EXAMPLE

■ **Add.**

$\frac{1}{3} + \frac{2}{5} = \frac{5}{15} + \frac{6}{15}$

$= \frac{11}{15}$

EXERCISES

Add or subtract. Write each answer in simplest form.

49. $\frac{3}{4} - \frac{1}{3}$

50. $\frac{1}{4} + \frac{3}{5}$

51. $\frac{4}{11} + \frac{4}{44}$

52. $\frac{4}{9} - \frac{1}{3}$

4-11 Adding and Subtracting Mixed Numbers (pp. 240–243)

EXAMPLE

■ **Add.**

$1\frac{1}{3} + 2\frac{1}{2} = 1\frac{2}{6} + 2\frac{3}{6}$

$= 3 + \frac{5}{6}$

$= 3\frac{5}{6}$

EXERCISES

Add or subtract. Write each answer in simplest form.

53. $3\frac{7}{8} + 2\frac{1}{3}$

54. $2\frac{1}{4} + 1\frac{1}{12}$

55. $8\frac{1}{2} - 2\frac{1}{4}$

56. $11\frac{3}{4} - 10\frac{1}{3}$

4-12 Solving Equations Containing Fractions (pp. 244–247)

EXAMPLE

■ **Solve. Write the answer in simplest form.**

$\frac{1}{4}x = \frac{1}{6}$

$\frac{4}{1} \cdot \frac{1}{4}x = \frac{1}{6} \cdot \frac{4}{1}$

$x = \frac{4}{6} = \frac{2}{3}$

EXERCISES

Solve. Write each answer in simplest form.

57. $\frac{1}{5}x = \frac{1}{3}$

58. $\frac{1}{3} + y = \frac{2}{5}$

59. $\frac{1}{6}x = \frac{2}{7}$

60. $\frac{2}{7} + x = \frac{3}{4}$

Study Guide and Review

Chapter Test

Estimate.

1. $19.95 + 21.36$
2. $49.17 - 5.88$
3. $3.21 \cdot 16.78$
4. $49.1 \div 5.6$

Add or subtract.

5. $3.086 + 6.152$
6. $5.91 + 12.8$
7. $3.1 - 2.076$
8. $14.75 - 6.926$

Multiply.

9. $3.25 \cdot 24$
10. $1.4 \cdot 2.5$
11. $-3.79 \cdot 0.9$
12. $-4.79 \cdot 7.2$

Divide.

13. $1.8 \div (-6)$
14. $3.2 \div 16$
15. $3.57 \div (-0.7)$
16. $5.88 \div 0.6$

Solve.

17. $w - 5.3 = 7.6$
18. $4.9 = c + 3.7$
19. $b \div 1.8 = 2.1$
20. $4.3h = 81.7$

Multiply. Write each answer in simplest form.

21. $4 \cdot \frac{5}{8}$
22. $5 \cdot 4\frac{1}{3}$
23. $2\frac{7}{10} \cdot 2\frac{2}{3}$
24. $\frac{3}{5} \cdot \frac{1}{2}$

Divide. Write each answer in simplest form.

25. $\frac{3}{10} \div \frac{4}{5}$
26. $2\frac{1}{5} \div 1\frac{5}{6}$
27. $\frac{1}{4} \div \frac{1}{4}$
28. $3 \div 1\frac{4}{5}$

Estimate each sum or difference.

29. $\frac{3}{4} + \frac{3}{8}$
30. $5\frac{7}{8} + 3\frac{3}{4}$
31. $6\frac{5}{7} - 2\frac{2}{9}$
32. $8\frac{1}{2} - 3\frac{9}{10}$

Add or subtract. Write each answer in simplest form.

33. $\frac{3}{10} + \frac{2}{5}$
34. $\frac{11}{16} - \frac{7}{8}$
35. $7\frac{1}{3} + 5\frac{11}{12}$
36. $9 - 3\frac{2}{5}$

Solve. Write each answer in simplest form.

37. $\frac{1}{5}a = \frac{1}{8}$
38. $\frac{1}{4}c = 980$
39. $-\frac{7}{9} + w = \frac{2}{3}$
40. $z - \frac{5}{13} = \frac{6}{7}$

41. Alan finished his homework in $1\frac{1}{2}$ hours. It took Jimmy $\frac{3}{4}$ of an hour longer than Alan to finish his homework. How long did it take Jimmy to finish his homework?

Chapter 4

Performance Assessment

Show What You Know

Create a portfolio of your best work from this chapter. Complete this page and include it with the four best pieces of your work from Chapter 4. Choose from your homework or lab assignments, mid-chapter quiz, or any journals you have done. Put them together using any design you want. Make your portfolio represent what you consider your best work.

Short Response

1. Amanda earns \$18.41 per hour. Tina earns \$12.07 per hour. Find the difference between their yearly salaries. Assume that they both work 40 hours each week for 52 weeks each year. Show your work.

2. A building proposal calls for 6 acres of land to be divided into $\frac{3}{4}$-acre lots. How many lots can be made? Explain your answer.

3. Mari bought 3 packages of colored paper. She used $\frac{3}{4}$ of a package to make greeting cards, used $1\frac{1}{6}$ packages for an art project, and gave $\frac{2}{3}$ of a package to her brother. How much colored paper does Mari have left? Show the steps that you used to find the answer.

Extended Problem Solving

4. A high school is hosting a triple-jump competition. In this event, athletes make three leaps in a row—a hop, a skip, and a jump—and try to cover the greatest distance.

   a. Tony's first two jumps were $11\frac{2}{3}$ ft and $11\frac{1}{2}$ ft. His total distance was 44 ft. Write and solve an equation to find the length of his final jump.

   b. Candice's three jumps were all the same length. Her total distance was 38 ft. What was the length of each of her jumps?

   c. The lengths of Davis's jumps were 11.6 ft, $11\frac{1}{4}$ ft, and $11\frac{2}{3}$ ft. Plot these distances on a number line. What is the average distance of the jumps?

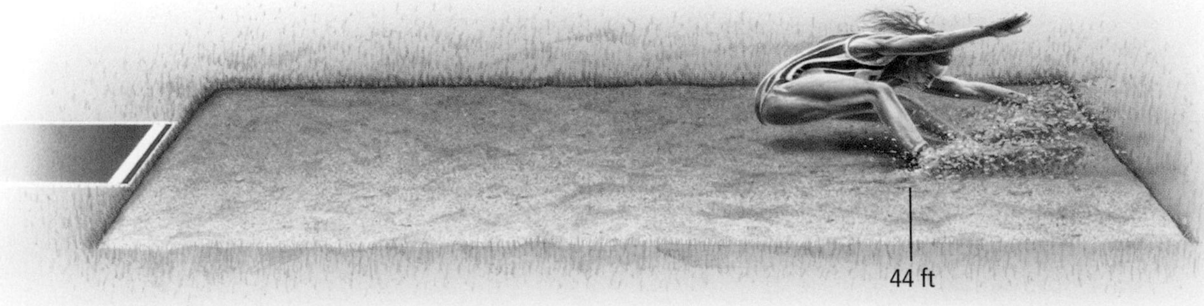

Performance Assessment

internet connect
EOG Practice Online
go.hrw.com Keyword: MS4 TestPrep

Getting Ready for EOG

Chapter 4

Cumulative Assessment, Chapters 1–4

1. On a baseball field, the distance from home plate to the pitcher's mound is $60\frac{1}{2}$ feet. The distance from home plate to second base is about $127\frac{7}{24}$ feet. What is the difference between the two distances?

   A $61\frac{1}{3}$ ft C $66\frac{19}{24}$ ft

   B $66\frac{5}{6}$ ft D $66\frac{5}{24}$ ft

2. Which number when rounded to the nearest whole number is 8?

   A $8\frac{7}{10}$ C $8\frac{11}{16}$

   B $8\frac{4}{9}$ D $8\frac{8}{15}$

3. Evaluate $3\frac{1}{2} \div \frac{1}{8}$.

   A $24\frac{1}{16}$ C $\frac{7}{16}$

   B 28 D $3\frac{1}{16}$

4. Evaluate $4 \cdot (3^2 + 7)$.

   A 64 C 43

   B 52 D 40

TEST TAKING TIP!
Eliminate choices by estimating the quotient.

5. What is $0.12 \div 0.03$?

   A 0.04 C 4

   B 0.4 D 40

6. Which is the GCF of 54 and 81?

   A 3 C 9

   B 6 D 27

7. Use compatible numbers to estimate $98.4 \div 19.2$.

   A 5.1 C 4

   B 5 D 4.75

8. Which word phrase best describes the expression $x + 2$?

   A a number plus itself

   B a number plus another number

   C a number more than two

   D twice a number

9. ***SHORT RESPONSE*** The graph shows the number of boys and the number of girls who tried out for a talent show. Write a statement giving a reasonable conclusion that could be drawn from the graph.

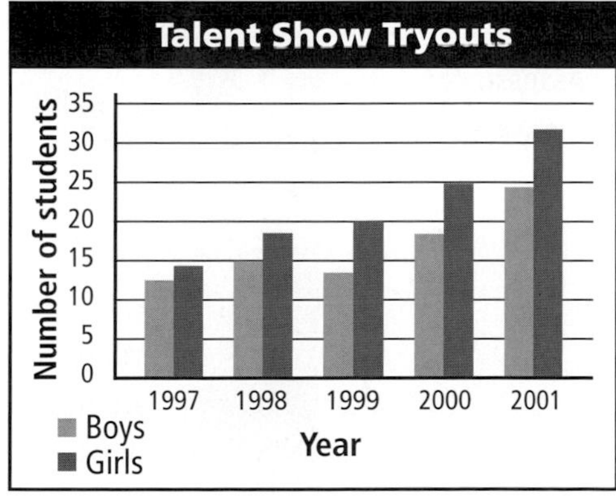

10. ***SHORT RESPONSE*** Write $\frac{3}{12}$ and $\frac{1}{3}$ as fractions with a common denominator. Explain your method. Then tell whether the fractions are equivalent.

Getting Ready for EOG

Chapter 5

Proportional Reasoning

TABLE OF CONTENTS

internet connect

Chapter Opener Online
go.hrw.com
KEYWORD: MS4 Ch5

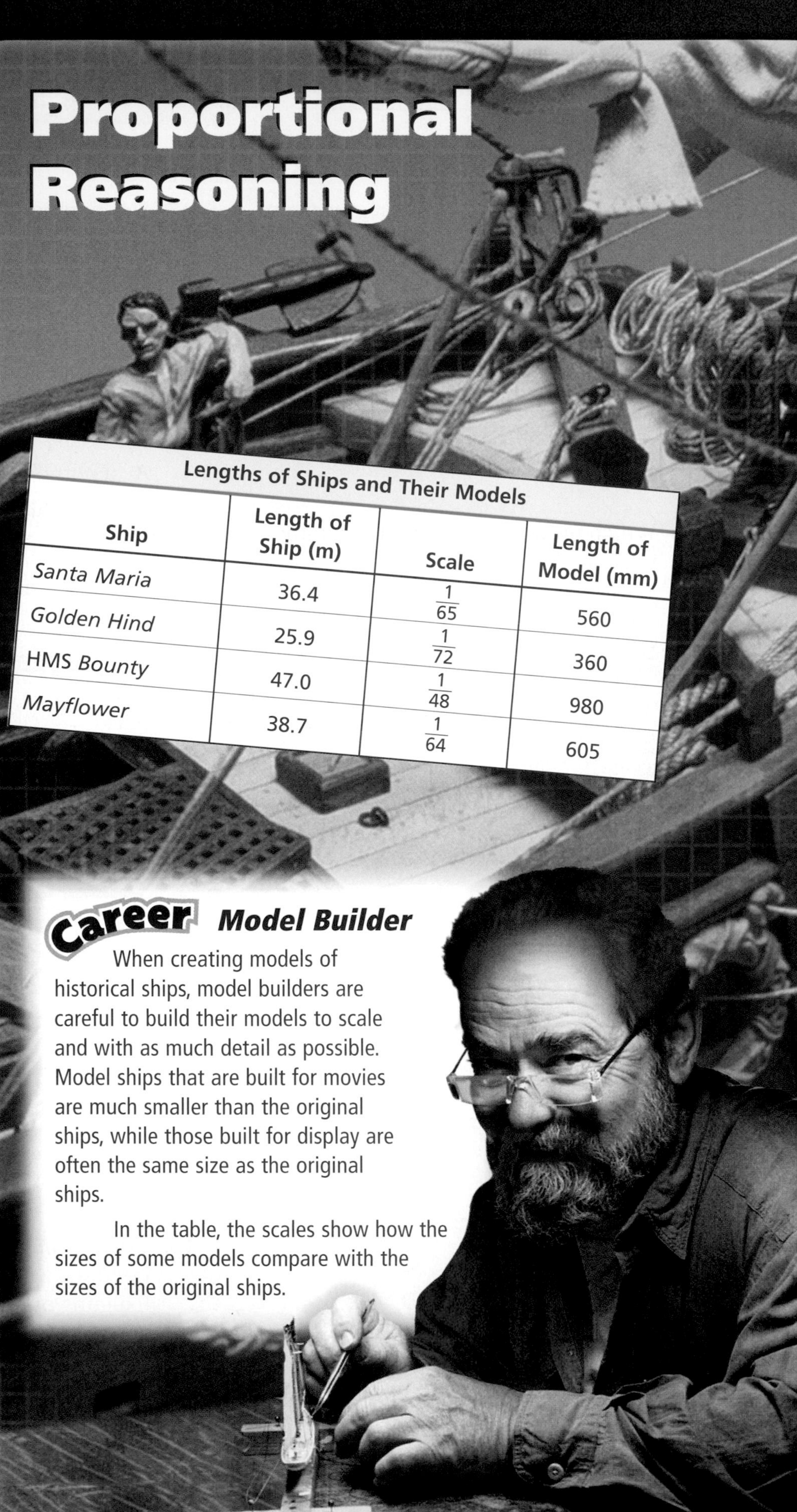

Lengths of Ships and Their Models

Ship	Length of Ship (m)	Scale	Length of Model (mm)
Santa Maria	36.4	$\frac{1}{65}$	560
Golden Hind	25.9	$\frac{1}{72}$	360
HMS *Bounty*	47.0	$\frac{1}{48}$	980
Mayflower	38.7	$\frac{1}{64}$	605

Career ***Model Builder***

When creating models of historical ships, model builders are careful to build their models to scale and with as much detail as possible. Model ships that are built for movies are much smaller than the original ships, while those built for display are often the same size as the original ships.

In the table, the scales show how the sizes of some models compare with the sizes of the original ships.

Are You Ready?

Choose the best term from the list to complete each sentence.

1. A(n) __?__ is a number that represents a part of a whole.	**common denominator**
2. A closed figure with three sides is called a(n) __?__.	**equivalent**
3. Two fractions are __?__ if they represent the same number.	**fraction**
	quadrilateral
4. One way to compare two fractions is to first find a(n) __?__ between them.	**triangle**

Complete these exercises to review skills you will need for this chapter.

✔ Write Equivalent Fractions

Find two fractions that are equivalent to each fraction.

5. $\frac{2}{5}$ **6.** $\frac{7}{11}$ **7.** $\frac{25}{100}$ **8.** $\frac{4}{6}$

9. $\frac{5}{17}$ **10.** $\frac{15}{23}$ **11.** $\frac{24}{78}$ **12.** $\frac{150}{325}$

✔ Compare Fractions

Compare. Write < or >.

13. $\frac{5}{6}$ ■ $\frac{2}{3}$ **14.** $\frac{3}{8}$ ■ $\frac{2}{5}$ **15.** $\frac{6}{11}$ ■ $\frac{1}{4}$ **16.** $\frac{5}{8}$ ■ $\frac{11}{12}$

17. $\frac{8}{9}$ ■ $\frac{12}{13}$ **18.** $\frac{5}{11}$ ■ $\frac{7}{21}$ **19.** $\frac{4}{10}$ ■ $\frac{3}{7}$ **20.** $\frac{3}{4}$ ■ $\frac{2}{9}$

✔ Solve Multiplication Equations

Solve each equation.

21. $3x = 12$ **22.** $15t = 75$ **23.** $2y = 14$ **24.** $7m = 84$

25. $25c = 125$ **26.** $16f = 320$ **27.** $11n = 121$ **28.** $53y = 318$

✔ Multiply Fractions

Solve. Write each answer in simplest form.

29. $\frac{2}{3} \cdot \frac{5}{7}$ **30.** $\frac{12}{16} \cdot \frac{3}{9}$ **31.** $\frac{4}{9} \cdot \frac{18}{24}$ **32.** $\frac{1}{56} \cdot \frac{50}{200}$

33. $\frac{1}{5} \cdot \frac{5}{9}$ **34.** $\frac{7}{8} \cdot \frac{4}{3}$ **35.** $\frac{25}{100} \cdot \frac{30}{90}$ **36.** $\frac{46}{91} \cdot \frac{3}{6}$

5-1 Ratios and Rates

Learn to identify, write, and compare ratios and rates.

Vocabulary

ratio

rate

unit rate

In basketball practice, Kathlene made 17 baskets in 25 attempts. She compared the number of baskets she made to the total number of attempts she made by using the *ratio* $\frac{17}{25}$. A **ratio** is a comparison of two quantities by division.

Kathlene can write her ratio of baskets made to attempts in three different ways.

$\frac{17}{25}$ 17 to 25 17:25

EXAMPLE 1 Writing Ratios

A recipe for homemade ice cream calls for 6 cups of cream, 1 cup of sugar, and 2 cups of fruit. Write each ratio in all three forms.

A cups of fruit to cups of cream

$\frac{2}{6}$, 2 to 6, 2:6 *For every 2 cups of fruit there are 6 cups of cream.*

The fraction $\frac{2}{6}$ can be simplified to $\frac{1}{3}$, so you can also write the following ratios:

$\frac{1}{3}$, 1 to 3, 1:3 *For each cup of fruit there are 3 cups of cream.*

B cups of sugar to total cups of ice cream mixture

$6 + 1 + 2 = 9$ *Find the total number of cups in the mixture.*

$\frac{1}{9}$, 1 to 9, 1:9 *For each cup of sugar there are 9 cups of mixture.*

A ratio that compares two quantities measured in different units is a **rate**. Suppose Ms. Latocki drove 75 miles in 3 hours. Her rate of travel was 75 miles in 3 hours, or $\frac{75 \text{ mi}}{3 \text{ hr}}$.

If the measure of the second quantity in a rate is one unit, then the rate is a **unit rate**. To change a rate to a unit rate, divide both the numerator and the denominator by the number in the denominator.

$$\frac{75 \text{ mi}}{3 \text{ hr}} = \frac{75 \text{ mi} \div 3}{3 \text{ hr} \div 3} = \frac{25 \text{ mi}}{1 \text{ hr}}$$

The unit rate 25 miles per hour expresses the average number of miles Ms. Latocki drove each hour.

Reading Math

The unit rate $\frac{25 \text{ miles}}{1 \text{ hour}}$ is read as "twenty-five miles per hour."

EXAMPLE 2 Writing Rates and Unit Rates

Find the unit rates and write them in both fraction and word form.

A **Belinda biked 36 miles in 4 hours.**

$\frac{36 \text{ mi}}{4 \text{ hr}}$ *Rate in fraction form*

$\frac{36 \text{ mi} \div 4}{4 \text{ hr} \div 4} = \frac{9 \text{ mi}}{1 \text{ hr}}$ *Unit rate in fraction form*

She rode 9 miles per hour. *Unit rate in word form*

B **A recipe for a garden weed killer involves mixing 11 tablespoons of liquid soap with 2 quarts of water.**

$\frac{11 \text{ tbsp}}{2 \text{ qt}}$ *Rate in fraction form*

$\frac{11 \text{ tbsp} \div 2}{2 \text{ qt} \div 2} = \frac{5.5 \text{ tbsp}}{1 \text{ qt}}$ *Unit rate in fraction form*

The recipe calls for 5.5 tablespoons of soap per quart of water. *Unit rate in word form*

It is often easy to compare ratios when they are written as fractions in simplest form—especially when they have a common denominator.

EXAMPLE 3 Simplifying Ratios to Make Comparisons

Tell which cookie has the greater ratio of protein to serving size.

	Fruit Cookie	Chocolate Cookie
Serving size	30 grams	45 grams
Protein	2 grams	3 grams
Fat	0 grams	9 grams
Energy	90 calories	200 calories

Fruit cookie: $\frac{\text{protein grams}}{\text{serving grams}} = \frac{2}{30} = \frac{1}{15}$ *Simplify the ratio.*

Chocolate cookie: $\frac{\text{protein grams}}{\text{serving grams}} = \frac{3}{45} = \frac{1}{15}$ *Simplify the ratio.*

Both cookies have the same ratio of protein to serving size.

Think and Discuss

1. **Tell** how to identify whether a ratio is a rate.
2. **Give an example** of a ratio that can be written as a unit rate.
3. **Explain** why the ratio in Example 1B is considered a "part-to-whole" ratio.

5-1 Exercises

FOR EOG PRACTICE
see page 674

1.01

GUIDED PRACTICE

See Example 1

Sun-Li has 10 blue marbles, 3 red marbles, and 17 white marbles. Write each ratio in all three forms.

1. blue marbles to red marbles
2. red marbles to total marbles

See Example 2

Find the unit rates and write them in both fraction and word form.

3. Helena runs 10 miles in 2 hours.

4. Geoff scores 96 points in 8 games.

See Example 3

5. A 5-lb box of soap powder costs $4.65. A 10-lb box of the same powder costs $8.90. Which is the better price? Explain.

INDEPENDENT PRACTICE

See Example 1

A soccer league has 25 sixth-graders, 30 seventh-graders, and 15 eighth-graders. Write each ratio in all three forms.

6. 6th-graders to 7th-graders
7. 6th-graders to total students
8. 7th-graders to 8th-graders
9. 7th- and 8th-graders to 6th-graders

See Example 2

Find the unit rates and write them in both fraction and word form.

10. Sam bikes 100 miles in 8 hours.

11. Kendra spends $21 for 3 CDs.

12. A recipe calls for 15 fl oz of punch mix per 3 qt of water.

See Example 3

13. Jonie paid $10.70 for 10 gallons of gasoline. André paid $15.60 for 12 gallons of gasoline. Who paid the higher price per gallon? Explain.

PRACTICE AND PROBLEM SOLVING

14. ***CONSUMER MATH*** Bottles of water are sold in various sizes. Write the ratios of price per volume for each size of bottled water. Which size is the best value?

$0.29
12 oz

$0.43
24.8 oz

$0.59
33.8 oz

$1.29
128 oz

Physical Science LINK

The pressure of water at different depths can be measured in *atmospheres,* or atm. The water pressure on a scuba diver increases as the diver descends below the surface. Use the table for Exercises 15–17.

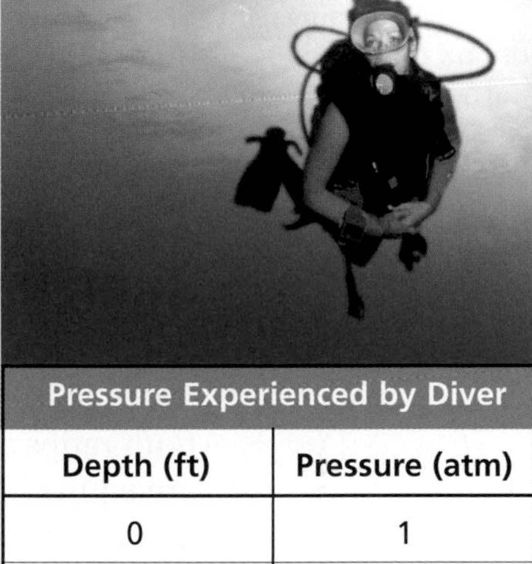

Pressure Experienced by Diver

Depth (ft)	Pressure (atm)
0	1
−33	2
−66	3
−99	4

15. Write each ratio in all three forms.

a. pressure at −33 ft to pressure at surface

b. pressure at −66 ft to pressure at surface

c. pressure at −99 ft to pressure at surface

d. pressure at −66 ft to pressure at −33 ft

e. pressure at −99 ft to pressure at −66 ft

16. A scuba diver descends 33 feet below the surface.

a. Write a rate in fraction form that shows the relation between the change in water pressure and the change in the diver's depth.

b. Tell how you would write this rate as a unit rate. Then write the unit rate in word form.

17. ***CHALLENGE*** The ratio of the beginning pressure and the new pressure when a scuba diver goes from −33 ft to −66 ft is less than the ratio of pressures when the diver goes from the surface to −33 ft. The ratio of pressures is even less when the diver goes from −66 ft to −99 ft. Use the ratios that you wrote in Exercise 15 to explain why this is true.

go.hrw.com
KEYWORD: MS4 Pressure
CNN Student News

Spiral Review

Solve each equation. (Lesson 2-12)

18. $3n = 18$ **19.** $15c = 225$ **20.** $8b = 120$ **21.** $19m = 95$

Find three fractions equivalent to each given fraction. (Lesson 3-8)

22. $\frac{1}{2}$ **23.** $\frac{3}{5}$ **24.** $\frac{1}{9}$ **25.** $\frac{4}{7}$

Write each fraction in simplest form. (Lesson 3-8)

26. $\frac{8}{12}$ **27.** $\frac{15}{18}$ **28.** $\frac{21}{24}$ **29.** $\frac{18}{36}$

30. **EOG PREP** Which is the product of 3.68 · 7.1? (Lesson 4-3)

A 2.6128 B 26.128 C 261.28 D 2,612.8

5-2 Identifying and Writing Proportions

Learn to find equivalent ratios and to identify proportions.

Vocabulary

equivalent ratios

proportion

Students in Mr. Howell's math class are measuring the width w and the length ℓ of their heads. The ratio of ℓ to w is 10 inches to 6 inches for Jean and 25 centimeters to 15 centimeters for Pat.

Calipers have adjustable arms that are used to measure the thickness of objects.

These ratios can be written as the fractions $\frac{10}{6}$ and $\frac{25}{15}$. Since both ratios simplify to $\frac{5}{3}$, they are equivalent. **Equivalent ratios** are ratios that name the same comparison.

An equation stating that two ratios are equivalent is called a **proportion** . The equation, or proportion, below states that the ratios $\frac{10}{6}$ and $\frac{25}{15}$ are equivalent.

Reading Math

Read the proportion $\frac{10}{6} = \frac{25}{15}$ by saying "ten is to six as twenty-five is to fifteen."

$$\frac{10}{6} = \frac{25}{15}$$

If two ratios are equivalent, they are said to be *proportional* to each other, or *in proportion.*

EXAMPLE 1 Comparing Ratios in Simplest Form

Determine whether the ratios are proportional.

A $\frac{2}{7}, \frac{6}{21}$

$\frac{2}{7}$ — *$\frac{2}{7}$ is already in simplest form.*

$\frac{6}{21} = \frac{6 \div 3}{21 \div 3} = \frac{2}{7}$ — *Simplify $\frac{6}{21}$.*

Since $\frac{2}{7} = \frac{2}{7}$, the ratios are proportional.

B $\frac{8}{24}, \frac{6}{20}$

$\frac{8}{24} = \frac{8 \div 8}{24 \div 8} = \frac{1}{3}$ — *Simplify $\frac{8}{24}$.*

$\frac{6}{20} = \frac{6 \div 2}{20 \div 2} = \frac{3}{10}$ — *Simplify $\frac{6}{20}$.*

Since $\frac{1}{3} \neq \frac{3}{10}$, the ratios are *not* proportional.

EXAMPLE 2 Comparing Ratios Using a Common Denominator

Use the data in the table to determine whether the ratios of oats to water are proportional for both servings of oatmeal.

Servings of Oatmeal	Cups of Oats	Cups of Water
8	2	4
12	3	6

Write the ratios of oats to water for 8 servings and for 12 servings.

Ratio of oats to water, 8 servings: $\frac{2}{4}$ *Write the ratio as a fraction.*

Ratio of oats to water, 12 servings: $\frac{3}{6}$ *Write the ratio as a fraction.*

$\frac{2}{4} = \frac{2 \cdot 6}{4 \cdot 6} = \frac{12}{24}$ *Write the ratios with a common denominator, such as 24.*

$\frac{3}{6} = \frac{3 \cdot 4}{6 \cdot 4} = \frac{12}{24}$

Since both ratios are equal to $\frac{12}{24}$, they are proportional.

You can find an equivalent ratio by multiplying or dividing the numerator and the denominator of a ratio by the same number.

EXAMPLE 3 Finding Equivalent Ratios and Writing Proportions

Find a ratio equivalent to each ratio. Then use the ratios to write a proportion.

A $\frac{8}{14}$

$\frac{8}{14} = \frac{8 \cdot 20}{14 \cdot 20} = \frac{160}{280}$ *Multiply both the numerator and denominator by any number, such as 20.*

$\frac{8}{14} = \frac{160}{280}$ *Write a proportion.*

B $\frac{4}{18}$

$\frac{4}{18} = \frac{4 \div 2}{18 \div 2} = \frac{2}{9}$ *Divide both the numerator and denominator by a common factor, such as 2.*

$\frac{4}{18} = \frac{2}{9}$ *Write a proportion.*

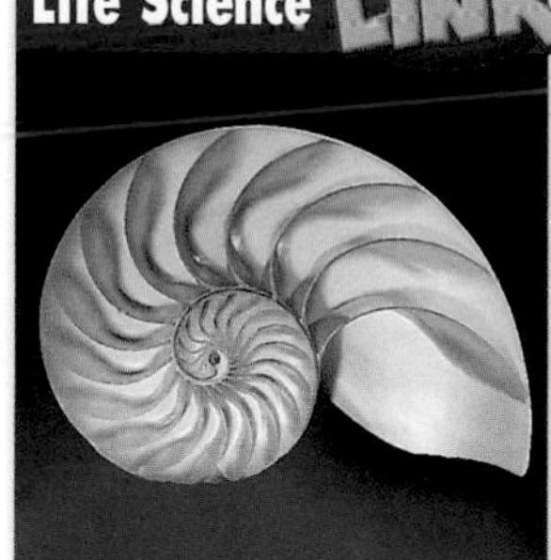

The ratios of the sizes of the segments of a nautilus shell are approximately equal to the *golden ratio*, 1.618.... This ratio can be found in many places in nature.

go.hrw.com
KEYWORD: MS4 Golden

CNN Student News

Think and Discuss

1. **Explain** why the ratios in Example 1B are not proportional.
2. **Describe** what it means for ratios to be proportional.
3. **Give an example** of a proportion. Then tell how you know it is a proportion.

5-2 Exercises

FOR EOG PRACTICE

see page 674

internet connect
Homework Help Online
go.hrw.com Keyword: MS4 5-2

1.01

GUIDED PRACTICE

See Example 1 **Determine whether the ratios are proportional.**

1. $\frac{2}{3}, \frac{4}{6}$ **2.** $\frac{5}{10}, \frac{8}{18}$ **3.** $\frac{9}{12}, \frac{15}{20}$ **4.** $\frac{3}{4}, \frac{8}{12}$

See Example 2 **5.** $\frac{10}{12}, \frac{15}{18}$ **6.** $\frac{6}{9}, \frac{8}{12}$ **7.** $\frac{3}{4}, \frac{5}{6}$ **8.** $\frac{4}{6}, \frac{6}{9}$

See Example 3 **Find a ratio equivalent to each ratio. Then use the ratios to write a proportion.**

9. $\frac{1}{3}$ **10.** $\frac{9}{21}$ **11.** $\frac{8}{3}$ **12.** $\frac{10}{4}$

INDEPENDENT PRACTICE

See Example 1 **Determine whether the ratios are proportional.**

13. $\frac{5}{8}, \frac{7}{14}$ **14.** $\frac{8}{24}, \frac{10}{30}$ **15.** $\frac{18}{20}, \frac{81}{180}$ **16.** $\frac{15}{20}, \frac{27}{35}$

See Example 2 **17.** $\frac{2}{3}, \frac{4}{9}$ **18.** $\frac{18}{12}, \frac{15}{10}$ **19.** $\frac{7}{8}, \frac{14}{24}$ **20.** $\frac{18}{54}, \frac{10}{30}$

See Example 3 **Find a ratio equivalent to each ratio. Then use the ratios to write a proportion.**

21. $\frac{5}{9}$ **22.** $\frac{27}{60}$ **23.** $\frac{6}{15}$ **24.** $\frac{121}{99}$

25. $\frac{11}{13}$ **26.** $\frac{5}{22}$ **27.** $\frac{78}{104}$ **28.** $\frac{27}{72}$

PRACTICE AND PROBLEM SOLVING

Complete each table of equivalent ratios.

29. 8 angelfish to 6 tiger fish

angelfish	4	■	■	20
tiger fish	■	6	18	■

30. 4 squares to 16 circles

squares	2	4	6	8
circles	■	■	■	■

Find two ratios equivalent to each given ratio.

31. 3 to 7 **32.** 6:2 **33.** $\frac{5}{12}$ **34.** 8:4

35. 6 to 9 **36.** $\frac{10}{50}$ **37.** 10:4 **38.** 1 to 10

39. ***ECOLOGY*** If you recycle one aluminum can, you save enough energy to run a TV for four hours.

a. Write the ratio of cans to hours.

b. Marti's class recycled enough aluminum cans to run a TV for 2,080 hours. Did the class recycle 545 cans? Justify your answer using equivalent ratios.

40. ***SOCIAL STUDIES*** Population density is the average number of people per unit of area.

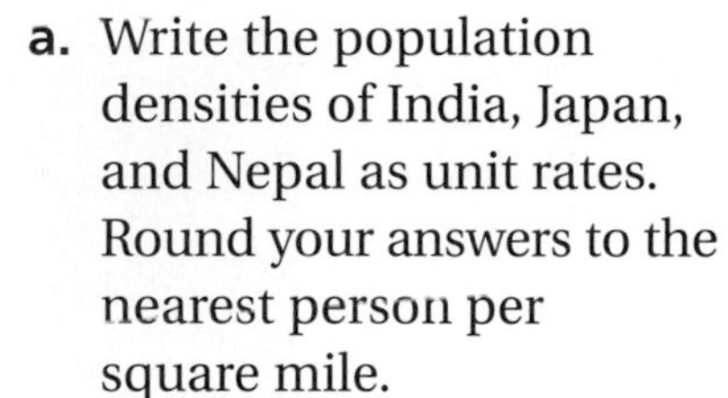

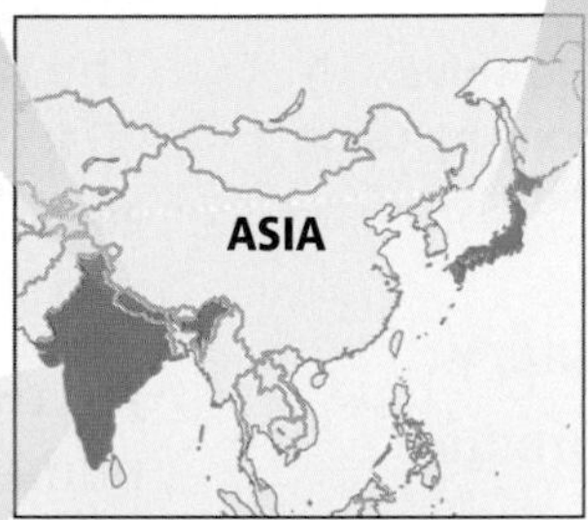

Source: 1999 World Almanac

a. Write the population densities of India, Japan, and Nepal as unit rates. Round your answers to the nearest person per square mile.

b. Which country has the greatest population density? Explain.

41. Last year in Kerry's school, the ratio of students to teachers was 22:1. Write an equivalent ratio to show how many students there were if the school had only 8 teachers.

42. Marcus earned \$230 for 40 hours of work. Phillip earned \$192 for 32 hours of work. Are these pay rates proportional? Explain.

43. ***WHAT'S THE ERROR?*** A student wrote the proportion $\frac{13}{20} = \frac{26}{60}$. What did the student do wrong?

44. ***WRITE ABOUT IT*** Explain two different ways to determine if two ratios are proportional.

45. ***CHALLENGE*** Write all possible proportions using only the numbers 1, 2, and 4.

Spiral Review

Write the prime factorization of each number. (Lesson 2-4)

46. 16 **47.** 120 **48.** 18 **49.** 48

Evaluate each algebraic expression for $x = 2$ and $y = 3$. (Lesson 2-7)

50. $2x$ **51.** $5 - y$ **52.** $3x + 2y$ **53.** $4y^2 - 3$

Add. (Lesson 3-3)

54. $8 + (-2)$ **55.** $-9 + 3$ **56.** $-7 + (-2)$ **57.** $11 + (-6)$

58. **EOG PREP** The best estimate of $\frac{1}{16} + \frac{1}{9}$ is __?__. (Lesson 4-9)

A 0 B $\frac{1}{2}$ C 1 D $1\frac{1}{2}$

5-3 Solving Proportions

Learn to solve proportions by using cross products.

Vocabulary

cross product

The tall stack of Jenga® blocks is 25.8 cm tall. How tall is the shorter stack of blocks? To find the answer, you will need to solve a proportion.

For two ratios, the product of the numerator in one ratio and the denominator in the other is a **cross product**. If the cross products of the ratios are equal, then the ratios form a proportion.

$$\frac{2}{5} = \frac{6}{15} \qquad 5 \cdot 6 = 30 \qquad 2 \cdot 15 = 30$$

CROSS PRODUCT RULE

In the proportion $\frac{a}{b} = \frac{c}{d}$, the cross products, $a \cdot d$ and $b \cdot c$, are equal.

You can use the cross product rule to solve proportions with variables.

EXAMPLE 1 **Solving Proportions Using Cross Products**

Use cross products to solve the proportion $\frac{p}{6} = \frac{10}{3}$.

$\frac{p}{6} = \frac{10}{3}$

$p \cdot 3 = 6 \cdot 10$ — *The cross products are equal.*

$3p = 60$ — *Multiply.*

$\frac{3p}{3} = \frac{60}{3}$ — *Divide each side by 3 to isolate the variable.*

$p = 20$

When setting up a proportion to solve a problem, use a variable to represent the number you want to find. In proportions that include different units of measurement, either the units in the numerators must be the same and the units in the denominators must be the same or the units within each ratio must be the same.

$$\frac{16\text{ mi}}{4\text{ hr}} = \frac{8\text{ mi}}{x\text{ hr}} \qquad \frac{16\text{ mi}}{8\text{ mi}} = \frac{4\text{ hr}}{x\text{ hr}}$$

EXAMPLE 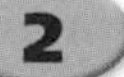 2 PROBLEM SOLVING APPLICATION

A stack of 18 Jenga blocks is about 25.8 cm tall. What is the height, to the nearest tenth of a centimeter, of a stack of 11 Jenga blocks?

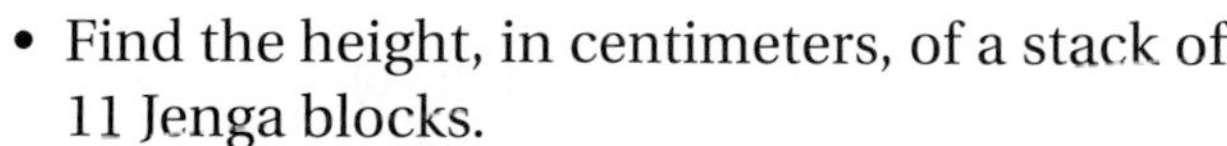

1 Understand the Problem

Rewrite the question as a statement.

- Find the height, in centimeters, of a stack of 11 Jenga blocks.

List the **important information:**

- A stack of 18 Jenga blocks is about 25.8 cm tall.

2 Make a Plan

Set up a proportion using the given information.

$$\frac{18 \text{ blocks}}{25.8 \text{ cm}} = \frac{11 \text{ blocks}}{h}$$ *Let h be the unknown height.*

3 Solve

$\frac{18}{25.8} = \frac{11}{h}$ *Write the proportion.*

$18 \cdot h = 25.8 \cdot 11$ *The cross products are equal.*

$18h = 283.8$ *Multiply.*

$\frac{18h}{18} = \frac{283.8}{18}$ *Divide each side by 18 to isolate the variable.*

$h \approx 15.7\overline{6}$

$h \approx 15.8$ *Round to the nearest tenth.*

The stack of 11 Jenga blocks is about 15.8 cm.

4 Look Back

Since the height of the smaller stack of blocks was rounded, the cross products of the proportion are not equal.

$\frac{18}{25.8} \neq \frac{11}{15.8}$ $\quad$ **$11 \cdot 25.8 = 283.8$**

$18 \cdot 15.8 = 284.4$

However, the cross products are close in value, so 15.8 cm is a reasonable answer.

Think and Discuss

1. **Describe** the error in these steps: $\frac{2}{3} = \frac{x}{12}$; $2x = 36$; $x = 18$.
2. **Show** how to use cross products to decide whether the ratios 6:45 and 2:15 are proportional.

5-3 Exercises

FOR EOG PRACTICE
see page 674

internet connect
Homework Help Online
go.hrw.com Keyword: MS4 5-3

1.01, 1.02d

GUIDED PRACTICE

See Example 1
Use cross products to solve each proportion.

1. $\frac{6}{10} = \frac{36}{x}$
2. $\frac{4}{7} = \frac{5}{p}$
3. $\frac{12.3}{m} = \frac{75}{100}$
4. $\frac{t}{42} = \frac{1.5}{3}$

See Example 2

5. A stack of 2,450 one dollar bills weighs 5 pounds. How much does a stack of 1,470 one dollar bills weigh?

INDEPENDENT PRACTICE

See Example 1
Use cross products to solve each proportion.

6. $\frac{4}{36} = \frac{x}{180}$
7. $\frac{7}{84} = \frac{12}{h}$
8. $\frac{3}{24} = \frac{r}{52}$
9. $\frac{5}{140} = \frac{12}{v}$
10. $\frac{45}{x} = \frac{15}{3}$
11. $\frac{t}{6} = \frac{96}{16}$
12. $\frac{2}{5} = \frac{s}{12}$
13. $\frac{14}{n} = \frac{5}{8}$

See Example 2

14. Euro coins come in eight denominations. One denomination is the one-euro coin, which is worth 100 cents. A stack of 10 one-euro coins is 21.25 millimeters tall. How tall would a stack of 45 one-euro coins be? Round your answer to the nearest hundredth of a millimeter.
15. There are 18.5 ounces of soup in a can. This is equivalent to 524 grams. If Jenna has 8 ounces of soup, how many grams does she have? Round your answer to the nearest whole gram.

PRACTICE AND PROBLEM SOLVING

Solve each proportion. Then find another equivalent ratio.

16. $\frac{4}{h} = \frac{12}{24}$
17. $\frac{x}{15} = \frac{12}{90}$
18. $\frac{39}{4} = \frac{t}{12}$
19. $\frac{5.5}{6} = \frac{16.5}{w}$
20. $\frac{1}{3} = \frac{y}{25.5}$
21. $\frac{18}{x} = \frac{1}{5}$
22. $\frac{m}{4} = \frac{175}{20}$
23. $\frac{8.7}{2} = \frac{q}{4}$
24. $\frac{r}{84} = \frac{32.5}{182}$
25. $\frac{76}{304} = \frac{81}{k}$
26. $\frac{9}{500} = \frac{p}{2,500}$
27. $\frac{5}{j} = \frac{6}{19.8}$

Arrange each set of numbers to form a proportion.

28. 10, 6, 30, 18
29. 4, 6, 10, 15
30. 12, 21, 7, 4
31. 75, 4, 3, 100
32. 30, 42, 5, 7
33. 5, 90, 108, 6
34. If you put an object that weighs 8 grams on one side of a balance scale, you would have to put about 20 paper clips on the other side to balance the weight. How many paper clips would balance the weight of a 10-gram object?
35. Sandra drove 126.2 miles in 2 hours at a constant speed. Use a proportion to find how long it would take her to drive 189.3 miles at the same speed.

36. ***LIFE SCIENCE*** On Monday a marine biologist took a random sample of 50 fish from a pond and tagged them. On Tuesday she took a new sample of 100 fish. Among them were 4 fish that had been tagged on Monday.

   a. What does the ratio $\frac{4}{100}$ show?

   b. How can you show the ratio of fish tagged on Monday to n, the estimated total number of fish in the pond?

   c. Use a proportion to estimate the number of fish in the pond.

37. ***CHEMISTRY*** The table shows the type and number of atoms in one molecule of citric acid. Use a proportion to find the number of oxygen atoms in 15 molecules of citric acid.

Composition of Citric Acid	
Type of Atom	**Number of Atoms**
Carbon	6
Hydrogen	8
Oxygen	7

38. A certain shade of paint is made by mixing 5 parts blue paint with 2 parts white paint. To get the correct shade, how many quarts of white paint should be mixed with 8.5 quarts of blue paint?

39. ***EARTH SCIENCE*** You can find your distance from a thunderstorm by counting the number of seconds between a lightning flash and the thunder. For example, if the time difference is 21 s, then the storm is 7 km away. How far away is a storm if the time difference is 9 s?

40. ***WHAT'S THE QUESTION?*** There are 20 grams of protein in 3 ounces of sautéed fish. If the answer is 9 ounces, what is the question?

41. ***WRITE ABOUT IT*** Give an example from your own life that can be described using a ratio. Then tell how a proportion can give you additional information.

42. ***CHALLENGE*** How can you use the proportion $\frac{a}{b} = \frac{c}{d}$ to show why the cross product rule works?

Spiral Review

Solve. (Lesson 3-6)

43. $y - 5 = -2$

44. $9 + n = -4$

45. $5 = x - 8$

Subtract. (Lesson 4-2)

46. $9.2 - 3.76$

47. $17 - 1.97$

48. $4.235 - 1.407$

49. **EOG PREP** What is the solution of $x + 1.6 = -4.87$? (Lesson 4-6)

   A 6.47
   B 3.27
   C −5.47
   D −6.47

50. **EOG PREP** What is the sum of $\frac{3}{8} + \frac{5}{24}$ in simplest form? (Lesson 4-10)

   A $\frac{7}{12}$
   B $\frac{14}{24}$
   C $\frac{1}{3}$
   D $\frac{1}{6}$

5-4 Dimensional Analysis

Learn to use dimensional analysis to make unit conversions.

Vocabulary

unit conversion factor

In 1999, NASA's *Climate Observer* was destroyed by the atmosphere of Mars. This happened because one NASA team used customary units (feet, pounds, etc.), while another team used metric units (meters, grams, etc.). The problem could have been avoided by changing the units of measurement used by one team to match those of the other team.

You can use a *unit conversion factor* to change, or convert, measurements from one unit to another. A **unit conversion factor** is a fraction in which the numerator and denominator represent the same quantity, but in different units. The fraction below is a unit conversion factor that can be used to convert miles to feet. Notice that it can be simplified to 1.

$$\frac{5{,}280 \text{ ft}}{1 \text{ mi}} = \frac{5{,}280 \text{ ft}}{5{,}280 \text{ ft}} = 1$$

Multiplying a quantity by a unit conversion factor changes only its units, not its value. The process of choosing an appropriate conversion factor is called dimensional analysis.

EXAMPLE 1 **Making Unit Conversions**

A bucket holds 16 quarts. How many gallons of water will fill the bucket? Use a unit conversion factor to convert the units.

One gallon equals 4 quarts, so use the unit conversion factor $\frac{1 \text{ gal}}{4 \text{ qt}}$ or $\frac{4 \text{ qt}}{1 \text{ gal}}$. Choose the first one so that the quart units will "cancel."

$$16 \cancel{\text{qt}} \cdot \frac{1 \text{ gal}}{4 \cancel{\text{qt}}} = \frac{16 \text{ gal}}{4} \quad \textit{Multiply.}$$

$$= 4 \text{ gal}$$

Four gallons will fill the 16-quart bucket.

Helpful Hint

When choosing a unit conversion factor, choose the one that cancels the units you want to change and replaces them with the units you want.

EXAMPLE 2 Making Rate Conversions

Use a unit conversion factor to convert the units within each rate.

A **Convert 80 miles per hour to feet per hour.**

There are 5,280 feet per mile, so use $\frac{5{,}280 \text{ ft}}{1 \text{ mi}}$ to cancel the miles.

$$\frac{80 \cancel{\text{mi}}}{1 \text{ hr}} \cdot \frac{5{,}280 \text{ ft}}{1 \cancel{\text{mi}}} = \frac{80 \cdot 5{,}280 \text{ ft}}{1 \text{ hr}} \qquad \textit{Multiply.}$$

$$= \frac{422{,}400 \text{ ft}}{1 \text{ hr}}$$

80 miles per hour is 422,400 feet per hour.

B **A phone service in the United States charges $1.99 per minute for a call to Australia. How many dollars per hour is the phone service charging?**

There are 60 minutes per hour, so use $\frac{60 \text{ min}}{1 \text{ hr}}$ to cancel the minutes.

$$\frac{\$1.99}{1 \cancel{\text{min}}} \cdot \frac{60 \cancel{\text{min}}}{1 \text{ hr}} = \frac{\$1.99 \cdot 60}{1 \text{ hr}} \qquad \textit{Multiply.}$$

$$= \frac{\$119.40}{1 \text{ hr}}$$

$1.99 per minute is $119.40 per hour.

EXAMPLE 3 *Measurement Application*

The *Climate Observer* missed its target altitude of about 145 km above Mars during its entry into orbit. Instead, its altitude was 57 km, where it was destroyed by the Martian atmosphere. How many feet above Mars is 145 km?

Use the unit conversion factors that convert kilometers to miles, and then miles to feet. One kilometer is equivalent to 0.62 mile.

$$145 \cancel{\text{km}} \cdot \frac{0.62 \cancel{\text{mi}}}{1 \cancel{\text{km}}} \cdot \frac{5{,}280 \text{ ft}}{1 \cancel{\text{mi}}} = \frac{145 \cdot 0.62 \cdot 5{,}280 \text{ ft}}{1}$$

$$= 474{,}672 \text{ ft}$$

145 km is 474,672 ft.

Think and Discuss

1. **Explain** why you cannot use the unit conversion factor $\frac{100 \text{ cm}}{1 \text{ m}}$ to convert 50 centimeters to meters.
2. **Tell** whether you get an equivalent rate when you multiply a rate by a conversion factor. Explain.
3. **Compare** the process of converting feet to inches with the process of converting feet per minute to inches per second.

5-4 Exercises

FOR EOG PRACTICE
see page 675

internet connect
Homework Help Online
go.hrw.com Keyword: MS4 5-4

1.01

GUIDED PRACTICE

See Example 1 **Use a unit conversion factor to convert the units.**

1. A bag of apples weighs 64 ounces. How many pounds does it weigh?
2. Dario drank 2 liters of water. How many milliliters of water did he drink?

See Example 2 **Use a unit conversion factor to convert the units within each rate.**

3. Convert 32 feet per second to inches per second.
4. A craft store charges $1.75 per foot for lace. How much per yard is this?

See Example 3

5. Pluto has a diameter of about 1,423 miles. What is the planet's diameter in meters? (There are about 1.61 kilometers in 1 mile.)

INDEPENDENT PRACTICE

See Example 1 **Use a unit conversion factor to convert the units.**

6. A soup recipe calls for 3.5 quarts of water. How many pints of water are needed?
7. You need 48 inches of ribbon. How many feet of ribbon do you need?

See Example 2 **Use a unit conversion factor to convert the units within each rate.**

8. Convert 63,360 feet per hour to miles per hour.
9. A company rents boats for $9 per hour. How much per minute is this?

See Example 3

10. Andy is 5.25 feet tall. What is Andy's height to the nearest centimeter? (There are about 2.54 centimeters in 1 inch.)

PRACTICE AND PROBLEM SOLVING

Write the appropriate unit conversion factor for each conversion.

11. inches to feet
12. meters to centimeters
13. minutes to hours
14. yards to feet

Convert each quantity to the given units.

15. 42 inches to feet
16. 1 hour to seconds
17. 2 kilometers to meters
18. 7 weeks to minutes
19. 36 pints to gallons
20. 500 milliliters to liters

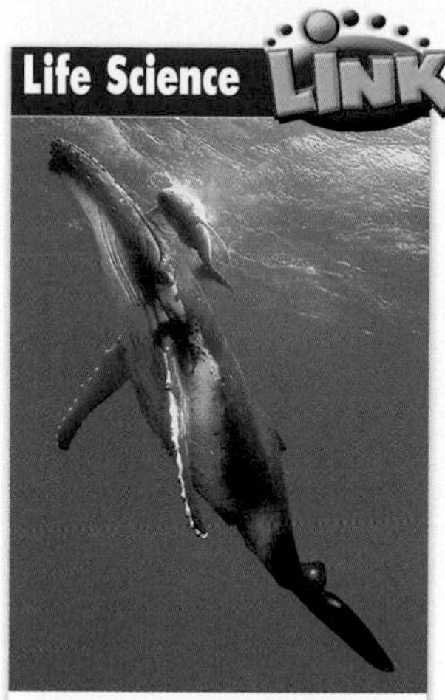

Whales communicate with each other by singing. Their songs are made up of moans, grunts, blasts, and shrieks, and the sound can travel up to 10,000 miles through the ocean.

21. ***PHYSICAL SCIENCE*** It takes 5 seconds for sound to travel 1 mile through air. It takes only 1 second for sound to travel 1 mile through water. How many minutes faster will a sound 90 miles away travel through water than through air?

22. ***EARTH SCIENCE*** The amount of time it takes for a planet to revolve around the Sun is called a period of revolution. The period of revolution for Earth is one Earth year, and the period of revolution for any other planet is one year on that planet.

    Periods of Revolution Compared to Earth's

Planet	One Revolution in Earth Years
Venus	0.615
Mars	1.88
Neptune	164.79

    a. How many Earth years does it take Venus to revolve around the Sun?

    b. Use a unit conversion factor to find the number of Venus years equivalent to three Earth years. Round to the nearest tenth.

    c. Find your age on each planet to the nearest year.

23. In England, a commonly used unit of measure is the stone. One stone is equivalent to 14 pounds. If Jo weighs 95 pounds, about how many stone does she weigh? Round your answer to the nearest tenth of a stone.

24. ***MONEY*** Fencing costs \$3.75 per foot. Harris wants to enclose his rectangular garden, which measures 6 yards by 4 yards. How much will fencing for the garden cost?

25. ***WRITE A PROBLEM*** Use units of time in a problem that can be solved by dimensional analysis.

26. ***WRITE ABOUT IT*** Explain how to determine which unit conversion factor to use when solving a problem using dimensional analysis.

27. ***CHALLENGE*** Your car gets 45 miles per gallon of gasoline. You have \$10, and gasoline costs \$1.50 per gallon. How far can you travel on \$10?

Spiral Review

Use compatible numbers to estimate. (Lesson 4-1)

28. $4.08 \cdot 31.17$ **29.** $-6.19 \cdot 88.5$ **30.** $53.7 \div 8.89$

Find each product. (Lesson 4-7)

31. $\frac{1}{3} \cdot 7$ **32.** $2 \cdot \frac{5}{6}$ **33.** $1\frac{4}{9} \cdot \frac{1}{3}$ **34.** $2\frac{1}{2} \cdot 9\frac{1}{3}$

35. **EOG PREP** What is the solution of $\frac{1}{8} \cdot x = 7$? (Lesson 4-12)

A $6\frac{7}{8}$ C 54

B 49 D 56

36. **EOG PREP** What is the solution of $w - \frac{4}{9} = \frac{2}{9}$? (Lesson 4-12)

A 6 C $\frac{2}{3}$

B 2 D $\frac{2}{9}$

Chapter 5

Mid-Chapter Quiz

LESSON 5-1 (pp. 260–263)

There are 16 boys and 14 girls on the bus. Write each ratio in all three forms.

1. boys to girls

2. girls to boys

3. girls to total students

4. total students to boys

Find the unit rates and write them in both fraction and word form.

5. Sean walked 7 miles in 2 hours.

6. Kara made juice using 12 fluid ounces of concentrate for 3 quarts of water.

7. Last month Mr. Oliver used 21 gallons of gasoline to drive 525 miles. This month he used 20 gallons to drive 480 miles. During which month did he get better gas mileage?

LESSON 5-2 (pp. 264–267)

Determine whether the ratios are proportional.

8. $\frac{6}{8}, \frac{9}{12}$

9. $\frac{16}{36}, \frac{20}{30}$

10. $\frac{7}{10}, \frac{21}{30}$

11. $\frac{21}{49}, \frac{15}{35}$

Find a ratio equivalent to each ratio. Then use the ratios to write a proportion.

12. $\frac{10}{16}$

13. $\frac{21}{28}$

14. $\frac{12}{25}$

15. $\frac{40}{48}$

LESSON  5-3 (pp. 268–271)

Use cross products to solve each proportion.

16. $\frac{n}{8} = \frac{15}{4}$

17. $\frac{20}{t} = \frac{2.5}{6}$

18. $\frac{6}{11} = \frac{0.12}{z}$

19. $\frac{15}{24} = \frac{x}{10}$

20. The average life expectancy of a dog is 12 to 14 years. One dog year is said to equal 7 human years. If Cliff's dog is 5.5 years old in dog years, what is his dog's age in human years?

LESSON  5-4 (pp. 272–275)

Use a unit conversion factor to convert the units.

21. 8 gallons to quarts

22. 7,920 feet to miles

23. 18 minutes to seconds

24. 450 grams to kilograms

Use a unit conversion factor to convert the units within each rate.

25. 15 miles per hour to feet per hour

26. \$720 per hour to dollars per minute

1.02a

Focus on Problem Solving

Make a Plan

• Choose a problem-solving strategy

The following are strategies that you might choose to help you solve a problem:

- Make a table
- Find a pattern
- Make an organized list
- Work backward
- Draw a diagram
- Guess and test
- Use logical reasoning
- Solve a simpler problem

Tell which strategy from the list above you would use to solve each problem. Explain your choice.

1. A recipe for blueberry muffins calls for 1 cup of milk and 1.5 cups of blueberries. Ashley wants to make more muffins than the recipe yields. In Ashley's muffin batter, there are 4.5 cups of blueberries. If she is using the recipe as a guide, how many cups of milk will she need?

2. One side of a triangle is 3 cm long. Another side is 1 cm longer than the first side, and the third side is 1 cm longer than the second side. On a similar triangle, the side that corresponds to the 3 cm side of the first triangle is 6 m long. How long is the third side of the similar triangle?

3. Jeremy is the oldest of four brothers. Each of the four boys gets an allowance for doing chores at home each week. The amount of money each boy receives depends on his age. Jeremy is 13 years old, and he gets $7.05. His 11-year-old brother gets $6.11, and his 9-year-old brother gets $5.17. How much money does his 7-year-old brother get?

4. Jorge lives in Santa Fe, New Mexico, and is going to see his grandparents in Tucson, Arizona. By using his map, he found that he will travel 375 miles. The scale of Jorge's map is 1 inch = 120 miles. What other information did Jorge use to find the distance in miles that he will travel?

Make Similar Figures

Use with Lesson 5-5

2.01

REMEMBER

- The ratios of the lengths of the corresponding sides of similar figures are proportional.
- All of the angles of a rectangle have the same measure.

internet connect
Lab Resources Online
go.hrw.com
KEYWORD: MS4 Lab5A

You can make similar figures by increasing or decreasing each dimension of a rectangle while keeping the ratios of the lengths of the corresponding sides proportional. Modeling similar figures using square tiles can help you solve proportions.

Activity

A rectangle made of square tiles measures 5 tiles long and 2 tiles wide. What is the length of a similar rectangle whose width is 6 tiles?

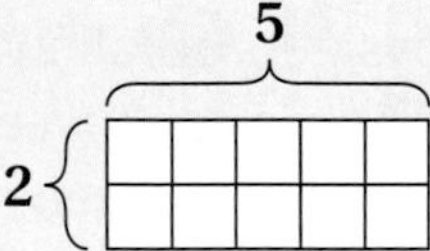

Use tiles to make a 5 × 2 rectangle.

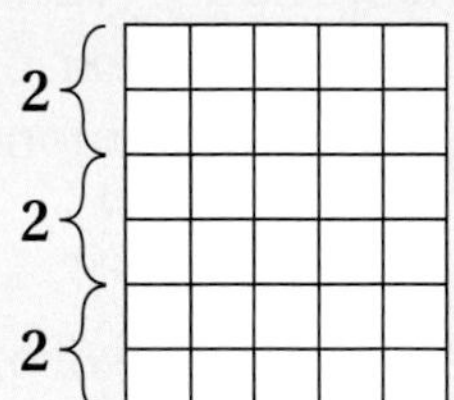

Add tiles to increase the width of the rectangle to 6 tiles.

Notice that there are now 3 sets of 2 tiles along the width of the rectangle because 2 × 3 = 6.

The width of the new rectangle is three times greater than the width of the original rectangle. To keep the ratios of the side measures proportional, the length must also be three times greater than the length of the original rectangle.

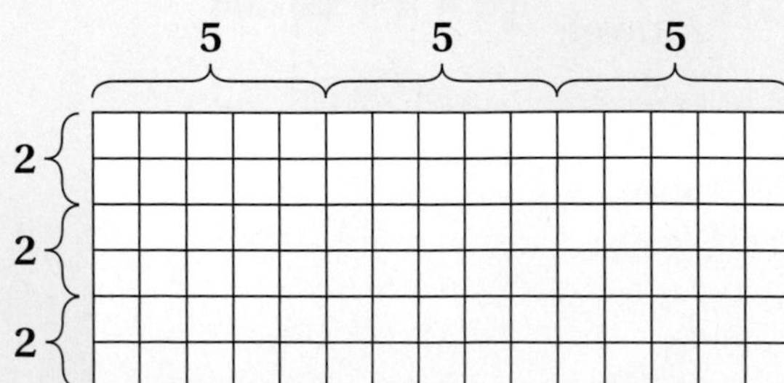

5 × 3 = 15

Add tiles to increase the length of the rectangle to 15 tiles.

The length of the similar rectangle is 15 tiles.

To check your answer, you can use ratios.

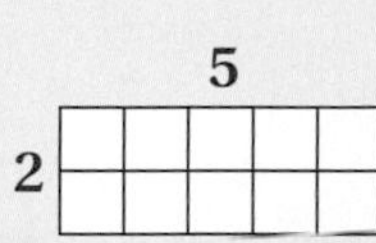

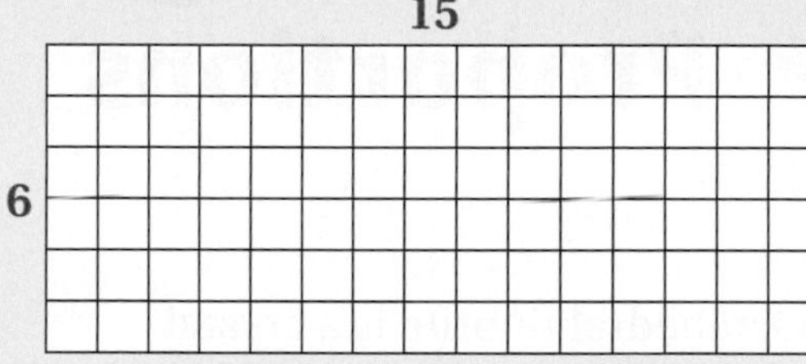

$\frac{2}{6} \stackrel{?}{=} \frac{5}{15}$ *Write ratios using the corresponding side lengths.*

$\frac{1}{3} \stackrel{?}{=} \frac{1}{3}$ *Simplify each ratio.*

1 Use square tiles to model similar figures with the given dimensions.

a. The original rectangle is 4 tiles wide by 3 tiles long. The similar rectangle is 8 tiles wide.

b. The original rectangle is 8 tiles wide by 10 tiles long. The similar rectangle is 15 tiles long.

Think and Discuss

1. In a backyard, a plot of land that is 5 yd × 8 yd is used to grow corn. The homeowner wants to decrease this plot to 5 yd × 4 yd. Will the new plot be similar to the original? Why or why not?
2. The homeowner wants to make the new plot similar to the original that is 5 yd × 8 yd by decreasing the longer side to 4 yd. What should the measure of the shorter side be?
3. What other dimensions could the homeowner have used for the new plot to make it similar to the old plot?

Try This

1. A rectangle is 3 feet long and 7 feet wide. What is the width of a similar rectangle whose length is 9 feet?
2. A rectangle is 6 feet long and 12 feet wide. What is the length of a similar rectangle whose width is 4 feet?

Use square tiles to model similar rectangles to solve each proportion.

3. $\frac{4}{5} = \frac{8}{x}$ **4.** $\frac{5}{9} = \frac{h}{18}$ **5.** $\frac{2}{y} = \frac{6}{18}$ **6.** $\frac{1}{t} = \frac{4}{16}$

7. $\frac{2}{3} = \frac{8}{m}$ **8.** $\frac{9}{12} = \frac{p}{4}$ **9.** $\frac{6}{r} = \frac{9}{15}$ **10.** $\frac{k}{12} = \frac{7}{6}$

5-5 Similar Figures and Proportions

Learn to use ratios to determine if two figures are similar.

Vocabulary

similar

corresponding sides

corresponding angles

Octahedral fluorite is a crystal found in nature. It grows in the shape of an octahedron, which is a solid figure with eight triangular faces. The triangles in different-sized fluorite crystals are *similar* figures. **Similar** figures have the same shape but not necessarily the same size.

Matching sides of two or more polygons are called **corresponding sides**, and matching angles are called **corresponding angles**.

SIMILAR FIGURES

If two figures are similar, then the measures of the corresponding angles are equal and the ratios of the lengths of the corresponding sides are proportional.

To find out if triangles are similar, determine whether the ratios of the lengths of their corresponding sides are proportional. If the ratios are proportional, then the corresponding angles must have equal measures.

EXAMPLE 1 Determining Whether Two Triangles Are Similar

Reading Math

A side of a figure can be named by its endpoints, with a bar above.

$\overline{AB}$

Without the bar, the letters indicate the *length* of the side.

Identify the corresponding sides in the pair of triangles. Then use ratios to determine whether the triangles are similar.

S
24 in.
36 in.
F
8 in.
12 in.
E 7 in. D
R
21 in.
Q

$\overline{DE}$ corresponds to $\overline{QR}$.
$\overline{EF}$ corresponds to $\overline{RS}$.
$\overline{DF}$ corresponds to $\overline{QS}$.

$\frac{DE}{QR} \stackrel{?}{=} \frac{EF}{RS} \stackrel{?}{=} \frac{DF}{QS}$ *Write ratios using the corresponding sides.*

$\frac{7}{21} \stackrel{?}{=} \frac{8}{24} \stackrel{?}{=} \frac{12}{36}$ *Substitute the lengths of the sides.*

$\frac{1}{3} \stackrel{?}{=} \frac{1}{3} \stackrel{?}{=} \frac{1}{3}$ *Simplify each ratio.*

Since the ratios of the corresponding sides are equivalent, the triangles are similar.

In figures with four or more sides, it is possible for the corresponding side lengths to be proportional and the figures to have different shapes. To find out if these figures are similar, first check that their corresponding angles have equal measures.

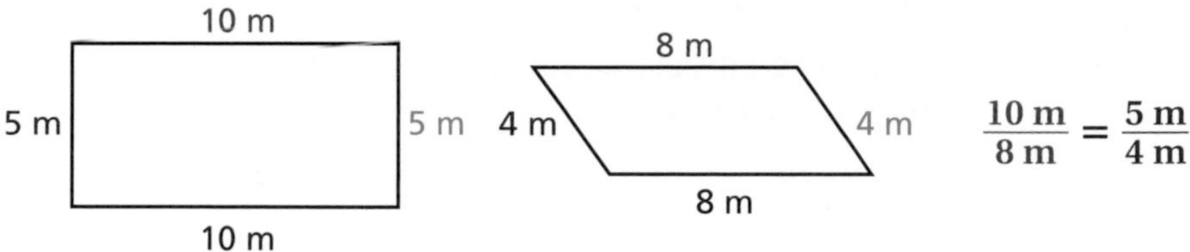

EXAMPLE 2 Determining Whether Two Four-Sided Figures Are Similar

Use the properties of similarity to determine whether the figures are similar.

The corresponding angles of the figures have equal measure. Write each set of corresponding sides as a ratio.

$\frac{EF}{LM}$ *$\overline{EF}$ corresponds to $\overline{LM}$.*

$\frac{FG}{MN}$ *$\overline{FG}$ corresponds to $\overline{MN}$.*

$\frac{GH}{NO}$ *$\overline{GH}$ corresponds to $\overline{NO}$.*

$\frac{EH}{LO}$ *$\overline{EH}$ corresponds to $\overline{LO}$.*

Determine whether the ratios of the lengths of the corresponding sides are proportional.

$\frac{EF}{LM} \stackrel{?}{=} \frac{FG}{MN} \stackrel{?}{=} \frac{GH}{NO} \stackrel{?}{=} \frac{EH}{LO}$ *Write ratios using the corresponding sides.*

$\frac{15}{6} \stackrel{?}{=} \frac{10}{4} \stackrel{?}{=} \frac{10}{4} \stackrel{?}{=} \frac{20}{8}$ *Substitute the lengths of the sides.*

$\frac{5}{2} \stackrel{?}{=} \frac{5}{2} \stackrel{?}{=} \frac{5}{2} \stackrel{?}{=} \frac{5}{2}$ *Simplify each ratio.*

Since the ratios of the corresponding sides are equivalent, the figures are similar.

Think and Discuss

1. **Identify** the corresponding angles of the triangles in Example 1.
2. **Explain** whether all rectangles are similar. Give specific examples to justify your answer.

5-5 Exercises

FOR EOG PRACTICE

see page 676

internet connect
Homework Help Online
go.hrw.com Keyword: MS4 5-5

1.01, 2.01, 3.02

GUIDED PRACTICE

See Example 1 **Identify the corresponding sides in each pair of triangles. Then use ratios to determine whether the triangles are similar.**

1. 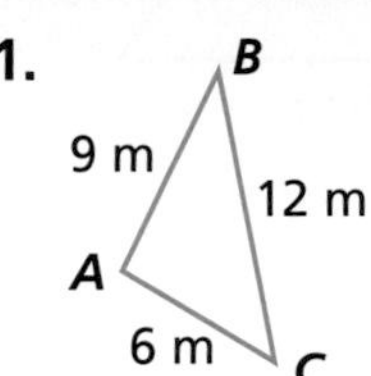

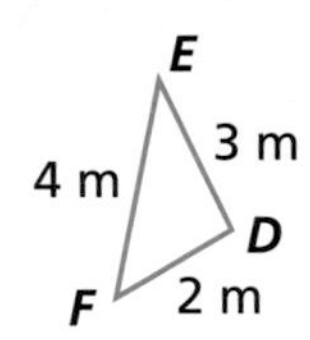

2. 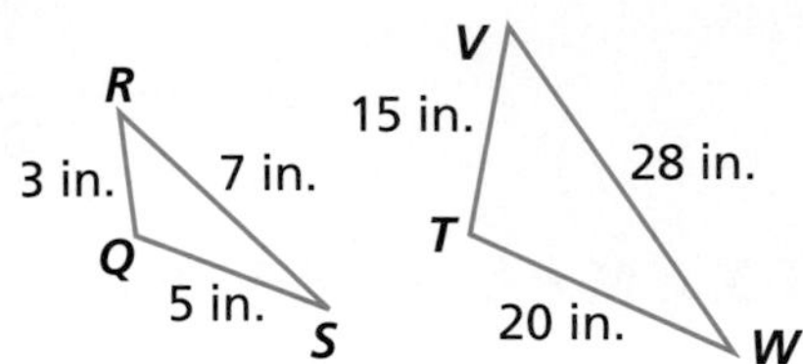

See Example 2 **Use the properties of similarity to determine whether the figures are similar.**

3. 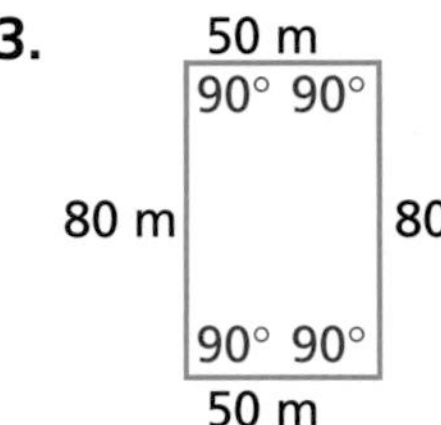

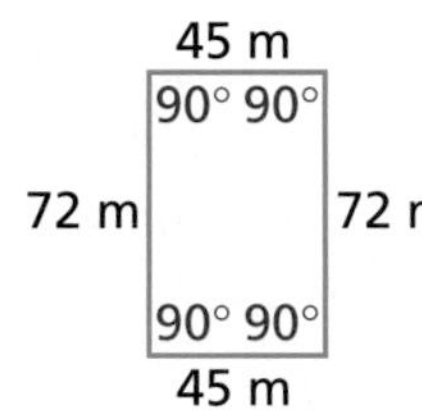

4. 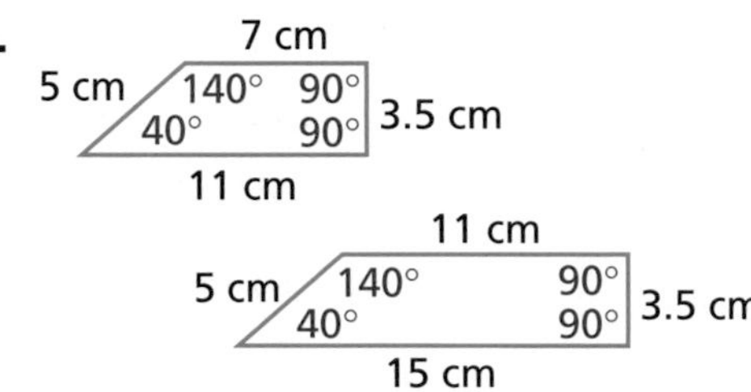

INDEPENDENT PRACTICE

See Example 1 **Identify the corresponding sides in each pair of triangles. Then use ratios to determine whether the triangles are similar.**

5. 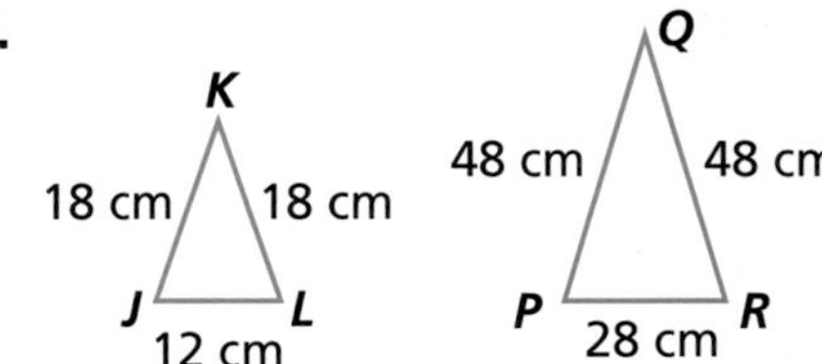

6. 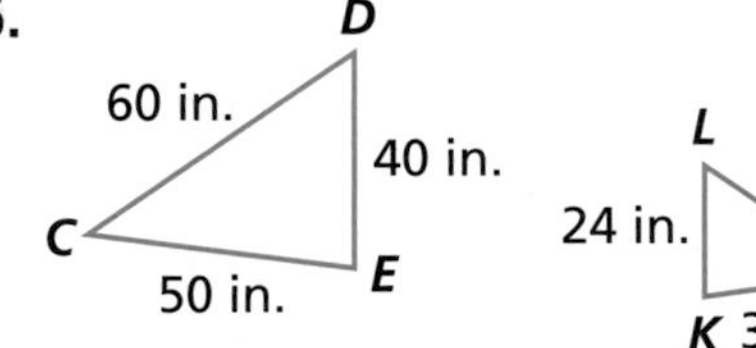

See Example 2 **Use the properties of similarity to determine whether the figures are similar.**

7. 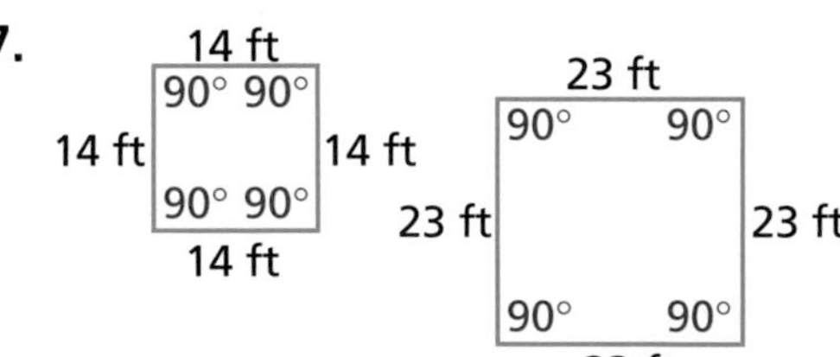

8. 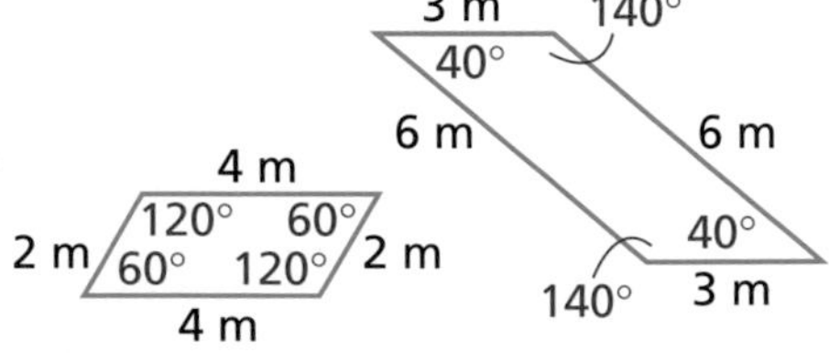

PRACTICE AND PROBLEM SOLVING

9. ***HOBBIES*** Michelle wants similar prints made in various sizes, both small and large, of a favorite photograph. The photo lab offers prints in these sizes: 3 in. × 5 in., 4 in. × 6 in., 8 in. × 18 in., 9 in. × 20 in., 16 in. × 24 in., and 20 in. × 30 in. Which could she order to get similar prints?

Identify the corresponding sides in each pair of triangles. Then use ratios to determine whether the triangles are similar.

10. A, B, C: 9 ft, 18 ft, 12 ft; L, M, N: 6 ft, 12 ft, 8 ft

11. 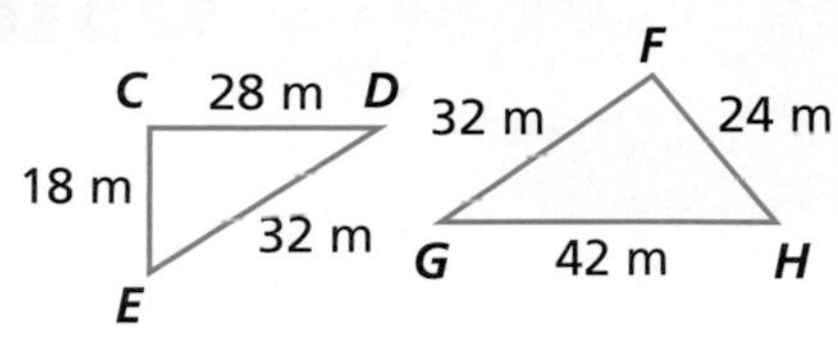

12. Tell whether the parallelogram and trapezoid could be similar. Explain your answer.

120° 60° 60° 120° | 120° 120° 60° 60°

13. The figure at right shows a rectangular piece of paper that has been cut into four rectangular parts. Tell whether the rectangles in each pair are similar. Explain your answers.

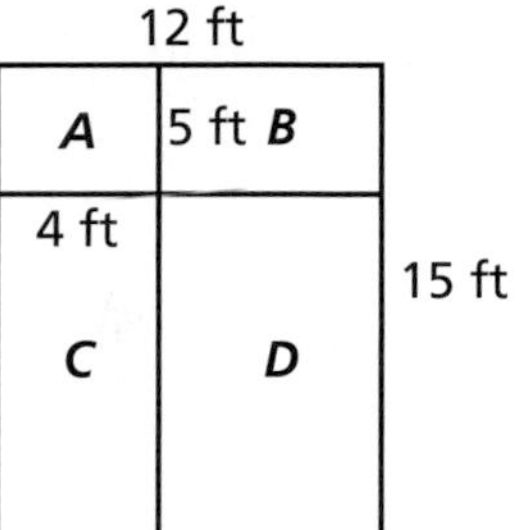

   a. rectangle A and the original rectangle
   b. rectangle C and rectangle B
   c. the original rectangle and rectangle D

For Exercises 14–16, justify your answers using words or drawings.

14. Are all squares similar?

15. Are all parallelograms similar?

16. Are all right triangles similar?

17. ***CHOOSE A STRATEGY*** What number gives the same result when multiplied by 6 as it does when 6 is added to it?

18. ***WRITE ABOUT IT*** Tell how to decide whether two triangles are similar.

19. ***CHALLENGE*** Two triangles are similar. The ratio of the lengths of the corresponding sides is $\frac{5}{4}$. If the length of one side of the larger triangle is 40 feet, what is the length of the corresponding side of the smaller triangle?

Spiral Review

Add or subtract. (Lesson 4-2)

20. $29 - 4.63$ **21.** $9.6 - 0.47$ **22.** $-14.7 + (-5.06) + (-0.9)$

Write each as a unit rate in fraction and word form. (Lesson 5-1)

23. $\frac{48 \text{ min}}{6 \text{ mi}}$ **24.** $\frac{78 \text{ m}}{4 \text{ s}}$ **25.** 8 cups per 4 servings

26. **EOG PREP** A store receives a delivery of 112 quarts of milk. How many gallons of milk does the store receive? (Lesson 5-4)

A 14 gal B 28 gal C 56 gal D 448 gal

5-6 Using Similar Figures

Learn to use similar figures to find unknown lengths.

Vocabulary

indirect measurement

Native Americans of the Northwest, such as the Tlingit tribe of Alaska, carved totem poles out of tree trunks. These poles, sometimes painted with bright colors, could stand up to 80 feet tall. Totem poles include carvings of animal figures, such as bears and eagles, which symbolize traits of the family or clan who built them.

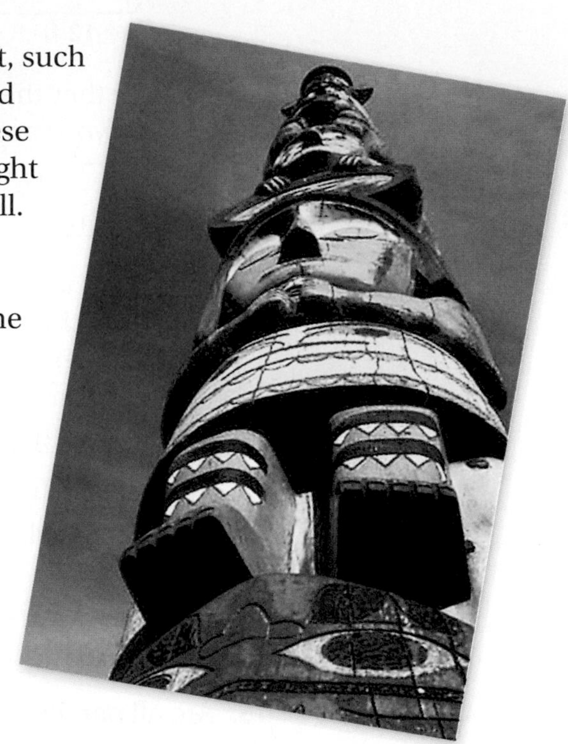

Measuring the heights of tall objects, like some totem poles, cannot be done by using a ruler or yardstick. Instead, you can use *indirect measurement*.

Indirect measurement is a method of using proportions to find an unknown length or distance in similar figures.

EXAMPLE 1 Finding Unknown Lengths in Similar Figures

Find the unknown length in the similar figures.

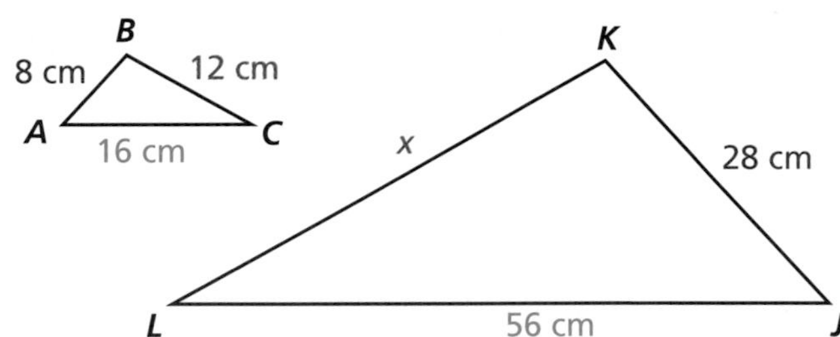

$\dfrac{AB}{JK} = \dfrac{BC}{KL}$ *Write a proportion using corresponding sides.*

$\dfrac{8}{28} = \dfrac{12}{x}$ *Substitute the lengths of the sides.*

$8 \cdot x = 28 \cdot 12$ *Find the cross products.*

$8x = 336$ *Multiply.*

$\dfrac{8x}{8} = \dfrac{336}{8}$ *Divide each side by 8 to isolate the variable.*

$x = 42$

KL is 42 centimeters.

EXAMPLE 2 *Measurement Application*

A volleyball court is a rectangle that is similar in shape to an Olympic-sized pool. Find the width of the pool.

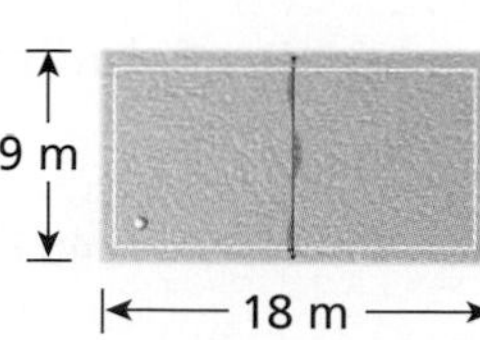

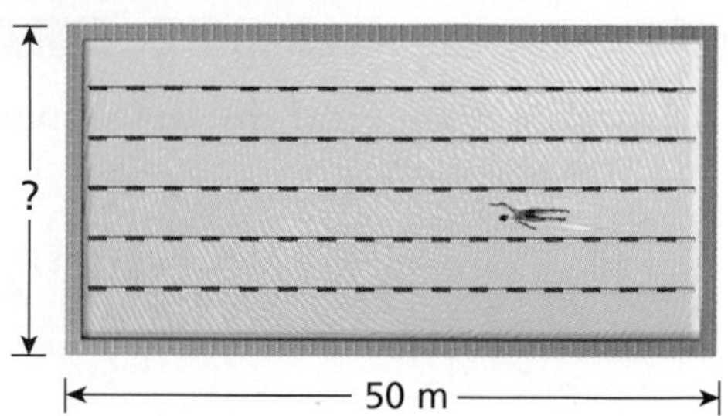

Let w = the width of the pool.

$\frac{18}{50} = \frac{9}{w}$	*Write a proportion using corresponding side lengths.*
$18 \cdot w = 50 \cdot 9$	*Find the cross products.*
$18w = 450$	*Multiply.*
$\frac{18w}{18} = \frac{450}{18}$	*Divide each side by 18 to isolate the variable.*
$w = 25$	

The pool is 25 meters wide.

EXAMPLE 3 Using Indirect Measurement

The birdhouse in Chantal's yard casts a shadow that is 13.5 ft long. Chantal casts a shadow that is 3.75 ft long. What is the height of the birdhouse?

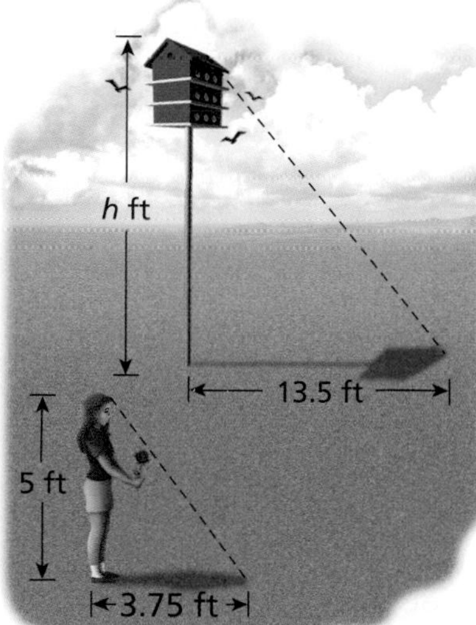

Let h = the height of the birdhouse.

$\frac{h}{5} = \frac{13.5}{3.75}$	*Write a proportion.*
$3.75h = 5 \cdot 13.5$	*Find the cross products.*
$3.75h = 67.5$	
$\frac{3.75h}{3.75} = \frac{67.5}{3.75}$	*Divide each side by 3.75 to isolate the variable.*
$h = 18$	

The birdhouse is 18 feet tall.

Think and Discuss

1. **Write** another proportion that could be used to find the value of x in Example 1.
2. **Name** two objects that would make sense to measure using indirect measurement.

5-6 Exercises

FOR EOG PRACTICE

see page 677

internet connect

Homework Help Online

go.hrw.com Keyword: MS4 5-6

1.01, 2.01, 3.03

GUIDED PRACTICE

See Example 1

Find the unknown length in each pair of similar figures.

1. 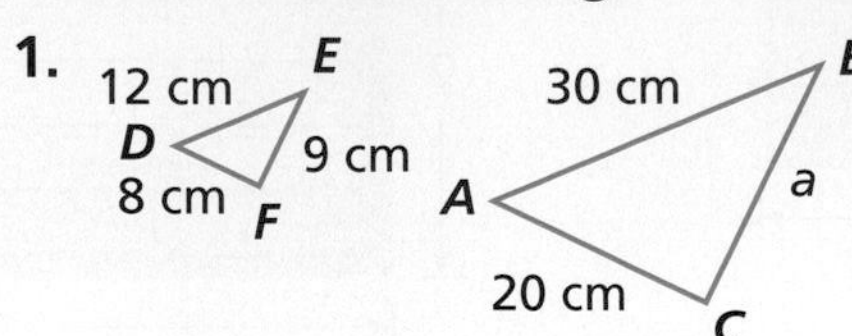

2. 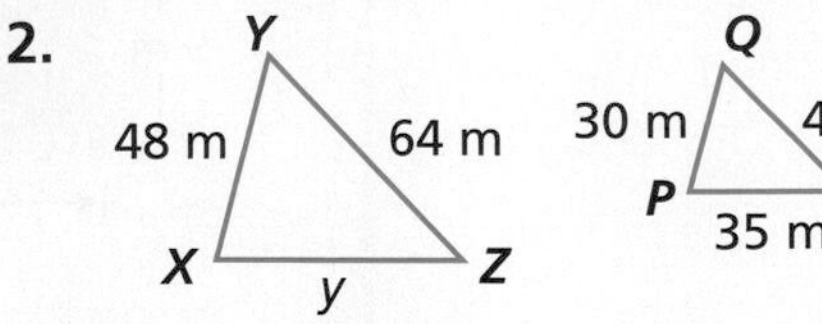

See Example 2

3. The rectangular gardens at right are similar in shape. How wide is the smaller garden?

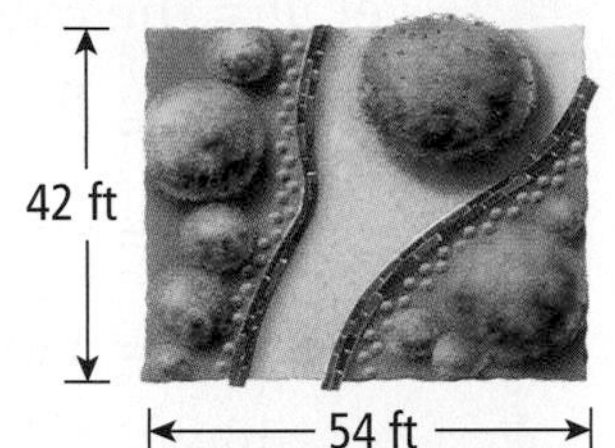

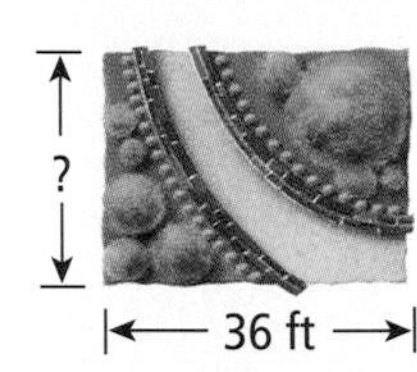

See Example 3

4. A water tower casts a shadow that is 20 ft long. A tree casts a shadow that is 8 ft long. What is the height of the water tower?

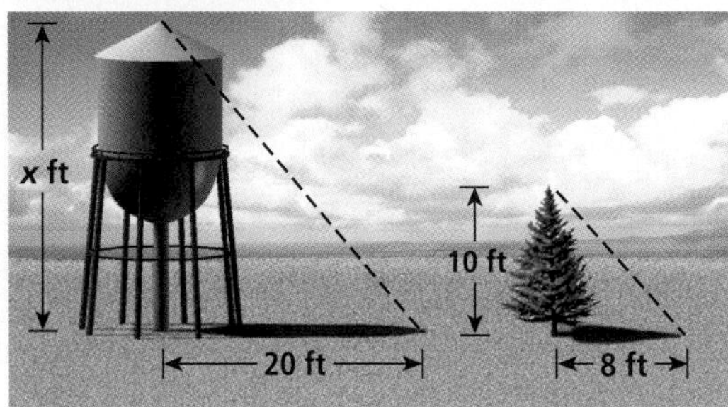

INDEPENDENT PRACTICE

See Example 1

Find the unknown length in each pair of similar figures.

5. 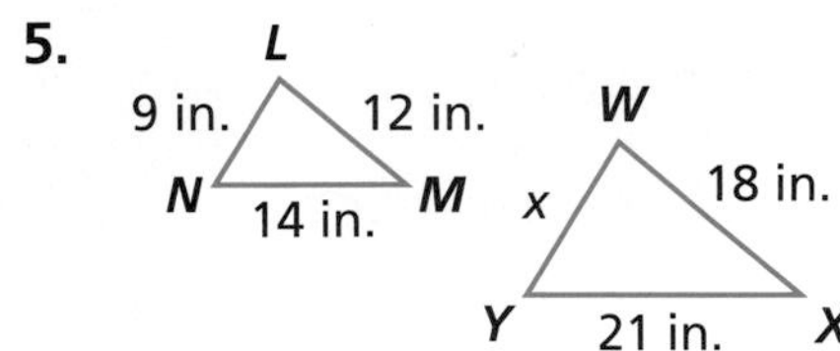

6. 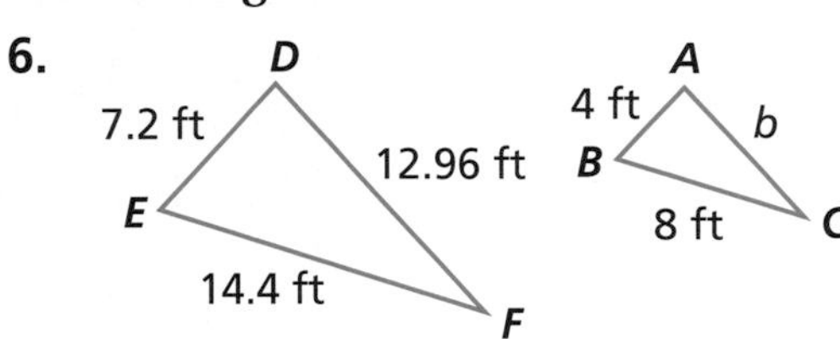

See Example 2

7. The two rectangular windows at right are similar. What is the height of the bigger window?

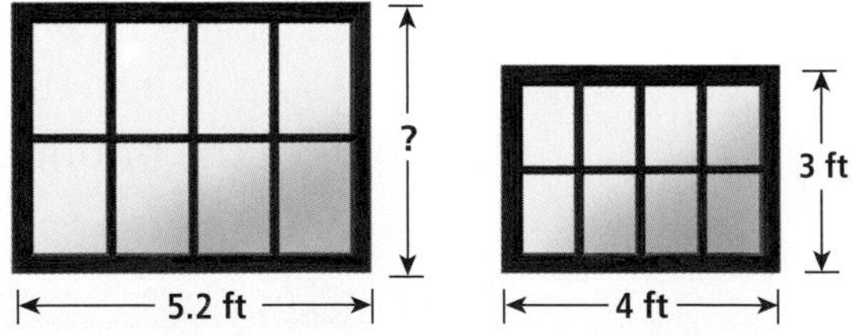

See Example 3

8. A cactus casts a shadow that is 11.25 ft long. A gate nearby casts a shadow that is 5 ft long. What is the height of the cactus?

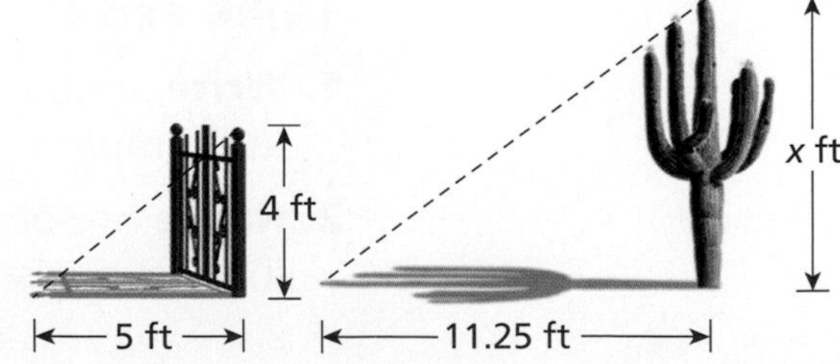

PRACTICE AND PROBLEM SOLVING

9. A building casts a shadow that is 16 m long while a taller building casts a 24 m long shadow. What is the height of the taller building?

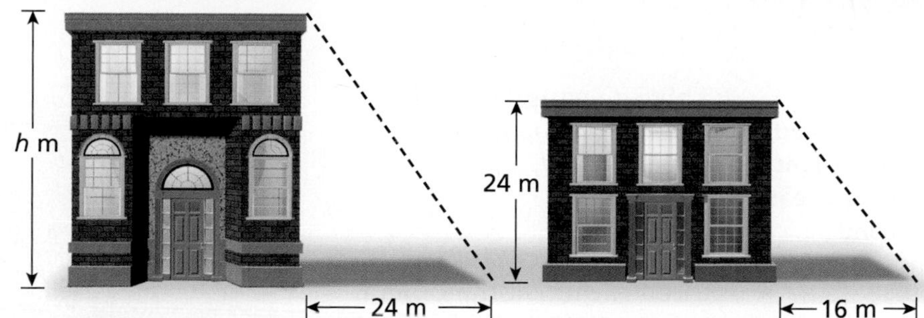

10. ***EARTH SCIENCE*** An art class is painting a wall mural of the solar system. The table shows the distances of three planets from the Sun in astronomical units (AU) and the corresponding distances planned for the mural. Find the missing distances in the table. Round your answers to the nearest tenth.

Planet	Distance from Sun (AU)	Distance on Mural (yd)
Venus	0.72	
Earth	1.0	2.5
Mars		3.8

11. ***WRITE A PROBLEM*** Write a problem that can be solved using indirect measurement.

12. ***WRITE ABOUT IT*** Assume you know the side lengths of one triangle and the length of one side of a second similar triangle. Explain how to use the rules of similar figures to find the unknown lengths in the second triangle.

13. ***CHALLENGE*** Triangle *ABE* is similar to triangle *ACD*. What is the value of y in the diagram?

Spiral Review

Solve. (Lesson 3-6)

14. $w + 8 = -5$ **15.** $y - 12 = -1$ **16.** $4 + m = 51$ **17.** $a - 23 = -23$

Multiply. (Lesson 4-3)

18. $8.6 \cdot 0.13$ **19.** $-5.12 \cdot 6$ **20.** $-7.2 \cdot (-1.9)$ **21.** $0.45 \cdot 0.23$

22. **EOG PREP** Andreas needs $\frac{2}{3}$ cup of cornmeal to make one batch of corn muffins. He needs to make six batches of muffins for a picnic. How many cups of cornmeal will he need? (Lesson 4-7)

A 4 cups B $4\frac{1}{3}$ cups C $6\frac{2}{3}$ cups D $12\frac{2}{3}$ cups

5-7 Scale Drawings and Scale Models

Learn to understand ratios and proportions in scale drawings. Learn to use ratios and proportions with scale.

Vocabulary
scale model
scale factor
scale
scale drawing

This HO gauge model train is a *scale model* of a historic train. A **scale model** is a proportional model of a three-dimensional object. Its dimensions are related to the dimensions of the actual object by a ratio called the **scale factor**. The scale factor of an HO gauge model train is $\frac{1}{87}$. This means that each dimension of the model is $\frac{1}{87}$ of the corresponding dimension of the actual train.

A **scale** is the ratio between two sets of measurements. Scales can use the same units or different units. The photograph shows a *scale drawing* of the model train. A **scale drawing** is a proportional drawing of an object. Both scale drawings and scale models can be smaller or larger than the objects they represent.

EXAMPLE 1 Finding a Scale Factor

Identify the scale factor.

	Race Car	Model
Length (in.)	132	11
Height (in.)	66	5.5

$$\frac{\text{model length}}{\text{race car length}} = \frac{11}{132}$$ *Write a ratio using one of the dimensions.*

$$= \frac{1}{12}$$ *Simplify.*

The scale factor is $\frac{1}{12}$.

EXAMPLE 2 Using Scale Factors to Find Unknown Lengths

A photograph of Vincent van Gogh's painting *Still Life with Irises Against a Yellow Background* has dimensions 6.13 cm and 4.90 cm. The scale factor is $\frac{1}{15}$. Find the size of the actual painting, to the nearest tenth of a centimeter.

Think: $\frac{\text{photo}}{\text{painting}} = \frac{1}{15}$

$\frac{6.13}{\ell} = \frac{1}{15}$ *Write a proportion to find the length ℓ.*

$\ell = 6.13 \cdot 15$ *Find the cross products.*

$\ell = 92.0$ *Multiply and round to the nearest tenth.*

$\frac{4.90}{w} = \frac{1}{15}$ *Write a proportion to find the width w.*

$w = 4.90 \cdot 15$ *Find the cross products.*

$w = 73.5$ *Multiply and round to the nearest tenth.*

The painting is 92.0 cm long and 73.5 cm wide.

EXAMPLE 3 *Measurement Application*

On a map of Florida, the distance between Hialeah and Tampa is 10.5 cm. What is the actual distance between the cities if the map scale is 3 cm = 80 mi?

Let d be the actual distance between the cities.

$\frac{3}{80} = \frac{10.5}{d}$ *Write a proportion.*

$3 \cdot d = 80 \cdot 10.5$ *Find the cross products.*

$3d = 840$

$\frac{3d}{3} = \frac{840}{3}$ *Divide.*

$d = 280$

The distance between the cities is 280 miles.

Think and Discuss

1. Given a scale factor of $\frac{5}{3}$, explain how you can tell whether a model is bigger or smaller than the original object.

2. Describe how to find the scale factor if an antenna is 60 feet long and a scale drawing shows the length as 1 foot long.

5-7 Exercises

FOR EOG PRACTICE
see page 677

internet connect
Homework Help Online
go.hrw.com Keyword: MS4 5-7

1.01, 1.02b, 2.01, 3.03

GUIDED PRACTICE

See Example 1 **Identify the scale factor.**

1.

	Grizzly Bear	Model
Height (in.)	84	6

See Example 

2. In a photograph, a sculpture is 4.2 cm tall and 2.5 cm wide. The scale factor is $\frac{1}{16}$. Find the size of the actual sculpture.

See Example 

3. Ms. Jackson is driving from South Bend to Indianapolis. She measures a distance of 4.3 cm between the cities on her Indiana road map. What is the actual distance between the cities if the map scale is 1 cm = 30 mi?

INDEPENDENT PRACTICE

See Example  **Identify the scale factor.**

4.

	Eagle	Model
Wingspan (in.)	90	6

See Example 

5. On a scale drawing, a tree is $6\frac{3}{4}$ inches tall. The scale factor is $\frac{1}{20}$. Find the height of the actual tree.

See Example 

6. On a road map of Virginia, the distance from Alexandria to Roanoke is 7.6 cm. What is the actual distance between the cities if the map scale is 2 cm = 50 mi?

PRACTICE AND PROBLEM SOLVING

The scale factor of each model is 1:12. Find the missing dimensions.

	Item	Actual Dimensions	Model Dimensions
7.	Lamp	Height: ▢	Height: $1\frac{1}{3}$ in.
8.	Grandfather clock	Height: 5 ft	Height: ▢
9.	Couch	Height: 32 in. Length: 69 in.	Height: ▢ Length: ▢
10.	Table	Height: ▢ Width: ▢ Length: ▢	Height: 6.25 cm Width: 11.75 cm Length: 20 cm
11.	Chair	Height: $51\frac{1}{2}$ in.	Height: ▢

History LINK

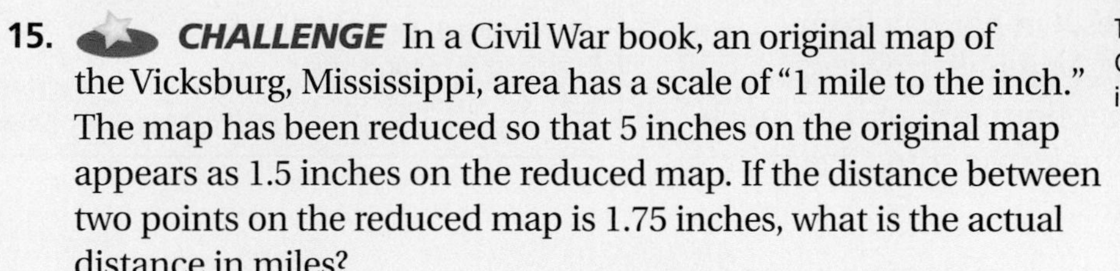

Use the map for Exercises 12 and 13. The map scale is 1 inch = 10 miles.

12. The battle at Gettysburg was the turning point of the American Civil War. It started when Confederate troops marched from Chambersburg into Gettysburg in search of badly needed shoes. There, they encountered Union troops. Use the ruler and the scale of the map to estimate how far the Confederate soldiers, many of whom were barefoot, marched.

13. Before the Civil War, the Mason-Dixon Line was considered the dividing line between the North and the South. If Gettysburg is about 8.1 miles north of the Mason-Dixon Line, how far apart in inches are Gettysburg and the Mason-Dixon Line on the map?

14. Uncle Toby is making a scale model of the battlefield at Fredericksburg. The area he wants to model measures about 11 mi by 7.5 mi. He plans to put the model on a 3.25 ft by 3.25 ft square table. On each side of the model he wants to leave at least 3 in. between the model and the table edges. What is the largest scale he can use?

This painting by H.A. Ogden depicts General Robert E. Lee at Fredericksburg in 1862.

15. ***CHALLENGE*** In a Civil War book, an original map of the Vicksburg, Mississippi, area has a scale of "1 mile to the inch." The map has been reduced so that 5 inches on the original map appears as 1.5 inches on the reduced map. If the distance between two points on the reduced map is 1.75 inches, what is the actual distance in miles?

Spiral Review

Write the numbers in order from least to greatest. (Lesson 3-10)

16. $\frac{4}{7}$, 0.41, 0.054 **17.** $\frac{1}{4}$, 0.2, −1.2 **18.** 0.7, $\frac{7}{9}$, $\frac{7}{11}$

Divide. (Lesson 4-4)

19. 0.32 ÷ 5 **20.** 78.57 ÷ 9 **21.** 40.5 ÷ 15 **22.** 29.68 ÷ 28

23. **EOG PREP** A spool has 21 meters of ribbon. The ribbon is cut into pieces that are each 0.6 meters long. How many pieces of ribbon are cut from the spool? (Lesson 4-5)

A 3.5 B 12.6 C 35 D 126

Problem Solving on Location

NORTH CAROLINA

North Carolina's Highways

With 77,400 miles of roads, North Carolina has the largest state-maintained highway system in the United States. Highway 178 is North Carolina's shortest highway, at 7 miles long. The longest, a 569-mile-long stretch of U.S. Highway 64, spans the state. It enters the state from Tennessee in the west, at Angelico Gap, and ends on the East Coast, in Nags Head.

For 1–3, use the map.

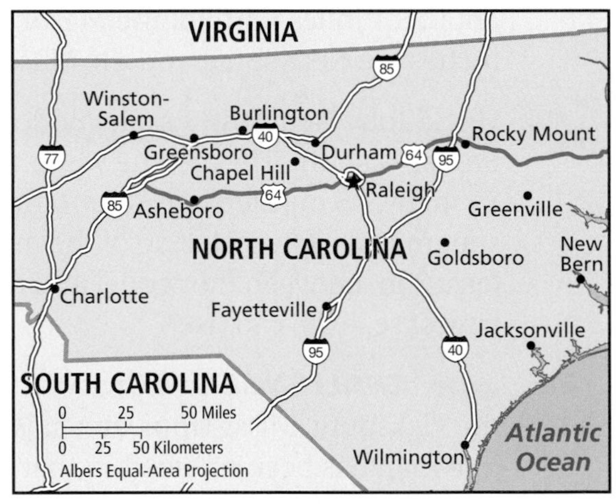

1. Nichole drove from Charlotte to Raleigh. She took Highways 85 and 40 so that she could stop in Greensboro on the way. Tell which of the following distances Nichole most likely drove from Charlotte to Raleigh: 110 miles, 170 miles, 235 miles.

2. The distance from Fayetteville to Rocky Mount along Highway 95 is about 85 miles. On his way from Fayetteville to Rocky Mount, Mitchell drove at an average speed of 50 miles per hour. About how much time did it take Mitchell to make the trip? Give your answer in hours and minutes.

3. The driving distance from Charlotte to Chapel Hill is about 148 miles. If it takes you 65 minutes to drive 50 miles, would it take you more or less than 3 hours to drive to Chapel Hill? Approximately how many minutes more or less would it take you?

Furniture

The furniture industry is huge in High Point, North Carolina, as is some of the furniture. In fact, an entire building has been constructed to look like a 40-foot-tall chest of drawers, and another 80-foot-tall chest of drawers is connected to a furniture store.

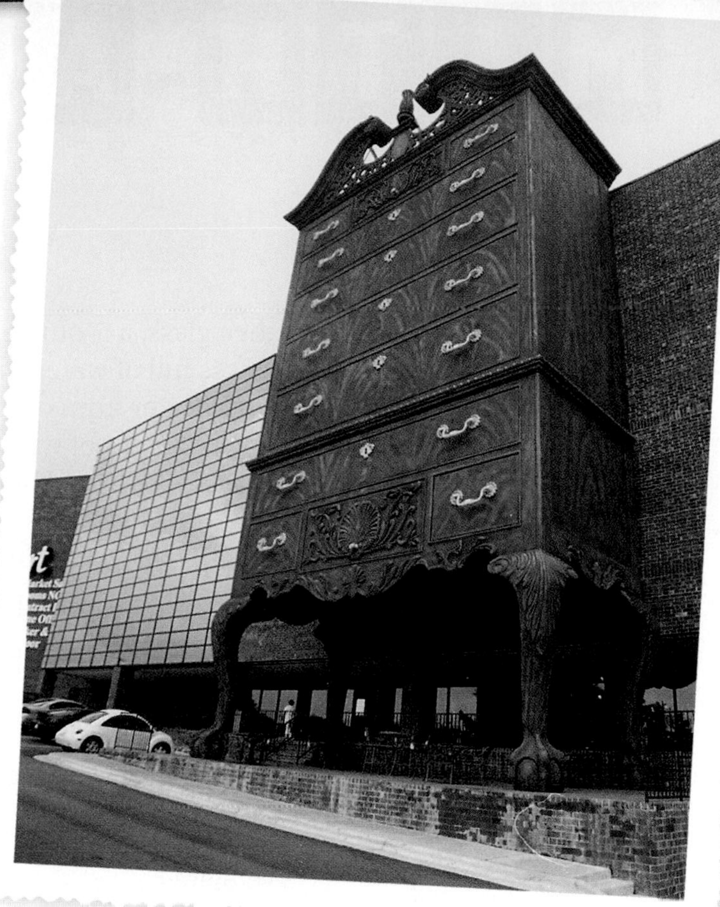

1. In 1920, a public office building was modeled after a chest of drawers. Then in 1996, the building was doubled in height to 40 feet and renovated to resemble a Goddard-Townsend block-front chest. A similar actual Goddard-Townsend chest is 42 inches tall. What scale did the builders use when they modeled the building from the piece of furniture?

2. The drawer pulls on the building in problem **1** are each 4 feet wide. How wide are the drawer pulls on the original piece of furniture, in inches?

3. Suppose the same scale were used to build an 80-foot-tall chest. How tall would the original chest be?

In Thomasville, a huge chair stands on a pedestal as a monument to North Carolina's furniture industry. The chair is an 18-foot-tall concrete-and-steel structure modeled after a Duncan Phyfe chair.

4. A similar Duncan Phyfe chair is commemorated in a photograph on a decorative pin. In the photograph, the chair is about 0.8 inch tall. What scale would have to be used to reduce the size of the chair in the monument to the size of the chair in the photograph?

MATH-ABLES

Water Works

You have three glasses: a 3-ounce glass, a 5-ounce glass, and an 8-ounce glass. The 8-ounce glass is full of water, and the other two glasses are empty. By pouring water from one glass to another, how can you get exactly 6 ounces of water in one of the glasses? The step-by-step solution is described below.

Step 1: Pour the water from the 8 oz glass into the 5 oz glass.

Step 2: Pour the water from the 5 oz glass into the 3 oz glass.

Step 3: Pour the water from the 3 oz glass into the 8 oz glass.

You now have 6 ounces of water in the 8-ounce glass.

Start again, but this time try to get exactly 4 ounces of water in one glass. (*Hint:* Find a way to get 1 ounce of water. Start by pouring water into the 3-ounce glass.)

Next, using 3-ounce, 8-ounce, and 11-ounce glasses, try to get exactly 9 ounces of water in one glass. Start with the 11-ounce glass full of water. (*Hint:* Start by pouring water into the 8-ounce glass.)

Look at the sizes of the glasses in the first and second problem and the sizes of the glasses in the third problem. The volume of the third glass is the sum of the volumes of the first two glasses:
$3 + 5 = 8$ and $3 + 8 = 11$.
Using any amounts for the two smaller glasses, and starting with the largest glass full, you can get any multiple of the smaller glass's volume. Try it and see.

Concentration

Each card in a deck of cards has a ratio printed on one side. The cards are laid out upside down. Each person or team takes a turn flipping over two cards. If the cards match, the person or team can keep the pair. If not, the next person or team flips two cards. After all the cards have been turned over, the person or team with the most pairs wins.

internet connect

Go to ***go.hrw.com*** for a complete set of rules and game pieces.
KEYWORD: MS4 Game5

Model Similar Triangles

Similar triangles can be constructed by drawing one line across a single triangle.

internet connect
Lab Resources Online
go.hrw.com
KEYWORD: MS4 TechLab5

Activity

1. Using geometry software, construct a triangle that resembles the one shown in the first window below. Label the vertices *A*, *B*, and *C*. Label a point *D* on $\overline{AB}$. Use the parallel line tool to draw a line parallel to $\overline{BC}$, through point *D*. Label point *E* where $\overline{AC}$ intersects the new line. Triangles *ABC* and *ADE* are similar.

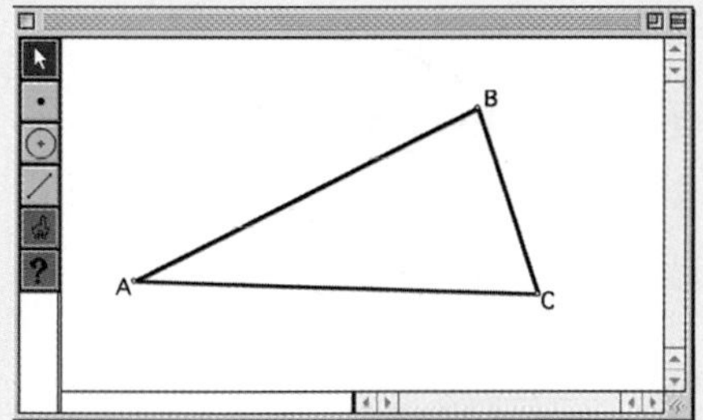

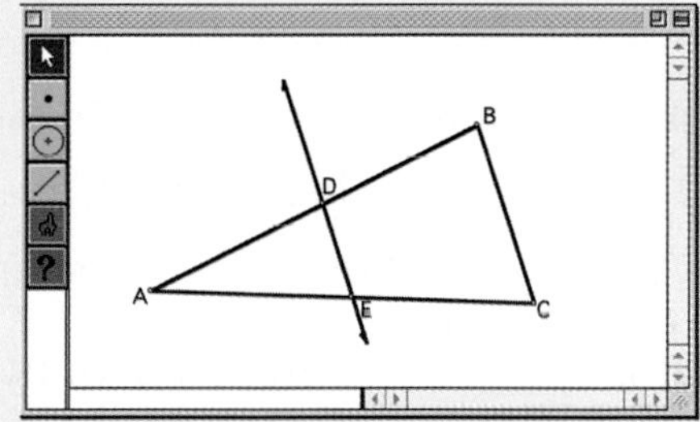

Select vertex *C* and drag it until your figure resembles the one shown in the first window below. Use your angle measure tool to measure $\angle ABC$ and $\angle ADE$. (Your measurements may differ from those shown.) Drag point *C* again. What do you notice about the measures of $\angle ABC$ and $\angle ADE$?

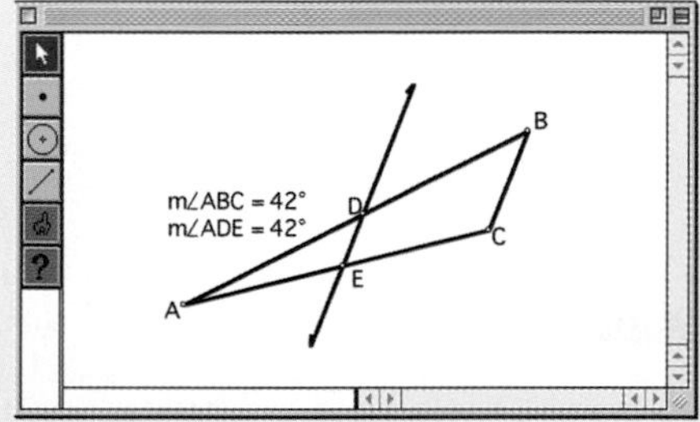

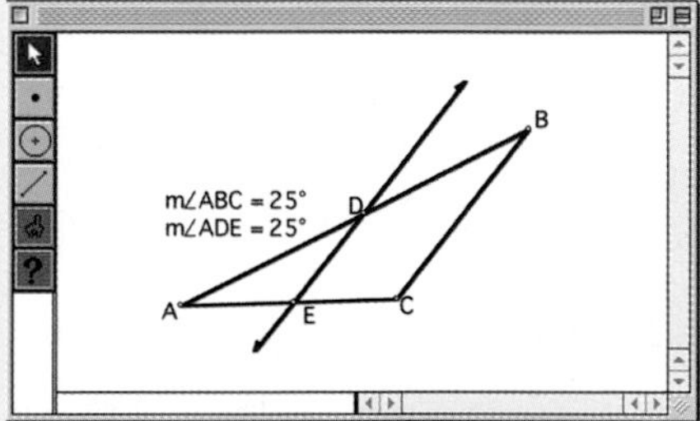

Think and Discuss

1. Notice quadrilateral *DBCE* that is formed when $\overline{DE}$ divides triangle *ABC*. Can such a quadrilateral ever be a rectangle? Explain.

Try This

1. Draw a quadrilateral *ABCD* that is *not* a rectangle. Connect vertices *A* and *C* with a line segment to form two triangles. Select and drag a point until $m\angle BAC$ is equal to $m\angle DCA$. Describe the new quadrilateral *ABCD*.

Chapter 5

Study Guide and Review

Vocabulary

Complete the sentences below with vocabulary words from the list above. Words may be used more than once.

1. ___?___ figures have the same shape but not necessarily the same size.
2. A(n) ___?___ is a comparison of two numbers, and a(n) ___?___ is a ratio that compares two quantities measured in different units.
3. The ratio used to enlarge or reduce similar figures is a(n) ___?___.
4. The process of using proportions to find an unknown length or distance in similar figures is called ___?___.

5-1 Ratios and Rates (pp. 260–263)

EXAMPLE

■ **Write the ratio of 2 servings of bread to 4 servings of vegetables in all three forms.**

$\frac{2}{4}$, 2 to 4, 2:4 — *For every 2 servings of bread there are 4 servings of vegetables.*

The fraction $\frac{2}{4}$ can be simplified to $\frac{1}{2}$.

$\frac{1}{2}$, 1 to 2, 1:2 — *For each serving of bread there are 2 servings of vegetables.*

EXERCISES

There are 6 cats and 10 dogs. Write each ratio in all three forms.

5. cats to dogs
6. dogs to total animals

Find the unit rates and write them in both fraction and word form.

7. 8 gallons in 4 minutes
8. 35 miles in 10 minutes

5-2 Identifying and Writing Proportions (pp. 264–267)

EXAMPLE

- **Determine if the ratios are proportional.**

$\frac{5}{12}, \frac{3}{9}$

$\frac{5}{12}$ — *$\frac{5}{12}$ is already in simplest form.*

$\frac{3}{9} = \frac{1}{3}$ — *Simplify $\frac{3}{9}$.*

Since $\frac{5}{12} \neq \frac{1}{3}$, the ratios are not proportional.

EXERCISES

Determine if the ratios are proportional.

9. $\frac{15}{20}, \frac{20}{21}$ **10.** $\frac{2}{3}, \frac{64}{90}$ **11.** $\frac{16}{32}, \frac{4}{8}$

12. $\frac{9}{27}, \frac{6}{20}$ **13.** $\frac{15}{25}, \frac{20}{30}$ **14.** $\frac{21}{14}, \frac{18}{12}$

Find a ratio equivalent to each ratio. Then use the ratios to write a proportion.

15. $\frac{10}{12}$ **16.** $\frac{45}{50}$ **17.** $\frac{9}{15}$

5-3 Solving Proportions (pp. 268–271)

EXAMPLE

- **Use cross products to solve $\frac{p}{8} = \frac{10}{12}$.**

$$\frac{p}{8} = \frac{10}{12}$$

$$p \cdot 12 = 8 \cdot 10$$

$$12p = 80$$

$$\frac{12p}{12} = \frac{80}{12}$$

$$p = \frac{20}{3}, \text{ or } 6\frac{2}{3}$$

EXERCISES

Use cross products to solve each proportion.

18. $\frac{4}{6} = \frac{n}{3}$ **19.** $\frac{2}{a} = \frac{5}{15}$

20. $\frac{b}{1.5} = \frac{8}{3}$ **21.** $\frac{16}{11} = \frac{96}{x}$

22. $\frac{2}{y} = \frac{1}{11}$ **23.** $\frac{7}{10} = \frac{70}{k}$

5-4 Dimensional Analysis (pp. 272–275)

EXAMPLE

- **Use a unit conversion factor to convert 14 gallons to quarts.**

1 gal = 4 qt, so use $\frac{1 \text{ gal}}{4 \text{ qt}}$ or $\frac{4 \text{ qt}}{1 \text{ gal}}$.

Choose the second conversion factor so that the gallon units cancel.

$$14 \cancel{\text{gal}} \cdot \frac{4 \text{ qt}}{1 \cancel{\text{gal}}} = 56 \text{ qt}$$

EXERCISES

Use a unit conversion factor to convert the units.

24. 10 gallons to quarts

25. 6 quarts to gallons

26. 11,616 feet to miles

27. How many feet are there in 15 yards?

Study Guide and Review

5-5 Similar Figures and Proportions (pp. 280–283)

EXAMPLE

■ **Determine if the triangles are similar.**

$\frac{AB}{DE} \stackrel{?}{=} \frac{BC}{EF} \stackrel{?}{=} \frac{AC}{DF}$

$\frac{2}{6} \stackrel{?}{=} \frac{4}{12} \stackrel{?}{=} \frac{3}{9}$

$\frac{1}{3} \stackrel{?}{=} \frac{1}{3} \stackrel{?}{=} \frac{1}{3}$

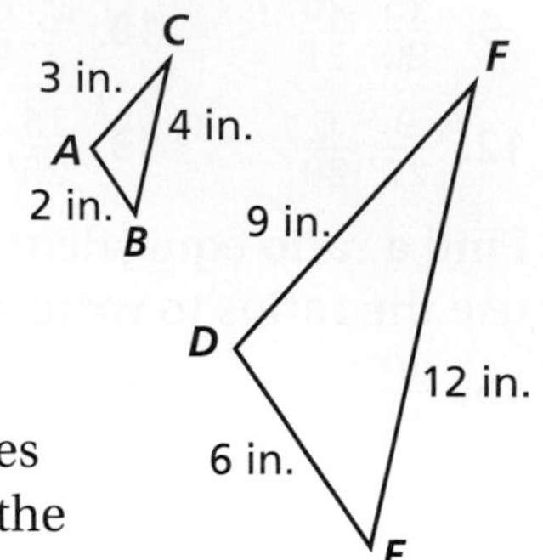

The ratios of the corresponding sides are equivalent, so the triangles are similar.

EXERCISES

Determine if the triangles are similar.

28.

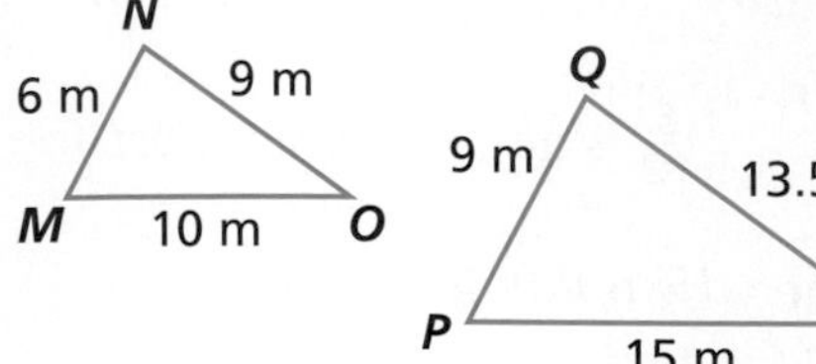

29.

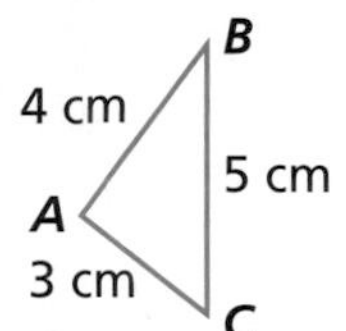

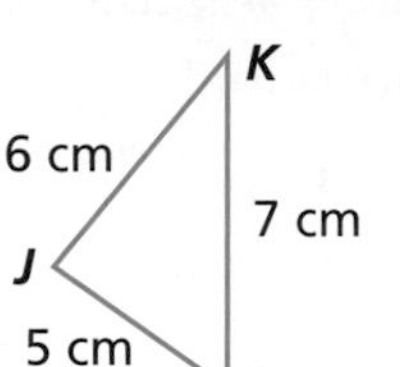

5-6 Using Similar Figures (pp. 284–287)

EXAMPLE

■ **Find the unknown length in the similar triangle.**

$\frac{AB}{LM} = \frac{AC}{LN}$

$\frac{8}{t} = \frac{11}{44}$

$8 \cdot 44 = t \cdot 11$

$352 = 11t$

$\frac{352}{11} = \frac{11t}{11}$

$32 \text{ in.} = t$

B, 8 in., 7 in., A, C, 11 in.

M, t, 28 in., L, N, 44 in.

EXERCISES

Find the unknown length in the similar triangle.

30.

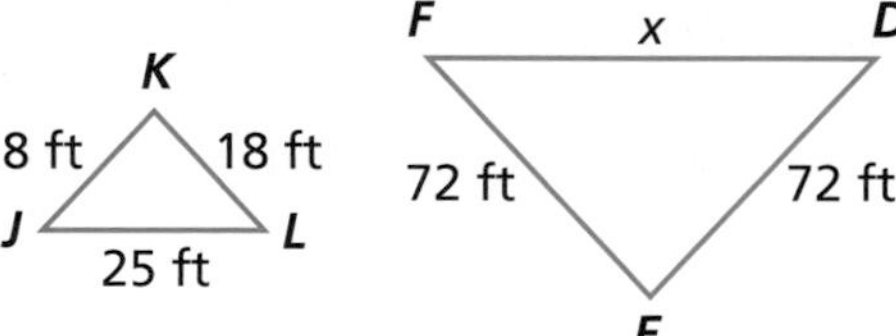

5-7 Scale Drawings and Scale Models (pp. 288–291)

EXAMPLE

■ **A model sailboat is 4 inches long. The scale factor is $\frac{1}{24}$. How long is the actual sailboat?**

$\frac{\text{model}}{\text{sailboat}} = \frac{1}{24}$

$\frac{4}{n} = \frac{1}{24}$ *Write a proportion.*

$4 \cdot 24 = n \cdot 1$ *Find the cross products.*

$96 = n$ *Solve.*

The sailboat is 96 inches long.

EXERCISES

Solve.

31. The Wright brothers' *Flyer* had a 484-inch wingspan. Carla bought a model of the plane with a scale factor of $\frac{1}{40}$. What is the model's wingspan?

32. Eduardo measured the distance from Asheville to Winston-Salem on a map to be 3.7 inches. The map scale is 1 inch = 40 miles. What is the actual distance?

Chapter Test

There are 20 wagons and 10 bicycles. Write each ratio in all three forms.

1. wagons to bicycles
2. bicycles to total vehicles
3. total vehicles to wagons

Determine whether the ratios are proportional.

4. $\frac{5}{6}, \frac{20}{24}$
5. $\frac{3}{4}, \frac{9}{16}$
6. $\frac{11}{33}, \frac{5}{15}$

Solve each proportion. Then write a ratio equivalent to the ratios in the proportion.

7. $\frac{9}{12} = \frac{m}{6}$
8. $\frac{x}{2} = \frac{18}{6}$
9. $\frac{3}{7} = \frac{21}{t}$

Use a unit conversion factor to convert the units.

10. 18 feet to inches
11. 6 gallons to quarts
12. 13,200 feet to miles

Use a unit conversion factor to convert the units within each rate.

13. 6 feet per second to feet per minute
14. $0.35 per ounce to dollars per pound

Determine whether the figures are similar.

15.

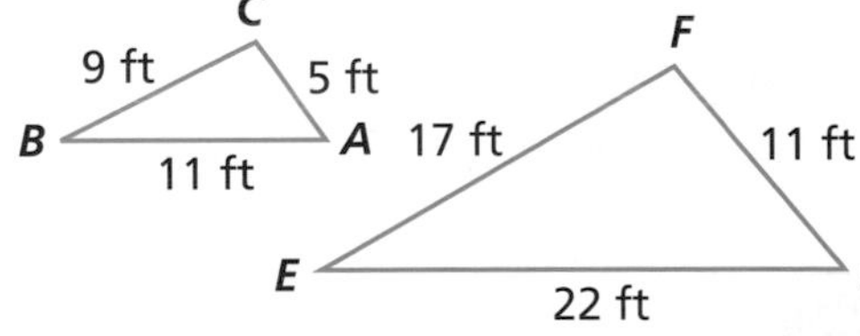

16.

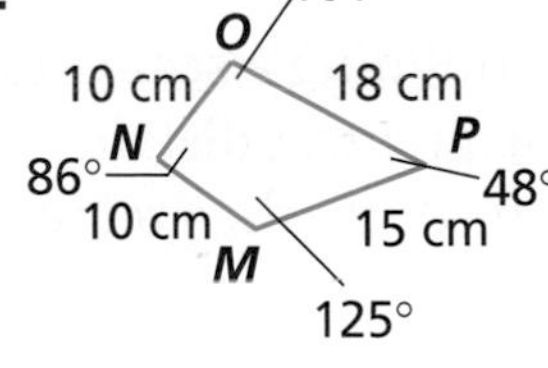

Find the unknown length in each pair of similar figures.

17.

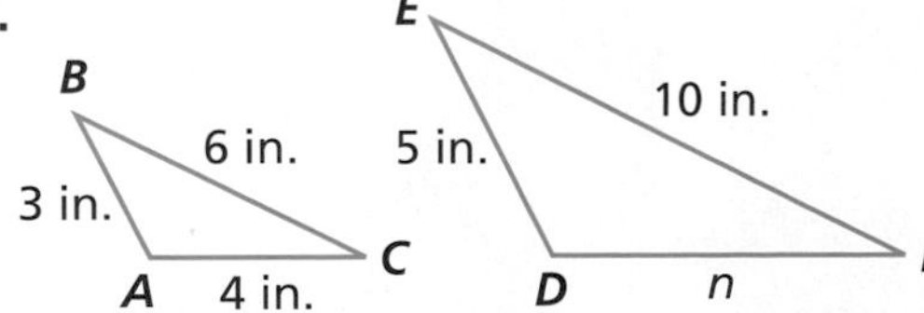

18.

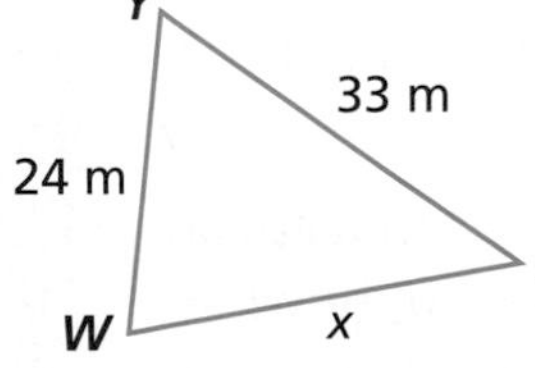

Solve.

19. A scale model of a building is 8 in. by 12 in. If the scale is 1 in. = 15 ft, what are the dimensions of the actual building?

20. The actual distance from Portland to Seaside is 75 mi. If the map scale is $1\frac{1}{4}$ in. = 25 mi, what is the distance in inches between the two towns on the map?

Chapter 5

Performance Assessment

Show What You Know

Create a portfolio of your best work from this chapter. Complete this page and include it with your four best pieces of work from Chapter 5. Choose from your homework or lab assignments, mid-chapter quiz, or any journal entries you have done. Put them together using any design you want. Make your portfolio represent what you consider your best work.

Short Response

1. A tree casts a shadow 20 feet long. Monica is 5 feet tall, and her shadow is 4 feet long. Describe in words how to find the height of the tree. Then find the height, and show your work.
2. In the past, $1 U.S. equaled $1.575 Canadian. Find the approximate value of $75 Canadian. Show the steps that you used to calculate the answer.
3. Adam is refilling his 10-gallon fish tank. He must add $\frac{1}{2}$ tsp of chemical solution for each quart of water. How much chemical solution will he need to treat the entire tank? Show your work.

Extended Problem Solving

4. Casey and Jacqui are building a scale model of the state of Virginia. From east to west, the greatest distance across the state is about 430 mi. From north to south, the greatest distance is about 200 mi.
   a. What scale factor are Casey and Jacqui using to build their model? Explain how you found your answer.
   b. How wide will their model be? Show your work.
   c. If an airplane travels at a speed of 880 feet per second, how long will it take for the plane to fly east to west across the widest part of Virginia?

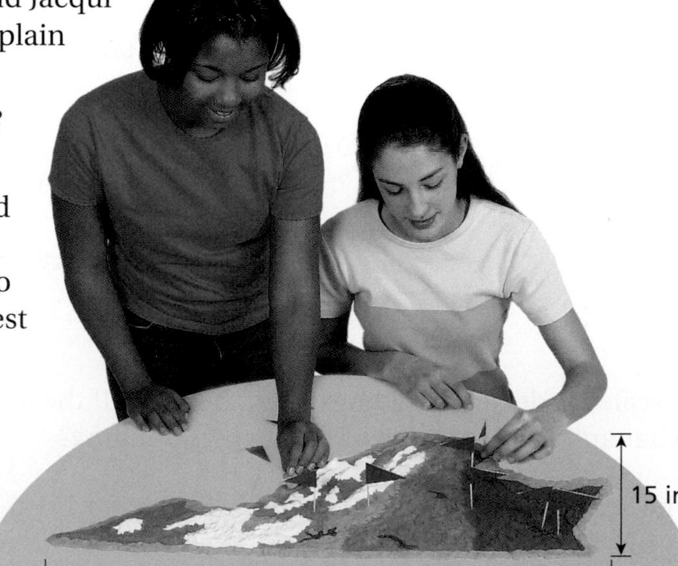

Performance Assessment

internet connect
EOG Practice Online
go.hrw.com Keyword: MS4 TestPrep

Getting Ready for EOG

Cumulative Assessment, Chapters 1–5

1. Find the difference: $-6 - (-3)$.
   A -9
   B -3
   C 3
   D 9

TEST TAKING TIP!
Eliminate choices by using estimation.

2. Multiply $5\frac{1}{3} \cdot 3\frac{3}{4}$. Write the answer in simplest form.
   A $15\frac{1}{4}$
   B $1\frac{19}{45}$
   C $9\frac{1}{12}$
   D 20

3. Find the mean of the numbers 6, 6, 8, 20, and 10.
   A 6
   B 8
   C 10
   D 14

4. Find the solution to the equation $y + (-8) = 12$.
   A $y = 20$
   B $y = -20$
   C $y = -96$
   D $y = 4$

5. Write 123,000 in scientific notation.
   A $123 \times 1{,}000$
   B 123×10^3
   C 0.123×10^6
   D 1.23×10^5

6. Which of the following numbers is irrational?
   A π
   B $3.\overline{8}$
   C $6\frac{1}{2}$
   D $\sqrt{4}$

7. Which is the LCM of 16 and 12?
   A 4
   B 24
   C 48
   D 192

8. Which is the GCF of 32 and 18?
   A 9
   B 6
   C 18
   D 2

9. Solve the proportion $\frac{m}{3} = \frac{18}{6}$.
   A $m = 36$
   B $m = 18$
   C $m = 1$
   D $m = 9$

10. ***SHORT RESPONSE*** Write an expression to describe the phrase "sixteen less than a number *m*." Can you use your expression to find the value of *m*? Why or why not?

11. ***SHORT RESPONSE*** The box-and-whisker plot shows the scores on a quiz students took in Ms. Santini's class. Can you find the mean quiz score from the box-and-whisker plot? If so, find it. If not, explain why not.

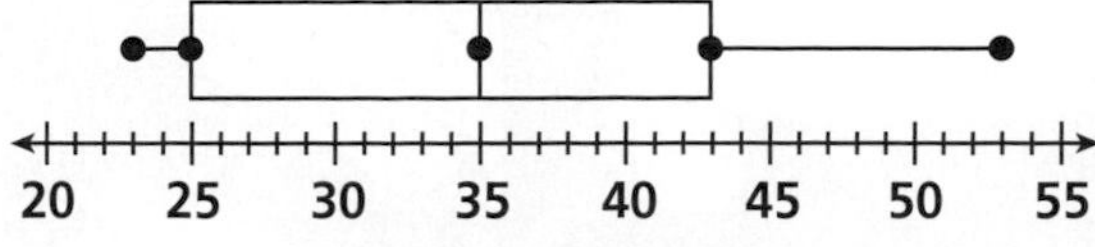

Chapter 6

Percents

TABLE OF CONTENTS

internet connect

Chapter Opener Online
go.hrw.com
KEYWORD: MS4 Ch6

Annual Urban Waste Deposited in U.S. Landfills (million tons)			
Soil	Wood	Concrete	Household Refuse
107.6	87.6	22.5	32.7

Career *Urban Archaeologist*

Have you ever wanted to study the lifestyles of people who lived long ago? If so, becoming an archaeologist might be for you. Archaeologists learn about past civilizations by excavating their cities and examining their artifacts. They even examine their garbage!

Since 1973, the urban archaeologists of the Garbage Project have been learning about the habits of present-day societies by excavating landfills and studying the things we throw away. They have found that over 80% of urban waste in the United States is deposited in landfills.

ARE YOU READY?

Choose the best term from the list to complete each sentence.

1. A statement that two ratios are equivalent is called a(n) __?__.

2. To write $\frac{2}{3}$ as a(n) __?__, divide the numerator by the denominator.

3. A(n) __?__ is a comparison by division of two quantities.

4. The __?__ of $\frac{9}{24}$ is $\frac{3}{8}$.

decimal
equation
fraction
proportion
ratio
simplest form

Complete these exercises to review skills you will need for this chapter.

✔ Write Fractions as Decimals

Write each fraction as a decimal.

5. $\frac{8}{10}$ **6.** $\frac{53}{100}$ **7.** $\frac{739}{1,000}$ **8.** $\frac{7}{100}$

9. $\frac{2}{5}$ **10.** $\frac{5}{8}$ **11.** $\frac{7}{12}$ **12.** $\frac{13}{20}$

✔ Write Decimals as Fractions

Write each decimal as a fraction in simplest form.

13. 0.05 **14.** 0.92 **15.** 0.013 **16.** 0.8

17. 0.006 **18.** 0.305 **19.** 0.0007 **20.** 1.04

✔ Solve Multiplication Equations

Solve each equation.

21. $100n = 300$ **22.** $38 = 0.4x$ **23.** $16p = 1,200$

24. $9 = 72y$ **25.** $0.07m = 56$ **26.** $25 = 100t$

✔ Solve Proportions

Solve each proportion.

27. $\frac{2}{3} = \frac{x}{12}$ **28.** $\frac{x}{20} = \frac{3}{4}$ **29.** $\frac{8}{15} = \frac{x}{45}$

30. $\frac{16}{28} = \frac{4}{n}$ **31.** $\frac{p}{100} = \frac{12}{36}$ **32.** $\frac{42}{12} = \frac{14}{n}$

6-1 Fractions, Decimals, and Percents

Learn to write equivalent fractions, decimals, and percents.

Vocabulary
percent

The students at Westview Middle School are collecting cans of food for the local food bank. Their goal is to collect 2,000 cans in one month. After 10 days, they have 800 cans of food. What *percent* of their goal have the students reached?

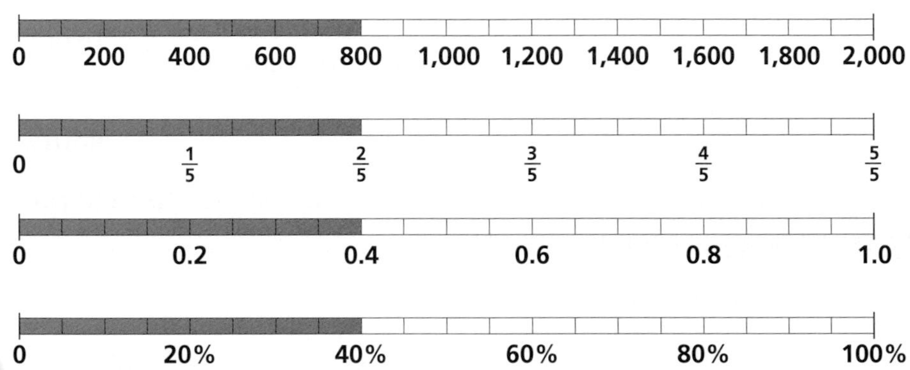

Reading Math

The word *percent* means "per hundred." So 40% means "40 out of 100."

The number lines show that the students have reached 40% of their goal.

A **percent** is a ratio of a number to 100. The symbol % is used to indicate that a number is a percent. For example, 40% is the ratio 40 to 100, or $\frac{40}{100}$. Percents can be written as fractions or decimals.

EXAMPLE 1 Writing Percents as Fractions

Write 35% as a fraction in simplest form.

$35\% = \frac{35}{100}$ *Write the percent as a fraction with a denominator of 100.*

$= \frac{7}{20}$ *Simplify.*

EXAMPLE 2 Writing Percents as Decimals

Write 43% as a decimal.

$43\% = \frac{43}{100}$ *Write the percent as a fraction with a denominator of 100.*

$= 0.43$ *Write the fraction as a decimal.*

Notice that both 43% and 0.43 mean "43 hundredths." Another way to write a percent as a decimal is to delete the percent sign and move the decimal point two places to the left.

$43.\% = 0.43$

EXAMPLE 3 Writing Decimals as Percents

Write each decimal as a percent.

A 0.07

$0.07 = \frac{7}{100}$ *Write the decimal as a fraction.*

$= 7\%$ *Write the fraction as a percent.*

B 0.2

$0.2 = \frac{2}{10}$ *Write the decimal as a fraction.*

$= \frac{20}{100}$ *Write an equivalent fraction with a denominator of 100.*

$= 20\%$ *Write the fraction as a percent.*

Both 0.07 and 7% mean "7 hundredths." You can write a decimal as a percent by moving the decimal point two places to the right and adding the percent sign.

$0.07 = 7.\%$

EXAMPLE 4 Writing Fractions as Percents

Helpful Hint

The method shown in Example 4A works only if the given denominator is a factor or multiple of 100. The method shown in Example 4B works for any denominator.

Write each fraction as a percent.

A $\frac{4}{5}$

$\frac{4}{5} = \frac{4 \cdot 20}{5 \cdot 20}$

$= \frac{80}{100}$ *Write an equivalent fraction with a denominator of 100.*

$= 80\%$ *Write the fraction as a percent.*

B $\frac{3}{8}$

$\frac{3}{8} = 3 \div 8$

$= 0.375$ *Use division to write the fraction as a decimal.*

$= 37.5\%$ *Write the decimal as a percent.*

Think and Discuss

1. **Explain** how to write 5% as a decimal.
2. **Describe** two methods you could use to write $\frac{3}{4}$ as a percent.
3. **Write** the ratio 25:100 as a fraction in simplest form, as a decimal, and as a percent.

6-1 Exercises

FOR EOG PRACTICE

see page 678

internet connect
Homework Help Online
go.hrw.com Keyword: MS4 6-1

1.01

GUIDED PRACTICE

See Example 1 Write each percent as a fraction in simplest form.

1. 10% **2.** 45% **3.** 60% **4.** 28%

See Example 2 Write each percent as a decimal.

5. 85% **6.** 30% **7.** 9% **8.** 100%

See Example 3 Write each decimal as a percent.

9. 0.18 **10.** 0.4 **11.** 0.75 **12.** 0.03

See Example 4 Write each fraction as a percent.

13. $\frac{2}{5}$ **14.** $\frac{1}{4}$ **15.** $\frac{7}{50}$ **16.** $\frac{1}{8}$

INDEPENDENT PRACTICE

See Example 1 Write each percent as a fraction in simplest form.

17. 30% **18.** 50% **19.** 88% **20.** 2%

See Example 2 Write each percent as a decimal.

21. 16% **22.** 32% **23.** 105% **24.** 0.1%

See Example 3 Write each decimal as a percent.

25. 0.21 **26.** 0.57 **27.** 0.08 **28.** 0.038

See Example 4 Write each fraction as a percent.

29. $\frac{3}{10}$ **30.** $\frac{7}{8}$ **31.** $\frac{11}{40}$ **32.** $\frac{9}{1,000}$

PRACTICE AND PROBLEM SOLVING

Write each percent as a fraction in simplest form and as a decimal.

33. 8% **34.** 120% **35.** 2.5% **36.** 300%

37. 0.75% **38.** 16% **39.** 70.3% **40.** 48%

Write as a percent. Round to the nearest tenth of a percent, if necessary.

41. $\frac{9}{10}$ **42.** 0.15 **43.** $\frac{1}{3}$ **44.** 5:8

45. 0.001 **46.** 4 to 9 **47.** 2.7 **48.** $\frac{5}{2}$

Compare. Write $<$, $>$, or $=$.

49. 45% ■ $\frac{2}{5}$ **50.** 9% ■ 0.9 **51.** $\frac{7}{12}$ ■ 60% **52.** 1.5 ■ 150%

Social Studies 

The table shows the results of a survey to determine the number of people who walk to work in five U.S. cities. Use the table for Exercises 53–55.

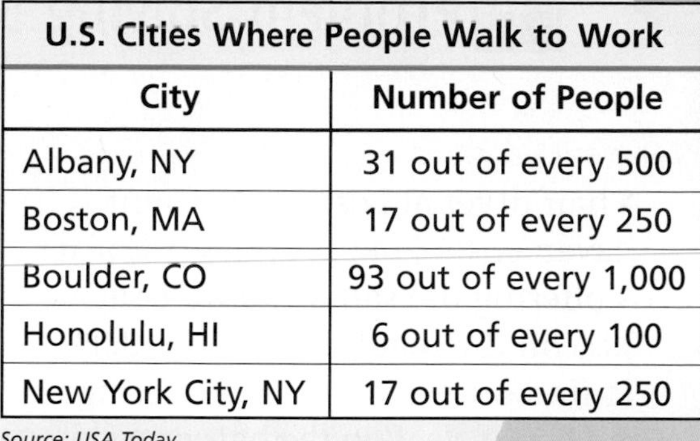

U.S. Cities Where People Walk to Work	
City	**Number of People**
Albany, NY	31 out of every 500
Boston, MA	17 out of every 250
Boulder, CO	93 out of every 1,000
Honolulu, HI	6 out of every 100
New York City, NY	17 out of every 250

Source: USA Today

53. What fraction of people walk to work in Honolulu? Write the fraction as a decimal and as a percent.

54. What percent of people walk to work in Boulder?

55. Compare the percent of people who walk to work in New York City with the percent of people who walk to work in Albany.

56. In a survey, 35% of downtown workers in Ann Arbor, Michigan, reported that they had walked to work at least once in the past 5 years. Write this percent as a fraction in simplest form and as a decimal.

57. ***CHALLENGE*** According to the 1990 U.S. Census, $\frac{530,857}{814,548}$ of people in Los Angeles drove alone to work each day and $\frac{17,881}{116,364}$ carpooled. What is the total percent of people in Los Angeles who used a car to get to work? Round your answer to the nearest tenth of a percent.

go.hrw.com
KEYWORD: MS4 Travel
CNN Student News

Spiral Review

Find each value. (Lesson 2-1)

58. 5^3 **59.** 12^2 **60.** 30^2 **61.** 8^3

62. **EOG PREP** A cookie recipe calls for $\frac{3}{4}$ cup brown sugar for 1 batch of cookies. If Jamie wants to make 3 batches of cookies, how much brown sugar does she need? (Lesson 4-7)

A 1 cup B $1\frac{1}{2}$ cups C $2\frac{1}{4}$ cups D $3\frac{1}{4}$ cups

63. **EOG PREP** What is the solution to the proportion $\frac{2}{3} = \frac{6}{a}$? (Lesson 5-3)

A 4 B 1 C 9 D 1.5

6-2 Estimate with Percents

 Problem Solving Skill

Learn to estimate percents.

A hair dryer at Lester's Discount Haven costs \$14.99. Carissa's Corner is offering the same hair dryer at 20% off the regular price of \$19.99. To find out which store is offering the better deal on the hair dryer, you can use estimation.

The table shows common percents and their fraction equivalents. You can use fractions to estimate the percent of a number by choosing a fraction that is close to a given percent.

Percent	10%	20%	25%	$33\frac{1}{3}\%$	50%	$66\frac{2}{3}\%$
Fraction	$\frac{1}{10}$	$\frac{1}{5}$	$\frac{1}{4}$	$\frac{1}{3}$	$\frac{1}{2}$	$\frac{2}{3}$

EXAMPLE  1 Using Fractions to Estimate Percents

Use a fraction to estimate 48% of 79.

$48\% \text{ of } 79 \approx \frac{1}{2} \cdot 79$ *Think: 48% is about 50% and 50% is equivalent to $\frac{1}{2}$.*

$\approx \frac{1}{2} \cdot 80$ *Change 79 to a compatible number.*

≈ 40 *Multiply.*

48% of 79 is about 40.

Remember!

Compatible numbers are close to the numbers in the problem and are used to do math mentally.

EXAMPLE 2 *Consumer Math Application*

Carissa's Corner is offering 20% off a hair dryer that costs \$19.99. If the same hair dryer costs \$14.99 at Lester's Discount Haven, which store offers the better deal?

First find the discount on the hair dryer at Carissa's Corner.

$20\% \text{ of } \$19.99 = \frac{1}{5} \cdot \$19.99$ *Think: 20% is equivalent to $\frac{1}{5}$.*

$\approx \frac{1}{5} \cdot \$20$ *Change \$19.99 to a compatible number.*

$\approx \$4$ *Multiply.*

The discount is approximately \$4. Since \$20 − \$4 = \$16, the \$14.99 hair dryer at Lester's Discount Haven is the better deal.

Another way to estimate percents is to find 1% or 10% of a number. You can do this by moving the decimal point in the number.

1% of 45 = .45.	10% of 45 = 4.5.
To find 1% of a number, move the decimal point two places to the left.	*To find 10% of a number, move the decimal point one place to the left.*

EXAMPLE 3 Estimating with Simple Percents

Use 1% or 10% to estimate the percent of each number.

A **3% of 59**

59 is about 60, so find 3% of 60.

1% of 60 = .60.

3% of 60 = 3 · 0.60 = 1.8 *3% equals 3 · 1%.*

3% of 59 is about 1.8.

B **18% of 45**

18% is about 20%, so find 20% of 45.

10% of 45 = 4.5.

20% of 45 = 2 · 4.5 = 9.0 *20% equals 2 · 10%.*

18% of 45 is about 9.

EXAMPLE 4 *Consumer Math Application*

Eric and Selena spent $25.85 for their meals at a restaurant. About how much money should they leave for a 15% tip?

Since $25.85 is about $26, find 15% of $26.

15% = 10% + 5% *Think: 15% is 10% plus 5%.*

10% of $26 = **$2.60**

5% of $26 = **$2.60** ÷ 2 = **$1.30** *5% is $\frac{1}{2}$ of 10%, so divide $2.60 by 2.*

$2.60 + $1.30 = $3.90 *Add the 10% and 5% estimates.*

Eric and Selena should leave about $3.90 for a 15% tip.

Think and Discuss

1. **Describe** two ways to estimate 51% of 88.
2. **Explain** why you might divide by 7 or multiply by $\frac{1}{7}$ to estimate a 15% tip.
3. **Give an example** of a situation in which an estimate of a percent is sufficient. Then give an example of a situation in which an exact percent is necessary.

6-2 Exercises

FOR EOG PRACTICE
see page 678

internet connect
Homework Help Online
go.hrw.com Keyword: MS4 6-2

1.01, 1.03

GUIDED PRACTICE

See Example 1 **Use a fraction to estimate the percent of each number.**

1. 30% of 86 **2.** 52% of 83 **3.** 10% of 48 **4.** 27% of 63

See Example 2 **5.** Darden has $35 to spend on a backpack. He finds one on sale for 35% off the regular price of $43.99. Does Darden have enough money to buy the backpack? Explain.

See Example 3 **Use 1% or 10% to estimate the percent of each number.**

6. 5% of 82 **7.** 39% of 19 **8.** 21% of 68 **9.** 7% of 109

See Example 4 **10.** Mrs. Coronado spent $23 on a manicure. About how much money should she leave for a 15% tip?

INDEPENDENT PRACTICE

See Example 1 **Use a fraction to estimate the percent of each number.**

11. 8% of 261 **12.** 34% of 93 **13.** 53% of 142 **14.** 23% of 98

15. 51% of 432 **16.** 18% of 42 **17.** 11% of 132 **18.** 54% of 39

See Example 2 **19.** A pair of shoes at The Value Store costs $20. Betty's Boutique has the same shoes on sale for 25% off the regular price of $23.99. Which store offers the better price on the shoes?

See Example 3 **Use 1% or 10% to estimate the percent of each number.**

20. 41% of 16 **21.** 8% of 310 **22.** 83% of 70 **23.** 2% of 634

24. 58% of 81 **25.** 24% of 49 **26.** 11% of 99 **27.** 63% of 39

See Example 4 **28.** Marc's lunch cost $8.92. He wants to leave a 15% tip for the service. About how much should his tip be?

PRACTICE AND PROBLEM SOLVING

Estimate.

29. 31% of 180 **30.** 18% of 150 **31.** 3% of 96 **32.** 2% of 198

33. 78% of 90 **34.** 52% of 234 **35.** 19% of 75 **36.** 4% of 311

37. The new package of Marti's Snacks contains 20% more snack mix than the old package. If there were 22 ounces of snack mix in the old package, about how many ounces are in the new package?

38. Frameworks charges $60.85 for framing. Including the 7% sales tax, about how much will it cost to have a painting framed?

39. ***SPORTS*** Last season, Ali had a hit 19.3% of the times he came to bat. If Ali batted 82 times last season, about how many hits did he have?

40. ***BUSINESS*** The table shows the results of a survey about the Internet. The number of people interviewed was 391.

    a. Estimate the number of people willing to give out their e-mail address.

    b. Estimate the number of people not willing to give out their credit card number.

Information People Are Willing to Give Out on the Internet

Information	Percent of People
E-mail address	78
Work phone number	53
Street address	49
Home phone number	35
Credit card number	33
Social Security number	11

Source: USA Today

41. ***SPORTS*** In a survey of 1,032 people, 58% said they got their information about the 2000 Summer Olympics by watching television. Estimate the number of people who got their information by watching television.

42. Sandi earns $43,000 per year. This year, she plans to spend about 27% of her income on rent.

    a. About how much does Sandi plan to spend on rent this year?

    b. About how much does she plan to spend on rent each month?

43. ***WRITE A PROBLEM*** Use information from the table in Exercise 40 to write a problem that can be solved by using estimation of a percent.

44. ***WRITE ABOUT IT*** Explain why it might be important to know whether your estimate of a percent is too high or too low. Give an example.

45. ***CHALLENGE*** Use the table from Exercise 40 to estimate how many more people will give out their work phone number than their Social Security number. Show your work using two different methods.

Spiral Review

Find each product. (Lesson 4-3)

46. 0.8 · 96 47. 30 · 0.04 48. 1.6 · 900 49. 0.005 · 75

50. **EOG PREP** Brandi spent $15 on a new CD. If this was $\frac{1}{3}$ of her monthly allowance, how much is Brandi's allowance? (Lesson 4-12)

A $5 B $30 C $35 D $45

51. **EOG PREP** What is the decimal equivalent of 4%? (Lesson 6-1)

A 4 B 0.4 C 0.04 D 0.004

6-3 Percent of a Number

Learn to find the percent of a number.

The human body is made up mostly of water. In fact, about 67% of a person's total (100%) body weight is water. If Cameron weighs 90 pounds, about how much of his weight is water?

Recall that a percent is a part of 100. Since you want to know the part of Cameron's body that is water, you can set up and solve a proportion to find the answer.

Part → $\frac{67}{100} = \frac{n}{90}$ ← *Part*

Whole → ← *Whole*

EXAMPLE 1 Using Proportions to Find Percents of Numbers

Find the percent of each number.

A 67% of 90

$$\frac{67}{100} = \frac{n}{90}$$ *Write a proportion.*

$$67 \cdot 90 = 100 \cdot n$$ *Set the cross products equal.*

$$6{,}030 = 100n$$ *Multiply.*

$$\frac{6{,}030}{100} = \frac{100n}{100}$$ *Divide each side by 100 to isolate the variable.*

$$60.3 = n$$

67% of 90 is 60.3.

B 145% of 210

$$\frac{145}{100} = \frac{n}{210}$$ *Write a proportion.*

$$145 \cdot 210 = 100 \cdot n$$ *Set the cross products equal.*

$$30{,}450 = 100n$$ *Multiply.*

$$\frac{30{,}450}{100} = \frac{100n}{100}$$ *Divide each side by 100 to isolate the variable.*

$$304.5 = n$$

145% of 210 is 304.5.

Helpful Hint

When solving a problem with a percent greater than 100%, the *part* will be greater than the *whole*.

In addition to using proportions, you can find the percent of a number by using decimal equivalents.

EXAMPLE 2 Using Decimal Equivalents to Find Percents of Numbers

Find the percent of each number. Estimate to check whether your answer is reasonable.

Helpful Hint

When you are solving problems with percents, *of* usually means "times."

A **8% of 50**

8% of 50 = 0.08 · 50 *Write the percent as a decimal.*

= 4 *Multiply.*

Estimate

10% of 50 = 5, so 8% of 50 is less than 5. Thus 4 is a reasonable answer.

B **0.5% of 36**

0.5% of 36 = 0.005 · 36 *Write the percent as a decimal.*

= 0.18 *Multiply.*

Estimate

1% of 36 = 0.36, so 0.5% of 36 is half of 0.36. Thus 0.18 is a reasonable answer.

EXAMPLE 3 *Geography Application*

Earth's total land area is about 57,308,738 mi^2. The land area of Asia is about 30% of this total. What is the approximate land area of Asia to the nearest square mile?

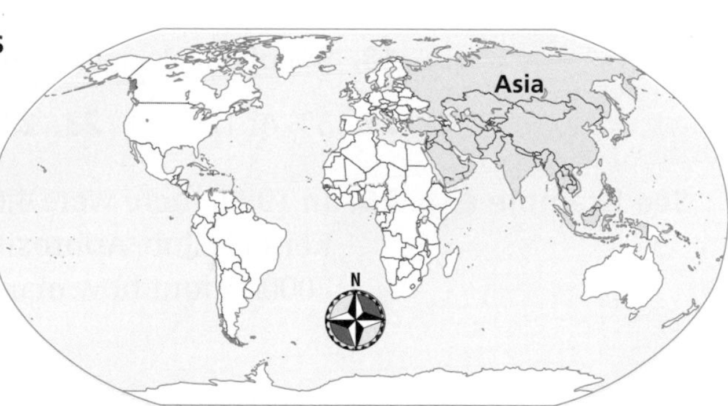

Find 30% of 57,308,738.

0.30 · 57,308,738 *Write the percent as a decimal.*

17,192,621.4 *Multiply.*

The land area of Asia is about 17,192,621 mi^2.

Think and Discuss

1. **Explain** how to set up a proportion to find 150% of a number. Describe how this proportion differs from the one used to find 50% of a number.
2. **Name** a situation in which you might need to find a percent of a number.

6-3 Exercises

FOR EOG PRACTICE

see page 678

internet connect
Homework Help Online
go.hrw.com Keyword: MS4 6-3

1.01, 1.03

GUIDED PRACTICE

See Example 1 **Find the percent of each number.**

1. 30% of 80
2. 38% of 400
3. 200% of 10
4. 180% of 90

See Example 2 **Find the percent of each number. Estimate to check whether your answer is reasonable.**

5. 16% of 50
6. 7% of 200
7. 47% of 900
8. 40% of 75

See Example 3

9. Of the 450 students at Miller Middle School, 38% ride the bus to school. How many students ride the bus to school?

INDEPENDENT PRACTICE

See Example 1 **Find the percent of each number.**

10. 80% of 35
11. 16% of 70
12. 150% of 80
13. 118% of 3,000
14. 5% of 58
15. 1% of 4
16. 103% of 50
17. 225% of 8

See Example 2 **Find the percent of each number. Estimate to check whether your answer is reasonable.**

18. 9% of 40
19. 20% of 65
20. 36% of 50
21. 2.9% of 60
22. 5% of 12
23. 220% of 18
24. 0.2% of 160
25. 155% of 8

See Example 3

26. In 1999, there were 4,652 Dalmatians registered by the American Kennel Club. Approximately 66% of this number were registered in 2000. About how many Dalmatians were registered in 2000?

PRACTICE AND PROBLEM SOLVING

Solve.

27. 60% of 10 is what number?
28. What number is 25% of 160?
29. What number is 15% of 30?
30. 10% of 84 is what number?
31. 25% of 47 is what number?
32. What number is 59% of 20?
33. What number is 125% of 4,100?
34. 150% of 150 is what number?

Find the percent of each number. If necessary, round to the nearest hundredth.

35. 160% of 50
36. 350% of 20
37. 480% of 25
38. 115% of 200
39. 18% of 3.4
40. 0.9% of 43
41. 98% of 4.3
42. 1.22% of 56

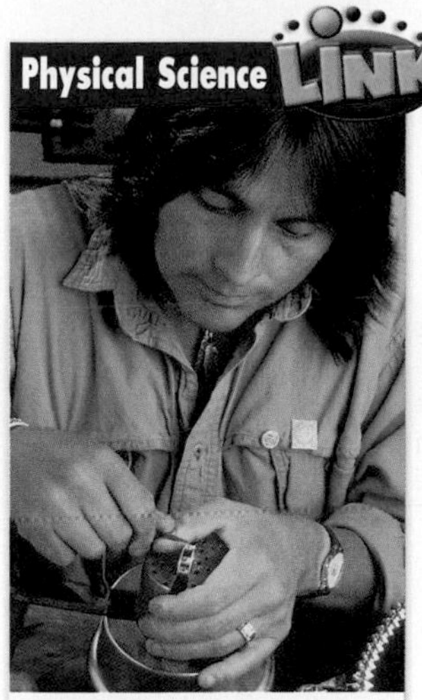

The mineral turquoise is often used as an ornamental stone in jewelry. Its color varies from greenish blue to sky-blue shades and can change from exposure to skin oils.

43. ***NUTRITION*** The United States Department of Agriculture recommends that women should eat 25 g of fiber each day. If a granola bar provides 9% of that amount, how many grams of fiber does it contain?

44. ***PHYSICAL SCIENCE*** The percent of pure gold in 14-karat gold is about 58.3%. If a 14-karat gold ring weighs 5.6 grams, about how many grams of pure gold are in the ring?

45. ***EARTH SCIENCE*** The apparent magnitude of the star Mimosa is 1.25. If Spica, another star, has an apparent magnitude that is 78.4% of Mimosa's, what is Spica's apparent magnitude?

46. ***CONSUMER MATH*** Trahn purchased a pair of slacks for $39.95 and a jacket for $64.00. The sales tax rate on his purchases was 5.5%.
    a. Find the amount of sales tax Trahn paid, to the nearest cent.
    b. Find the total cost of the purchases, including sales tax.

47. The graph shows the results of a student survey about computers. Use the graph to predict how many students in your class have a computer at home.

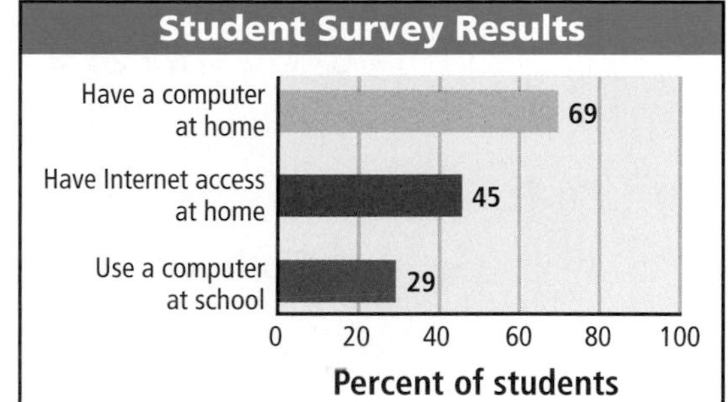

48. ***WHAT'S THE ERROR?*** A student used the proportion $\frac{n}{100} = \frac{5}{26}$ to find 5% of 26. What did the student do wrong?

49. ***WRITE ABOUT IT*** Describe two ways to find 18% of 40.

50. ***CHALLENGE*** François's starting pay was $6.25 per hour. During his annual review, he received a 5% raise. Find François's pay raise to the nearest cent and the amount he will earn with his raise. Then find 105% of $6.25. What can you conclude?

Spiral Review

Solve each equation. (Lesson 4-6)

51. $0.4x = 16$ **52.** $c \div 1.25 = 4$ **53.** $2.2b = 2.42$ **54.** $\frac{x}{2.5} = 20$

Find the unit rates and write them in both fraction and word form. (Lesson 5-1)

55. Monica buys 3 greeting cards for $5.25.

56. Kevin types 295 words in 5 minutes.

57. **EOG PREP** The decimal 0.0125 is equivalent to which percent? (Lesson 6-1)

A 0.125% C 12.5%
B 1.25% D 125%

58. **EOG PREP** Which is the closest estimate of 150% of 83? (Lesson 6-2)

A 40 C 120
B 80 D 160

6-4 Solving Equations Containing Percents

Learn to solve one-step equations containing percents.

Sloths may seem lazy, but their extremely slow movement helps to make them almost invisible to predators. Sloths sleep an average of 16.5 hours per day. To find out what percent of a 24-hour day 16.5 hours is, you can use a proportion or an equation.

Proportion method

$$\begin{array}{l}\textit{Part} \rightarrow \\ \textit{Whole} \rightarrow\end{array} \frac{n}{100} = \frac{16.5}{24} \begin{array}{l}\leftarrow \textit{Part} \\ \leftarrow \textit{Whole}\end{array}$$

$$n \cdot 24 = 100 \cdot 16.5$$
$$24n = 1{,}650$$
$$n = 68.75$$

Equation method

What **percent** of **24** is **16.5**?

$$n \cdot 24 = 16.5$$
$$n = 0.6875$$
$$n = 68.75\%$$

Sloths spend about **69%** of the day sleeping!

EXAMPLE 1 Using Proportions to Solve Problems with Percents

Solve.

A **45 is what percent of 90?**

$$\frac{n}{100} = \frac{45}{90}$$ *Write a proportion.*

$$n \cdot 90 = 100 \cdot 45$$ *Set the cross products equal.*

$$90n = 4{,}500$$ *Multiply.*

$$\frac{90n}{90} = \frac{4{,}500}{90}$$ *Divide each side by 90 to isolate the variable.*

$$n = 50$$

45 is 50% of 90.

B **12 is 8% of what number?**

$$\frac{8}{100} = \frac{12}{n}$$ *Write a proportion.*

$$8 \cdot n = 100 \cdot 12$$ *Set the cross products equal.*

$$8n = 1{,}200$$ *Multiply.*

$$\frac{8n}{8} = \frac{1{,}200}{8}$$ *Divide each side by 8 to isolate the variable.*

$$n = 150$$

12 is 8% of 150.

EXAMPLE 2 Using Equations to Solve Problems with Percents

Solve.

A **48 is 20% of what number?**

$48 = 20\% \cdot n$ *Write an equation.*

$48 = 0.2 \cdot n$ *Write 20% as a decimal.*

$\frac{48}{0.2} = \frac{0.2 \cdot n}{0.2}$ *Divide each side by 0.2 to isolate the variable.*

$240 = n$

48 is 20% of 240.

B **105 is what percent of 75?**

$105 = n \cdot 75$ *Write an equation.*

$\frac{105}{75} = \frac{n \cdot 75}{75}$ *Divide each side by 75 to isolate the variable.*

$1.4 = n$

$140\% = n$ *Write the decimal as a percent.*

105 is 140% of 75.

EXAMPLE 3 *Social Studies Application*

The table shows the average number of annual vacation days earned by workers with one year of service. What percent of the vacation days that workers in Finland earn do workers in Mexico earn?

Country	Vacation Days
Mexico	6
United States	10
Canada	10
Britain	20
Finland	30

Workers in Mexico earn 6 vacation days, and workers in Finland earn 30. Restate the question: What percent of 30 is 6?

$n \cdot 30 = 6$ *Write an equation.*

$\frac{n \cdot 30}{30} = \frac{6}{30}$ *Divide each side by 30 to isolate the variable.*

$n = 0.2$

$n = 20\%$ *Write the decimal as a percent.*

Workers in Mexico earn 20% of the number of vacation days that workers in Finland earn.

Think and Discuss

1. **Tell** what number is always used when you use a proportion to solve a percent problem. Explain.
2. **Write** and solve an equation to find the number that is 40% of 65.

6-4 Exercises

FOR EOG PRACTICE
see page 679

internet connect
Homework Help Online
go.hrw.com Keyword: MS4 6-4

1.01, 5.03

GUIDED PRACTICE

See Example 1

Solve.

1. 25 is what percent of 100?

2. 8 is 20% of what number?

3. 9 is what percent of 50?

4. 30 is 15% of what number?

See Example 2

5. 6 is 10% of what number?

6. 4 is what percent of 5?

7. 7 is 14% of what number?

8. 27 is what percent of 30?

See Example 3

9. Each week, Sandler spends \$10 of his allowance on school lunches. If his allowance is \$25, what percent of his allowance does he spend on school lunches?

INDEPENDENT PRACTICE

See Example 1

Solve.

10. 56 is 140% of what number?

11. 16 is what percent of 48?

12. 9 is what percent of 45?

13. 9 is 30% of what number?

14. 210% of what number is 147?

15. 12.4 is what percent of 12.4?

See Example 2

16. 40 is what percent of 60?

17. 45 is 20% of what number?

18. 18 is 15% of what number?

19. 18 is what percent of 6?

20. What percent of 80 is 10?

21. 8.8 is 40% of what number?

See Example 3

22. On average, teens spend 4 hours a week using the Internet and 4 hours doing chores. They spend 10 hours listening to the radio. What percent of the total time teens spend using the Internet and doing chores is the time they spend listening to the radio?

23. Montrell saves 8% of each paycheck for his college fund. If Montrell saved \$18.80 from his last paycheck, how much was he paid?

PRACTICE AND PROBLEM SOLVING

Solve. Round answers to the nearest tenth, if necessary.

24. 5 is what percent of 9?

25. What is 45% of 39?

26. 55 is 80% of what number?

27. 12 is what percent of 19?

28. What is 155% of 50?

29. 5.8 is 0.9% of what number?

30. 36% of what number is 57?

31. What percent of 64 is 40?

32. Every night, Sean reads 12 pages of a 286-page novel. What percent of the book is this? Round to the nearest tenth of a percent.

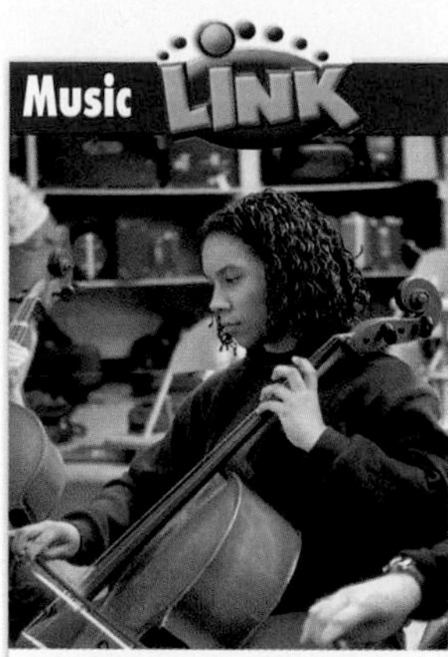

The viola family is made up of the cello, violin, and viola. Of the three instruments, the cello is the largest. Originally, the cello was named *violoncello*, which means little big viola.

33. ***MUSIC*** Beethoven wrote 9 trios for the piano, violin, and cello. If 20% of the chamber music pieces Beethoven wrote are trios, how many pieces of chamber music did he write?

34. ***HEALTH*** The circle graph shows the approximate distribution of blood types among people in the United States.

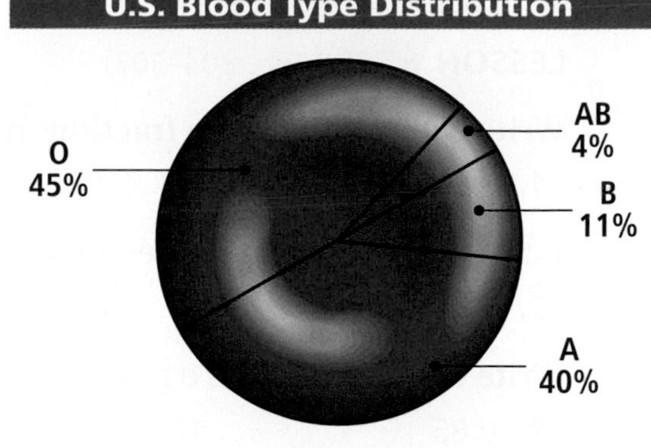

a. In a survey, 126 people had type O blood. Predict how many people were surveyed.

b. How many of the people surveyed had type AB blood?

35. ***HISTORY*** The length of Abraham Lincoln's first inaugural speech was 3,635 words. The length of his second inaugural speech was about 19.3% of the length of his first speech. About how long was Lincoln's second speech?

36. ***WHAT'S THE QUESTION?*** The first lap of an auto race is 2,500 m. This is 10% of the total race distance. The answer is 10. What is the question?

37. ***WRITE ABOUT IT*** If 35 is 110% of a number, is the number greater than or less than 35? Explain.

38. ***CHALLENGE*** Kayleen has been offered two jobs. The first job offers an annual salary of $32,000. The second job offers an annual salary of $10,000 plus 8% commission on all of her sales. How much money per month would Kayleen need to make in sales to earn enough commission to make more money at the second job?

Spiral Review

39. What is the range of the data in the stem-and-leaf plot? (Lesson 1-2)

Stems	Leaves
6	1 2 2 5 9
7	0 4 6 7 8
8	3 3 3 5 6

Key: 7 | 0 means 70

Write each fraction as a mixed number. (Lesson 3-8)

40. $\frac{28}{3}$ 41. $\frac{40}{7}$ 42. $\frac{19}{5}$

43. **EOG PREP** The speed limit on a stretch of highway is 70 mi/h. What is this in miles per minute? (Lesson 5-4)

A $1\frac{1}{6}$ mi/min B 6,160 mi/min C 369,600 mi/min D 70 mi/min

44. **EOG PREP** Which proportion would you use to find 38% of 90? (Lesson 6-3)

A $\frac{38}{100} = \frac{n}{90}$ B $\frac{100}{38} = \frac{n}{90}$ C $\frac{38}{100} = \frac{90}{n}$ D $\frac{38}{n} = \frac{90}{100}$

Chapter 6

Mid-Chapter Quiz

LESSON 6-1 (pp. 304–307)

Write each percent as a fraction in simplest form.

1. 9% **2.** 43% **3.** 5% **4.** 18%

Write each percent as a decimal.

5. 22% **6.** 90% **7.** 29% **8.** 5%

Write each decimal as a percent.

9. 0.85 **10.** 0.026 **11.** 0.1111 **12.** 0.56

Write each fraction as a percent. Round to the nearest tenth of a percent, if necessary.

13. $\frac{14}{81}$ **14.** $\frac{25}{52}$ **15.** $\frac{55}{78}$ **16.** $\frac{13}{32}$

LESSON 6-2 (pp. 308–311)

Estimate.

17. 49% of 46 **18.** 9% of 25 **19.** 36% of 150 **20.** 5% of 60

21. 18% of 80 **22.** 26% of 115 **23.** 91% of 300 **24.** 42% of 197

25. The Carsons find a new video game system selling for $230.00 on the Internet. A local electronics store is selling the same system for 25% off the regular price of $299.99. Which is the better offer for the system?

LESSON 

(pp. 312–315)

Find the percent of each number.

26. 25% of 84 **27.** 52% of 300 **28.** 0.5% of 40 **29.** 160% of 450

30. 41% of 122 **31.** 178% of 35 **32.** 29% of 88 **33.** 80% of 176

LESSON 

(pp. 316–319)

Solve. Round to the nearest tenth, if necessary.

34. 14 is what percent of 280? **35.** 8 is 32% of what number?

36. 14 is 44% of what number? **37.** 22 is what percent of 900?

38. 99 is what percent of 396? **39.** 75 is 24% of what number?

40. 36 is 18% of what number? **41.** 16 is what percent of 34?

1.03

Focus on Problem Solving

Make a Plan

- **Estimate or find an exact answer**

Sometimes an estimate is sufficient when you are solving a problem. Other times you need to find an exact answer. Before you try to solve a problem, you should decide whether an estimate will be sufficient. Usually if a problem includes the word *about,* then you can estimate the answer.

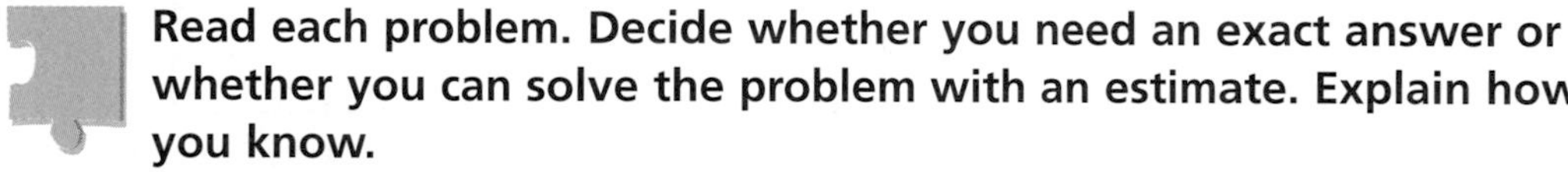

Read each problem. Decide whether you need an exact answer or whether you can solve the problem with an estimate. Explain how you know.

1. Barry has $21.50 left from his allowance. He wants to buy a book for $5.85 and a CD for $14.99. Assuming these prices include tax, does Barry have enough money left to buy both the book and the CD?

2. Last weekend Valerie practiced playing the drums for 3 hours. This is 40% of the total time she spent practicing last week. How much time did Valerie spend practicing last week?

3. Amber is shopping for a winter coat. She finds one that costs $157. The coat is on sale and is discounted 25% today only. About how much money will Amber save if she buys the coat today?

4. Marcus is planning a budget. He plans to spend less than 35% of his allowance each week on entertainment. Last week Marcus spent $7.42 on entertainment. If Marcus gets $20.00 each week, did he stay within his budget?

5. An upright piano is on sale for 20% off the original price. If the original price is $9,840, what is the sale price?

6. The Mapleton Middle School band has 41 students. If 6 of the students in the band play percussion instruments, do more than 15% of the students play percussion instruments?

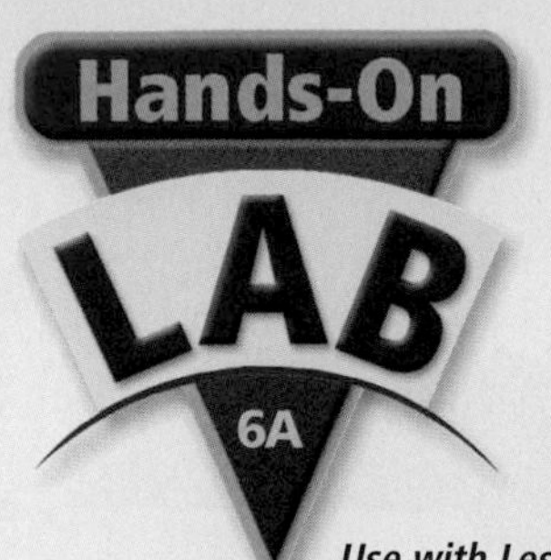

Percent of Increase and Decrease

Use with Lesson 6-5

You can use a geoboard or dot paper to help you understand the concepts of percent of increase and percent of decrease.

internet connect
Lab Resources Online
go.hrw.com
KEYWORD: MS4 Lab6A

Activity

1. Follow the steps to model a percent decrease.

   a. Make a 4-by-4 square like the one shown at right.

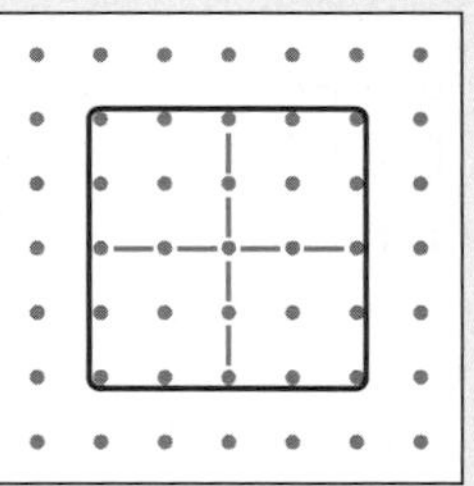

   b. To decrease the area of the square by 25%, first divide the square into four equal parts. (Recall that 25% is equivalent to $\frac{1}{4}$.)

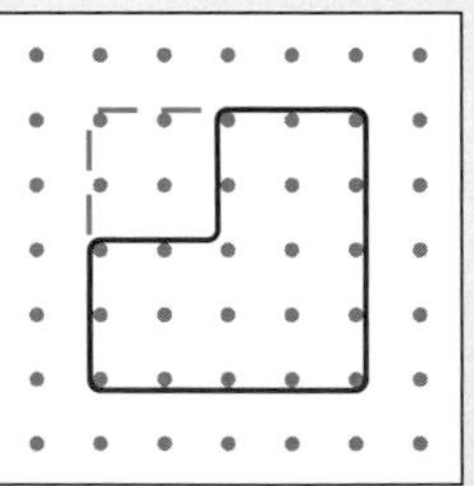

   c. Remove one part, which represents 25%, or $\frac{1}{4}$, from the original square to model a decrease of 25%.

2. Follow the steps to model a percent increase.

   a. Make a 3-by-3 square like the one shown at right.

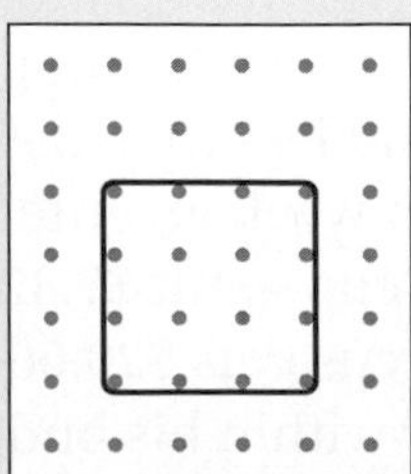

b. To increase the area of the square by $33\frac{1}{3}\%$, first divide the square into three equal parts. (Recall that $33\frac{1}{3}\%$ is equivalent to $\frac{1}{3}$.)

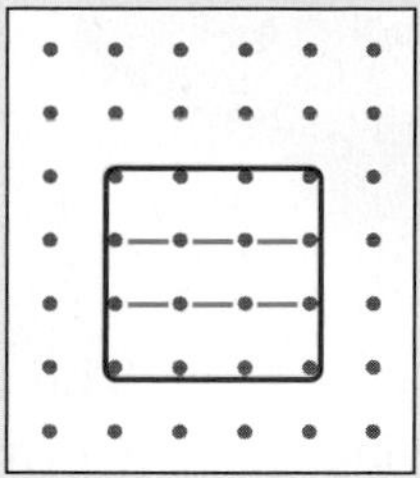

c. Add one part, which represents $33\frac{1}{3}\%$, or $\frac{1}{3}$, from the original square to model an increase of $33\frac{1}{3}\%$.

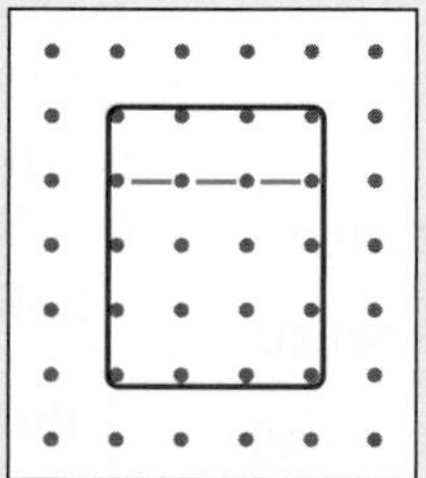

Think and Discuss

1. Explain why you divided the square in the first activity into four equal parts.
2. What percent of the 4-by-4 square remains after the 25% decrease?
3. Give an example of a figure that you might use to model a 40% increase.
4. In the second activity, what is the ratio of the area of the new rectangle to the area of the original square? Write this ratio as a percent. Then explain what this percent means.

Try This

1. Use a 2-by-2 square to model a 25% decrease. Then use the square to model a 25% increase.
2. Use a 3-by-3 square to model a $33\frac{1}{3}\%$ decrease.
3. Use a 4-by-4 square to model a 50% increase.
4. Use the figure at right to model a 50% decrease.

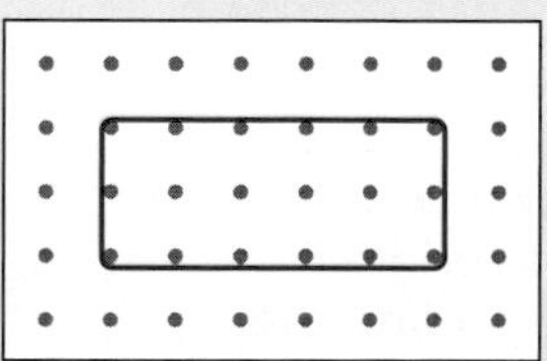

5. Choose a figure to use to model an increase of 10%. Explain why you chose that figure.
6. Look at the figures on the geoboard below. By what percent must you increase the area of the figure on the left to get the figure on the right?

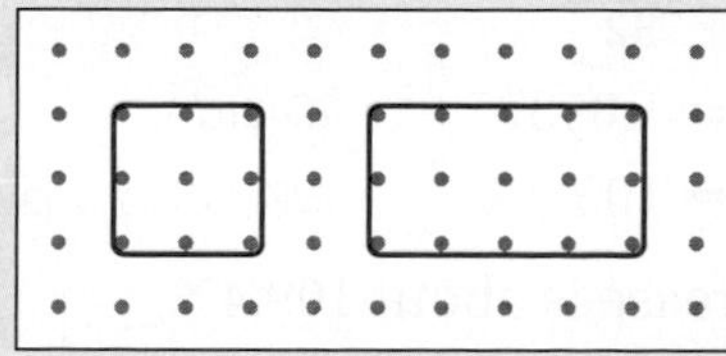

6-5 Percent of Change

Learn to solve problems involving percent of change.

Vocabulary

percent of change

percent of increase

percent of decrease

The U.S. Consumer Product Safety Commission has reported that, in 2000, 4,390 injuries related to motorized scooters were treated in hospital emergency rooms. This was a 230% increase from 1999's report of 1,330 injuries.

A percent can be used to describe an amount of change. The **percent of change** is the amount, stated as a percent, that a number increases or decreases. If the amount goes up, it is a **percent of increase**. If the amount goes down, it is a **percent of decrease**.

You can find the percent of change by using the following formula.

$$\text{percent of change} = \frac{\textbf{amount of change}}{\textbf{original amount}}$$

EXAMPLE 1 **Finding Percent of Change**

Find each percent of change. Round answers to the nearest tenth of a percent, if necessary.

Helpful Hint

When a number is decreased, subtract the new amount from the original amount to find the amount of change. When a number is increased, subtract the original amount from the new amount.

A **27 is decreased to 20.**

$27 - 20 = 7$ *Find the amount of change.*

$\text{percent of change} = \frac{7}{27}$ *Substitute values into formula.*

≈ 0.259259 *Divide.*

$\approx 25.9\%$ *Write as a percent. Round.*

The percent of decrease is about 25.9%.

B **32 is increased to 67.**

$67 - 32 = 35$ *Find the amount of change.*

$\text{percent of change} = \frac{35}{32}$ *Substitute values into formula.*

$= 1.09375$ *Divide.*

$\approx 109.4\%$ *Write as a percent. Round.*

The percent of increase is about 109.4%.

When you know the percent of change, you can use an equation to find the actual amount of change.

EXAMPLE 2 Using Percent of Change

The regular price of a portable CD player at Edwin's Electronics is $31.99. This week the CD player is on sale for 25% off.

A Find the actual amount of the discount.

$25\% \cdot 31.99 = d$	*Think: 25% of $31.99 is what number?*
$0.25 \cdot 31.99 = d$	*Write the percent as a decimal.*
$7.9975 = d$	*Multiply.*
$\$8.00 \approx d$	*Round to the nearest cent.*

The discount is $8.00.

B Find the sale price.

$\$31.99 - \$8.00 = \$23.99$	*Subtract the discount from the regular price.*

The sale price is $23.99.

EXAMPLE 3 *Business Application*

Helpful Hint

The retail price is the wholesale price from the manufacturer plus the amount of increase.

Winter Wonders buys snow globes from a manufacturer for $9.20 each and sells them at a 95% increase in price. What is the retail price of the snow globes?

First find the actual amount of increase.

Think: 95% of $9.20 is what number?

$95\% \cdot 9.20 = n$	
$0.95 \cdot 9.20 = n$	*Write the percent as a decimal.*
$8.74 = n$	*Multiply.*

The amount of increase is $8.74. Now find the retail price.

Think: retail price = wholesale price + amount of increase

$r = \$9.20 + \$8.74$

$r = \$17.94$

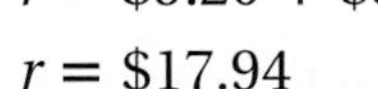

The retail price of the snow globes is $17.94 each.

Think and Discuss

1. **Explain** what is meant by a 100% decrease.
2. **Give an example** in which the amount of increase is greater than the original amount. What do you know about the percent of increase?

6-5 Exercises

FOR EOG PRACTICE
see page 680

1.01

GUIDED PRACTICE

See Example 1 

Find each percent of change. Round answers to the nearest tenth of a percent, if necessary.

1. 25 is decreased to 18.

2. 36 is increased to 84.

3. 62 is decreased to 52.

4. 28 is increased to 96.

See Example 2

5. The regular price of a sweater is $42.99. It is on sale for 20% off. Find the amount of the discount and the sale price.

See Example 3

6. The retail price of a pair of shoes is a 98% increase from its wholesale price. If the wholesale price of the shoes is $12.50, what is the retail price?

INDEPENDENT PRACTICE

See Example 1 

Find each percent of change. Round answers to the nearest tenth of a percent, if necessary.

7. 72 is decreased to 45.

8. 40 is increased to 95.

9. 12 is increased to 56.

10. 90 is decreased to 55.

11. 180 is decreased to 140.

12. 230 is increased to 250.

See Example 2 

13. A skateboard that sells for $65 is on sale for 15% off. Find the amount of the discount and the sale price.

14. A store is closing out its stock of sunglasses. The original price of the sunglasses was $44.95. The closeout price is 40% off the original price. Find the amount of the discount and the sale price.

See Example 3 

15. A jeweler buys a ring from an artisan for $85. He sells the ring in his store at a 135% increase in price. What is the retail price of the ring?

16. A water tank holds 45 gallons of water. A new water tank can hold 25% more water. What is the capacity of the new water tank?

PRACTICE AND PROBLEM SOLVING

Find each percent of change, amount of increase, or amount of decrease. Round answers to the nearest tenth, if necessary.

17. $8.80 is increased to $17.60.

18. 6.2 is decreased to 5.9.

19. 39.2 is increased to 56.3.

20. 28 is increased by 150%.

21. 75 is decreased by 40%.

22. $325 is decreased by 100%.

Economics 

23. The information at right shows the expenses for the Kramer family for one year.

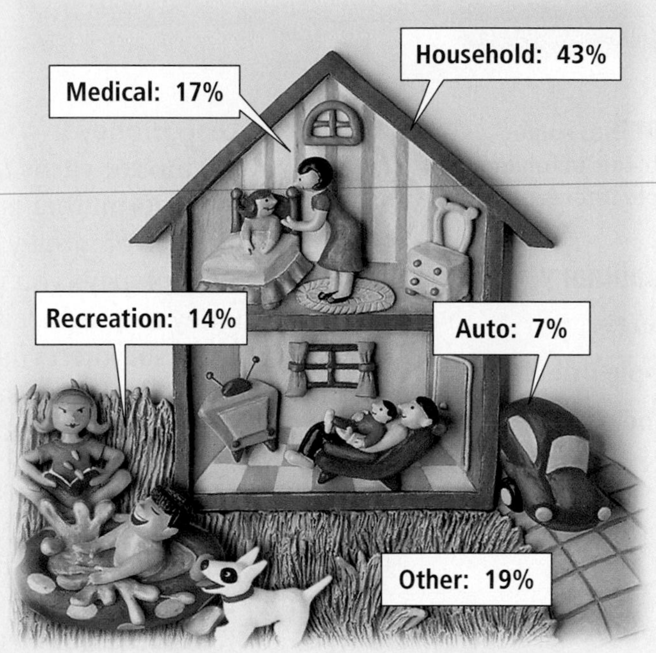

  a. If the Kramers spent $2,905 on auto expenses, what was their income for the year?

  b. How much money was spent on household expenses?

  c. The Kramers pay $14,400 per year on their mortgage. What percent of their household expenses is this? Round your answer to the nearest tenth.

24. United States health expenses were $428.7 billion in 1985 and $991.4 billion in 1995. What was the percent of increase in health expenses during this ten-year period? Round your answer to the nearest tenth of a percent.

25. In 1990, the total amount of energy consumed for transportation in the United States was 22,540 trillion British thermal units (Btu).

  a. From 1990 to 1991, there was a 1.8% decrease in energy consumed. About how many Btu of energy were consumed in 1991?

  b. From 1950 to 1990, there was a 165% increase in energy consumed for transportation. About how many Btu of energy were consumed in 1950?

26. ***CHALLENGE*** In 1960, 21.5% of U.S. households did not have a telephone. This statistic decreased by 75.8% between 1960 and 1990. In 1990, what percent of U.S. households had a telephone?

Spiral Review

Convert using dimensional analysis. (Lesson 5-4)

27. 8 gallons to pints **28.** 3 weeks to minutes **29.** 120 feet to yards

Write each percent as a fraction in simplest form. (Lesson 6-1)

30. 80% **31.** 2% **32.** 65% **33.** 48%

34. **EOG PREP** Of the 765 people registered to run in a race, only 80% ran. How many people ran in the race? (Lesson 6-3)

A 61 B 612 C 676 D 685

6-6 Simple Interest

Learn to solve problems involving simple interest.

Vocabulary

interest

simple interest

principal

When you keep money in a savings account, your money earns *interest.* **Interest** is an amount that is collected or paid for the use of money. For example, the bank pays you interest to use your money to conduct its business. Likewise, when you borrow money from the bank, the bank collects interest on its loan to you.

One type of interest, called **simple interest**, is money paid only on the *principal.* The **principal** is the amount of money deposited or borrowed. To solve problems involving simple interest, you can use the following formula.

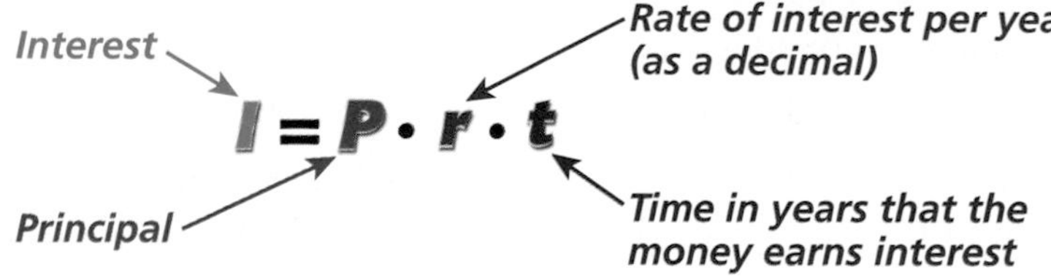

EXAMPLE 1 Using the Simple Interest Formula

Find each missing value.

A $I = \square$, $P = \$225$, $r = 3\%$, $t = 2$ years

$I = P \cdot r \cdot t$

$I = 225 \cdot 0.03 \cdot 2$ *Substitute. Use 0.03 for 3%.*

$I = 13.5$ *Multiply.*

The simple interest is $13.50.

B $I = \$300$, $P = \$1{,}000$, $r = \square$, $t = 5$ years

$I = P \cdot r \cdot t$

$300 = 1{,}000 \cdot r \cdot 5$ *Substitute.*

$300 = 5{,}000r$ *Multiply.*

$\frac{300}{5{,}000} = \frac{5{,}000r}{5{,}000}$ *Divide by 5,000 to isolate the variable.*

$0.06 = r$

The interest rate is 6%.

EXAMPLE  PROBLEM SOLVING APPLICATION

Olivia deposits $7,000 in an account that earns 7% simple interest. About how long will it take for her account balance to reach $8,000?

1 Understand the Problem

Rewrite the question as a statement:

- Find the number of years it will take for Olivia's account balance to reach $8,000.

List the **important information:**

- The principal is $7,000.
- The interest rate is 7%.
- Her account balance will be $8,000.

2 Make a Plan

Olivia's account balance A includes the principal plus the interest: $A = P + I$. Once you solve for I, you can use $I = P \cdot r \cdot t$ to find the time.

3 Solve

$$A = P + I$$

$$8{,}000 = 7{,}000 + I \qquad \textit{Substitute.}$$

$$\underline{-7{,}000} \quad \underline{-7{,}000} \qquad \textit{Subtract to isolate the variable.}$$

$$1{,}000 = I$$

$$I = P \cdot r \cdot t$$

$$1{,}000 = 7{,}000 \cdot 0.07 \cdot t \qquad \textit{Substitute. Use 0.07 for 7\%.}$$

$$1{,}000 = 490t \qquad \textit{Multiply.}$$

$$\frac{1{,}000}{\mathbf{490}} = \frac{490t}{\mathbf{490}} \qquad \textit{Divide to isolate the variable.}$$

$$2.04 \approx t \qquad \textit{Round to the nearest hundredth.}$$

It will take just over 2 years.

4 Look Back

After exactly 2 years, Olivia's money will have earned $980 in simple interest and her account balance will be $7,980.

$$I = 7{,}000 \cdot 0.07 \cdot 2 = 980$$

So it will take just over 2 years for her account to reach $8,000.

Think and Discuss

1. Write the value of t for a time period of 6 months.

2. Show how to find r if $I = \$10$, $P = \$100$, and $t = 2$ years.

6-6 Exercises

FOR EOG PRACTICE
see page 681

internet connect
Homework Help Online
go.hrw.com Keyword: MS4 6-6

5.04

GUIDED PRACTICE

See Example 1

Find each missing value.

1. $I =$ ■, $P = \$300$, $r = 4\%$, $t = 2$ years
2. $I =$ ■, $P = \$500$, $r = 2\%$, $t = 1$ year
3. $I = \$120$, $P =$ ■, $r = 6\%$, $t = 5$ years
4. $I = \$240$, $P = \$4{,}000$, $r =$ ■, $t = 2$ years

See Example 2

5. Scott deposits $8,000 in an account that earns 6% simple interest. How long will it be before the total amount is $10,000?

INDEPENDENT PRACTICE

See Example 1

Find each missing value.

6. $I =$ ■, $P = \$600$, $r = 7\%$, $t = 2$ years
7. $I =$ ■, $P = \$12{,}000$, $r = 3\%$, $t = 9$ years
8. $I = \$364$, $P = \$1{,}300$, $r =$ ■, $t = 7$ years
9. $I = \$440$, $P =$ ■, $r = 5\%$, $t = 4$ years
10. $I = \$455$, $P =$ ■, $r = 7\%$, $t = 5$ years
11. $I = \$231$, $P = \$700$, $r =$ ■, $t = 3$ years

See Example 2

12. Broderick deposits $6,000 in an account that earns 5.5% simple interest. How long will it be before the total amount is $9,000?
13. Teresa deposits $4,000 in an account that earns 7% simple interest. How long will it be before the total amount is $6,500?

PRACTICE AND PROBLEM SOLVING

Complete the table.

	Principal	Interest Rate	Time	Simple Interest
14.	$2,455	3%	■	$441.90
15.	■	4.25%	3 years	$663
16.	$18,500	■	42 months	$1,942.50
17.	$425.50	5%	10 years	■
18.	■	6%	3 years	$2,952

19. ***FINANCE*** How many years will it take for $4,000 to double at a simple interest rate of 5%?

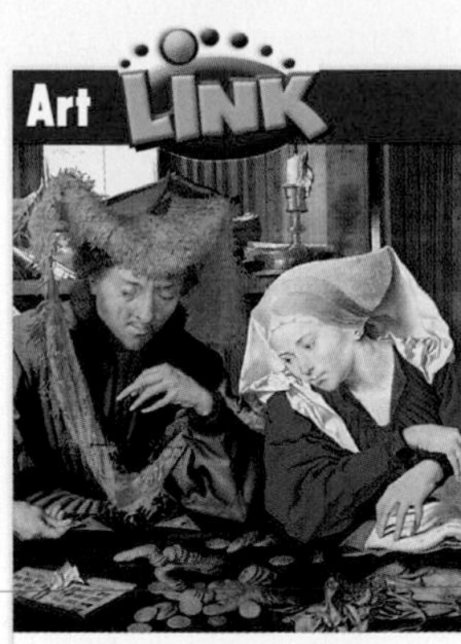

The Flemish painter Marinus van Reymerswaele painted several portraits of bankers and tax collectors. His *The Money-Changer and His Wife* was finished in 1539 and is now in a museum in Madrid, Spain.

go.hrw.com KEYWORD: MS4 Reserve

20. ***BANKING*** The graph shows interest rate returns for different types of investment as of July 2001.

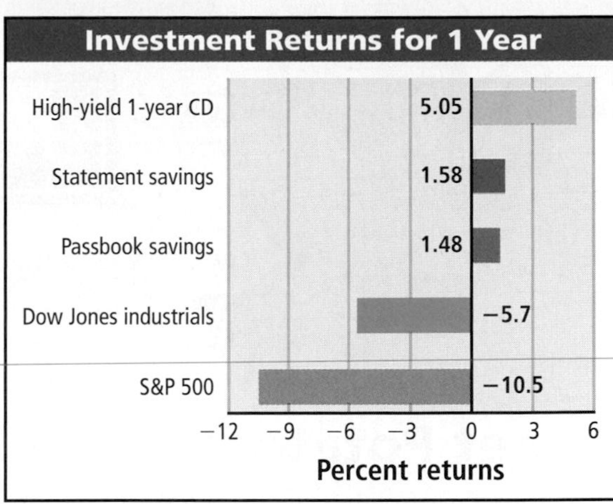

Source: USA Today

a. How much more interest was earned on $8,000 deposited for 6 months in a statement savings account than in a passbook savings account?

b. How much money was lost on $5,000 invested in S&P 500 stocks for one year?

c. Compare the returns on $12,000 invested in the high-yield 1-year CD and the Dow Jones industrials for one year.

21. ***BANKING*** Alexis put $400 in an account that earns 2% simple interest. At the end of the first year, the bank deposited the interest earned on the principal into her account. After that, the money in Alexis's account included both the principal and the interest, and the entire sum earned interest. How much will Alexis have in the account at the end of 2 years?

22. ***WRITE A PROBLEM*** Use the graph in Exercise 20 to write a problem that can be solved by using the simple interest formula.

23. ***WRITE ABOUT IT*** Explain whether you would pay more simple interest on a loan if you used plan A or plan B.

Plan A: $1,500 for 8 years at 6%

Plan B: $1,500 for 6 years at 8%

24. ***CHALLENGE*** The Jacksons are opening a savings account for their child's college education. In 18 years, they will need about $134,000. If the account earns 6% simple interest, how much money must the Jacksons invest now to cover the cost of the college education?

Spiral Review

Write each fraction as a percent. (Lesson 6-1)

25. $\frac{1}{20}$ **26.** $\frac{3}{5}$ **27.** $\frac{7}{50}$ **28.** $\frac{11}{25}$

29. **EOG PREP** What percent of 150 is 48? (Lesson 6-4)

A 32% B 50% C 72% D 312.5%

30. **EOG PREP** The regular price of a helmet is $36.50. It is on sale for 20% off. What is the sale price of the helmet? (Lesson 6-5)

A $7.30 B $16.50 C $29.20 D $35.77

Problem Solving on Location

NORTH CAROLINA

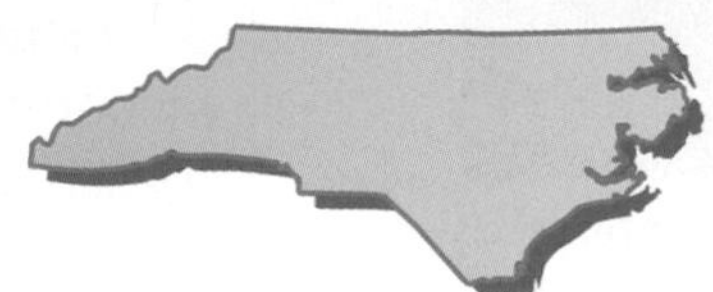

Sweet Potatoes

The sweet potato is a root vegetable that originated in the South and Central Americas. Many of the sweet potatoes produced in the United States are grown in North Carolina. For this reason, the sweet potato has been named North Carolina's state vegetable.

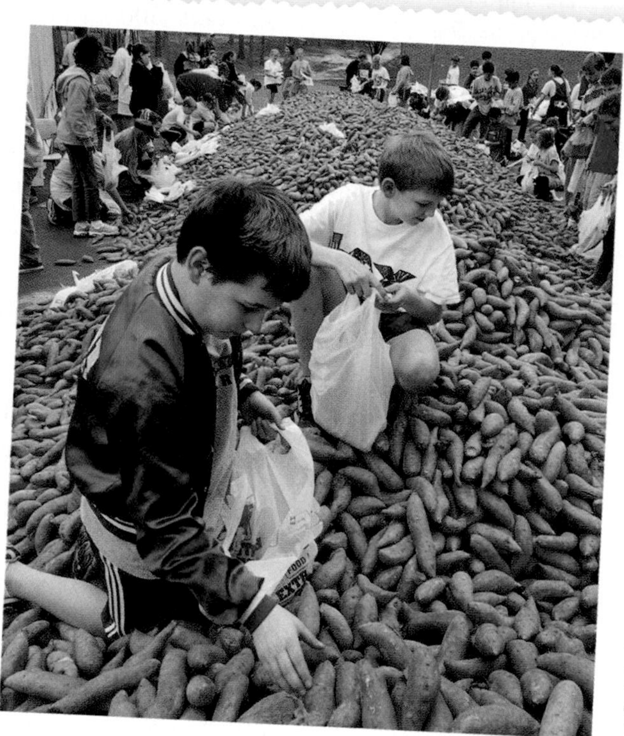

1. One year, 108,200 acres of sweet potatoes were planted in North Carolina. If the farmers were able to harvest 95.5% of the acres that year, how many acres did they harvest? Round your answer to the nearest hundred acres.

2. The United States produced about 1,238,200,000 pounds of sweet potatoes one year. About 544,000,000 pounds of the crop were produced in North Carolina alone. About what percent of the total weight of sweet potatoes did North Carolina produce that year?

3. North Carolina produced about 416,000,000 pounds of sweet potatoes one year and about 527,000,000 pounds the next. What was the percent of increase in sweet-potato production from the first year to the second? Round your answer to the nearest percent.

4. The circle graph at right shows the percent of the sweet-potato production in the United States that is supplied by North Carolina. What percent should be labeled for the remainder of the United States?

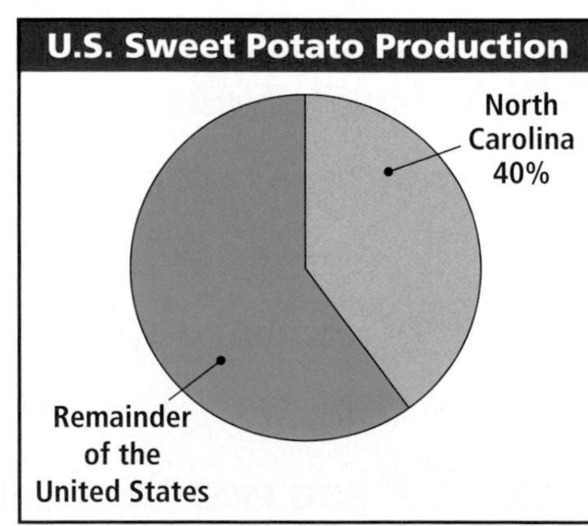

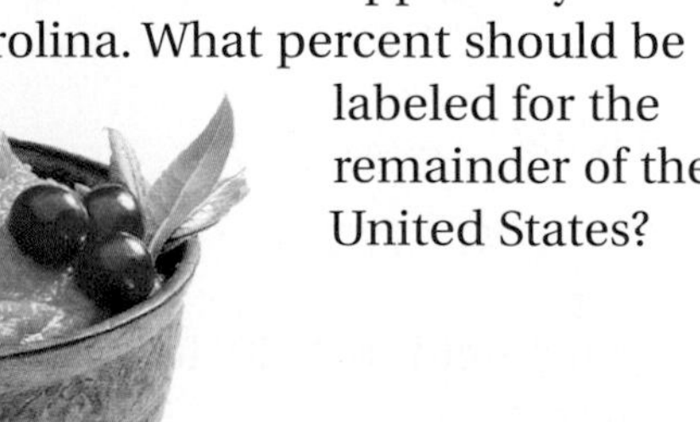

Azaleas

From late March to late June, azaleas can be seen in full bloom across the North Carolina landscape. The azalea plant is one of the most popular decorative shrubs in North Carolina and can be grown in all parts of the state. Some of the most common azalea plants grown in the area are known as Kurume hybrids. These come in a wide range of beautiful colors. A term that is often used in describing azaleas is *hose-in-hose*. This means that the azalea flower looks like a flower inside another flower.

Kurume Hybrids	
Hybrid	**Color**
Appleblossom	Light pink
Christmas Cheer	Hose-in-hose red
Coral Bells	Hose-in-hose pink
Delaware Valley White	White
Flame	Orange red
Hershey's Red	Semi-double bright red
Hexe	Hose-in-hose violet purple
Hinodegiri	Red
Salmon Beauty	Hose-in-hose salmon pink
Sherwood Red	Orange red
Snow	Hose-in-hose white

For 1–2, use the table.

1. What percent of the Kurume hybrids in the table are white or hose-in-hose white? Round to the nearest percent.

2. Helen planted 12 azalea shrubs in her yard. Of these, 4 shrubs grew orange red flowers, 3 grew light pink flowers, and 5 grew flowers that were hose-in-hose salmon pink. What percent of Helen's azalea shrubs were Appleblossoms?

3. A local nursery sells Kurume azalea shrubs every fall, which is the ideal time for planting them. Last year, the nursery sold 80 azalea shrubs. This year, the nursery sold 15% fewer shrubs than it sold last year. How many azalea shrubs did the nursery sell this year?

MATH-ABLES

Lighten Up

On a digital clock, up to seven light bulbs make up each digit on the display. You can label each light bulb as shown below.

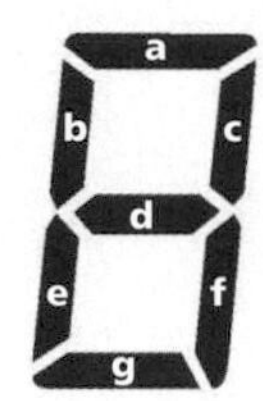

If each number were lit up for the same amount of time, you could find out which light bulb is lit the greatest percent of the time. You could also find out which light bulb is lit the least percent of the time.

For each number 0–9, list the letters of the light bulbs that are used when that number is showing. The first few numbers have been done for you.

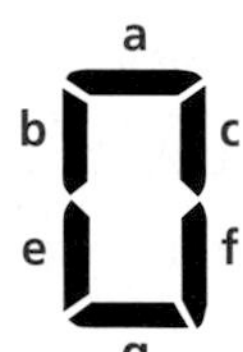

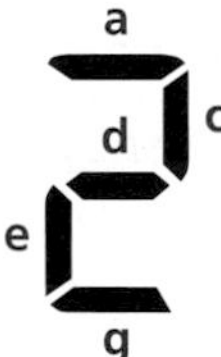

Once you have determined which bulbs are lit for each number, count how many times each bulb is lit. What percent of the time is each bulb lit? What does this tell you about which bulb will burn out first?

Percent Bingo

Each bingo card has numbers and percents on it. The caller has a collection of percent problems. The caller reads a problem. Then the players solve the problem, and the solution is a number or a percent. If players have the solution on their card, they mark it off. Normal bingo rules apply. You can win with a horizontal, vertical, or diagonal row.

internet connect

Go to ***go.hrw.com*** for a complete set of rules and game pieces.
KEYWORD: MS4 Game6

Create Circle Graphs

internet connect
Lab Resources Online
go.hrw.com
KEYWORD: MS4 TechLab6

A circle graph shows the relationship of parts to a whole. You can use a spreadsheet to create circle graphs.

Activity

1 The table shows the number of videos rented in one week. Use a spreadsheet to create a circle graph of the data.

Videos Rented	
Type	**Number of Rentals**
Classics	200
Comedy	250
Drama	250
Kids	300

Enter the data in the spreadsheet as shown below. Highlight the types and numbers of rentals by clicking on cell **A2** and dragging down to cell **B5.** Select the Graph Wizard tool and select the **Pie** option from the **Chart Type** list. Then click Finish to choose the first type of circle graph.

	A	B
1	Type	Number of Rentals
2	Classics	200
3	Comedy	250
4	Drama	250
5	Kids	300

Classics 20%
Comedy 25%
Drama 25%
Kids 30%

Classics
Comedy
Drama
Kids

The graph shown uses an option that displays the types of movies and the percent each type is of the total.

Think and Discuss

1. Explain the advantages of using a circle graph rather than a bar graph to represent the data in the Activity.

Try This

1. The table shows the results of a survey to determine students' favorite subjects. Use a spreadsheet to make a circle graph of the data.

Subject	English	Health	History	Math	Science
Number of Students	179	108	234	242	184

Chapter 6

Study Guide and Review

Vocabulary

Complete the sentences below with vocabulary words from the list above. Words may be used more than once.

1. __?__ is an amount that is collected or paid for the use of money. The equation $I = P \cdot r \cdot t$ is used for calculating __?__. The letter P represents the __?__ and the letter r represents the annual rate.

2. The ratio of an amount of increase to the original amount is the __?__.

3. The ratio of an amount of decrease to the original amount is the __?__.

4. A(n) __?__ is a ratio whose denominator is 100.

6-1 Fractions, Decimals, and Percents (pp. 304–307)

EXAMPLE

- **Write 12% as a decimal.**

$$12\% = \frac{12}{100}$$
$$= 0.12$$

- **Write $\frac{7}{8}$ as a percent.**

$$\frac{7}{8} = 7 \div 8$$
$$= 0.875$$
$$= 87.5\%$$

EXERCISES

Write each percent as a decimal.

5. 78% **6.** 40%

7. 5% **8.** 119%

Write each fraction as a percent. Round to the nearest tenth of a percent, if necessary.

9. $\frac{3}{5}$ **10.** $\frac{1}{6}$

11. $\frac{2}{3}$ **12.** $\frac{2}{25}$

6-2 Estimate with Percents (pp. 308–311)

EXAMPLE

■ **Estimate 26% of 77.**

$$26\% \text{ of } 77 \approx \frac{1}{4} \cdot 77$$
$$\approx \frac{1}{4} \cdot 80$$
$$\approx 20$$

26% of 77 is about 20.

EXERCISES

Estimate.

13. 22% of 44

14. 43% of 64

15. 49% of 82

16. 74% of 120

17. 31% of 97

18. 6% of 53

6-3 Percent of a Number (pp. 312–315)

EXAMPLE

■ **Find the percent of the number.**

125% of 610

$$\frac{125}{100} = \frac{n}{610}$$
$$125 \cdot 610 = 100 \cdot n$$
$$76{,}250 = 100n$$
$$\frac{76{,}250}{100} = \frac{100n}{100}$$
$$762.5 = n$$

125% of 610 is 762.5.

EXERCISES

Find the percent of each number.

19. 16% of 425

20. 48% of 50

21. 7% of 63

22. 96% of 125

23. 130% of 21

24. 72% of 75

6-4 Solving Equations Containing Percents (pp. 316–319)

EXAMPLE

■ **Solve.**

80 is 32% of what number?

$$80 = 32\% \cdot n$$
$$80 = 0.32 \cdot n$$
$$\frac{80}{0.32} = \frac{0.32 \cdot n}{0.32}$$
$$250 = n$$

80 is 32% of 250.

EXERCISES

Solve.

25. 20% of what number is 25?

26. 4 is what percent of 50?

27. 30 is 250% of what number?

28. What percent of 96 is 36?

29. 6 is 75% of what number?

30. 200 is what percent of 720?

6-5 Percent of Change (pp. 324–327)

EXAMPLE

Find each percent of change. Round answers to the nearest tenth, if necessary.

- **25 is decreased to 16.**

$25 - 16 = 9$

$\text{percent of change} = \frac{9}{25}$

$= 0.36$

$= 36\%$

The percent of decrease is 36%.

- **13.5 is increased to 27.**

$27 - 13.5 = 13.5$

$\text{percent of change} = \frac{13.5}{13.5}$

$= 1$

$= 100\%$

The percent of increase is 100%.

EXERCISES

Find each percent of change. Round answers to the nearest tenth, if necessary.

31. 54 is increased to 81.
32. 14 is decreased to 12.
33. 110 is increased to 143.
34. 90 is decreased to 15.2.
35. 26 is increased to 32.
36. 84 is decreased to 21.
37. The regular price of a new pair of skis is $245. This week the skis are on sale for 15% off. Find the amount of the discount and the sale price.

6-6 Simple Interest (pp. 328–331)

EXAMPLE

Find each missing value.

- $I = \square$, $P = \$545$, $r = 1.5\%$, $t = 2$ **years**

$I = P \cdot r \cdot t$

$I = 545 \cdot 0.015 \cdot 2$

$I = 16.35$

The simple interest is $16.35.

- $I = \$825$, $P = \square$, $r = 6\%$, $t = 11$ **years**

$I = P \cdot r \cdot t$

$825 = P \cdot 0.06 \cdot 11$

$825 = P \cdot 0.66$

$\frac{825}{0.66} = \frac{P \cdot 0.66}{0.66}$

$1{,}250 = P$

The principal is $1,250.

EXERCISES

Find each missing value.

38. $I = \square$, $P = \$1{,}000$, $r = 3\%$, $t = 6$ months
39. $I = \$452.16$, $P = \$1{,}256$, $r = 12\%$, $t = \square$
40. $I = \square$, $P = \$675$, $r = 4.5\%$, $t = 8$ years
41. $I = \$555.75$, $P = \$950$, $r = \square$, $t = 15$ years
42. $I = \$172.50$, $P = \square$, $r = 5\%$, $t = 18$ months
43. Craig deposits $1,000 in a savings account that earns 5% simple interest. How long will it take for the total amount in his account to reach $1,350?

Chapter Test

Write each percent as a fraction in simplest form and as a decimal.

1. 95% **2.** 37.5% **3.** 4% **4.** 0.01%

Write as a percent.

5. 0.75 **6.** 0.06 **7.** 0.8 **8.** 0.0039

9. $\frac{3}{10}$ **10.** $\frac{9}{20}$ **11.** $\frac{5}{16}$ **12.** $\frac{21}{7}$

Estimate.

13. 48% of 8 **14.** 3% of 119 **15.** 26% of 32 **16.** 76% of 280

17. The Pattersons spent $47.89 for a meal at a restaurant. About how much should they leave for a 15% tip?

Find the percent of each number.

18. 90% of 200 **19.** 35% of 210 **20.** 16% of 85

21. 250% of 30 **22.** 38% of 11 **23.** 5% of 145

Solve.

24. 36 is what percent of 150? **25.** 29 is what percent of 145?

26. 51 is what percent of 340? **27.** 36 is 40% of what number?

28. 70 is 14% of what number? **29.** 25 is 20% of what number?

30. Hampton Middle School is expecting 376 seventh-graders next year. This is 40% of the expected school enrollment. How many students are expected to enroll in the school next year?

Find each percent of change.

31. 30 is increased to 45. **32.** 115 is decreased to 46.

33. 116 is increased to 145. **34.** 128 is decreased to 32.

35. A community theater sold 8,500 tickets to performances during its first year. By its tenth year, ticket sales had increased by 34%. How many tickets did the theater sell during its tenth year?

Find each missing value.

36. $I =$ ■, $P = \$500$, $r = 5\%$, $t = 1$ year **37.** $I = \$702$, $P = \$1{,}200$, $r = 3.9\%$, $t =$ ■

38. $I = \$468$, $P = \$900$, $r =$ ■, $t = 8$ years **39.** $I = \$37.50$, $P =$ ■, $r = 10\%$, $t = 6$ months

40. Kate invested $3,500 at a 5% simple interest rate. How many years will it take for the original amount to double?

Chapter 6

Performance Assessment

Show What You Know

Create a portfolio of your work from this chapter. Complete this page and include it with your four best pieces of work from Chapter 6. Choose from your homework or lab assignments, mid-chapter quiz, or any journal entries you have done. Put them together using any design you want. Make your portfolio represent what you consider your best work.

Short Response

1. A bank advertises that a deposit of $150.00 will earn $2.75 in 6 months. What simple interest rate does the bank pay? Show the work that you used to determine your answer.

2. You have a coupon for 25% off the price of a mystery novel that regularly sells for $7.00. The sales tax is 7.25%. If you gave the cashier $10.00, how much change would you expect in return? Show your work.

3. Amber's meal cost $15.70. She paid her server $18.76. Approximately what percent tip did Amber give the server? Show your calculations.

Extended Problem Solving

4. The table shows the results of a survey about favorite types of restaurants.
   a. Make a bar graph of the data in the table.
   b. Write each number as a percent. Show your work.
   c. What is the total of the four percents? How can you find this answer without adding the four percents?
   d. Explain how you can use the percents from part **b** to find how many more people prefer steak houses than prefer formal-dining restaurants. Then find this number.

Type of Restaurant	Number of People
Cafeteria	32
Fast food	63
Formal dining	38
Steak house	47

Performance Assessment

internet connect
EOG Practice Online
go.hrw.com Keyword: MS4 TestPrep

Chapter 6

Getting Ready for EOG

Cumulative Assessment, Chapters 1–6

1. You deposit \$750 in an account that earns 5% simple interest. How much interest will your money earn in 6 months?

   A \$375.00 C \$37.50
   B \$187.50 D \$18.75

2. What are all the factors of 15?

   A 3, 5 C 3, 5, 15
   B 1, 3, 5, 15 D 3, 5, 10, 15

TEST TAKING TIP!

Try breaking a complex problem into smaller parts.

3. This year, 592 students said they recycle. Last year, only $\frac{5}{8}$ of that number said they recycled. How many more students recycle this year than last year?

   A 74 students C 370 students
   B 222 students D 974 students

4. Which number is **not** equivalent to the other three?

   A $\frac{27}{8}$ C $3\frac{3}{8}$
   B 3.375 D $\frac{61}{24}$

5. Which expression has the greatest value?

   A $-3(3) - 9$ C $-3(-3) - 9$
   B $-3(3) + 9$ D $-3(-3) + 9$

6. In a survey, 28 students said that pizza is their favorite lunch. If 28 is 16% of the class, how many students are in the class?

   A 4 students C 32 students
   B 175 students D 448 students

7. Divide $\frac{7}{10}$ by $2\frac{1}{2}$.

   A $1\frac{3}{4}$ C $\frac{14}{25}$
   B $1\frac{4}{10}$ D $\frac{7}{25}$

8. The graph shows the results of a survey. Use the graph to determine which statement is false.

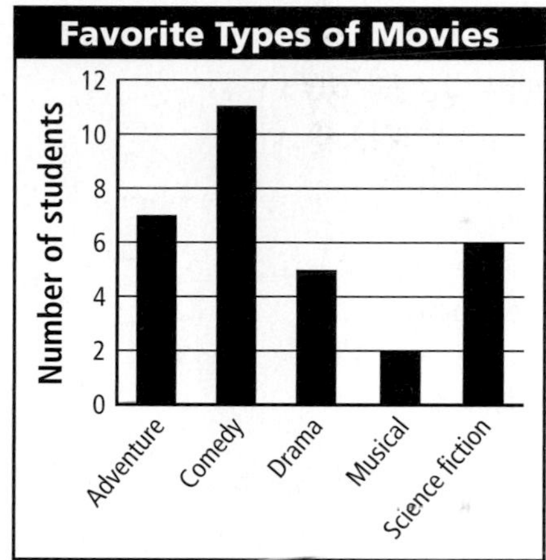

   A More than half the students chose comedies.
   B Musicals are less popular than dramas.
   C Fewer than $\frac{1}{5}$ of the students chose dramas.
   D Fewer than $\frac{1}{10}$ of the students chose musicals.

9. ***SHORT RESPONSE*** Set up a proportion to represent the following question: 50 is 125% of what number? Explain each term in your proportion. Then solve it, and show all of your steps.

10. ***SHORT RESPONSE*** Write and simplify an expression that represents the phrase "four times the difference of ten and three." Explain how you used the order of operations to simplify your expression.

Chapter

7

Plane Figures

TABLE OF CONTENTS

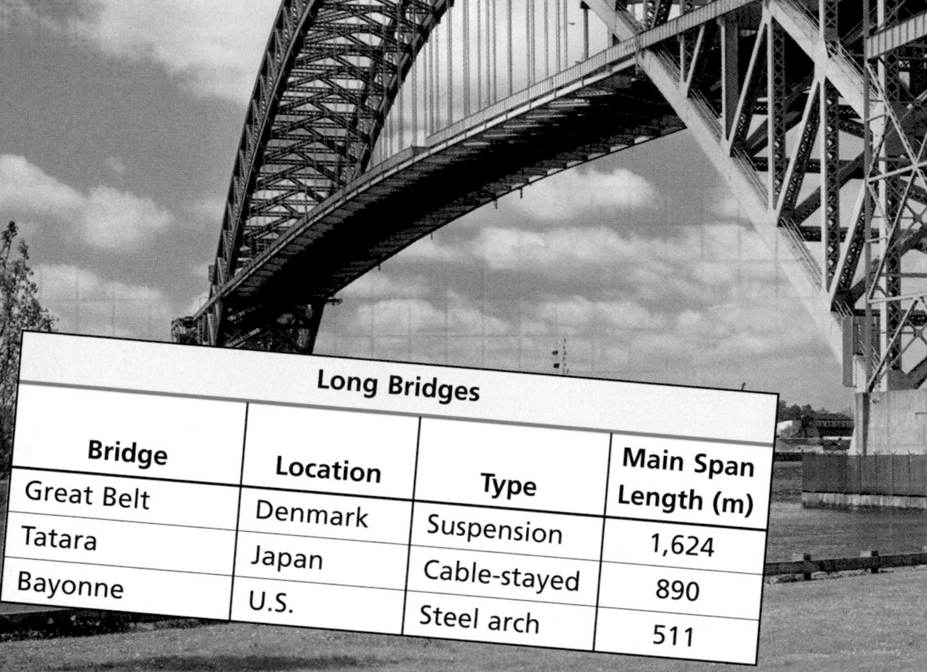

Long Bridges

Bridge	Location	Type	Main Span Length (m)
Great Belt	Denmark	Suspension	1,624
Tatara	Japan	Cable-stayed	890
Bayonne	U.S.	Steel arch	511

Career Bridge Designer

Many factors influence the way a bridge is constructed. A bridge must be able to withstand winds, snow, and the weight of traffic while supporting its own weight.

Bridge designers also have to consider the distance that a bridge must cross (called the *span*), the nature of the land, and the look of the structure. Bridge designers often combine technological know-how with artistry to create structures that are both functional and beautiful.

internet connect

Chapter Opener Online
go.hrw.com
KEYWORD: MS4 Ch7

ARE YOU READY?

Choose the best term from the list to complete each sentence.

1. An equation showing that two ratios are equal is a(n) __?__.
2. The coordinates of a point on a grid are written as a(n) __?__ .
3. A(n) __?__ is a special ratio that compares a number to 100 and uses the symbol %.
4. The number –3 is a(n) __?__.
5. The __?__ between two numbers is found by subtracting one of the numbers from the other.

decimal
difference
integer
percent
proportion
ordered pair
sum

Complete these exercises to review skills you will need for this chapter.

✔ Write Percents and Decimals

Write each decimal as a percent.

6. 0.77 **7.** 0.06 **8.** 0.9 **9.** 1.04

Write each percent as a decimal.

10. 42% **11.** 80% **12.** 1% **13.** 131%

✔ Multiply with Fractions and Decimals

Find the percent of each number. Round to the nearest hundredth, if necessary.

14. 10% of 40 **15.** 12% of 100 **16.** 6% of 18 **17.** 99% of 60

18. 100% of 81 **19.** 150% of 20 **20.** 45% of 360 **21.** 55% of 1,024

✔ Solve Proportions

Solve each proportion.

22. $\frac{n}{30} = \frac{3}{15}$ **23.** $\frac{x}{50} = \frac{3}{75}$ **24.** $\frac{51}{17} = \frac{k}{3}$ **25.** $\frac{3}{45} = \frac{4}{y}$

26. $\frac{s}{12} = \frac{4}{16}$ **27.** $\frac{8}{56} = \frac{t}{14}$ **28.** $\frac{5}{h} = \frac{4}{8}$ **29.** $\frac{9}{57} = \frac{3}{p}$

✔ Graph Ordered Pairs

Graph each ordered pair on a coordinate plane.

30. $A(0, 2)$ **31.** $B(-6, 7)$ **32.** $C(4, -6)$ **33.** $D(-3, -5)$

34. $E(5, 0)$ **35.** $F(-5, 5)$ **36.** $G(6, 1)$ **37.** $H(0, -7)$

7-1 Points, Lines, and Planes

Learn to identify and describe geometric figures.

Vocabulary

point
line
ray
line segment
plane
congruent

Artists often use basic geometric figures when creating their works. For example, Wassily Kandinsky used *line segments* in his painting called *Red Circle*, which is shown at right.

A **point** is an exact location in space. It is usually represented as a dot, but it has no size at all.	• A	point A *Use a capital letter to name a point.*
A **line** is a straight path that extends without end in opposite directions.	X Y	$\overleftrightarrow{XY}$, or $\overleftrightarrow{YX}$ *Use two points on the line to name a line.*
A **ray** is a part of a line. It has one endpoint and extends without end in one direction.	G H	$\overrightarrow{GH}$ *Name the endpoint first when naming a ray.*
A **line segment** is a part of a line or a ray that extends from one endpoint to another.	L M	$\overline{LM}$, or $\overline{ML}$ *Use the endpoints to name a line segment.*
A **plane** is a perfectly flat surface that extends infinitely in all directions.	Q S R	plane QRS *Use three points in any order, not on the same line, to name a plane.*

Helpful Hint

A number line is an example of a line, and a coordinate plane is an example of a plane.

EXAMPLE 1 Identifying Points, Lines, and Planes

Identify the figures in the diagram.

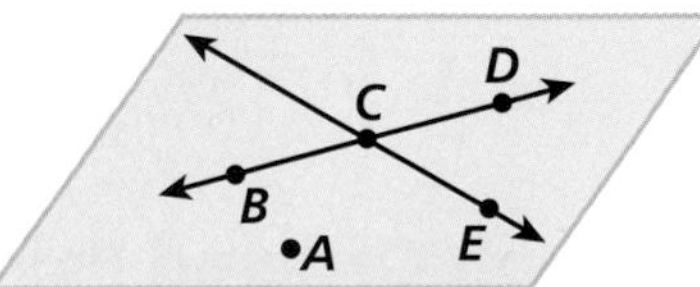

A three points

A, B, and C

B two lines

$\overleftrightarrow{BD}$, $\overleftrightarrow{CE}$ *Choose any two points on a line to name the line.*

C a plane

plane ABC *Choose any three points on a plane to name the plane.*

EXAMPLE 2 Identifying Line Segments and Rays

Identify the figures in the diagram.

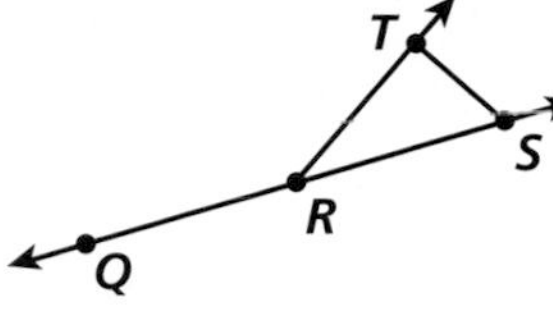

A three rays
$\overrightarrow{RQ}$, $\overrightarrow{RT}$, and $\overrightarrow{SQ}$

Name the endpoint of a ray first.

B three line segments
$\overline{RQ}$, $\overline{QS}$, and $\overline{ST}$

Use the endpoints in any order to name a segment.

Figures are **congruent** if they have the same shape and size. If you place one on top of the other, they match exactly. Line segments are congruent if they have the same length.

Tick marks are used to indicate congruent line segments. In the illustration below, segments that have the same number of tick marks are congruent. Line segments AB and BC are congruent (one tick mark), and line segments MN and OP are congruent (two tick marks).

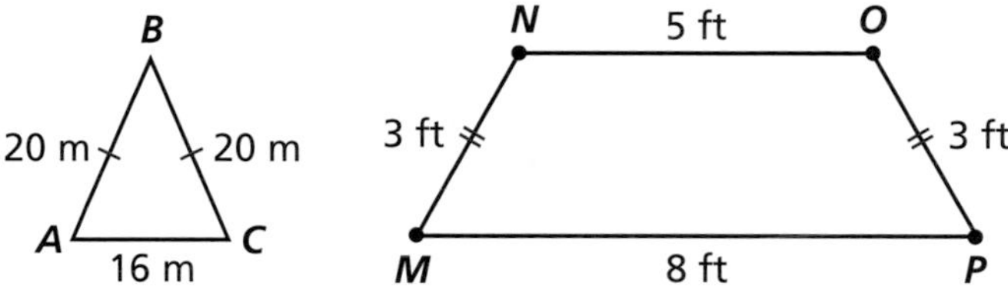

EXAMPLE 3 Identifying Congruent Line Segments

Reading Math

The symbol ≅ means "is congruent to."

Identify the line segments that are congruent.

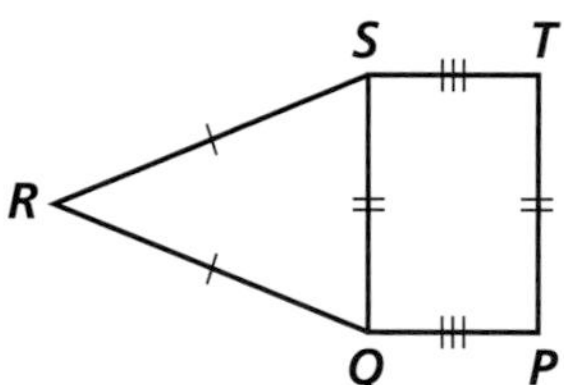

$\overline{QR} \cong \overline{SR}$ *One tick mark*
$\overline{QS} \cong \overline{PT}$ *Two tick marks*
$\overline{QP} \cong \overline{ST}$ *Three tick marks*

Think and Discuss

1. **Explain** why a line and a plane can be named in more than two ways. How many ways can a line segment be named?
2. **Explain** why it is important to choose three points that are not on the same line when naming a plane.
3. **Tell** how you can determine whether two line segments are congruent without knowing their actual measures.

7-1 Exercises

FOR EOG PRACTICE
see page 682

3.02

GUIDED PRACTICE

See Example 1 **Identify the figures in the diagram.**

1. three points
2. two lines
3. a plane

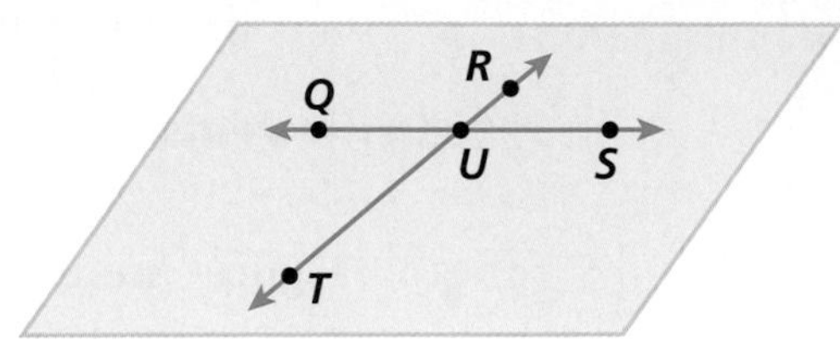

See Example 2

4. three rays
5. three line segments

See Example 3

6. Identify the line segments that are congruent.

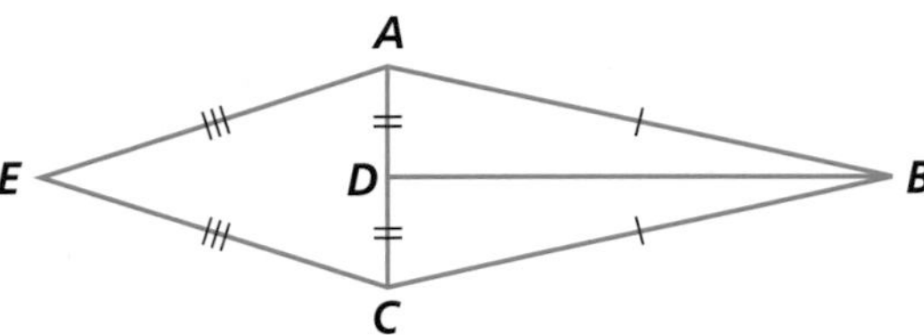

INDEPENDENT PRACTICE

See Example 1 **Identify the figures in the diagram.**

7. three points
8. two lines
9. a plane

See Example 2

10. three rays
11. three line segments

See Example 3

12. Identify the line segments that are congruent.

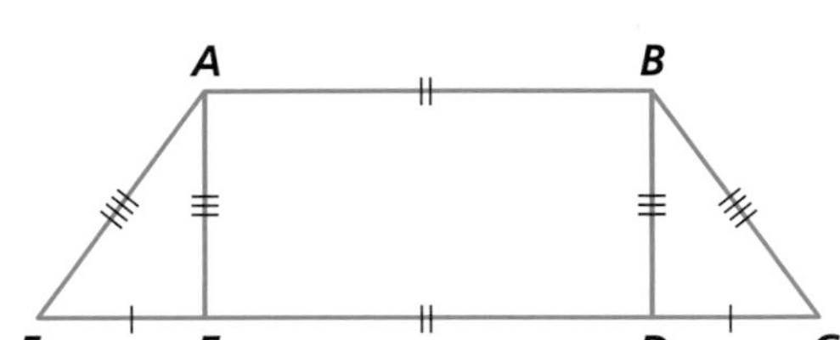

PRACTICE AND PROBLEM SOLVING

13. Identify the points, lines, line segments, and rays that are represented in the illustration, and tell what plane each is in. Some figures may be in more than one plane.

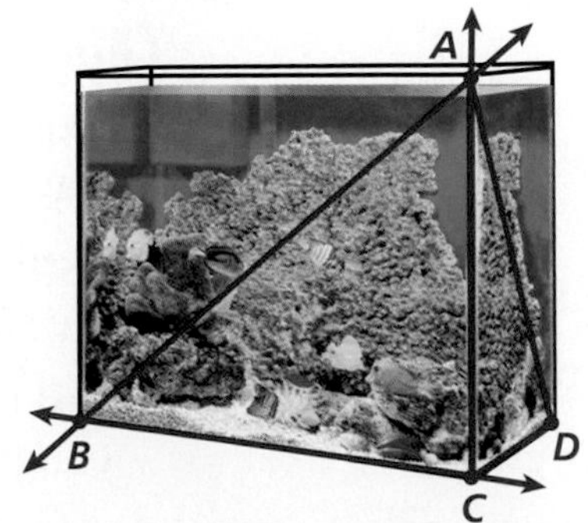

Art LINK

14. The painting at right, by Piet Mondrian, is called *Composition with Red, Yellow, and Blue.*
    a. Copy the line segments in the painting.
    b. Add tick marks to show congruent line segments.
    c. Label the endpoints of the segments, including the points of intersection. Then name five pairs of congruent line segments.
15. For each of the following, make an illustration.
    a. Show at least three sets of congruent line segments. Label the endpoints and use notation to tell which line segments are congruent.
    b. Show how one point could be shared by more than one line segment.
    c. Show how one point could be shared by more than one plane.
16. Can two endpoints be shared by two different line segments? Make a drawing to illustrate your answer.
17. Draw a diagram in which a plane, 5 points, 4 rays, and 2 lines can be identified. Then identify these figures.
18. ***WRITE ABOUT IT*** Explain the difference between a line, a line segment, and a ray. Is it possible to estimate the length of any of these figures? If so, tell which ones and why.
19. ***CHALLENGE*** The wooden sculpture at right, by Vantongerloo, is called *Interrelation of Volumes.* Explain whether two separate faces on the front of the sculpture could be on the same plane.

Spiral Review

Write each percent as a fraction in simplest form. (Lesson 6-1)

20. 24% **21.** 7% **22.** 95% **23.** 62%

Find the percent of each number. (Lesson 6-3)

24. 12% of 30 **25.** 9% of 50 **26.** 0.5% of 8

27. **EOG PREP** Sixty people were expected at a reception. Only 45 people came. What percent of the people who were expected at the reception actually came? (Lesson 6-4)

A 7.5% B 13.3% C 75% D 133.3%

7-2 Angles

Learn to identify angles and parts of angles.

Vocabulary

angle
vertex
right angle
acute angle
obtuse angle
straight angle
complementary angles
supplementary angles

When riding down a ramp on a skateboard, the speed you gain depends partly on the *angle* that the ramp makes with the ground.

An **angle** is formed by two rays with a common endpoint. The two rays are the sides of the angle. The common endpoint is the **vertex**.

You can name an angle in three ways:

- with the capital letter at the vertex: $\angle B$,
- with the number inside the angle: $\angle 1$,
- with three capital letters so that the letter at the vertex is in the middle: $\angle ABC$ or $\angle CBA$

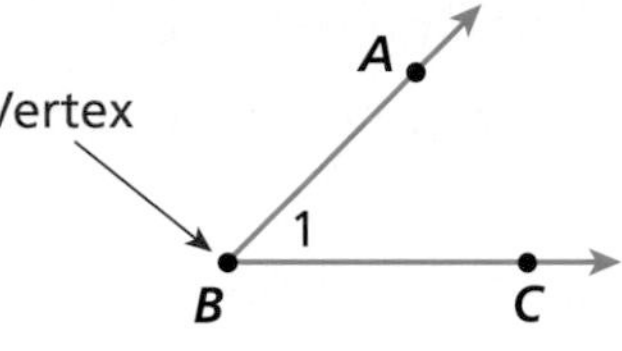

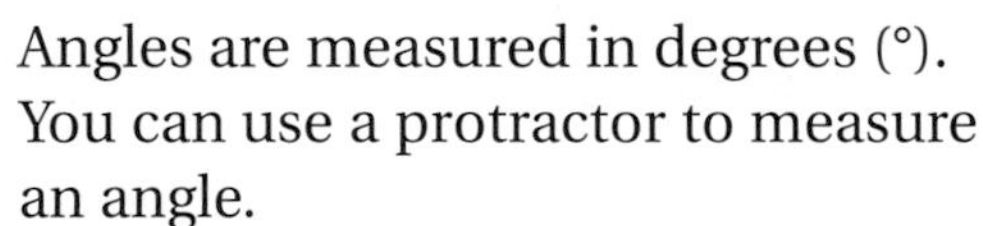

Angles are measured in degrees (°). You can use a protractor to measure an angle.

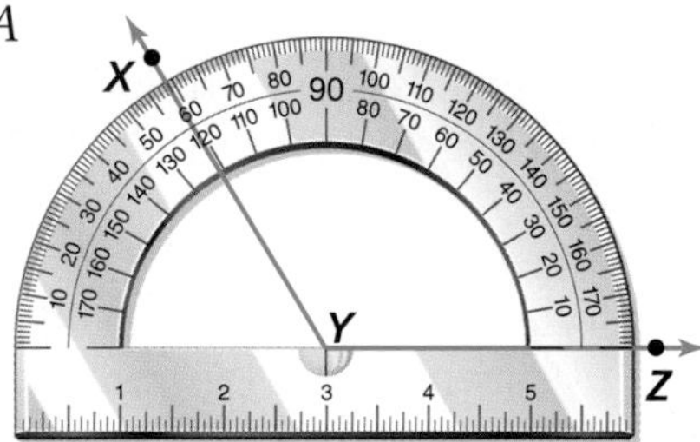

The measure of $\angle XYZ$ is 122°, or $m\angle XYZ = 122°$.

EXAMPLE 1 Identifying Angle Measures

Give the measure of each angle.

A $\angle EAD$

$m\angle EAD = 105°$

B $\angle CAB$

$m\angle CAB = 45°$

C $\angle DAB$

$m\angle DAB = 45° + 30°$

$m\angle DAB = 75°$

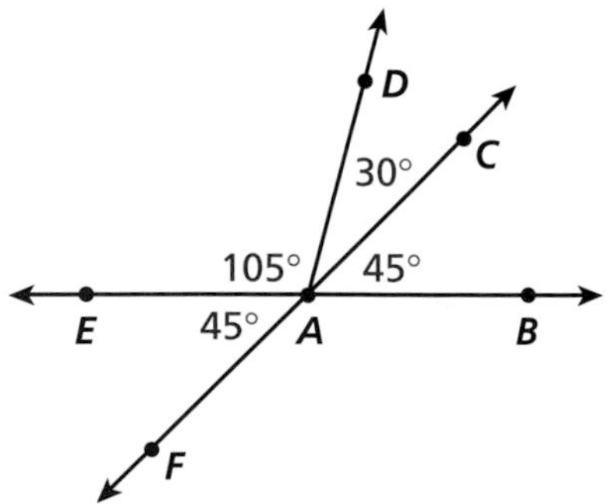

D $\angle FAC$

$m\angle FAC = 30° + 105° + 45°$

$m\angle FAC = 180°$

An angle's measure determines the type of angle it is.

A **right angle** is an angle that measures exactly 90°. The symbol ⌉ indicates a right angle.

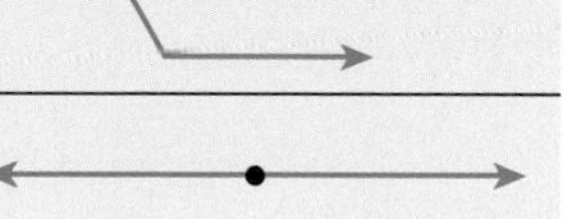

An **acute angle** is an angle that measures less than 90°.

An **obtuse angle** is an angle that measures more than 90° but less than 180°.

A **straight angle** is an angle that measures exactly 180°.

EXAMPLE **2** **Classifying Angles**

Tell whether each angle is acute, right, obtuse, or straight.

Right angle

Obtuse angle

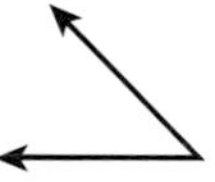

Acute angle

If the sum of the measures of two angles is 90°, then the angles are **complementary angles**. If the sum of the measures of two angles is 180°, then the angles are **supplementary angles**.

EXAMPLE **3** **Identifying Complementary and Supplementary Angles**

Use the figure to name the following.

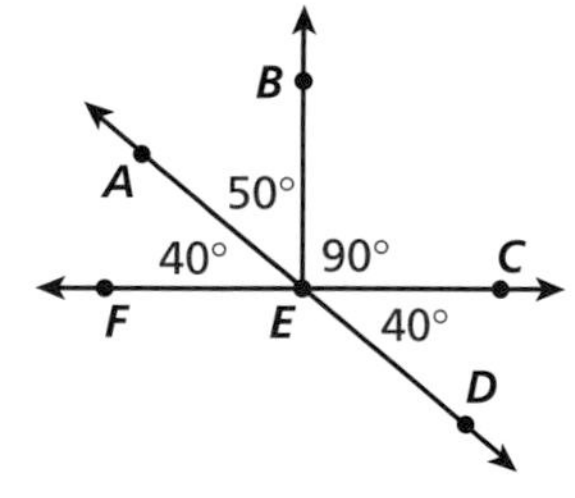

A **one pair of complementary angles**
$m\angle AEF = 40°$ and $m\angle AEB = 50°$

Since $40° + 50° = 90°$, $\angle AEF$ and $\angle AEB$ are complementary.

B **one pair of supplementary angles**
$m\angle AEB = 50°$ and $m\angle BED = 90° + 40° = 130°$

Since $50° + 130° = 180°$, $\angle AEB$ and $\angle BED$ are supplementary.

Think and Discuss

1. **Describe** three different ways to name an angle.
2. **Explain** how to find the measure of $\angle P$ if $\angle P$ and $\angle Q$ are complementary angles and $m\angle Q = 25°$.

7-2 Exercises

FOR EOG PRACTICE
see page 682

internet connect
Homework Help Online
go.hrw.com Keyword: MS4 7-2

3.02

GUIDED PRACTICE

See Example 1 **Give the measure of each angle.**

1. ∠*CAD*
2. ∠*EAF*
3. ∠*DAF*
4. ∠*BAE*

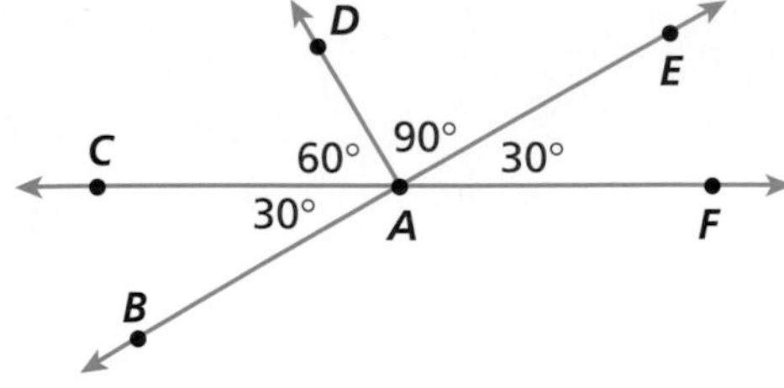

See Example 2 **Tell whether each angle is acute, right, obtuse, or straight.**

5.

6. 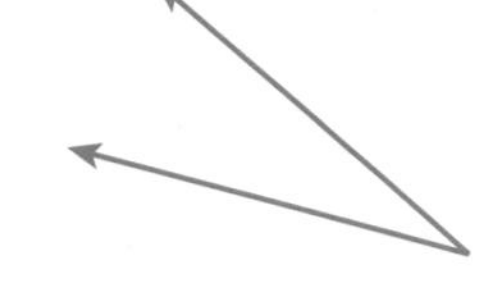

7. 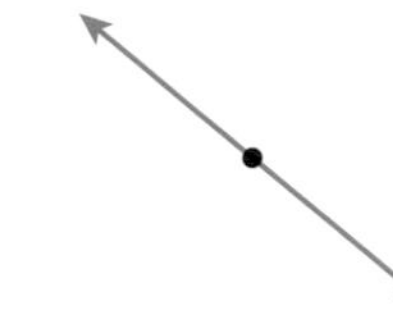

See Example 3 **Use the figure to name the following.**

8. two pairs of complementary angles
9. two pairs of supplementary angles

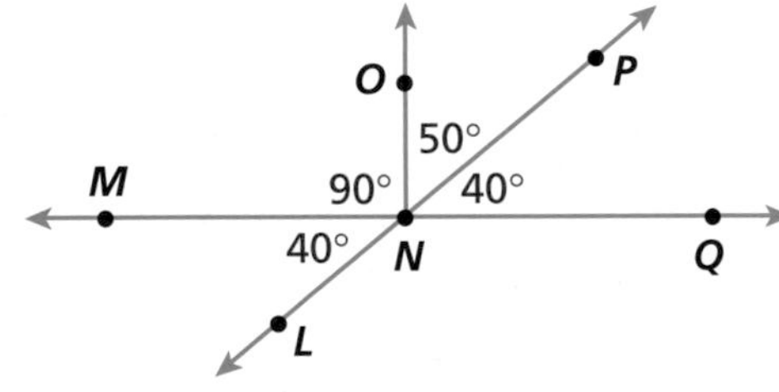

INDEPENDENT PRACTICE

See Example 1 **Give the measure of each angle.**

10. ∠*MJN*
11. ∠*LJK*
12. ∠*MJK*
13. ∠*MJO*

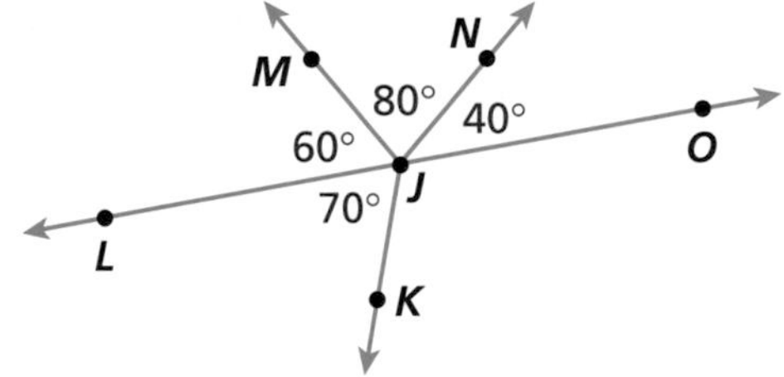

See Example 2 **Tell whether each angle is acute, right, obtuse, or straight.**

14.

15.

16. 

See Example 3 **Use the figure to name the following.**

17. two pairs of complementary angles
18. two pairs of supplementary angles

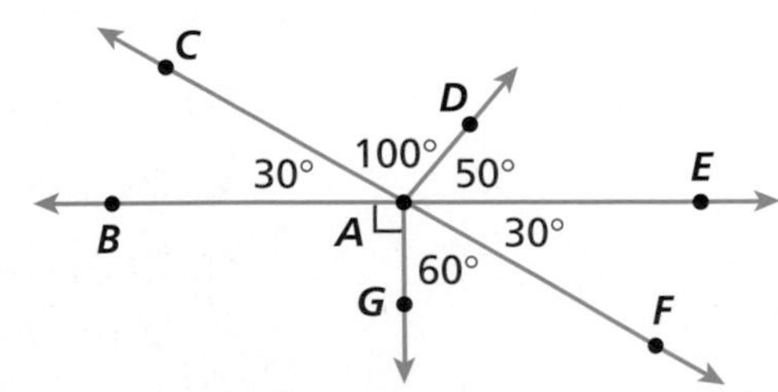

PRACTICE AND PROBLEM SOLVING

Classify each pair of angles as complementary or supplementary. Then find the missing angle measure.

19.

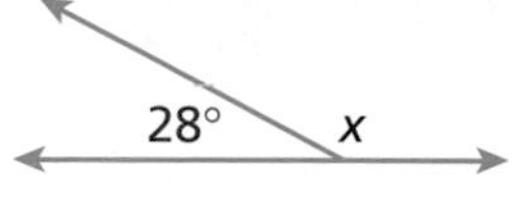

20.

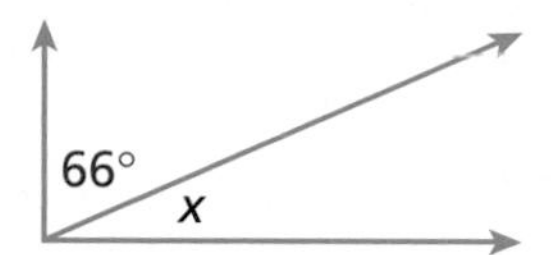

21.

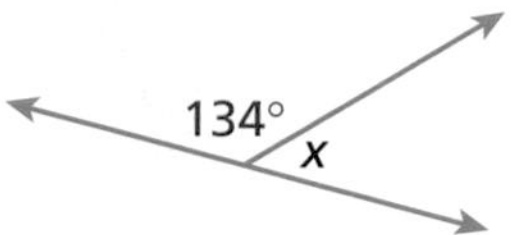

22. The hands of a clock form an acute angle at 1:00. What type of angle is formed at 6:00? at 3:00? at 5:00?

23. ***GEOGRAPHY*** Imaginary curves around Earth show distances in degrees from the equator and Prime Meridian. On a flat map, these curves are displayed as horizontal lines (latitude) and vertical lines (longitude).

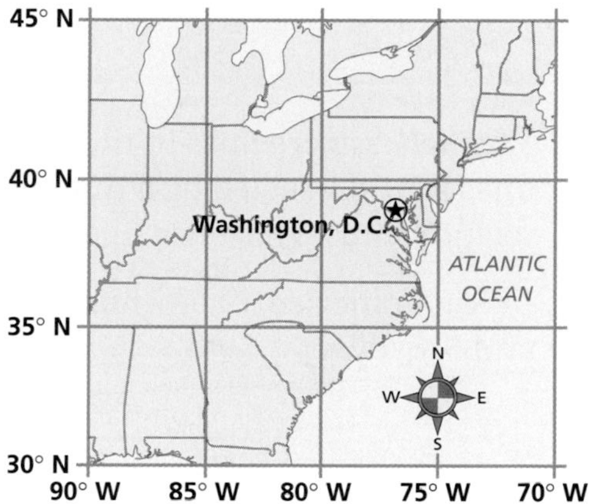

a. What type of angle is formed where a line of latitude and a line of longitude cross?

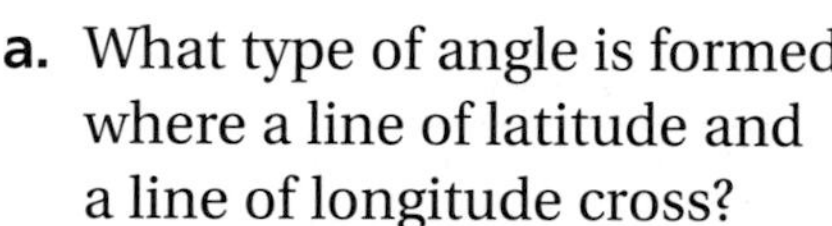

b. Estimate the latitude and longitude of Washington, D.C.

24. ***WHAT'S THE ERROR?*** Jamal drew the pair of complementary angles at right. His classmate says that the angles are not complementary since they do not share a ray. Why is his classmate wrong?

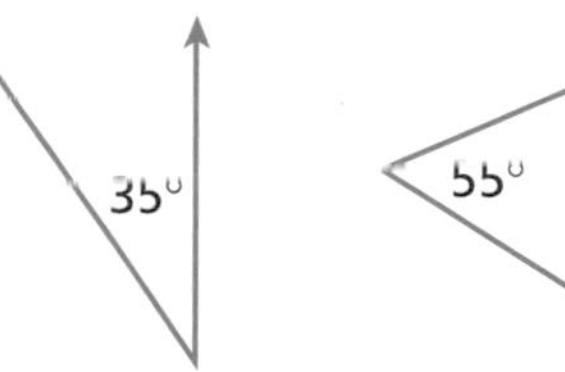

25. ***WRITE ABOUT IT*** Describe the relationship between complementary angles and supplementary angles.

26. ***CHALLENGE*** Find m$\angle BAC$ in the figure.

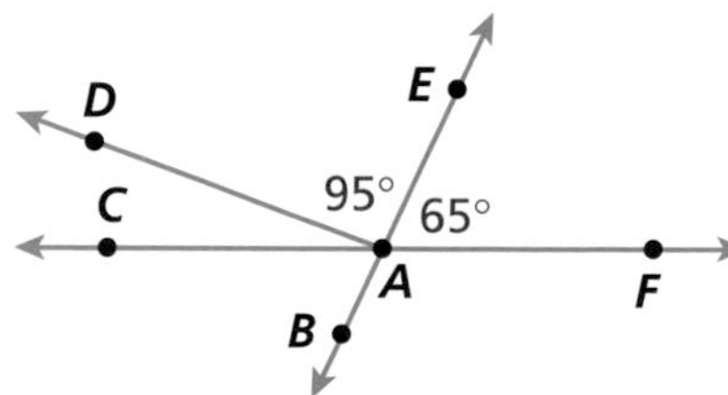

Spiral Review

Write each decimal as a percent. (Lesson 6-1)

27. 0.7 **28.** 0.09 **29.** 1.45 **30.** 0.234

Find each percent of change. Round your answer to the nearest tenth. (Lesson 6-5)

31. 35 is decreased to 15. **32.** 42 is increased to 98. **33.** 150 is decreased to 80.

34. **EOG PREP** Which geometric figure has one endpoint and extends infinitely in one direction? (Lesson 7-1)

A Line B Ray C Line segment D Plane

Explore Parallel Lines and Transversals

Use with Lesson 7-3

REMEMBER
- Two angles are supplementary if the sum of their measures is 180°.
- Angles with measures less than 90° are acute.
- Angles with measures greater than 90° but less than 180° are obtuse.

Parallel lines are lines in the same plane that never cross. When two parallel lines are intersected by a third line, the angles formed have special relationships. This third line is called a *transversal.*

In San Francisco, California, many streets are parallel such as Lombard St. and Broadway.

Columbus Ave. is a transversal that runs diagonally across them. The eight angles that are formed are labeled on the diagram below.

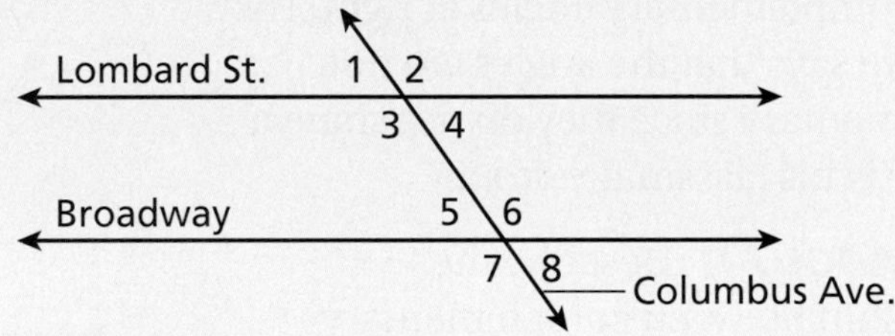

Activity

1. Copy the table below. Then measure angles 1–8 in the diagram below. Write these measures in your table.

Angle Number	Angle Measure
1	
2	
3	
4	
5	
6	
7	
8	

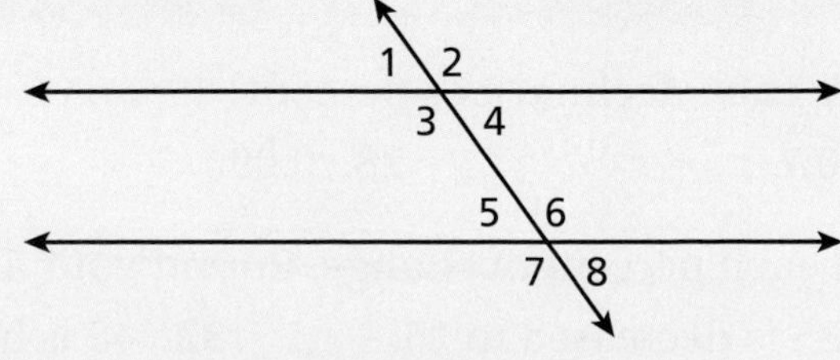

2 Use the table you completed and the corresponding diagram for the following problems.

a. Angles inside the parallel lines are *interior angles.* Name them.

b. Angles outside the parallel lines are *exterior angles.* Name them.

c. Angles 3 and 6 and angles 4 and 5 are *alternate interior angles.* What do you notice about angles 3 and 6? What do you notice about angles 4 and 5?

d. Angles 2 and 7 and angles 1 and 8 are *alternate exterior angles.* How do the measures of each pair of alternate exterior angles compare?

e. Angles 1 and 5 are *corresponding angles* because they are in the same position on each of the parallel lines. How do the measures of angles 1 and 5 compare? Name another set of corresponding angles.

f. Add the measures of angles 1 and 2. Now add the measures of angles 3 and 8. What can you say about the relationship of the angles in each of these sets? Name two other angles that have the same relationship.

Think and Discuss

1. $\overleftrightarrow{FG}$ and $\overleftrightarrow{LO}$ are parallel. Tell what you know about the angles that are labeled 1 through 8.

2. If angle 2 measures 125°, what are the measures of angles 1, 3, 4, 5, 6, 7, and 8?

3. If a transversal intersects two parallel lines and one of the angles formed measures 90°, discuss the relationship between all the angles.

Try This

Use a protractor to measure one angle in each diagram. Then find the measures of all the other angles without using a protractor. Tell how to find each angle measure.

1.

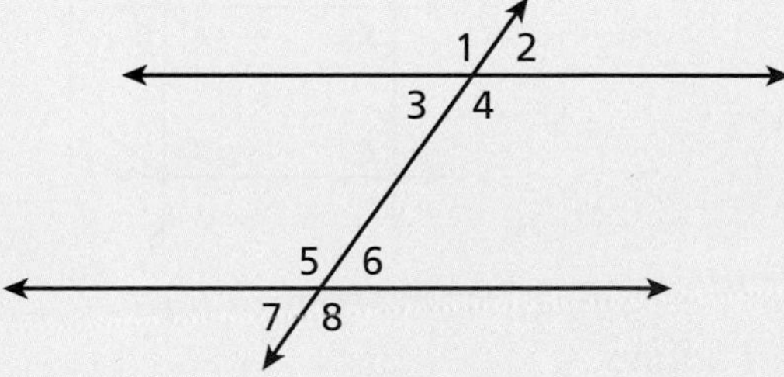

2.

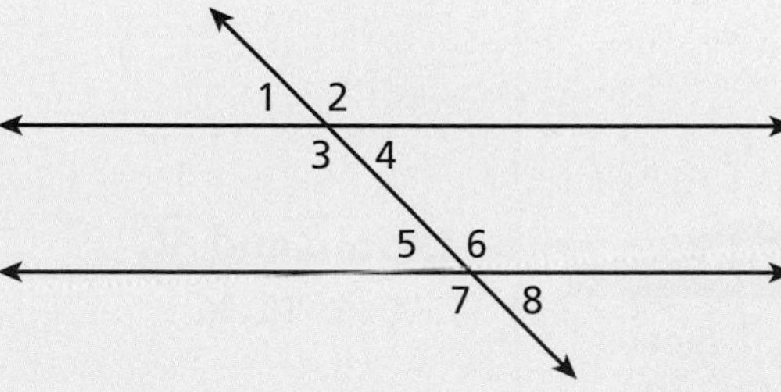

3.

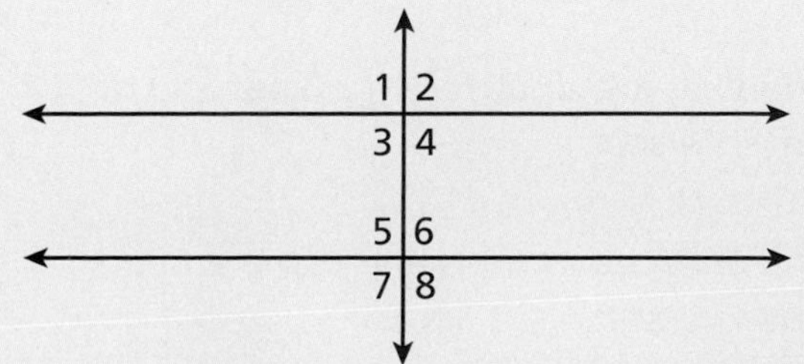

4.

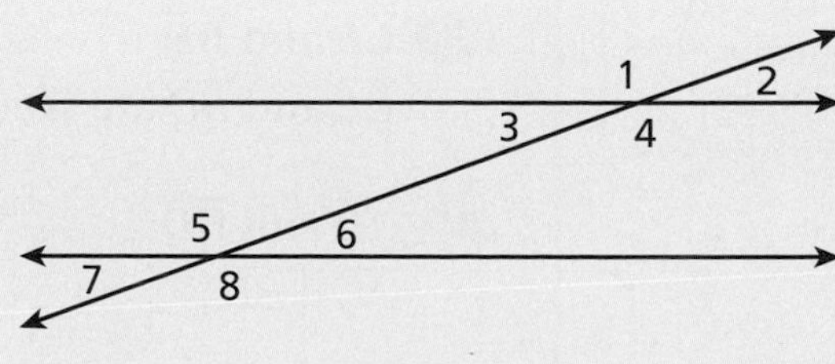

7-3 Parallel and Perpendicular Lines

Learn to identify parallel, perpendicular, and skew lines, and angles formed by a transversal.

Vocabulary

perpendicular lines

parallel lines

skew lines

vertical angles

transversal

When lines, segments, or rays intersect, they form angles. If the angles formed by two intersecting lines are equal to 90°, the lines are **perpendicular lines**. The red and yellow line segments in the photograph of the skyscraper are perpendicular because they form 90° angles.

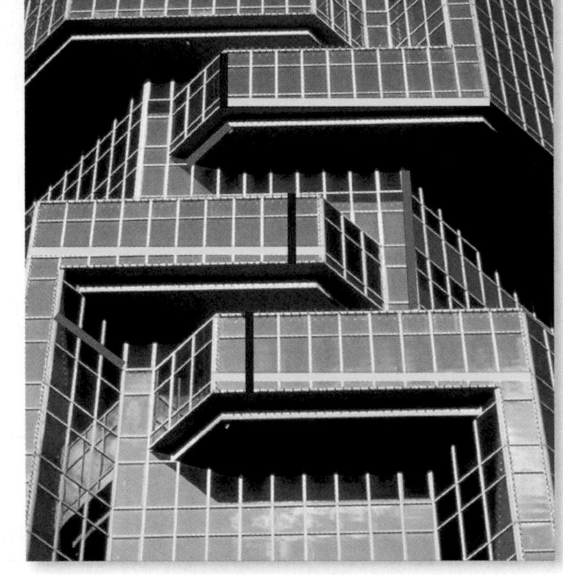

Some lines in the same plane do not intersect at all. These lines are **parallel lines**. Segments and rays that are parts of parallel lines are also parallel. All of the vertical line segments in the photo appear to be parallel. Even if you extend the line segments, it seems that they will never cross. The same is true of the horizontal line segments.

Skew lines do not intersect, and yet they are also not parallel. They lie in different planes. The orange line segments in the photograph are skew.

EXAMPLE 1 Identifying Parallel, Perpendicular, and Skew Lines

Tell whether the lines appear parallel, perpendicular, or skew.

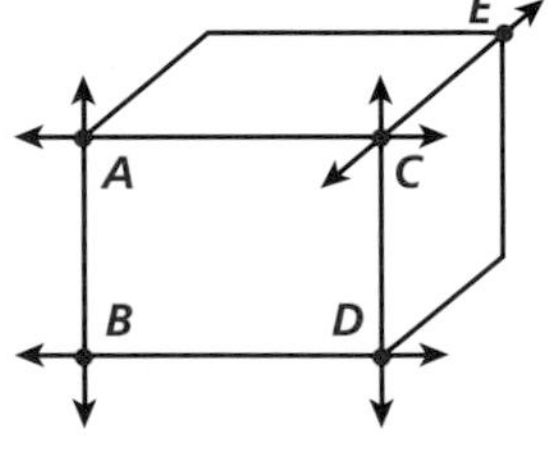

Reading Math

The symbol || means "is parallel to." The symbol ⊥ means "is perpendicular to."

A $\overleftrightarrow{AB}$ and $\overleftrightarrow{AC}$

$\overleftrightarrow{AB} \perp \overleftrightarrow{AC}$ *The lines appear to intersect to form right angles.*

B $\overleftrightarrow{CE}$ and $\overleftrightarrow{BD}$

$\overleftrightarrow{CE}$ and $\overleftrightarrow{BD}$ are skew. *The lines are in different planes and do not intersect.*

C $\overleftrightarrow{AC}$ and $\overleftrightarrow{BD}$

$\overleftrightarrow{AC} \parallel \overleftrightarrow{BD}$ *The lines are in the same plane and do not intersect.*

Reading Math

Angles with the same number of tick marks are congruent. The tick marks are placed in arcs drawn inside the angles.

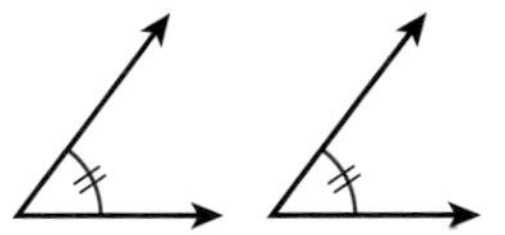

Vertical angles are the opposite angles formed by two intersecting lines. When two lines intersect, two pairs of vertical angles are formed. Vertical angles have the same measure, so they are congruent.

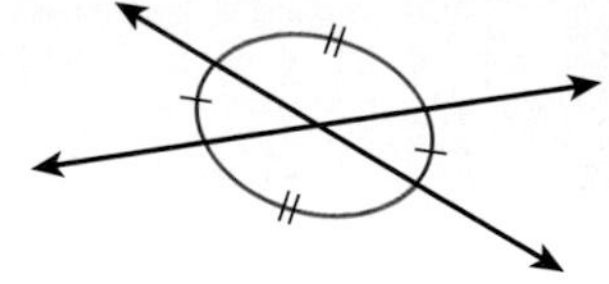

A **transversal** is a line that intersects two or more lines. Eight angles are formed when a transversal intersects two lines. When those two lines are parallel, all of the **acute** angles formed are congruent, and all of the **obtuse** angles formed are congruent. These obtuse and acute angles are supplementary.

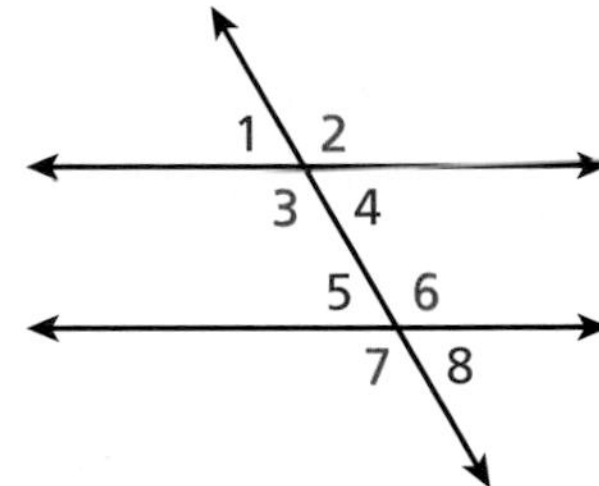

EXAMPLE 2 Using Angle Relationships to Find Angle Measures

Line $n \parallel$ line p. Find the measure of each angle.

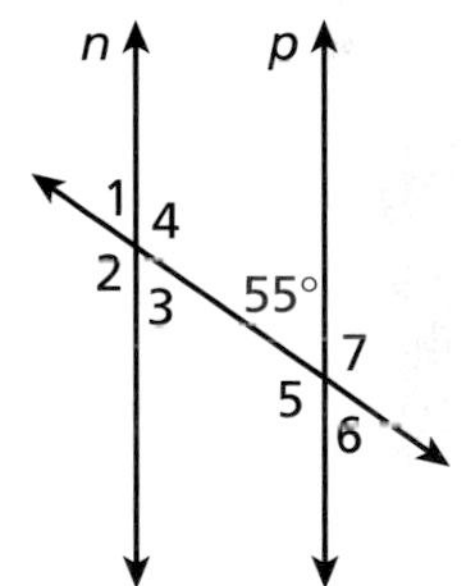

A $\angle 6$

$\angle 6$ and the 55° angle are vertical angles. Since vertical angles are congruent, $m\angle 6 = 55°$

B $\angle 1$

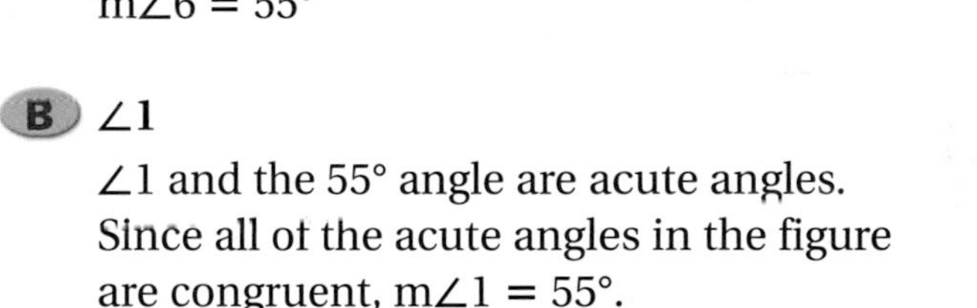

$\angle 1$ and the 55° angle are acute angles. Since all of the acute angles in the figure are congruent, $m\angle 1 = 55°$.

C $\angle 2$

$\angle 2$ is an obtuse angle.

$m\angle 1 + m\angle 2 = 180°$ *In the figure, the acute and obtuse angles are supplementary.*

$55° + m\angle 2 = 180°$ *Substitute for $m\angle 1$.*

$\underline{-55°} \qquad \underline{-55°}$ *Subtract 55° to isolate $m\angle 2$.*

$m\angle 2 = 125°$

Think and Discuss

1. **Draw** a pair of parallel lines intersected by a transversal. Use tick marks to indicate the congruent angles.
2. **Give** some examples in which parallel, perpendicular, and skew relationships can be seen in the real world.

7-3 Exercises

FOR EOG PRACTICE
see page 683

internet connect
Homework Help Online
go.hrw.com Keyword: MS4 7-3

3.02

GUIDED PRACTICE

See Example 1 Tell whether the lines appear parallel, perpendicular, or skew.

1. $\overleftrightarrow{JL}$ and $\overleftrightarrow{KM}$
2. $\overleftrightarrow{LM}$ and $\overleftrightarrow{KN}$
3. $\overleftrightarrow{LM}$ and $\overleftrightarrow{KM}$

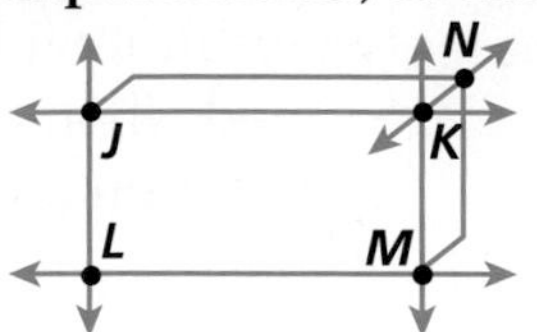

See Example 2 Line $r \parallel$ line s. Find the measure of each angle.

4. $\angle 5$
5. $\angle 2$
6. $\angle 7$

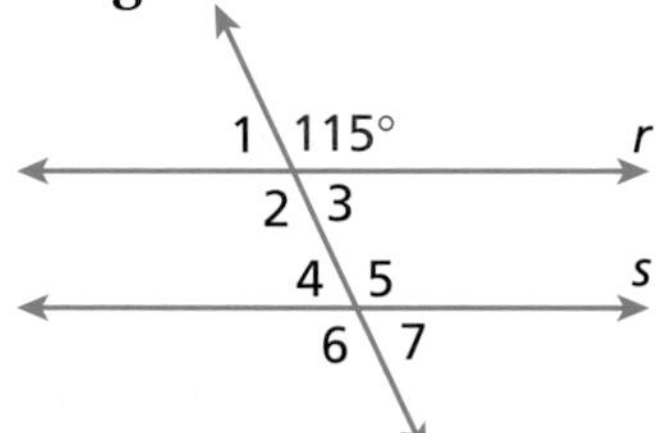

INDEPENDENT PRACTICE

See Example 1 Tell whether the lines appear parallel, perpendicular, or skew.

7. $\overleftrightarrow{UX}$ and $\overleftrightarrow{YZ}$
8. $\overleftrightarrow{YZ}$ and $\overleftrightarrow{XY}$
9. $\overleftrightarrow{UX}$ and $\overleftrightarrow{VW}$

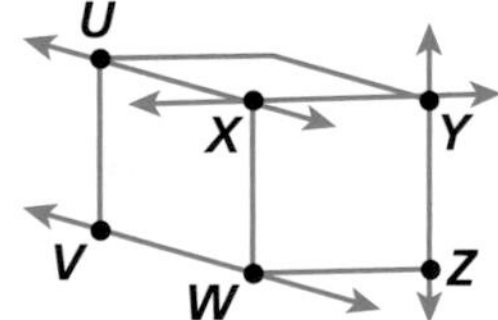

See Example 2 Line $k \parallel$ line m. Find the measure of each angle.

10. $\angle 1$
11. $\angle 3$
12. $\angle 6$

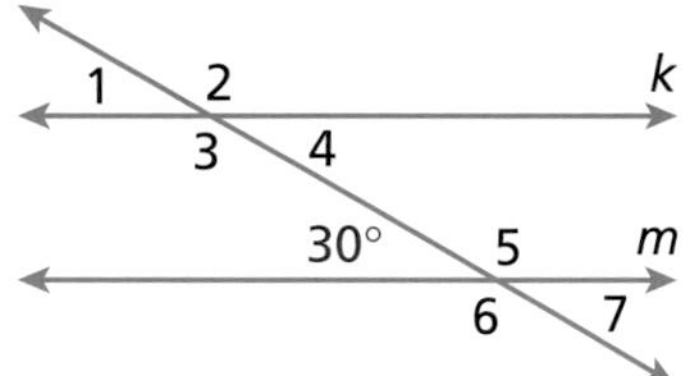

PRACTICE AND PROBLEM SOLVING

For Exercises 13–16, use the figure to complete each statement.

13. Lines x and y are __?__.
14. $\angle 3$ and $\angle 4$ are __?__.
15. Lines u and x are __?__.
16. $\angle 2$ and $\angle 6$ are __?__. They are also __?__.

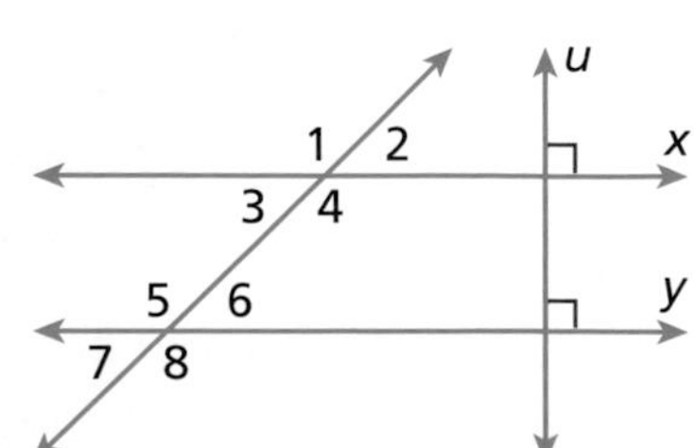

17. A pair of complementary angles are congruent. What is the measure of each angle?

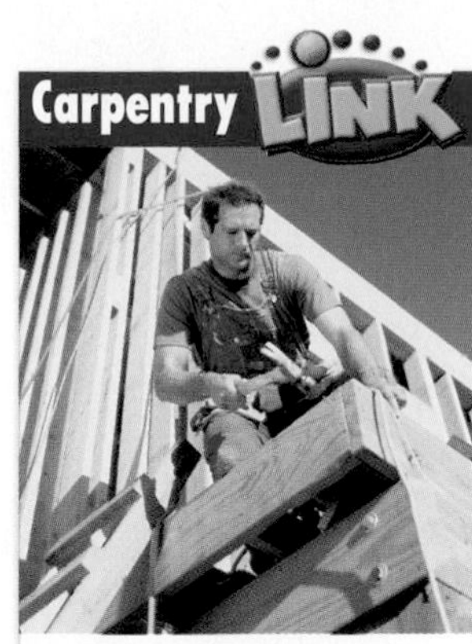

Construction workers use a special tool called a *square* to make sure that boards are perpendicular.

18. ***CARPENTRY*** In the diagram of a partial wall frame, the *ceiling joist,* $\overline{PQ}$, and the *soleplate,* $\overline{RS}$, are parallel.

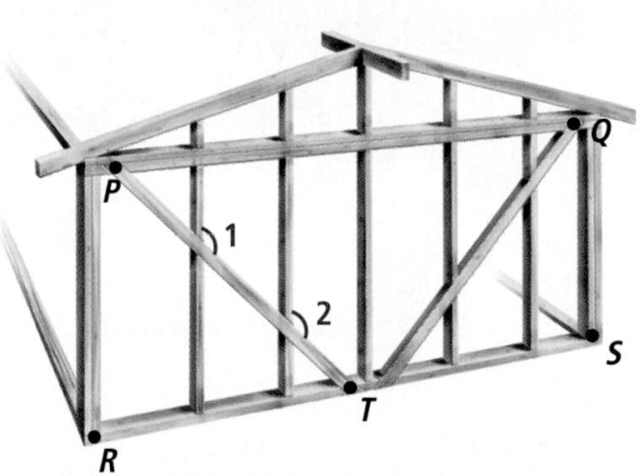

a. How are $\angle RTP$ and $\angle TPQ$ related?

b. How is $\overline{PT}$ related to the vertical beams that it crosses?

c. How are $\angle 1$ and $\angle 2$ related?

19. Tell whether the lines in each pair appear parallel, perpendicular, skew, or none of these.

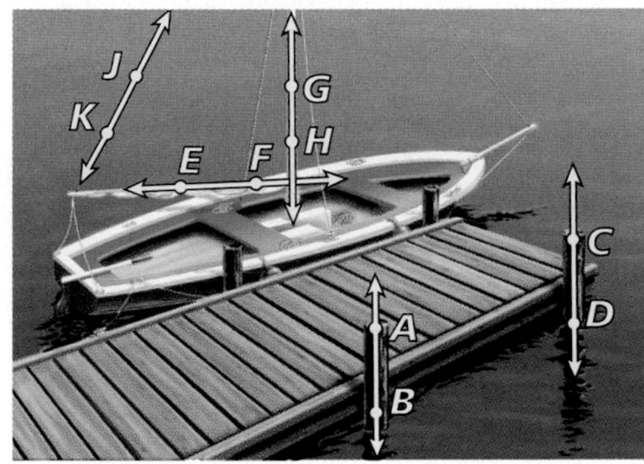

a. $\overleftrightarrow{AB}$ and $\overleftrightarrow{CD}$

b. $\overleftrightarrow{AB}$ and $\overleftrightarrow{JK}$

c. $\overleftrightarrow{GH}$ and $\overleftrightarrow{EF}$

d. $\overleftrightarrow{EF}$ and $\overleftrightarrow{JK}$

20. ***CHOOSE A STRATEGY*** Trace the dots in the figure. Draw all the lines that connect three dots. How many pairs of perpendicular lines have you drawn?

A 8 C 10

B 9 D 14

21. ***WRITE ABOUT IT*** Describe the relationship between the acute and obtuse angles formed when a transversal intersects two parallel lines.

22. ***CHALLENGE*** The lines in the parking lot appear to be parallel. How could you check that the lines are parallel? What would you have to assume to do this?

Spiral Review

Find the simple interest for each principal, interest rate, and time. (Lesson 6-6)

23. $550, 3%, 2 years

24. $900, 7%, 6 months

25. $1,200, 2.5%, 18 months

Give a real-world example of each type of line segment pair. (Lesson 7-1)

26. intersecting

27. parallel

28. skew

29. congruent

30. **EOG PREP** $\angle 1$ and $\angle 2$ are *complementary* angles. What is the sum of their measures? (Lesson 7-2)

A 45° B 90° C 180° D 360°

Construct Bisectors and Congruent Angles

Use with Lesson 7-3

REMEMBER
- To bisect a segment or an angle is to divide it into two congruent parts.
- Congruent angles have the same measure, and congruent segments are the same length.

You can bisect segments and angles, and construct congruent angles without using a protractor or ruler. Instead, you can use a compass and a straightedge.

Activity

1 Bisect a line segment.

a. Draw a line segment $\overline{JS}$ on a piece of paper.

b. Place your compass on endpoint J and, using an opening that is greater than half the length of $\overline{JS}$, draw an arc that intersects $\overline{JS}$.

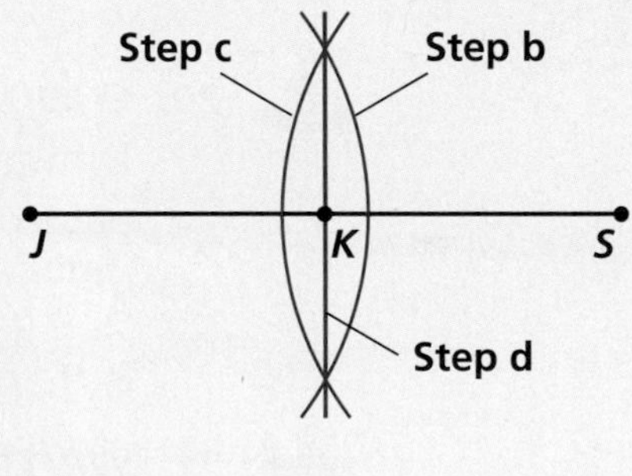

c. Place your compass on endpoint S and draw an arc using the same opening as you did in part **b.** The arc should intersect the first arc at both ends.

d. Draw a line to connect the intersections of the arcs. Label the intersection of $\overline{JS}$ and the line point K.

Measure $\overline{JS}$, $\overline{JK}$, and $\overline{KS}$. What do you notice?

The bisector of $\overline{JS}$ is a *perpendicular bisector* because all of the angles it forms with $\overline{JS}$ measure 90°.

2 Bisect an angle.

a. Draw an acute angle GHE on a piece of paper. Label the vertex H.

b. Place the point of your compass on H and draw an arc through both sides of the angle. Label points G and E where the arc crosses each side of the angle.

c. Without changing your compass opening, draw intersecting arcs from point G and point E. Label the point of intersection D.

d. Draw $\overrightarrow{HD}$.

Use your protractor to measure angles GHE, GHD, and DHE. What do you notice?

3 Construct congruent angles.

a. Draw angle *ABM* on your paper.

b. To construct an angle congruent to angle *ABM*, begin by drawing a ray, and label its endpoint *C*.

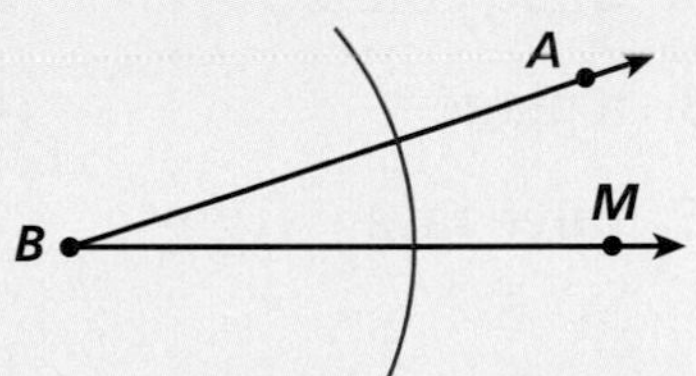

c. With your compass point on *B*, draw an arc through angle *ABM*.

d. With the same compass opening, place the compass point on *C* and draw an arc through the ray. Label point *D* where the arc crosses the ray.

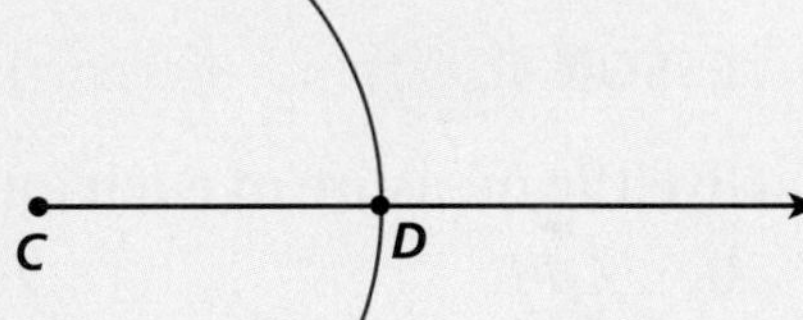

e. With your compass, measure the arc in angle *ABM*.

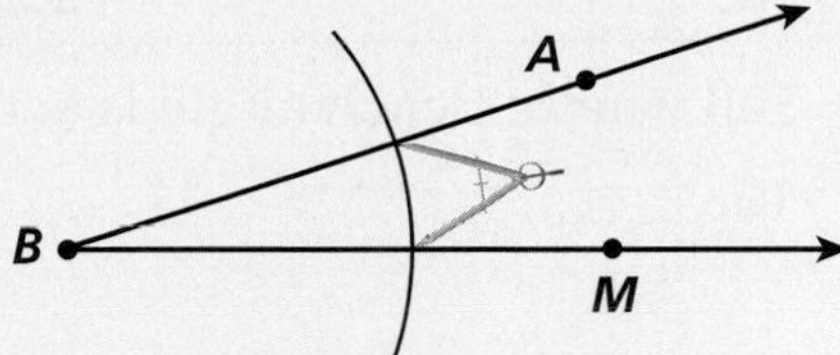

f. With the same opening, place your compass point on *D*, and draw another arc intersecting the first one. Label the intersection *F*. Draw $\overrightarrow{CF}$.

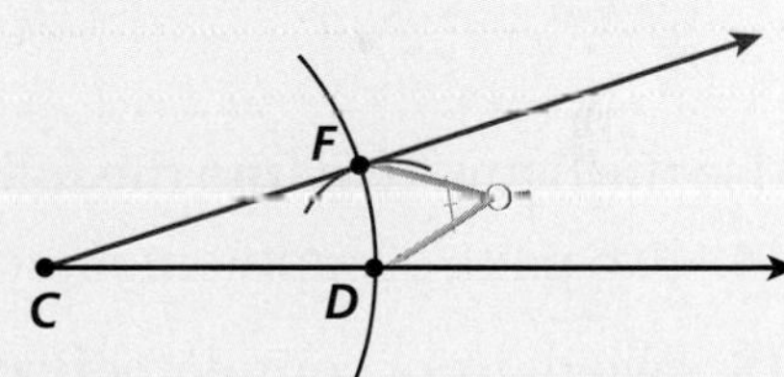

Use your protractor to measure angle *ABM* and angle *FCD*. What do you find?

Think and Discuss

1. How many bisectors would you use to divide an angle into four equal parts?

2. An 88° angle is bisected, and then each of the two angles formed are bisected. What is the measure of each of the smaller angles formed?

Try This

Use a compass and a straightedge to perform each construction.

1. Bisect a line segment.

2. Bisect an angle like angle *GOB*.

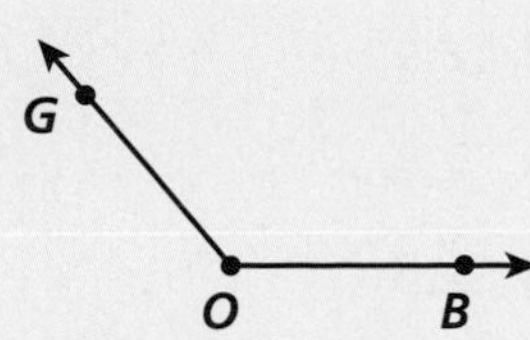

3. Draw an angle congruent to the angle you bisected in problem 2.

Chapter 7

Mid-Chapter Quiz

LESSON 7-1 (pp. 344–347)

Identify the figures in the diagram.

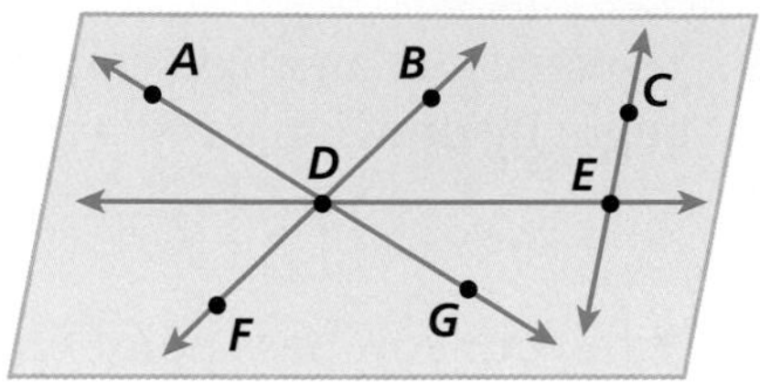

1. three points
2. three lines
3. a plane
4. three line segments
5. three rays

LESSON 7-2 (pp. 348–351)

Give the measure of each angle.

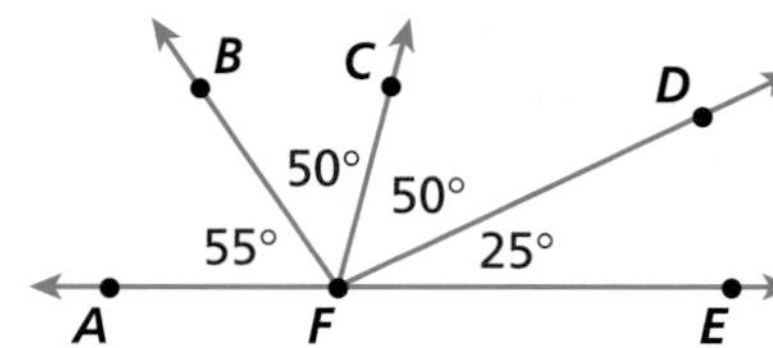

6. ∠*DFE*
7. ∠*BFC*
8. ∠*AFE*
9. ∠*AFD*

Tell whether each angle is acute, right, obtuse, or straight.

10. 
11. 
12. 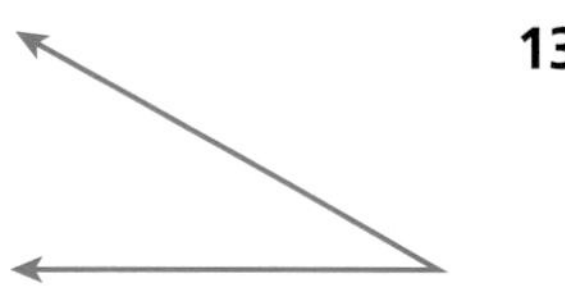
13. 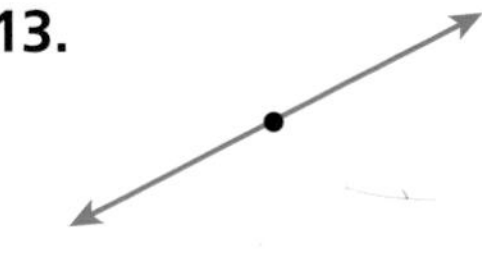

Use the figure to name the following in 14 and 15.

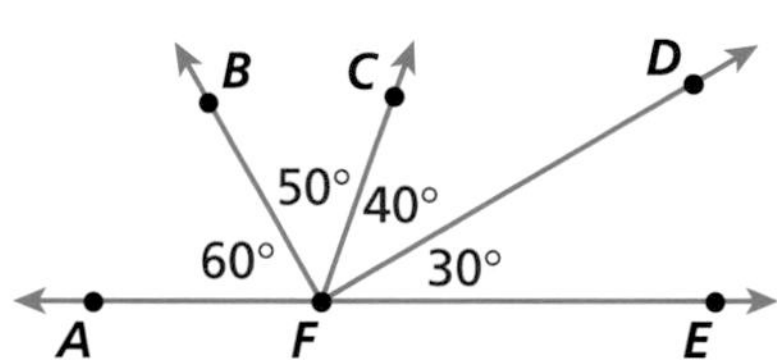

14. two pairs of supplementary angles
15. two pairs of complementary angles
16. Two supplementary angles are congruent. What is their measure?

LESSON 7-3 (pp. 354–357)

Tell whether the lines appear parallel, perpendicular, or skew.

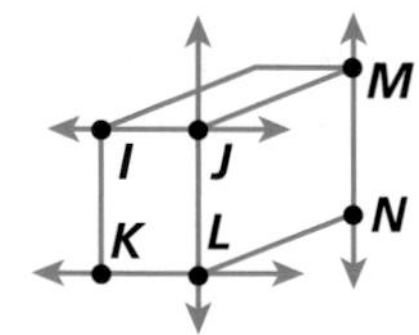

17. $\overleftrightarrow{KL}$ and $\overleftrightarrow{MN}$
18. $\overleftrightarrow{JL}$ and $\overleftrightarrow{MN}$
19. $\overleftrightarrow{KL}$ and $\overleftrightarrow{JL}$
20. $\overleftrightarrow{IJ}$ and $\overleftrightarrow{MN}$

Line *a* || line *b*. Find the measure of each angle.

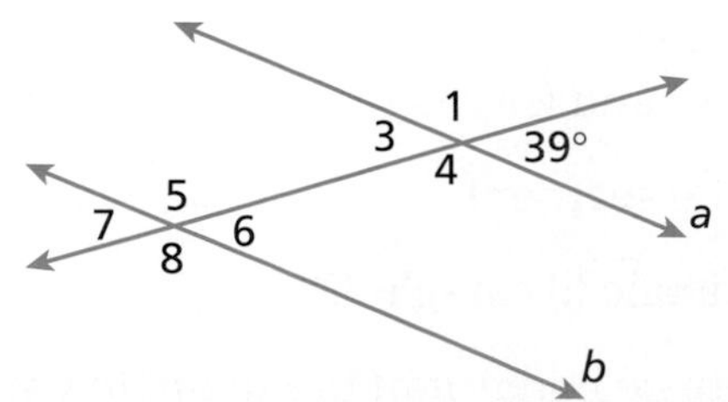

21. ∠3
22. ∠4
23. ∠8
24. ∠6
25. ∠1

Mid-Chapter Quiz

Construct Perpendicular and Parallel Lines

Use with Lesson 7-3

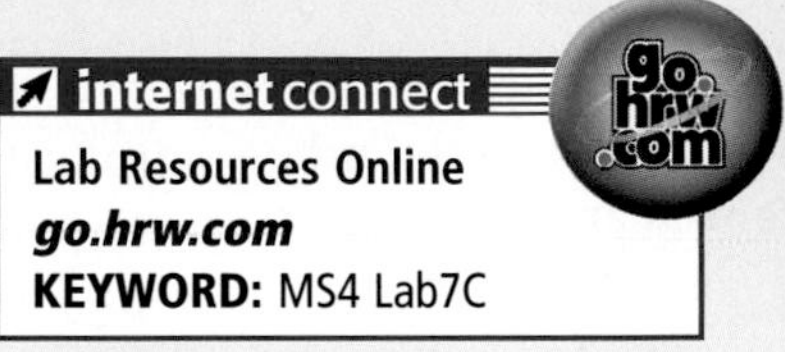

Activity

1 Construct perpendicular lines.

a. Draw $\overleftrightarrow{NO}$ and a point P above or below it.

b. Put your compass point at P and draw an arc intersecting $\overleftrightarrow{NO}$. Label points U and R.

c. Using the same compass opening, draw intersecting arcs from points U and R. Label the intersection H.

d. Draw $\overleftrightarrow{PH}$.

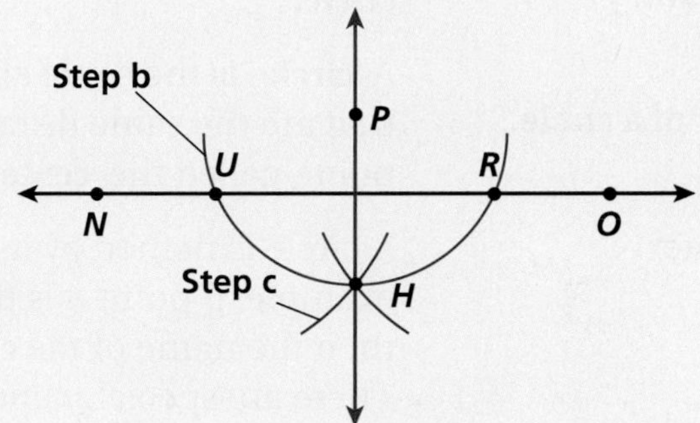

What angle does $\overleftrightarrow{NO}$ make with $\overleftrightarrow{PH}$?

2 Construct parallel lines.

a. Draw $\overleftrightarrow{LR}$ and a point S above or below it.

b. Draw a line that intersects point S and $\overleftrightarrow{LR}$. Label the intersection T.

c. Use your compass to construct an angle USW at point S that is congruent to angle STR.

d. Draw $\overleftrightarrow{SW}$.

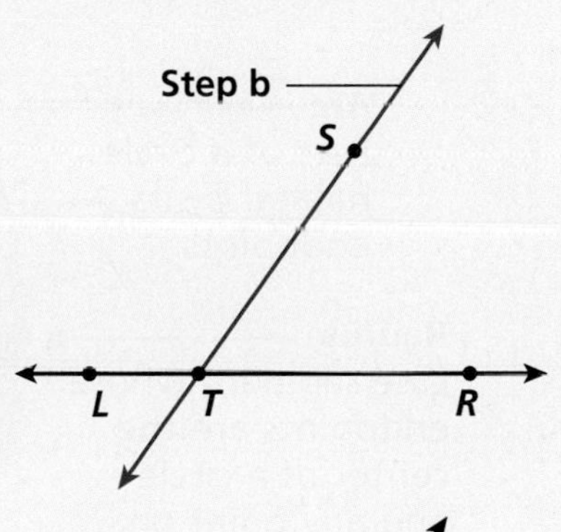

How do $\overleftrightarrow{SW}$ and $\overleftrightarrow{LR}$ relate to each other?

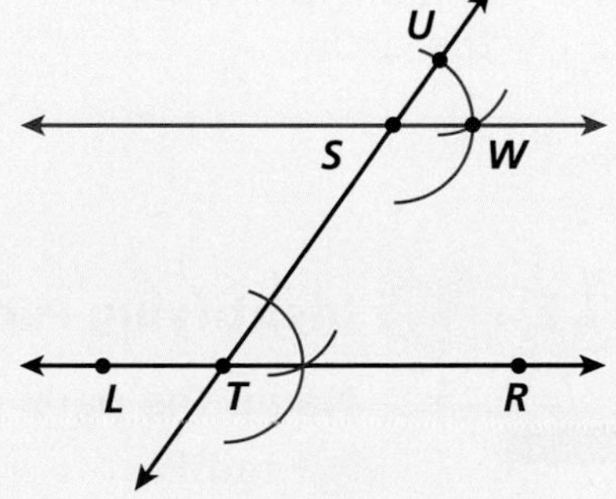

Think and Discuss

1. In **1**, how would you confirm that $\overleftrightarrow{PH}$ and $\overleftrightarrow{NO}$ are perpendicular?

2. In **2**, if $\overleftrightarrow{ST}$ is perpendicular to $\overleftrightarrow{LR}$, what can you say about angle USW?

Try This

1. Starting with $\overleftrightarrow{MT}$, construct a line segment parallel to $\overleftrightarrow{MT}$.

2. Starting with $\overleftrightarrow{PS}$, construct a line segment perpendicular to $\overleftrightarrow{PS}$.

7-4 Circles

Learn to identify parts of a circle and to find central angle measures.

Vocabulary

circle

center of a circle

radius

diameter

chord

arc

central angle

sector

The wheel is one of the most important inventions of all time. Vehicles with wheels—from ancient chariots to modern bicycles and cars—rely on the idea of a *circle*.

A **circle** is the set of all points in a plane that are the same distance from a given point, called the **center of a circle**.

A circle is named by its center. For example, if point A is the center of a circle, then the name of the circle is circle A. There are special names for the different parts of a circle.

This relief sculpture was made around 645 A.D., and shows King Ashurbanipal of Nineveh riding on his chariot.

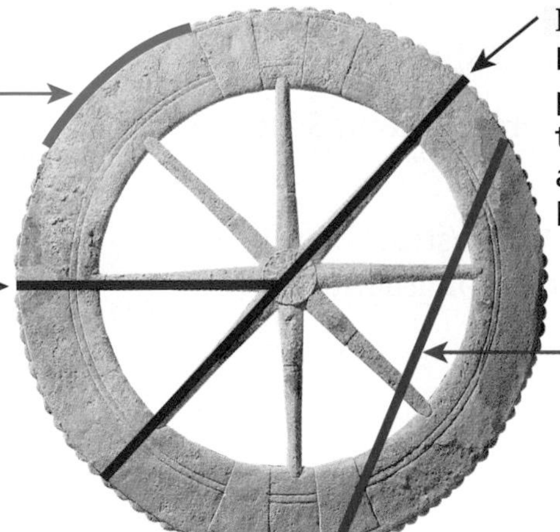

EXAMPLE 1 Identifying Parts of Circles

Reading Math

Radii is the plural form of *radius.*

Name the parts of circle *P*.

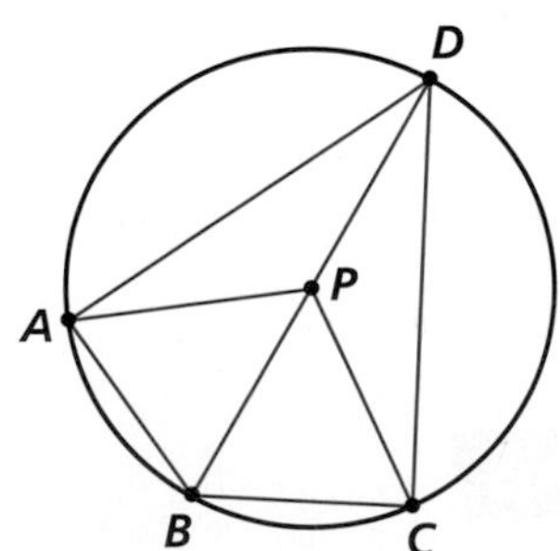

A radii

$\overline{PA}, \overline{PB}, \overline{PC}, \overline{PD}$

B diameter

$\overline{BD}$

C chords

$\overline{AD}, \overline{DC}, \overline{AB}, \overline{BC}, \overline{BD}$

A **central angle** of a circle is an angle formed by two radii. A **sector** of a circle is the part of the circle enclosed by two radii and an arc connecting them.

The sum of the measures of all of the central angles in a circle is 360°. We say that there are 360° in a circle.

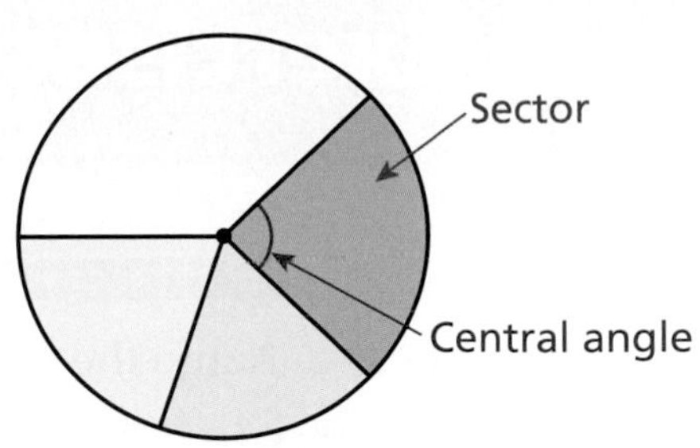

EXAMPLE 2 **PROBLEM SOLVING APPLICATION**

The circle graph shows the results of a survey to determine how people feel about keeping the penny. Find the central angle measure of the sector that shows the percent of people who are against keeping the penny.

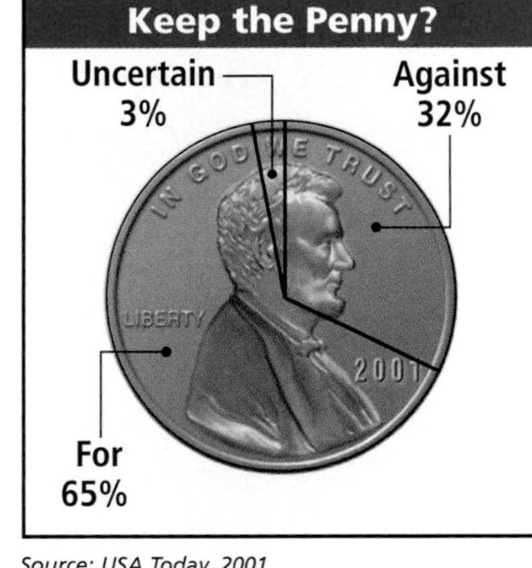

Source: USA Today, 2001

1 Understand the Problem

List the **important information:**

- The percent of people who are against keeping the penny is 32%.

2 Make a Plan

The central angle measure of the sector that represents those people against keeping the penny is 32% of the angle measure of the whole circle. The angle measure of a circle is 360°. Since the sector is 32% of the circle graph, the central angle measure is 32% of 360°.

$32\% \text{ of } 360° = 0.32 \cdot 360°$

3 Solve

$0.32 \cdot 360° = 115.2°$ *Multiply.*

The central angle of the sector measures 115.2°.

4 Look Back

The 32% sector is about one-third of the graph, and 120° is one-third of 360°. Since 115.2° is close to 120°, the answer is reasonable.

Think and Discuss

1. Explain why a diameter is a chord but a radius is not.

2. Draw a circle with a central angle of 90°.

7-4 Exercises

FOR EOG PRACTICE

see page 684

internet connect

Homework Help Online

go.hrw.com Keyword: MS4 7-4

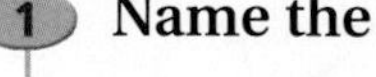

1.02d

GUIDED PRACTICE

See Example 1

Name the parts of circle *O*.

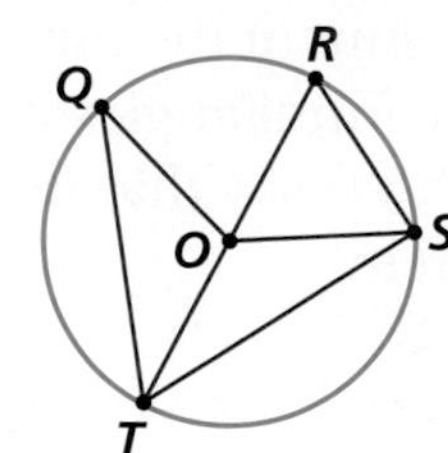

1. radii
2. diameter
3. chords

See Example 2

4. The circle graph shows the results of a 2001 survey in which the following question was asked: "If you had to describe your office environment as a type of television show, which would it be?" Find the central angle measure of the sector that shows the percent of people who described their workplace as a courtroom drama.

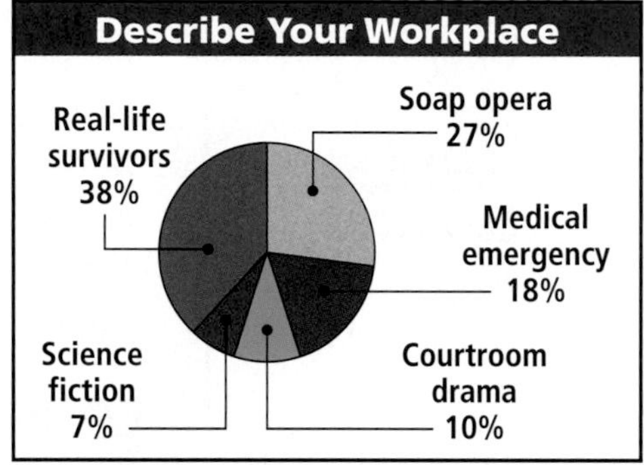

Source: USA Today

INDEPENDENT PRACTICE

See Example 1

Name the parts of circle *C*.

5. radii
6. diameters
7. chords

See Example 2

8. The circle graph shows the areas from which the United States imports bananas. Find the central angle measure of the sector that shows the percent of banana imports from South America.

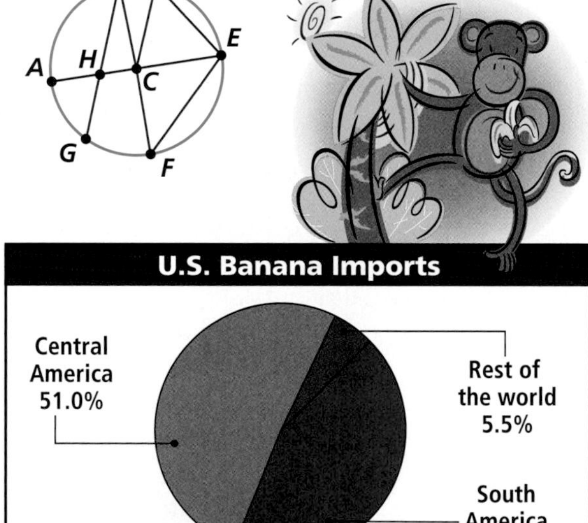

Source: US Bureau of the Census Trade Data

PRACTICE AND PROBLEM SOLVING

9. What is the distance between the centers of the circles?

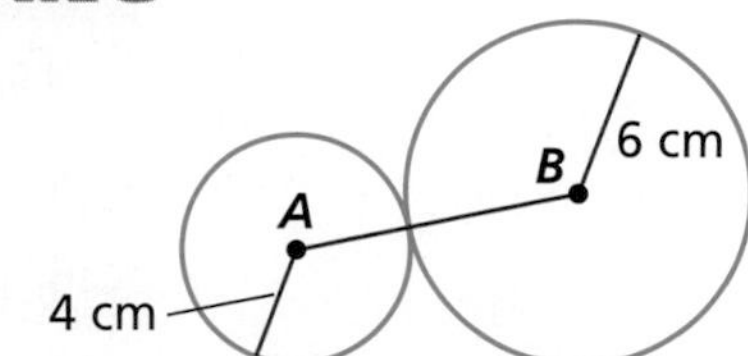

10. a. Name all of the chords of the circle at right.

b. If $\overline{AB} \parallel \overline{CD}$, what is the measure of $\angle 1$? Explain your answer.

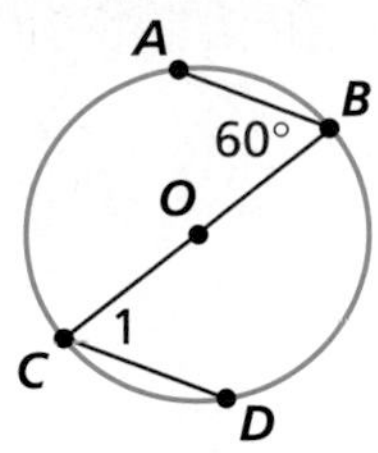

11. A circle is divided into five equal sectors. Find the measure of the central angle of each sector.

12. What is the measure of the central angle of a quarter of a circle?

13. ***MUSIC*** The circle graphs show the results of a music survey.

a. Find the central angle measure of the sector that shows the percent of people who chose *The Star-Spangled Banner* as the song they prefer.

b. Find the central angle measure of the sector that shows the percent of people who prefer *God Bless America* as the national anthem.

c. What point is made by showing the two circle graphs together?

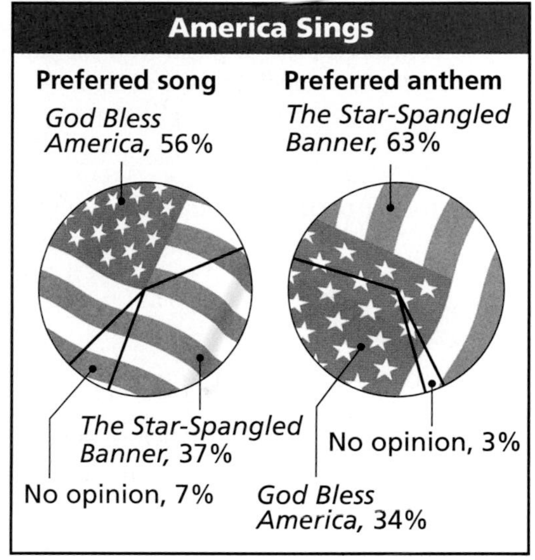

Source: USA Today

14. ***WRITE A PROBLEM*** Find a circle graph in your science or social studies textbook. Use the graph to write a problem that can be solved by finding the central angle measure of one of the sectors of the circle.

15. ***WRITE ABOUT IT*** Compare central angles of a circle with sectors of a circle.

16. ***CHALLENGE*** Find the angle measure between the minute and hour hands on the clock at right.

Spiral Review

Solve. (Lesson 4-6)

17. $w - 7.3 = 12$ **18.** $x + 0.5 = 3.1$ **19.** $y - 1.06 = 0.9$ **20.** $a + 9.6 = 10$

Tell whether each angle is acute, obtuse, right, or straight. (Lesson 7-2)

21.  **22.** 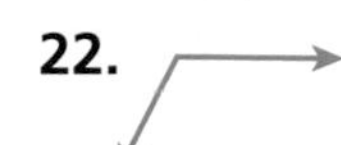 **23.**  **24.** 

25. **EOG PREP** Which best describes the relationship between the obtuse angles that are formed when a transversal intersects two parallel lines? (Lesson 7-3)

A Congruent B Intersecting C Complementary D Supplementary

Construct Circle Graphs

Use with Lesson 7-4

REMEMBER
- There are 360° in a circle.
- A radius is a line segment with one endpoint at the center of a circle and the other endpoint on the circle.

A circle graph can be used to compare data that are parts of a whole.

Activity

You can make a circle graph using information from a table.

At Booker Middle School, a survey was conducted to find the percent of students who favor certain types of books. The results are shown in the table below.

To make a circle graph, you need to find the size of each part of your graph. Each part is a *sector*.

To find the size of a sector, you must find the measure of its angle. You do this by finding what percent of the whole circle that sector represents.

1 Find the size of each sector.

a. Copy the table at right.

b. Find a decimal equivalent for each percent given, and fill in the decimal column of your table.

c. Find the fraction equivalent for each percent given, and fill in the fraction column of your table.

d. Find the angle measure of each sector by multiplying each fraction or decimal by 360°. Fill in the last column of your table.

Students Favorite Types of Books

Type of Book	Percent	Decimal	Fraction	Degrees
Mysteries	35%			
Science Fiction	25%			
Sports	20%			
Biographies	15%			
Humor	5%			

2 Follow the steps below to draw a circle graph.

a. Using a compass, draw a circle. Using a straightedge, draw one radius.

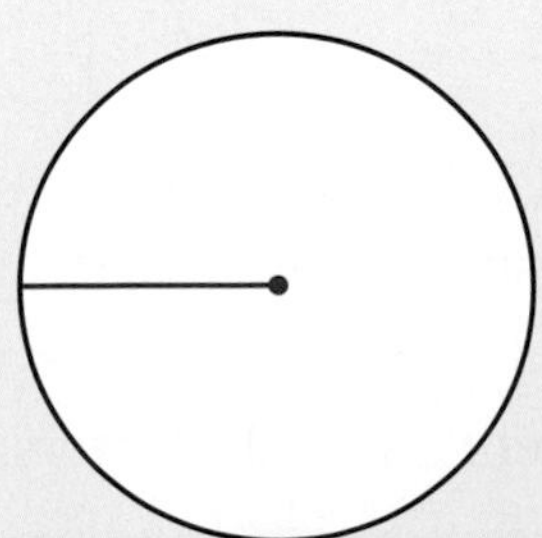

b. Use a protractor to measure the angle of the first sector. Draw the angle.

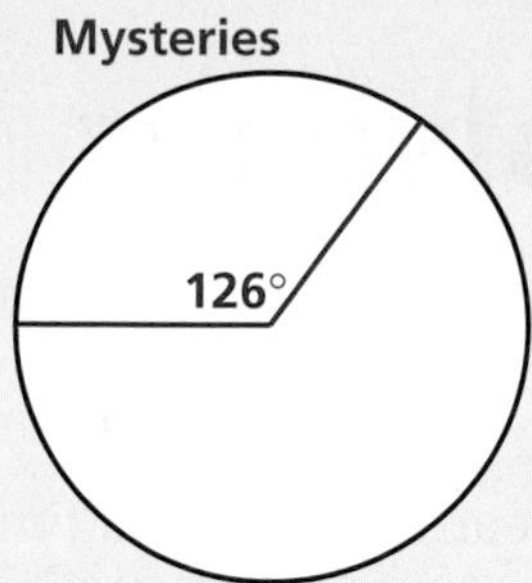

c. Use a protractor to measure the angle of the next sector. Draw the angle.

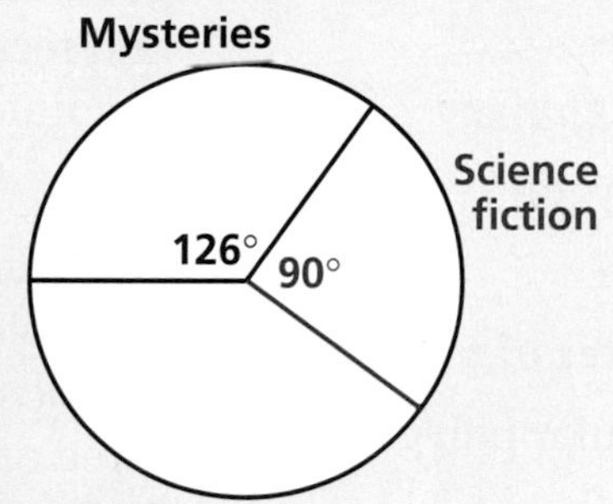

d. Continue until your graph is complete. Label each sector with its name and percent.

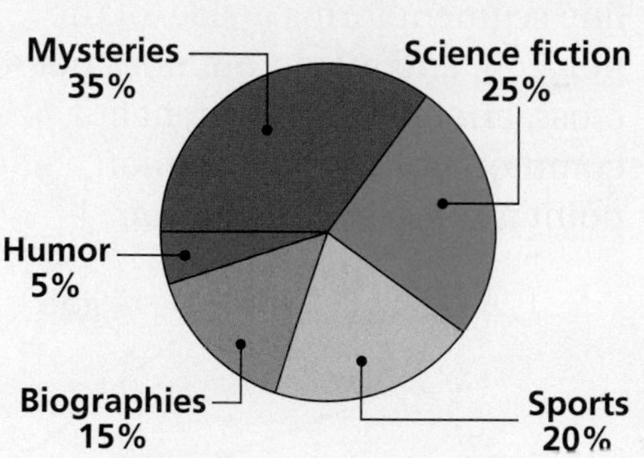

Think and Discuss

1. Total each column in the table from the beginning of the activity. What do you notice?
2. What type of data would you want to display using a circle graph?
3. How does the size of each sector of your circle graph relate to the percent, the decimal, and the fraction in your table?

Try This

1. Complete the table below and use the information to make a circle graph.

   On a typical Saturday, Alan divides his leisure time and spends it in the following ways:

Time Spent for Leisure				
Activity	**Percent**	**Decimal**	**Fraction**	**Degrees**
Reading	30%			
Playing sports	25%			
Working on computer	40%			
Watching TV	5%			

7-5 Polygons

Learn to identify and name polygons.

Vocabulary

polygon
side
vertex of a polygon
regular polygon

From the earliest recorded time, geometric shapes, such as triangles and rectangles, have been used to decorate buildings and works of art.

Triangles and rectangles are examples of *polygons*. A **polygon** is a closed plane figure formed by three or more line segments. Each line segment forms a **side** of the polygon, and meets, but does not cross, another line segment at a common point. This common point is a **vertex of a polygon**.

The Paracas were an ancient native culture of Peru. Among the items that have been excavated from their lands are color tapestries, such as this one.

Remember!

Vertices is plural for *vertex*.

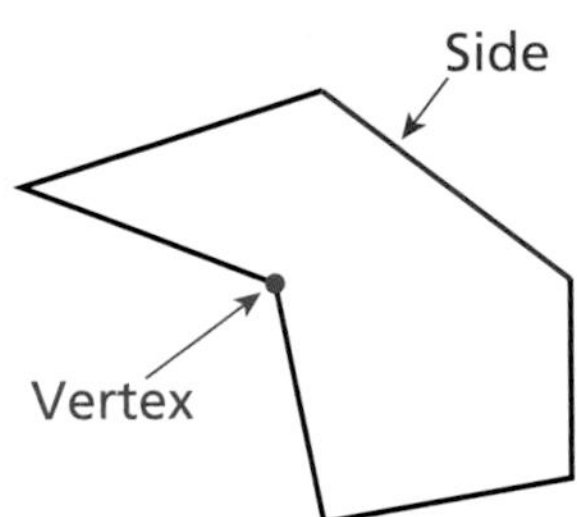

The polygon at left has six sides and six vertices.

EXAMPLE 1 Identifying Polygons

Determine whether each figure is a polygon. If it is not, explain why not.

A

The figure is a polygon.
It is a closed figure with 5 sides.

B

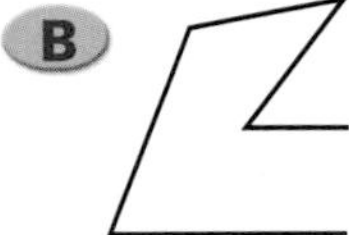

The figure is not a polygon.
It is not a closed figure.

C

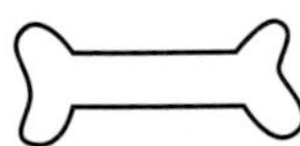

The figure is not a polygon.
Not all of the sides of the figure are line segments.

D

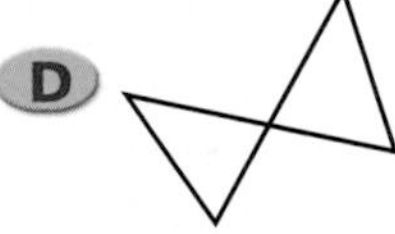

The figure is not a polygon.
There are line segments in the figure that cross.

Polygons are classified by the number of sides and angles they have.

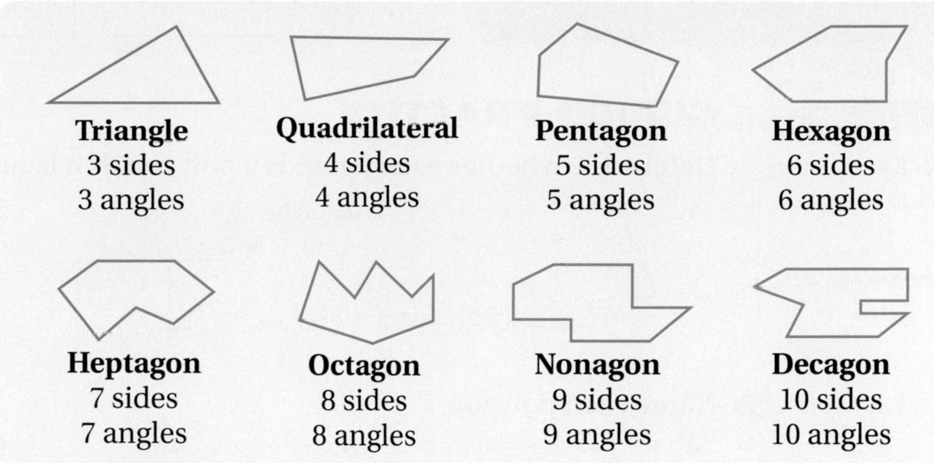

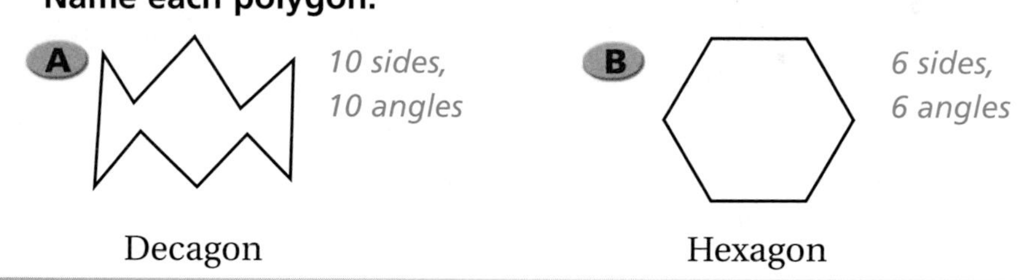

EXAMPLE 2 Classifying Polygons

Name each polygon.

A

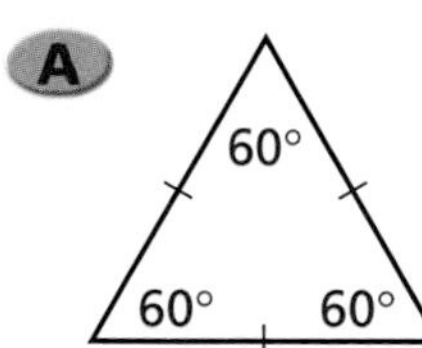

10 sides, 10 angles

Decagon

B

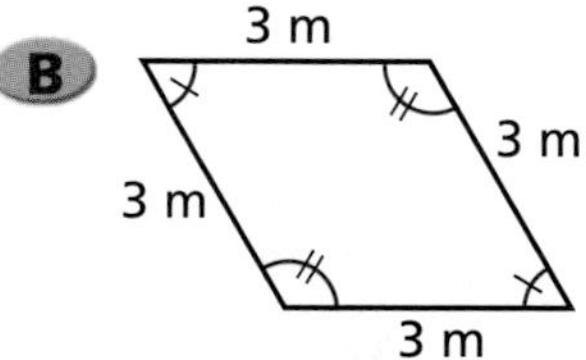

6 sides, 6 angles

Hexagon

A **regular polygon** is a polygon in which all sides are congruent and all angles are congruent.

EXAMPLE 3 Identifying and Classifying Regular Polygons

Name each polygon, and tell whether it is a regular polygon. If it is not, explain why not.

A

60° 60° 60°

The figure is a regular triangle.

B

3 m 3 m 3 m 3 m

The figure is a quadrilateral. It is not a regular polygon because all of the angles are not congruent.

Think and Discuss

1. **Explain** why a circle is not a polygon.
2. **Draw** a dodecagon, which is a polygon with 12 sides and 12 angles.

7-5 Exercises

FOR EOG PRACTICE
see page 684

internet connect
Homework Help Online
go.hrw.com Keyword: MS4 7-5

3.02

GUIDED PRACTICE

See Example 1 Determine whether each figure is a polygon. If it is not, explain why not.

1. 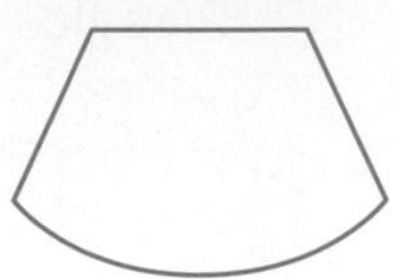

2. 

3. 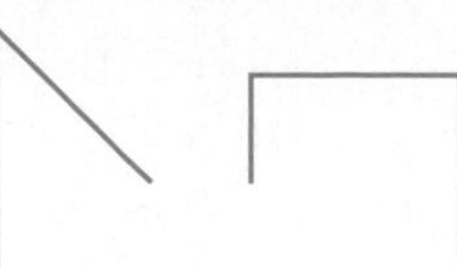

See Example 2 Name each polygon.

4. 

5. 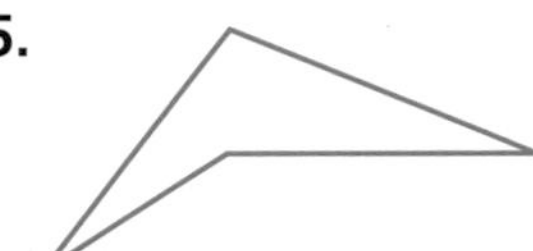

6. 

See Example 3 Name each polygon, and tell whether it is a regular polygon. If it is not, explain why not.

7. 24 in. / 24 in. / 24 in. / 24 in.

8. 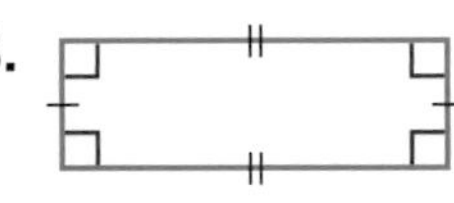

9. 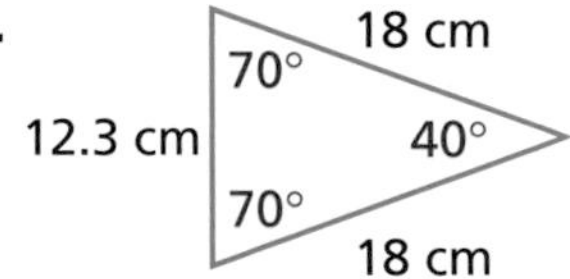

INDEPENDENT PRACTICE

See Example 1 Determine whether each figure is a polygon. If it is not, explain why not.

10. 

11. 

12. 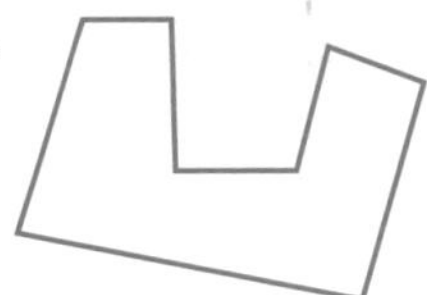

See Example 2 Name each polygon.

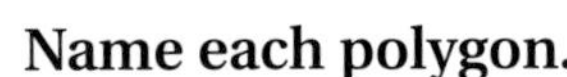

13. 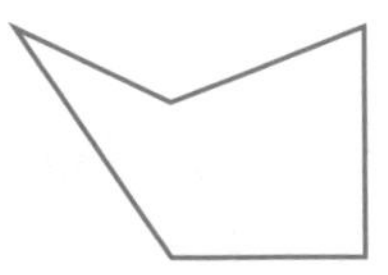

14. 

15. 

See Example 3 Name each polygon, and tell whether it is a regular polygon. If it is not, explain why not.

16. 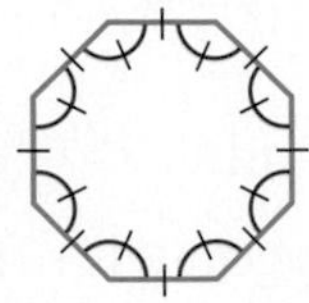

17. 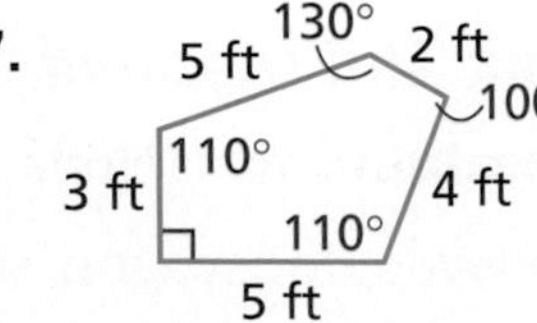

18. 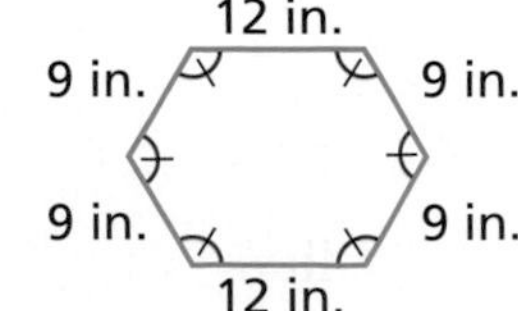

Art LINK

19. The design of the quilt at right is made up of triangles.
    a. What other polygons can you find in the pattern?
    b. Which of the polygons in the pattern appear to be regular?

Use the photograph of the star quilt for Exercises 20 and 21.

20. The large star in the quilt pattern is made of smaller shapes stitched together. These smaller shapes are all the same type of polygon. What type of polygon are the smaller shapes?

21. A polygon can be named by the number of its sides followed by *-gon*. For example, a polygon with 14 sides is called a 14-gon. What is the name of the large star-shaped polygon on the quilt?

22. The pattern on the quilt at right is called a bow tie pattern.
    a. Name the polygons that are used in the bow tie pattern.
    b. Do regular polygons appear to be used in the pattern? Explain.

23. ***CHALLENGE*** The quilt at right has a modern design. Copy one of each type of polygon, from a triangle to a decagon, onto your paper from the design. Write the name of each polygon next to its drawing.

go.hrw.com
KEYWORD: MS4 Quilt
CNN Student News

Spiral Review

$\angle 1$ and $\angle 2$ are complementary angles. Find m$\angle 2$. (Lesson 7-2)

24. m$\angle 1 = 50°$ 25. m$\angle 1 = 25°$ 26. m$\angle 1 = 12°$ 27. m$\angle 1 = 66°$

Line $a \parallel$ line b. Use the diagram to find each angle measure. (Lesson 7-3)

28. $\angle 1$ 29. $\angle 2$ 30. $\angle 3$

31. **EOG PREP** Which is a chord of a circle? (Lesson 7-4)

A Radius B Diameter C Sector D Central angle

Constructions with Polygons and Circles

Use with Lesson 7-5

REMEMBER
- An equilateral triangle has three congruent sides and three congruent angles.
- A square has four congruent sides and four congruent angles.
- A regular hexagon has six congruent sides and six congruent angles.
- A perpendicular bisector of a line segment is a line that intersects the segment at its midpoint at a right angle.
- The diameter of a circle is a line segment that passes through the center of the circle and whose endpoints are on the circle.

internet connect
Lab Resources Online
go.hrw.com
KEYWORD: MS4 Lab7E

A regular polygon is inscribed in a circle when all of its vertices lie on a circle that surrounds it.

Activity

1 Inscribe an equilateral triangle in a circle.

a. Draw a circle using a compass. Make the circle at least an inch in diameter.

b. Draw a diameter anywhere inside the circle.

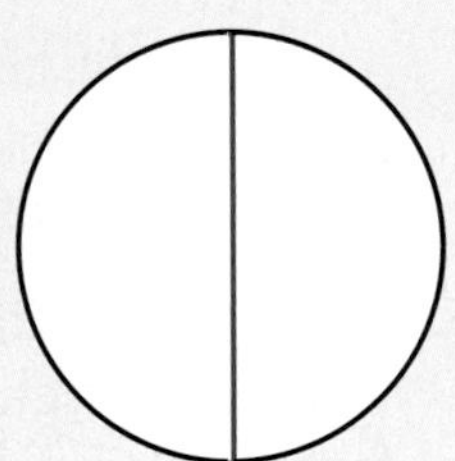

c. Place the point of your compass at one end of the diameter. Using the same compass opening that was used to make the circle, draw two arcs, one on each side of the point, that intersect the circle.

d. Label the points where the arcs intersect the circle as *A* and *B*. Label point *C* where the diameter meets the circle farthest from the arcs.

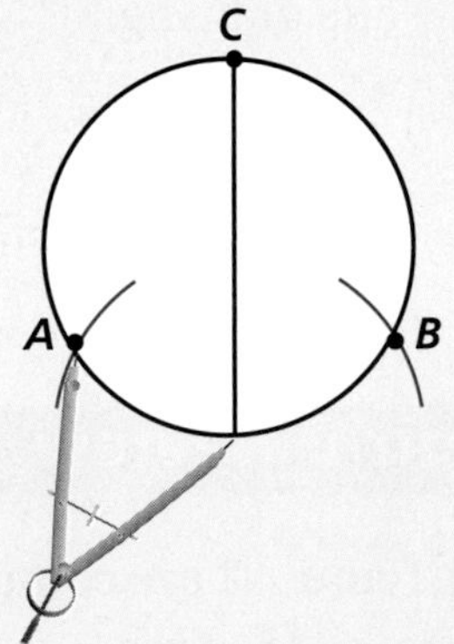

e. Draw line segments to connect the points to form an equilateral triangle.

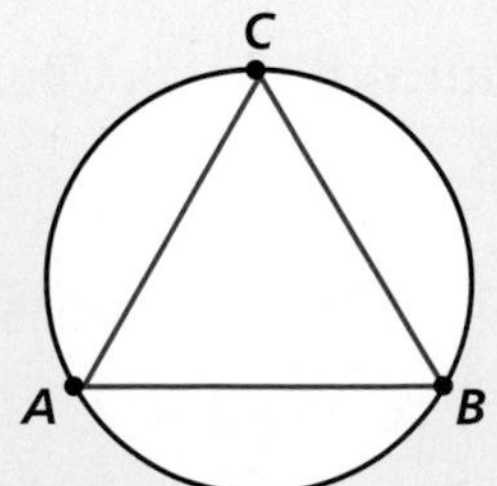

2 Inscribe a square in a circle.

a. Repeat steps **a** and **b** from **1**.

b. Construct a perpendicular bisector of the diameter. Then label points *D, E, F,* and *G* where the diameter and its bisector intersect the circle.

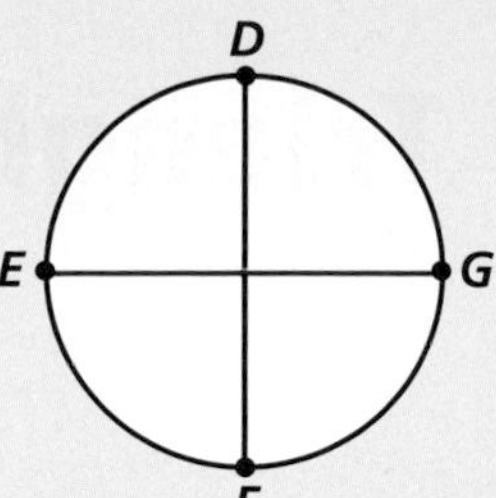

c. Draw line segments to connect the points to form a square.

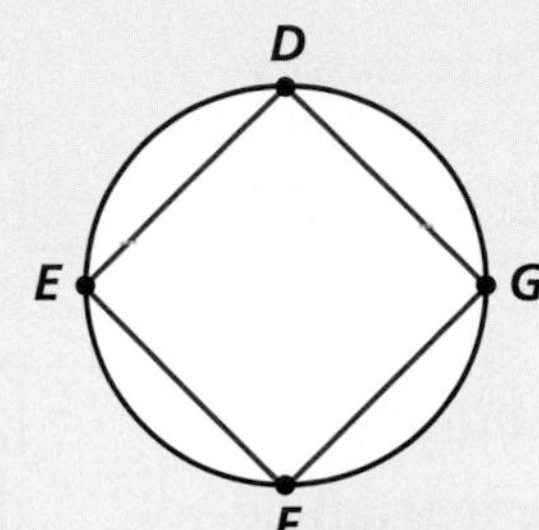

3 Inscribe a regular hexagon in a circle.

a. Repeat steps **a, b,** and **c** from **1**.

b. Draw two diameters that each include one of the points where the arcs intersect the circle. Then label the points *L, M, N, O, P,* and *Q* where the diameters meet the circle.

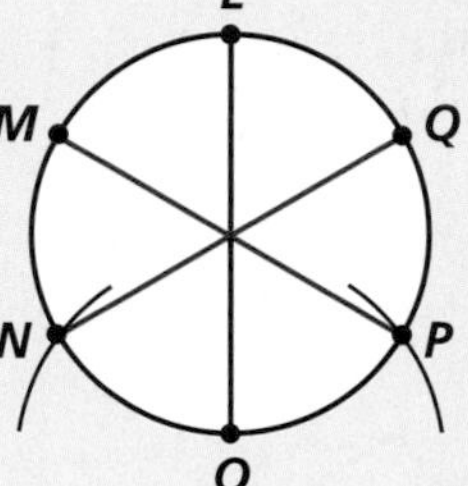

c. Draw line segments to connect the points to form a regular hexagon.

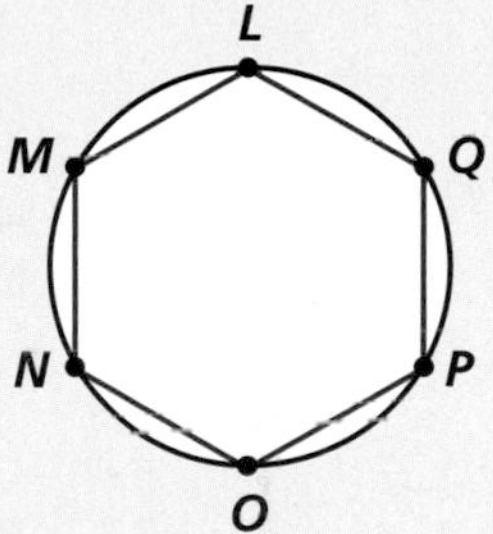

Think and Discuss

1. Do you have to keep the compass opening the same at all times when inscribing an equilateral triangle in a circle? Explain.
2. Do you have to keep the compass opening the same at all times when inscribing a square in a circle? Explain.
3. When you are inscribing a regular hexagon in a circle, how can you tell that all of the sides of the hexagon are congruent? How does the length of each side of the hexagon compare with the radius of the circle?
4. When you are inscribing a square in a circle, do the sides of the square have the same length as the radius of the circle? Explain.

Try This

1. Inscribe an equilateral triangle in a circle with a 3-inch diameter.
2. Inscribe a square in a circle with a 2-inch diameter.
3. Inscribe a regular hexagon in a circle with a 1.5-inch diameter.
4. Inscribe an equilateral triangle, a square, and a regular hexagon in the same circle with a 4-inch diameter.

7-6 Triangles

Learn to classify triangles by their side lengths and angle measures.

Vocabulary
scalene triangle
isosceles triangle
equilateral triangle
acute triangle
obtuse triangle
right triangle

How many different kinds of triangles can you find in the origami butterfly at right? To answer this question, you must first have ways of classifying the triangles. One way to classify them is by the lengths of their sides. Another way is by the measures of their angles.

Triangles classified by sides

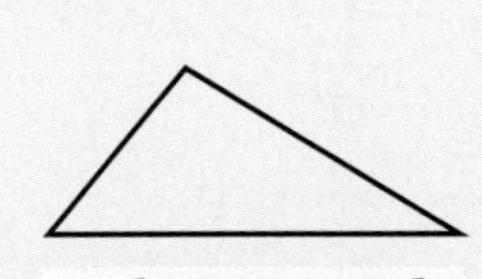

A **scalene triangle** has no congruent sides.

An **isosceles triangle** has at least 2 congruent sides.

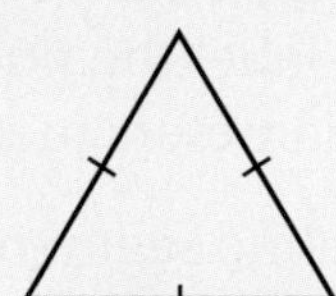

In an **equilateral triangle**, all of the sides are congruent.

Triangles classified by angles

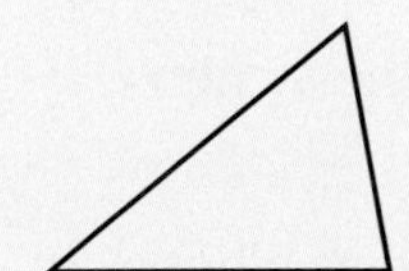

In an **acute triangle**, all of the angles are acute.

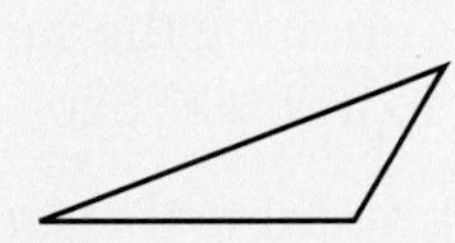

An **obtuse triangle** has one obtuse angle.

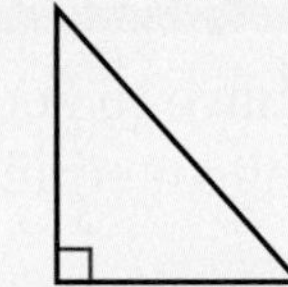

A **right triangle** has one right angle.

EXAMPLE 1 Classifying Triangles

Writing Math

When classifying a triangle according to its sides and angles, use the word that describes the sides first, and use the word that describes the angles second.

Classify each triangle according to its sides and angles.

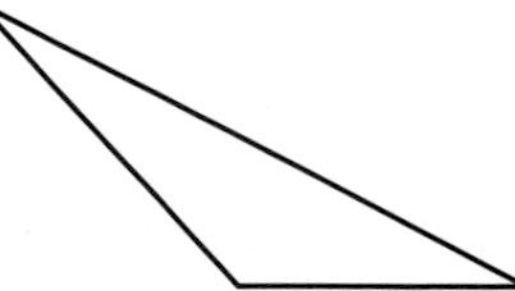

scalene *No congruent sides*
obtuse *One obtuse angle*

This is a scalene obtuse triangle.

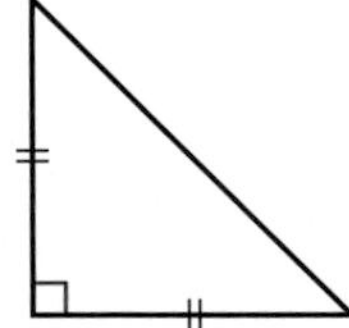

isosceles *Two congruent sides*
right *One right angle*

This is an isosceles right triangle.

Classify each triangle according to its sides and angles.

scalene *No congruent sides*
right *One right angle*

This is a scalene right triangle.

D

isosceles *Two congruent sides*
obtuse *One obtuse angle*

This is an isosceles obtuse triangle.

EXAMPLE 2 Identifying Triangles

Identify the different types of triangles in the figure, and determine how many of each there are.

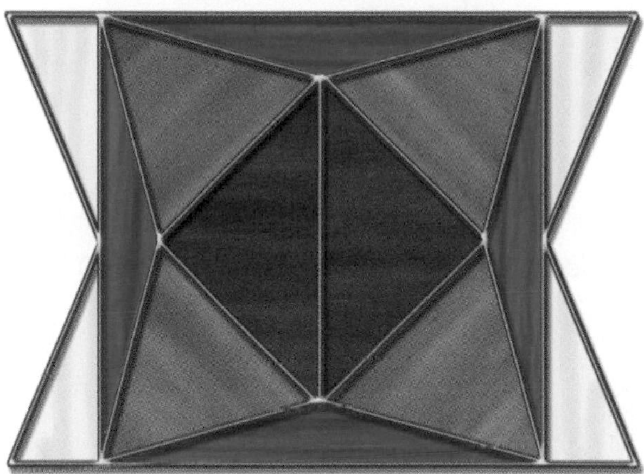

Type	How Many	Colors	Type	How Many	Colors
Scalene	4	Yellow	Right	6	Purple, yellow
Isosceles	10	Green, pink, purple	Obtuse	4	Green
Equilateral	4	Pink	Acute	4	Pink

Think and Discuss

1. Draw an isosceles acute triangle and an isosceles obtuse triangle.

2. Draw a triangle that is right and scalene.

3. Explain why an equilateral triangle is also an isosceles triangle but an isosceles triangle is not always an equilateral triangle.

7-6 Exercises

FOR EOG PRACTICE

see page 685

internet connect

Homework Help Online

go.hrw.com Keyword: MS4 7-6

3.02

GUIDED PRACTICE

See Example 1

Classify each triangle according to its sides and angles.

1.

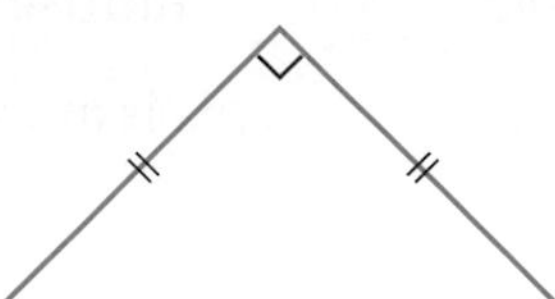

2.

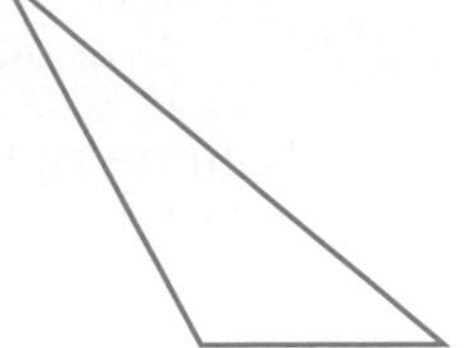

3.

See Example 2

4. Identify the different types of triangles in the figure, and determine how many of each there are.

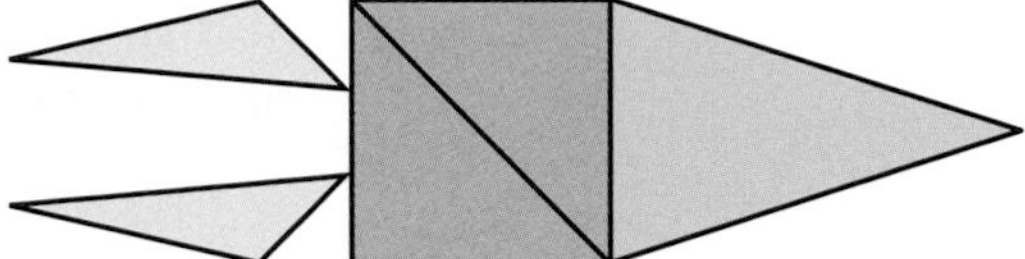

INDEPENDENT PRACTICE

See Example 1

Classify each triangle according to its sides and angles.

5.

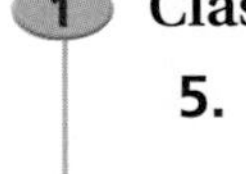

6.

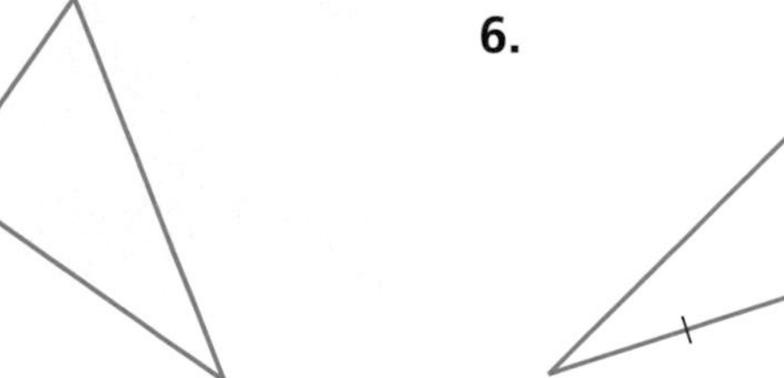

7.

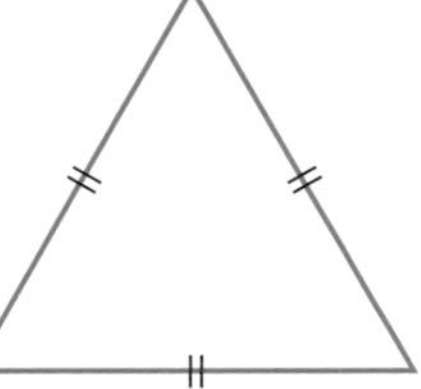

See Example 2

8. Identify the different types of triangles in the figure, and determine how many of each there are.

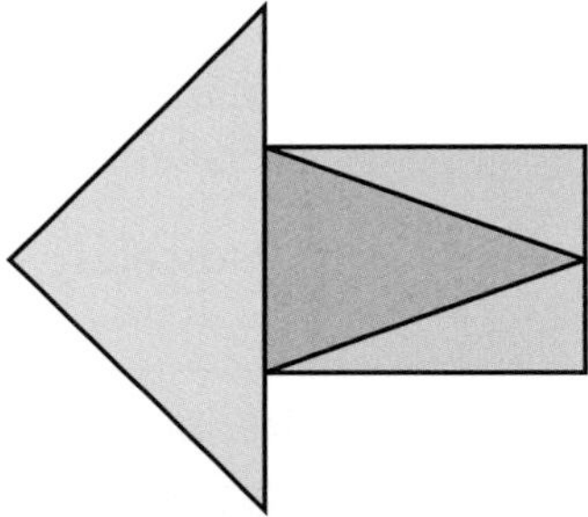

PRACTICE AND PROBLEM SOLVING

Classify each triangle according to the lengths of its sides.

9. 3 cm, 5 cm, 3 cm
10. 3 ft, 4 ft, 5 ft
11. 4 m, 4 m, 4 m
12. 6 ft, 9 ft, 12 ft
13. 2 in., 2 in., 2 in.
14. 7.4 mi, 7.4 mi, 4 mi

Classify each triangle according to the measures of its angles.

15. 60°, 60°, 60°
16. 105°, 38°, 37°
17. 79°, 11°, 90°
18. 30°, 120°, 30°
19. 45°, 90°, 45°
20. 40°, 60°, 80°

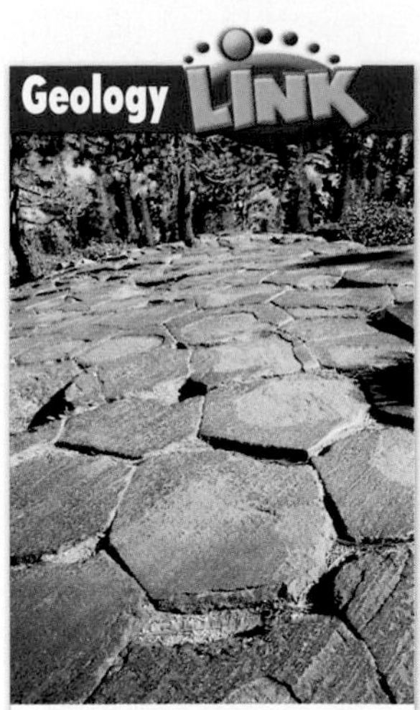

The Devil's Postpile in California is an area where lava once cooled and solidified to form hexagonal structures.

21. The sum of the lengths of the sides of triangle ABC is 25 in. The lengths of sides $\overline{AB}$ and $\overline{BC}$ are 9 inches and 8 inches. Find the length of side $\overline{AC}$ and classify the triangle.

22. Draw a square. Divide it into two triangles. Describe the triangles.

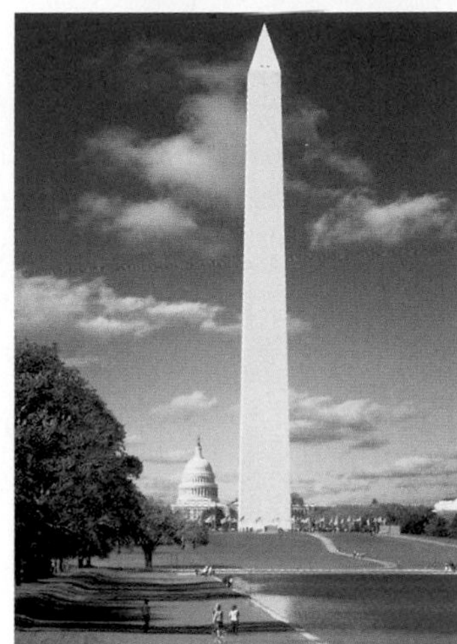

23. ***GEOLOGY*** Each face of a topaz crystal is a triangle whose sides are all different lengths. What kind of triangle is each face of a topaz crystal?

24. ***ARCHITECTURE*** The Washington Monument is an obelisk, the top of which is a pyramid. The pyramid has four triangular faces. The bottom edge of each face measures 10.5 m. The other edges measure 17.0 m. What kind of triangle is each face of the pyramid?

25. ***CHOOSE A STRATEGY*** How many triangles are in the figure?

A 6 **C** 10

B 9 **D** 13

26. ***WRITE ABOUT IT*** Is it possible for an equilateral triangle to be obtuse? Explain your answer.

27. ***CHALLENGE*** The centers of circles A, B, C, D, and E are connected by line segments as shown. Classify each triangle that is formed in the figure according to its side lengths, given the following information:

$DE = 5$ $BD = 6$

$CB = 8$ $AC = 8$

The diameter of circle D is 4.

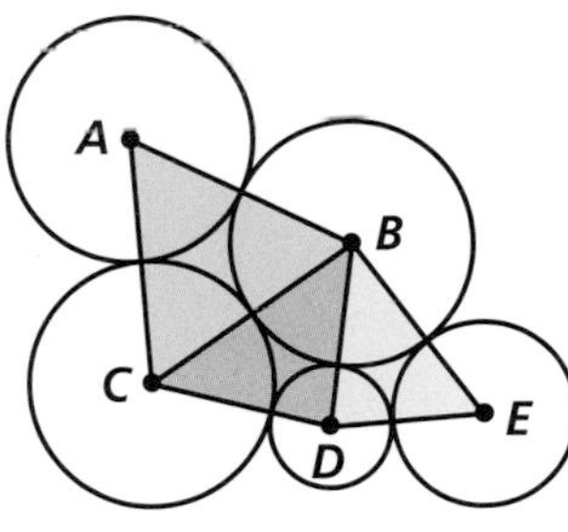

Spiral Review

Multiply. (Lesson 4-3)

28. $0.2 \cdot 10^3$ **29.** $3.84 \cdot 10^2$ **30.** $0.006 \cdot 10^4$ **31.** $0.0056 \cdot 10^2$

Solve. (Lesson 5-3)

32. $\frac{6}{a} = \frac{15}{10}$ **33.** $\frac{h}{9} = \frac{10}{15}$ **34.** $\frac{4}{17} = \frac{k}{34}$ **35.** $\frac{12}{5} = \frac{24}{m}$

36. **EOG PREP** What percent of 15 is 5? (Lesson 6-4)

A 300% B 3% C $\frac{1}{3}$% D $33\frac{1}{3}$%

7-7 Quadrilaterals

Learn to name and identify types of quadrilaterals.

Vocabulary

parallelogram

rhombus

rectangle

square

trapezoid

kite

College campuses are often built around an open space called a "quad" or "quadrangle." A quadrangle is a four-sided enclosure, or a quadrilateral.

Some quadrilaterals have properties that classify them as *special quadrilaterals.* The figures below are six major special quadrilaterals.

A quad at Cornell University in Ithaca, NY

A **parallelogram** has two pairs of parallel sides.	A **rhombus** has four congruent sides.
A **rectangle** has four right angles.	A **square** has four congruent sides and four right angles.
A **trapezoid** has exactly one pair of parallel sides.	A **kite** has exactly two pairs of congruent, adjacent sides.

Quadrilaterals can have more than one name because the special quadrilaterals sometimes share properties.

EXAMPLE 1 Identifying Types of Quadrilaterals

Give all of the names that apply to each quadrilateral.

A

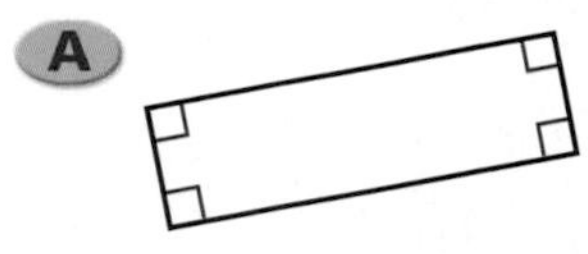

The figure has two pairs of parallel sides, so it is a **parallelogram**. It has four right angles, so it is also a **rectangle**.

Give all of the names that apply to each quadrilateral.

B

The figure has exactly one pair of parallel sides, so it is a **trapezoid**. It does not fit the definitions of any of the other special quadrilaterals.

C

5 cm 5 cm 5 cm 5 cm

The figure has two pairs of parallel sides, so it is a **parallelogram**. It has four right angles, so it is also a **rectangle**. It has four congruent sides, so it is also a **rhombus** and a **square**.

EXAMPLE 2 Recognizing Relationships Between Quadrilaterals

Tell whether each statement is true or false. Explain your answer.

A All squares are rhombuses.

Any quadrilateral that has four congruent sides is a rhombus. Because a square has four congruent sides, it is a rhombus. The statement is true.

B All kites are parallelograms.

The opposite sides of a kite are not parallel, but the opposite sides of a parallelogram are parallel. The statement is false.

C All rhombuses are rectangles.

A rhombus may or may not have right angles, but a rectangle must have four right angles. The statement is false.

Think and Discuss

1. **Compare** a parallelogram with a trapezoid.
2. **Describe** how you can decide whether a rhombus is also a square. Use drawings to justify your answer.
3. **Discuss** why a rhombus is not a kite.

7-7 Exercises

FOR EOG PRACTICE
see page 685

internet connect
Homework Help Online
go.hrw.com Keyword: MS4 7-7

3.02

GUIDED PRACTICE

See Example 1 **Give all of the names that apply to each quadrilateral.**

1. 6 yd, 4.5 yd, 4.5 yd, 6 yd

2.

3. 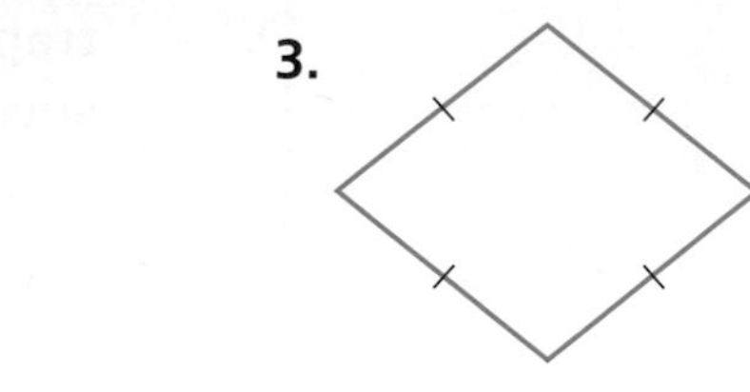

See Example 2 **Tell whether each statement is true or false. Explain your answer.**

4. All rhombuses are squares.

5. All rectangles are parallelograms.

INDEPENDENT PRACTICE

See Example 1 **Give all of the names that apply to each quadrilateral.**

6. 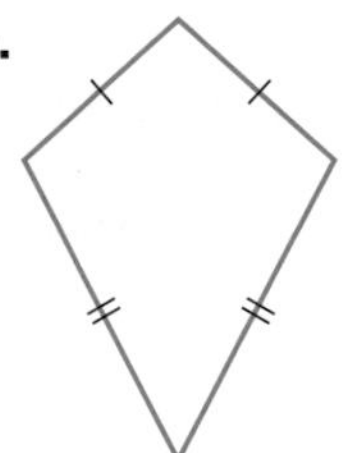

7. 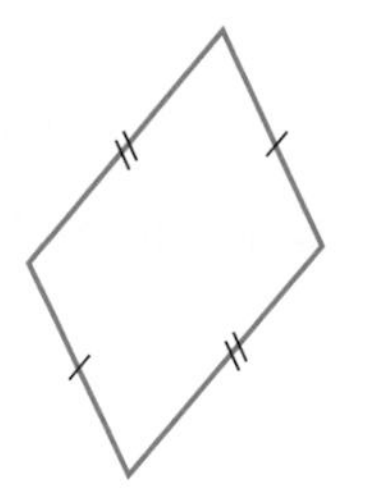

8. 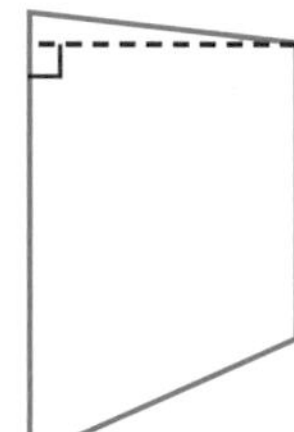

9. 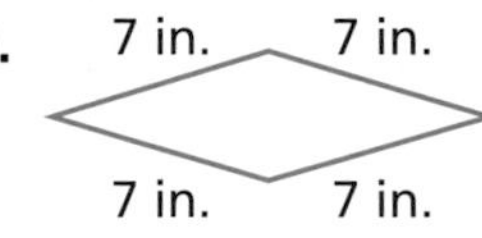

10. 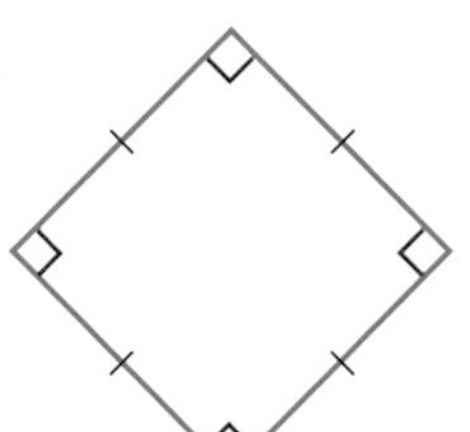

11. 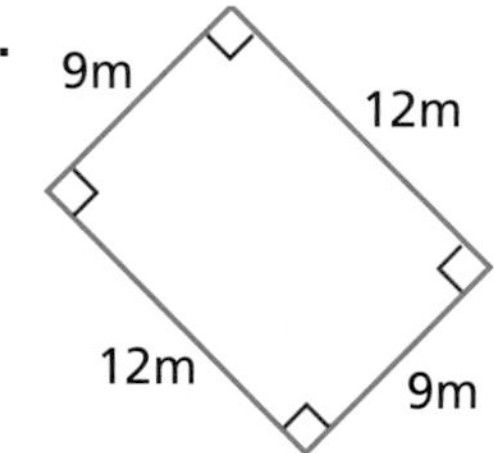

See Example 2 **Tell whether each statement is true or false. Explain your answer.**

12. All squares are rectangles.

13. All rectangles are squares.

14. Some rectangles are squares.

15. Some trapezoids are squares.

PRACTICE AND PROBLEM SOLVING

Name the types of quadrilaterals with each feature.

16. four right angles

17. two pairs of opposite, parallel sides

18. four congruent sides

19. opposite sides are congruent

20. Graph the points $A(-2, -2)$, $B(4, 1)$, $C(3, 4)$, and $D(-1, 2)$, and draw line segments to connect the points. What kind of quadrilateral did you draw?

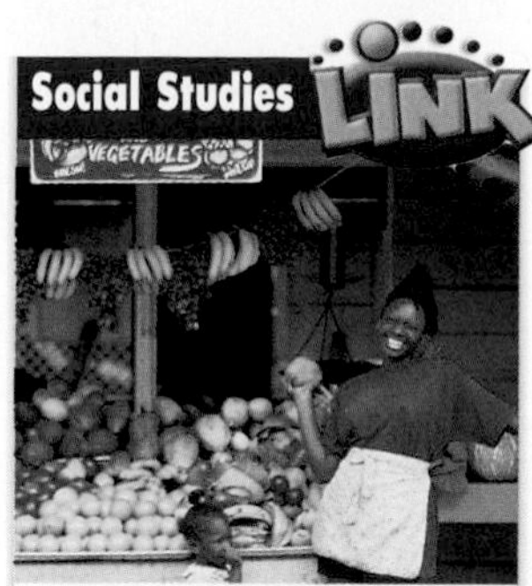

Native fruits of the Bahamas include guavas, pineapples, coconuts, bananas, and plantains.

go.hrw.com
KEYWORD: MS4 Bahamas

21. ***SOCIAL STUDIES*** Name the polygons made by each color in the flag of the Bahamas. Give the specific names of any quadrilaterals you find.

22. Describe how to construct a parallelogram from the figure at right, then complete the construction.

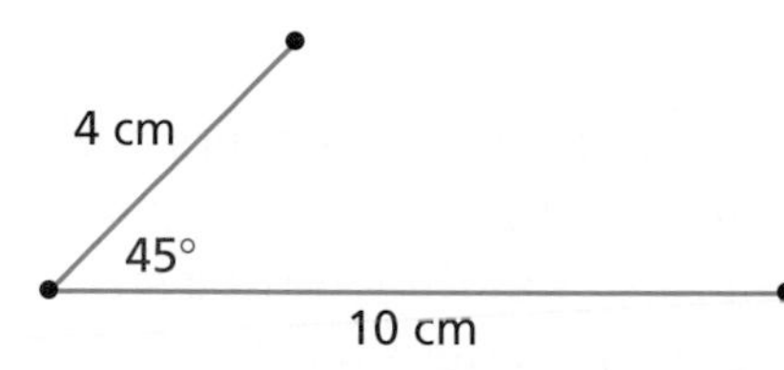

23. Bandon Highway is being built perpendicular to Avenue A and Avenue B, which are parallel. What kinds of polygons could be made by adding a fourth road?

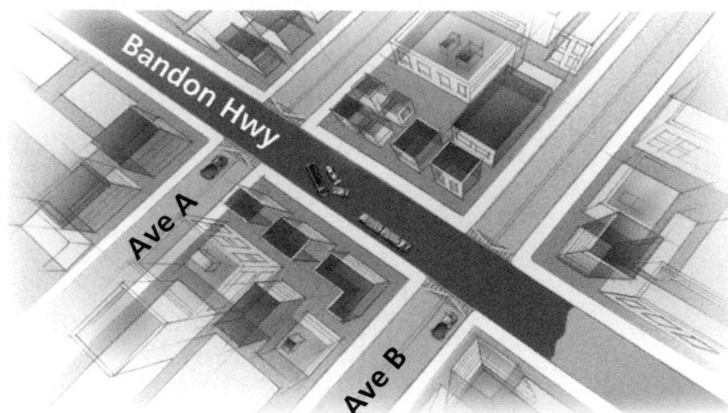

24. ***WRITE A PROBLEM*** Draw a design, or find one in a book, and then write a problem about the design that involves identifying quadrilaterals.

25. ***WRITE ABOUT IT*** Compare a kite with a parallelogram.

26. ***CHALLENGE*** The diagonals of a parallelogram bisect each other, and the diagonals of a rectangle are congruent. Given this information, what type of quadrilateral is figure *BCFD* if *ABFE* is a rectangle? Be specific, and explain your answer.

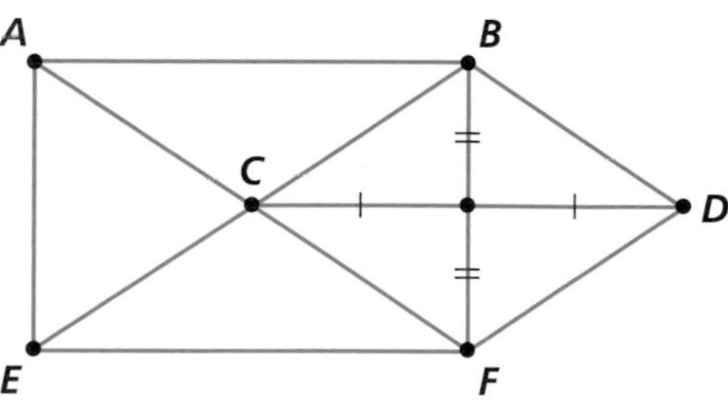

Spiral Review

Find the GCF of each set of numbers. (Lesson 2-5)

27. 16, 32 **28.** 8, 28 **29.** 15, 35 **30.** 19, 21

Find each sum or difference. (Lessons 3-3, 3-4)

31. $5 + (-8)$ **32.** $-9 - 2$ **33.** $17 - 22$ **34.** $-4 + 6$

35. **EOG PREP** What is the solution to $45 \div 0.009$? (Lesson 4-5)

A 5,000 B 500 C 50 D 5

7-8 Angles in Polygons

Learn to find the measures of angles in polygons.

Vocabulary
diagonal

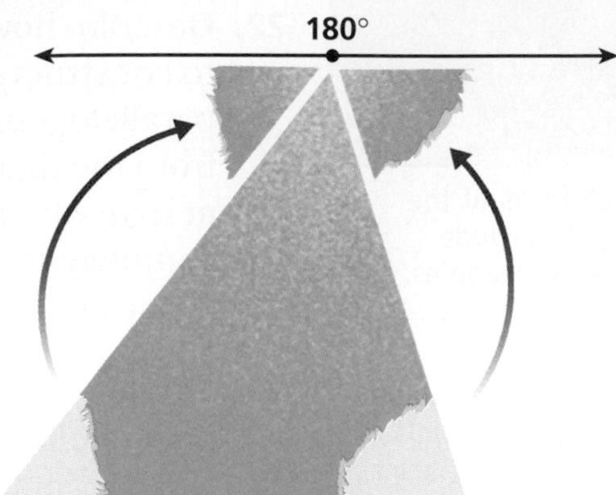

If you tear off the corners of a triangle and put them together, you will find that they form a straight angle. This suggests that the sum of the measures of the angles in a triangle is 180°.

You can prove mathematically that the angle measures in a triangle add up to 180° by drawing a diagram using the following steps.

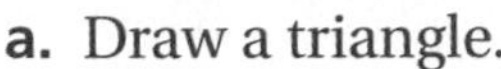

a. Draw a triangle.

b. Extend the sides of the triangle.

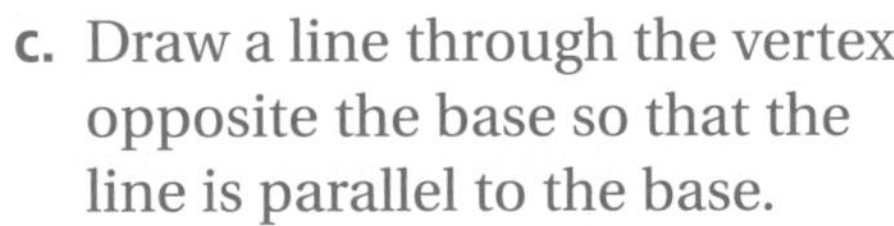

c. Draw a line through the vertex opposite the base so that the line is parallel to the base.

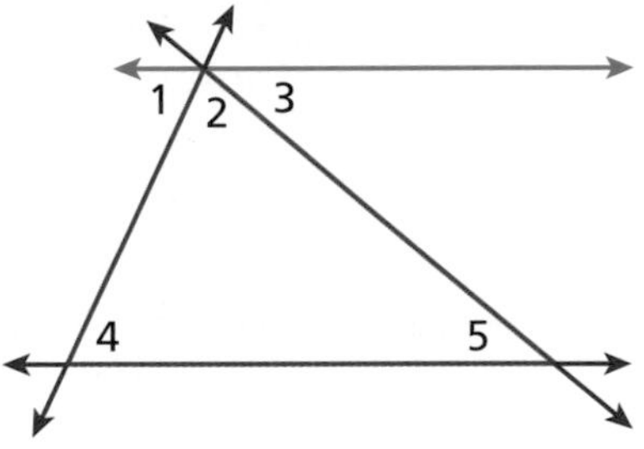

Notice that ∠1, ∠2, and ∠3 together form a straight angle. That is, the sum of their measures is 180°.

Notice also that the figure you have drawn consists of two parallel lines cut by two transversals. So if you were to tear off ∠4 and ∠5 from the triangle, they would fit exactly over ∠1 and ∠3. This shows that the sum of the measures of the angles in the triangle is 180°.

EXAMPLE 1 Determining the Measure of an Unknown Interior Angle

Find the measure of the unknown angle.

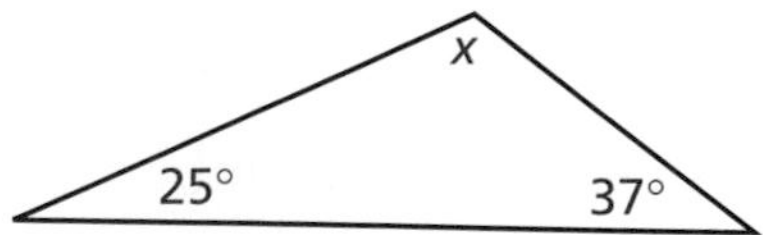

$25° + 37° + x = 180°$ *The sum of the measures of the angles is 180°.*

$62° + x = 180°$ *Combine like terms.*

$\underline{-62°} \quad \underline{-62°}$ *Subtract 62° from both sides.*

$x = 118°$

The measure of the unknown angle is 118°.

The sum of the angle measures in other polygons can be found by dividing the polygons into triangles. A polygon can be divided into triangles by drawing all of the *diagonals* from one of its vertices.

A **diagonal** of a polygon is a segment that is drawn from one vertex to another and is not one of the sides of the polygon. You can divide a polygon into triangles by using diagonals only if all of the diagonals of that polygon are inside the polygon. The sum of the angle measures in the polygon is then found by combining the sums of the angle measures in the triangles.

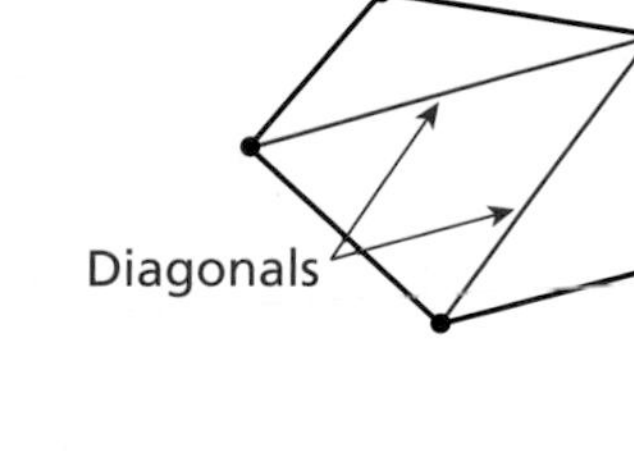

Number of triangles in pentagon		Sum of angle measures in each triangle		Sum of angle measures in pentagon
3	•	180°	=	540°

EXAMPLE 2 **Drawing Triangles to Find the Sum of Interior Angles**

Divide each polygon into triangles to find the sum of its angle measures.

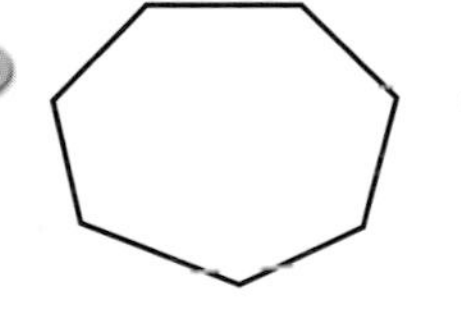

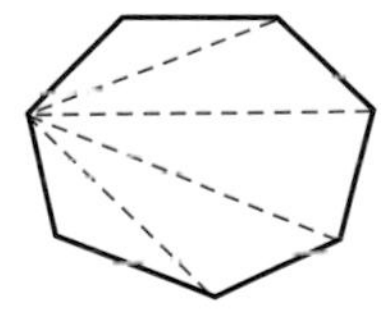

There are 5 triangles.

$5 \cdot 180° = 900°$

The sum of the angle measures of a heptagon is 900°.

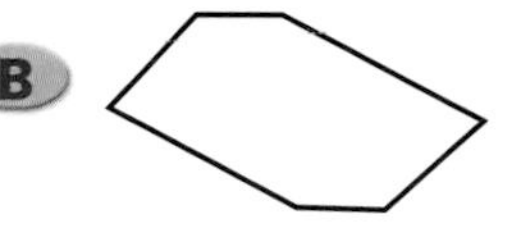

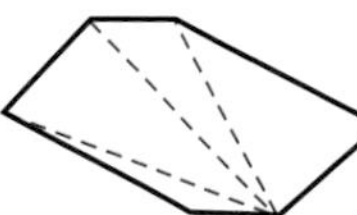

There are 4 triangles.

$4 \cdot 180° = 720°$

The sum of the angle measures of a hexagon is 720°.

Think and Discuss

1. **Explain** how to find the measure of an angle in a triangle when the measures of the two other angles are known.
2. **Determine** for which polygon the sum of the angle measures is greater, a pentagon or an octagon.
3. **Explain** how the measure of each angle in a regular polygon changes as the number of sides increases.

7-8 Exercises

FOR EOG PRACTICE
see page 685

internet connect
Homework Help Online
go.hrw.com Keyword: MS4 7-8

GUIDED PRACTICE

See Example 1 Find the measure of each unknown angle.

1.
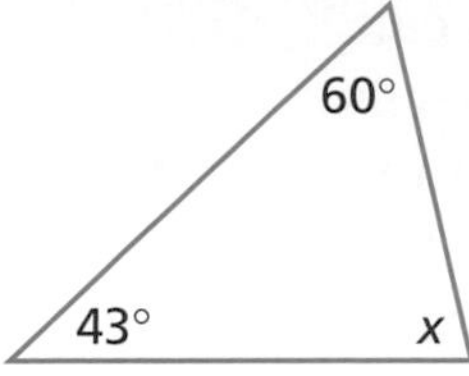

2.
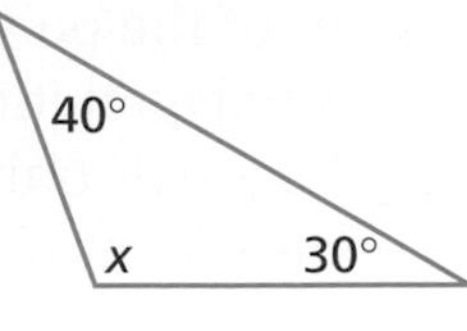

3.
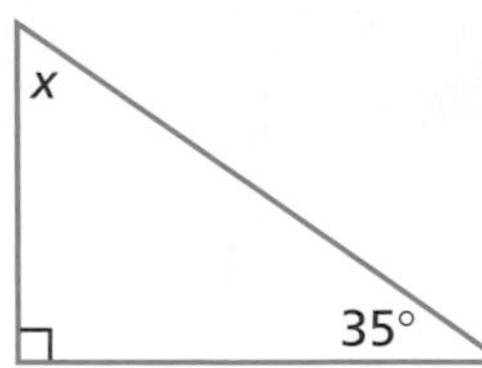

See Example 2 Divide each polygon into triangles to find the sum of its angle measures.

4.
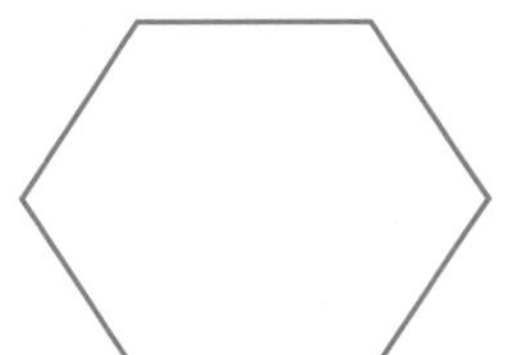

5.
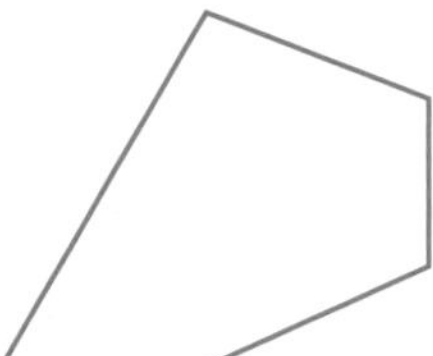

6.

INDEPENDENT PRACTICE

See Example  Find the measure of each unknown angle.

7.

8.
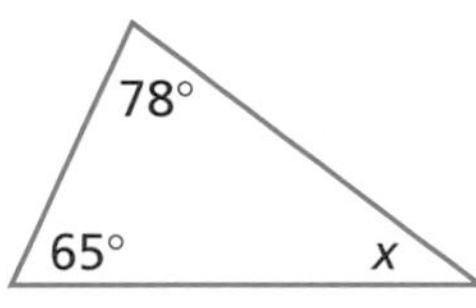

9.
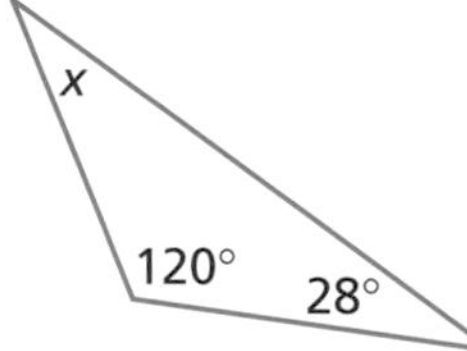

See Example 2 Divide each polygon into triangles to find the sum of its angle measures.

10.
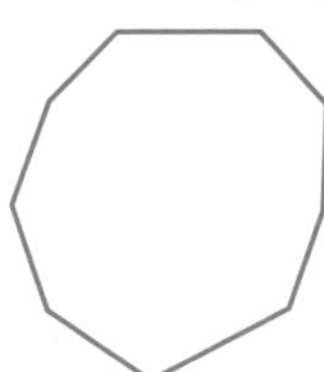

11.
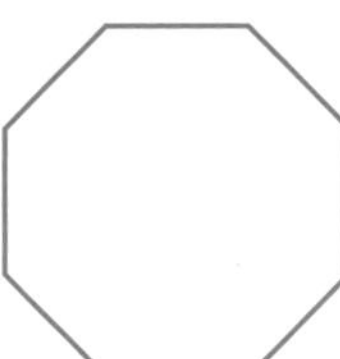

12.
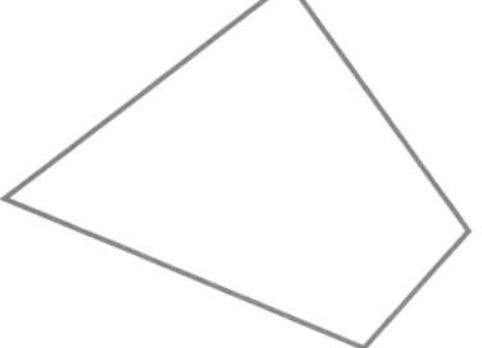

PRACTICE AND PROBLEM SOLVING

Find the measure of the third angle in each triangle, given two angle measures.

13. 56°, 101° **14.** 18°, 63° **15.** 62°, 58°

Decide whether the angle measures can form a triangle. If they can, classify the triangle.

16. 54°, 68°, 58° **17.** 120°, 59°, 30° **18.** 110°, 30°, 40°

19. ***ARCHITECTURE*** Each outer wall of the Pentagon in Washington, D.C., measures 921 feet.

a. What is the sum of the angle measures in the shape made by the Pentagon's outer walls?

b. What is the measure of each angle in this shape?

20. ***EARTH SCIENCE*** A sundial consists of a circular base and a right triangle mounted upright on the base. If one acute angle in the right triangle is 52°, what is the measure of the other acute angle?

21. ***NAVIGATION*** The angle between the lines of sight from a lighthouse to a tugboat and to a cargo ship is 27°. The angle between the lines of sight at the cargo ship is twice the angle between the lines of sight at the tugboat. What are the angles at the tugboat and at the cargo ship?

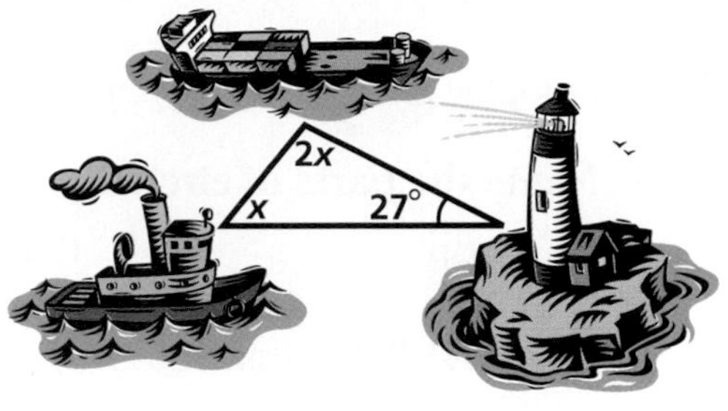

22. ***CONSTRUCTION*** A truss bridge is supported by frames called trusses. Each truss consists of steel or wooden parts connected to form triangles. If every triangle in a truss bridge is an isosceles right triangle, what is the measure of each angle in one of the triangles?

23. ***WHAT'S THE ERROR?*** If you try to find the sum of the angle measures in an octagon by multiplying $7 \cdot 180°$, what is your error?

24. ***WRITE ABOUT IT*** Describe how you would find the sum of the angle measures in a polygon with diagonals that are all inside the polygon.

25. ***CHALLENGE*** Write a formula for the measure of each angle in a regular polygon with n sides.

Spiral Review

Use the graph for Exercises 26–28. (Lesson 1-4)

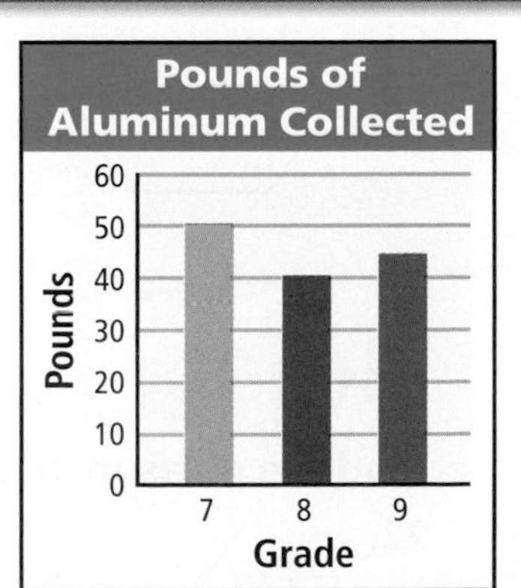

26. Which grade collected the most aluminum?

27. About how many more pounds of aluminum did the ninth grade collect than the eighth grade?

28. About how many pounds were collected by all three grades?

Find each sum. (Lesson 4-2)

29. 4.2 + 3.68

30. 16.8 + 7.02

31. 431.025 + 14.86

32. **EOG PREP** What is 0.4% written as a fraction? (Lesson 6-1)

A 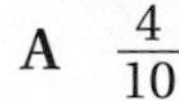 $\frac{4}{10}$

B 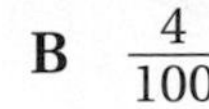 $\frac{4}{100}$

C $\frac{4}{1,000}$

D 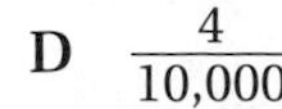 $\frac{4}{10,000}$

Chapter 7

Mid-Chapter Quiz

LESSONS 7-1, 7-2, AND 7-3 (pp. 344–351, and 354–357)

Identify the figures in the diagram.

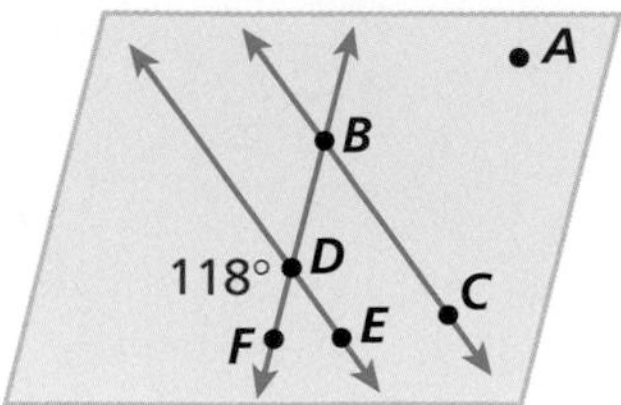

1. a plane
2. three line segments
3. three rays

Line *BC* || line *DE*. Find the measure of each angle, and tell whether the angle is acute, right, obtuse, or straight.

4. $\angle BDE$
5. $\angle CBD$
6. $\angle FDB$

LESSON 7-4 (pp. 362–365)

Name the parts of circle *B*.

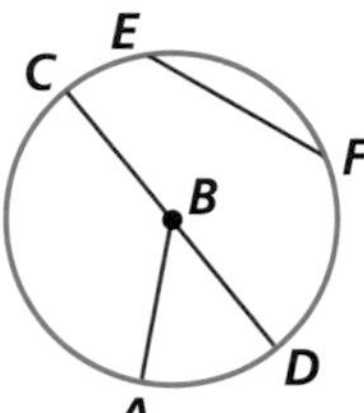

7. radii
8. diameter
9. chords

LESSON 7-5 (pp. 368–371)

Determine whether each figure is a polygon. If it is, name it. If it is not, explain why not.

10. 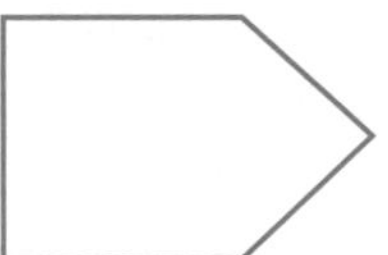

11. 

12. 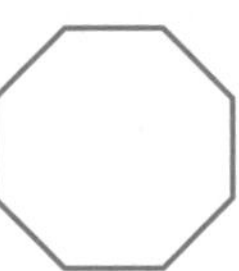

13. 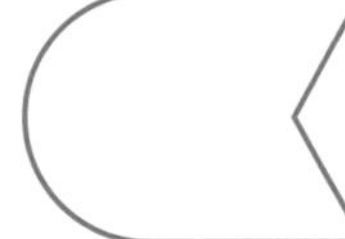

LESSON 7-6 (pp. 374–377)

Classify each triangle according to its sides and angles.

14. 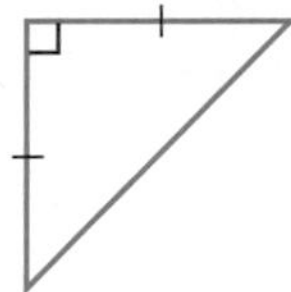

15. 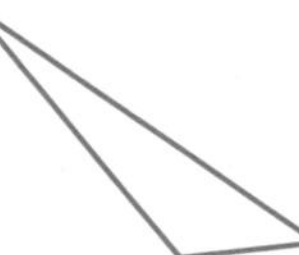

16. 

17. 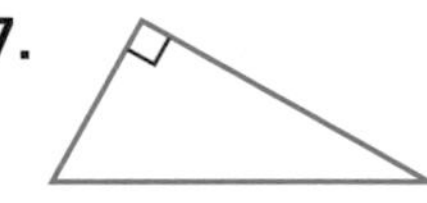

LESSON 7-7 (pp. 378–381)

Give all of the names that apply to each quadrilateral.

18. 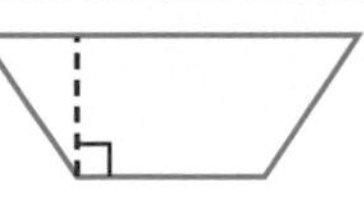

19. 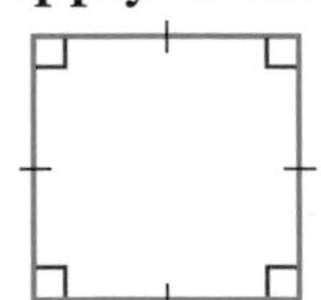

20. 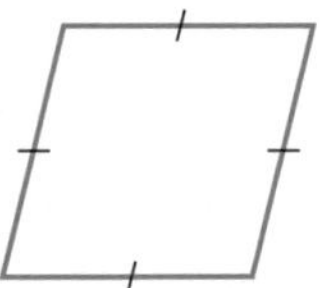

21. 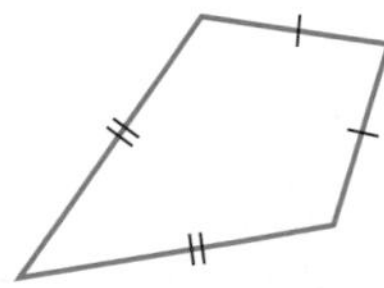

LESSON 7-8 (pp. 382–385)

Find the measure of each unknown angle.

22. 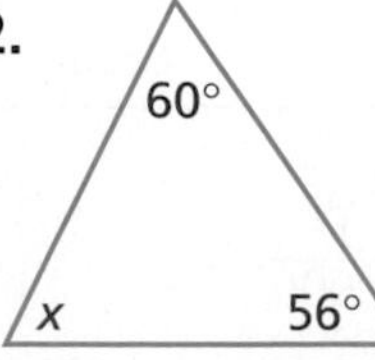

23. 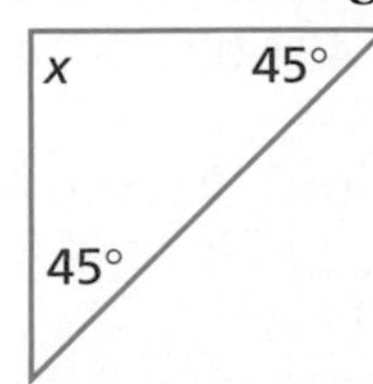

24. 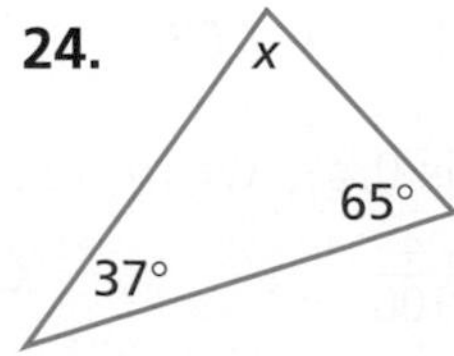

25. 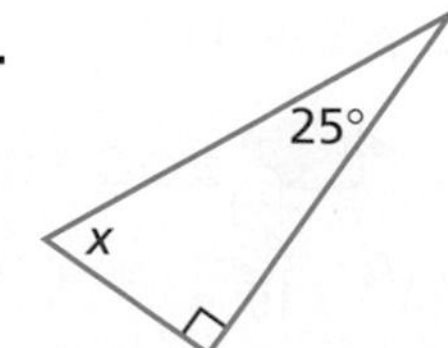

Mid-Chapter Quiz

Focus on Problem Solving

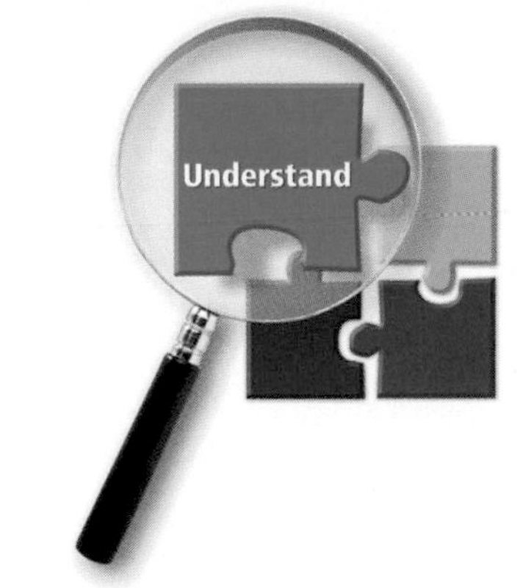

Understand the Problem

• Understand the words in the problem

Words that you do not understand can sometimes make a simple problem seem difficult. Some of those words, such as the names of things or persons, may not even be necessary to solve the problem. If a problem contains an unfamiliar name, or one that you cannot pronounce, you can substitute another word for it. If a word that you don't understand is necessary to solve the problem, look the word up to find its meaning.

Read each problem, and make a list of unusual or unfamiliar words. If a word is not necessary to solve the problem, replace it with a familiar one. If a word is necessary, look up the word and write its meaning.

1. Using a pair of calipers, Mr. Papadimitriou measures the diameter of an ancient Greek amphora to be 17.8 cm at its widest point. What is the radius of the amphora at this point?

2. Joseph wants to plant gloxinia and hydrangeas in two similar rectangular gardens. The length of one garden is 5 ft, and the width is 4 ft. The other garden's length is 20 ft. What is the width of the second garden?

3. Mr. Manityche is sailing his catamaran from Kaua'i to Ni'ihau, a distance of about 12 nautical miles. If his speed averages 10 knots, how long will the trip take him?

4. Aimee's lepidoptera collection includes a butterfly with dots that appear to form a scalene triangle on each wing. What is the sum of the angles of each triangle on the butterfly's wings?

5. Students in a physics class use wire and resistors to build a Wheatstone bridge. Each side of their rhombus-shaped design is 2 cm long. What angle measures would the design have to have for its shape to be a square?

7-9 Congruent Figures

Learn to identify congruent figures and to use congruence to solve problems.

Vocabulary

Side-Side-Side Rule

Look at the two patterns. Which center circle do you think is bigger? In spite of appearances, the two center circles are congruent. Their apparent differences are optical illusions. One way to determine whether figures are congruent is to see if one figure will fit exactly over the other one.

EXAMPLE 1 Identifying Congruent Figures in the Real World

Identify any congruent figures.

A

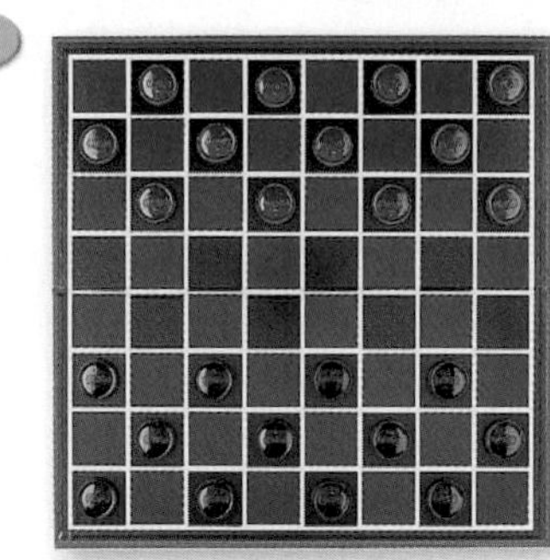

The squares on a checkerboard are congruent. The checkers are also congruent.

B

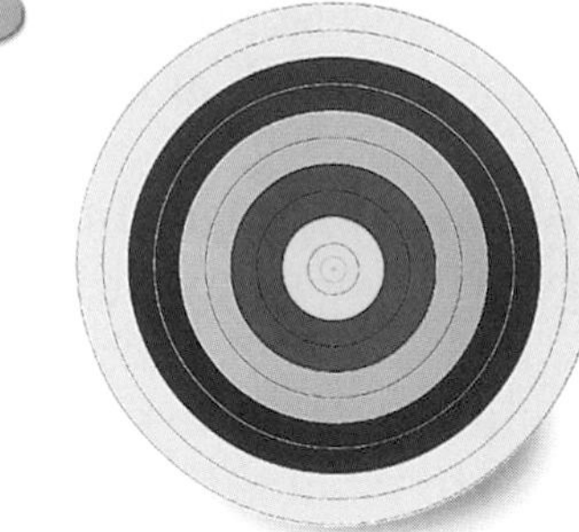

The rings on a target are not congruent. Each ring is larger than the one inside of it.

If all of the corresponding sides and angles of two polygons are congruent, then the polygons are congruent. For triangles in particular, the corresponding angles will always be congruent if the corresponding sides are congruent. This is called the **Side-Side-Side Rule**. Because of this rule, when determining whether triangles are congruent, you only need to determine whether the sides are congruent.

EXAMPLE 2 Identifying Congruent Triangles

Determine whether the triangles are congruent.

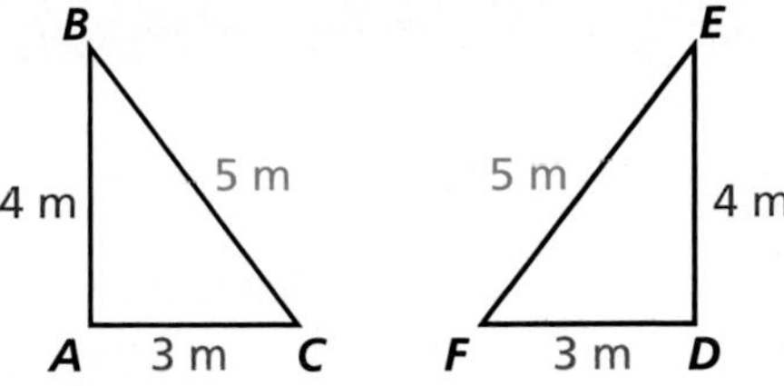

$AC = 3$ m $DF = 3$ m
$AB = 4$ m $DE = 4$ m
$BC = 5$ m $EF = 5$ m

By the Side-Side-Side Rule, $\triangle ABC$ is congruent to $\triangle DEF$, or $\triangle ABC \cong \triangle DEF$. If you flip one triangle, it will fit exactly over the other.

Reading Math

The notation $\triangle ABC$ is read "triangle *ABC*."

For polygons with more than three sides, it is not enough to compare the measures of their sides. For example, the corresponding sides of the figures below are congruent, but the figures are not congruent.

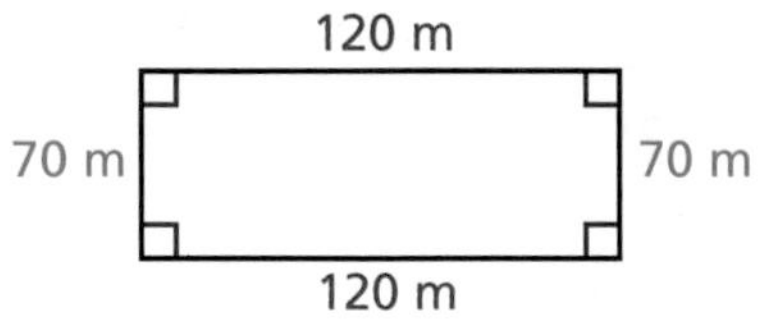

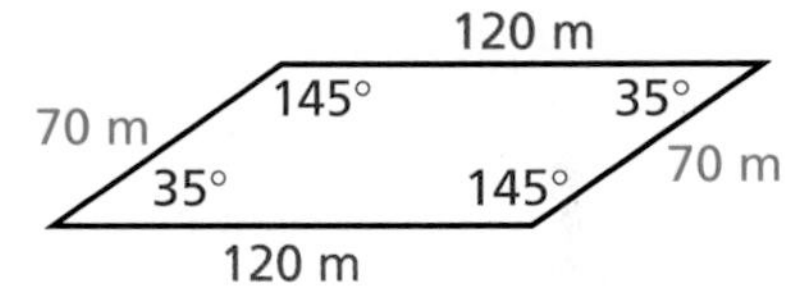

If you know that two figures are congruent, you can find missing measures in the figures.

EXAMPLE 3 Using Congruence to Find Missing Measures

Determine the missing measure in each set of congruent polygons.

A

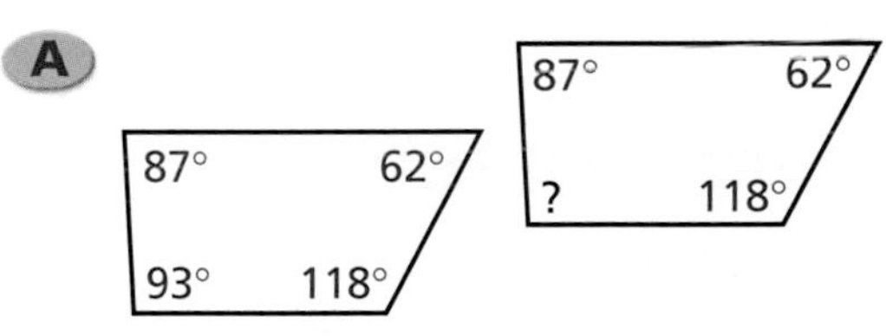

The corresponding angles of congruent polygons are congruent.

The missing angle measure is 93°.

B

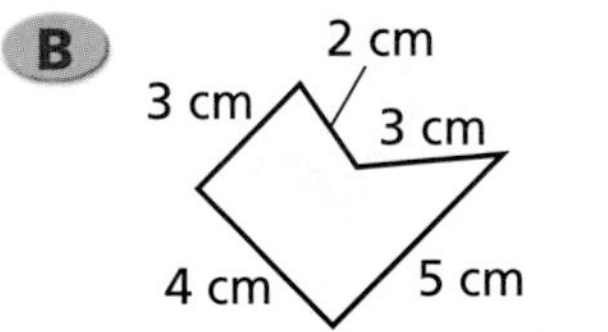

2 cm ?
3 cm
5 cm
4 cm

The corresponding sides of congruent polygons are congruent.

The missing side measure is 3 cm.

Think and Discuss

1. **Draw** an illustration to explain whether an isosceles triangle can be congruent to a right triangle.
2. **Explain** whether an isosceles triangle can be congruent to a scalene triangle.

7-9 Exercises

FOR EOG PRACTICE
see page 686

internet connect
Homework Help Online
go.hrw.com Keyword: MS4 7-9

3.02

GUIDED PRACTICE

See Example 1 **Identify any congruent figures.**

1. 

2. 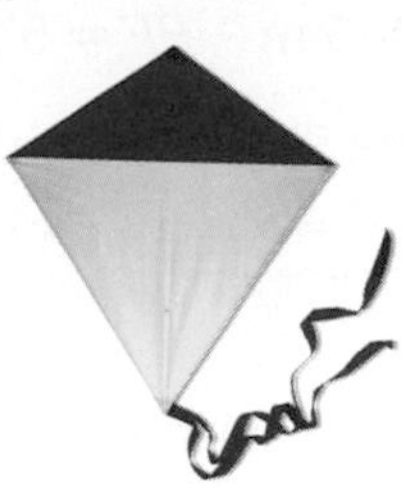

3. 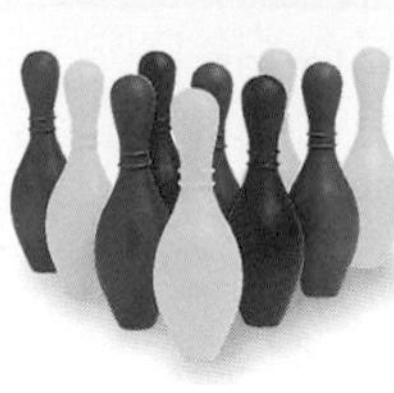

See Example 2 **Determine whether the triangles are congruent.**

4. 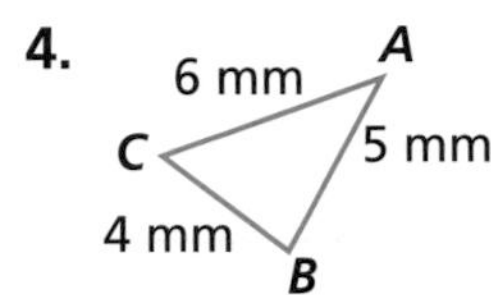

5. 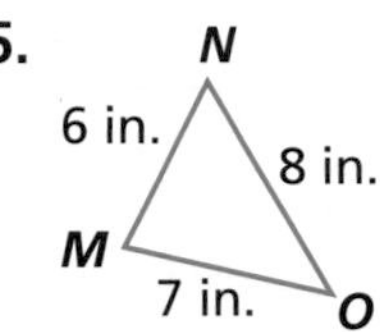

See Example 3 **Determine the missing measure in each set of congruent polygons.**

6. 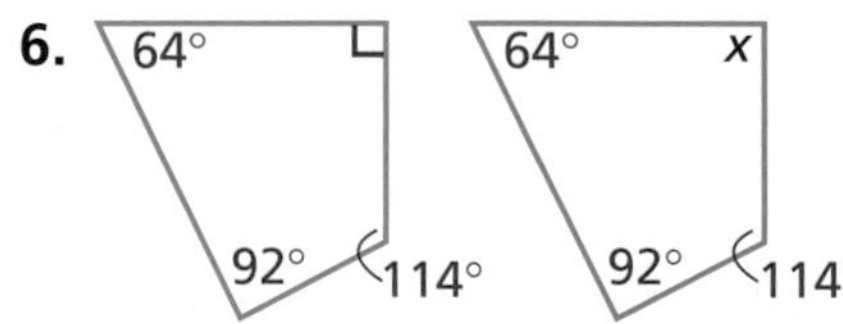

7. 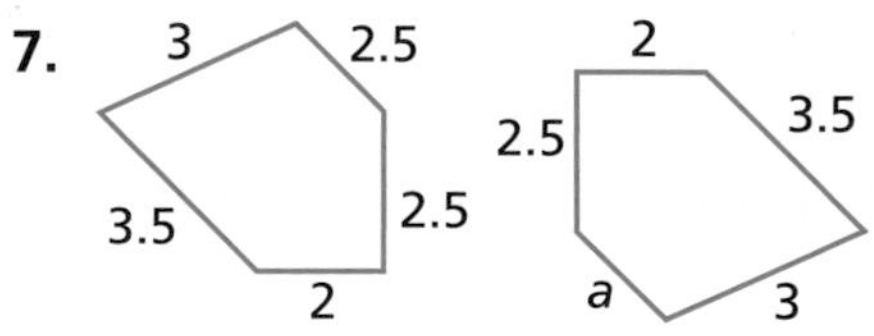

INDEPENDENT PRACTICE

See Example 1 **Identify any congruent figures.**

8. 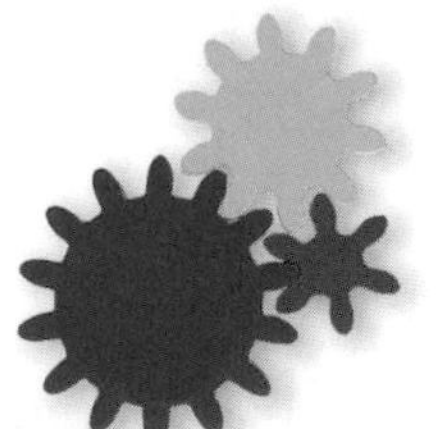

9. 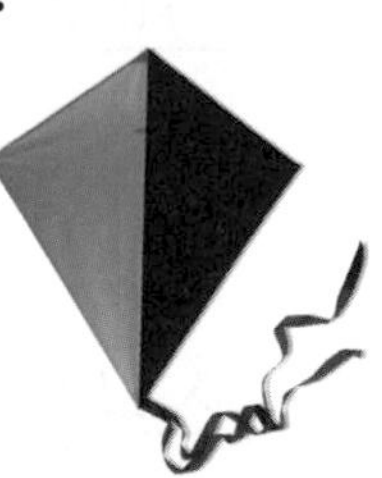

10. 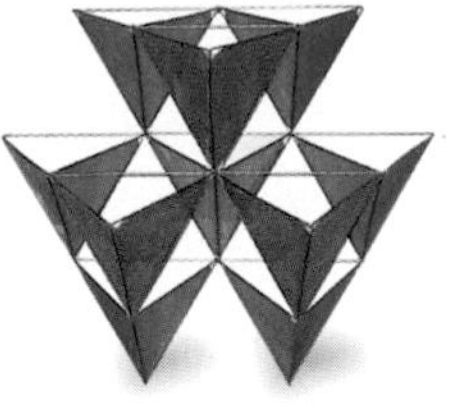

See Example 2 **Determine whether the triangles are congruent.**

11. 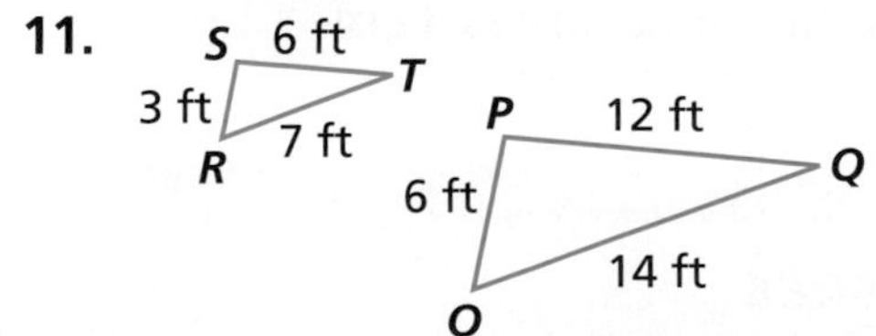

12. 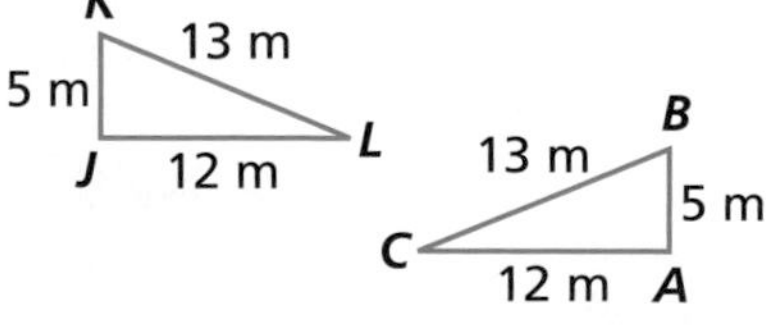

See Example 3 **Determine the missing measures in each set of congruent polygons.**

13. 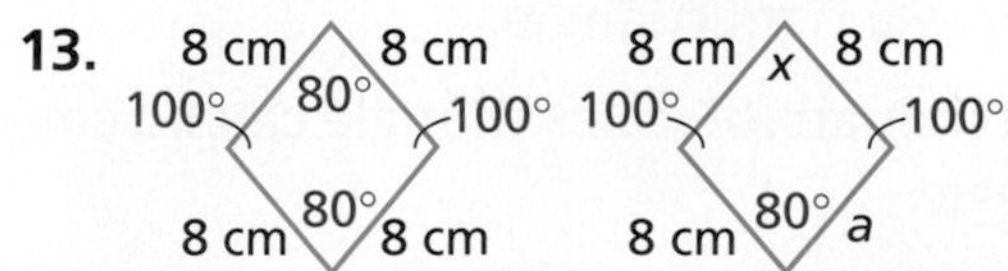

14. 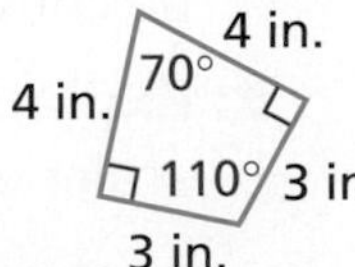

PRACTICE AND PROBLEM SOLVING

Tell the minimum amount of information needed to determine whether the figures are congruent.

15. two triangles **16.** two squares **17.** two rectangles **18.** two pentagons

19. The diagonals of rectangle *ABCD* bisect each other at point *E*. Explain why $\triangle CED \cong \triangle AEB$.

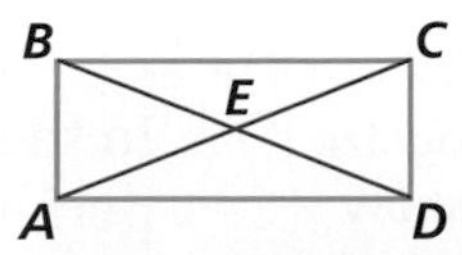

20. ***SURVEYING*** In the figure, trees *A* and *B* are on opposite sides of the stream. Jamil wants to string a rope from one tree to the other. If triangles *ABC* and *DEC* are congruent, what is the distance between the trees?

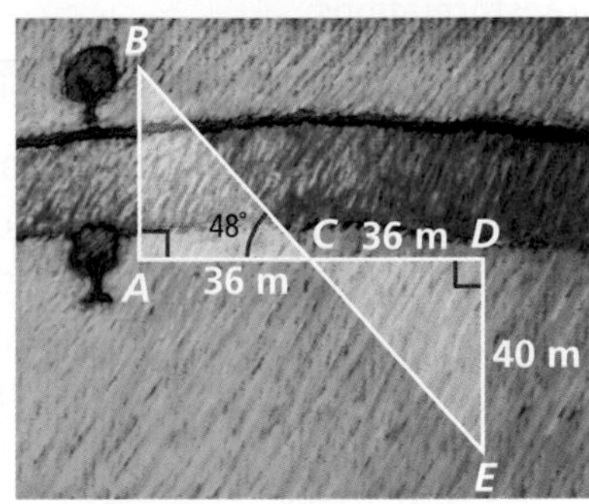

21. ***HOBBIES*** In the quilt block, which figures appear congruent?

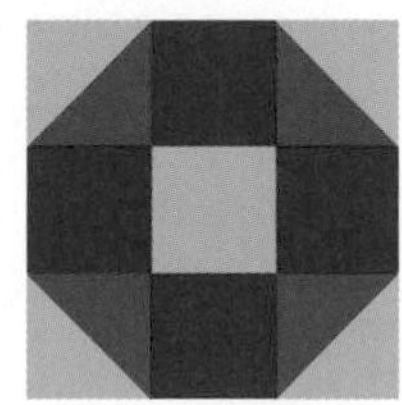

22. ***CHOOSE A STRATEGY*** Brittany and her brother Art walk to school every day along the routes in the figure. They start at the same time and walk at the same rate. Who gets to school first?

A Brittany

B Art

C They arrive at the same time.

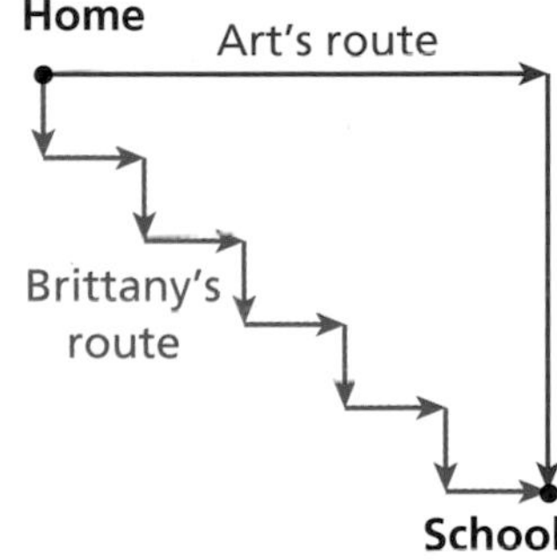

23. ***WRITE ABOUT IT*** Explain how you can determine whether two triangles are congruent.

24. ***CHALLENGE*** If all of the angles in two triangles have the same measure, are the triangles necessarily congruent?

Spiral Review

Round each decimal to the nearest tenth. (Lesson 4-1)

25. 8.032 **26.** 0.985 **27.** 1.246 **28.** 18.872

Find each value. (Lesson 6-4)

29. 28 is what percent of 40? **30.** 30% of what number is 12? **31.** 8% of what number is 4?

32. **EOG PREP** An angle that measures 148° is what kind of angle? (Lesson 7-2)

A Obtuse B Right C Congruent D Acute

7-10 Transformations

Learn to recognize, describe, and show transformations.

Vocabulary

transformation
image
translation
rotation
reflection
line of reflection

In the photograph, Kristi Yamaguchi is performing a *layback spin.* She is holding her body in one position while she rotates. This is an example of a *transformation.*

In mathematics, a **transformation** changes the position or orientation of a figure. The resulting figure is the **image** of the original. Images resulting from the transformations described below are congruent to the original figures.

Types of Transformations

Translation
The figure slides along a straight line without turning.

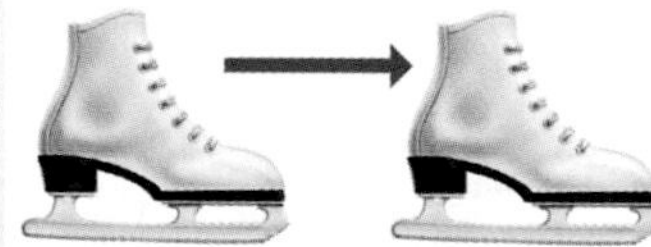

Rotation
The figure turns around a fixed point.

Reflection
The figure flips across a **line of reflection**, creating a mirror image.

EXAMPLE 1 Identifying Types of Transformations

Helpful Hint

The point that a figure rotates around may be on the figure or away from the figure.

Identify each type of transformation.

A

Translation

B

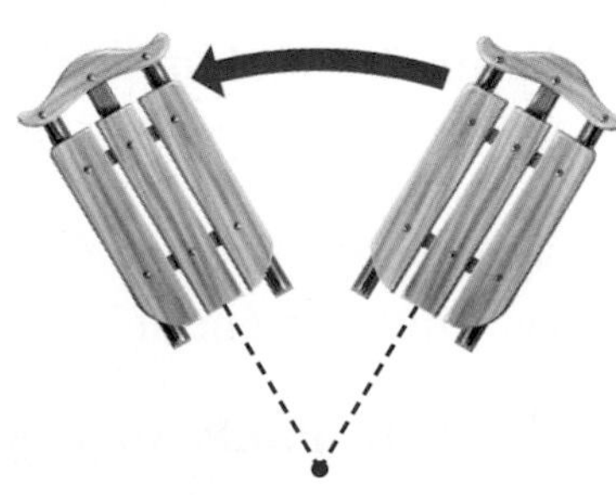

Rotation

EXAMPLE 2 Graphing Transformations on a Coordinate Plane

Graph each transformation.

Reading Math

A′ is read "A prime" and is used to represent the point on the image that corresponds to point *A* of the original figure.

A Translate $\triangle ABC$ 6 units right and 4 units down.

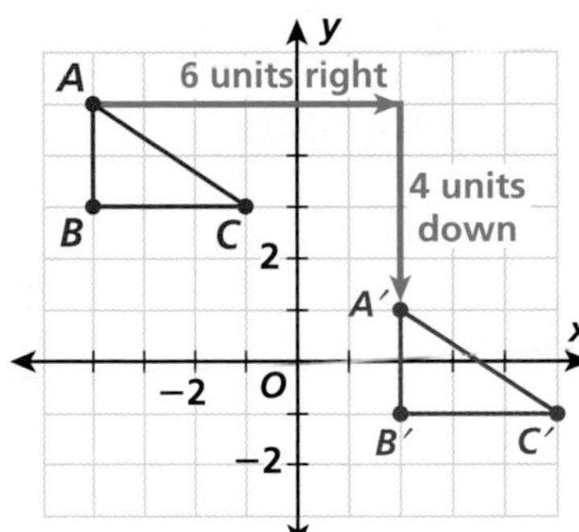

Each vertex is moved 6 units right and 4 units down.

B Rotate $\triangle JKL$ 90° counterclockwise around the vertex *J*.

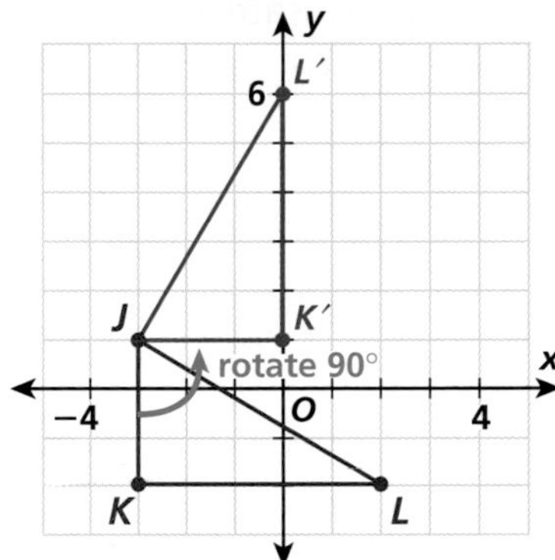

The corresponding sides, $\overline{JK}$ and $\overline{JK'}$, make a 90° angle.

Notice that vertex K is 3 units below vertex J, and vertex K′ is 3 units to the right of vertex J.

C Reflect the figure across the *y*-axis.

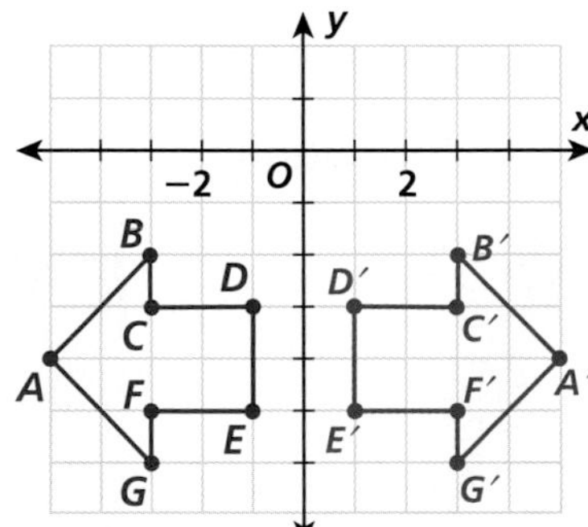

The y-coordinates of the corresponding vertices are the same, and the x-coordinates of the corresponding vertices are opposites.

Think and Discuss

1. **Describe** a classroom situation that illustrates a translation.
2. **Explain** how a figure skater might perform a translation and a rotation at the same time.
3. **Draw** a rectangle, and place a dot in one corner. Show the rectangle after a translation, a reflection, and a rotation.

7-10 Exercises

FOR EOG PRACTICE
see page 686

internet connect
Homework Help Online
go.hrw.com Keyword: MS4 7-10

GUIDED PRACTICE

See Example 1 **Identify each type of transformation.**

1.

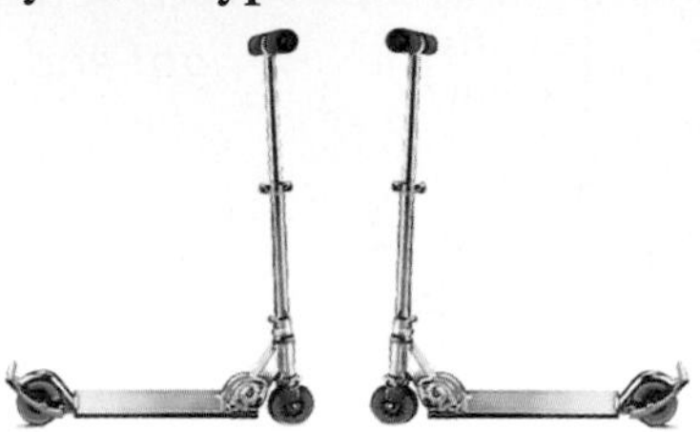

2.

See Example 2 **Graph each transformation.**

3. Translate $\triangle ABC$ 2 units left and 3 units up.

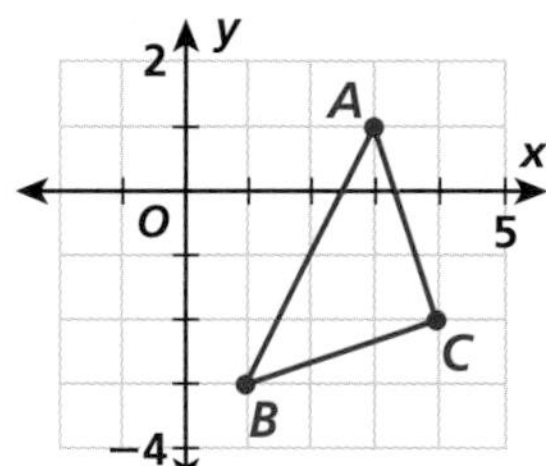

4. Rotate $\triangle LMN$ 180° around the vertex L.

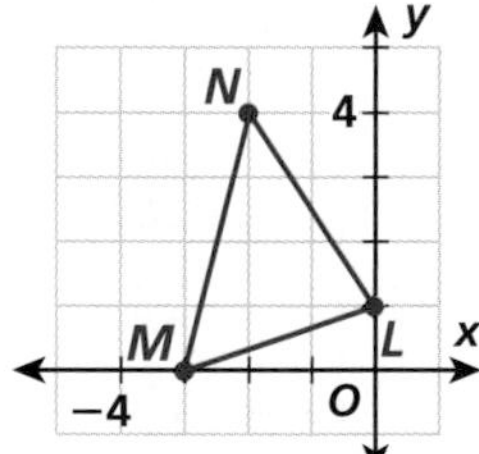

5. Reflect the figure across the x-axis.

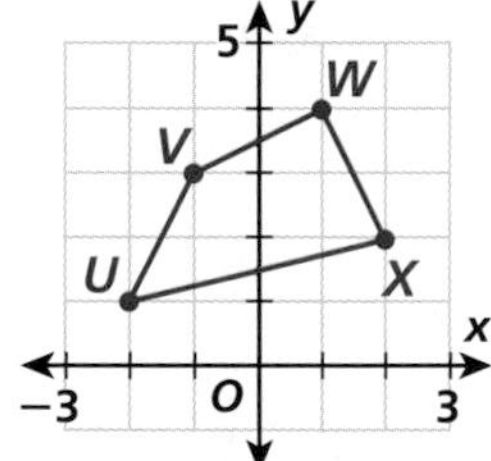

INDEPENDENT PRACTICE

See Example  **Identify each type of transformation.**

6.

7.

See Example  **Graph each transformation.**

8. Rotate $\triangle MNL$ 90° counterclockwise around the vertex L.

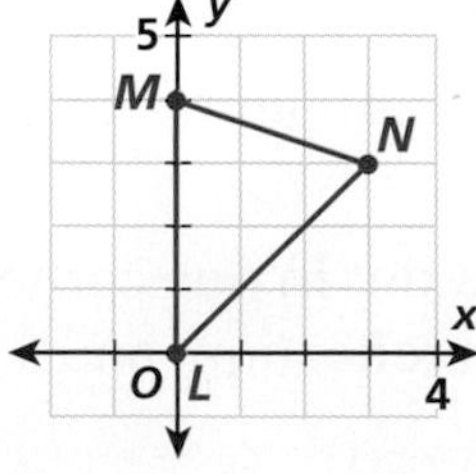

9. Translate $\triangle XYZ$ 5 units right and 1 unit down.

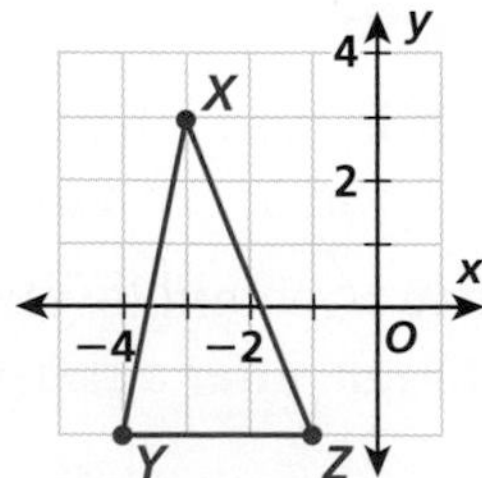

10. Reflect the figure across the y-axis.

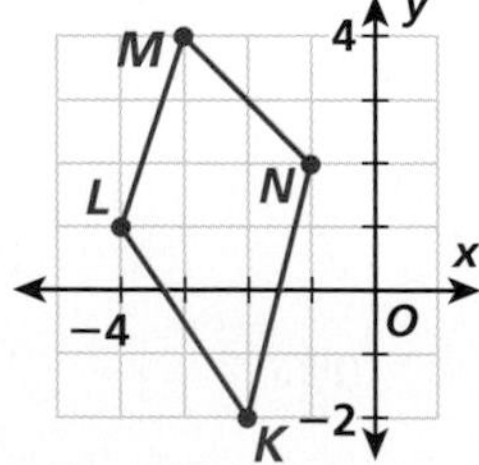

Social Studies LINK

Native Americans often use transformations in their art. The art pieces in the photos show combinations of transformations. Use the photos for Exercises 11–13.

11. The bowl at right was made by Maria Martinez, a Native American potter from the San Ildefonso pueblo in New Mexico. Are translations, rotations, and reflections all shown in the design? If not, which transformations are not shown?

12. ***WRITE ABOUT IT*** The Navajo blanket in the photo has a design based on a sand painting. The two people in the design are standing next to a stalk of corn, which the Native Americans called *maize*. The red, white, and black stripes represent a rainbow.

a. Tell how the figure shows reflections. Also explain what parts of the design do not show reflections.

b. A *glide reflection* is a combination of a reflection and a translation. Tell where the design shows a glide reflection.

13. ***CHALLENGE*** The bead design in the saddle bag at right shows different types of transformations. What part of the design can be described as three separate transformations? Draw diagrams to illustrate your answer.

Spiral Review

Find the mode of each data set. (Lesson 1-2)

14. 1, 3, 8, 2, 7, 1, 8, 4, 5

15. 6.2, 3.8, 4.1, 9.5, 6.3, 5.4

Solve each equation. (Lesson 2-12)

16. $2x = 8$

17. $\frac{y}{2} = 12$

18. $5m = 170$

19. **EOG PREP** What is the percent of increase from 120 to 300? (Lesson 6-5)

A 150%

B 40%

C 250%

D 60%

7-11 Symmetry

Learn to identify symmetry in figures.

Vocabulary
line symmetry
line of symmetry
asymmetry
rotational symmetry
center of rotation

Many architects and artists use symmetry in their buildings and artwork because symmetry is pleasing to the eye.

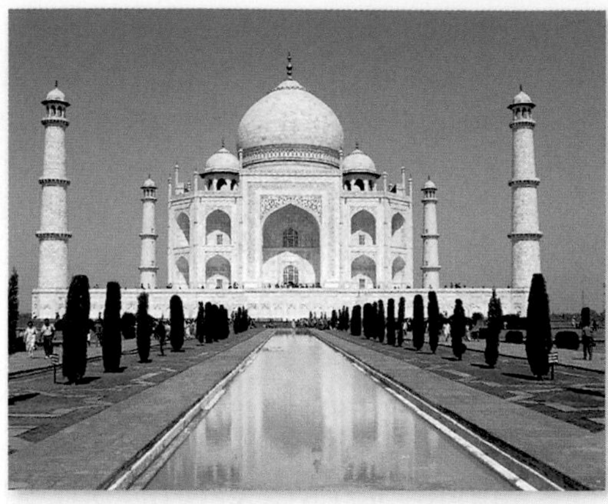

The Taj Mahal in Agra, India is an example of Mughal architecture.

When you can draw a line through a plane figure so that the two halves are mirror images of each other, the figure has **line symmetry**, or is symmetrical. The line along which the figure is divided is called the **line of symmetry**.

When a figure is not symmetrical, then it has **asymmetry**, or is asymmetrical.

The image of the Taj Mahal is symmetrical. You can draw a line of symmetry down the center of the building. Also, each window in the image has its own line of symmetry.

EXAMPLE 1 Identifying Line Symmetry

Decide whether each figure has line symmetry. If it does, draw all the lines of symmetry.

A

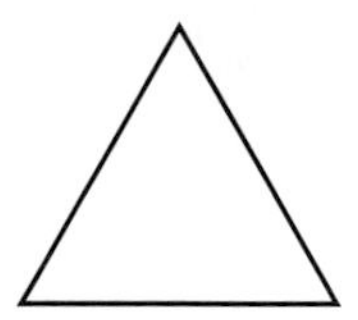

3 lines of symmetry

B

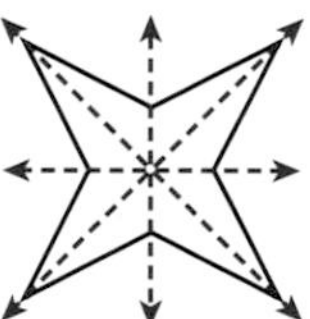

4 lines of symmetry

EXAMPLE 2 *Social Studies Application*

Find all the lines of symmetry in each flag.

A
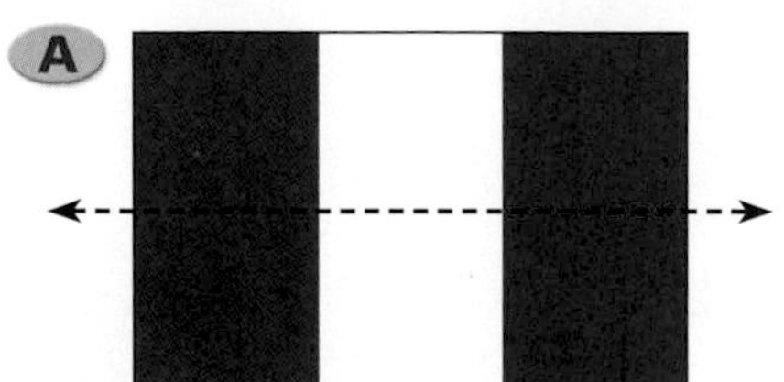

There is 1 line of symmetry.

B

There are no lines of symmetry.

A figure has **rotational symmetry** if, when it is rotated less than 360° around a central point, it coincides with itself. The central point is called the **center of rotation**.

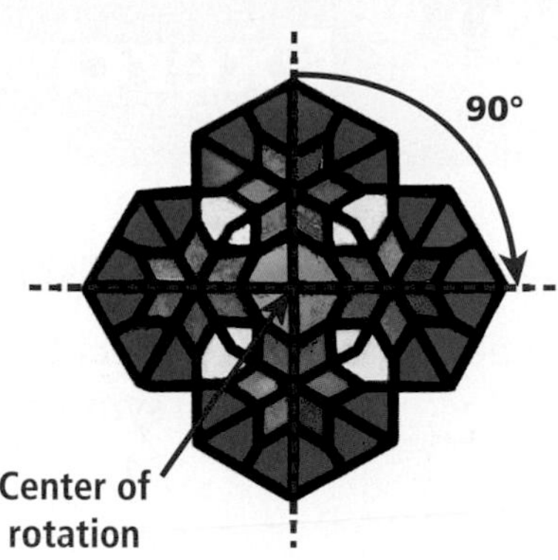

If the stained glass window at right is rotated 90°, as shown, the image looks the same as the original stained glass window. Therefore the window has rotational symmetry.

EXAMPLE 3 **Identifying Rotational Symmetry**

Tell how many times each figure will show rotational symmetry within one full rotation.

A

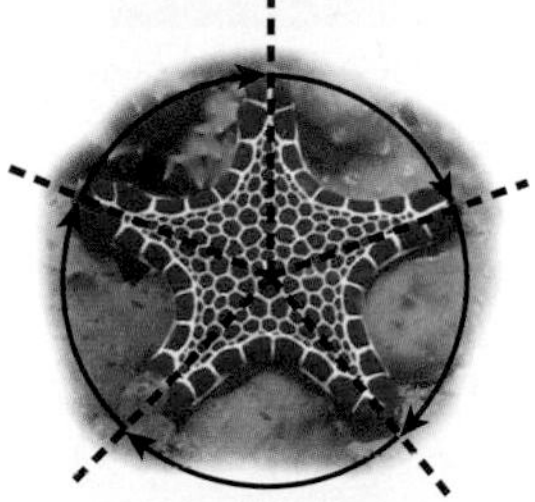

Draw lines from the center of the figure out through identical places in the figure.

The starfish will show rotational symmetry 5 times within a 360° rotation.

Count the number of lines drawn.

B

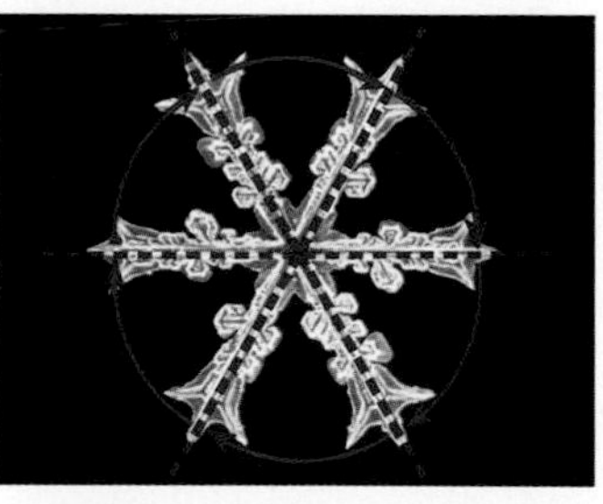

Draw lines from the center of the figure out through identical places in the figure.

The snowflake will show rotational symmetry 6 times within a 360° rotation.

Count the number of lines drawn.

Think and Discuss

1. **Draw** a figure that has no rotational symmetry.
2. **Determine** whether an equilateral triangle has rotational symmetry. If so, tell how many times it shows rotational symmetry within one full rotation.

7-11 Exercises

FOR EOG PRACTICE
see page 687

internet connect
Homework Help Online
go.hrw.com Keyword: MS4 7-11

GUIDED PRACTICE

See Example 1

Decide whether each figure has line symmetry. If it does, draw all the lines of symmetry.

1. 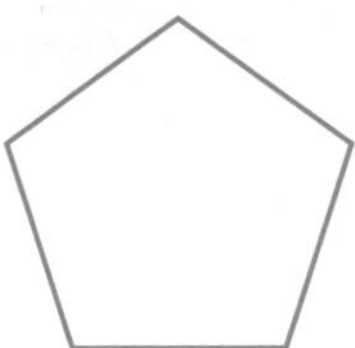

2. 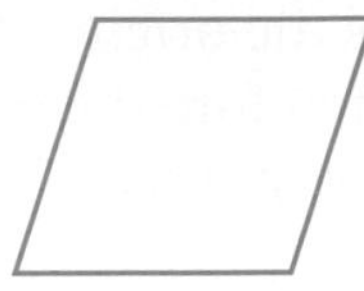

3. 

See Example 2

Find all the lines of symmetry in each flag.

4. 

5. 

6. 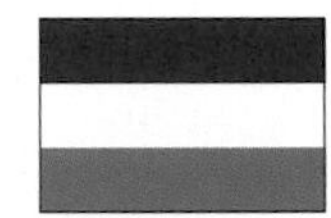

See Example 3

Tell how many times each figure will show rotational symmetry within one full rotation.

7. 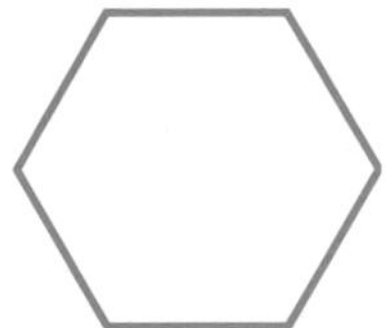

8. 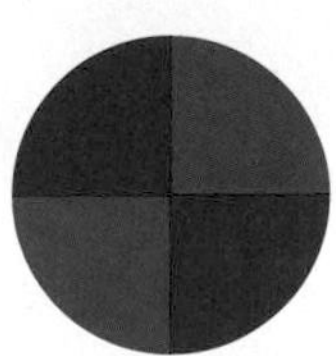

9. 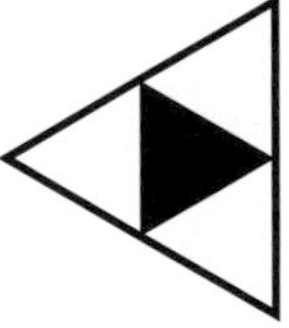

INDEPENDENT PRACTICE

See Example 1

Decide whether each figure has line symmetry. If it does, draw all the lines of symmetry.

10. 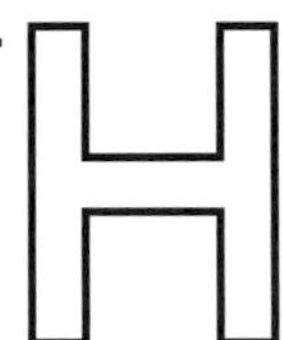

11. 

12. 

See Example 2

Find all the lines of symmetry in each flag.

13. 

14. 

15. 

See Example 3

Tell how many times each figure will show rotational symmetry within one full rotation.

16. 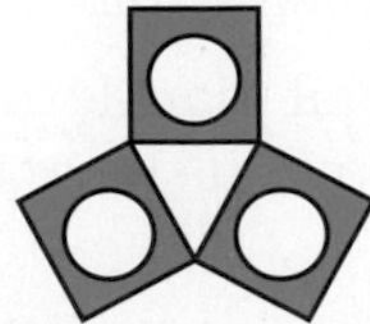

17. 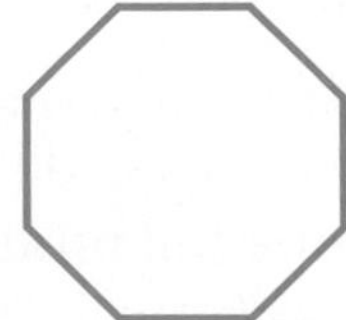

18. 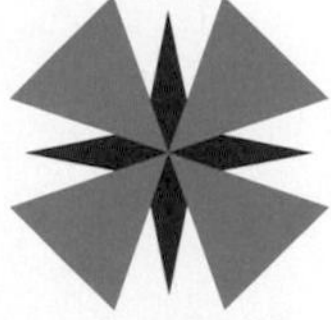

PRACTICE AND PROBLEM SOLVING

19. Draw a figure with at least two lines of symmetry.

20. Which regular polygon shows rotational symmetry 9 times within one full rotation?

21. ***LIFE SCIENCE*** How many lines of symmetry does the image of the moth have?

22. Fold a piece of paper in half vertically and then in half horizontally. Cut or tear a design into one of the folded edges. Then unfold the paper. Does the design have a vertical or horizontal line of symmetry? rotational symmetry? Explain your answer.

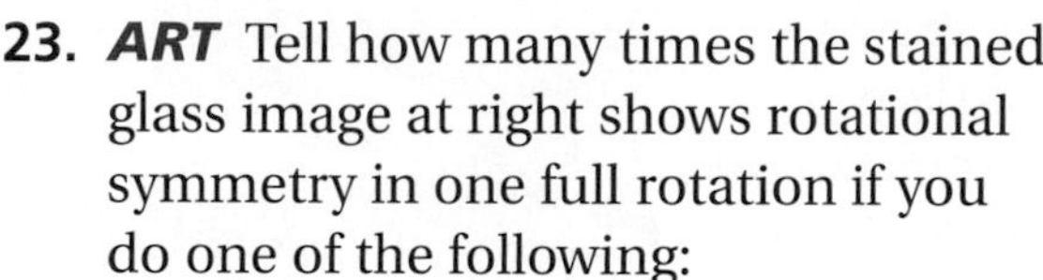

23. ***ART*** Tell how many times the stained glass image at right shows rotational symmetry in one full rotation if you do one of the following:
    - **a.** consider only the shape of the design and
    - **b.** consider both the shape and the colors in the design.

24. ***WHAT'S THE QUESTION?*** Marla drew a square on the chalkboard. As an answer to Marla's question about symmetry, Rob said "90°." What question did Marla ask?

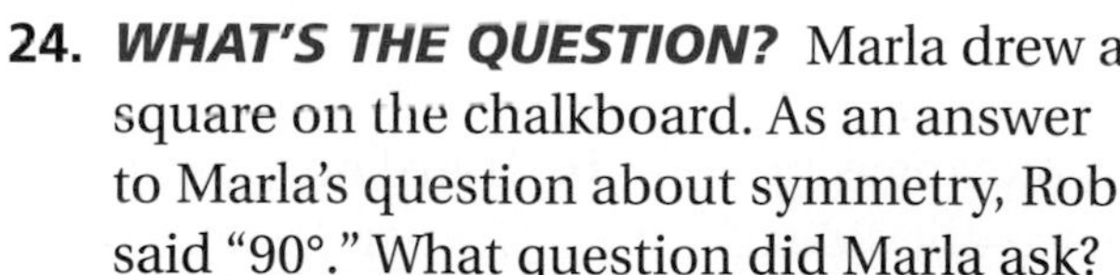

25. ***WRITE ABOUT IT*** Explain why an angle of rotation must be less than 360° for a figure to have rotational symmetry.

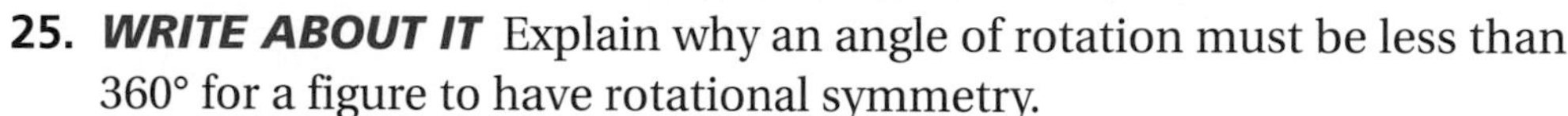

26. ***CHALLENGE*** Print a word in capital letters, using only letters that have horizontal lines of symmetry. Print another word with capital letters that have vertical lines of symmetry.

Spiral Review

Find each sum. (Lesson 4-10)

27. $\frac{3}{8} + \frac{3}{8}$ **28.** $\frac{5}{9} + \frac{2}{9}$ **29.** $\frac{7}{10} + \frac{3}{10}$ **30.** $\frac{5}{6} + \frac{5}{6}$

Solve each equation. (Lesson 4-12)

31. $\frac{2}{3}x = 12$ **32.** $y - \frac{3}{4} = \frac{3}{8}$ **33.** $\frac{5}{9} + m = \frac{2}{3}$

34. **EOG PREP** What is the interest after 6 months on $8,000 at 4.5% interest? (Lesson 6-6)

A $3,600 **B** $720 **C** $360 **D** $180

Create Tessellations

Use with Lesson 7-10

Tessellations are patterns of identical shapes that completely cover a plane with no gaps or overlaps. The artist M. C. Escher created many fascinating tessellations.

Activity

1 Create a translation tessellation.

The tessellation by M. C. Escher shown at right is an example of a *translation tessellation*. To create your own translation tessellation, follow the steps below.

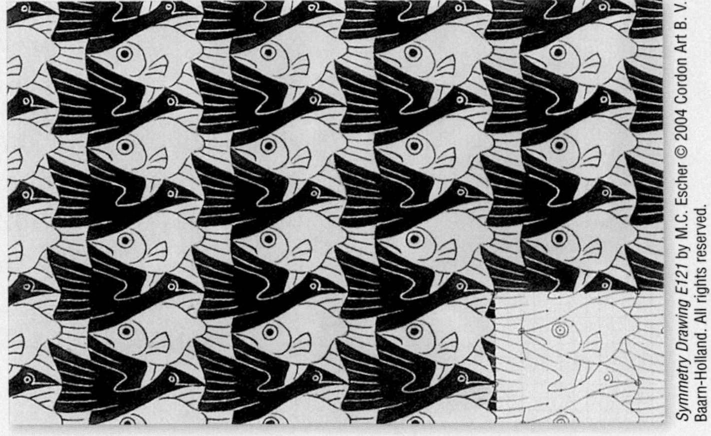

a. Start by drawing a square, rectangle, or other parallelogram. Replace one side of the parallelogram with a curve, as shown.

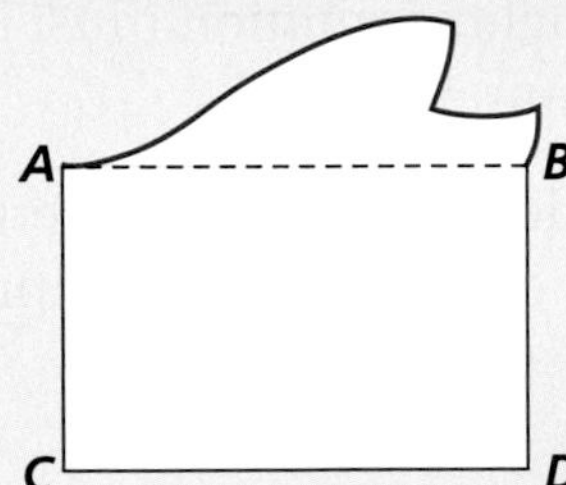

b. Translate the curve to the opposite side of the parallelogram.

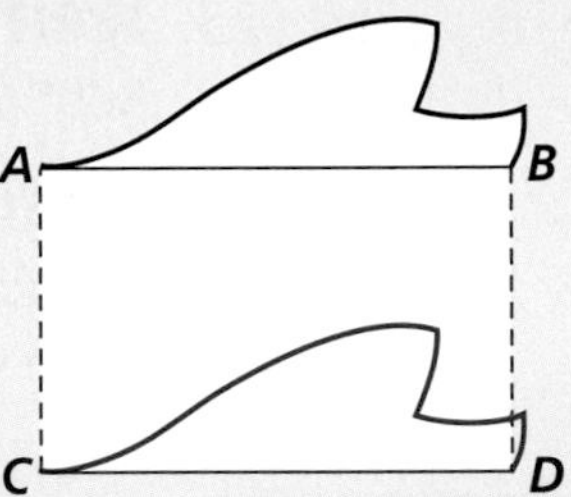

c. Repeat steps **a** and **b** for the other two sides of your parallelogram.

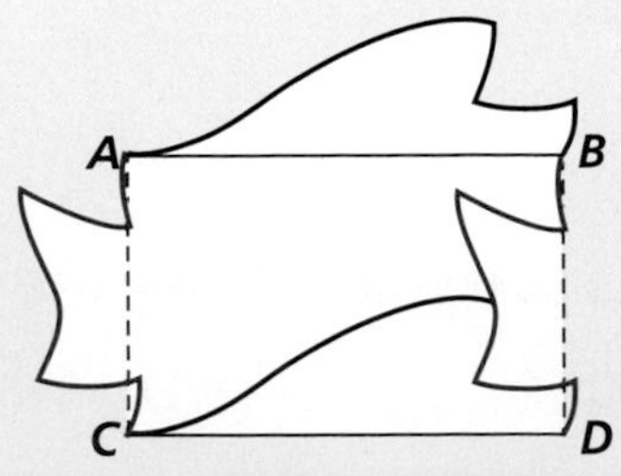

d. The figure can be translated to create an interlocking design, or tessellation. You can add details to your figure or divide it into two or more parts, as shown below.

2 Create a rotation tessellation.

The tessellation by M. C. Escher shown at right is an example of a *rotation tessellation*. To create your own rotation tessellation, follow the steps below.

a. Start with a regular hexagon. Replace one side of the hexagon with a curve. Rotate the curve about point B so that the endpoint at point A is moved to point C.

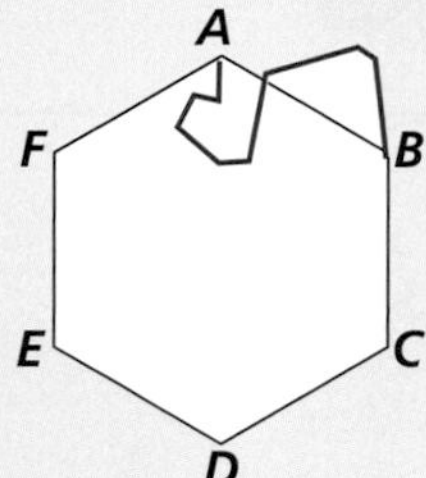

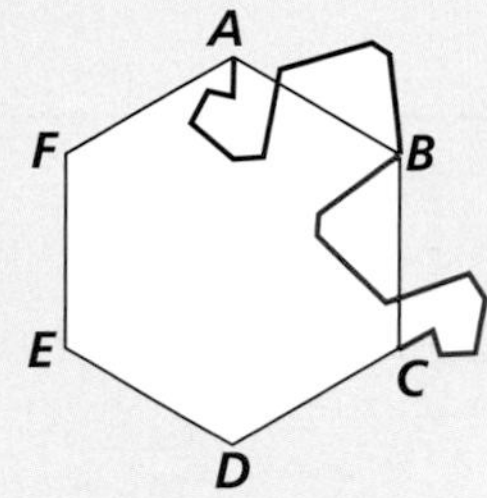

b. Replace side $\overline{CD}$ with a new curve, and rotate it about point D to replace side $\overline{DE}$.

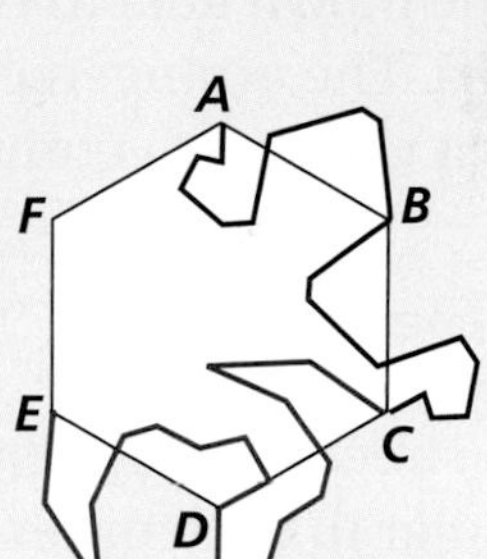

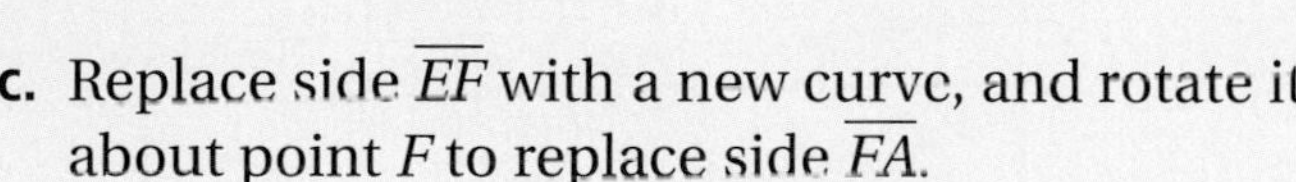

c. Replace side $\overline{EF}$ with a new curve, and rotate it about point F to replace side $\overline{FA}$.

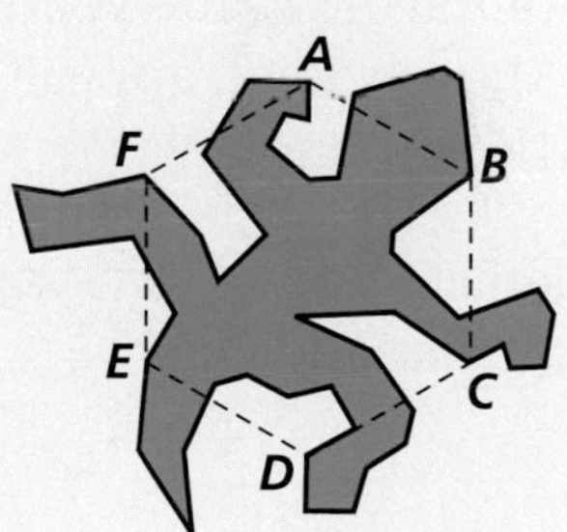

The figure can be rotated and fitted together with copies of itself to create an interlocking design, or tessellation. You can add details to your figure, if desired.

Think and Discuss

1. Explain why the two types of tessellations in this activity are known as translation and rotation tessellations.

Try This

1. Create your own design for a translation or rotation tessellation.

2. Cut out copies of your design from **1** and fit them together to fill a space with your pattern.

Problem Solving on Location

NORTH CAROLINA

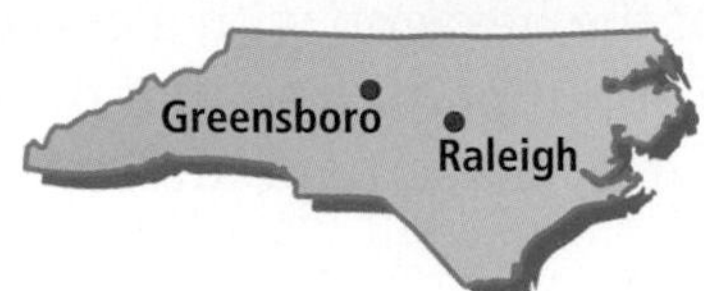

Downtown Greensboro

Engineers, architects, city planners, and designers use geometry to help them develop drawings and models of streets and buildings before they are built. Below is a map of the downtown area in Greensboro, North Carolina. The arrangement of the streets shows a variety of geometric patterns.

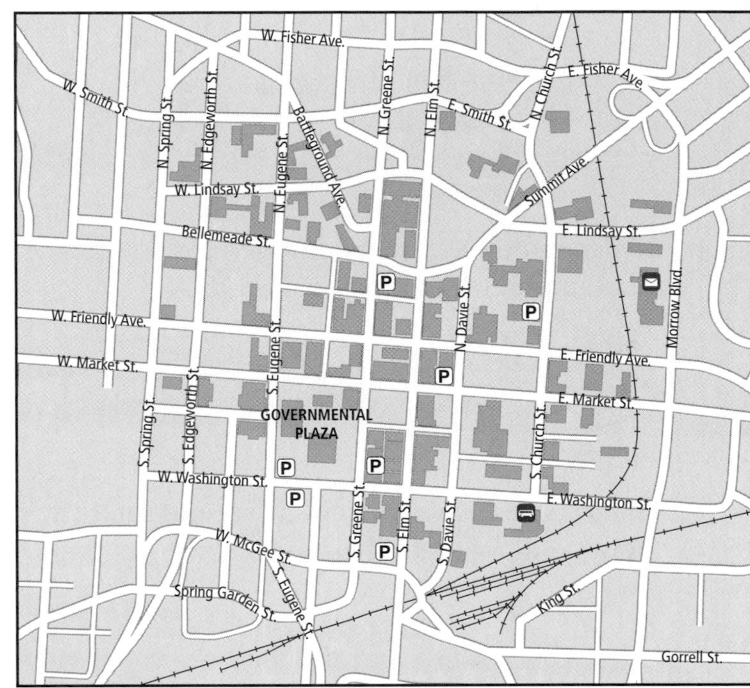

For 1–6, use the map.

1. Name two streets that appear to be perpendicular to each other.
2. Do W. Friendly Ave. and W. Market St. west of Morrow Blvd. appear to be parallel or perpendicular?
3. Do E. Market St. and Elm St. appear to be parallel or perpendicular?

4. Describe the shape of the area of the map that includes Governmental Plaza, which is bound by W. Market St., W. Washington St., S. Eugene St., and S. Greene St.
5. Name two streets whose intersection appears to form a right angle.
6. Describe the figure formed by W. Smith St. to the south, Battleground Ave. to the north, N. Edgeworth St. to the west, and N. Eugene St. to the east.

International Festival

The mission of the annual International Festival held in Raleigh is to promote awareness and appreciation of the cultural diversity found along the East Coast. The festival features many booths, demonstrations, cafés, and exhibits. There is also a stage where traditional dances from cultures around the world are performed. For the children, there are storytellers who hand down stories and folk tales from a variety of far-away places.

For 1–7, use the poster.

1. Of the five birds depicted on the poster, how many of the birds have tails that appear to have line symmetry?
2. Describe a set of parallel lines from the poster.
3. Do the triangles that make up the frame of the poster appear to be equilateral or scalene?
4. Name two types of polygons in the picture.
5. Describe a pair of figures in the poster that appear to be congruent.
6. Find two other congruent figures on the poster.
7. Does the poster from the International Festival appear to have a line of symmetry? Explain.

MATH-ABLES

Networks

A network is a figure that uses vertices and segments to show how objects are connected. You can use a network to show distances between cities. In the network at right, the vertices identify four cities in North Carolina, and the segments show the distances in miles between the cities.

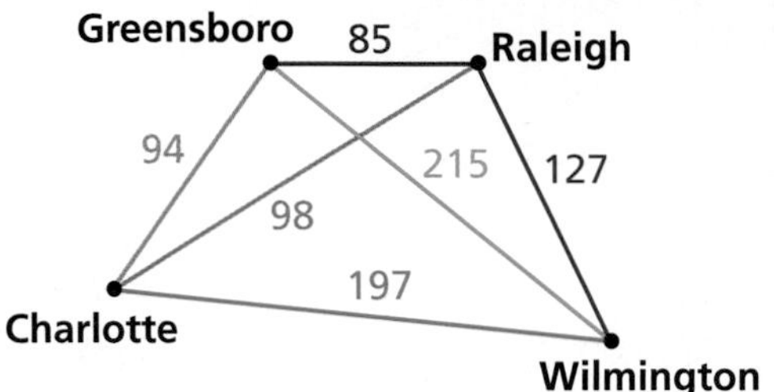

You can use the network to find the shortest route from Charlotte to the other three cities and back to Charlotte. First find all the possible routes. Then find the distance in miles for each route. One route has been identified below.

CGRWC $94 + 85 + 127 + 197 = 503$

Which is the shortest route, and what is the distance?

Color Craze

You can use rhombus-shaped tiles to build a variety of polygons. Each side of a tile is a different color. Build each design by matching the same-colored sides of tiles. Then see if you can create your own designs with the tiles. Try to make designs that have line or rotational symmetry.

internet connect
Go to ***go.hrw.com*** for a complete set of tiles.
KEYWORD: MS4 Game7

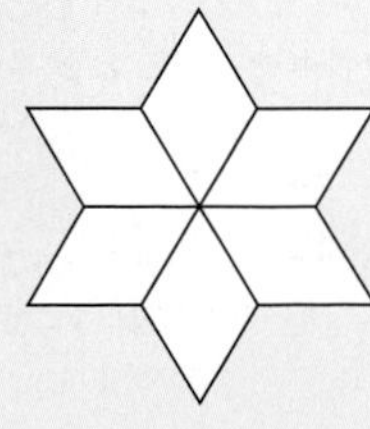

Transformations

You can use geometry software to perform transformations of geometric figures.

Activity

1. Use your dynamic geometry software to construct a 5-sided polygon like the one below. Label the vertices *A, B, C, D,* and *E.* Use the translation tool to translate the polygon 2 units right and $\frac{1}{2}$ unit up.

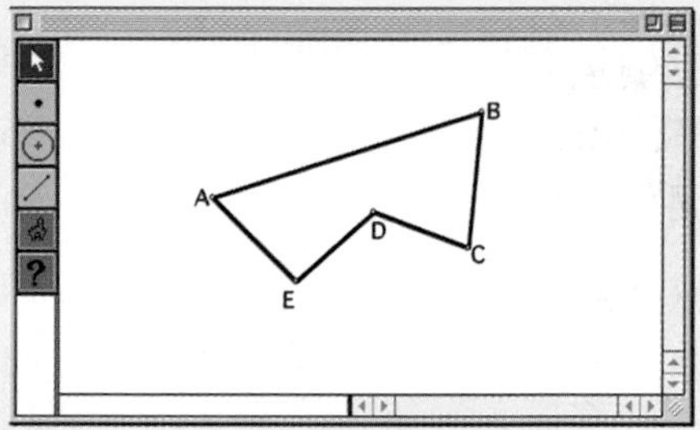

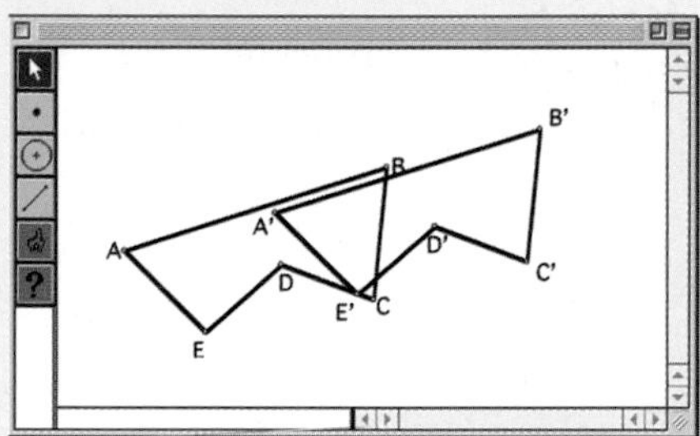

2. Start with the polygon from 1. Use the rotation tool to rotate the polygon 30° and then 150°, both about the vertex *C*.

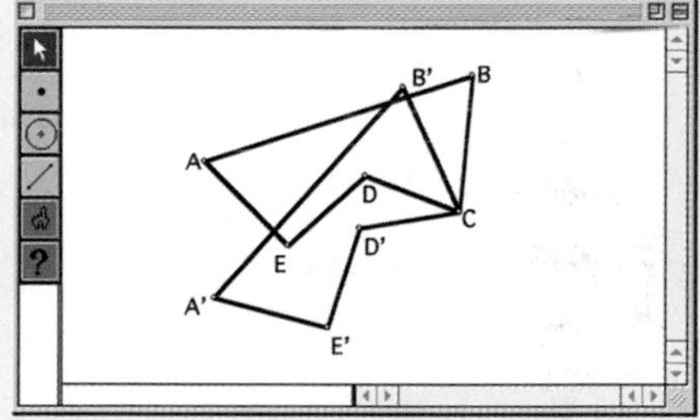

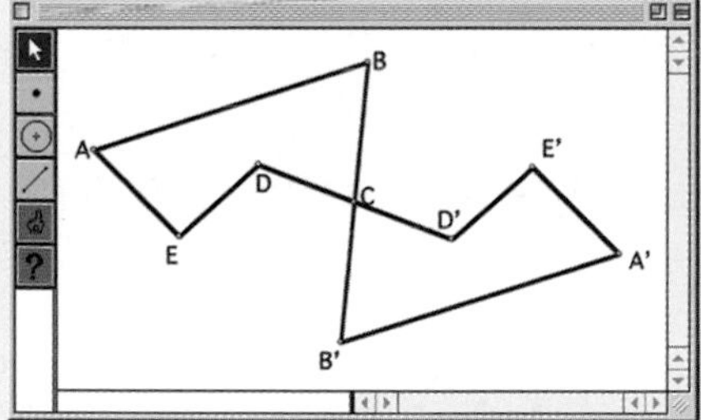

Think and Discuss

1. Rotate a triangle 30° about a point outside the triangle. Can this image be found by combining a vertical translation (slide up or down) and a horizontal translation (slide left or right) of the original triangle?
2. After what angle of rotation will the rotated image of a figure have the same orientation as the original figure?

Try This

1. Construct a quadrilateral *ABCD* using the geometry software.
   a. Translate the figure 2 units right and 1 unit up.
   b. Rotate the figure 30°, 45°, and 60°.

Chapter 7

Study Guide and Review

Vocabulary

Complete the sentences below with vocabulary words from the list above.

1. Every equilateral triangle is also a(n) __?__ triangle.
2. Lines in the same plane that do not intersect are __?__.

7-1 Points, Lines, and Planes (pp. 344–347)

EXAMPLE

Identify each figure in the diagram.

- points: A, B, C
- lines: $\overleftrightarrow{AB}$
- planes: ABC
- rays: $\overrightarrow{BA}$; $\overrightarrow{AB}$
- line segments: $\overline{AB}$; $\overline{BC}$

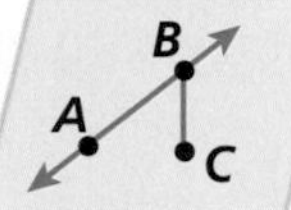

EXERCISES

Identify each figure in the diagram.

3. points
4. lines
5. planes
6. rays
7. line segments

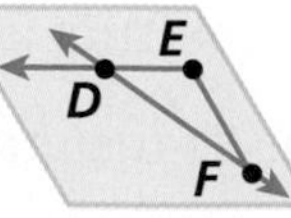

7-2 Angles (pp. 348–351)

EXAMPLE

- **Tell whether the angle is acute, right, obtuse, or straight.**

The angle is a right angle.

EXERCISES

Tell whether each angle is acute, right, obtuse, or straight.

8. 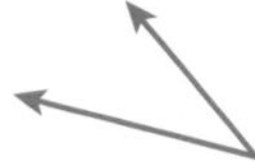

9. 

7-3 Parallel and Perpendicular Lines (pp. 354–357)

EXAMPLE

- **Tell whether the lines appear parallel, perpendicular, or skew.**

perpendicular

EXERCISES

Tell whether the lines appear parallel, perpendicular, or skew.

10. 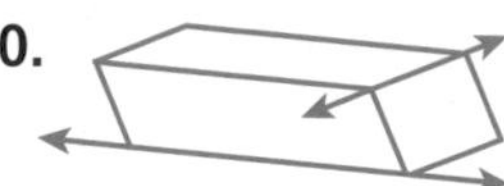

7-4 Circles (pp. 362–365)

EXAMPLE

Name the parts of circle D.

- radii: $\overline{DB}, \overline{DC}, \overline{DE}$
- diameter: $\overline{EB}$
- chords: $\overline{AB}, \overline{EB}, \overline{EF}$

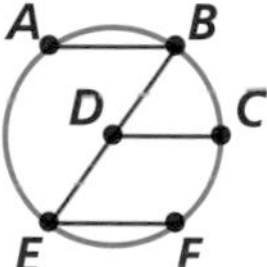

EXERCISES

Name the parts of circle F.

11. radii

12. diameter

13. chords

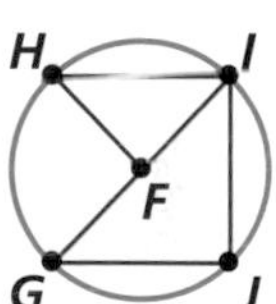

7-5 Polygons (pp. 368–371)

EXAMPLE

- **Tell whether the figure is a regular polygon. If it is not, explain why not.**
  No, all the angles in the polygon are not congruent.

EXERCISES

Tell whether each figure is a regular polygon. If it is not, explain why not.

14. 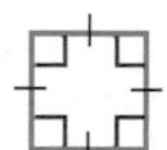

15. 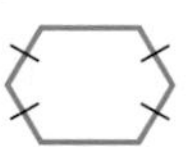

7-6 Triangles (pp. 374–377)

EXAMPLE

- **Classify the triangle according to its sides and angles.**

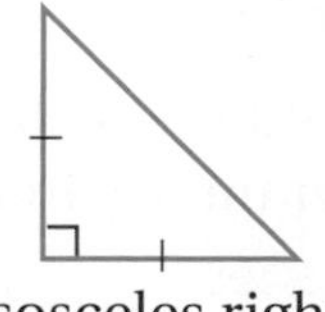

Isosceles right

EXERCISES

Classify each triangle according to its sides and angles.

16. 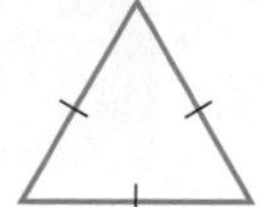

17. 

7-7 Quadrilaterals (pp. 378–381)

EXAMPLE

- **Give all of the names that apply to the quadrilateral.**

trapezoid

EXERCISES

Give all of the names that apply to each quadrilateral.

18.

19.

7-8 Angles in Polygons (pp. 382–385)

EXAMPLE

- **Find the measure of the unknown angle.**

$62° + 45° + x = 180°$

$107° + x = 180°$

$x = 73°$

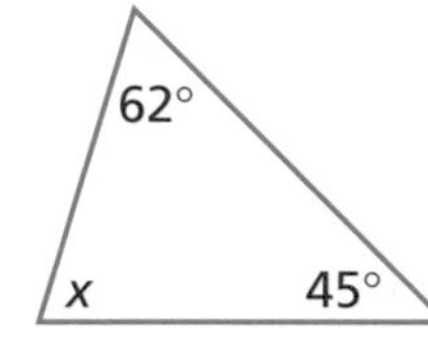

EXERCISES

Find the measure of the unknown angle.

20.

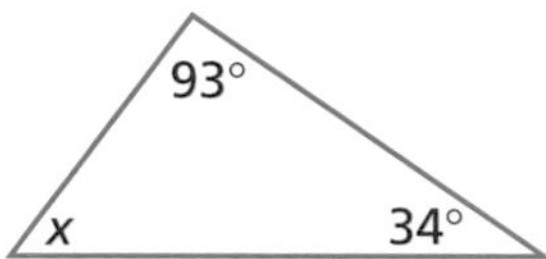

7-9 Congruent Figures (pp. 388–391)

EXAMPLE

- **Determine the missing measure in the set of congruent polygons.**

The angle measures 53°.

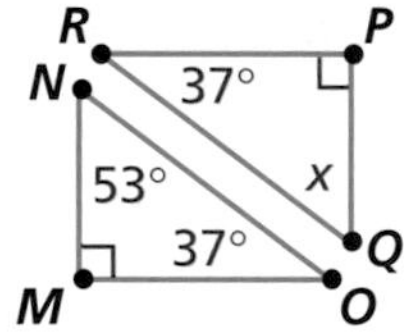

EXERCISES

21. Determine the missing measure in the set of congruent polygons.

133° 47° 47° 133° | 133° 47° 47° x

7-10 Transformations (pp. 392–395)

EXAMPLE

- **Graph the translation.**

Translate $\triangle ABC$ 1 unit right and 3 units down.

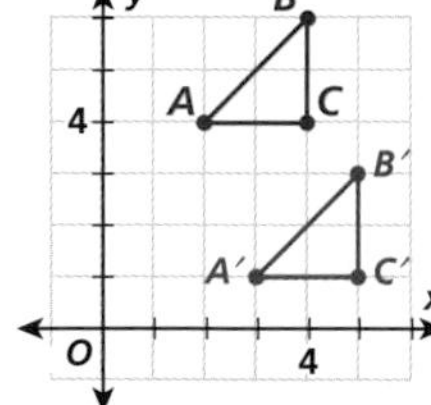

EXERCISES

Graph the translation.

22. Translate $\triangle BCD$ 2 units left and 4 units down.

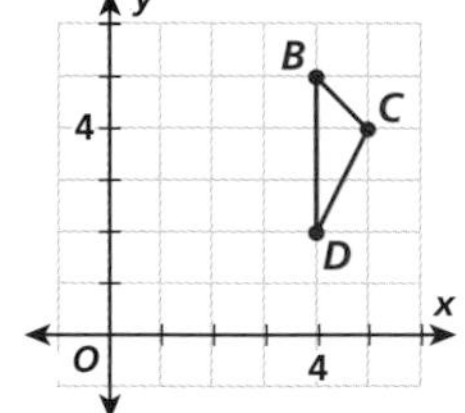

7-11 Symmetry (pp. 396–399)

EXAMPLE

- **Find all the lines of symmetry in the flag.**

The flag has four lines of symmetry.

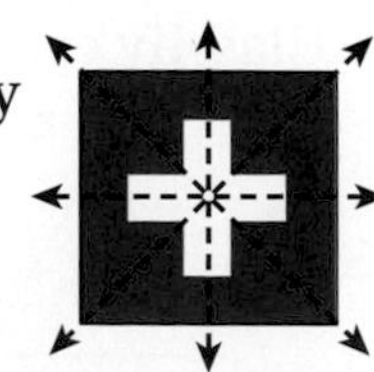

EXERCISES

23. Find all the lines of symmetry in the flag.

Chapter Test

Chapter 7

Identify the figures in the diagram.

1. 5 points **2.** 3 lines **3.** a plane

4. 5 line segments **5.** 6 rays

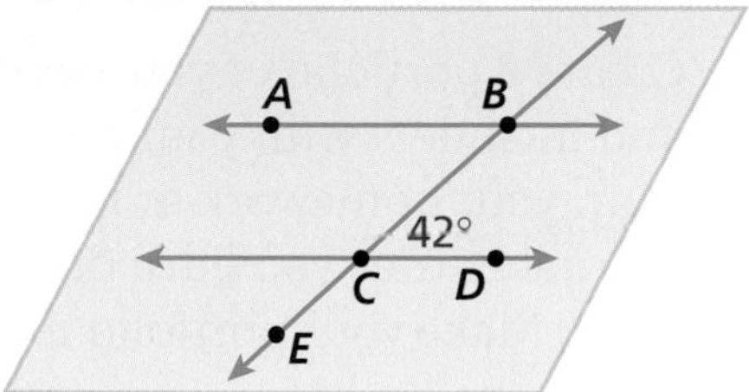

Line *AB* || line *CD*. Find the measure of each angle and tell whether the angle is acute, right, obtuse, or straight.

6. $\angle ABC$ **7.** $\angle BCE$ **8.** $\angle DCE$

Tell whether the lines appear parallel, perpendicular, or skew.

9. $\overleftrightarrow{MN}$ and $\overleftrightarrow{PO}$ **10.** $\overleftrightarrow{LM}$ and $\overleftrightarrow{PO}$ **11.** $\overleftrightarrow{NO}$ and $\overleftrightarrow{MN}$

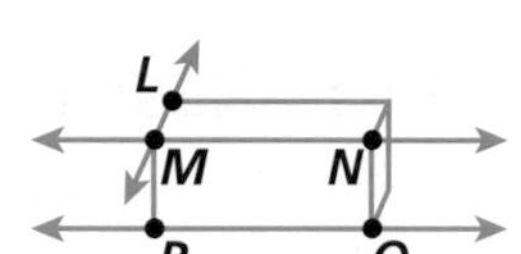

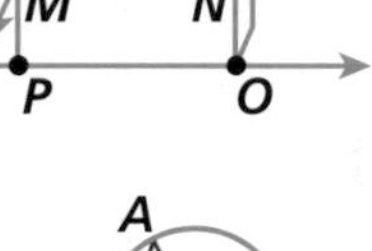

Name the parts of circle *E*.

12. radii **13.** chords **14.** diameter

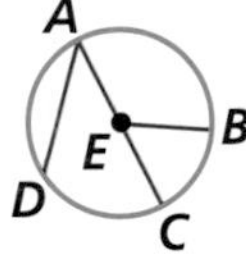

Tell whether each figure is a regular polygon. If it is not, explain why not.

15. **16.**  **17.**

Classify each triangle according to its sides and angles.

18. **19.** **20.**

Give all the names that apply to each quadrilateral.

21. 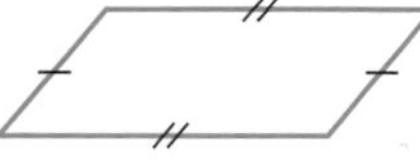 **22.** 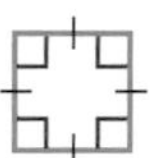 **23.** 

Find the measure of each unknown angle.

24. 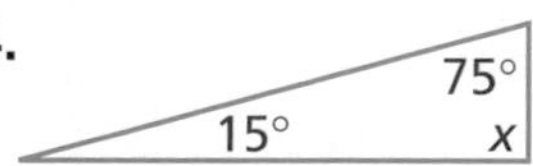

25. 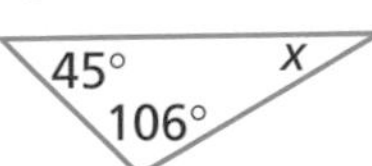

26. 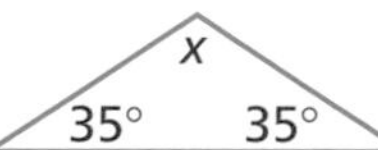

27. Determine the missing measure in the congruent polygons.

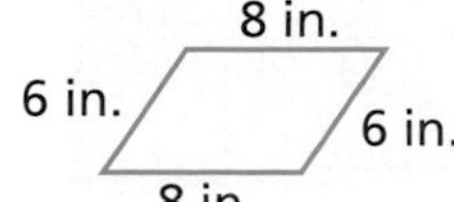

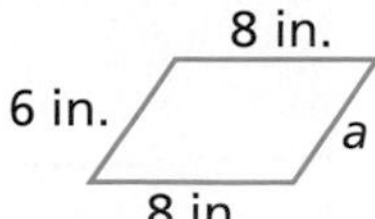

28. The vertices of a triangle have the coordinates $(-1, -3)$, $(-4, -1)$, and $(-1, -1)$. Graph the triangle after a translation 3 units left.

Find all the lines of symmetry in each flag.

29. 

30. 

Chapter Test

Chapter 7 Performance Assessment

Show What You Know

Create a portfolio of your work from this chapter. Complete this page and include it with your four best pieces of work from Chapter 7. Choose from your homework or lab assignments, mid-chapter quizzes, or any journal entries you have done. Put them together using any design you want. Make your portfolio represent what you consider your best work.

Short Response

1. The vertices of a triangle are $A(6, 4)$, $B(6, 2)$, and $C(8, 2)$. Translate triangle ABC left 7 units, and list the coordinates of the translated image.
2. According to a circle graph, 55% of the people who responded to a survey eat cereal for breakfast. Find the central angle measure of the sector that shows this percent. Show your work.
3. Jacob drew a pair of parallel lines intersected by a transversal. One of the angles formed by one of the parallel lines and the transversal is a right angle. What can you conclude about the other angles? Explain your answer.
4. Barry drew a quadrilateral. One angle measures 80°, and another angle measures 130°. The fourth angle is twice as large as the third angle. Write and solve an equation to find the measures of the third and fourth angles. Show your work.

Extended Problem Solving

5. Use the tapestry for the following problems.
   a. Identify as many places as possible where the line symmetry is broken.
   b. Identify a rotation in the tapestry.
   c. Identify a reflection in the tapestry.
   d. Identify a combination of a rotation and a translation in the tapestry.

Performance Assessment

internet connect

EOG Practice Online

go.hrw.com Keyword: MS4 TestPrep

Chapter 7

Getting Ready for EOG

Cumulative Assessment, Chapters 1–7

1. Which statement about the figure is true?

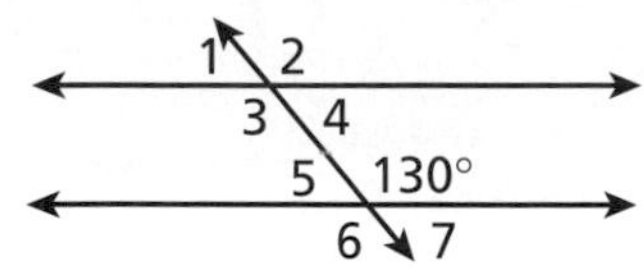

A $\angle 1$ and $\angle 6$ are congruent.

B The measure of $\angle 7$ is 85°.

C $\angle 4$ and $\angle 6$ are complementary angles.

D $\angle 2$ and $\angle 3$ are vertical angles.

2. Find the median of -47, 25, -10, 7, -24, and -29.

A 17 C -17

B -10 D -24

3. A pair of shoes that normally cost \$18.50 are on sale for \$14.80. What is the percent of decrease?

A 125% C 20%

B 80% D 10%

4. Which number is **not** expressed in scientific notation?

A 1.7×10^{34} C 2.354×10^{1}

B 3×10^{2} D 10.3×10^{5}

5. The mean of a set of five numbers is 8.8. If the number 7 is added to the set, what is the new mean?

A 8 C 9

B 8.5 D 9.5

6. What is 65 mi/h expressed in ft/min?

A 0.74 ft/min C 4,874 ft/min

B 1,907 ft/min D 5,720 ft/min

7. Solve $\frac{2}{3}x = \frac{5}{6}$.

A $x = \frac{5}{4}$ C $x = \frac{4}{5}$

B $x = \frac{5}{9}$ D $x = \frac{9}{5}$

8. What is the prime factorization of 1,000?

A $2^3 \cdot 3 \cdot 5^2$ C $2^3 \cdot 5^3$

B $2^2 \cdot 5^4$ D $2^2 \cdot 5^3 \cdot 7$

9. ***SHORT RESPONSE*** The triangles are congruent. What is the value of x? How do you know?

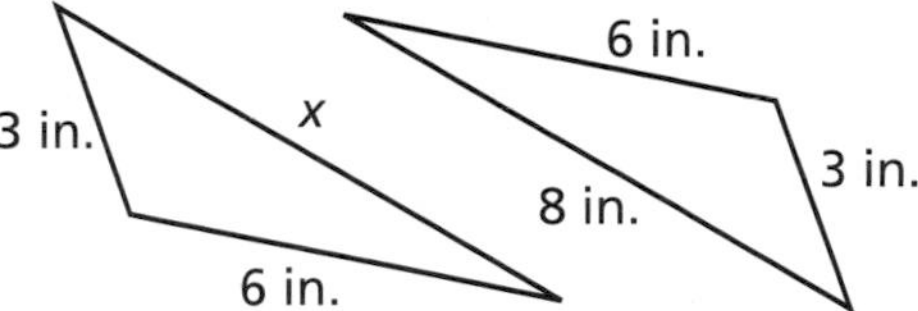

TEST TAKING TIP!

When solving a geometry word problem, draw a diagram to illustrate the problem.

10. ***SHORT RESPONSE*** Four of the angles in a pentagon measure 81°, 115°, 139°, and 90°. What is the measure of the fifth angle? Explain how you determined your answer.

Chapter 8

Perimeter, Circumference, and Area

TABLE OF CONTENTS

internet connect

Chapter Opener Online
go.hrw.com
KEYWORD: MS4 Ch8

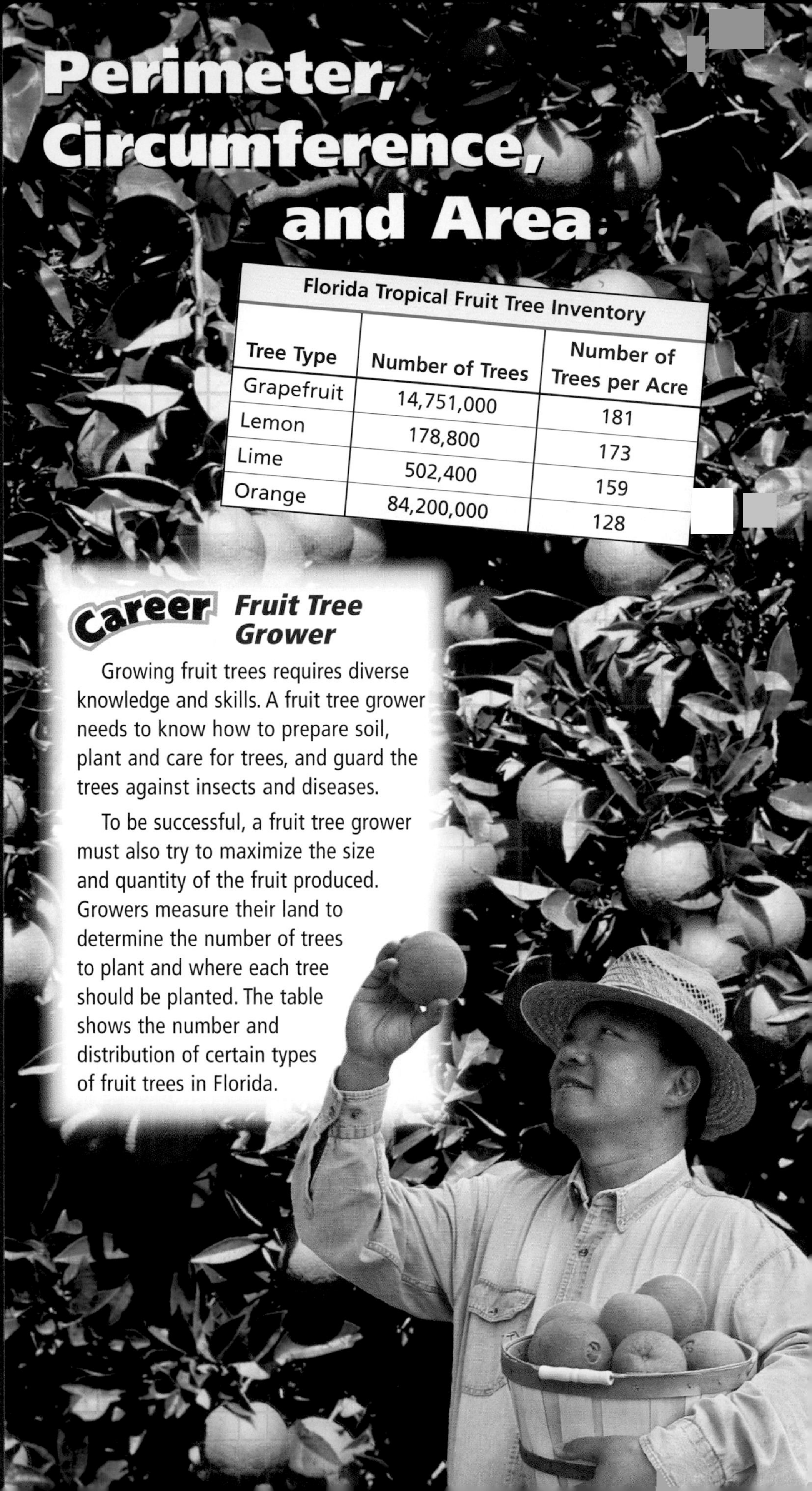

Florida Tropical Fruit Tree Inventory

Tree Type	Number of Trees	Number of Trees per Acre
Grapefruit	14,751,000	181
Lemon	178,800	173
Lime	502,400	159
Orange	84,200,000	128

Career *Fruit Tree Grower*

Growing fruit trees requires diverse knowledge and skills. A fruit tree grower needs to know how to prepare soil, plant and care for trees, and guard the trees against insects and diseases.

To be successful, a fruit tree grower must also try to maximize the size and quantity of the fruit produced. Growers measure their land to determine the number of trees to plant and where each tree should be planted. The table shows the number and distribution of certain types of fruit trees in Florida.

ARE YOU READY?

Choose the best term from the list to complete each sentence.

1. A(n) __?__ is a quadrilateral with exactly one pair of parallel sides.
2. A(n) __?__ is a four-sided figure with opposite sides that are congruent and parallel.
3. The __?__ of a circle is one-half the __?__ of the circle.

diameter
parallelogram
radius
right triangle
trapezoid

Complete these exercises to review skills you will need for this chapter.

✔ Round Whole Numbers

Round each number to the nearest ten and nearest hundred.

4. 1,535 **5.** 294 **6.** 30,758 **7.** 497

✔ Round Decimals

Round each number to the nearest whole number and nearest tenth.

8. 6.18 **9.** 10.50 **10.** 513.93 **11.** 29.06

✔ Multiply with Decimals

Multiply.

12. $5.63 \cdot 8$ **13.** $9.67 \cdot 4.3$ **14.** $8.34 \cdot 16$ **15.** $6.08 \cdot 0.56$

✔ Order of Operations

Evaluate each expression.

16. $2 \cdot 9 + 2 \cdot 6$ **17.** $2(15 + 8)$ **18.** $4 \cdot 6.8 + 7 \cdot 9.3$ **19.** $14(25.9 + 13.6)$

✔ Identify Polygons

Name each figure.

20. 

21. 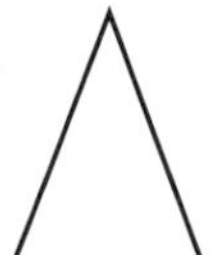

22.

8-1 Customary and Metric Measurements

Learn **to convert measurements within the customary and metric systems.**

To conserve energy and to avoid the heat of the day, lions spend much of their time resting—usually up to 20 hours a day. This means that in one year, a lion spends about 7,300 hours resting.

You can use *unit conversion*, which you explored in Lesson 5-4, to find out how many days there are in 7,300 hours. Recall that a *unit conversion factor* is a fraction whose numerator and denominator represent the same quantity but in different units.

EXAMPLE 1 Converting Between Units of Time

Remember!

60 s = 1 min
60 min = 1 hr
24 hr = 1 day
365 days = 1 yr

About how many days are there in 7,300 hours?

There are 24 hours in 1 day, so the possible conversion factors are $\frac{1 \text{ day}}{24 \text{ hr}}$ or $\frac{24 \text{ hr}}{1 \text{ day}}$. Choose the first factor so that hour units will cancel.

$$7{,}300 \cancel{\text{hr}} \cdot \frac{1 \text{ day}}{24 \cancel{\text{hr}}} = \frac{7{,}300 \text{ days}}{24}$$

$$\approx 304 \text{ days}$$

There are about 304 days in 7,300 hours.

We use two systems of measurement: customary and metric. The customary system is based on measurements from old English law, and the metric system is based on powers of ten. Each system has its own units.

Measure	Customary System	Metric System
Length/ distance	12 inches (in.) = 1 foot (ft) 3 feet = 1 yard (yd) 5,280 feet = 1 mile (mi)	10 millimeters (mm) = 1 centimeter (cm) 100 centimeters = 1 meter (m) 1,000 meters = 1 kilometer (km)
Volume/ capacity	8 fluid ounces (fl oz) = 1 cup (c) 2 cups = 1 pint (pt) 2 pints = 1 quart (qt) 4 quarts = 1 gallon (gal)	1,000 milliliters (mL) = 1 liter (L)
Weight/ mass	16 ounces (oz) = 1 pound (lb) 2,000 pounds = 1 ton	1,000 milligrams (mg) = 1 gram (g) 1,000 grams = 1 kilogram (kg)

EXAMPLE 2 Making One-Step Unit Conversions

Convert 25 kilometers to meters.

There are 1,000 meters in 1 kilometer.

$$25\ \cancel{km} \cdot \frac{1{,}000\ m}{1\ \cancel{km}} = \frac{25 \cdot 1{,}000\ m}{1}$$ *Use a unit conversion factor.*

$$= 25{,}000\ m$$

So 25 kilometers is equal to 25,000 meters.

EXAMPLE 3 PROBLEM SOLVING APPLICATION

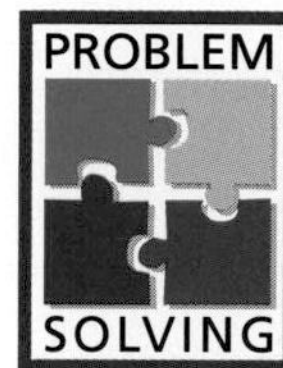

Selena is selling juice during the intermission of a school play. If she has 6 quarts of juice, how many 1-cup servings of juice can she pour?

1 Understand the Problem

Rewrite the question as a statement:

- Find the number of cups in 6 quarts.

List the **important information:**

- Selena has 6 quarts of juice.

2 Make a Plan

Use unit conversion factors to convert quarts to pints and then pints to cups.

3 Solve

$$6\ \cancel{qt} \cdot \frac{2\ \cancel{pt}}{1\ \cancel{qt}} \cdot \frac{2\ c}{1\ \cancel{pt}} = 6 \cdot 2 \cdot 2\ c$$ *Use two conversion factors.*

$$= 24\ c$$

Since 6 quarts is equal to 24 cups, Selena can pour 24 one-cup servings of juice.

4 Look Back

A cup is a smaller unit than a quart, so the answer should be greater than 6. So 24 is a reasonable answer.

Think and Discuss

1. **Describe** how you could use mental math to convert measurements within the metric system.
2. **Write** a unit conversion factor that relates ounces to tons.

8-1 Exercises

FOR EOG PRACTICE

see page 688

internet connect
Homework Help Online
go.hrw.com Keyword: MS4 8-1

1.01, 1.02d, 1.03

GUIDED PRACTICE

See Example 1

1. How many hours are there in 24 days?
2. How many minutes are there in 450 seconds?

See Example 2

Convert.

3. 5 gal = ■ qt
4. 600 min = ■ hr
5. 3,700 mg = ■ g
6. 100 m = ■ cm
7. 8 mi = ■ ft
8. 72 oz = ■ lb

See Example 3

9. Roger has 48 pints of strawberries. How many gallon containers would he need to hold all of the strawberries?

INDEPENDENT PRACTICE

See Example 1

10. How many days are there in 9 years?
11. How many seconds are there in 45 minutes?
12. How many years are there in 292 days?
13. How many minutes are there in 8.3 hours?

See Example 2

Convert.

14. 4.2 kg = ■ g
15. 0.5 ton = ■ lb
16. 3.8 mm = ■ cm
17. 217 cm = ■ m
18. 48 oz = ■ lb
19. 10 qt = ■ gal
20. 3.5 qt = ■ pt
21. $8\frac{1}{3}$ yd = ■ ft
22. 2.2 mi = ■ ft

See Example 3

23. Cassie is planning a 2-day nonstop Valentine's Day tribute on her radio show. If each person can give a 1-minute tribute, how many tributes can Cassie play?
24. In 1998, a 2,505-gallon ice cream float was made in Atlanta, Georgia. How many 1-pint servings did the float contain?

PRACTICE AND PROBLEM SOLVING

Convert.

25. 64 oz = ■ lb
26. 7.9 m = ■ km
27. 2 mi = ■ yd
28. 4 gal = ■ c
29. 1.6 m = ■ mm
30. 2,160 min = ■ days

Compare. Write $<$, $>$, or $=$.

31. 6 yd ■ 12 ft
32. 70 s ■ 1 min
33. 80 oz ■ 5 lb
34. 40 cm ■ 40 mm
35. 25 m ■ 250 cm
36. 18 in. ■ 3 ft

37. ***EARTH SCIENCE*** The average depth of the Pacific Ocean is 12,925 feet. How deep is this, rounded to the nearest tenth of a mile?

The owner of this 1,009.6 lb pumpkin won the New England Giant Pumpkin Weigh-Off in September 2000.

38. ***AGRICULTURE*** In one year, the United States produced nearly 895 million pounds of pumpkins. How many ounces were produced by the state with the lowest production shown in the table?

U.S. Pumpkin Production

State	Pumpkins (million pounds)
California	180
Illinois	364
New York	114
Pennsylvania	109

39. How many milliliters are there in a 2.2-liter container of juice?

40. ***MUSIC*** The shortest published opera is *The Sands of Time*, which lasted 4 minutes 9 seconds when it was first performed in 1993. How many seconds long was the performance?

41. ***LITERATURE*** The novel *Twenty Thousand Leagues Under the Sea* was written by Jules Verne in 1873. One league is approximately 3.45 miles. How many miles are in 20,000 leagues?

42. ***SPORTS*** A marathon is a race that is 26 miles 385 yards long.
    a. Write a unit conversion factor that can be used to convert miles to yards.
    b. What is the length of a marathon in yards?

43. ***WHAT'S THE ERROR?*** A student used the following method to convert 85 centimeters to meters. What did the student do wrong?

$$85 \text{ cm} \cdot \frac{100 \text{ cm}}{1 \text{ m}} = 8{,}500 \text{ m}$$

44. ***WRITE ABOUT IT*** Explain how to convert 12,000 milligrams to kilograms.

45. ***CHALLENGE*** Chen's heart beats once per second. How many times does it beat in a year?

Spiral Review

Find each sum or difference. (Lesson 4-2)

46. 3.8 + 4.02 **47.** 8 − 6.18 **48.** 18.4 − 13.61 **49.** 15.07 + 3.08

50. **EOG PREP** Eighty-four is 12% of what number? (Lesson 6-4)

A 700 C 10.08
B 70 D 14.3

51. **EOG PREP** Which transformation is shown? (Lesson 7-10)

A Rotation C Reflection
B Translation D Glide reflection

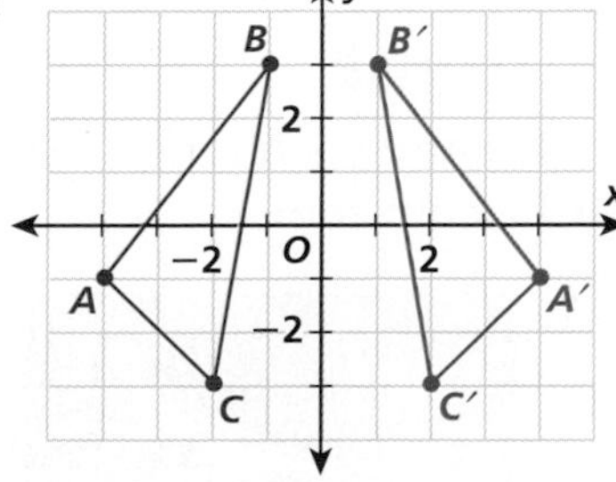

Measure Objects

Use with Lesson 8-2

To measure the length in inches of an object, you can use a standard ruler. To measure the length in centimeters of an object, you can use a metric ruler.

Activity 1

1 Use a metric ruler to measure the length of the pencil to the nearest centimeter and to the nearest tenth of a centimeter.

On a metric ruler, the numbered divisions represent centimeters. Since the tip of the pencil is between the 18 cm mark and the 19 cm mark, the pencil is between 18 cm and 19 cm. The tip of the pencil is closest to the 18 cm mark, so to the nearest centimeter it is 18 cm long.

Each centimeter on the ruler is divided into tenths. Since the tip of the pencil is between the 18.1 cm mark and the 18.2 cm mark, the pencil is between 18.1 cm and 18.2 cm. The tip of the pencil is closest to the 18.1 cm mark, so to the nearest tenth of a centimeter it is 18.1 cm long.

Think and Discuss

1. Which measurement, 18 cm or 18.1 cm, best represents the actual length of the pencil? Explain.
2. Suppose you used a standard ruler to measure the pencil in inches. Would this measurement better represent the length of the pencil? Explain.

Try This

Measure each length to the nearest centimeter and to the nearest tenth of a centimeter.

1. 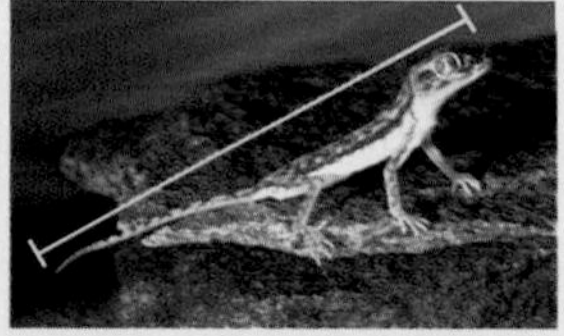

2. 

Activity 2

1 Use a standard ruler to measure the lengths of the sides of the rectangle.

A standard ruler is 12 inches long, and each inch is divided into sixteenths. Therefore, each division on a ruler represents $\frac{1}{16}$ inch. The length of one of the longer sides of the rectangle is halfway between the 3-inch and 4-inch marks. So the length is $3\frac{8}{16}$ in., or $3\frac{1}{2}$ in. The length of one of the shorter sides is just over 1 inch. Since the side ends at the second mark after the 1-inch mark, the length is $1\frac{2}{16}$ in., or $1\frac{1}{8}$ in.

2 Find the distance around the rectangle.

$3\frac{8}{16} + 3\frac{8}{16} + 1\frac{2}{16} + 1\frac{2}{16} = 8 + \frac{20}{16}$ *Add the whole numbers. Then add the fractions.*

$= 8 + 1\frac{4}{16}$ *Write the fraction as a mixed number.*

$= 9\frac{4}{16}$ *Add the whole numbers.*

$= 9\frac{1}{4}$ *Write the fraction in simplest form.*

The distance around the rectangle is $9\frac{1}{4}$ inches.

Think and Discuss

1. Use a metric ruler to measure the length of each side of the rectangle in **1** to the nearest tenth of a centimeter. What can you conclude about the relationship between an inch and a centimeter?

Try This

Use a standard ruler to measure the lengths of the sides of each rectangle. Then find the distance around each rectangle.

1.

2.

Use a metric ruler to measure the lengths of the sides of each rectangle. Then find the distance around each rectangle.

3.

4.

8-2 Accuracy and Precision

Learn to compare the precision of measurements and to determine acceptable levels of accuracy.

Vocabulary

precision

accuracy

significant digits

Ancient Greeks used measurements taken during lunar eclipses to determine that the Moon was 240,000 miles from Earth. In 1969, the distance was measured as 221,463 miles.

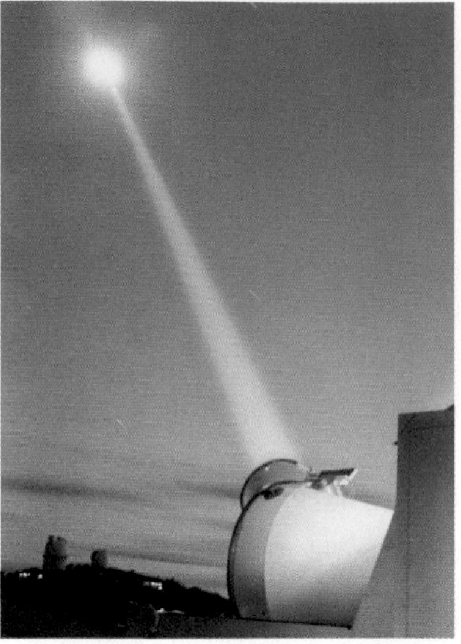

At the University of Texas McDonald Observatory, a laser is used to measure the distance from Earth to the Moon.

There is a difference between these measurements because modern scientists conducted the measurement with greater *precision*. **Precision** is the level of detail an instrument can measure.

The smaller the unit an instrument can measure, the more precise its measurements will be. For example, a millimeter ruler has greater precision than a centimeter ruler because it can measure smaller units.

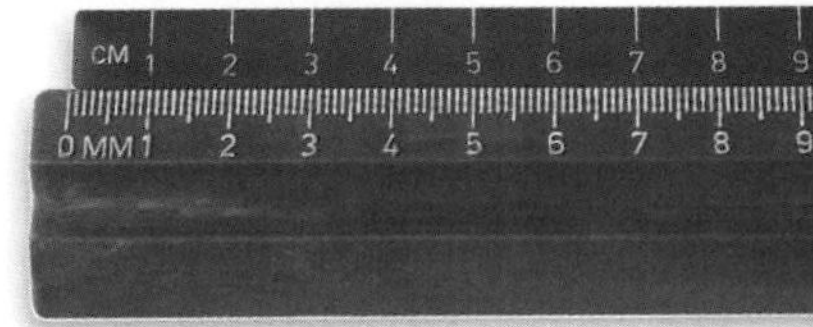

EXAMPLE 1 Judging Precision of Measurements

Choose the more precise measurement in each pair.

A **37 in., 3 ft**

Since an inch is a smaller unit than a foot, 37 in. is more precise.

B **5 km, 5.8 km**

Since tenths are smaller than ones, 5.8 km is more precise.

In the real world, no measurement is exact. The relative exactness of a measurement is its **accuracy**. In a measured value, all the digits that are known with certainty are called **significant digits**. Zeros at the end of a whole number are assumed to be nonsignificant. The table shows the rules for identifying significant digits.

Rule	Example	Number of Significant Digits
• Nonzero digits	45.7	3 significant digits
• Zeros between significant digits	78,002	5 significant digits
• Zeros after the last nonzero digit and to the right of a decimal point	0.0040	2 significant digits

EXAMPLE 2 Identifying Significant Digits

Determine the number of significant digits in each measurement.

A **120.1 mi**

The digits 1 and 2 are nonzero digits, and 0 is between two nonzero digits.

So 120.1 mi has 4 significant digits.

B **0.0350 kg**

The digits 3 and 5 are nonzero digits, and 0 is to the right of the decimal after the last nonzero digit.

So 0.0350 kg has 3 significant digits.

When you are adding and subtracting measurements, the answer should have the same number of digits to the right of the decimal point as the measurement with the least number of digits to the right of the decimal point.

EXAMPLE 3 Using Significant Digits in Addition or Subtraction

Calculate 45 mi − 0.9 mi. Use the correct number of significant digits in the answer.

45	*0 digits to the right of the decimal point*
− 0.9	*1 digit to the right of the decimal point*
44.1 ≈ 44 mi	*Round the difference so that it has no digits to the right of the decimal point.*

When you are multiplying and dividing measurements, the answer must have the same number of significant digits as the measurement with the least number of significant digits.

EXAMPLE 4 Using Significant Digits in Multiplication or Division

Calculate 32.8 m · 1.5 m. Use the correct number of significant digits in the answer.

32.8	*3 significant digits*
× 1.5	*2 significant digits*
49.2 ≈ 49 m	*Round the product so that it has 2 significant digits.*

Think and Discuss

1. **Tell** how many significant digits there are in 380.102.
2. **Choose** the more precise measurement: 18 oz or 1 lb. Explain.

8-2 Exercises

1.03

FOR EOG PRACTICE
see page 688

internet connect
Homework Help Online
go.hrw.com Keyword: MS4 8-2

GUIDED PRACTICE

See Example 1 **Choose the more precise measurement in each pair.**

1. 4 ft, 1 yd **2.** 2 cm, 21 mm **3.** $5\frac{1}{2}$ in., $5\frac{1}{4}$ in.

See Example 2 **Determine the number of significant digits in each measurement.**

4. 2.703 g **5.** 0.02 km **6.** 28,000 lb

See Example 3 **Calculate. Use the correct number of significant digits in each answer.**

7. 16 − 3.8 **8.** 3.5 + 0.66 **9.** 11.3 − 4

See Example 4 **10.** 47.9 • 3.8 **11.** 7.0 • 3.6 **12.** 50.2 ÷ 8.0

INDEPENDENT PRACTICE

See Example 1 **Choose the more precise measurement in each pair.**

13. 11 in., 1 ft **14.** 7.2 m, 6.2 cm **15.** 14.2 km, 14 km

16. $4\frac{3}{8}$ in., $4\frac{7}{16}$ in. **17.** 2.8 m, 3 m **18.** 37 g, 37.0 g

See Example 2 **Determine the number of significant digits in each measurement.**

19. 0.00002 kg **20.** 10,000,000 lb **21.** 200.060 m

22. 4.003 L **23.** 0.230 cm **24.** 940.0 ft

See Example 3 **Calculate. Use the correct number of significant digits in each answer.**

25. 6.2 + 8.93 **26.** 7.02 + 15 **27.** 8 − 6.6

28. 29.1 − 13.204 **29.** 8.6 + 9.43 **30.** 43.5 + 876.23

See Example 4 **31.** 17 • 104 **32.** 21.8 • 10.9 **33.** 7.0 ÷ 3.11

34. 1,680 ÷ 5.025 **35.** 14.2 ÷ 0.05 **36.** 5.22 • 6.3

PRACTICE AND PROBLEM SOLVING

Which unit is more precise?

37. foot or mile **38.** centimeter or millimeter

39. liter or milliliter **40.** minute or second

Calculate. Use the correct number of significant digits in each answer.

41. 38,000 • 4.8 **42.** 2.879 + 113.6 **43.** 290 − 6.1

44. 5.6 ÷ 0.6 **45.** 40.29 − 18.5 **46.** 24 ÷ 6.02

Health LINK

The food labels at right give information about two types of soup: cream of tomato and minestrone. Use the labels for Exercises 47–49.

Cream of Tomato

Nutrition Facts	
Serving size 1 cup (240mL)	
Servings per container about 2	
Amount per Serving	
Calories 100	Calories from Fat 20
	% Daily Value*
Total Fat 2 g	3%
Saturated Fat 1.5 g	6%
Cholesterol 10 mg	3%
Sodium 690 mg	29%
Total Carbohydrate 17 g	6%
Dietary Fiber 4 g	18%
Sugars 11 g	
Protein 2 g	
Vitamin A 20%	Vitamin C 20%
Calcium 0%	Iron 8%

*Percent daily values are based on a 2,000 calorie diet.

Minestrone

Nutrition Facts	
Serving size 1 cup (240mL)	
Servings per container about 2	
Amount per Serving	
Calories 90	Calories from Fat 10
	% Daily Value*
Total Fat 1.5 g	2%
Saturated Fat 0 g	0%
Cholesterol 0 mg	0%
Sodium 540 mg	22%
Total Carbohydrate 17 g	6%
Dietary Fiber 3 g	14%
Sugars 5 g	
Protein 3 g	
Vitamin A 30%	Vitamin C 10%
Calcium 2%	Iron 6%

*Percent daily values are based on a 2,000 calorie diet.

47. In one serving of minestrone, how many significant digits are there in the number of milligrams of sodium?

48. Which measurement is more precise, the total amount of fat in cream of tomato soup or the total amount in minestrone? Explain.

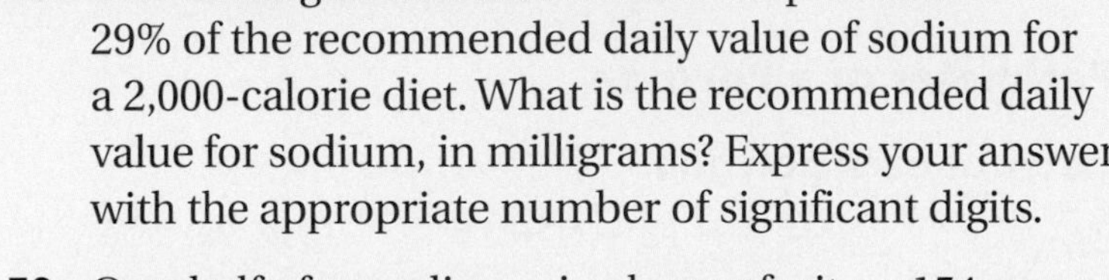

49. One serving of cream of tomato soup contains 29% of the recommended daily value of sodium for a 2,000-calorie diet. What is the recommended daily value for sodium, in milligrams? Express your answer with the appropriate number of significant digits.

50. One-half of a medium-sized grapefruit, or 154 grams, counts as one serving of fruit. How many servings of fruit are in 1 kilogram of grapefruit? Express your answer with the appropriate number of significant digits.

51. ***CHALLENGE*** The greatest possible error of any measurement is half of the smallest unit used in the measurement. For example, 1 pt of juice may actually measure between $\frac{1}{2}$ pt and $1\frac{1}{2}$ pt. What is the range of possible actual weights for a watermelon that was weighed at $19\frac{1}{4}$ lb?

The food pyramid can be used as a guide for meeting daily balanced nutrition needs.

Spiral Review

Write each mixed number as an improper fraction. (Lesson 3-8)

52. $3\frac{1}{2}$ **53.** $7\frac{3}{4}$ **54.** $16\frac{1}{3}$ **55.** $10\frac{2}{5}$

Solve. (Lesson 4-6)

56. $2.7 + y = 9.3$ **57.** $\frac{m}{0.2} = 16$ **58.** $0.5t = 28$ **59.** $n - 8.01 = 16.2$

60. TEST PREP What is 0.2 written as a percent? (Lesson 6-1)

A 20% **C** 0.02%

B 2% **D** 0.002%

61. TEST PREP What is 137% of 52? (Lesson 6-3)

F 7.124 **H** 71.24

G 19.24 **J** 123.24

8-3 Perimeter and Circumference

Learn to find the perimeter of a polygon and the circumference of a circle.

Vocabulary

perimeter

circumference

The distance around a geometric figure is its **perimeter**. To find the perimeter P of a polygon, you can add the lengths of its sides.

If a ball hits the perimeter of a tennis court with chalked lines, a mark will show that it landed in the court. To find the perimeter of a tennis court, add the lengths of each baseline and each sideline.

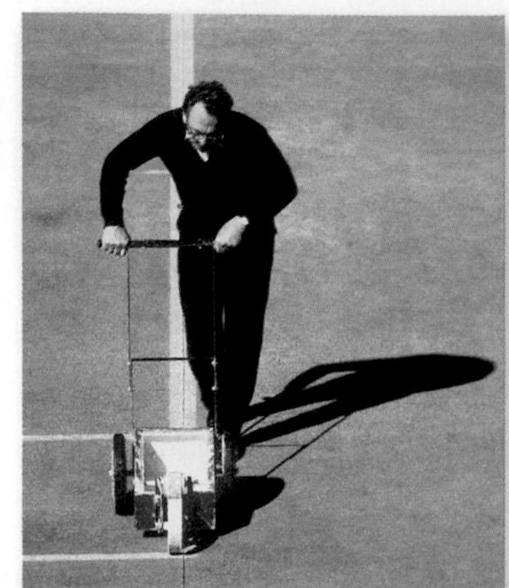

EXAMPLE 1 Finding the Perimeter of a Polygon

Find the perimeter of each polygon.

A

12 in., 8 in., 9 in., 16 in.

$P = 8 + 12 + 9 + 16$ *Use the side lengths.*

$P = 45$ *Add.*

The perimeter of the trapezoid is 45 in.

B

9 cm, 12 cm, 11 cm

$P = 9 + 12 + 11$ *Use the side lengths.*

$P = 32$ *Add.*

The perimeter of the triangle is 32 cm.

You can find the perimeter of a rectangle by adding the lengths of its sides. Or, since opposite sides of a rectangle are equal in length, you can find the perimeter by using the formula $P = 2\ell + 2w$.

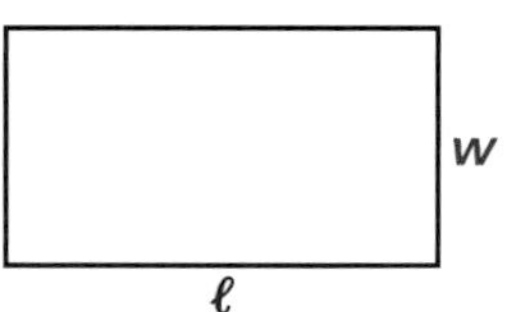

EXAMPLE 2 Using Properties of a Rectangle to Find Perimeter

Find the perimeter of the rectangle.

15 m, 32 m

$P = 2\ell + 2w$ *Use the formula.*

$P = (2 \cdot \mathbf{32}) + (2 \cdot \mathbf{15})$ *Substitute for ℓ and w.*

$P = 64 + 30$ *Multiply.*

$P = 94$ *Add.*

The perimeter of the rectangle is 94 m.

The distance around a circle is called **circumference**. For every circle, the ratio of circumference C to diameter d is the same. This ratio, $\frac{C}{d}$, is represented by the symbol π, called *pi. Pi* is approximately equal to 3.14 or $\frac{22}{7}$. By multiplying both sides of the equation $\frac{C}{d} = \pi$ by d, you get the formula for circumference, $C = \pi d$, or $C = 2\pi r$.

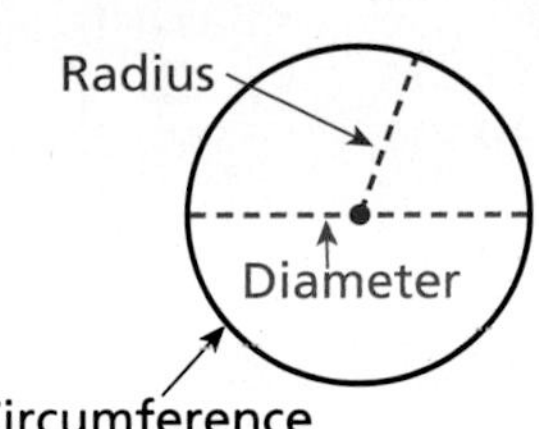

EXAMPLE 3 Finding the Circumference of a Circle

Find the circumference of each circle to the nearest tenth. Use 3.14 for π.

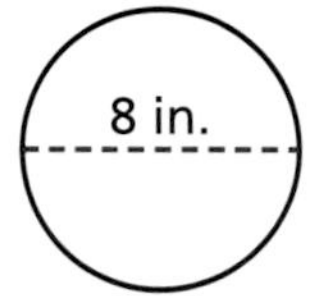

$C = \pi d$	*You know the diameter.*
$C \approx 3.14 \cdot 8$	*Substitute for π and d.*
$C \approx 25.12$	*Multiply.*

The circumference of the circle is about 25.1 in.

B

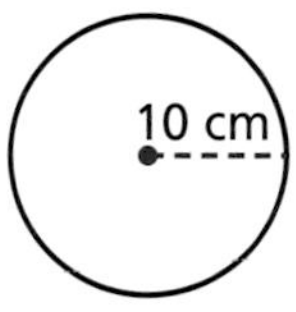

$C = 2\pi r$	*You know the radius.*
$C \approx 2 \cdot 3.14 \cdot 10$	*Substitute for π and r.*
$C \approx 62.8$	*Multiply.*

The circumference of the circle is about 62.8 cm.

EXAMPLE 4 *Art Application*

Melanie is drawing plans for a circular fountain. If the diameter of the fountain is 21 ft, what is its circumference? Use $\frac{22}{7}$ for π.

$C = \pi d$	*You know the diameter.*
$C \approx \frac{22}{7} \cdot 21$	*Substitute $\frac{22}{7}$ for π and 21 for d.*
$C \approx \frac{22}{7} \cdot \frac{21}{1}$	*Write 21 as a fraction.*
$C \approx \frac{22}{\cancel{7}_1} \cdot \frac{\cancel{21}^3}{1}$	*Simplify.*
$C \approx 66$	*Multiply.*

The circumference of the fountain is about 66 ft.

Think and Discuss

1. **Describe** two ways to find the perimeter of a rectangle.
2. **Explain** how to use the formula $C = \pi d$ to find the circumference of a circle if you know the radius.

8-3 Exercises

5.04

FOR EOG PRACTICE
see page 689

internet connect
Homework Help Online
go.hrw.com Keyword: MS4 8-3

GUIDED PRACTICE

See Example 1 **Find the perimeter of each polygon.**

1.
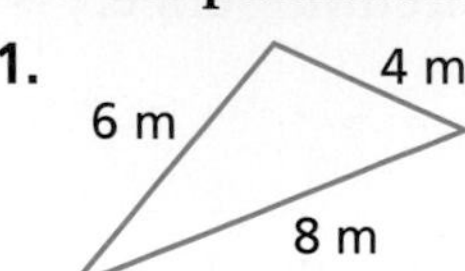

2.
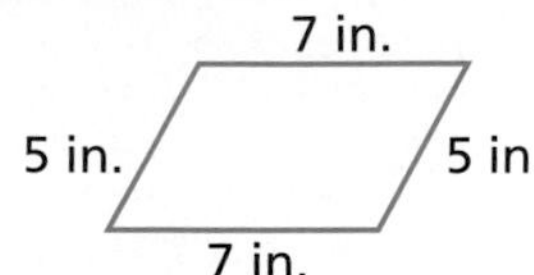

3.
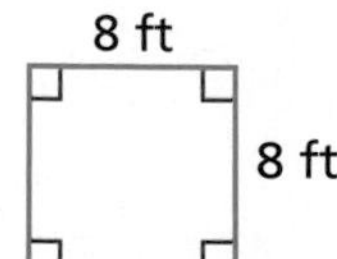

See Example 2 **Find the perimeter of each rectangle.**

4.

5.
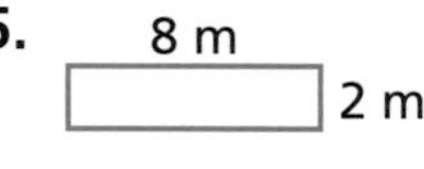

6.
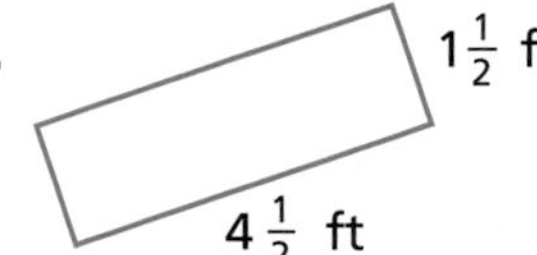

See Example 3 **Find the circumference of each circle to the nearest tenth. Use 3.14 for π.**

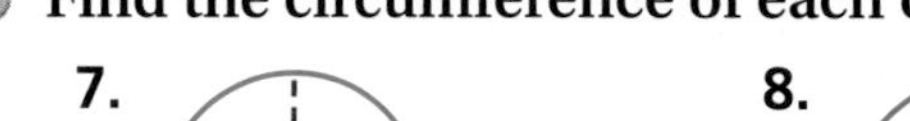

7.
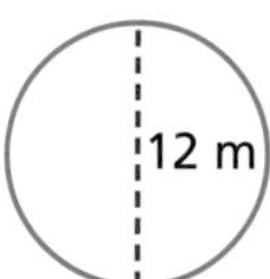

8.
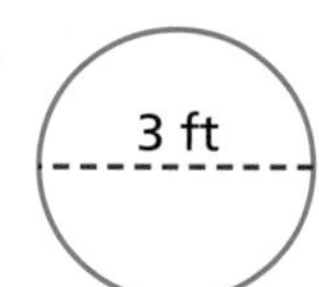

9.

See Example 4 

10. A Ferris wheel has a diameter of 140 feet. What is the circumference of the Ferris wheel? Use $\frac{22}{7}$ for π.

INDEPENDENT PRACTICE

See Example 1  **Find the perimeter of each polygon.**

11.
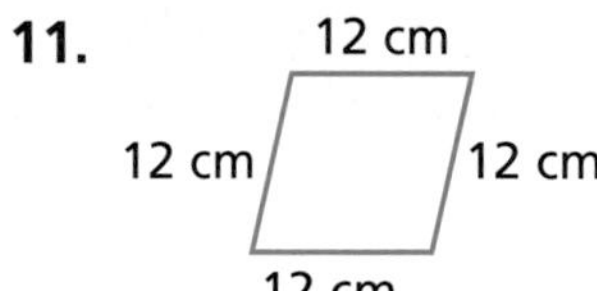

12.
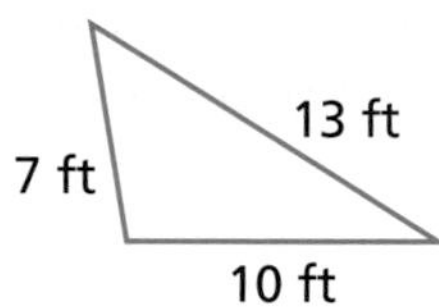

13.
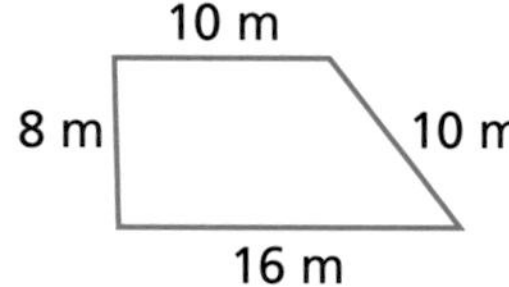

See Example 2  **Find the perimeter of each rectangle.**

14.
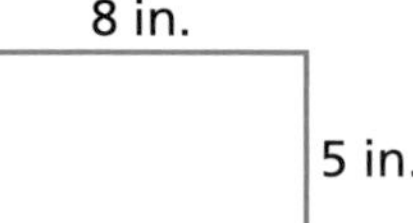

15.
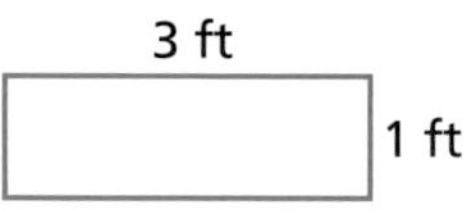

16.
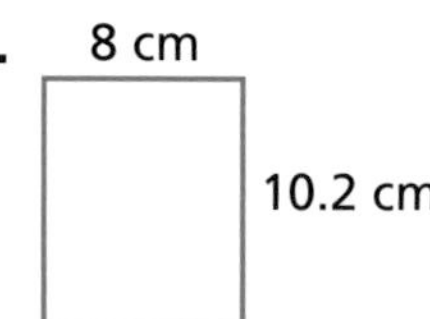

See Example 3  **Find the circumference of each circle to the nearest tenth. Use 3.14 for π.**

17.
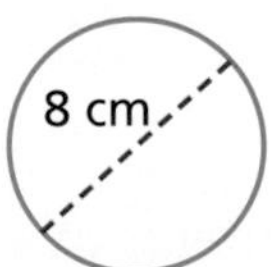
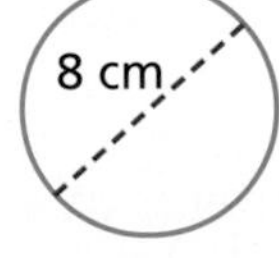

18.
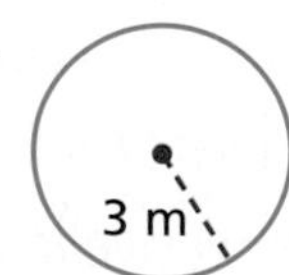

19.

See Example 4 

20. The diameter of Kayla's bicycle wheel is 28 inches. What is the circumference of her bicycle wheel? Use $\frac{22}{7}$ for π.

PRACTICE AND PROBLEM SOLVING

Find each missing measurement to the nearest tenth. Use 3.14 for π.

21. $r = \square$; $d = \square$; $C = 17.8$ m

22. $r = 6.7$ yd; $d = \square$; $C = \square$

23. $r = \square$; $d = 10.6$ in.; $C = \square$

24. $r = \square$; $d = \square$; $C = \pi$

Find the perimeter of each polygon.

25. a parallelogram with side lengths 0.23 m and 0.76 m

26. a regular hexagon with side lengths $4\frac{2}{3}$ in.

27. ***GEOGRAPHY*** The map shows the distances in miles between the airports on the Big Island of Hawaii. If Dembe, a pilot, flies from Kailua-Kona to Waimea to Hilo and back to Kailua-Kona, how far does he travel?

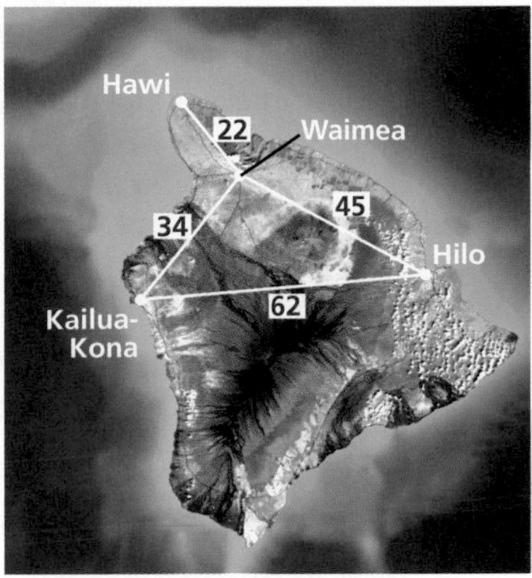

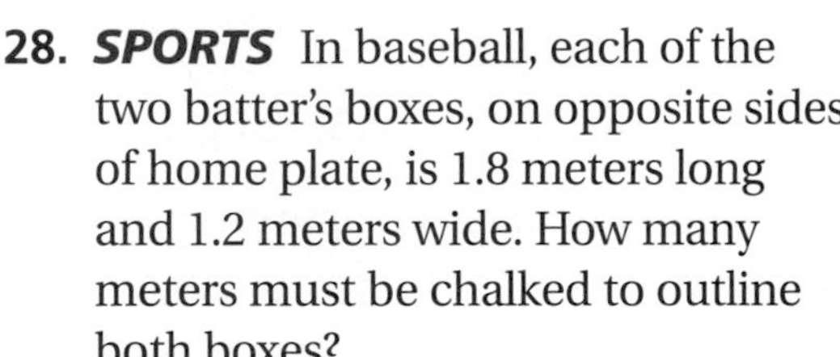

28. ***SPORTS*** In baseball, each of the two batter's boxes, on opposite sides of home plate, is 1.8 meters long and 1.2 meters wide. How many meters must be chalked to outline both boxes?

29. ***ARCHITECTURE*** The Capitol Rotunda connects the House and Senate sides of the U.S. Capitol. The rotunda is 180 feet tall and has a circumference of about 301.5 feet. What is its approximate diameter, to the nearest foot?

30. ***WRITE A PROBLEM*** Write a problem about finding the perimeter or circumference of an object in your school or classroom.

31. ***WRITE ABOUT IT*** Explain how to find the width of a rectangle if you know its perimeter and length.

32. ***CHALLENGE*** The perimeter of a regular nonagon is $25\frac{1}{2}$ in. What is the length of one side of the nonagon?

Spiral Review

Multiply or divide. (Lesson 3-5)

33. $4 \cdot (-3)$ **34.** $-18 \div 9$ **35.** $6 \cdot 9$ **36.** $45 \div (-5)$

Divide. (Lesson 4-5)

37. $9.8 \div 0.4$ **38.** $0.02 \div 0.5$ **39.** $30 \div 2.5$ **40.** $1.752 \div 1.2$

41. **EOG PREP** How many millimeters are there in 4.2 cm? (Lesson 8-1)

A 0.042 mm B 0.42 mm C 42 mm D 420 mm

Chapter 8 Mid-Chapter Quiz

LESSON 8-1 (pp. 414–417)

Convert.

1. 336 hr = ■ days
2. 3,500 cm = ■ m
3. 1,500 s = ■ min
4. 47 kg = ■ g
5. 21,120 ft = ■ mi
6. 324 in. = ■ yd
7. 248 oz = ■ lb
8. 675 km = ■ mm
9. 12 gal = ■ pt
10. How many ounces are there in 16 pounds?

LESSON 8-2 (pp. 420–423)

Choose the more precise measurement in each pair.

11. 5 in., 56 ft
12. 46 cm, 46.2 cm
13. 24 g, 2 kg

Determine the number of significant digits in each measurement.

14. 305.7 km
15. 0.0840 g
16. 6,030.0 mi

Calculate. Use the correct number of significant digits in each answer.

17. 13 + 2.5
18. $5.6 \cdot 2.59$
19. 27.1 − 4
20. $82.5 \div 16$
21. 329 + 640
22. $205.0 \cdot 0.009$

LESSON  **8-3** (pp. 424–427)

Find the perimeter of each polygon.

23.
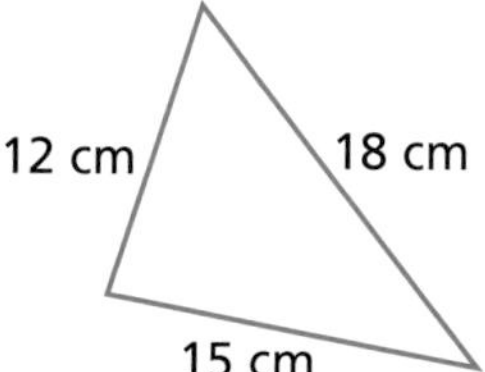

24.
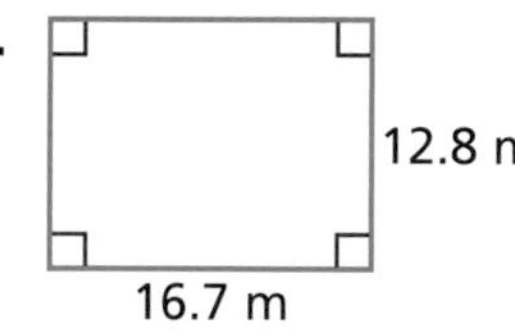

25.
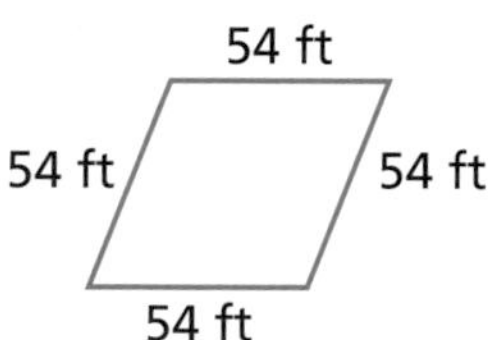

Find the circumference of each circle to the nearest tenth. Use 3.14 for π.

26.
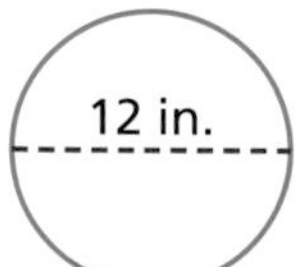

27.

28.
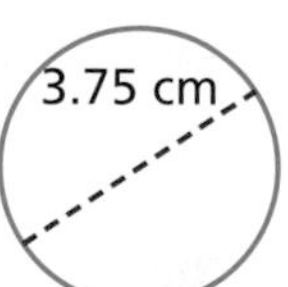

29. The diameter of a circle is 56 centimeters. What is the circumference of the circle? Use $\frac{22}{7}$ for π.

Focus on Problem Solving

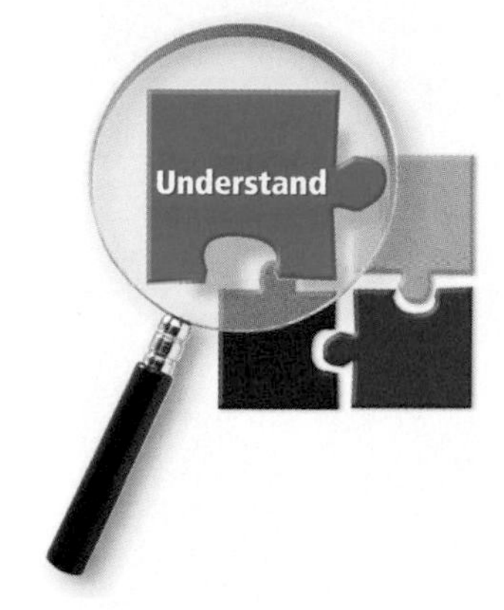

Understand the Problem

- **Restate the problem in your own words**

By writing a problem in your own words, you may understand it better. Before writing the problem, you may need to reread it several times, perhaps aloud so that you can hear yourself saying the words.

Once you have written the problem in your own words, check to make sure you included all of the necessary information to solve it.

Write each problem in your own words. Check to make sure you have included all of the information needed to solve the problem.

1. College basketball is played on a court that measures 94 feet in length and 50 feet in width. The gym crew needs to paint a stripe around the court. What is the distance around the court?

2. The tallest living tree in the world is a redwood in Montgomery State Reserve in California. The tree is 112 meters tall and has a diameter of 3.14 meters. It is estimated to be over 1,000 years old. What is the circumference of this tree?

3. The shape of a raindrop varies, depending on its size and the air resistance as it falls. The smallest raindrop produced during a drizzle has a mass of about 0.004 milligrams. The largest raindrop produced during a heavy storm has a mass of about 300 milligrams. How many significant digits are in the mass of the smallest raindrop?

4. The volume of blood in an average human adult is between 4.7 and 5 liters. People who live at high altitudes, where the air contains less oxygen, may have up to 1.9 liters more blood than people who live at low altitudes. The extra blood delivers additional oxygen to body cells. How many more milliliters of blood do people at high altitudes have?

8-4 Area of Parallelograms

Learn to find the area of rectangles and other parallelograms.

Vocabulary
area

The **area** of a figure is the number of unit squares needed to cover the figure. Area is measured in square units. For example, the area of a chessboard can be measured in square inches. The area of a lawn chessboard is much larger, so it can be measured in square feet or square yards.

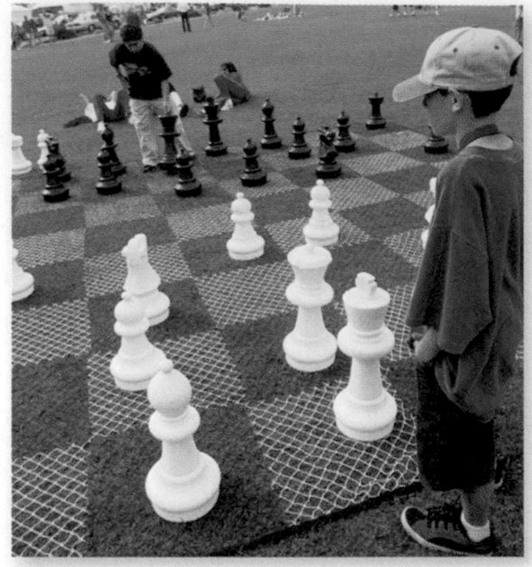

AREA OF A RECTANGLE		
The area A of a rectangle is the product of its length ℓ and its width w.	$A = \ell w$	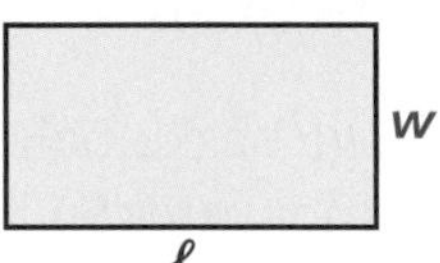

EXAMPLE 1 Finding the Area of a Rectangle

Find the area of the rectangle.

7.5 ft
10 ft

$A = \ell w$ *Use the formula.*

$A = 10 \cdot 7.5$ *Substitute for ℓ and w.*

$A = 75$ *Multiply.*

The area of the rectangle is 75 ft^2.

Helpful Hint

The *base* of the parallelogram is the length of the rectangle. The *height* of the parallelogram is the width of the rectangle.

For any parallelogram that is not a rectangle, you can cut a right triangle-shaped piece from one side and move it to the other side to form a rectangle.

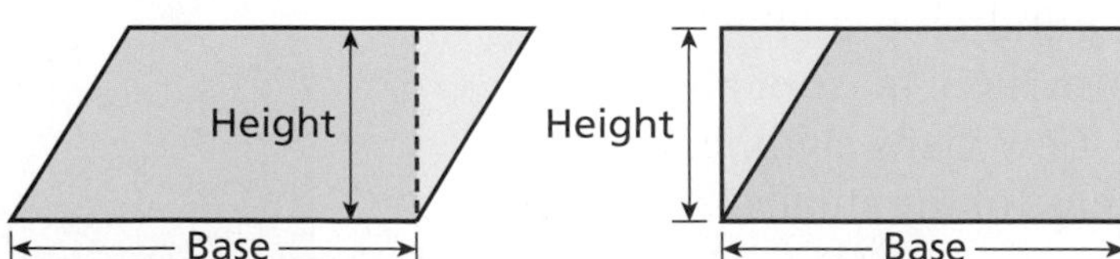

The base of a parallelogram is the length of one side. The height of a parallelogram is the perpendicular distance from the base to the opposite side.

AREA OF A PARALLELOGRAM		
The area A of a parallelogram is the product of its base b and its height h.	$A = bh$	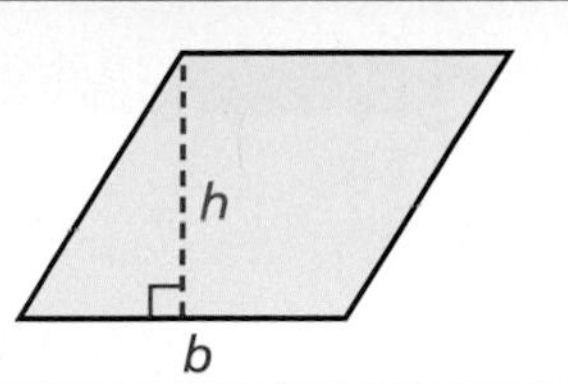

EXAMPLE 2 Finding the Area of a Parallelogram

Find the area of the parallelogram.

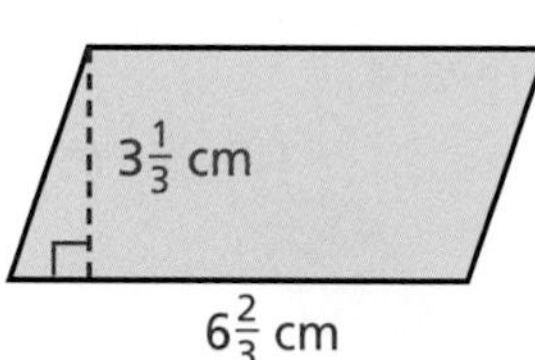

$A = bh$ *Use the formula.*

$A = 6\frac{2}{3} \cdot 3\frac{1}{3}$ *Substitute for b and h.*

$A = \frac{20}{3} \cdot \frac{10}{3}$

$A = \frac{200}{9}$ or $22\frac{2}{9}$

The area of the parallelogram is $22\frac{2}{9}$ cm^2.

EXAMPLE 3 *Measurement Application*

Teesha and Justin are using 4 ft by 8 ft plywood sheets to build a rectangular stage for an outdoor concert. If the area of the stage is 200 ft^2, what is the least number of sheets they need?

First find the area of each sheet of plywood.

$A = \ell w$ *Use the formula for the area of a rectangle.*

$A = 8 \cdot 4$ *Substitute 8 for ℓ and 4 for w.*

$A = 32$ *Multiply.*

The area of each sheet of plywood is 32 ft^2.

To find the number of plywood sheets needed, divide the area of the stage by the area of one sheet.

$\frac{200 \text{ ft}^2}{32 \text{ ft}^2} = 6.25$

Since building the stage requires more than 6 sheets of plywood, Teesha and Justin need at least 7 sheets.

Think and Discuss

1. Write a formula for the area of a square, using an exponent. Explain your answer.

2. Explain why the area of a nonrectangular parallelogram with side lengths 5 in. and 3 in. is not 15 in^2.

8-4 Exercises

FOR EOG PRACTICE

see page 690

internet connect
Homework Help Online
go.hrw.com Keyword: MS4 8-4

5.04

GUIDED PRACTICE

See Example 1

Find the area of each rectangle.

1.

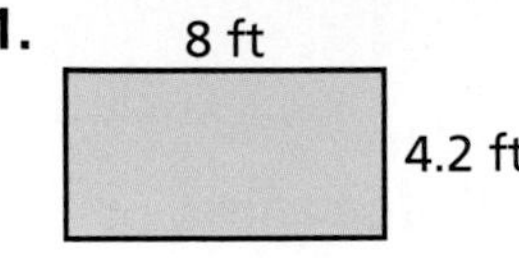

2. 3 m, 7 m

3. 16.4 cm, 9 cm

See Example 2

Find the area of each parallelogram.

4.

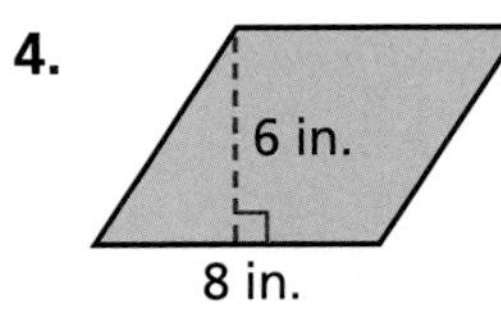

5.

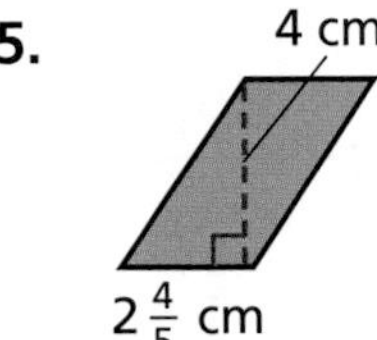

6.

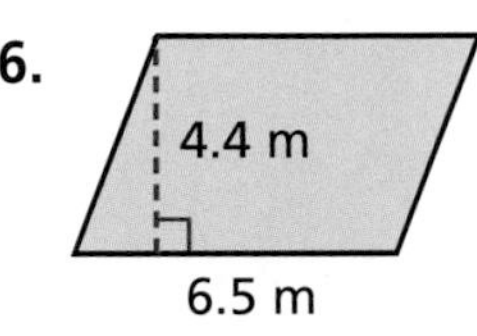

See Example 3

7. Leanne is using 1.5 ft by 1 ft tiles to tile her kitchen floor. If her floor is 10 ft by 6 ft, what is the least number of tiles she will need?

INDEPENDENT PRACTICE

See Example 1

Find the area of each rectangle.

8.

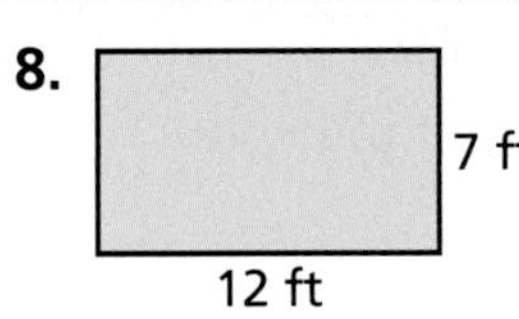

9.

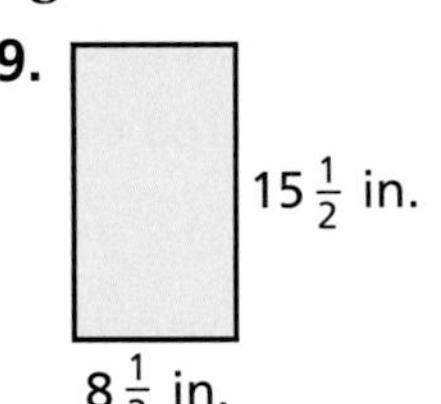

10.

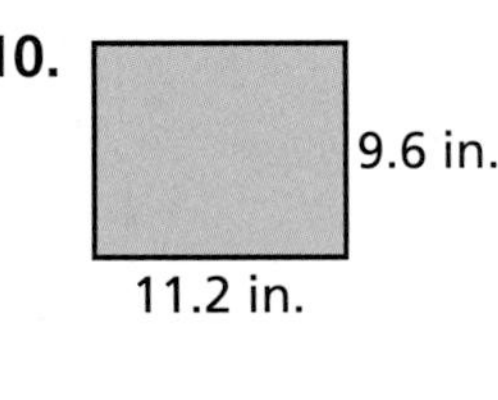

See Example 2

Find the area of each parallelogram.

11.

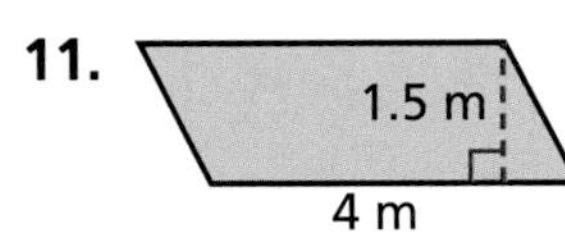

12. $2\frac{1}{3}$ ft, $7\frac{1}{2}$ ft

13.

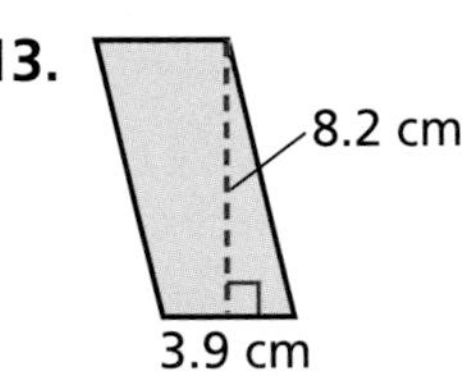

See Example 3

14. Roberto has four 4 ft by 6 ft carpet remnants that he will use to cover a game room floor. If the floor is 9 ft by 12 ft, does he have enough carpet to cover the floor? Explain.

PRACTICE AND PROBLEM SOLVING

Find the area of each polygon.

15. rectangle: $\ell = 9$ yd; $w = 8$ yd

16. parallelogram: $b = 7$ m; $h = 4.2$ m

17. rectangle: $\ell = 16$ cm; $w = 12$ cm

18. parallelogram: $b = 2\frac{1}{2}$ ft; $h = \frac{2}{5}$ ft

Graph the polygon with the given vertices. Then find the area of the polygon.

19. (2, 0), (2, −2), (9, 0), (9, −2)

20. (−3, 1), (−3, 6), (1, 1), (1, 6)

21. (1, 2), (3, 5), (7, 2), (9, 5)

22. (4, 1), (4, 7), (8, 4), (8, 10)

23. What is the height of a parallelogram with an area of 66 in^2 and a base of 11 in.?

24. What is the width of a rectangle with an area of 105 cm^2 and a length of 7.5 cm?

25. ***ART*** Without the frame, the painting *Girl of Tehuantepec* by Diego Rivera measures about 23 in. by 31 in. The width of the frame is 3 in.

a. What is the area of the painting?

b. What is the perimeter of the painting?

c. What is the total area covered by the painting and the frame?

Girl of Tehuantepec by Diego Rivera

26. A local grocery store has diagonal parking spaces that are shaped like parallelograms. If a space is 9 ft wide and 24 ft long, what is its area?

27. ***CHOOSE A STRATEGY*** The area of a parallelogram is 84 cm^2. If the base is 5 cm longer than the height, what is the length of the base?

A 5 cm　**B** 7 cm　**C** 12 cm　**D** 14 cm

28. ***WRITE ABOUT IT*** A rectangle and a parallelogram have sides that measure 3 m, 4 m, 3 m, and 4 m. Do the figures have the same area? Explain.

29. ***CHALLENGE*** Two parallelograms have the same base length, but the height of the first is half that of the second. What is the ratio of the area of the first parallelogram to that of the second? What would the ratio be if both the height and the base of the first parallelogram were half those of the second?

Spiral Review

Solve each equation. (Lesson 3-6)

30. $n - 8 = 16$

31. $-12d = -96$

32. $\frac{t}{-6} = 5$

33. $5 + b = 1$

Find the percent of each number. (Lesson 6-3)

34. 25% of 48

35. 72% of 60

36. 4% of 35

37. 30% of 115

38. **EOG PREP** What is an angle that measures less than 90° called? (Lesson 7-2)

A Right　B Acute　C Obtuse　D Straight

8-5 Area of Triangles and Trapezoids

Learn to find the area of triangles and trapezoids.

The Bermuda Triangle is a triangular region between Bermuda, Florida, and Puerto Rico. To find the area of this region, you could use the formula for the area of a triangle, which is related to the formula for the area of a parallelogram.

A diagonal of a parallelogram divides the parallelogram into two congruent triangles. So the area of each triangle is half the area of the parallelogram.

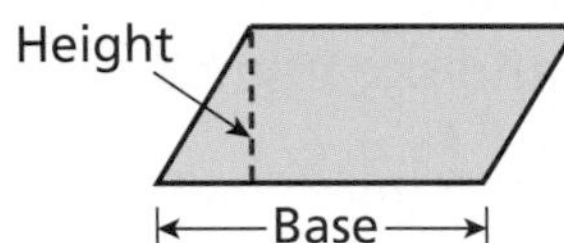

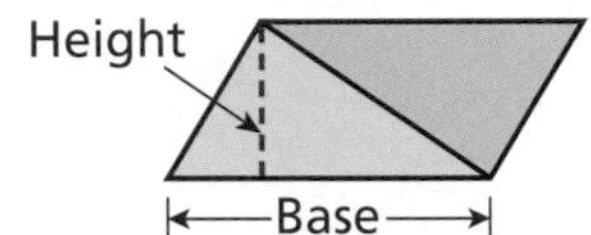

The base of a triangle can be any side. The height of a triangle is the perpendicular distance from the base to the opposite vertex.

AREA OF A TRIANGLE		
The area A of a triangle is half the product of its base b and its height h.	$A = \frac{1}{2}bh$	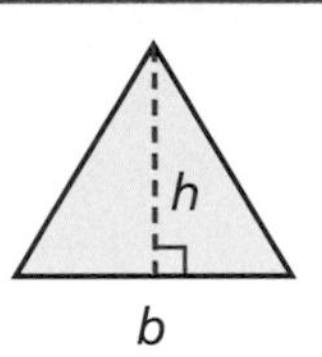

EXAMPLE 1 Finding the Area of a Triangle

Find the area of each triangle.

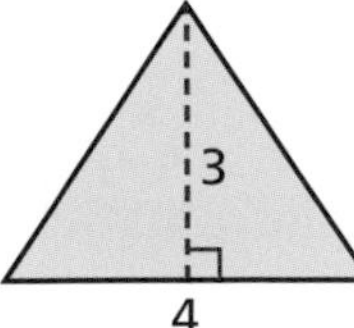

$A = \frac{1}{2}bh$ *Use the formula.*

$A = \frac{1}{2}(4 \cdot 3)$ *Substitute 4 for b and 3 for h.*

$A = 6$

The area of the triangle is 6 square units.

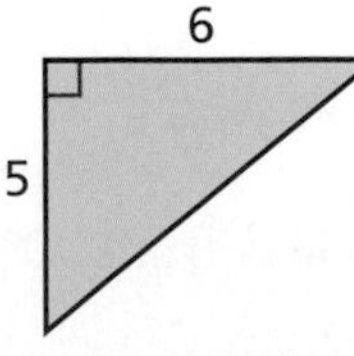

$A = \frac{1}{2}bh$ *Use the formula.*

$A = \frac{1}{2}(6 \cdot 5)$ *Substitute 6 for b and 5 for h.*

$A = 15$

The area of the triangle is 15 square units.

A parallelogram can be divided into two congruent trapezoids. The area of each trapezoid is one-half the area of the parallelogram.

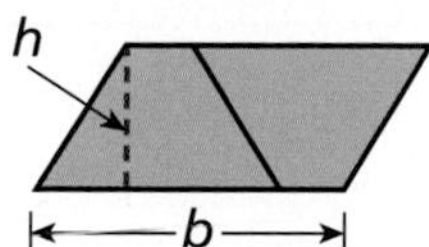

Area of trapezoid $= \frac{1}{2}$*(base of parallelogram)(height)*

The two parallel sides of a trapezoid are its bases. If we call the longer side b_1 and the shorter side b_2, then the base of the parallelogram is $b_1 + b_2$.

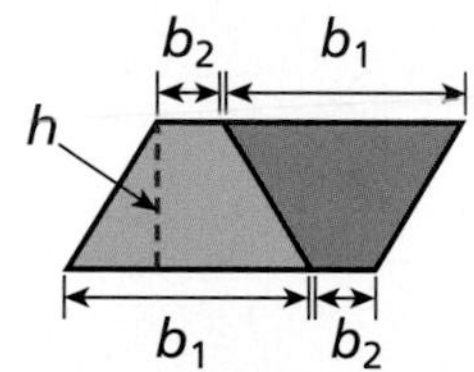

Area of trapezoid $= \frac{1}{2}$*(base 1 + base 2)(height)*

AREA OF A TRAPEZOID		
The area of a trapezoid is half its height multiplied by the sum of the lengths of its two bases.	$A = \frac{1}{2}h(b_1 + b_2)$	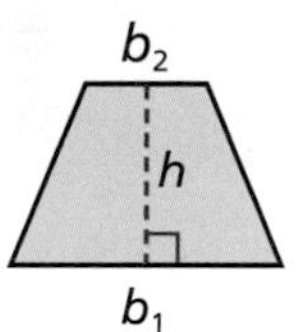

EXAMPLE 2 Finding the Area of a Trapezoid

Find the area of each trapezoid.

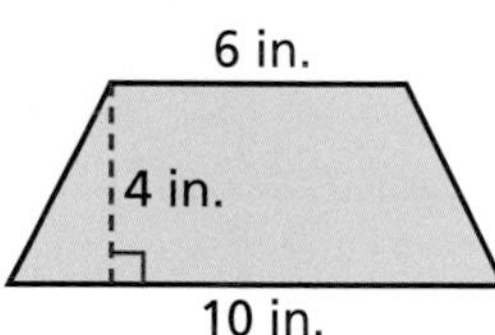

$A = \frac{1}{2}h(b_1 + b_2)$	*Use the formula.*
$A = \frac{1}{2} \cdot 4(10 + 6)$	*Substitute.*
$A = \frac{1}{2} \cdot 4(16)$	*Add.*
$A = 32$	*Multiply.*

The area of the trapezoid is 32 in^2.

B

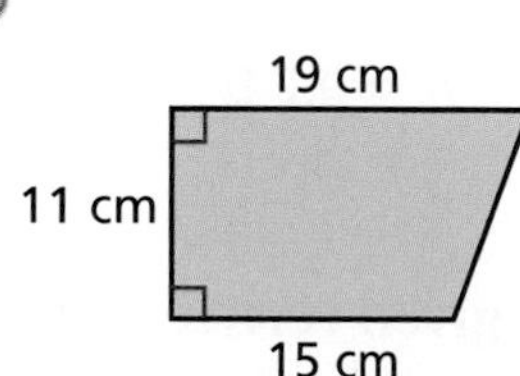

$A = \frac{1}{2}h(b_1 + b_2)$	*Use the formula.*
$A = \frac{1}{2} \cdot 11(15 + 19)$	*Substitute.*
$A = \frac{1}{2} \cdot 11(34)$	*Add.*
$A = 187$	*Multiply.*

The area of the trapezoid is 187 cm^2.

Think and Discuss

1. Tell how to use the sides of a right triangle to find its area.

2. Explain how to find the area of a trapezoid.

8-5 Exercises

FOR EOG PRACTICE
see page 691

internet connect
Homework Help Online
go.hrw.com Keyword: MS4 8-5

5.04

GUIDED PRACTICE

See Example 1 **Find the area of each triangle.**

1. 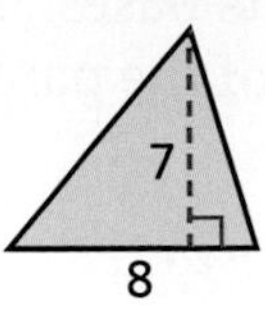

2. 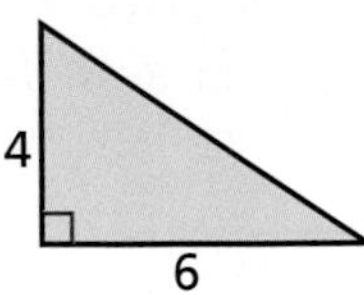

3. 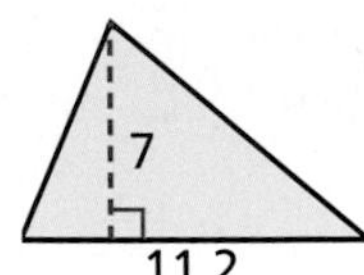

See Example 2 **Find the area of each trapezoid.**

4. 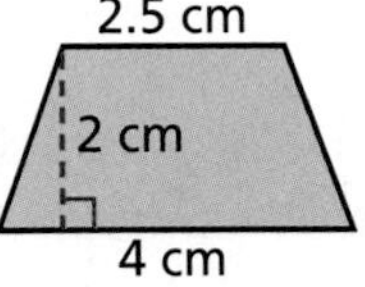

5. 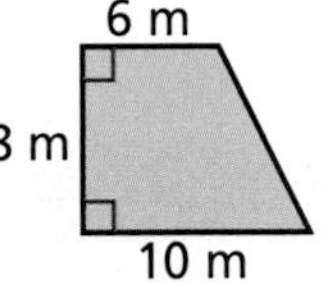

6. 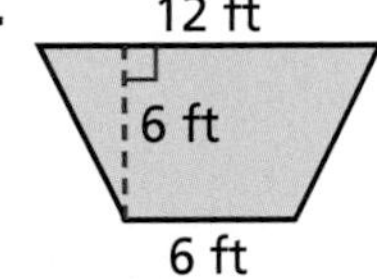

INDEPENDENT PRACTICE

See Example 1 **Find the area of each triangle.**

7. 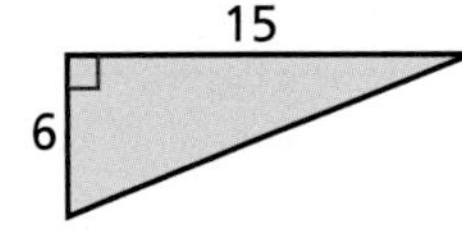

8. 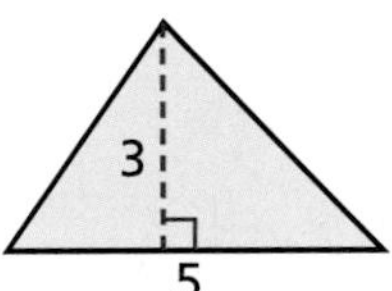

9. 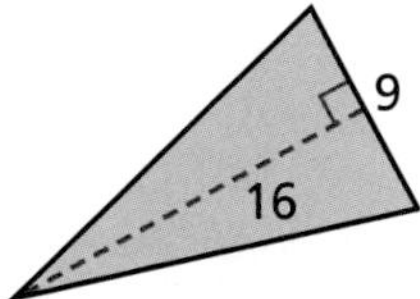

See Example 2 **Find the area of each trapezoid.**

10. 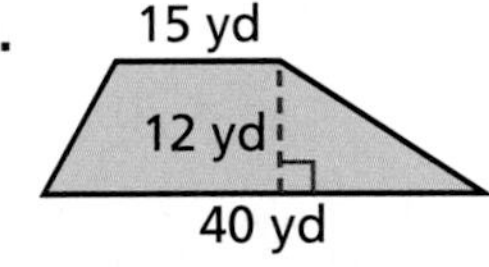

11. 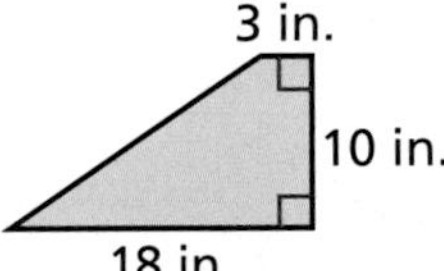

12. 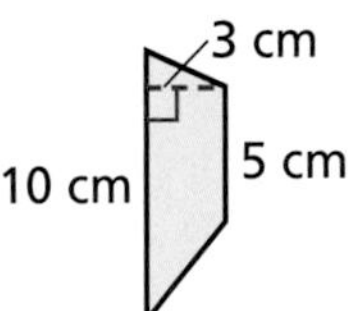

PRACTICE AND PROBLEM SOLVING

Find the missing measurement of each triangle.

13. $b = 8$ cm
$h = \blacksquare$
$A = 18$ cm^2

14. $b = 16$ ft
$h = 0.7$ ft
$A = \blacksquare$

15. $b = \blacksquare$
$h = 95$ in.
$A = 1{,}045$ in^2

Graph the polygon with the given vertices. Then find the area of the polygon.

16. (1, 2), (4, 5), (8, 2), (8, 5)

17. (1, −6), (5, −1), (7, −6)

18. (2, 3), (2, 10), (7, 6), (7, 8)

19. (3, 0), (3, 4), (−3, 0)

20. When the Erie Canal opened, it was 42 ft wide at the top, 28 ft wide at the bottom, and 4 ft deep. Find the area of a trapezoidal cross section of the canal.

21. What is the height of a trapezoid with an area of 9 m^2 and bases that measure 2.4 m and 3.6 m?

Geography LINK

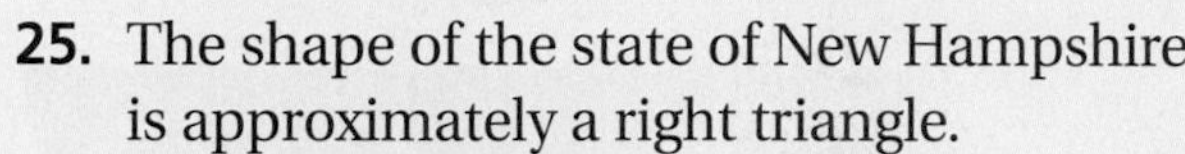

22. The state of Tennessee is shaped somewhat like a parallelogram. Find the approximate area of Tennessee.

23. The state of Nevada is shaped somewhat like a trapezoid. What is the approximate area of Nevada?

24. The state of Colorado is somewhat rectangular in shape. Estimate the perimeter and area of Colorado.

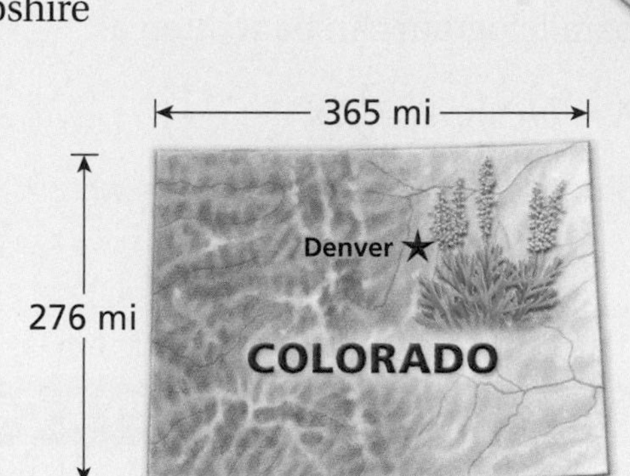

25. The shape of the state of New Hampshire is approximately a right triangle.
    a. Estimate the area of New Hampshire.
    b. In 2000, the population of New Hampshire was 1,235,786. Estimate the population per square mile.

26. ***WRITE ABOUT IT*** Explain how to use the formulas for the area of a rectangle and the area of a triangle to estimate the area of Nevada.

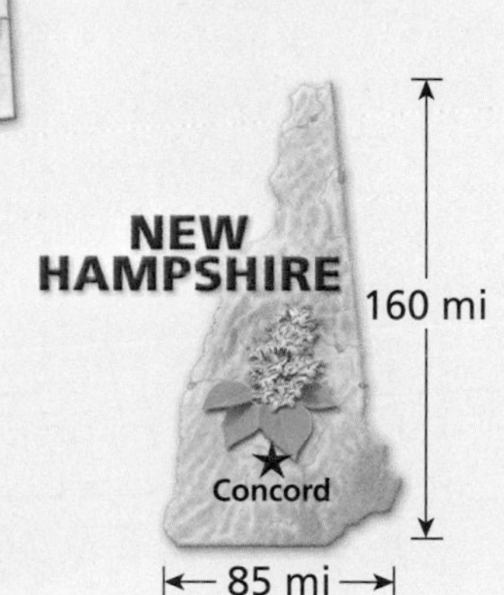

27. ***CHALLENGE*** The state of North Dakota is trapezoidal in shape and has an area of 70,704 mi^2. If the southern border is 359 mi and the distance between the northern border and the southern border is 210 mi, what is the approximate length of the northern border?

Spiral Review

Identify each number as prime or composite. (Lesson 2-4)

28. 16　　**29.** 17　　**30.** 111　　**31.** 29

Write each fraction as a decimal. (Lesson 3-9)

32. $\frac{1}{4}$　　**33.** $\frac{3}{5}$　　**34.** $\frac{5}{8}$　　**35.** $\frac{9}{6}$

36. **EOG PREP** What is the approximate radius of a circle with circumference 36? (Lesson 8-3)

A 3.14　　B 5.73　　C 11.46　　D 113.68

8-6 Area of Circles

Learn to find the area of circles.

A circle can be cut into equal-sized sectors and arranged to resemble a parallelogram. The height h of the parallelogram is equal to the radius r of the circle, and the base b of the parallelogram is equal to one-half the circumference C of the circle. So the area of the parallelogram can be written as

$A = bh$, or $A = \frac{1}{2}Cr$.

Since $C = 2\pi r$, $A = \frac{1}{2}(2\pi r)r = \pi r^2$.

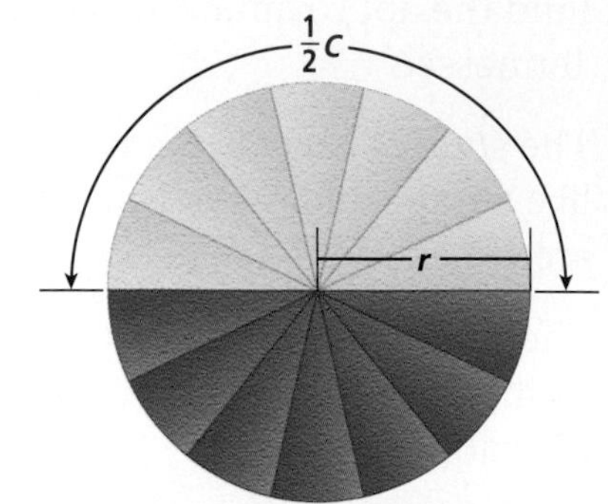

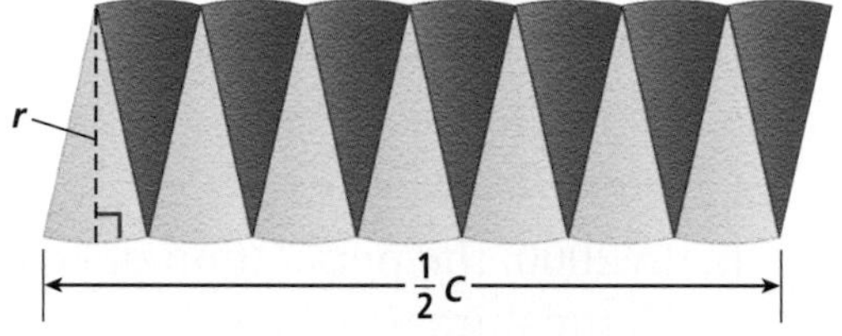

AREA OF A CIRCLE		
The area A of a circle is the product of π and the square of the circle's radius r.	$A = \pi r^2$	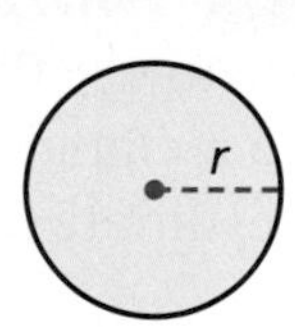

EXAMPLE 1 Finding the Area of a Circle

Remember!

The order of operations calls for evaluating the exponents before multiplying.

Find the area of each circle to the nearest tenth. Use 3.14 for π.

A

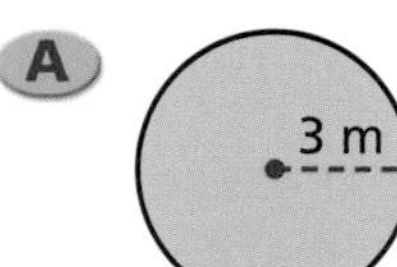

$A = \pi r^2$	*Use the formula.*
$A \approx 3.14 \cdot 3^2$	*Substitute. Use 3 for r.*
$A \approx 3.14 \cdot 9$	*Evaluate the power.*
$A \approx 28.26$	*Multiply.*

The area of the circle is about 28.3 m^2.

B

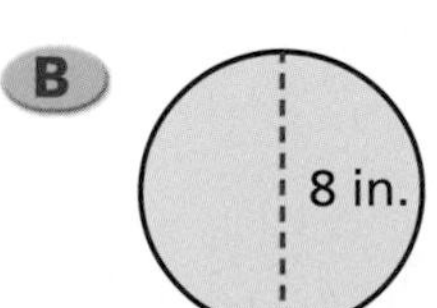

$A = \pi r^2$	*Use the formula.*
$A \approx 3.14 \cdot 4^2$	*Substitute. Use 4 for r.*
$A \approx 3.14 \cdot 16$	*Evaluate the power.*
$A \approx 50.24$	*Multiply.*

The area of the circle is about 50.2 in^2.

EXAMPLE 2 Social Studies Application

Social Studies LINK

Nomads in Mongolia carried their homes wherever they roamed. These homes, called *yurts*, were made of wood and felt.

A group of historians are building a yurt to display at a local multicultural fair. The yurt has a height of 8 feet 9 inches at its center, and it has a circular floor of radius 7 feet. What is the area of the floor of the yurt? Use $\frac{22}{7}$ for π.

$A = \pi r^2$ — *Use the formula for the area of a circle.*

$A \approx \frac{22}{7} \cdot 7^2$ — *Substitute. Use 7 for r.*

$A \approx \frac{22}{\cancel{7}_1} \cdot \cancel{49}^7$ — *Evaluate the power.*

$A \approx 22 \cdot 7$

$A \approx 154$ — *Multiply.*

The area of the floor of the yurt is about 154 ft^2.

EXAMPLE 3 Agriculture Application

These fields are located in the Wadi Rum Desert in Jordan. Here crops are grown on circular patches of irrigated land. If the radius of one irrigated field is 60 feet, what is the area of the field? Round your answer to the nearest whole number.

$A = \pi r^2$ — *Use the formula for the area of a circle.*

$A = \pi \cdot 60^2$ — *Substitute. Use 60 for r.*

$A \approx 11309.73355$ — *Use a calculator.*   60 

$A \approx 11{,}310$ — *Round.*

The sprinkler covers about 11,310 ft^2.

Think and Discuss

1. **Compare** finding the area of a circle when given the radius with finding the area when given the diameter.
2. **Explain** how to find the area of a circle with a diameter of 3 feet.
3. **Give an example** of a circular object in your classroom. Tell how you could estimate the area of the object, and then estimate.

8-6 Exercises

FOR EOG PRACTICE
see page 691

internet connect
Homework Help Online
go.hrw.com Keyword: MS4 8-6

5.04

GUIDED PRACTICE

See Example 1 **Find the area of each circle to the nearest tenth. Use 3.14 for π.**

1. 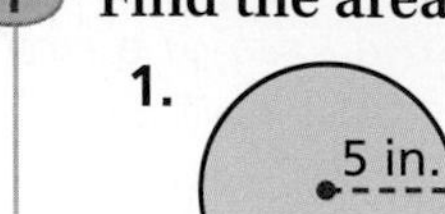

2. 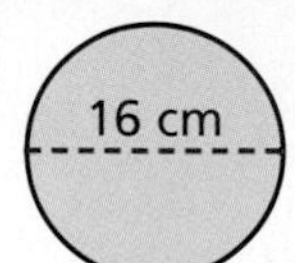

3. 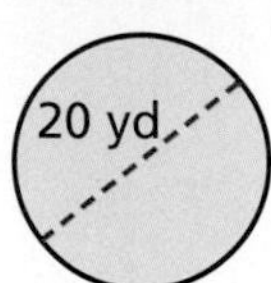

See Example  **4.** The most popular pizza at Sam's Pizza is the 14-inch pepperoni pizza. What is the area of a pizza with a diameter of 14 inches? Use $\frac{22}{7}$ for π.

See Example  **5.** A radio station broadcasts a signal over an area with a 75-mile radius. What is the area of the region that receives the radio signal?

INDEPENDENT PRACTICE

See Example 1 **Find the area of each circle to the nearest tenth. Use 3.14 for π.**

6. 

7. 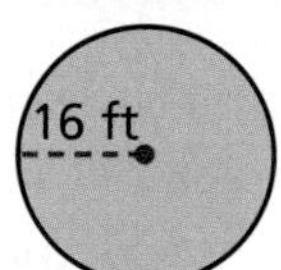

8. 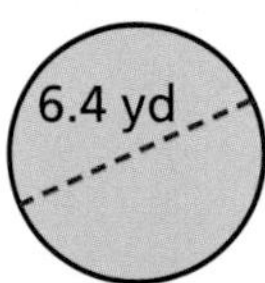

9. 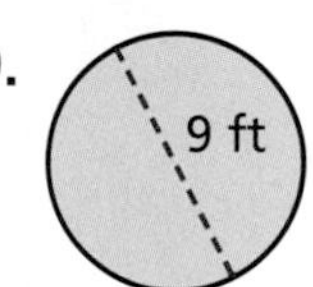

10. 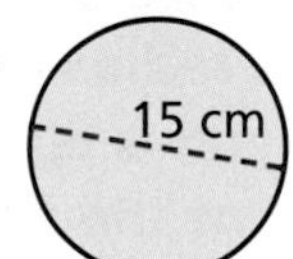

11. 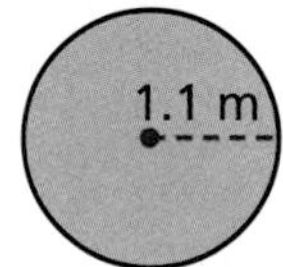

See Example  **12.** A circle has a radius of 14 centimeters. What is the area of the circle? Use $\frac{22}{7}$ for π.

See Example  **13.** A circular flower bed in Kay's backyard has a diameter of 8 feet. What is the area of the flower bed? Round your answer to the nearest tenth.

14. A company is manufacturing aluminum lids. If the radius of each lid is 3 centimeters, what is the area of one lid? Round your answer to the nearest tenth.

PRACTICE AND PROBLEM SOLVING

Given the radius or diameter, find the circumference and area of each circle to the nearest tenth. Use 3.14 for π.

15. $r = 7$ m **16.** $d = 18$ in. **17.** $d = 24$ ft

18. $r = 11$ m **19.** $r = 6.4$ cm **20.** $d = 19$ in.

Given the area, find the radius of each circle. Use 3.14 for π.

21. $A = 113.04\ \text{cm}^2$ **22.** $A = 3.14\ \text{ft}^2$ **23.** $A = 28.26\ \text{in}^2$

24. ***PHYSICAL SCIENCE*** The tower of a wind turbine is about the height of a 20-story building, and each turbine can produce 24 megawatt-hours of electricity in one day. Find the area covered by the turbine when it is rotating. Use 3.14 for π. Round your answer to the nearest tenth.

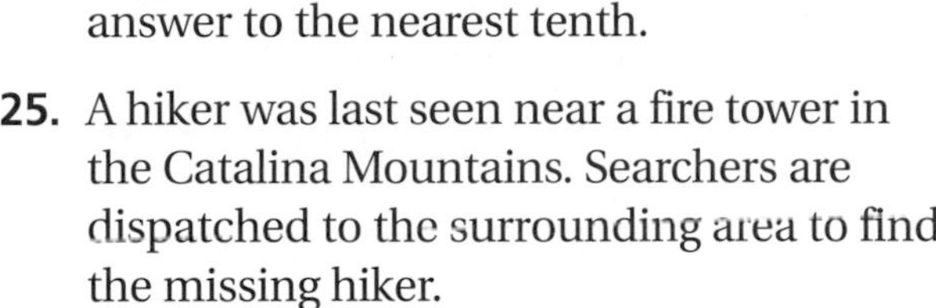

25. A hiker was last seen near a fire tower in the Catalina Mountains. Searchers are dispatched to the surrounding area to find the missing hiker.

a. Assume the hiker could walk in any direction at a rate of 3 miles per hour. How large an area would searchers have to cover if the hiker was last seen 2 hours ago? Use 3.14 for π. Round your answer to the nearest square mile.

b. How much additional area would the searchers have to cover if the hiker was last seen 3 hours ago?

26. Two sprinklers are installed on opposite sides of a square lawn. Each sprinkler waters a semicircular area as shown. Find the area of the lawn not watered by the sprinklers. Use 3.14 for π. Round your answer to the nearest square yard.

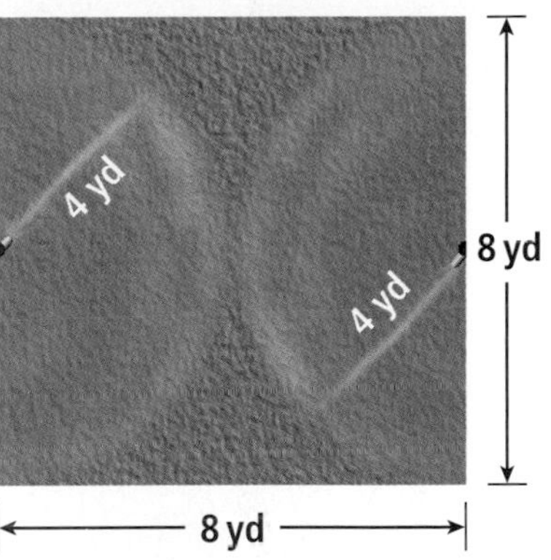

27. ***WHAT'S THE QUESTION?*** Chang painted half of a free-throw circle that had a diameter of 12 ft. The answer is 56.52 ft^2. What is the question?

28. ***WRITE ABOUT IT*** Two circles have the same radius. Is the combined area of the two circles the same as the area of one circle with twice the radius? Explain your answer.

29. ***CHALLENGE*** How does the area of a circle change if you multiply the radius by a factor of n, where n is a whole number?

Spiral Review

Simplify. (Lesson 2-1)

30. 2^6 **31.** 5^3 **32.** 8^2 **33.** 3^5

Add or subtract. (Lessons 3-4, 3-5)

34. $9 + (-3)$ **35.** $-3 - 8$ **36.** $-2 + (-9)$ **37.** $-12 - 6$

38. **EOG PREP** Which type of triangle is ΔABC? (Lesson 7-6)

A Acute B Right C Scalene D Obtuse

A
B C

Chapter 8

Mid-Chapter Quiz

LESSON 8-1 (pp. 414–417)

Convert.

1. 132 hr = ▬ days
2. 748 m = ▬ km
3. 12.5 gal = ▬ pt

LESSON 8-2 (pp. 420–423)

Choose the more precise measurement in each pair.

4. 12 yd, 36 ft
5. 27 cm, 30.2 m
6. 150 g, 150.0 g

Calculate. Use the correct number of significant digits in each answer.

7. $1.5 + 1.33$
8. $4.9 \cdot 2.70$
9. $2.2 - 1.03$

LESSON 8-3 (pp. 424–427)

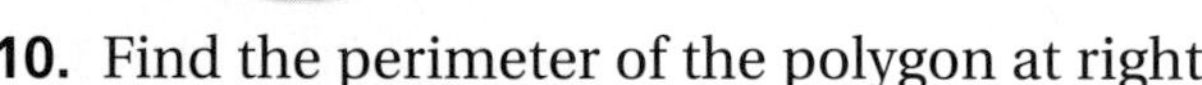

10. Find the perimeter of the polygon at right.

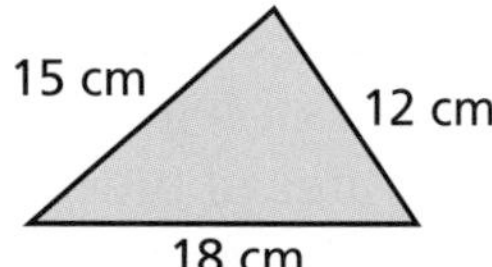

11. Find the circumference of a circle with a diameter of 3.75 cm. Use 3.14 for π.

LESSONS  8-4 AND  8-5 (pp. 430–437)

Find the area of each figure.

12. 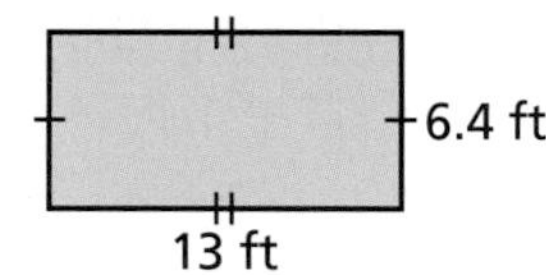

13. 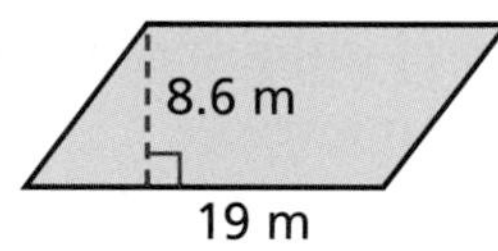

14. 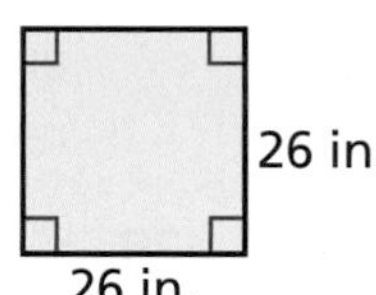

15. 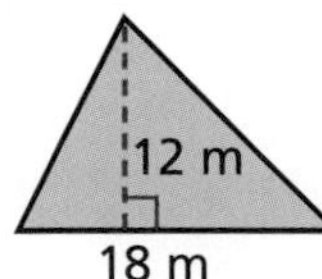

16. 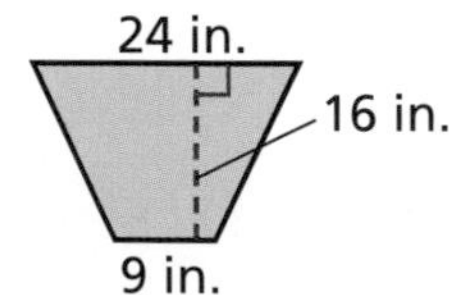

17. 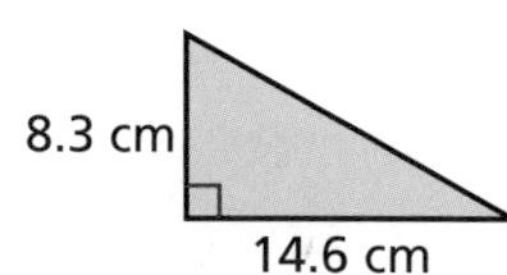

LESSON 8-6 (pp. 438–441)

Find the area of each circle to the nearest tenth. Use 3.14 for π.

18. 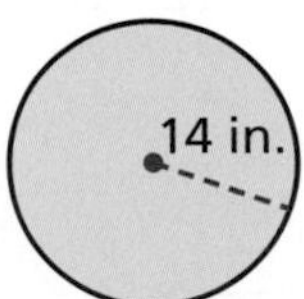

19. 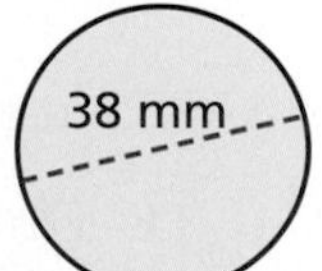

20. 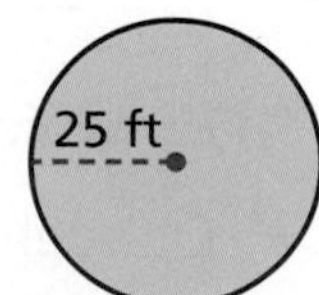

Mid-Chapter Quiz

Focus on Problem Solving

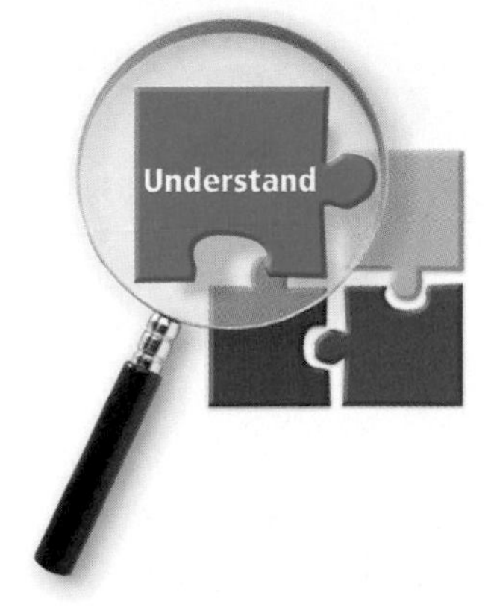

Understand the Problem

• Identify too much or too little information

Problems involving real-world situations sometimes give too much or too little information. Before solving these types of problems, you must decide what information is necessary and whether you have all the necessary information.

If the problem gives too much information, identify which of the facts are really needed to solve the problem. If the problem gives too little information, determine what additional information is required to solve the problem.

Copy each problem and underline the information you need to solve it. If necessary information is missing, write down what additional information is required.

1. Mrs. Wong wants to put a fence around her garden. One side of her garden measures 8 feet. Another side measures 5 feet. What length of fencing does Mrs. Wong need to enclose her garden?

2. Two sides of a triangle measure 17 inches and 13 inches. The perimeter of the triangle is 45 inches. What is the length in feet of the third side of the triangle? (There are 12 inches in 1 foot.)

3. During swim practice, Peggy swims 2 laps each of freestyle and backstroke. The dimensions of the pool are 25 meters by 50 meters. What is the area of the pool?

4. Each afternoon, Curtis walks his dog two times around the park. The park is a rectangle that is 315 yards long. How far does Curtis walk his dog each afternoon?

5. A trapezoid has bases that measure 12 meters and 18 meters and one side that measures 9 meters. The trapezoid has no right angles. What is the area of the trapezoid?

8-7 Powers and Roots

Learn to express and evaluate numbers using powers and roots.

Vocabulary

perfect square

square root

radical sign

Recall that a power is a number represented by a base and an exponent. The exponent tells you how many times to use the base as a repeated factor.

A square with sides that measure 3 units each has an area of $3 \cdot 3$, or 3^2. Notice that the area of the square is represented by a power in which the base is the side length and the exponent is 2. A power in which the exponent is 2 is called a *square.*

EXAMPLE 1 Finding Squares

Model each power using a square. Then evaluate the power.

A 7^2

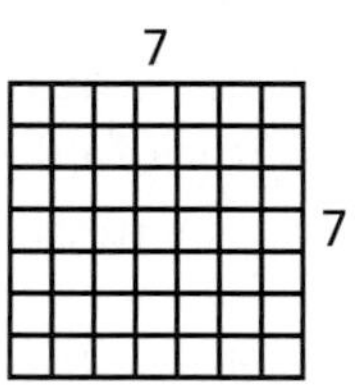

$A = \ell w$

$A = 7 \cdot 7$ *Substitute.*

$A = 49$ *Multiply.*

The square of 7 is 49.

B $(2.5)^2$

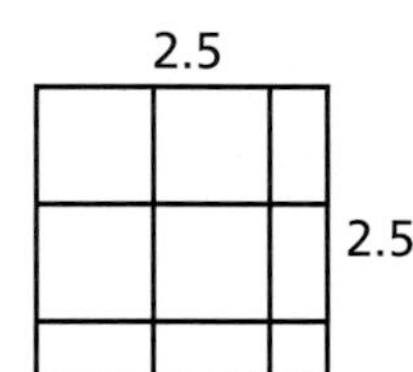

$A = \ell w$

$A = 2.5 \cdot 2.5$ *Substitute.*

$A = 6.25$ *Multiply.*

The square of 2.5 is 6.25.

A **perfect square** is the square of a whole number. The number 49 is a perfect square because $49 = 7^2$ and 7 is a whole number. The number 6.25 is not a perfect square.

Reading Math

$\sqrt{16} = 4$ is read as "The square root of 16 is 4."

The **square root** of a number is one of the two equal factors of the number. Four is a square root of 16 because $4 \cdot 4 = 16$. The symbol for a square root is $\sqrt{\ }$, which is called a **radical sign**.

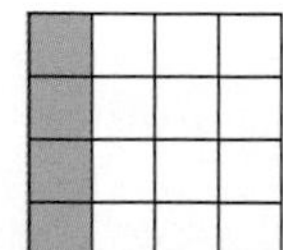

Most calculators have square-root keys that you can use to quickly find approximate square roots of nonperfect squares. You can also use perfect squares to estimate the square roots of nonperfect squares.

EXAMPLE 2 Estimating Square Roots

Estimate each square root to the nearest whole number. Use a calculator to check your answer.

A $\sqrt{30}$

$25 < 30 < 36$ *Find the perfect squares nearest 30.*

$\sqrt{25} < \sqrt{30} < \sqrt{36}$

$5 < \sqrt{30} < 6$ *Find the square roots of 25 and 36.*

$\sqrt{30} \approx 5$ *30 is nearer in value to 25 than to 36.*

Check

$\sqrt{30} \approx 5.477225575$ *Use a calculator to approximate $\sqrt{30}$. 5 is a reasonable estimate.*

B $\sqrt{99}$

$81 < 99 < 100$ *Find the perfect squares nearest 99.*

$\sqrt{81} < \sqrt{99} < \sqrt{100}$

$9 < \sqrt{99} < 10$ *Find the square roots of 81 and 100.*

$\sqrt{99} \approx 10$ *99 is nearer in value to 100 than to 81.*

Check

$\sqrt{99} \approx 9.949874371$ *Use a calculator to approximate $\sqrt{99}$. 10 is a reasonable estimate.*

EXAMPLE 3 *Recreation Application*

While searching for a lost hiker, a helicopter pilot covers a square area of 150 mi^2. What is the approximate length of each side of the square area? Round your answer to the nearest mile.

The length of each side of the square is $\sqrt{150}$.

$144 < 150 < 169$ *Find the perfect squares nearest 150.*

$\sqrt{144} < \sqrt{150} < \sqrt{169}$

$12 < \sqrt{150} < 13$ *Find the square roots of 144 and 169.*

$\sqrt{150} \approx 12$ *150 is nearer in value to 144 than to 169.*

Each side of the search area is about 12 miles long.

Think and Discuss

1. **Explain** how to estimate $\sqrt{75}$.
2. **Explain** how you might find the square root of 3^2.

8-7 Exercises

FOR EOG PRACTICE

see page 692

internet connect

Homework Help Online
go.hrw.com Keyword: MS4 8-7

1.03

GUIDED PRACTICE

See Example 1 **Model each power using a square. Then evaluate the power.**

1. 4^2 2. $(1.5)^2$ 3. 9^2 4. 6^2

See Example 2 **Estimate each square root to the nearest whole number. Use a calculator to check your answer.**

5. $\sqrt{20}$ 6. $\sqrt{45}$ 7. $\sqrt{84}$ 8. $\sqrt{58}$

See Example 3

9. A Coast Guard ship patrols an area of 125 square miles. If the area the ship patrols is a square, about how long is each side of the area? Round your answer to the nearest mile.

INDEPENDENT PRACTICE

See Example 1 **Model each power using a square. Then evaluate the power.**

10. 3^2 11. 8^2 12. 11^2 13. $(4.5)^2$

See Example 2 **Estimate each square root to the nearest whole number. Use a calculator to check your answer.**

14. $\sqrt{12}$ 15. $\sqrt{39}$ 16. $\sqrt{73}$ 17. $\sqrt{109}$

18. $\sqrt{6}$ 19. $\sqrt{180}$ 20. $\sqrt{145}$ 21. $\sqrt{216}$

See Example 3

22. The area of a square field is 200 ft^2. What is the approximate length of each side of the field? Round your answer to the nearest foot.

23. A square bandanna has an area of 1,000 cm^2. About how long is each edge of the bandanna? Round your answer to the nearest centimeter.

PRACTICE AND PROBLEM SOLVING

Evaluate each power.

24. 20^2 25. $\left(\frac{3}{8}\right)^2$ 26. $(0.16)^2$ 27. $\left(\frac{1}{3}\right)^2$

Estimate each square root to the nearest whole number. Use a calculator to check your answer.

28. $\sqrt{300}$ 29. $\sqrt{420}$ 30. $\sqrt{700}$ 31. $\sqrt{1,500}$

Estimate each sum or difference to the nearest whole number.

32. $\sqrt{18} + \sqrt{9}$ 33. $\sqrt{34} + \sqrt{35}$ 34. $\sqrt{50} - \sqrt{10}$

35. $\sqrt{98} - \sqrt{89}$ 36. $\sqrt{8} + 8^2$ 37. $14^2 - \sqrt{14}$

38. Order π, $\sqrt{15}$, 2.9, $\sqrt{20}$, and $\frac{25}{6}$ from greatest to least.

39. Find the perimeter of a square whose area is 49 square inches.

To find the distance at which an object becomes visible, you can use your distance to the horizon and the object's distance to the horizon.

40. ***EARTH SCIENCE*** The formula $D = 3.56 \cdot \sqrt{A}$ gives the distance D in kilometers to the horizon from an airplane flying at an altitude A in meters. If a pilot is flying at an altitude of 1,800 m, about how far away is the horizon? Round your answer to the nearest kilometer.

41. ***ART*** An artist is making two square stained-glass windows. One window has a perimeter of 48 inches. The other has an area of 110 square inches. Which window is bigger? Explain.

42. The Ricci family is building a new house on a square foundation. The floor area of the house is 1,316 ft^2. What is the length of one wall of the house, to the nearest foot?

43. Darien will need 154 ft^2 of wall-to-wall carpeting to cover the floor of his new square bedroom. What is the length of each wall of his room? Round your answer to the nearest foot.

44. For his new room, Darien's grandmother gave him a handmade quilt. The quilt is made up of 16 squares set in 4 rows of 4. If the area of each square is 324 in^2, what are the dimensions of the quilt in inches?

45. ***CHOOSE A STRATEGY*** The figure shows how two squares can be formed by drawing only seven lines. Show how two squares can be formed by drawing only six lines.

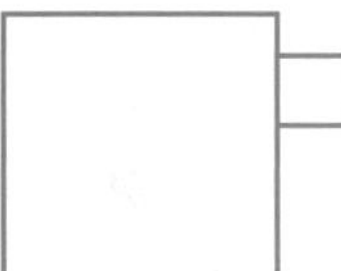

46. ***WRITE ABOUT IT*** Explain how to estimate the square root of a nonperfect square.

47. ***CHALLENGE*** Find the value of $\sqrt{5^2 + 12^2}$.

Spiral Review

Given the radius or diameter, find the circumference of each circle to the nearest tenth. Use 3.14 for π. (Lesson 8-3)

48. $d = 6$ ft **49.** $d = 9$ cm **50.** $r = 4$ in. **51.** $r = 5$ m

Find the area of each parallelogram. (Lesson 8-4)

52. $b = 5$ cm; $h = 3$ cm **53.** $b = 12$ m; $h = 9$ m **54.** $b = 4.8$ ft; $h = 2$ ft

55.  **EOG PREP** What is the area of the trapezoid? (Lesson 8-5)

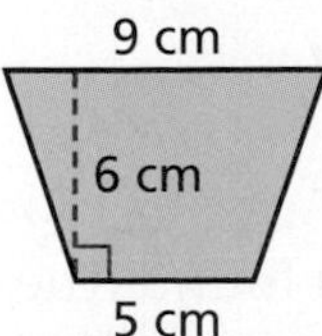

A $42\ cm^2$

B $84\ cm^2$

C $135\ cm^2$

D $270\ cm^2$

Explore the Pythagorean Theorem

Use with Lesson 8-8

An important and famous relationship in mathematics, known as the Pythagorean Theorem, involves the three sides of a right triangle. Recall that a right triangle is a triangle that has one right angle. If you know the lengths of two sides of a right triangle, you can find the length of the third side.

Activity 1

1. The drawing at right shows an isosceles right triangle and three squares. Make your own drawing similar to the one shown. (Recall that an isosceles right triangle has two congruent sides and a right angle.)

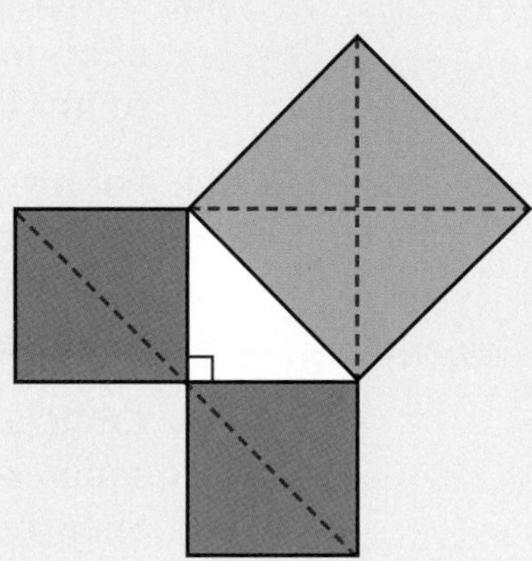

Cut out the two smaller squares of your drawing, then cut those squares in half along a diagonal. Fit the pieces of the smaller squares on top of the blue square.

Think and Discuss

1. What can you tell about the relationship between the areas of the squares?
2. a. How does the side length of a square relate to the area of the square?

   b. How do the side lengths of the triangle in your drawing relate to the areas of the squares around it?

   c. Write an equation that shows the relationship between the lengths of the sides of the triangle in your drawing. Use the variables a and b to represent the lengths of the two shorter sides of your triangle, and c to represent the length of the longest side.

Try This

1. Repeat Activity 1 for other isosceles right triangles. Is the relationship that you found true for the areas of the squares around each triangle?

Activity 2

1 On graph paper, draw a segment that is 3 units long. At one end of this segment, draw a perpendicular segment that is 4 units long. Draw a third segment to form a triangle. Cut out the triangle.

Cut out a 3-by-3 square and a 4-by-4 square from the same graph paper. Place the edges of the squares against the corresponding sides of the right triangle.

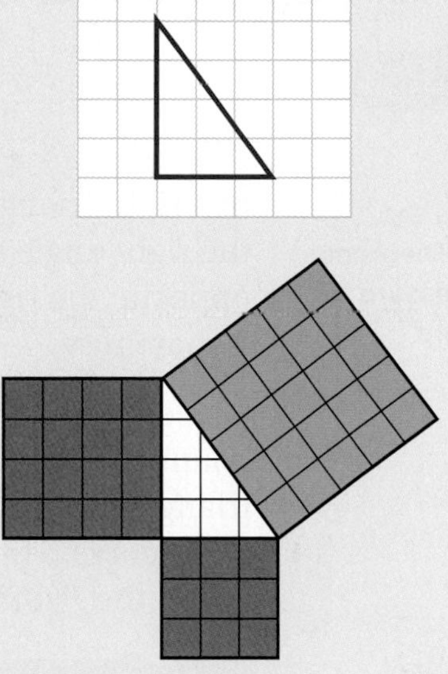

Cut the two squares into individual squares or strips. Arrange the squares into a large square along the third side of the triangle.

Think and Discuss

1. What is the area of each of the three squares? What relationship is there between the areas of the small squares and the area of the large square?
2. What is the length of the third side of the triangle?
3. Substitute the side lengths of your triangle into the equation you wrote in Think and Discuss problem **2c** in Activity 1. What do you find?
4. Do you think the relationship is true for triangles that are not right triangles?

Try This

1. Use graph paper to cut out three squares with sides that are 3 units, 4 units, and 6 units long. Fit the squares together to form a triangle as shown at right. Is the relationship between the areas of the red squares and the area of the blue square the same as the relationship shown in Activity 2? Explain.

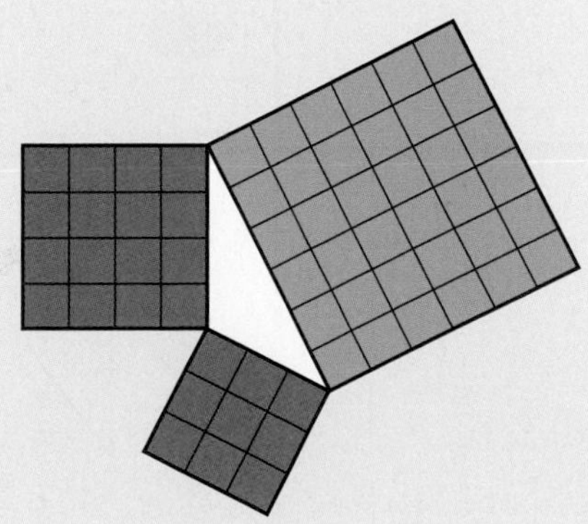

2. If you know the lengths of the two short sides of a right triangle are 9 and 12, can you find the length of the longest side? Show your work.
3. If you know the length of the longest side of a right triangle and the length of one of the shorter sides, how would you find the length of the third side?

8-8 The Pythagorean Theorem

Learn to use the Pythagorean Theorem to find the measure of a side of a right triangle.

Vocabulary

leg

hypotenuse

Pythagorean Theorem

In a right triangle, the two sides that form the right angle are called **legs**. The side opposite the right angle is called the **hypotenuse**.

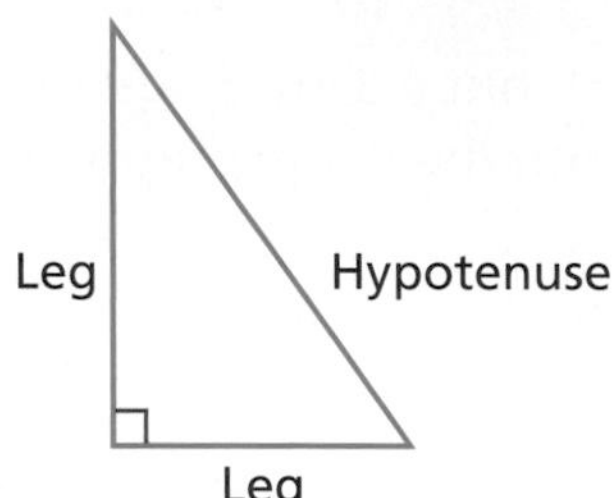

One of the first people to recognize the relationship between the sides of a right triangle was the Greek mathematician Pythagoras. This special relationship is called the *Pythagorean Theorem.*

PYTHAGOREAN THEOREM		
In a right triangle, the sum of the squares of the lengths of the legs is equal to the square of the length of the hypotenuse.	$a^2 + b^2 = c^2$	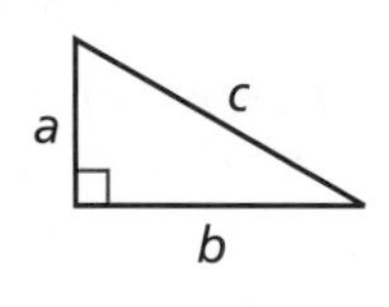

EXAMPLE 1 Calculating the Length of a Side of a Right Triangle

Use the Pythagorean Theorem to find each missing measure.

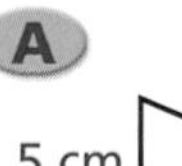

A

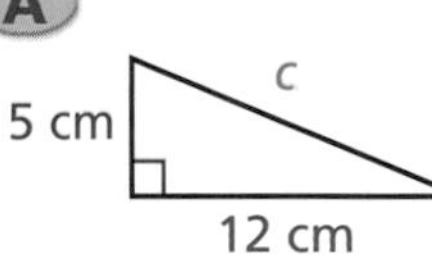

$a^2 + b^2 = c^2$ *Use the Pythagorean Theorem.*

$5^2 + 12^2 = c^2$ *Substitute for a and b.*

$25 + 144 = c^2$ *Evaluate the powers.*

$169 = c^2$ *Add.*

$\sqrt{169} = \sqrt{c^2}$ *Take the square root of both sides.*

$13 = c$

The length of the hypotenuse is 13 cm.

B

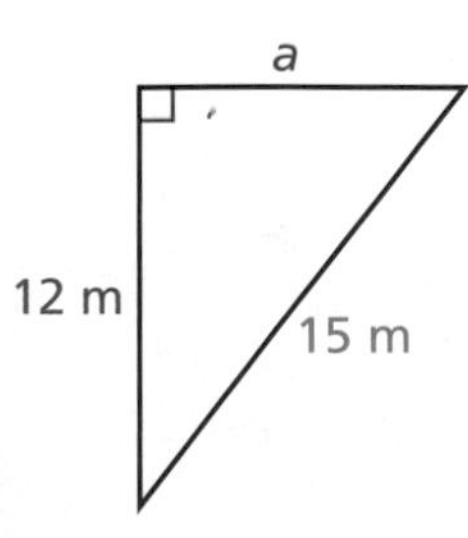

$a^2 + b^2 = c^2$ *Use the Pythagorean Theorem.*

$a^2 + 12^2 = 15^2$ *Substitute for b and c.*

$a^2 + 144 = 225$ *Evaluate the powers.*

$\underline{\quad - 144} \quad \underline{- 144}$

$a^2 = 81$ *Subtract 144 from both sides.*

$\sqrt{a^2} = \sqrt{81}$ *Take the square root of both sides.*

$a = 9$

The length of the missing leg is 9 m.

EXAMPLE 2 PROBLEM SOLVING APPLICATION

A regulation baseball diamond is a square with sides that measure 90 feet. About how far is it from home plate to second base? Round your answer to the nearest tenth.

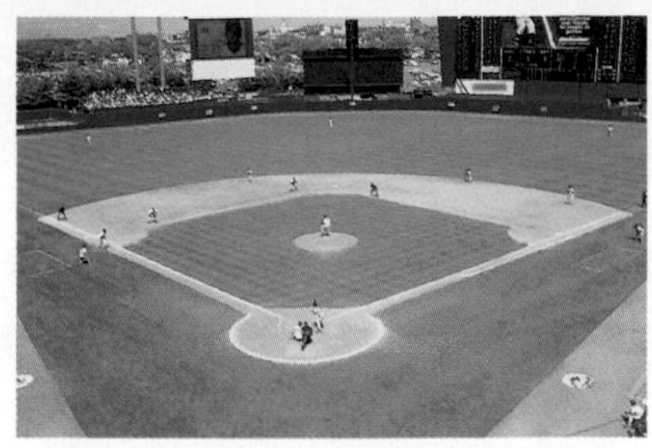

1 Understand the Problem

Rewrite the question as a statement.

- Find the distance from home plate to second base.

List the **important information:**

- Drawing a segment between home plate and second base divides the diamond into two right triangles.
- The segment between home and second base is the hypotenuse.
- The base lines are legs, and they are each 90 feet long.

2 Make a Plan

You can use the Pythagorean Theorem to write an equation.

3 Solve

$a^2 + b^2 = c^2$	*Use the Pythagorean Theorem.*
$90^2 + 90^2 = c^2$	*Substitute for the known variables.*
$8{,}100 + 8{,}100 = c^2$	*Evaluate the powers.*
$16{,}200 = c^2$	*Add.*
$127.279 \approx c$	*Take the square root of both sides.*
$127.3 \approx c$	*Round.*

The distance from home plate to second base is about 127.3 ft.

4 Look Back

The hypotenuse is the longest side of a right triangle. Since the distance from home plate to second base is greater than the distance between the bases, the answer is reasonable.

Think and Discuss

1. **Explain** how to use the Pythagorean Theorem to determine whether a triangle with side lengths 3, 4, and 5 could be a right triangle.
2. **Demonstrate** whether a leg of a right triangle can be longer than the hypotenuse.

8-8 Exercises

FOR EOG PRACTICE
see page 692

internet connect
Homework Help Online
go.hrw.com Keyword: MS4 8-8

1.02d, 5.04

GUIDED PRACTICE

See Example 1

Use the Pythagorean Theorem to find each missing measure.

1.

12 m
c
16 m

2.

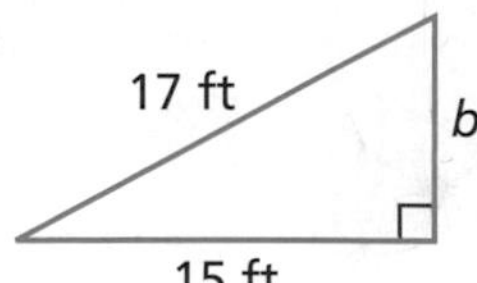

See Example 2

3. A 10 ft ladder is leaning against a wall. If the ladder is 5 ft from the base of the wall, how far above the ground does the ladder touch the wall? Round your answer to the nearest tenth.

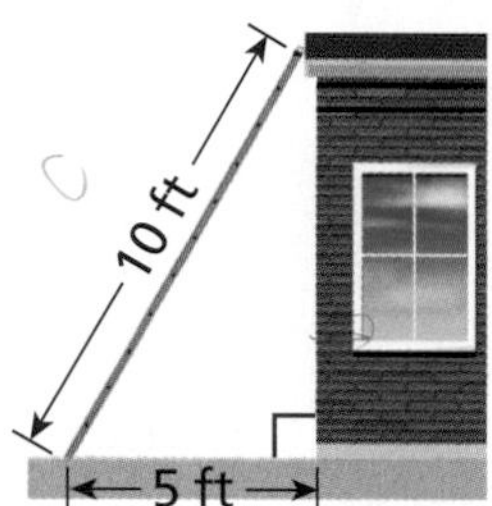

INDEPENDENT PRACTICE

See Example 1

Use the Pythagorean Theorem to find each missing measure.

4.

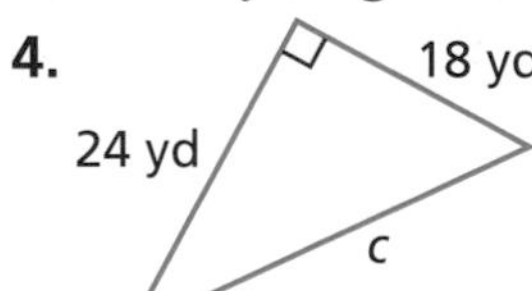

5.

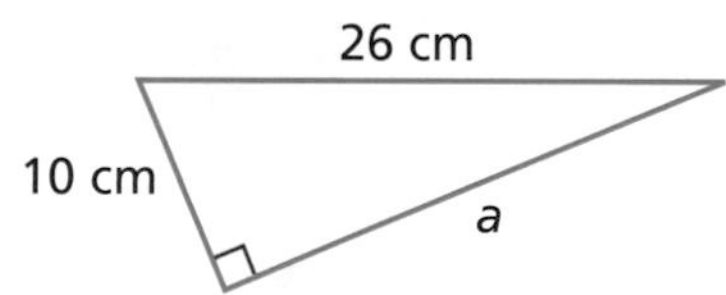

6.

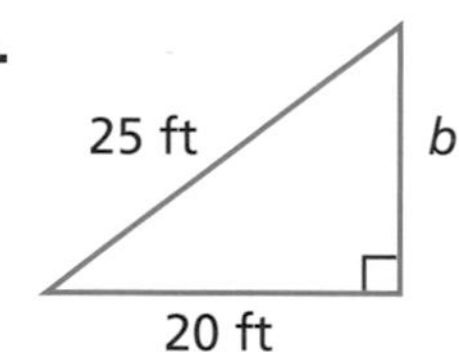

7.

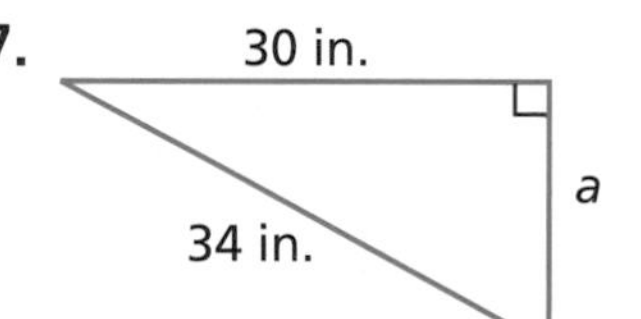

See Example 2

8. James rides his bike 15 miles west. Then he turns north and rides another 15 miles before he stops to rest. How far is James from his starting point when he stops to rest? Round your answer to the nearest tenth.

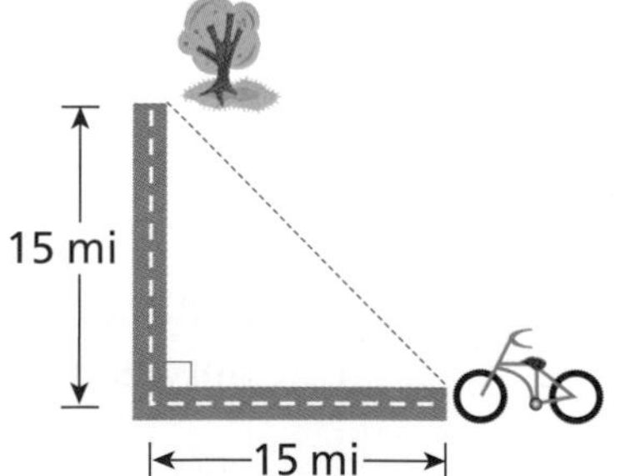

PRACTICE AND PROBLEM SOLVING

Find the missing length to the nearest tenth for each right triangle.

9. $a = 5$; $b = 8$; $c = \square$

10. $a = 10$; $b = \square$; $c = 15$

11. $a = \square$; $b = 13$; $c = 18$

12. $a = 21$; $b = 20$; $c = \square$

Determine whether each set of lengths forms a right triangle.

13. $a = 11$; $b = 60$; $c = 61$

14. $a = 17$; $b = 20$; $c = 25$

History LINK

15. Ancient Egyptians built pyramids to serve as tombs for their kings. One pyramid, called Menkaure, has a square base with an area of about 12,100 m^2.

a. What is the length of each side of the base?

b. What is the length of a diagonal of the base? Round your answer to the nearest tenth.

16. The photograph shows the Pyramid of Khafre in Egypt. Each side of its square base is about 214 meters long. Each triangular side is an isosceles triangle with a height of about 179 meters. What is the area of one side of the pyramid?

17. Use the Pythagorean Theorem to find the distance from one corner of the Pyramid of Khafre to its peak. Round your answer to the nearest tenth.

18. The pyramids were constructed using a unit of measurement called a cubit. There are about 21 inches in 1 cubit. If the height of a pyramid is 471 feet, what is its height in cubits?

19. ***WRITE ABOUT IT*** To determine the boundaries of their fields after the Nile flooded each spring, ancient Egyptians used a loop of rope that was knotted at 12 equal intervals. They stretched the rope around three stakes to form a triangle so that the measures of the sides were 3, 4, and 5. Explain why the triangle formed by the knotted rope could be a right triangle.

20. ***CHALLENGE*** The pyramid at right has a square base. Find the height of the pyramid to the nearest tenth.

Spiral Review

Choose the more precise measurement in each pair. (Lesson 8-2)

21. 12 in., 1 ft

22. 5 m, 8 km

23. 3.5 yd, 22 ft

Calculate. Use the correct number of significant digits in each answer. (Lesson 8-2)

24. $12.6 + 7.32$

25. $19 - 5.7$

26. $4.3 \cdot 2.5$

27. **EOG PREP** A circular mat has a radius of 6 inches. What is the area of the mat? (Lesson 8-6)

A 18.84 in^2

B 37.68 in^2

C 113.04 in^2

D 226.08 in^2

Graph Irrational Numbers

Use with Lesson 8-7

A rational number is a number that can be expressed as a ratio of two integers. All rational numbers can be written as either terminating or repeating decimals. An irrational number is a number that cannot be expressed as a ratio of two integers or as a terminating or repeating decimal.

Every point on the number line corresponds to a real number, either a rational number or an irrational number. Between every two real numbers there is always another real number.

One way to find an approximate value of an irrational number is to locate it between two rational numbers on the number line. The number line below shows the location of several rational numbers.

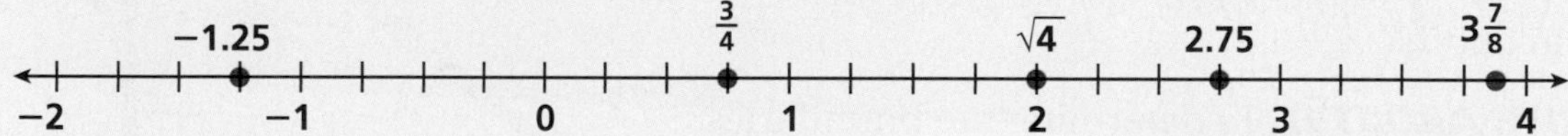

Activity

1 Copy the number line below. Locate $\sqrt{2}$ on the number line.

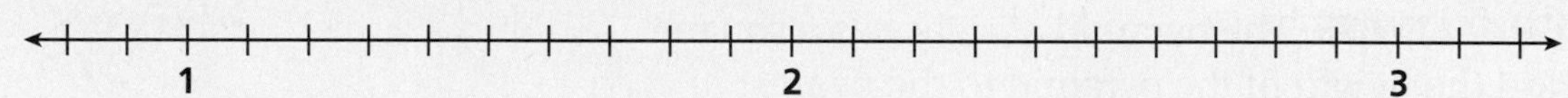

Since $\sqrt{2}$ is an irrational number, you must find an approximate value.

Think: 2 is between the perfect squares 1 and 4. Therefore, $\sqrt{2}$ is between $\sqrt{1}$ and $\sqrt{4}$, or between 1 and 2.

To find a closer approximation, you can use decimals rounded to the tenths place.

Think:
$1.1^2 = 1.21$
$1.2^2 = 1.44$
$1.3^2 = 1.69$
$\mathbf{1.4^2 = 1.96}$
$\mathbf{1.5^2 = 2.25}$

Since $1.4^2 = 1.96$ and $1.5^2 = 2.25$, $\sqrt{2}$ is between 1.4 and 1.5.

To find an even closer approximation, you can use decimals rounded to the hundredths place or the thousandths place.

Think: $\mathbf{1.41^2 = 1.9881}$
$\mathbf{1.42^2 = 2.0164}$

Since $1.41^2 = 1.9881$ and $1.42^2 = 2.0164$, $\sqrt{2}$ is between 1.41 and 1.42.

Think: $1.411^2 = 1.990921$
$1.412^2 = 1.993744$
$1.413^2 = 1.996569$
$\mathbf{1.414^2 = 1.999396}$
$\mathbf{1.415^2 = 2.002225}$

So $\sqrt{2}$ is between 1.414 and 1.415.

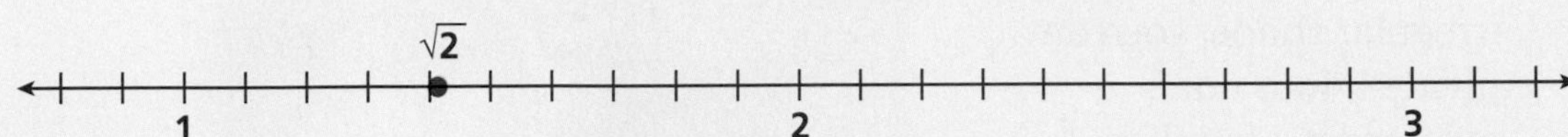

2 Locate $\sqrt{5}$ on a number line.

Think: 5 is between the perfect squares 4 and 9. Therefore, $\sqrt{5}$ is between $\sqrt{4}$ and $\sqrt{9}$, or between 2 and 3.

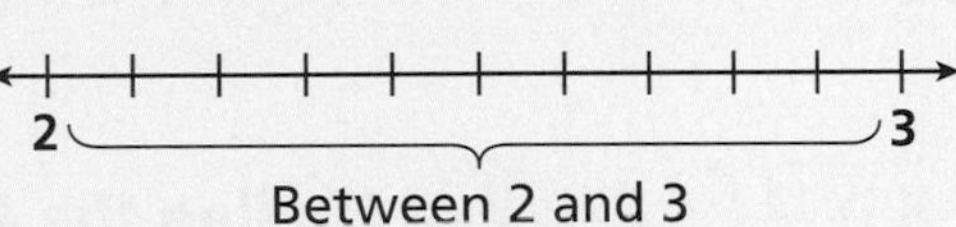

$2.1^2 = 4.41$
$\mathbf{2.2^2 = 4.84}$
$\mathbf{2.3^2 = 5.29}$

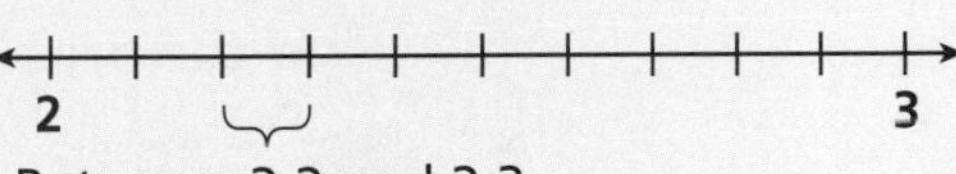

$2.21^2 = 4.8841$
$2.22^2 = 4.9284$
$\mathbf{2.23^2 = 4.9729}$
$\mathbf{2.24^2 = 5.0176}$

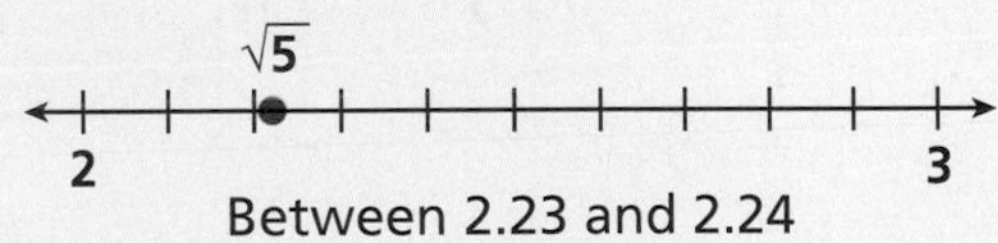

So $\sqrt{5}$ is between 2.23 and 2.24.

Think and Discuss

1. Use a calculator to find $\sqrt{5}$. How is this answer similar to the one you found? How is it different? Which is more precise? Why?
2. Is π a rational number or an irrational number? Explain.

Try This

Draw a number line from 2 to 4 that is marked in tenths. Locate each number on the number line.

1. $3\frac{3}{5}$ **2.** $\sqrt{8}$ **3.** 2.1 **4.** $\sqrt{10}$

5. $2\frac{1}{2}$ **6.** $\sqrt{11}$ **7.** $\sqrt{6}$ **8.** 2.25

EXTENSION Area of Irregular Figures

Learn to find the areas of irregular shapes.

A group of Eagle Scouts are constructing a new path through a local park. The section Martin and Carl are working on is irregularly shaped.

To find the area of an irregular shape, you can divide it into non-overlapping familiar shapes. The sum of these areas is the area of the irregular shape.

EXAMPLE 1 Finding the Area of an Irregular Figure

Find the area of each figure.

A

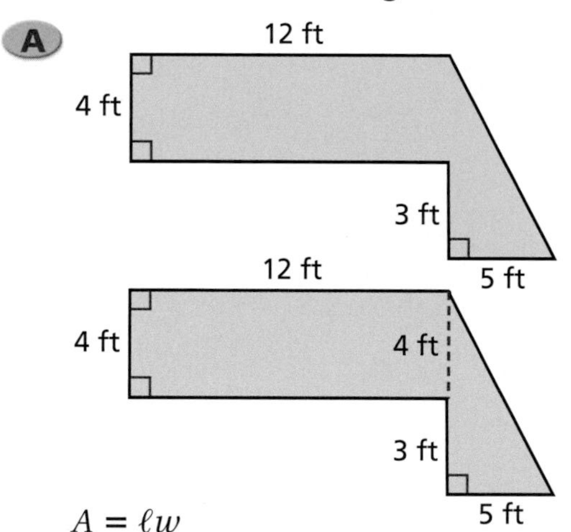

Divide the figure into a rectangle and a triangle.

$A = \ell w$

$A = 12 \cdot 4$ *Find the area of the rectangle.*

$A = 48$

$A = \frac{1}{2}bh$

$A = \frac{1}{2}(5)(3 + 4)$ *Find the area of the triangle.*

$A = \frac{1}{2}(5)(7)$

$A = 17.5$

$A = 48 + 17.5$ *Total area = area of rectangle + area of triangle.*

$A = 65.5$ *Add the areas to find the total area.*

The area of the figure is 65.5 ft^2.

Find the area of each figure. Use 3.14 for π.

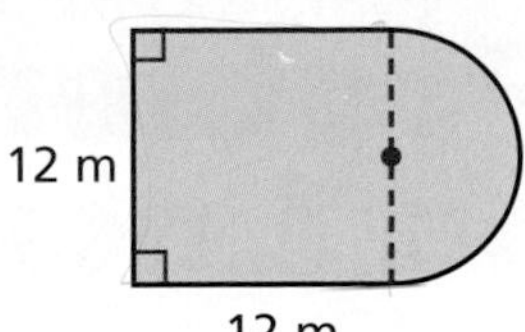

$A = s^2$ $\quad$ $A = \frac{1}{2}(\pi r^2)$

$A = 12^2$ $\quad$ $A \approx \frac{1}{2}(3.14 \cdot 6^2)$

$A = 144$ $\quad$ $A \approx \frac{1}{2}(113.04)$

$A \approx 56.52$

Find the area of the square and the semicircle. The area of a semicircle is one-half the area of a circle.

Total area = area of square + area of semicircle

$\approx 144 + 56.52$

≈ 200.52

Add the areas to find the total area.

The area of the figure is about 200.52 m^2.

EXTENSION Exercises

5.04

Find the area of each figure. Use 3.14 for π.

1. 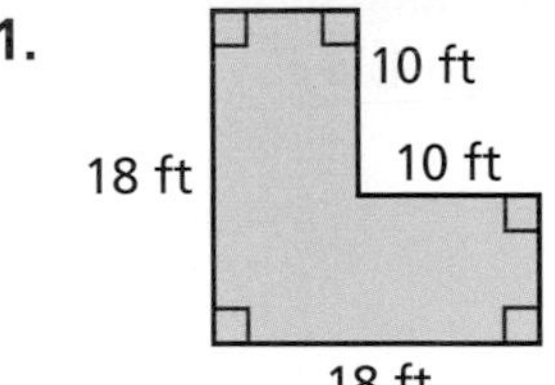

2. 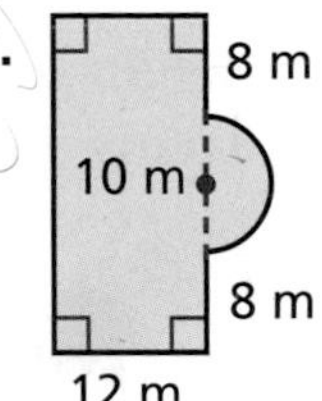

3. 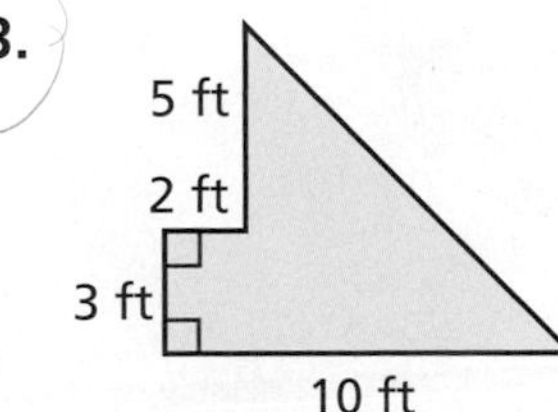

4. 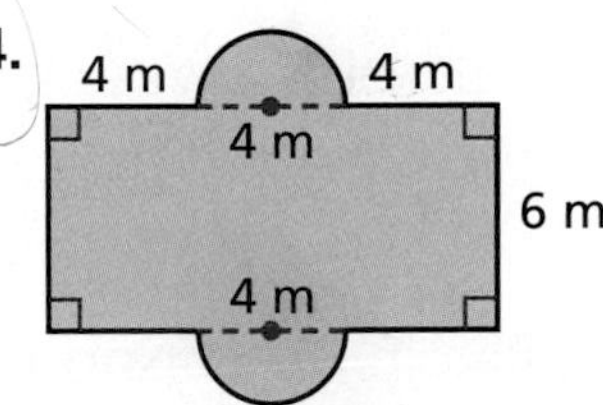

5. 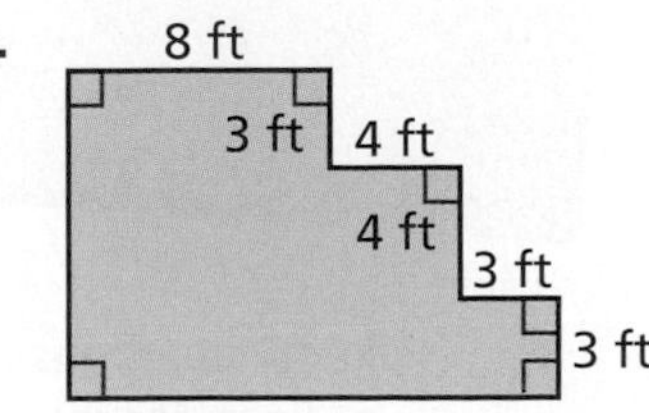

6. 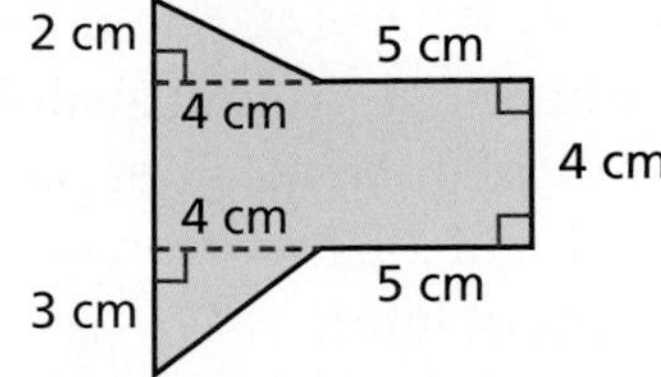

Problem Solving on Location

NORTH CAROLINA

Extreme Dimensions

North Carolina has some contrasts in physical geography. Elevations range from sea level, along the Atlantic Ocean, to 6,684 feet, atop Mt. Mitchell to the west. The Coastal Plain, which is the eastern part of the state, is characterized by flatland with some gently rolling hills. The midsection of the state, the Piedmont Providence, is composed of hills and ridges. The Blue Ridge, to the west, boasts steep mountain ridges and valleys, giving this area a rugged mountain character. The state's dimensions are about 505 miles from east to west and about 190 miles from north to south, with a mean elevation of 700 feet. Its area is 53,821 square miles, making it the twenty-eighth largest state.

A fishing boat heads for shore at sunset near Cedar Island.

1. What is the difference in height between the highest point in North Carolina and the state's mean elevation?
2. If you wanted to find the approximate perimeter of North Carolina, which measurements would you use?
3. What is the perimeter of a rectangle that has the same length and width as North Carolina?
4. About 5,103 square miles of North Carolina are covered by inland water. How many square miles are not covered by inland water?
5. Approximately 8,320,146 people live in North Carolina. Use your answer from **4** to give the approximate area per person in North Carolina if each person occupied the same amount of land.

Rhododendrons bloom in the high elevations of Roan Mountain.

State Capitol

The North Carolina State Capitol, in Raleigh, is the second capitol to stand on Union Square. The State House, the first capitol, burned in 1831 when workmen were attempting to fireproof the roof by pouring melted zinc onto the shingles. Only a few furnishings were saved, including the Senate Speaker's chair, a senate desk chair, and a painting of George Washington. The building was rebuilt and was completed in 1840.

The North Carolina State Capitol was built in the Greek revival style of architecture and is designated as a National Historic Landmark. The new capitol is a cross-shaped building with a domed rotunda in the center. It is 160 feet long from north to south by 140 feet wide from east to west.

1. How much longer is the capitol building than it is wide?
2. The height of the capitol building is 97.5 feet from the base of the rotunda to the crown atop the dome. What is the height in inches?
3. There is a road around the capitol. If the road is 10 feet from the outside of the building on all sides, what is the perimeter of the property along the side of the road that is closest to the capitol building?
4. Suppose you could draw a rectangle around the cross-shaped capitol building. What would the approximate area of a rectangle that has the same length and width as the capitol building?

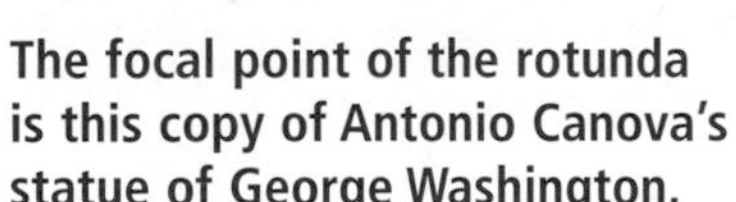

The focal point of the rotunda is this copy of Antonio Canova's statue of George Washington.

MATH-ABLES

Shape Up

Rectangles

The square below has been divided into four rectangles. The areas of two of the rectangles are given. If the length of each of the segments in the diagram is an integer, what is the area of the original square?

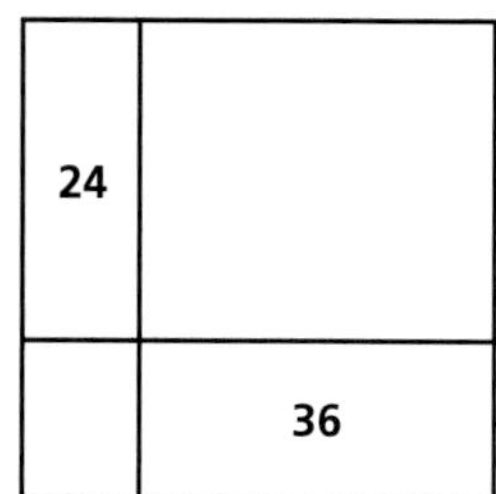

(*Hint:* Remember $a + c = b + d$.)

Use different lengths and a different answer to create your own version of this puzzle.

Circles

What is the maximum number of times that six circles of the same size can intersect? To find the answer, start by drawing two circles that are the same size. What is the greatest number of times they can intersect? Add another circle, and another, and so on.

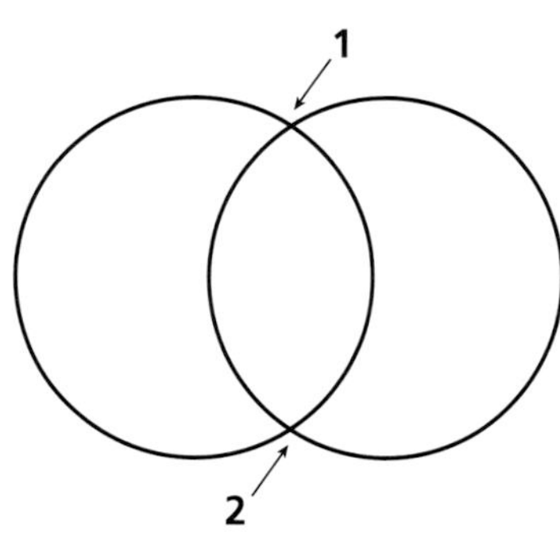

Circles and Squares

Two players start with a sequence of circles and squares. Before beginning the game, each player chooses whether to be a "circle" or a "square." The goal of the game is to have the final remaining shape be the shape you chose to be. Shapes are removed from the sequence according to the following rules: On each move, a player selects two shapes. If the shapes are identical, they are replaced with one square. If the shapes are different, they are replaced with one circle.

internet connect

Go to ***go.hrw.com*** for a complete set of rules and game pieces.
KEYWORD: MS4 Game8

Technology LAB

Area and Perimeter

You can use geometry software to explore the area and perimeter of geometric figures.

Activity

1. Construct and label a quadrilateral that resembles the one shown in the first window below. Then select the interior of the quadrilateral.

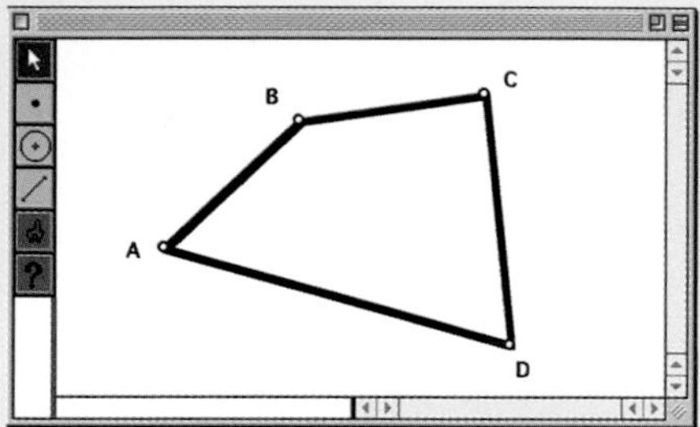

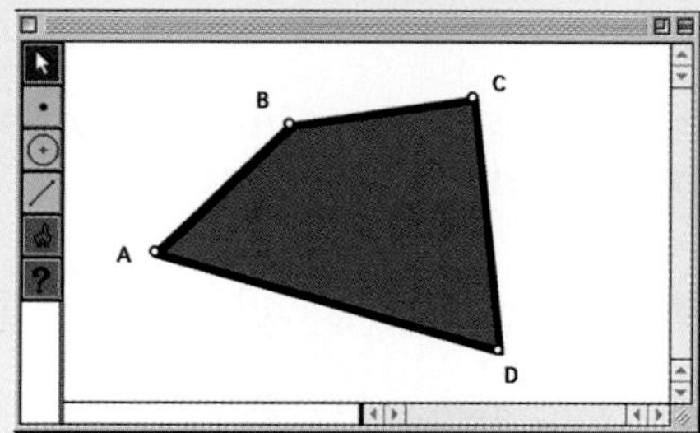

Use the software tools to measure the area and the perimeter of *ABCD*. Then select vertex *D* and drag it around. Notice that the measures of the area and the perimeter change as the shape of the quadrilateral changes.

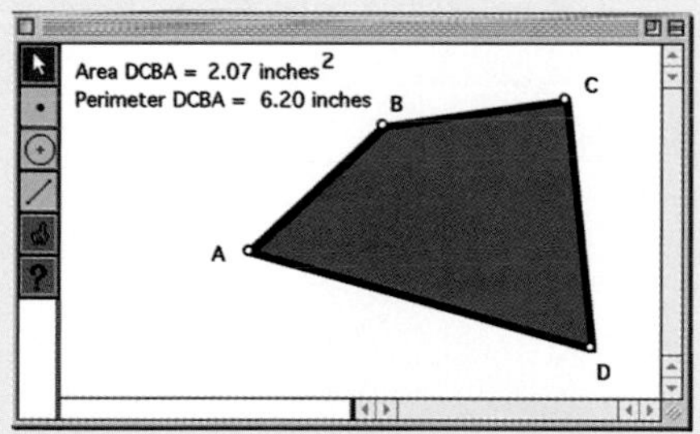

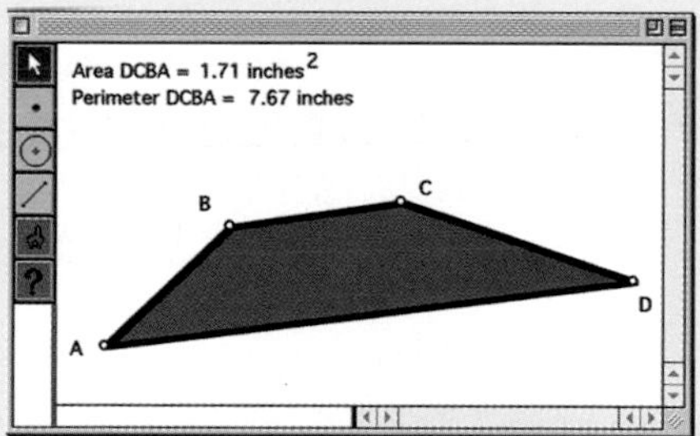

Think and Discuss

1. Will the area and perimeter of a translated polygon be the same as the area and perimeter of the original polygon? Explain.

Try This

1. Use geometry software to construct triangle *ABC*. Measure the area and perimeter of *ABC*. Then construct triangle *DEF* that has *twice the area* of *ABC*. Is the perimeter of *DEF* twice the perimeter of *ABC*? Explain.

2. Use geometry software to construct a 5-sided polygon *ABCDE*. Measure the area of *ABCDE*. Identify three triangles that make up polygon *ABCDE*. Measure the area of each triangle and find the sum of the areas. Compare this sum with the area of *ABCDE*.

Chapter 8

Study Guide and Review

Vocabulary

Complete the sentences below with vocabulary words from the list above. Words may be used more than once.

1. The longest side of a right triangle is called the ___?___.
2. The ___?___ is the distance around a circle.
3. ___?___ is the level of detail an instrument can measure.
4. A(n) ___?___ is one of the two equal factors of a number.

8-1 Customary and Metric Measurements (pp. 414–417)

EXAMPLE

- **Convert 54 meters to centimeters.**

  There are 100 centimeters in 1 meter.

  $54 \cancel{m} \cdot \frac{100 \text{ cm}}{1 \cancel{m}} = 5{,}400 \text{ cm}$

  54 meters is equal to 5,400 centimeters.

EXERCISES

Convert.

5. 312 hr = ▢ days
6. 3,400 g = ▢ kg
7. 8 yd = ▢ in.
8. 458 km = ▢ m

8-2 Accuracy and Precision (pp. 420–423)

EXAMPLE

- **Determine the number of significant digits in 705.4 mL.**

  The digits 7, 5, and 4 are nonzero digits, and 0 is between two nonzero digits. So 705.4 mL has 4 significant digits.

EXERCISES

Determine the number of significant digits in each measurement.

9. 0.450 kg
10. 6,703.0 ft
11. 30,000 lb
12. 0.00078 g
13. 900.5 cm
14. 1,204 gal

8-3 Perimeter and Circumference (pp. 424–427)

EXAMPLE

- **Find the perimeter of the triangle.**

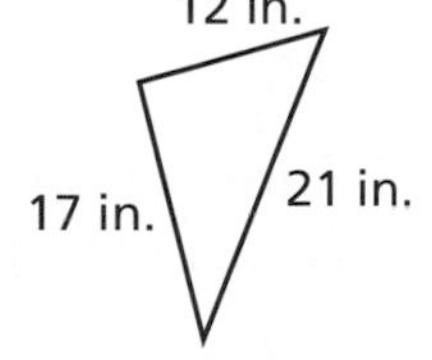

$P = 12 + 17 + 21$
$P = 50$
The perimeter of the triangle is 50 in.

- **Find the circumference of the circle. Use 3.14 for π.**

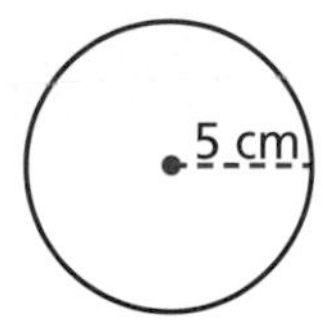

$C = 2\pi r$
$C \approx 2 \cdot 3.14 \cdot 5$
$C \approx 31.4$
The circumference of the circle is about 31.4 cm.

EXERCISES

Find the perimeter of each polygon.

15.

16.

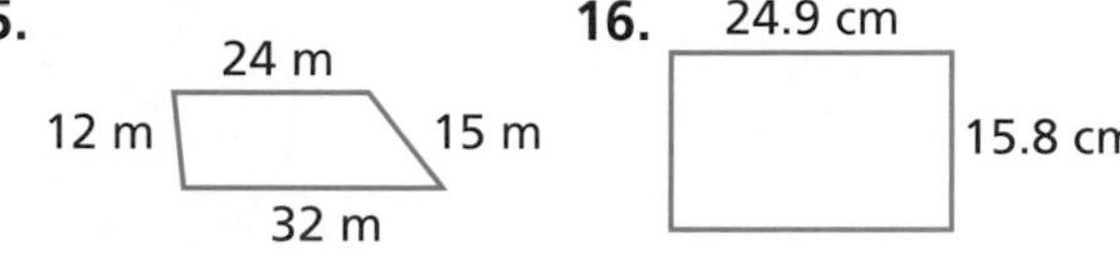

Find the circumference of each circle to the nearest tenth. Use 3.14 for π.

17.

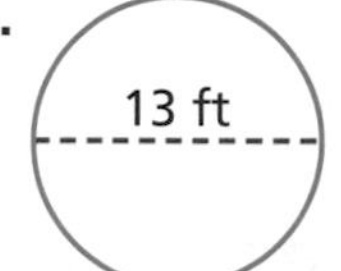

18.

8-4 Area of Parallelograms (pp. 430–433)

EXAMPLE

- **Find the area of the rectangle.**

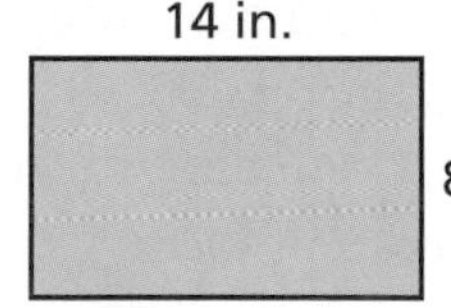

$A = \ell w$
$A = 14 \cdot 8.6$
$A = 120.4$
The area of the rectangle is 120.4 in^2.

EXERCISES

Find the area of each figure.

19.

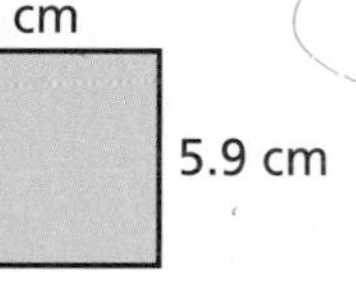

20.

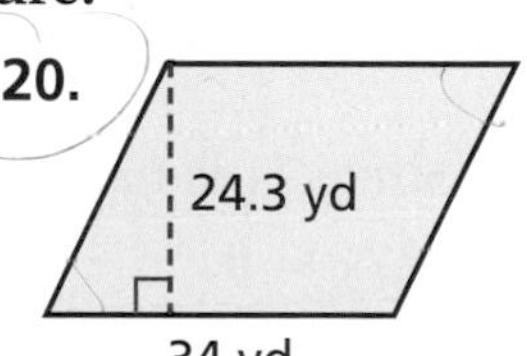

8-5 Area of Triangles and Trapezoids (pp. 434–437)

EXAMPLE

- **Find the area of the triangle.**

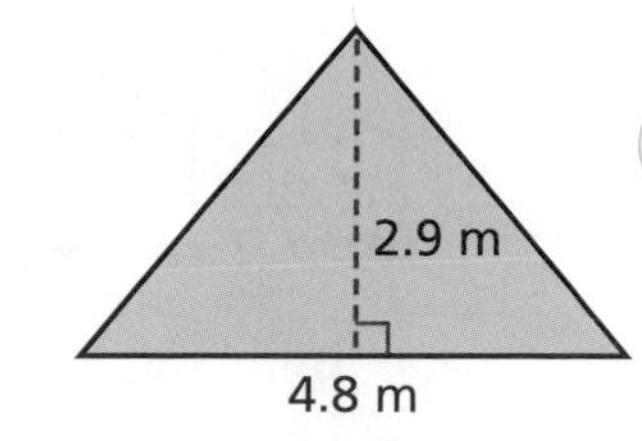

$A = \frac{1}{2}bh$
$A = \frac{1}{2}(4.8 \cdot 2.9)$
$A = \frac{1}{2}(13.92)$
$A = 6.96$
The area of the triangle is 6.96 m^2.

EXERCISES

Find the area of each polygon.

21.

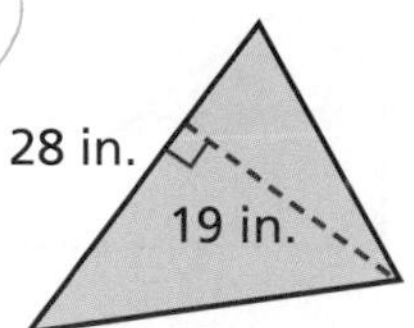

22.

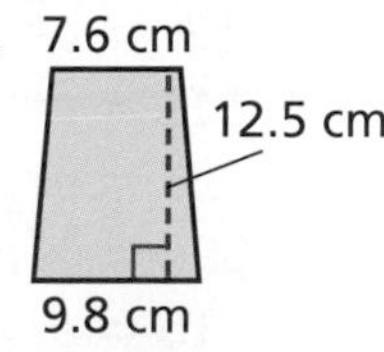

23.

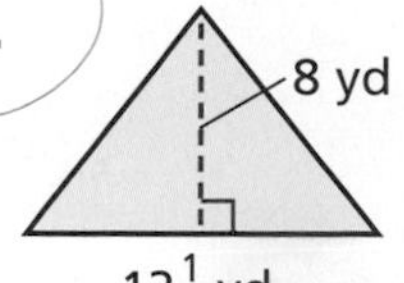

24.

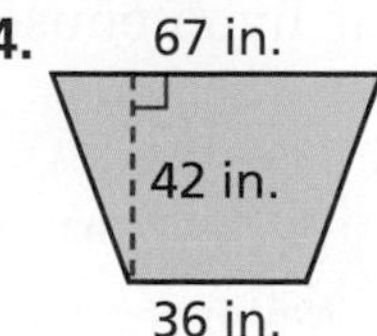

8-6 Area of Circles (pp. 438–441)

EXAMPLE

- **Find the area of the circle to the nearest tenth. Use 3.14 for π.**

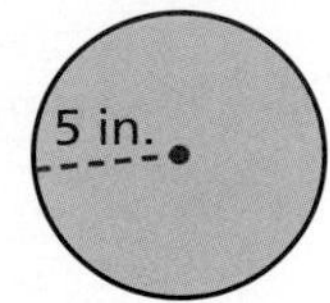

$A = \pi r^2$
$A \approx 3.14 \cdot 5^2$
$A \approx 3.14 \cdot 25$
$A \approx 78.5$
The area of the circle is about 78.5 in^2.

EXERCISES

Find the area of each circle to the nearest tenth. Use 3.14 for π.

25.

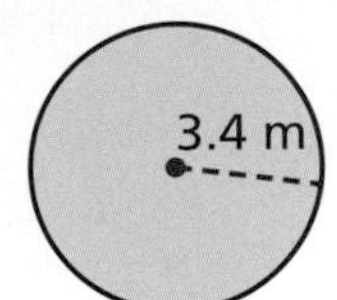

26.

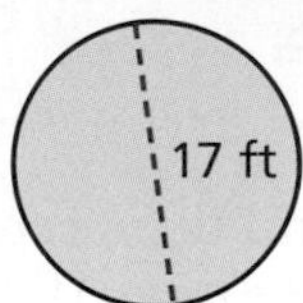

8-7 Powers and Roots (pp. 444–447)

EXAMPLE

- **Estimate $\sqrt{71}$ to the nearest whole number.**

$64 < \quad 71 < 81$ *Find the perfect squares nearest 71.*

$\sqrt{64} < \sqrt{71} < \sqrt{81}$
$8 < \sqrt{71} < 9$ *Find the square roots of 64 and 81.*

Since 71 is nearer in value to 64 than to 81, $\sqrt{71} \approx 8$.

EXERCISES

Estimate each square root to the nearest whole number.

27. $\sqrt{29}$ **28.** $\sqrt{92}$
29. $\sqrt{106}$ **30.** $\sqrt{150}$
31. The area of Rita's square vegetable garden is 265 ft^2. What is the length of each side of the garden to the nearest foot?

8-8 Pythagorean Theorem (pp. 450–453)

EXAMPLE

- **Use the Pythagorean Theorem to find the missing measure.**

$a^2 + b^2 = c^2$
$9^2 + 12^2 = c^2$
$81 + 144 = c^2$
$225 = c^2$
$\sqrt{225} = \sqrt{c^2}$
$15 = c$

9 in.
12 in.

The hypotenuse is 15 in.

EXERCISES

Use the Pythagorean Theorem to find each missing measure.

32.

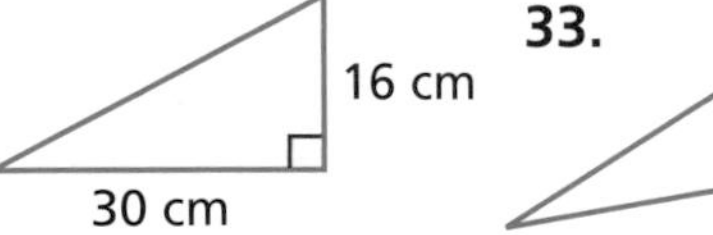

33.

34. **35.**

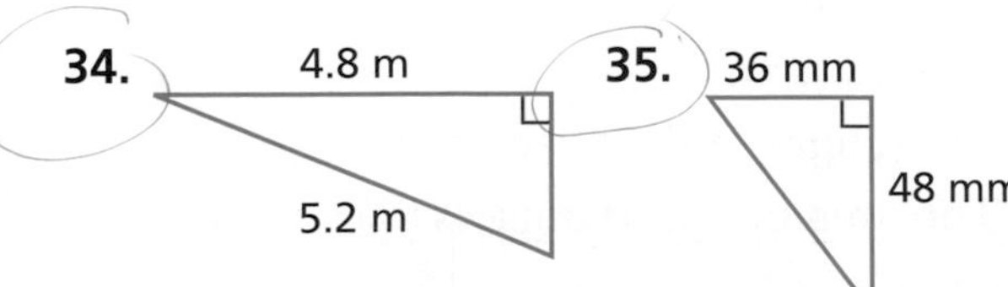

Chapter Test

Convert.

1. 13 lb = ▢ oz
2. 450 m = ▢ km
3. 6 hr = ▢ s

Choose the more precise measurement in each pair.

4. 80 m, 7.9 cm
5. 18 yd, 5 mi
6. 500 lb, 18 oz

Calculate. Use the correct number of significant digits in each answer.

7. 5.882 in. + 5.17 in.
8. 5.6 g ÷ 2.59
9. 3.14 · 125 cm

Find the perimeter of each polygon.

10. 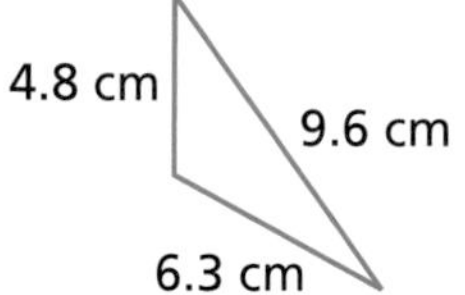

11. 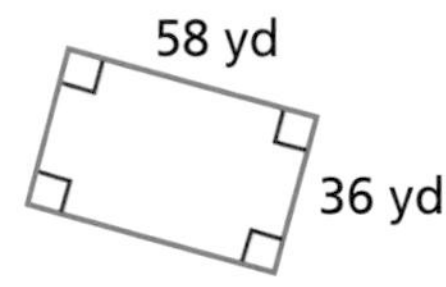

12. 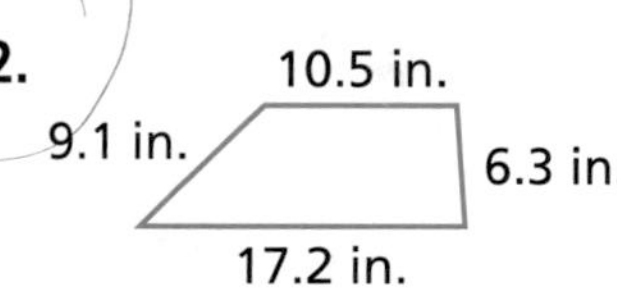

Find the area of each figure.

13. 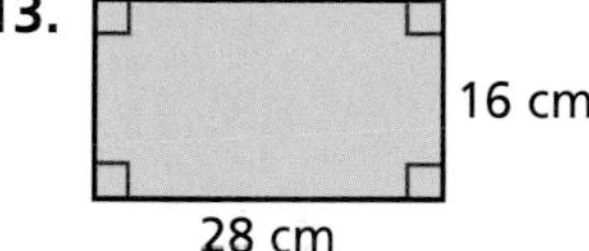

14. 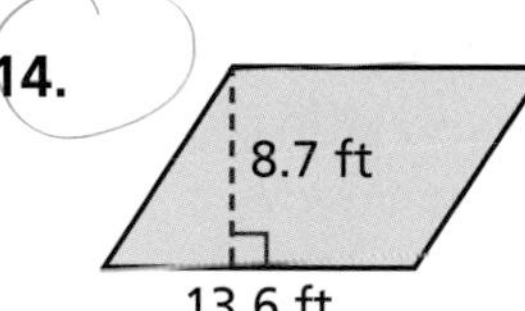

15. 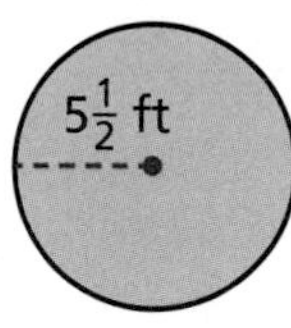

Use the circle at right for items 16 and 17. Use 3.14 for π.

16. Find the circumference of the circle to the nearest tenth.
17. Find the area of the circle to the nearest tenth.
18. What is the radius of a circle with circumference 51.496 in.? Use 3.14 for π.

$5\frac{1}{2}$ ft

Estimate each square root to the nearest whole number.

19. $\sqrt{42}$
20. $\sqrt{78}$
21. $\sqrt{115}$
22. The area of a square chessboard is 212 in^2. What is the length to the nearest inch of each side of the chessboard?

Use the triangle at right for items 23 and 24.

23. Use the Pythagorean Theorem to find the missing measure of the triangle.
24. Find the area of the triangle.

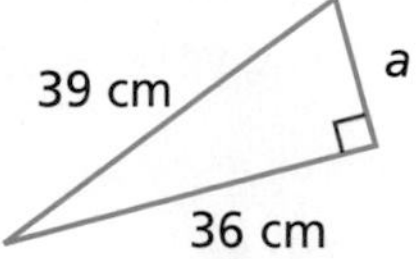

25. A triangle has side lengths of 18 cm, 20 cm, and 29 cm. Could the triangle be a right triangle? Explain your answer.

Chapter Test

Chapter 8

Performance Assessment

Show What You Know

Create a portfolio of your work from this chapter. Complete this page and include it with your four best pieces of work from Chapter 8. Choose from your homework or lab assignments, mid-chapter quizzes, or any journal entries you have done. Put them together using any design you want. Make your portfolio represent what you consider your best work.

Short Response

1. The bases of a trapezoid are 6.4 m and 8.2 m, and the height is 3.25 m. Show the steps necessary to find the area of the trapezoid. Write the area using the rules for significant digits.
2. Jeanette wants to mount a circular photo onto a rectangular piece of cardboard. The area of the photo is 50.24 in^2. What are the smallest possible dimensions the piece of cardboard can have and still hold the entire photo? Use 3.14 for π and explain your answer.
3. Find the perimeter and area of a rectangle with length 12 m and width 7 m. Then find the length of each side of a square with the same area as the rectangle. Round your answers to the nearest meter, and show your work.

Extended Problem Solving

4. Washington, D.C., occupies about 68% of the area of the square shown in the center of the map. The length of each side of the square is 10 miles.
   a. Find the area of the square. Show your work.
   b. Explain how to use your answer from part **a** to calculate the area of Washington, D.C. Then find the area of the city.
   c. Explain how you could find the diagonal distance across the square. Find this distance to the nearest tenth. Show your work.

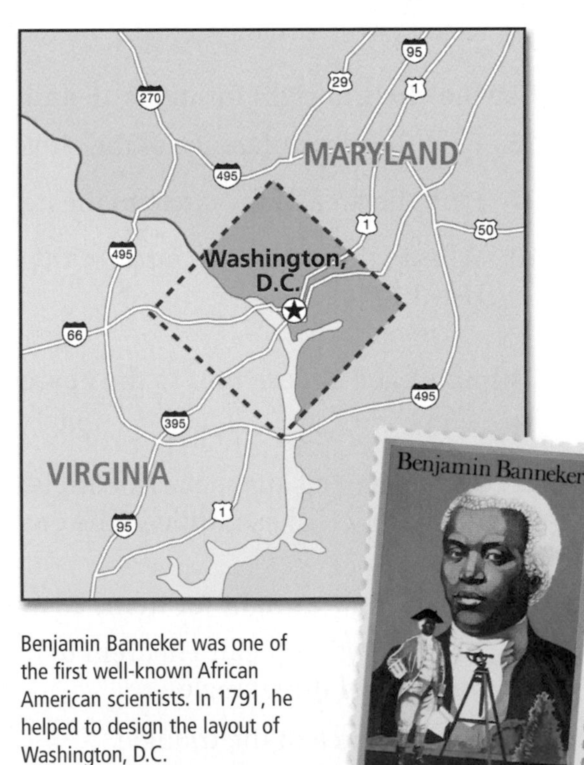

Benjamin Banneker was one of the first well-known African American scientists. In 1791, he helped to design the layout of Washington, D.C.

internet connect
EOG Test Practice Online
go.hrw.com Keyword: MS4 TestPrep

Getting Ready for EOG

Chapter 8

Cumulative Assessment, Chapters 1–8

1. Which is a solution to the equation $78 = a + 49$.

   A $a = 29$ C $a = 31$

   B $a = 30$ D $a = 127$

2. Which statement is false?

   A $\frac{85}{34}$ is equivalent to 2.5.

   B $\frac{7}{80}$ is equivalent to 0.875.

   C $\frac{3}{500}$ is equivalent to 0.006.

   D $\frac{7}{35}$ is equivalent to 0.20.

3. What is the area of the trapezoid?

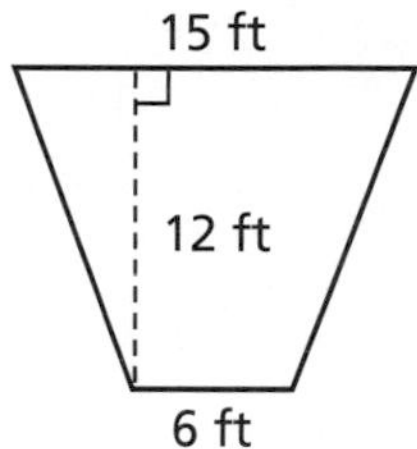

   A 81 cm^2 C 135 cm^2

   B 126 cm^2 D 252 cm^2

4. The distance from Chattanooga to Nashville, Tennessee, is 128 mi. If a map scale is $\frac{1}{2}$ in. = 10 mi, what is the distance between the cities on the map?

   A 6.4 in. C 12.8 in.

   B 10.8 in. D 25.6 in.

TEST TAKING TIP!

Eliminate choices by using 3 for π.

5. What is the circumference of a circle with diameter 18 m? Use 3.14 for π.

   A 28.26 m C 113.03 m

   B 56.52 m D 254.34 m

6. A histogram looks most like which type of graph?

   A Bar graph

   B Line graph

   C Circle graph

   D Box-and-whisker plot

7. The track team jogged 2.5 miles east from the school and then 1.7 miles north. At this point, about how far from the school was the team?

   A 4.2 miles C 2.2 miles

   B 3.0 miles D 0.8 miles

8. What type of triangle is triangle ABC?

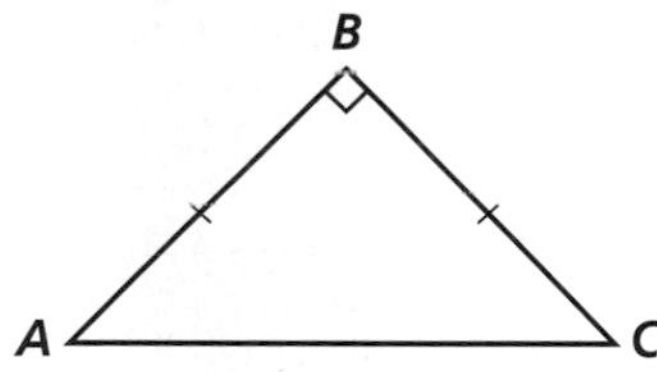

   A Isosceles acute

   B Scalene obtuse

   C Isosceles right

   D Scalene acute

9. ***SHORT RESPONSE*** 120 is 80% of what number? Show your work.

10. ***SHORT RESPONSE*** Libby bought 4.4 pounds of tomatoes for $6.60. How much did she pay per pound? Explain how you determined your answer.

Getting Ready for EOG

Chapter 9

Volume and Surface Area

TABLE OF CONTENTS

internet connect

Chapter Opener Online
go.hrw.com
KEYWORD: MS4 Ch9

Pyramid	Location	Height (m)	Base Length (m)
El Castillo	Chichén Itzá, Mexico	55.5	79.0
Tikal	Tikal, Guatemala	30.0	80.0
Pyramid of the Sun	Teotihuácan, Mexico	63.0	225.0

Archaeological Architect

Did you ever wonder how the pyramids were built? Archaeologists who are also architects combine a love of the past with the skills of a building designer to study the construction of ancient buildings.

In recent years, archaeological architects have built machines like those used in ancient times to demonstrate how the pyramids might have been constructed. The table shows the dimensions of a few famous pyramids.

ARE YOU READY?

Choose the best term from the list to complete each sentence.

1. A polygon with six sides is called a(n) __?__.
2. __?__ figures are the same size and shape.
3. A(n) __?__ is a ratio that relates the dimensions of two similar objects.
4. The formula for the __?__ of a circle can be written as πd or $2\pi r$.
5. __?__ figures are the same shape but not necessarily the same size.
6. A polygon with five sides is called a(n) __?__.

area
circumference
congruent
hexagon
pentagon
scale factor
similar

Complete these exercises to review skills you will need for this chapter.

✔ Area of Rectangles and Triangles

Find the area of each figure.

7.

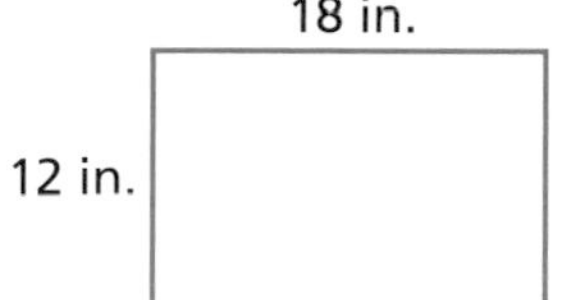

8.

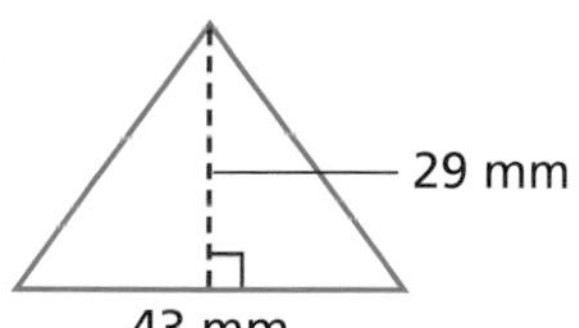

9.

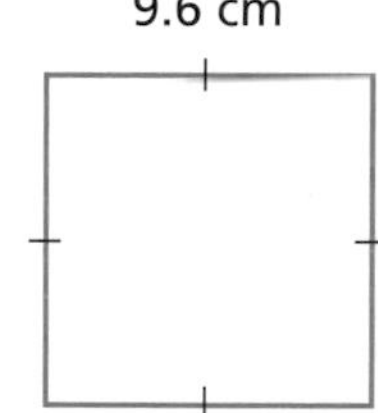

✔ Area of Circles

Find the area of each circle to the nearest tenth. Use 3.14 for π.

10.

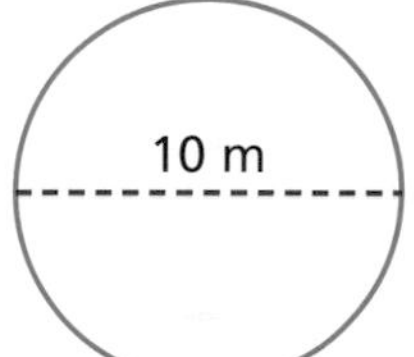

11.

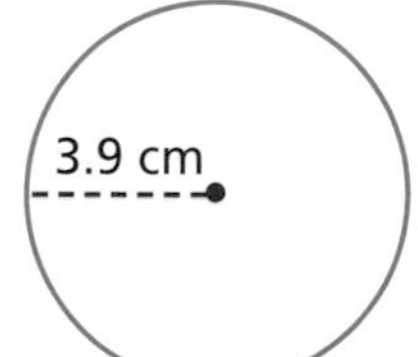

12.

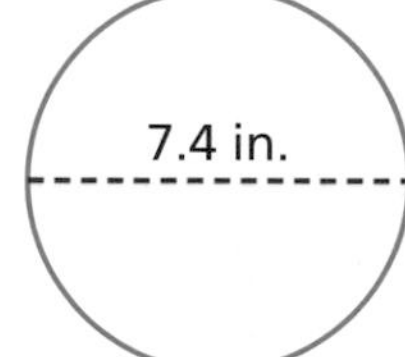

✔ Find the Cube of a Number

Find each value.

13. 3^3 **14.** 8^3 **15.** 2.5^3 **16.** 6.2^3

17. 10^3 **18.** 5.9^3 **19.** 800^3 **20.** 98^3

Draw Three-Dimensional Figures

Use with Lesson 9-1

2.01, 3.01b, 3.01c

Three-dimensional figures have length, width, and height. To draw a three-dimensional figure, you first have to choose which view of the figure you will draw.

Activity

1 Draw the side view of a cone with a radius of 1.5 cm and a height of 2.5 cm.

a. From the side, the circular base of a cone looks like an oval. Since the diameter of the cone is 3 cm, draw an oval with a length of 3 cm. Make the top half of the oval dashed.

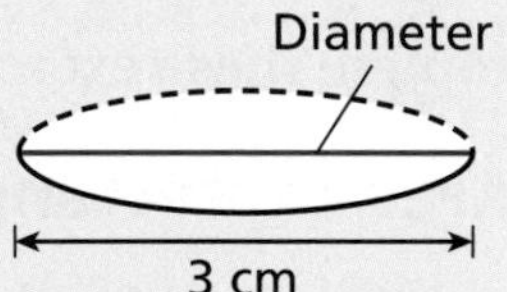

b. Recall that a *perpendicular bisector* bisects a segment at 90° angles. Draw a perpendicular bisector of the diameter, so that one of its endpoints is on the diameter. The length of the bisector is 2.5 cm. This is the height of the cone.

c. Draw two line segments, one from each endpoint of the diameter to the *vertex* of the cone.

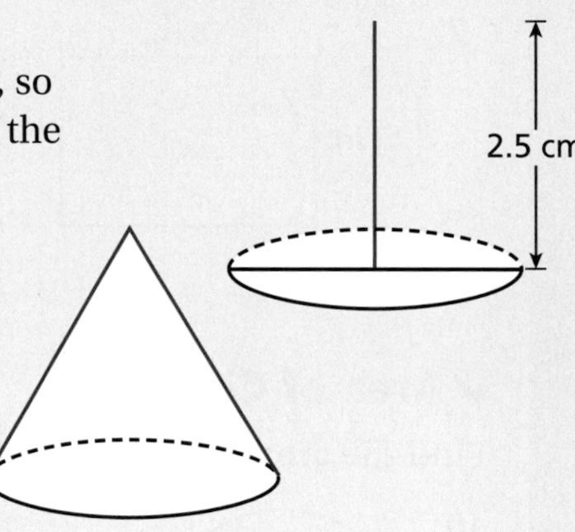

2 Draw the side view of a pyramid with a height of 2.5 cm and a square base that is 2 cm on each side.

a. From the side, the square base looks like a parallelogram with no right angles. Draw a parallelogram like the one shown. The top and bottom are each 2 cm long. Make the top and one side dashed as shown.

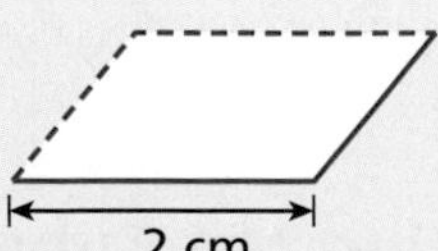

b. Find the midpoint of each side of the parallelogram. Draw two line segments to connect these points as shown.

c. From the point where the two segments meet, draw a 2.5 cm line segment perpendicular to the horizontal segment. The length of this vertical segment is the height of the pyramid.

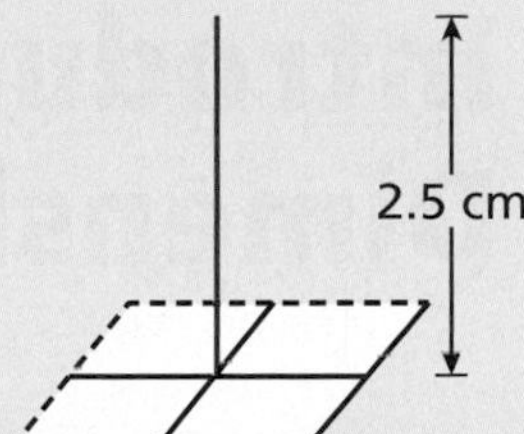

d. Draw four line segments, one from each vertex of the parallelogram to the *vertex* of the pyramid. The segment that meets the two dashed segments should also be dashed.

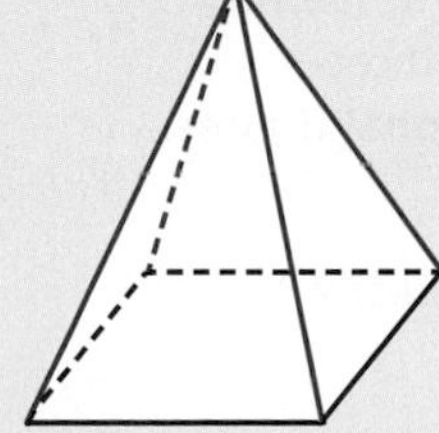

3 Draw the side view of a hemisphere and a sphere with a radius of 0.5 in.

a. From the side, the circular base of a hemisphere looks like an oval. Since the diameter of the hemisphere is 1 in., draw an oval with a length of 1 in. Make the top half of the oval dashed.

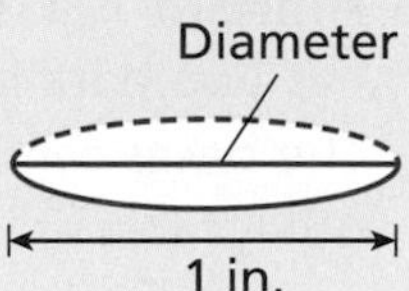

b. Draw a half of a circle, so that the endpoints of the arc meet with the endpoints of the diameter. This completes the drawing of the hemisphere.

c. Now draw the other half of the circle. This completes the drawing of the sphere.

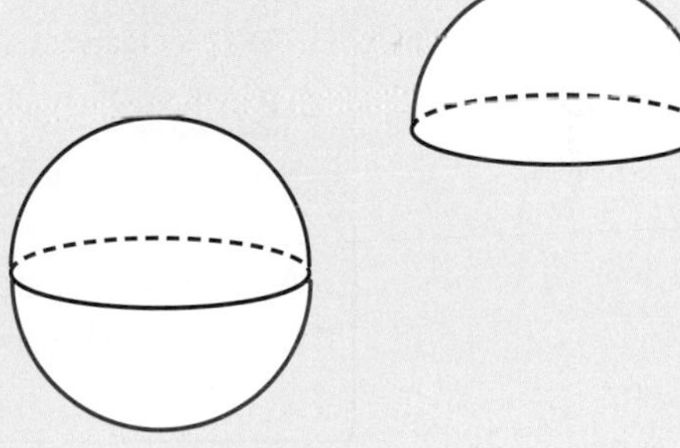

Think and Discuss

1. How is drawing a cone similar to drawing a pyramid?
2. How is drawing a cone similar to drawing a hemisphere?
3. How does drawing the side view of a cone or pyramid affect the appearance of the base?

Try This

1. Draw the side view of a cone with a radius of 3 cm and a height of 3 cm.
2. Draw the side view of a pyramid with a height of 3.5 cm and a square base that is 2 cm on each side.
3. Draw the side view of a hemisphere and a sphere with a radius of 2 cm.

9-1 Introduction to Three-Dimensional Figures

Learn to identify various three-dimensional figures.

Vocabulary

face
edge
vertex
base
polyhedron
prism
pyramid
lateral surface
cylinder
cone
sphere
hemisphere

Three-dimensional figures, or solids, have length, width, and height. A flat surface of a solid is a **face**. An **edge** is where two faces meet, and a **vertex** is where three or more edges meet. The face that is used to classify a solid is a **base**.

The surfaces of a three-dimensional figure determine the type of solid it is. A **polyhedron** is a three-dimensional figure whose surfaces, or faces, are all polygons. *Prisms* and *pyramids* are two types of polyhedrons.

Prisms	Pyramids
A **prism** is a polyhedron that has two parallel congruent bases. The bases can be any polygon. The other faces are parallelograms. Vertex, 2 bases, Edge	A **pyramid** is a polyhedron that has one base. The base can be any polygon. The other faces are triangles. Vertex, 1 base, Edge
A *cube* is a special prism whose faces are all congruent squares.	A *regular tetrahedron* is a special pyramid whose faces are all congruent equilateral triangles.

Prisms and pyramids are named by the shapes of their bases.

EXAMPLE 1 Naming Prisms and Pyramids

Identify the base or bases of each solid. Then name the solid.

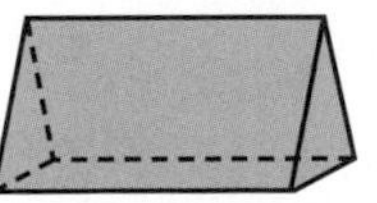

There are two bases, and they are both triangles.
The other faces are parallelograms.
The figure is a triangular prism.

B

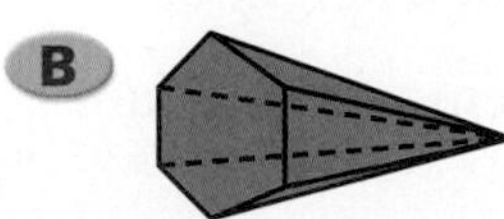

There is one base, and it is a hexagon.
The other faces are triangles.
The figure is a hexagonal pyramid.

Remember!

A polygon with six sides is called a hexagon.

Other three-dimensional figures include *cylinders, cones,* and *spheres.* These figures are different from polyhedrons because they each have a curved surface and their bases are not polygons. The curved surface of a cylinder or a cone is called a **lateral surface** .

A **cylinder** has two parallel, congruent circular bases connected by a lateral surface.

A **cone** has one circular base and a lateral surface. The lateral surface of a cone comes to a point called its vertex.

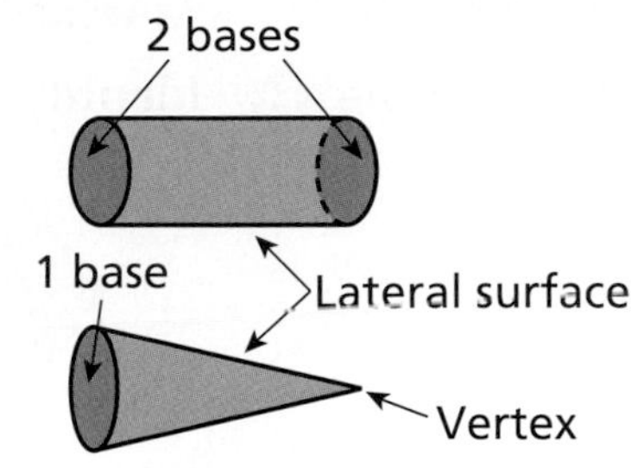

A **sphere** has only one surface, which is curved, and has no base. All of the points on the surface are the same distance from the center of the sphere.

A plane that intersects a sphere through its center divides the sphere into two halves, or **hemispheres** .

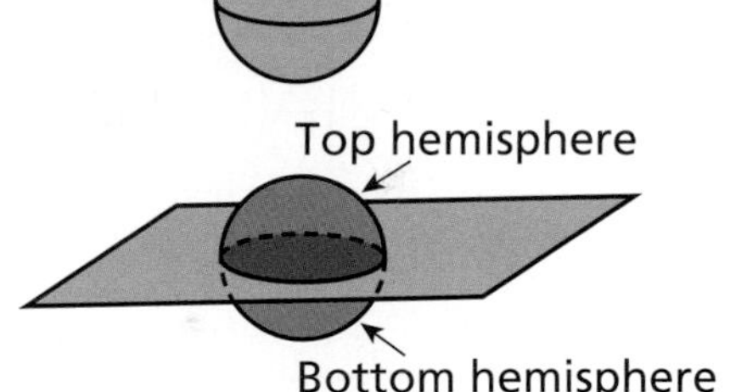

EXAMPLE 2 Identifying Combinations of Solids

Tell what solids make up each figure.

A

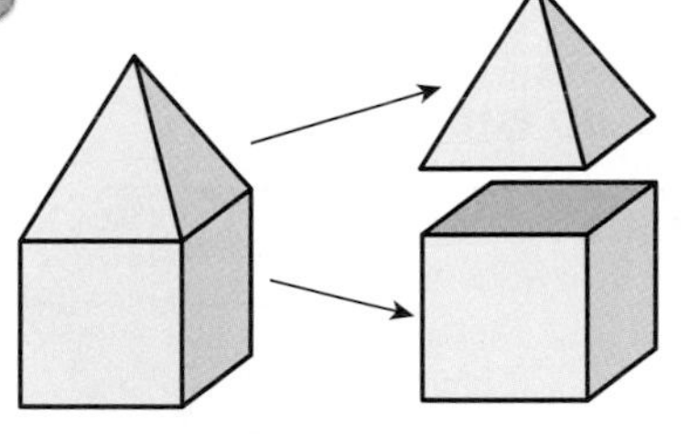

The figure is made up of a rectangular pyramid and a rectangular prism.

B

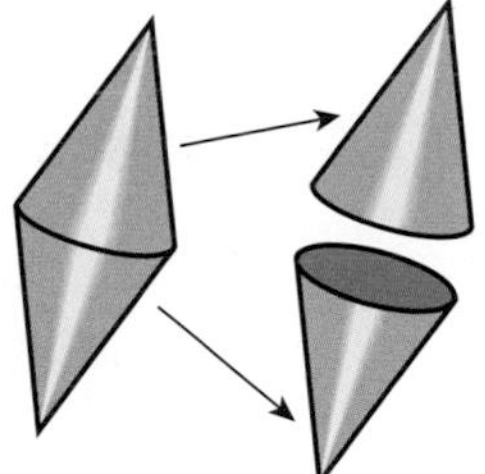

The figure is made up of two cones.

Think and Discuss

1. **Explain** how to identify the bases of a prism and the base of a pyramid. With what type of pyramid could any of the faces be considered the pyramid's base?
2. **Describe** what kinds of figures are made by cutting through a cylinder parallel to its bases.

9-1 Exercises

3.01a, 3.01c

FOR EOG PRACTICE

see page 694

internet connect

Homework Help Online
go.hrw.com Keyword: MS4 9-1

GUIDED PRACTICE

See Example 1 **Identify the base or bases of each solid. Then name the solid.**

1. 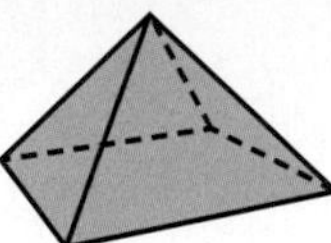

2. 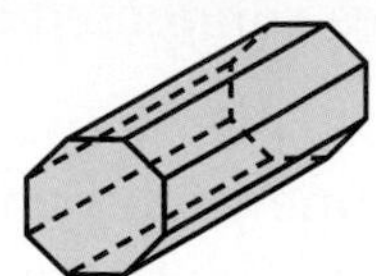

3. 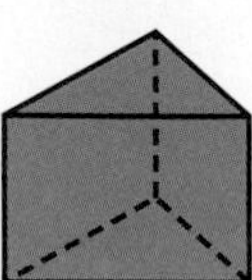

See Example 2 **Tell what solids make up each figure.**

4. 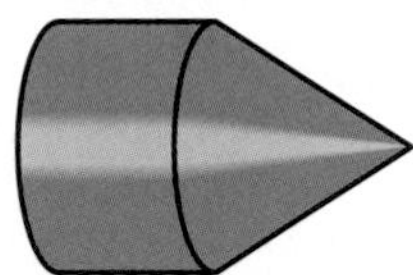

5. 

6. 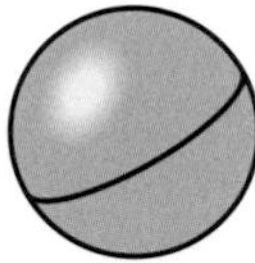

INDEPENDENT PRACTICE

See Example 1 **Identify the base or bases of each solid. Then name the solid.**

7. 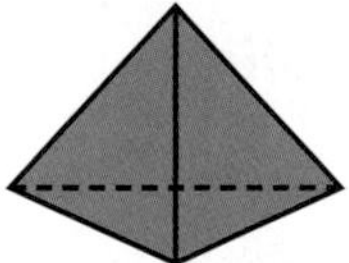

8. 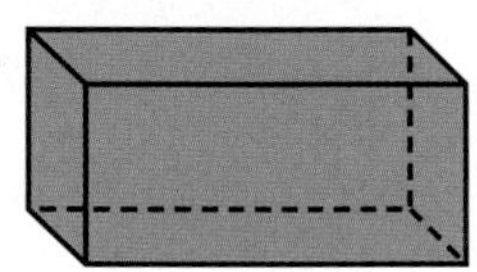

9. 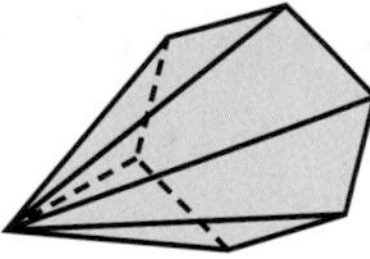

See Example 2 **Tell what solids make up each figure.**

10. 

11. 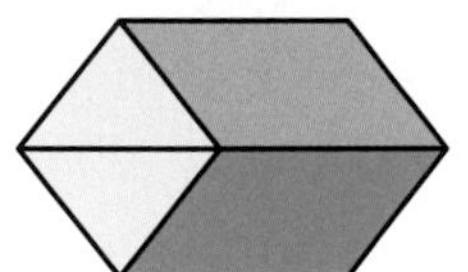

12. 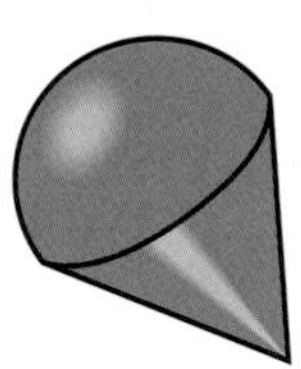

PRACTICE AND PROBLEM SOLVING

Identify the three-dimensional figure described.

13. six congruent square faces

14. two circular bases, curved lateral surface

15. triangular base, three triangular lateral faces

16. All points on the surface are the same distance from a given point.

History LINK

17. The structures in the photo at right are tombs of ancient Egyptian kings. No one knows exactly when the tombs were built, but some archaeologists think the first one might have been built around 2780 B.C. Name the shape of the ancient Egyptian structures.

18. The Parthenon was built around 440 B.C. by the ancient Greeks. Its purpose was to house a statue of Athena, the Greek goddess of wisdom. Describe the three-dimensional shapes you see in the structure.

440 B.C. Parthenon

19. The Leaning Tower of Pisa began to lean as it was being built. To keep the tower from falling over, the upper sections (floors) were built slightly off center so that the tower would curve away from the way it was leaning. What shape is each section of the tower?

20. ***CHALLENGE*** The stainless steel structure at right, called the Unisphere, became the symbol of the New York World's Fair of 1964–1965. Explain why the structure is not a true representation of a sphere.

1964 Unisphere

2600 B.C. Ancient Egyptian structures at Giza

1173 Leaning Tower of Pisa

go.hrw.com
KEYWORD: MS4 Structures
CNN Student News

Spiral Review

Divide. (Lesson 4-4)

21. $4.24 \div 4$ **22.** $3.5 \div 5$ **23.** $28.53 \div 9$ **24.** $0.04 \div 8$

Estimate each sum. (Lesson 4-9)

25. $\frac{2}{5} + \frac{3}{8}$ **26.** $\frac{1}{16} + \frac{4}{9}$ **27.** $\frac{7}{9} + \frac{11}{12}$ **28.** $\frac{1}{10} + \frac{1}{16}$

29. **EOG PREP** Which of the following is equivalent to 36 feet? (Lesson 8-1)

A 432 inches B 108 yards C 0.06 miles D 3 yards

30. **EOG PREP** Which of the following could be the lengths of the sides of a right triangle? (Lesson 8-8)

A 5, 11, 12 B 5, 7, 12 C 3, 4, 7 D 5, 12, 13

9-2 Volume of Prisms and Cylinders

Learn to find the volume of prisms and cylinders.

Vocabulary

volume

Any solid figure can be filled completely with congruent cubes and parts of cubes. The **volume** of a solid is the number of cubes it can hold. Each cube represents a unit of measure called a cubic unit.

EXAMPLE 1 Using Cubes to Find the Volume of a Rectangular Prism

Find how many cubes the prism holds. Then give the prism's volume.

You can find the volume of this prism by counting how many cubes tall, long, and wide the prism is and then multiplying.

$2 \cdot 4 \cdot 2 = 16$

There are 16 cubes in the prism, so the volume is 16 cubic units.

Reading Math

Any unit of measurement with an exponent of 3 is a cubic unit. For example, cm^3 means "cubic centimeter" and in^3 means "cubic inch."

A cube that measures one centimeter on each side represents one cubic centimeter of volume. Suppose the cubes in the prism in Example 1 measure one centimeter on each side. The volume of the prism would be 16 cm^3.

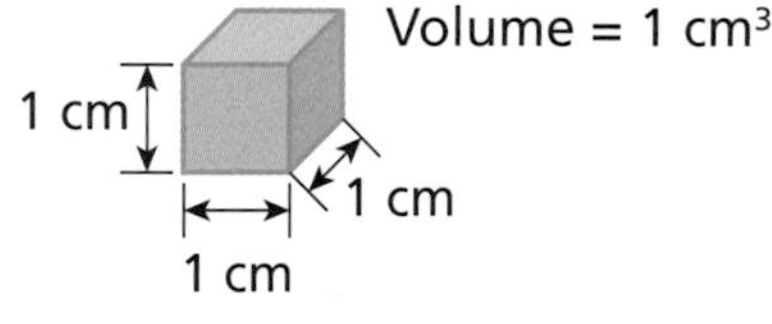

This volume is found by multiplying the prism's length times its width times its height.

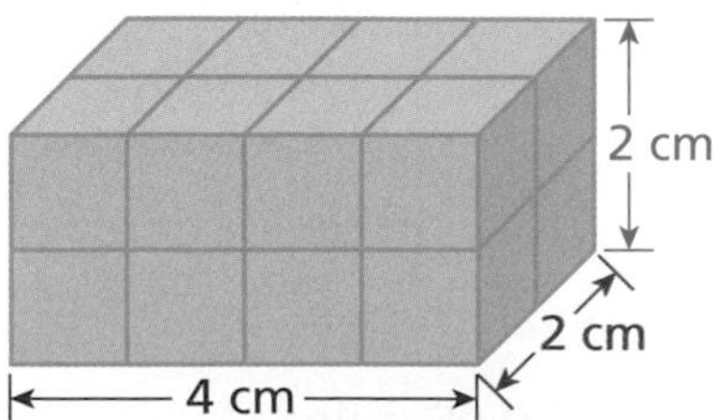

4 cm · **2 cm** · 2 cm = 16 cm^3

length · **width** · height = volume

area of base · height = volume

Notice that for the rectangular prism, the volume is found by multiplying the area of its base times its height. This method can be used for finding the volume of any prism.

VOLUME OF A PRISM

The volume V of a prism is the area of its base B times its height h.

$$V = Bh$$

EXAMPLE 2 Using a Formula to Find the Volume of a Prism

Find the volume of the prism to the nearest tenth.

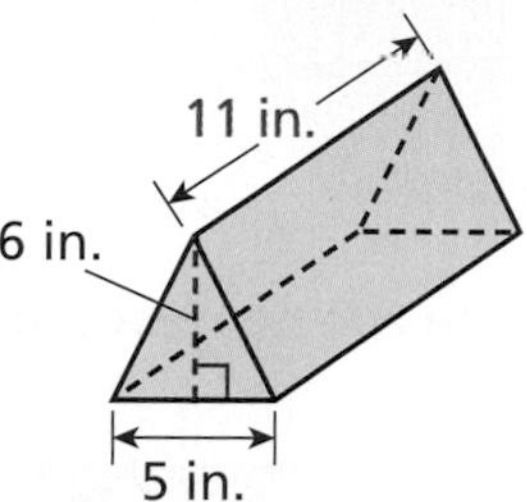

$V = Bh$ *Use the formula.*

The bases are triangles. The base of each triangle is 5 in., and the height is 6 in. The area of each triangular base is $\frac{1}{2}bh$, or $\frac{1}{2} \cdot 5 \cdot 6 = 15$.

$V = 15 \cdot 11$ *Substitute for B and h.*

$V = 165$ *Multiply.*

The volume is 165.0 in^3.

Finding the volume of a cylinder is similar to finding the volume of a prism.

VOLUME OF A CYLINDER

The volume V of a cylinder is the area of its base, πr^2, times its height h.

$$V = \pi r^2 h$$

EXAMPLE 3 Using a Formula to Find the Volume of a Cylinder

Find the volume of the cylinder to the nearest tenth. Use 3.14 for π.

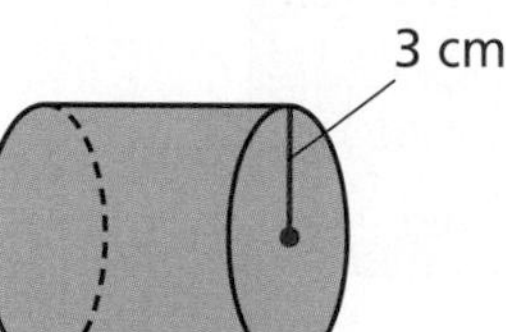

$V = \pi r^2 h$ *Use the formula.*

The radius of the cylinder is 3 cm, and the height is 6.1 cm.

$V \approx 3.14 \cdot 3^2 \cdot 6.1$ *Substitute for r and h.*

$V \approx 172.386$ *Multiply.*

$V \approx 172.4$ *Round.*

The volume is about 172.4 cm^3.

Think and Discuss

1. **Explain** what a cubic unit is.
2. **Tell** what units you would use to express the volume of a cylinder if the radius and the height are given in yards.

9-2 Exercises

FOR EOG PRACTICE

see page 694

2.02, 3.01a, 5.04

GUIDED PRACTICE

See Example 1 

Find how many cubes each prism holds. Then give the prism's volume.

1. 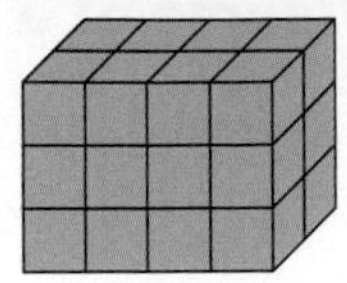

2. 

3. 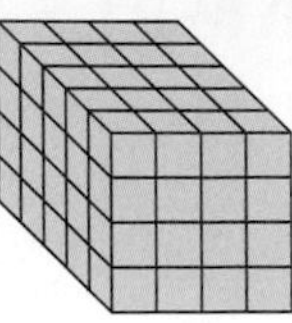

See Example 2 

Find the volume of each prism to the nearest tenth.

4. 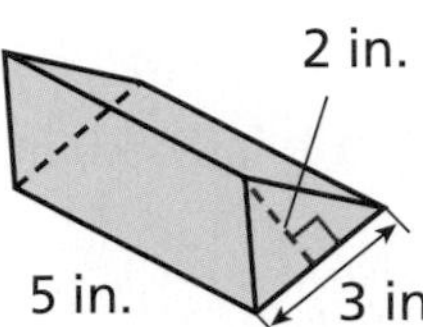

5. 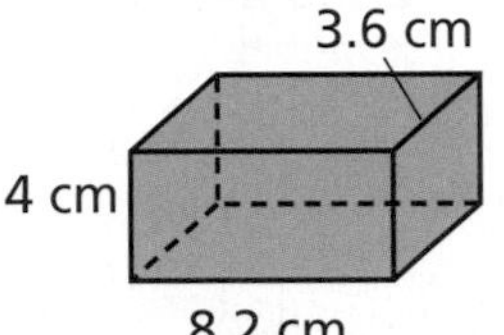

6. 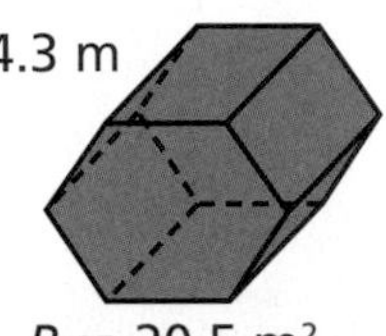

See Example 3 

Find the volume of each cylinder to the nearest tenth. Use 3.14 for π.

7. 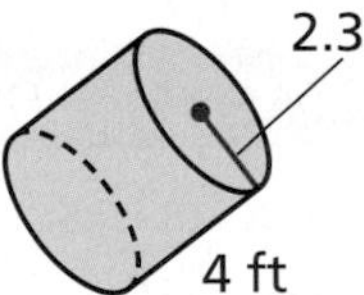

8. 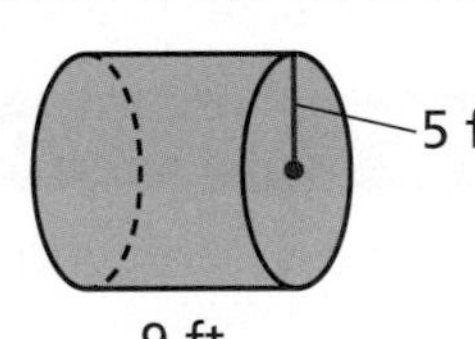

9. 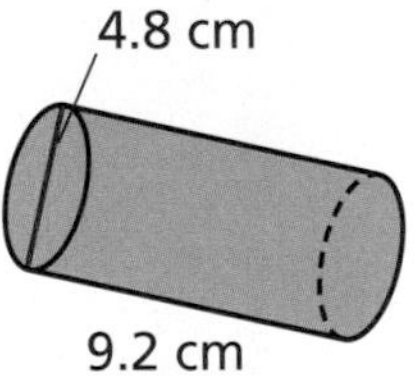

INDEPENDENT PRACTICE

See Example 1

Find how many cubes each prism holds. Then give the prism's volume.

10. 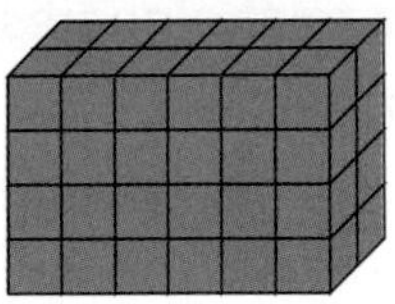

11. 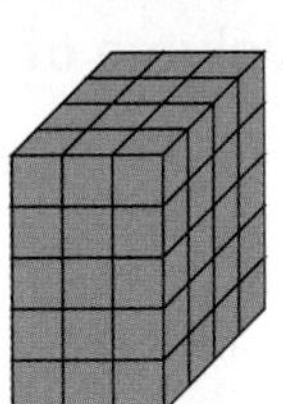

12. 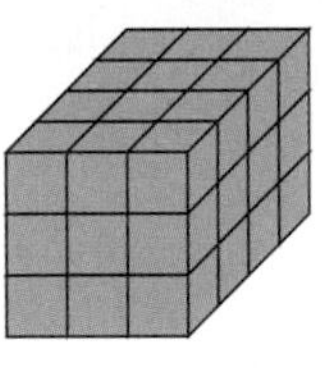

See Example 2 

Find the volume of each prism to the nearest tenth.

13. 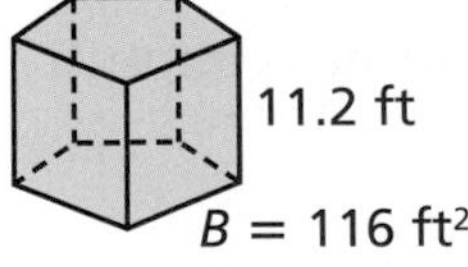

14. 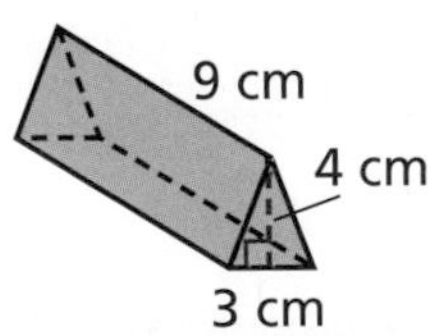

15. 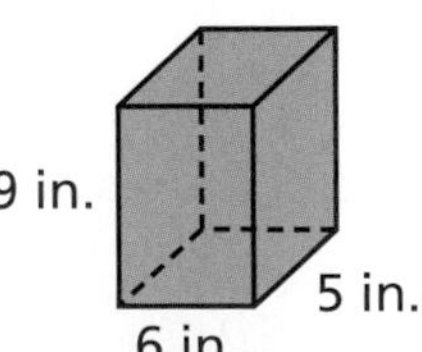

See Example 3 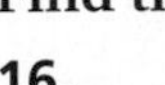

Find the volume of each cylinder to the nearest tenth. Use 3.14 for π.

16. 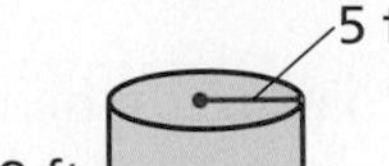 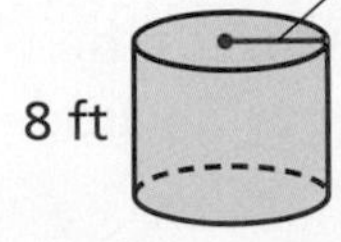

17. 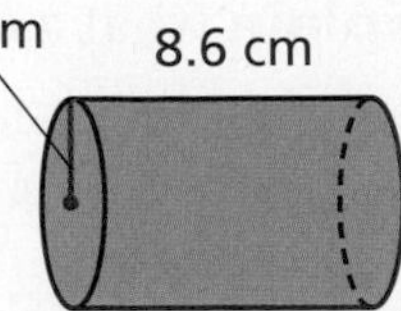 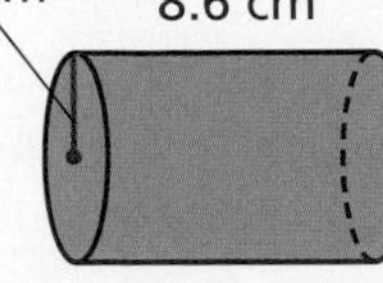

18. 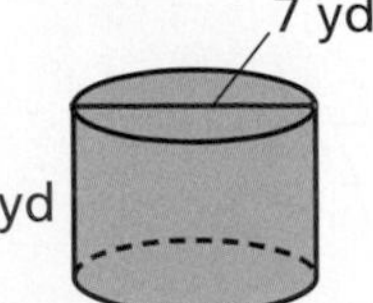

PRACTICE AND PROBLEM SOLVING

Find the volume of each solid to the nearest tenth. Use 3.14 for π.

19. cylinder: $r = 10$ m, $h = 7$ m

20. rectangular prism: $\ell = 8$ cm, $w = 7.3$ cm, $h = 10.7$ cm

21. triangular prism: $B = 15\ \text{ft}^2$, $h = 8$ ft

22. cylinder: $d = 10$ m, $h = 9.3$ m

Life Science LINK

Marine biologists insert tags containing tiny microchips into the bellies of salmon to study the migration patterns of these fish.

go.hrw.com
KEYWORD: MS4 Tags

CNN Student News

23. The base of a triangular prism is a right triangle with hypotenuse 10 m long and one leg 6 m long. If the height of the prism is 12 m, what is the volume of the prism?

24. ***ENTERTAINMENT*** A compact-disc case, or jewel case, is 14 cm long, 12.5 cm wide, and 1 cm tall. What is the volume of the jewel case?

25. An ID tag containing a microchip can be injected into a pet, such as a dog or cat. These microchips are cylindrical and can be as small as 12 mm in length and 2.1 mm in diameter. What is the volume, to the nearest tenth, of one of these microchips? Use 3.14 for π.

26. ***RECREATION*** The tent shown is in the shape of a triangular prism. How many cubic feet of space are in the tent?

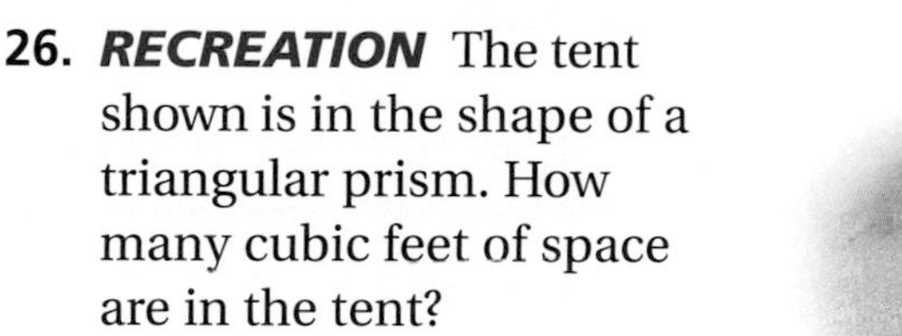

27. ***WHAT'S THE ERROR?*** A student said the volume of a cylinder with a 3-inch diameter is two times the volume of a cylinder with the same height and a 1.5-inch radius. What is the error?

28. ***WRITE ABOUT IT*** Explain the similarities and differences between finding the volume of a cylinder and finding the volume of a triangular prism.

29. ***CHALLENGE*** Find the volume, to the nearest tenth, of the material that makes up the pipe shown. Use 3.14 for π.

6 cm
15 cm
8.4 cm

Spiral Review

Write each fraction in decimal form. (Lesson 3-7)

30. $\frac{1}{2}$ **31.** $\frac{3}{5}$ **32.** $\frac{7}{8}$ **33.** $\frac{5}{16}$

Find the simple interest. (Lesson 6-6)

34. $P = \$3{,}600$; $r = 5\%$; $t = 1.5$ years **35.** $P = \$10{,}000$; $r = 3.2\%$; $t = 2$ years

36.  **EOG PREP** Which is the equivalent in meters of 100.6 centimeters? (Lesson 8-1)

A 1.006 B 10.06 C 1,006 D 10,060

9-3 Volume of Pyramids, Cones, and Spheres

Learn to find the volume of pyramids, cones, and spheres.

If you pour sand from a pyramid-shaped container into a prism-shaped container with the same height, base shape, and base size, you will discover an interesting relationship. The prism-shaped container appears to hold three times as much sand as the pyramid-shaped container.

In fact, the volume of a pyramid is exactly one-third the volume of a prism that has the same height, base shape, and base size as the pyramid. The height of a pyramid is the perpendicular distance from the pyramid's base to its vertex.

VOLUME OF A PYRAMID

The volume V of a pyramid is one-third the area of its base B times its height h.

$$V = \frac{1}{3}Bh$$

EXAMPLE 1 Finding the Volume of a Pyramid

Find the volume of each pyramid.

A

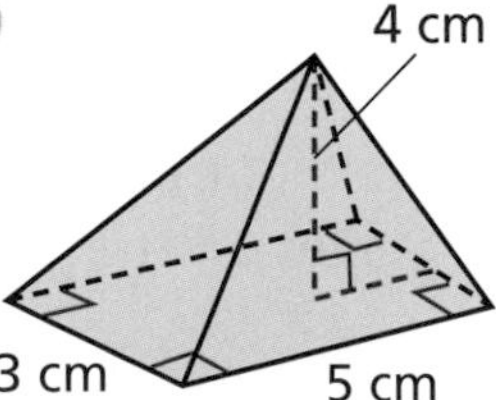

$V = \frac{1}{3}Bh$	*Use the formula.*
$B = 3 \cdot 5 = \mathbf{15}$	*Find the area of the rectangular base.*
$V = \frac{1}{3} \cdot \mathbf{15} \cdot \mathbf{4}$	*Substitute for B and h.*
$V = 20$	*Multiply.*

The volume is 20 cm^3.

B

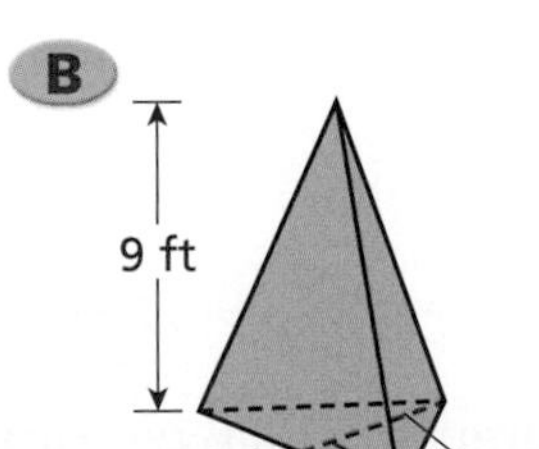

$V = \frac{1}{3}Bh$	*Use the formula.*
$B = \frac{1}{2} \cdot 6 \cdot 4 = \mathbf{12}$	*Find the area of the triangular base.*
$V = \frac{1}{3} \cdot \mathbf{12} \cdot \mathbf{9}$	*Substitute for B and h.*
$V = 36$	*Multiply.*

The volume is 36 ft^3.

The volume of a cone is one-third the volume of a cylinder with the same height and a congruent base. The height of a cone is the perpendicular distance from the cone's base to its vertex.

VOLUME OF A CONE

The volume V of a cone is one-third the area of its base, πr^2, times its height h.

$$V = \frac{1}{3}\pi r^2 h$$

EXAMPLE 2 Finding the Volume of a Cone

Find the volume of the cone to the nearest tenth. Use 3.14 for π.

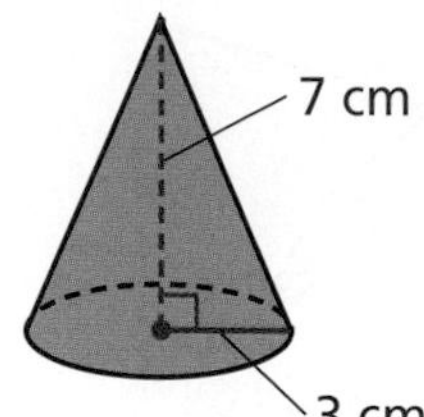

$V = \frac{1}{3}\pi r^2 h$ — *Use the formula.*

$V \approx \frac{1}{3} \cdot 3.14 \cdot 3^2 \cdot 7$ — *Substitute.*

$V \approx 65.94$ — *Multiply.*

$V \approx 65.9$ — *Round.*

The volume is about 65.9 cm^3.

VOLUME OF A SPHERE

The volume V of a sphere is $\frac{4}{3}$ times π times the radius r cubed.

$$V = \frac{4}{3}\pi r^3$$

EXAMPLE 3 Finding the Volume of a Sphere

Find the volume of the sphere to the nearest tenth. Use 3.14 for π.

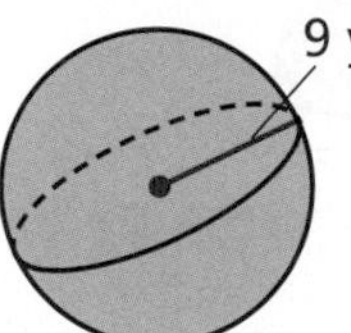

$V = \frac{4}{3}\pi r^3$ — *Use the formula.*

$V \approx \frac{4}{3} \cdot 3.14 \cdot 9^3$ — *Substitute.*

$V \approx 3{,}052.08$ — *Multiply.*

$V \approx 3{,}052.1$ — *Round.*

The volume is about 3,052.1 yd^3.

Think and Discuss

1. **Compare** the formulas for the volume of a pyramid and the volume of a prism.
2. **Describe** how to find the volume of a sphere with diameter 8 m.

9-3 Exercises

FOR EOG PRACTICE

see page 695

internet connect

Homework Help Online

go.hrw.com Keyword: MS4 9-3

2.02, 5.04

GUIDED PRACTICE

See Example 1 Find the volume of each pyramid.

1. 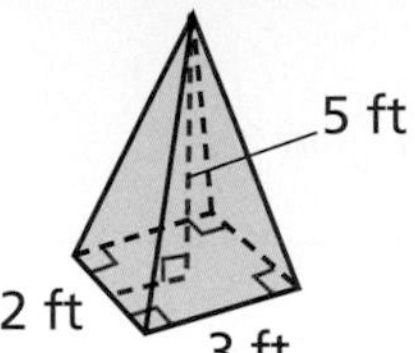

2. 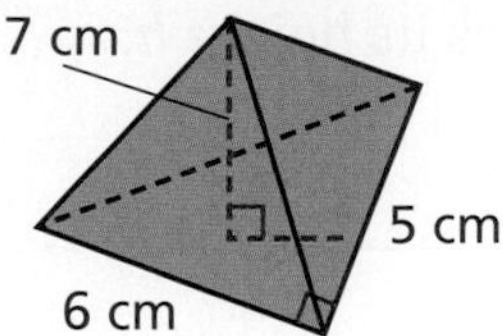

3. 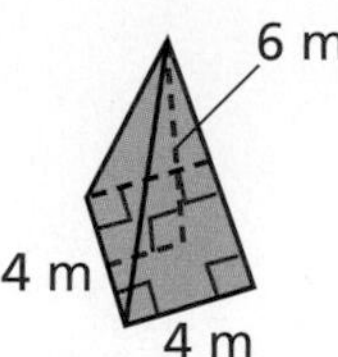

See Example 2 Find the volume of each cone to the nearest tenth. Use 3.14 for π.

4. 10 ft, 6 ft

5. 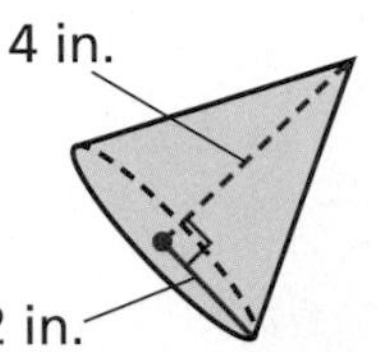

6. 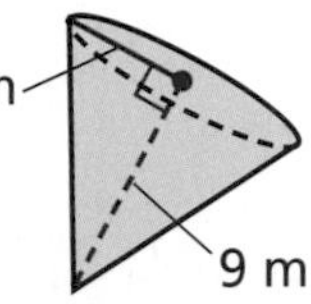

See Example 3 Find the volume of each sphere to the nearest tenth. Use 3.14 for π.

7. 6 cm

8. 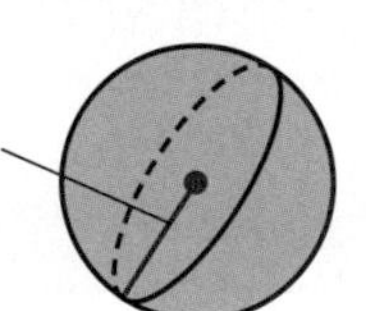

9. 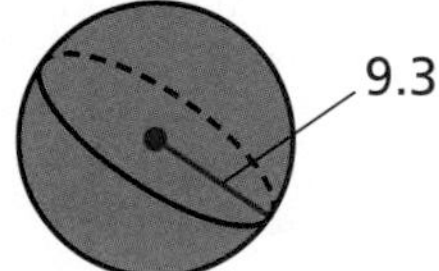

INDEPENDENT PRACTICE

See Example 1 Find the volume of each pyramid.

10. 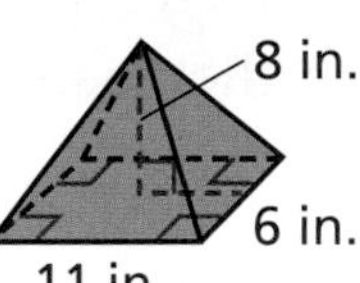

11. 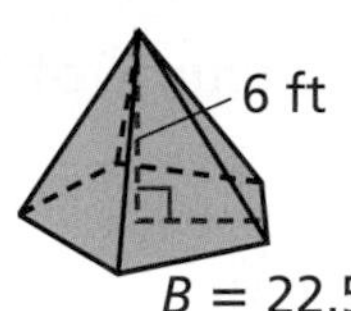

12. 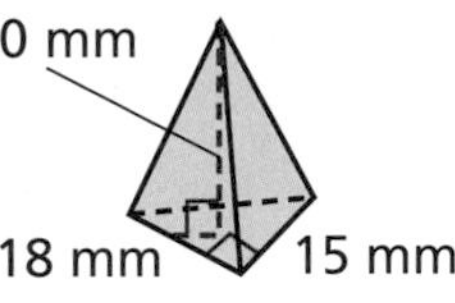

See Example 2 Find the volume of each cone to the nearest tenth. Use 3.14 for π.

13. 5 in., 3 in.

14. 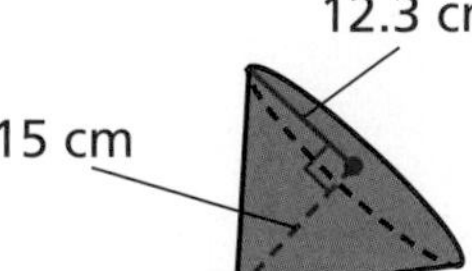

15. 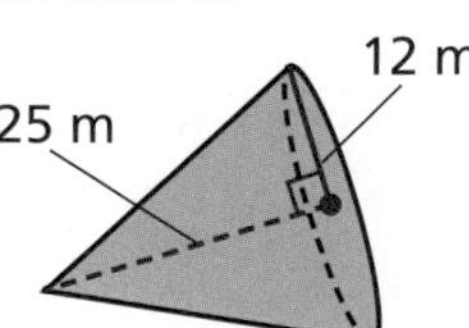

See Example 3 Find the volume of each sphere to the nearest tenth. Use 3.14 for π.

16. 8 ft

17. 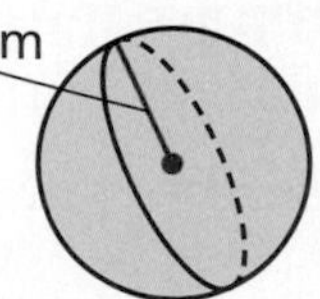

18. 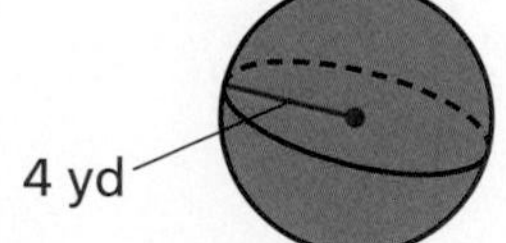

PRACTICE AND PROBLEM SOLVING

Find the volume of each solid to the nearest tenth. Use 3.14 for π.

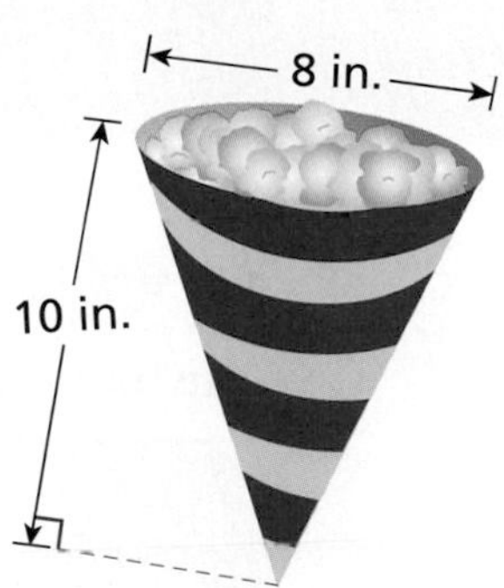

19. a 7 ft tall rectangular pyramid with base 4 ft by 5 ft

20. a cone with radius 8 yd and height 12 yd

21. a sphere with diameter 5 m

22. ***BUSINESS*** A snack bar sells popcorn in the containers shown at right.

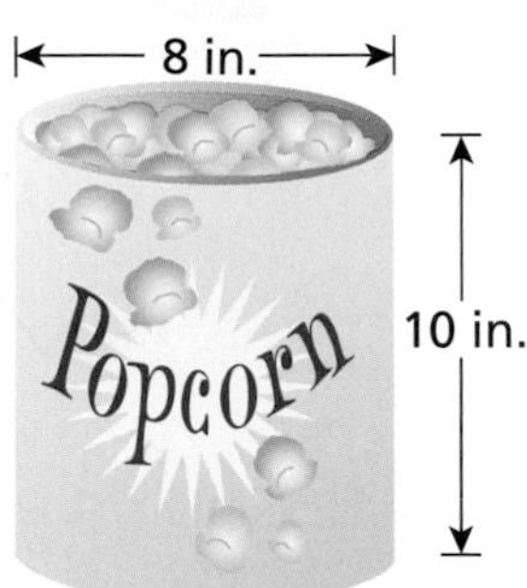

a. How many cubic inches of popcorn, to the nearest tenth, does the cone-shaped container hold? Use 3.14 for π.

b. How many cubic inches of popcorn does the cylinder-shaped container hold? Use 3.14 for π.

c. About how many times as much popcorn does the larger container hold?

23. ***ARCHITECTURE*** The steeple on a building is a square pyramid with base area 12 square feet and height 15 feet. How many cubic feet of concrete was used to make the pyramid?

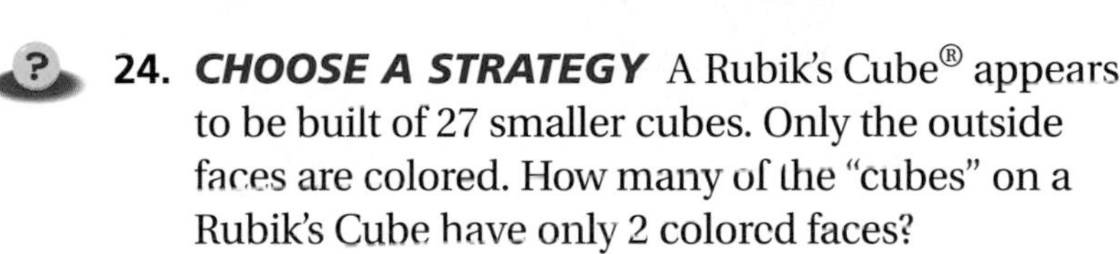

24. ***CHOOSE A STRATEGY*** A Rubik's Cube® appears to be built of 27 smaller cubes. Only the outside faces are colored. How many of the "cubes" on a Rubik's Cube have only 2 colored faces?

A 3 **B** 8 **C** 9 **D** 12

25. ***WRITE ABOUT IT*** Compare finding the volume of a cylinder with finding the volume of a cone that has the same height and base.

26. ***CHALLENGE*** What effect does doubling the radius of a sphere have on the sphere's volume?

Spiral Review

Solve by using cross products. (Lesson 5-3)

27. $\frac{5}{6} = \frac{a}{12}$ **28.** $\frac{c}{16} = \frac{5}{8}$ **29.** $\frac{b}{9} = \frac{21}{27}$ **30.** $\frac{3}{5} = \frac{m}{25}$

The two quadrilaterals are congruent. Give the measure of each angle. (Lesson 7-9)

31. $\angle GFE$ **32.** $\angle FGH$

33. **EOG PREP** What is the volume of a cylinder with diameter 4 m and height 10 m? (Lesson 9-2)

A 62.8 m^3 B 125.6 m^3 C 160 m^3 D 502.4 m^3

Chapter 9

Mid-Chapter Quiz

LESSON 9-1 (pp. 472–475)

Give the name of each prism or pyramid.

1. 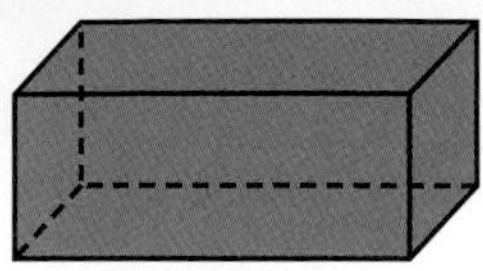

2. 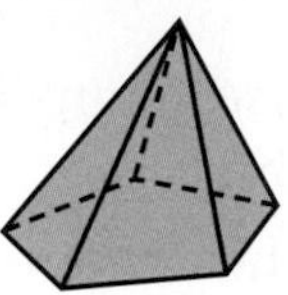

3. 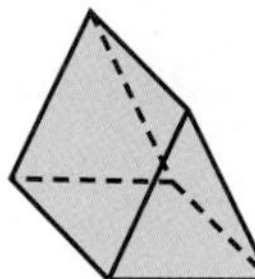

4. 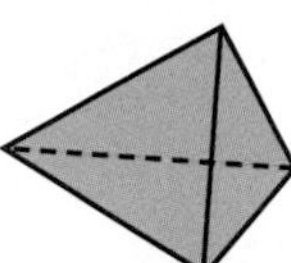

5. 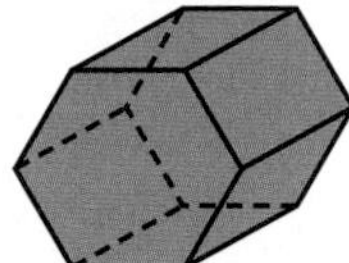

6. 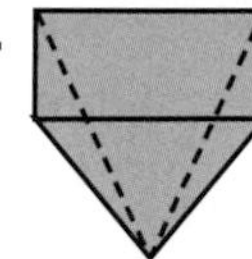

LESSON  9-2 (pp. 476–479)

Find the volume of each figure to the nearest tenth. Use 3.14 for π.

7. 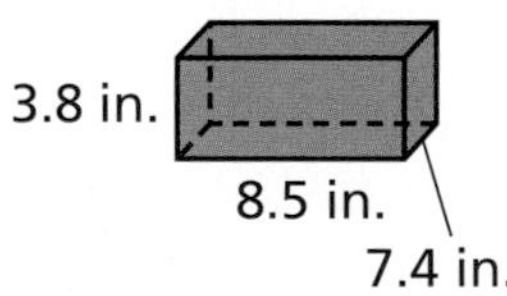

8. 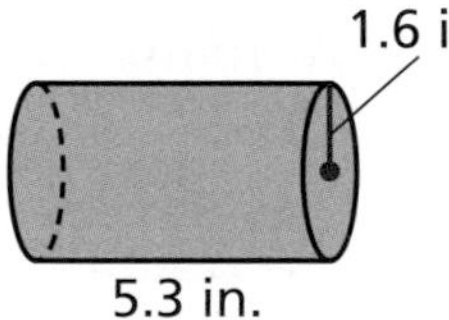

9. 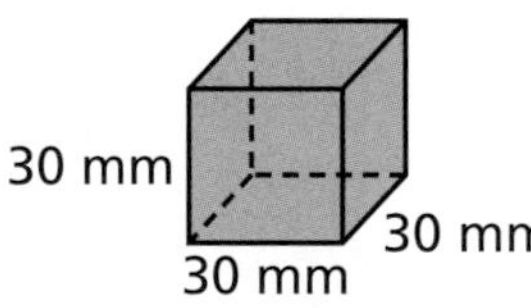

10. 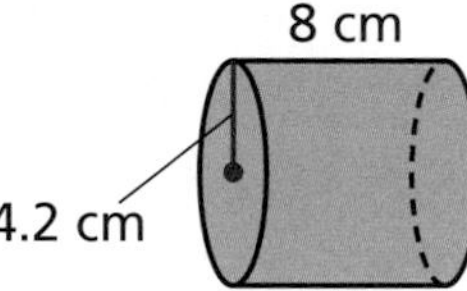

11. 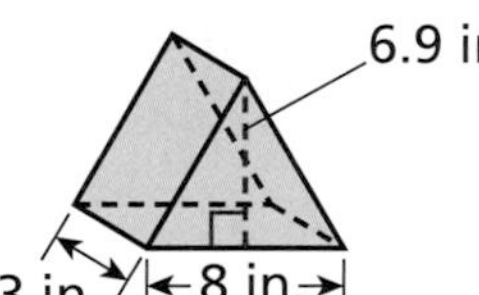

12. 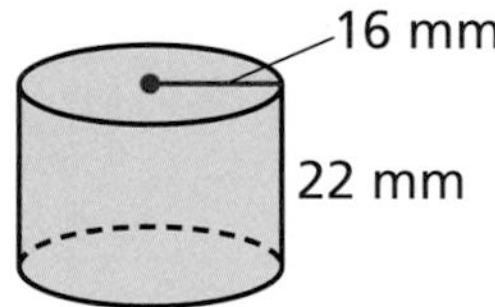

LESSON  9-3 (pp. 480–483)

Find the volume of each figure to the nearest tenth. Use 3.14 for π.

13. 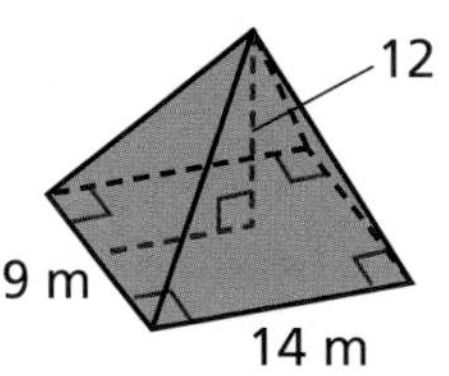

14. 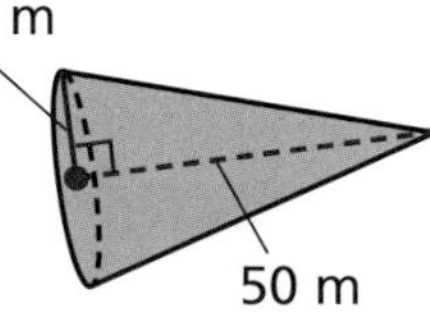

15. 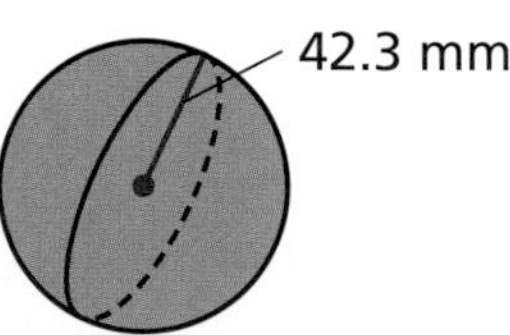

16. 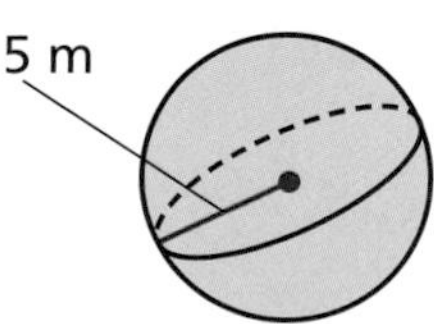

17. 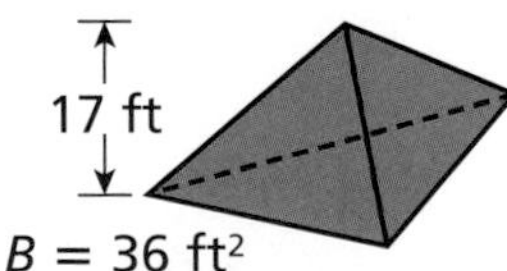

18. 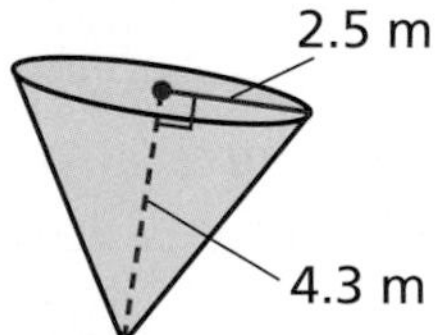

19. The diameter of a volleyball is about 21 cm. About how much air, to the nearest cubic centimeter, will fill the volleyball? Use 3.14 for π.

20. A cone has a radius of 2.5 cm and a height of 14 cm. What is the volume of the cone to the nearest hundredth? Use 3.14 for π.

1.02a

Focus on Problem Solving

Solve

• **Choose an operation**

When choosing an operation to use when solving a problem, you need to decide which action the problem is asking you to take. If you are asked to combine numbers, then you need to add. If you are asked to take away numbers or to find the difference between two numbers, then you need to subtract. You need to use multiplication when you put equal parts together and division when you separate something into equal parts.

Determine the action in each problem. Then tell which operation should be used to solve the problem. Explain your choice.

1. Jeremy filled a sugar cone completely full of frozen yogurt and then put one scoop of frozen yogurt on top. The volume of Jeremy's cone is about 20.93 in^3, and the volume of the scoop that Jeremy used is about 16.75 in^3. About how much frozen yogurt, in cubic inches, did Jeremy use?

2. The volume of a cylinder equals the combined volumes of three cones that each have the same height and base size as the cylinder. What is the volume of a cylinder if a cone of the same height and base size has a volume of 45.2 cm^3?

3. The biology class at Jefferson High School takes care of a family of turtles that is kept in a glass tank with water, rocks, and plants. The volume of the tank is 2.75 cubic feet. At the end of the year, the baby turtles will have grown and will be moved into a tank that is 6.15 cubic feet. How much greater will the volume of the new tank be than that of the old tank?

4. Brianna is adding a second section to her hamster cage. The two sections will be connected by a tunnel that is made of 4 cylindrical parts, all the same size. If the volume of the tunnel is 56.52 cubic inches, what is the volume of each part of the tunnel?

9-4 Surface Area of Prisms, Cylinders, and Spheres

Learn to find the surface area of prisms, cylinders, and spheres.

Vocabulary

net

surface area

If you remove the surface from a three-dimensional figure and lay it out flat, the pattern you make is called a **net**. You can construct nets to cover almost any geometric solid.

Since nets allow you to see all the surfaces of a solid at one time, you can use them to help find the *surface area* of a three-dimensional figure. **Surface area** is the sum of the areas of all of the surfaces of a figure.

SURFACE AREA OF A POLYHEDRON

The surface area of a polyhedron is found by adding the areas of each face of the polyhedron.

You can use nets to write formulas for the surface area of prisms. The surface area S is the sum of the areas of the faces of the prism. For the rectangular prism shown, $S = \ell w + \ell h + wh + \ell w + \ell h + wh$

$= 2\ell w + 2\ell h + 2wh.$

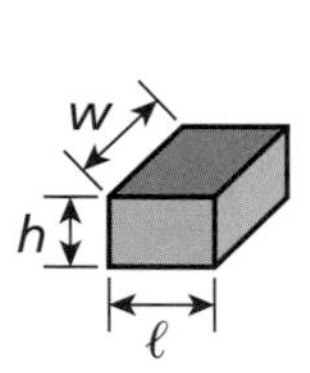

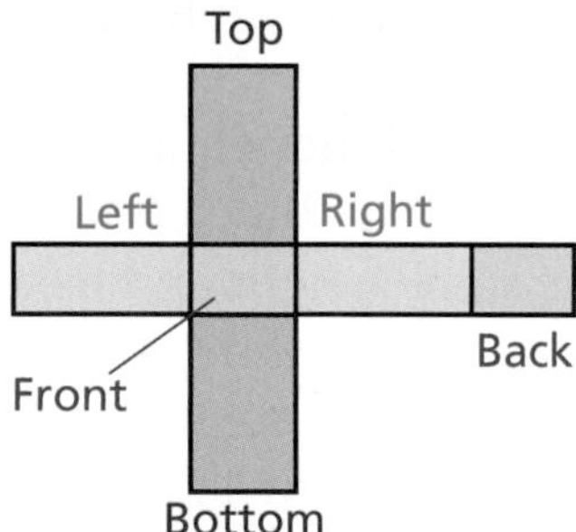

EXAMPLE 1 **Finding the Surface Area of a Prism**

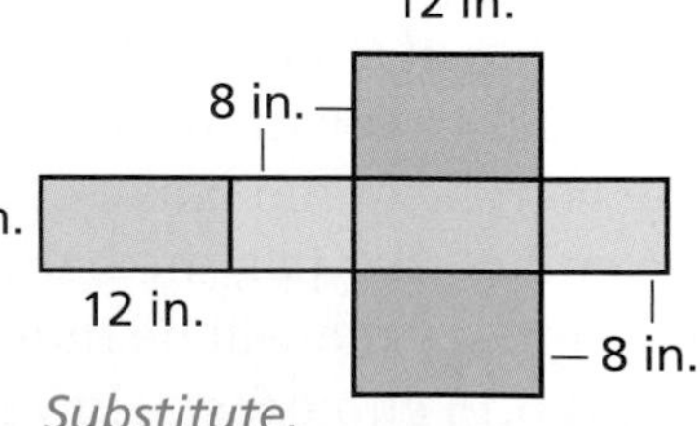

Find the surface area of the prism formed by the net.

$S = 2\ell w + 2\ell h + 2wh$

$S = (2 \cdot 12 \cdot 8) + (2 \cdot 12 \cdot 6) + (2 \cdot 8 \cdot 6)$ *Substitute.*

$S = 192 + 144 + 96$ *Multiply.*

$S = 432$ *Add.*

The surface area of the prism is 432 in^2.

If you could remove the lateral surface from a cylinder, like peeling a label from a can, you would see that it has the shape of a rectangle when flattened out.

You can draw a net for a cylinder by drawing the circular bases (like the ends of a can) and the rectangular lateral surface as shown below. The length of the rectangle is the circumference, $2\pi r$, of the cylinder. So the area of the lateral surface is $2\pi rh$. The area of each base is πr^2.

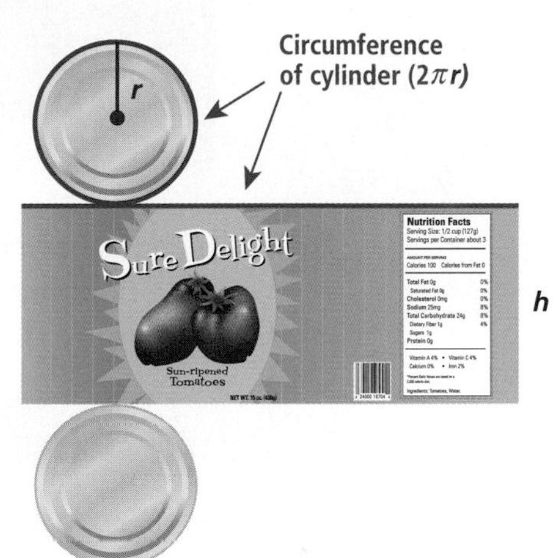

SURFACE AREA OF A CYLINDER

The surface area S of a cylinder is the sum of the areas of its bases, $2\pi r^2$, plus the area of its lateral surface, $2\pi rh$.

$$S = 2\pi r^2 + 2\pi rh$$

EXAMPLE 2 Finding the Surface Area of a Cylinder

Find the surface area of the cylinder formed by the net to the nearest tenth. Use 3.14 for π.

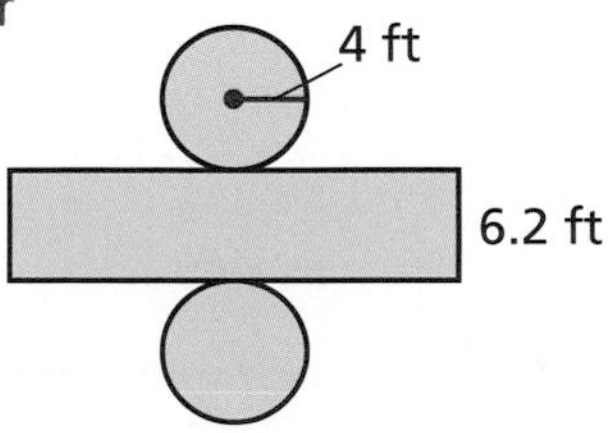

$S = 2\pi r^2 + 2\pi rh$	*Use the formula.*
$S \approx (2 \cdot 3.14 \cdot 4^2) + (2 \cdot 3.14 \cdot 4 \cdot 6.2)$	*Substitute.*
$S \approx 100.48 + 155.744$	*Multiply.*
$S \approx 256.224$	*Add.*
$S \approx 256.2$	*Round.*

The surface area of the cylinder is about 256.2 ft^2.

Unlike the surface of a prism or a cylinder, the surface of a sphere cannot be flattened without stretching or shrinking. This is why it is not possible to draw a perfectly accurate flat map of Earth. In the *Mercator projection* of Earth, for example, Greenland appears much too large.

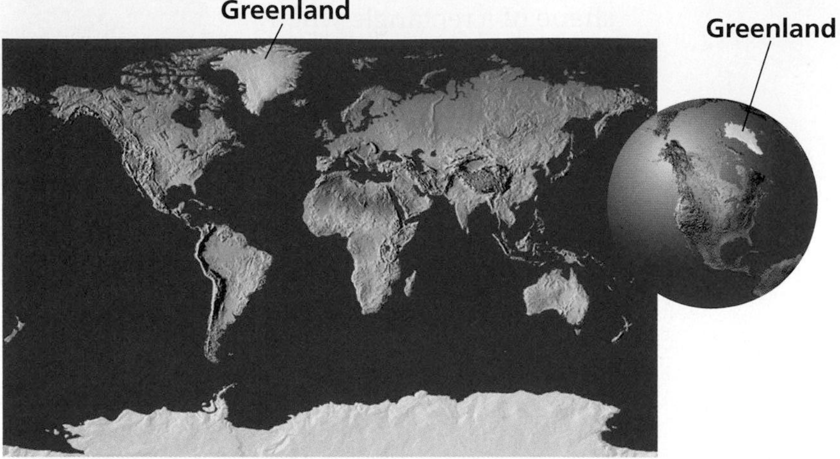

Because the surface of a sphere cannot be flattened out, it is impossible to make a net for a sphere. However, there is an exact formula for the area of a sphere.

SURFACE AREA OF A SPHERE

The surface area S of a sphere is 4 times π times the radius r squared.

$$S = 4\pi r^2$$

EXAMPLE 3 Finding the Surface Area of a Sphere

Find the surface area of the sphere to the nearest tenth. Use 3.14 for π.

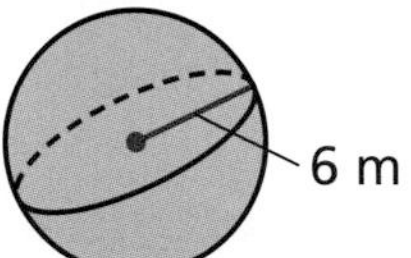

$S = 4\pi r^2$	*Use the formula.*
$S \approx 4 \cdot 3.14 \cdot 6^2$	*Substitute.*
$S \approx 452.16$	*Multiply.*
$S \approx 452.2$	*Round.*

The surface area of the sphere is about 452.2 m^2.

Think and Discuss

1. **Explain** how you would find the surface area of a rectangular prism.
2. **Describe** how to find the surface area of a cylinder 2 inches tall with a radius of 3 inches.
3. **Describe** the relationship between the area of a circle and the surface area of a sphere with the same radius.

9-4 Exercises

2.02, 3.01a

FOR EOG PRACTICE
see page 696

internet connect
Homework Help Online
go.hrw.com Keyword: MS4 9-4

GUIDED PRACTICE

See Example 1 **Find the surface area of the prism formed by each net.**

1\. 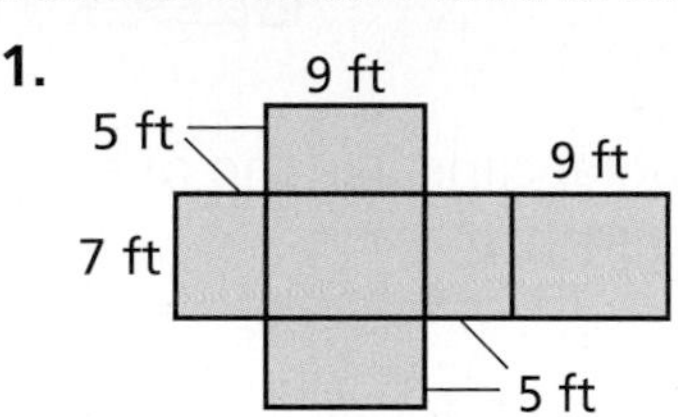

2\. 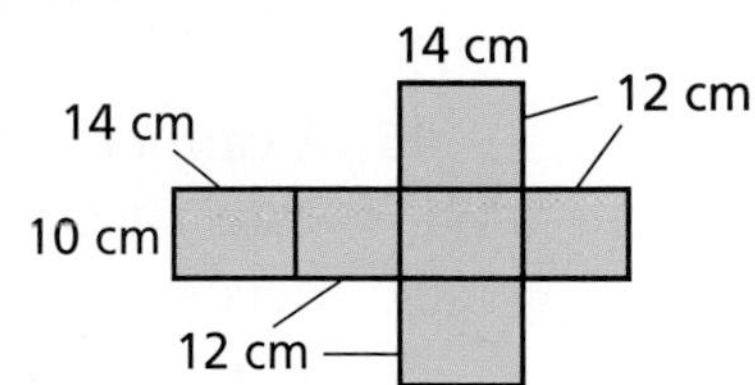

See Example 2 **Find the surface area of the cylinder formed by each net to the nearest tenth. Use 3.14 for π.**

3\. 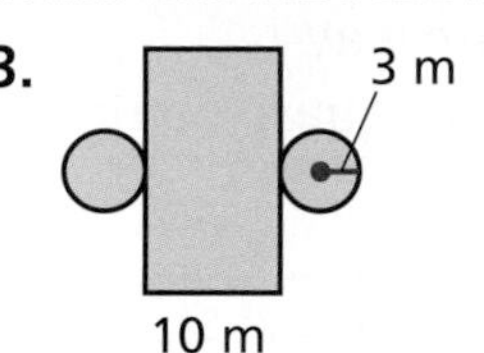

4\. 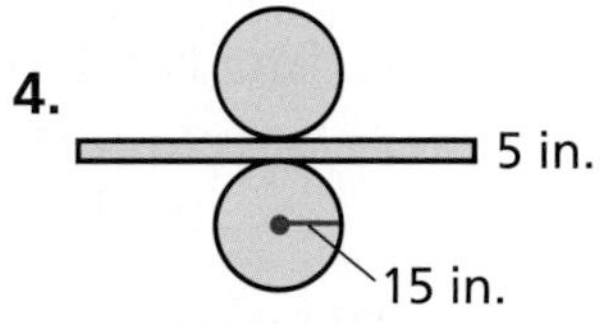

See Example 3 **Find the surface area of each sphere to the nearest tenth. Use 3.14 for π.**

5\. 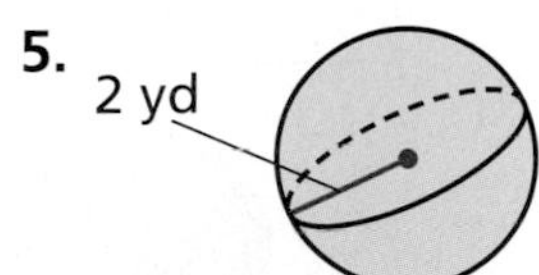

6\. 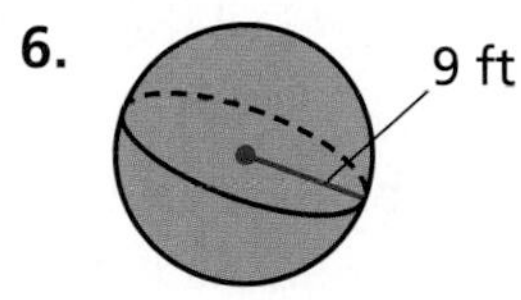

INDEPENDENT PRACTICE

See Example 1 **Find the surface area of the prism formed by each net.**

7\. 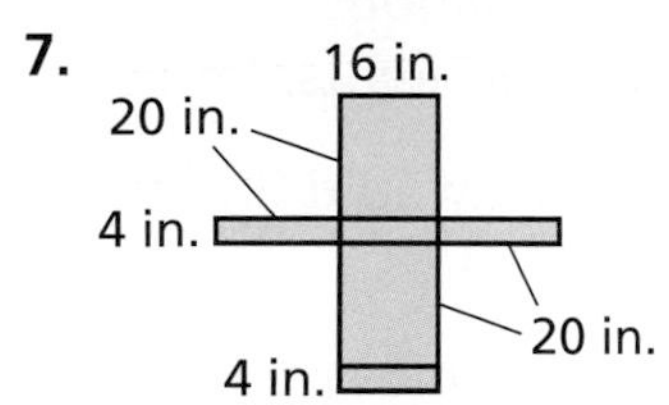

8\. 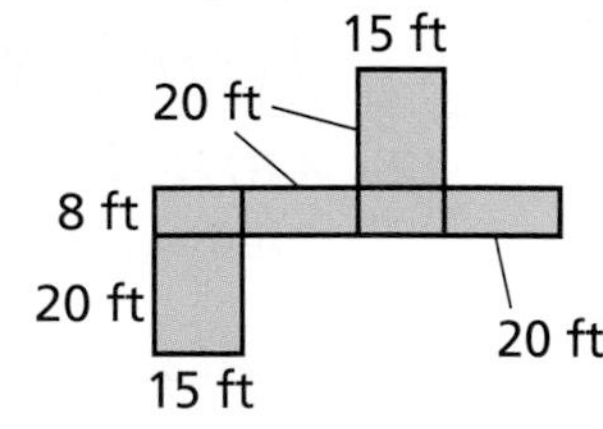

See Example 2 **Find the surface area of the cylinder formed by each net to the nearest tenth. Use 3.14 for π.**

9\. 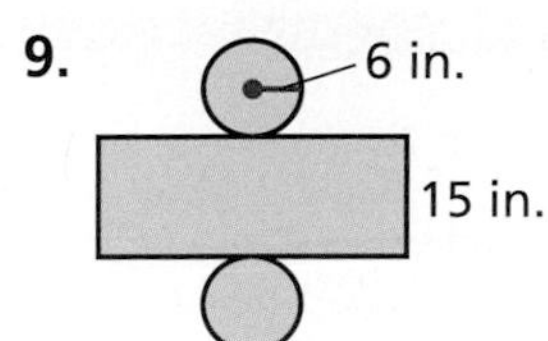

10\. 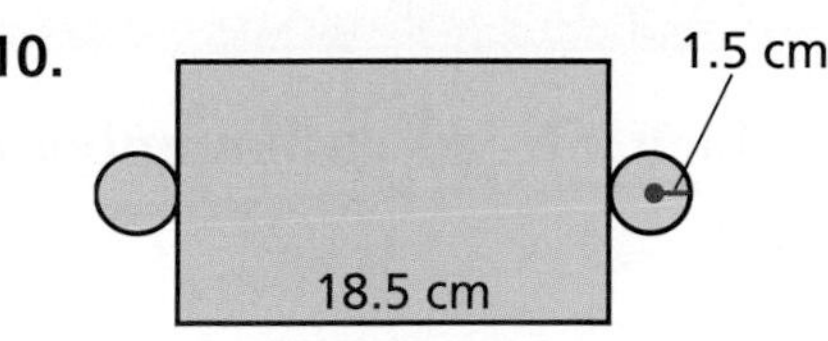

See Example 3 **Find the surface area of each sphere to the nearest tenth. Use 3.14 for π.**

11\. 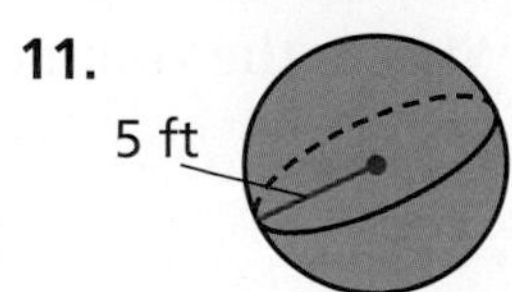

12\. 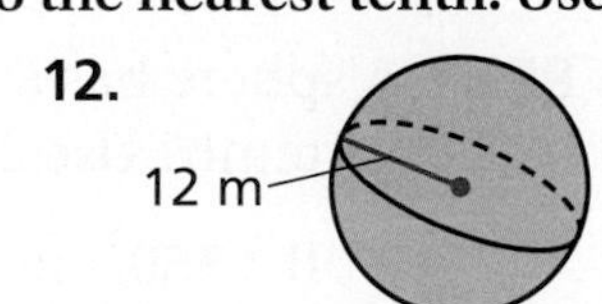

PRACTICE AND PROBLEM SOLVING

Find the surface area of each figure to the nearest tenth. Use 3.14 for π.

13.

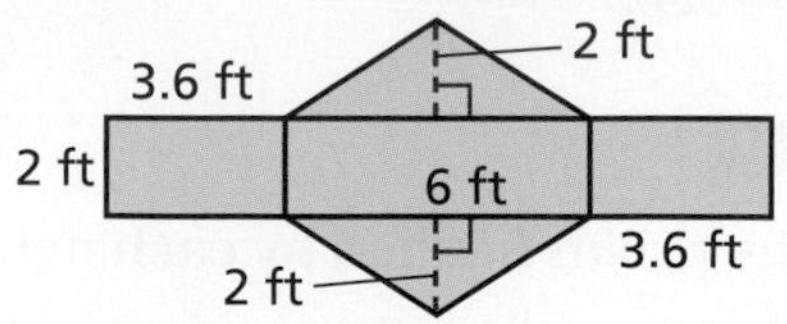

14.

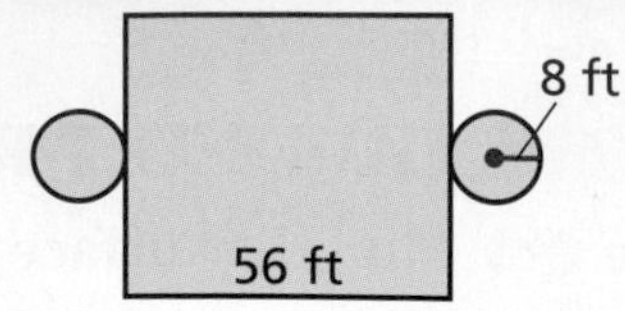

15. A cannery packs tuna into metal cans like the one shown. Round your answers to the nearest tenth, if necessary. Use 3.14 for π.

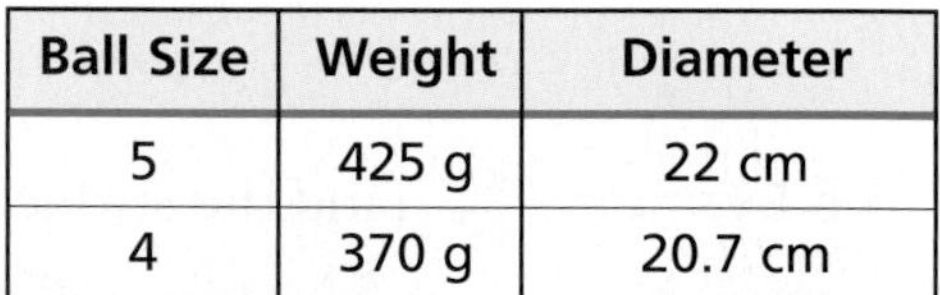

a. Draw and label a net of the cylinder.

b. About how many square centimeters of metal are used to make each can?

c. The label for each can goes all the way around the can. About how many square centimeters of paper are needed for each label?

16. The table shows the approximate weight and diameter of size 4 and size 5 soccer balls.

Ball Size	Weight	Diameter
5	425 g	22 cm
4	370 g	20.7 cm

a. Find the surface area of each ball to the nearest tenth. Use 3.14 for π.

b. Find the difference between the surface areas.

Size 5

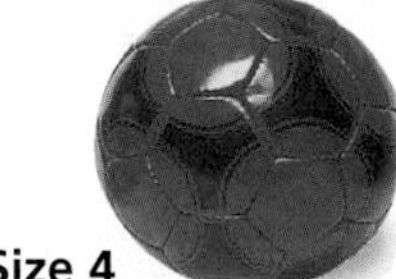
Size 4

17. ***WRITE A PROBLEM*** Write a problem about finding the surface area of an object whose shape is some combination of prisms, cylinders, and spheres (or hemispheres).

18. ***WRITE ABOUT IT*** Explain how you would find the surface area of the figure you described in Exercise 17.

19. ***CHALLENGE*** Write a formula using r for radius and the symbol π to express the surface area of a hemisphere. Include in your formula the area of a base of the hemisphere.

Spiral Review

State which integers each number lies between. (Lesson 3-7)

20. −6.4 **21.** 5.3 **22.** −2.1 **23.** −0.7

Divide. (Lesson 4-4)

24. 109.8 ÷ 6 **25.** −28.24 ÷ 4 **26.** −27.6 ÷ 23 **27.** 132.8 ÷ 16

28. **EOG PREP** A sphere has a radius of 6 inches. What is the volume of the sphere to the nearest tenth? Use 3.14 for π. (Lesson 9-3)

A 75.4 in^3 B 150.7 in^3 C 452.2 in^3 D 904.3 in^3

9-5 Changing Dimensions

Learn to find the volume and surface area of similar three-dimensional figures.

Recall that similar figures are proportional. The surface areas of similar three-dimensional figures are also proportional. To see this relationship, you can compare the areas of corresponding faces of similar rectangular prisms.

Remember!

You can multiply numbers in any order. So $(3 \cdot 2) \cdot (5 \cdot 2)$ is the same as $(3 \cdot 5) \cdot (2 \cdot 2)$.

Area of front of smaller prism	Area of front of larger prism	
$3 \cdot 5$	$6 \cdot 10$	
15	$(3 \cdot 2) \cdot (5 \cdot 2)$	← Each dimension has a scale factor of 2.
	$(3 \cdot 5) \cdot (2 \cdot 2)$	
	$15 \cdot 2^2$	

The area of the front face of the larger prism is 2^2 times the area of the front face of the smaller prism. This is true for all of the corresponding faces. Thus it is also true for the entire surface area of the prisms.

SURFACE AREA OF SIMILAR FIGURES

The surface area of a three-dimensional figure *A* is equal to the surface area of a similar figure *B* times the square of the scale factor of figure *A*.

$$\text{surface area of figure } A = \text{surface area of figure } B \cdot (\text{scale factor of figure } A)^2$$

EXAMPLE 1 Finding the Surface Area of a Similar Figure

A The surface area of a box is 27 in². What is the surface area of a larger, similarly shaped box that has a scale factor of 5?

$S = 27 \cdot 5^2$ *Use the surface area of the smaller box and the square of the scale factor.*

$S = 27 \cdot 25$ *Evaluate the power.*

$S = 675$ *Multiply.*

The surface area of the larger box is 675 in².

B **The surface area of the Great Pyramid was originally 1,160,280 ft^2. Find the surface area, to the nearest tenth, of a model of the pyramid that has a scale factor of $\frac{1}{500}$.**

$S = 1{,}160{,}280 \cdot \left(\frac{1}{500}\right)^2$ — *Use the surface area of the actual pyramid and the square of the scale factor.*

$S = 1{,}160{,}280 \cdot \frac{1}{250{,}000}$ — *Evaluate the power.*

$S = 4.64112$ — *Multiply.*

The surface area of the model is about 4.6 ft^2.

The volumes of similar three-dimensional figures are also related.

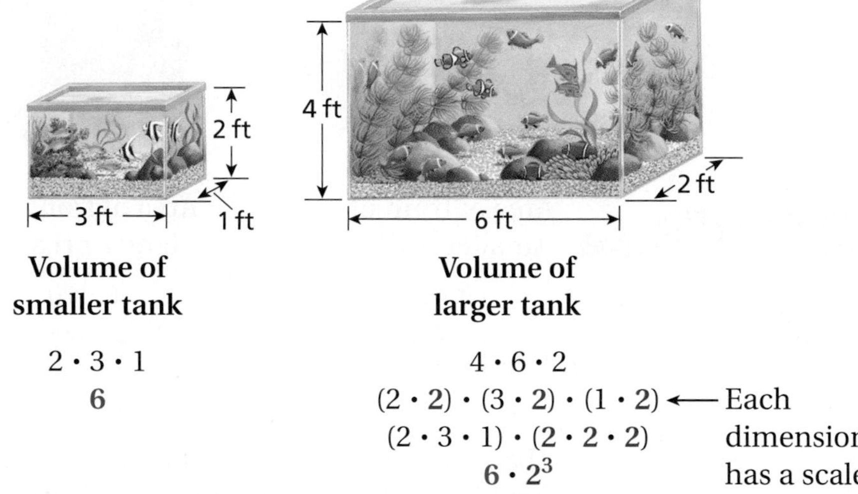

Volume of smaller tank	**Volume of larger tank**
$2 \cdot 3 \cdot 1$	$4 \cdot 6 \cdot 2$
6	$(2 \cdot 2) \cdot (3 \cdot 2) \cdot (1 \cdot 2)$ ← Each dimension has a scale factor of 2.
	$(2 \cdot 3 \cdot 1) \cdot (2 \cdot 2 \cdot 2)$
	$6 \cdot 2^3$

Remember!

$2 \cdot 2 \cdot 2 = 2^3$

The volume of the larger tank is 2^3 times the volume of the smaller tank.

VOLUME OF SIMILAR FIGURES

The volume of a three-dimensional figure *A* is equal to the volume of a similar figure *B* times the cube of the scale factor of figure *A*.

volume of figure *A* = volume of figure *B* · (scale factor of figure *A*)^3

EXAMPLE 2 Finding Volume Using Similar Figures

The volume of a bucket is 231 in^3. What is the volume of a larger, similarly shaped bucket that has a scale factor of 3?

$V = 231 \cdot 3^3$ — *Use the volume of the smaller bucket and the cube of the scale factor.*

$V = 231 \cdot 27$ — *Evaluate the power.*

$V = 6{,}237$ — *Multiply.*

The volume of the larger bucket is 6,237 in^3.

When similar three-dimensional figures are made of the same material, their weights compare in the same way as their volumes. The weight of a three-dimensional figure is the weight of a similar three-dimensional figure multiplied by the cube of the scale factor of the original figure.

EXAMPLE 3 **PROBLEM SOLVING APPLICATION**

PROBLEM SOLVING

Elise's fish tank weighs 167 pounds when full. She bought a larger tank that has a similar shape with a scale factor of 2. How much does the new tank weigh when full?

1 Understand the Problem

You can find the weight of the larger fish tank using the weight of the smaller tank.

List the **important information:**

- The smaller tank weighs 167 pounds when full.
- The scale factor of the larger tank is 2.

2 Make a Plan

You can write an equation that relates the weight of the smaller fish tank to the weight of the larger fish tank.

weight of large tank = weight of small tank · (scale factor)^3

3 Solve

weight of large tank = weight of small tank · (scale factor)^3

$= 167 \cdot 2^3$ *Substitute.*

$= 167 \cdot 8$ *Evaluate the power.*

$= 1{,}336$ *Multiply.*

The larger tank weighs 1,336 pounds when full.

4 Look Back

By estimation, the smaller tank weighs about 150 pounds when full, and 150 pounds · 8 = 1,200 pounds. So the answer is reasonable.

Think and Discuss

1. Tell whether an object's surface area has increased or decreased if the size of the object is changed by a factor of $\frac{1}{3}$. Explain.

2. Compare finding the volume of a similar figure with finding the weight of a similar figure.

9-5 Exercises

FOR EOG PRACTICE

see page 697

internet connect
Homework Help Online
go.hrw.com Keyword: MS4 9-5

1.02d, 2.02

GUIDED PRACTICE

See Example 1

1. The surface area of a box is 10.4 cm^2. What is the surface area of a larger, similarly shaped box that has a scale factor of 3?
2. The surface area of a ship's hull is about 11,000 m^2. Find the surface area, to the nearest tenth, of the hull of a model ship that has a scale factor of $\frac{1}{150}$.

See Example 2

3. The volume of an ice chest is 2,160 in^3. What is the volume of a larger, similarly shaped ice chest that has a scale factor of 2.5?

See Example 3

4. A fish tank weighs 18 pounds when it is full. A larger, similarly shaped fish tank has a scale factor of 3. How much does the larger tank weigh when full?

INDEPENDENT PRACTICE

See Example 1

5. The surface area of a triangular prism is 13.99 in^2. What is the surface area of a larger, similarly shaped prism that has a scale factor of 4?
6. The surface area of a car frame is about 200 ft^2. Find the surface area, to the nearest tenth, of a square foot, of a model of the car that has a scale factor of $\frac{1}{12}$.

See Example 2

7. The volume of a sphere is about 523 cm^3. What is the volume, to the nearest tenth, of a smaller sphere that has a scale factor of $\frac{1}{4}$?

See Example 3

8. A small steel anchor weighs 17 pounds. A larger steel anchor that is similar in shape has a scale factor of 5. How much does the larger anchor weigh?

PRACTICE AND PROBLEM SOLVING

For each figure shown, find the surface area and volume of a similar figure with a scale factor of 25. Use 3.14 for π.

9.
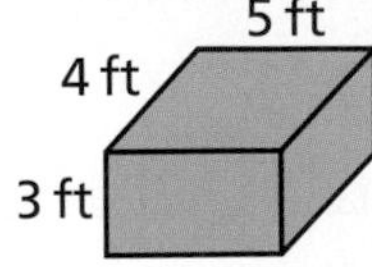

10.
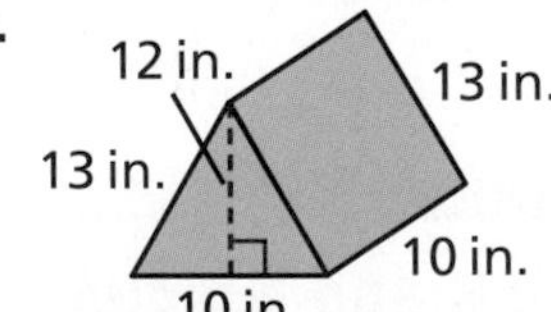

11.
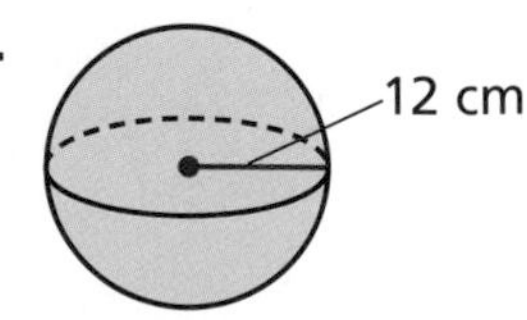

12. The surface area of a cylinder is 1,620 m^2. Its volume is about 1,130 m^3. What are the surface area and volume of a similarly shaped cylinder that has a scale factor of $\frac{1}{9}$? Round to the nearest tenth, if necessary.

History LINK

Natalie and Rebecca are making a scale model of the *Titanic* for a history class project. Their model has a scale factor of $\frac{1}{100}$. For Exercises 13–16, express your answers in both centimeters and meters. Use the conversion chart at right if needed.

METRIC CONVERSIONS	
1 m = 100 cm	1 cm = 0.01 m
1 m^2 = 10,000 cm^2	1 cm^2 = 0.0001 m^2
1 m^3 = 1,000,000 cm^3	1 cm^3 = 0.000001 m^3

13. The length and height of the *Titanic* are shown in the drawing above. What are the length and height of the students' scale model?

14. On the students' model, the diameter of the outer propellers is 7.16 cm. What was the diameter of these propellers on the ship?

These are propellers from the *Olympic*, the *Titanic*'s sister ship. They are identical to those that were on the *Titanic*.

15. The surface area of the deck of the students' model is 4,156.75 cm^2. What was the surface area of the deck of the ship?

16. The volume of the students' model is about 127,426 cm^3. What was the volume of the ship?

17. ***CHALLENGE*** The weight of the *Titanic* was 38,760 tons. If the students could make a detailed model using the same materials as the ship, what would the model weigh in pounds? (*Hint:* 1 ton = 2,000 pounds.)

Some people have made very detailed models of the *Titanic*.

Spiral Review

Determine whether the ratios are proportional. (Lesson 5-2)

18. $\frac{7}{56}, \frac{35}{280}$ **19.** $\frac{12}{20}, \frac{60}{140}$ **20.** $\frac{9}{45}, \frac{45}{225}$ **21.** $\frac{5}{82}, \frac{65}{1,054}$

Write each percent as a fraction in simplest form. (Lesson 6-1)

22. 75% **23.** 60% **24.** 52% **25.** 18%

26. **EOG PREP** Which polygon has ten angles and ten sides? (Lesson 7-5)

A Pentagon **B** Octagon **C** Decagon **D** Dodecagon

Build Polyhedrons, Cylinders, and Cones

Use with Lesson 9-4

internet connect
Lab Resources Online
go.hrw.com
KEYWORD: MS4 Lab9B

You can build geometric solids using nets or cubes.

Activity

1 Using a net, construct a cylinder with a radius of 1 inch and a height of 3.5 inches.

a. Draw a rectangle that is 6.28 inches by 3.5 inches. Then draw two circles as shown, one on each side of the rectangle. Each circle should have a radius of 1 inch. This is the net of a cylinder.

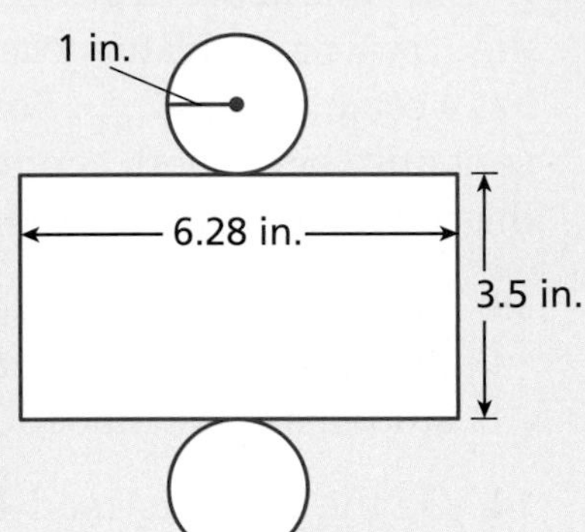

b. Cut out the net. Fold the net as shown to make a cylinder. Tape the edges of the paper to hold them in place.

2 Using a net, construct a cone with a radius of 1 inch.

a. Draw a 115° angle. Make each side of the angle 3.13 inches long.

b. Place your compass on the vertex of the angle, and open it to the length of one side. Draw an arc to connect the two sides as shown.

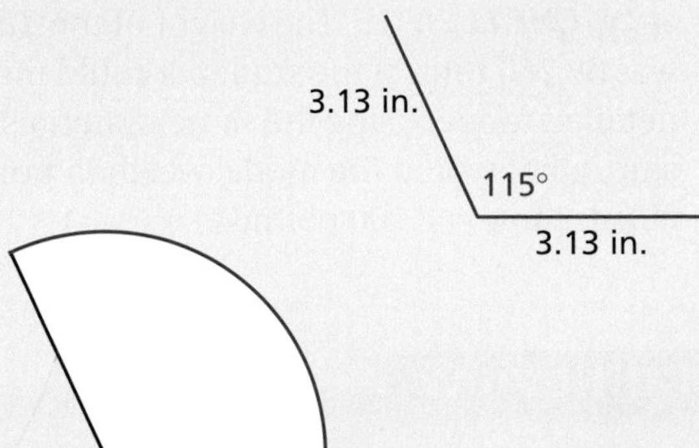

c. Draw a circle with a 1-inch radius, touching the figure from part **b** as shown. This completes the net of a cone.

d. Cut out the net. Fold the net as shown to make a cone. Tape the edges of the paper to hold them in place.

3 Construct a rectangular prism using a net.

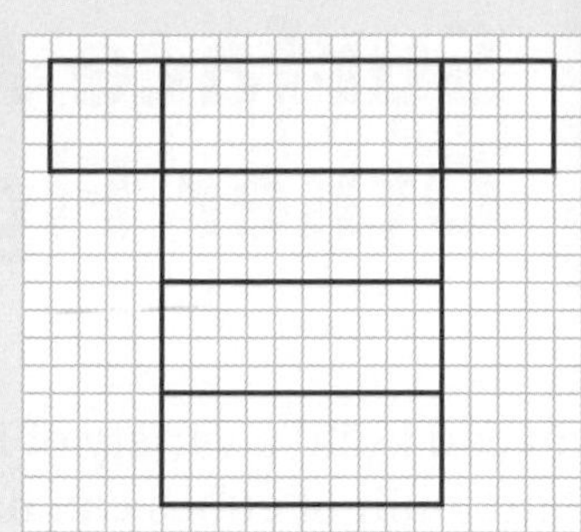

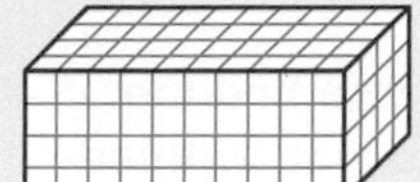

a. Draw the net at right on a piece of graph paper. Each rectangle should be 10 squares by 4 squares. Each of the two large squares should be 4 small squares on each side.

b. Cut out the net. Fold the net along the edges of each rectangle to make a rectangular prism. Tape the edges of the paper to hold them in place.

4 Use centimeter cubes to build a rectangular prism.

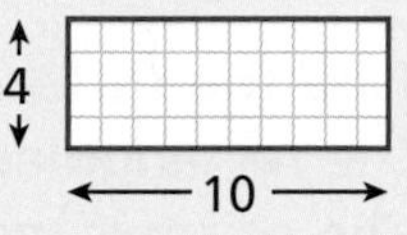

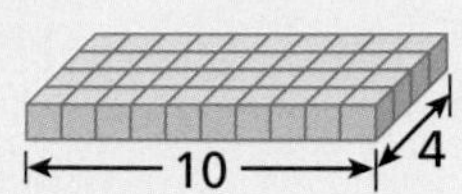

a. Using the prism that you built in **3**, look at the graph paper squares on one of the large faces. Arrange centimeter cubes to match the view of the squares.

b. Now look at the graph paper squares on one of the smaller faces of the paper prism. Stack more cubes on top of the first layer of cubes until the stack's height matches the paper prism's height.

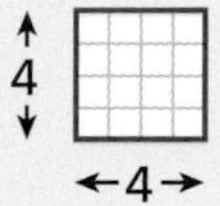

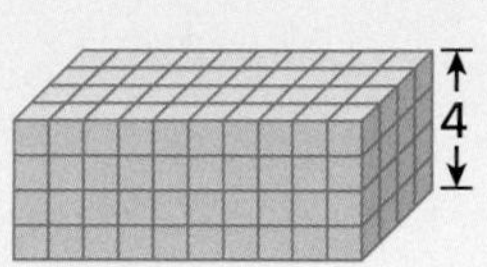

Think and Discuss

1. Find the volume in cubic units of the prism that you built in **4**.
2. Find the surface area of the prism that you built in **4** by counting the exposed cube faces. Express your answer in square units.
3. On the net of the cone, describe the shape that forms the lateral surface of the cone. How is that shape related to the base of the cone?
4. In **1**, the width of the rectangle in the net is 6.28 inches. How does the width of the rectangle relate to the radius of the cylinder?

Try This

1. Use a net to construct a cylinder with a radius of 2 centimeters and a height of 4 centimeters.
2. Use a net to construct a cone whose base has a radius of 2 centimeters. To make the lateral surface, use an angle of 90° and side lengths of 8 centimeters.
3. Use a net to construct a rectangular prism that is 2 inches by 3 inches by 4 inches.
4. Use cubes to construct the rectangular prism in problem 3.

EXTENSION Surface Area of Other Figures

Learn to find the surface area of three-dimensional figures built from cubes.

You can build a variety of figures with cubes. The surface area of a figure built of cubes can be found by adding the areas of the exposed cube faces.

EXAMPLE 1 Finding the Surface Area of a Prism Built of Cubes

Find the surface area of the rectangular prism. The prism is made up of congruent cubes.

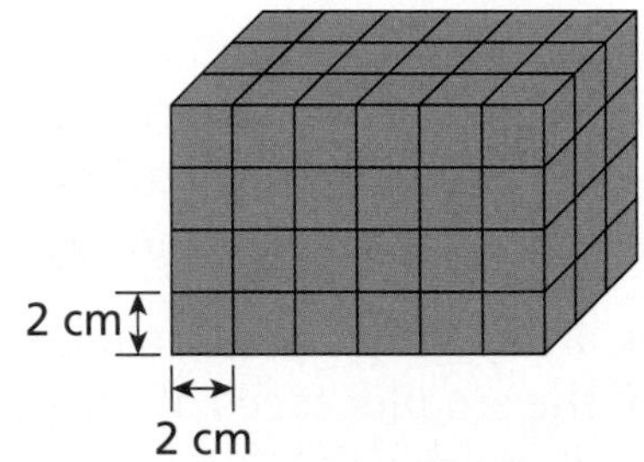

First, draw each view of the prism.

Top view

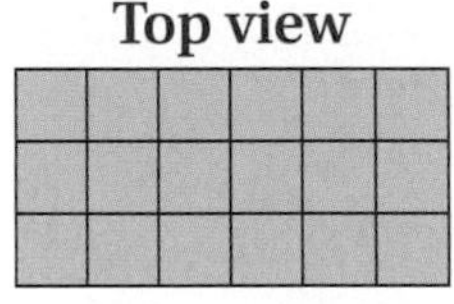

Front view

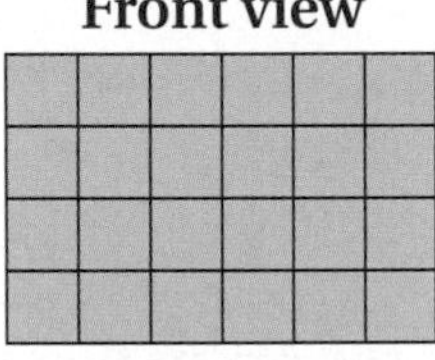

Left side view

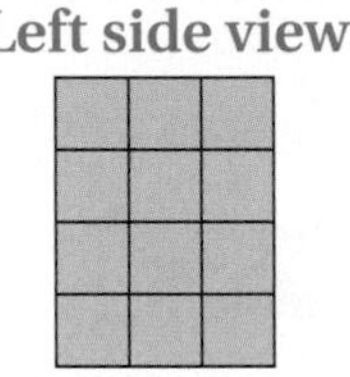

Bottom view

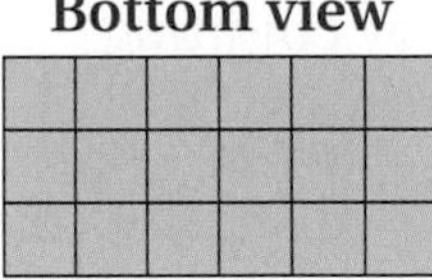

Back view

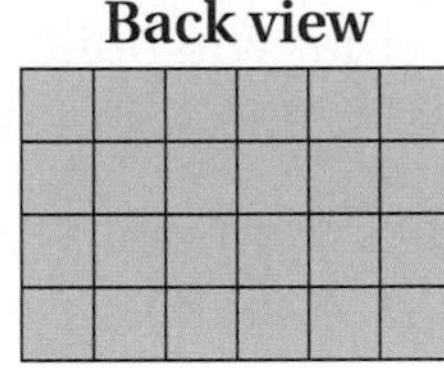

Right side view

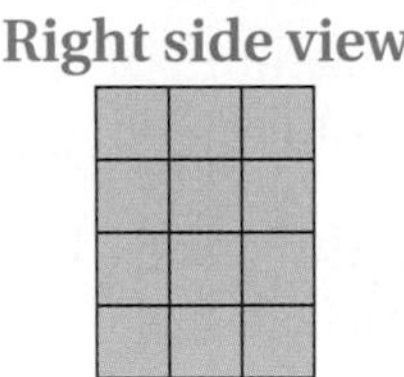

Since the cubes are congruent, you can count all of the cube faces that show and multiply this number by the area of one cube face. This will give you the total surface area of the prism.

Top view: 18 cube faces	**Front view:** 24 cube faces	**Left view:** 12 cube faces
Bottom view: 18 cube faces	**Back view:** 24 cube faces	**Right view:** 12 cube faces

$18 + 18 + 24 + 24 + 12 + 12 = 108$ *Find the total number of cube faces that show.*

Each edge of a cube is 2 cm, so the area of each face is 4 cm^2.

$108 \cdot 4 = 432$ *Multiply by the area of each face.*

The surface area of the prism is 432 cm^2.

EXAMPLE 2 Finding the Surface Area of a Complex Figure Built of Cubes

Find the surface area of the figure. The figure is made up of congruent cubes.

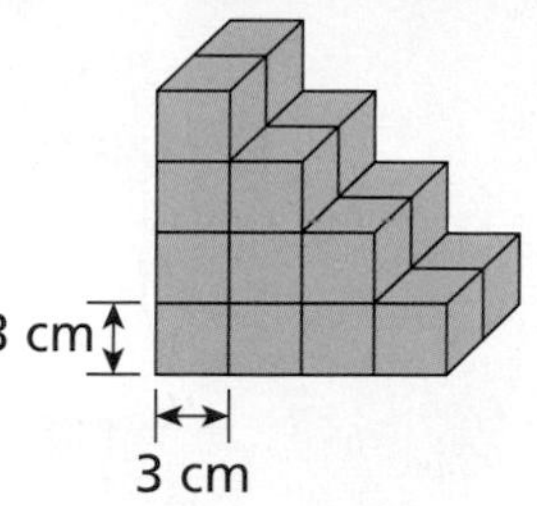

Draw each view.

Top view

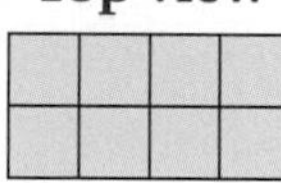

Front view

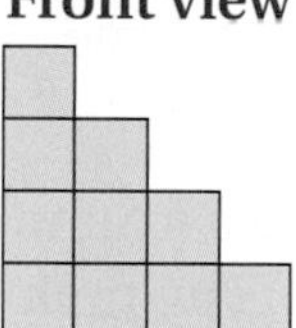

Left side view

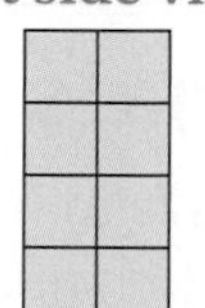

Bottom view

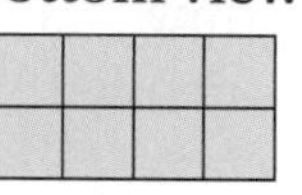

Back view

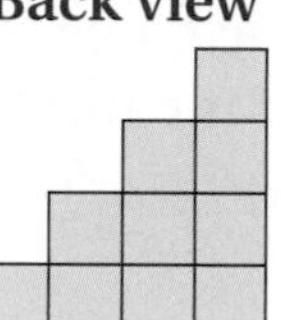

Right side view

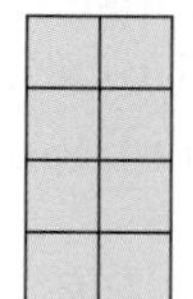

$8 + 8 + 10 + 10 + 8 + 8 = 52$ *Find the total number of cube faces that show.*

Each edge of a cube is 3 cm, so the area of each face is 9 cm^2.

$52 \cdot 9 = 468$ *Multiply by the area of each face.*

The surface area of the figure is 468 cm^2.

EXTENSION Exercises

3.01a, 3.01b

Find the surface area of each figure. Each figure is made up of congruent cubes.

1.

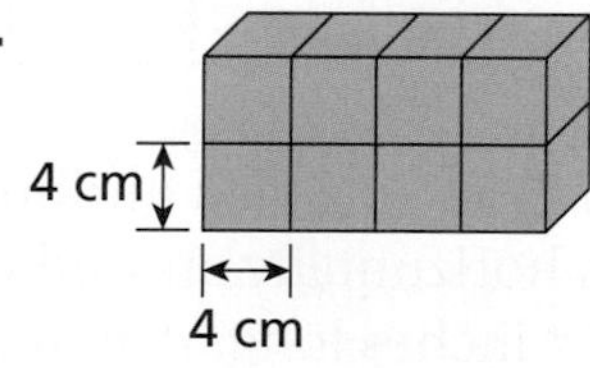

2.

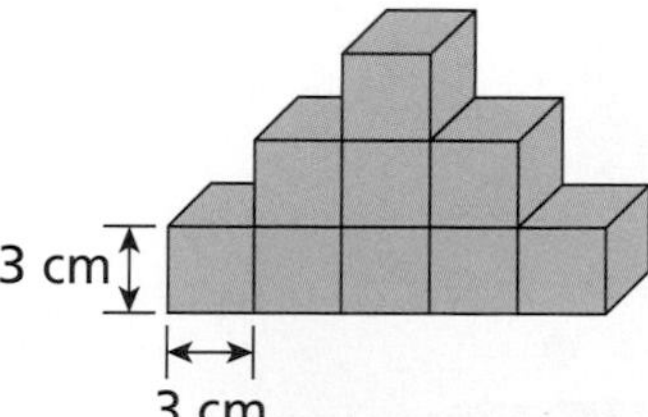

3.

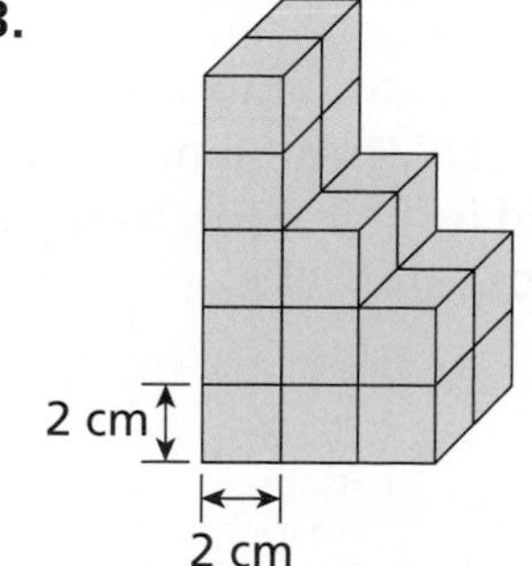

4.

5 cm

5 cm

NORTH CAROLINA

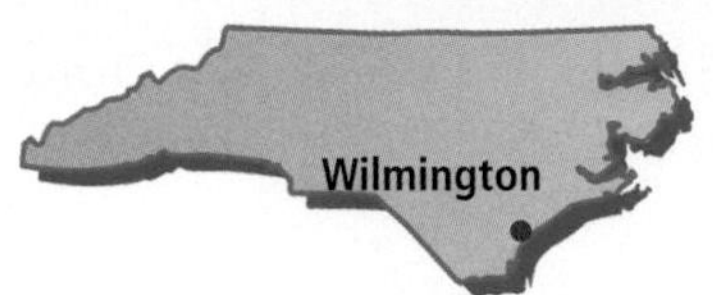

Cardinals

Cardinals live in North Carolina year-round and are one of the most common bird species in the state. The male cardinal is mostly red, but one area of its throat and bill is black. The female cardinal has less red, which is confined to its crest, wings, and tail. People sometimes make birdhouses to attract specific types of birds, including cardinals. The table gives some of the minimum recommended dimensions of some birdhouse designs for specific species.

For 1–5, use the table.

1. Eileen wants to paint her chickadee birdhouse. If the chickadee house is a rectangular prism, how would Eileen find the surface area of the outside walls of the birdhouse? What is its surface area?

2. One quart of paint covers about 14,400 in^2 of surface. Using your answer from **1**, estimate about how many chickadee birdhouses could be painted with 2 quarts of paint.

3. What is the area of the bottom of the Carolina wren birdhouse?

Bird Species	Length (in.)	Width (in.)	Height (in.)
Bluebird	4	4	9
Cardinal	5	5	8
Carolina wren	4	5	9
Chickadee	4	4	9
Purple martin	6	6	9

4. Suppose you wanted to give a birdhouse as a gift and wanted to put a ribbon horizontally around the outside of the house. The ribbon is 14 inches long. Would it fit around the outside of the chickadee house?

5. Suppose a family of cardinals is nesting in the cardinal birdhouse and each cardinal needs 5 square inches to sufficiently build its nest. How many cardinals would fit inside the birdhouse?

North Carolina Studios

Wilmington, North Carolina, is home to the largest motion-picture facility in the United States outside of California. EUE/Screen Gems Studios in North Carolina has been involved with over 300 film and television productions, and television commercials. There are nine different stages on the 32-acre film lot. These stages range in size from 7,200 to 35,000 square feet.

For 1–6, use the table.

1. What is the area of stage 1?
2. If a movie company needs 10,000 square feet to make a scene for a movie, which stage should be used, 2 or 6?
3. What is the difference in area between stages 3 and 4?
4. Which stage has the greatest area?
5. Explain how you would find the surface area of the outside walls of one of the enclosed stages at Screen Gems Studios.
6. Use a net to find the surface area of stage 6, not including the floor.

Approximate Dimensions of Stages (ft)			
Stage	**Length**	**Width**	**Height**
1	60	120	35
2	60	120	35
3	60	120	35
4	100	200	45
5	80	140	45
6	80	140	45
7	100	200	45
8	100	155	45
9	100	150	45

MATH-ABLES

Blooming Minds

Students in the Agriculture Club at Carter Middle School are designing a flower bed for the front of the school. The flower bed will be in the shape of the letter *C*. After considering the two designs shown below, the students decided to build the flower bed that required the least amount of peat moss. Which design did the students choose? (*Hint:* Find the volume of each flower bed.)

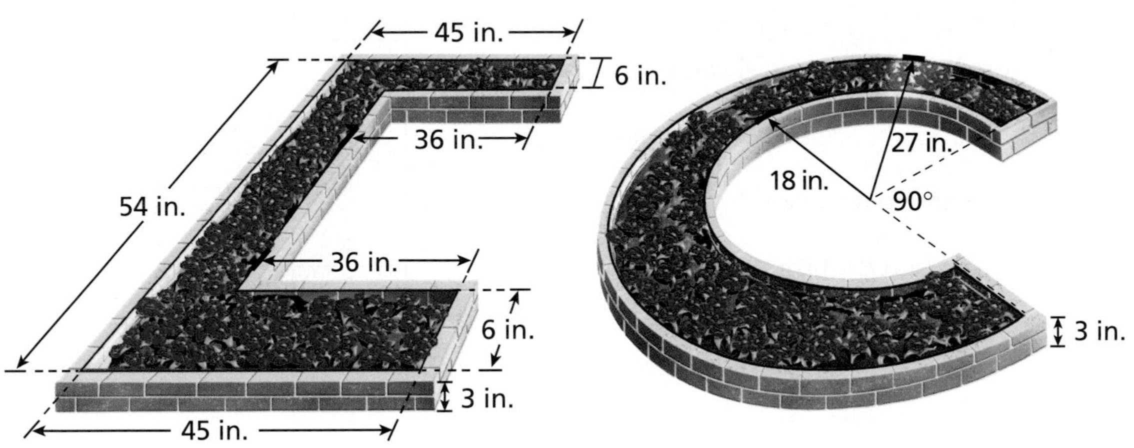

Magic Cubes

Four magic cubes are used in this fun puzzle. Each side of the four cubes has the number 1, 2, 3, or 4 written on it. The object of the game is to stack the cubes so that the numbers along each side of the stack add up to 10. No number can be repeated along any side of the stack.

internet connect

Go to ***go.hrw.com*** for a complete set of rules and to print out nets for the cubes.
KEYWORD: MS4 Game9

Explore Volume of Solids

You can use a spreadsheet to explore how changing the dimensions of a rectangular pyramid affects the volume of the pyramid.

Activity

1. On a spreadsheet, enter the following headings:
Base Length in cell A1,
Base Width in cell B1,
Height in cell C1, and
Volume in cell D1.

   In row 2, enter the numbers 15, 7, and 22, as shown.

H9 =

	A	B	C	D
1	Base Length	Base Width	Height	Volume
2	15	7	22	

2. Then enter the formula for the volume of a pyramid in cell D2. To do this, enter **=(1/3)*A2*B2*C2.** Press **ENTER** and notice that the volume is 770.

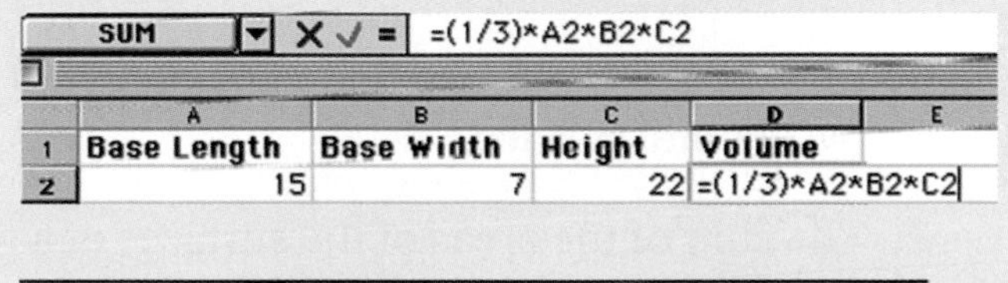

SUM =(1/3)*A2*B2*C2

	A	B	C	D	E
1	Base Length	Base Width	Height	Volume	
2	15	7	22	=(1/3)*A2*B2*C2	

3. Enter 30 in cell A2 and 11 in cell C2 to find out what happens to the volume when you double the base length and halve the height.

C23 =

	A	B	C	D
1	Base Length	Base Width	Height	Volume
2	30	7	11	770

Think and Discuss

1. Explain why doubling the base length and halving the height does not change the volume of the pyramid.
2. What other ways could you change the dimensions of the pyramid without changing its volume?

Try This

1. Use a spreadsheet to compute the volume of each cone. Use 3.14 for π.
   a. radius = 2.75 inches; height = 8.5 inches
   b. radius = 7.5 inches; height = 14.5 inches
2. What would the volumes in problem 1 be if the radii were doubled?

Chapter 9

Study Guide and Review

Vocabulary

Study Guide and Review

Complete the sentences below with vocabulary words from the list above. Words may be used more than once.

1. A(n) ___?___ has two parallel, congruent circular bases connected by a lateral surface.
2. The sum of the areas of the surfaces of a three-dimensional figure is called the ___?___.
3. The two parallel congruent faces of a prism are the ___?___, and they can be any polygon.
4. A(n) ___?___ has one circular face and a curved lateral surface.

9-1 Introduction to Three-Dimensional Figures (pp. 472–475)

EXAMPLE

- **Give the name of the prism or pyramid.**

hexagonal prism

EXERCISES

Give the name of each prism or pyramid.

5.

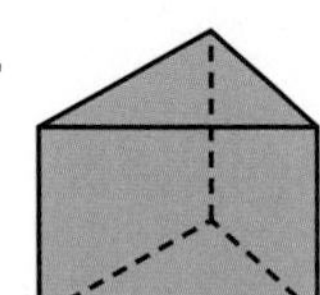

6.

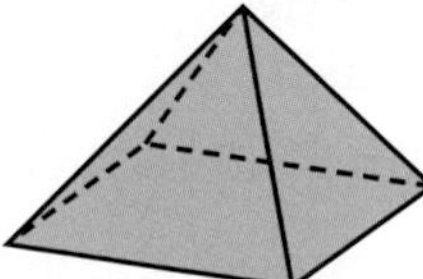

9-2 Volume of Prisms and Cylinders (pp. 476–479)

EXAMPLE

■ **Find the volume of the prism.**

$V = Bh$

$V = (15 \cdot 4) \cdot 9$

$V = 540$

The volume of the prism is 540 ft^3.

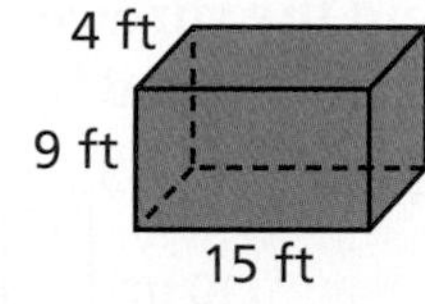

■ **Find the volume of the cylinder to the nearest tenth. Use 3.14 for π.**

$V = \pi r^2 h$

$V \approx 3.14 \cdot 3^2 \cdot 4$

$V \approx 113.04$

The volume is about 113.0 cm^3.

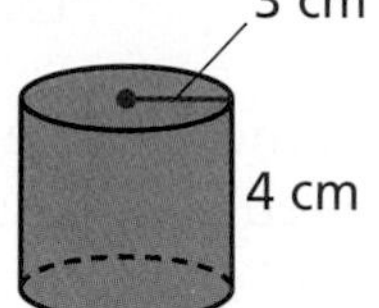

EXERCISES

Find the volume of the prism.

7.

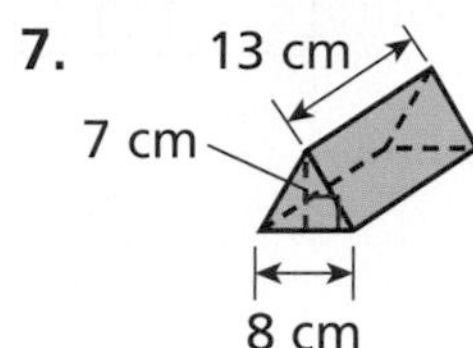

Find the volume of each cylinder to the nearest tenth. Use 3.14 for π.

8.

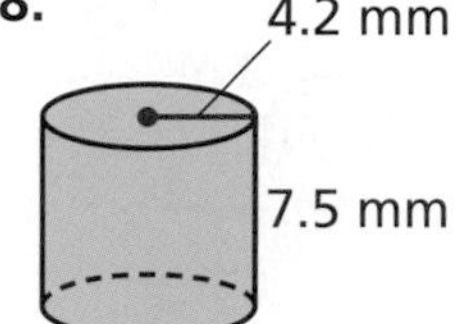

9.

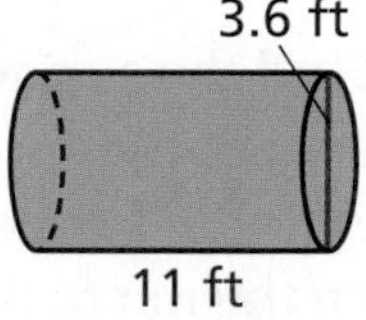

9-3 Volume of Pyramids, Cones, and Spheres (pp. 480–483)

EXAMPLE

■ **Find the volume of the pyramid.**

$V = \frac{1}{3} Bh$

$V = \frac{1}{3} \cdot (5 \cdot 6) \cdot 7$

$V = 70$

The volume is 70 m^3.

■ **Find the volume of the cone to the nearest tenth. Use 3.14 for π.**

$V = \frac{1}{3} \pi r^2 h$

$V \approx \frac{1}{3} \cdot 3.14 \cdot 4^2 \cdot 9$

$V \approx 150.72$

The volume is about 150.7 ft^3.

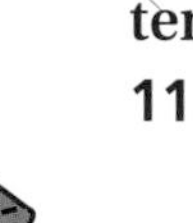

■ **Find the volume of the sphere to the nearest tenth. Use 3.14 for π.**

$V = \frac{4}{3} \pi r^3$

$V \approx \frac{4}{3} \cdot 3.14 \cdot 8.5^3$

$V \approx 2{,}571.13\overline{6}$

The volume is about 2,571.1 ft^3.

EXERCISES

Find the volume of the pyramid.

10.

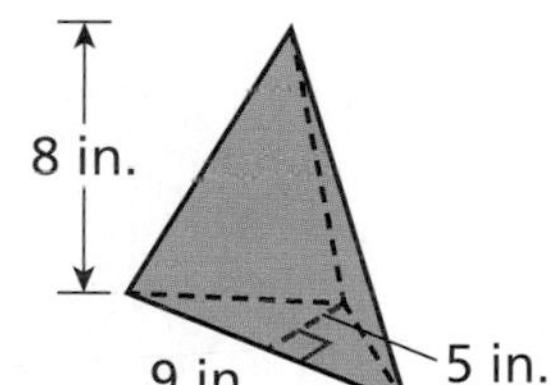

Find the volume of the cone to the nearest tenth. Use 3.14 for π.

11.

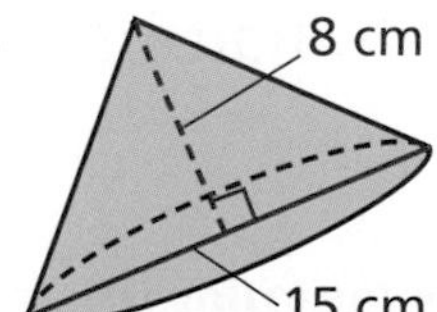
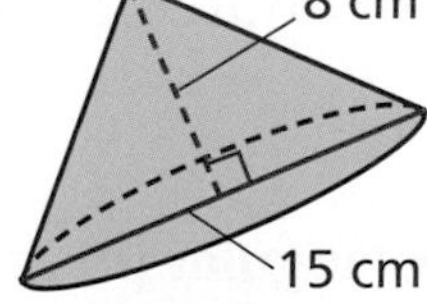

Find the volume of each sphere to the nearest tenth. Use 3.14 for π.

12.

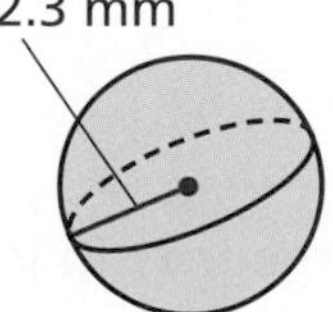

13.

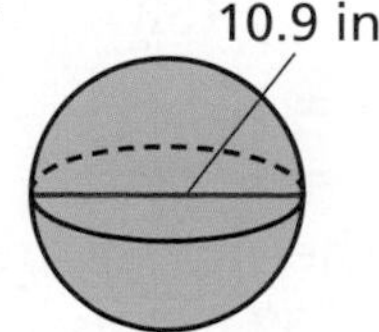

9-4 Surface Area of Prisms, Cylinders, and Spheres (pp. 486–490)

EXAMPLE

- **Find the surface area of the rectangular prism formed by the net.**

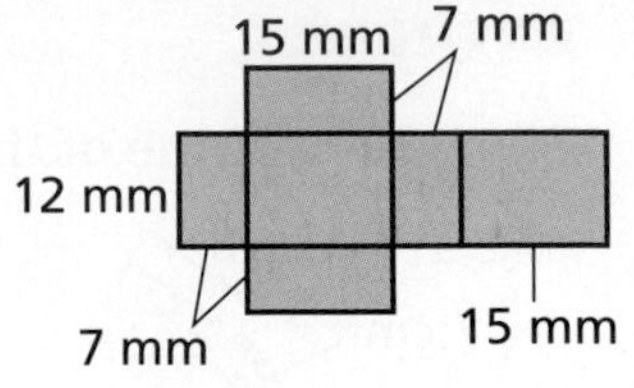

$S = 2\ell w + 2\ell h + 2wh$

$S = (2 \cdot 15 \cdot 7) + (2 \cdot 15 \cdot 12) + (2 \cdot 7 \cdot 12)$

$S = 738$

The surface area is 738 mm^2.

- **Find the surface area of the cylinder formed by the net to the nearest tenth. Use 3.14 for π.**

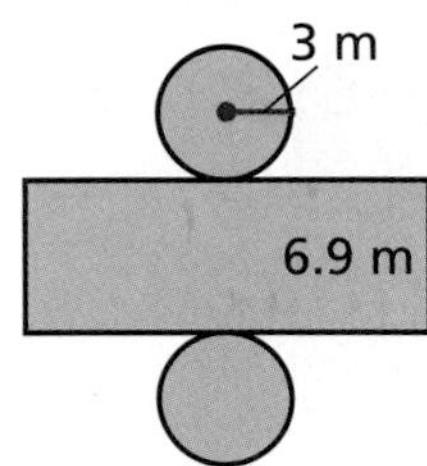

$S = 2\pi r^2 + 2\pi rh$

$S \approx (2 \cdot 3.14 \cdot 3^2) + (2 \cdot 3.14 \cdot 3 \cdot 6.9)$

$S \approx 186.516$

The surface area is about 186.5 m^2.

- **Find the surface area of the sphere to the nearest tenth. Use 3.14 for π.**

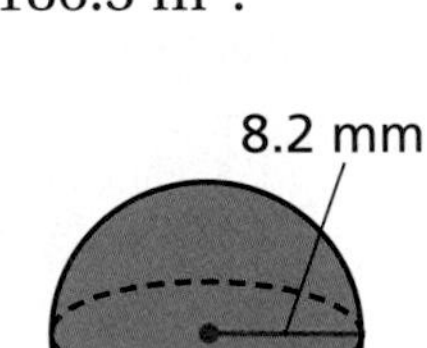

$S = 4\pi r^2$

$S \approx 4 \cdot 3.14 \cdot 8.2^2$

$S \approx 844.5344$

The surface area is about 844.5 mm^2.

EXERCISES

Find the surface area of the rectangular prism formed by the net.

14.

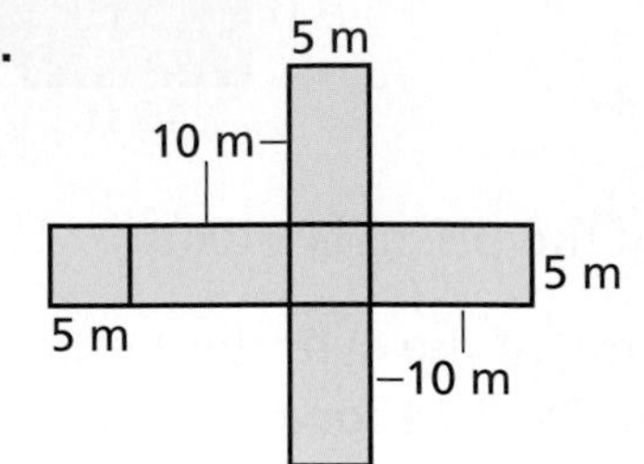

Find the surface area of the cylinder formed by the net to the nearest tenth. Use 3.14 for π.

15.

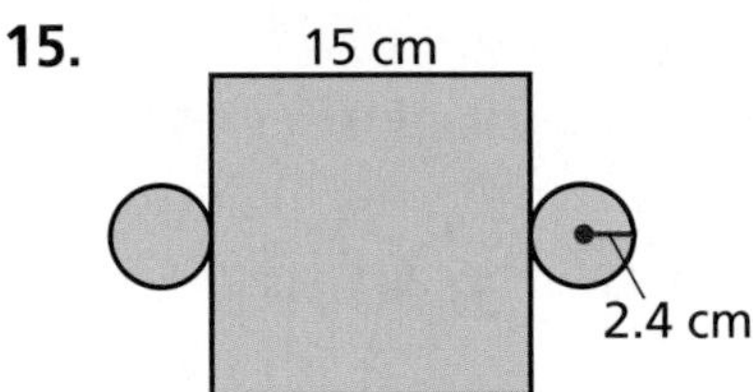

Find the surface area of each sphere to the nearest tenth. Use 3.14 for π.

16. **17.**

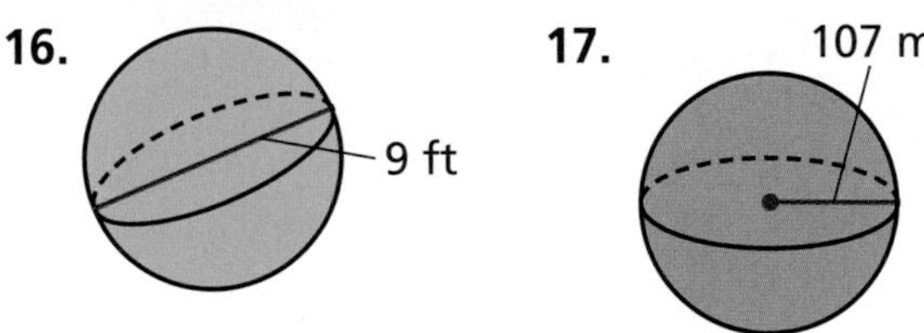

9-5 Changing Dimensions (pp. 491–495)

EXAMPLE

- **The surface area of a rectangular prism is 32 m^2, and its volume is 12 m^3. What are the surface area and volume of a similar rectangular prism with a scale factor of 6?**

$S = 32 \cdot 6^2$
$= 1{,}152$

$V = 12 \cdot 6^3$
$= 2{,}592$

The surface area of the larger prism is 1,152 m^2. Its volume is 2,592 m^3.

EXERCISES

18. A cylinder has a surface area of about 13.2 in^2. What is the surface area of a similar cylinder that has a scale factor of 15?

19. A refrigerator has a volume of 14 ft^3. What is the volume, to the nearest tenth, of a similarly shaped refrigerator that has a scale factor of $\frac{2}{3}$?

Chapter Test

Chapter 9

Give the name of each prism or pyramid.

1. 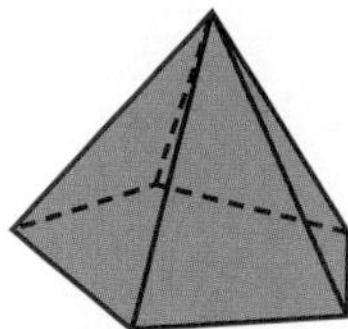

2. 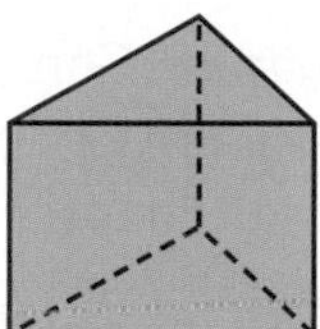

3. 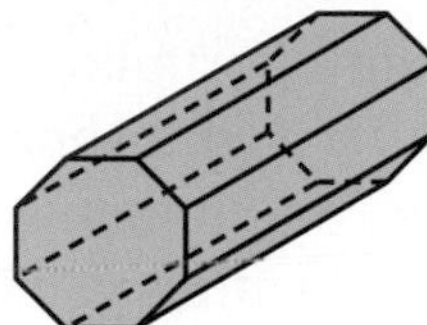

4. 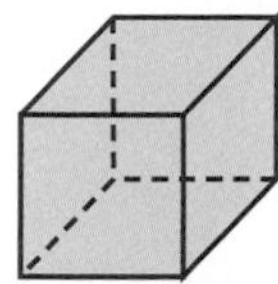

5. 

6. 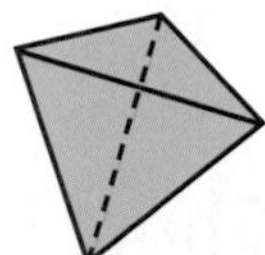

Find the volume of each figure to the nearest tenth. Use 3.14 for π.

7. 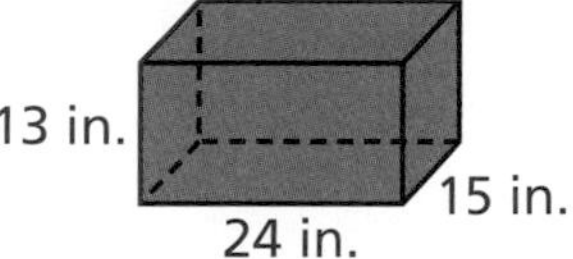

8. 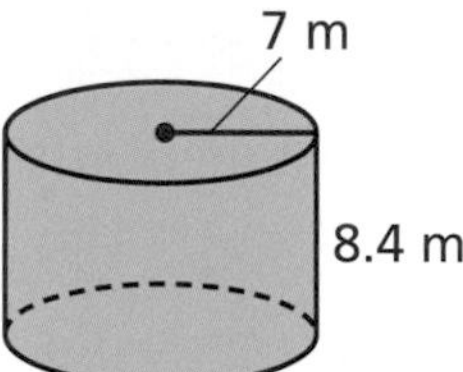

9. 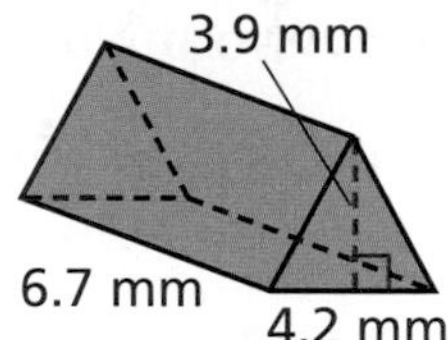

10. 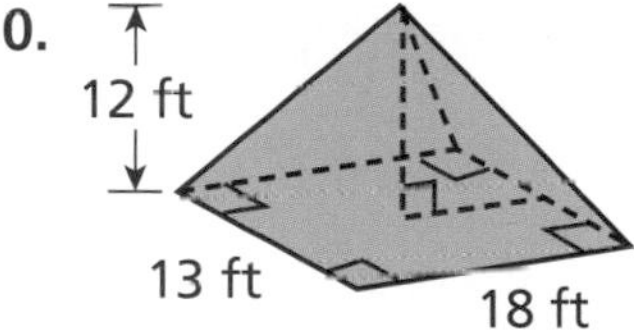

11. 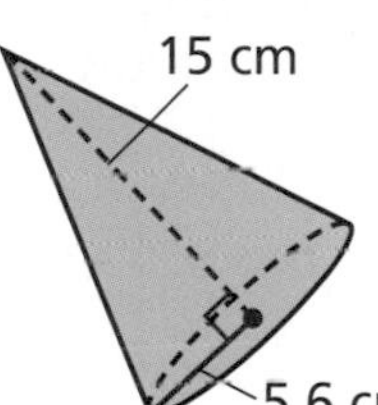

12. 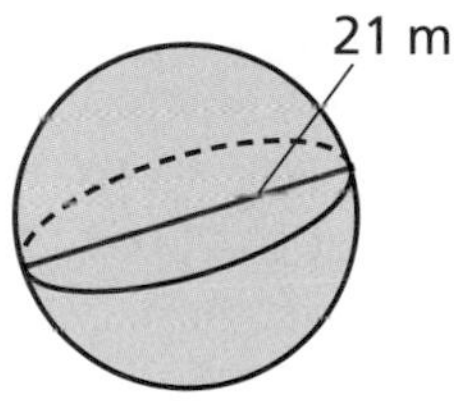

Find the surface area of each figure to the nearest tenth. Use 3.14 for π.

13. 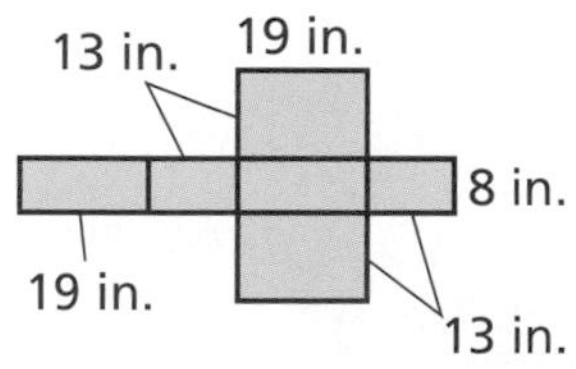

14. 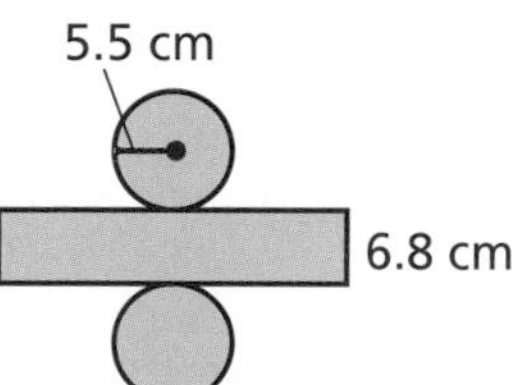

15. 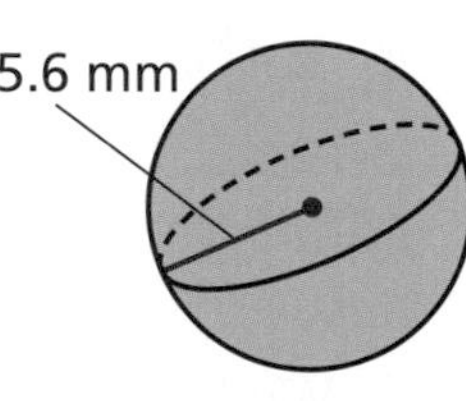

16. The surface area of a rectangular prism is 52 ft^2. What is the surface area of a similar prism that has a scale factor of 7?

17. The surface area of a table is 8.4 m^2. What is the surface area to the nearest hundredth of a similarly shaped table that has a scale factor of $\frac{1}{8}$?

18. The volume of a cube is 35 mm^3. What is the volume of a larger cube that has a scale factor of 9?

19. The volume of a flowerpot is 7.5 cm^3. What is the volume to the nearest hundredth of a similarly shaped flowerpot that has a scale factor of $\frac{1}{2}$?

20. A round balloon has a volume of about 104.7 in^3. If the balloon is inflated by a scale factor of 2, what is its new volume?

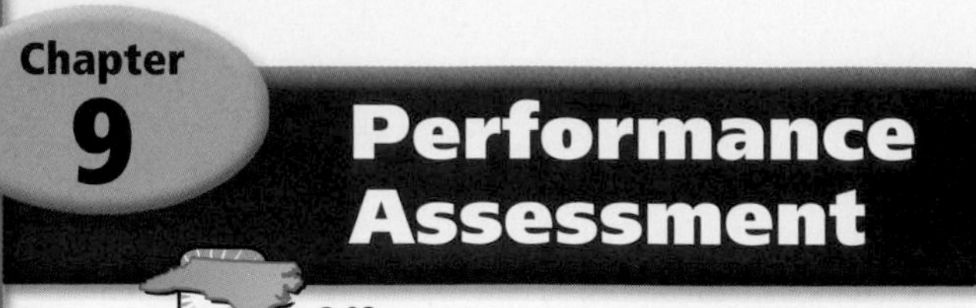

Chapter 9 Performance Assessment

2.02

Show What You Know

Create a portfolio of your work from this chapter. Complete this page and include it with your four best pieces of work from Chapter 9. Choose from your homework or lab assignments, mid-chapter quiz, or any journal entries you have done. Put them together using any design you want. Make your portfolio represent what you consider your best work.

Short Response

1. A polyhedron has two parallel square bases with edges 6 inches long. Its height is 12 inches. Identify the figure and find its volume. Show your work.
2. Find the volume and surface area of a sphere with radius 3 feet. Show your work.
3. The volume of a triangular prism is 60 m^3. A similar prism has a scale factor of $\frac{5}{3}$. Which prism is larger? Explain your answer. Then find the volume of the second prism, and show your work.
4. A three-dimensional figure has one vertex and a circular base. Tell whether the figure is a polyhedron. If it is, name the figure. If it is not, explain why not.

Extended Problem Solving

5. Use the diagram for the following problems. Round your answers to the nearest tenth, if necessary. Use 3.14 for π.
   a. What three-dimensional shapes make up the silo?
   b. What is the volume of the silo? Show your work.
   c. Approximate the surface area of the silo. Show your work.
   d. The bales of hay in the diagram are similarly shaped. The scale factor of the larger bale is 2, and it contains 60 cubic feet of hay. How much hay is in the smaller bale? Explain your answer.

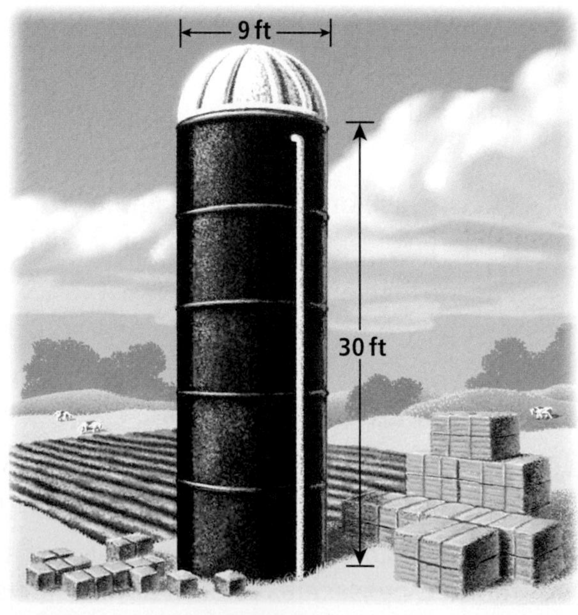

Performance Assessment

internet connect
EOG Test Practice Online
go.hrw.com Keyword: MS4 TestPrep

Getting Ready for EOG

Chapter 9

Cumulative Assessment, Chapters 1–9

1. What is the measure of the unknown angle?

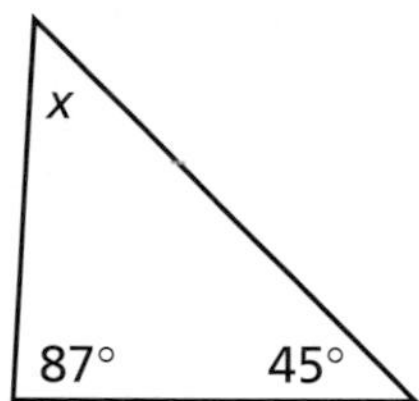

A 48°
B 228°
C 132°
D 52°

2. What is the mean of the data set 23, 15, 21, 23, 18, 20, 19, 13?

A 23
B 19.5
C 10
D 19

3. Which set of numbers is in order from least to greatest?

A −3, −7, 0, 1, 4
B −7, −3, 0, 1, 4
C 4, 1, 0, −3, −7
D 4, 1, 0, −7, −3

4. What is the simple interest on $8,500 invested at 6% for 18 months?

A $9,180
B $765
C $7,650
D $1,020

5. What is the volume of the cylinder to the nearest tenth? Use 3.14 for π.

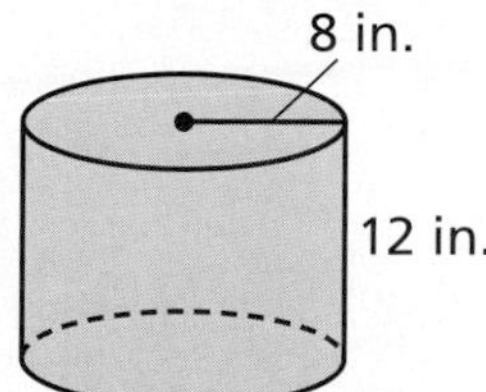

A 602.9 in^3
B 3,215.4 in^3
C 1,205.8 in^3
D 2,411.5 in^3

6. What is the surface area of the rectangular prism?

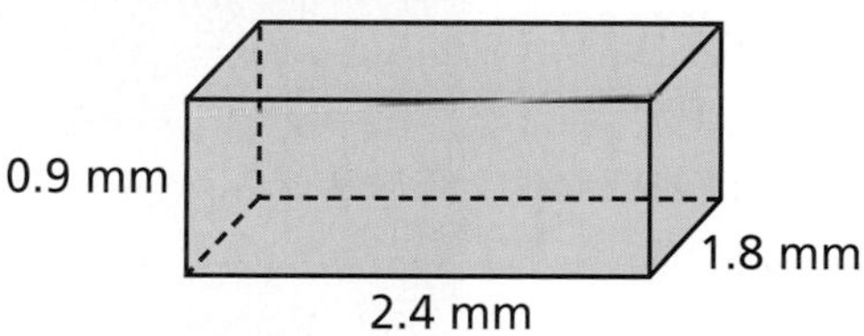

A 8.1 mm^2
B 16.2 mm^2
C 15.84 mm^2
D 3.888 mm^2

TEST TAKING TIP!

When finding the prime factorization of a number, you can eliminate choices by looking for nonprime factors.

7. Which is the prime factorization of 180?

A $4 \cdot 9 \cdot 5$
B $10 \cdot 18$
C $2^2 \cdot 3^2 \cdot 5$
D $2^3 \cdot 3 \cdot 5$

8. What is the sum of $3\frac{2}{3} + 1\frac{3}{5}$ written in simplest form?

A $5\frac{4}{15}$
B $2\frac{1}{15}$
C $4\frac{5}{8}$
D $3\frac{6}{15}$

9. ***SHORT RESPONSE*** The surface area of a sphere is about 86 ft^2.

   a. Find the surface area of a larger sphere that has a scale factor of 4.

   b. How does the surface area change when the dimensions are increased by a factor of 4?

10. ***SHORT RESPONSE*** What is the base length of a parallelogram with a height of 8 in. and an area of 56 in^2?

Chapter

10

Probability

Country	Mobile Phones per 100 People
Finland	67.8
Norway	62.7
Sweden	59.0
Italy	52.2

Countries with Highest per Capita Mobile-Phone Use

TABLE OF CONTENTS

internet connect

Chapter Opener Online
go.hrw.com
KEYWORD: MS4 Ch10

Career *Demographer*

Demographers study people: their numbers, how old they are, where they live, with whom they live, where they are moving, and more. They examine how age affects buying habits, which occupations are most popular, how people are affecting the natural environment, and many other behavioral data.

Businesses use demographers to analyze how products are used. The table lists the countries with the highest per capita mobile-phone use. Who might be interested in this kind of demographic information?

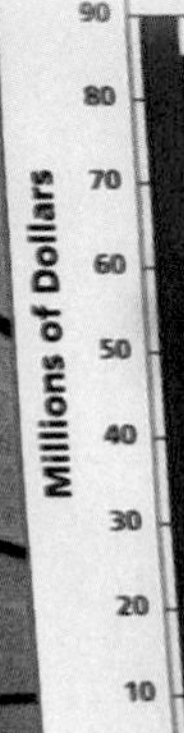

ARE YOU READY?

Choose the best term from the list to complete each sentence.

1. A(n) __?__ is a comparison of two quantities by division.
2. A(n) __?__ is an integer that is divisible by 2.
3. A(n) __?__ is a ratio that compares a number to 100.
4. A(n) __?__ is a number greater than 1 that has more than two whole number factors.
5. A(n) __?__ is an integer that is not divisible by 2.

composite number
even number
odd number
percent
prime number
ratio

Complete these exercises to review skills you will need for this chapter.

✔ Simplify Fractions

Simplify.

6. $\frac{6}{9}$ **7.** $\frac{12}{15}$ **8.** $\frac{8}{10}$ **9.** $\frac{20}{24}$

10. $\frac{2}{4}$ **11.** $\frac{7}{35}$ **12.** $\frac{12}{22}$ **13.** $\frac{72}{81}$

✔ Write Fractions as Decimals

Write each fraction as a decimal.

14. $\frac{3}{5}$ **15.** $\frac{9}{20}$ **16.** $\frac{57}{100}$ **17.** $\frac{12}{25}$

18. $\frac{3}{25}$ **19.** $\frac{1}{2}$ **20.** $\frac{7}{10}$ **21.** $\frac{9}{5}$

✔ Percents and Decimals

Write each decimal as a percent.

22. 0.14 **23.** 0.08 **24.** 0.75 **25.** 0.38

26. 0.27 **27.** 1.89 **28.** 0.234 **29.** 0.0025

✔ Multiply Fractions

Multiply. Write each answer in simplest form.

30. $\frac{1}{2} \cdot \frac{1}{4}$ **31.** $\frac{2}{3} \cdot \frac{3}{5}$ **32.** $\frac{3}{10} \cdot \frac{1}{2}$ **33.** $\frac{5}{6} \cdot \frac{3}{4}$

34. $\frac{5}{14} \cdot \frac{7}{17}$ **35.** $-\frac{1}{8} \cdot \frac{3}{8}$ **36.** $-\frac{2}{15} \cdot \left(-\frac{2}{3}\right)$ **37.** $\frac{1}{4} \cdot \left(-\frac{1}{6}\right)$

10-1 Probability

Learn to use informal measures of probability.

Vocabulary

experiment
outcome
event
probability
equally likely
impossible
certain

Suppose you rolled one of these dice. The blue one is equally likely to land on any of the six numbers. The red one is more likely to land on one of the larger faces. So the likelihood is greater that you would roll a 5 with the red die than with the blue one.

Any activity involving chance, such as the roll of a die, is an **experiment**. The result of an experiment is an **outcome**. An **event** is a set of one or more outcomes.

Events that have the same probability are **equally likely**. **Probability** is the measure of how likely an event is to occur. The more likely an event is to occur, the higher its probability. The less likely an event is to occur, the lower its probability.

EXAMPLE 1 Determining the Likelihood of an Event

A bag contains 6 blue marbles, 6 red marbles, 3 green marbles, and 1 yellow marble. All of the marbles are the same size and weight.

A Would you be more likely to pull a red marble or a green marble from the bag?

Since there are more red marbles than green marbles, it is more likely that you would pull a red marble than a green marble.

B Would you be more likely to pull a red marble or a blue marble from the bag?

Since the number of red marbles equals the number of blue marbles, it is just as likely that you would pull a red marble as a blue marble. The events are equally likely.

Every event is either impossible, certain, or somewhere between these extremes. An event is mathematically **impossible** if it can never happen and mathematically **certain** if it will always happen. If an event is as likely as not, the probability that it will happen is the same as the probability that it will not happen.

Impossible — **Unlikely** — **As likely as not** — **Likely** — **Certain**

EXAMPLE 2 Classifying Likelihood

Determine whether each event is impossible, unlikely, as likely as not, likely, or certain.

A **All of the students in Ms. Caro's class are in school today. Jamie is in Ms. Caro's class. How likely is it that Jamie is in school today?**

It is certain that Jamie is in school today.

B **Karl has soccer practice on Monday afternoons. How likely is it that Karl is playing basketball on Monday afternoon?**

Soccer practice could have been canceled, and Karl could be playing basketball. However, it is unlikely that Karl is playing basketball.

C **There are four 2's and four 6's in a set of eight cards. If you draw a card, how likely is it that you will randomly draw a 6?**

Since the number of 2's equals the number of 6's, it is as likely as not that you will draw a 6.

D **Rita's family is visiting the Smithville Zoo. Her mother is out of town on a business trip. How likely is it that Rita's mother is at the Smithville Zoo?**

It is impossible that Rita's mother is at the Smithville Zoo.

EXAMPLE 3 *School Application*

Eric's math teacher almost always gives a pop quiz if the class did not ask many questions about the lesson on the previous class day. If it is Monday and no one asked questions during class on Friday, should Eric expect a pop quiz? Explain.

Since Eric's teacher often gives quizzes on days after few questions were asked, a quiz on this day is likely.

Think and Discuss

1. **Determine** whether you would be more likely to roll a 4 on the red die or on the blue die, as shown at the beginning of the lesson. Explain your answer.
2. **Give an example** of a certain event and of an impossible event.

10-1 Exercises

FOR EOG PRACTICE

see page 698

internet connect

Homework Help Online
go.hrw.com Keyword: MS4 10-1

GUIDED PRACTICE

See Example 1

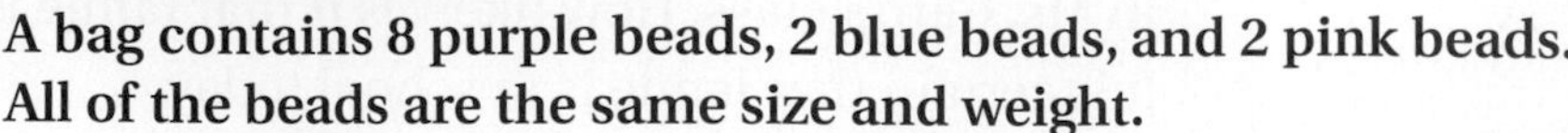

A bag contains 8 purple beads, 2 blue beads, and 2 pink beads. All of the beads are the same size and weight.

1. Would you be more likely to pull a purple bead or a pink bead from the bag?

2. Would you be more likely to pull a blue bead or a pink bead from the bag?

See Example 2

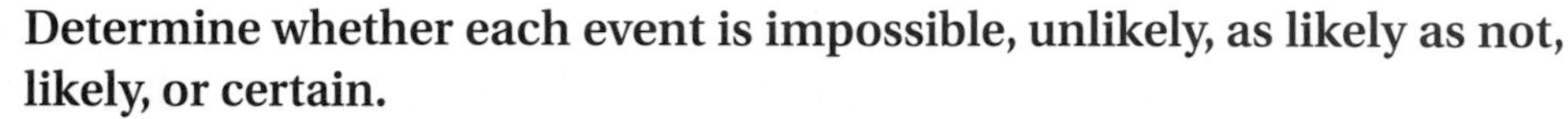

Determine whether each event is impossible, unlikely, as likely as not, likely, or certain.

3. Natalie has dance lessons on Saturday morning. How likely is it that Natalie is attending dance class on Saturday morning?

4. There are three 2's, three 4's, and three 6's in a set of nine cards. If you draw a card, how likely is it that you will randomly draw a 3?

See Example 3

5. Timothy went to a theater to see a movie that started at 3 P.M. and lasts for 2 hours 10 minutes. If it is now 5 P.M., would you expect Timothy to be in the theater? Explain.

INDEPENDENT PRACTICE

See Example 1

A bag contains 5 red markers, 4 blue markers, 4 black markers, and 2 yellow markers. All of the markers are the same size and weight.

6. Would you be more likely to pull a blue marker or a black marker from the bag?

7. Would you be more likely to pull a black marker or a yellow marker from the bag?

See Example 2

Determine whether each event is impossible, unlikely, as likely as not, likely, or certain.

8. A bag contains 12 red checkers and 12 black checkers. How likely is it that you will pull a white checker from the bag?

9. Sixth-grade students have a 30 minute lunch break in the cafeteria at 11:30 A.M., and seventh-graders eat at noon. How likely is it that a sixth-grader is in the cafeteria at 12:15 P.M.?

See Example 3

10. The planetarium opens weekdays at noon. On Saturday, it is open from 10 A.M. until 10 P.M. Trisha wants to go to the planetarium on Wednesday morning. Should Trisha expect the planetarium to be open? Explain.

PRACTICE AND PROBLEM SOLVING

11. ***LIFE SCIENCE*** In a scientist's garden, there are 700 sweet pea plants with purple flowers and 200 with white flowers. If one plant from the garden were selected at random, would you expect the plant to have purple or white flowers? Explain.

12. ***LIFE SCIENCE*** Sharks belong to a class of fishes that have skeletons made of cartilage. Bony fishes, which account for 95% of all species of fish, have skeletons made of bone.

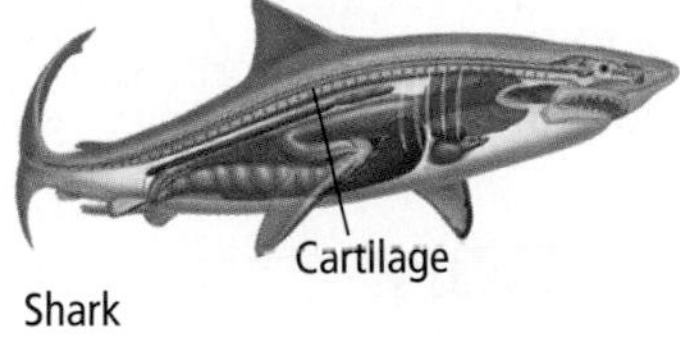

Shark

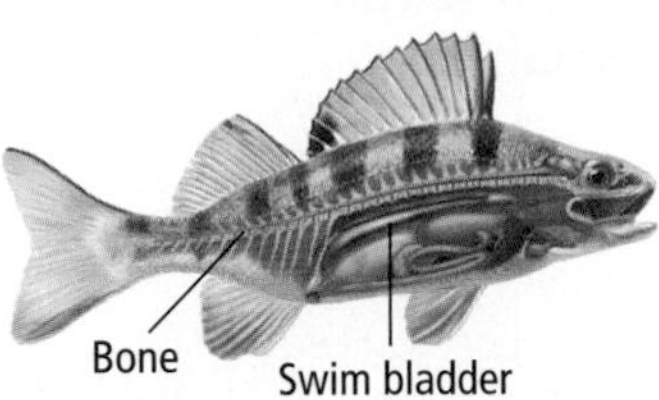

Bony fish

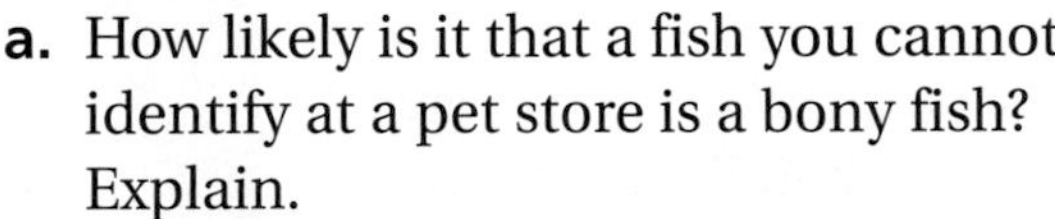

   a. How likely is it that a fish you cannot identify at a pet store is a bony fish? Explain.

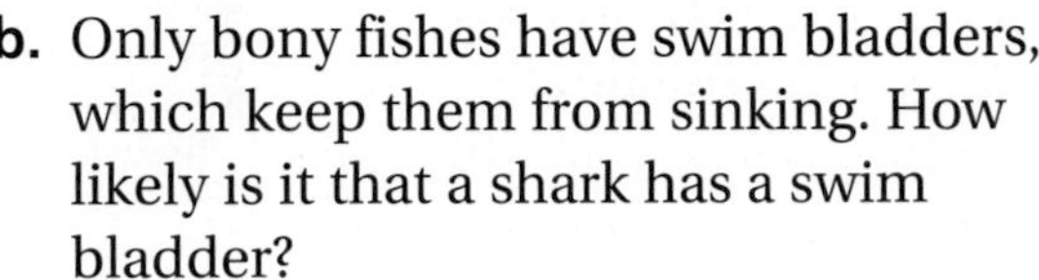

   b. Only bony fishes have swim bladders, which keep them from sinking. How likely is it that a shark has a swim bladder?

13. ***EARTH SCIENCE*** The graph shows the carbon dioxide levels in the atmosphere from 1958 to 1994. How likely is it that the level of carbon dioxide fell from 1994 to 2000? Explain.

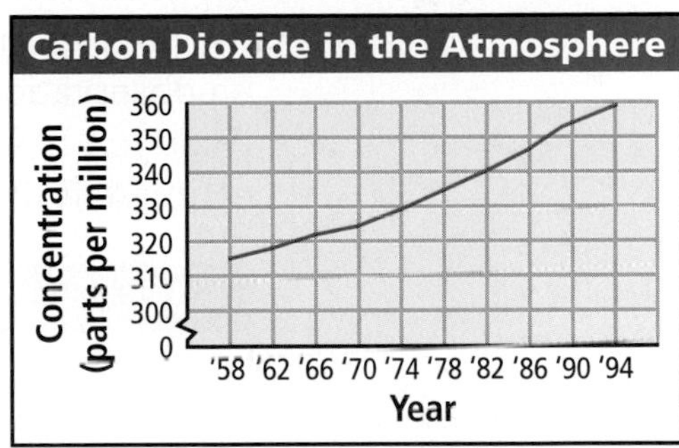

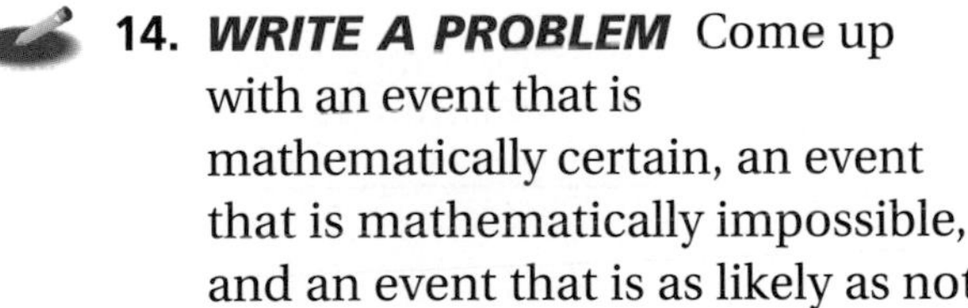

14. ***WRITE A PROBLEM*** Come up with an event that is mathematically certain, an event that is mathematically impossible, and an event that is as likely as not.

15. ***WRITE ABOUT IT*** Explain how to tell if an event is as likely as not.

16. ***CHALLENGE*** A bag contains 10 red marbles and 8 blue marbles, all the same size and weight. Keiko draws 2 red marbles from the bag and does not replace them. Will Keiko be more likely to draw a red marble than a blue marble on her next draw? Explain.

Spiral Review

Solve. (Lesson 6-4)

17. 27 is what percent of 54?

18. 13 is 40% of what number?

19. **EOG PREP** What is the volume of the prism? (Lesson 9-2)

   A 19 in^3 B 240 in^3 C 480 in^3 D $1{,}200 \text{ in}^3$

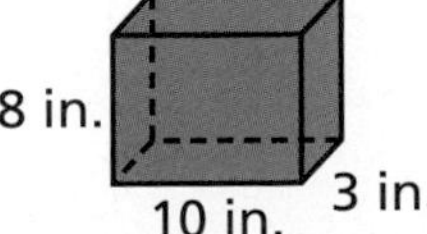

20. **EOG PREP** What is the surface area of the prism? (Lesson 9-4)

   A 70 in^2 B 118 in^2 C 240 in^2 D 268 in^2

10-2 Experimental Probability

Learn to find experimental probability.

Vocabulary

experimental probability

trial

During hockey practice, Tanya made saves on 18 out of 25 shots. Based on these numbers, what is the probability that Tanya will make a save on the next shot?

Experimental probability is one way of estimating the probability of an event. It is based on actual experiments or observations. **Experimental probability** is found by comparing the number of times an event occurs to the total number of **trials**, the times that an experiment is carried out or an observation is made. The more trials you have, the more accurate the estimate is likely to be.

EXPERIMENTAL PROBABILITY

$$\text{probability} \approx \frac{\text{number of times an event occurs}}{\text{total number of trials}}$$

EXAMPLE 1 ***Sports Application***

If Tanya made saves on 18 out of 25 shots, what is the experimental probability that she will make a save on the next shot?

$$P \approx \frac{\text{number of times an event occurs}}{\text{total number of trials}}$$

$$P(\text{save}) \approx \frac{\text{number of saves made}}{\text{number of shots attempted}}$$

$$\approx \frac{18}{25} \qquad \textit{Substitute.}$$

The experimental probability that Tanya will make a save on the next shot is approximately $\frac{18}{25}$.

Writing Math

"*P*(event)" represents the probability that an event will occur. For example, the probability of a flipped coin landing heads up could be written as "*P*(heads)."

Recall that an outcome of an experiment can be impossible, certain, or in between. If an event is impossible, it will never happen in any trial.

$$P(\text{impossible event}) = \frac{0}{\text{total number of trials}}$$

$$= 0 \qquad \textit{0 divided by any number except 0 equals 0.}$$

If an event is certain, it will always happen in every trial. This means that the number of times the event happens is equal to the total number of trials.

$$P(\text{certain event}) = \frac{\text{total number of trials}}{\text{total number of trials}}$$

$$= 1 \qquad \textit{Any number except 0 divided by itself equals 1.}$$

All probabilities can be expressed numerically on a scale from 0 to 1.

Impossible	Unlikely	As likely as not	Likely	Certain
0		$\frac{1}{2}$		1

EXAMPLE 2 ***Weather Application***

For the past three weeks, Karl has been recording the daily high temperatures for a science project. During that time, the high temperature was above 75°F for 14 out of 21 days.

A What is the experimental probability that the high temperature will be above 75°F on the next day?

$$P(\text{warmer than } 75°\text{F}) \approx \frac{\text{number of days warmer than } 75°\text{F}}{\text{total number of days}}$$

$$\approx \frac{14}{21} \qquad \textit{Substitute.}$$

$$\approx \frac{2}{3} \qquad \textit{Write in simplest form.}$$

The experimental probability that the temperature will be above 75°F on the next day is approximately $\frac{2}{3}$.

Helpful Hint

The events "The temperature is above 75°F" and "The temperature is 75°F or below" are called complements.

B What is the experimental probability that the high temperature will be 75°F or below on the next day?

$$P(75°\text{F or cooler}) \approx \frac{\text{number of days } 75°\text{F or cooler}}{\text{total number of days}}$$

$$\approx \frac{7}{21} \qquad \textit{Substitute.}$$

$$\approx \frac{1}{3} \qquad \textit{Write in simplest form.}$$

The experimental probability that the high temperature will be 75°F or below on the next day is approximately $\frac{1}{3}$.

Think and Discuss

1. **Find** the sum of the probabilities in Examples 2A and 2B. Why do you think adding the probabilities gives this sum?

2. **Tell** whether the experimental probability of an event is always the same. Explain.

10-2 Exercises

FOR EOG PRACTICE
see page 698

internet connect
Homework Help Online
go.hrw.com Keyword: MS4 10-2

GUIDED PRACTICE

See Example 

1. Terry's PE class is practicing archery. If Terry hit the target on 12 out of 20 tries, what is the experimental probability that she will hit the target on her next try?

See Example 

2. A reporter surveys 75 people to determine whether they plan to vote for or against a proposed bill. Of these people, 65 say they plan to vote for the bill.
   a. What is the experimental probability that the next person surveyed would say he or she plans to vote for it?
   b. What is the experimental probability that the next person surveyed would say he or she plans to vote against it?

INDEPENDENT PRACTICE

See Example 

3. If Jack hit a baseball on 13 out of 30 tries during practice, what is the experimental probability that he will hit the ball on his next try?
4. While playing darts, Cam hit the bull's-eye 8 times out of 15 throws. What is the experimental probability that Cam's next throw will hit the bull's-eye?

See Example 

5. For the past two weeks, Benita has been recording the number of people at Eastside Park at lunchtime. During that time, there were 50 or more people at the park 9 out of 14 days.
   a. What is the experimental probability that there will be 50 or more people at the park during lunchtime on the fifteenth day?
   b. What is the experimental probability that there will be fewer than 50 people at the park during lunchtime on the fifteenth day?

PRACTICE AND PROBLEM SOLVING

6. George is watching cars drive by. Of the first 46 cars he sees, 18 are red, and the rest are other colors. What is the experimental probability that the next car he sees will be red?
7. While bowling with friends, Alexis rolls a strike in 4 out of the 10 frames. What is the experimental probability that Alexis will roll a strike in the first frame of the next game?
8. Jeremiah is standing in line at a music store. Of the first 25 people he sees enter the store, 16 are wearing jackets and 9 are not. What is the experimental probability that the next person to enter the store will be wearing a jacket?

Earth Science 

9. The stem-and-leaf plot shows the depth of snow in inches recorded in Buffalo, New York, over a 10-day period.

Stems	Leaves
7	9 9
8	
9	1 1 1 1 8 8
10	
11	8
12	
13	0

Key: 7|9 means 7.9

   a. What is the median depth of snow for the 10-day period?
   b. What is the experimental probability that the snow will be less than 6 in. deep on the eleventh day?
   c. What is the experimental probability that the snow will be more than 10 in. deep on the eleventh day?

10. Monty's class has been recording the daily low temperatures for 30 days. During that time, the low temperature was 48°F on 11 out of 30 days. What is the experimental probability that the low temperature will be 48°F on the next day?

11. The table shows the high temperatures recorded on July 4 in Orlando, Florida, over an eight-year period.
   a. What is the experimental probability that the high temperature on July 4, 2002 is below 90°F?
   b. What is the experimental probability that the high temperature on July 4, 2002 is above 100°F?

Year	Temp (°F)	Year	Temp (°F)
1994	86.0	1998	96.8
1995	95.0	1999	89.1
1996	78.8	2000	90.0
1997	98.6	2001	91.0

Source: Old Farmers' Almanac

12. ***CHALLENGE*** A weather forecaster says that the probability of rain is 30%. What is the probability that it will not rain?

Spiral Review

Estimate each sum or difference. (Lesson 4-9)

13. $\frac{5}{9} + \frac{2}{5}$ **14.** $\frac{5}{6} - \frac{1}{8}$ **15.** $1\frac{4}{9} + 3\frac{7}{15}$ **16.** $6\frac{1}{6} - 2\frac{7}{8}$

17. **EOG PREP** How many days are equal to 360 hours? (Lesson 8-1)

A 12 days B 15 days C 30 days D 4,320 days

18. **EOG PREP** Rita's dog is named Ally. How likely is it that Rita has a pet? (Lesson 10-1)

A Certain B As likely as not C Unlikely D Impossible

10-3 Make a List to Find Sample Spaces

Problem Solving Skill

Learn to use counting methods to determine possible outcomes.

Vocabulary

sample space

Fundamental Counting Principle

Because you can roll the numbers 1, 2, 3, 4, 5, and 6 on a number cube, there are 6 possible outcomes. Together, all the possible outcomes of an experiment make up the **sample space**.

You can make an organized list to show all possible outcomes of an experiment.

EXAMPLE 1 PROBLEM SOLVING APPLICATION

Lucia flips two quarters at the same time. What are all the possible outcomes? How large is the sample space?

1 Understand the Problem

Rewrite the question as a statement.

- Find all the possible outcomes of flipping two quarters, and determine the size of the sample space.

List the **important information:**

- There are two quarters.
- Each quarter can land heads up or tails up.

2 Make a Plan

You can make an organized list to show all the possible outcomes.

3 Solve

Let H = heads and T = tails.

Quarter 1	Quarter 2
H	H
H	T
T	H
T	T

Record each possible outcome.

The possible outcomes are HH, HT, TH, and TT. There are four possible outcomes in the sample space.

4 Look Back

Each possible outcome that is recorded in the list is different.

When the number of possible outcomes of an experiment increases, it may be easier to track all the possible outcomes on a tree diagram.

EXAMPLE 2 Using a Tree Diagram to Find a Sample Space

Claudia spins each of the spinners. What are all the possible outcomes? How large is the sample space?

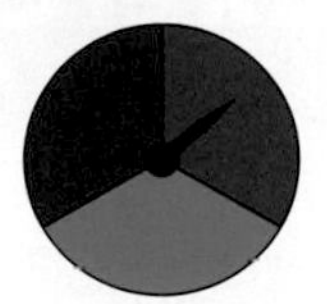

You can make a tree diagram to show the sample space.

List each color on spinner 1. Then list each number on spinner 2 for each color on spinner 1.

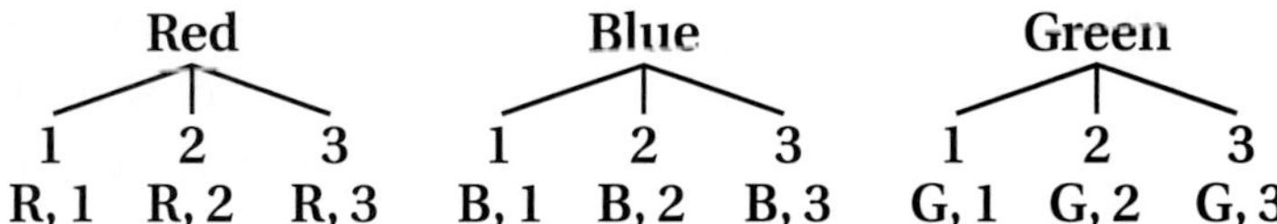

There are nine possible outcomes in the sample space.

In Example 1, there are two outcomes for each coin.

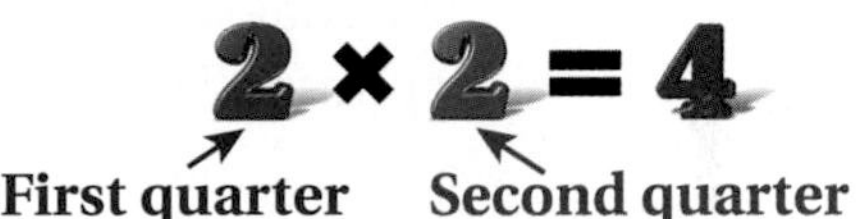

In Example 2, there are three outcomes for each spinner.

3 × 3 = 9

First spinner   Second spinner

The **Fundamental Counting Principle** states that you can find the total number of ways that two or more separate tasks can happen by multiplying the number of ways each task can happen separately.

EXAMPLE 3 *Recreation Application*

In a game each player rolls a 1–6 number cube and spins a spinner. The spinner is divided into thirds, numbered 1, 2, and 3. How many outcomes are possible during one player's turn?

Use the Fundamental Counting Principle.

Number of ways the number cube can land: 6

Number of ways the spinner can land: 3

$6 \cdot 3 = 18$ *Multiply the number of ways each task can happen.*

There are 18 possible outcomes during one player's turn.

Think and Discuss

1. **Find** the size of the sample space for flipping 5 coins.
2. **Explain** how to use the Fundamental Counting Principle to find the number of possible outcomes in Example 1.

10-3 Exercises

FOR EOG PRACTICE
see page 699

internet connect
Homework Help Online
go.hrw.com Keyword: MS4 10-3

GUIDED PRACTICE

See Example 

1. Enrique flips a dime and spins the spinner at right. What are all the possible outcomes? How large is the sample space?

See Example 

2. An ice cream stand offers sugar cones, cake cones, waffle cones, or cups to hold ice cream. You can get vanilla, chocolate, strawberry, pistachio, or coffee flavored ice cream. If you order a single scoop, what are all the possible options you have? How large is the sample space?

See Example 

3. A game includes a 1–6 number cube and a spinner divided into 4 equal sectors. Each player rolls the number cube and spins the spinner. How many outcomes are possible?

INDEPENDENT PRACTICE

See Example 

4. At noon, Aretha can watch a football game, a basketball game, or a documentary about horses on TV. At 3:00, she can watch a second football game, a movie, or a concert. What are all the possible outcomes? How large is the sample space?

5. A spinner is divided into fourths and numbered 1 through 4. Jory spins the spinner and tosses a nickel. What are all the possible outcomes? How large is the sample space?

See Example 

6. Berto tosses a coin and spins the spinner at right. What are all the possible outcomes? How large is the sample space?

7. For breakfast, Clarissa can choose from oatmeal, corn flakes, or scrambled eggs. She can drink milk, orange juice, apple juice, or hot chocolate. What are all the possible outcomes? How large is the sample space?

See Example 

8. A pizza shop offers thick crust, thin crust, or stuffed crust. The choices of toppings are pepperoni, cheese, hamburger, Italian sausage, Canadian bacon, onions, bell peppers, mushrooms, and pineapple. How many different one-topping pizzas could you order?

PRACTICE AND PROBLEM SOLVING

9. Andie has a blue sweater, a red sweater, and a purple sweater. She has a white shirt and a tan shirt. How many different ways can she wear a sweater and a shirt together?

Health LINK

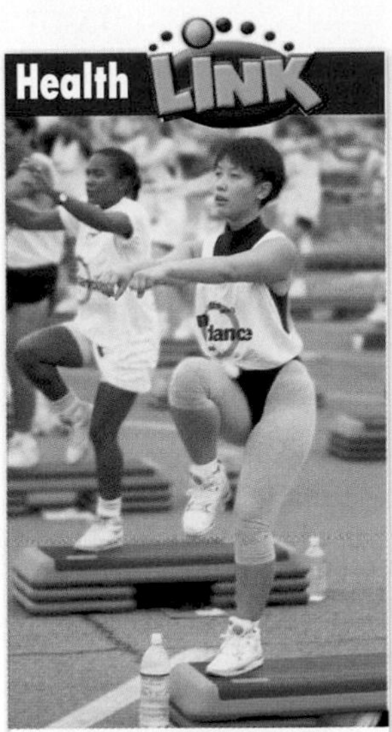

The American Heart Association recommends that people exercise for 30–60 minutes three or four times a week to maintain healthy hearts.

10. ***HEALTH*** The graph shows the kinds of classes that health club members would like to see offered.

    a. If the health club offers the four most popular classes on one day, how many ways could they be arranged?

    b. If the health club offers each of the five classes on a different day of the week, how many ways could they be arranged?

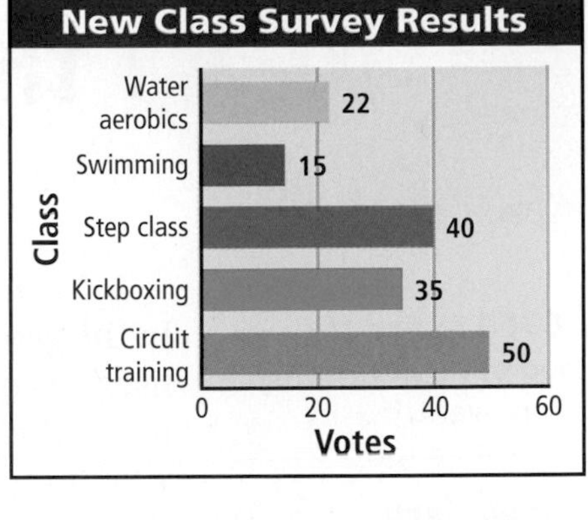

11. ***HEALTH*** For each pair of food groups to the right, give the number of possible outcomes if one item is chosen from each group.

    a. Group A and Group B

    b. Group B and Group D

    c. Group A and Group C

Group A	Group B	Group C	Group D
milk cheese yogurt	beef fish poultry	bread cereal pasta rice	vegetables fruit

12. ***RECREATION*** There are 3 trails from the South Canyon trail head to Lake Solitude. There are 4 trails from Lake Solitude to Hidden Lake. How many possible routes could you take to hike from the South Canyon trail head to Hidden Lake that pass Lake Solitude?

13. ***WHAT'S THE QUESTION?*** Dan has 4 face cards and 5 number cards. He shuffles the cards separately and places each set in a separate pile. The answer is 20 possible outcomes. What is the question?

14. ***WRITE ABOUT IT*** Explain how to determine the size of the sample space when you toss two 1–6 number cubes at the same time.

15. ***CHALLENGE*** Suppose you flip a penny, a nickel, and a dime at the same time. What are all the possible outcomes?

Spiral Review

Write each fraction as a percent. (Lesson 6-1)

16. $\frac{1}{8}$ **17.** $\frac{3}{4}$ **18.** $\frac{2}{5}$ **19.** $\frac{3}{10}$

20. **EOG PREP** What is the volume of a sphere with diameter 8 in.? Use 3.14 for π. (Lesson 9-3)

A 2143.6 in^3 B 267.9 in^3 C 50.24 in^3 D 200.96 in^3

21. **EOG PREP** During basketball practice, Owen made 15 out of 24 free throws. What is the experimental probability that he will make his next free throw? (Lesson 10-2)

A $\frac{3}{8}$ B $\frac{1}{2}$ C $\frac{5}{8}$ D $\frac{5}{6}$

10-4 Theoretical Probability

Learn to find the theoretical probability of an event.

Vocabulary

- favorable outcome
- theoretical probability
- fair

In the game of Scrabble®, players use tiles bearing the letters of the alphabet to form words. Of the 100 tiles used in a Scrabble game, 12 have the letter *E* on them. What is the probability of drawing an *E* from a bag of 100 Scrabble tiles?

In this case, pulling an *E* from the bag is called a *favorable outcome.* A **favorable outcome** is an outcome that you are looking for when you conduct an experiment.

To find the probability of drawing an *E,* you can draw tiles from a bag and record your results, or you can find the *theoretical probability.* **Theoretical probability** is used to estimate the probability of an event when all outcomes are equally likely.

THEORETICAL PROBABILITY

$$\text{probability} = \frac{\text{number of favorable outcomes}}{\text{total number of possible outcomes}}$$

If each possible outcome of an experiment is equally likely, then the experiment is said to be **fair**. Experiments involving number cubes and coins are usually assumed to be fair.

You can write probability as a fraction, a decimal, or a percent.

EXAMPLE 1 Finding Theoretical Probability

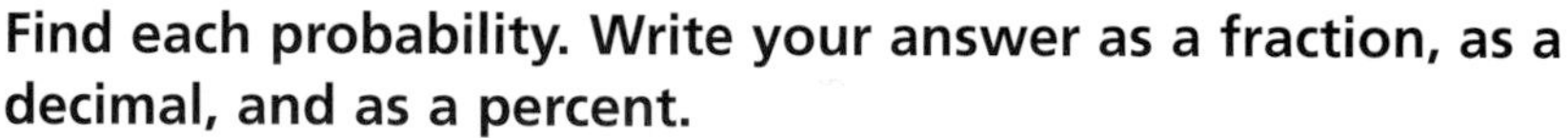

Find each probability. Write your answer as a fraction, as a decimal, and as a percent.

A **What is the probability of drawing one of the 12 *E*'s from a bag of 100 Scrabble tiles?**

$$P = \frac{\text{number of favorable outcomes}}{\text{total number of possible outcomes}}$$

$P(E) = \frac{\text{number of } E\text{'s}}{\text{total number of tiles}}$ — *Write the ratio.*

$= \frac{12}{100}$ — *Substitute.*

$= \frac{3}{25}$ — *Write in simplest form.*

$= 0.12 = 12\%$ — *Write as a decimal and as a percent.*

The theoretical probability of drawing an *E* is $\frac{3}{25}$, 0.12, or 12%.

Find each probability. Write your answer as a fraction, as a decimal, and as a percent.

B **What is the probability of rolling a number greater than 2 on a fair number cube?**

For a fair number cube, each of the six possible outcomes is equally likely. There are 4 ways to roll a number greater than 2: 3, 4, 5, or 6.

$$P = \frac{\text{number of favorable outcomes}}{\text{total number of possible outcomes}}$$

$$P = \frac{\text{4 numbers greater than 2}}{\text{6 possible outcomes}}$$

$$P = \frac{4}{6}$$

$$P = \frac{2}{3}$$

$$P \approx 0.667 \approx 66.7\%$$

EXAMPLE 2 ***School Application***

Mr. Ashley has written the names of each of his students on a craft stick. He draws randomly from these sticks to choose a student to answer a question, and then replaces the stick in the pile.

A **If there are 11 boys and 16 girls in Mr. Ashley's class, what is the theoretical probability that a boy's name will be drawn?**

$P(\text{boy}) = \frac{\text{number of boys in class}}{\text{number of students in class}}$ *Find the theoretical probability.*

$= \frac{11}{27}$ *Substitute.*

B **What is the theoretical probability that a girl's name will be drawn?**

$P(\text{girl}) = \frac{\text{number of girls in class}}{\text{number of students in class}}$ *Find the theoretical probability.*

$= \frac{16}{27}$ *Substitute.*

Think and Discuss

1. Give an example of an experiment in which all of the outcomes are not equally likely. Explain.

2. Find the theoretical probability that a number cube will land with 6 up. If you roll a number cube 5 times without its landing 6 up, what is the theoretical probability that the cube will land 6 up on the sixth roll? Explain your answer.

10-4 Exercises

FOR EOG PRACTICE

see page 699

GUIDED PRACTICE

See Example 1

Find each probability. Write your answer as a fraction, as a decimal, and as a percent.

1. In Chinese checkers, 15 red marbles, 15 blue marbles, 15 green marbles, 15 yellow marbles, 15 black marbles, and 15 white marbles are used. What is the probability of randomly choosing a red marble from a bag of Chinese checkers marbles?

2. If you toss 2 fair pennies, what is the probability that both will land heads up?

See Example 2

3. A set of cards includes 15 yellow cards, 10 green cards, and 10 blue cards.
   a. What is the probability that the card chosen at random will be yellow?
   b. What is the probability that the card chosen at random will be green?
   c. What is the probability that the card chosen at random will not be yellow or green?

INDEPENDENT PRACTICE

See Example 1

Find each probability. Write your answer as a fraction, as a decimal, and as a percent.

4. A standard deck of playing cards has 52 cards. These cards are divided into four 13-card suits: diamonds, hearts, clubs, and spades. Find the probability of drawing a heart or a club at random from a deck of shuffled cards.

5. A game requires 13 red disks, 13 purple disks, 13 orange disks, and 13 white disks. All of the disks are the same size and shape. What is the probability of randomly drawing a purple disk?

6. In Scrabble, 2 of the 100 tiles are blank. Find the probability of drawing a blank tile from an entire set of Scrabble tiles.

See Example 

7. Sifu gives each student in his karate class a different-colored slip of paper. He puts corresponding slips of paper in a bag and randomly selects a slip. The student with the slip of paper that matches the slip drawn must demonstrate a self-defense technique.
   a. If there are 6 girls and 8 boys in the class, what is the probability that a girl will be selected?
   b. What is the probability that a boy will be selected?

PRACTICE AND PROBLEM SOLVING

8. ***RECREATION*** The table shows the approximate number of visitors to five different amusement parks in the United States.

   a. What is the probability that a randomly selected visitor to one of the amusement parks visited Disney World? Write your answer as a percent.

   b. What is the probability that a randomly selected visitor to one of the amusement parks visited a park in California? Write your answer as a decimal.

Amusement Parks	Number of Visitors
Disney World, FL	15,640,000
Disneyland, CA	13,680,000
SeaWorld, FL	4,900,000
Busch Gardens, FL	4,200,000
SeaWorld, CA	3,700,000

Source: The Top 10 of Everything 2000

9. ***GARDENING*** A package of mixed lettuce seeds contains seeds for green lettuce and red lettuce.

   a. Of the first 50 seeds planted, 18 are red lettuce seeds, and the rest are green lettuce seeds. Based on these results, what is the experimental probability that the next seed planted will be a green lettuce seed? Write your answer as a percent.

   b. If the package contains 150 green lettuce seeds and 50 red lettuce seeds, what is the probability that a randomly selected seed will be a red lettuce seed? Write your answer as a percent.

10. ***CHOOSE A STRATEGY*** Francis, Amanda, Raymond, and Albert wore different-colored T-shirts. The colors were tan, orange, purple, and aqua. Neither Raymond nor Amanda ever wears orange, and neither Francis nor Raymond ever wears aqua. Albert wore purple. What color was each person's T-shirt?

11. ***WRITE ABOUT IT*** Suppose the probability of an event happening is $\frac{3}{8}$. Explain what each number in the ratio represents.

12. ***CHALLENGE*** A spinner is divided into three sectors. Half of the spinner is red, $\frac{1}{3}$ is blue, and $\frac{1}{6}$ is green. What is the probability that the spinner will land on either red or green?

Spiral Review

Solve. (Lesson 3-6)

13. $n + 12 = -9$ **14.** $x - 25 = -6$ **15.** $7 + s = -34$ **16.** $y - 52 = -19$

17. **EOG PREP** There are 3 red cards and 9 blue cards in a bag. If you draw a card, how likely is it that you will randomly draw a red card? (Lesson 10-1)

A Impossible B Unlikely C As likely as not D Likely

Chapter 10 Mid-Chapter Quiz

LESSON 10-1 (pp. 512–515)

Determine whether each event is impossible, unlikely, as likely as not, likely, or certain.

1. Ramona went to school today. If her class went on a field trip, how likely is it that Ramona went on the field trip?
2. Neil has blue, black, tan, and gray pants. How likely is it that he will wear green pants to school today?

LESSON 10-2 (pp. 516–519)

3. Carl is conducting a survey for the school paper. Of the 31 students he has interviewed, 10 have only a dog, 9 have only a cat, and 5 have both a dog and a cat. What is the experimental probability that the next person Carl surveys will not have a dog or a cat?
4. During her ride home from school, Dana sees 15 cars driven by men and 34 cars driven by women. What is the experimental probability that the next car Dana sees will be driven by a man?

LESSON 10-3 (pp. 520–523)

5. Shelly and Anthony are playing a game using a number cube numbered 1–6 and a nickel. Each player rolls the number cube and flips the coin. What are all the possible outcomes during one turn? How large is the sample space?

6. A yogurt shop offers 4 different flavors of yogurt and 3 different fruit toppings. How many different desserts can you have if you choose one flavor of yogurt and one topping?

LESSON 10-4 (pp. 524–527)

A spinner with 10 equal sections numbered 1 through 10 is spun. Find each probability. Write your answer as a fraction, as a decimal, and as a percent.

7. $P(5)$
8. $P(\text{prime number})$
9. $P(\text{even number})$
10. $P(20)$

Focus on Problem Solving

Understand the Problem

- **Identify important details**

When you are solving word problems, you need to identify information that is important to the problem. Read the problem several times to find all the important details. Sometimes it is helpful to read the problem aloud so that you can hear the words. Highlight the facts that are needed to solve the problem. Then list any other information that is necessary.

Highlight the important information in each problem, and then list any other important details.

1. A bag of bubble gum has 25 pink pieces, 20 blue pieces, and 15 green pieces. If Lauren selects 1 piece of bubble gum without looking, what is the probability that it is not blue?

2. Regina has a bag of marbles that contains 6 red marbles, 3 green marbles, and 4 blue marbles. Regina pulls 1 marble from the bag without looking. What is the probability that the marble is red?

3. Marco is counting the cars he sees on his ride home from school. Of 20 cars, 10 are white, 6 are red, 2 are blue, and 2 are green. What is the experimental probability that the next car Marco sees will be red?

4. Frederica has 8 red socks, 6 blue socks, 10 white socks, and 4 yellow socks in a drawer. What is the probability that she will randomly pull a brown sock from the drawer?

5. During the first 20 minutes of lunch, 5 male students, 7 female students, and 3 teachers went through the lunch line. What is the experimental probability that the next person through the line will be a teacher?

10-5 Probability of Independent and Dependent Events

Learn to find the probability of independent and dependent events.

Vocabulary

independent events

dependent events

Raji and Kara must each choose a topic from a list of topics to research for their class. If Raji's choice has no effect on Kara's choice and vice versa, the events are *independent.* For **independent events**, the occurrence of one event has no effect on the probability that a second event will occur.

If once Raji chooses a topic, Kara must choose from the remaining topics, then the events are *dependent.* For **dependent events**, the occurrence of one event *does* have an effect on the probability that a second event will occur.

EXAMPLE 1 Determining Whether Events Are Independent or Dependent

Decide whether each set of events is independent or dependent. Explain your answer.

A Erika rolls a 3 on one number cube and a 2 on another number cube.

Since the outcome of rolling one number cube does not affect the outcome of rolling the second number cube, the events are independent.

B Tomoko chooses a seventh-grader for her team from a group of seventh- and eighth-graders, and then Juan chooses a different seventh-grader from the remaining students.

Since Juan cannot pick the same student that Tomoko picked, and since there are fewer students for Juan to choose from after Tomoko chooses, the events are dependent.

To find the probability that two independent events will happen, multiply the probabilities of the two events.

Probability of Two Independent Events

$$P(A \text{ and } B) = P(A) \cdot P(B)$$

Probability of both events — *Probability of first event* — *Probability of second event*

EXAMPLE 2 Finding the Probability of Independent Events

Find the probability of flipping a coin and getting heads and then rolling a 6 on a number cube numbered 1 through 6.

The outcome of flipping the coin does not affect the outcome of rolling the number cube, so the events are independent.

$P(\text{heads and } 6) = P(\text{heads}) \cdot P(6)$

$= \frac{1}{2} \cdot \frac{1}{6}$ *There are 2 ways a coin can land and 6 ways a number cube can land.*

$= \frac{1}{12}$ *Multiply.*

The probability of getting heads and a 6 is $\frac{1}{12}$.

To find the probability of two dependent events, you must determine the effect that the first event has on the probability of the second event.

Probability of Two Dependent Events

$P(A \text{ and } B) = P(A) \cdot P(B \text{ after } A)$

Probability of both events — *Probability of first event* — *Probability of second event*

EXAMPLE 3 Finding the Probability of Dependent Events

Mica has five \$1 bills, three \$10 bills, and two \$20 bills in her wallet. If she picks two bills at random, what is the probability of her picking the two \$20 bills?

The first draw changes the number of bills left, and may change the number of \$20 bills left, so the events are dependent.

$P(\text{first } \$20) = \frac{2}{10} = \frac{1}{5}$ *There are two \$20 bills out of ten bills.*

$P(\text{second } \$20) = \frac{1}{9}$ *There is one \$20 bill left out of nine bills.*

$P(\text{first } \$20, \text{ then second } \$20) = P(A) \cdot P(B \text{ after } A)$

$= \frac{1}{5} \cdot \frac{1}{9}$

$= \frac{1}{45}$ *Multiply.*

The probability of Mica picking two \$20 bills is $\frac{1}{45}$.

Think and Discuss

1. **Compare** probabilities of independent and dependent events.
2. **Explain** whether the probability of two events is greater or less than the probability of each individual event.

10-5 Exercises

FOR EOG PRACTICE

see page 700

GUIDED PRACTICE

See Example 1

Decide whether each set of events is independent or dependent. Explain your answer.

1. A student flips heads on one coin and tails on a second coin.
2. A student chooses a red marble from a bag of marbles and then chooses another red marble without replacing the first.

See Example 2

Find the probability of each set of independent events.

3. a flipped coin landing heads up and rolling a 5 or a 6 on a number cube numbered 1 through 6
4. drawing a 5 from 10 cards numbered 1 through 10 and rolling a 2 on a number cube numbered 1 through 6

See Example 3

5. Each day, Mr. Samms randomly chooses 2 students from his class to serve as helpers. If there are 15 boys and 10 girls in the class, what is the probability that Mr. Samms will choose 2 girls to be helpers?

INDEPENDENT PRACTICE

See Example 1

Decide whether each set of events is independent or dependent. Explain your answer.

6. A student chooses a fiction book at random from a list of books and then chooses a second fiction book from those remaining.
7. A woman chooses a lily from one bunch of flowers and then chooses a tulip from a different bunch.

See Example 2

Find the probability of each set of independent events.

8. drawing a red marble from a bag of 6 red and 4 blue marbles, replacing it, and then drawing a blue marble
9. rolling an even number on a number cube numbered 1 through 6 and rolling an odd number on a second roll of the same cube

See Example 3

10. Francisco has 7 quarters in his pocket. Of these, 3 depict the state of Delaware, 2 depict Georgia, 1 depicts Connecticut, and 1 depicts Pennsylvania. Francisco removes 1 quarter from his pocket and then removes a second quarter without replacing the first. What is the probability that both will be Delaware quarters?

PRACTICE AND PROBLEM SOLVING

11. An even number is chosen randomly from a set of cards labeled with the numbers 1 through 8. A second even number is chosen without the first card's being replaced. Are these independent or dependent events? What is the probability of both events occurring?

12. A school cafeteria has 3 containers of white milk, 5 containers of chocolate milk, and 2 containers of apple juice left. Ilana is first in line and Vishal is second. If the drinks are given out randomly, what is the probability that Ilana will get chocolate milk and Vishal will get apple juice?

13. On a multiple-choice test, each question has five possible answers. A student does not know the answers to two questions, so he guesses. What is the probability that the student will get both answers wrong?

14. ***BUSINESS*** The graph shows the dogs bathed at a dog-grooming business one day. What is the probability that the first two dogs bathed were large dogs?

15. ***WRITE A PROBLEM*** Describe two events that are either independent or dependent, and make up a probability problem about them.

16. ***WRITE ABOUT IT*** At the beginning of a game of Scrabble, players take turns drawing 7 tiles. Is drawing an *A* on the first two tiles dependent or independent events? Explain.

17. ***CHALLENGE*** Weather forecasters have accurately predicted rain in one community $\frac{4}{5}$ of the time. What is the probability that they will accurately predict rain two days in a row?

Spiral Review

Estimate each square root to the nearest whole number. (Lesson 8-7)

18. $\sqrt{134}$ **19.** $\sqrt{11}$ **20.** $\sqrt{175}$ **21.** $\sqrt{217}$

22. **EOG PREP** Fritz jogged $1\frac{3}{4}$ mi on Monday, $2\frac{1}{2}$ mi on Wednesday, and 3 mi on Friday. How many miles did he jog altogether on these days? (Lesson 4-11)

A $6\frac{3}{4}$ mi B $7\frac{1}{4}$ mi C $8\frac{1}{4}$ mi D $6\frac{2}{3}$ mi

23. **EOG PREP** There are 15 boys and 20 girls in a class. If you pick one student at random, what is the probability of choosing a girl? (Lesson 10-4)

A $\frac{4}{7}$ B $\frac{3}{4}$ C $\frac{4}{3}$ D $\frac{3}{7}$

Pascal's Triangle

Use with Lesson 10-6

REMEMBER
- Probability is the likelihood of an event occurring.
- A combination is an arrangement of items or events in which order is not important.

The triangular arrangement of numbers below is called **Pascal's Triangle.** Each row starts and ends with 1. Each other number in the triangle is the sum of the two numbers above it.

1

1 1

1 2 1

1 3 3 1

1 4 6 4 1 $4 = 3 + 1$

1 5 10 10 5 1

You can use Pascal's Triangle to solve problems involving probability.

Activity

1. Geri, Jan, Kathy, Annie, and Mia are on a women's bobsled team. Only two women can race at a time. How many pairings of bobsledders are possible?

   Write each name on a separate card. You will need four cards for each name. Show all of the possible pairings of bobsledders.

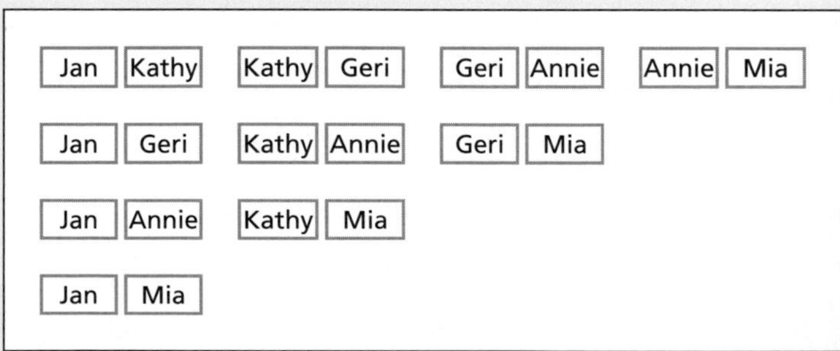

   Each pair of bobsledders shown is different. So there are 10 possible pairings of bobsledders.

Each pairing is a **combination,** because the order is not important. The number of possible combinations can be expressed using the following notation:

2 **Total number of people** → $_5C_2$ ← Number of people in each combination

You can use Pascal's Triangle to find the value of $_5C_2$ or any other combination.

a. Copy Pascal's Triangle. Label the rows and columns as shown.

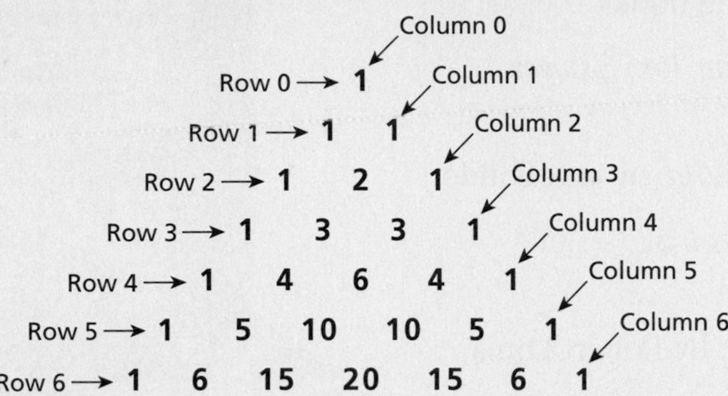

b. To find the value of $_5C_2$, look at where row 5 and column 2 intersect.

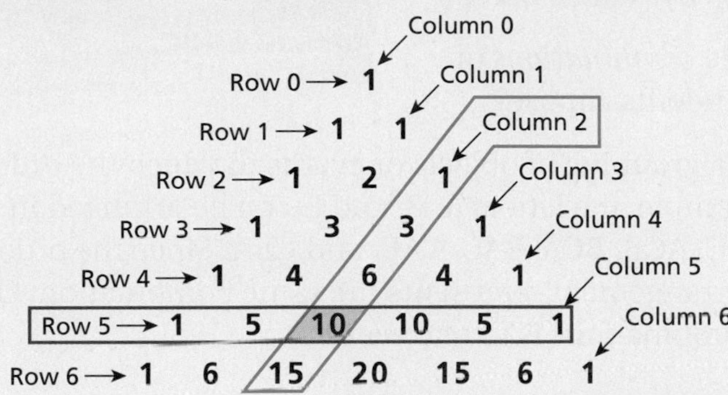

Therefore, $_5C_2 = 10$.

Think and Discuss

1. In **1**, why is the order of the bobsledders not important?
2. Complete rows 7 and 8 in Pascal's Triangle.
3. What are some patterns that you see in Pascal's Triangle?
4. Find $_3C_3$, $_2C_2$, $_6C_6$, and $_4C_4$. What do you notice? Explain.
5. Ten students are on the basketball team, but only 5 can play at a time. Use Pascal's Triangle to find the number of possible combinations of players.

Try This

Use Pascal's Triangle to find the number of combinations.

1. $_3C_2$	**2.** $_3C_1$	**3.** $_6C_4$	**4.** $_5C_5$
5. $_7C_4$	**6.** $_6C_3$	**7.** $_6C_1$	**8.** $_4C_2$

10-6 Combinations

Learn to find the number of possible combinations.

Vocabulary

combination

Mrs. Logan's students have to read any two of the following books.

1. *The Adventures of Tom Sawyer,* by Mark Twain
2. *The Call of the Wild,* by Jack London
3. *A Christmas Carol,* by Charles Dickens
4. *Treasure Island,* by Robert Louis Stevenson
5. *Tuck Everlasting,* by Natalie Babbit

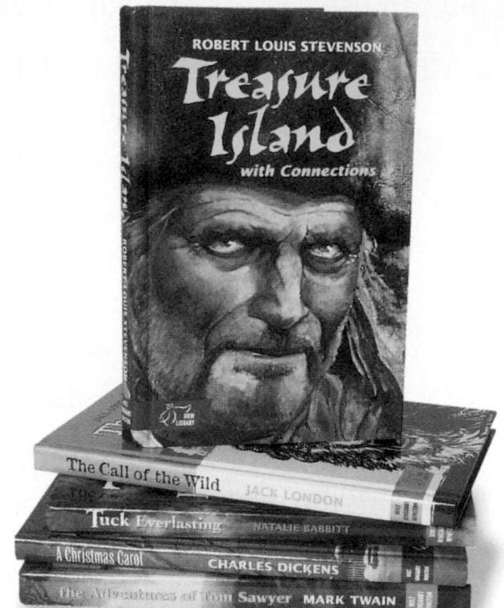

How many possible *combinations* of books could the students choose?

A **combination** is a grouping of objects or events in which the order does not matter. For example, the letters A, B, and C can be arranged in 6 different ways: ABC, ACB, BCA, BAC, CAB, and CBA. Since the order does not matter, each arrangement represents the same combination. One way to find possible combinations is to make a list.

EXAMPLE 1 Using a List to Find Combinations

How many different combinations of two books are possible from Mrs. Logan's list of five books?

Begin by listing all of the possible groupings of books taken two at a time.

1, 2	2, 1	3, 1	4, 1	5, 1
1, 3	2, 3	3, 2	4, 2	5, 2
1, 4	2, 4	3, 4	4, 3	5, 3
1, 5	2, 5	3, 5	4, 5	5, 4

Because order does not matter, you can eliminate repeated pairs. For example, 1, 2 is already listed, so 2, 1 can be eliminated.

1, 2	~~2, 1~~	~~3, 1~~	~~4, 1~~	~~5, 1~~
1, 3	2, 3	~~3, 2~~	~~4, 2~~	~~5, 2~~
1, 4	2, 4	3, 4	~~4, 3~~	~~5, 3~~
1, 5	2, 5	3, 5	4, 5	~~5, 4~~

There are 10 different combinations of two books on Mrs. Logan's list of five books.

You can also use a tree diagram to find possible combinations.

EXAMPLE 2 PROBLEM SOLVING APPLICATION

As a caterer, Cuong offers four vegetable choices: potatoes, corn, peas, and carrots. Each person can choose two vegetables. How many different combinations of two vegetables can a person choose?

1 Understand the Problem

Rewrite the question as a statement.

- Find the number of possible combinations of two vegetables a person can choose.

List the **important information:**

- There are four vegetable choices in all.

2 Make a Plan

You can make a tree diagram to show the possible combinations.

3 Solve

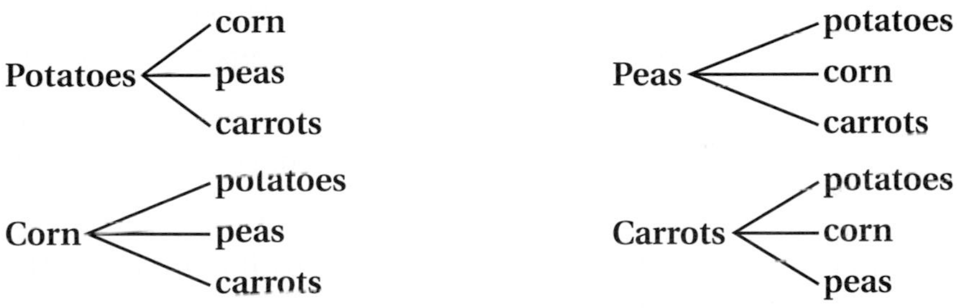

The tree diagram shows 12 possible ways to combine two vegetables, but each combination is listed twice. So there are $12 \div 2 = 6$ possible combinations.

4 Look Back

You can also check by making a list. The potatoes can be paired with three other vegetables, corn with two, and peas with one. The total number of possible pairs is $3 + 2 + 1 = 6$.

Think and Discuss

1. **Find** the number of combinations there would be in Example 1 if the students had to choose three books.
2. **Describe** how you can use combinations to find the probability of an event.

10-6 Exercises

FOR EOG PRACTICE
see page 700

internet connect
Homework Help Online
go.hrw.com Keyword: MS4 10-6

GUIDED PRACTICE

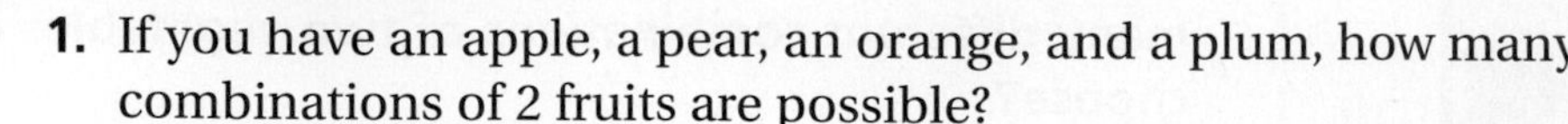

See Example 1

1. If you have an apple, a pear, an orange, and a plum, how many combinations of 2 fruits are possible?
2. How many 3-letter combinations are possible from *A, E, I, O,* and *U*?

See Example 2

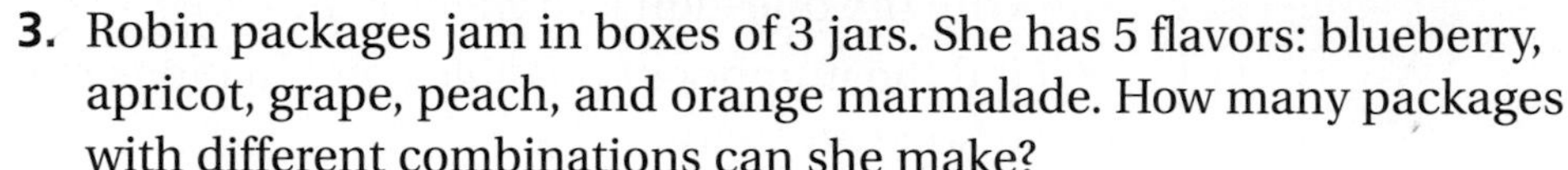

3. Robin packages jam in boxes of 3 jars. She has 5 flavors: blueberry, apricot, grape, peach, and orange marmalade. How many packages with different combinations can she make?
4. Eduardo has 6 colors of fabric: red, blue, green, yellow, orange, and white. He plans to make flags using 2 colors. How many possible combinations of 2 colors can he choose?

INDEPENDENT PRACTICE

See Example 1

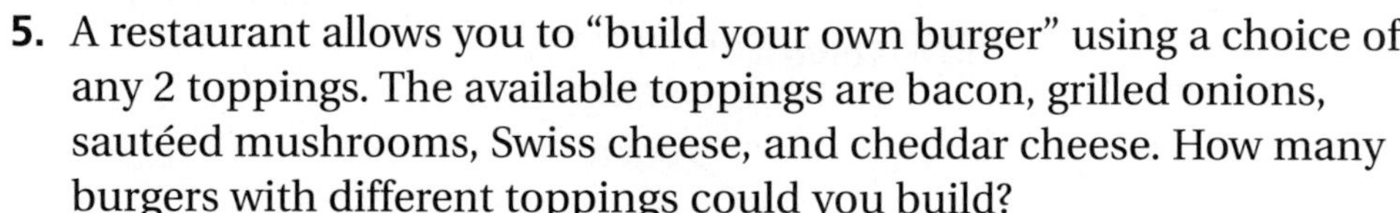

5. A restaurant allows you to "build your own burger" using a choice of any 2 toppings. The available toppings are bacon, grilled onions, sautéed mushrooms, Swiss cheese, and cheddar cheese. How many burgers with different toppings could you build?
6. Jamil has to do reports on 3 cities. He can choose from Paris, New York, Moscow, and London. How many different combinations of cities are possible?

See Example 2

7. A florist can choose from 6 different types of flowers to make a bouquet: carnations, roses, lilies, daisies, irises, and tulips. How many different combinations of 3 types of flowers can he choose from?
8. How many different 2-member tennis teams can be made from 7 students?

PRACTICE AND PROBLEM SOLVING

Find the number of combinations.

9. 7 things taken 2 at a time
10. 8 items taken 3 at a time
11. 6 things taken 5 at a time
12. 6 things taken 1 at a time
13. Rob, Caryn, and Sari are pairing up to play a series of chess matches.
    a. In how many different ways can they pair up?
    b. What is the probability that Sari will play Rob in the first match?
14. Gary has to write short biographies about 2 historical figures. He can choose from Winston Churchill, Martin Luther King, Jr., and Nelson Mandela. How many different combinations of biographies can Gary write?

Art LINK

15. Ms. Frennelle is teaching her class about famous impressionist painters. She asked her students to choose 2 artists from among Renoir, Monet, Manet, Degas, Pissarro, and Cassatt, and to find information about a painting made by each artist.

The White Water Lilies, 1899, by Claude Monet

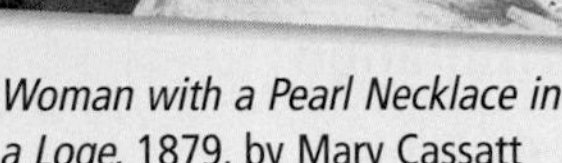

Woman with a Pearl Necklace in a Loge, 1879, by Mary Cassatt

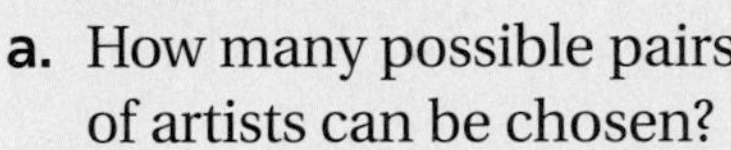

a. How many possible pairs of artists can be chosen?

b. What is the probability that a student selecting a pair at random will select a pair including Renoir?

16. Trina wants to select 3 of Ansel Adams's 5 "surf sequence" photos to hang on her wall.

a. How many combinations of 3 photos are possible?

b. What is the probability that the three photos she chooses will include surf sequence 1 and surf sequence 5?

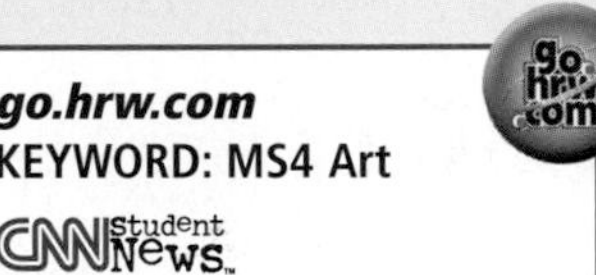

17. The graph shows the number of paintings by artists of different nationalities that Matt found in an art history book. In how many ways can Matt select 4 paintings by Chinese artists?

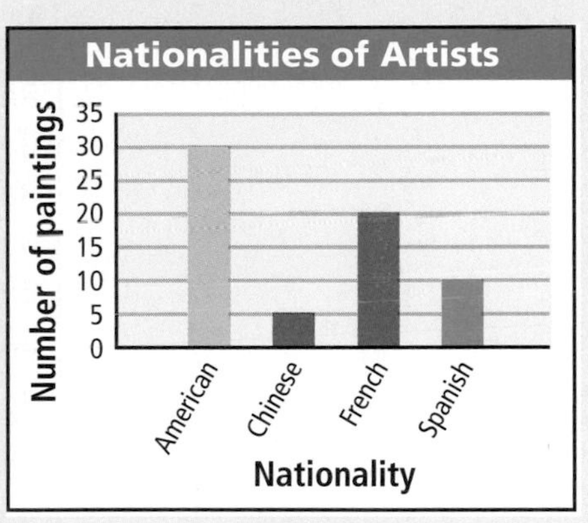

18. ***CHALLENGE*** A gallery is preparing a show by a new artist. The gallery has enough wall space to display 7 pieces of art. If the artist has prepared 4 paintings and 5 sculptures, how many distinct combinations of the artist's works are possible?

Spiral Review

Find the area of each circle to the nearest tenth. Use 3.14 for π. (Lesson 8-6)

19. $r = 8$ cm **20.** $r = 2$ ft **21.** $d = 10$ in. **22.** $d = 5$ m

Identify each figure. (Lesson 9-1)

23.

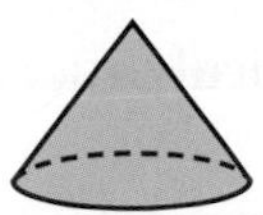

24.

25.

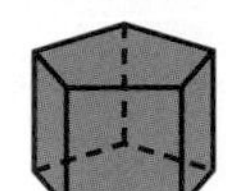

26. **EOG PREP** Which of the following describes the probability of selecting a red marble from a bag containing 4 blue marbles, 3 green marbles, and 3 white marbles? (Lesson 10-1)

A Certain B Unlikely C Impossible D As likely as not

10-7 Permutations

Learn to find the number of possible permutations and the probability that a specific permutation will occur.

Vocabulary

permutation

factorial

The conductor of a symphony orchestra is planning a concert titled "An Evening with the Killer B's." The concert will feature music by Bach, Beethoven, Brahms, and Bartok. In how many different ways can the conductor program each composer's music?

An arrangement of objects or events in which the order is important is called a **permutation**. You can use a list to find the number of permutations of a group of objects.

EXAMPLE 1 Using a List to Find Permutations

A **In how many different ways can the conductor order pieces composed by Bach, Beethoven, Brahms, and Bartok?**

Use a list to find the possible permutations.

Bach, Beethoven, Brahms, Bartok	Brahms, Bach, Beethoven, Bartok
Bach, Beethoven, Bartok, Brahms	Brahms, Bach, Bartok, Beethoven
Bach, Bartok, Beethoven, Brahms	Brahms, Beethoven, Bach, Bartok
Bach, Bartok, Brahms, Beethoven	Brahms, Beethoven, Bartok, Bach
Bach, Brahms, Beethoven, Bartok	Brahms, Bartok, Beethoven, Bach
Bach, Brahms, Bartok, Beethoven	Brahms, Bartok, Bach, Beethoven
Beethoven, Bach, Brahms, Bartok	Bartok, Bach, Beethoven, Brahms
Beethoven, Bach, Bartok, Brahms	Bartok, Bach, Brahms, Beethoven
Beethoven, Brahms, Bach, Bartok	Bartok, Beethoven, Bach, Brahms
Beethoven, Brahms, Bartok, Bach	Bartok, Beethoven, Brahms, Bach
Beethoven, Bartok, Brahms, Bach	Bartok, Brahms, Beethoven, Bach
Beethoven, Bartok, Bach, Brahms	Bartok, Brahms, Bach, Beethoven

There are 24 ways to order the music.

B **If the order is chosen randomly, what is the probability that the composers will be featured in alphabetical order?**

Of the 24 possible permutations, only 1 is in alphabetical order (Bach, Bartok, Beethoven, Brahms).

$$P(\text{alphabetical order}) = \frac{\text{number of alphabetical arrangements}}{\text{total number of arrangements}}$$

$$= \frac{1}{24}$$

The probability that the composers will be featured in alphabetical order is $\frac{1}{24}$.

By making an organized list, you can find the possible permutations as well as the number of permutations. You can use the Fundamental Counting Principle to find only the number of permutations.

EXAMPLE 2 Using the Fundamental Counting Principle to Find the Number of Permutations

Ed, Emily, and Lila have agreed to be president, vice president, and secretary of the Yearbook Club. In how many different ways can the students fill the positions? What is the probability that any two of the permutations will be chosen?

Once you fill a position, you have one less choice for the next position.

There are three choices for the first position.

There are two remaining choices for the second position.

There is one choice for the third position.

$3 \cdot 2 \cdot 1 = 6$ *Multiply.*

There are 6 different ways that the students can fill the positions. The probability that any two of the permutations will be chosen is $\frac{2}{6}$, or $\frac{1}{3}$.

Multiplying $3 \cdot 2 \cdot 1$ is called 3 *factorial* and is written as "3!" You can find the **factorial** of a whole number by multiplying all the whole numbers except zero that are less than or equal to the number.

$3! = 3 \cdot 2 \cdot 1 = 6$ $6! = 6 \cdot 5 \cdot 4 \cdot 3 \cdot 2 \cdot 1 = 720$

You can use factorials to find the number of permutations in a given situation.

EXAMPLE 3 Using Factorials to Find the Number of Permutations

There are nine players in a baseball lineup. How many different batting orders are possible for these players?

$$\begin{aligned}\text{Number of permutations} &= 9! \\ &= 9 \cdot 8 \cdot 7 \cdot 6 \cdot 5 \cdot 4 \cdot 3 \cdot 2 \cdot 1 \\ &= 362{,}880\end{aligned}$$

There are 362,880 different batting orders possible for 9 players.

Think and Discuss

1. **Give an example** that shows the difference between permutations and combinations.
2. **Explain** why 8! gives the number of permutations of 8 objects.

10-7 Exercises

FOR EOG PRACTICE
see page 701

internet connect
Homework Help Online
go.hrw.com Keyword: MS4 10-7

GUIDED PRACTICE

See Example 1
1. **a.** In how many ways can you arrange the numbers 1, 2, 3, and 4 to make a four-digit number?
   b. If you choose one of the four-digit numbers at random, what is the probability that the number will be less than 2,000?

See Example 2
2. **a.** Find the number of permutations of the letters in the word *quiet.*
   b. What is the probability that any one of the permutations will be chosen?

See Example 3

3. Sam wants to call 6 friends to invite them to a party. In how many possible orders can he make the calls?
4. Seven people are waiting to audition for a play. In how many different orders can the auditions be done?

INDEPENDENT PRACTICE

See Example 1
5. **a.** In how many ways can Eric, Meera, and Roger stand in line?
   b. If you choose one of the orders at random, what is the probability that Meera will be first in line?

See Example 2

6. **a.** Find the number of ways you can arrange the letters in the word *art.*
   b. What is the probability that any one of the permutations will be chosen?

See Example 3
7. How many permutations of the letters *A* through *I* are there?
8. In how many different ways can 8 riders be matched up with 8 horses?

PRACTICE AND PROBLEM SOLVING

Determine whether each problem involves combinations or permutations. Explain your answer.

9. Choose five books to read from a group of ten.
10. Decide how many ways five people can be assigned to sit in five chairs.
11. Choose a 4-digit code from the digits 3, 7, 1, and 8.
12. Carl, Melba, Sean, and Ricki are going to present individual reports in their Spanish class. What is the probability that Melba will present her report first?
13. Using the digits 1 through 7, Pima County is assigning new seven-digit numbers to all households. How many possible numbers can the county assign without repeating any of the digits in a number?

Since its initial publication in 1868, *Little Women*, by Louisa May Alcott, has never been out of print. Over the years, the story has been translated into at least 27 languages.

14. ***LITERATURE*** The school library has 13 books by Louisa May Alcott. Merina wants to read all 13 of them one after another. Write an expression to show the number of ways she can do that.

15. ***HEALTH*** A survey was taken to find out how 200 people age 40 and older rate their memory now compared to 10 years ago. In how many different orders could interviews be conducted of people who think their memory is the same?

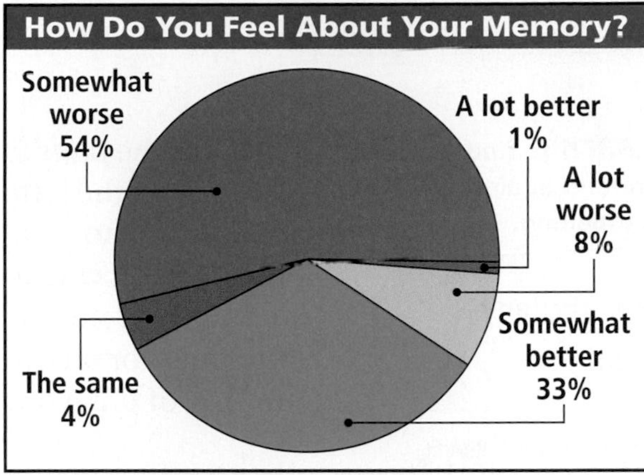

16. Use the letters *A, D, E, R*.
    a. How many permutations of the letters are there?
    b. How many arrangements form English words?
    c. What is the probability that a random arrangement of the letters shown will form an English word?

17. ***SPORTS*** Ten golfers on a team are playing in a tournament. How many different lineups can the golf coach make?

18. ***WHAT'S THE ERROR?*** A student was trying to find 5! and wrote the equation $5 + 4 + 3 + 2 + 1 = 15$. Why is this student incorrect?

19. ***WRITE ABOUT IT*** Explain the difference between combinations of objects and permutations of objects. Give examples of each.

20. ***CHALLENGE*** How many permutations are there of four objects taken two at a time?

Spiral Review

Find each volume. (Lesson 9-2)

21. cylinder with $r = 3$ m and $h = 5$ m
22. rectangular prism with $\ell = 4$ ft, $w = 3$ ft, and $h = 9$ ft
23. triangular prism with $B = 4.5\text{ cm}^2$ and $h = 13$ cm

Determine the experimental probability. (Lesson 10-2)

24. Kaisia made 3 hits in 8 at bats. What is the probability that she will make a hit the next time at bat?

25. **EOG PREP** Bertram has four \$1 bills, three \$5 bills, and three \$10 bills in his pocket. He draws a bill at random and then draws a second one without replacing the first. What is the probability that Bertram pulls out a \$5 bill and then a \$1 bill? (Lesson 10-5)

A $\frac{7}{10}$ B 0.12 C $\frac{2}{15}$ D 0.1

EXTENSION

Computing Odds

Learn to find the odds for and against events happening.

Vocabulary

odds

What are the *odds* that the spinner will stop on the letter *e*? **Odds** are another way to express the likelihood that an event will occur. Odds compare the number of favorable outcomes for an event with the number of unfavorable outcomes.

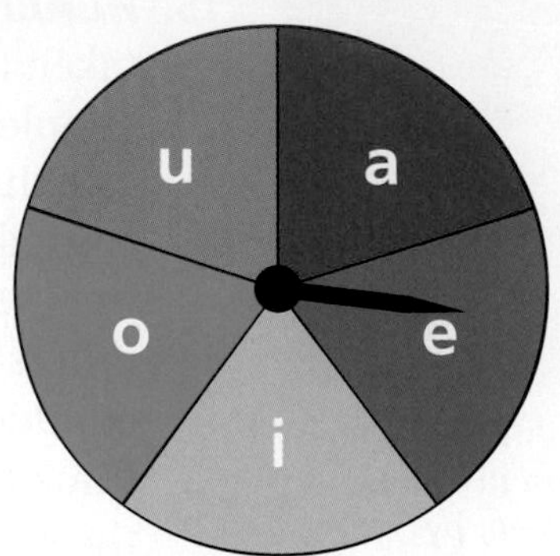

Reading Math

Odds are read as the number of favorable outcomes *to* the number of unfavorable outcomes. The odds of the spinner landing on e are 1 to 4, or 1:4.

ODDS OF AN EVENT

$$\text{odds} = \frac{\text{number of favorable outcomes}}{\text{number of unfavorable outcomes}}$$

There is only one way the spinner can land on *e*, but there are four equal ways for it not to land on *e*, so the odds of landing on *e* are $\frac{1}{4}$. In contrast, the probability that the spinner will land on *e* is $\frac{1}{5}$.

EXAMPLE 1 Computing the Odds of an Event

Suppose you choose a letter at random from the word *Indiana*.

A **What are the odds that the letter you choose will be a vowel?**

$$\text{odds of choosing a vowel} = \frac{\text{number of vowels}}{\text{number of consonants}} = \frac{4}{3}$$

The odds of choosing a vowel are 4:3.

B **What are the odds that the letter you choose will not be a vowel?**

$$\text{odds of not choosing a vowel} = \frac{\text{number of consonants}}{\text{number of vowels}} = \frac{3}{4}$$

The odds of not choosing a vowel are 3:4.

C **What are the odds that the letter you choose will be an *a*?**

$$\text{odds of choosing an } a = \frac{\text{number of } a\text{'s}}{\text{number of letters that are not } a} = \frac{2}{5}$$

The odds of choosing an *a* are 2:5.

EXTENSION

Exercises

Each card in a set of 7 cards has one letter from the word *Florida*.

1. What are the odds of choosing a card with a vowel?
2. What are the odds of choosing a card that does not have a vowel?
3. What are the odds of choosing a card with an *F*?
4. What are the odds of choosing a card that does not have an *F*?

A student is chosen at random from a class of 10 boys and 20 girls. Find the odds of each event.

5. A girl is chosen.
6. A boy is chosen.
7. A girl is not chosen.
8. A boy is not chosen.

Casey spins the spinner shown. Find the odds of each event.

9. The pointer lands on red.
10. The pointer lands on blue.
11. The pointer lands on yellow.
12. The pointer lands on a color other than red.
13. Find the odds of a coin landing heads up when you toss a coin.
14. What are the odds against rolling a 1 on a 1–6 number cube?
15. Each letter of the alphabet is written on one craft stick, for a total of 26 sticks. You choose one stick at random.
    a. What are the odds of choosing a stick with a vowel?
    b. What are the odds against choosing a stick with a vowel?
16. The letters in *Montgomery, Alabama* are written on slips of paper, one letter per slip.
    a. What are the odds of drawing an *A*?
    b. What are the odds of drawing an *M*?
17. The probability of choosing a red marble from a bag of marbles is $\frac{2}{5}$.
    a. What are the odds of choosing a red marble from the bag?
    b. What are the odds against choosing a red marble from the bag?

Problem Solving on Location

NORTH CAROLINA

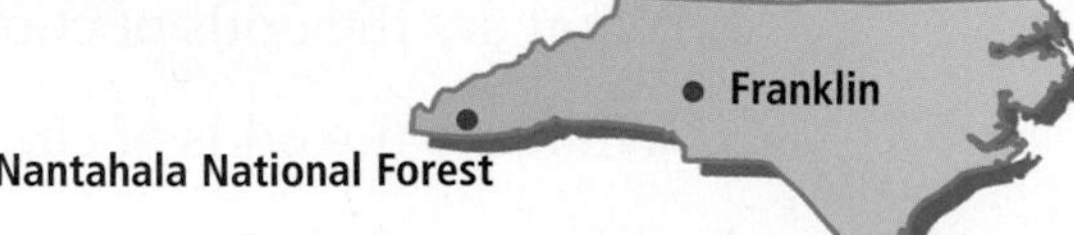

Gems and Minerals

North Carolina has a variety of mineral deposits where gemstones such as rubies, topaz, sapphires, garnets, emeralds, and sometimes even diamonds can be found. Franklin, North Carolina, is well known for its rubies, and you can pan for them at several "pan for a fee" locations.

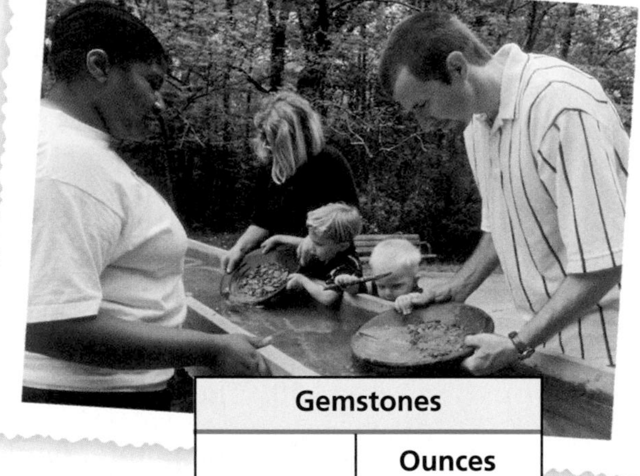

A miner panned for 5 years and kept track of the gemstones he found during this time period.

For 1–5, use the table.

Gemstones	
Gem	Ounces Collected
Ruby	2
Sapphire	1.5
Garnet	8
Emerald	1
Diamond	0.5

1. Which gemstone did the miner find the most ounces of?
2. How many more ounces of rubies, emeralds, and diamonds combined did the miner find than ounces of sapphires?
3. If all of the gemstones in the table weighed the same, what is the approximate probability as a percent that the next gemstone the miner finds will be a ruby?
4. What is the probability the miner will find no diamonds in the next 5 years? Write your answer as a decimal, rounded to the nearest hundredth.

5. Suppose all of the gemstones in the table weigh the same. What is the probability that the miner will find sapphires while panning? Write your answer as a fraction.
6. In addition to finding the gemstones in the table, the miner also found 2.5 ounces of gold during this time period. If he panned for gold for the next 25 years with the same probability of finding the same amount of gold, predict how much gold the miner would find.

Mountain Biking

The Tsali Recreation Area, near Lake Fontana in the Nantahala National Forest, boasts 42 miles of challenging trails for mountain biking. The four trails—Right Loop, Left Loop, Mouse Branch, and Thompson Loop—wind around scenic landscapes covered by tall pines.

1. As Celeste is taking a break on the Left Loop trail, she counts 50 cyclists riding by. Of the cyclists, 22 are female and 28 are male. Based on this count, what is the probability that the next person Celeste sees ride by will be male?

2. If it doesn't rain on Saturday, a group of friends plan to ride the Thompson Loop. The weather forecast calls for a sunny day with a slight chance of showers. Is it impossible, unlikely, as likely as not, likely, or certain that the group will make their ride?

For 3–4, use the map.

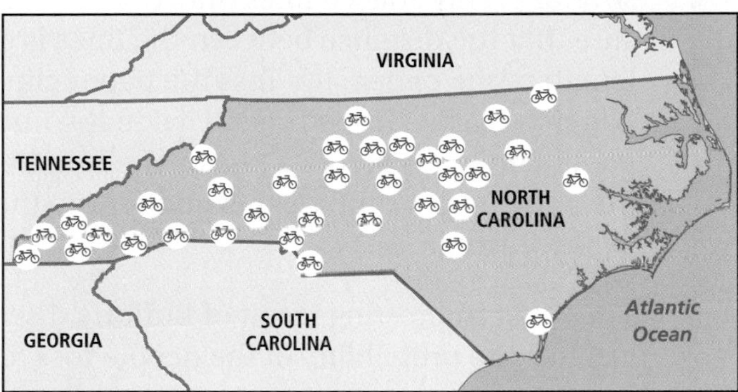

3. The map shows the locations of some cities in North Carolina that have mountain biking trails. Would you be more likely to find a town with trails for mountain biking in eastern North Carolina or in western North Carolina?

4. The 35 cities shown on the map have trails for mountain biking available in or around them. Bryson City has 6 trails. Three other towns on the map also have names that start with the letter *B*. If you were to choose one of the towns on the map at random, what is the probability that you would select a town whose name begins with *B*? Write your answer as a fraction, as a decimal, and as a percent.

MATH-ABLES

4.01

Buffon's Needle

If you drop a needle of a given length onto a wooden floor with evenly spaced cracks, what is the probability that it will land across a crack?

Compte de Buffon (1707–1788) asked this problem of geometric probability. To answer his question, Buffon developed a formula using ℓ to represent the length of the needle and d to represent the distance between the cracks.

$$\text{probability} = \frac{2\ell}{\pi d}$$

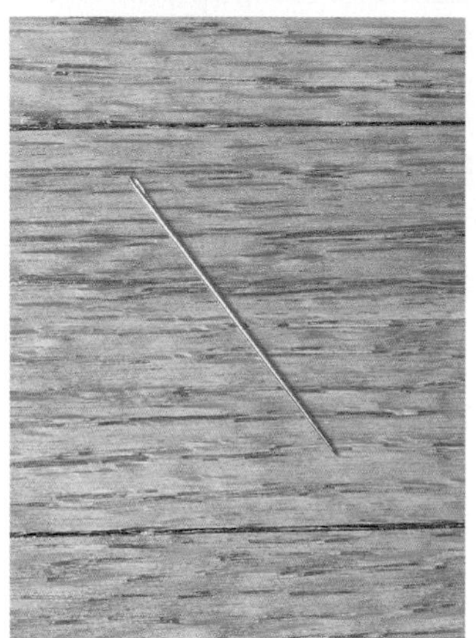

To re-create this experiment, you need a paper clip and several evenly spaced lines drawn on a piece of paper. Make sure that the distance between the lines is greater than the length of the paper clip. Toss the paper clip onto the piece of paper at least a dozen times. Divide the number of times the paper clip lands across a line by the number of times you toss the paper clip. Compare this quotient to the probability given by the formula.

The other interesting result of Buffon's discovery is that you can use the probability of the needle toss to estimate *pi.*

$$\pi = \frac{2\ell}{\text{probability} \cdot d}$$

Toss the paper clip 20 times to find the experimental probability. Use this probability in the formula above, and compare the result to 3.14.

Pattern Match

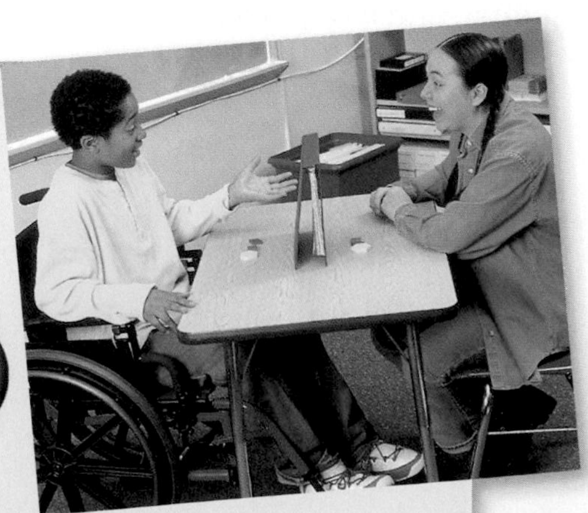

This game is for two players. Player A arranges four pattern blocks in a row out of the view of player B. Player B then tries to guess the arrangement. After each guess, player A reveals how many of the blocks are in the correct position without telling which blocks they are. The round ends when player B correctly guesses the arrangement.

internet connect

Go to ***go.hrw.com*** for a complete set of game pieces.
KEYWORD: MS4 Game10

Technology LAB

Factorials and Permutations

internet connect
Lab Resources Online
go.hrw.com
KEYWORD: MS4 TechLab10

You can use a graphing calculator to help compute factorials and permutations.

Activity

1 To compute 7! you can do the computation $7 \cdot 6 \cdot 5 \cdot 4 \cdot 3 \cdot 2 \cdot 1$ on a calculator as shown at right.

You can also use the factorial command, **!**, on a graphing calculator. This command is found in the **MATH** menu in the **PRB** submenu. It provides a faster method of computation.

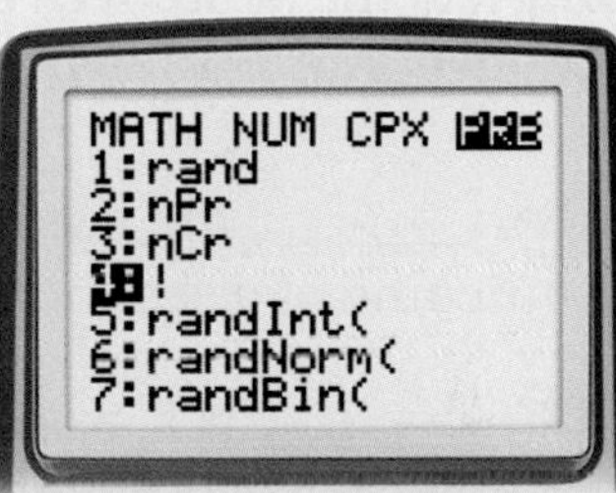

2 To compute 7! on a graphing calculator press

7 MATH ▶ ▶ ▶ 4 ENTER.

Think and Discuss

1. What is the greatest factorial you can compute on your calculator without getting an answer in scientific notation?
2. Is $m! \cdot n! = (m \cdot n)!$ a true statement? Check by substituting values for the variables.

Try This

Compute each factorial by multiplication and by using the graphing calculator factorial command.

1. 5!
2. 8!
3. 10!
4. 12!
5. $11! \div 4!$
6. $15! \div 10!$

Chapter 10

Study Guide and Review

Vocabulary

certain	512	Fundamental Counting Principle	521
combination	536	impossible	512
dependent events	530	independent events	530
equally likely	512	outcome	512
event	512	permutation	540
experiment	512	probability	512
experimental probability	516	sample space	520
factorial	541	theoretical probability	524
fair	524	trial	516
favorable outcome	524		

Complete the sentences below with vocabulary words from the list above. Words may be used more than once.

1. For ___?___ , the outcome of one event has no effect on the outcome of a second event.
2. A(n) ___?___ is a grouping of objects or events in which order does not matter.
3. All the possible outcomes to an experiment make up the ___?___.
4. A(n) ___?___ is a result of an experiment.

10-1 Probability (pp. 512–515)

EXAMPLE

- **Sergio has math class during first period. Is it impossible, unlikely, as likely as not, likely, or certain that Sergio is in math class during second period?**

  It is unlikely that Sergio is in math class during second period because he has math during first period.

EXERCISES

Suppose you draw a card from a set of cards that is numbered 2–16. Determine whether each event is impossible, unlikely, as likely as not, likely, or certain.

5. How likely is it that you will draw a composite number?
6. How likely is it that you will draw a number less than 4?
7. How likely is it that you will draw a number greater than 20?

10-2 Experimental Probability (pp. 516–519)

EXAMPLE

■ **Of the first 50 people surveyed, 21 said they liked mysteries better than comedies. What is the probability that the next person surveyed will favor mysteries?**

$$P(\text{mysteries}) = \frac{\text{number who like mysteries}}{\text{total number surveyed}}$$

$$P(\text{mysteries}) = \frac{21}{50}$$

The probability that the next person surveyed will favor mysteries is $\frac{21}{50}$.

EXERCISES

Sami has been keeping a record of her math grades. Of her first 15 grades, 10 have been above 82.

8. What is the probability that her next grade will be above 82?

9. What is the probability that her next grade will be 82 or below?

10-3 Make a List to Find Sample Spaces (pp. 520–523)

EXAMPLE

■ **Anita tosses a coin and a 1–6 number cube. How many outcomes are possible?**

Use the Fundamental Counting Principle.
Number of ways the coin can land: 2
Number of ways the cube can land: 6
$2 \cdot 6 = 12$
There are 12 possible outcomes.

EXERCISES

Chen spins each of the spinners once.

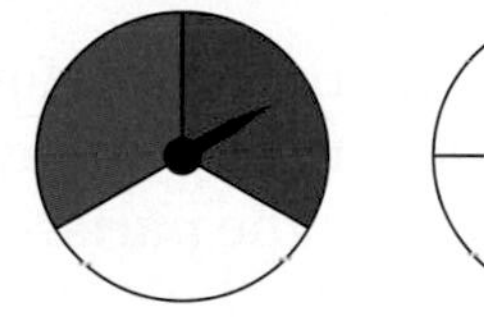

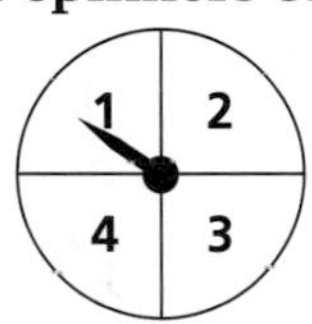

10. What are all the possible outcomes?

11. How large is the sample space?

10-4 Theoretical Probability (pp. 524–527)

EXAMPLE

■ **Find the probability of drawing a 4 from a standard deck of 52 playing cards. Four cards in the deck are numbered 4. Write your answer as a fraction, as a decimal, and as a percent.**

$$P(4) = \frac{\text{number of 4s in deck}}{\text{number of cards in deck}}$$

$$= \frac{4}{52}$$

$$= \frac{1}{13}$$

$$\approx 0.077 \approx 7.7\%$$

EXERCISES

Find each probability. Write your answer as a fraction, as a decimal, and as a percent.

12. There are 9 girls and 12 boys on the student council. What is the probability that a girl will be chosen as president?

13. Anita tosses 3 coins. What is the probability that each coin will land tails up?

10-5 Probability of Independent and Dependent Events (pp. 530–533)

EXAMPLE

- **There are 4 red marbles, 3 green marbles, 6 blue marbles, and 2 black marbles in a bag. What is the probability that Angie will pick a green marble and then a black marble without replacing the first marble?**

$P(\text{green marble}) = \frac{3}{15} = \frac{1}{5}$

$P(\text{black marble}) = \frac{2}{14} = \frac{1}{7}$

$P(\text{green, then black}) = \frac{1}{5} \cdot \frac{1}{7} = \frac{1}{35}$

The probability of picking a green marble and then a black marble is $\frac{1}{35}$.

EXERCISES

14. There are 40 tags numbered 1 through 40 in a bag. What is the probability that Glenn will pick a multiple of 5 and then a multiple of 9 without replacing the first tag?

15. Each letter of the word *probability* is written on a card and put in a bag. What is the probability of picking a vowel on the first try and again on the second try if the first card is replaced?

10-6 Combinations (pp. 536–539)

EXAMPLE

- **Tina, Sam, and Jo are trying out for the 2 lead parts in a play. In how many ways can they be chosen for the parts?**

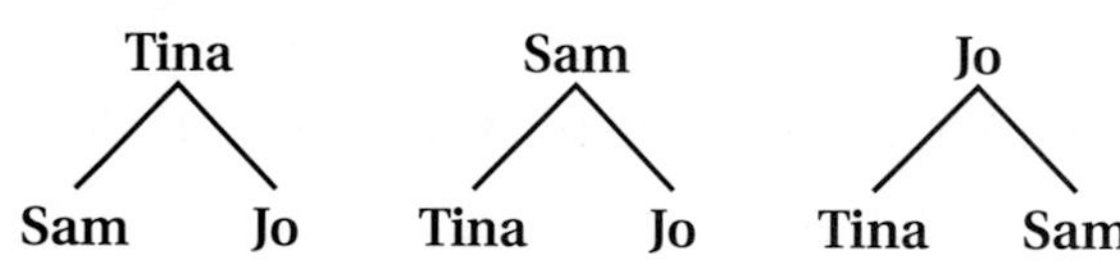

There are 3 possible ways the students can be chosen for the parts.

EXERCISES

16. How many ways can you select 3 pieces of fruit from a basket of 5 pieces?

17. How many 4-person committees can be chosen from 7 people?

18. How many combinations of 3 balloons can be chosen from 9 balloons?

10-7 Permutations (pp. 540–543)

EXAMPLE

- **How many different four-digit numbers can you make from the numbers 2, 4, 6, and 8 using each just once?**

There are 4 choices for the first digit, 3 choices for the second, 2 choices for the third, and 1 for the fourth.

$4 \cdot 3 \cdot 2 \cdot 1 = 24$

There are 24 different four-digit numbers possible.

EXERCISES

19. How many different batting orders are possible for 10 players on a softball team?

20. How many different ways can you arrange the letters in the word *number*?

21. In how many ways can Tanya, Rika, Andy, Evan, and Tanisha line up for lunch?

Chapter Test

A box contains 3 orange cubes, 2 white cubes, 3 black cubes, and 4 blue cubes.

1. Would you be more likely to pick a white cube or a blue cube?
2. Would you be more likely to pick an orange cube or a black cube?
3. Simon tosses a coin 20 times. The coin lands heads up 7 times. Based on these results, how many times can Simon expect the coin to land heads up in the next 100 tosses?
4. Emilio spins a 1–8 spinner 10 times. In his first three spins, the spinner lands on 8. What is the experimental probability that Emilio will spin a 10 on his fourth spin?
5. A brand of jeans comes in 8 different waist sizes: 28, 30, 32, 34, 36, 38, 40, and 42. The jeans also come in three different colors: blue, black, and tan. How many different combinations of waist sizes and colors are possible?
6. Greg is planning his vacation. He can choose from 3 ways to travel—train, bus, or plane—and four different activities—skiing, skating, snowboarding, or hiking. What are all the possible outcomes? How many different vacations could Greg plan?

Rachel spins a spinner that is divided into 10 equal sectors and numbered 1 through 10. Find each probability. Write your answer as a fraction, as a decimal, and as a percent.

7. *P*(odd number)
8. *P*(composite number)
9. *P*(number greater than 10)

Find the probability of each set of events.

10. spinning red on a spinner with equally sized red, blue, yellow, and green sections, and flipping a coin that lands tails up
11. choosing a card labeled *vanilla* from a group of cards labeled *vanilla, chocolate, strawberry,* and *swirl,* and then choosing a card labeled *chocolate* without replacing the first card
12. How many ways can 3 students be chosen from 10 students?
13. How many different 5-member basketball teams can be made from 7 students?
14. Timothy wants to arrange his 6 model cars on a shelf. How many ways could he arrange them? How many ways could he arrange 7 model cars?

Chapter Test

Chapter 10

Performance Assessment

Show What You Know

Create a portfolio of your work from this chapter. Complete this page and include it with your four best pieces of work from Chapter 10. Choose from your homework or lab assignments, mid-chapter quiz, or any journal entries you have done. Put them together using any design you want. Make your portfolio represent what you consider your best work.

Short Response

1. A coin is tossed 3 times. List the sample space. What is the probability that the outcomes of all 3 tosses are the same, either all heads or all tails? Explain how you found your answer.

2. Rhonda has 3 different-colored T-shirts—red, blue, and green—and a pair of blue jeans and a pair of white jeans. She randomly chooses a T-shirt and a pair of jeans. What is the probability that she will pair the red T-shirt with the white jeans? Show how you found your answer.

3. Chandra, Elias, Kenia, and Rob line up for lunch. In how many different orders can they line up? Are the orders permutations or combinations? Explain your answer.

4. The chart shows the results that Gregor Mendel obtained when he crossed two tall pea plants, each carrying a dominant (D) and a recessive (d) gene for tallness. What is the probability that an offspring will be a dwarf (dd) plant? Explain your answer.

	D	*d*
D	*DD*	*Dd*
d	*Dd*	*dd*

Gregor Mendel

Extended Problem Solving

5. A bag contains 5 blue blocks, 3 red blocks, and 2 yellow blocks.
   a. What is the probability that Tip will draw a red block and then a blue block at random if the first block is replaced before the second is drawn? Show the steps necessary to find your answer.
   b. What is the probability that Tip will draw a red block and then a blue block at random if the first block is not replaced before the second is drawn? Show your work.
   c. Explain why replacing the first block affects the answers to parts **a** and **b**.

internet connect

EOG Test Practice Online
go.hrw.com Keyword: MS4 TestPrep

Getting Ready for EOG

Chapter 10

Cumulative Assessment, Chapters 1–10

1. What is 50 meters written as kilometers?

A 50,000 km   C 500 km
B 0.05 km   D 0.5 km

2. What is the perimeter of the rectangle?

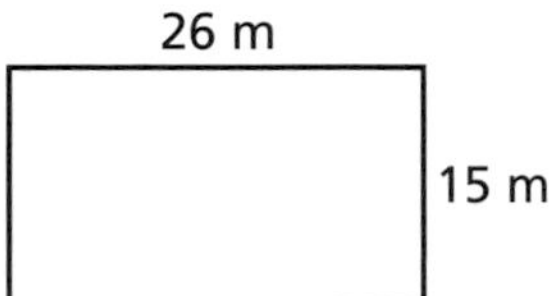

A 41 m   C 82 m
B 67 m   D 390 m

3. What is the probability that a phone number selected at random will end in a digit that is *less* than 5?

A As likely as not   C Unlikely
B Impossible   D Likely

4. Marissa made 3 out of 10 foul shots. What is the probability that she will make the next foul shot?

A 30%   C 3%
B 70%   D 7%

5. Manheim Middle School is ordering school caps. The caps come in 4 colors with 3 different designs. How many different caps could the school order?

A 7   C 24
B 12   D 14

6. Rena, Nicole, Akira, and Rosa are going to the movies. In how many ways can they sit in four adjacent seats?

A 12   C 7
B 16   D 24

7. What is the number of combinations of 8 things taken 3 at a time?

A 336   C 56
B 24   D 28

8. The graph shows a town's high temperatures over a 5-day period. What was the average high temperature over these 5 days?

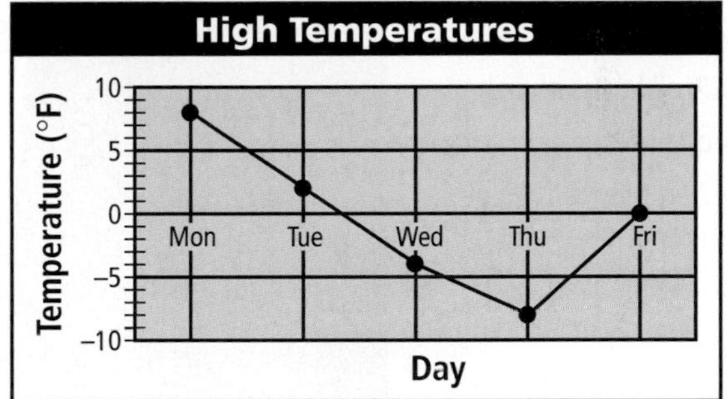

A −0.4°F   C 0.4°F
B 4.4°F   D −4.4°F

TEST TAKING TIP!

Eliminate choices by estimating the ratio of the two numbers.

9. 18 is what percent of 75?

A 76%   C 24%
B 42%   D 416%

10. ***SHORT RESPONSE*** Find the measure of the third angle of a triangle whose other angles measure 59° and 22°. Explain your work.

11. ***SHORT RESPONSE*** Evaluate $72 \div 4 \cdot 3^2 - 13$. Explain how you got your answer, and show your steps.

Getting Ready for EOG

Multistep Equations and Inequalities

TABLE OF CONTENTS

internet connect
Chapter Opener Online
go.hrw.com
KEYWORD: MS4 Ch11

Altitudes of Artificial Satellites	
Satellite	Altitude (km)
Sputnik	245
Skylab	270
Mir	390
International Space Station	420

Career *Satellite Engineer*

Artificial satellites were born with the launch of *Sputnik* on October 4, 1957. The 84 kg ball with a 56 cm diameter circled Earth every 35 minutes and signified the beginning of changes in the way people live. Today, there are over 2,500 satellites orbiting Earth.

Satellite engineers work on satellite design, construction, orbit determination, launch, tracking, and orbit adjustment. Satellites can monitor weather, crop growth, and natural resources and communicate this information using television, radio, and other communication signals. Satellites can even directionally guide people who have GPS (Global Positioning System) devices.

ARE YOU READY?

Choose the best term from the list to complete each sentence.

1. __?__ are mathematical operations that undo each other.
2. To solve an equation you need to __?__.
3. A(n) __?__ is a mathematical statement that two expressions are equivalent.
4. A(n) __?__ is a mathematical statement that two ratios are equivalent.

isolate the variable

equation

proportion

inverse operations

expression

Complete these exercises to review skills you will need for this chapter.

✔ Add Whole Numbers, Decimals, Fractions, and Integers

Add.

5. $-24 + 16$ **6.** $-34 + (-47)$ **7.** $35 + (-61)$ **8.** $-12 + (-29) + 53$

9. $2.7 + 3.5$ **10.** $\frac{2}{3} + \frac{1}{2}$ **11.** $-5.87 + 10.6$ **12.** $\frac{8}{9} + \left(-\frac{9}{11}\right)$

✔ Evaluate Expressions

Evaluate each expression for $a = 7$ and $b = -2$.

13. $a - b$ **14.** $b - a$ **15.** $\frac{b}{a}$ **16.** $2a + 3b$

17. $\frac{-4a}{b}$ **18.** $3a - \frac{8}{b}$ **19.** $1.2a + 2.3b$ **20.** $-5a - (-6b)$

✔ Solve Multiplication Equations

Solve.

21. $8x = -72$ **22.** $-12a = -60$ **23.** $\frac{2}{3}y = 16$ **24.** $-12b = 9$

25. $12 = -4x$ **26.** $13 = \frac{1}{2}c$ **27.** $-2.4 = -0.8p$ **28.** $\frac{3}{4} = 6x$

✔ Solve Proportions

Solve.

29. $\frac{3}{4} = \frac{x}{24}$ **30.** $\frac{8}{9} = \frac{4}{a}$ **31.** $-\frac{12}{5} = \frac{15}{c}$ **32.** $\frac{y}{50} = \frac{35}{20}$

33. $\frac{2}{3} = \frac{18}{w}$ **34.** $\frac{35}{21} = \frac{d}{3}$ **35.** $\frac{7}{13} = \frac{h}{195}$ **36.** $\frac{9}{-15} = \frac{-27}{p}$

Model Two-Step Equations

Use with Lesson 11-1

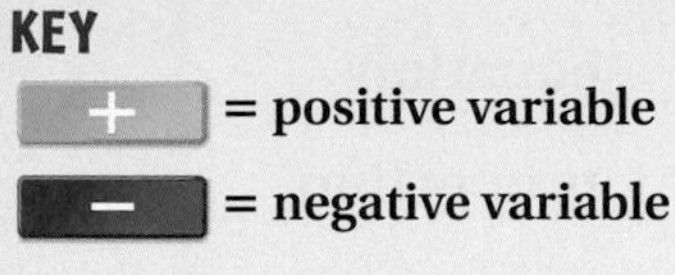

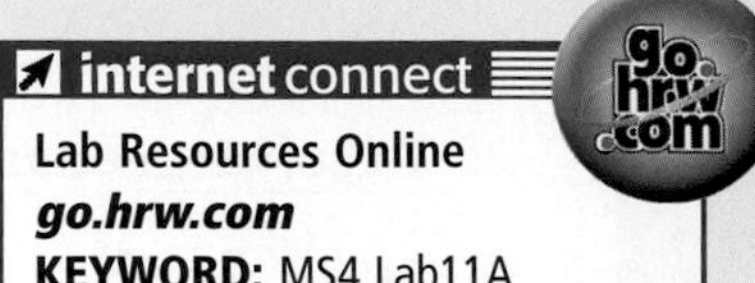

Lab Resources Online
go.hrw.com
KEYWORD: MS4 Lab11A

KEY

[+] = positive variable

[−] = negative variable

[+] = 1 [−] = −1

REMEMBER

- [+] + [−] = 0
- [+] + [−] = 0
- In an equation, the expressions on both sides of the equal sign are equivalent.

In Lab 3C, you learned how to solve one-step equations using algebra tiles. For example, to solve the equation $x + 2 = 6$, you need to isolate the variable by removing tiles.

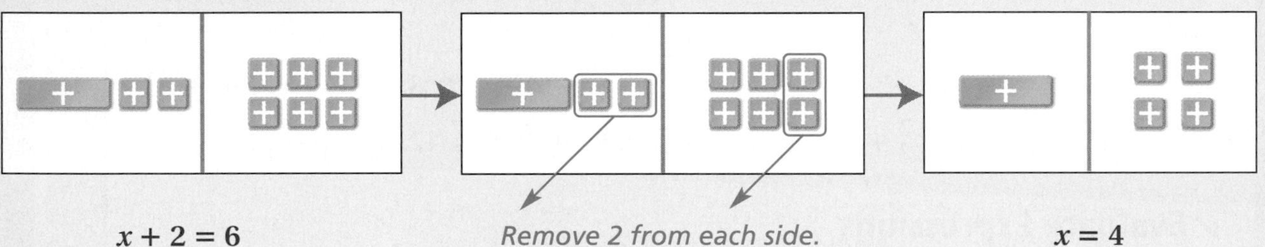

$x + 2 = 6$ *Remove 2 from each side.* $x = 4$

You can also use algebra tiles to solve two-step equations.

Activity

When solving a two-step equation, such as $2p + 2 = 10$, it is easiest to perform addition and subtraction before multiplication and division.

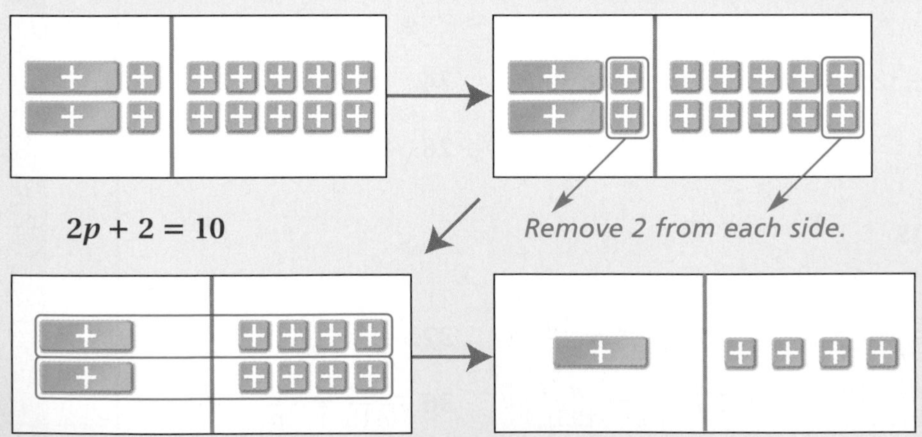

$2p + 2 = 10$ *Remove 2 from each side.*

Divide each side into 2 equal groups. $p = 4$

1. Use algebra tiles to model and solve each equation.

   a. $3x + 5 = 14$ **b.** $2n - 1 = -5$ **c.** $4w + 3 = 7$ **d.** $3n - 6 = -18$

To model solving the equation $3n + 6 = -15$, you must add neutral pairs to isolate the variable term.

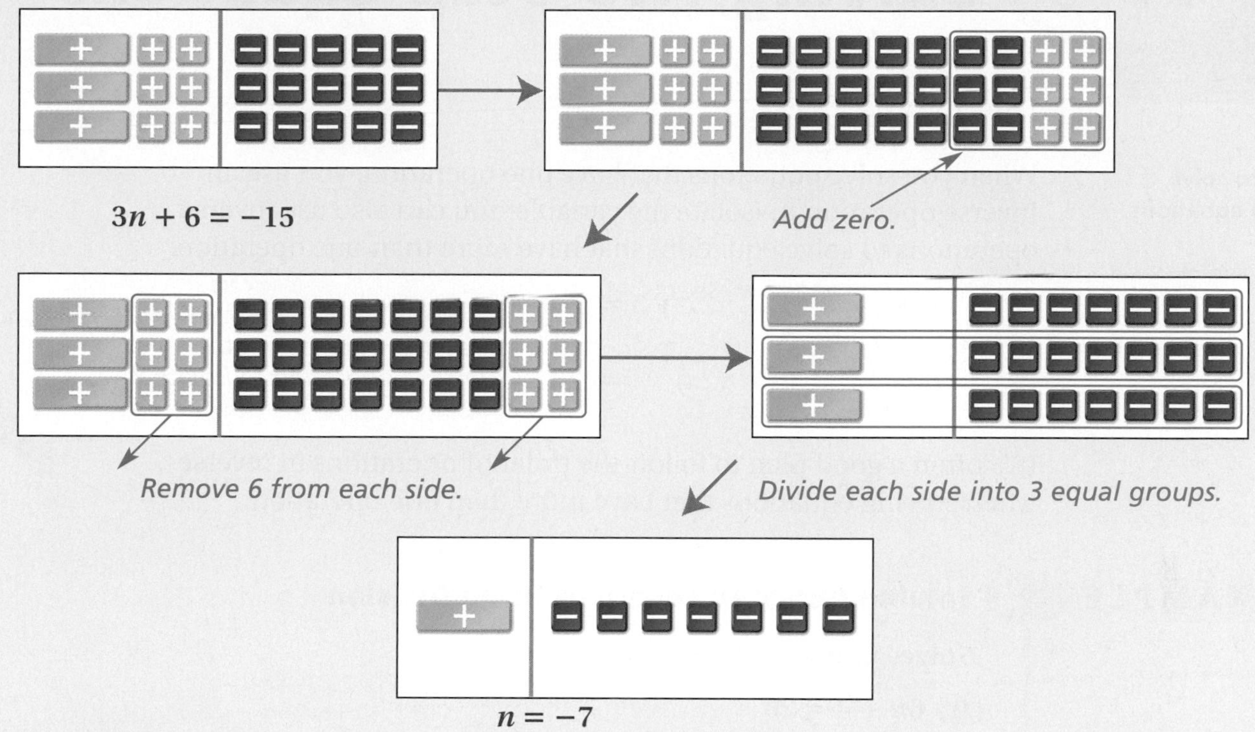

2 Use algebra tiles to model and solve each equation.

a. $2y - 4 = 10$ **b.** $3k + 3 = -12$ **c.** $-1 + 5m = 9$ **d.** $5 + 2k = -3$

Think and Discuss

1. When you subtract a value from one side of an equation, why do you also have to subtract the same value from the other side?
2. When you solved $3n + 6 = -15$ in the activity, why didn't you have to add six yellow unit tiles and six red unit tiles to the left side of the equation when you added them to the right side?
3. Model and solve $3x - 5 = 10$. Explain each step.
4. How would you check the solution to $3n + 6 = -15$ using algebra tiles?

Try This

Use algebra tiles to model and solve each equation.

1. $4 + 2x = 20$
2. $3r + 7 = -8$
3. $-4m + 3 = -25$
4. $-2n - 5 = 17$
5. $10 = 2j - 4$
6. $5 + r = 7$
7. $4h + 2h + 3 = 15$
8. $-3g = 9$
9. $5k + (-7) = 13$

11-1 Solving Two-Step Equations

Learn to solve two-step equations.

When you solve equations that have one operation, you use an inverse operation to isolate the variable. You can also use inverse operations to solve equations that have more than one operation.

$$\begin{array}{rcl} n + 7 &=& 15 \\ -7 & & -7 \\ \hline n &=& 8 \end{array} \qquad \begin{array}{rcl} 2x + 3 &=& 23 \\ -3 & & -3 \\ \hline \boxed{2x} &=& 20 \end{array}$$

You need to use another inverse operation to isolate x.

It is often a good plan to follow the order of operations in reverse when solving equations that have more than one operation.

EXAMPLE 1 Solving Two-Step Equations Using Division

Solve. Check each answer.

A $6n + 4 = 28$

$$\begin{array}{rcl} 6n + 4 &=& 28 \\ -4 & & -4 \\ \hline 6n &=& 24 \end{array}$$ *Subtract 4 from both sides.*

$$\frac{6n}{6} = \frac{24}{6}$$ *Divide both sides by 6.*

$$n = 4$$

Check

$$6n + 4 = 28$$

$$6(4) + 4 \stackrel{?}{=} 28$$ *Substitute 4 for n.*

$$24 + 4 \stackrel{?}{=} 28$$

$$28 \stackrel{?}{=} 28 \checkmark$$ *4 is a solution.*

B $-3p - 8 = 19$

$$\begin{array}{rcl} -3p - 8 &=& 19 \\ +8 & & +8 \\ \hline -3p &=& 27 \end{array}$$ *Add 8 to both sides.*

$$\frac{-3p}{-3} = \frac{27}{-3}$$ *Divide both sides by −3.*

$$p = -9$$

Check

$$-3p - 8 = 19$$

$$-3(-9) - 8 \stackrel{?}{=} 19$$ *Substitute −9 for p.*

$$27 - 8 \stackrel{?}{=} 19$$

$$19 \stackrel{?}{=} 19 \checkmark$$ *−9 is a solution.*

EXAMPLE 2 Solving Two-Step Equations Using Multiplication

Solve.

A $8 + \frac{j}{4} = 17$

$$\begin{aligned} 8 + \frac{j}{4} &= 17 \\ -8 \quad & \quad -8 \\ \frac{j}{4} &= 9 \\ (4)\frac{j}{4} &= (4)9 \\ j &= 36 \end{aligned}$$

Subtract 8 from both sides.

Multiply both sides by 4.

B $\frac{u}{6} - 12 = 3$

$$\begin{aligned} \frac{u}{6} - 12 &= 3 \\ +12 \quad & \quad +12 \\ \frac{u}{6} \qquad &= 15 \\ (6)\frac{u}{6} &= (6)15 \\ u &= 90 \end{aligned}$$

Add 12 to both sides.

Multiply both sides by 6.

EXAMPLE 3 *Consumer Math Application*

A new one-year membership at Workout Nation costs \$630. A registration fee of \$150 is paid up front, and the rest is paid monthly. How much do new members pay each month?

Let m represent the monthly cost.

$$\begin{aligned} 12m + 150 &= 630 \\ -150 \quad & \quad -150 \\ 12m \qquad &= 480 \\ \frac{12m}{12} &= \frac{480}{12} \\ m &= 40 \end{aligned}$$

Subtract 150 from both sides.

Divide both sides by 12.

New members pay \$40 per month for a one-year membership.

Think and Discuss

1. **Explain** how you decide which inverse operation to use first when solving a two-step equation.
2. **Tell** how each term in the equation in Example 3 relates to the word problem.

11-1 Exercises

5.03

FOR EOG PRACTICE
see page 702

internet connect
Homework Help Online
go.hrw.com Keyword: MS4 11-1

GUIDED PRACTICE

See Example 1 **Solve. Check each answer.**

1. $3n + 8 = 29$
2. $-4m - 7 = 17$
3. $-6x + 4 = 2$

See Example 2 **Solve.**

4. $12 + \frac{b}{6} = 16$
5. $\frac{y}{8} - 15 = 2$
6. $-8 + \frac{n}{4} = 10$

See Example 3

7. A coffee shop offers a souvenir ceramic mug filled with coffee for \$8.95. After that, refills cost \$1.50. Sandra spent \$26.95 on a mug and refills last month. How many refills did she buy?

INDEPENDENT PRACTICE

See Example 1 **Solve. Check each answer.**

8. $5x + 6 = 41$
9. $-9p - 15 = 93$
10. $-2m + 14 = 10$
11. $7d - 8 = -7$
12. $-3c + 14 = -7$
13. $12y - 11 = 49$

See Example 2 **Solve.**

14. $24 + \frac{h}{4} = 10$
15. $\frac{k}{5} - 13 = 4$
16. $-17 + \frac{q}{8} = 13$
17. $\frac{m}{10} + 32 = 24$
18. $15 + \frac{v}{3} = -9$
19. $\frac{m}{-7} - 14 = 2$

See Example 3

20. Every Saturday, Workout Nation holds a 45-minute aerobics class. Weekday aerobics classes last 30 minutes. The number of weekday classes varies. Last week there were a total of 165 minutes of aerobics classes available. How many weekday aerobics classes were there?

PRACTICE AND PROBLEM SOLVING

Translate each equation into words, and then solve the equation.

21. $6 + \frac{m}{3} = 18$
22. $3x + 15 = 27$
23. $\frac{n}{5} - 4 = 2$

Solve.

24. $18 + \frac{y}{4} = 12$
25. $5x + 30 = 40$
26. $\frac{s}{12} - 7 = 8$
27. $-10 + 6g = 110$
28. $\frac{z}{7} + 2 = -8$
29. $-6w - 8 = 46$
30. $-7 + \frac{r}{3} = 15$
31. $-4p - 12 = -20$
32. $\frac{1}{2} + \frac{r}{7} = \frac{5}{14}$

33. A long-distance phone company charges \$1.01 for the first 25 minutes of a call, and then \$0.09 for each additional minute. If a call cost \$9.56, how long did it last?

Health LINK

As a service to health-conscious customers, many grocery stores have installed scanners that calculate the total number of calories purchased.

34. If you double the number of calories per day that the U.S. Department of Agriculture recommends for children who are 1 to 3 years old and then subtract 100, you get the number of calories per day recommended for teenage boys. If 2,500 calories are recommended for teenage boys, how many calories should children consume?

35. According to the U.S. Department of Agriculture, children who are 4 to 6 years old need about 1,800 calories per day. This is 700 calories more than half the recommended calories for teenage girls. How many calories does a teenage girl need per day?

36. Hector consumed 2,130 calories from food in one day. Of these, he consumed 350 calories at breakfast and 400 calories having a snack. He also ate 2 portions of one of the items shown in the table for lunch and the same for dinner. What did Hector eat for lunch and dinner?

Calorie Counter

Food	Portion	Calories
Stir-fry	1 cup	250
Enchilada	1 whole	310
Pizza	1 slice	345
Tomato soup	1 cup	160

37. The U.S. Department of Agriculture recommends that the total amount of saturated fat in a diet not exceed 10% of calories. If a teenage girl consumes 2,200 calories per day, and a gram of fat has 9 calories, how many grams of fat can she consume without exceeding the recommended amounts?

38. ***CHALLENGE*** There are 30 mg of cholesterol in a box of macaroni and cheese. This is 77 mg minus $\frac{1}{10}$ the number of milligrams of sodium it contains. How many milligrams of sodium are in a box of macaroni and cheese?

Spiral Review

39. Caleb flips a dime and spins a spinner that is divided into fourths and numbered 1 through 4. How many outcomes are possible? (Lesson 10-3)

40. **EOG PREP** The volume of a cylinder with height 6 inches and radius 2 inches is approximately __?__ cubic inches. (Lesson 9-2)

A 18.84 B 37.68 C 75.36 D 226.08

41. **EOG PREP** The surface area of a cube with a side length of 4 cm is __?__ cm^2. (Lesson 9-4)

A 256 B 96 C 64 D 24

11-2 Solving Multistep Equations

Learn to solve multistep equations.

Levi has half as many comic books as Jamal has. If you add 6 to the number of comic books Jamal has and then divide by 7, you get the number of comic books Brooke has. If Brooke has 30 comic books, how many comic books does Levi have? To answer this question, you need to set up an equation that requires more than two steps to solve.

EXAMPLE 1 Solving Equations That Contain Like Terms

Solve $7n - 1 - 2n = 14$.

$$7n - 1 - 2n = 14$$

$$5n - 1 = 14 \quad \textit{Combine like terms.}$$

$$\underline{+1 \quad +1} \quad \textit{Add 1 to both sides.}$$

$$5n = 15$$

$$\frac{5n}{5} = \frac{15}{5} \quad \textit{Divide both sides by 5.}$$

$$n = 3$$

Sometimes one side of an equation has a variable expression as the numerator of a fraction. With this type of equation, it may help to first multiply both sides of the equation by the denominator.

EXAMPLE 2 Solving Equations That Contain Fractions

Solve $\frac{4x - 8}{9} = 12$.

$$\frac{4x - 8}{9} = 12$$

$$(9)\frac{4x - 8}{9} = (9)12 \quad \textit{Multiply both sides by 9.}$$

$$4x - 8 = 108$$

$$\underline{+8 \quad +8} \quad \textit{Add 8 to both sides.}$$

$$4x = 116$$

$$\frac{4x}{4} = \frac{116}{4} \quad \textit{Divide both sides by 4.}$$

$$x = 29$$

EXAMPLE 3 PROBLEM SOLVING APPLICATION

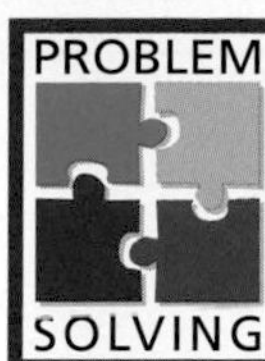

Jamal owns twice as many comic books as Levi owns. Adding 6 to the number of comic books Jamal owns and then dividing by 7 gives the number Brooke owns. If Brooke owns 30 comic books, how many does Levi own?

1 Understand the Problem

Rewrite the question as a statement.

- Find the number of comic books that Levi owns.

List the **important information:**

- Jamal owns 2 times as many comic books as Levi owns.
- The number of comic books Jamal owns added to 6 and then divided by 7 equals the number Brooke owns.
- Brooke owns 30 comic books.

2 Make a Plan

Let c represent the number of comic books Levi owns. Then $2c$ represents the number Jamal owns, and $\frac{2c+6}{7}$ represents the number Brooke owns, which equals 30. Solve the equation $\frac{2c+6}{7} = 30$ for c to find the number of comic books that Levi owns.

3 Solve

$$\frac{2c+6}{7} = 30$$

$$(7)\frac{2c+6}{7} = (7)30 \quad \textit{Multiply both sides by 7.}$$

$$2c + 6 - 6 = 210 - 6 \quad \textit{Subtract 6 from both sides.}$$

$$2c = 204$$

$$\frac{2c}{2} = \frac{204}{2} \quad \textit{Divide both sides by 2.}$$

$$c = 102$$

Levi owns 102 comic books.

4 Look Back

Check your answer by substituting 102 in the equation. $\frac{2(102)+6}{7} = 30$ ✔

Think and Discuss

1. **List** the steps required to solve $-n + 5n + 3 = 27$.
2. **Describe** how to solve the equations $\frac{2}{3}x + 7 = 4$ and $\frac{2x+7}{3} = 4$. Are the solutions the same or different? Explain.

11-2 Exercises

FOR EOG PRACTICE
see page 702

internet connect
Homework Help Online
go.hrw.com Keyword: MS4 11-2

5.03

GUIDED PRACTICE

See Example 1 **Solve.**

1. $14n + 2 - 7n = 37$ **2.** $10x - 11 - 4x = 43$ **3.** $-3 + 4p - 2p = 1$

See Example 2 **4.** $\frac{3m + 6}{4} = 33$ **5.** $\frac{5y - 6.6}{8} = 4.8$ **6.** $\frac{4m - 6}{6} = -\frac{2}{3}$

See Example 3 **7.** Keisha read twice as many books this year as Ben read. Subtracting 4 from the number of books Keisha read and dividing by 2 gives the number of books Sheldon read. If Sheldon read 10 books this year, how many books did Ben read?

INDEPENDENT PRACTICE

See Example 1 **Solve.**

8. $b + 18 + 3b = 74$ **9.** $10x - 3 - 2x = 4$

10. $18w - 10 - 6w = 50$ **11.** $5n + 7 - 3n = 19$

12. $-3p + 15 - 3p = -27$ **13.** $-x - 8 + 14x = -34$

See Example 2 **14.** $\frac{2x + 4}{12} = \frac{5}{6}$ **15.** $\frac{3n + 5}{8} = 4$

16. $\frac{9p - 5.2}{4} = 9.95$ **17.** $\frac{4r - 8}{10} = 48$

18. $\frac{7k + 13}{-18} = \frac{4}{9}$ **19.** $\frac{19 - 11t}{13} = 4$

See Example 3 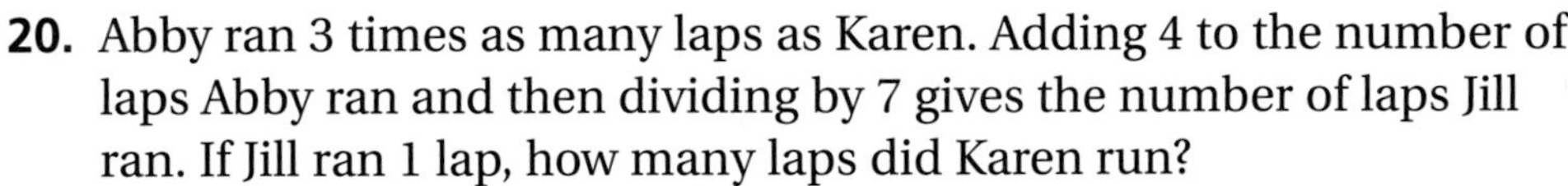
20. Abby ran 3 times as many laps as Karen. Adding 4 to the number of laps Abby ran and then dividing by 7 gives the number of laps Jill ran. If Jill ran 1 lap, how many laps did Karen run?

PRACTICE AND PROBLEM SOLVING

Solve.

21. $\frac{0.5x + 7}{8} = 5$ **22.** $\frac{8t - 4}{5} = 6$ **23.** $8w + 2.6 - 3.6 = 63$

24. $\frac{3w + 0.5}{2} = 1$ **25.** $\frac{\frac{1}{4}a - 12}{8} = 4$ **26.** $1.8 + 6n - 3.2 = 7.6$

27. $\frac{2b - 3.4}{0.6} = -29$ **28.** $\frac{34.6 + 4h}{5} = 8.44$ **29.** $-2.5x + 18 - 1.6x = 5.7$

30. Three friends ate dinner at a restaurant. The total bill for their dinner was $27.00. The friends decided to add a 15% tip and then split the bill evenly. How much did each friend pay?

31. Ann earns 1.5 times her normal hourly pay for each hour that she works over 40 hours in a week. Last week she worked 51 hours and earned $378.55. What is her normal hourly pay?

32. ***CONSUMER MATH*** Patrice used a \$15 gift certificate when she purchased a pair of sandals. After 8% sales tax was applied to the price of the sandals, the \$15 was deducted. If Patrice had to pay an additional \$12, how much did the sandals cost before tax?

33. ***PHYSICAL SCIENCE*** To convert temperatures between degrees Celsius and degrees Fahrenheit, you can use the formula $F = \frac{9}{5}C + 32$. The table shows the melting points of various elements.

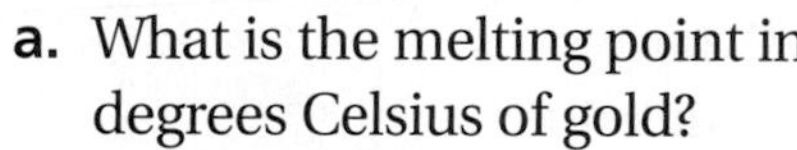

   a. What is the melting point in degrees Celsius of gold?

   b. What is the melting point in degrees Celsius of hydrogen?

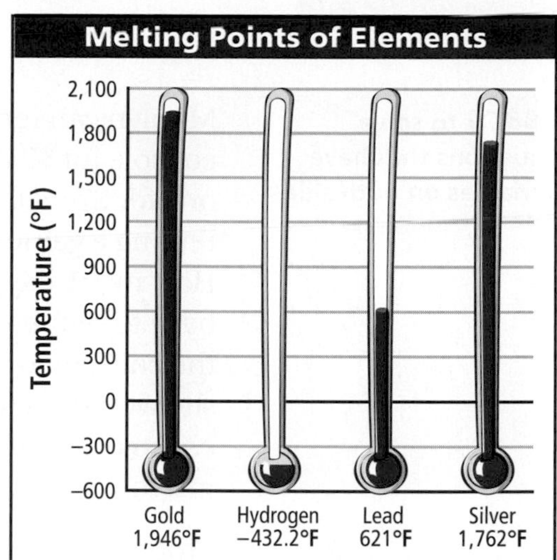

34. On his first two social studies tests, Billy made an 86 and a 93. What grade must Billy make on the third test to have an average of 90 for all three tests?

35. ***WHAT'S THE QUESTION?*** Three friends shared a taxi ride from the airport to their hotel. After adding a \$7.00 tip, the friends divided the cost of the ride evenly. If solving the equation $\frac{c + \$7.00}{3} = \$11.25$ gives the answer, what is the question?

36. ***WRITE ABOUT IT*** Explain why multiplying first in the equation $\frac{2x - 6}{5} = 2$ makes finding the solution easier than adding first does.

37. ***CHALLENGE*** Are the solutions to the following equations the same? Explain.

$$\frac{3y}{4} + 2 = 4 \text{ and } 3y + 8 = 16$$

Spiral Review

Solve. Write each answer in simplest form. (Lesson 4-12)

38. $\frac{3}{4}x = 9$ **39.** $y + \frac{7}{11} = \frac{2}{3}$ **40.** $\frac{5}{6}k = \frac{5}{9}$ **41.** $\frac{3}{8} + n = -\frac{11}{24}$

Find the number of permutations for each situation. (Lesson 10-7)

42. There are 10 students lining up for lunch.

43. Dan is putting 8 books on a shelf.

44. **EOG PREP** Which of the ratios are proportional? (Lesson 5-2)

A $\frac{2}{6}, \frac{1}{12}$ B $\frac{15}{36}, \frac{18}{45}$ C $\frac{3}{12}, \frac{12}{36}$ D $\frac{12}{36}, \frac{60}{180}$

45. **EOG PREP** A spinner that is divided into ten sections numbered 1–10 is spun. What is the probability of landing on a prime number? (Lesson 10-2)

A 40% B 0.1 C $\frac{1}{4}$ D 0.06

11-3 Solving Equations with Variables on Both Sides

Learn to solve equations that have variables on both sides.

Marilyn can rent a video game console for \$2.20 per day or buy one for \$74.80. The cost of renting a game is \$1.99 per day. How many days would Marilyn have to rent both the game and the console to pay as much as she would if she had bought the console and rented the game instead?

Problems such as this require you to solve equations that have the same variable on both sides of the equal sign. To solve this kind of problem, you need to get the terms with variables on one side of the equal sign.

EXAMPLE 1 Using Inverse Operations to Group Terms with Variables

Group the terms with variables on one side of the equal sign, and simplify.

A $\mathbf{6m = 4m + 12}$

$6m = 4m + 12$

$6m - \mathbf{4m} = 4m - \mathbf{4m} + 12$ — *Subtract 4m from both sides.*

$2m = 12$ — *Simplify.*

B $\mathbf{-7x - 198 = 5x}$

$-7x - 198 = 5x$

$-7x + \mathbf{7x} - 198 = 5x + \mathbf{7x}$ — *Add 7x to both sides.*

$-198 = 12x$ — *Simplify.*

EXAMPLE 2 Solving Equations with Variables on Both Sides

Solve.

A $\mathbf{5n = 3n + 26}$

$5n = 3n + 26$

$5n - \mathbf{3n} = 3n - \mathbf{3n} + 26$ — *Subtract 3n from both sides.*

$2n = 26$ — *Simplify.*

$\frac{2n}{2} = \frac{26}{2}$ — *Divide both sides by 2.*

$n = 13$

Solve.

B $19 + 7n = -2n + 37$

$19 + 7n = -2n + 37$	
$19 + 7n + \mathbf{2n} = -2n + \mathbf{2n} + 37$	*Add 2n to both sides.*
$19 + 9n = 37$	*Simplify.*
$19 + 9n - \mathbf{19} = 37 - \mathbf{19}$	*Subtract 19 from both sides.*
$9n = 18$	*Simplify.*
$\frac{9n}{\mathbf{9}} = \frac{18}{\mathbf{9}}$	*Divide both sides by 9.*
$n = 2$	

C $\frac{5}{9}x = \frac{4}{9}x + 9$

$\frac{5}{9}x = \frac{4}{9}x + 9$	
$\frac{5}{9}x - \mathbf{\frac{4}{9}x} = \frac{4}{9}x - \mathbf{\frac{4}{9}x} + 9$	*Subtract $\frac{4}{9}x$ from both sides.*
$\frac{1}{9}x = 9$	*Simplify.*
$(\mathbf{9})\frac{1}{9}x = (\mathbf{9})9$	*Multiply both sides by 9.*
$x = 81$	

EXAMPLE 3 *Consumer Math Application*

Marilyn can buy a video game console for $74.80 and rent a game for $1.99 per day, or she can rent a console and the same game for a total of $4.19 per day. How many days would Marilyn have to rent both the video game and the console to pay as much as she would if she had bought the console and rented the game instead?

Let d represent the number of days.

$4.19d = 74.80 + 1.99d$	
$4.19d - \mathbf{1.99d} = 74.80 + 1.99d - \mathbf{1.99d}$	*Subtract 1.99d from both sides.*
$2.20d = 74.80$	*Simplify.*
$\frac{2.20d}{\mathbf{2.20}} = \frac{74.80}{\mathbf{2.20}}$	*Divide both sides by 2.20.*
$d = 34$	

Marilyn would need to rent both the video game and the console for 34 days to pay as much as she would have if she had bought the console.

Think and Discuss

1. **Explain** how you would solve $\frac{1}{2}x + 7 = \frac{2}{3}x - 2$.
2. **Describe** how you would decide which variable term to add or subtract on both sides of the equation $-3x + 7 = 4x - 9$.

11-3 Exercises

FOR EOG PRACTICE

see page 703

internet connect

Homework Help Online
go.hrw.com Keyword: MS4 11-3

5.03

GUIDED PRACTICE

See Example 1 **Group the terms with variables on one side of the equal sign, and simplify.**

1. $5n = 4n + 32$ **2.** $-6x - 28 = 4x$ **3.** $8w = 32 - 4w$

See Example 2 **Solve.**

4. $4y = 2y + 40$ **5.** $8 + 6a = -2a + 24$ **6.** $\frac{3}{4}d + 4 = \frac{1}{4}d + 18$

See Example 3 **7.** Members at the Star Theater pay \$30.00 per month plus \$1.95 for each movie. Nonmembers pay the regular \$7.95 admission fee. How many movies would both a member and a nonmember have to see in a month to pay the same amount?

INDEPENDENT PRACTICE

See Example 1 **Group the terms with variables on one side of the equal sign, and simplify.**

8. $12h = 9h + 84$ **9.** $-10p - 8 = 2p$ **10.** $6q = 18 - 2q$

11. $-4c - 6 = -2c$ **12.** $-7s + 12 = -9s$ **13.** $6 + \frac{4}{5}a = \frac{9}{10}a$

See Example 2 **Solve.**

14. $9t = 4t + 120$ **15.** $42 + 3b = -4b - 14$

16. $\frac{6}{11}x + 4 = \frac{2}{11}x + 16$ **17.** $1.5a + 6 = 9a + 12$

18. $32 - \frac{3}{8}y = \frac{3}{4}y + 5$ **19.** $-6 - 8c = 3c + 16$

See Example 3 **20.** Members at a swim club pay \$5 per lesson plus a one-time membership fee of \$60. Nonmembers pay \$11 per lesson. How many lessons would both a member and a nonmember have to take to pay the same amount?

PRACTICE AND PROBLEM SOLVING

Solve. Check each answer.

21. $3y + 7 = -6y - 56$ **22.** $-\frac{7}{8}x - 6 = -\frac{3}{8}x - 14$

23. $5r + 6 - 2r = 7r - 10$ **24.** $-10p + 8 = 7p + 12$

25. $9 + 5r = -17 - 8r$ **26.** $0.8k + 7 = -0.7k + 1$

27. A choir is performing for the school fall festival. On the first night, 12 choir members were absent, so the choir stood in 5 equal rows. On the second night, only 1 choir member was absent, so the choir stood in 6 equal rows. How many members are in the choir?

The figures in each pair have the same perimeter. Find the value of each variable.

28. Rectangle: x, $x + 4$; Triangle: x, $x + 9$, $x + 5$

29. Rectangle: $s + 7$, $3s$; Triangle: $2s + 12$, $2s + 12$, $2s + 12$

Recreation LINK

The first indoor rock-climbing gym was built in 1987. Today, hundreds of these kinds of gyms exist around the country.

30. ***RECREATION*** A rock-climbing gym charges nonmembers \$18 per day to use the wall plus \$7 per day for equipment rental. Members pay an annual fee of \$400 plus \$5 per day for equipment rental. How many days must both a member and a nonmember use the wall in one year so that both pay the same amount?

31. ***SOCIAL STUDIES*** Two families drove from Denver to Cincinnati. After driving 582 miles the first day, the Smiths spread the rest of the trip equally over the next 3 days. The Chows spread their trip equally over 6 days. The distance the Chows drove each day was equal to the distance the Smiths drove each of the three days.

a. How many miles did the Chows drive each day?

b. How far is it from Denver to Cincinnati?

32. ***WHAT'S THE ERROR?*** To combine terms in the equation $-8a - 4 = 2a + 34$, a student wrote $-6a = 38$. What is the error?

33. ***WRITE ABOUT IT*** If the same variable is on both sides of an equation, must it have the same value on each side? Explain your answer.

34. ***CHALLENGE*** Combine terms before solving the equation $12x - 4 - 12 = 4x + 8 + 8x - 24$. Do you think there is just one solution to the equation? Why or why not?

Spiral Review

Estimate each percent. (Lesson 6-2)

35. 52% of 62 **36.** 31% of 47 **37.** 9% of 87 **38.** 23% of 79

Convert. (Lesson 8-1)

39. 24 feet to inches **40.** 442 milliliters to liters **41.** 4 quarts to cups

42. **EOG PREP** Which number is the solution to the equation $-13 = 14 + x$? (Lesson 2-10)

A 1 B -27 C -1 D 27

43. **EOG PREP** For a circle with diameter 5 inches, what is the area to the nearest tenth? Use 3.14 for π. (Lesson 8-6)

A 15.7 in^2 B 78.5 in^2 C 19.6 in^2 D 7.9 in^2

Chapter 11

Mid-Chapter Quiz

LESSON 11-1 (pp. 560–563)

Solve.

1. $-4x + 6 = 54$
2. $15 + \frac{y}{3} = 6$
3. $\frac{z}{8} - 5 = -3$
4. $-7a - 5 = -33$
5. $\frac{r}{12} - 19 = -27$
6. $11 - 2n = -13$
7. $3x + 13 = 37$
8. $\frac{p}{-8} - 7 = 12$
9. $\frac{u}{7} + 45 = -60$
10. A taxi service charges an initial fee of \$1.50 plus \$1.50 for every mile traveled. If a taxi ride costs \$21.00, how many miles did the taxi travel?

LESSON 11-2 (pp. 564–567)

Solve.

11. $\frac{3x - 4}{5} = 7$
12. $9b + 6 - 8 = -38$
13. $\frac{15c + 3}{6} = -12$
14. $\frac{24.6 + 3a}{4} = 9.54$
15. $\frac{2b + 9}{11} = 18$
16. $2c + 3 + 5c = 13$
17. $\frac{16w - 12}{4} = 17$
18. $\frac{1.2s + 3.69}{0.3} = 47.9$
19. $\frac{5p - 8}{12} = \frac{1}{2}$
20. Peter used a \$5.00 gift certificate to help pay for his lunch. After adding a 15% tip to the cost of his meal, Peter still had to pay \$2.36 in cash. How much did Peter's meal cost?
21. A group of 10 friends had dinner together at a restaurant. The meal cost a total of \$99.50, including tax. After a 15% tip was added, they split the cost evenly. How much did each person pay? Round your answer to the nearest cent.

LESSON 11-3 (pp. 568–571)

Solve.

22. $12m = 3m + 108$
23. $\frac{7}{8}n - 3 = \frac{5}{8}n + 12$
24. $1.2x + 3.7 = 2.2x - 4.5$
25. $-7 - 7p = 3p + 23$
26. $-2.3q + 16 = -5q - 38$
27. $\frac{3}{5}k + \frac{7}{10} = \frac{11}{15}k - \frac{2}{5}$
28. $-19m + 12 = -14m - 8$
29. $\frac{2}{3}v + \frac{1}{6} = \frac{7}{9}v - \frac{5}{6}$
30. $8.9 - 3.3j = -2.2j + 2.3$
31. $4a - 7 = -6a + 12$
32. Nine more than 3 times a number is equal to 8 times the number decreased by 16. What is the number?

Focus on Problem Solving

Solve

• **Write an equation**

When you are asked to solve a problem, be sure to read the entire problem before you begin solving it. Sometimes you will need to perform several steps to solve the problem, and you will need to know all of the information in the problem before you decide which steps to take.

Read each problem and determine what steps are needed to solve it. Then write an equation that can be used to solve the problem.

1. Martin can buy a pair of inline skates and safety equipment for \$49.50. At a roller rink, Martin can rent a pair of inline skates for \$2.50 per day, but he still needs to buy safety equipment for \$19.50. How many days would Martin have to skate to pay as much to rent skates and buy safety equipment as he would have to buy both?

2. Christopher draws caricatures at the local mall. He charges \$5 for a simple sketch and \$15 for a larger drawing. In one day, Christopher earned \$175. If he drew 20 simple sketches that day, how many larger drawings did he make?

3. Coach Willis has won 150 games during his career as a baseball coach. This is 10 more than $\frac{1}{2}$ as many games as Coach Gentry has won. How many games has Coach Gentry won?

4. Book-club members are required to buy a minimum number of books each year. Leslee bought 3 times the minimum. Denise bought 7 more than the minimum. Together, they bought 23 books. What is the minimum number of books?

5. The perimeter of an isosceles triangle is 4 times the length of the shortest side. The longer sides are 4.5 ft longer than the shortest side. What is the length of each side of the triangle?

6. Miss Rankin's class is raising money for a class trip. The class has \$100.00 so far and needs to collect a total of \$225.00. How many \$0.50 carnations must the class sell to reach its goal?

11-4 Inequalities

Learn to read and write inequalities and graph them on a number line.

Vocabulary

inequality
algebraic inequality
solution set
compound inequality

An **inequality** states that two quantities either are not equal or may not be equal. An inequality uses one of the following symbols:

Symbol	Meaning	Word Phrases
$<$	Is less than	Fewer than, below
$>$	Is greater than	More than, above
$\leq$	Is less than or equal to	At most, no more than
$\geq$	Is greater than or equal to	At least, no less than

EXAMPLE 1 Writing Inequalities

Write an inequality for each situation.

A **There are at least 25 students in the auditorium.**

number of students ≥ 25 — *"At least" means greater than or equal to.*

B **No more than 150 people can occupy the room.**

room capacity ≤ 150 — *"No more than" means less than or equal to.*

An inequality that contains a variable is an **algebraic inequality**. A value of the variable that makes the inequality true is a solution of the inequality.

An inequality may have more than one solution. Together, all of the solutions are called the **solution set**.

You can graph the solutions of an inequality on a number line. If the variable is "greater than" or "less than" a number, then that number is indicated with an open circle.

This open circle shows that 5 is not a solution.

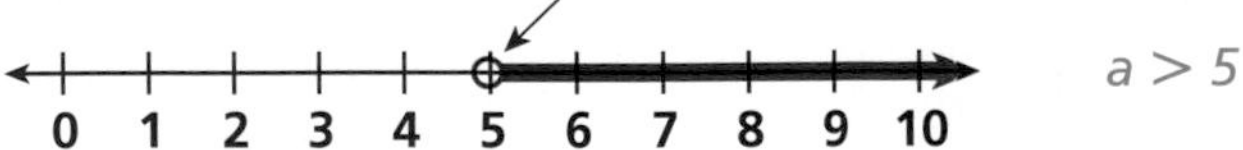

$a > 5$

If the variable is "greater than or equal to" or "less than or equal to" a number, that number is indicated with a closed circle.

This closed circle shows that 3 is a solution.

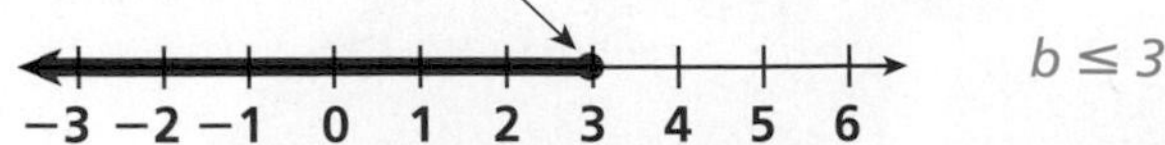

$b \leq 3$

EXAMPLE 2 Graphing Simple Inequalities

Graph each inequality.

A $x > -2$

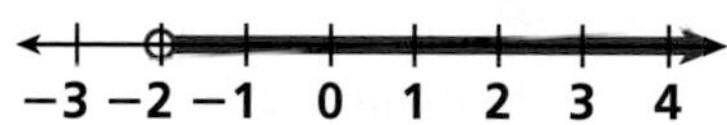

−2 is not a solution, so draw an open circle at −2. Shade the line to the right of −2.

B $y \leq -1$

−5 −4 −3 −2 −1 0 1 2 3

−1 is a solution, so draw a closed circle at −1. Shade the line to the left of −1.

> **Writing Math**
>
> The compound inequality $-2 < y$ and $y < 4$ can be written as $-2 < y < 4$.

A **compound inequality** is the result of combining two inequalities. The words *and* and *or* are used to describe how the two parts are related.

$x > 3$ or $x < -1$	$-2 < y$ and $y < 4$
x is either greater than 3 or less than −1.	*y is both greater than −2 and less than 4. y is between −2 and 4.*

EXAMPLE 3 Graphing Compound Inequalities

Graph each compound inequality.

A $s \geq 0$ or $s < -3$

First graph each inequality separately.

$s \geq 0$

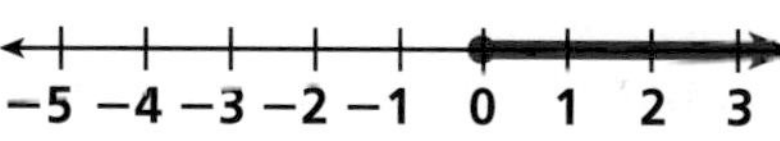

$s < -3$

−5 −4 −3 −2 −1 0 1 2 3

Then combine the graphs.

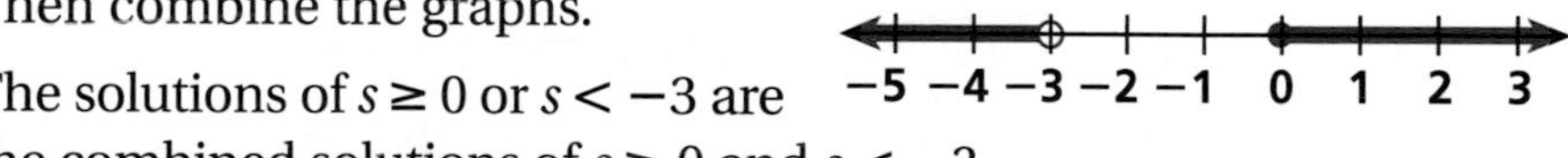

The solutions of $s \geq 0$ or $s < -3$ are the combined solutions of $s \geq 0$ and $s < -3$.

B $1 < p \leq 5$

$1 < p \leq 5$ can be written as $1 < p$ and $p \leq 5$. Graph each inequality.

$1 < p$

−2 −1 0 1 2 3 4 5 6

$p \leq 5$

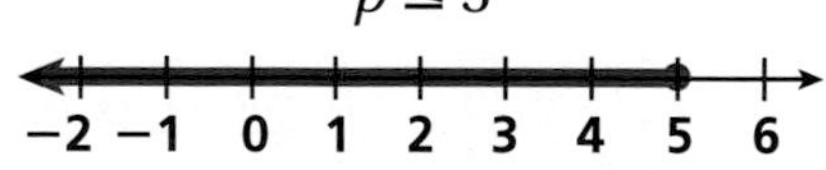

Then combine the graphs.

−2 −1 0 1 2 3 4 5 6

The solutions of $1 < p \leq 5$ are the solutions common to $1 < p$ and $p \leq 5$.

Think and Discuss

1. **Compare** the graphs of the inequalities $y > 2$ and $y \geq 2$.
2. **Explain** how to graph each type of compound inequality.

11-4 Exercises

FOR EOG PRACTICE
see page 704

internet connect
Homework Help Online
go.hrw.com Keyword: MS4 11-4

5.02, 5.03

GUIDED PRACTICE

See Example 1 **Write an inequality for each situation.**

1. No more than 18 people are allowed in the gallery at one time.
2. There are fewer than 8 fish in the aquarium.
3. The water level is above 45 inches.

See Example 2 **Graph each inequality.**

4. $x < 3$ 5. $r \leq \frac{1}{2}$ 6. $w > 2.8$ 7. $y \geq -4$

See Example 3 **Graph each compound inequality.**

8. $a > 2$ or $a \leq -1$ 9. $-4 < p \leq 6$ 10. $-2 \leq n < 0$

INDEPENDENT PRACTICE

See Example 1 **Write an inequality for each situation.**

11. The temperature is below 40°F.
12. There are at least 24 pictures on the roll of film.
13. No more than 35 tables are in the cafeteria.
14. Fewer than 250 people attended the rally.

See Example 2 **Graph each inequality.**

15. $s \geq -1$ 16. $y < 0$ 17. $n \leq -3$
18. $x > 2$ 19. $b \geq -6$ 20. $m < -4$

See Example 3 **Graph each compound inequality.**

21. $p > 3$ or $p < 0$ 22. $1 \leq x \leq 4$ 23. $-3 < y < -1$
24. $k > 0$ or $k \leq -2$ 25. $n \geq 1$ or $n \leq -1$ 26. $-2 < w \leq 2$

PRACTICE AND PROBLEM SOLVING

Write each statement using inequality symbols.

27. The number c is between -2 and 3.
28. The number y is greater than -10.

Write an inequality shown by each graph.

29. −4 −3 −2 −1 0 1 2 3 4

30. 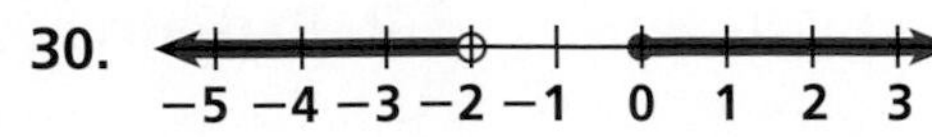

Earth Science 

A continental margin is divided into the continental shelf, the continental slope, and the continental rise.

31. The continental shelf begins at the shoreline and slopes toward the open ocean. The depth of the continental shelf can reach 200 meters. Write a compound inequality for the depth of the continental shelf.

32. The continental slope begins at the edge of the continental shelf and continues down to the flattest part of the ocean floor. The depth of the continental slope ranges from about 200 meters to about 4,000 meters. Write a compound inequality for the depth of the continental slope.

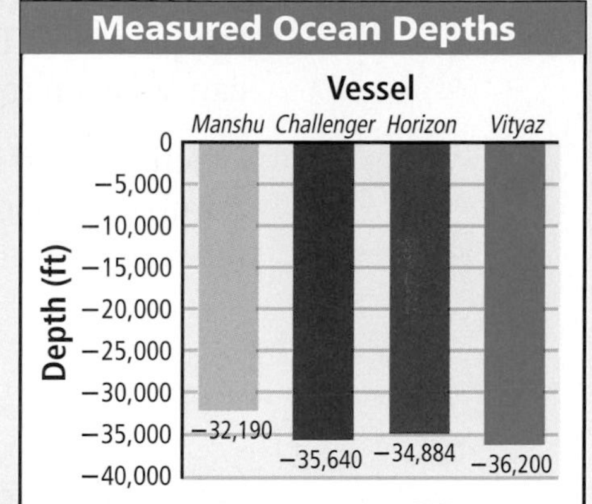

33. The bar graph shows the depth of the ocean in various locations as measured by different research vessels. Write a compound inequality that shows the ranges of depth measured by each vessel.

34. A submarine's *crush depth* is the depth at which water pressure will cause the submarine to collapse. A certain submarine has a crush depth of 2,250 feet. Write a compound inequality that shows at what depths this submarine can travel.

35. ***CHALLENGE*** Water freezes at 32°F and boils at 212°F. Write three inequalities to show the ranges of temperatures for which water is a solid, a liquid, and a gas.

Deep Flight is designed to explore the ocean in underwater flights.

Spiral Review

A 1–6 number cube is rolled. Find each probability. Write your answer as a fraction, as a decimal, and as a percent. (Lesson 10-4)

36. $P(3)$ **37.** $P(\text{even number})$ **38.** $P(12)$ **39.** $P(2 \text{ or } 3)$

40. **EOG PREP** What is the volume of a cylinder with radius 2 m and height 5 m? Use 3.14 for π. (Lesson 9-2)

A 62.8 m^3 B 98.596 m^3 C 157 m^3 D 31.4 m^3

41. **EOG PREP** Which is a solution of $2x + 7 = 91$? (Lesson 11-1)

A $x = 49$ B $x = 42$ C $x = 168$ D $x = 196$

11-5 Solving Inequalities by Adding or Subtracting

Learn to solve one-step inequalities by adding or subtracting.

Monday's high temperature was 32°F. The weather forecaster predicted a high of at least 71°F on Tuesday. At least how many degrees warmer is Tuesday's forecasted high temperature than Monday's high temperature? To find the answer, you can solve an inequality.

When you add or subtract the same number on both sides of an inequality, the resulting statement will still be true.

$$\begin{array}{rcr} -2 & < & 5 \\ \underline{+\,7} & & \underline{+\,7} \\ 5 & < & 12 \end{array}$$

You can find solution sets of inequalities the same way you find solutions of equations, by isolating the variable.

EXAMPLE 1 Solving Inequalities by Adding

Solve. Then graph each solution set on a number line.

A $x - 12 > 32$

$$\begin{array}{rcr} x - 12 & > & 32 \\ \underline{+\,12} & & \underline{+\,12} \\ x & > & 44 \end{array}$$

Add 12 to both sides.

0 11 22 33 44 55 66 77 88

44 is not a solution, so draw an open circle at 44. Then shade the line to include values greater than 44.

B $y - 8 \le -14$

$$\begin{array}{rcr} y - 8 & \le & -14 \\ \underline{+\,8} & & \underline{+\,8} \\ y & \le & -6 \end{array}$$

Add 8 to both sides.

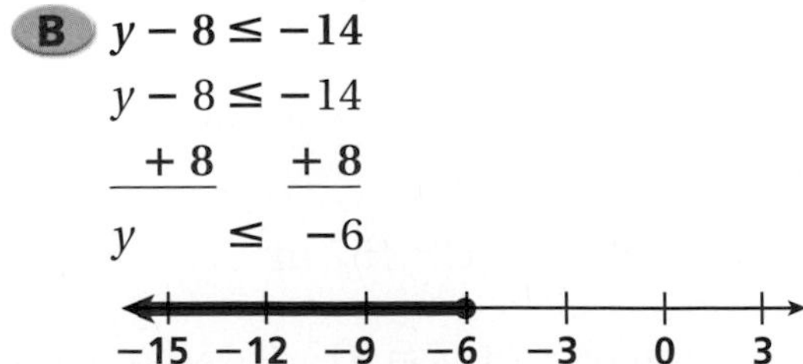

−6 is a solution, so draw a closed circle at −6. Then shade the line to include values less than −6.

You can see if the solution to an inequality is true by choosing any number in the solution set and substituting it into the original inequality.

EXAMPLE 2 Solving Inequalities by Subtracting

Solve. Check each answer.

A $c + 9 < 20$

$$\begin{aligned} c + 9 &< 20 \\ \underline{-9} &\quad \underline{-9} \\ c &< 11 \end{aligned}$$ *Subtract 9 from both sides.*

Check

$$c + 9 < 20$$
$$10 + 9 \overset{?}{<} 20$$ *10 is less than 11. Substitute 10 for c.*
$$19 \overset{?}{<} 20 \checkmark$$

B $x + 16 > -2$

$$\begin{aligned} x + 16 &> -2 \\ \underline{-16} &\quad \underline{-16} \\ x &> -18 \end{aligned}$$ *Subtract 16 from both sides.*

Check

$$x + 16 > -2$$
$$0 + 16 \overset{?}{>} -2$$ *0 is greater than −18. Substitute 0 for x.*
$$16 \overset{?}{>} -2 \checkmark$$

Helpful Hint

When checking your solution, choose a number in the solution set that is easy to work with.

EXAMPLE 3 *Meteorology Application*

Monday's high temperature was 32°F. The weather forecast for Tuesday includes a high temperature of at least 71°F. At least how many degrees warmer is the forecasted temperature for Tuesday than Monday's high temperature?

Let t represent the temperature increase from Monday to Tuesday.

$$\begin{aligned} 32 + t &\geq 71 \\ \underline{-32} &\quad \underline{-32} \\ t &\geq 39 \end{aligned}$$ *Subtract 32 from both sides.*

The temperature forecasted for Tuesday is at least 39°F warmer than the high temperature on Monday.

Think and Discuss

1. **Compare** solving addition and subtraction equations with solving addition and subtraction inequalities.
2. **Describe** how to check whether −36 is a solution of $s - 5 > 1$.

11-5 Exercises

FOR EOG PRACTICE

see page 704

internet connect

Homework Help Online
go.hrw.com Keyword: MS4 11-5

5.03

GUIDED PRACTICE

See Example 1 **Solve. Then graph each solution set on a number line.**

1. $x - 9 < 18$
2. $y - 11 \geq -7$
3. $p - 3 \leq 4$

See Example 2 **Solve. Check each answer.**

4. $n + 5 > 26$
5. $b + 21 \leq -3$
6. $12 + k \geq 9$

See Example 3

7. Yesterday's high temperature was 30°F. Tomorrow's weather forecast includes a high temperature of at most 42°F. At most, how many degrees warmer is the high-temperature forecast for tomorrow than yesterday's high temperature?

INDEPENDENT PRACTICE

See Example 1 **Solve. Then graph each solution set on a number line.**

8. $s - 2 > 14$
9. $m - 14 < -3$
10. $b - 25 > -30$
11. $c - 17 \leq -6$
12. $y - 53 < -25$
13. $x - 9 \geq 71$

See Example 2 **Solve. Check each answer.**

14. $w + 16 < 4$
15. $z + 9 > -3$
16. $p + 21 \leq -4$
17. $f + 32 > 26$
18. $k + 54 < 65$
19. $n + 29 \geq 25$

See Example 3

20. Clark scored at least 12 points more than Josh scored. If Josh scored 15 points, at least how many points did Clark score?
21. Adriana counted 8 fewer birds on Tuesday than on Thursday. She counted at most 32 birds on Thursday. At most, how many birds did she count on Tuesday?

PRACTICE AND PROBLEM SOLVING

Solve.

22. $k + 3.2 \geq 8$
23. $a - 1.3 > -1$
24. $c - 6\frac{1}{2} < -1\frac{1}{4}$
25. $18 + m \leq -20$
26. $x + 7.02 > 4$
27. $g + 3\frac{2}{3} < 10$
28. $r - 58 < -109$
29. $5.9 + w \leq 21.6$
30. $n - 21.6 > 26$
31. $t + 92 \geq -150$
32. $y + 4\frac{3}{4} \geq 1\frac{1}{8}$
33. $v - 0.9 \leq -1.5$
34. ***CONSUMER MATH*** To get a group discount for baseball tickets, Marco's group must have at least 20 people. So far, 13 people have signed up for tickets. At least how many more people must sign up for the group to get a discount?

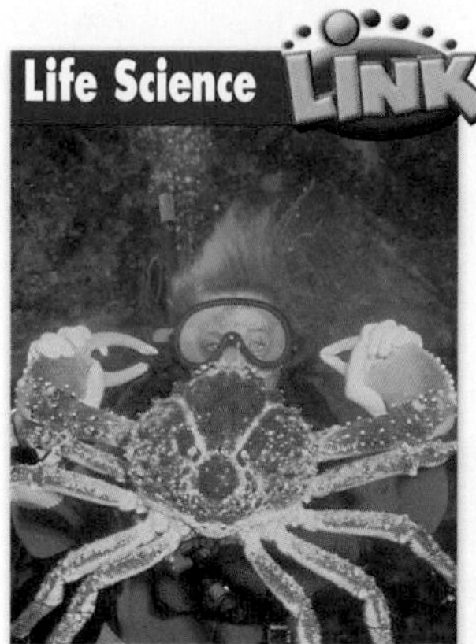

One of the largest giant spider crabs on record weighed 41 lb and had a claw span of 12 ft 2 in.

go.hrw.com
KEYWORD: MS4 Crab
CNN Student News

35. ***LIFE SCIENCE*** The giant spider crab, the world's largest crab, lives off the southeastern coast of Japan. Giant spider crabs can grow to be more than 3.6 meters across. A scientist measures a giant spider crab that is 3.1 meters across. At least how many more meters might the crab grow?

36. The *shinkansen*, or bullet train, of Japan travels at an average speed of 162.3 miles per hour. It has a top speed of 186 miles per hour. At most, how many more miles per hour can the train travel beyond its average speed before it reaches its maximum speed?

37. The line graph shows the number of miles Amelia rode her bike in each of the last four months. She wants to ride at least 5 miles more in May than she did in April. At least how many miles does Amelia want to ride in May?

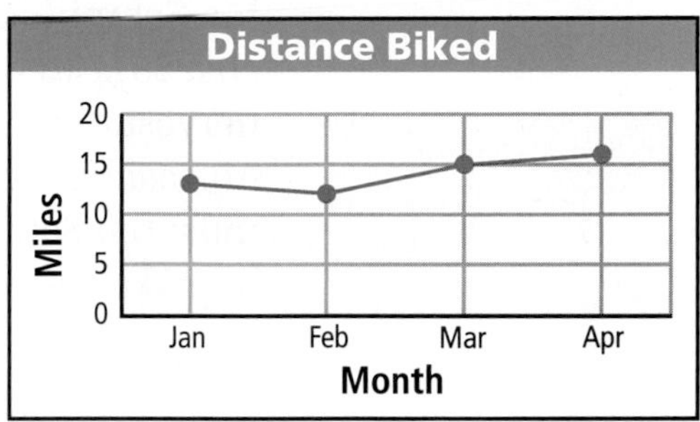

38. ***PHYSICAL SCIENCE*** The average human ear can detect sounds that have frequencies between 20 hertz and 20,000 hertz. The average dog ear can detect sounds with frequencies of up to 30,000 hertz greater than those a human ear can detect. Up to how many hertz can a dog hear?

39. ***LIFE SCIENCE*** Cheetahs have been known to run at speeds of more than 105 km/h for short bursts. A cheetah in a nature preserve is clocked running at 92.6 km/h. How many more kilometers per hour might an even faster cheetah run?

40. ***CHOOSE A STRATEGY*** If five days ago was the day after Saturday, what was the day before yesterday?

41. ***WRITE ABOUT IT*** Explain how to solve and check the inequality $n - 9 < -15$.

42. ***CHALLENGE*** Solve the inequality $x + (4^2 - 2^3)^2 > -1$.

Spiral Review

The surface area of a prism is 16 in^2. Find the surface area of a larger, similarly shaped prism that has each scale factor. (Lesson 9-5)

43. scale factor = 3
44. scale factor = 8
45. scale factor = 10

46. **EOG PREP** Amber rolls two different-colored 1–6 number cubes at the same time. How many outcomes are possible? (Lesson 10-3)

A 6 B 12 C 24 D 36

47. **EOG PREP** Which graph is a solution of $-3 < x \leq 1$? (Lesson 11-4)

A −4 −2 0 2
B −4 −2 0 2
C −2 0 2
D −4 −2 0 2

11-6 Solving Inequalities by Multiplying or Dividing

Learn to solve one-step inequalities by multiplying or dividing.

During the spring, the Schmidt family sells watermelons at a roadside stand for $5 apiece. Mr. Schmidt calculated that it cost $517 to plant, grow, and harvest the melons this year. At least how many melons must the Schmidts sell in order to make a profit for the year?

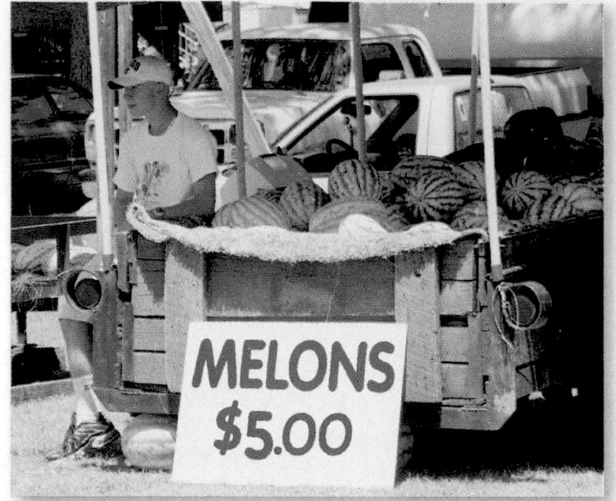

Problems like this require you to multiply or divide to solve an inequality.

When you multiply or divide both sides of an inequality by the same positive number, the statement will still be true. However, when you multiply or divide both sides by the same *negative* number, you need to reverse the direction of the inequality symbol for the statement to be true.

$-4 < 2$	$-4 < 2$
$(3)(-4) < (3)(2)$	$(-3)(-4) > (-3)(2)$
$-12 < 6$	$12 > -6$

EXAMPLE 1 Solving Inequalities by Multiplying

Solve.

A $\frac{x}{11} < 3$

$\frac{x}{11} < 3$

$(11)\frac{x}{11} < (11)3$ *Multiply both sides by 11.*

$x < 33$

B $\frac{r}{-6} \geq 4.8$

$\frac{r}{-6} \geq 4.8$

$(-6)\frac{r}{-6} \leq (-6)4.8$ *Multiply both sides by −6, and reverse the inequality symbol.*

$r \leq -28.8$

EXAMPLE 2 Solving Inequalities by Dividing

Solve. Check each answer.

A $\mathbf{4x > 9}$

$4x > 9$

$\frac{4x}{4} > \frac{9}{4}$ *Divide both sides by 4.*

$x > \frac{9}{4}$, or $2\frac{1}{4}$

Check

$4x > 9$

$4(3) \overset{?}{>} 9$ *3 is greater than $2\frac{1}{4}$. Substitute 3 for x.*

$12 \overset{?}{>} 9$ ✔

B $\mathbf{-12y \leq -60}$

$-12y \leq -60$

$\frac{-12y}{-12} \geq \frac{-60}{-12}$ *Divide both sides by −12, and reverse the inequality symbol.*

$y \geq 5$

Check

$-12y \leq -60$

$-12(10) \overset{?}{\leq} -60$ *10 is greater than 5. Substitute 10 for y.*

$-120 \overset{?}{\leq} -60$ ✔

EXAMPLE  3 *Agriculture Application*

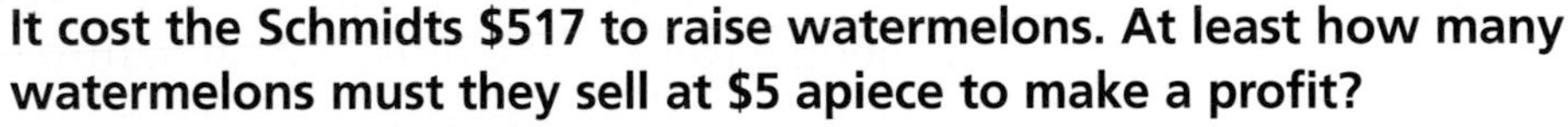

It cost the Schmidts $517 to raise watermelons. At least how many watermelons must they sell at $5 apiece to make a profit?

Since profit is the amount earned minus the amount spent, the Schmidts need to earn more than $517.

Let w represent the number of watermelons they must sell.

$5w > 517$ *Write an inequality.*

$\frac{5w}{5} > \frac{517}{5}$ *Divide both sides by 5.*

$w > 103.4$

The Schmidts cannot sell 0.4 watermelon, so they need to sell at least 104 watermelons to earn a profit.

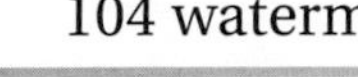

Think and Discuss

1. **Compare** solving multiplication and division equations with solving multiplication and division inequalities.
2. **Explain** how you would solve the inequality $0.5y > 4.5$.

11-6 Exercises

FOR EOG PRACTICE

see page 705

internet connect

Homework Help Online
go.hrw.com Keyword: MS4 11-6

5.03

GUIDED PRACTICE

See Example 1 **Solve.**

1. $\frac{w}{8} < -4$
2. $\frac{z}{-6} \geq 7$
3. $\frac{p}{-12} > -4$

See Example 2 **Solve. Check each answer.**

4. $3m > -15$
5. $-8y < 11$
6. $25c \leq 200$

See Example 3

7. It cost Deirdre $212 to make candles this month. At least how many candles must she sell at $8 apiece to make a profit?

INDEPENDENT PRACTICE

See Example 1 **Solve.**

8. $\frac{s}{5} > 1.4$
9. $\frac{m}{-4} < -13$
10. $\frac{b}{6} > -30$
11. $\frac{c}{-10} \leq 12$
12. $\frac{y}{9} < 2.5$
13. $\frac{x}{1.1} \geq -1$

See Example 2 **Solve. Check each answer.**

14. $6w < 4$
15. $-5z > -3$
16. $15p \leq -45$
17. $-9f > 27$
18. $20k < 30$
19. $-18n \geq 180$

See Example 3

20. Attendance at a museum more than tripled from Monday to Saturday. On Monday, 186 people went to the museum. At least how many people went to the museum on Saturday?
21. It cost George $678 to make wreaths. At least how many wreaths must he sell at $15 apiece to make a profit?

PRACTICE AND PROBLEM SOLVING

Solve.

22. $\frac{a}{65} \leq -10$
23. $0.4p > 1.6$
24. $-\frac{m}{5} < -20$
25. $\frac{2}{3}y \geq 12$
26. $\frac{x}{-9} \leq \frac{3}{5}$
27. $\frac{g}{2.1} > 0.3$
28. $\frac{r}{6} \geq \frac{2}{3}$
29. $4w \leq 1\frac{1}{2}$
30. $-10n < 10^2$
31. $-1\frac{3}{5}t > -4$
32. $-\frac{y}{12} < 3\frac{1}{2}$
33. $5.6v \geq -14$
34. A community theater group produced 8 plays over the last two years. The group's goal for the next two years is to produce at least $1\frac{1}{2}$ times as many plays as they did in the two previous years. At least how many plays does the group want to produce in the next two years?

35. ***SOCIAL STUDIES*** Of the total U.S. population, about 874,000 people are Pacific Islanders. The graph shows where most of these Americans live.

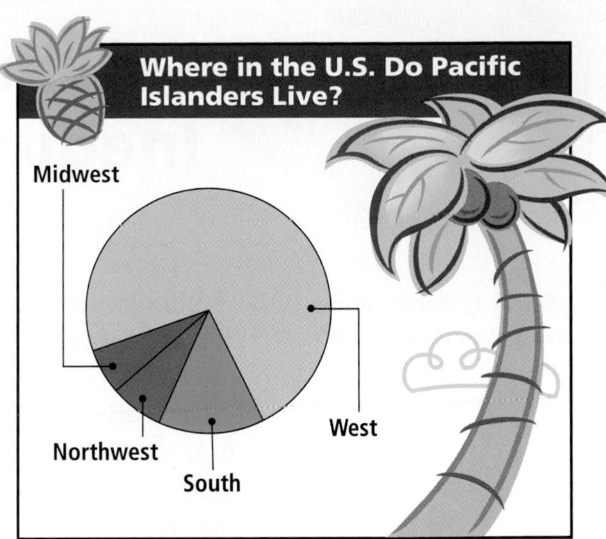

Source: USA Today

a. According to the graph, more than 25% of Pacific Islanders do not live in the West. More than how many Pacific Islanders do not live in the West?

b. According to the graph, less than 75% of Pacific Islanders live in the West. Fewer than how many Pacific Islanders live in the West?

36. Seventh-graders at Mountain Middle School have sold 360 subscriptions to magazines. This is $\frac{3}{4}$ of the number of subscriptions that they need to sell to reach their goal and to beat the eighth grade's sales. At least how many total subscriptions must they sell to reach their goal?

37. ***RECREATION*** Malcolm has saved \$362 to spend on his vacation. He wants to have at least \$35 a day available to spend. How many days of vacation does Malcolm have enough money for?

38. ***WRITE A PROBLEM*** Write a word problem that can be solved using the inequality $\frac{x}{2} \geq 7$. Solve the inequality.

39. ***WRITE ABOUT IT*** Explain how to solve the inequality $\frac{n}{-8} < -40$.

40. ***CHALLENGE*** Use what you have learned about solving multistep equations to solve the inequality $4x - 5 \leq 7x + 4$.

Spiral Review

Solve. (Lesson 11-1)

41. $3x - 5 = 28$ **42.** $\frac{t}{2} + 8 = 9$ **43.** $5r - 12 = 18$ **44.** $\frac{y}{6} + 30 = 10$

45. **EOG PREP** Allana is painting a wall that is 11 feet tall and 23 feet long. What is the area of the wall? (Lesson 8-4)

A 34 ft^2 B 253 ft^2 C 153 ft^2 D 353 ft^2

46. **EOG PREP** Which is the solution of the inequality $w + 8 < -5$? (Lesson 11-5)

A $w < -13$ B $w > -3$ C $w < 3$ D $w > 13$

11-7 Solving Two-Step Inequalities

Learn to solve simple two-step inequalities.

The band students at Newman Middle School are trying to raise at least \$5,000 to buy new percussion instruments. They already have raised \$850. How much should each of the 83 band students still raise, on average, to meet their goal?

When you solve two-step equations, you can use the order of operations in reverse to isolate the variable. You can use the same process when solving two-step inequalities.

Remember!

Draw a closed circle when the inequality includes the point and an open circle when it does not include the point.

EXAMPLE 1 Solving Two-Step Inequalities

Solve. Then graph each solution set on a number line.

A $\frac{x}{5} - 15 < 10$

$\frac{x}{5} - 15 < 10$

$\underline{\quad + 15} \quad \underline{+ 15}$ *Add 15 to both sides.*

$\frac{x}{5} < 25$

$(5)\frac{x}{5} < (5)25$ *Multiply both sides by 5.*

$x < 125$

−25 0 25 50 75 100 125 150 175

B $\frac{y}{-9} + 30 \geq 42$

$\frac{y}{-9} + 30 \geq 42$

$\underline{\quad - 30} \quad \underline{- 30}$ *Subtract 30 from both sides.*

$\frac{y}{-9} \geq 12$

$(-9)\frac{y}{-9} \leq (-9)12$ *Multiply both sides by −9, and reverse the inequality symbol.*

$y \leq -108$

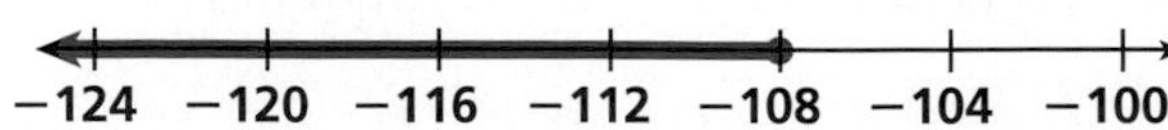

Solve. Then graph each solution set on a number line.

C $3x - 12 \geq 9$

$$\begin{aligned} 3x - 12 &\geq 9 \\ +12 &\quad +12 \\ \hline 3x &\geq 21 \end{aligned}$$ *Add 12 to both sides.*

$$\frac{3x}{3} \geq \frac{21}{3}$$ *Divide both sides by 3.*

$$x \geq 7$$

2 3 4 5 6 7 8 9 10 11 12

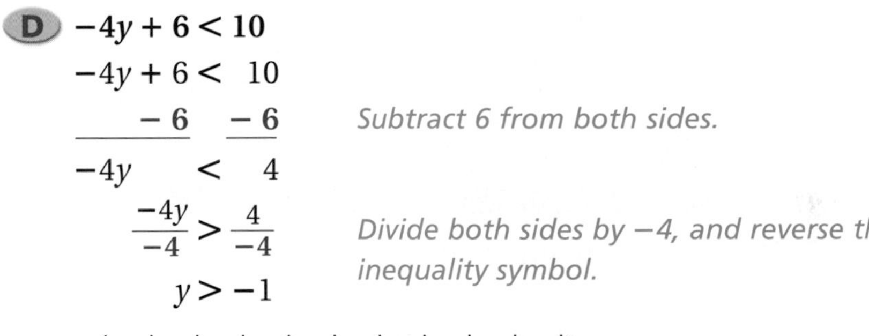

D $-4y + 6 < 10$

$$\begin{aligned} -4y + 6 &< 10 \\ -6 &\quad -6 \\ \hline -4y &< 4 \end{aligned}$$ *Subtract 6 from both sides.*

$$\frac{-4y}{-4} > \frac{4}{-4}$$ *Divide both sides by −4, and reverse the inequality symbol.*

$$y > -1$$

−5 −4 −3 −2 −1 0 1 2 3 4 5

EXAMPLE 2 *School Application*

The 83 members of the Newman Middle School Band are trying to raise at least $5,000 to buy new percussion instruments. If they have already raised $850, how much should each student still raise, on average, to meet the goal?

Let d represent the average amount each student should still raise.

$$83d + 850 \geq 5{,}000$$ *Write an inequality.*

$$\begin{aligned} -850 &\quad -850 \\ \hline 83d &\geq 4{,}150 \end{aligned}$$ *Subtract 850 from both sides.*

$$\frac{83d}{83} \geq \frac{4{,}150}{83}$$ *Divide both sides by 83.*

$$d \geq 50$$

On average, each band member should raise at least $50.

Think and Discuss

1. **Tell** how you would solve the inequality $8x + 5 < 20$.
2. **Explain** why the *greater than or equal to* symbol was used in the inequality in Example 2.

11-7 Exercises

5.02, 5.03

FOR EOG PRACTICE see page 705

internet connect Homework Help Online go.hrw.com Keyword: MS4 11-7

GUIDED PRACTICE

See Example 1 **Solve. Then graph each solution set on a number line.**

1. $5x + 3 < 18$
2. $\frac{z}{7} + 23 \leq -19$
3. $3y - 4 \geq 14$
4. $\frac{m}{4} - 2 > -3$
5. $-11p - 13 \geq 42$
6. $\frac{n}{-3} - 4 > 4$

See Example 2

7. Three students collected more than \$93 washing cars. They used \$15 to reimburse their parents for cleaning supplies. Then they divided the remaining money equally. How much did each student earn?

INDEPENDENT PRACTICE

See Example 1 **Solve. Then graph each solution set on a number line.**

8. $5s - 7 > -42$
9. $\frac{b}{2} + 3 < 9$
10. $-2q + 5 \geq 19$
11. $-8c - 11 \leq 13$
12. $\frac{y}{-4} + 6 > 10$
13. $\frac{x}{9} - 5 \leq -8$
14. $\frac{r}{-2} - 9 > -14$
15. $13j + 18 \leq 44$
16. $\frac{d}{13} - 12 > 27$

See Example 2

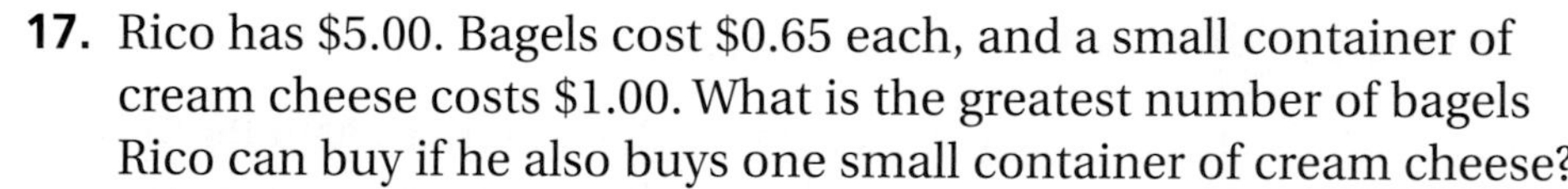

17. Rico has \$5.00. Bagels cost \$0.65 each, and a small container of cream cheese costs \$1.00. What is the greatest number of bagels Rico can buy if he also buys one small container of cream cheese?
18. The 35 members of the Tigers drill team are trying to raise at least \$1,200 to cover travel costs to a training camp. If they have already raised \$500, at least how much should each member still raise, on average, to meet the goal?

PRACTICE AND PROBLEM SOLVING

Solve.

19. $9 - 2y < 15$
20. $-3q + 10 < -2$
21. $\frac{a}{-6} - 5 \geq 4$
22. $-4x + 8 \leq 32$
23. $0.5 + \frac{n}{5} > -0.5$
24. $1.4 + \frac{c}{3} < 2$
25. $-\frac{3}{4}b - 2.2 > -1$
26. $12 + 2w - 8 \leq 20$
27. $5k + 6 - k \geq -14$
28. $\frac{s}{2} + 9 > 12 - 15$
29. $4t - 3 - 10t < 15$
30. $\frac{d}{2} + 1 + \frac{d}{2} \leq 5$
31. Mr. Monroe keeps a bag of small prizes to distribute to his students. He likes to keep at least twice as many prizes in the bag as he has students. The bag currently has 79 prizes in it. If Mr. Monroe has 117 students, at least how many more prizes does he need to buy?

32. The bar graph shows how many students from Warren Middle School participated in a reading challenge each of the past four years. This year, the goal is for at least 10 more students to participate than the average number of participants from the past four years. What is the goal for this year?

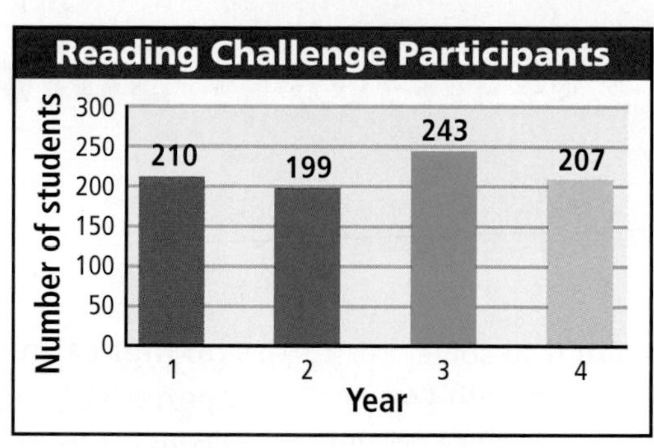

33. ***BUSINESS*** Darcy earns a salary of $1,400 per month, plus a commission of 4% of her sales. If she wants to earn a total of at least $1,600 this month, what is the least amount of sales she needs?

34. ***CONSUMER MATH*** Michael wants to buy a belt that costs $18. He also wants to buy some shirts that are on sale for $14 each. He has $70. At most, how many shirts can Michael buy together with the belt?

35. ***EARTH SCIENCE*** A granite rock contains the minerals feldspar, quartz, and biotite mica. The rock has $\frac{1}{3}$ as much biotite mica as quartz. If the rock is at least 30% quartz, what percent of the rock is feldspar?

36. ***WHAT'S THE ERROR?*** A student's solution to the inequality $\frac{x}{-9} - 5 > 2$ was $x > 63$. What error did the student make in the solution?

37. ***WRITE ABOUT IT*** Explain how to solve the inequality $4y + 6 < -2$.

38. ***CHALLENGE*** A student scored 92, 87, 90, and 85 on four tests. She wants her average score for five tests to be at least 90. What is the lowest score the student can get on her fifth test?

Spiral Review

39. Four friends are standing in line. In how many possible orders can they be standing? (Lesson 10-7)

40. There are 6 jars lined up on a shelf. In how many possible orders can the jars be lined up? (Lesson 10-7)

Solve. (Lesson 11-3)

41. $-8x - 210 = 6x$ **42.** $50 + 9y = -3y + 14$ **43.** $n + 27 = -n + 35$

44. **EOG PREP** Which is the solution of the inequality $-6x > -54$? (Lesson 11-7)

A $x < -9$ B $x > -9$ C $x < 9$ D $x > 9$

EXTENSION Solving for a Variable

Learn **to solve formulas with two or more variables for one of the variables.**

The highest recorded speed of a magnetically elevated vehicle was achieved by the MLX01 on the Yamanashi Maglev Test Line in Japan. At its top speed, the MLX01 could travel the 229 miles from Tokyo to Kyoto in less than an hour.

The formula *distance* = *rate* · *time* ($d = rt$) tells how far an object travels at a certain rate over a certain time. In an equation or a formula that contains more than one variable, you can isolate one of the variables by using inverse operations. Recall that you cannot divide by a variable if it represents 0.

The MLX01 attained the record speed of 343 miles per hour in January 1998.

EXAMPLE  1 Solving for Variables in Formulas

Solve $d = rt$ for r.

$$d = rt$$

$$\frac{d}{t} = \frac{rt}{t} \qquad \textit{Divide both sides by } t.$$

$$\frac{d}{t} = r$$

EXAMPLE 2 *Physical Science Application*

How long would it take the MLX01 to travel 1,029 mi if it travels at a speed of 343 mi/h?

First solve the distance formula for t, since you want to find the time. Then use the given values to find t.

$$d = rt$$

$$\frac{d}{r} = \frac{rt}{r} \qquad \textit{Divide both sides by } r.$$

$$\frac{d}{r} = t$$

$$\frac{1{,}029}{343} = t \qquad \textit{Substitute 1,029 for } d \textit{ and 343 for } r.$$

$$3 = t$$

It would take the MLX01 3 hours to travel 1,029 miles.

EXTENSION Exercises

5.03

Solve each equation for the given variable.

1. $A = bh$ for h
2. $A = bh$ for b
3. $C = \pi d$ for d
4. $P = 4s$ for s
5. $V = Bh$ for B
6. $d = 2r$ for r
7. $xy = k$ for y
8. $A = \ell w$ for w
9. $W = Fd$ for F
10. $I = Prt$ for P
11. $C = 2\pi r$ for r
12. $A = \frac{1}{2}bh$ for h
13. $V = \frac{1}{3}Bh$ for h
14. $K = C + 273$ for C
15. $E = Pt$ for t
16. $D = \frac{m}{v}$ for v
17. $F = ma$ for a
18. $P = VI$ for I
19. $r = \frac{V}{I}$ for V
20. $I = Prt$ for r
21. $P = 2\ell + 2w$ for ℓ
22. $V = \pi r^2 h$ for h

23. ***PHYSICAL SCIENCE*** The formula $E = mc^2$ tells the amount of energy an object at rest has. In the equation, E stands for the amount of energy in joules, m stands for the rest mass in kilograms of the object, and c is the speed of light (approximately 300,000,000 meters per second). What is the rest mass of an object that has 90,000,000,000,000 joules of energy?

THE FAR SIDE® BY GARY LARSON

"*Now* that desk looks better. Everything's squared away, yessir, squaaaaaared away."

24. ***PHYSICAL SCIENCE*** The Kelvin scale is a temperature scale. To convert between the Celsius temperature scale and the Kelvin temperature scale, use the formula $C = K - 273$, where C represents the temperature in degrees Celsius and K represents the temperature in kelvins. Use the formula to convert 38°C to an equivalent Kelvin temperature.

25. ***PHYSICAL SCIENCE*** Density is mass per unit volume. The formula for density is $D = \frac{m}{v}$, where D represents density, m represents mass, and v represents volume. Find the mass of a gear with a density of 3.75 g/cm^3 and a volume of 20 cm^3.

26. What is the height of the cone if its volume is 8,138.88 ft^3? Use 3.14 for π.

12 ft

Problem Solving on Location

NORTH CAROLINA

Great Dismal Swamp National Wildlife Refuge

Stretching through parts of northern North Carolina and southern Virginia is an area known as the Great Dismal Swamp. The Great Dismal Swamp is 37 miles long and 12 miles wide at its widest point. The refuge consists of 107,000 acres of forested wetlands and surrounds Lake Drummond, a 3,100-acre natural lake on the Virginia side. Many animals make the refuge their home.

1. Falls Lake is in North Carolina. One-fourth the size of Falls Lake plus 100 equals the size of Lake Drummond. Write and solve an equation to find the size of Falls Lake.
2. Of the 200 species of birds found in the refuge, 96 have been seen nesting. What is the percent of birds that have been seen nesting in the refuge? 
3. Sixty percent of the swamp is in four North Carolina counties, and the remaining portion is in Virginia. Write and solve an equation to find the percent of the Great Dismal Swamp that is in Virginia. Then make a circle graph to show how much of the swamp is located in each state.
4. Three species of poisonous snakes can be found in the Great Dismal Swamp. The total number of snake species in the swamp is 21. Write and solve an equation to find how many nonpoisonous snake species are in the swamp.

State Parks

North Carolina has 29 state parks, four recreation areas, and a number of natural areas. The educational and recreational opportunities at North Carolina's state parks include camping, boating, picnicking, canoeing, hiking, and more. The map below shows several of the state parks in eastern North Carolina as well as some of the popular activities at those parks.

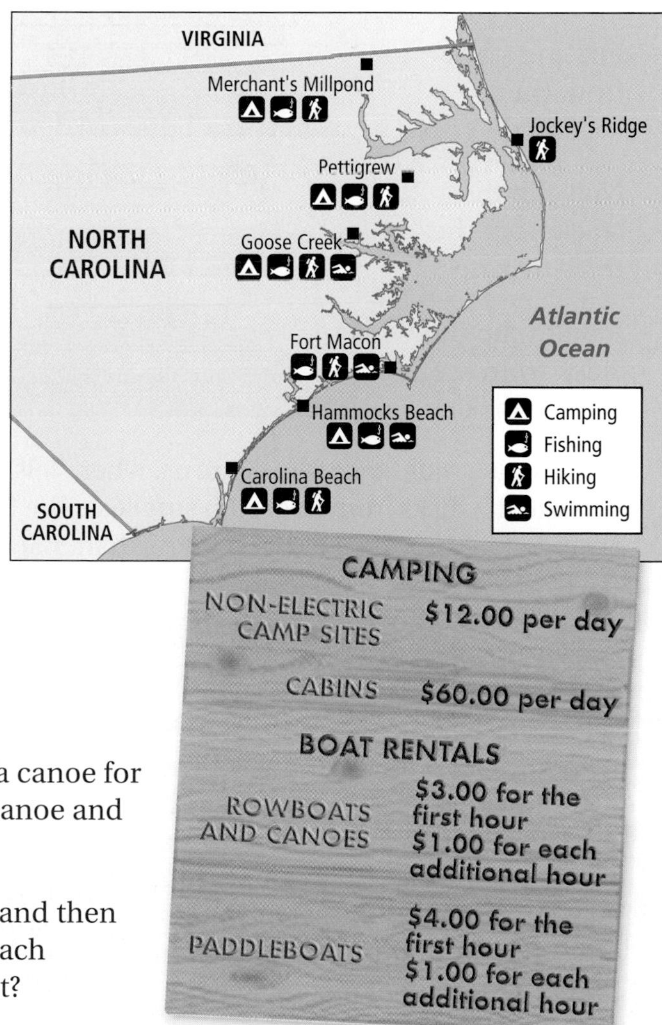

For 1–2, use the map.

1. Carolina Beach State Park has about 6 miles of hiking trails. If you double the number of miles of hiking trails at Carolina Beach State Park and then subtract 3 miles, the result is the number of miles of hiking trails at Merchant's Millpond State Park. Write and solve an equation to find the number of miles of hiking trails at Merchant's Millpond State Park.

2. Using s for the number of state parks with swimming areas, write an inequality that shows the least number of state parks in North Carolina that have swimming areas.

For 3–4, use the sign.

3. Suppose you camp in a tent for 5 days and rent a canoe for 2 hours each day. What is the total cost of your canoe and camping trip?

4. A group of 5 friends camped for 2 days in a tent and then rented a cabin for 2 more days. How much did each person have to pay if each paid an equal amount?

MATH-ABLES

Flapjacks

Five pancakes of different sizes are stacked in a random order. How can you get the pancakes in order from largest to smallest by flipping portions of the stack?

To find the answer, stack five disks of different sizes in no particular order. Arrange the disks from largest to smallest in the fewest number of moves possible. Move disks by choosing a disk and flipping the whole stack over from that disk up.

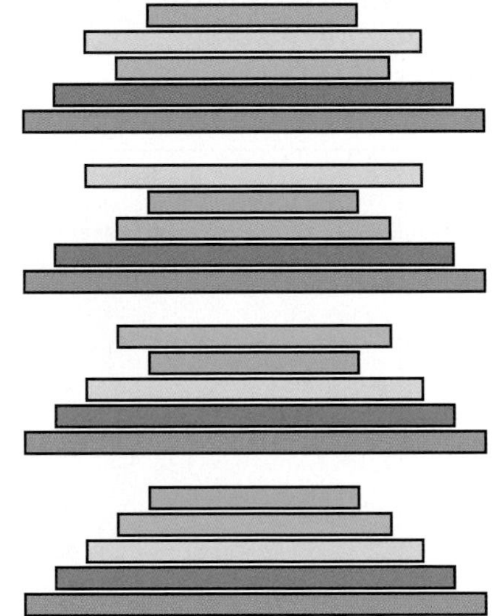

Start with a stack of five.

Flip the stack from the second disk up.

Now flip the stack from the third disk up.

Finally, flip the stack from the second disk up.

At most, it should take $3n - 2$ turns, where n is the number of disks, to arrange the disks from largest to smallest. The five disks above were arranged in three turns, which is less than $3(5) - 2 = 13$. Try it on your own.

Leaping Counters

Remove all but one of the counters from the board by jumping over each counter with another and removing the jumped counter. The game is over when you can no longer jump a counter. A perfect game would result in one counter being left in the center of the board.

internet connect

Go to ***go.hrw.com*** for a complete set of rules and a game board.
KEYWORD: MS4 Game11

Solve Multistep Equations

You can use the graph function on your graphing calculator to find the solutions to multistep equations.

Activity

1 Use a graphing calculator to solve the equation $3x - 7 = 26$.

On your calculator, enter the expressions that are from each side of the equation separately. Set the size of the window as shown.

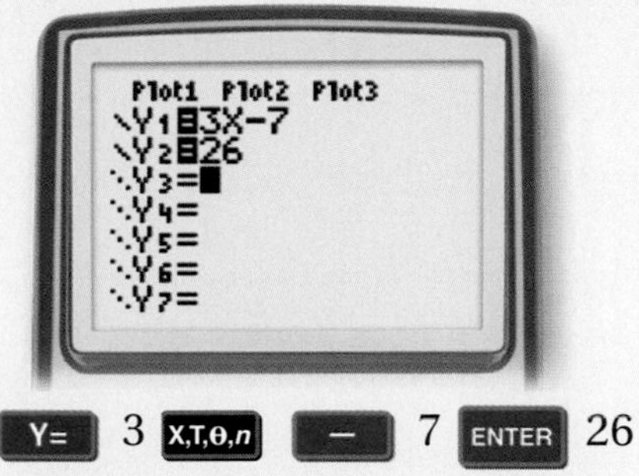

Y= 3 X,T,θ,n − 7 ENTER 26

WINDOW (−) 20 ENTER 20 ENTER 2 ENTER (−) 50 ENTER 50 ENTER 5 ENTER

To see the graph, press GRAPH.

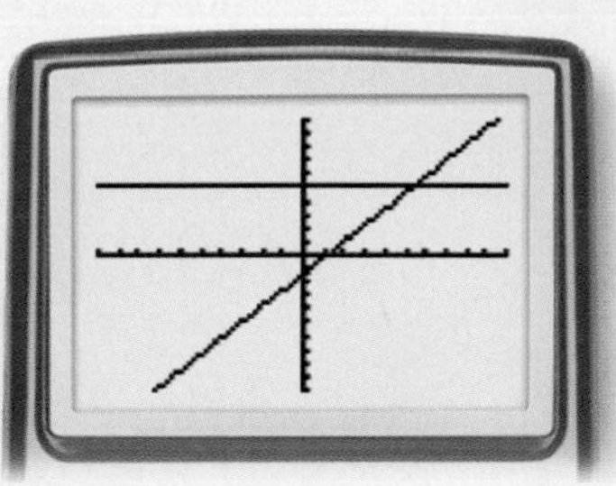

Use the **CALC** function to find the intersection.

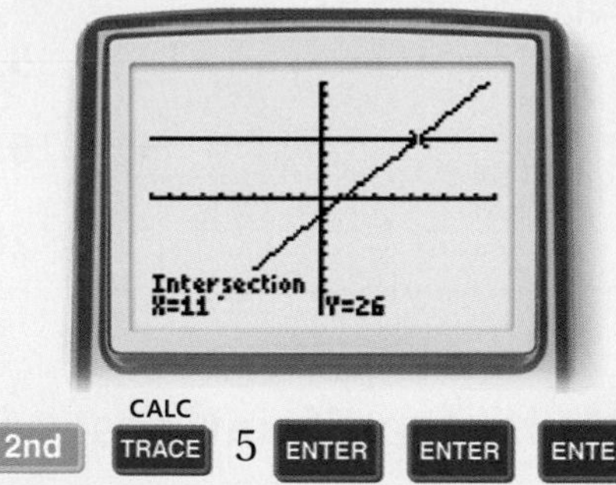

2nd CALC TRACE 5 ENTER ENTER ENTER

The solution to the equation is the value shown for x. For $3x - 7 = 26$, $x = 11$.

Think and Discuss

1. Why does the intersection of the two lines give the solution to the equation?

2. How would you enter the expression $\frac{2x+6}{3}$ into the calculator?

Try This

Use a graphing calculator to solve each equation.

1. $4x + 9 = 17$ **2.** $\frac{x-8}{5} = 3$ **3.** $-3x - 2 = -14$ **4.** $\frac{2x+1}{7} = 3$

Chapter 11

Study Guide and Review

Vocabulary

Complete the sentences below with vocabulary words from the list above.

1. A(n) ___?___ states that two quantities either are not equal or may not be equal.
2. A(n) ___?___ is a combination of more than one inequality.
3. Together, the solutions of an inequality are called the ___?___.

11-1 Solving Two-Step Equations (pp. 560–563)

EXAMPLE

■ **Solve $6a - 3 = 15$.**

$$6a - 3 = 15$$
$$6a - 3 + 3 = 15 + 3$$
$$6a = 18$$
$$\frac{6a}{6} = \frac{18}{6}$$
$$a = 3$$

EXERCISES

Solve.

4. $-5y + 6 = -34$
5. $9 + \frac{z}{6} = 14$
6. $\frac{w}{-7} + 13 = -8$

11-2 Solving Multistep Equations (pp. 564–567)

EXAMPLE

■ **Solve $\frac{4x - 3}{7} = 3$.**

$$\frac{4x - 3}{7} = 3$$
$$(7)\frac{4x - 3}{7} = (7)3$$
$$4x - 3 = 21$$
$$4x - 3 + 3 = 21 + 3$$
$$4x = 24$$
$$\frac{4x}{4} = \frac{24}{4}$$
$$x = 6$$

EXERCISES

Solve.

7. $7a + 4 - 13a = 46$
8. $\frac{8b - 5}{3} = 9$
9. $\frac{6j - 18}{4} = 9$
10. $-9 + 16y - 19 = 52$
11. $\frac{12x + 15}{3} = 53$

11-3 Solving Equations with Variables on Both Sides (pp. 568–571)

EXAMPLE

- **Solve $8a = 3a + 25$.**

$$8a = 3a + 25$$
$$8a - 3a = 3a - 3a + 25$$
$$5a = 25$$
$$\frac{5a}{5} = \frac{25}{5}$$
$$a = 5$$

EXERCISES

Solve.

12. $-6b + 9 = 12b$
13. $5 - 7c = -3c - 19$
14. $18m - 14 = 12m + 2$
15. $4 - \frac{2}{5}x = \frac{1}{5}x - 8$

11-4 Inequalities (pp. 574–577)

EXAMPLE

Write an inequality for each situation.

- **You have to be at least 17 years old to drive a car in New Jersey.**
  age of driver ≥ 17
- **There can be at most 60 people on the bus.**
  number of people ≤ 60

EXERCISES

Write an inequality for each situation.

16. You must have an average of at least 65 to pass math class.
17. A bridge's load limit is at most 9 tons.
18. The large tree in the park is more than 200 years old.
19. It is less than 2 miles from home to the grocery store.

11-5 Solving Inequalities by Adding or Subtracting (pp. 578–581)

EXAMPLE

Solve. Graph each solution set.

- $b + 6 > -10$

$$b + 6 > -10$$
$$b + 6 - 6 > -10 - 6$$
$$b > -16$$

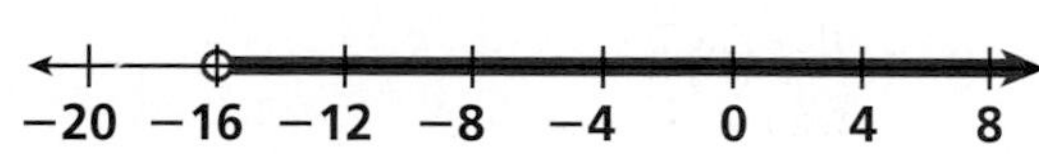

- $p - 17 \leq 25$

$$p - 17 \leq 25$$
$$p - 17 + 17 \leq 25 + 17$$
$$p \leq 42$$

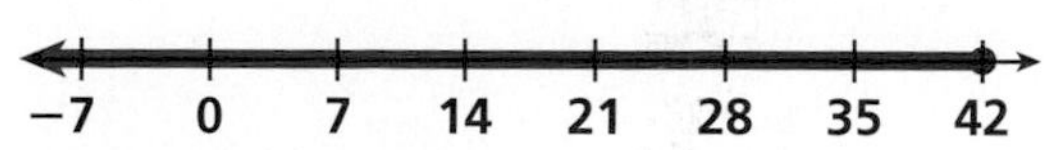

EXERCISES

Solve. Graph each solution set.

20. $r - 16 > 9$
21. $s + 7 \geq 21$
22. $12 + x \leq -14$
23. $\frac{3}{4} + g < 8\frac{3}{4}$
24. $\frac{2}{3} + t \leq \frac{5}{6}$
25. $7.46 > r - 1.54$
26. $z + 17 < -13$
27. $u - 57.7 \geq -123.7$

Study Guide and Review

11-6 Solving Inequalities by Multiplying or Dividing (pp. 582–585)

EXAMPLE

Solve.

■ $\frac{m}{-4} \geq 3.8$

$\frac{m}{-4} \geq 3.8$

$(-4)\frac{m}{-4} \leq (-4)3.8$

$m \leq -15.2$

■ $8b < -48$

$8b < -48$

$\frac{8b}{8} < \frac{-48}{8}$

$b < -6$

EXERCISES

Solve.

28. $\frac{n}{-8} > 6.9$

29. $4x \leq 24$

30. $-3p \geq -18$

31. $\frac{k}{13} < -10$

32. $-5p > -25$

33. $\frac{v}{1.2} \geq 2.3$

34. $\frac{c}{-11} < -3$

35. $1.3y \leq 3.51$

11-7 Solving Two-Step Inequalities (pp. 586–589)

EXAMPLE

Solve. Graph each solution set.

■ $\frac{k}{3} - 18 > 24$

$\frac{k}{3} - 18 > 24$

$\frac{k}{3} - 18 + 18 > 24 + 18$

$\frac{k}{3} > 42$

$(3)\frac{k}{3} > (3)42$

$k > 126$

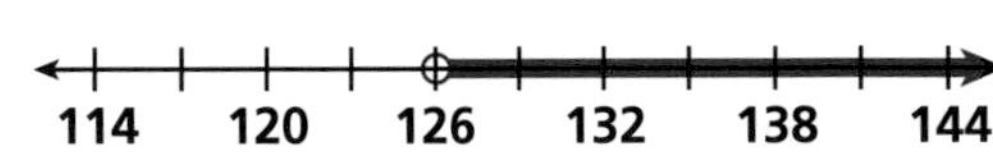

■ $-5b + 11 \leq -4$

$-5b + 11 \leq -4$

$-5b + 11 - 11 \leq -4 - 11$

$-5b \leq -15$

$\frac{-5b}{-5} \geq \frac{-15}{-5}$

$b \geq 3$

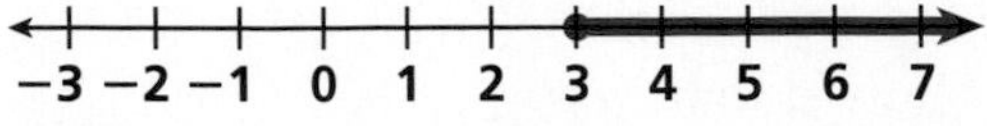

EXERCISES

Solve. Graph each solution set.

36. $-7b - 16 > -2$

37. $3.8 + \frac{d}{5} < 2.6$

38. $15 - 4n + 9 \leq 40$

39. $\frac{y}{-3} + 18 \geq 12$

40. $\frac{c}{3} + 7 > -11$

41. $4x - 8 \leq 32$

42. $12 - 7m < 47$

43. $18 + \frac{h}{6} \geq -8$

44. $14 > -2t - 6$

45. $-3 < \frac{w}{-4} + 10$

46. $\frac{y}{7} + 3.9 \leq 8.9$

Chapter Test

Solve.

1. $3y - 8 = 16$
2. $\frac{x}{3} + 12 = -4$
3. $\frac{a}{6} - 7 = -4$
4. $-7b + 5 = -51$
5. $\frac{5y - 4}{3} = 7$
6. $8r + 7 - 13 = 58$
7. $\frac{12s - 6}{5} = 6$
8. $\frac{19.8 - 4t}{3} = 8.7$
9. $-14q = 4q - 126$
10. $\frac{5}{6}p + 4 = \frac{1}{6}p - 16$
11. $9 - 6k = 3k - 54$
12. $-3.6l = -7l + 34$
13. The bill for the repair of a computer was \$179. The cost of the parts was \$44, and the labor charge was \$45 per hour. How many hours did it take to repair the computer?
14. Members of the choir are baking cookies for a fund-raiser. It costs \$2.25 to make a dozen cookies, and the choir's initial expenses were \$15.75. If they sell the cookies for \$4.50 a dozen, how many dozen do they have to sell to cover their costs?

Write an inequality for each situation.

15. You must be more than 4 ft tall to go on the ride.
16. You cannot go more than 65 miles per hour on Route 18.

Graph each inequality.

17. $a < -2$
18. $b \geq 3$
19. $c > -1$ or $c < -5$
20. $-5 < d \leq 2$

Solve. Then graph each solution on a number line.

21. $n + 8 < -9$
22. $n - 124 > -59$
23. $\frac{x}{32} < -40$
24. $-\frac{3}{4}y \leq -12$
25. Rosa wants to save at least \$125 to buy a new skateboard. She has already saved \$46. How much more does Rosa need to save?
26. Gasoline costs \$1.25 a gallon. At most, how many gallons can be bought for \$15.00?

Solve.

27. $m - 7.8 \leq 23.7$
28. $6z > -2\frac{2}{3}$
29. $\frac{w}{-4.9} \leq 3.4$
30. $4a + 9 > -15$
31. $2.8 - \frac{c}{4} \geq 7.4$
32. $\frac{d}{5} - 8 > -4$
33. The seventh-grade students at Fulmore Middle School are trying to raise at least \$7,500 for the local public library. So far, each of the 198 students has raised an average of \$20. At least how much more money must the seventh-graders collect to reach their goal?

Chapter 11 Performance Assessment

Show What You Know

Create a portfolio of your work from this chapter. Complete this page and include it with your four best pieces of work from Chapter 11. Choose from your homework or lab assignments, mid-chapter quiz, or any journal entries you have done. Put them together using any design you want. Make your portfolio represent what you consider your best work.

Short Response

1. Write an equation for the sentence "Two more than three times a number is seventeen." Then solve the equation and show your work.

2. Solve the inequality $-5y < 25$, and then graph the solution set on a number line.

3. Jumping Jack's Fitness Facility charges members $50 per month to work out. Sweat Dog Fitness Center charges members $100 to join and $40 per month. After how many months of being a member of one place would you have paid as much as if you were a member of the other? Show how you found your answer.

4. The maximum weight that the school elevator can carry is 1,500 pounds. If a group of 35 seventh-graders, with an average weight of 90 pounds each, wants to use the elevator, how many trips will be needed to take the students to the fourth floor? Explain your answer.

Extended Problem Solving

5. Tim and his crew trim trees. They charge a service fee of $40 for each job, plus an hourly rate.
   a. Use the graph to determine the crew's hourly rate. Explain how you found your answer.
   b. Write an equation to find y, their income for x hours of work.
   c. How many hours did Tim's crew work if they earned $490? Show your work.

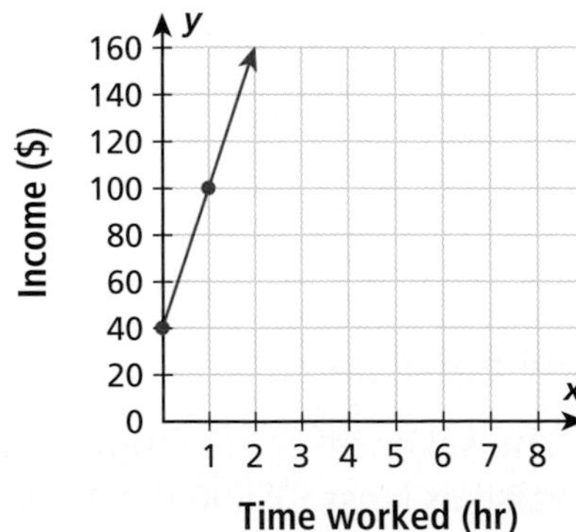

Performance Assessment

internet connect
EOG Test Practice Online
go.hrw.com Keyword: MS4 TestPrep

Getting Ready for EOG

Chapter 11

Cumulative Assessment, Chapters 1–11

1. What is $\sqrt{55}$ to the nearest whole number?

A 8 C 7.4
B 7 D 3,025

2. What is 26 m − 0.42 m, given with the correct number of significant digits?

A 25.58 m C 26.6 m
B 30 m D 26 m

3. Meera has 5 yards of beaded ribbon with which to make necklaces. How many 15-inch necklaces can she make?

A 12 C 75
B 4 D 3

4. Of 4,500 books, $\frac{3}{5}$ are fiction. Of the remaining books, $\frac{1}{2}$ are nonfiction. The rest are children's books. How many of the books are children's books?

A 500 C 1,800
B 900 D 2,700

5. What is a solution of the inequality $\frac{y}{8} > -12$?

A $y < -96$ C $y < -1.5$
B $y < 96$ D $y > -96$

6. If 6 is added to 3 times a number and then that result is divided by 2, the answer is 12. What is the number?

A 2 C 6
B 10 D 18

TEST TAKING TIP!
Eliminate choices by considering the meanings of words such as *increase* and *decrease*.

7. If the number of students in seventh grade at Madison Middle School is going to increase by 15% from year three to year four, what will enrollment be?

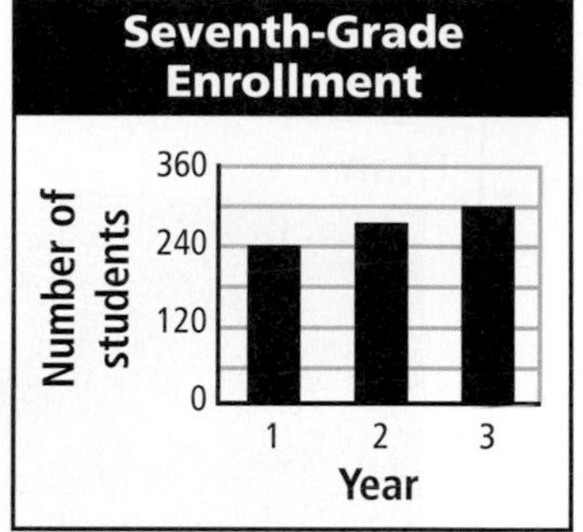

A 42 C 345
B 295 D 238

8. What is a solution of the equation $x + -9 = -20$.

A $x = -29$ C $x = 11$
B $x = 29$ D $x = -11$

9. ***SHORT RESPONSE*** A spinner is divided into ten equal sectors numbered 1–10. Monica spins the spinner. List all of the outcomes that are a multiple of 3. What is the probability that the spinner will land on a multiple of 3?

10. ***SHORT RESPONSE*** Nine less than four times a number is the same as twice the number increased by 11. What is the number?

a. Write the above statement as an equation.

b. Solve the equation.

Chapter 12

Graphs and Functions

TABLE OF CONTENTS

Fastest U.S. Roller Coasters

Roller Coaster	Speed (mi/h)
Superman the Escape	100
Millennium Force	92
Goliath	85
Titan	85

Roller Coaster Designer

Traditional roller-coaster designs rely on gravity for a coaster to gain speed. Some of these designs include loops and turns to make rides more exciting.

Jim Seay is a roller-coaster designer who uses high-tech methods to create exhilarating rides. His designs include a system that can propel a coaster from 0 to 70 miles per hour in less than four seconds!

Chapter Opener Online
go.hrw.com
KEYWORD: MS4 Ch12

ARE YOU READY?

Choose the best term from the list to complete each sentence.

1. A(n) __?__ states that two expressions are equivalent.

2. To __?__ an expression is to substitute a number for the variable.

3. The __?__ is the horizontal number line on a coordinate plane.

4. A(n) __?__ is a number that can be written as a ratio of two integers.

equation
evaluate
irrational number
rational number
***x*-axis**
***y*-axis**

Complete these exercises to review skills you will need for this chapter.

✔ Evaluate Expressions

Evaluate each expression.

5. $x + 5$ for $x = -18$

6. $-9y$ for $y = 13$

7. $\frac{z}{-6}$ for $z = 96$

8. $w - 9$ for $w = -13$

9. $-3z + 1$ for $z = 4$

10. $3w + 9$ for $w = 7$

11. $5 - \frac{y}{3}$ for $y = -3$

12. $x^2 + 1$ for $x = -2$

✔ Solve Equations

Solve each equation.

13. $13 + y = -3$

14. $-4y = -56$

15. $3y = -12$

16. $25 - y = 7$

17. $3y + 8 = 3$

18. $5 - 3y = 7$

19. $5y - 4 = 16$

20. $\frac{y}{3} = -9$

✔ Write Ordered Pairs

Write the ordered pair for each point.

21. point A

22. point B

23. point C

24. point D

25. point E

26. point F

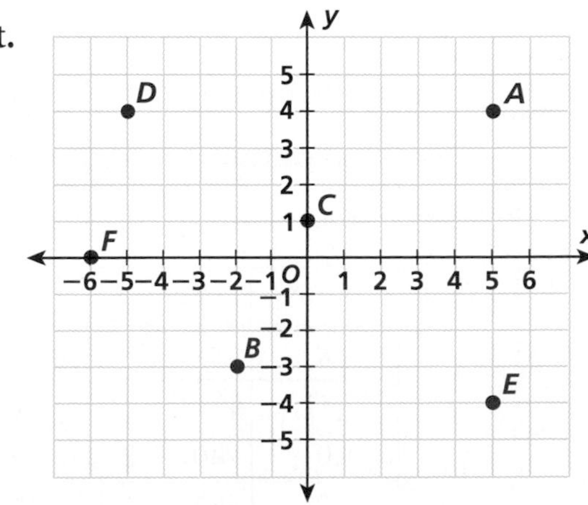

12-1 Introduction to Functions

Learn to use function tables to generate and graph ordered pairs.

Vocabulary

function

Rube Goldberg, a famous cartoonist, invented machines that perform ordinary tasks in extraordinary ways. Each machine operates according to a rule, or a set of steps, to produce a particular output.

As you raise spoon of soup (A) to your mouth, it pulls string (B), thereby jerking ladle (C), which throws cracker (D) past parrot (E). Parrot jumps after cracker, and perch (F) tilts, upsetting seeds (G) into pail (H). Extra weight in pail pulls cord (I), which opens and lights automatic cigar lighter (J), setting off sky rocket (K), which causes sickle (L) to cut string (M) and allow pendulum with attached napkin to swing back and forth, thereby wiping off your chin.

In mathematics, a **function** operates according to a rule to produce a single output value for each input value.

A function can be represented as a rule written in words, such as **"double the number and then add nine to the result."**

A function can also be represented by an equation with two variables. One variable represents the input, and the other represents the output.

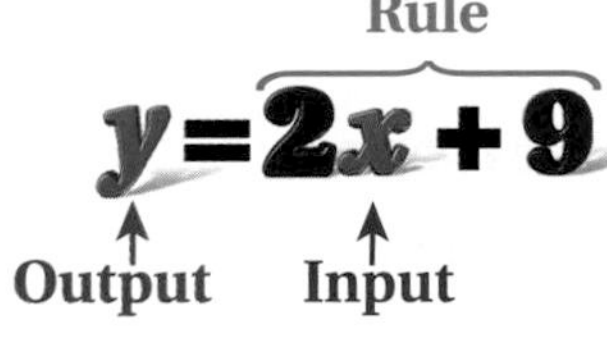

You can use a table to organize the input and output values of a function. Your table may show as many possible input and output values as you choose.

EXAMPLE 1 Completing a Function Table

Find the output for each input.

A $y = 4x - 2$

Input	Rule	Output
x	$4x - 2$	y
-1	$4(-1) - 2$	-6
0	$4(0) - 2$	-2
3	$4(3) - 2$	10

Substitute −1 for x and simplify.
Substitute 0 for x and simplify.
Substitute 3 for x and simplify.

Find the output for each input.

B $y = 6x^2$

Input	Rule	Output
x	$6x^2$	y
−5	$6(-5)^2$	150
0	$6(0)^2$	0
5	$6(5)^2$	150

Substitute −5 for x and simplify.
Substitute 0 for x and simplify.
Substitute 5 for x and simplify.

Remember!

An ordered pair is a pair of numbers that represents a point on a graph.

You can also use a graph to represent a function. The corresponding input and output values together form unique ordered pairs.

EXAMPLE 2 Graphing Functions Using Ordered Pairs

Make a function table, and graph the resulting ordered pairs.

Helpful Hint

When writing an ordered pair, write the input value first and then the output value.

A $y = 2x$

Input	Rule	Output	Ordered Pair
x	$2x$	y	(x, y)
−2	2(−2)	−4	(−2, −4)
−1	2(−1)	−2	(−1, −2)
0	2(0)	0	(0, 0)
1	2(1)	2	(1, 2)
2	2(2)	4	(2, 4)

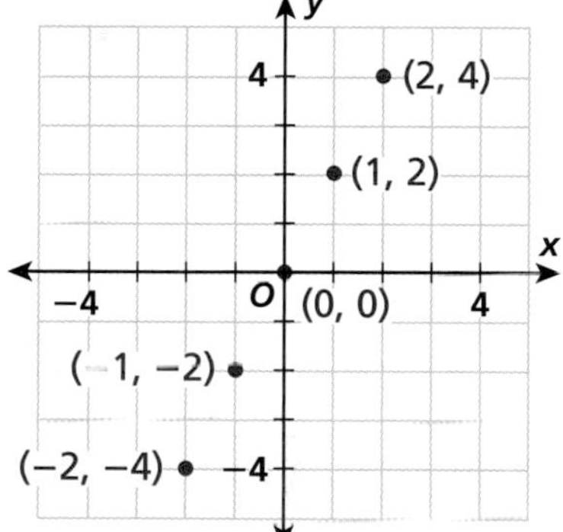

B $y = x^2$

Input	Rule	Output	Ordered Pair
x	x^2	y	(x, y)
−2	$(-2)^2$	4	(−2, 4)
−1	$(-1)^2$	1	(−1, 1)
0	$(0)^2$	0	(0, 0)
1	$(1)^2$	1	(1, 1)
2	$(2)^2$	4	(2, 4)

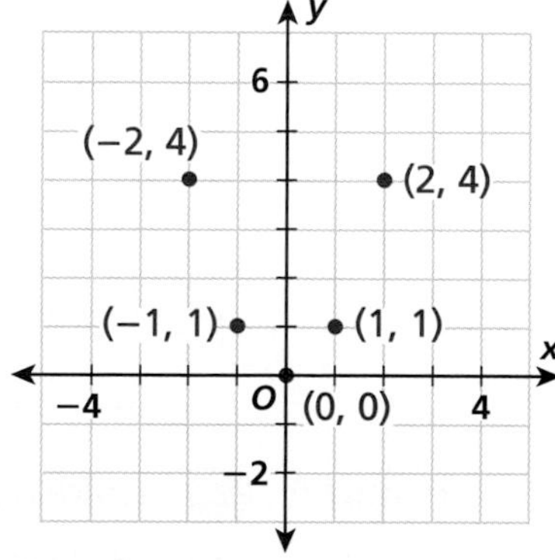

Think and Discuss

1. **Describe** how a function works like a machine.
2. **Give an example** of a rule that takes an input value of 4 and produces an output value of 10.

12-1 Exercises

5.01

FOR EOG PRACTICE
see page 706

internet connect
Homework Help Online
go.hrw.com Keyword: MS4 12-1

GUIDED PRACTICE

See Example 1 Find the output for each input.

1. $y = 2x + 1$

Input	Rule	Output
x	$2x + 1$	y
−3		
0		
1		

2. $y = -x + 3$

Input	Rule	Output
x	$-x + 3$	y
−2		
0		
2		

3. $y = 2x^2$

Input	Rule	Output
x	$2x^2$	y
−5		
1		
3		

See Example 2 Make a function table, and graph the resulting ordered pairs.

4. $y = 3x - 2$

Input	Rule	Output	Ordered Pair
x	$3x - 2$	y	(x, y)
−1			
0			
1			
2			

5. $y = x^2 + 2$

Input	Rule	Output	Ordered Pair
x	$x^2 + 2$	y	(x, y)
−1			
0			
1			
2			

INDEPENDENT PRACTICE

See Example 1 Find the output for each input.

6. $y = -2x$

Input	Rule	Output
x	$-2x$	y
−2		
0		
4		

7. $y = 3x + 2$

Input	Rule	Output
x	$3x + 2$	y
−3		
−1		
2		

8. $y = 3x^2$

Input	Rule	Output
x	$3x^2$	y
−10		
−6		
−2		

See Example 2 Make a function table, and graph the resulting ordered pairs.

9. $y = x \div 2$

Input	Rule	Output	Ordered Pair
x	$x \div 2$	y	(x, y)
−1			
0			
1			
2			

10. $y = x^2 - 4$

Input	Rule	Output	Ordered Pair
x	$x^2 - 4$	y	(x, y)
−1			
0			
1			
2			

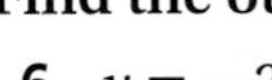

PRACTICE AND PROBLEM SOLVING

11. ***PHYSICAL SCIENCE*** The equation $F = \frac{9}{5}C + 32$ gives the Fahrenheit temperature F for a given Celsius temperature C. Make a function table for the values $C = -20°, -5°, 0°, 20°$, and $100°$.

12. ***HEALTH*** You burn about 3 calories a minute paddling a canoe. The equation $y = 3x$, where y is the number of calories burned and x is the number of minutes, describes your calorie use. Make a function table using the values $x = 1, 2, 3, 4$, and 5. Then graph the ordered pairs.

13. ***CONSUMER MATH*** Sharlyn pays \$15 annually for a pool membership and \$2 each time she swims. The equation $y = 2x + 15$ gives her cost y to swim x times per year. Make a function table using $x = 0, 1, 5$, and 10. Then graph the ordered pairs.

Weather LINK

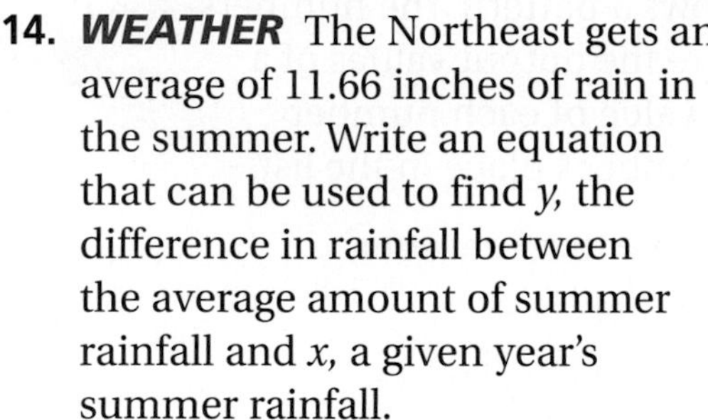

Most of Jordan is desert, but parts of the country have a rainy season. The heavy rains in Jordan flooded the steps where this photo was taken.

go.hrw.com KEYWORD: MS4 Weather

CNN Student News

14. ***WEATHER*** The Northeast gets an average of 11.66 inches of rain in the summer. Write an equation that can be used to find y, the difference in rainfall between the average amount of summer rainfall and x, a given year's summer rainfall.

a. Make a function table using each year's summer rainfall data.

b. What do the values of y in the table tell you?

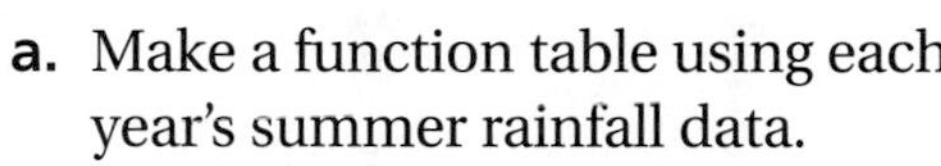

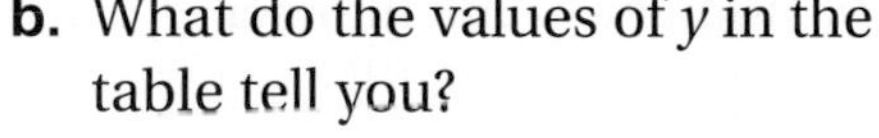

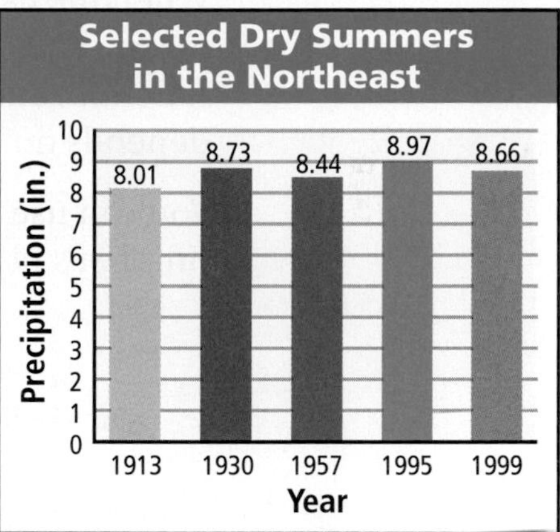

Source: *USA Today*, August 17, 2001

15. ***WHAT'S THE ERROR?*** What is the error in the function table at right?

x	$y = x^3$	y
−2	$y = (-2)^3$	8
−1	$y = (-1)^3$	1
0	$y = 0^3$	0
1	$y = 1^3$	1

16. ***WRITE ABOUT IT*** Explain how to make a function table for $y = 2x + 11$.

17. ***CHALLENGE*** Mountain Rental charges a \$25 deposit plus \$10 per hour to rent a bicycle. Write an equation that gives the cost y to rent a bike for x hours, and then write the ordered pairs for $x = \frac{1}{2}$, 5, and $8\frac{1}{2}$.

Spiral Review

Find each probability if you roll a number cube labeled 1–6. (Lesson 10-4)

18. $P(1)$ **19.** $P(2 \text{ or } 3)$ **20.** $P(\text{even number})$

Solve each equation. (Lesson 11-1)

21. $3x + 5 = 14$ **22.** $\frac{y}{2} - 7 = 5$ **23.** $5x - 3x = 11$

24. **EOG PREP** Which of the following is the solution to $4x + 3 < 11$? (Lesson 11-7)

A $x > 3.5$ B $x > -3.5$ C $x < 2$ D $x > -2$

12-2 Find a Pattern in Sequences

 Problem Solving Skill

Learn to find patterns to complete sequences using function tables.

Vocabulary
sequence

A **sequence** is an ordered list of numbers. One of the most well-known sequences is the Fibonacci sequence. In this sequence, each term after the second term is the sum of the two terms before it.

1, 1, 2, 3, 5, 8, 13, . . .

Many natural things, such as the arrangement of seeds in the head of a sunflower, follow the pattern of the Fibonacci sequence.

When the list follows a pattern, the numbers in the sequence are the output values of a function, and the value of each number depends on the number's place in the list.

You can use a variable to represent a number's place in a sequence.

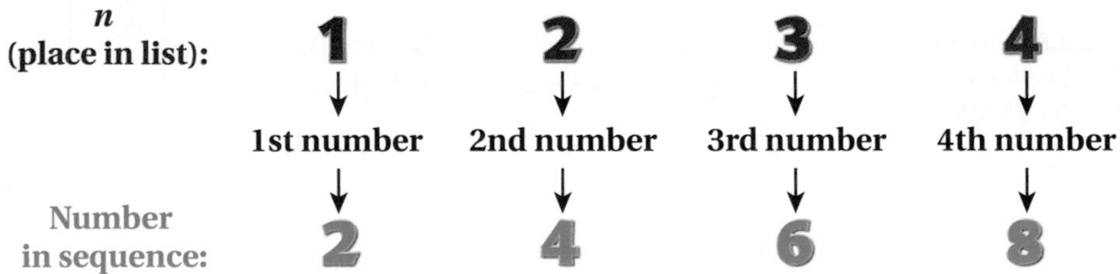

You can use a function table to help identify the pattern in a sequence.

EXAMPLE 1 Identifying Functions in Sequences

Find a function that describes each sequence. Use *y* for the term in the sequence and *n* for its place in the list. Then use the function to find the next three terms in the sequence.

A 2, 4, 6, 8, . . .

n	Rule	y
1	1 · 2	2
2	2 · 2	4
3	3 · 2	6
4	4 · 2	8

Multiply n by 2.

$y = 2n$ *Write the function.*
$y = 2(5) = 10$ *Substitute for n*
$y = 2(6) = 12$ *to find the next*
$y = 2(7) = 14$ *three terms.*

B 4, 5, 6, 7, . . .

n	Rule	y
1	1 + 3	4
2	2 + 3	5
3	3 + 3	6
4	4 + 3	7

Add 3 to n.

$y = n + 3$ *Write the function.*
$y = 5 + 3 = 8$ *Substitute for n*
$y = 6 + 3 = 9$ *to find the next*
$y = 7 + 3 = 10$ *three terms.*

EXAMPLE 2 PROBLEM SOLVING APPLICATION

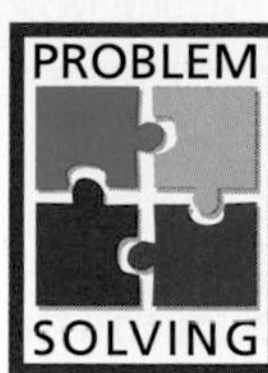

Kevin has \$20.00. Every day, he spends \$1.80 on bus fare. Write a sequence to show how much money Kevin will have left after 1, 2, 3, and 4 days.

1 Understand the Problem

List the **important information:**

- Kevin has \$20.00.
- He spends \$1.80 each day.

The **answer** will be the amount of money he has left after 1, 2, 3, and 4 days.

2 Make a Plan

You can find a pattern, then write a rule that can be used to find a sequence.

After the first day, Kevin will have \$20.00 − \$1.80.

After the second day, Kevin will have \$20.00 − 2 · \$1.80.

The rule \$20.00 − n · \$1.80 can be used to find the terms in the sequence.

3 Solve

n	Rule	y
1	20 − 1 · 1.80	18.20
2	20 − 2 · 1.80	16.40
3	20 − 3 · 1.80	14.60
4	20 − 4 · 1.80	12.80

After 1, 2, 3, and 4 days, Kevin will have \$18.20, \$16.40, \$14.60, and \$12.80 left, respectively.

4 Look Back

If Kevin spends about \$2 per day, he will have about \$18, \$16, \$14, and \$12 left after each of the 4 days, respectively. The answer is reasonable.

Think and Discuss

1. **Give an example** of a sequence involving addition, and give the rule you used.
2. **Describe** how to find a pattern in the sequence 1, 4, 16, 64, ...

12-2 Exercises

FOR EOG PRACTICE

see page 706

internet connect
Homework Help Online
go.hrw.com Keyword: MS4 12-2

1.02d, 5.01

GUIDED PRACTICE

See Example 1

Find a function that describes each sequence. Use *y* for the term in the sequence and *n* for its place in the list. Then use the function to find the next three terms in the sequence.

1. 3, 6, 9, 12, . . . **2.** 3, 4, 5, 6, . . . **3.** 0, 1, 2, 3, . . . **4.** 5, 10, 15, 20, . . .

See Example 2

5. Sara ran 20 miles last week. This week, she ran 3 miles each day for five days. Write a sequence to show the total distance, including the first 20 miles, that Sara had run after each of the last 5 days.

INDEPENDENT PRACTICE

See Example 1

Find a function that describes each sequence. Use *y* for the term in the sequence and *n* for its place in the list. Then use the function to find the next three terms in the sequence.

6. 1.5, 2.5, 3.5, 4.5, . . .

7. $\frac{1}{2}, 1, \frac{3}{2}, 2, \ldots$

8. 7, 14, 21, 28, . . .

9. −2, −1, 0, 1, . . .

10. 20, 40, 60, 80, . . .

11. 3, 5, 7, 9, . . .

12. 5, 6, 7, 8, . . .

13. 1, 4, 9, 16, . . .

See Example 2

14. Macy purchased a box of cereal that contained 567 grams of cereal. Each day, he ate 52 grams for breakfast. Write a sequence to show how much cereal he had left after each of the 4 days.

15. Shaundra opened a savings account with $50.00. Every week, she added $2.50. Write a sequence to show how much money she had in her account after 1, 2, 3, 4, 5, and 6 weeks.

PRACTICE AND PROBLEM SOLVING

Write the rule for each sequence in words. Then find the next three terms.

16. 35, 70, 105, 140, . . .

17. 0.7, 1.7, 2.7, 3.7, . . .

18. $\frac{3}{2}, \frac{5}{2}, \frac{7}{2}, \frac{9}{2}, \ldots$

19. −1, 0, 1, 2, . . .

20. $\frac{1}{3}, \frac{2}{3}, 1, \frac{4}{3}, \ldots$

21. 6, 11, 16, 21, . . .

Find a function that describes each sequence. Use the function to find the tenth term in the sequence.

22. 0.5, 1.5, 2.5, 3.5, . . .

23. 0, 2, 4, 6, . . .

24. 5, 8, 11, 14, . . .

25. 3, 8, 13, 18, . . .

26. 1, 3, 5, 7, . . .

27. 6, 10, 14, 18, . . .

28. Dominique ran 40 minutes on Saturday, 1 hour 20 minutes on Monday, and 2 hours on Wednesday. Use the sequence to predict the next day that Dominique ran and how long he ran on that day.

Computer Science 

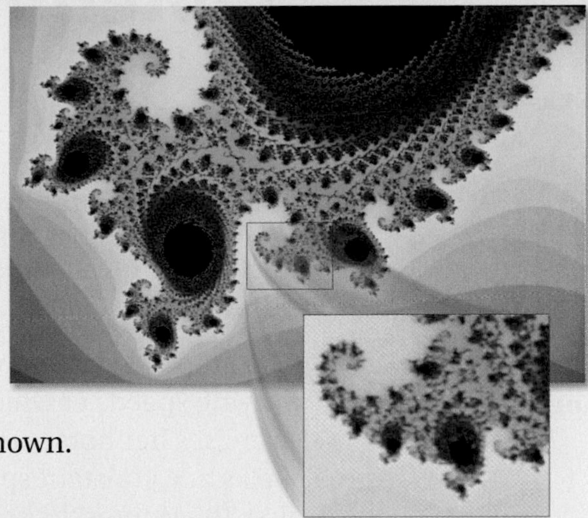

Computer programmers use functions to create beautiful designs known as *fractals.* A fractal is a *self-similar* pattern, which means that each part of the pattern is similar to the whole pattern. Fractals are created by repeating a set of steps, called *iterations.* Use this information for Exercises 29–31.

29. Below is part of a famous fractal called Cantor dust. In each iteration, part of a line segment is removed, resulting in twice as many segments as before. The table lists the number of line segments that result from the iterations shown. Find a function that describes the sequence.

Iteration (n)	Number of Segments (y)
1	2
2	4
3	8

30. These are the first three iterations of another famous fractal called the Sierpinski triangle. In each iteration, a certain number of smaller triangles are cut out of the larger triangle in the pattern shown.

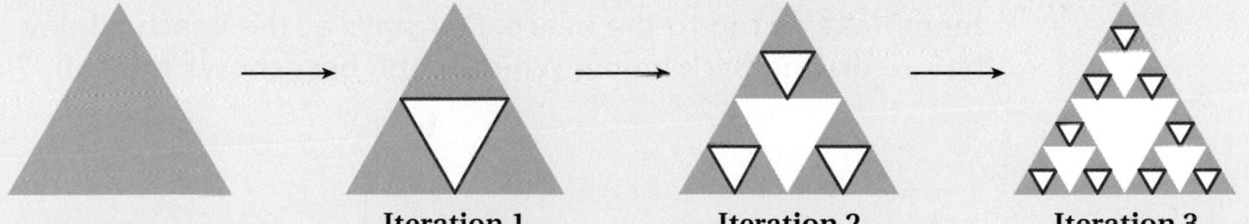

Iteration 1 1 triangle removed — **Iteration 2** 3 more triangles removed — **Iteration 3** 9 more triangles removed

Create a table to list the total number of yellow triangles that exist after each iteration. Then find a function that describes the sequence.

31. ***CHALLENGE*** Find a function that describes the number of triangles removed in each iteration of the Sierpinski triangle.

go.hrw.com
KEYWORD: MS4 Fractals
CNN Student News

Spiral Review

Determine the number of combinations. (Lesson 10-6)

32. 6 objects taken 2 at a time

33. 8 objects taken 7 at a time

34. **EOG PREP** Martin spent $26 for 4 equally priced CDs. Which equation could be used to find how much each CD cost? (Lesson 2-12)

A $4 \cdot 26 = n$ B $n = 26 - 4$ C $4 \cdot n = 26$ D $n \cdot 26 = 4$

12-3 Interpreting Graphs

Learn to relate graphs to situations.

Vocabulary
domain
range

You can use a graph to show the relationship between speed and time, time and distance, or speed and distance.

The graph at right shows the varying speeds at which Emma exercises her horse. The horse walks at a constant speed for the first 10 minutes. Its speed increases over the next 7 minutes, and then it gallops at a constant rate for 20 minutes. Then it slows down over the next 3 minutes and then walks at a constant pace for 10 minutes.

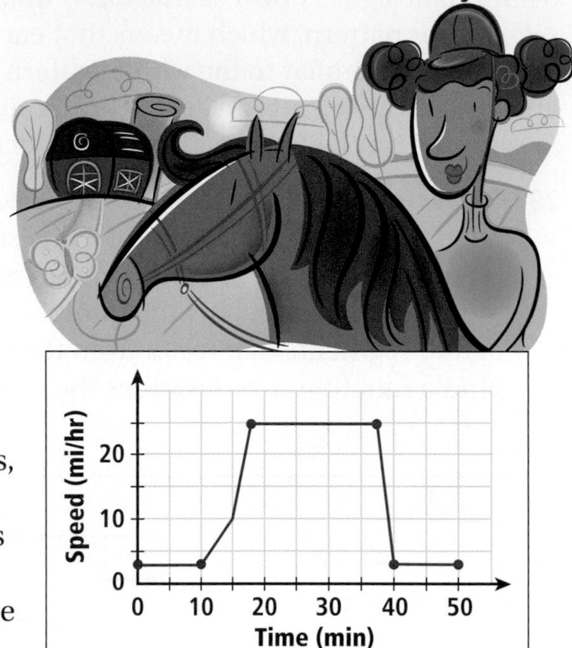

EXAMPLE 1 Relating Graphs to Stories

Jenny takes a trip to the beach. She stays at the beach all day before driving back home. Which graph best shows the story?

a.

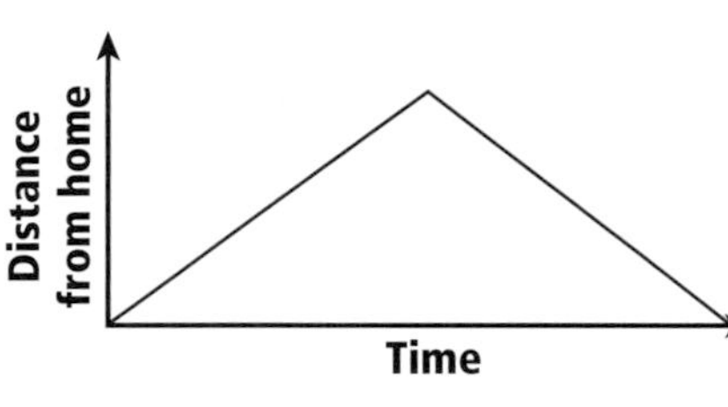

b.

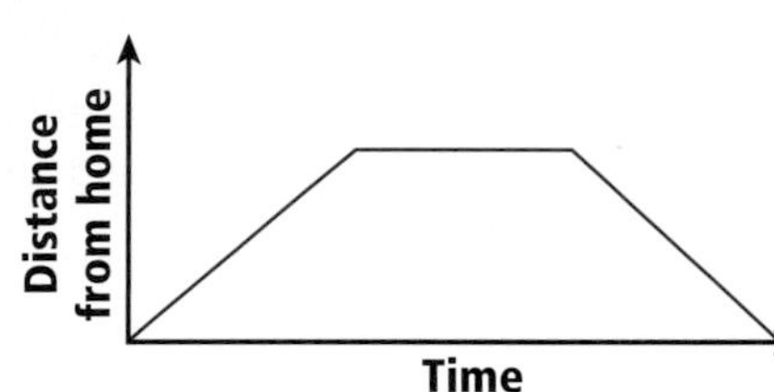

c.

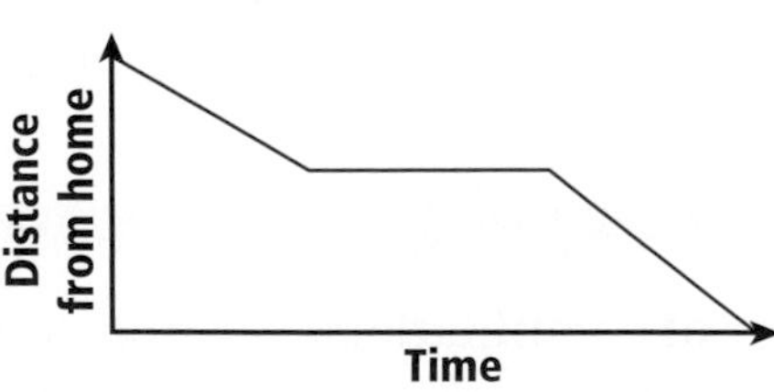

As Jenny drives to the beach, her distance from home *increases*. While she is at the beach, her distance from home is *constant*. As she returns home, her distance from home *decreases*. The answer is graph b.

EXAMPLE 2 Using a Graph to Tell a Story

Draw a graph for each situation.

> **Helpful Hint**
>
> Some graphs are curves or lines that show continuous data. Other graphs may show only points.

A **Maria is selling boxes of greeting cards for $5 a box. Draw a graph to show her possible income from sales.**

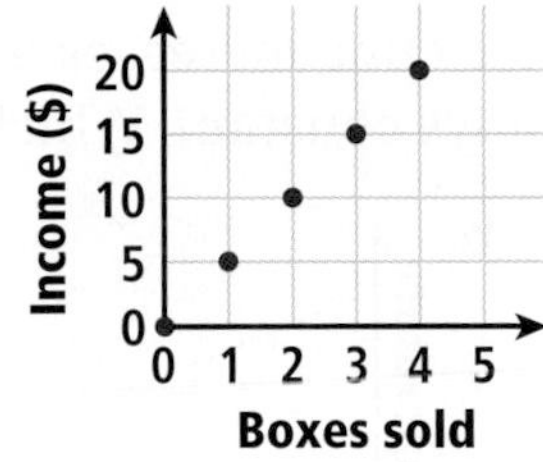

Each box is $5, so the amount of money she can make is a multiple of $5. This graph is only points.

B **Pecans cost $3 per pound, and they can be sold in fractions of a pound. Draw a graph to show how much the pecans can cost.**

One pound costs $3, two pounds cost $6, and so on. However, you can buy any weight of pecans. This graph is continuous.

The set of input values of a function is the **domain**, and the set of resulting output values is the **range**. In the graph of boxes sold, the domain is the possible number of boxes sold, or the set of whole numbers. The range in this graph is Maria's possible income, or all nonnegative multiples of $5.

EXAMPLE 3 Finding Domain and Range

Find the domain and range of the graph.

> **Helpful Hint**
>
> When a graph is continuous, its domain and range contain all real numbers or those within a certain interval.

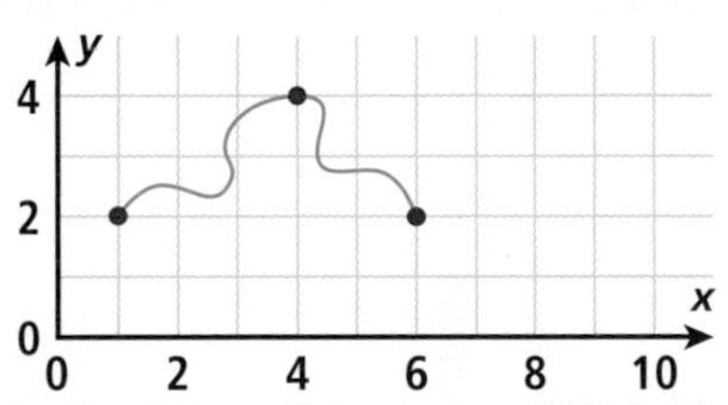

The graph goes from 1 to 6 on the x-axis and from 2 to 4 on the y-axis.

Domain: all real numbers from 1 through 6; $1 \leq x \leq 6$

Range: all real numbers from 2 through 4; $2 \leq y \leq 4$

Think and Discuss

1. **Give an example** of a situation that could be represented by a graph that shows the range increasing as the domain increases.
2. **Compare** domain and range of a function.

12-3 Exercises

FOR EOG PRACTICE
see page 706

internet connect
Homework Help Online
go.hrw.com Keyword: MS4 12-3

5.01

GUIDED PRACTICE

See Example 1

1. The temperature of an ice cube increases until it starts to melt. While it melts, its temperature stays constant. Which graph best shows the story?

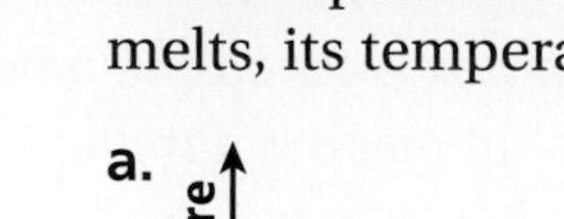

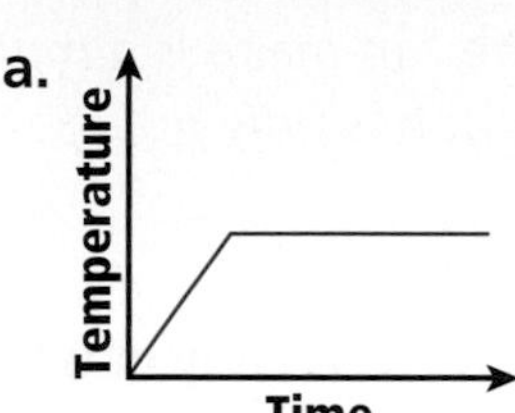

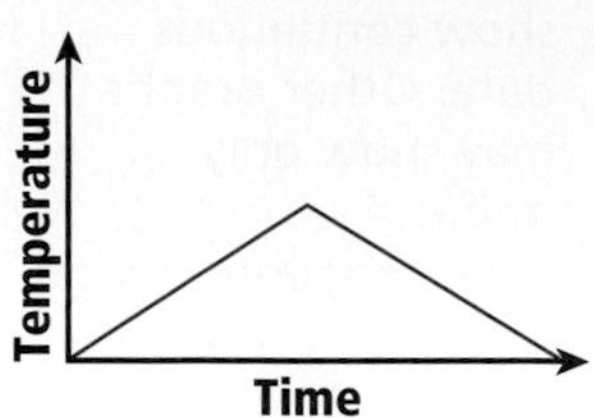

a. b. c.

See Example 2

2. Kim sells blueberries for $3 a box. Draw a graph to show how much he can make.

See Example 3

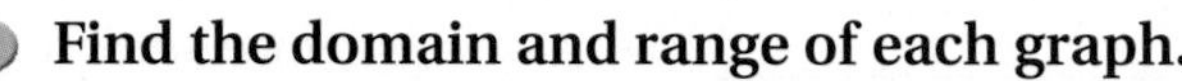

Find the domain and range of each graph.

3.
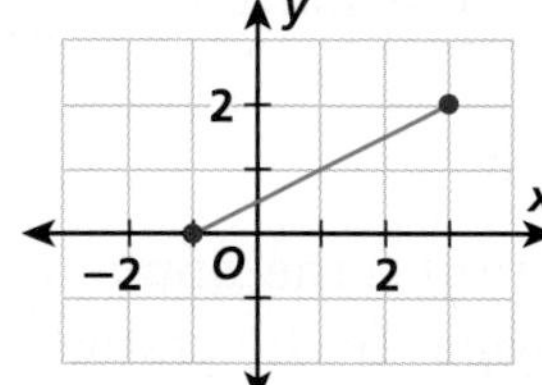

4.
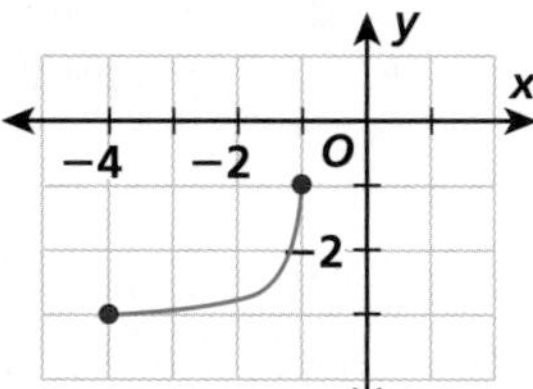

5.
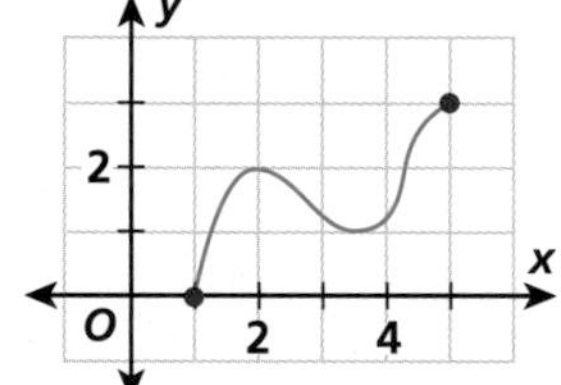

INDEPENDENT PRACTICE

See Example 1

6. The ink in a printer is used until the ink cartridge is empty. The cartridge is refilled, and the ink is used up again. Which graph best shows the story?

a.
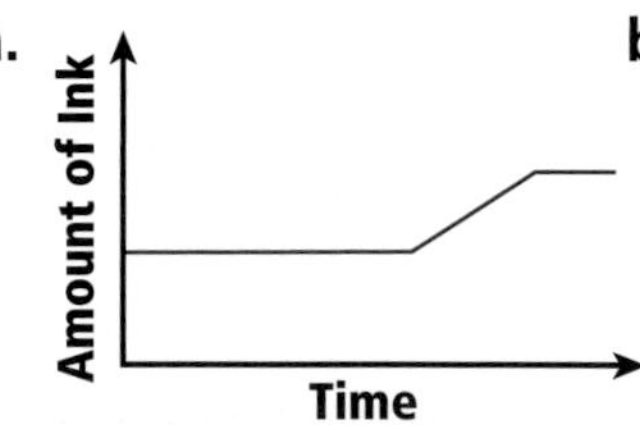

b.
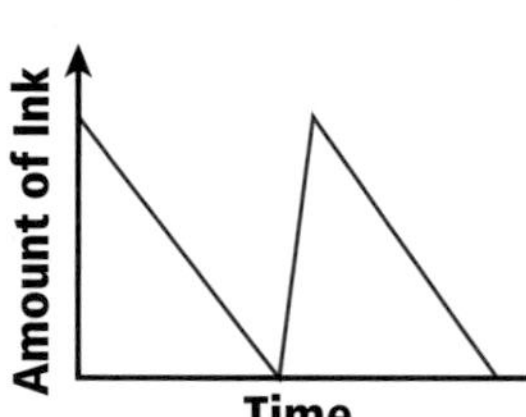

c.
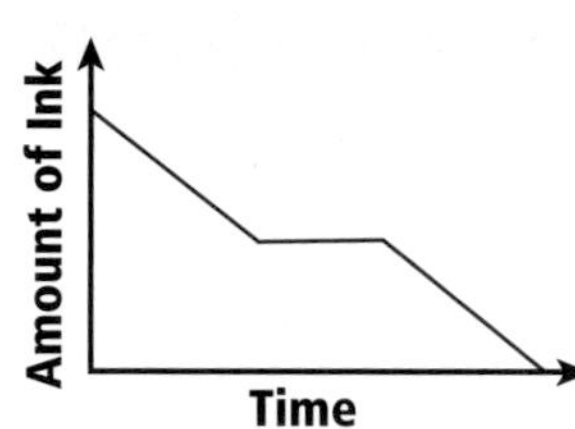

See Example 2

7. Elena uses 4 gallons of water per minute to fill plastic bottles. Draw a graph to show how much water she uses.

See Example 3

Find the domain and range of each graph.

8.
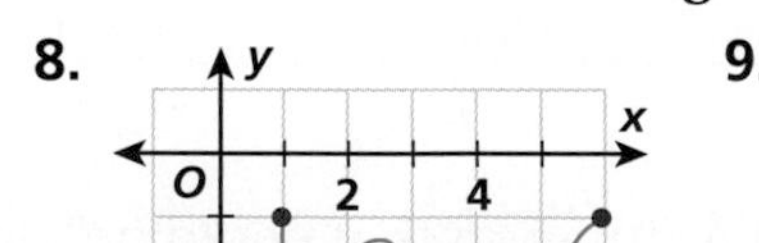

9.
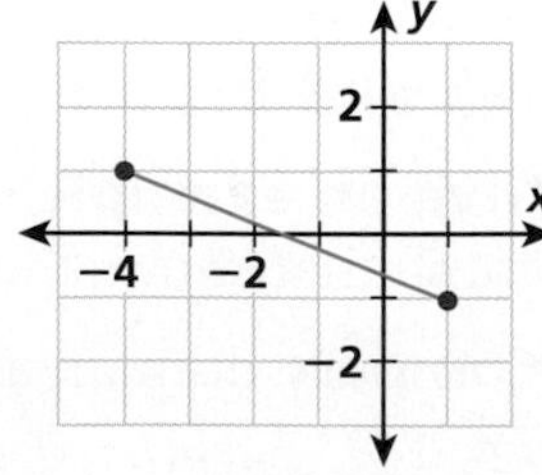

10.
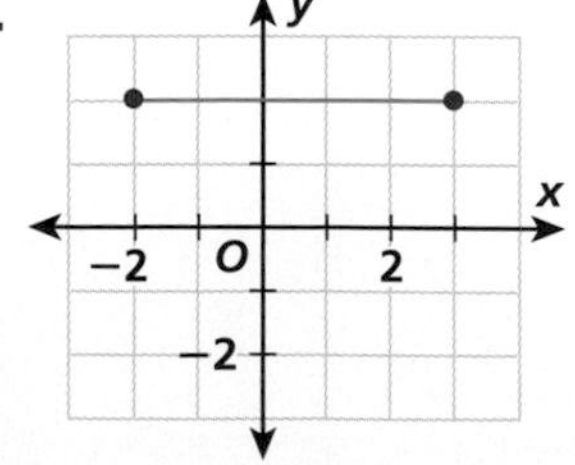

PRACTICE AND PROBLEM SOLVING

11. Tell a story that fits the graph at right.

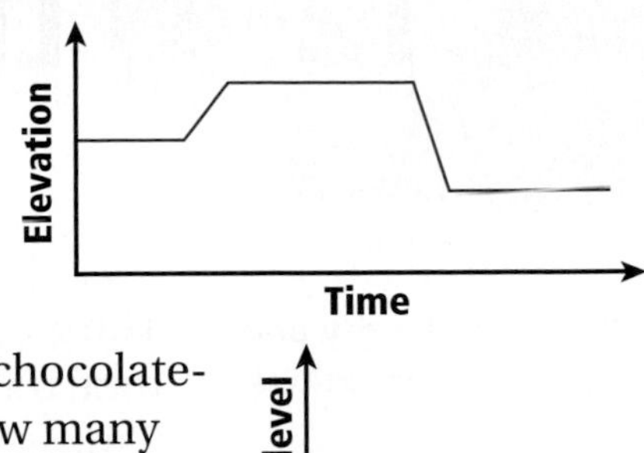

12. ***SPORTS*** A basketball team plays 2 games a week. Make a graph showing the total number of games the team can play in a 12-week season.

13. ***NUTRITION*** There are 60 calories in each chocolate-crunch rice cake. Make a graph to show how many calories you can consume eating these rice cakes.

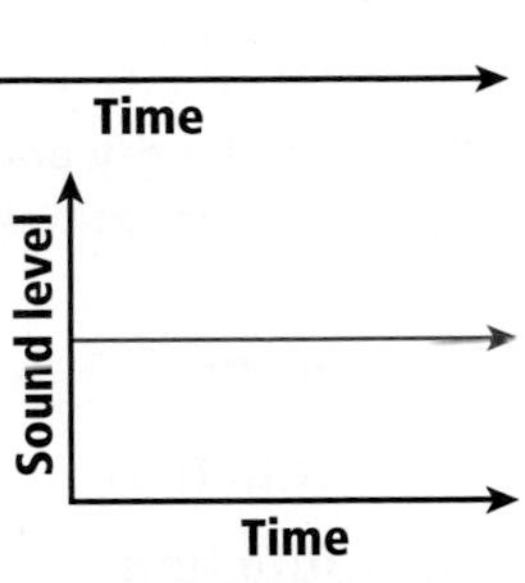

14. The graph at right shows sound being recorded in a studio. Interpret the graph.

15. The graph at right shows high school enrollment, including future projections.
    a. Give the domain of the graph.
    b. Give the range of the graph.

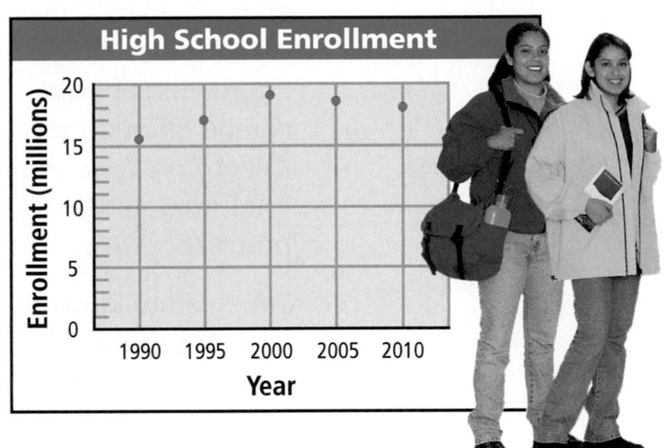

16. ***CHOOSE A STRATEGY*** Three bananas were given to two mothers who were with their daughters. Each person had a banana to eat. How is that possible?

17. ***WRITE ABOUT IT*** Explain the difference between a graph made up of single points and a continuous graph.

18. ***CHALLENGE*** The graph of a line segment represents the function $y = 3x$, where the domain is all real numbers from -2 to 4, including -2 and 4. What is the range of the graph?

Spiral Review

Estimate each value. (Lesson 6-2)

19. 18% of 21 **20.** 35% of 88 **21.** 51% of 39

Determine the number of lines of symmetry in each figure. (Lesson 7-11)

22.

23.

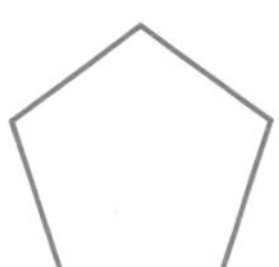

24.

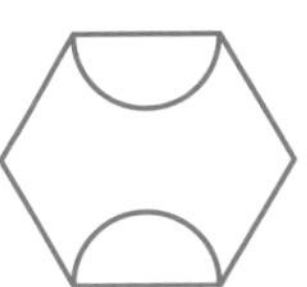

25. **EOG PREP** By what will the volume of a rectangular prism be multiplied if you double each dimension? (Lesson 9-5)

A 8 B 4 C 2 D 1

12-4 Linear Functions

Learn to identify and graph linear equations.

Vocabulary

linear equation

linear function

The graph below shows how far an inner tube travels down a river if the current flows 2 miles per hour. The graph is linear because all of the points fall on a line. It is part of the graph of a *linear equation.*

A **linear equation** is an equation whose graph is a line. The solutions of a linear equation are the points that make up its graph. Linear equations and linear graphs can be different representations of *linear functions.* A **linear function** is a function whose graph is a nonvertical line.

You need to know only two points to draw the graph of a linear function. However, graphing a third point serves as a check. You can use a function table to find each ordered pair.

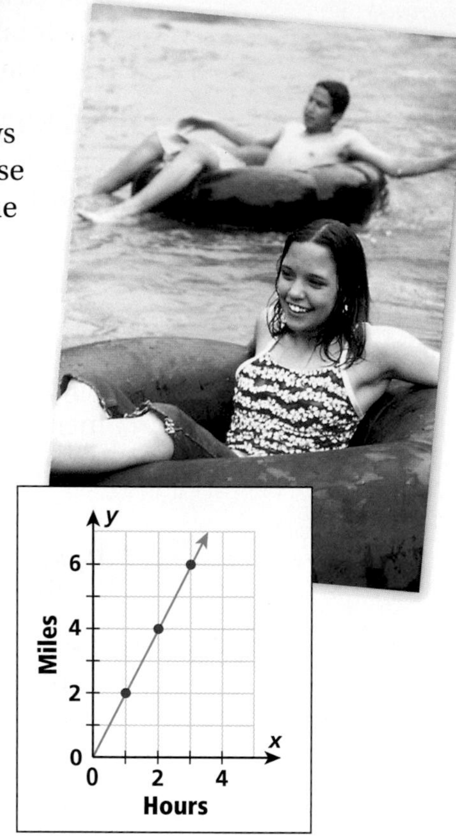

EXAMPLE 1 **Graphing Linear Functions**

Graph each linear function.

A $y = 2x + 1$

Input	Rule	Output	Ordered Pair
x	$2x + 1$	y	(x, y)
0	2(0) + 1	1	(0, 1)
2	2(2) + 1	5	(2, 5)
−2	2(−2) + 1	−3	(−2, −3)

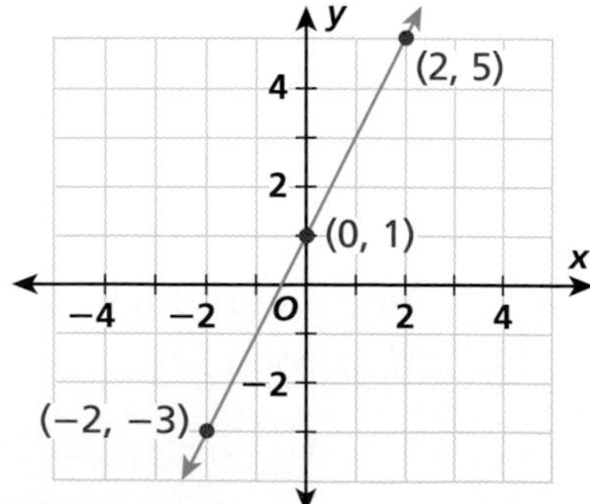

Place each ordered pair on the coordinate grid, and then connect the points with a line.

Graph each linear function.

B $y = 4$

The equation $y = 4$ is the same equation as $y = 0x + 4$.

Input	Rule	Output	Ordered Pair
x	$0x + 4$	y	(x, y)
0	0(0) + 4	4	(0, 4)
3	0(3) + 4	4	(3, 4)
−1	0(−1) + 4	4	(−1, 4)

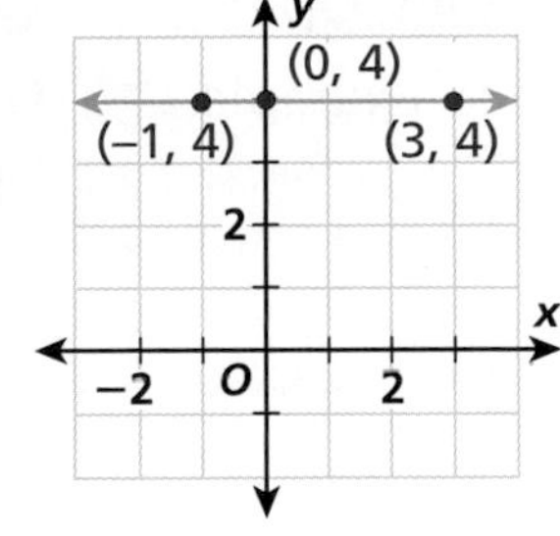

EXAMPLE 2 *Earth Science Application*

The Atlantic Ocean floor is spreading by 4 cm each year. Scientists began studying two parts of the ocean floor when they were 10 cm apart. Write a linear function that describes the spread of the ocean floor over time. Then make a graph to show how far the ocean floor will spread over the next 10 years.

The function is $y = 4x + 10$, where x is the number of years and y is the spread in centimeters.

The dark red region in the center of the map is the midocean ridge, where the ocean floor is spreading.

Input	Rule	Output
x	$4x + 10$	y
0	4(0) + 10	10
5	4(5) + 10	30
10	4(10) + 10	50

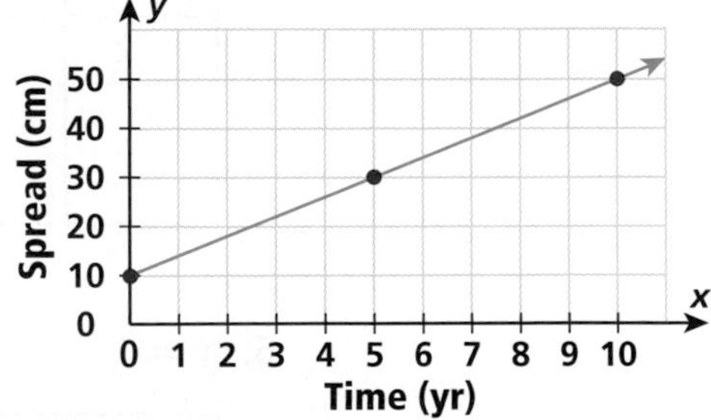

Since each output y depends on the input x, y is called the *dependent variable* and x is called the *independent variable.*

Think and Discuss

1. **Describe** how a linear equation is related to a linear graph.
2. **Explain** what $x = -1$ would mean in Example 2.

12-4 Exercises

FOR EOG PRACTICE
see page 707

internet connect
Homework Help Online
go.hrw.com Keyword: MS4 12-4

5.01

GUIDED PRACTICE

See Example 1 **Graph each linear function.**

1. $y = x + 3$

Input	Rule	Output	Ordered Pair
x	$x + 3$	y	(x, y)
−2			
0			
2			

2. $y = 2x - 2$

Input	Rule	Output	Ordered Pair
x	$2x - 2$	y	(x, y)
−2			
0			
2			

See Example 2

3. A water tanker is used to fill a community pool. The tanker pumps 750 gallons of water per hour. Write a linear function that describes the amount of water in the pool over time. Then make a graph to show the amount of water in the pool over the first 6 hours.

INDEPENDENT PRACTICE

See Example 1 **Graph each linear function.**

4. $y = -x + 2$

Input	Rule	Output	Ordered Pair
x	$-x + 2$	y	(x, y)
−2			
0			
2			

5. $y = \frac{x}{2} + 1$

Input	Rule	Output	Ordered Pair
x	$\frac{x}{2} + 1$	y	(x, y)
−2			
0			
2			

6. $y = 1$

Input	Rule	Output	Ordered Pair
x	$0x + 1$	y	(x, y)
−2			
0			
2			

7. $y = 2x + 3$

Input	Rule	Output	Ordered Pair
x	$2x + 3$	y	(x, y)
−2			
0			
2			

See Example 

8. The temperature of a liquid is increasing at the rate of 3°C per hour. When Joe begins measuring the temperature, it is 40°C. Write a linear function that describes the temperature of the liquid over time. Then make a graph to show the temperature over the first 12 hours.

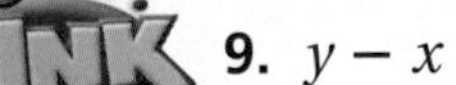

PRACTICE AND PROBLEM SOLVING

Solve each equation for *y*. Then graph the function.

9. $y - x = 0$ **10.** $2x + y = 1$ **11.** $6y = -12$

12. $2y - 5x + 2 = 0$ **13.** $\frac{y}{2} - 3 = 0$ **14.** $2y + x + 6 = 0$

Environment LINK

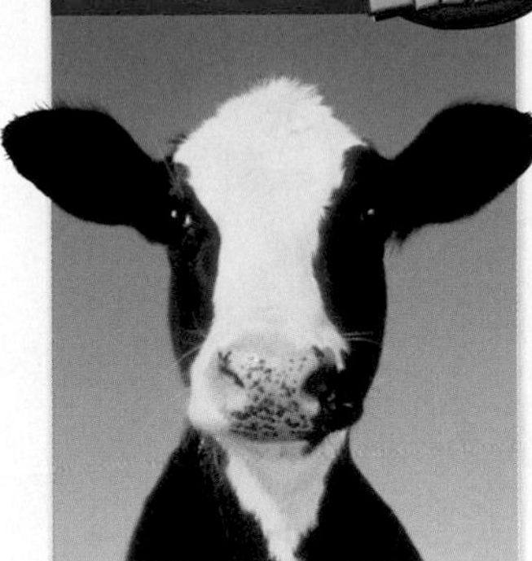

About 15% of the methane gas in the atmosphere comes from farm animals such as cows and sheep.

15. ***ENVIRONMENT*** The graph shows the concentration of carbon dioxide in the atmosphere from 1958 to 1994.

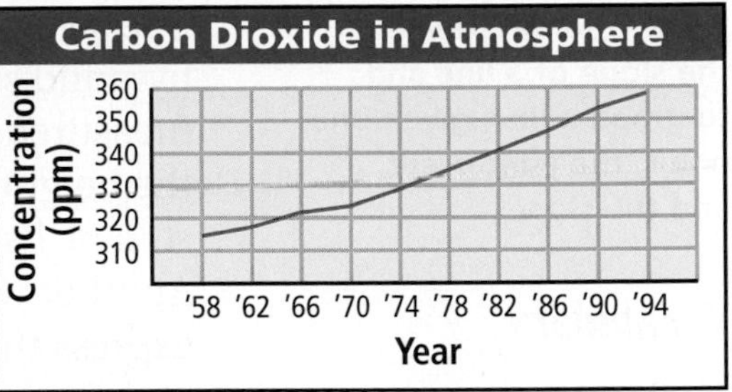

a. The graph is approximately linear. About how many parts per million (ppm) were added each 4-year period?

b. Given the parts per million in 1994 shown on the graph, about how many parts per million do you predict there will be after four more 4-year periods, or in 2010?

16. ***EARTH SCIENCE*** Water is seeping into an underground water supply at a rate of 10 centimeters per year. It is being pumped out at a rate of 2 meters per year. If the depth of the water in the water supply is 100 meters deep, how deep will it be in 10 years?

17. ***WHAT'S THE QUESTION?*** Tron used the equation $y = 100 + 25x$ to track his savings *y* after *x* months. If the answer is $250, what is the question?

18. ***WRITE ABOUT IT*** Explain how to graph $y = 2x - 5$.

19. ***CHALLENGE*** Bacteria of a certain species divide every 30 minutes. To find how many bacteria there are after each half-hour period, starting with one bacterium, you can use the function $y = 2^x$, where *x* is the number of half-hour periods. Make a table of values for $x = 1, 2, 3, 4$, and 5. Graph the points. How does the graph differ from those you have seen so far in this lesson?

Spiral Review

Name the polygon with the given number of sides. (Lesson 7-5)

20. five **21.** eight **22.** six **23.** three

Find the area of each parallelogram. (Lesson 8-4)

24.

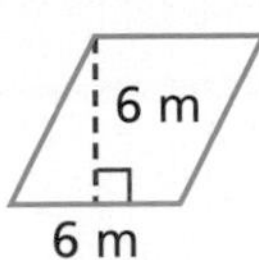

25.

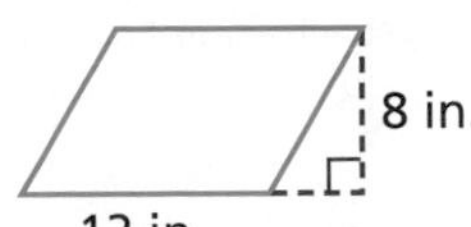

26.

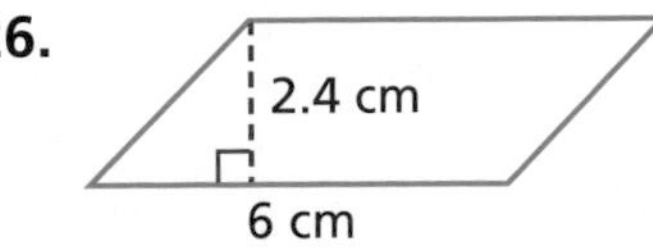

27. **EOG PREP** Using the pattern in the sequence 3, 2, 0, −3, . . . , which is the next integer in the sequence? (Lesson 12-2)

A 0 B −7 C −6 D −4

12-5 Slope

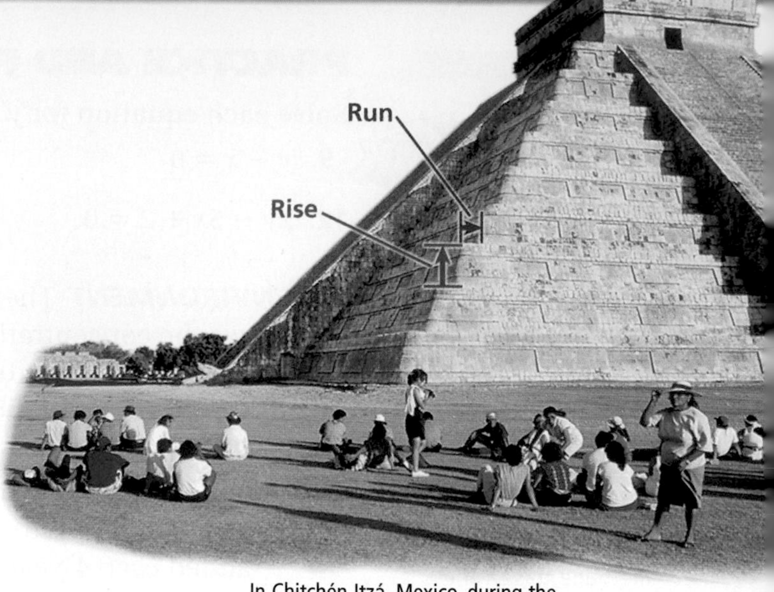

In Chitchén Itzá, Mexico, during the spring and fall equinoxes, shadows fall on the pyramid El Castillo, giving the illusion of a snake crawling down the steps.

Learn to determine the slope of a line and to graph a line, given one point, the *y*-intercept, and the slope.

Vocabulary

slope

***y*-intercept**

slope-intercept form

The steepness of the pyramid steps is measured by dividing the height of each step by its depth. Another way to express the height and depth is with the words *rise* and *run*.

The **slope** of a line is a measure of its steepness and is the ratio of rise to run:

$$\textbf{slope} = \frac{\textbf{rise}}{\textbf{run}}$$

The *y*-coordinate of the point where a line crosses the *y*-axis is called the ***y*-intercept**.

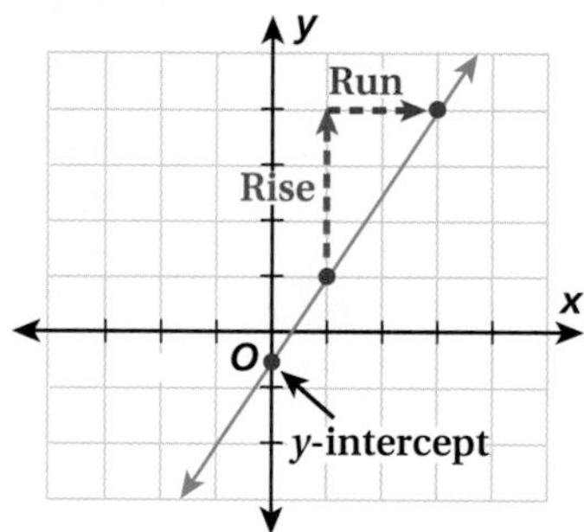

EXAMPLE 1 Identifying Slope and *y*-intercept of a Line

Give the slope and *y*-intercept of each line.

A

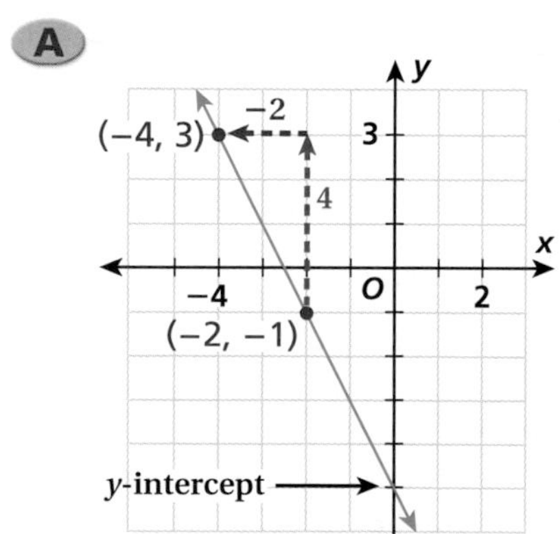

The rise is 4.

The run is –2.

slope $= \frac{\text{rise}}{\text{run}} = \frac{4}{-2} = -2$

The *y*-intercept is -5.

B

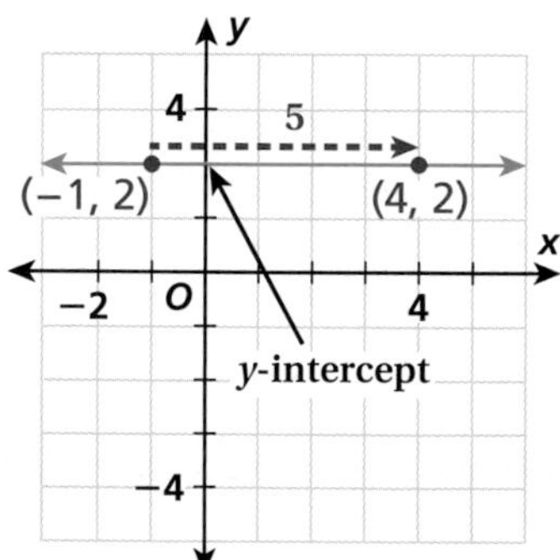

The rise is 0.

The run is 5.

slope $= \frac{\text{rise}}{\text{run}} = \frac{0}{5} = 0$

The *y*-intercept is 2.

A linear equation with the variable y isolated on the left-hand side of the equation is in **slope-intercept form**.

$$y = mx + b$$

Slope (m) y-intercept (b)

EXAMPLE 2 Identifying Slope and y-intercept in a Linear Equation

Give the slope and y-intercept of each line, given the equation.

A $y = 12x - 9$

$y = 12x - 9$

The slope is 12.

The y-intercept is -9.

B $y = \frac{4}{5}x$

$y = \frac{4}{5}x + 0$

The slope is $\frac{4}{5}$.

The y-intercept is 0.

When you know the slope and any one point on a line, you can graph the line.

EXAMPLE 3 Graphing Linear Equations Using Slope and y-intercept

Graph each line, given the equation.

A $y = -\frac{1}{3}x + 2$

$y = -\frac{1}{3}x + 2$

The slope is $-\frac{1}{3}$ or $\frac{1}{-3}$. The y-intercept is 2. From point (0, 2), move 1 unit up (rise) and 3 units left (run) to find another point on the line. Draw a line to connect the points.

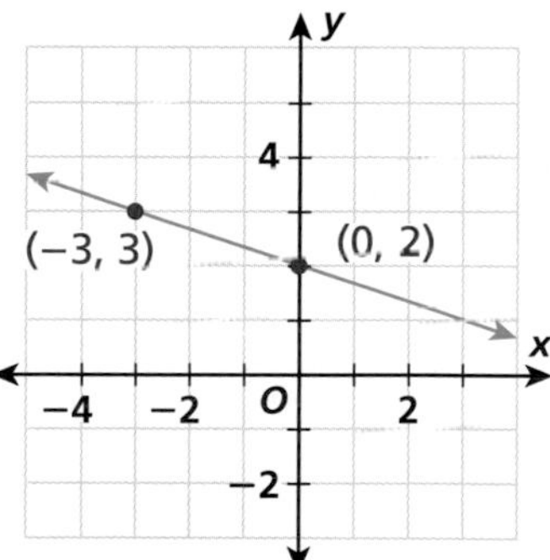

B $y = -1$

$y = 0x - 1$

The slope is 0. The y-intercept is -1. Locate the point $(0, -1)$. When the slope is zero, the line is horizontal. Draw a horizontal line through point $(0, -1)$.

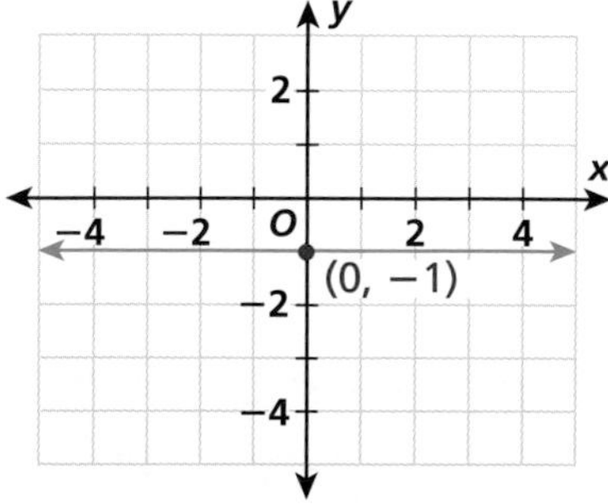

Think and Discuss

1. **Explain** why a horizontal line has a slope of 0.
2. **Describe** a line with a negative value for slope.
3. **Explain** how to write $-x + 2y = 8$ in slope-intercept form.

12-5 Exercises

FOR EOG PRACTICE
see page 707

internet connect
Homework Help Online
go.hrw.com Keyword: MS4 12-5

5.01

GUIDED PRACTICE

See Example 1 Give the slope and y-intercept of each line.

1.

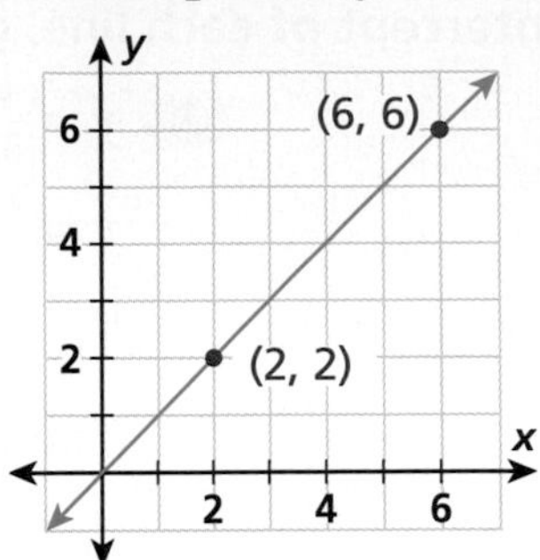

2.

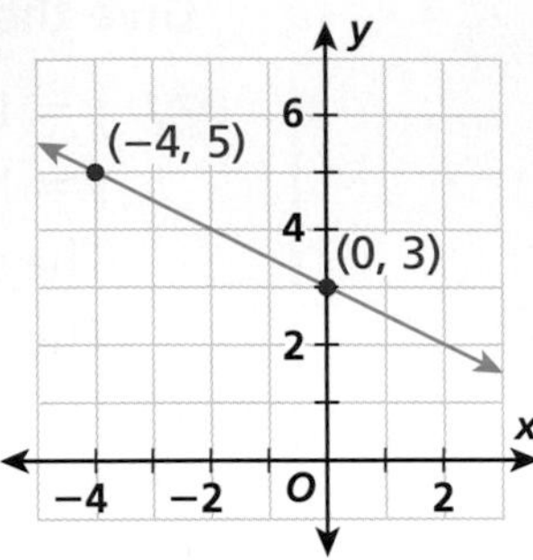

See Example 2 Give the slope and y-intercept of each line, given the equation.

3. $y = 3x - 4$
4. $y = \frac{1}{3}x$
5. $y = -3x + 8$

See Example 3 Graph each line, given the equation.

6. $y = -\frac{1}{4}x + 5$
7. $y = -5$
8. $y = \frac{2}{5}x - 2$

INDEPENDENT PRACTICE

See Example 1 Give the slope and y-intercept of each line.

9.

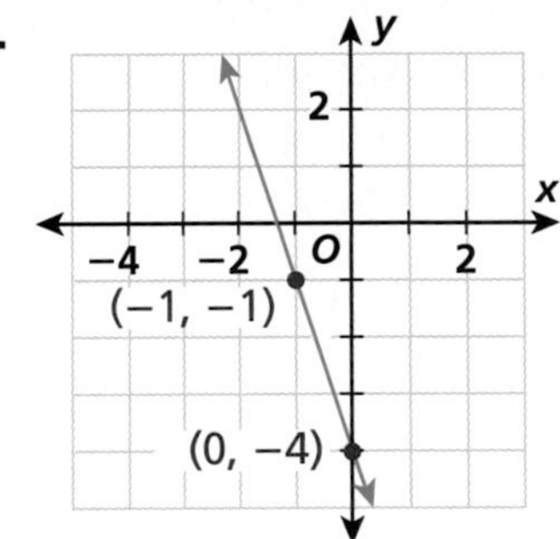

10.

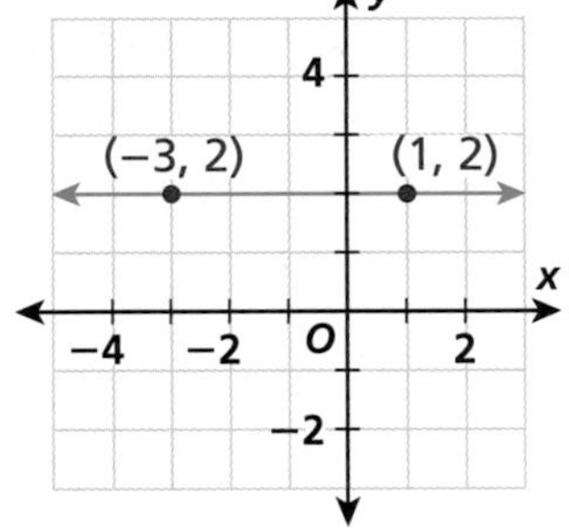

See Example 2 Give the slope and y-intercept of each line, given the equation.

11. $y = x - 2$
12. $y = \frac{2}{5}x$
13. $y = 4x + 2$

See Example 3 Graph each line, given the equation.

14. $y = -\frac{5}{6}x + 4$
15. $y = 3$
16. $y = \frac{1}{5}x - 1$

PRACTICE AND PROBLEM SOLVING

Write each equation in slope-intercept form. Then tell whether its graph has a positive or a negative slope.

17. $-3x + y = 8$
18. $3 + y = \frac{7}{8}x$
19. $y + 5x = 1$

Graph each line, given the equation. Solve for y when necessary.

20. $2x + y = -1$ **21.** $2x + 5y = 0$ **22.** $3y - 2x = 3$

Graph each pair of points. Find the slope and y-intercept of each line that connects the points in each pair.

23. (3, 2), (−2, −3) **24.** (−4, −3), (0, −4) **25.** (−3, 3), (3, 5)

Graph each pair of points. Find the slope and y-intercept of the line containing each pair of points. Then write an equation for the line in slope-intercept form.

26. (3, 2), (4, 2) **27.** (−2, −6), (2, 2) **28.** (−2, 13), (1, 4)

29. (−2, −3), (2, 3) **30.** (2, −3), (3, −5) **31.** (−5, −3), (4, −3)

32. A carpenter is building a staircase leading up to a loft. The slope of the staircase is $\frac{1}{2}$. If the staircase has 30 steps and the depth of each step is 8 inches (from its edge to the bottom of the next step), how high up is the loft?

33. ***CONSUMER MATH*** The table shows the value of a used car over time.

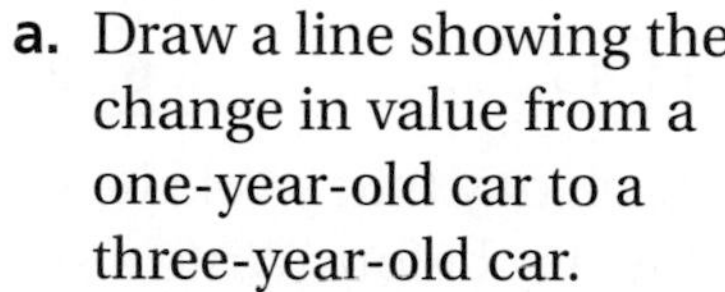

a. Draw a line showing the change in value from a one-year-old car to a three-year-old car.

b. Find the slope of the line.

c. Tell whether the slope is positive or negative.

d. What does the slope show about the value of the car?

Value of a Used Ford Mustang to the Nearest Thousand Dollars

Age	Value
1 year	$14,000
2 years	$13,000
3 years	$12,000

Source: Kelley Blue Book

34. ***WHAT'S THE QUESTION?*** A line has point A at (3, 2) and point B at (4, 3). What question could you answer about this line?

35. ***WRITE ABOUT IT*** Explain how you can tell whether the slope of a linear graph is positive or negative.

36. ***CHALLENGE*** Describe the slope of a line that contains points (3, 4) and (3, 6).

Spiral Review

Solve. (Lesson 11-6)

37. $\frac{b}{4} < 8$ **38.** $\frac{s}{-8} \geq 1.6$ **39.** $5m > 1.5$ **40.** $-9y \leq 108$

41. **EOG PREP** Using the pattern in the sequence −2, ▇, 0, 1, 2, . . . what is the missing integer? (Lesson 12-2)

A 3 B −3 C −1 D 1

42. **EOG PREP** What is the value of the expression $x - y$ for $x = 6$ and $y = -4$? (Lesson 2-7)

A −10 B −2 C 2 D 10

Chapter 12

Mid-Chapter Quiz

LESSON 12-1 (pp. 604–607)

Make a function table, and graph the resulting ordered pairs.

1. $y = 3x - 1$

Input	Rule	Output
x	$3x - 1$	y
−3		
−2		
−1		

2. $y = x \div (-3)$

Input	Rule	Output
x	$x \div (-3)$	y
−6		
−3		
0		

3. $y = x^2 + 2$

Input	Rule	Output
x	$x^2 + 2$	y
−2		
0		
2		

LESSON 12-2 (pp. 608–611)

Find a function that describes each sequence. Use the function to find the next three terms in the sequence.

4. 99, 199, 299, 399, . . .

5. 12, 13, 14, 15, . . .

6. 21, 41, 61, 81, . . .

7. Jamie put 15 cents in a jar on Sunday. Each day that week, he put double the amount that he had put in the jar the day before. Write a sequence to find how much money Jamie put in the jar on Friday.

LESSON 12-3 (pp. 612–615)

Find the domain and range of each graph.

8.

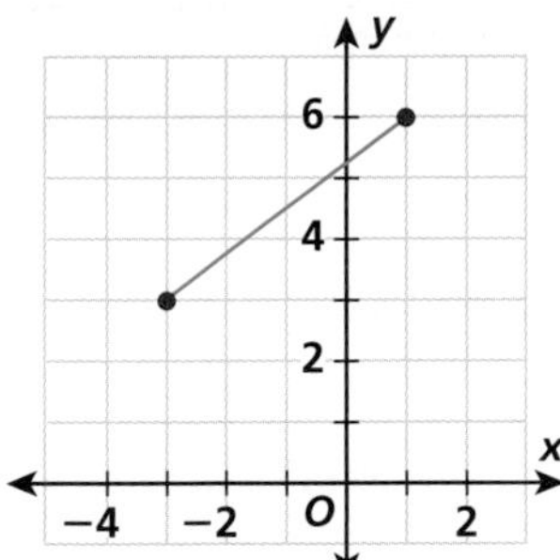

9.

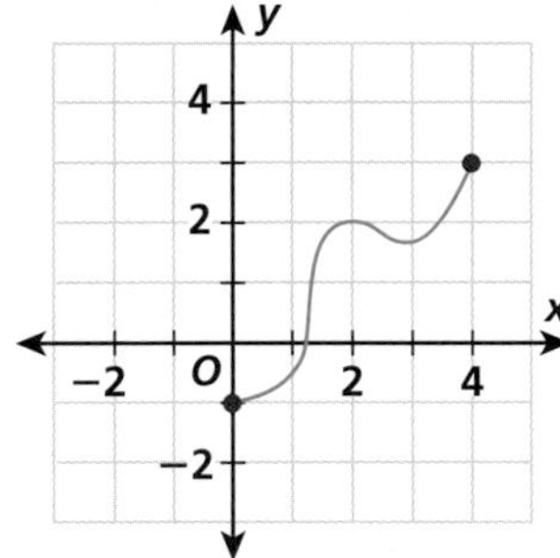

10.

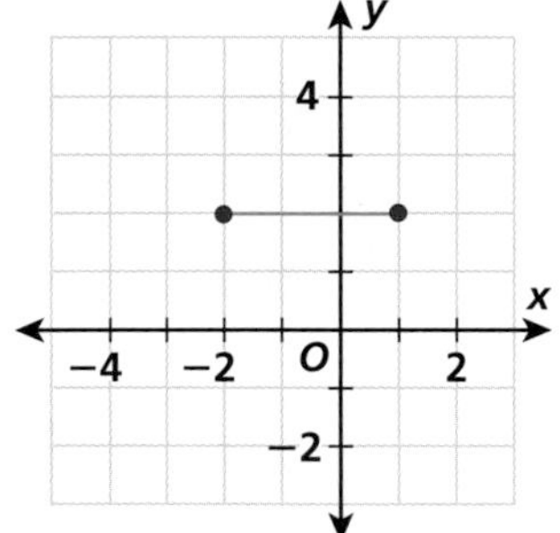

LESSON 12-4 (pp. 616–619)

Solve each equation for *y*. Then graph the linear function.

11. $3x + y = 2$

12. $4y = -20x$

13. $3y - 4x + 6 = 0$

LESSON 12-5 (pp. 620–623)

Give the slope and *y*-intercept of each line, given the equation.

14. $y = 4x + 1$

15. $y = \frac{3}{4}x$

16. $y + 2x = 6$

Mid-Chapter Quiz

Focus on Problem Solving

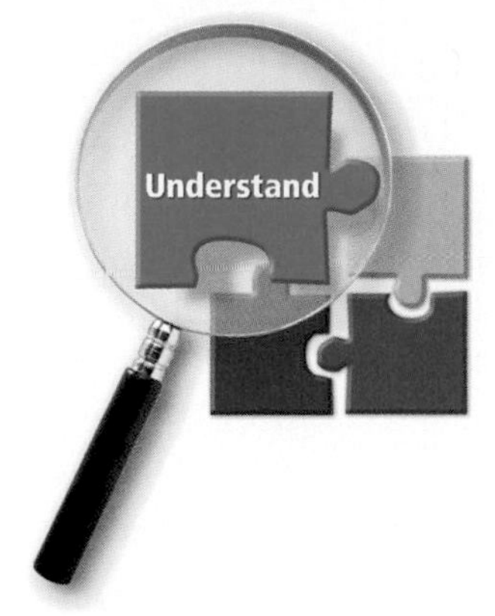

Understand the Problem

• **Sequence and prioritize information**

When you are reading a math problem, putting events in order, or in *sequence,* can help you understand the problem better. It helps to *prioritize* the information when you put it in sequence. To prioritize, you decide which of the information in your list is most important. The most important information has highest priority.

Use the information in the list or table to answer each question.

1. The list at right shows all of the things that Roderick has to do on Saturday. He starts the day without any money.

   a. Which two activities on Roderick's list must be done before any of the other activities? Do these two activities have higher or lower priority?

   b. Is there more than one way that he can order his activities? Explain.

   c. List the order in which Roderick's activities could occur on Saturday.

2. Tara and her family will visit Ocean World Park from 9:30 to 4:00. They want to see the waterskiing show at 10:00. Each show in the park is 50 minutes long. The time they choose to eat lunch will depend on the schedule they choose for seeing the shows.

   a. Which of the information given in the paragraph above has the highest priority? Which has the lowest priority?

   b. List the order in which they can see all of the shows, including the time they will see each.

   c. At what time should they plan to have lunch?

Show Times at Ocean World Park	
9:00, 12:00	Underwater acrobats
9:00, 3:00	Whale acts
10:00, 2:00	Dolphin acts
10:00, 1:00	Waterskiing
11:00, 4:00	Aquarium tour

Explore Nonlinear Functions

Use with Lesson 12-6

REMEMBER
- Linear functions can be represented by straight lines on a coordinate plane.
- The graphs of nonlinear functions are not straight lines.

You can use patterns involving squares and rectangles to explore the difference between linear and nonlinear functions.

Activity

1 The perimeter of a 1-inch-long square tile is 4 inches. Place 2 tiles together side by side. The perimeter of this figure is 6 inches.

1 in. / 1 in. / 1 in. / 1 in.   2 in. / 1 in. / 1 in. / 2 in.

a. Complete the table at right by adding tiles side by side and finding the perimeter of each new figure.

b. If x equals the number of tiles, what is the difference between consecutive x-values? If y equals the perimeter, what is the difference between consecutive y-values? How do these differences compare?

c. Graph the ordered pairs from your table on a coordinate plane. Is the graph linear or nonlinear? What does the table indicate about this type of function?

Number of Tiles	Perimeter (in.)
1	4
2	6
3	■
4	■
5	■

2 Draw the pattern at right and complete the next two sets of dots in the pattern.

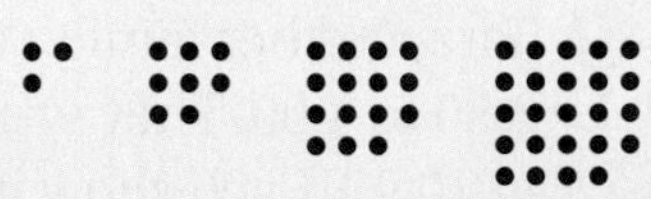

a. Complete the table at right. Let x equal the number of dots in the top row of each set, and let y equal the total number of dots in the set.

b. What is the difference between consecutive x-values? What is the difference between consecutive y-values? How do these differences compare?

c. Graph the ordered pairs on a coordinate plane. Is the graph linear or nonlinear? What does the table indicate about this type of function?

x	y
2	3
3	■
4	■
5	■
6	■

3 Use square tiles to model rectangles of the following sizes: 1×2, 2×3, 3×4, 4×5, and 5×6. An example of the first three rectangles is shown.

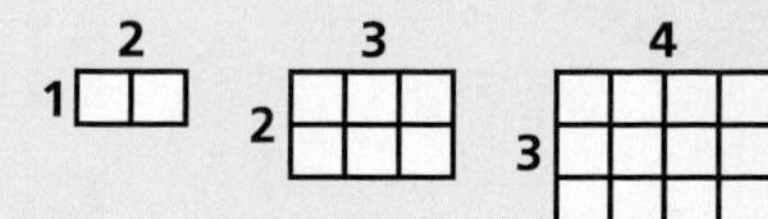

a. Find the perimeter and area of each rectangle. Complete the table at right. Let x = perimeter and y = area.

b. What is the difference between consecutive x-values? What is the difference between consecutive y-values? How do these differences compare?

c. Using what you have observed in **1** and **2**, tell whether the relationship between x and y in the table is linear or nonlinear.

d. Graph the ordered pairs from your table on a coordinate plane. Does the shape of your graph agree with the type of function you gave in **c**?

Rectangle	x	y
1×2		
2×3		
3×4		
4×5		
5×6		

Think and Discuss

1. Refer to **1** and **2** to explain how linear and nonlinear functions compare.
2. Is $y = x^2$ linear or nonlinear? Explain your answer.
3. If $x = 1$ and keeps increasing by 1, are $y = x^2$ and $x = y^2$ the same graph? Does $x = y^2$ represent a linear or nonlinear function? Explain your answer.

Try This

1. Use square tiles to model each pattern.
2. Model the next two sets in each pattern using square tiles.
3. Complete each table.
4. Explain whether each function is linear or nonlinear.

Pattern 1

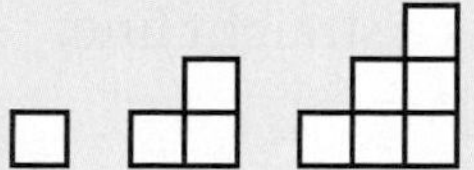

Perimeter	Area
4	
8	
12	

Pattern 2

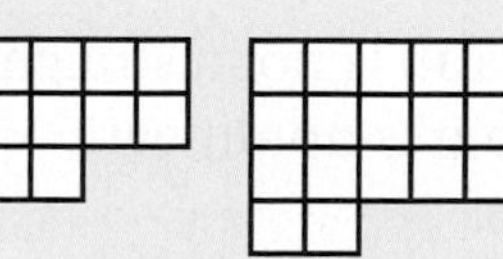

Perimeter	Area
10	
14	
18	

Pattern 3

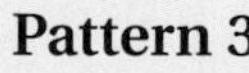

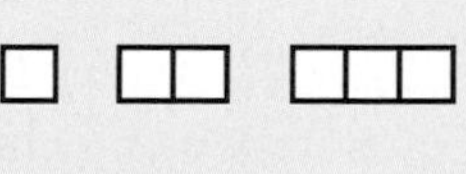

Perimeter	Area
4	
6	
8	

12-6 Nonlinear Functions

Learn to identify nonlinear functions.

Vocabulary

nonlinear function

As you inflate a balloon, its volume increases. The table at right shows the increase in volume of a round balloon as its radius changes. Do you think a graph of the data would or would not be a straight line? You can make a graph to find out.

Radius (in.)	Volume (in^3)
1	4.19
2	33.52
3	113.13
4	268.16
5	523.75

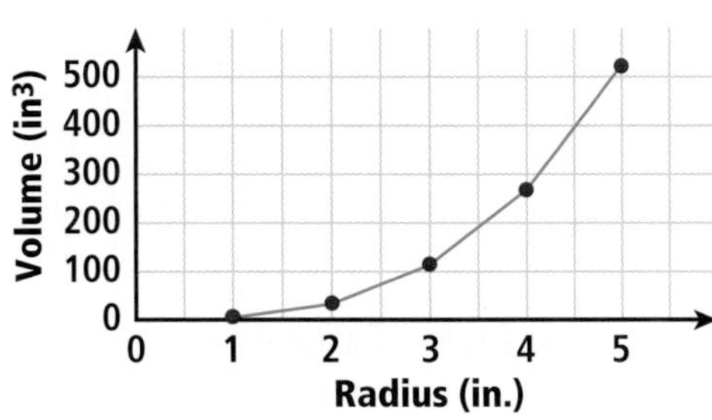

A **nonlinear function** is a function whose graph is not a straight line.

EXAMPLE 1 Identifying Graphs of Nonlinear Functions

Tell whether the graph is linear or nonlinear.

A

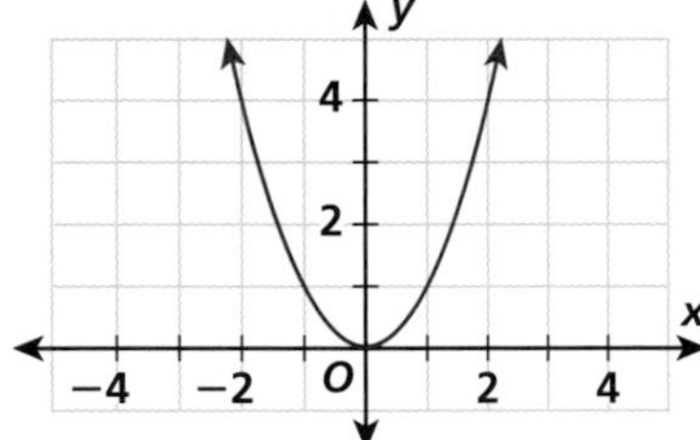

The graph is not a straight line, so it is nonlinear.

B

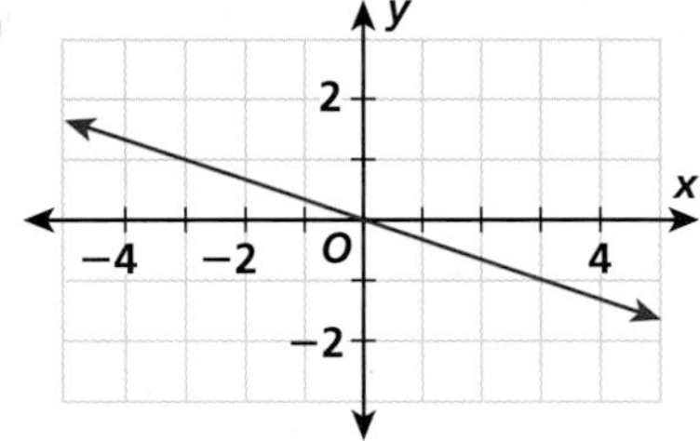
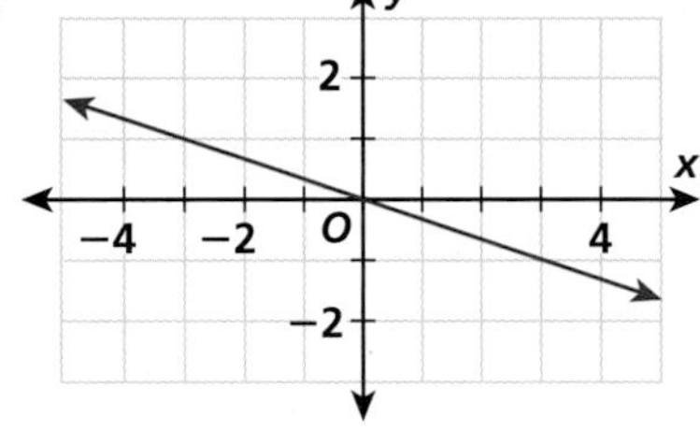

The graph is a straight line, so it is linear.

C

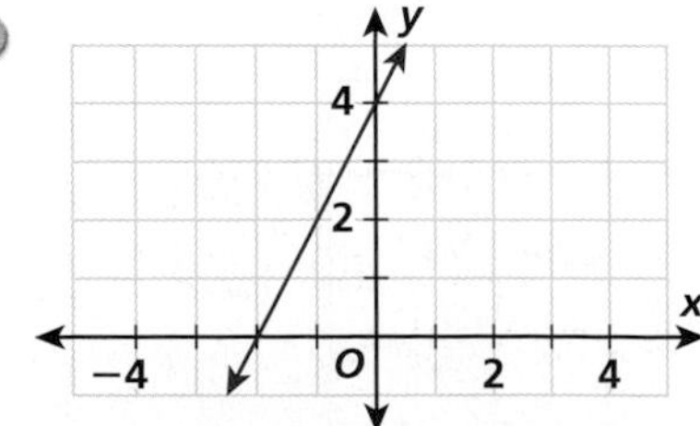

The graph is a straight line, so it is linear.

D

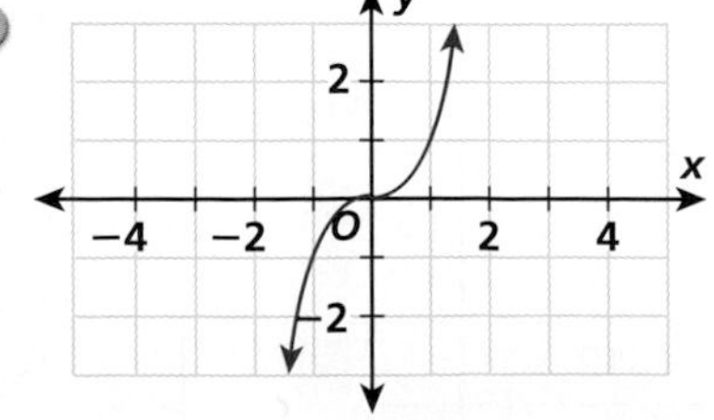

The graph is not a straight line, so it is nonlinear.

You can use a function table to determine whether ordered pairs describe a linear or a nonlinear relationship.

For a function that has a linear relationship, when the difference between each successive input value is constant, the difference between each corresponding output value is *constant.*

For a function that has a nonlinear relationship, when the difference between each successive input value is constant, the difference between each corresponding output value *varies.*

EXAMPLE 2 Identifying Nonlinear Relationships in Function Tables

Tell whether the function represented in each table has a linear or nonlinear relationship.

A

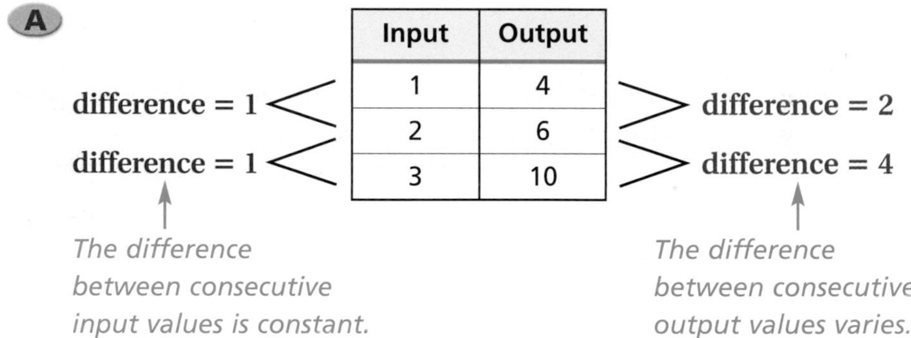

Input	Output
1	4
2	6
3	10

The function represented in the table has a nonlinear relationship.

B

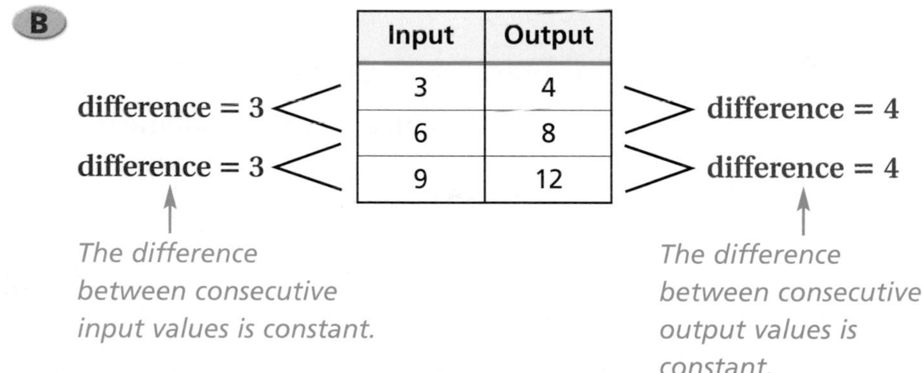

Input	Output
3	4
6	8
9	12

The function represented in the table has a linear relationship.

Think and Discuss

1. Tell whether the relationship between the area of a square and the length of its sides is linear or nonlinear. Explain your answer.

2. Give an example of a nonlinear relationship between two sets of data and a linear relationship between two sets of data.

12-6 Exercises

FOR EOG PRACTICE
see page 708

internet connect
Homework Help Online
go.hrw.com Keyword: MS4 12-6

5.01

GUIDED PRACTICE

See Example 1 Tell whether the graph is linear or nonlinear.

1. 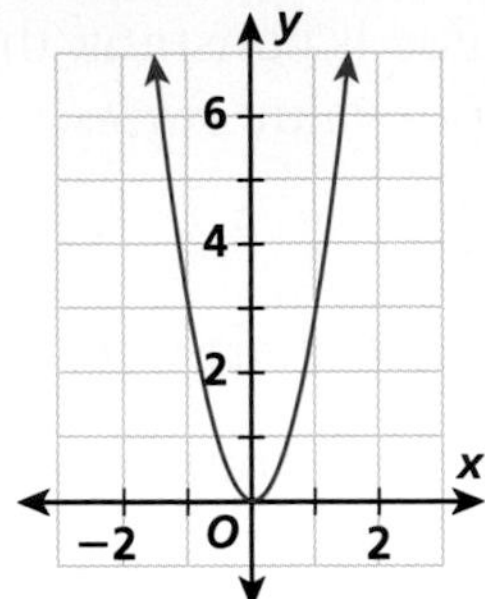

2. 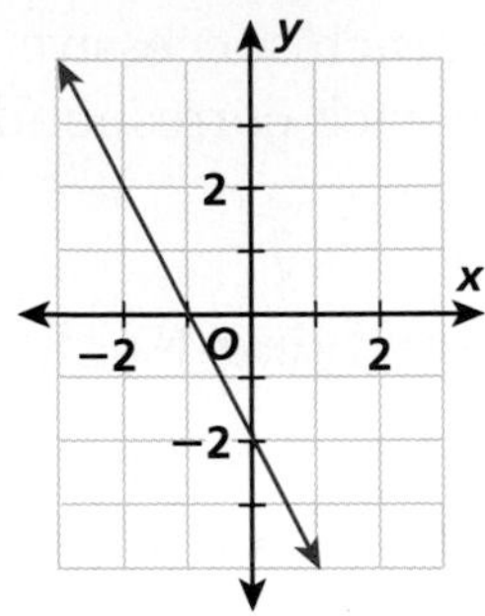

3. 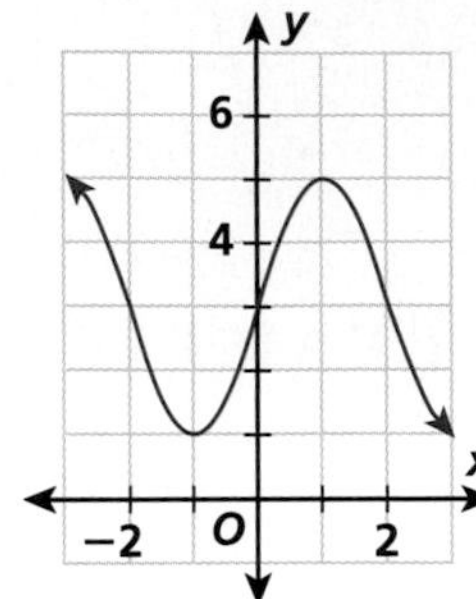

See Example 2 Tell whether the function represented in each table has a linear or nonlinear relationship.

4.

x	y
2	5
4	7
6	9

5.

x	y
10	100
20	400
30	900

6.

x	y
1	6
2	9
3	14

INDEPENDENT PRACTICE

See Example  1 Tell whether the graph is linear or nonlinear.

7. 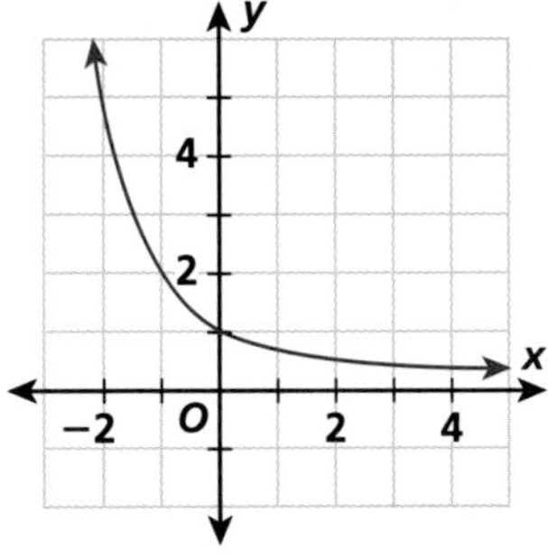

8. 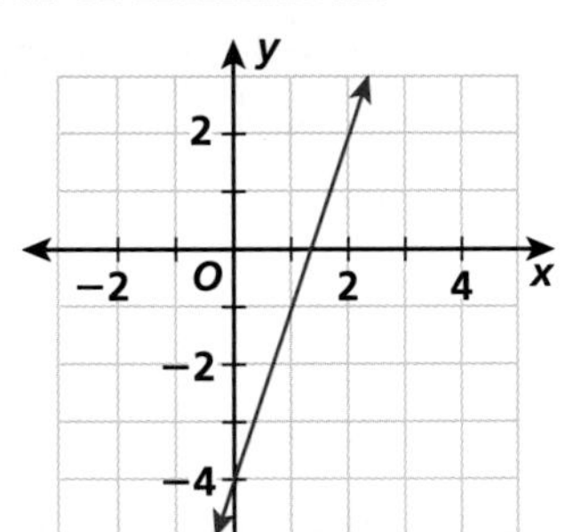

9. 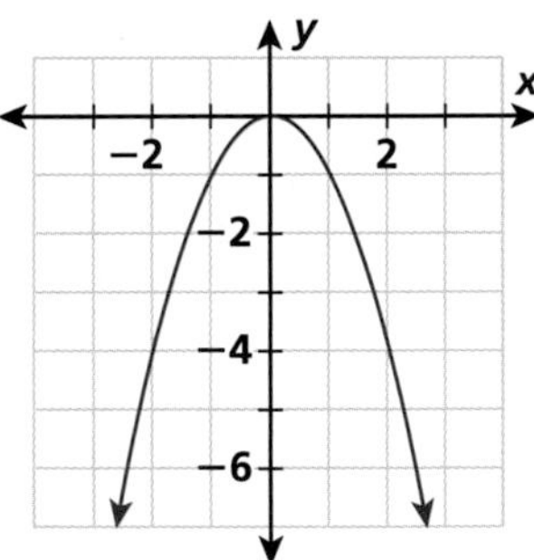

See Example 2 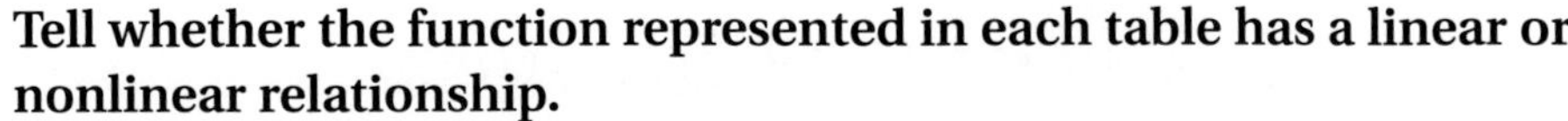
Tell whether the function represented in each table has a linear or nonlinear relationship.

10.

x	y
4	25
8	36
12	49

11.

x	y
25	125
50	2,000
75	5,125

12.

x	y
1	822
2	824
3	826

Physical Science 

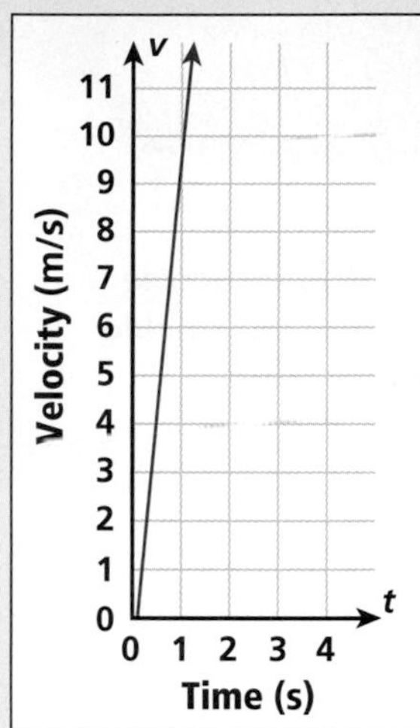

13. Galileo showed that falling objects increase their speed by about 9.8 meters per second for each second they fall. The speed of an object falling from rest is represented by the function $v = g \cdot t$, where v stands for velocity or speed, g stands for the acceleration due to gravity, or 9.8 m/s^2, and t stands for time.

a. Explain what the graph shows about time and speed.

b. Is the graph linear or nonlinear?

c. What speed would an object have after falling for 10 seconds?

14. Galileo also developed the following rule to calculate the distance an object falls from rest in a given amount of time, on Earth.

$$d = \frac{1}{2} \cdot \frac{32 \text{ ft}}{\text{s}^2} \cdot t^2$$

The graph shows the distance in feet that an object fell from 0 to 4 seconds. Make a table to show the data on the graph. Is the graph linear or nonlinear?

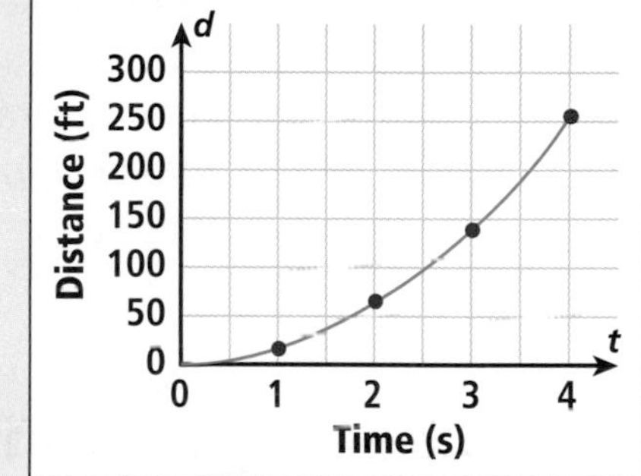

15. ***WRITE ABOUT IT*** Explain what the graph in Exercise 14 shows about the relationship between distance and time for falling objects.

16. ***CHALLENGE*** Pretend you have discovered a new planet. Create a rule for the distance that an object falls in a given amount of time on your planet. Use the rule to write a function that has a nonlinear relationship. Show a table for your rule.

Spiral Review

Solve each proportion using cross products. (Lesson 5-3)

17. $\frac{1}{p} = \frac{3}{12}$ **18.** $\frac{4}{7} = \frac{x}{14}$ **19.** $\frac{m}{5} = \frac{16}{20}$ **20.** $\frac{7}{9} = \frac{35}{s}$ **21.** $\frac{6}{7} = \frac{24}{a}$

Solve. (Lesson 11-3)

22. $5x + 2 = 2x - 1$ **23.** $4m + 14 = 11m$ **24.** $3y - 2 = 2 - y$

25. **EOG PREP** Which is a ratio equivalent to $\frac{10}{15}$? (Lesson 5-2)

A $\frac{2}{5}$ B $\frac{20}{45}$ C $\frac{5}{30}$ D $\frac{20}{30}$

12-7 Rates of Change

Learn to recognize constant and variable rates of change.

If you dropped a bowling ball and a baseball from the same height at the same time, which do you think would hit the ground first? Without air resistance, both would hit the ground at the same time. Gravity pulls all objects to the ground at the same rate, no matter what their sizes or weights.

As each ball drops, it falls faster and faster due to gravity. The *rate of change,* the change in the ball's speed as it falls, is a change that can be measured.

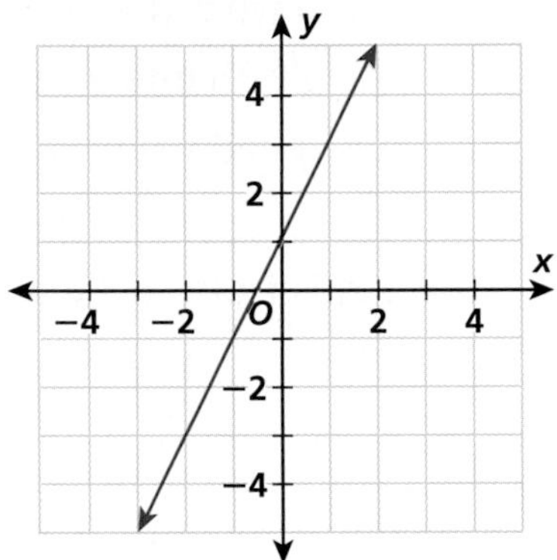

A *constant rate of change* means that something changes by the same amount during equal intervals. A graph that has a constant rate of change is a line, and the *rate of change* is the same as the *slope* of the line.

EXAMPLE 1 Identifying Constant Rates of Change

Tell in which intervals of x the graph shows constant rates of change.

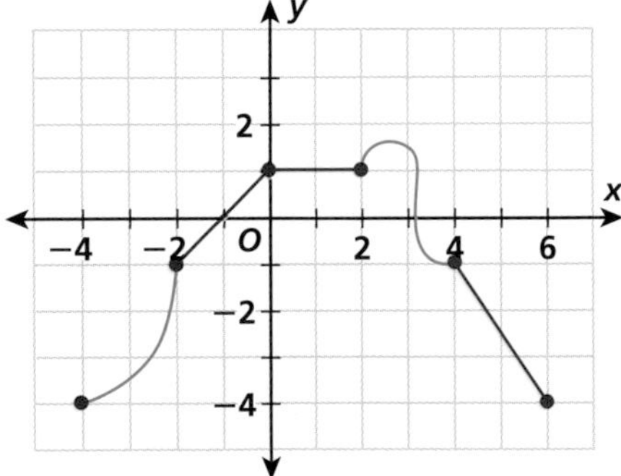

Look for parts of the graph that are line segments.

Identify the intervals of x for each line segment.

The graph shows constant rates of change in the intervals from $x = -2$ to $x = 0$, $x = 0$ to $x = 2$, and $x = 4$ to $x = 6$.

A *variable rate of change* means that something changes by a different amount during equal intervals. Variable rates of change are also called nonlinear rates of change, because a graph that represents a variable rate of change is not a line.

EXAMPLE 2 Identifying Variable Rates of Change

Tell whether each graph shows a constant or variable rate of change.

A

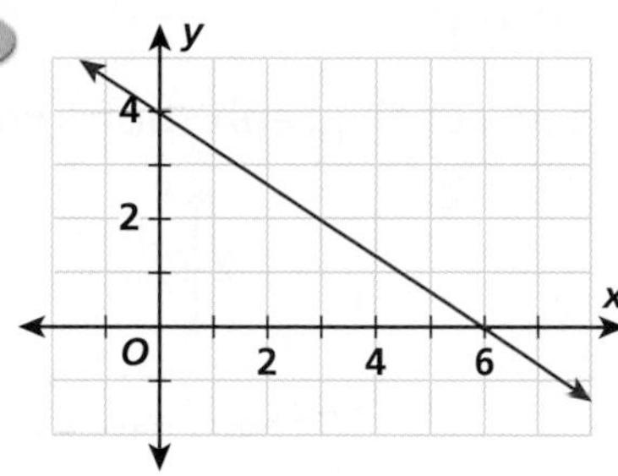

The graph is linear, so the rate of change is constant.

B

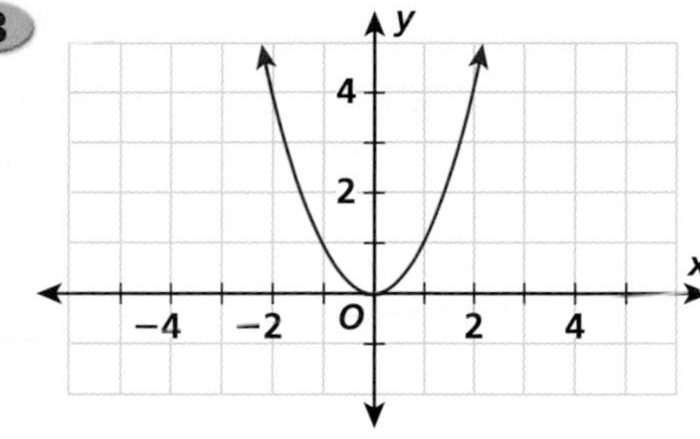

The graph is nonlinear, so the rate of change is variable.

EXAMPLE 3 Using Rate of Change to Solve Problems

Remember!

slope $= \frac{\text{rise}}{\text{run}}$

The graph shows the distance a bicyclist travels over time. Does the bicyclist travel at a constant or variable speed? How fast does the bicyclist travel?

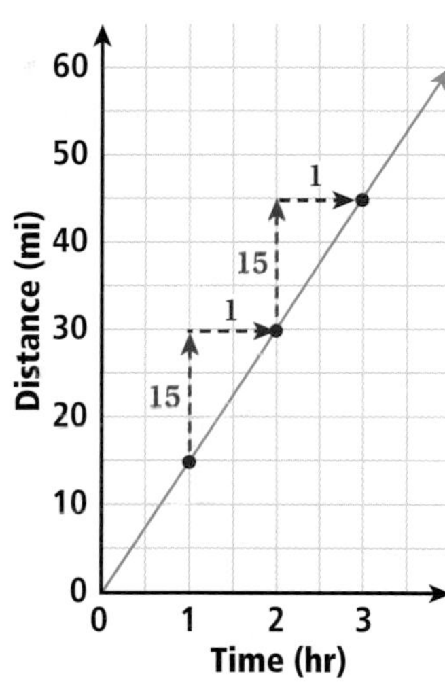

Use two points on the line, such as (1, 15) and (2, 30), to find the slope.

The graph is linear. You can see that the same amount of distance (rise) is traveled during equal time intervals (run), so the bicyclist is traveling at a constant speed.

The slope of the graph is rise ÷ run, or distance ÷ time.

Since speed = distance ÷ time, you can find the speed by finding the slope.

slope $= \frac{\text{rise}}{\text{run}} = \frac{15 \text{ mi}}{1 \text{ hr}}$

The bicyclist travels at 15 miles per hour.

Think and Discuss

1. **Compare** constant and variable rates of change.
2. **Explain** how you can use the appearance of a graph to decide whether the rate of change is constant or variable.
3. **Give an example** of a real-world situation that would involve more than one constant rate of change.

12-7 Exercises

5.01

FOR EOG PRACTICE

see page 709

internet connect

Homework Help Online
go.hrw.com Keyword: MS4 12-7

GUIDED PRACTICE

See Example 1 **Tell in which intervals of x each graph shows constant rates of change.**

1.

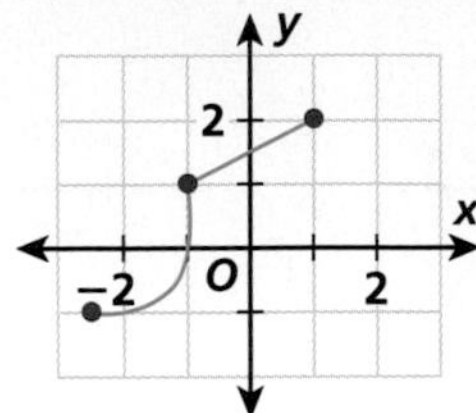

2.

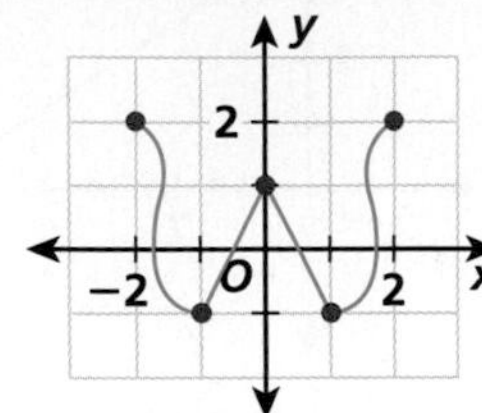

3.

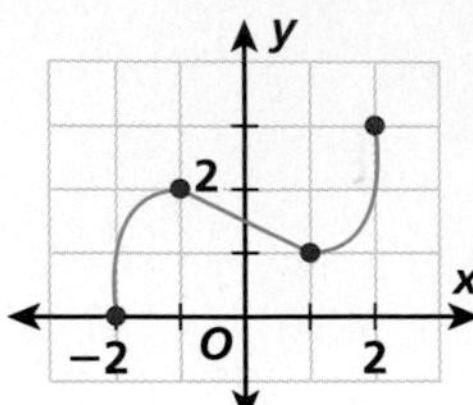

See Example 2 **Tell whether each graph shows a constant or variable rate of change.**

4.

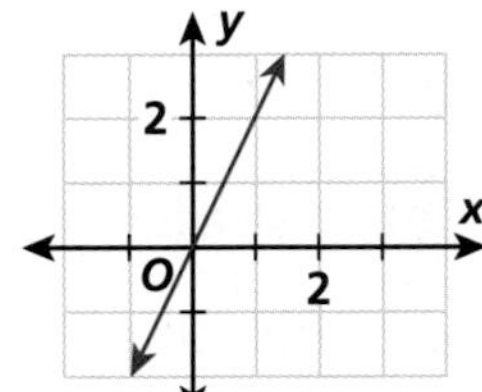

5.

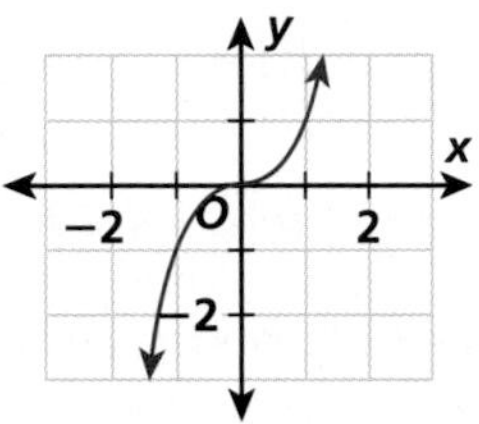

6.

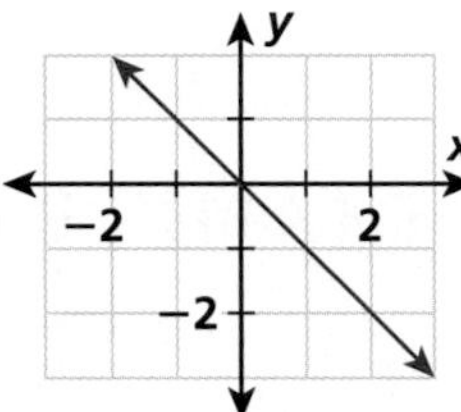

See Example 3 **7.** The graph shows the distance a trout swims over time. Does the trout swim at a constant or variable speed? How fast does the trout swim?

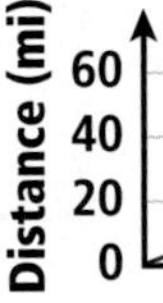

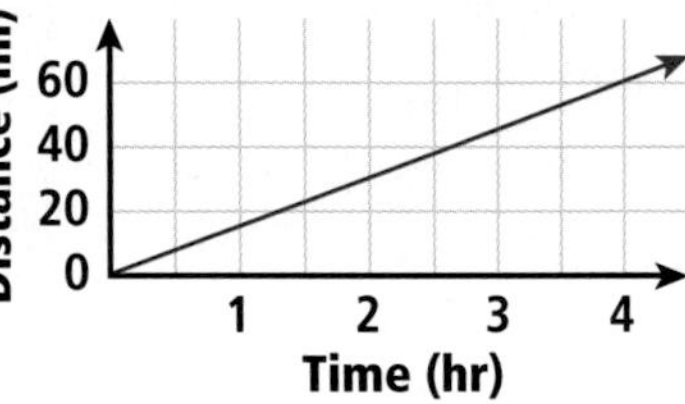

INDEPENDENT PRACTICE

See Example 1 **Tell in which intervals of x each graph shows constant rates of change.**

8.

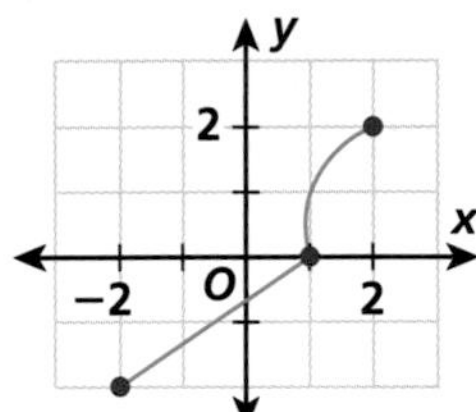

9.

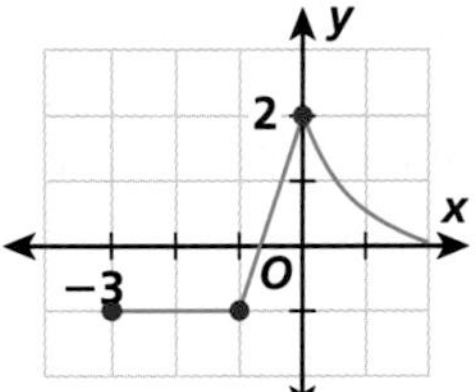

10.

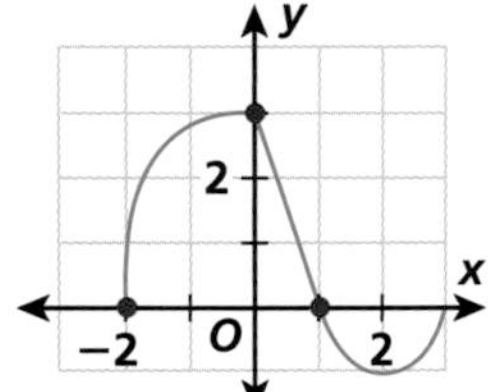

See Example 2 **Tell whether each graph shows a constant or variable rate of change.**

11.

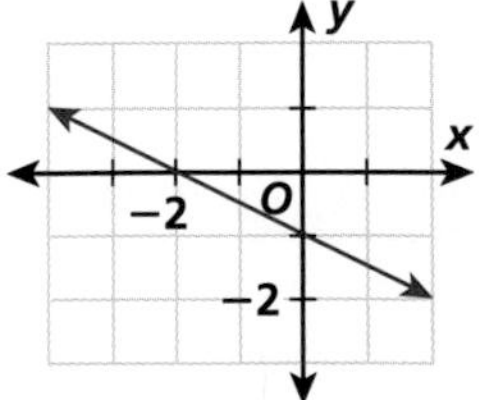

12.

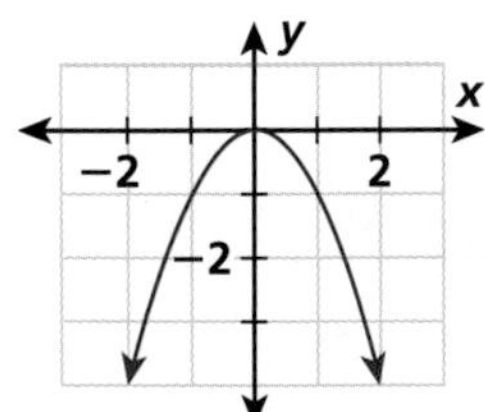

13.

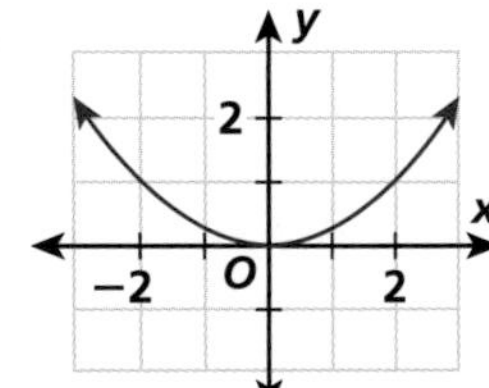

See Example 3 **14.** The graph shows an amount of rain that falls over time. Does the rain fall at a constant or variable rate? How much rain falls per hour?

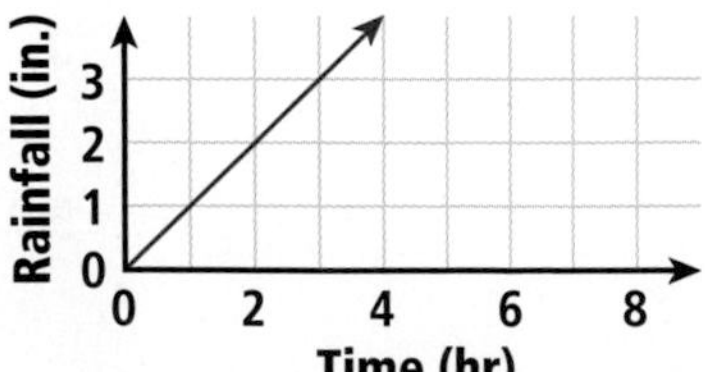

PRACTICE AND PROBLEM SOLVING

Agriculture LINK

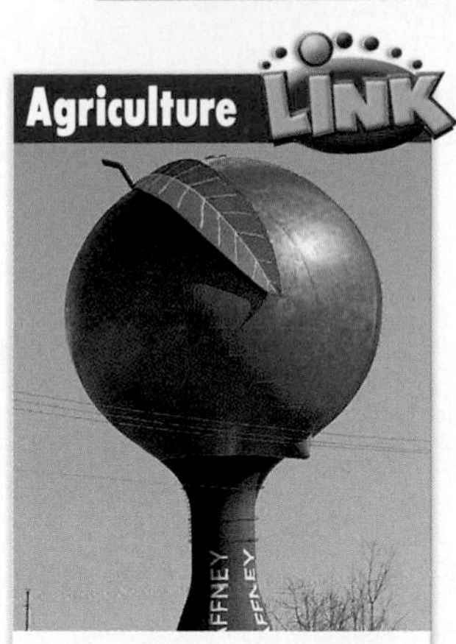

This water tower can be seen in Gaffney, South Carolina, where the South Carolina Peach Festival is held every year.

15. ***AGRICULTURE*** The graph at right shows the cost of buying peaches when the cost is the same per pound no matter how much you buy.

   a. How much does a pound of peaches cost?

   b. Write an equation that represents the cost of the peaches.

16. ***SPORTS*** The graph at right shows a runner's progress during a race.

   a. During which interval was the runner going at a constant speed?

   b. During which interval was the runner's speed changing at a constant rate?

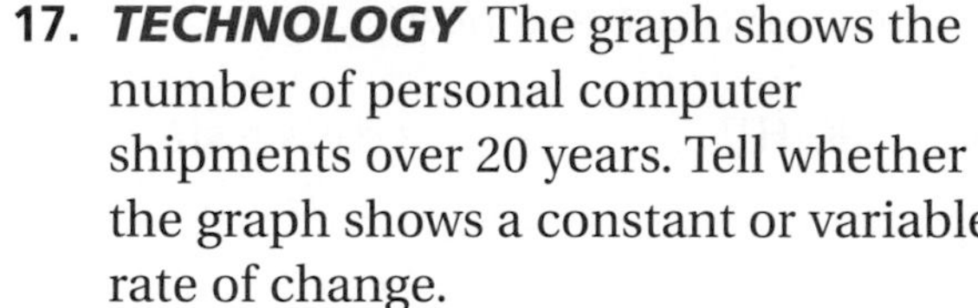

17. ***TECHNOLOGY*** The graph shows the number of personal computer shipments over 20 years. Tell whether the graph shows a constant or variable rate of change.

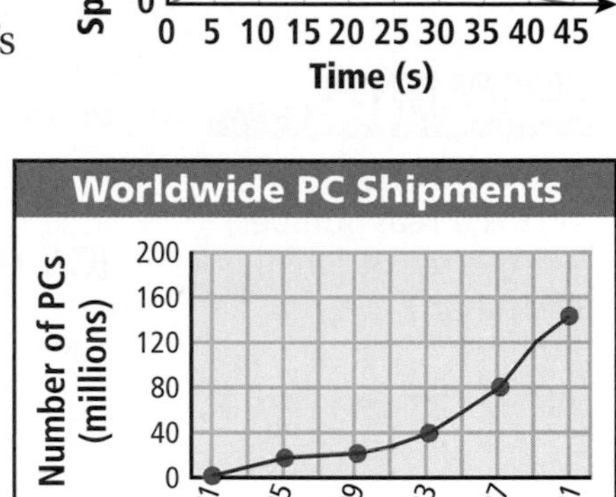

18. ***WRITE A PROBLEM*** Write a problem that can be solved by deciding whether the graph illustrating the problem shows a constant or variable rate of change.

19. ***WRITE ABOUT IT*** Describe how you can use three points shown on a graph to determine whether the rate of change is constant or variable.

20. ***CHALLENGE*** The population of prairie dogs in a park doubles every year. Does this population show a constant or variable rate of change? Explain.

Spiral Review

Solve each inequality. (Lesson 11-6)

21. $x + 9 < 12$ **22.** $x - 3 > -2$ **23.** $-11 < 4 + x$ **24.** $x + 8 > -5$

Graph each linear function. (Lesson 12-4)

25. $y = x - 3$ **26.** $y = 2x + 1$ **27.** $y = 4$ **28.** $y = -x + 1$

29. **EOG PREP** Which of the following points is not on the graph of $y = -x^2 + 1$? (Lesson 12-5)

A (1, 0) B (−1, 0) C (0, 0) D (0, 1)

12-8 Set Theory

Learn to use set notation and to identify relationships between sets.

Vocabulary

set

element

Reading Math

Read the notation {x|x is a real number} as "the set of all x such that x is a real number."

The states that are bordered by the Pacific Ocean form a *set*. The *set* of states in the United States that are bordered by the Pacific Ocean includes California, Oregon, Washington, Alaska, and Hawaii.

A **set** is a collection of objects. Each object in a set is an **element** of the set. The special set that contains no elements is the *null set,* or *empty set.* When describing sets using symbols, you can use either *roster notation* or *set-builder notation.*

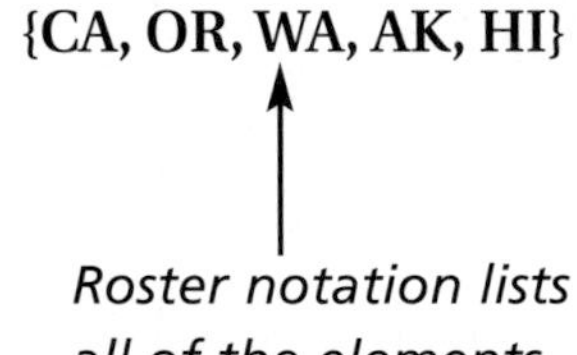

{CA, OR, WA, AK, HI}

Roster notation lists all of the elements.

{x|x is a state that is bordered by the Pacific Ocean}

Set-builder notation gives a rule.

EXAMPLE 1 Writing in Set Notation

Write each set using roster notation or set-builder notation.

A the set of all even numbers from 1 to 10, including 10

{2, 4, 6, 8, 10} — *The numbers can be listed conveniently. Use roster notation.*

B the set of all integers greater than zero

{x|x is a counting number} — *The numbers follow a rule but cannot all be listed conveniently. Use set-builder notation.*

The following notations show how different sets can relate to each other.

Subset		
Words	**Notation**	**Example**
A is a *subset* of B if all the elements of set A are in set B.	$A \subset B$	A = {Mary, Bill} B = {Mary, Bill, Chim, Selena} $A \subset B$ All of the elements of A are in B.

Intersection		
Words	**Notation**	**Example**
The *intersection* of sets A and B is the set of all elements that are in both sets A and B.	$A \cap B$	A = {Des Moines, Kent, Seattle} B = {Seattle, Portland, San Francisco} $A \cap B$ = {Seattle} Seattle is the only city in both sets.

Union		
Words	**Notation**	**Example**
The *union* of sets A and B is the set of all elements that are in either set A or set B.	$A \cup B$	$A = \{1, 2, 4, 7\}$; $B = \{5, 6, 7\}$ $A \cup B = \{1, 2, 4, 5, 6, 7\}$ All of the numbers in either set are included.

EXAMPLE 2 Identifying Relationships Between Sets

Find $A \cap B$ and $A \cup B$. Tell whether $A \subset B$, $B \subset A$, or neither.

A $A = \{2, 4, 6, 8\}$; $B = \{2, 4, 8\}$

$A \cap B = \{2, 4, 8\}$ — *2, 4, and 8 are in both A and B.*

$A \cup B = \{2, 4, 6, 8\}$ — *List the elements of both sets.*

$B \subset A$ — *All of the elements of B are in A.*

B $A = \{0, 3, 6\}$; $B = \{0, 2, 3, 5, 6\}$

$A \cap B = \{0, 3, 6\}$ — *0, 3, and 6 are in both A and B.*

$A \cup B = \{0, 2, 3, 5, 6\}$ — *List the elements of both sets.*

$A \subset B$ — *All the elements of A are in B.*

C $A = \{x|x$ **is a rational number**$\}$; $B = \{x|x$ **is an irrational number**$\}$

$A \cap B = \{\ \} = \varnothing$ — *No number is both rational and irrational.*

$A \cup B = \{x|x \text{ is a real number}\}$ — *Together, the set of rational numbers and the set of irrational numbers make up the set of real numbers.*

Neither set is a subset of the other.

Writing Math

The null, or empty, set is written as { } or ∅.

Think and Discuss

1. **Give an example** of two sets whose intersection is the null set.
2. **Tell** what is true if the intersection of sets A and B is set A.

12-8 Exercises

FOR EOG PRACTICE

see page 709

internet connect

Homework Help Online

go.hrw.com Keyword: MS4 12-8

GUIDED PRACTICE

See Example 1

Write each set using roster notation or set-builder notation.

1. the set of all odd numbers from 1 through 11
2. the set of all negative integers
3. the set of all integers from −4 through −6
4. the set of all even numbers greater than 10

See Example 2

Find $A \cap B$ and $A \cup B$. Tell whether $A \subset B$, $B \subset A$, or neither.

5. $A = \{11, 13, 15, 19\}$; $B = \{5, 10, 15, 20\}$
6. $A = \{100, 200, 300\}$; $B = \{100, 200, 300, 400\}$
7. $A = \{x|x \text{ is an even integer}\}$; $B = \{x|x \text{ is an odd integer}\}$
8. $A = \{x|x \text{ is a real number}\}$; $B = \{x|x = 0\}$

INDEPENDENT PRACTICE

See Example 1

Write each set using roster notation or set-builder notation.

9. the set of positive multiples of 3, less than 20
10. the set of even multiples of 5 from 10 through 20
11. the set of even numbers
12. the set of all two-word states in the United States

See Example 2

Find $A \cap B$ and $A \cup B$. Tell whether $A \subset B$, $B \subset A$, or neither.

13. $A = \{-6, -4, -2, 0\}$; $B = \{-2, 0, 4, 6\}$
14. $A = \{x|x \text{ is a vowel}\}$; $B = \{x|x \text{ is a consonant}\}$
15. $A = \{1, 2, 3, 4, 5\}$; $B = \{1, 2, 3\}$
16. $A = \{1, 3, 12, 15\}$; $B = \{1, 3\}$
17. $A = \{x|x \text{ is a European country}\}$; $B = \{\text{Germany, France}\}$
18. $A = \{x|x \text{ is a multiple of } 3\}$; $B = \{x|x \text{ is a multiple of } 2\}$

PRACTICE AND PROBLEM SOLVING

Find the missing set of values.

19. $A = \{1, 2, 3\}$; $B = \blacksquare$
    $A \cup B = \{1, 2, 3, 4, 5\}$
    $A \cap B = \varnothing$

20. $A = \blacksquare$; $B = \{0, 5, 10, 15\}$
    $A \subset B$
    $A \cap B = \{0, 5, 10\}$

21. The graph shows the number of branches of several libraries serving 1,000,000 or more people.

a. Give the set A of libraries with 70 or more branches.

b. Give the set B of all libraries with a number of branches that is a multiple of 20.

c. Find $A \cap B$.

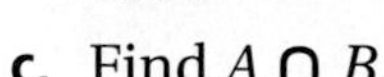

22. ***GEOGRAPHY*** Find $A \cap B$ if $A = \{\text{all U.S. states} \mid \text{the state name starts with } W\}$ and $B = \{\text{all U.S. states} \mid \text{the state borders Oregon}\}$.

23. ***SPORTS*** Given that $A = \{\text{all basketball players}\}$ and $B = \{\text{all people who are more than 80 inches tall}\}$, describe the members of the intersection of A and B.

24. ***SOCIAL STUDIES*** Let $M = \{x \mid x \text{ is a person who lives in Illinois}\}$ and $N = \{x \mid x \text{ is a person who lives in Chicago}\}$.

a. Determine $M \cup N$ and $M \cap N$.

b. Is either set a subset of the other?

25. ***WHAT'S THE ERROR?*** If $A = \{1, 2, 3, 4\}$ and $B = \{2, 4\}$, what is the error if you write $A \cup B = B$?

26. ***WRITE ABOUT IT*** Explain how you can determine the members of the intersection of two sets A and B.

27. ***CHALLENGE*** If a is in $A \cap B$, is a in $A \cup B$? Explain. If a is in $A \cup B$, is a in $A \cap B$? Explain.

Spiral Review

Determine the number of permutations in each case. (Lesson 10-7)

28. the numbers 1–5

29. the letters A, B, C, and D

30. the number of ways six people can line up to go into the auditorium

Solve each equation. (Lesson 11-3)

31. $2x + 5 = 3x$

32. $4 - x = x$

33. $3n + 16 = 7n$

34. $11x + 1 = 8x + 10$

35. **EOG PREP** What is the slope of the line that passes through (3, 5) and (6, 10)? (Lesson 12-5)

A $\frac{5}{3}$ B $\frac{3}{5}$ C $-\frac{5}{3}$ D $-\frac{3}{5}$

EXTENSION Venn Diagrams

Learn to create and analyze Venn diagrams.

Vocabulary

Venn diagram

How are spaghetti and earthworms alike? In more ways than you might think. You can organize their similarities and differences in a *Venn diagram.*

A **Venn diagram** is a drawing that shows the relationships between two or more sets.

Characteristics that are shared are in the overlapping region. This is the intersection. Characteristics that appear in any region are part of the union.

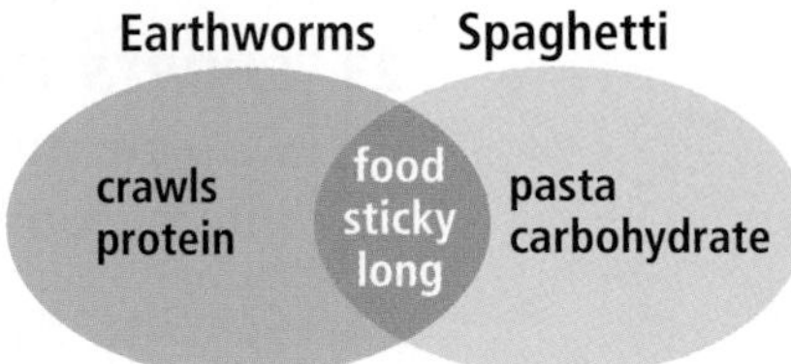

Since all of the characteristics of a subset are shared by the larger set, the subset is drawn completely inside the larger set.

EXAMPLE 1 Drawing Venn Diagrams

Draw a Venn diagram to show the relationships between the sets.

A

Set	Elements
School band	Bass drums, flutes, clarinets, trumpets, tubas
School orchestra	violins, flutes, clarinets, trumpets, timpani, french horns

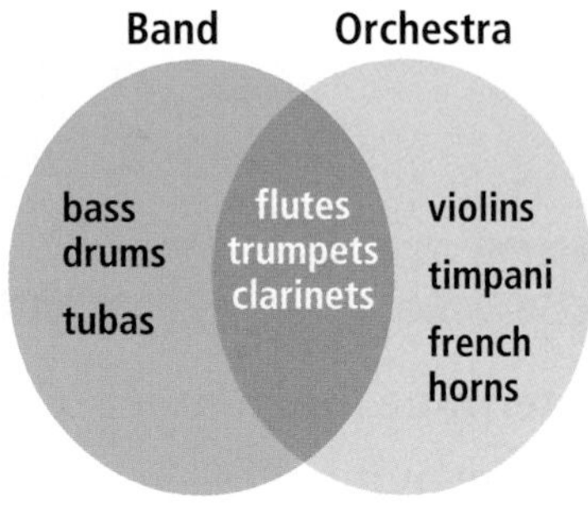

The two sets have an intersection of flutes, clarinets, and trumpets.

B

Set	Elements
Multiples of 4	4, 8, 12, 16, 20, 24, 28, 32, 36, 40
Multiples of 8	8, 16, 24, 32, 40

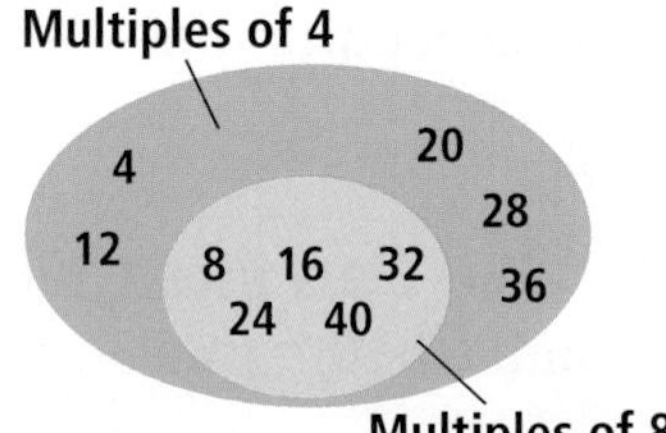

The set "Multiples of 8" is a subset of the set "Multiples of 4."

EXAMPLE 2 Analyzing Venn Diagrams

Use the Venn diagrams to identify intersections, unions, and subsets.

A

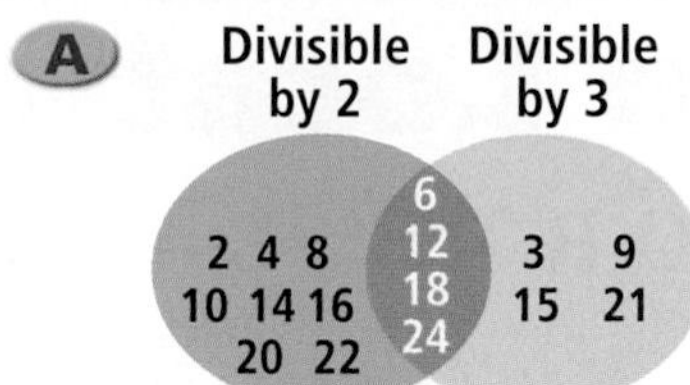

Intersection: 6, 12, 18, 24

Union: 2, 3, 4, 6, 8, 9, 10, 12, 14, 15, 16, 18, 20, 21, 22, 24

B

Fruit
strawberries
apples
bananas
grapes
pears
oranges
lemons
Citrus fruit

Intersection: oranges, lemons

Union: apples, pears, bananas, grapes, strawberries, oranges, lemons

Subset: citrus fruit

EXTENSION Exercises

Draw a Venn diagram to show the relationships between the sets.

1.

Set	Elements
Math club members	Harry, Jason, Juan, Doris, Keisha, Elden
Band members	Tyrone, Cara, Harry, Megan, Doris, Ricky

2.

Set	Elements
Multiples of 7 between 0 and 50	7, 14, 21, 28, 35, 42, 49
Odd numbers between 0 and 50	1, 3, 5, 7, 9, 11, 13, 15, 17, 19, 21, 23, 25, 27, 29, 31, 33, 35, 37, 39, 41, 43, 45, 47, 49

Use the Venn diagrams to identify intersections, unions, and subsets.

3. 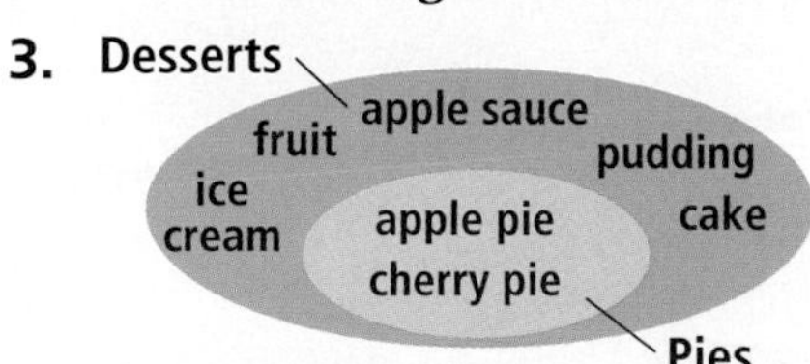

4. 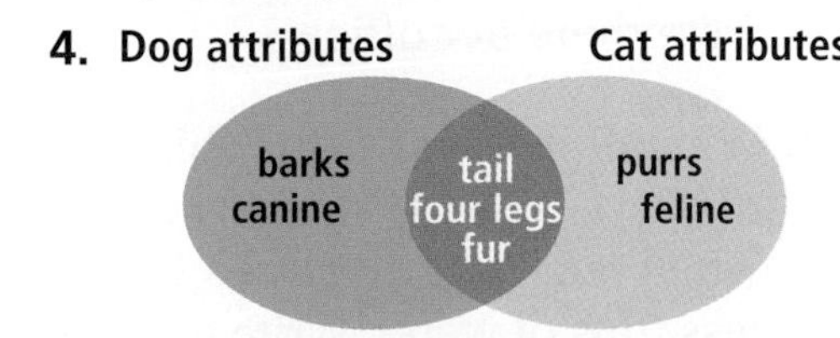

5. ***CHALLENGE*** Draw a Venn diagram to show the relationships between different kinds of quadrilaterals.

Problem Solving on Location

NORTH CAROLINA

Battleship *North Carolina*

The USS *North Carolina* was the first modern American battleship built after World War I. At that time, the *North Carolina* set a standard for shipbuilding technology. Currently, the battleship is located on the west bank of the Cape Fear River in Wilmington, North Carolina. Visitors on self-guided tours can see portions of the battleship's nine decks and a number of interesting exhibits.

The USS *North Carolina* is about 728 feet long and 108 feet wide and once held a crew of over 2,300. The ship holds 7,167 tons of fuel and can reach a maximum speed of 28 knots. On one journey, the ship used 7,167 tons of fuel. Below is a table that shows how much fuel would have been used in five trips of the same length.

Trip Number	1	2	3	4	5
Fuel (tons)	7,167	14,334	21,501	28,668	35,835

For 1–2, use the table.

1. How much fuel would have been used in 6 trips?

2. Write and solve an expression to find how many tons of fuel would have been used during 12 trips, where x is the number of trips.

3. On one trip, the battleship held 2,339 people, consisting of 144 officers and 2,195 enlisted sailors. About how many enlisted sailors were there per officer?

4. When fully loaded, the ship displaces 44,800 tons of water; when empty, it displaces 11,200 less tons of water. Write and solve an equation to find the amount of water the ship displaces when it is empty.

The Andy Griffith Show made its debut on October 3, 1960, and ran for 249 episodes, 90 of which were in color.

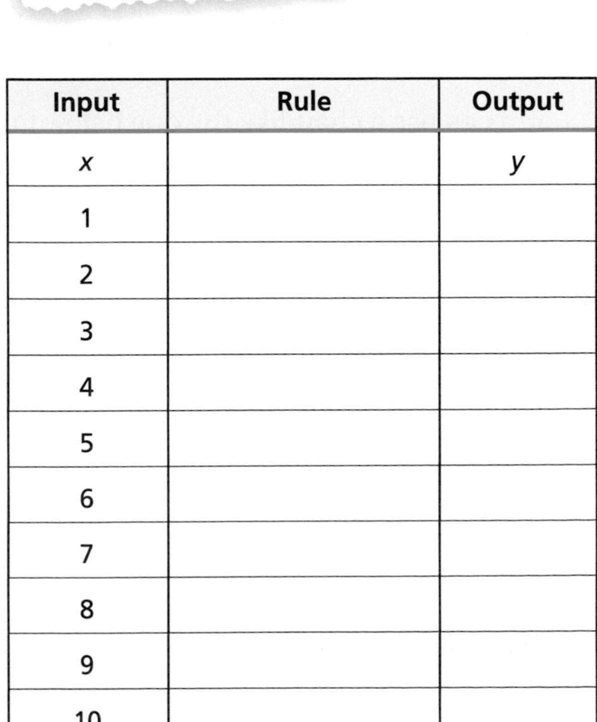

Surry County

Actor Andy Griffith was born and raised in Mount Airy, a small town in Surry County, North Carolina. Mount Airy eventually became the inspiration for the town of Mayberry, the setting of the popular 1960s TV series *The Andy Griffith Show*.

The U.S. Census Bureau statistics show that the population of Surry County in the year 2000 was 71,219. Suppose that the population of Surry County is projected to grow by an average of 911 people per year for the next 10 years.

1. Copy and complete the table at right to find a pattern in Surry County's population increase over the next ten years. Let x equal the number of years after 2000. Let y equal the total population after x years.

2. What is Surry County's population projected to be in the year 2010?

3. Does the function represented in the table have a linear or a nonlinear relationship?

4. Write the function that describes the sequence of y-values.

5. Find the slope and y-intercept of the graph of the function.

Input	Rule	Output
x		y
1		
2		
3		
4		
5		
6		
7		
8		
9		
10		

MATH-ABLES

Clothes Encounters

Five students from the same math class met to study for an upcoming test. They sat around a circular table with seat 1 and seat 5 next to each other. No two students were wearing the same color of shirt or the same type of shoes. From the clues provided, determine where each student sat, each student's shirt color, and what type of shoes each student was wearing.

1. The girls' shoes were sandals, flip-flops, and boots.
2. Robin, wearing a blue shirt, was sitting next to the person wearing the green shirt. She was not sitting next to the person wearing the orange shirt.
3. 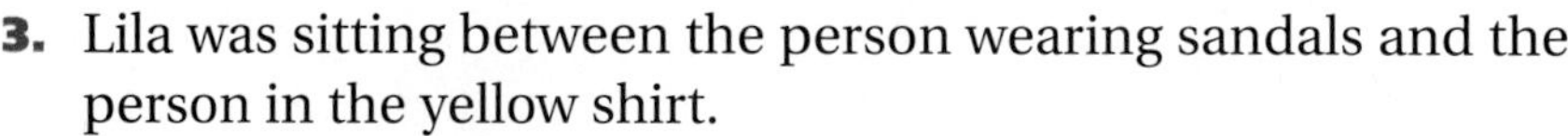Lila was sitting between the person wearing sandals and the person in the yellow shirt.
4. The boy who was wearing the tennis shoes was wearing the orange shirt.
5. April had on flip-flops and was sitting between Lila and Charles.
6. Glenn was wearing loafers, but his shirt was not brown.
7. Robin sat in seat 1.

You can use a chart like the one below to organize the information given. Put *X*'s in the spaces where the information is false and *O*'s in the spaces where the information is true. Some of the information from the first two clues has been included on the chart already. You will need to read through the clues several times and use logic to complete the chart.

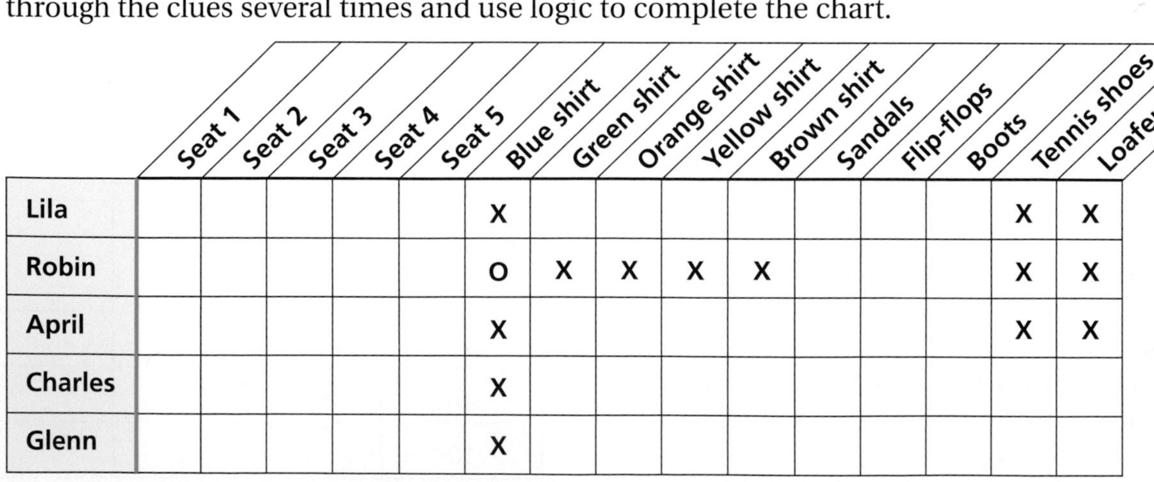

	Seat 1	Seat 2	Seat 3	Seat 4	Seat 5	Blue shirt	Green shirt	Orange shirt	Yellow shirt	Brown shirt	Sandals	Flip-flops	Boots	Tennis shoes	Loafers
Lila						X								X	X
Robin						O	X	X	X	X				X	X
April						X								X	X
Charles						X									
Glenn						X									

Graph Linear Equations

internet connect
Lab Resources Online
go.hrw.com
KEYWORD: MS4 TechLab12

A graphing calculator can be used to graph linear equations.

Activity

1 Graph the equation $y = 3x - 5$.

a. Start by pressing [Y=] 3 [X,T,θ,n] [−] 5 [ENTER]. You will get the screen shown at right.

b. Now select the standard viewing window by pressing [ZOOM] 6. The graph will immediately be displayed as shown.

You can check the graph by finding its slope and y-intercept. The graph crosses the y-axis at (0, −5), so the y-intercept is **−5**. The graph also crosses the x-axis at (2, 0), which gives a slope of **3**. This agrees with the equation $y = \mathbf{3}x - \mathbf{5}$, which is in slope-intercept form.

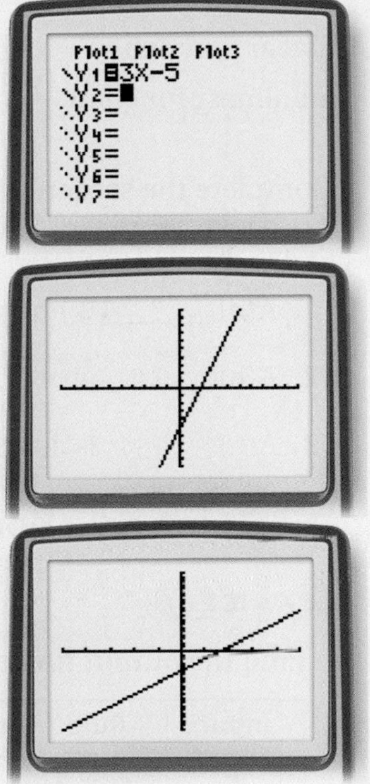

2 Change the window for the graph.

Change the window for the graphed equation by pressing [WINDOW], and edit the entries by pressing [(−)] 5 [ENTER] 5 [ENTER] [ENTER] [(−)] 20 [ENTER] 20. Then press [GRAPH].

The slope of the graph looks like it has changed, but the mathematical slope is still 3.

Think and Discuss

1. Explain why the slope appears to have changed in **2**. Is the graph incorrect? Does the graph still represent the original equation? Explain.

Try This

Use a graphing calculator to graph each equation. Specify an appropriate viewing window. Check the graph by finding the slope and y-intercept of the equation to see whether they agree with the graph.

1. $y = x + 3$

2. $y = 4 - x$

3. $y = 4 - 2x$

4. $y = 4x + 8$

5. $y = 10x + 15$

6. $x - y = 2$

Chapter 12

Study Guide and Review

Vocabulary

domain	613	range	613
element	636	sequence	608
function	604	set	636
linear equation	616	slope	620
linear function	616	slope-intercept form	621
nonlinear function	628	*y*-intercept	620

Complete the sentences below with vocabulary words from the list above. Words may be used more than once.

1. A(n) ___?___ is an ordered list of numbers.
2. A(n) ___?___ gives exactly one output for every input.
3. A(n) ___?___ is a function whose graph is a nonvertical line.

12-1 Introduction to Functions (pp. 604–607)

EXAMPLE

■ Find the output for each input.

Input	Rule	Output
x	$3x + 4$	y
−1	3(−1) + 4	1
0	3(0) + 4	4
2	3(2) + 4	10

EXERCISES

Find the output for each input.

4.

Input	Rule	Output
x	$x^2 - 1$	y
−2		
3		
5		

12-2 Find a Pattern in Sequences (pp. 608–611)

EXAMPLE

■ Find a function that describes the sequence. Then use it to find the next three terms in the sequence.

3, 6, 9, 12, . . .

Function: $y = 3n$

The next three terms are 15, 18, and 21.

n	Rule	y
1	1 · 3	3
2	2 · 3	6
3	3 · 3	9
4	4 · 3	12

EXERCISES

Find a function that describes each sequence. Then use it to find the next three terms in the sequence.

5. 25, 50, 75, 100, . . .
6. −3, −2, −1, 0, . . .
7. −4, −1, 2, 5, . . .
8. 4, 6, 8, 10, . . .

12-3 Interpreting Graphs (pp. 612–615)

EXAMPLE

- Find the domain and range of the graph.

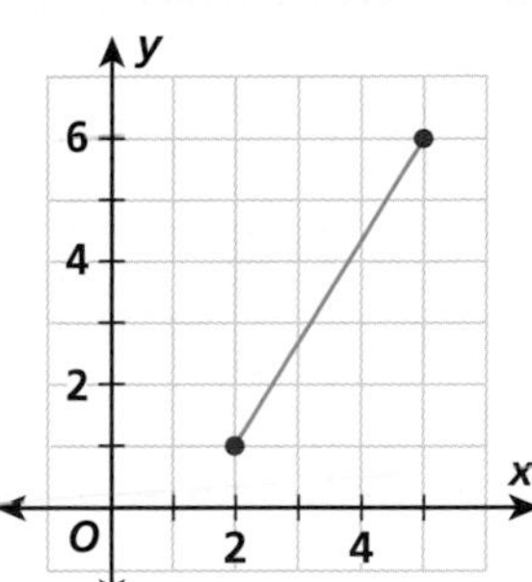

Domain: $2 \le x \le 5$

Range: $1 \le y \le 6$

EXERCISES

Find the domain and range of the graph.

9.

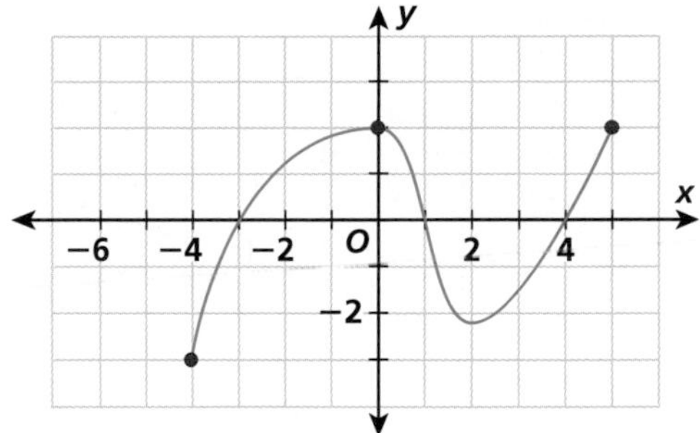

12-4 Linear Functions (pp. 616–619)

EXAMPLE

- Graph $y = -x + 2$.

Input	Rule	Output	Ordered Pair
x	$-x + 2$	y	(x, y)
0	$-(0) + 2$	2	(0, 2)
2	$-(2) + 2$	0	(2, 0)
−1	$-(-1) + 2$	3	(−1, 3)

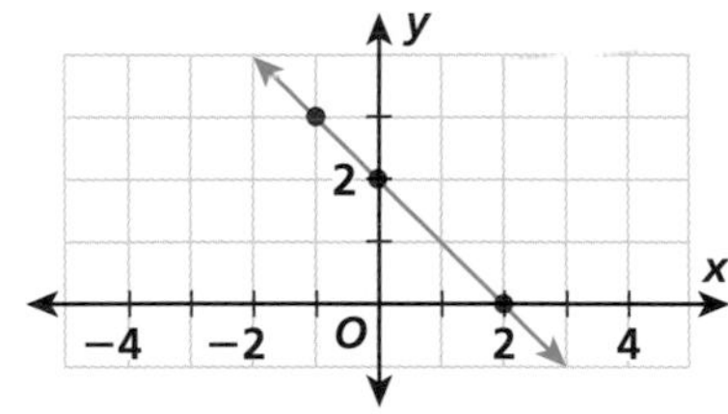

EXERCISES

Graph each linear function.

10. $y = 2x - 1$
11. $y = -3x$
12. $x + y = -3$
13. $y - 2x = 4$
14. $2y = x - 6$
15. $x - 3y = -9$

12-5 Slope (pp. 620–623)

EXAMPLE

- Graph $y = x - 2$.

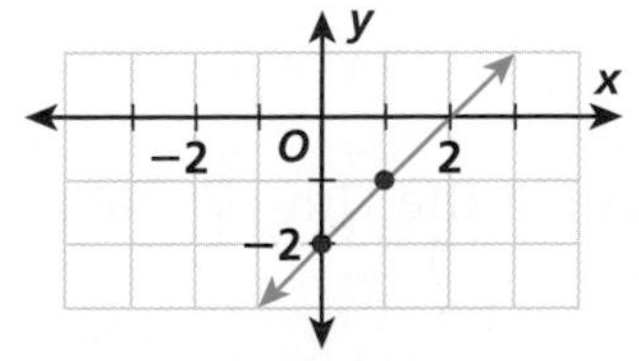

The slope is 1. The y-intercept is –2.

EXERCISES

Graph each line, given the equation.

16. $y = \frac{1}{3}x + 1$
17. $y = -x + 4$
18. $y = -1$

12-6 Nonlinear Functions (pp. 628–631)

EXAMPLE

■ **Tell whether the function represented in the table has a linear or nonlinear relationship.**

Difference is constant. 1, 1

x	y
2	5
3	10
4	17

Difference varies. 5, 7

The function has a nonlinear relationship.

EXERCISES

Tell whether the function represented in each table has a linear or nonlinear relationship.

19.

x	y
0	0
1	3
2	12

20.

x	y
0	0
1	−3
2	−6

12-7 Rates of Change (pp. 632–635)

EXAMPLE

■ **Tell whether the graph shows a constant or variable rate of change.**

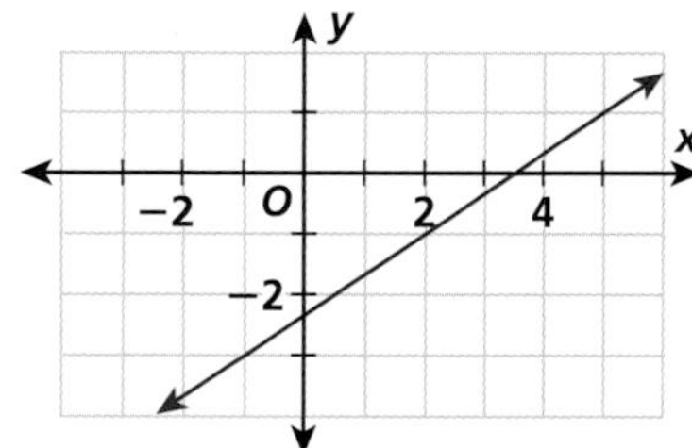

The rate of change is constant.

EXERCISES

Tell whether the graph shows a constant or variable rate of change.

21.

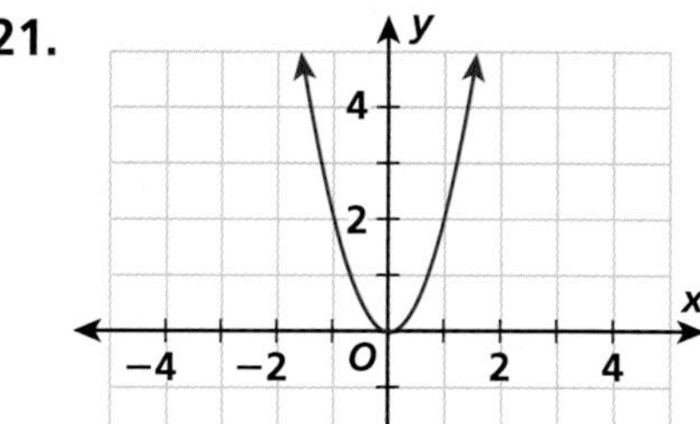

12-8 Set Theory (pp. 636–639)

EXAMPLE

■ **Find $A \cap B$ and $A \cup B$, and tell whether one of the sets is a subset of the other.**

$A = \{1, 3, 5, 7\}; B = \{3, 5\}$

$A \cap B = \{3, 5\}$
$A \cup B = \{1, 3, 5, 7\}$
$B \subset A$

EXERCISES

Find $A \cap B$ and $A \cup B$, and tell whether one of the sets is a subset of the other.

22. $A = \{x|x$ is a prime number less than $10\}$; $B = \{x|x$ is an odd number less than $10\}$

23. $A = \{x|x$ is divisible by $2\}$; $B = \{x|x$ is divisible by $8\}$

24. $A = \{x|x$ is a U.S. state beginning with the word *New*$\}$; $B = \{x|x$ is a state in the United States$\}$

Chapter Test

Chapter 12

Find the output for each input.

1.

Input	Rule	Output
x	$-2x + 5$	y
−1		
0		
1		

2.

Input	Rule	Output
x	$x \div 4$	y
−8		
0		
4		

Find a function that describes each sequence. Use the function to find the next three terms in the sequence.

3. 4, 6, 8, 10, . . .

4. 11, 21, 31, 41, . . .

5. −2, −1, 0, 1, . . .

6. 15, 30, 45, 60, . . .

Find the domain and range of each graph.

7. 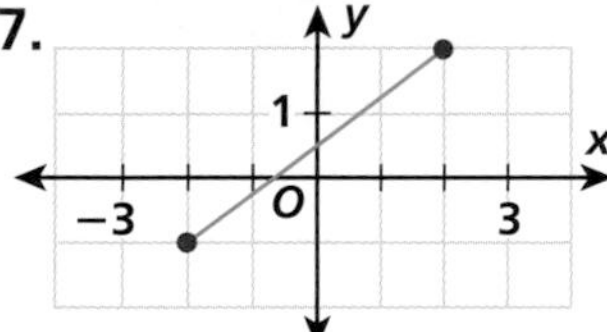

8. 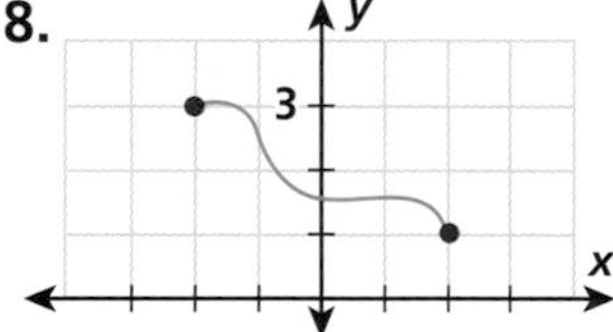

9. 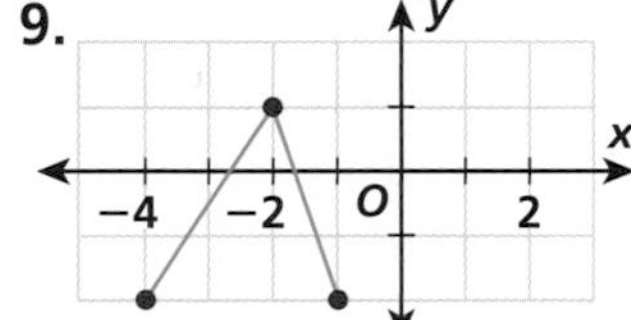

10. June deposited \$1,500 into a savings account. Because of the interest earned, the amount will double every 10 years. Write a sequence to show how much money will be in the account at the end of 40 years.

Use the slope and *y*-intercept to graph each linear function.

11. $y = 2x - 5$

12. $y = 4$

13. $2y - 5x - 4 = 0$

14. $-x + 2y = 10$

Tell whether the function represented in each table has a linear or nonlinear relationship.

15.

x	y
1	−3
2	−12
3	−27

16.

x	y
1	4
2	1
3	0

17.

x	y
1	6
2	10
3	14

18. Tell whether the graph shows a constant or variable rate of change.

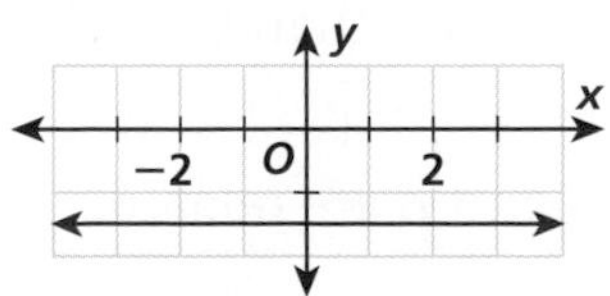

Find $A \cap B$ and $A \cup B$, and tell whether one of the sets is a subset of the other.

19. $A = \{-5, -3, -1, 1\}$; $B = \{-3, -1\}$

20. $A = \{x|x \text{ is a cat}\}$; $B = \{x|x \text{ is a dog}\}$

Chapter Test

Chapter 12

Performance Assessment

Show What You Know

Create a portfolio of your work from this chapter. Complete this page and include it with your four best pieces of work from Chapter 12. Choose from your homework or lab assignments, mid-chapter quiz, or any journal entries you have done. Put them together using any design you want. Make your portfolio represent what you consider your best work.

Short Response

1. The graph shows the population growth of a species of paramecium, a microscopic organism, in a test tube. Describe the growth. Is it linear or variable? Explain your answer.

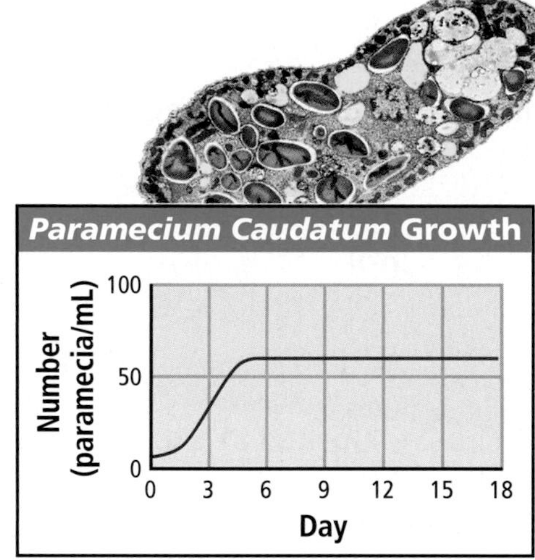

2. Make a function table for $y = 2x - 3$. Is the function linear or nonlinear? Explain your answer.
3. Find a pattern in the sequence 17, 34, 51, 68, ... Write a function for the sequence, and give the next three terms.
4. Graph the function $y = x^2 - 1$. Give the domain and range of the function.
5. Find the slope and y-intercept of the line that represents the equation $y = 3x + 1$. Explain in words how you found your answer. Then use the slope and intercept to graph the line.

Extended Problem Solving

6. The graph shows speed limits on rural interstates in six states.
   a. List the members of set A, the set of all states with a speed limit of 70 mi/h.
   b. List the members of set B, the set of all states with a speed limit greater than 65 mi/h.
   c. Find $A \cap B$ and $A \cup B$.
   d. Explain how sets A and B are related.

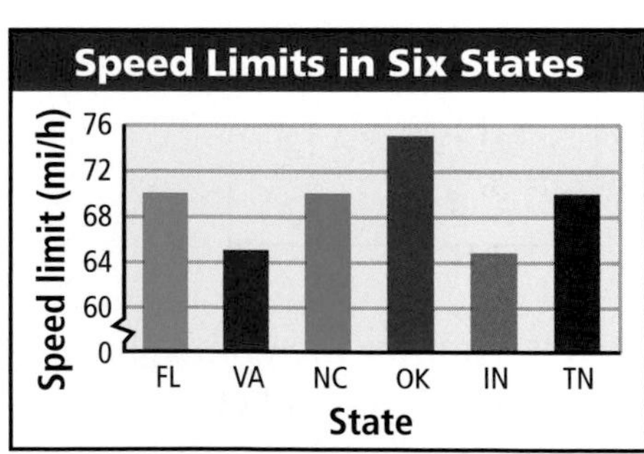

Performance Assessment

internet connect
EOG Test Practice Online
go.hrw.com Keyword: MS4 TestPrep

Getting Ready for EOG

Chapter 12

Cumulative Assessment, Chapters 1–12

1. What is the sum of $\sqrt{21} + \sqrt{68}$ to the nearest whole number?

   A 11 C 13
   B 40 D 14

2. What is the range of the data set?
   13, 22, 21, 18, 12, 21, 20, 25

   A 21 C 12
   B 13 D 19

3. What is the least common multiple of 18, 24, and 30?

   A 360 C 12,960
   B 240 D 6,480

4. What is the value of the expression $3a - b$ for $a = -5$ and $b = 9$?

   A -6 C -21
   B 21 D -24

5. What is the area of the circle to the nearest tenth? Use 3.14 for π.

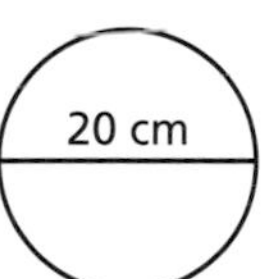

   A 1,256 cm^2 C 62.8 cm^2
   B 314 cm^2 D 125.6 cm^2

6. Stewart spent $12.86 at a restaurant. About how much should he leave for a 15% tip?

   A $1.95 C $6.00
   B $3.00 D $2.50

7. Which rule can describe the sequence 5, 7, 9, 11, . . . ?

   A $n + 2$ C $n - 2$
   B $2n - 3$ D $2n + 3$

TEST TAKING TIP!

When finding the percent of a number, you can eliminate choices by deciding whether the answer should be greater or less than the original number.

8. A seventh-grade class is going to increase in size by 15% next year. There are 280 seventh graders now. What will be the enrollment next year?

   A 42 C 322
   B 295 D 238

9. This is a graph of which linear equation?

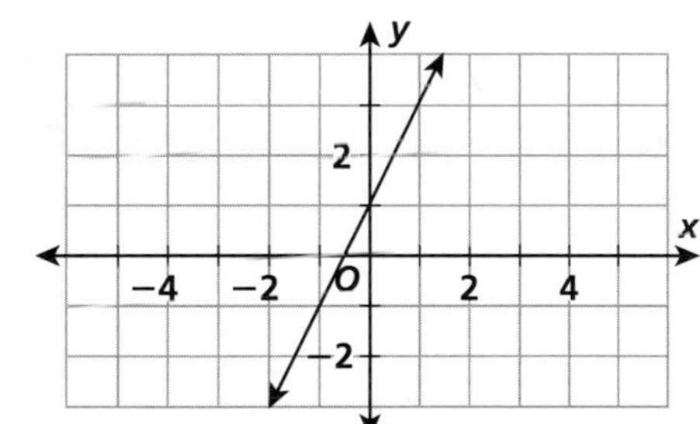

   A $y = 2x + 1$ C $y = 2x$
   B $y = -2x - 1$ D $y = 2x - 1$

10. ***SHORT RESPONSE*** A right triangle has side lengths of 7 inches and 24 inches. Use the Pythagorean Theorem to find the length of the hypotenuse. Show your work.

11. ***SHORT RESPONSE*** Identify the sample space for tossing three coins, and give the probability that when three coins are tossed, all three would land on heads.

Student Handbook

Student Handbook

Preview Skills

Science Skills

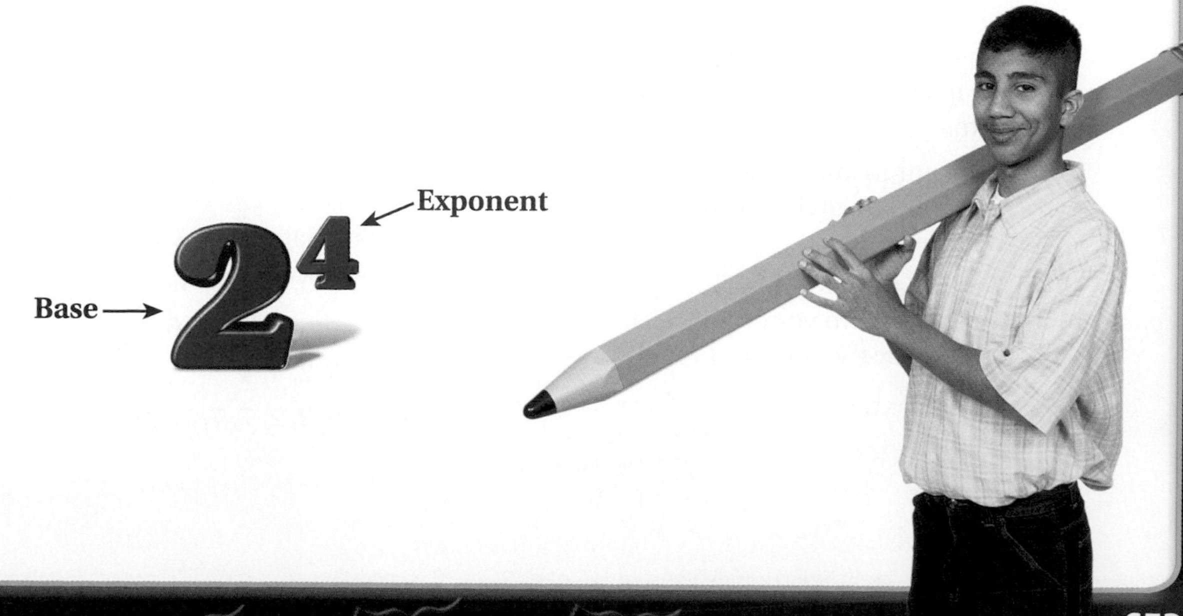

EOG Practice ▪ Chapter 1

1A Organizing Data

LESSON 1-1

1. Researchers poll 2,000 middle school students from across the country to find out about the kinds of music middle school students like. Which of the following accurately represents the population for this situation?

   A the 2,000 middle school students who are polled

   B the middle school students from Rhode Island

   C all middle school students in the United States

   D all seventh grade students

2. Scientists tag the ears of 85 deer across the country to get information on the migratory patterns of deer in the wild. Which of the following accurately represents the sample for this situation?

   A all deer in North America

   B the 85 deer that are tagged

   C all deer in the United States

   D 1,000 deer that are tagged

3. In which situation should you survey the population instead of using a sample?

   A You want to know the types of computers used in all of the schools across the country.

   B You want to know the favorite music group of all of the teenagers in your state.

   C You want to know how many hours per week the males in your city surf the Internet.

   D You want to know how many hours the students in your science class studied for the quiz.

4. In which of the following situations would it *not* be logical to sample the population?

   A You want to know how many of your cousins have graphing calculators.

   B You want to know what type of cars your teachers drive.

   C You want to know the average salaries of veterinarians across the country.

   D You want to know the number of people who are in the student council at your school.

5. ***SHORT RESPONSE*** A reporter is gathering responses from customers at a local grocery store about what kind of cereal they prefer. He surveys all of the people whose last names begin with the letters *A*-*G*. Explain whether the sampling method is random. If the sampling method is not random, give one that is.

LESSON  1-2

6. Which number represents the median of the following data set?

   13, 8, 40, 19, 5, 8

   A 7.5 C 15.5

   B 10.5 D 35

7. Which number represents the mean of the following data set?

   61, 89, 93, 102, 47, 93, 61

   A 55 C 78

   B 61 D 89

8. Which data set has a mode of 2?

   A 1, 7, 5, 2, 3, 3, 6

   B 5, 2, 6, 9, 11, 2, 6, 2

   C 5, 9, 2, 4, 4

   D 9, 9, 2, 2, 7, 7, 7, 10

EOG Practice ■ Chapter 1

1A Organizing Data

9. ***SHORT RESPONSE*** The ages of the participants in a charity walk from Elmore Middle School are 12, 13, 35, 41, 13, 13, 29, 12, 11, 14. Find the mean, median, and mode of the ages. Explain the steps you used to find your answers.

10. ***SHORT RESPONSE*** Identify the outlier in the following data set and determine how it affects the mean, median, and mode.

    36, 37, 35, 39, 35, 39, 10, 31

LESSON 1-3

Use the following stem-and-leaf plot for problems 11–15.

Stems	Leaves
0	3 5
1	0 1 3
2	0 3 3 6
3	
4	8

Key: 4|8 means 48

11. What is the range of the numbers in the data set?
    - A 16.5
    - B 18.2
    - C 45
    - D 48

12. What is the mode of the data in the stem-and-leaf plot?
    - A 11
    - B 16.5
    - C 18
    - D 23

13. What is the median of the data in the stem-and-leaf plot?
    - A 16.5
    - B 23
    - C 45
    - D 48

14. Which of the following numbers is *not* part of the data set?
    - A 1
    - B 3
    - C 10
    - D 26

15. Which of the following is *most* likely the source of the data in the stem-and-leaf plot?
    - A shoe sizes of 10 basketball players
    - B numbers of calories eaten in a day by 10 people
    - C temperatures for 10 days in Alaska
    - D number of pets per household on Dee Street

The table shows the number of points a player scored during the last ten games of the season. Use the following table for problems 16–17.

Points Scored Per Game			
Game Date	Points	Game Date	Points
Feb 7	36	Feb 25	18
Feb 14	34	Feb 27	31
Feb 18	27	Mar 1	43
Feb 20	46	Mar 3	42
Feb 23	32	Mar 4	28

16. Which set of numbers accurately represents the mean, median, mode, and range of the data in the table?
    - A mean = 37; median = 34; mode = 18; range = 28
    - B mean = 33.7; median = 33; no mode; range = 28
    - C mean = 33; median = 33; no mode; range = 34
    - D mean = 28; median = 33; mode = 36; range =34

17. ***SHORT RESPONSE*** Make a cumulative frequency table of the data. Explain how you determined the intervals you chose.

EOG Practice ■ Chapter 1

1B Displaying Data

LESSON 1-4

Use the following table for problems 1–4.

Populations of Selected Countries in 1998 and 2001		
Country	1998 Population (millions)	2001 Population (millions)
Algeria	30.1	31.7
Syria	15.3	16.7
Tunisia	9.3	9.7
Turkey	64.5	66.5

1. If you were making a double-bar graph of this data, which scale, in millions, would be *best* to use for the *y*-axis?
   - A 0 to 500, intervals of 100
   - B 0 to 70, intervals of 10
   - C 9 to 31, intervals of 2
   - D 1 to 10, intervals of 1

2. Which country's population was the *greatest* in 2001?
   - A Tunisia
   - B Syria
   - C Turkey
   - D Algeria

3. Which country's population was the *least* in 1998?
   - A Tunisia
   - B Syria
   - C Turkey
   - D Algeria

4. Which country's population increased the *most* from 1998 to 2001?
   - A Tunisia
   - B Syria
   - C Turkey
   - D Algeria

Use the following histogram for problems 5–7.

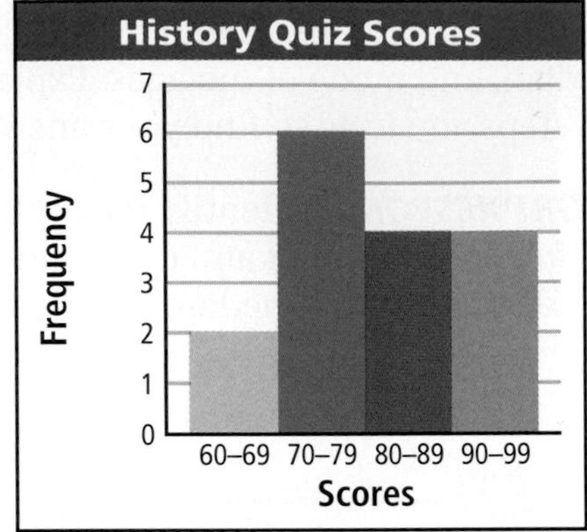

5. How many students scored between 70 and 79 on their history quiz?
   - A 2
   - B 6
   - C 8
   - D 16

6. If a grade of 70 is considered a passing score, how many students passed their history quiz?
   - A 2
   - B 8
   - C 14
   - D 16

7. Eight students scored within which range of values?
   - A 60–89
   - B 70–89
   - C 80–99
   - D 90–99

8. ***SHORT RESPONSE*** The table below shows the results of a survey to determine the types of pets most students have. Make a bar graph of the data.

Types of Pets Owned by Students	
Cats	~~||||~~ ~~||||~~ ~~||||~~
Dogs	~~||||~~ ~~||||~~ |
Fish	~~||||~~
Hamsters	||
Rabbits	|
Snakes	||

EOG Practice ▪ Chapter 1

1B Displaying Data

LESSON 1-5

The following circle graph shows the results of a survey of 100 people from Iran who were asked about their ethnic backgrounds. Use the circle graph for problems 9–12.

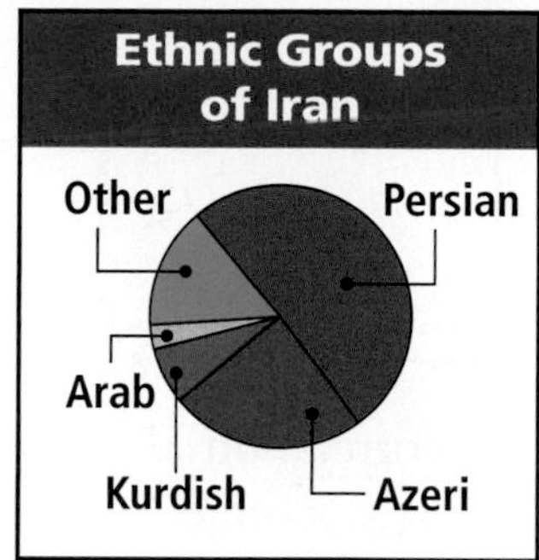

9. Which ethnic group is the *second largest*?

A Persian

B Arab

C Kurdish

D None of the above

10. Approximately what percent of the people are Persian?

A 100% C 50%

B 75% D 25%

11. According to the survey, 3% of the people are Arab. How many of the people surveyed are Arab?

A 3 people

B 12 people

C 97 people

D 100 people

12. About what percent of the people surveyed said they were Kurdish or Azeri?

A 32% C 12%

B 64% D 50%

13. ***SHORT RESPONSE*** Decide whether a bar graph or a circle graph would *best* display the number of guitars sold compared with the number of drum sets sold in one year. Explain why that type of graph would be best.

14. ***SHORT RESPONSE*** Decide whether a bar graph or a circle graph would *best* display the average temperature for each day of one week. Explain why that type of graph would be *best*.

LESSON 1-6

Use the following box-and-whisker plot for problems 15–18.

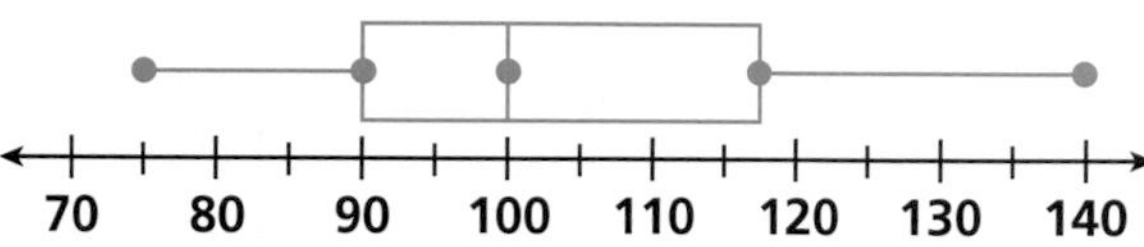

15. What is the first quartile?

A 65 C 87

B 75 D 90

16. What is the range of the data?

A 140 C 65

B 100 D none of these

17. About what fraction of the data are *less* than 90?

A $\frac{2}{11}$ C $\frac{1}{4}$

B $\frac{1}{2}$ D $\frac{7}{8}$

18. What does the point at 140 represent on the box-and-whisker plot?

A median

B upper extreme

C lower extreme

D third quartile

19. What is the difference between the minimum and maximum value of a data set called?

A mean

B mode

C median

D range

EOG Practice

EOG Practice  ■ Chapter 1

1C Trends and Relations in Graphs

LESSON 1-7

The table shows the number of students Karen tutored during certain months of the year. Use the following table for problems 1–4.

Karen's Tutoring	
Month	Number of Students
Jan	5
Mar	8
May	9
Jul	12
Sep	14
Nov	18

1. ***SHORT RESPONSE*** Make a line graph of the data in the table. Between which two months did the number of students increase the *most?* Use your graph to justify your answer.

2. When did the number of students increase the *least*?
   A from January to March
   B from March to May
   C from May to July
   D from July to September

3. Which is the *best* estimate of the number of students Karen tutored during the month of October?
   A about 12 students
   B about 14 students
   C about 16 students
   D about 18 students

4. Which month can you predict that Karen tutored the most students?
   A April
   B June
   C February
   D October

LESSON 1-8

5. What type of correlation is shown by the scatter plot?

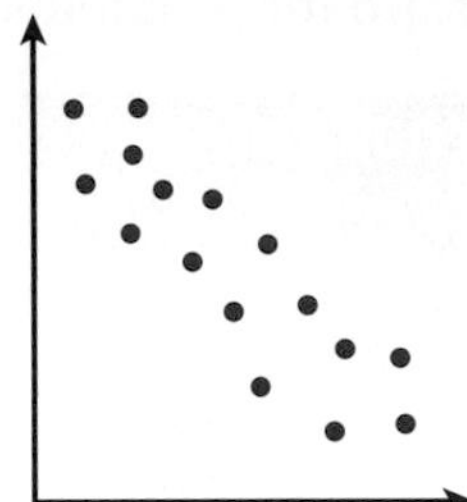

   A positive correlation
   B negative correlation
   C no correlation
   D perfect correlation

6. What type of correlation could be used to describe the relationship between the number of students at a school and the number of teachers at a school?
   A positive correlation
   B negative correlation
   C no correlation
   D cannot be determined

7. What type of correlation could be used to describe the relationship between the length of a person's hair and the size of a person's shoes?
   A positive correlation
   B negative correlation
   C no correlation
   D cannot be determined

8. What type of correlation could be used to describe the relationship between the number of people in a movie theater and the number of empty seats in the movie theater?
   A positive correlation
   B negative correlation
   C no correlation
   D cannot be determined

EOG Practice

1C Trends and Relations in Graphs

9. The table shows the average number of points per game that Michael Jordan scored during each season with the Chicago Bulls. Which of the scatter plots below accurately represents the data from the table?

Average Points Scored Per Game			
Year	**Points**	**Year**	**Points**
1990	33.6	1994	26.9
1991	31.5	1995	30.4
1992	30.1	1996	29.6
1993	32.6	1997	28.7

A

Points: 0, 5, 10, 15, 20, 25, 30
Year: 1990, 1992, 1994, 1996, 1998

B

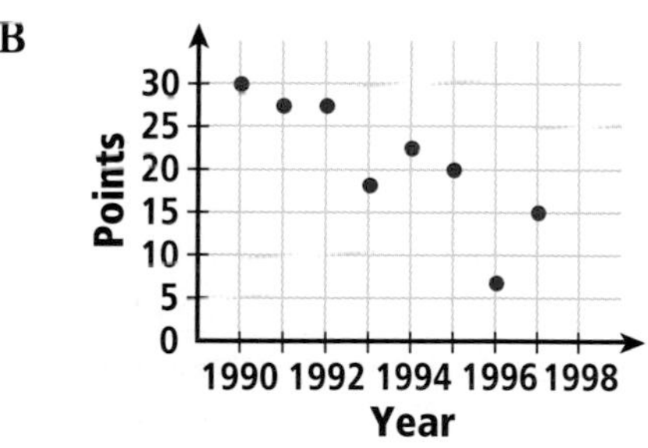

C

Points: 0, 5, 10, 15, 20, 25, 30
Year: 1990, 1992, 1994, 1996, 1998

D

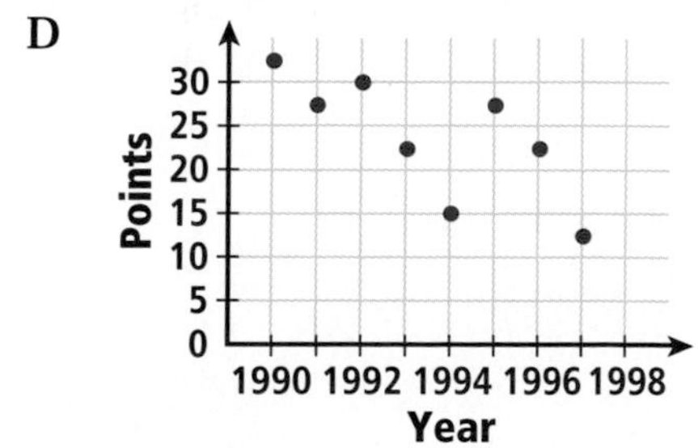

LESSON 1-9

SHORT RESPONSE **Explain why each of the following graphs could be considered misleading. What incorrect conclusions could be drawn from each graph?**

10.

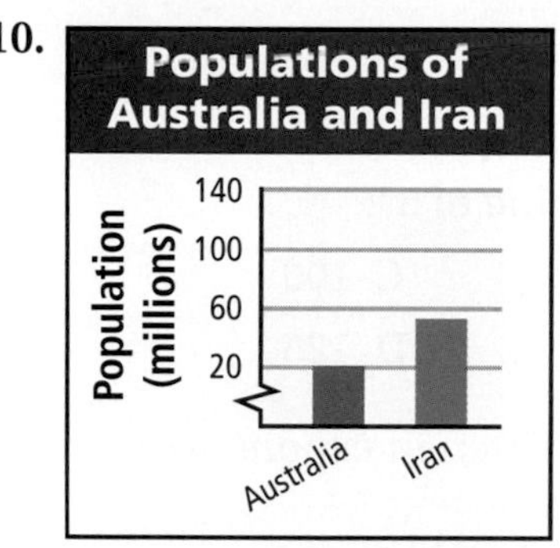

11.

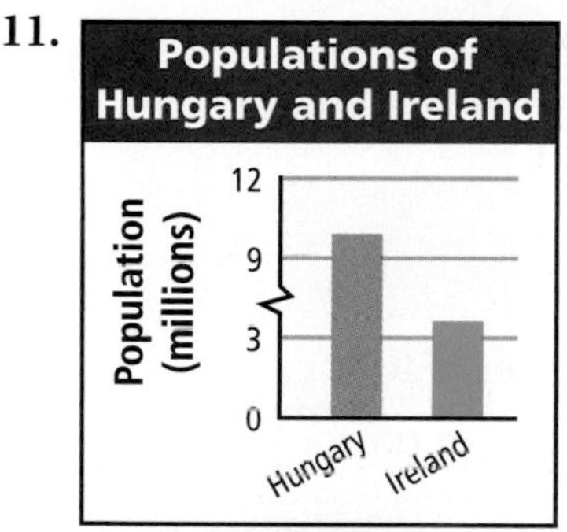

12.

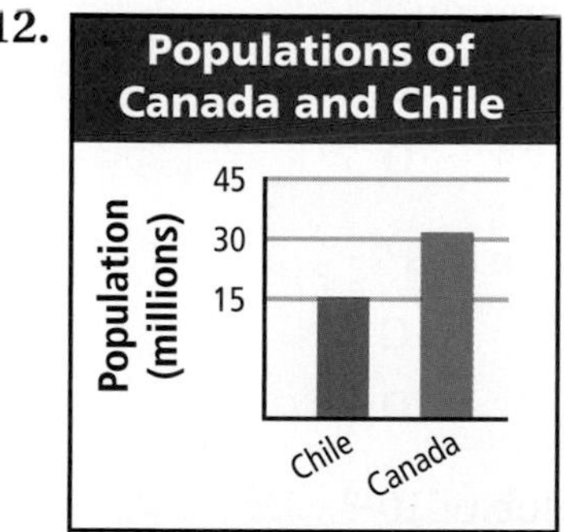

13.

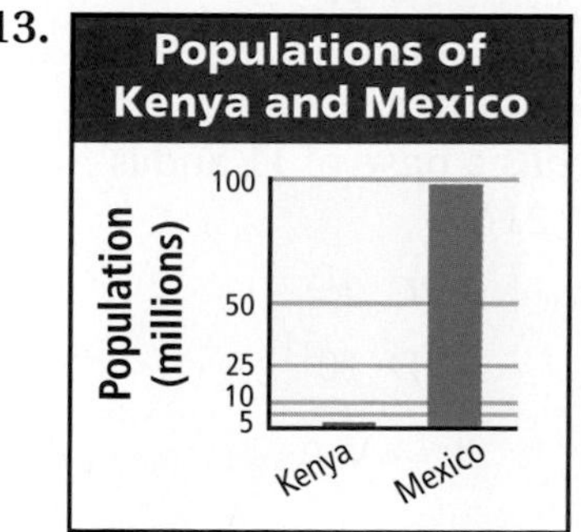

EOG Practice

EOG Practice ▪ Chapter 2

2A Exponents

LESSON 2-1

1. Which is another way of expressing 6^4?
   - A $4 \times 4 \times 4 \times 4 \times 4 \times 4$
   - B $6 \times 6 \times 6 \times 6$
   - C 6×4
   - D $6 \div 4$

2. What is the value of 5^3?
   - A 15
   - B 25
   - C 100
   - D 125

3. What is 729 written in exponential notation?
   - A 24^3
   - B 200^3
   - C 9^3
   - D 43^2

4. How is 5^4 read?
   - A nine
   - B five cubed
   - C the fifth power of four
   - D five to the fourth power

5. What is $7 \cdot 7 \cdot 7 \cdot 7 \cdot 7 \cdot 7$ written in exponential notation?
   - A 7^7
   - B 7^6
   - C 6^7
   - D 7^0

6. What is the value of 5^0?
   - A 0
   - B 1
   - C 5
   - D 50

7. What is the value of 10^1?
   - A 0
   - B 1
   - C 10
   - D 11

8. Which power has a base of 11 and is equivalent to 121?
   - A 11^2
   - B 11^{11}
   - C 2^{11}
   - D 10^{11}

9. Which power has a base of 20 and is equivalent to 8,000?
   - A 20^8
   - B 20^3
   - C 3^{20}
   - D 8000^{20}

10. Trey decided to save $2.00 one week and to double the amount he saves each week after that. How much will he have saved after 7 weeks?
    - A $14.00
    - B $7.00
    - C $128.00
    - D $49.99

LESSON 2-2

11. What is the value of $24 \cdot 10^3$?
    - A 243
    - B 2,400
    - C 24,000
    - D 24,103

12. What is the product of 525 and 10^5?
    - A 525,000
    - B 525,105
    - C 52,500,000
    - D 525,000,000

13. Which of the following expressions does *not* represent 100?
    - A 10^2
    - B 100^1
    - C 100^0
    - D $1,000 - 900$

14. To mentally multiply $281 \cdot 10^5$, how many times would you need to move the decimal point and in which direction?
    - A 5 places to the right
    - B 5 places to the left
    - C 10 places to the right
    - D 2 places to the right

EOG Practice

2A Exponents

15. The population of a city doubles every 10 years. What will the city's population be in 2033 if its population in 2003 was 354,230?

A 177,115

B 1,416,920

C 2,833,840

D 5,667,680

16. Which is 6,800,000 written in scientific notation?

A 6.8×10^8

B 68×10^4

C 6.8×10^6

D 680×10^6

17. Which number does 26×10^3 represent?

A 260

B 26,000

C 260,000

D 2,600,000

18. Which is one million, three hundred fifty thousand written in scientific notation?

A 1.35×10^5

B 1.35×10^6

C 135×10^6

D 1,350,000

19. Which is two hundred seventy-seven thousand, six hundred and two written in standard form?

A 207,672

B 2,276,021

C 2.77602×10^5

D 277,602

20. ***SHORT RESPONSE*** When asked to write 81 in exponential notation, a student wrote 9×2. What did this student do wrong? What is the correct way to write 81 in exponential notation?

LESSON 2-3

21. Which is the value of $9 \div 3 + 6 \cdot 5$?

A 0.2 C 5

B 3 D 33

22. Which part of the following expression should be worked first?

$$16 \cdot 2 + (20 \div 5) - 3^2 \div 3 + 1$$

A $3^2 \div 3$

B $20 \div 5$

C $16 \cdot 2$

D $3 + 1$

23. Charlotte bought 4 shirts and 3 pairs of pants. She got the pants at a discount. The expression $4 \cdot 32 + 3 \cdot 25 - (3 \cdot 25) \div 5$ can be used to find out how much she paid for the clothes. What is the value of the expression?

A \$69.00 C \$188.00

B \$158.00 D \$348.50

24. Rebecca paid a \$10 basic fee plus \$3.50 per hour to rent a canoe. The expression $10 + 3 \cdot 3.50$ can be used to find out how much it cost her to rent the canoe for 3 hours. What is the value of the expression?

A \$45.50 C \$22.00

B \$20.50 D \$16.50

25. According to the order of operations, which of the following should be evaluated first in a numerical expression?

A subtraction

B exponents

C division

D multiplication

26. ***SHORT RESPONSE*** Describe in what order you would perform the operations to find the value of $(4 \cdot 4 - 6)^2 - (5 \cdot 7)$. How could you get a different answer if you did not follow the order of operations?

EOG Practice

EOG Practice ■ Chapter 2

2B Factors and Multiples

LESSON 2-4

1. Which is an example of prime factorization?

   A $2 \times 3 \times 5 \times 7$

   B $3 \times 6 \times 9 \times 12$

   C 12^2

   D $2 \times 2 \times 4$

2. What is the prime factorization of 156?

   A $2 \times 2 \times 13$

   B $2 \times 2 \times 3 \times 13$

   C $3 \times 4 \times 13$

   D 12×13

3. Which of the following shows the prime factorization of 144?

   A $2^3 \cdot 3^2$

   B 14^4

   C $2^4 \cdot 3^2$

   D $2 \cdot 2 \cdot 2 \cdot 2 \cdot 2 \cdot 2 \cdot 2$

4. This factor tree is not complete. Which set of prime numbers should appear at the bottom of the factor tree?

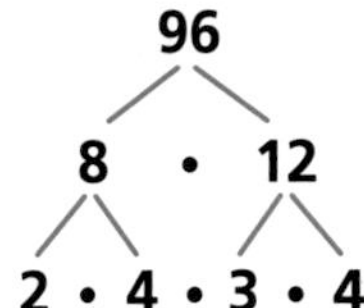

   A $2 \times 2 \times 3 \times 2 \times 2$

   B $2 \times 2 \times 2 \times 2 \times 2 \times 2$

   C $2 \times 4 \times 3 \times 2 \times 2$

   D $2 \times 2 \times 2 \times 3 \times 2 \times 2$

5. The expression $2^2 \times 3^3 \times 7$ is the prime factorization of which number?

   A 27

   B 252

   C 378

   D 756

6. The step diagram used for finding the prime factorization of 675 is not complete. What number is missing from the step diagram?

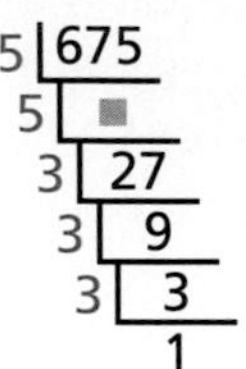

   A 54   C 135

   B 75   D 675

7. Which of the following numbers is divisible by 2, 3, 5, and 10?

   A 30   C 40

   B 35   D 100

8. ***SHORT RESPONSE*** One way to write 45 as a product of two factors is 15×3. What other ways can 45 be written as a product of two factors? Explain how you can use the prime factorization of 45 to find your answer.

LESSON  2-5

9. What is the greatest common factor of 18 and 27?

   A 3   C 18

   B 9   D 486

10. What is the greatest common factor of 27 and 189?

    A 3   C 27

    B 9   D 89

11. What is the greatest common factor of 64, 84, and 120?

    A 2   C 8

    B 4   D 12

12. Which of the following pairs of numbers is 12 the greatest common factor of?

    A 12, 18   C 48, 100

    B 24, 38   D 24, 36

EOG Practice

2B Factors and Multiples

13. Jose is making grab bags to sell at his concert. He has 51 CDs and 34 copies of his book. If he uses the same number of CDs and the same number of books in each bag, what is the *greatest* number of grab bags Jose can make if he uses all of the CDs and all of the books?

   A 34 grab bags
   B 25 grab bags
   C 17 grab bags
   D 2 grab bags

14. ***SHORT RESPONSE*** Use a list to find the greatest common factor of 36 and 120. Explain your steps in words.

LESSON 2-6

15. What is the least common multiple of 12 and 15?

   A 180   C 15
   B 60   D 3

16. What is the least common multiple of 12 and 36?

   A 36   C 4
   B 12   D 3

17. What is the least common multiple of 15, 22, and 30?

   A 3   C 660
   B 330   D 9,900

18. Marita completes a lap around the track in 5 minutes. Joselle completes a lap around the track in 7 minutes. They start running at the same time. How many minutes will have passed when they start another lap at the same time?

   A 10 minutes
   B 14 minutes
   C 35 minutes
   D They will never start another lap at the same time.

19. Ahmad can paint a sign in 4 minutes. Hoang can paint a sign in 6 minutes. They start painting at the same time. How many minutes will have passed when they first finish a sign at the same time?

   A 24 minutes
   B 12 minutes
   C 10 minutes
   D They will never finish a sign at the same time.

20. Kanisha shoots a basket every 7 seconds. Thomas shoots a basket every 12 seconds. They start out together. How many seconds will have passed when they next shoot a basket at the same time?

   A 19 seconds
   B 60 seconds
   C 84 seconds
   D They will never shoot a basket at the same time.

21. At an electronic store's grand opening, for the first 1000 people, every 50th person through the door receives a free CD and every 100th person receives a free television. How many people will receive both a CD and a television?

   A 10
   B 50
   C 100
   D 5,000

22. Darius washes his car every two weeks and bathes his dog every three weeks. If he does both today, how many weeks will pass before he does them both on the same day again?

   A 4
   B 5
   C 6
   D 12

EOG Practice ■ Chapter 2

2C Beginning Algebra

LESSON 2-7

1. If $r = 5$, what is the value of the expression $3r - 20 \div r$?

   A -1 C 11

   B 5 D 19

2. You can factor $10y^2$ as $2 \cdot 5 \cdot y \cdot y$. How is $15z^3x^4$ factored in the same manner?

   A $15 \cdot y \cdot y$

   B $15 \cdot z \cdot z \cdot x \cdot x \cdot x \cdot x$

   C $3 \cdot 5 \cdot z \cdot z \cdot z \cdot x \cdot x \cdot x \cdot x$

   D $3 \cdot 5 \cdot z \cdot x$

3. What is the value of $5x^2 + 3x$ for $x = 3$?

   A 8 C 48

   B 18 D 54

LESSON 2-8

4. A music store sells packages of guitar strings. David bought s strings for \$24. Which of the following algebraic expressions represents the cost of one string?

   A $24 \div s$

   B $24s$

   C $s \div 24$

   D $s + 24$

5. Which word phrase represents $6(13 + n)$?

   A six plus thirteen plus a number

   B six times the sum of thirteen and a number

   C a number times six plus thirteen

   D the sum of thirteen and six plus a number

6. What is "the quotient of a number and eight" written as an algebraic expression?

   A $8n$ C $n - 8$

   B $8 + n$ D $n \div 8$

7. Which of the following words is *not* used to mean subtraction?

   A difference

   B take away

   C product

   D minus

8. ***SHORT RESPONSE*** Marc spends \$78 for n shirts. Write an algebraic expression to represent the cost of one shirt. Evaluate your expression to find out the cost of each shirt if Marc bought 4 shirts. Explain how you determined your answer in words.

LESSON 2-9

9. When you combine the like terms in the expression $3u + 6 + 5k + u$, which expression is the result?

   A $4u + 5k + 6$

   B $14u$

   C $9uk + 6$

   D $9u + 5k$

10. Which expression has all of the like terms combined?

    A $z + 5z + 11$

    B $4p + 5p + 4 + 7p$

    C $7x + 8 + x^2$

    D $9s + 5s^2 + 3s$

11. ***SHORT RESPONSE*** To find the perimeter of a figure, add the lengths of the sides. Write an expression for the perimeter of the given figure. Then find the perimeter when $s = 1$, $s = 2$, and $s = 3$.

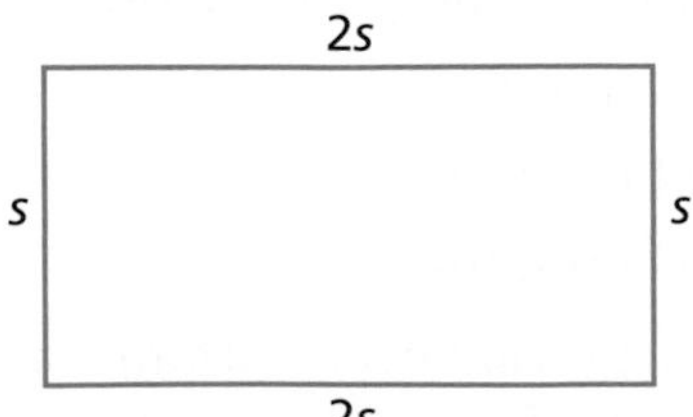

EOG Practice ■ Chapter 2

2C Beginning Algebra

LESSON 2-10

12. Which of the following numbers is a solution of $17 = 45 - j$?

A 14 C 28

B 22 D 31

13. Which of the following numbers is a solution of $x + 23 = 51$?

A 19 C 31

B 28 D 42

14. Dano has 87 CDs. This is 12 more than Megan has. The equation $87 = c + 12$ can be used to represent the number of CDs that Megan has. How many CDs does Megan have?

A 99 C 75

B 85 D 51

15. ***SHORT RESPONSE*** Jessica wants to buy a new pair of inline skates. She has $28, which is $19 *less* than she needs. Do the skates cost $42, $47, or $49? Show your work.

LESSON 2-11

16. Which value of n is a solution of $n - 22 = 16$?

A 6 C 34

B 8 D 38

17. Which value of y is a solution of $y + 27 = 42$?

A 15 C 28

B 16 D 69

18. Which equation has the solution $x = 95$?

A $x - 4 = 95$

B $4 + x = 99$

C $91 = x + 4$

D $4 - x = 91$

19. Which equation does NOT have the solution $t = 32$?

A $11 + t = 43$

B $t - 9 = 23$

C $t + 6 = 26$

D $7 = t - 25$

20. ***SHORT RESPONSE*** After Ella deposited a check for $39, her new account balance was $248. Write and solve an equation to find the amount that was in her account before the deposit. Show your work.

LESSON 2-12

21. It costs $18 per ticket for groups of 20 or more people to enter an amusement park. If Celia's group paid a total of $414 to enter, how many people were in the group?

A 29 C 23

B 25 D 18

22. Mario rents his apartment at a rate of $1.25 per square foot per month. If he pays $1562.50 in rent per month, how many square feet is his apartment?

A 1,200 C 1,350

B 1,250 D 1,500

SHORT RESPONSE **For each equation, write a word problem that could be solved using the equation. Then solve the equation to find the answer to your problem. Show your work.**

23. $20 = s \div 3$

24. $12y = 84$

25. $15 = \frac{n}{9}$

26. $\frac{m}{36} = 12$

27. $144 = 3p$

28. $72j = 360$

EOG Practice ■ Chapter 3

3A Integers

LESSON 3-1

1. On which of the following number lines are 5, −3, −1, 2, and 0 accurately graphed?

A 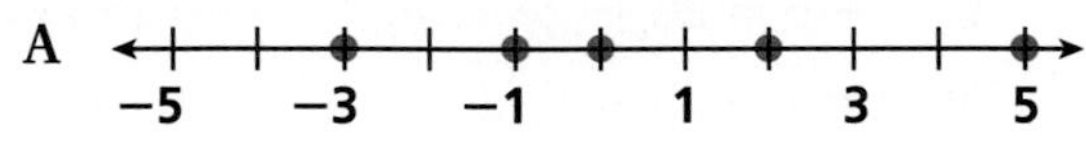

B 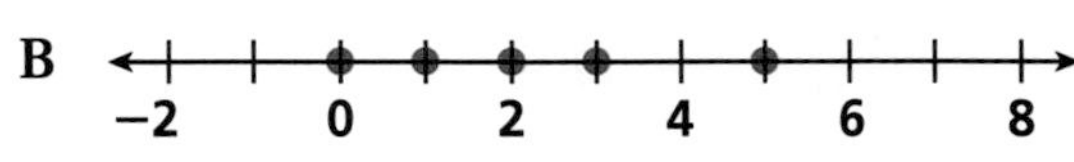

C

D 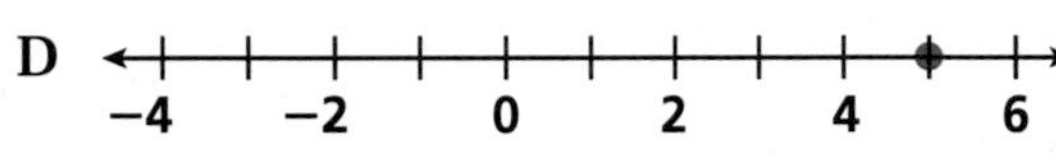

2. Which of the following sets of numbers is correctly ordered from least to greatest?

A 0, 2, −3, 4, −5

B 4, 2, 0, −3, −5

C −5, −3, 0, 2, 4

D 0, 2, 4, −3, −5

3. What integers are graphed on the following number line?

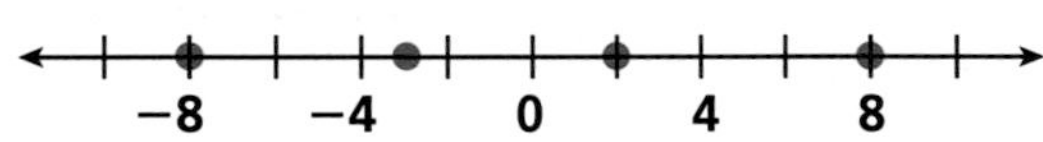

A −4, −2.5, 1, 4

B −9, 1, −1, 9

C −8, −3, 2, 9

D 8, −3, 2, −8

4. What is the absolute value of −20?

A −20 C 20

B 0 D 400

5. What is the opposite of $|-13|$?

A 26 C 0

B 13 D −13

6. ***SHORT RESPONSE*** What values can m have if $|m| = 9$? Explain how you determined your answer.

LESSON 3-2

7. On a coordinate plane, in which direction do the numbers on the x-axis increase?

A left C up

B right D down

Use the coordinate plane for Exercises 8–10.

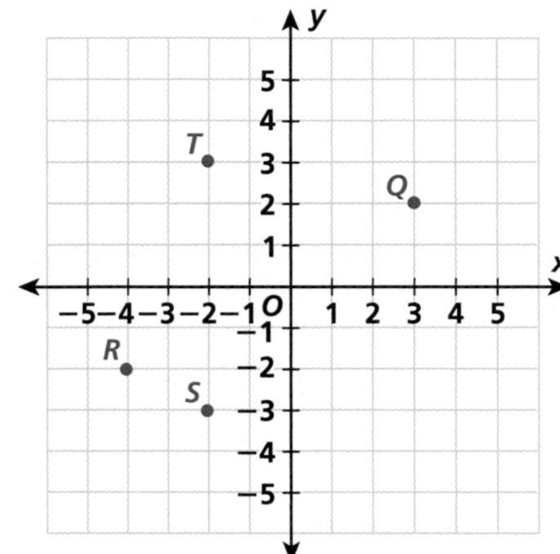

8. Which quadrant has no points plotted in it?

A Quadrant I

B Quadrant II

C Quadrant III

D Quadrant IV

9. What are the coordinates of point R?

A (−2, −4) C (−4, 2)

B (−4, −2) D (−2, 4)

10. What are the coordinates of point T, and what quadrant is it in?

A (−2, 3); Quadrant I

B (3, −2); Quadrant II

C (−2, 3); Quadrant II

D (−3, −2); Quadrant II

SHORT RESPONSE **Plot each point on a coordinate plane.**

11. $M(-1, 1)$

12. $N(4, 4)$

13. $Q(3, 1)$

EOG Practice ▪ Chapter 3

3A Integers

LESSON 3-3

14. A football team loses 7 yards on one play and then gains 16 yards on the next. Which of the following integer addition problems and solutions correctly represents the team's total yardage?

    A $7 + (-16) = -9$

    B $-7 + 16 = 9$

    C $-7 + 16 = -9$

    D $7 + 16 = 23$

SHORT RESPONSE **Evaluate $c + d$ for the given values. Explain how you determined whether your answer for each of the following would be positive or negative.**

15. $c = 5, d = 9$
16. $c = 12, d = -9$
17. $c = 7, d = -2$
18. $c = -16, d = -8$

LESSON 3-4

19. If the temperature rises from −2°F to 16°F, how much does the temperature increase?

    A −16°F

    B 14°F

    C 16°F

    D 18°F

SHORT RESPONSE **Evaluate $a - b$ for the given values. If $a - b$ is positive, what can you say about the values of a and b? What if $a - b$ is negative?**

20. $a = 5, b = -8$
21. $a = -12, b = -6$
22. $a = 6, b = 13$
23. $a = 9, b = -17$

LESSON 3-4

24. One negative number is divided by another negative number. Which of the following describes the result?

    A The quotient is a negative number.

    B The quotient is a positive number.

    C The quotient is 0.

    D The quotient is 1.

SHORT RESPONSE **Without solving, tell whether each product or quotient is positive or negative. Explain in words how you determined your answer.**

25. $-9 \div 3$
26. $-2 \cdot -9$
27. $8 \cdot (-3)$
28. $15 \div (-5)$
29. $16 \div 4$
30. $6 \cdot 7$
31. $-7 \cdot 3$
32. $-72 \div (-12)$

LESSON 3-6

33. Tim runs the same distance every day. In March and April he ran a total of 183 miles. How many miles did he run each day? (*Hint:* March has 31 days and April has 30 days.)

    A 1

    B 2

    C 3

    D 4

34. ***SHORT RESPONSE*** If $n - 25 = -18$, is n *greater* than or *less* than −18? Solve the equation and show your work, or explain in words how you found your answer.

35. ***SHORT RESPONSE*** If $y + (-13) = 61$, is y *greater* than or *less* than 61? Solve the equation and show your work, or explain in words how you found your answer.

36. ***SHORT RESPONSE*** If $\frac{k}{-18} = 2$, is k positive or negative? Solve the equation and show your work, or explain in words how you found your answer.

EOG Practice ■ Chapter 3

3B Rational Numbers

LESSON 3-7

1. On which number line is −0.7 graphed correctly?

A 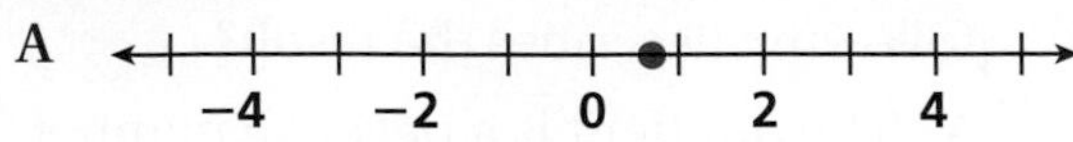

B 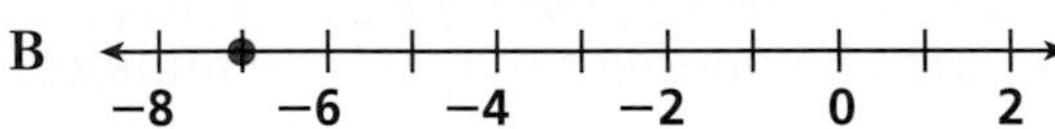

C 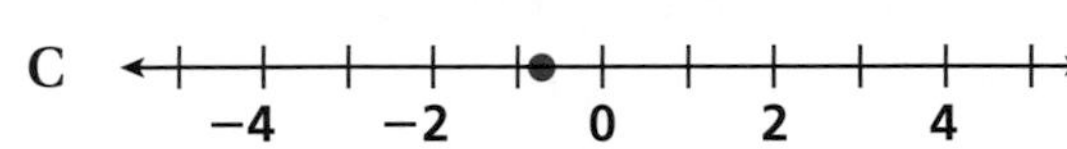

D 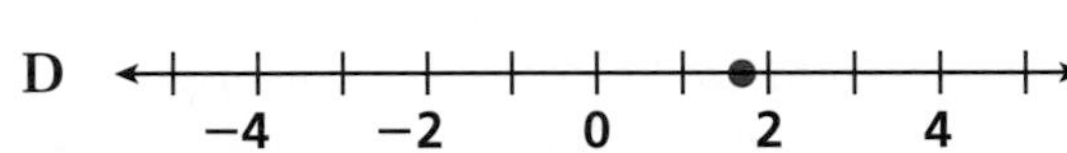

2. On which number line is $\frac{6}{7}$ graphed correctly?

A 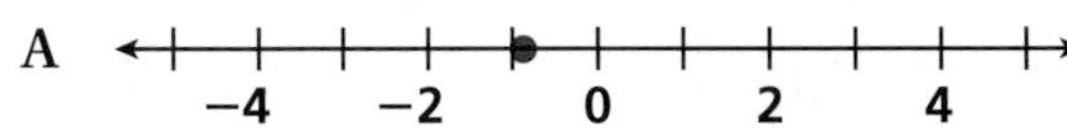

B −4 −2 0 2 4

C −8 −6 −4 −2 0 2

D −4 −2 0 2 4 6

3. Which number is graphed on the following number line?

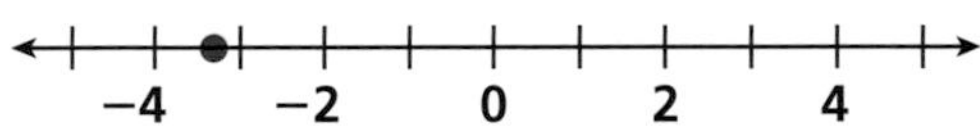

A −3.3　　C 3
B −3　　D 3.3

SHORT RESPONSE **Show that each number is a rational number by writing it as a fraction. Show your work, or explain in words how you found each answer.**

4. −12
5. 1.25
6. 0.50
7. −0.25

LESSON 3-8

8. Which fraction is written in simplest form?

A $\frac{5}{10}$　　C $\frac{2}{10}$
B $\frac{2}{5}$　　D $\frac{3}{21}$

9. Which fraction is equivalent to $7\frac{2}{3}$?

A $\frac{14}{3}$　　C $\frac{70}{3}$
B $\frac{23}{3}$　　D $\frac{72}{3}$

10. Which fraction is *not* equivalent to $\frac{8}{9}$?

A $\frac{16}{18}$　　C $\frac{40}{45}$
B $\frac{32}{36}$　　D $\frac{48}{63}$

11. Which mixed number is equivalent to $\frac{50}{13}$?

A $2\frac{1}{13}$　　C $2\frac{11}{13}$
B $2\frac{2}{13}$　　D $3\frac{11}{13}$

12. Which of the following is a common denominator of $\frac{2}{7}$ and $\frac{3}{4}$?

A 7　　C 16
B 14　　D 28

13. In which pair are the fractions equivalent?

A $\frac{12}{18}, \frac{4}{6}$　　C $\frac{6}{10}, \frac{1}{3}$
B $\frac{4}{18}, \frac{2}{3}$　　D $\frac{12}{24}, \frac{1}{4}$

14. In which pair are the fractions *not* equivalent?

A $\frac{14}{16}, \frac{7}{8}$　　C $\frac{49}{56}, \frac{7}{8}$
B $\frac{21}{24}, \frac{3}{4}$　　D $\frac{14}{15}, \frac{28}{30}$

15. Which mixed number is equivalent to $\frac{19}{5}$?

A $3\frac{4}{5}$　　C $9\frac{1}{5}$
B $3\frac{9}{10}$　　D $18\frac{1}{5}$

3B Rational Numbers

SHORT RESPONSE **Write each mixed number as an improper fraction. Show your work.**

16. $3\frac{1}{8}$

17. $9\frac{2}{7}$

18. $1\frac{4}{5}$

LESSON 3-9

19. Brianna sold 84 of the 96 CDs that she brought to sell at her concert. What portion of the CDs did she sell?

    A 0.12 C 0.875

    B 0.75 D 1.14

20. Jacob used 44 pages out of the 60-page journal that he bought. About what portion of the pages did he use?

    A 0.16 C 0.73

    B 0.44 D 1.36

21. Brent washed 4 out of the 10 cars at a school car wash on Saturday. What portion of the cars did Brent wash?

    A 0.04 C 0.40

    B 0.10 D 0.60

22. A survey states that 9 out of 10 students like to use the Internet. According to this survey, what portion of the students do *not* like to use the Internet?

    A 0.10 C 0.90

    B 0.45 D 10.0

SHORT RESPONSE **Write each fraction as a decimal. Show your work.**

23. $\frac{4}{5}$

24. $\frac{6}{8}$

25. $\frac{57}{15}$

26. $-\frac{75}{10}$

27. $\frac{2}{3}$

SHORT RESPONSE **Write each decimal as a fraction in simplest form. Then, explain how to use place value to write each decimal as a fraction in simplest form.**

28. 0.85

29. −0.04

30. 0.875

31. 2.6

32. 1.375

33. −0.22

LESSON 3-10

34. Which numbers are ordered from *least* to *greatest?*

    A 2.7, 2.59, $2\frac{7}{12}$

    B −0.61, −0.55, $-\frac{9}{15}$

    C $\frac{6}{13}$, 0.5, 0.58

    D $\frac{1}{4}$, 0.2, 3.2

35. Which number is the *greatest*?

    A $\frac{3}{5}$

    B 0.87

    C −2.1

    D $1\frac{1}{8}$

36. ***SHORT RESPONSE*** A recipe calls for $\frac{3}{4}$ cup flour and $\frac{2}{3}$ cup butter. If you double the recipe, would it call for a greater amount of flour or butter? Explain how you determined your answer.

37. ***SHORT RESPONSE*** Explain how to determine which is greater, 0.35 or $\frac{4}{9}$.

EOG Practice ▪ Chapter 4

4A Decimal Operations

LESSON 4-1

1. Helen saves $7.50 each week. She wants to buy a TV that costs $179.00. For about how many weeks will she have to save her money before she can buy the TV?

   A about 15 weeks

   B about 18 weeks

   C about 24 weeks

   D about 104 weeks

2. ***SHORT RESPONSE*** Joey has a piece of rope that is 63.25 feet long. He wants to cut it into smaller pieces that are each 4 feet long. He thinks he will get about 15 smaller pieces of rope. Use estimation to check whether his assumption is reasonable.

LESSON 4-2

3. Pedro runs his first lap around the track in 2.94 minutes, his second lap in 2.63 minutes, his third lap in 2.54 minutes, and his fourth lap in 2.78 minutes. What is his total time for the 4 laps?

   A 8.11 minutes

   B 10.35 minutes

   C 10.89 minutes

   D 11 minutes

4. ***SHORT RESPONSE*** When adding or subtracting decimals, what should you do if the two numbers do not have the same number of decimal places? Use an example in your explanation.

5. ***SHORT RESPONSE*** When adding or subtracting, what should you do when one number is a decimal and the other is a whole number? Use an example in your explanation.

LESSON 4-3

6. Casey and her friends hike an average of 4.8 miles per hour. How far could they hike in 2.5 hours?

   A 1.92 miles

   B 2.3 miles

   C 11 miles

   D 12 miles

7. Emilio can drive his car 21.9 miles on 1 gallon of gasoline. How far is he able to drive on 14.1 gallons?

   A 36 miles

   B 155 miles

   C 306.6 miles

   D 308.79 miles

SHORT RESPONSE **Find each product. Show your work.**

8. $4.3 \cdot 2.8$
9. $-3.38 \cdot 0.8$
10. $-8 \cdot (-0.07)$
11. $7.59 \cdot (-36)$
12. $-67.4 \cdot (-8.7)$
13. $5.66 \cdot (-16.34)$
14. $-43.9 \cdot (-4.7)$
15. $73.3 \cdot 6.85$

LESSON 4-4

16. The total cost of a baseball team's uniforms is $716.25. If there are 25 people on the team, how much does each player owe?

   A $25.00

   B $28.65

   C $34.90

   D $71.62

EOG Practice ■ Chapter 4

4A Decimal Operations

SHORT RESPONSE **Find each quotient. Show your work.**

17. $32.8 \div (-4)$
18. $-10.5 \div 4$
19. $-25.6 \div 8$
20. $-69.6 \div (-6)$
21. $63.5 \div (-2)$
22. $36.6 \div 6$
23. $-62.8 \div 8$
24. $56.05 \div 2$

LESSON 4-5

25. What division problem is modeled by the decimal grid below?

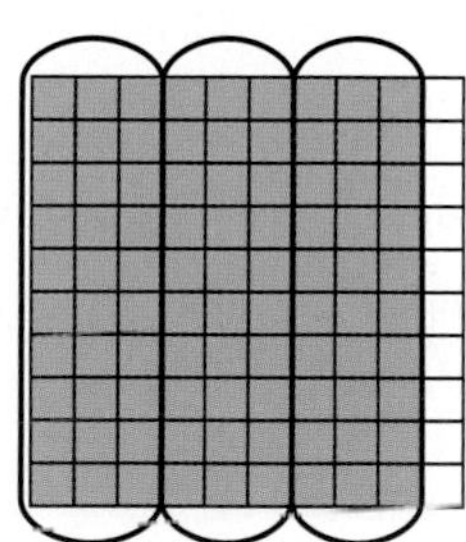

A $9 \div 3 = 3$

B $0.9 \div 3 = 0.3$

C $0.3 \div 9 = 0.033$

D $3 \div 9 = 0.33$

26. Brandi used 16.5 gallons of gasoline to drive her car 404.25 miles. What was her car's gas mileage?

A 40.8 mi/gal

B 25.5 mi/gal

C 24.5 mi/gal

D 23.9 mi/gal

SHORT RESPONSE **Tell the least power of ten that you could use to multiply each number by to make the divisor an integer.**

27. $16.9 \div (-1.3)$
28. $74.25 \div 6.6$
29. $-4.8 \div 0.12$
30. $-0.63 \div (-0.7)$
31. $-36.04 \div 4.24$
32. $34.672 \div (-4.402)$
33. $-128.685 \div 37.379$
34. $-231.28 \div (-41.31)$

LESSON 4-6

35. Billy worked 7.5 hours and earned $56.70. What is Billy's hourly wage?

A $6.75

B $7.00

C $7.56

D $8.00

36. A single movie ticket costs $7.25. The Brown family consists of Mr. and Mrs. Brown, Amy, and her two brothers. What does it cost the Brown family to go to the movies together?

A $14.50

B $21.75

C $29.00

D $36.25

37. The same cereal costs $3.99 per box at one store, $3.25 per box at another store, and $3.59 per box at a third store. What is the average price per box of the cereal?

A $3.61

B $3.75

C $3.92

D $10.83

SHORT RESPONSE **Solve each equation. Show your work, or explain in words how you found each answer.**

38. $4.7 + s = 9$
39. $t - 1.35 = -22$
40. $-4.8 = -6x$
41. $9.6 = \frac{v}{8}$
42. $-6.5 + n = 5.9$

EOG Practice ■ Chapter 4

4B Fraction Operations

LESSON 4-7

1. In a survey, 400 families were asked on what day of the week they went out to dinner the most often. Of the families, $\frac{5}{8}$ said Friday. How many families said they went out to dinner most often on Fridays?

   A 62.5   C 300

   B 250   D 350

2. A recipe calls for $\frac{2}{3}$ cup sugar. If Jill wants to triple the recipe, how much sugar does she need?

   A $\frac{3}{4}$ cup

   B $1\frac{2}{3}$ cups

   C 2 cups

   D 3 cups

3. What number is missing from the multiplication sentence?

$$\frac{2}{3} \cdot \frac{\blacksquare}{7} = \frac{8}{21}$$

   A 3

   B 4

   C 5

   D 6

SHORT RESPONSE **Find each product. Give the improper fraction you converted each mixed number to in order to multiply, and show the steps you used to get each product.**

4. $\frac{2}{3} \cdot 12\frac{3}{4}$
5. $3\frac{2}{9} \cdot \frac{1}{2}$
6. $\frac{5}{7} \cdot 4\frac{3}{8}$
7. $5\frac{2}{3} \cdot \frac{7}{12}$
8. $4\frac{3}{5} \cdot 3\frac{2}{3}$
9. $3\frac{1}{3} \cdot 2\frac{5}{6}$

LESSON 4-8

10. Jennifer's water bottle holds $80\frac{3}{4}$ ounces of water. How many $8\frac{1}{2}$-ounce glasses of water can be filled with that much water?

    A $8\frac{1}{2}$ glasses

    B 9 glasses

    C $9\frac{1}{2}$ glasses

    D $10\frac{1}{2}$ glasses

11. Kristi, Matt, and Monica are driving round-trip to a basketball game that is 190 miles from their town. If each of them drives the same distance, how far will each person drive?

    A $15\frac{7}{8}$ miles

    B $63\frac{1}{3}$ miles

    C $126\frac{2}{3}$ miles

    D 190 miles

LESSON 4-9

12. If salmon costs $5.99 a pound, about how much does $1\frac{1}{2}$ pounds of salmon cost?

    A $6.00

    B $8.00

    C $9.00

    D $12.00

SHORT RESPONSE **Estimate each sum or difference. Explain how you determined each answer.**

13. $\frac{3}{8} + \frac{5}{6}$
14. $\frac{7}{16} - 2\frac{3}{4}$
15. $4\frac{7}{12} + 2\frac{3}{8}$
16. $\frac{7}{8} - \frac{2}{3}$

EOG Practice

4B Fraction Operations

LESSON 4-10

17. Angelica lives $\frac{1}{4}$ mile from the bus stop. If she has already walked $\frac{1}{10}$ mile, how much farther must she walk to get to the bus stop?

A $\frac{1}{6}$ mile

B $\frac{3}{20}$ mile

C $\frac{1}{14}$ mile

D $\frac{3}{15}$ mile

18. On Tuesday, Brant spent $\frac{1}{2}$ hour on his math homework, $\frac{1}{4}$ hour on his science homework, and $\frac{2}{3}$ hour reading his book for English class. How long did Brant spend Tuesday on his homework?

A $1\frac{5}{12}$ hours

B 1 hour

C $1\frac{1}{4}$ hours

D $1\frac{1}{2}$ hours

LESSON 4-11

19. Madison walked 2 miles on Monday, $1\frac{1}{4}$ miles on Wednesday, and $2\frac{1}{2}$ miles on Friday. How many miles did she walk this week?

A $3\frac{4}{5}$ miles

B 5 miles

C $5\frac{1}{4}$ miles

D $5\frac{3}{4}$ miles

20. Yolande and Ashley ran in a 55-meter dash. Yolande finished in $6\frac{3}{5}$ seconds, and Ashley finished in $6\frac{79}{100}$ seconds. Who had the fastest time and by how much?

A Ashley, by $\frac{19}{100}$ second

B Yolande, by $\frac{19}{100}$ second

C Ashley, by $1\frac{39}{100}$ seconds

D Yolande, by $1\frac{39}{100}$ seconds

SHORT RESPONSE **Find each sum or difference. Show your work, and write each answer in simplest form.**

21. $9\frac{7}{8} - 4\frac{1}{4}$

22. $3\frac{1}{2} + 2\frac{3}{4}$

23. $9\frac{5}{6} - 6\frac{1}{3}$

24. $5\frac{7}{12} + 2\frac{5}{8}$

LESSON 4-12

25. Jorge owns $1\frac{3}{4}$ acres of land. Juanita, his neighbor, owns $2\frac{2}{3}$ acres. How many acres do they own in all?

A $\frac{7}{4}$

B $2\frac{5}{7}$

C $3\frac{5}{7}$

D $4\frac{5}{12}$

26. Kyra uses $2\frac{1}{3}$ yards of ribbon to wrap each of the identical gift baskets of fruit that she sells. How many baskets can she wrap with a 50-yard roll of ribbon?

A 16 baskets

B 21 baskets

C 25 baskets

D 57 baskets

SHORT RESPONSE **Solve each equation. Show your work, and write each answer in simplest form.**

27. $\frac{1}{3} + s = \frac{2}{5}$

28. $t - \frac{3}{8} = -\frac{5}{6}$

29. $\frac{r}{6} = \frac{1}{8}$

30. $-\frac{2}{3}y = -\frac{3}{4}$

EOG Practice ▪ Chapter 5

5A Numerical Proportions

LESSON 5-1

1. One day, a veterinarian saw 20 cats and 30 dogs. Which is *not* a way to write the ratio of dogs to cats she saw that day?

   A 30 to 20

   B 20:30

   C $\frac{30}{20}$

   D 30:20

2. For which of the following rates is $\frac{3 \text{ gallons}}{1 \text{ minute}}$ the unit rate?

   A 12 gallons in 3 minutes

   B 12 gallons in 4 minutes

   C 28 gallons in 7 minutes

   D 10 gallons in 4 minutes

3. A compact car gets 135 miles per 5 gallons of gas. A midsize car gets 210 miles per 10 gallons of gas. An SUV gets 18 miles per gallon of gas. Which automobile gets the most miles per gallon?

   A the compact car

   B the midsize car

   C the SUV

   D All cars get the same gas mileage.

SHORT RESPONSE **An art class has 7 sixth-graders, 12 seventh-graders, and 9 eighth-graders. Write each ratio in all three forms.**

4. sixth-graders to seventh-graders
5. seventh-graders to eighth-graders
6. sixth- and seventh-graders to eighth-graders
7. sixth-graders to eighth-graders
8. sixth-graders to seventh- and eighth-graders

LESSON 5-2

9. Which of the following ratios is equivalent to $\frac{8}{12}$?

   A $\frac{2}{6}$ C $\frac{8}{12}$

   B $\frac{4}{12}$ D $\frac{2}{3}$

10. Which of the following ratios is *not* equivalent to $\frac{15}{40}$?

    A $\frac{3}{8}$ C $\frac{45}{120}$

    B $\frac{30}{80}$ D $\frac{5}{8}$

11. Which ratio is equivalent to $\frac{35}{56}$?

    A $\frac{7}{8}$ C $\frac{5}{8}$

    B $\frac{3}{5}$ D $\frac{3}{8}$

12. ***SHORT RESPONSE*** Are the ratios $\frac{25}{40}$ and $\frac{30}{48}$ proportional? How do you know?

13. ***SHORT RESPONSE*** Find a ratio that is equivalent to $\frac{7}{3}$, and then use the ratios to write a proportion. Show your work.

LESSON 5-3

14. What is the value of p in the proportion?

$$\frac{7}{p} = \frac{42}{18}$$

    A $p = 3$

    B $p = 6$

    C $p = 14$

    D $p = 126$

15. The ratio of a person's weight on Earth to his or her weight on the Moon is 6 to 1. If Rafael weighs 90 pounds on Earth, how much would he weigh on the Moon?

    A 15 pounds

    B 18 pounds

    C 45 pounds

    D 60 pounds

EOG Practice ■ Chapter 5

5A Numerical Proportions

SHORT RESPONSE **Solve each proportion. Show your work or explain in words the method you used to find your answer.**

16. $\frac{8}{n} = \frac{12}{18}$

17. $\frac{4}{7} = \frac{p}{28}$

18. $\frac{u}{14} = -\frac{21}{28}$

19. $\frac{3}{21} = \frac{t}{49}$

20. $\frac{y}{35} = \frac{63}{45}$

21. $-\frac{6}{n} = -\frac{48}{12}$

22. $\frac{30}{x} = \frac{52}{117}$

23. $\frac{56}{80} = \frac{105}{m}$

LESSON 5-4

24. Which conversion factor can be used to convert 10 miles to feet?

    A $\frac{5{,}280 \text{ feet}}{1 \text{ mile}}$ C $\frac{10 \text{ miles}}{5{,}280 \text{ feet}}$

    B $\frac{1 \text{ mile}}{5{,}280 \text{ feet}}$ D $\frac{5{,}280 \text{ feet}}{10 \text{ miles}}$

25. How many gallons is equivalent to 16 quarts?

    A 4 gallons C 16 gallons

    B 6 gallons D 32 gallons

26. A bag of potatoes weighs 15 pounds. How many ounces does the bag of potatoes weigh?

    A 240 ounces

    B 128 ounces

    C 60 ounces

    D 30 ounces

27. Ivan is 6.1 feet tall. What is Ivan's height to the nearest centimeter? (*Hint:* There are 2.54 centimeters in 1 inch.)

    A 15 centimeters

    B 73 centimeters

    C 168 centimeters

    D 186 centimeters

28. Which of the following is equivalent to 88 feet per second?

    A 80 miles per hour

    B 60 miles per hour

    C 44 miles per hour

    D 28 miles per hour

29. Jimmy drinks 64 fluid ounces of water each day. How many gallons of water does Jimmy drink per week?

    A 5.5 gallons per week

    B 3.5 gallons per week

    C 3.2 gallons per week

    D 1 gallon per week

30. A CD that is played from beginning to end plays for 42 minutes and 34 seconds. How many hours, to the nearest tenth, does the CD play?

    A 0.3 hour

    B 0.7 hour

    C 0.9 hour

    D 1.9 hours

31. An 18-wheeler truck weighs 14 tons. How many pounds is equivalent to 14 tons?

    A 2,800 pounds

    B 14,000 pounds

    C 25,200 pounds

    D 28,000 pounds

32. ***SHORT RESPONSE*** You have $20, and gasoline costs $1.59 per gallon. How far can you travel on $20 worth of gasoline if your car gets 28 miles to the gallon? Show your work, or explain in words how you found your answer.

33. ***SHORT RESPONSE*** The speed of light is about 300,000,000 meters per second. If Earth is about 150,000,000,000 meters from the Sun, does it take *greater* or *less* than 1 minute for the Sun's light to reach Earth? Explain how you determined your answer.

5B Geometric Proportions

LESSON 5-5

1. The figures in which of the following pairs are similar?

A
9 cm, 5 cm, 8 cm
27 cm, 15 cm, 24 cm

B
130°, 6 in., 50°, 10 in.
130°, 4 in., 6 in., 50°

C
12 ft, 5 ft, 9 ft
36 ft, 20 ft, 50 ft

D
6 m, 10 m, 8 m
10 m, 11 m, 12 m

2. The sides of triangle *ABC* are 3 inches, 4 inches, and 5 inches long. Two sides of triangle *XYZ* are 9 inches and 12 inches long. What is the length of the third side of triangle *XYZ* if the two triangles are similar?

A 5 inches

B 10 inches

C 15 inches

D 25 inches

Use the following similar triangles for problems 3–4.

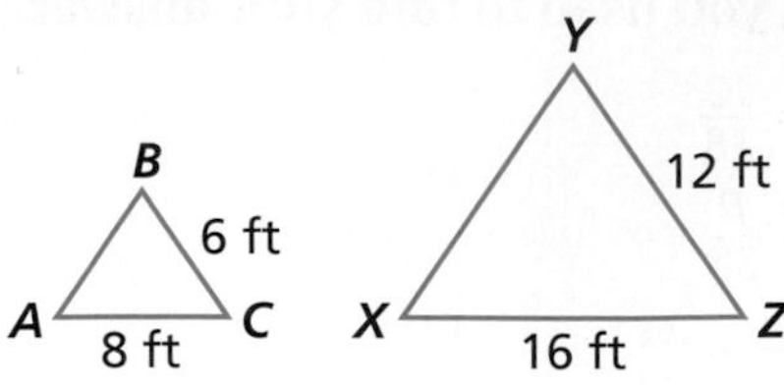

3. Which side corresponds to $\overline{AB}$?

A $\overline{XY}$

B $\overline{YZ}$

C $\overline{XZ}$

D $\overline{AC}$

4. Which angle corresponds to $\angle XYZ$?

A $\angle XZY$

B $\angle ABC$

C $\angle BAC$

D $\angle BCA$

5. Which size photograph is similar to a photograph that measured 8 × 10?

A 3 × 5

B 4 × 5

C 5 × 7

D $8\frac{1}{2} \times 11$

6. Which size photograph is *not* similar to a photograph that measures 5 × 7?

A 10 × 14

B 4 × 5

C 15 × 21

D 20 × 28

5B Geometric Proportions

LESSON 5-6

7. What is the length of x in the pair of similar figures?

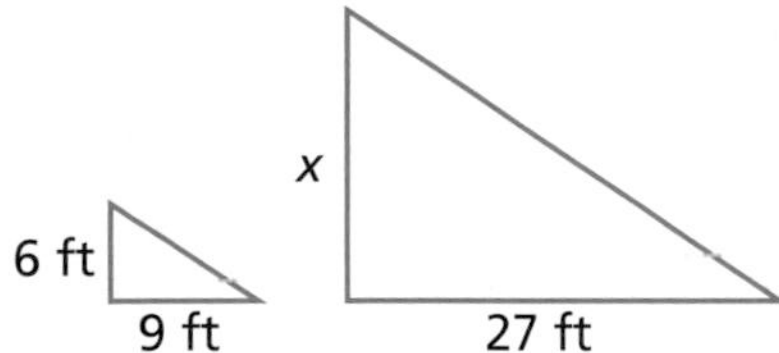

A 3 feet

B 4.5 feet

C 18 feet

D 21 feet

8. What is the length of x in the pair of similar figures?

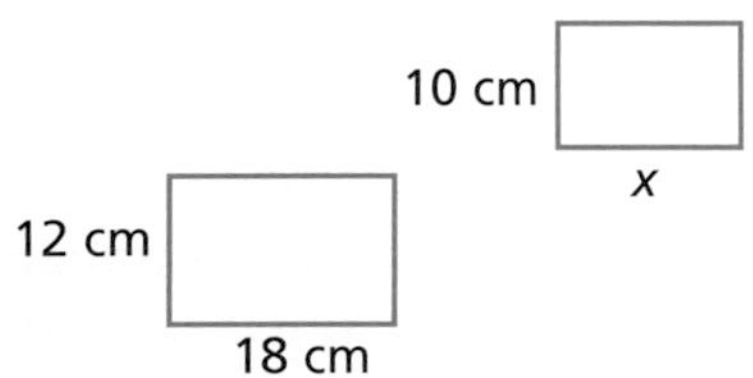

A 8 centimeters

B 12 centimeters

C 15 centimeters

D 18 centimeters

9. A 5-foot-tall girl casts a 7-foot-long shadow. A nearby telephone pole casts a 35-foot-long shadow. What is the height of the telephone pole?

A 7 feet

B 25 feet

C 28 feet

D 35 feet

10. A 24-foot-tall tree casts a 30-foot-long shadow. A 4-foot-tall child is standing nearby. How long is the child's shadow?

A 5 feet

B 6 feet

C 7.5 feet

D 8 feet

LESSON 5-7

11. A scale model of the Empire State Building is 3.125 feet tall with a scale factor of $\frac{1}{400}$. What is the height of the actual Empire State Building?

A 128 feet

B 1,250 feet

C 1,500 feet

D 10,000 feet

12. A dining-room table has a height of 30 inches. If the scale factor of a model of the dining-room table is $\frac{1}{12}$, what is the height of the model?

A 0.4 inch

B 2.5 inches

C 4 inches

D 25 inches

13. On a map, the distance between Branchburg and Trunktown is 4.3 centimeters. If the map scale is 1 centimeter = 25 kilometers, what is the actual distance between the cities?

A 5.8 kilometers

B 69.7 kilometers

C 107.5 kilometers

D 430 kilometers

14. Kira is drawing a map with a scale of 1 inch = 30 miles. The actual distance from Park City to Gatesville is 80 miles. How far from the dot representing Gatesville should Kira draw the dot representing Park City?

A 0.375 inch

B 1 inch

C $2\frac{2}{3}$ inches

D $3\frac{3}{4}$ inches

EOG Practice ▪ Chapter 6

6A Introduction to Percent

LESSON 6-1

1. Which fraction is equivalent to 14% and is written in simplest form?

   A $\frac{14}{100}$ C $\frac{7}{14}$

   B $\frac{7}{50}$ D $\frac{1}{4}$

2. Which fraction or mixed number is equivalent to 110% written in simplest form?

   A $1\frac{1}{10}$ C $\frac{110}{10}$

   B $1\frac{10}{100}$ D $1\frac{1}{5}$

3. How is 7% written as a decimal?

   A 0.07   C 0.77

   B 0.7   D 7.0

4. How is 125% written as a decimal?

   A 0.125   C 12.5

   B 1.25   D 125.0

5. What percent is equivalent to 0.06?

   A 0.06%   C 6%

   B 0.6%   D 60%

6. Which of the following is 42.0 written as a percent?

   A 0.42%

   B 42%

   C 420%

   D 4,200%

SHORT RESPONSE **Write the following fractions as percents, and round to the nearest tenth of a percent if necessary. Show your work.**

7. $\frac{15}{34}$
8. $\frac{29}{86}$
9. $\frac{33}{44}$
10. $\frac{61}{91}$

LESSON 6-2

11. Which of the following is the best estimate of 48% of 34?

   A about 48

   B about 34

   C about 17

   D about 10

12. Tyler wants to donate 45% of his 33 stuffed animals to the children's hospital. About how many stuffed animals will he donate?

   A about 4 stuffed animals

   B about 15 stuffed animals

   C about 33 stuffed animals

   D about 45 stuffed animals

SHORT RESPONSE **Use 1% or 10% to estimate the percent of each number. Explain how you determined each answer.**

13. 3% of 70
14. 28% of 125
15. 89% of 175
16. 53% of 84

LESSON 6-3

17. Last year, Maria's retirement fund lost 19%. If the fund was worth $18,000 at the beginning of the year, how much money did she lose?

   A $1,900   C $8,000

   B $3,420   D $14,580

18. Every year, about 300 movies are made. Only 13% are considered to be hits. About how many movies are considered hits in a year?

   A about 13 movies

   B about 39 movies

   C about 150 movies

   D about 287 movies

EOG Practice ■ Chapter 6

6A Introduction to Percent

SHORT RESPONSE **Write a word problem that could be solved by finding each percent. Then find the percent to solve your problem.**

19. 25% of 64

20. 48% of 200

21. 120% of 35

22. 0.5% of 90

23. 27% of 76

24. 65% of 300

25. 150% of 84

26. 15% of 15

LESSON 6-4

27. What percent of 150 is 60?

A 60%

B 50%

C 40%

D $\frac{2}{5}$%

28. Nine is 15% of what number?

A 18

B 50

C 60

D 100

29. Sixteen percent of 120 is what number?

A 1.9

B 19.2

C 50

D 52

30. Sixty is what percent of 400?

A $6\frac{2}{3}$%

B 15%

C 25%

D 55%

31. What is 8% of 125?

A 5%

B 10%

C 25%

D 54%

32. Twenty is 30% of a number. Which equation could be used to find the number?

A $\frac{x}{30} = \frac{20}{100}$

B $\frac{30}{x} = \frac{20}{100}$

C $\frac{x}{30} = \frac{100}{20}$

D $\frac{30}{100} = \frac{20}{x}$

33. Ryan bought a new CD holder for his car. He could fit only 60 of his CDs in the holder. This represents 60% of his collection. How many CDs does Ryan have?

A 36 CDs

B 60 CDs

C 75 CDs

D 100 CDs

34. Each week, Kelli spends $7.50 of her allowance on new books. If her allowance is $30 per week, what percent does she spend on books?

A 25%

B 29%

C 35%

D 75%

35. Yasmine saves 10% of each of her paychecks for her college fund. If Yasmine saved $25 from her last paycheck, how much was she paid?

A $100

B $250

C $400

D $450

36. ***SHORT RESPONSE*** If 40 is 150% of a number, is the number greater than or less than 30? Explain in words how you found your answer.

6B Using Percents

LESSON 6-5

1. If 54 is increased to 68, what is the percent of change rounded to the nearest tenth of a percent?
   A 25.9%
   B 29.5%
   C 35%
   D 79.4%

2. If 90 is decreased to 82, what is the percent of change rounded to the nearest tenth of a percent?
   A 8%
   B 8.9%
   C 9.1%
   D 110%

3. What is the percent of change, rounded to the nearest tenth of a percent, if 60 is increased to 80?
   A 25%
   B 33.3%
   C 66.6%
   D 80%

4. If 76 is decreased to 55, what is the percent of change rounded to the nearest tenth of a percent?
   A 21.0%
   B 27.6%
   C 72.3%
   D 99.2%

5. What is the percent of change, rounded to the nearest tenth of a percent, if 1 is increased to 3?
   A 100%
   B 133%
   C 200%
   D 300%

6. Abby's Appliances sells DVD players at 7% above the wholesale cost of $89. How much does the store charge for a DVD player?
   A $199.00
   B $108.95
   C $99.99
   D $95.23

7. The Nut Shop is having a sale on nutcrackers. For two days only, nutcrackers are marked down by 35% from their regular price of $49. What is the sale price?
   A $15.99
   B $31.85
   C $39.95
   D $43.55

8. A market's old parking lot held 48 cars. The new lot holds 66 cars. What is the percent of increase in the number of parking spaces?
   A 37.5%
   B 72%
   C 100%
   D 114%

9. A regular bag of potato chips contains 12 ounces. A jumbo bag of chips contains 32 ounces. What is the percent of increase in the amount of chips, rounded to the nearest tenth of a percent?
   A 200%
   B 166.7%
   C 113%
   D 37.5%

10. ***SHORT RESPONSE*** The regular price of a computer monitor at the electronics store is $499. This month the monitor is on sale for 15% off. Find the sale price of the computer monitor. Show your work.

EOG Practice

EOG Practice ■ Chapter 6

6B Using Percents

LESSON 6-6

11. Using the equation $I = prt$, what is the value of I, in dollars, if $p = \$70$, $r = 1.5\%$, and $t = 2$ years?

    A $2.10

    B $21.00

    C $73.50

    D $210.00

12. Austin deposits $8,500 in an account that earns 4.5% in simple interest per year. How long will it be before the total amount is $12,000?

    A 4.5 years

    B 8.25 years

    C 9.15 years

    D 31.37 years

13. Liz deposits $5,000 in an account that earns 5% simple interest per year. How long will it be before the total amount is $7,500?

    A 10 years

    B 15 years

    C 30 years

    D 32 years

14. How many years will it take $3,000 to triple at a yearly simple interest rate of 4%?

    A 7 years

    B 25 years

    C 50 years

    D 100 years

15. Shane deposited $9,000 into a savings account. After $3\frac{1}{2}$ years, he closed his account with the bank. His balance was $10,500. What was his yearly simple interest rate?

    A 3.3%

    B 4.8%

    C 8%

    D 19.5%

16. Nikolas put $2,500 into a savings account that earns 4% yearly simple interest. How many years will it take him to double his initial deposit?

    A 9 years

    B 15 years

    C 25 years

    D 35 years

17. Michelle put $5,000 into a money-market account and left it there for 9 months. She then closed the account and received a check for $5,300. What was the yearly simple interest rate on the account?

    A 19%

    B 11%

    C 8%

    D 0.7%

18. Ping is buying a new car for $17,500. His bank will lend him the money for the car now, but Ping must repay it at the end of 4 years. He must also pay 5.5% simple interest each year until the loan is due. How much interest will he have paid to the bank after 4 years?

    A $4,500

    B $3,896

    C $3,850

    D $3,125

19. ***SHORT RESPONSE*** Copy and complete the following table and explain how you found each answer.

Principal	Interest Rate	Time	Simple Interest
$500	5%	1 year	■
$800	■	3 years	$168
■	6%	2 years	$30
$3,500	5%	■	$262.50

EOG Practice

EOG Practice ▪ Chapter 7

7A Lines and Angles

LESSON 7-1

1. Point D is *not* part of which line segment?

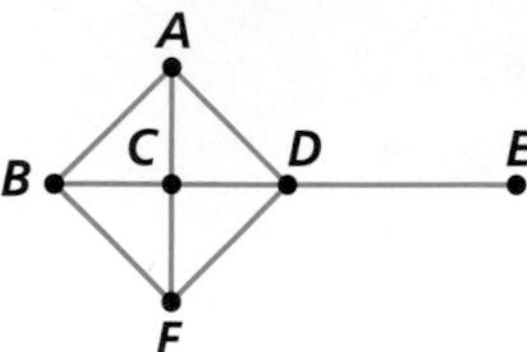

A $\overline{BC}$
C $\overline{CD}$
B $\overline{BE}$
D $\overline{FD}$

2. Which ray is part of line $\overleftrightarrow{SW}$?

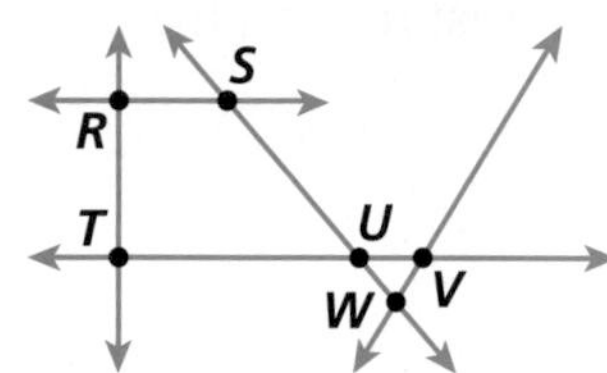

A $\overrightarrow{SR}$
C $\overrightarrow{UW}$
B $\overrightarrow{UV}$
D $\overrightarrow{WV}$

3. Which statement is false?

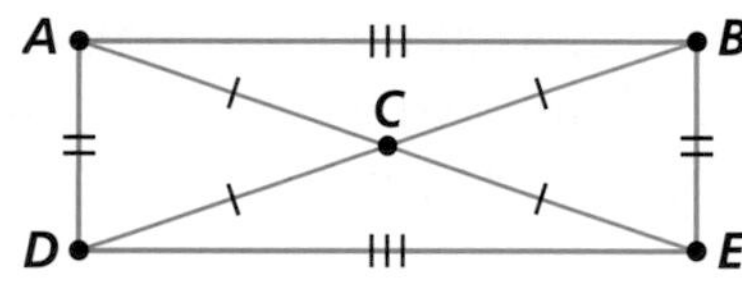

A $\overline{AD} \cong \overline{BE}$
B $\overline{AE} \cong \overline{DB}$
C $\overline{AC} \cong \overline{CB}$
D $\overline{AB} \cong \overline{BE}$

4. Which statement is true?

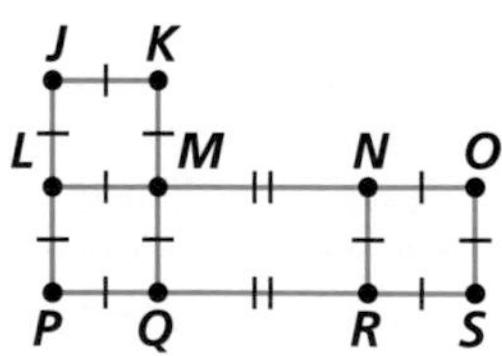

A $\overline{JP} \cong \overline{LN}$
B $\overline{QR} \cong \overline{MO}$
C $\overline{LN} \cong \overline{QS}$
D $\overline{MN} \cong \overline{JK}$

5. ***SHORT RESPONSE*** Draw a line and label three points on it. Then identify three rays in your drawing.

6. ***SHORT RESPONSE*** Draw two intersecting line segments. Label the endpoints and the point where the segments intersect, and then identify all the line segments in your drawing.

LESSON 7-2

7. What is the measure of $\angle BFC$?

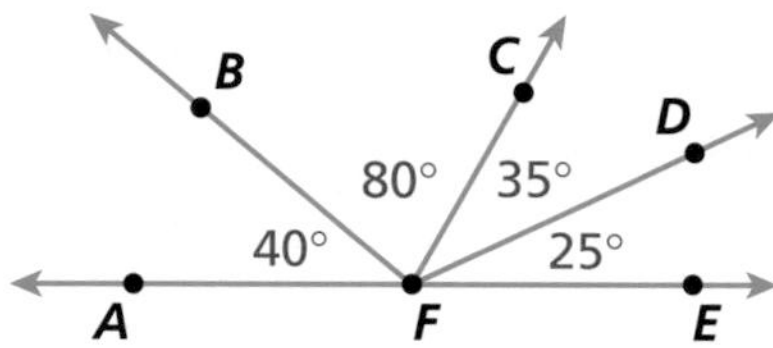

A 60°
B 80°
C 90°
D 120°

8. The measures of three angles add up to 180°. One of the angles is a right angle. What word describes the relationship between the other two angles?

A adjacent
B complementary
C supplementary
D vertical

9. What type of angle is $\angle HML$?

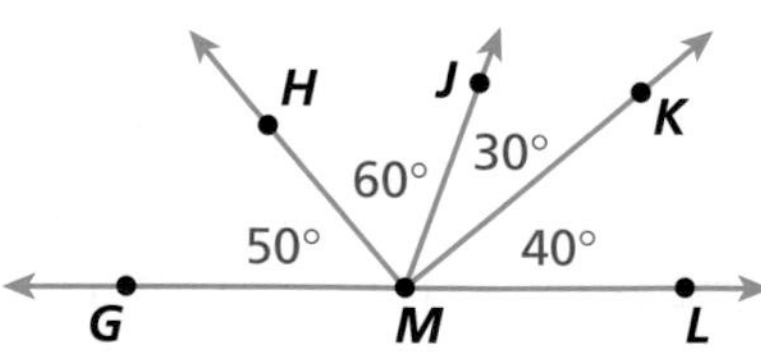

A acute
B obtuse
C right
D straight

EOG Practice

7A Lines and Angles

10. Which statement is *true*?

    A The sum of the measures of two supplementary angles is 160°.

    B The measure of an acute angle is less than or equal to 90°.

    C The measure of a straight angle is always 180°.

    D Complementary angles are always adjacent.

11. ***SHORT RESPONSE*** What is the measure of $\angle GLH$? Show your work, and explain in words how you determined your answer.

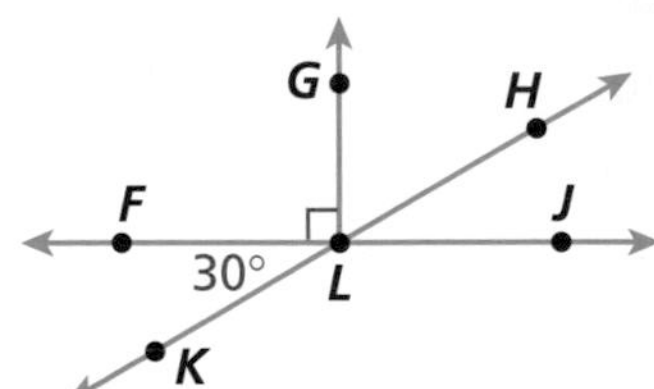

12. ***SHORT RESPONSE*** If $\angle P$ and $\angle L$ are complementary angles and $m\angle P = 53°$, what is $m\angle L$? Show your work.

LESSON 7-3

All of the angles in the figure below are right angles.

13. Which word describes the relationship between $\overleftrightarrow{PN}$ and $\overleftrightarrow{QR}$?

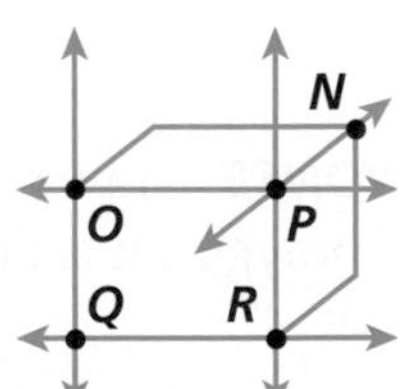

    A intersecting

    B parallel

    C perpendicular

    D skew

14. Two line segments intersect to form four angles. One of the angles measures 43°. What are the measures of the other three angles?

    A 43°, 137°, 137°

    B 43°, 47°, 47°

    C 43°, 137°, 47°

    D 43°, 43°, 43°

15. Two lines are in different planes and do not intersect. What word describes the relationship between these two lines?

    A intersecting

    B parallel

    C perpendicular

    D skew

16. A transversal intersects 3 parallel lines. How many angles are formed?

    A 3

    B 8

    C 12

    D 16

17. Line $j \parallel$ line k. What is the measure of $\angle 3$?

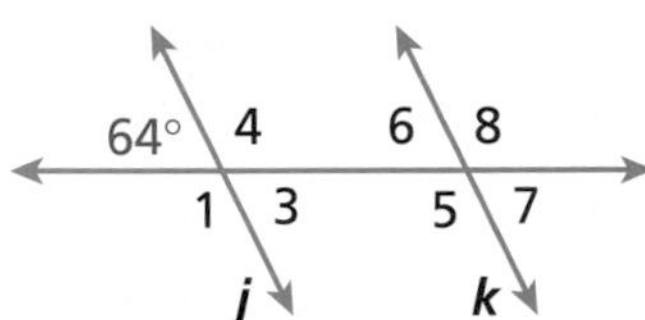

    A 26°

    B 64°

    C 116°

    D 154°

18. ***SHORT RESPONSE*** Draw two intersecting lines and label the four angles that are made. Then tell which angles are supplementary and which angles are vertical.

19. ***SHORT RESPONSE*** Draw two parallel lines and a transversal. Label the angles 1–8. Explain which angles are congruent and why.

EOG Practice ▪ Chapter 7

7B Closed Figures

LESSON 7-4

1. How many radii are shown in the figure?

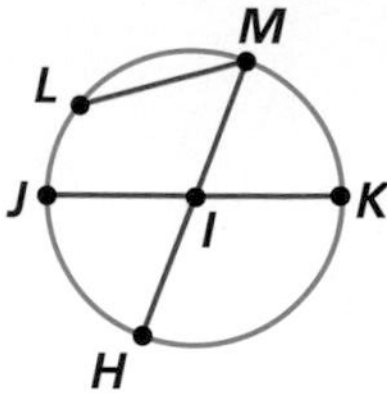

A 1
B 3
C 4
D 5

2. Which statement is *false*?
   A A diameter is a chord.
   B A diameter is always twice as long as a radius.
   C A radius is always half as long as a chord.
   D The endpoints of a chord are points on a circle.

The circle graph below shows the results of a survey in which people were asked to choose their preferred form of aerobic activity.

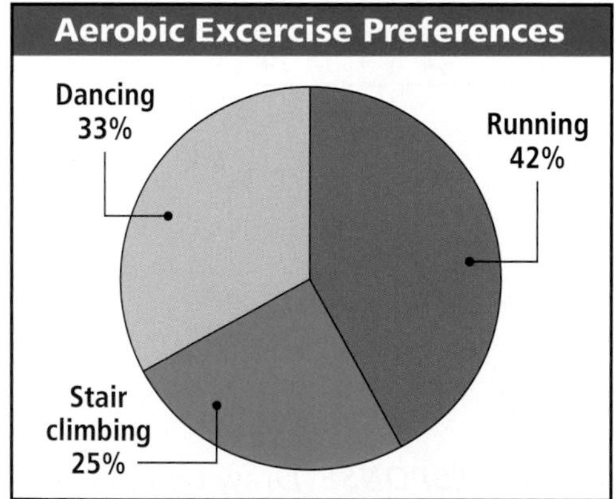

3. ***SHORT RESPONSE*** Find the central angle measure of the sector that shows the percent of people who prefer running. Show your work.

LESSON 7-5

4. What is the name of the polygon below?

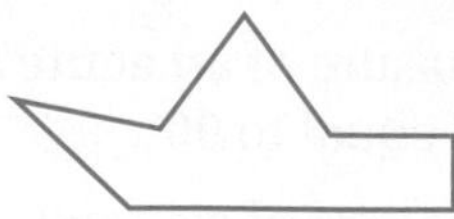

A heptagon
B hexagon
C pentagon
D quadrilateral

5. Which of the following is *not* a polygon?
   A square
   B circle
   C triangle
   D hexagon

6. How many angles are in a quadrilateral?
   A 10
   B 8
   C 6
   D 4

7. How many more sides does a decagon have than a heptagon?
   A 3
   B 4
   C 1
   D 5

8. ***SHORT RESPONSE*** Draw a polygon. Tell what type of polygon it is and explain how you know.

9. ***SHORT RESPONSE*** Draw a figure that is *not* a polygon. Then explain why it is not a polygon.

10. ***SHORT RESPONSE*** A figure has seven sides and seven angles. All but one of the angles in the figure are congruent. All of the sides of the figure are congruent. Name the polygon, and explain whether it is a regular polygon.

EOG Practice

EOG Practice ■ Chapter 7

7B Closed Figures

LESSON 7-6

11. The lengths of the sides of a triangle are 1 meter, 5 meters, and 7 meters. The measures of the angles in the triangle are 50°, 116°, and 14°. What type of triangle is it?

A scalene obtuse

B isosceles right

C scalene right

D isosceles obtuse

12. A line segment is drawn from one corner of a square to another to form two triangles. What type of triangles are they?

A scalene right

B equilateral

C isosceles right

D scalene obtuse

13. ***SHORT RESPONSE*** Two sides of a triangle are 3 inches and 4 inches long. The angle between them is a right angle. Draw the triangle to determine the length of the third side.

LESSON 7-7

14. Which of the following does *not* describe the figure below?

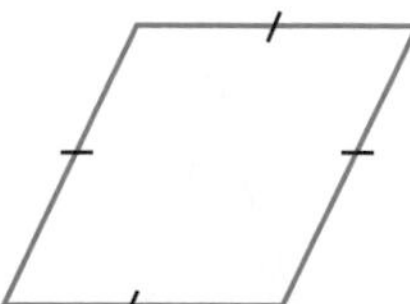

A parallelogram

B square

C rhombus

D quadrilateral

15. Two opposite sides of a quadrilateral are parallel but not congruent. What type of quadrilateral is the figure?

A rhombus

B kite

C trapezoid

D rectangle

16. ***SHORT RESPONSE*** While playing a trivia game, Joey was asked the following question: Is it true or false that a square is a rectangle? Joey's answer was "True." Was Joey correct? Explain.

17. ***SHORT RESPONSE*** A quadrilateral has two sides that are parallel. Exactly two of the angles in the quadrilateral are right angles. Make a sketch of a figure that matches this description, and give all of the names that apply to the quadrilateral.

LESSON 7-8

18. What is the value of x?

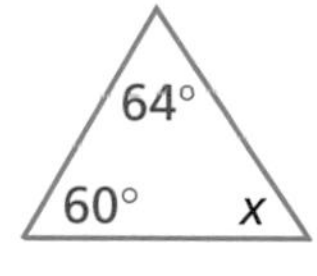

A 56° C 64°

B 60° D 66°

19. What is the value of y?

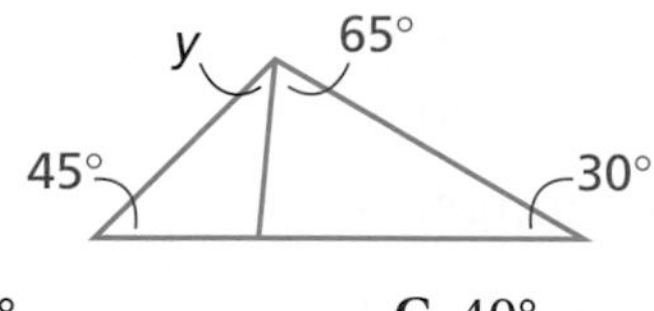

A 65° C 40°

B 45° D 25°

20. ***SHORT RESPONSE*** Draw a hexagon. Divide the figure into triangles, and find the sum of the figure's angle measures. Show your work.

7C Closed Figure Relationships

LESSON 7-9

1. Which of the triangles below are congruent?

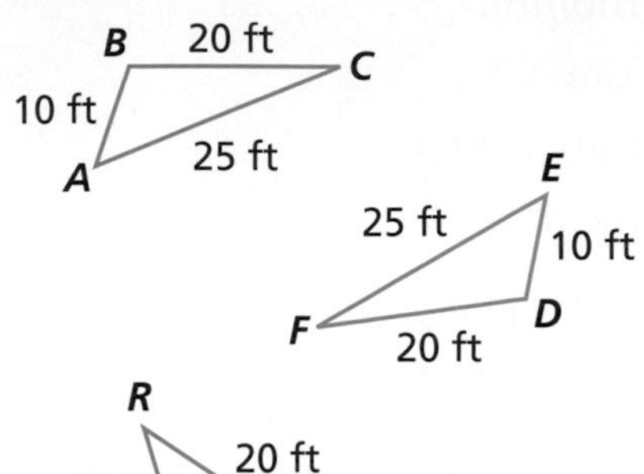

A △*ABC* and △*DEF*

B △*ABC* and △*PQR*

C △*DEF* and △*PQR*

D All three triangles are congruent.

2. What is the missing measure in the set of congruent polygons?

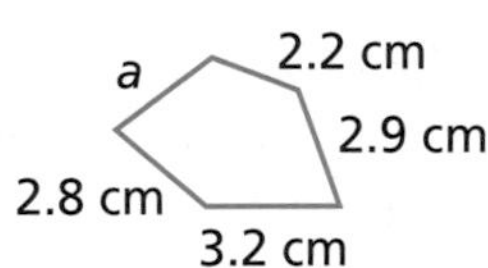

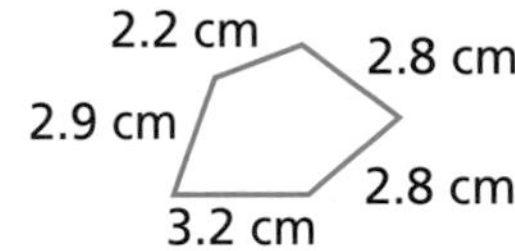

A 2.2 centimeters

B 2.8 centimeters

C 2.9 centimeters

D 3.2 centimeters

3. A rectangle that measures 4 inches long and 7 inches wide is divided into three parts. One part is 4 inches long and 3.5 inches wide. The other two parts are congruent. What is the width of one of the two other parts?

A 1.25 inches

B 1.75 inches

C 3.5 inches

D 4 inches

4. ***SHORT RESPONSE*** Explain whether the triangles below are congruent.

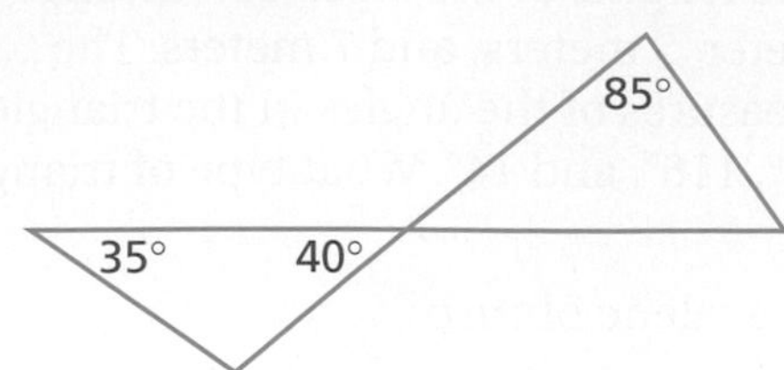

5. ***SHORT RESPONSE*** What is the minimum amount of information needed to determine whether two parallelograms are congruent?

LESSON 7-10

6. What is the transformation illustrated below?

A reflection

B rotation

C tessellation

D translation

7. Which figure is the resulting image if the figure below is reflected across a vertical line that passes through point *B* and then rotated 180°?

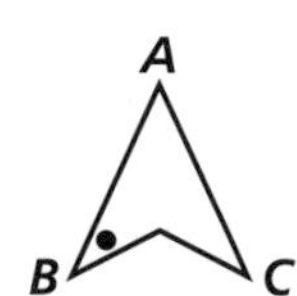

A 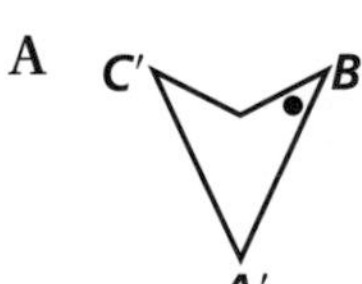

C 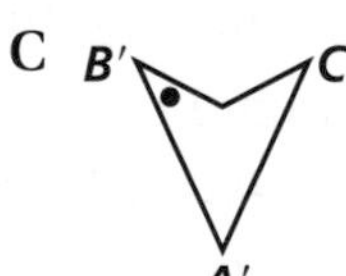

B 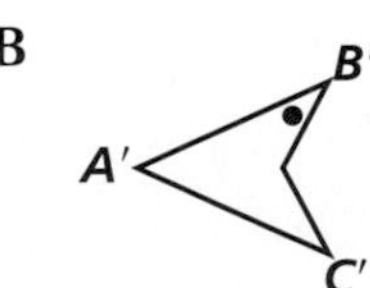

D 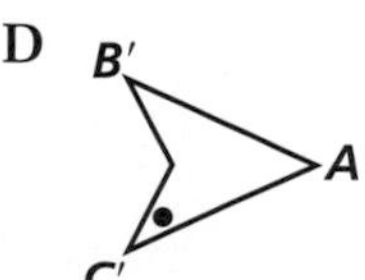

7C Closed Figure Relationships

8. Which of the following is illustrated in the figure below?

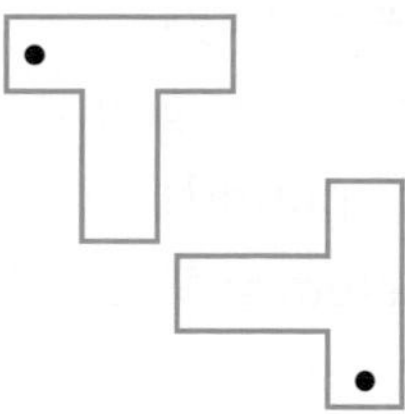

A reflection and rotation

B rotation and translation

C translation and reflection

D rotation and tessellation

9. ***SHORT RESPONSE*** Make a drawing to show how two congruent triangles can form a rectangle. What type of transformation would one of the triangles have to make to match the other triangle?

10. ***SHORT RESPONSE*** The figure below is rotated 90° clockwise about point E and translated 3 units right. What are the coordinates of the vertices of the image? Explain in words how you determined each coordinate.

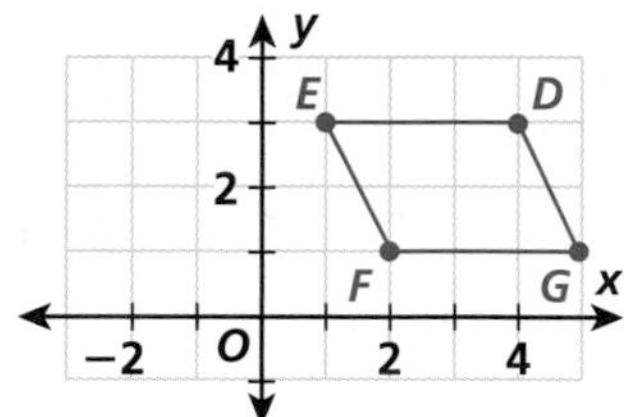

11. ***SHORT RESPONSE*** Draw a rectangle and place a dot in one of its corners. Then draw the image of that figure that results when the original figure is rotated 90° clockwise and reflected across one of its sides. Indicate which side the rectangle has been reflected across. Show each transformation in a separate step.

LESSON 7-11

12. How many lines of symmetry does the figure below have?

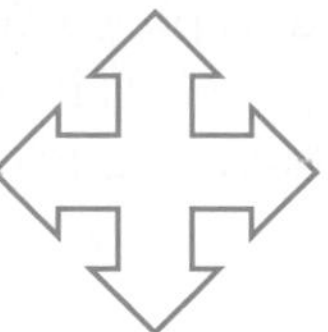

A 2 C 6

B 4 D 8

13. How many times does the figure below show rotational symmetry within one full rotation?

A 12 C 6

B 9 D 3

14. ***SHORT RESPONSE*** Trace the figure below, and then draw all the lines of symmetry.

15. ***SHORT RESPONSE*** Draw a figure that has two lines of symmetry, and draw the lines of symmetry on your figure.

16. ***SHORT RESPONSE*** Draw a figure that can show rotational symmetry three times within one full rotation. Include curved arrows in your drawing to illustrate how far the figure must turn each time to show rotational symmetry.

EOG Practice ▪ Chapter 8

8A Measurement, Perimeter, and Circumference

LESSON 8-1

1. Byron and his band are touring for 6 weeks along the East Coast. If they have 12 nights off and play a concert on each of the remaining nights, how many concerts will the band play while on tour?

   A 6 C 30

   B 18 D 72

2. Carol Ann ran a marathon, which is 26.2 miles. How many feet did Carol Ann run?

   A 0.005 foot

   B 202 feet

   C 26,200 feet

   D 138,336 feet

3. How many quarts are there in 53 gallons?

   A 424 quarts

   B 212 quarts

   C 13.25 quarts

   D 6.625 quarts

4. The driving distance from Clearwater, Florida, to Bradenton is 0.97 kilometer. If there is about 0.62 mile in 1 kilometer, how many miles is the driving distance from Clearwater to Bradenton?

   A 0.35 mile

   B 0.60 mile

   C 0.64 mile

   D 1.6 miles

5. Ollie's goal was to jog a total of 8 hours from Monday through Friday. On Monday, he jogged for 90 minutes. Each day following, he jogged 15 minutes longer than he did on the day before. How much time was Ollie over or under his initial goal?

   A 1 hour and 20 minutes under

   B 40 minutes under

   C 1 hour and 25 minutes over

   D 2 hours over

SHORT RESPONSE Make each conversion. Show your work.

6. 192 hours = ▪ days
7. 8.9 kilograms = ▪ grams
8. 0.6 ton = ▪ pounds
9. 338 millimeters = ▪ centimeters
10. 420 minutes = ▪ hours

LESSON 8-2

11. Which of the following is the most precise measurement?

    A 20 feet

    B 22 feet

    C 260 inches

    D 264 inches

12. How many significant digits are in the measurement 360.840 liters?

    A 2 C 5

    B 4 D 6

13. Which of the following has 3 significant digits?

    A 4,005 pounds

    B 4,050 pounds

    C 4,055 pounds

    D 4,505 pounds

14. Which of the following does *not* have 4 significant digits?

    A 98.05 yards

    B 98.50 yards

    C 985.0 yards

    D 9,850 yards

15. How many significant digits does the sum of 7.02 and 6.9 have?

    A 1 C 3

    B 2 D 4

8A Measurement, Perimeter, and Circumference

16. ***SHORT RESPONSE*** Find the difference of 12 and 5.88. Use the correct number of significant digits in your answer. Show your work.

17. ***SHORT RESPONSE*** Find the product of 3.6 and 1.8. Use the correct number of significant digits in your answer. Show your work.

LESSON 8-3

18. What is the perimeter of the figure below?

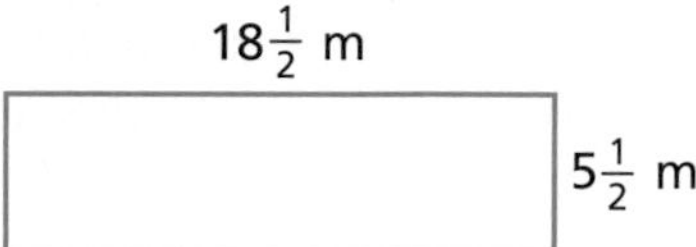

A 24 meters
C 90 meters
B 48 meters
D 101.75 meters

19. The perimeter of the figure below is 18.5 centimeters. What is the value of *a*?

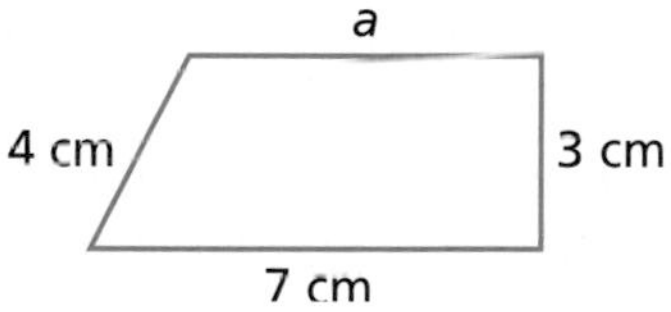

A 4.5 centimeters
B 5.5 centimeters
C 7 centimeters
D 14 centimeters

20. What is the circumference of the circle to the nearest tenth of an inch? Use 3.14 for π.

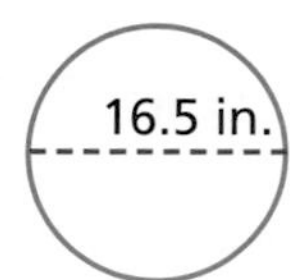

A 213.7 inches
B 103.6 inches
C 51.8 inches
D 41.6 inches

In the figure below, *b* is the distance between the centers of the circles.

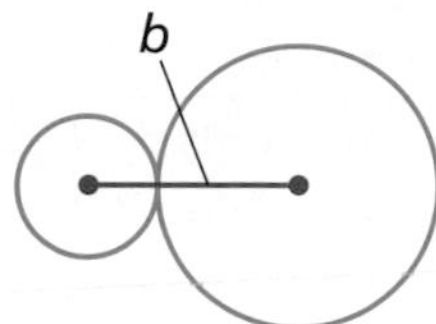

21. The radius of the smaller circle is half as long as the radius of the larger circle. If $b = 12$ feet, what is the circumference of the smaller circle? Round to the nearest foot. Use 3.14 for π.

A 50 feet
B 25 feet
C 13 feet
D 4 feet

22. ***SHORT RESPONSE*** The figure below is made up of an isosceles triangle and a square. The perimeter of the entire figure is 44 feet. What is the value of *x*? Show your work.

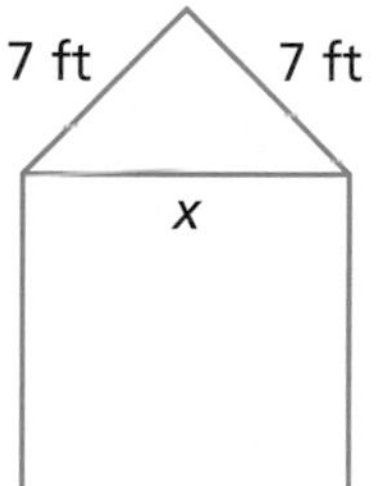

23. ***SHORT RESPONSE*** Renee is making a decorative jar filled with jellybeans for a gift. She wants to put a circular piece of material under the lid so that when the lid is closed, the material hangs out of the jar about 1 inch. The circumference of the lid is 10.5 inches, which is 6.28 inches less than the circumference of the material. What is the circumference of the material? Show your work.

8B Area

LESSON 

1. What is the area of the rectangle?

11 cm

3.3 cm

A 36.3 square centimeters

B 8.6 square centimeters

C 14.3 square centimeters

D 3.3 square centimeters

2. What is the area of the parallelogram?

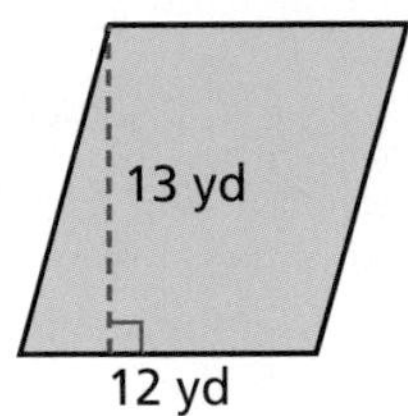

A 25 square yards

B 50 square yards

C 78 square yards

D 156 square yards

3. Harry is using 16 Japanese *tatami* mats to cover a floor. Each mat measures 3 feet by 2 feet. What is the total area that will be covered by the mats?

A 576 square feet

B 96 square feet

C 21 square feet

D 12 square feet

4. A parallelogram has a height of 15 meters and a base length that is $\frac{1}{3}$ of the height. What is the area of the parallelogram?

A 5 square meters

B $15\frac{1}{3}$ square meters

C 45 square meters

D 75 square meters

5. The area of a parallelogram is 720 square miles. What is the height of the parallelogram if the base length is 30 miles?

A 24 miles

B 690 miles

C 750 miles

D 21,600 miles

6. Megan and Mikayla are helping their father paint the garage. They are asked to paint a part of the wall that is 240 inches wide and 60 inches tall. If Megan and Mikayla each paint the same amount of space on the wall, how many square inches of wall does each sister paint?

A 1,800 square inches

B 3,600 square inches

C 7,200 square inches

D 14,400 square inches

7. ***SHORT RESPONSE*** Higinio is planting a tomato garden. The area he is planting in is 4 meters wide. Manda is also planting a garden. Her garden is 6 meters long. The shape of Manda's garden is congruent to the shape of Higinio's garden. If both Higinio's and Manda's gardens are rectangular, what is the total area of their gardens combined?

The figure below is made up of a parallelogram and a square.

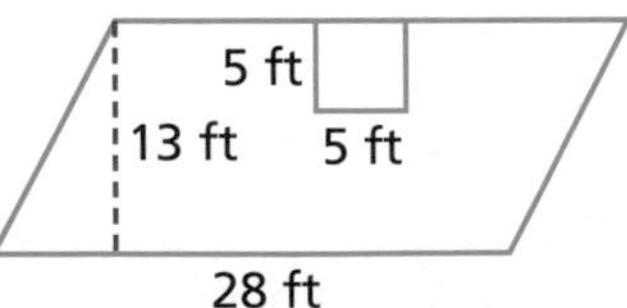

8. ***SHORT RESPONSE*** The figure above is the floor plan of a room in Angela's house. She wants to cover the square entrance with tile and the rest of the floor with carpet. How many square feet of carpet does Angela need?

EOG Practice ▪ Chapter 8

8B Area

LESSON 8-5

9. What is the area of the triangle?

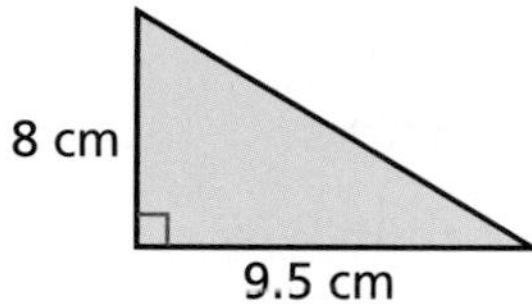

A 8.75 square centimeters

B 17.5 square centimeters

C 38 square centimeters

D 76 square centimeters

10. What is the area of the trapezoid?

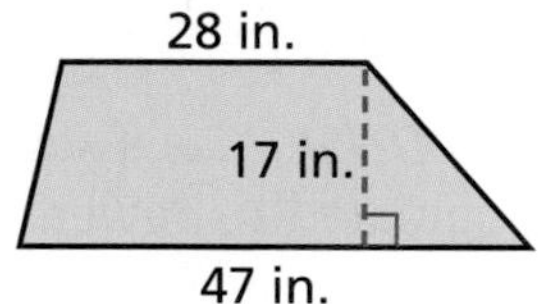

A 427.5 square inches

B 637.5 square inches

C 999 square inches

D 1,275 square inches

11. The area of a triangle is 144 square yards. What is the height of the triangle if the base length is 72 yards?

A 4 yards

B 36 yards

C 72 yards

D 216 yards

12. ***SHORT RESPONSE*** A barn door is shaped like a trapezoid with bases of 5 meters and 6.5 meters and height of 4 meters. One gallon of paint will cover 8 square meters. How many gallons of paint are required to paint the front of the barn door? Show your work.

13. ***SHORT RESPONSE*** The base of a triangle is 7 feet long. The height of the triangle is half the length of the base. What is the area of the triangle? Show your work.

LESSON 8-6

14. What is the area of the circle to the nearest hundredth? Use 3.14 for π.

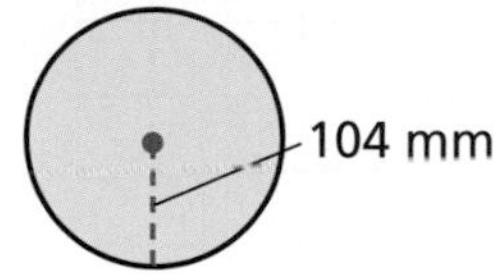

A 326.56 square millimeters

B 653.12 square millimeters

C 1,025.40 square millimeters

D 33,962.24 square millimeters

15. A circular wading pool has a diameter of 4.5 yards. What is the area to the nearest tenth of the surface of the water in the pool? Use 3.14 for π.

A 7.1 square yards

B 14.1 square yards

C 15.9 square yards

D 63.6 square yards

16. The area of a circle is 30 square feet. A second circle has a radius that is 2 feet shorter than that of the first circle. What is the area, to the nearest tenth, of the second circle? Use 3.14 for π.

A 3.7 square feet

B 10.0 square feet

C 38.0 square feet

D 179.2 square feet

17. The area of a circle is 393.9 square meters. What is the diameter of the circle, to the nearest tenth?

A 11.2 meters

B 22.4 meters

C 125.2 meters

D 250.3 meters

18. ***SHORT RESPONSE*** If the area of a circle is twice as great as the area of another circle, explain how the radius of the smaller circle is related to the radius of the larger circle?

EOG Practice ■ Chapter 8

8C The Pythagorean Theorem

LESSON 8-7

1. The figure below models which power?

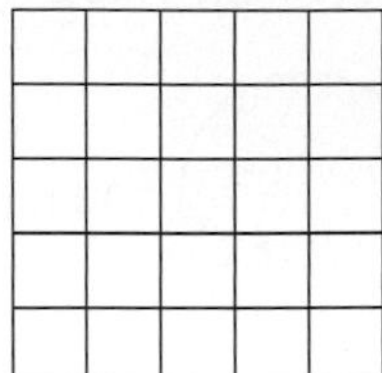

A 25^2
B 16^2
C 10^2
D 5^2

2. Which figure models the power 4^2?

A

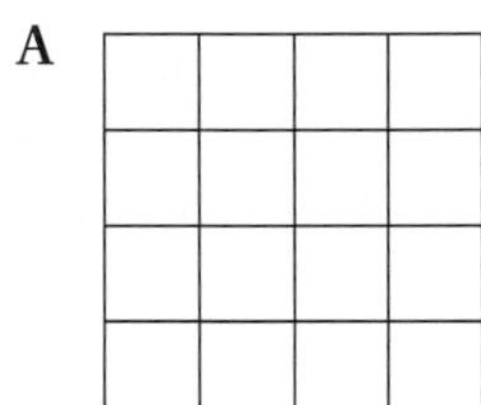

B

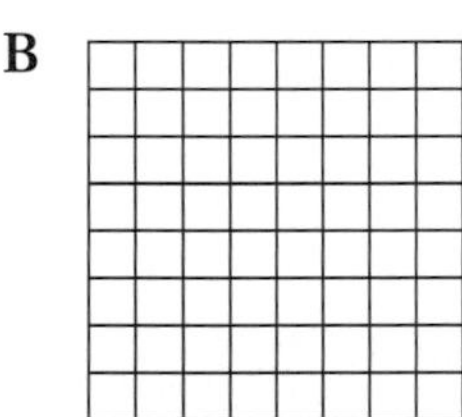

C

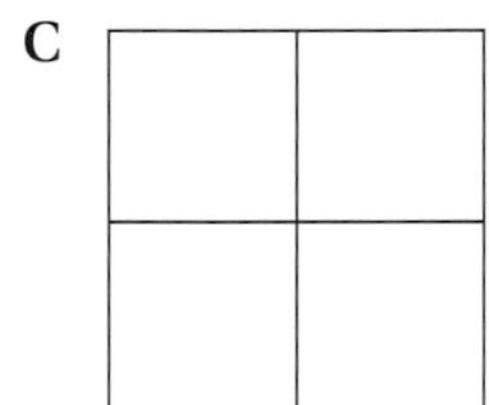

D

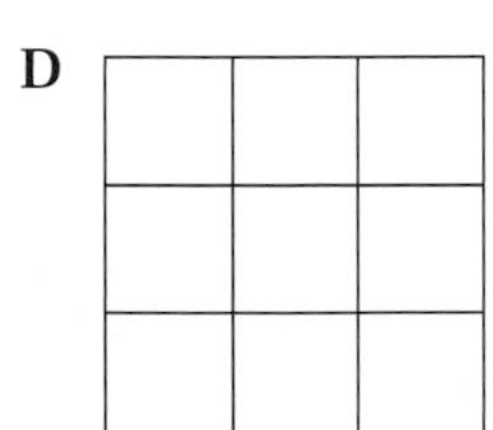

3. The area of a square room is 264 square feet. What is the approximate length of each side of the room?

A 132 feet
B 130 feet
C 17 feet
D 16 feet

4. A square painting has an area of 2,728 square centimeters. About how long is each side of the painting?

A 6 centimeters
B 52 centimeters
C 1,364 centimeters
D 5,456 centimeters

5. Which is the *best* estimate of the square root of 109, to the nearest whole number?

A 10
B 11
C 54
D 55

6. ***SHORT RESPONSE*** Use a square to model 6^2. Then evaluate the power.

7. ***SHORT RESPONSE*** Estimate the square root of 217 to the nearest whole number. Show your work.

8. ***SHORT RESPONSE*** James had enough rye grass seed to effectively cover 145 square feet of his back yard. If James spread the rye seed in an area of the yard that is square shaped, about how wide was that area of the yard?

LESSON 8-8

9. What is the value of x?

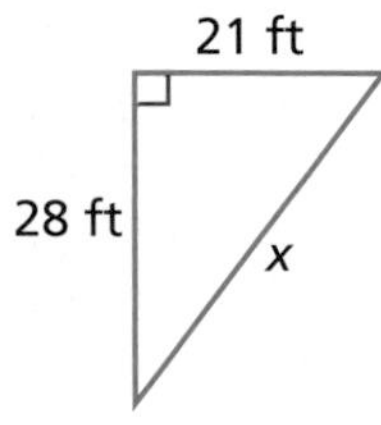

A 400 yards
B 200 yards
C 35 yards
D 20 yards

EOG Practice ■ Chapter 8

8C The Pythagorean Theorem

10. What is the value of a?

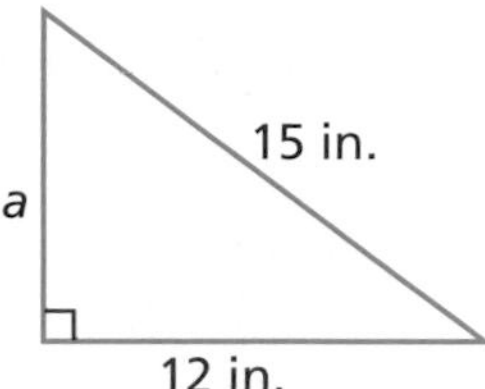

A 9 inches

B 19.2 inches

C 184.5 inches

D 364 inches

11. What is the value of t?

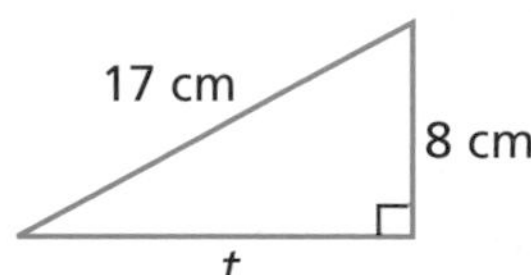

A 353 centimeters

B 176.5 centimeters

C 15 centimeters

D 9 centimeters

12. The legs of a right triangle are 13 inches long and 54 inches long. How many inches long is the hypotenuse of the triangle?

A 55.5 inches

B 52.4 inches

C 20.5 inches

D 6 inches

13. Ricky rides his bike 25 miles south and then turns east and rides another 25 miles before he stops to rest. How far is Ricky from his starting point?

A 50 miles

B 35.4 miles

C 25 miles

D 0 miles

14. One of the legs of a right triangle measures 3.5 millimeters. The other leg is 4 millimeters longer. What is the length of the hypotenuse to the nearest tenth of a millimeter?

A 28.2 millimeters

B 8.3 millimeters

C 5.3 millimeters

D 2.7 millimeters

15. Which equation can be used to find the value of x in the figure?

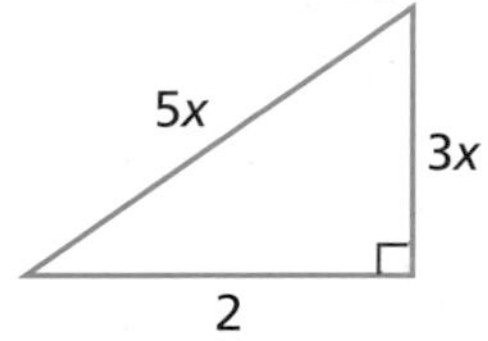

A $3x^2 + 5x^2 = 4$

B $(3x)^2 + (5x)^2 = 4$

C $3x^2 + 2 = 5x^2$

D $(3x)^2 + 4 = (5x)^2$

16. One leg of a right triangle is twice as long as the other leg. Which equation shows the relationship between the shorter leg and the hypotenuse, where ℓ is the length of the shorter leg and h is the length of the hypotenuse?

A $5\ell^2 = h^2$ C $\ell^2 = h^2$

B $3\ell^2 + \ell^2 = h^2$ D $\ell^2 = 2h^2$

17. ***SHORT RESPONSE*** Marissa walks 3.7 miles north and then turns west and walks another 2.8 miles. How far is Marissa from her starting point? Round your answer to the nearest tenth of a mile. Show your work.

18. ***SHORT RESPONSE*** The hypotenuse of a right triangle measures 5 meters. The legs are congruent. What is the length of one of the legs? Round your answer to the nearest tenth. Show your work.

EOG Practice ■ Chapter 9

9A Volume

LESSON 

1. What is the name of the figure below?

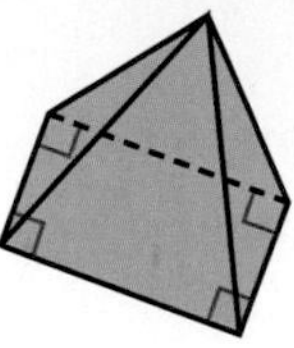

A rectangular prism
B triangular prism
C rectangular pyramid
D triangular pyramid

2. What polygon is the base of the figure below?

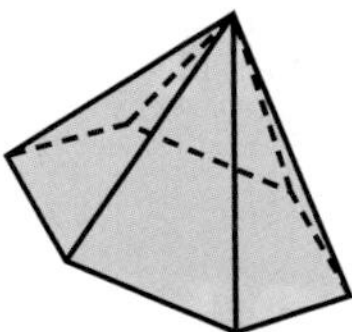

A hexagon
B triangle
C octagon
D pentagon

3. How many bases does the figure below have?

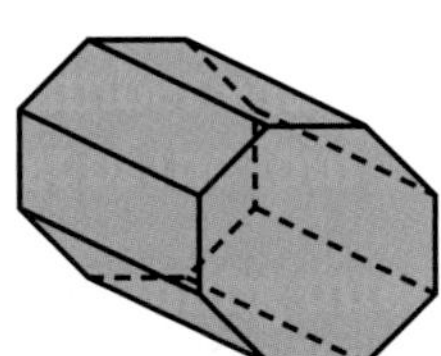

A 10
B 8
C 5
D 2

4. ***SHORT RESPONSE*** How many faces does a rectangular prism have? How many faces does a triangular prism have? Explain why there is a difference between the number of faces on these two types of prisms.

LESSON 

5. What is the volume of the figure below?

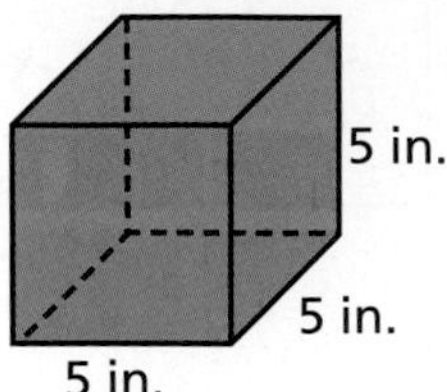

A 15 cubic inches
B 30 cubic inches
C 50 cubic inches
D 125 cubic inches

6. What is the volume of the figure below?

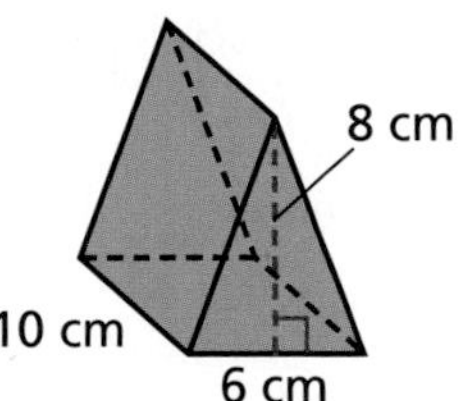

A 480 cubic centimeters
B 240 cubic centimeters
C 148 cubic centimeters
D 68 cubic centimeters

7. The volume of the figure below is 2,289.06 cubic inches. What is the value of x?

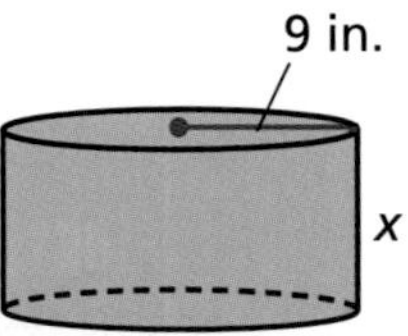

A 254.34 inches
B 81 inches
C 9 inches
D 4.5 inches

EOG Practice ■ Chapter 9

9A Volume

8. ***SHORT RESPONSE*** Danny is filling a sandbox for his little brother. The sandbox is shaped like a rectangular prism that is 12 inches deep and has a base with an area of 10 square feet. The sand is sold in bags, and each bag contains 4 cubic feet of sand. How many bags of sand does Danny need to buy to fill the sandbox? Show your work.

9. ***SHORT RESPONSE*** On a triangular prism, each triangular face has a height of 7 centimeters and a base length of 5.3 centimeters. The volume of the prism is 185.5 cubic centimeters. What is the distance between the bases of the prism? Show your work.

LESSON 9-3

10. What is the volume of the figure below?

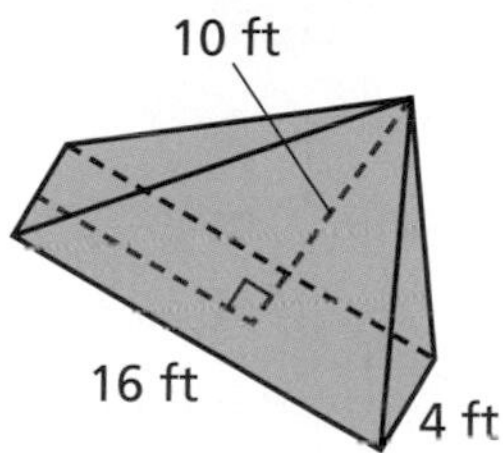

A 106.7 cubic feet
B 213.3 cubic feet
C 320 cubic feet
D 640 cubic feet

11. The base of a hexagonal pyramid has an area of 51.25 square meters. The height of the pyramid is 29 meters. What is the volume of the pyramid?

A 495.4 cubic meters
B 743.1 cubic meters
C 1,486.3 cubic meters
D 76,170 cubic meters

12. What is the volume of the figure below?

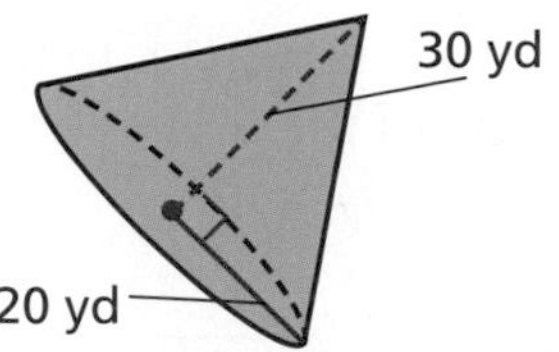

A 600 cubic yards
B 1,884 cubic yards
C 12,560 cubic yards
D 37,680 cubic yards

13. What is the volume of the figure below?

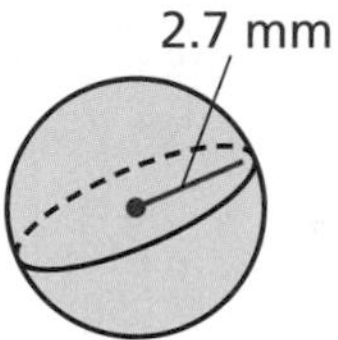

A 30.5 cubic millimeters
B 82.4 cubic millimeters
C 247.2 cubic millimeters
D 1,444.4 cubic millimeters

14. ***SHORT RESPONSE*** The radius of a large sphere is 5 times greater than the radius of a smaller sphere. How many times greater is the volume of the larger sphere than the volume of the smaller sphere? Show your work.

15. ***SHORT RESPONSE*** Find the volume of a pyramid that is 50 feet tall and has an octagonal base with an area of 473 square feet. Show your work.

16. ***SHORT RESPONSE*** A heptagonal pyramid has a volume of 32 cubic inches. What is the area of the pyramid's base if the height is 8 inches? Show your work.

EOG Practice ■ Chapter 9

9B Surface Area

LESSON 9-4

1. What is the surface area, to the nearest square inch, of the prism formed by the net below?

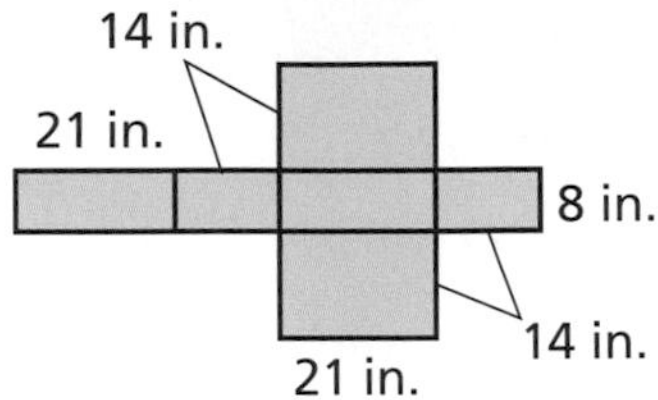

A 78 square inches
B 574 square inches
C 1,148 square inches
D 691,488 square inches

2. What is the surface area, to the nearest square yard, of the cylinder formed by the net below?

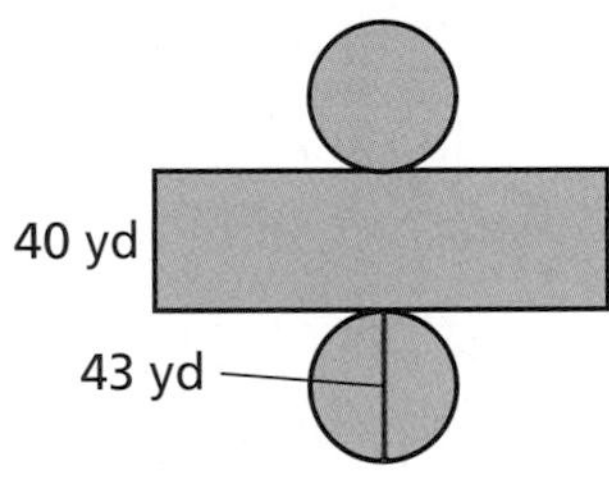

A 22,413 square yards
B 8,304 square yards
C 6,852 square yards
D 1,720 square yards

3. A can of beans is 12.5 centimeters tall and has a diameter of 9 centimeters. The label completely covers the lateral surface of the can with no overlap. About how much paper was used to make the label?

A 177 square centimeters
B 353 square centimeters
C 707 square centimeters
D 795 square centimeters

4. What is the surface area, to the nearest tenth, of the triangular prism formed by the net below?

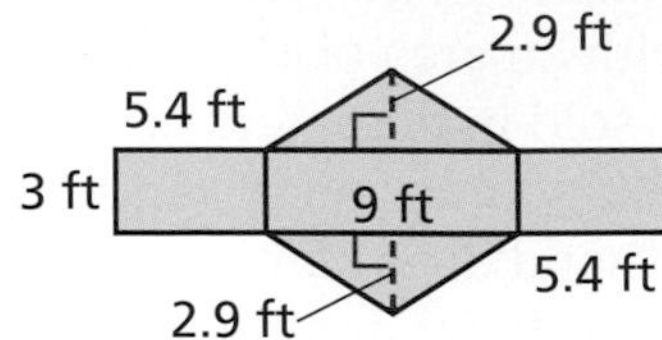

A 56.3 square feet
B 85.5 square feet
C 112.5 square feet
D 1,300.9 square feet

5. What is the surface area, to the nearest tenth, of the sphere? Use 3.14 for π.

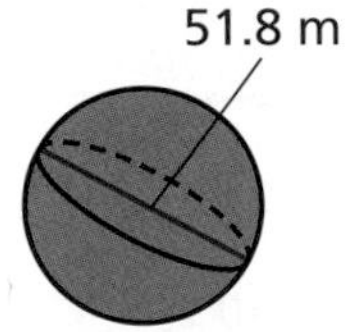

A 650.6 square meters
B 8,425.4 square meters
C 11,233.8 square meters
D 33,701.5 square meters

6. ***SHORT RESPONSE*** Identify the figure formed by the net below. Then find the surface area of the figure to the nearest tenth. Show your work.

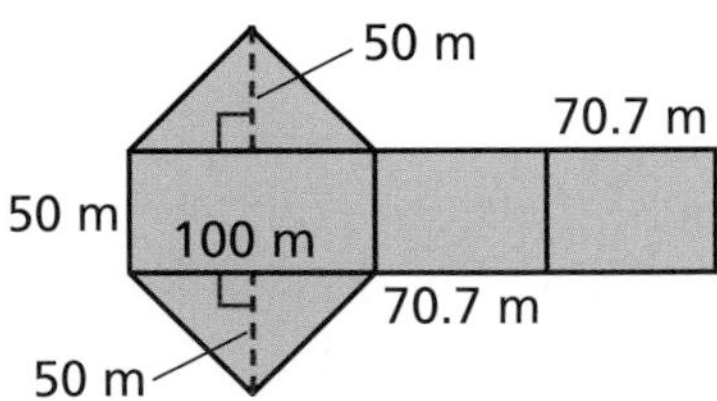

7. ***SHORT RESPONSE*** Draw a net of a rectangular prism that has a height of 1 centimeter, a width of 2 centimeters, and a length of 3 centimeters. Label the dimensions on the net.

EOG Practice ■ Chapter 9

9B Surface Area

LESSON 9-5

8. The surface area of a cylinder is 49 square meters. What is the surface area of a similar cylinder that has a scale factor of 6?

   A 294 square meters

   B 400 square meters

   C 1,764 square meters

   D 14,406 square meters

9. The painted surface of a cabinet is 36 square feet. Earl is painting a similarly shaped cabinet that has a scale factor of $\frac{1}{4}$. How much surface does Earl need to paint?

   A 324 square feet

   B 18 square feet

   C 9 square feet

   D 2.25 square feet

10. The surface area of a hexagonal prism is 65 square inches. What is the surface area of a similar prism that has a scale factor of 8?

   A 1 square inch

   B 8 square inches

   C 520 square inches

   D 4,160 square inches

11. The volume of a cube is 50 cubic meters. What is the volume of a similar cube that has a scale factor of 7?

   A 350 cubic meters

   B 2,450 cubic meters

   C 17,150 cubic meters

   D 17,500 cubic meters

12. An oil drum has a volume of 513 cubic centimeters. What is the volume of a similarly shaped oil drum that has a scale factor of $\frac{1}{3}$?

   A 19 cubic centimeters

   B 57 cubic centimeters

   C 171 cubic centimeters

   D 13,851 cubic centimeters

13. A block of wood weighs 3 pounds. What does a similarly shaped block of the same type of wood weigh if its scale factor is 14?

   A 42 pounds

   B 378 pounds

   C 588 pounds

   D 8,232 pounds

14. What is the surface area of a cylinder that is similar to the cylinder formed by the net below and has a scale factor of 10?

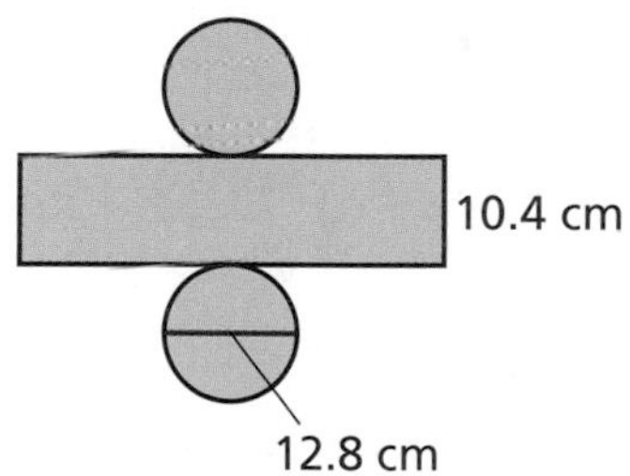

   A 546.6 square meters

   B 675.2 square meters

   C 54,661.1 square meters

   D 67,522.6 square meters

15. ***SHORT RESPONSE*** A rectangular prism is 3.5 feet wide, 2.6 feet long, and 5.7 feet tall. What is the surface area, to the nearest tenth, of a similar prism that has a scale factor of $\frac{1}{5}$? Show your work.

EOG Practice

EOG Practice ■ Chapter 10

10A Introduction to Probability

LESSON 10-1

1. Ian's soccer team has a game every evening in April and May. If it is a Monday evening in May, how likely is it that Ian is playing soccer?

   A unlikely

   B as likely as not

   C likely

   D certain

2. A pouch contains 9 white stones and 9 red stones. If you pull out a stone without looking, how likely is it that it will be white?

   A impossible

   B unlikely

   C as likely as not

   D likely

3. The newscaster said that Tuesday would be a sunny day. How likely is it that it will rain on Tuesday?

   A impossible C likely

   B unlikely D certain

4. Jennifer has just enough gas in her car to drive 50 miles. How likely is it that Jennifer will drive 100 miles in her car before stopping to fill up her gas tank?

   A impossible C likely

   B unlikely D certain

5. ***SHORT RESPONSE*** Louise has a bag of 8 blue marbles, 7 yellow marbles, 3 white marbles, and 10 green marbles. If Louise takes a marble out of the bag without looking, which color would that marble more likely be? Explain your answer.

6. ***SHORT RESPONSE*** A spinner has 4 sections numbered 1–4. How likely is it that when spun, the spinner will land on a section that is numbered either 1, 2, 3, or 4? Explain your answer.

LESSON 10-2

7. Drivers for a courier service made 24 trips last week. Eighteen of those trips were across the Delaware River. What is the experimental probability that the next trip will be across the Delaware River?

   A $\frac{1}{18}$ C $\frac{3}{4}$

   B $\frac{1}{4}$ D $\frac{4}{3}$

8. Out of the last 1,000 games Bo bowled, he scored 200 points or better in 85 of them. What is the experimental probability that Bo will score less than 200 the next time he bowls?

   A $\frac{17}{200}$ C $\frac{200}{183}$

   B $\frac{183}{200}$ D $\frac{200}{17}$

9. A quarterback completed 16 of 24 passes during the first half of a football game. What is the experimental probability that he will complete the first pass he throws in the second half of the game?

   A $\frac{1}{8}$ C $\frac{3}{2}$

   B $\frac{2}{3}$ D $\frac{8}{1}$

10. Three of the last 5 times that someone called on the telephone, the person was calling to talk to Audry. What is the experimental probability that the next person who calls will *not* be calling for Audry?

    A $\frac{2}{5}$ C $\frac{5}{3}$

    B $\frac{3}{5}$ D $\frac{5}{2}$

11. ***SHORT RESPONSE*** A bubble-gum-ball machine has blue, yellow, red, green, and white gum balls in it. There are an equal number of each color and a total of 200 gum balls. What is the probability that the first person who buys a gum ball will get a yellow gum ball? Explain in words how you determined your answer.

EOG Practice ■ Chapter 10

10A Introduction to Probability

LESSON 10-3

12. The Heavenly Dairy makes 23 flavors of ice cream. You can get each flavor in a waffle cone, a sugar cone, a cake cone, or a cup. How large is the sample space?

A 23 C 80

B 27 D 92

13. There are 3 bridges and 2 tunnels that each connect New Jersey and New York City. There are 3 highways that each connect New York City and Connecticut. How many possible routes are there by bridge or tunnel and by highway between New Jersey and Connecticut that go through New York City?

A 8 C 15

B 9 D 18

14. For her new car, Mrs. Chen has a choice of black, white, tan, silver, blue, or red. She also must choose the interior color. The choices are black or ivory leather or gray or tan fabric. How large is the sample space?

A 10 C 24

B 12 D 36

15. Joley is buying tacos. She has a choice of soft flour tortillas or crunchy corn shells, a choice of chicken or beef filling, and a choice between two different kinds of cheeses. How large is the sample space?

A 2 C 6

B 4 D 8

16. ***SHORT RESPONSE*** A lamp store sells 3 styles of small lamp bases, 2 different styles of lamp shades to fit those bases, and 3 colors of light bulbs. Make a diagram to find the sample space.

LESSON 10-4

17. A spinner with 12 equal sections numbered 1–12 is spun. What is the probability that the spinner will land on an odd number?

A $\frac{1}{12}$ C $\frac{1}{2}$

B $\frac{5}{12}$ D $\frac{7}{12}$

18. A spinner with 10 equal sections numbered 1–10 is spun. Each section is equally divided into one red, one blue, and one yellow subsection. What is the probability that the spinner will land in the blue subsection of section number 3?

A $\frac{1}{30}$ C $\frac{1}{10}$

B $\frac{1}{13}$ D $\frac{3}{10}$

19. Janet has different colors of paper stacked in a random order. Four pieces of the paper are blue, 10 are white, and 7 are pink. What is the probability that the first piece of paper that Janet takes from the top of the stack will be white or blue?

A $\frac{2}{3}$ C $\frac{2}{5}$

B $\frac{7}{11}$ D $\frac{2}{11}$

20. While fishing, Mr. Watson catches 4 catfish and 6 bass and puts them in a basket. If he removes a fish from the basket at random, what is the probability that he will pull out a bass?

A $\frac{2}{5}$ C $\frac{3}{2}$

B $\frac{3}{5}$ D $\frac{5}{2}$

21. ***SHORT RESPONSE*** Karen has an equal number of peach, apple, lemon, and berry flavored tea bags in a small box. If there are 20 bags in the box, what is the probability that Karen will randomly pull out a peach flavored tea bag? Show your work.

EOG Practice

EOG Practice ■ Chapter 10

10B Applications of Probability

LESSON 10-5

1. What is the probability of 2 coins being flipped at the same time and both landing heads up?

   A $\frac{1}{4}$

   B $\frac{1}{2}$

   C $\frac{2}{3}$

   D 1

2. What is the probability of drawing a 3 from 5 cards numbered 1 through 5 and rolling an even number on a number cube numbered from 1 through 6?

   A $\frac{1}{2}$

   B $\frac{1}{5}$

   C $\frac{1}{6}$

   D $\frac{1}{10}$

3. What is the probability of drawing a 4 from 5 cards numbered 1 through 5 and then drawing a 3 without replacing the first card?

   A $\frac{1}{20}$

   B $\frac{1}{5}$

   C $\frac{1}{4}$

   D $\frac{9}{20}$

4. Charissa has 4 pennies, 2 dimes, and 3 quarters in her pocket. She reaches in her pocket and pulls out a penny. If she pulls out another coin at random, what is the probability that the coin she chooses will *not* be a penny?

   A $\frac{3}{8}$

   B $\frac{4}{9}$

   C $\frac{5}{9}$

   D $\frac{5}{8}$

5. ***SHORT RESPONSE*** Explain whether the following events are independent or dependent.

   Alex draws a card out of a stack of 52 cards and holds onto it while Suzi draws a card from the same stack.

6. ***SHORT RESPONSE*** Explain whether the following events are independent or dependent.

   There are 14 boys and 16 girls in Mr. Fernandez's class. Mr. Fernandez randomly picks a boy and then a girl to represent the class at the school spelling bee.

LESSON 10-6

7. Venus has decided to have a 2-color paint job done on her car. There are 6 paint colors from which to choose. How many combinations are possible?

   A 6   C 15

   B 12   D 36

8. Philip has 5 different coins. How many combinations of 3 coins can he make from those?

   A 3   C 10

   B 8   D 15

9. A juice bar offers 8 different juice blends. You and a friend want to each try a different blend. How many combinations are possible?

   A 64   C 16

   B 28   D 8

10. Wagger's hot-dog stand offers the following choices of toppings: mustard, ketchup, sauerkraut, relish, onion, sour pickles, and chili. If you can choose 3 toppings, how many combinations are possible?

    A 64   C 24

    B 35   D 11

10B Applications of Probability

11. Lara used red, yellow, purple, black, and white beads to make necklaces. She made as many necklaces as possible such that each necklace had a different combination of 3 colors of beads. How many necklaces did she make?

    A 3 C 10

    B 8 D 15

12. ***SHORT RESPONSE*** Sara made a bag of party favors for each guest at her birthday party. She had 6 different types of party favors to choose from and put 4 different items in each bag. Make a list to determine how many different combinations of party favors Sara could put in the bags.

LESSON 10-7

13. Ralph, Randy, and Robert are getting tickets to see a movie. In how many different ways can they stand in line at the movie theater?

    A 3 C 9

    B 6 D 12

14. Roseanne and Rita join Ralph, Randy, and Robert at the movie theater. In how many different ways could they all stand in line?

    A 5 C 25

    B 6 D 120

15. Doris has a $1 bill, a $2 bill, a $5 bill, a $10 bill, a $20 bill, and a $50 bill. In how many different ways can she arrange them in a stack?

    A 720 C 12

    B 360 D 6

16. George has chosen 8 books to read over summer vacation as required by his book club. In how many different orders can George read the books?

    A 40,320 C 640

    B 5,040 D 64

17. Mila joined a fitness club. She has a choice of weight lifting, jogging, swimming, bike riding, stair climbing, aerobics, or racquetball. Mila decided to visit the fitness club every day her first week and to do a different activity during each visit. In how many different orders can Mila choose the activities?

    A 10,920

    B 5,040

    C 1,260

    D 49

18. Mrs. Kerbopple is writing math problems for a quiz. She is including a multiplication problem, a division problem, an addition problem, and a subtraction problem. In how many different orders can Mrs. Kerbopple arrange the problems?

    A 4

    B 16

    C 24

    D 48

19. ***SHORT RESPONSE*** Nancy is arranging 30 vacation photos in a photo album according to the day the photos were taken. Ten of the photos were taken on the first day, 8 of them were taken on the second day, and the rest were taken on the third day. In how many different orders can Nancy arrange the photos that were taken on the third day? Show your work.

20. ***SHORT RESPONSE*** Use a list to find the possible permutations of the letters in the word *cat*. In how many different ways can the letters be arranged?

21. ***SHORT RESPONSE*** A train engine is hauling 9 freight cars. How many different ways can the cars be arranged on the track? Explain in words how you determined your answer.

EOG Practice ▪ Chapter 11

11A Multistep Equations

LESSON 11-1

1. Which value of j is the solution to $-5j - 13 = 22$?

   A $j = -7$ C $j = 4$

   B $j = -1\frac{4}{5}$ D $j = 40$

2. Which value of e is the solution to $\frac{e}{7} + 2 = 5$?

   A $e = 49$ C $e = 1$

   B $e = 21$ D $e = \frac{3}{7}$

3. For which equation is $c = 7$ the solution?

   A $4c - 15 = -13$

   B $4c + 15 = 13$

   C $4c + 13 = 15$

   D $4c - 13 = 15$

4. For which equation is $m = 24$ the solution?

   A $\frac{m}{3} - 1 = -3$ C $\frac{m}{6} - 3 = 1$

   B $\frac{m}{3} - 6 = 1$ D $6m - 3 = 1$

5. For which equation is $h = 5$ *not* the solution?

   A $12 + 8 = 4h$ C $3h + 8 = 23$

   B $7 + 2h = -3$ D $3h + 14 = 29$

6. At the grocery store, apples cost \$0.99 per pound. The small bag of apples weighs 5 pounds. The large bag of apples weighs 3 pounds more than the small bag. How much does the large bag of apples cost?

   A \$1.98

   B \$7.92

   C \$7.95

   D \$7.97

7. ***SHORT RESPONSE*** If you multiply the number of DVDs Sarah has by 6 and then add 5, you get 41. How many DVDs does Sarah have? Show your work.

SHORT RESPONSE **Solve each equation. Show your work, or explain in words how you determined the answer.**

8. $\frac{x}{3} + 5 = 13$
9. $(3 + t)4 = 16$
10. $7y - 12 = 30$
11. $\frac{m}{5} + 18 = 23$
12. $4.5 + 9g = 67.5$
13. $1.1 - 8k = -42.7$

LESSON 11-2

14. Which value of w is the solution to $2w - 11 + 4w = 7$?

   A $w = -2$ C $w = 3$

   B $w = -\frac{2}{3}$ D $w = 9$

15. Which value of a is the solution to $12a - 3 - 8a = -1$?

   A $a = -5$ C $a = \frac{1}{2}$

   B $a = -1$ D $a = 8$

16. For which equation is $r = 10$ *not* the solution?

   A $2r + r + 8 = 10$

   B $r + 10 + r = 30$

   C $2r - 1 + 8r = 99$

   D $r + 27 - 3r = 7$

17. Erika received scores of 82, 87, 93, 95, 88, and 90 on her math quizzes. What score must Erika get on her next quiz to have an average of 90?

   A 89 C 95

   B 90 D 100

18. Earnest made oatmeal squares. He kept 4 of them for himself and equally divided the rest among Uma and Ethan. Uma ate 2 of her oatmeal squares and then had 3 left. How many oatmeal squares did Earnest make?

   A 10 C 15

   B 14 D 40

EOG Practice

11A Multistep Equations

19. ***SHORT RESPONSE*** A number plus 4 is divided by 17 and subtracted from 9. The answer is 4. Write and solve the equation. Show your work.

SHORT RESPONSE **Solve each equation. Show your work, or explain in words how you determined the answer.**

20. $-7z + 4 - z = -12$

21. $2t - 7 - 5t = 11$

22. $\frac{5x - 7}{3} = 15$

23. $\frac{10 - 4t}{8} = -12$

LESSON 11-3

24. Which value of m is the solution to $7 + 5m = 2 - m$?

   A $m = -1\frac{1}{4}$
   B $m = -\frac{5}{6}$
   C $m = 1\frac{1}{4}$
   D $m = 2\frac{1}{4}$

25. Which value of g is the solution to $\frac{2}{5}g + 9 = -6 - \frac{6}{10}g$?

   A $g = 15$
   B $g = 3$
   C $g = -3$
   D $g = -15$

26. For which equation is $a = 2$ the solution?

   A $6a - 6 = 3a$
   B $2a - 6 = a$
   C $6a - 8 = 4a$
   D $6a = 4a - 8$

27. For which equation is $p = 12$ *not* the solution?

   A $p + 3 = 2p - 14$
   B $6 + 11p = 9p + 30$
   C $1 + 29p = 13 + 28p$
   D $18p - 7 = 233 - 2p$

28. A number times 4 is half the difference of the number and 21. What is the number?

   A 3
   B $\frac{6}{7}$
   C $-\frac{6}{7}$
   D -3

29. Roberta and Stanley are collecting signatures for a petition. So far, Roberta has twice as many signatures as Stanley has. If she collects 30 more signatures, she will have 4 times as many signatures as Stanley currently has. How many signatures has Stanley collected?

   A 5
   B 10
   C 15
   D 30

30. Ben has b pieces of gum. June has half as many pieces of gum as Ben does, and Eunice has 5 more than $\frac{1}{3}$ the number of pieces of gum that Ben has. June and Eunice have the same number of pieces. Which equation can be used to find the number of pieces of gum that Ben has?

   A $\frac{1}{2}b + 5 = \frac{1}{3}b$
   B $\frac{1}{2}b = 5 + \frac{1}{3}b$
   C $2b = 5 + 3b$
   D $2b + 5 = 3b$

31. ***SHORT RESPONSE*** The quotient of a number and 9 is added to 12. The answer is 52. What is the number? Show your work.

32. ***SHORT RESPONSE*** Olivia bought a souvenir for \$4.50 plus tax. Her total cost was $17\frac{2}{3}$ times the amount she paid in tax. How much money did Olivia spend? Show your work.

SHORT RESPONSE **Solve each equation. Show your work, or explain in words how you determined the answer.**

33. $3d - 5 = 7d - 9$

34. $7y - 9 = -2y$

35. $2c - 13 = 5c + 11$

36. $7d + 4 = 8 - d$

37. $1.2k + 2.3 = -0.5k + 7.4$

EOG Practice ■ Chapter 11

11B Inequalities

LESSON 11-4

1. The cafeteria could hold no more than 50 people. Which inequality represents this situation?
   - A $50 > x$
   - B $50 \geq x$
   - C $50 < x$
   - D $50 \leq x$

2. Which situation could $y < 20$ represent?
   - A There were at least 20 ducks on the pond.
   - B There were fewer than 20 boats in the marina.
   - C There were 20 flowers in the garden.
   - D There were at most 20 people at the show.

3. Which inequality is represented by the graph below?

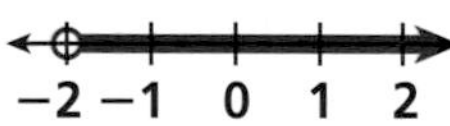

   - A $x > -2$
   - B $-2 \leq x$
   - C $x \leq -2$
   - D $-2 > x$

4. Which compound inequality is represented by the graph below?

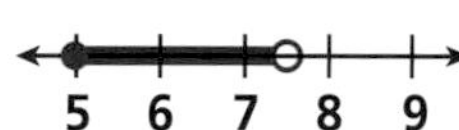

   - A $5 < x < 7.5$
   - B $5 \leq x < 7.5$
   - C $5 < x \leq 7.5$
   - D $5 \leq x \leq 7.5$

5. ***SHORT RESPONSE*** The spin-o-rama ride at the amusement park can hold a maximum of 15 people at one time. The ride requires a minimum of 5 riders. Write an inequality that represents this situation, and then graph the inequality.

LESSON 11-5

6. Which is the solution to $c - 6 > 5$?
   - A $c > -1$
   - B $c < -1$
   - C $c > 11$
   - D $c < -11$

7. For which inequality is $b < 14$ the solution?
   - A $b - 7 < 7$
   - B $b + 4 < 7$
   - C $b - 4 > 10$
   - D $b - 4 > -18$

8. The graph below represents the solution to which inequality?

   - A $m + 1 > 0$
   - B $m - 1 > 0$
   - C $m + 1 < 0$
   - D $m - 1 < 0$

9. By Saturday night, 3 inches of rain had fallen in Happy Valley. The weekend forecast predicted at least 8 inches of rain. How much more rain must fall on Sunday for this forecast to be correct?
   - A at least 3 inches
   - B at least 5 inches
   - C at least 8 inches
   - D at least 11 inches

SHORT RESPONSE **Solve each inequality. Show your work, or explain in words how you determined the answer.**

10. $q + 3 \leq 5$
11. $v - 3 \geq 1$
12. $p + 7 \leq 4$
13. $a - 2 \leq 5$

EOG Practice ■ Chapter 11

11B Inequalities

LESSON 11-6

14. Which is the solution to $\frac{a}{5} \leq 4.5$?

A $a \leq 0.9$

B $a \geq 22.5$

C $a \geq 0.9$

D $a \leq 22.5$

15. For which inequality is $s \leq 6$ a solution?

A $-11 - 5 \leq -s$

B $5 + s \leq 11$

C $6 - s \geq 6$

D $12 + s \leq 6$

16. For which inequality is $y \geq 15$ *not* a solution?

A $2y - 13 \geq y + 2$

B $100y - 25 \geq 10y + 1{,}325$

C $10y \geq y + 3$

D $7y + 4 + 9y \geq 15y + 19$

17. A seed store buys seeds in bulk and then sells them by the pound. If the store owner spends \$215 on sunflower seeds and then sells them for \$11.45 per pound, how many pounds must the owner sell to make a profit?

A at least 18 pounds

B at least 19 pounds

C at least 203 pounds

D at least 215 pounds

18. ***SHORT RESPONSE*** Haley has 6 days to make a minimum of 90 paper snowflakes to decorate the gym for the prom. Write and solve an inequality to determine at least how many paper snowflakes Haley needs to make each day.

SHORT RESPONSE **Solve each inequality. Show your work, or explain in words how you determined the answer.**

19. $-\frac{v}{2} > 2$

20. $-7r < 56$

LESSON 11-7

21. Which is the solution to $4j - 6 > 16$?

A $j > \frac{5}{2}$ C $j > \frac{11}{2}$

B $j < \frac{5}{2}$ D $j < \frac{11}{2}$

22. For which inequality is $q \geq 4.8$ a solution?

A $7.2q - 4.8 \geq 24$

B $11 + 2q > 5.6$

C $-8q + 3.2 \leq -35.2$

D $-q - 5.2 \leq 4$

23. The graph below represents the solution to which inequality?

A $2x - 4 \geq 12$

B $-2x + 4 \leq 12$

C $-2x - 4 \geq 12$

D $2x + 4 \geq 12$

24. Jill, Serena, and Erin are trying to earn enough money to rent a beach house for a week. They estimate that it will cost at least \$1,650. Jill has already earned \$600. If the other two girls plan to earn about the same amount of money, at least how much must Serena earn?

A at least \$1,050

B at least \$1,000

C at least \$525

D at least \$500

SHORT RESPONSE **Solve each inequality and graph each solution. Show your work.**

25. $\frac{m}{3} - 1 \leq 2$

26. $-1 - \frac{s}{3.5} \leq 1$

27. $5 - 2u < 15$

EOG Practice

EOG Practice ▪ Chapter 12

12A Introducing Functions

LESSON 12-1

1. For $y = 7.5x - 12$, what is the output if the input is 2.5?

   A -2 C 1.93
   B -1.27 D 6.75

2. Which is the missing ordered pair in the function table below?

Input	Rule	Output	Ordered Pair
x	2*x* − 5	*y*	(*x*, *y*)
0	2(0) + 5	5	■
1	2(1) + 5	7	(1, 7)
2	2(2) + 5	9	(2, 9)

   A (2, 5) C (5, 0)
   B (5, 2) D (0, 5)

3. The ordered pairs in the graph below are solutions to which function?

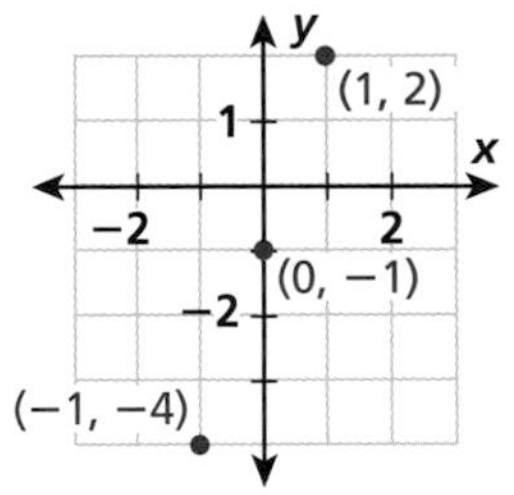

   A $y = 3x$ C $y = 3x + 1$
   B $y = 3x - 1$ D $y = x + 3$

4. Which of the points in the graph below represents an ordered-pair solution to $y = 3x^2$?

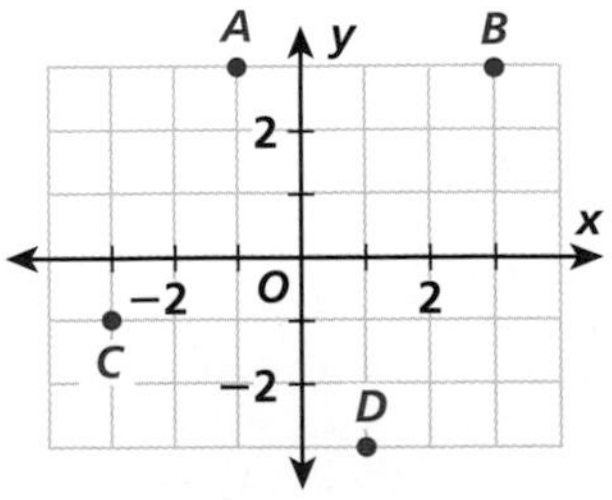

   A *A* C *C*
   B *B* D *D*

5. ***SHORT RESPONSE*** Make a function table to find the output values of $y = 3x + 1$ for the input values -1, 0, 1, and 2. Graph the resulting ordered pairs.

LESSON 12-2

6. Which function describes the sequence 5, 6, 7, 8, ...?

   A $y = n + 1$ C $y = 2n + 1$
   B $y = n + 4$ D $y = 5n + 1$

7. The function $y = n - 5$ describes which sequence?

   A 5, 6, 7, 8, . . . C −4, −3, −2, −1, . . .
   B 1, 2, 3, 4, . . . D −5, −6, −7, −8, . . .

8. What is the eighth term in the sequence that is described by $y = 7n - 2$?

   A 54 C 8
   B 42 D $\frac{10}{7}$

9. ***SHORT RESPONSE*** Andre has $0.50 in a jar. Each week he adds $0.50 to the jar. Write a sequence to find out how much money Andre will have in the jar after each of the first 4 weeks. Use the sequence to predict the amount of money Andre will have in the jar after 7 weeks.

LESSON 12-3

10. What is the domain of the function graphed below?

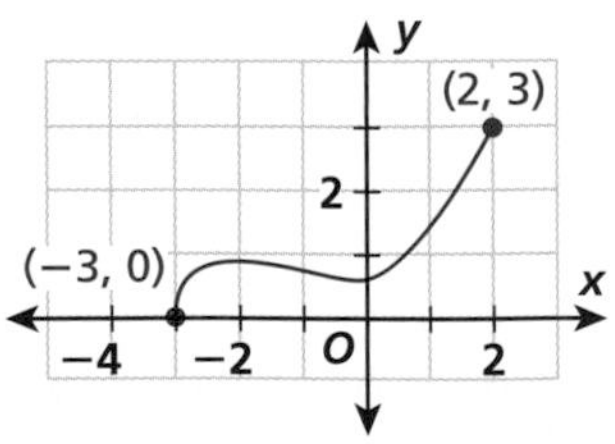

   A $0 \le y \le 3$ C $-3 \le y \le 2$
   B $0 \le x \le 3$ D $-3 \le x \le 2$

EOG Practice ■ Chapter 12

12A Introducing Functions

11. What is the range of the function graphed below?

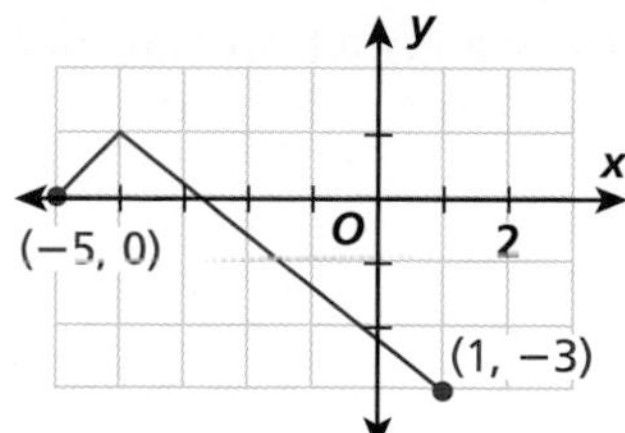

A $-3 \le y \le 1$

B $-3 \le x \le 1$

C $-3 \le y \le 0$

D $-5 \le x \le 1$

LESSON 12-4

12. The function $y = x + \frac{1}{2}$ is represented by which graph?

A

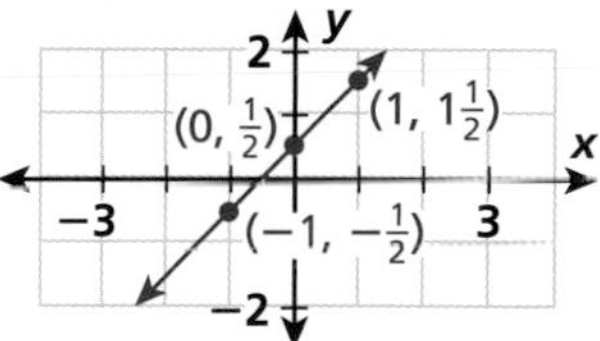

B

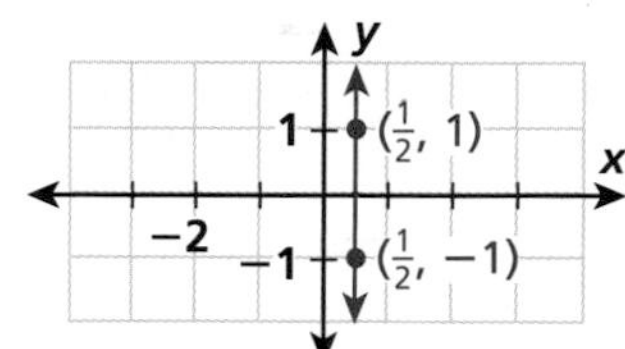

C

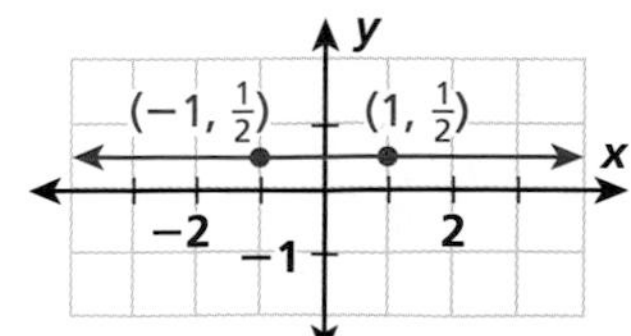

D

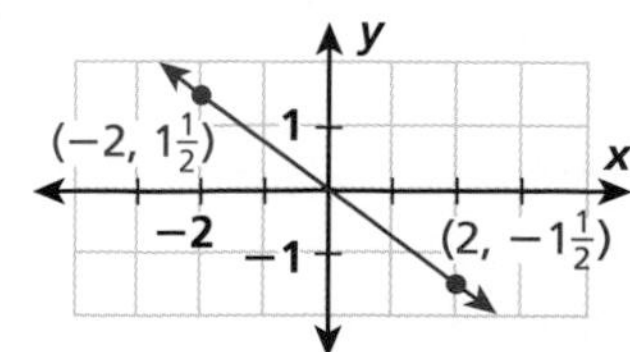

13. Which linear function is represented by the graph below?

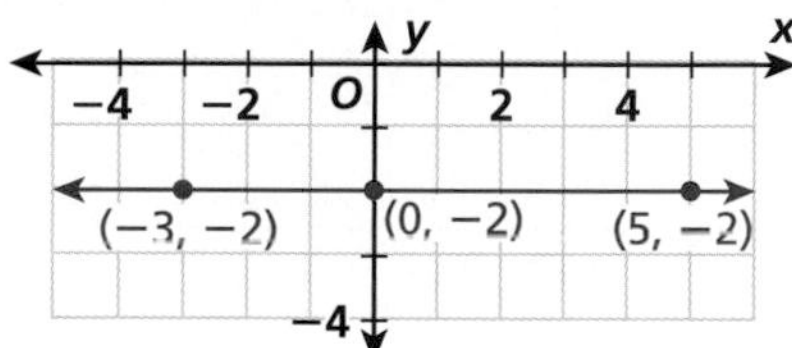

A $y = -2x - 2$

B $y = -x - 2$

C $y = -2x$

D $y = -2$

14. ***SHORT RESPONSE*** Jo measured the outside temperature as 52°F right before a cold front blew in. He noticed that the temperature dropped 2° every 30 minutes for the next 5 hours. Write a linear function that describes the drop in temperature over time. Make a graph to show the temperature decrease that Jo measured.

LESSON 12-5

15. What is the slope of the line graphed below?

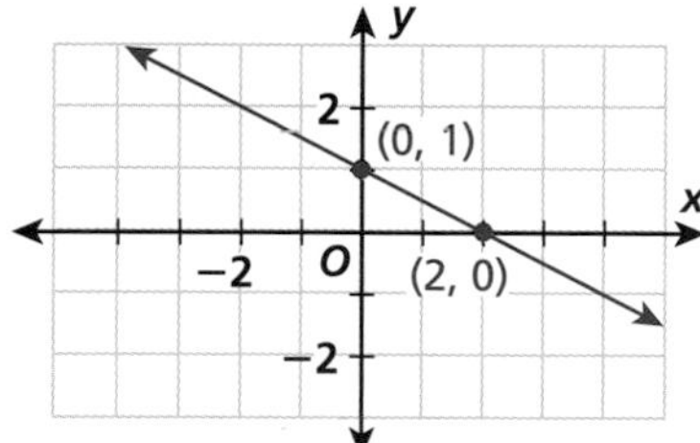

A -2

B -1

C $-\frac{1}{2}$

D 2

16. What is the slope of the line given by the equation $y = 9x - 4$?

A 9

B $\frac{9}{4}$

C $-\frac{9}{4}$

D -4

17. What is the y-intercept of the line given by the equation $y = -\frac{x}{3} + 1$?

A -3

B $-\frac{1}{3}$

C $\frac{1}{3}$

D 1

EOG Practice ▪ Chapter 12

12B Nonlinear Functions and Set Theory

LESSON 12-6

1. Which graph is linear?

A

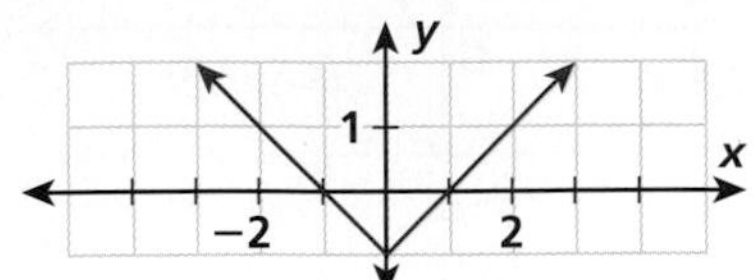

B

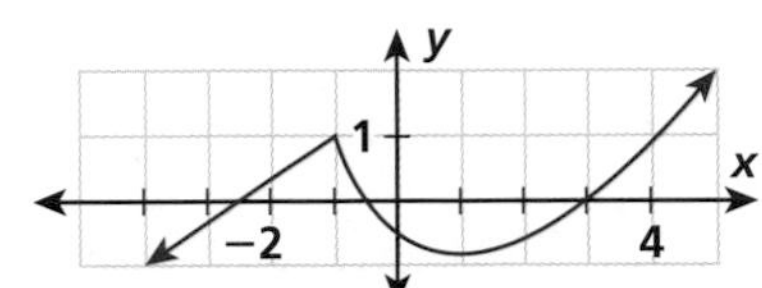

C

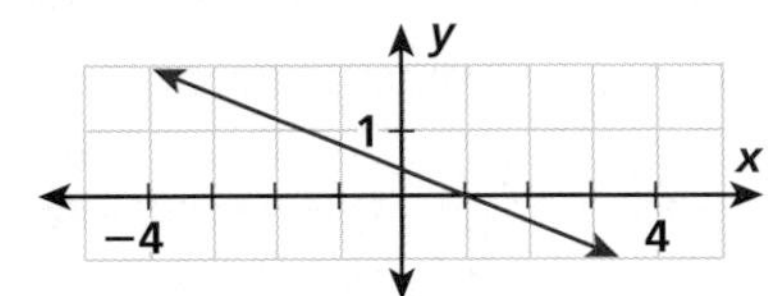

D

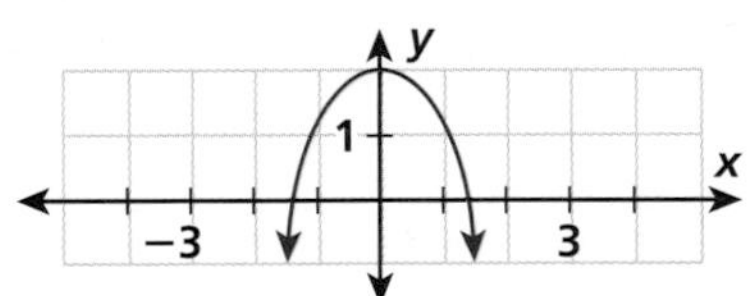

2. Which table represents a function that has a nonlinear relationship?

A

x	y
1	3
4	7
7	11

C

x	y
8	3
13	15
18	27

B

x	y
4	0
3	5
2	7

D

x	y
6	0
15	2
24	4

3. ***SHORT RESPONSE*** Make a function table to determine whether $y = 7x$ is linear or nonlinear. Graph the function to check your answer.

LESSON 12-7

4. In which interval of x does the graph below show a constant rate of change?

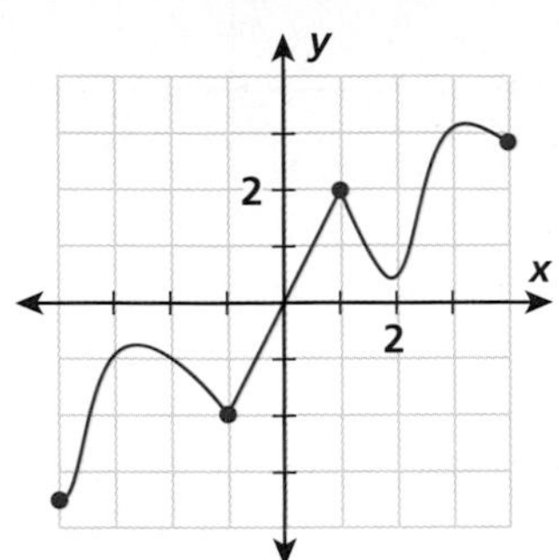

A $-4 \leq x \leq -1$

B $-1 \leq x \leq 1$

C $1 \leq x \leq 4$

D $-4 \leq x \leq 4$

5. Which of the following graphs shows a variable rate of change?

A

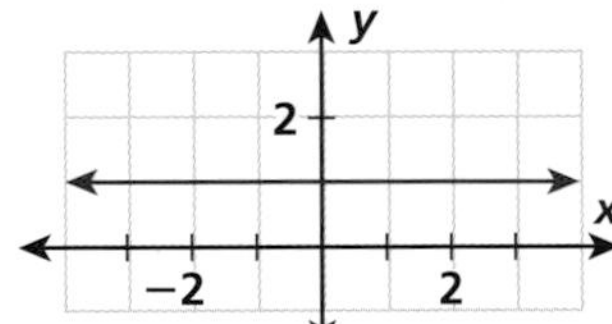

B

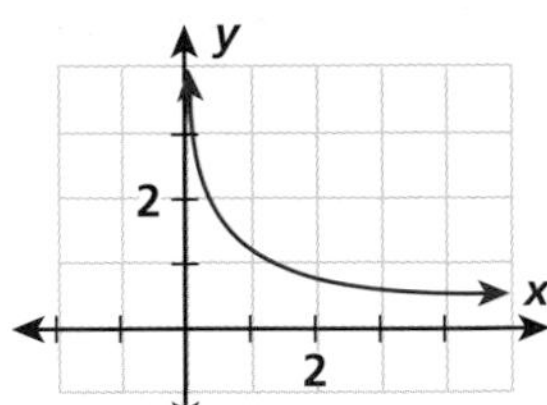

C

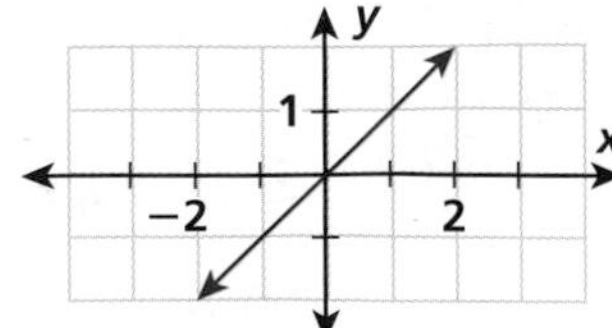

D

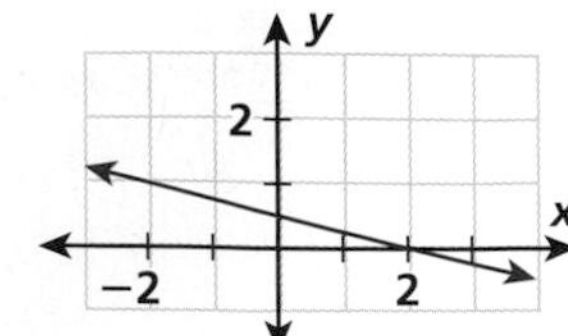

EOG Practice

EOG Practice ■ Chapter 12

12B Nonlinear Functions and Set Theory

LESSON 12-8

6. Which of the following sets is written in roster notation?
   - A the set of all elements that are noble gases
   - B helium, argon, neon, krypton, xenon, radon
   - C {He, Ar, Ne, Kr, Xe, Rn}
   - D $\{x|x \text{ is a noble gas}\}$

7. Which of the following sets is written in set-builder notation?
   - A the set of all regular solids
   - B cube, tetrahedron, octahedron, dodecahedron, icosahedron
   - C {cube, tetrahedron, octahedron, dodecahedron, icosahedron}
   - D $\{x|x \text{ is a regular solid}\}$

8. Which of the following describes the set {1, 3, 5, 7, 11, 13, 17, 19}?
   - A the set of all odd numbers between 0 and 20
   - B the set of all prime numbers between 0 and 20
   - C the set of all composite numbers between 0 and 20
   - D the set of all numbers between 0 and 20 that are divisible by 1

9. Which of the following would be most appropriately written in roster notation?
   - A the set of all integers that are greater than 6
   - B the set of all natural numbers that are less than 15
   - C the set of all prime numbers between 1 and 500
   - D the set of all odd numbers

10. Which of the following notations means "A is a subset of B"?
    - A $A \subset B$
    - B $A \supset B$
    - C $A \cup B$
    - D $A \cap B$

11. Which of the following notations means "the union of sets A and B"?
    - A $A \subset B$
    - B $A \supset B$
    - C $A \cup B$
    - D $A \cap B$

12. Which of the following notations means "the intersection of sets A and B"?
    - A $A \subset B$
    - B $A \supset B$
    - C $A \cup B$
    - D $A \cap B$

13. Which of the following is $A \cup B$ for $A = \{4, 8, 12, 16\}$ and $B = \{8, 16, 24, 32\}$?
    - A {4, 8, 12, 16, 24, 32}
    - B {4, 12, 24, 32}
    - C {4, 8, 12, 16}
    - D {8, 16}

14. Which of the following is $A \cap B$ for $A = \{x|x \text{ is an even number between 1 and 11}\}$ and $B = \{x|x \text{ is a multiple of 4 between 1 and 10}\}$?
    - A {2, 4, 6, 8, 10}
    - B {2, 6, 10}
    - C {4, 6, 8}
    - D {4, 8}

15. For which of the following pairs of sets is $A \cap B = \varnothing$?
    - A $A = \{0, 7, 9, 11\}$; $B = \{2, 3, 4, 5\}$
    - B $A = \{-1, 0, 4, 7, 13\}$; $B = \{-5, -2, 0\}$
    - C $A = \{0\}$; $B = \{0\}$
    - D $A = \{0\}$; $B = \{0, 1, 2\}$

Skills Bank Review Skills

Place Value

You can use a place-value chart to help you read and write numbers.
The number 213,867 is shown.

Hundred Thousands	Ten Thousands	Thousands	Hundreds	Tens	Ones
2	1	3	8	6	7

EXAMPLE

Use the chart to determine the place value of each digit.

A 2

The 2 is in the hundred thousands place.

B 8

The 8 is in the hundreds place.

PRACTICE

Determine the place value of each underlined digit.

1. 543,201 **2.** 239,487 **3.** 730,432 **4.** 4,382,121

Compare and Order Whole Numbers

You can use place values from left to right to compare and order numbers.

EXAMPLE

Compare and order from least to greatest: 42,810; 142,997; 42,729; 42,638.

Start at the leftmost place value.

There is one number with a digit in the greatest place. It is the greatest of the four numbers.

Compare the remaining three numbers. All values in the next two places, the ten thousands and thousands, are the same.

In the hundreds place, the values are different. Use this digit to order the remaining numbers.

42,810
142,997
42,729
42,638

42,638; 42,729; 42,810; 142,997

PRACTICE

Compare and order the numbers in each set from least to greatest.

1. 2,564; 2,546; 2,465; 2,654

2. 6,237; 6,372; 6,273; 6,327

3. 132,957; 232,795; 32,975; 31,999

4. 9,614; 29,461; 129,164; 129,146

Read and Write Decimals

When reading and writing a decimal, you need to know the place value of the digit in the last decimal place. Also, when writing a decimal in word form remember the following:

- "and" goes in place of the decimal for numbers greater than one.
- a hyphen is used in two-digit numbers, such as twenty-five.
- a hyphen is used in two-word place values, such as ten-thousandths.

EXAMPLE

Write 728.34 in words.

The 4 is in the hundredths place, so 728.34 is written as "seven hundred twenty-eight and thirty-four hundredths."

PRACTICE

Write each decimal in words.

1. 17.238 **2.** 9.0023 **3.** 534.01972 **4.** 33.00084 **5.** 4,356.67

Rules for Rounding

To round a number to a certain place value, locate the digit with that place value, and look at the digit to the right of it.

- If the digit to the right is 5 or greater, increase the number you are rounding by 1.
- If the digit to the right is 4 or less, leave the number you are rounding as it is.

EXAMPLE

A Round 765.48201 to the nearest hundredth.

765.4<u>8</u>201 *Locate the hundredths place.*

The digit to the right is less than 5, so the digit in the rounding place stays the same.

765.48

B Round 765.48201 to the nearest tenth.

765.<u>4</u>8201 *Locate the tenths place.*

The digit to the right is greater than 5, so the digit in the rounding place increases by 1.

765.5

PRACTICE

Round 203.94587 to the place indicated.

1. hundreds place **2.** hundredths place **3.** thousandths place

4. tens place **5.** ones place **6.** tenths place

Properties

Addition and multiplication follow certain rules. The tables show basic properties of addition and multiplication.

Addition Properties	
Commutative:	$a + b = b + a$
Associative:	$(a + b) + c = a + (b + c)$
Identity Property of Zero:	$a + 0 = a$
Inverse Property:	$a + (-a) = 0$
Closure Property:	The sum of two real numbers is a real number.

Multiplication Properties	
Commutative:	$a \times b = b \times a$
Associative:	$(a \times b) \times c = a \times (b \times c)$
Identity Property of One:	$a \times 1 = a$
Inverse Property:	$a \times \frac{1}{a} = 1$ if $a \neq 0$
Property of Zero:	$a \times 0 = 0$
Closure Property:	The product of two real numbers is a real number.
Distributive:	$a(b + c) = a \times b + a \times c$

The following properties are true when a, b, and c are real numbers.

Substitution Property: If $a = b$, then a can be substituted for b in any expression.

Transitive Property: If $a = b$ and $b = c$, then $a = c$.

PRACTICE

Name the property represented by each equation.

1. $8 + 0 = 8$

2. $(9 \times 3) \times 7 = 9 \times (3 \times 7)$

3. 3×5 is a real number

4. $7 \times 345 = 345 \times 7$

5. $2(3 + 5) = 2 \times 3 + 2 \times 5$

6. $15 \times \frac{1}{15} = 1$

7. $3.6 + 4.4 = 4.4 + 3.6$

8. $\frac{3}{4} \times \frac{4}{4} = \frac{3}{4}$

9. $18 + (-18) = 0$

10. $(5 + 17) + 23 = 5 + (17 + 23)$

Overestimates and Underestimates

An **overestimate** is an estimate that is greater than the actual answer. An **underestimate** is an estimate that is less than the actual answer.

EXAMPLE 1

Give an overestimate for each expression.

A $124 + 371$

$124 + 371 \approx 130 + 380$
≈ 510

B $316 \div 12$

$316 \div 12 \approx 320 \div 10$
≈ 32

EXAMPLE 2

Give an underestimate for each expression.

A $64 - 12$

$64 - 12 \approx 60 - 15$
≈ 45

B $28 \cdot 8$

$28 \cdot 8 \approx 25 \cdot 8$
≈ 200

PRACTICE

Give an overestimate and underestimate for each expression.

1. $224 + 545$ **2.** $756 + 142$ **3.** $643 - 104$ **4.** $2,456 - 435$

5. 13×17 **6.** 7×85 **7.** $261 \div 9$ **8.** $85 \div 34$

Skills Bank

Compatible Numbers

You can use compatible numbers to estimate products and quotients. Compatible numbers are close to the numbers in the problem and can help you do math mentally.

EXAMPLES

Estimate each product or quotient.

A $327 \cdot 28$

Compatible numbers

$327 \cdot 28 \approx 300 \cdot 30$
$\approx 9,000$ ← *Estimate*

B $637 \div 8$

Compatible numbers

$637 \div 8 \approx 640 \div 8$
≈ 80 ← *Estimate*

PRACTICE

Use compatible numbers to estimate each product or quotient.

1. $42 \cdot 7$ **2.** $3,957 \div 23$ **3.** $5,169 \cdot 21$ **4.** $813 \div 8$ **5.** $78 \cdot 42$

6. $1,443 \div 7$ **7.** $98 \cdot 48$ **8.** $3,372 \div 415$ **9.** $58 \cdot 9$ **10.** $27,657 \div 67$

Multiply and Divide by Powers of Ten

When you *multiply* by powers of ten, move the decimal point one place to the right for each zero in the power of ten. When you *divide* by powers of ten, move the decimal point one place to the left for each zero in the power of ten.

EXAMPLE

Find each product or quotient.

A **0.37 · 100**

0.37 · 100 = 0.37

= 37

B **43 · 1,000**

43 · 1,000 = 43.000

= 43,000

C **0.24 ÷ 10**

0.24 ÷ 10 = 0.24

= 0.024

D **1,467 ÷ 100**

1,467 ÷ 100 = 1467.

= 14.67

PRACTICE

Find each product or quotient.

1. 10 × 8.53 **2.** 0.55×10^4 **3.** 48.6 × 1,000 **4.** 2.487 ÷ 1,000 **5.** $6.03 \div 10^3$

Multiply Whole Numbers

When you multiply two whole numbers, think of the second number's expanded form, and multiply by each value.

EXAMPLE

Find the product of 621 · 485.

Step 1: Think of 485 as 4 hundreds, 8 tens, and 5 ones. Multiply 621 by 5 ones.	**Step 2:** Multiply 621 by 8 tens.	**Step 3:** Multiply 621 by 4 hundreds.	**Step 4:** Add the partial products.
621 × 485 3,105 ←5 × 621	621 × 485 3,105 49,680 ←80 × 621	621 × 485 3,105 49,680 248,400 ←400 × 621	621 × 485 3,105 49,680 + 248,400 301,185

621 · 485 = 301,185

PRACTICE

Multiply.

1. 493 × 37 **2.** 539 × 82 **3.** 134 × 145 **4.** 857 × 662

5. 1,872 × 43 **6.** 5,849 × 67 **7.** 36,735 × 28 **8.** 121,614 × 58

Divide Whole Numbers

EXAMPLE

Find the quotient of 5,712 ÷ 28.

Step 1: Write the first number inside the long division symbol, and write the second number to the left of the symbol. Divide by the number outside the symbol.	**Step 2:** Multiply 28 by 2, and place the product under 57. Subtract and bring down the next digit of the dividend.	**Step 3:** Divide 112 by 28. Multiply 28 by 4 and place the product under 112. Subtract.
$\begin{array}{r} 2 \\ 28\overline{)5712} \end{array}$ *28 cannot go into 5, so try 57.*	$\begin{array}{r} 20 \\ 28\overline{)5712} \\ -56\downarrow \\ \hline 11 \\ -0\downarrow \\ \hline 112 \end{array}$ *28 cannot go into 11, so put a 0 in the quotient, and bring down the 2.*	$\begin{array}{r} 204 \\ 28\overline{)5712} \\ -56 \\ \hline 11 \\ -0 \\ \hline 112 \\ -112 \\ \hline 0 \end{array}$

PRACTICE

Divide.

1. 23,148 ÷ 18 **2.** 5,772 ÷ 37 **3.** 56,088 ÷ 41 **4.** 34,540 ÷ 55

5. 68,894 ÷ 74 **6.** 143,296 ÷ 32 **7.** 398,736 ÷ 72 **8.** 566,746 ÷ 79

Divisibility Rules

A number is divisible by another number if the quotient is a whole number with no remainder.

A number is divisible by . . .	Divisible	Not Divisible
2 if the last digit is an even number.	13,776	4,221
3 if the sum of the digits is divisible by 3.	327	97
4 if the last two digits form a number divisible by 4.	3,128	526
5 if the last digit is 0 or 5.	9,415	50,501
6 if the number is divisible by 2 and 3.	762	62
9 if the sum of the digits is divisible by 9.	21,222	96
10 if the last digit is 0.	1,680	8,255

PRACTICE

Determine whether each number is divisible by 2, 3, 4, 5, 6, 9, or 10.

1. 324 **2.** 501 **3.** 200 **4.** 812 **5.** 60

6. 784 **7.** 351 **8.** 3,009 **9.** 2,345 **10.** 555,555

Factors

A **factor** of a number is any number that divides into it without leaving a remainder.

EXAMPLE

List all the factors of 28.

The possible factors are whole numbers from 1 to 28.

$1 \cdot 28 = 28$ *The numbers 1 and 28 are factors of 28.*

$2 \cdot 14 = 28$ *The numbers 2 and 14 are factors of 28.*

$3 \cdot ? = 28$ *No whole number multiplied by 3 equals 28, so 3 is not a factor of 28.*

$4 \cdot 7 = 28$ *The numbers 4 and 7 are factors of 28.*

$5 \cdot ? = 28$ *No whole number multiplied by 5 equals 28, so 5 is not a factor of 28.*

$6 \cdot ? = 28$ *No whole number multiplied by 6 equals 28, so 6 is not a factor of 28.*

The factors of 28 are 1, 2, 4, 7, 14, and 28.

PRACTICE

List all the factors of each number.

1. 10 **2.** 8 **3.** 18 **4.** 54 **5.** 27 **6.** 36

7. 19 **8.** 24 **9.** 50 **10.** 32 **11.** 49 **12.** 39

Prime and Composite Numbers

A **prime number** has exactly two factors, 1 and itself.
A **composite number** has more than two factors.

EXAMPLE

Determine whether each number is prime or composite.

A **19**
Factors: 1, 19
So 19 is prime.

B **20**
Factors: 1, 2, 4, 5, 10, 20
So 20 is composite.

PRACTICE

Determine whether each number is prime or composite.

1. 7 **2.** 15 **3.** 18 **4.** 8

5. 113 **6.** 31 **7.** 12 **8.** 49

9. 77 **10.** 67 **11.** 9 **12.** 79

Simplest Form of Fractions

A fraction is in simplest form when the only common factor of the numerator and denominator is 1.

EXAMPLE

Simplify.

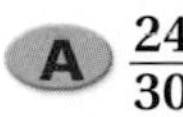 $\frac{24}{30}$

24: 1, 2, 3, 4, **6**, 8, 12, 24 *Find the greatest*
30: 1, 2, 3, 5, **6**, 10, 15, 30 *common factor of 24 and 30.*

$\frac{24 \div 6}{30 \div 6} = \frac{4}{5}$ *Divide both the numerator and the denominator by 6.*

B $\frac{18}{28}$

18: 1, **2**, 3, 6, 9, 18 *Find the greatest*
28: 1, **2**, 4, 7, 14, 28 *common factor of 18 and 28.*

$\frac{18 \div 2}{28 \div 2} = \frac{9}{14}$ *Divide both the numerator and the denominator by 2.*

PRACTICE

Simplify.

1. $\frac{15}{20}$ **2.** $\frac{32}{40}$ **3.** $\frac{14}{35}$ **4.** $\frac{30}{75}$ **5.** $\frac{17}{51}$

6. $\frac{18}{42}$ **7.** $\frac{19}{38}$ **8.** $\frac{22}{121}$ **9.** $\frac{10}{32}$ **10.** $\frac{39}{91}$

Roman Numerals

In the Roman numeral system, numbers do not have place values to show what they represent. Instead, numbers are represented by letters.

I = 1 V = 5 X = 10 L = 50 C = 100 D = 500 M = 1,000

The values of the letters do not change based on their place in a number.

If a numeral is to the right of an equal or greater numeral, add the two numerals' values. If a numeral is immediately to the left of a greater numeral, subtract the numeral's value from the greater numeral.

EXAMPLE

 **Write CLIV as a decimal number.**

CLIV = C + L + (V − I)
= 100 + 50 + (5 − 1)
= 154

B Write 1,109 as a Roman numeral.

1,109 = 1,000 + 100 + 9
= M + C + (X − I)
= MCIX

PRACTICE

Write each decimal number as a Roman numeral and each Roman numeral as a decimal number.

1. XXVI **2.** 29 **3.** MCMLII **4.** 224 **5.** DCCCVI

6. 8 **7.** XLIV **8.** 1,557 **9.** XCIX **10.** 2,004

Binary Numbers

Computers use the **binary number system.** In the binary, or base-2, system of numbers, numbers are formed using the digits 0 and 1. Each place in a binary number is associated with a power of 2. Binary numbers are written with the subscript *two* so that they are not confused with numbers in the decimal system.

The binary number 1101_{two} can be thought of as $(1 \cdot 2^3) + (1 \cdot 2^2) + (0 \cdot 2^1) + (1 \cdot 2^0)$.

Binary Place Value

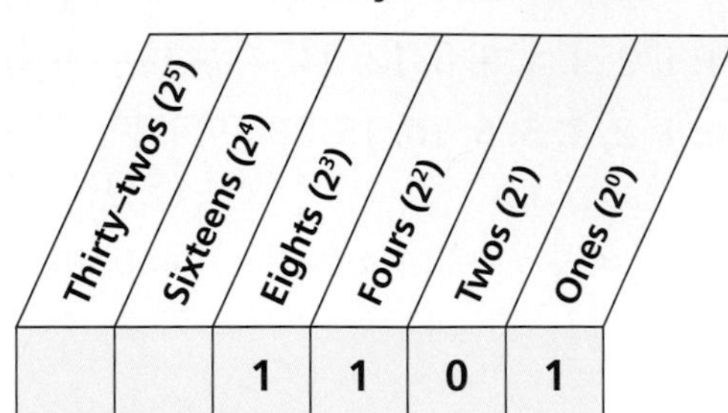

You can use the expanded form of 1101_{two} to find the value of the number as a decimal, or base-10, number.

$$\begin{aligned}(1 \cdot 2^3) + (1 \cdot 2^2) + (0 \cdot 2^1) + (1 \cdot 2^0) &= (1 \cdot 8) + (1 \cdot 4) + (0 \cdot 2) + (1 \cdot 1)\\ &= 8 + 4 + 0 + 1\\ &= 13\end{aligned}$$

So $1101_{two} = 13_{ten}$.

EXAMPLE

Write each binary number as a decimal number.

A 101110_{two}

$$\begin{aligned}101110_{two} &= (1 \cdot 2^5) + (0 \cdot 2^4) + (1 \cdot 2^3) + (1 \cdot 2^2) + (1 \cdot 2^1) + (0 \cdot 1)\\ &= 32 + 0 + 8 + 4 + 2 + 0\\ &= 46\end{aligned}$$

B 10001_{two}

$$\begin{aligned}10001_{two} &= (1 \cdot 2^4) + (0 \cdot 2^3) + (0 \cdot 2^2) + (0 \cdot 2^1) + (1 \cdot 1)\\ &= 16 + 0 + 0 + 0 + 1\\ &= 17\end{aligned}$$

PRACTICE

Write each binary number as a decimal number.

1. 100_{two} **2.** 110_{two} **3.** 101_{two} **4.** 1100_{two} **5.** 1011_{two}

6. 11011_{two} **7.** 11110_{two} **8.** 101010_{two} **9.** 111111_{two} **10.** 100111_{two}

Estimate Measurements

You can use benchmarks to estimate with metric and customary units.

1 meter (m)	Width of a doorway	**1 centimeter (cm)**	Width of a large paper clip
1 liter (L)	Water in a 1-quart bottle	**1 milliliter (mL)**	Water in an eyedropper
1 gram (g)	Mass of a dollar bill	**1 kilogram (kg)**	Mass of 8 rolls of pennies
30°C (Celsius)	Temperature on a hot day	**0°C (Celsius)**	Temperature on a freezing day

EXAMPLE 1

Choose the most reasonable estimate of the height of the ceiling in your classroom.

A 30 cm B 3 m C 30 m D 30,000 cm

The most reasonable estimate is 3 m.

Length	Temperature	Capacity
1 inch (in.)—about the length of a small paper clip **1 foot (ft)**—about the length of a standard sheet of paper **1 yard (yd)**—about the width of a doorway	**32°F** (Fahrenheit)—water freezes **70°F**—air on a comfortably warm day **90°F**—air on a hot day **212°F**—boiling water	**1 fluid ounce (fl oz)**—amount of water in two tablespoons **1 cup (c)**—amount of water held in a standard measuring cup **1 pint (pt), 1 quart (qt), 1 gallon (gal)**—Think about containers of water at a store.

EXAMPLE 2

Choose the most appropriate estimate.

A the length of a classroom

A 30 in. B 30 ft C 30 yd

The most appropriate estimate is 30 ft.

B a temperature for wearing a T-shirt

A 20°F B 40°F C 80°F

The most appropriate estimate is 80°F.

PRACTICE

Choose the most reasonable estimate.

1. the temperature on a warm day

   A −22°C B 22°C C 68°C

2. the capacity of a kitchen sink

   A 12 mL B 1,200 mL C 12 L

Choose the most appropriate estimate.

3. the capacity of a tall drinking glass

   A 1 pt B 4 qt C $\frac{1}{2}$ gal

4. a temperature for wearing a warm coat

   A 20°F B 60°F C 80°F

5. the temperature of a cup of hot cocoa

   A 32°F B 120°F C 250°F

6. the width of a pizza box

   A 18 in. B 8 ft C 2 yd

Relate Metric Units of Length, Mass, and Capacity

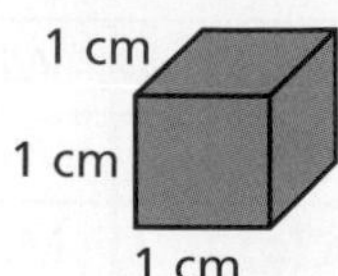

Within the metric system, measures of length, mass, and capacity are related. A cube that has a volume of 1 cm^3 has a capacity of 1 mL. If the cube were filled with water, the mass of the water would be 1 g.

EXAMPLE

Find the capacity of a 50 cm × 60 cm × 30 cm rectangular box. Then find the mass of the water that would fill the box.

Volume: 50 cm × 60 cm × 30 cm = 90,000 cm^3

Capacity: 1 cm^3 = 1 mL, so 90,000 cm^3 = 90,000 mL, or 90 L.

Mass: 1 mL of water has a mass of 1 g, so 90,000 mL of water has a mass of 90,000 g, or 90 kg.

PRACTICE

Find the capacity of each rectangular box. Then find the mass of the water that would fill the box.

1. 2 cm × 5 cm × 8 cm
2. 10 cm × 18 cm × 4 cm
3. 8 cm × 8 cm × 8 cm
4. 10 cm × 10 cm × 10 cm
5. 15 cm × 18 cm × 16 cm
6. 23 cm × 19 cm × 11 cm

Negative Exponents

A positive exponent tells how many times the base of a power is used as a factor. The table below shows one way you can determine the values of powers with negative exponents.

Power	8^3	8^2	8^1	8^0	8^{-1}	8^{-2}
Value	512	64	8	1	$\frac{1}{8}$	$\frac{1}{64}$

÷ 8 ÷ 8 ÷ 8 ÷ 8 ÷ 8

Each time the exponent is decreased by 1, the value of the power is divided by the base. There is another pattern in the table.

$8^{-1} = \frac{1}{8^1} = \frac{1}{8}$ $8^{-2} = \frac{1}{8^2} = \frac{1}{64}$ $8^{-3} = \frac{1}{8^3} = \frac{1}{512}$

A number raised to a negative exponent equals 1 divided by that number raised to the opposite (positive) exponent. You can write the general rule for this property as $a^{-b} = \frac{1}{a^b}$.

PRACTICE

Evaluate.

1. 2^{-3}
2. 7^{-2}
3. 10^{-4}
4. 8^{-4}
5. 1^{-20}
6. 5^{-1}

Skills Bank Preview Skills

Probability of Two Disjoint Events

In probability, two events are considered to be **disjoint** if they cannot happen at the same time. Examples of disjoint events are getting a 5 or a 6 on a single roll of a 1–6 number cube. To find the probability that either one or the other of two disjoint events will occur, add the probabilities of each event occurring separately.

EXAMPLE

Find the probability of each set of disjoint events.

A rolling either a 5 or a 6 on a 1–6 number cube

$$\begin{aligned} P(5 \text{ or } 6) &= P(5) + P(6) \\ &= \frac{1}{6} + \frac{1}{6} \\ &= \frac{2}{6} \\ &= \frac{1}{3} \end{aligned}$$

The probability of rolling a 5 or a 6 on a 1–6 number cube is $\frac{1}{3}$.

B choosing either an *A* or an *E* from the letters in the word *mathematics*

$$\begin{aligned} P(A \text{ or } E) &= P(A) + P(E) \\ &= \frac{2}{11} + \frac{1}{11} \\ &= \frac{3}{11} \end{aligned}$$

The probability of choosing an *A* or an *E* is $\frac{3}{11}$.

PRACTICE

Find the probability of each set of disjoint events.

1. tossing a coin and getting heads or tails
2. spinning red or green on a spinner that has four equal sectors colored red, green, blue, and yellow
3. drawing a black marble or a red marble from a bag that contains 4 white marbles, 3 black marbles, and 2 red marbles
4. choosing either a boy or a girl from a class of 13 boys and 17 girls
5. choosing either *A* or *E* from a list of the five vowels
6. choosing either a number less than 3 or a number greater than 12 from a set of 20 cards numbered 1–20

Inductive and Deductive Reasoning

You use **inductive reasoning** when you look for a pattern in individual cases to draw conclusions. Conclusions drawn using inductive reasoning are sometimes like predictions. They may be proven false.

You use **deductive reasoning** when you use given facts to draw conclusions. A conclusion based on facts must be true.

EXAMPLE

Identify the type of reasoning used. Explain your answers.

A ***Statement:*** **A number pattern begins with 2, 5, 8, 11, . . .**

Conclusion: **The next number in the pattern will be 14.**

This is inductive reasoning. The conclusion is based on the pattern established by the first four terms in the sequence.

B ***Statement:*** **It has rained for the past three days.**

Conclusion: **It will rain tomorrow.**

This is inductive reasoning. The conclusion is based on the weather pattern over the past three days.

C ***Statement:*** **The measures of two angles of a triangle are 30° and 70°.**

Conclusion: **The measure of the third angle is 80°.**

This is deductive reasoning. Since you know that the measures of the angles of a triangle have a sum of 180°, the third angle of this triangle must measure 80° ($30° + 70° + 80° = 180°$).

PRACTICE

Identify the type of reasoning used. Explain your answers.

1. *Statement:* Shawna has received a score of 100 on the last five math tests.
   Conclusion: Shawna will receive a score of 100 on the next math test.
2. *Statement:* The mail has arrived late every Monday for the past 4 weeks.
   Conclusion: The mail will arrive late next Monday.
3. *Statement:* Three angles of a quadrilateral measure 100°, 90°, and 70°.
   Conclusion: The measure of the fourth angle is 100°.
4. *Statement:* Perpendicular lines AB and CD intersect at point E.
   Conclusion: Angle AED is a right angle.
5. *Statement:* A pattern of numbers begins 1, 2, 4, . . .
   Conclusion: The next number in the pattern is 8.
6. *Statement:* Ten of the first ten seventh-grade students surveyed listed soccer as their favorite sport.
   Conclusion: Soccer is the favorite sport of all seventh-graders.

Make Conjectures

Conjecture is another word for conclusion. Conjectures in math are based on observations and in some cases have not yet been proven to be true. To prove that a conjecture is false, you need to find just one case, or *counterexample,* for which the conclusion does not hold true.

EXAMPLE 1

Test each conjecture to decide whether it is true or false. If the conjecture is false, give a counterexample.

A **The sum of two even numbers is always an even number.**

An even number is divisible by 2. The sum of two even numbers can be written as $2m + 2n = 2(m + n)$, which is divisible by 2, so it is even. The conjecture is true.

B **All prime numbers are odd.**

The first prime number is 2, which is an even number. The conjecture is false.

EXAMPLE 2

Formulate a conjecture based on the given information. Then test your conjecture.

$1 \cdot 3 = 3$ $3 \cdot 5 = 15$ $5 \cdot 7 = 35$ $7 \cdot 9 = 63$

Conjecture: The product of two odd numbers is always an odd number.

An odd number does not have 2 as a factor, so the product of two odd numbers also does not have 2 as a factor. The conjecture is true.

PRACTICE

Test each conjecture to decide whether it is true or false. If the conjecture is false, give a counterexample.

1. The sum of two odd numbers is always an odd number.
2. The product of two even numbers is always an even number.
3. The sum of twice a whole number and 1 is always an odd number.
4. If you subtract a whole number from another whole number, the result will always be a whole number.
5. If you multiply two fractions, the product will always be greater than either fraction.

Formulate a conjecture based on the given information. Then test your conjecture.

6. $12 + 21 = 33$ $13 + 31 = 44$ $23 + 32 = 55$ $17 + 71 = 88$
7. $15 \times 15 = 225$ $25 \times 25 = 625$ $35 \times 35 = 1{,}225$

Trigonometric Ratios

You can use ratios to find information about the sides and angles of a right triangle. These ratios are called *trigonometric ratios*, and they have names, such as **sine** (abbreviated *sin*), **cosine** (abbreviated *cos*), and **tangent** (abbreviated *tan*).

The **sine** of $\angle 1 = \sin \angle 1 = \frac{\text{length of side opposite } \angle 1}{\text{length of hypotenuse}} = \frac{a}{c}$.

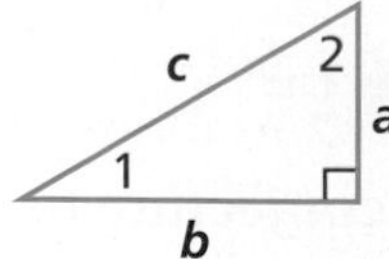

The **cosine** of $\angle 1 = \cos \angle 1 = \frac{\text{length of side adjacent to } \angle 1}{\text{length of hypotenuse}} = \frac{b}{c}$.

The **tangent** of $\angle 1 = \tan \angle 1 = \frac{\text{length of side opposite } \angle 1}{\text{length of side adjacent to } \angle 1} = \frac{a}{b}$.

EXAMPLE 1

Find the sine, cosine, and tangent of $\angle J$.

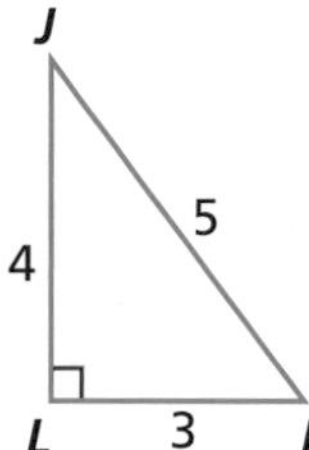

$\sin \angle J = \frac{LK}{JK} = \frac{3}{5}$

$\cos \angle J = \frac{JL}{JK} = \frac{4}{5}$

$\tan \angle J = \frac{LK}{JL} = \frac{3}{4}$

EXAMPLE 2

Use your calculator to find the measure of side $\overline{MN}$ to the nearest tenth.

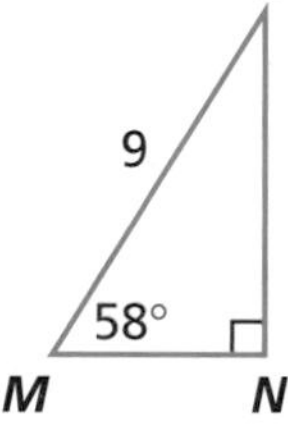

Side $\overline{MN}$ is adjacent to the 58° angle. The length of the hypotenuse is given. The ratio that uses the lengths of the adjacent side and the hypotenuse is cosine.

$\cos(58°) = \frac{MN}{9}$ — *Write the ratio that is equal to the cosine of 58°.*

$9 \cdot \cos(58°) = MN$ — *Multiply both sides by 9.*

9 

58 ENTER — *Use your calculator.*

$MN = 4.8$

PRACTICE

Find the sine, cosine, and tangent of each angle.

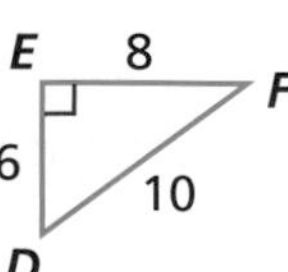

1. $\angle D$
2. $\angle F$

Use your calculator to find the measure of each side, to the nearest tenth.

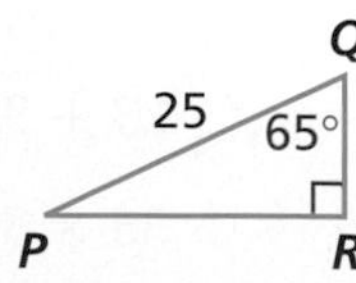

3. $\overline{QR}$
4. $\overline{PR}$

Science Skills

Half-life

Some atoms give off energy by emitting particles from their centers, or nuclei. The ability of these atoms to release nuclear radiation is called *radioactivity,* and the process is called *radioactive decay.*

Half-life is the amount of time it takes for one-half of the nuclei of a radioactive sample to decay. The half-life of an element can range from less than a second to millions of years.

EXAMPLE 1

The half-life of sodium-24 is 15 hours. If a sample of sodium-24 contains $\frac{1}{8}$ of its original amount, how old is the sample?

Every 15 hours, $\frac{1}{2}$ the sample decays.

Fraction of Sample	1	$\frac{1}{2}$	$\frac{1}{4}$	$\frac{1}{8}$
Time	0 hours	15 hours	30 hours	45 hours

The sample is 45 hours old.

EXAMPLE 2

The half-life of phosphorous-24 is 14.3 days. How much of a 6-gram sample will remain after 42.9 days?

Time	0 days	14.3 days	28.6 days	42.9 days
Amount of Sample (g)	6	3	1.5	0.75

After 42.9 days, 0.75 g of phosphorous-24 will remain.

PRACTICE

1. The half-life of cobalt-60 is 5.26 years. If a sample of cobalt-60 contains $\frac{1}{4}$ of its original amount, how old is the sample?
2. The half-life of sodium-24 is 15 hours. How much of a 9.6-gram sample will remain after 60 hours?
3. Iodine-131 has a half-life of 8.07 days. How much of a 4.4 g sample will there be after 40.35 days?
4. A sample of bismuth-212 decayed from 18 g to 1.125 g in 242 minutes. What is the half-life of bismuth-212?

pH Scale

An acid is a compound that produces hydrogen ions in solution. A base is a compound that produces hydroxide ions in solution. Chemists use the **pH scale** to measure how acidic or basic a solution is.

The range of the pH scale for a solution is 0 to 14. A solution with a pH below 7 is acidic. A solution with a pH above 7 is basic. A solution with a pH of 7 is neutral—that is, it has an equal number of hydrogen and hydroxide ions.

The pH numbers are related by powers of 10.

A pH of 6 is 10 times more acidic than a pH of 7.

A pH of 8 is 10 times more basic than a pH of 7.

0 Strong acids   Weak acids   7   Weak bases   Strong bases 14

EXAMPLE 1

Solution A and solution B have the same volume. Solution A has a pH of 2, and solution B has a pH of 4. How much more acidic is solution A than solution B?

Since $4 - 2 = 2$ and $10^2 = 100$ solution A is 100 times more acidic than solution B.

EXAMPLE 2

Solution C and solution D have the same volume. Solution C has a pH of 13, and solution D has a pH of 8. How much more basic is solution C than solution D?

Since $13 - 8 = 5$ and $10^5 = 100{,}000$ solution C is 100,000 times more basic than solution D.

PRACTICE

1. Solution E and solution F have the same volume. Solution E has a pH of 5, and solution F has a pH of 1. How much more basic is solution E than solution F?

2. Solution G and solution H have the same volume. Solution G has a pH of 9, and solution H has a pH of 8. How much more basic is solution G than solution H?

3. Solution K and solution J have the same volume. Solution J has a pH of 7, and solution K has a pH of 5. How much more acidic is solution K than solution J?

4. Solution M and solution L have the same volume. Solution L has a pH of 14, and solution M has a pH of 7. How much more basic is solution L than solution M?

Richter Scale

The magnitude of an earthquake is a measure of the amount of energy the earthquake releases. The **Richter scale** is used to express the magnitude of earthquakes. This scale uses the counting numbers. Each number represents a magnitude that is 10 times stronger than the magnitude represented by the previous number.

You can relate Richter scale numbers to the exponents in powers of 10.

$10^1 = 10$ $\quad$ $10^2 = 100$ $\quad$ $10^3 = 1{,}000$ $\quad$ $10^4 = 10{,}000$ $\quad$ $10^5 = 100{,}000$

Just as 10^2 is 10 times 10^1, an earthquake with a magnitude of 2 on the Richter scale is 10 times stronger than an earthquake with a magnitude of 1.

EXAMPLE 1

An earthquake has a magnitude of 5 on the Richter scale. How much stronger is it than an earthquake with a magnitude of 2?

$10^5 = 100{,}000$ and $10^2 = 100$
Since 100,000 is 1,000 times 100, 10^5 is 1,000 times 10^2.

An earthquake with a magnitude of 5 is 1,000 times stronger than an earthquake with a magnitude of 2.

EXAMPLE 2

An earthquake had a magnitude of 3. If the earthquake had been 10,000 times stronger, what would its magnitude have been?

$10^3 = 1{,}000$
$1{,}000 \cdot 10{,}000 = 10{,}000{,}000 = 10^7$

The earthquake would have had a magnitude of 7.

PRACTICE

1. How many times stronger is an earthquake with a magnitude of 3 than an earthquake with a magnitude of 1?
2. How many times stronger is an earthquake with a magnitude of 6 than an earthquake with a magnitude of 3?
3. An earthquake has a magnitude of 2. How many times stronger would the earthquake have to be to have a magnitude of 6?
4. An earthquake has a magnitude of 3. How many times stronger would the earthquake have to be to have a magnitude of 9?
5. An earthquake had a magnitude of 5. If the earthquake had been 1,000 times stronger, what would its magnitude have been?
6. An earthquake had a magnitude of 4. If the earthquake had been 100,000 times stronger, what would its magnitude have been?

Surface Area to Volume Ratios

A surface area to volume ratio is a ratio that compares the surface area and volume of a solid. You can use the surface area to volume ratio of a solid to find the surface area of the solid if you know its volume, or to find the volume of the solid if you know its surface area.

EXAMPLE 1

Find the surface area to volume ratio of the cube.

5 ft
5 ft
5 ft

$$\begin{aligned} \text{surface area} &= 2\ell w + 2\ell h + 2wh \\ &= (2 \cdot 5 \cdot 5) + (2 \cdot 5 \cdot 5) + (2 \cdot 5 \cdot 5) \\ &= 150 \end{aligned}$$

$$\begin{aligned} \text{volume} &= \ell wh \\ &= 5 \cdot 5 \cdot 5 \\ &= 125 \end{aligned}$$

$$\frac{\text{surface area}}{\text{volume}} = \frac{150}{125}$$

$$= \frac{6}{5} \qquad \textit{Simplify.}$$

The surface area to volume ratio is $\frac{6}{5}$.

EXAMPLE 2

Find the surface area of a cube that has a volume of 64 cubic units and a surface area to volume ratio of $\frac{3}{2}$.

$\frac{3}{2} = \frac{\text{surface area}}{64}$	*Write a proportion.*
$2 \cdot \text{surface area} = 3 \cdot 64$	*Find the cross products.*
$2(\text{surface area}) = 192$	*Multiply.*
$\text{surface area} = 96$	*Divide both sides by 2.*

The surface area of the cube is 96 square units.

PRACTICE

Find the surface area to volume ratio of each solid.

1.

25 in.
5 in.
10 in.

2.

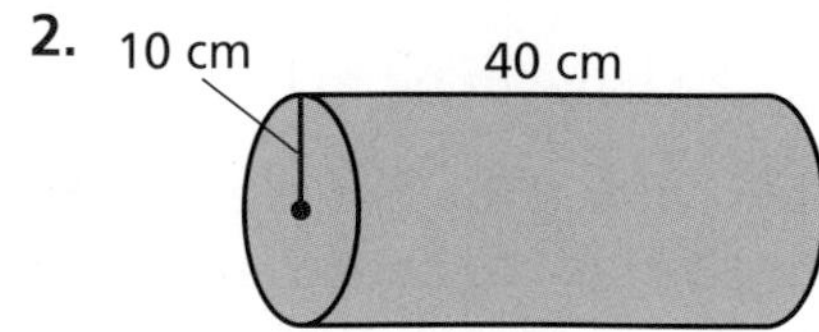

3. Find the surface area of a cylinder that has a volume of 5,001 cubic meters and a surface area to volume ratio of $\frac{1}{3}$.

4. Find the volume of a cylinder that has a surface area of 11,781 square feet and a surface area to volume ratio of $\frac{231}{500}$.

Skills Bank

Quadratic Relationships

Quadratic relationships involve one squared value related to another value. An example of a quadratic relationship is shown in the equation $a = x^2 + 5$. If you know the value of one variable, you can substitute for it in the equation and then solve to find the second variable.

EXAMPLE

The distance d in feet that an object falls is related to the amount of time t in seconds that it falls. This relationship is given by the equation $d = 16t^2$.

What distance will an object fall in 3 seconds?

$d = 16t^2$ *Write the equation.*
$d = 16 \cdot (3)^2$ *Substitute 3 for t.*
$= 144$ *Simplify.*

The object will fall 144 feet in 3 seconds.

PRACTICE

A small rocket is shot vertically upward from the ground. The distance d in feet between the rocket and the ground as the rocket goes up can be found by using the equation $d = 128t - 16t^2$, where t is the amount of time in seconds that the rocket has been flying upward.

1. How far above the ground is the rocket at 1 second and at 2 seconds?
2. Did the rocket's distance change by the same amount in each of the first 2 seconds? Explain.
3. When the rocket is returning to the ground, the distance that the rocket falls is given by the equation $d = 16t^2$. If the rocket hits the ground 4 seconds after it starts to return, how far up did it go?
4. As the rocket falls to the ground, does it fall the same distance each second? Explain.

The graph for $y = x^2$ is shown at right. Use the graph for problems 5–7.

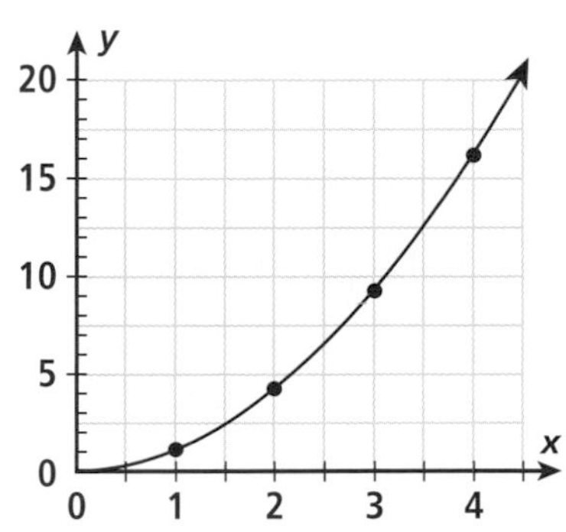

5. Find the value of y for $x = 1, 2, 3, 4$, and 5.
6. Does y increase by the same amount for each value of x? Explain.
7. How would the part of the graph from $x = 5$ to $x = 6$ compare to the part of the graph from $x = 4$ to $x = 5$?

Skills Bank

Selected Answers

Chapter 1

1-1 Exercises

1. population: all humpback whales; sample: the pod of humpback whales being studied **2.** population: all seventh-grade students; sample: the 25 students who are asked their ideas **3.** not random **4.** random **5.** population: all wild moose; sample: the 50 moose that are tagged **7.** population: all voters, sample: the voters who are polled **9.** not random **11.** use a sample **13.** survey the population **17. a.** students in grade 7 at your school who watch TV or who use a computer **19.** not random **23.** 88 **25.** 63 **27.** 11 **29.** 7 **31.** B

1-2 Exercises

1. 20; 20; 5 and 20; 30 **2.** 55; 54; 48; 24 **3.** 12; Adding the outlier increases the mean by 1. The median and the mode did not change. **5.** 14; 11; 11; 14 **7.** 151; Adding the outlier increased the mean by 10 and the median by 7.5. The mode did not change because there was no mode. **9.** $791; $1,822; $1,793 **11.** 9; 8; 12 **17.** 100 **19.** 820 **21.** 1,000 **23.** 6,100

1-3 Exercises

1.

Ages of American Presidents		
Age	Frequency	Cumulative Frequency
40–49	3	3
50–59	10	13
60–69	5	18

2. Ages of American Presidents

Stems	Leaves
4	6 8 9
5	0 1 2 4 6 7 7 7 7 8
6	1 1 4 5 8

Key: 5|2 means 52

3.

States with Drive-ins in 2000		
Age	Frequency	Cumulative Frequency
20–29	6	6
30–39	1	7
40–49	1	8
50–59	3	11

5. 4; 31; 27 **7.** b **9.** Chile and Surinam; Brazil

11. 6 Endangered Species in Each Country of South America

Stems	Leaves
0	5 5 6 6 6
1	1 3
2	0 1 3 4
3	5
4	
5	
6	4

Key: 2|3 means 23

13. 64; Adding the outlier increases the mean by approximately 3.8. **17.** no **19.** yes **21.** D

1-4 Exercises

1. grapes **2.** about 15 pounds

3.

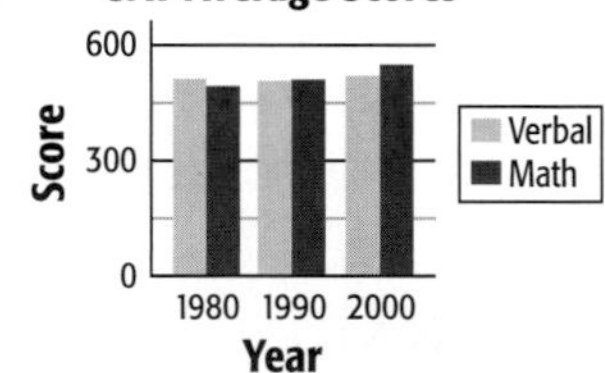

4.

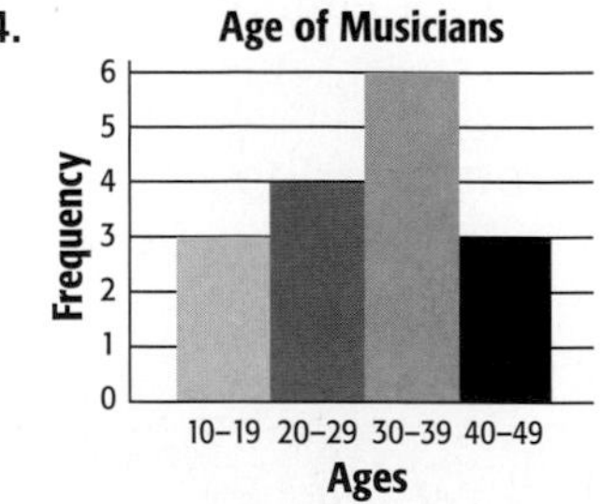

5. Florida

7.

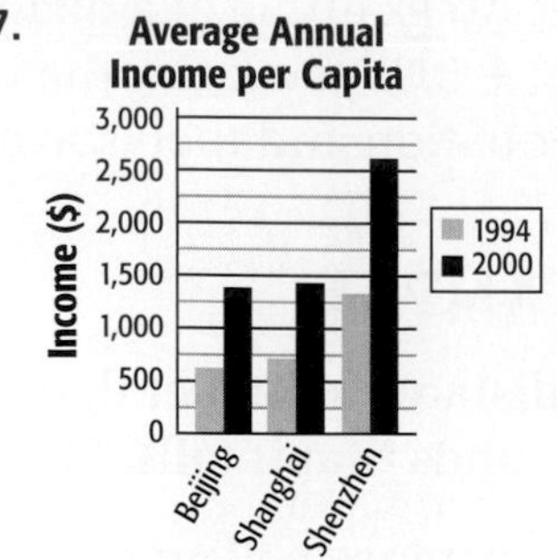

9.

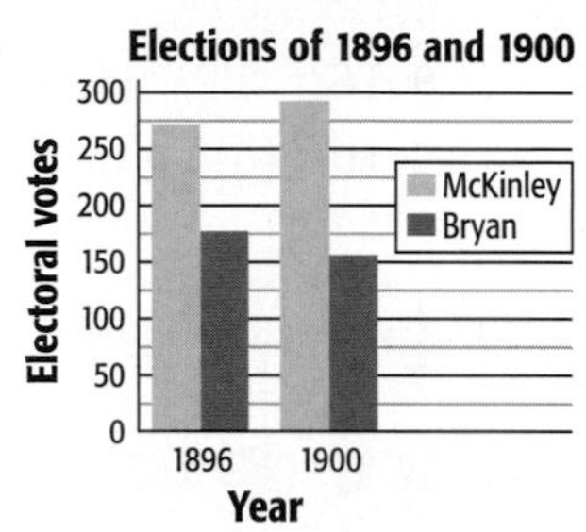

11. about 100

13.

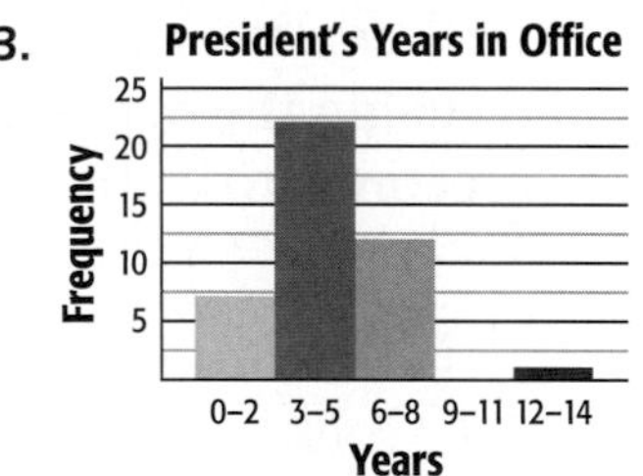

17. < **19.** 398, 402, 410, 417 **21.** 7°F, 8°F, 14°F, 41°F, 78°F

1-5 Exercises

1. outdoor **2.** 25% **3.** $50,000 **4.** bar graph **5.** circle graph **7.** 30% **9.** circle graph **11.** Asia, Africa, North America, South America, Antarctica, Europe, Australia **13.** about 25% **19.** 62 **21.** 4 **23.** 423

1-6 Exercises

1. 175 in. **2.** 45 in. **3.** $\frac{1}{4}$

4.

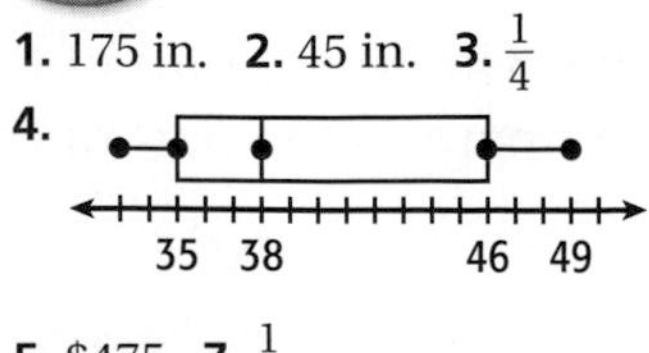

5. $475 **7.** $\frac{1}{2}$

9. a.

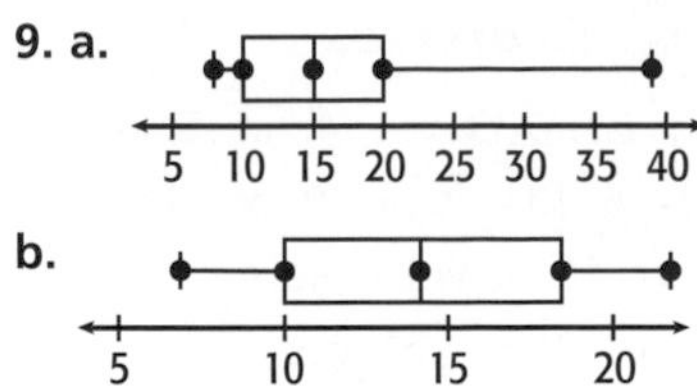

11. a. 60 in.
b.

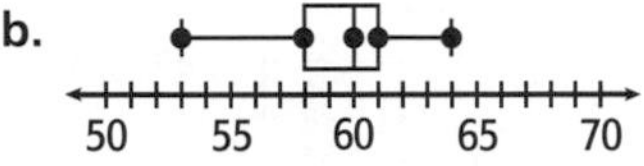

15. 730 **17.** 378 **19.** C

1-7 Exercises

1. 1990–1995
2. about $4.90
3.

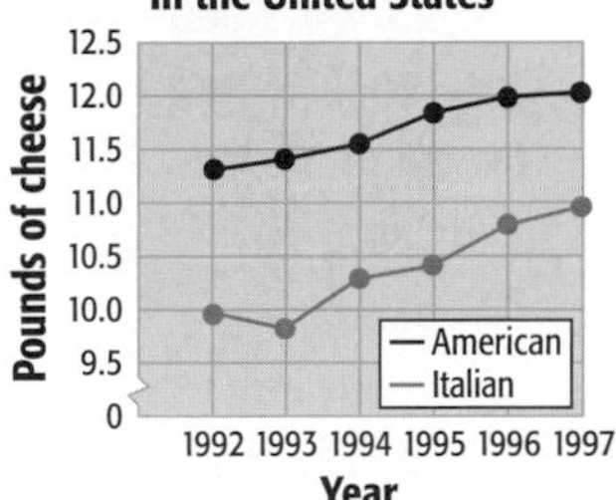

5. 1990–1995
7.

Normal Daily Temperature

Temperature (°F) 80 70 60 50 40 30 20

Peoria Portland

Jul Aug Sep Oct Nov Dec

Month

9. a. 0.9 million **13.** 168
15. 44,062 **17.** 5,000 + 40 + 7
19. 100,000 + 9,000 + 200 + 40 + 4

1-8 Exercises

1. The heart rate tends to decrease as the size increases. **2.** no correlation **3.** positive correlation
4. negative correlation
5. The scatter plot shows that the capacity increases with time, a positive correlation.
7. negative correlation
9. negative correlation
11. positive correlation
13. As latitude increases, temperature decreases.
17. 3 **19.** 3 **21.** A

1-9 Exercises

1. graph A **2.** The vertical axis is broken, so differences in weight appear greater. **3.** The vertical axis does not begin with zero, so differences in scale appear greater. **5.** The scale of the graph is not divided into equal intervals, so differences in sales appear less than they actually are.
7. a.

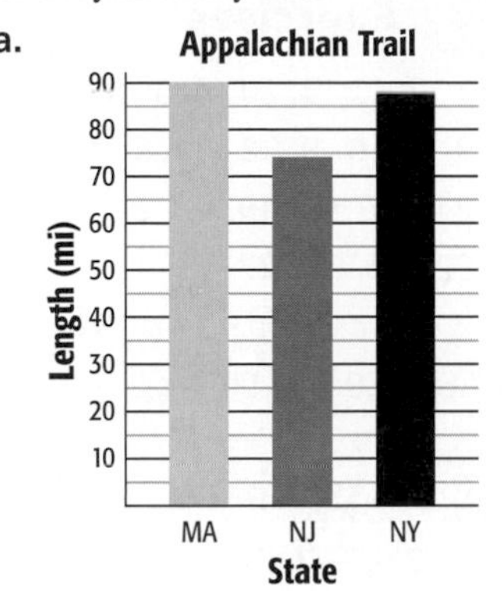

13. 120 **15.** 33 **17.** 606

Chapter 1 Study Guide and Review

1. population; sample
2. mean **3.** histogram; bar graph **4.** negative correlation
5. population: the students in the school; sample: the first 5 students **6.** 4; 4; 2 and 5; 8
7. 302; 311.5; 233 and 324; 166
8. 43; 46; none; 25
9.

	Frequency	Cumulative Frequency
0–9	1	1
10–19	3	4
20–29	3	7
30–39	2	9

10.

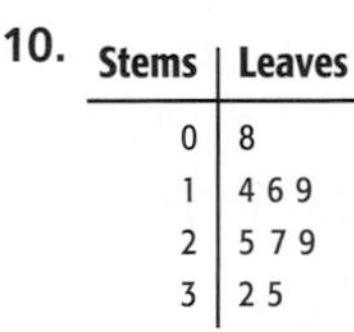

11.

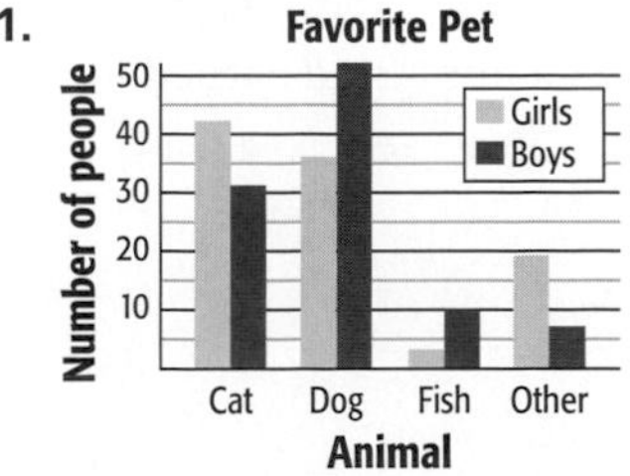

12. yellow **13.** 35 people
14.

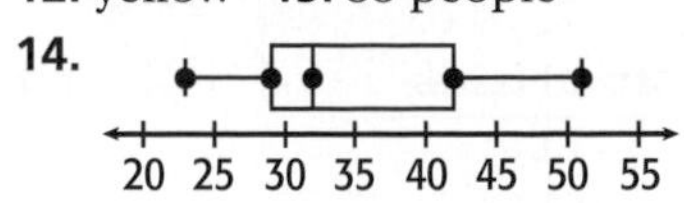

15.

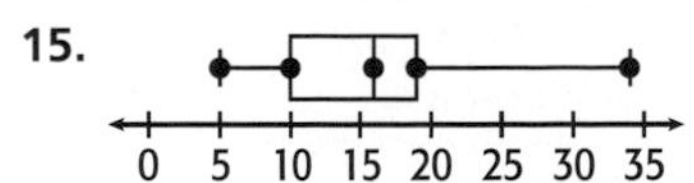

16.

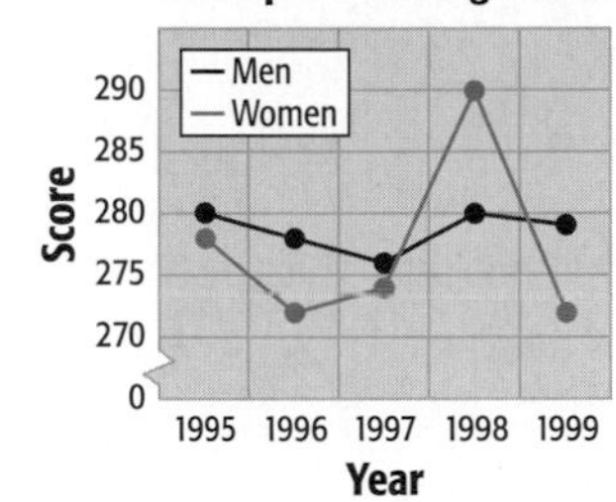

17. positive correlation
18. The temperature did not change that much, though the graph makes the changes look significant.

Chapter 2

2-1 Exercises

1. 32 **2.** 27 **3.** 36 **4.** 1 **5.** 5^2
6. 4^2 **7.** 3^3 **8.** 10^2 **9.** 100 times stronger **11.** 243 **13.** 1 **15.** 81
17. 5 **19.** 125 **21.** 10,000 **23.** 4^1
25. 7^0 **27.** 2^7 **29.** 50^2 **31.** 3^4 = 81 points **33.** 2^4 or 4^2 **35.** 9^3 or 3^6
37. 25 **39.** 108 **41.** 126 **43.** 3
45. 216 cubic inches **47.** Yuma: 484,152; Phoenix: 9,272,112 **49.** 4, 16, and 64 **53.** a **55.** 4.3 **57.** A

2-2 Exercises

1. 1,500 **2.** 18,000 **3.** 11
4. 120,000 **5.** 208,000
6. 1,130,000,000 **7.** 472,000
8. 362.2 **9.** 3.6×10^6 **10.** 2.14×10^5 **11.** 8.0×10^9 **12.** 4.2×10^4
13. 2,000,000,000,000,000,000,000
15. 80,000 **17.** 400,000 **19.** 19,000
21. 140 **23.** 105 **25.** 216.4
27. 1,020 **29.** 700,300 **31.** 1.61×10^6 **33.** 6.01×10^4 **35.** 2.98×10^8
37. 5×10^5 **39.** 9.8×10^8 feet per second **41.** 1.68 **43.** 5 **45.** 6.0; 7

47. 2.44; 8 **49.** 1.83×10^8 years

53.

Stems	Leaves
4	3 5 8
5	1 7
6	0 2 2

Key: 4|5 means 45

55. C

2-3 Exercises

1. 47 **2.** 30 **3.** 23 **4.** 7 **5.** 4 **6.** 40 **7.** $280 **9.** 42 **11.** 15 **13.** 73 **15.** 588 **17.** $139 **19.** $>$ **21.** $>$ **23.** $=$ **25.** $4 \cdot (8 - 3) = 20$ **27.** $(12 - 2)^2 \div 5 = 20$ **29.** $(4 + 6 - 3) \div 7 = 1$ **31.** $82 **33. a.** $4 \cdot 15$ **b.** $2 \cdot 30$ **c.** $126 **39.** C

2-4 Exercises

1. 2^4 **2.** $2 \cdot 3^3$ **3.** 3^4 **4.** $3 \cdot 5 \cdot 7$ **5.** $2 \cdot 3^2$ **6.** $2 \cdot 13$ **7.** $3^2 \cdot 5$ **8.** $2^4 \cdot 5$ **9.** $2 \cdot 5^2$ **10.** $2 \cdot 3^2 \cdot 5$ **11.** $2^2 \cdot 5^2$ **12.** $2^2 \cdot 3 \cdot 5$ **13.** $3^2 \cdot 7$ **14.** $2 \cdot 7$ **15.** $2^3 \cdot 5^3$ **16.** $2^2 \cdot 5 \cdot 7$ **17.** $2^2 \cdot 17$ **19.** $2^3 \cdot 3 \cdot 5$ **21.** $3^3 \cdot 5$ **23.** $2 \cdot 7 \cdot 11$ **25.** $2^5 \cdot 5^2$ **27.** 5^4 **29.** $3^2 \cdot 5 \cdot 7$ **31.** $3^3 \cdot 7$ **33.** $2 \cdot 11^2$ **35.** $11 \cdot 17$ **37.** $5^2 \cdot 7^2$ **39.** $2^3 \cdot 3^2 \cdot 5$ **41.** $2^4 \cdot 3^2$ **43.** $2^3 \cdot 5^2$ **45.** $2 \cdot 32$, $4 \cdot 16$, $8 \cdot 8$ **47.** 180 **49.** 448 **51.** 462 **53.** 117 **55.** 210 **57.** the chicken coop and the sheep pen **63.** 5.5 **65.** 23 **67.** 343 **69.** C

2-5 Exercises

1. 6 **2.** 9 **3.** 12 **4.** 3 **5.** 4 **6.** 10 **7.** 12 kits **9.** 12 **11.** 11 **13.** 38 **15.** 2 **17.** 26 **19.** 3 **21.** 1 **23.** 2 **25.** 22 **27.** 40 **29.** 1 **31.** 7 **33.** 3 **35.** 13 **37.** $2^2 \cdot 3 \cdot 11$ **39.** 4 sections

45.

Stems	Leaves
1	0 7 9
2	2 3 5 7 9
3	1 9

Key: 2|5 means 25

47. 13 **49.** C

2-6 Exercises

1. 28 **2.** 84 **3.** 48 **4.** 240 **5.** 45 **6.** 200 **7.** 24 minutes **9.** 24 **11.** 42 **13.** 120 **15.** 80 **17.** 180 **19.** 360 **21.** 60 minutes **23.** 10 **25.** 36 **27.** 108 **29.** 90 **31.** 12 **33.** 144 **35.** 60 **37.** 42 days **39.** no **43.** 1 **45.** 1,000 **47.** $1.0245 \cdot 10^2$ **49.** 769 **51.** D

2-7 Exercises

1. 12 **2.** 11 **3.** 20 **4.** 5 **5.** 8 **6.** 26 **7.** 19 **8.** 16 **9.** 22 **11.** 5 **13.** 11 **15.** 24 **17.** 12 **19.** 41 **21.** 10 **23.** 22 **25.** 24 **27.** 13 **29.** $2 \cdot 5 \cdot c \cdot c \cdot c \cdot d \cdot d$ **31.** $4.50 **33.** 86°F **39.** $3^2 \cdot 11$ **41.** $2^3 \cdot 7$ **43.** C

2-8 Exercises

1. $7p$ **2.** $n - 3$ **3.** $3(n + 5)$ **4.** $\$5 \div n$ or $\frac{5}{n}$ **5.** $46 + 21m$ **7.** $y - 2$ **9.** $9n$ **11.** $45 \div v$ **13.** $m + 6n$ **15.** $6k - 14$ **17.** $100 \div (6 + w)$ **19–23.** Possible answers given. **19.** h plus 3 **21.** s minus 405 **23.** the difference between 4 times p and 10 **25. a.** $17n$ **b.** $17 \cdot 24d$ **27.** $8m$ **33.** 93 **35.** Possible answer: 8^4 **37.** 43,800,000 **39.** B

2-9 Exercises

1. $6b$ and $\frac{b}{2}$, $5x^2$ and x^2 **2.** $12a^2$ and $4a^2$, $4x^3$ and $3.5x^3$, b and $\frac{5}{6}b$ **3.** $8x$ **4.** $5a^2 + 16$ **5.** There are no like terms. **6.** $10n + 12b$ **7.** b^6 and $3b^6$, $2b$ and b **9.** m and $2m$, 3^3 and 2 **11.** $8a + 2b$ **13.** $12b + 10$ **15.** $18 + 2d^3 + 4d$ **17.** $11a + 4n$ **19.** $27y$ **21.** $2d^2 + d$ **23.** no like terms **25.** no like terms **29.** $4n + 5n + 6n = 15n$; 15, 30, 45, 60, 75 **31. a.** $21.5d + 23d + 15.5d + 19d$; $79d$ **b.** $750.50 **c.** the amount Brad earned in June **37.** 3^5 **39.** 5^3 **41.** 9 **43.** 44

2-10 Exercises

1. no **2.** no **3.** yes **4.** no **5.** 88 cards **6.** $34 **7.** no **9.** no **11.** no **13.** no **15.** 53 video games **17.** no **19.** yes **21.** yes **23.** no **25.** no **27.** Possible answer: $6 - j = 3$ **29.** 318 mi/h **33.** 4 **35.** 142 **37.** $2^3 \cdot 7$ **39.** $2^2 \cdot 3^3$ **41.** C

2-11 Exercises

1. $r = 176$ **2.** $v = 168$ **3.** $x = 88$ **4.** $d = 9$ **5.** $f = 9$ **6.** $m = 971$ **7.** 14 yd **9.** $t = 82$ **11.** $m = 74$ **13.** $w = 43$ **15.** $x = 35$ **17.** $q = 99$ **19.** 2 books **21.** $n = 81$ **23.** $a = 45$ **25.** $f = 1{,}000$ **27.** $s = 159$ **29.** $z = 766$ **31.** $f = 14$ **33.** $m = \$39$ **37.** 262,144 **39.** 1,024 **41.** 14,700 **43.** C

2-12 Exercises

1. $s = 847$ **2.** $b = 100$ **3.** $y = 40$ **4.** $x = 9$ **5.** $c = 32$ **6.** $x = 1$ **7.** 9 people **9.** $k = 1{,}296$ **11.** $c = 175$ **13.** $n = 306$ **15.** $p = 21$ **17.** $a = 2$ **19.** $d = 45$ **21.** $g = 27$ **23.** $r = 12$ **25.** $b = 62$ **27.** $c = 716$ **29.** $d = 42$ **31.** $d = 12$ **33.** $x = \$20$ **35.** 12 toys **37.** $13,300 **41.** 10 **43.** $7x^2 + 6$ **45.** D

Chapter 2 Study Guide and Review

1. exponent; base **2.** numerical expression **3.** composite number **4.** algebraic expression **5.** 81 **6.** 10 **7.** 128 **8.** 1 **9.** 121 **10.** 14,400 **11.** 1,320 **12.** 220,000,000 **13.** 340 **14.** 560,000 **15.** 780 **16.** 3 **17.** 103 **18.** $2 \cdot 2 \cdot 2 \cdot 11$ **19.** $3 \cdot 3 \cdot 3$ **20.** $2 \cdot 3 \cdot 3 \cdot 3 \cdot 3$ **21.** $2 \cdot 2 \cdot 2 \cdot 2 \cdot 2 \cdot 3$ **22.** 30 **23.** 3 **24.** 12 **25.** 220 **26.** 60 **27.** 32 **28.** 27 **29.** 90 **30.** 12 **31.** 315 **32.** 19 **33.** 524 **34.** $4 \div (n + 12)$ **35.** $2(t - 11)$

36. $10b^2 + 8$ **37.** $15a^2 + 2$
38. $x^4 + x^3 + 6x^2$ **39.** yes **40.** no
41. no **42.** $b = 8$ **43.** $n = 32$
44. $c = 18$ **45.** $t = 112$ **46.** $n = 72$
47. $p = 9$ **48.** $d = 98$ **49.** $x = 13$

Chapter 3

3-1 Exercises

1. 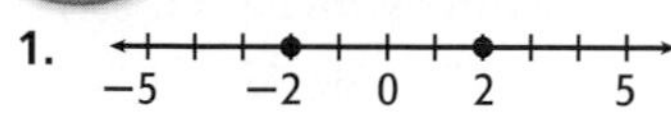

2. −10 −5 0 5 10

3. −5 −3 −1 1 3 5

4. −10 −5 0 5 10

5. −5, −3, −1, 4, 6 **6.** −8, −2, 1, 7, 8 **7.** −6, −4, 0, 1, 3 **8.** 2 **9.** 8 **10.** 7 **11.** 10

13. −10 −5 0 5 10

15. −10 −5 0 5 10

17. −9, −7, −5, −2, 0 **19.** 16 **21.** 20 **23.** < **25.** = **27.** = **29.** > **31.** > **33.** < **35.** 294 **37.** 45 **39.** −32 **41.** −282 **43. a.** baseball **b.** decreased by about 9% **c.** increased by about 27% **45.** second quarter **49.** 64 **51.** 1 **53.** 64 **55.** 256 **57.** 75 **59.** 4 **61.** 585 **63.** 0 **65.** C

3-2 Exercises

1. II **2.** IV **3.** III **4.** I
5–8.

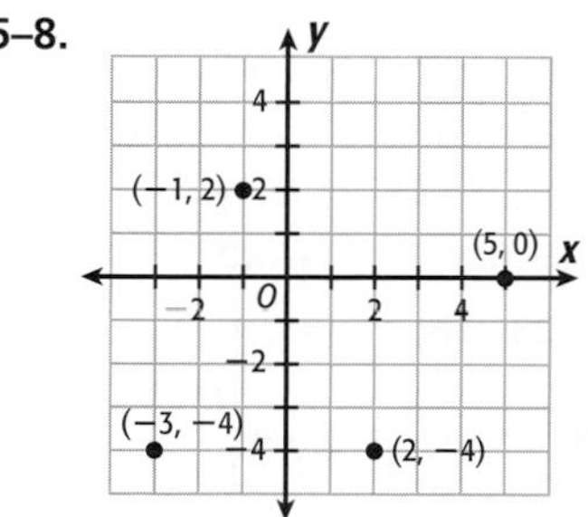

9. (6, −3) **10.** (−3, 2) **11.** (−4, 0) **12.** (5, 0) **13.** I **15.** IV **21.** (−4, 4) **23.** (−5, −4) **25.** (5, 6)

27.

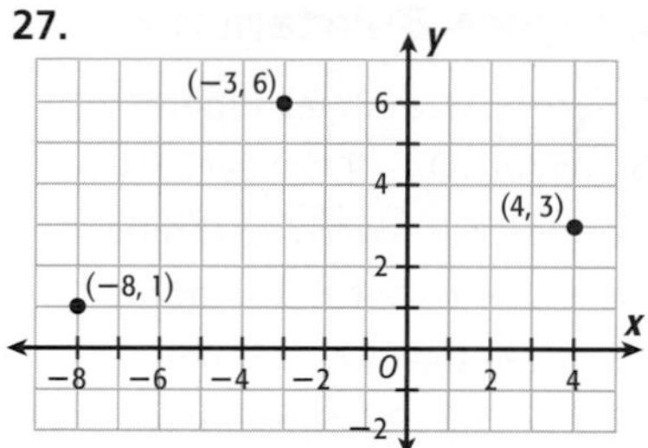

29. triangle; Quadrants I and II **31.** III **33.** II **35. a.** (−68°, 26°) **b.** (−80°, 26°) **c.** (−91°, 32°) **39.** 15 **41.** 87 **43.** C

3-3 Exercises

1. 12 **2.** −6 **3.** −2 **4.** 3 **5.** 15 **6.** −13 **7.** −15 **8.** 11 **9.** −12 **10.** 0 **11.** −20 **12.** −5 yards **13.** −9 **15.** 13 **17.** 7 **19.** −17 **21.** −19 **23.** −16 **25.** −88 **27.** −55 **29.** −14 **31.** −13 **33.** −13 **35.** −15 **37.** 27 **39.** −18 **41.** −29 **43.** = **45.** < **47.** = **49.** 7 **51.** 3 **53.** −9 **55.** 4,150 ft **57.** $4 **61.** 8.39×10^5 **63.** 4.023×10^9 **65.** A

3-4 Exercises

1. −3 **2.** −11 **3.** 6 **4.** −6 **5.** −4 **6.** 5 **7.** −10 **8.** −10 **9.** 7 **10.** −14 **11.** −14 **12.** 47°F **13.** −5 **15.** 8 **17.** 12 **19.** 16 **21.** −17 **23.** 8 **25.** 50 **27.** 18 **29.** 16 **31.** −5 **33.** −20 **35.** 83°F **37.** −14 **39.** −2 **41.** 2 **43.** 16 **45.** 0 **47.** 16 **49.** −27 **51.** −17 **53.** 1,266°F **55.** 531°F **57.** 16,500 ft **59.** 8 **61.** 22 **63.** 0 **65.** B

3-5 Exercises

1. −15 **2.** −10 **3.** −15 **4.** 15 **5.** 10 **6.** 15 **7.** −8 **8.** −6 **9.** 4 **10.** −7 **11.** 3 ft **13.** −10 **15.** 24 **17.** 32 **19.** 7 **21.** −8 **23.** −40 **25.** −3 **27.** 50 **29.** −120 **31.** 90 **33.** 16 **35.** −1 **37.** −24 **39.** 5 **41.** −26 **43.** 10 **45.** −60 ft **47.** The average change was $55 per day. **51.** $2x + 2y$ **53.** $8a^2$ **55.** $2y + 6$ **57.** D

3-6 Exercises

1. $w = 4$ **2.** $x = -12$ **3.** $k = -7$ **4.** $n = -8$ **5.** $y = -30$ **6.** $a = -60$ **7.** This year's loss is $57 million. **9.** $k = -3$ **11.** $v = -4$ **13.** $a = 20$ **15.** $t = -32$ **17.** $n = 150$ **19.** $l = -144$ **21.** $y = 100$ **23.** $j = -63$ **25.** $c = 17$ **27.** $y = -11$ **29.** $w = -41$ **31.** $x = -58$ **33.** $x = 4$ **35.** $t = 9$ **37.** $h = 20$ **39.** 60° **41. a.** $280 **b.** $240 **c.** $15 **43.** oceans or beaches **45.** C **49.** −10, −8, −3, 0, 7 **51.** −3, −1, 0, 2, 4 **53.** −9, −2, −1, 7, 10 **55.** B

3-7 Exercises

1. 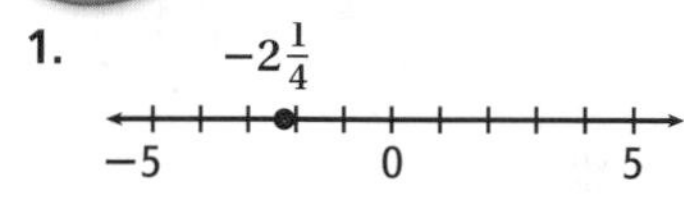

2. −5 0 3.25 5

3. −5 −1.5 0 5

4. $\frac{1}{2}$ −5 0 5

5. $-\frac{1}{4}$ **6.** $\frac{3}{4}$ **7.** $-\frac{1}{2}$ **8. a.** $-\frac{1}{2}$ in. **b.** 1.75 in.

9. −1 $-\frac{2}{3}$ $-\frac{1}{3}$ 0 $\frac{1}{3}$ $\frac{2}{3}$ 1

11. −1.1 −2.5 −1.5 −0.5 0 0.5 1.5 2.5

13. −1 $-\frac{3}{5}$ $-\frac{1}{5}$ 0 $\frac{1}{5}$ $\frac{2}{5}$ $\frac{3}{5}$ $\frac{4}{5}$ 1

15. $-\frac{3}{4}$ −5 0 5

17. $\frac{5}{4}$ **19.** $-\frac{1}{1}$ **21.** $\frac{0}{1}$ **23. a.** about $\frac{3}{2}$°C **b.** about −3.5°C

25. 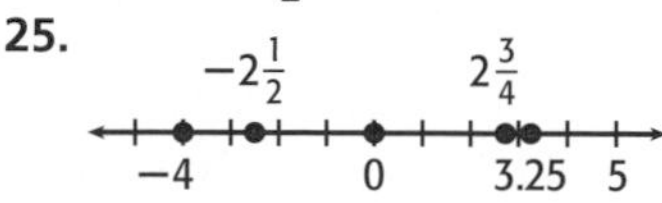

27–31. Possible answers given. **27.** $-\frac{1}{4}$ **29.** $\frac{3}{4}$ **31.** $\frac{1}{2}$ **37.** 4.5 **39.** 71 **41.** D

3-8 Exercises

1–25. Possible answers given. **1.** $\frac{2}{4}$ **2.** $\frac{12}{20}$ **3.** $\frac{30}{36}$ **4.** $\frac{45}{120}$ **5.** $\frac{24}{72}, \frac{54}{72}$; no **6.** $\frac{5}{6}, \frac{5}{6}$; yes **7.** $\frac{4}{3}, \frac{4}{3}$; yes **8.** $\frac{45}{24}, \frac{38}{24}$; no **9.** $3\frac{3}{4}$ **10.** $4\frac{2}{5}$ **11.** $1\frac{4}{13}$ **12.** $4\frac{2}{3}$ **13.** $\frac{31}{5}$ **14.** $\frac{23}{12}$ **15.** $\frac{38}{5}$ **16.** $\frac{39}{16}$ **17.** $\frac{3}{9}$ **19.** $\frac{9}{10}$ **21.** $\frac{9}{12}$ **23.** $\frac{18}{30}$ **25.** $\frac{1}{2}, \frac{1}{2}$; yes
27–31. Possible answers given. **27.** $\frac{5}{4}, \frac{5}{4}$; yes **29.** $\frac{2}{3}, \frac{2}{3}$; yes **31.** $\frac{6}{11}, \frac{7}{11}$; no **33.** $6\frac{1}{3}$ **35.** $7\frac{4}{11}$ **37.** $\frac{128}{5}$ **39.** $\frac{29}{3}$ **41–51.** Possible answers given. **41.** $\frac{10}{2}$ **43.** $\frac{52}{8}$ **45.** $\frac{32}{84}$ **47.** $\frac{11}{2}$ **49.** $\frac{202}{2}$ **51.** $\frac{19}{3}$ **53.** $\frac{21}{35}, \frac{3}{5}$ **55.** $\frac{2}{3}, \frac{20}{30}$ **57.** $2\frac{1}{12}$ ft **59.** $30\frac{2}{12}$ or $30\frac{1}{6}$ ft **61.** $80\frac{5}{12}$ ft **63.** $\frac{150}{4}$ **69.** 36 **71.** 100,000 **73.** 4.75×10^5 **75.** 8.8×10^1 **77.** D

3-9 Exercises

1. 0.6 **2.** 2.63 **3.** 1.83 **4.** 0.78 **5.** $\frac{1}{125}$ **6.** $-\frac{3}{5}$ **7.** $-2\frac{1}{20}$ **8.** $3\frac{3}{4}$ **9.** 0.720 **11.** 6.4 **13.** 0.88 **15.** 1 **17.** 1.92 **19.** $\frac{1}{100}$ **21.** $-\frac{2}{25}$ **23.** $\frac{61}{4}$, $15\frac{1}{4}$ **25.** $\frac{67}{8}, 8\frac{3}{8}$ **27–33.** Possible answers given. **27.** 8.75, $8\frac{6}{8}$ **29.** $5\frac{1}{20}, 5\frac{5}{100}$ **31.** $15\frac{7}{20}, \frac{307}{20}$ **33.** 4.003, $\frac{4,003}{1,000}$ **35.** yes **37.** no **39.** yes **41.** no **43.** $17\frac{9}{10}, 18\frac{1}{20}, 18\frac{1}{25}, 18\frac{11}{20}$ **47.** 32 **49.** 36 **51.** 9 **53.** 8 **55.** B

3-10 Exercises

1. < **2.** > **3.** < **4.** > **5.** < **6.** > **7.** < **8.** > **9.** 0.505, 0.55, $\frac{3}{4}$ **10.** 2.05, 2.5, $\frac{13}{5}$ **11.** −0.875, $\frac{5}{8}$, 0.877 **13.** > **15.** > **17.** < **19.** < **21.** < **23.** < **25.** < **27.** 1.6, $1\frac{4}{5}$, 1.82 **29.** −3.02, −3, $1\frac{1}{2}$ **31.** $\frac{4}{5}$, 0.82, $\frac{5}{6}$ **33.** $-1\frac{2}{5}$, −1.20, −1.02 **35.** $\frac{3}{4}$ **37.** $\frac{7}{8}$ **39.** 0.32 **41.** $-\frac{7}{8}$ **43.** $\frac{22}{24}$, 0.917 **45.** Saturn, Jupiter and Uranus, Neptune, Pluto, Mars, Venus, Mercury, Earth **51.** 1.08×10^2 **53.** 1×10^7 **55.** $a = 28.4$ **57.** $c = 320$

Chapter 3 Extension

1. terminates **3.** terminates **5.** repeats, $0.\overline{1}$ **7.** repeats, $0.\overline{01}$ **9.** repeats, $0.\overline{142857}$ **11.** repeats, no visible pattern **13.** rational **15.** irrational **17.** rational **19.** rational **21.** irrational **23.** irrational

Chapter 3 Study Guide and Review

1. rational number; integer; terminating decimal **2.** ordered pair; x-axis **3.** −6, −2, 0, 4, 5 **4.** 0 **5.** 17 **6.** 6 **7.** I **8.** III **9.** II **10.** IV
7–10.

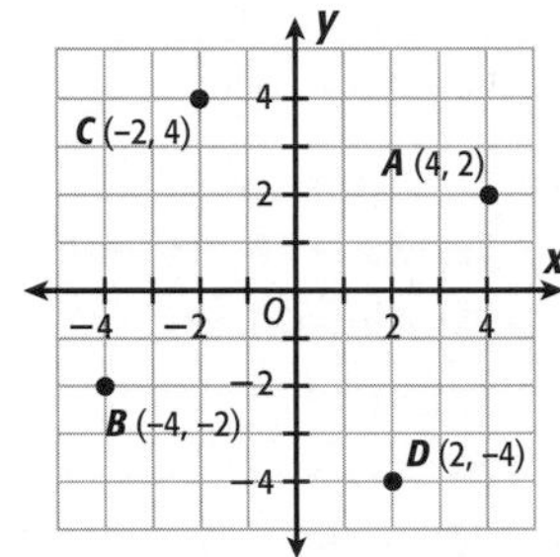

11. −3 **12.** 1 **13.** −56 **14.** 9 **15.** 14 **16.** −6 **17.** 6 **18.** −9 **19.** −1 **20.** −9 **21.** −50 **22.** 3 **23.** 16 **24.** −2 **25.** −12 **26.** −3 **27.** $y = 10$ **28.** $d = 14$ **29.** $j = -26$ **30.** $n = 72$ **31.** $c = 13$ **32.** $m = -4$
33–35. −4.25 5.5 −10 −3.25 0 10

36. $\frac{12}{1}$ **37.** $\frac{1}{4}$ **38.** $-\frac{37}{10}$ **39.** $\frac{21}{5}$ **40.** $\frac{19}{6}$ **41.** $\frac{43}{4}$ **42.** $3\frac{1}{3}$ **43.** $2\frac{1}{2}$ **44.** $2\frac{3}{7}$ **45.** $\frac{1}{4}$ **46.** $-\frac{1}{250}$ **47.** $\frac{1}{20}$ **48.** 3.5 **49.** −0.6 **50.** 0.667 **51.** < **52.** > **53.** >

Chapter 4

4-1 Exercises

1–6. Possible answers given. **1.** 63 **2.** 12 **3.** 2 **4.** 240 **5.** −225 **6.** 7 **7.** no, 36 • 10 = 360 **9.** 92 **11.** 8 **13.** 55
15–35. Possible answers given. **15.** 5 **17.** −120 **19.** 9 **21.** −7 **23.** −59 **25.** −90 **27.** −36 **29.** 11 **31.** −98 **33.** 225 **35.** 13 **37.** about 8 weeks **39. a.** English **b.** 36% **41.** yes **45.** −17, −12, −3, 6, 9 **47.** −5, −3, −1, 2, 4 **49.** 10 **51.** −35 **53.** A

4-2 Exercises

1. 21.82 **2.** −18.52 **3.** 12.826 **4.** 1.98 **5.** 9.65 **6.** 1.77 **7.** \$37.2 billion **9.** 18.97 **11.** 15.33 **13.** −25.52 **15.** 9.01 **17.** 4.47 **19.** 31.15 **21.** 77.13 g **23.** −4.883 **25.** −125.55 **27.** 30.12 **29.** 661.902 **31.** −3.457 **33.** 0.9 g/mL **39.** 7 **41.** −6 **43.** $w = 19$ **45.** $p = -66$ **47.** A

4-3 Exercises

1. −3.6 **2.** 0.6 **3.** 0.18 **4.** 2.04 **5.** 1.04 **6.** −0.315 **7.** 334.7379 miles **9.** 0.35 **11.** 29.4 **13.** 90.73 **15.** 4.48 **17.** −0.1542 **19.** −40.945 **21.** 5.445 mi **23.** 0.0021 **25.** −0.0876 **27.** 1.911 **29.** −0.7728 **31.** 21.5124 **33.** −83.538 **35.** −6,692.985 **37. a.** about 47 million people **b.** about 60 million people **43.** 35 **45.** −25 **47.** −24 **49.** C

4-4 Exercises

1. 6.14 **2.** 3.06 **3.** −3.09 **4.** −2.56 **5.** 0.017 **6.** 0.234 **7.** \$5.54 **9.** −8.9 **11.** −8.92 **13.** −4.8 **15.** 2.04 **17.** 1.13 **19.** −3.07 **21.** \$9.75 **23.** 1.56 **25.** 4.19 **27.** −2.8 **29.** −1.91 **31.** −0.019 **33.** 6.49 **35.** 1.44 **37.** −262.113 **39.** 93.295 **41.** 14.53 million people **43.** \$72.33 **47.** −5 **49.** 38 **51.** −23 **53.** 31 **55.** A

4-5 Exercises

1. 0.9 **2.** −35 **3.** 4.6 **4.** 5.3 **5.** −3.2 **6.** −12.24 **7.** 2.5 **8.** 35 **9.** −16 **10.** −20 **11.** −4.8 **12.** 22.5 **13.** 28 mi/gal **15.** −0.12 **17.** −14 **19.** 4.2 **21.** 47.5 **23.** 4 **25.** −48.75 **27.** 2.4 min **29.** 22.5

31. −0.4 **33.** 25 **35.** 5.9 **37.** 16 **39.** −5.8 **41.** 253.2 **43.** −27 **45.** 240 days **47.** 202.228 m **51.** $2^3 \cdot 11$ **53.** $3^2 \cdot 5^2$ **55.** −3.328 **57.** 0.4806 **59.** D

4-6 Exercises

1. $w = 7$ **2.** $x = 7.85$ **3.** $k = 24.09$ **4.** $n = 21.25$ **5.** $b = 5.04$ **6.** $x = 5.76$ **7.** $t = 9$ **8.** $y = 1.5$ **9.** \$4.25 **11.** $c = 44.56$ **13.** $a = 5.08$ **15.** $p = -53.21$ **17.** $z = 16$ **19.** $w = 11.76$ **21.** $a = -74.305$ **23.** \$7.50 **25.** $n = -4.92$ **27.** $r = 0.72$ **29.** $m = -0.15$ **31.** $k = 0.9$ **33.** $t = 0.936$ **35.** $v = -2$ **37.** $n = 12.254$ **39. a.** 148.1 million **b.** between Italian and English **41.** 7.14 g/cm^3 **45.** 6 **47.** 141 **49.** $-3, \frac{3}{5}, 3.5$ **51.** $-2, -1\frac{1}{4}, 1.4$ **53.** A

4-7 Exercises

1. $2\frac{1}{2}$ hr **2.** −6 **3.** $\frac{2}{5}$ **4.** $-\frac{1}{6}$ **5.** 14 **6.** $\frac{12}{5}$ or $2\frac{2}{5}$ **7.** $\frac{7}{3}$ or $2\frac{1}{3}$ **9.** $1\frac{2}{3}$ tsp **11.** $\frac{1}{2}$ **13.** 4 **15.** $\frac{1}{4}$ **17.** $-\frac{5}{9}$ **19.** $\frac{222}{5}$ or $44\frac{2}{5}$ **21.** $\frac{35}{2}$ or $17\frac{1}{2}$ **23.** $\frac{7}{3}$ or $2\frac{1}{3}$ **25.** $\frac{33}{4}$ or $8\frac{1}{4}$ **27.** $\frac{155}{42}$ or $3\frac{29}{42}$ **29.** $\frac{1}{3}$ **31.** $-\frac{1}{6}$ **33.** $\frac{1}{12}$ **35.** $\frac{1}{5}$ **37.** $\frac{7}{10}$ **39.** $\frac{1}{5}$ **41.** 1 **43.** 5 **45.** $2\frac{1}{12}$ lb **47.** $11\frac{1}{3}$ mi **53.** 3.43×10^6 **55.** −27 **57.** 60

4-8 Exercises

1. 18 **2.** $\frac{4}{5}$ **3.** $\frac{3}{32}$ **4.** $\frac{1}{4}$ **5.** $\frac{5}{4}$ or $1\frac{1}{4}$ **6.** 2 **7.** 3 capes **9.** 18 **11.** $\frac{32}{9}$ or $3\frac{5}{9}$ **13.** $\frac{3}{20}$ **15.** $\frac{5}{14}$ **17.** $\frac{13}{32}$ **19.** $\frac{77}{24}$ or $3\frac{5}{24}$ **21.** 6 pieces **23.** $\frac{3}{4}$ **25.** $\frac{60}{7}$ or $8\frac{4}{7}$ **27.** $\frac{8}{147}$ **29.** $\frac{16}{25}$ **31.** $\frac{18}{25}$ **33.** $\frac{21}{2}$ or $10\frac{1}{2}$ **35.** $-\frac{4}{9}$ **37.** $34\frac{1}{4}$ mi **39.** 42 side pieces **41.** $11\frac{3}{4}$ hours **45.** $2^6 \cdot 5$ **47.** B

4-9 Exercises

1. about 4 feet **2.** $1\frac{1}{2}$ **3.** 0 **4.** $5\frac{1}{2}$ **5.** exact **6.** estimate **7.** 3 feet **9.** $4\frac{1}{2}$ **11.** $\frac{1}{2}$ **13.** $11\frac{1}{2}$ **15.** exact **17.** $1\frac{1}{2}$ **19.** 4 **21.** 1 **23.** $18\frac{1}{2}$ **25.** 10 **27.** 5 **29.** $10\frac{1}{2}$ miles **35.** $5b + 2a$ **37.** $6z - 5y$ **39.** B

4-10 Exercises

1. $\frac{1}{3}$ **2.** $\frac{1}{6}$ **3.** $\frac{3}{7}$ **4.** $\frac{1}{2}$ **5.** $\frac{3}{20}$ **6.** $\frac{19}{24}$ **7.** $\frac{1}{12}$ **9.** $\frac{1}{2}$ **11.** $\frac{1}{6}$ **13.** $\frac{1}{3}$ **15.** $\frac{1}{4}$ **17.** $\frac{3}{8}$ **19.** $\frac{3}{14}$ **21.** $\frac{1}{6}$ mi **23.** $\frac{13}{18}$ **25.** $\frac{22}{21}$ or $1\frac{1}{21}$ **27.** $\frac{23}{20}$ or $1\frac{3}{20}$ **29.** $-\frac{1}{20}$ **31.** $\frac{9}{56}$ **33.** 0 **35.** 0 **37.** $\frac{3}{8}$ cup **39.** $\frac{3}{8}$ measure **41. a.** $\frac{1}{16}$ lb **b.** 1 oz **47.** 8 **49.** $y = 7$

4-11 Exercises

1. $5\frac{1}{6}$ ft **2.** $7\frac{3}{5}$ **3.** $6\frac{5}{8}$ **4.** $6\frac{1}{3}$ **5.** $1\frac{1}{3}$ **6.** $5\frac{1}{3}$ **7.** $\frac{11}{12}$ **9.** 15 **11.** $5\frac{2}{3}$ **13.** $15\frac{1}{40}$ **15.** $\frac{6}{7}$ **17.** $5\frac{1}{4}$ **19.** $\frac{17}{24}$ **21.** $15\frac{8}{15}$ **23.** $13\frac{5}{6}$ **25.** $29\frac{1}{6}$ **27.** $8\frac{3}{4}$ **29.** $4\frac{1}{6}$ **31.** $4\frac{1}{3}$ **33.** $<$ **35.** $>$ **37.** $\frac{97}{100}$ **39.** the waterfall trail; $\frac{1}{6}$ mi **43.** 5 **45.** −4 **47.** 0.625 **49.** $0.\overline{2}$ **51.** B

4-12 Exercises

1. $a = \frac{3}{4}$ **2.** $m = \frac{2}{3}$ **3.** $p = \frac{3}{2}$ or $1\frac{1}{2}$ **4.** $x = 40$ **5.** $r = \frac{9}{10}$ **6.** $w = \frac{1}{7}$ **7.** $\frac{9}{8}$ c or $1\frac{1}{8}$ c **9.** $t = \frac{5}{8}$ **11.** $x = \frac{53}{24}$ or $2\frac{5}{24}$ **13.** $y = \frac{7}{60}$ **15.** $w = \frac{1}{2}$ **17.** $z = \frac{1}{12}$ **19.** $n = \frac{48}{25}$ or $1\frac{23}{25}$ **21.** $t = \frac{1}{4}$ **23.** $w = 6$ **25.** $x = \frac{3}{5}$ **27.** $n = \frac{12}{5}$ or $2\frac{2}{5}$ **29.** $y = \frac{1}{2}$ **31.** $r = \frac{1}{77}$ **33.** $h = -\frac{1}{12}$ **35.** $v = \frac{3}{4}$ **37.** $d = \frac{577}{40}$ or $14\frac{17}{40}$ **39.** $11\frac{3}{16}$ lb **41.** 15 million species **43.** 48 stories **49.** $2\frac{1}{5}$ **51.** $4\frac{1}{3}$ **53.** $1\frac{4}{7}$

Chapter 4 Study Guide and Review

1. compatible numbers **2.** reciprocals **3.** 110 **4.** 5 **5.** 75 **6.** 4 **7.** 27.88 **8.** −51.2 **9.** 6.22 **10.** 52.902 **11.** 14.095 **12.** 35.88 **13.** 3.5 **14.** −38.7 **15.** 40.495 **16.** 60.282 **17.** 77.348 **18.** −18.81 **19.** 2.3 **20.** −4.9 **21.** 0.08 **22.** −5.8 **23.** −1.65 **24.** 3.4 **25.** 4.5 **26.** −1.09 **27.** −15.4 **28.** −500 **29.** 2 **30.** 4 **31.** $x = -10.44$ **32.** $s = 107$ **33.** $n = 0.007$ **34.** $k = 8.64$ **35.** $7\frac{1}{2}$ **36.** $1\frac{21}{25}$ **37.** $17\frac{17}{63}$ **38.** $6\frac{1}{4}$ **39.** $\frac{4}{75}$ **40.** $\frac{2}{15}$ **41.** 1 **42.** $1\frac{11}{12}$ **43.** 24 **44.** $2\frac{1}{2}$ **45.** −8 **46.** $22\frac{1}{2}$ **47.** 3 **48.** 1 **49.** $\frac{5}{12}$ **50.** $\frac{17}{20}$ **51.** $\frac{5}{11}$ **52.** $\frac{1}{9}$ **53.** $6\frac{5}{24}$ **54.** $3\frac{1}{3}$ **55.** $6\frac{1}{4}$ **56.** $1\frac{5}{12}$ **57.** $1\frac{2}{3}$ **58.** $\frac{1}{15}$ **59.** $1\frac{5}{7}$ **60.** $\frac{13}{28}$

Chapter 5

5-1 Exercises

1. $\frac{10}{3}$, 10 to 3, 10:3 **2.** Possible answer: $\frac{3}{30}$, 3 to 30, 3:30 **3.** $\frac{5 \text{ mi}}{1 \text{ h}}$; 5 miles per hour **4.** $\frac{12 \text{ points}}{1 \text{ game}}$; 12 points per game **5.** the 10-lb box **7.** $\frac{25}{70}$, 25 to 70, 25:70 **9.** $\frac{45}{25}$, 45 to 25, 45:25 **11.** $\frac{\$7}{1 \text{ CD}}$; \$7 per CD **13.** Andre **15.** 2:1, $\frac{2}{1}$, 2 to 1; 3:1, $\frac{3}{1}$, 3 to 1; 4:1, $\frac{4}{1}$, 4 to 1; 3:2, $\frac{3}{2}$, 3 to 2; 4:3, $\frac{4}{3}$, 4 to 3 **19.** $c = 15$ **21.** $m = 5$ **23.** Possible answer: $\frac{6}{10}, \frac{9}{15}, \frac{12}{20}$ **25.** Possible answer: $\frac{8}{14}, \frac{12}{21}, \frac{16}{28}$ **27.** $\frac{5}{6}$ **29.** $\frac{1}{2}$

5-2 Exercises

1. yes **2.** no **3.** yes **4.** no **5.** yes **6.** yes **7.** no **8.** yes **9–12.** Possible answers given. **9.** $\frac{1}{3} = \frac{2}{6}$ **10.** $\frac{9}{21} = \frac{3}{7}$ **11.** $\frac{8}{3} = \frac{16}{6}$ **12.** $\frac{10}{4} = \frac{5}{2}$ **13.** no **15.** no **17.** no **19.** no **21–27.** Possible answers given. **21.** $\frac{5}{9} = \frac{10}{18}$ **23.** $\frac{6}{15} = \frac{2}{5}$ **25.** $\frac{11}{13} = \frac{55}{65}$ **27.** $\frac{78}{104} = \frac{39}{52}$ **29.** 3, 8, 24, 15 **31–37.** Possible answers given. **31.** 6 to 14, 9 to 21 **33.** $\frac{10}{24}, \frac{15}{36}$ **35.** 2 to 3, 12 to 18 **37.** 5:2, 20:8 **39. a.** $\frac{1 \text{ can}}{4 \text{ hours}}$ **b.** no **41.** 176 students **47.** $2^3 \cdot 3 \cdot 5$ **49.** $2^4 \cdot 3$ **51.** 2 **53.** 33 **55.** −6 **57.** 5

5-3 Exercises

1. $x = 60$ **2.** $p = 8.75$ **3.** $m = 16.4$ **4.** $t = 21$ **5.** 3 lb **7.** $h = 144$ **9.** $v = 336$ **11.** $t = 36$ **13.** $n = 22.4$ **15.** 227 grams **17–33.** Possible answers given. **17.** $x = 2; \frac{4}{30}$ **19.** $w = 18; \frac{11}{12}$ **21.** $x = 90; \frac{2}{10}$ **23.** $q = 17.4; \frac{26.1}{6}$ **25.** $k = 324; \frac{1}{4}$ **27.** $j = 16.5; \frac{10}{33}$ **29.** $\frac{4}{10} = \frac{6}{15}$ **31.** $\frac{3}{75} = \frac{4}{100}$ **33.** $\frac{5}{6} = \frac{90}{108}$ **35.** 3 hours **37.** 105 oxygen atoms **39.** 3 km **43.** $y = 3$ **45.** $x = 13$ **47.** 15.03 **49.** D

5-4 Exercises

1. 4 lb **2.** 2,000 mL **3.** $\frac{384 \text{ in.}}{1 \text{ s}}$ **4.** $\frac{\$5.25}{1 \text{ yd}}$ **5.** 2,291,030 m **7.** 4 ft **9.** $\frac{\$0.15}{1 \text{ min}}$ **11.** $\frac{1 \text{ ft}}{12 \text{ in.}}$ **13.** $\frac{1 \text{ h}}{60 \text{ min}}$ **15.** 3.5 ft **17.** 2,000 m **19.** 4.5 gallons **21.** 6 minutes faster **23.** 6.8 stone **29.** Possible answer: −540 **31.** $\frac{7}{3}$ or $2\frac{1}{3}$ **33.** $\frac{13}{27}$ **35.** D

5-5 Exercises

1. similar **2.** not similar **3.** similar **4.** not similar **5.** not similar **7.** similar **9.** 4 in. × 6 in., 16 in. × 24 in., 20 in. × 30 in. **11.** not similar **13. a.** similar **b.** not similar **c.** similar **15.** no **21.** 9.13 **23.** $\frac{8 \text{ min}}{1 \text{ mi}}$; 8 minutes per mile **25.** $\frac{2 \text{ c}}{1 \text{ serving}}$; 2 cups per serving

5-6 Exercises

1. $a = 22.5$ cm **2.** $y = 56$ m **3.** 28 ft **4.** 25 ft **5.** $x = 13.5$ in. **7.** 3.9 ft **9.** 36 m **15.** $y = 11$ **17.** $a = 0$ **19.** −30.72 **21.** 0.1035

5-7 Exercises

1. $\frac{1}{14}$ **2.** 67.2 cm tall, 40 cm wide **3.** 129 mi **5.** 135 in. **7.** 16 in. **9.** height: $2\frac{2}{3}$ in., length: $5\frac{3}{4}$ in. **11.** $4\frac{7}{24}$ in. **13.** 0.81 in., or about $\frac{13}{16}$ in. **17.** −1.2, 0.2, $\frac{1}{4}$ **19.** 0.064 **21.** 2.7 **23.** C

Chapter 5 Study Guide and Review

1. similar **2.** ratio; rate or unit rate **3.** scale factor **4.** indirect measurement **5.** $\frac{6}{10}$, 6 to 10, 6:10 **6.** $\frac{10}{16}$, 10 to 16, 10:16 **7.** $\frac{2 \text{ gal}}{1 \text{ min}}$; 2 gallons per minute **8.** $\frac{3.5 \text{ mi}}{1 \text{ min}}$; 3.5 miles per minute **9.** no **10.** no **11.** yes **12.** no **13.** no **14.** yes **15–17.** Possible answers given. **15.** $\frac{10}{12} = \frac{5}{6}$ **16.** $\frac{45}{50} = \frac{9}{10}$ **17.** $\frac{9}{15} = \frac{3}{5}$ **18.** $n = 2$ **19.** $a = 6$ **20.** $b = 4$ **21.** $x = 66$ **22.** $y = 22$ **23.** $k = 100$ **24.** 40 qt **25.** 1.5 gal **26.** 2.2 mi **27.** 45 ft **28.** similar **29.** not similar **30.** $x = 100$ ft **31.** 12.1 in. **32.** 148 mi

Chapter 6

6-1 Exercises

1. $\frac{1}{10}$ **2.** $\frac{9}{20}$ **3.** $\frac{3}{5}$ **4.** $\frac{7}{25}$ **5.** 0.85 **6.** 0.3 **7.** 0.09 **8.** 1.0 **9.** 18% **10.** 40% **11.** 75% **12.** 3% **13.** 40% **14.** 25% **15.** 14% **16.** 12.5% **17.** $\frac{3}{10}$ **19.** $\frac{22}{25}$ **21.** 0.16 **23.** 1.05 **25.** 21% **27.** 8% **29.** 30% **31.** 27.5% **33.** $\frac{2}{25}$, 0.08 **35.** $\frac{1}{40}$, 0.025 **37.** $\frac{3}{400}$, 0.0075 **39.** $\frac{703}{1,000}$, 0.703 **41.** 90% **43.** 33.3% **45.** 0.1% **47.** 270% **49.** $>$ **51.** $<$ **53.** $\frac{3}{50}$, 0.06, 6% **55.** 6.8% > 6.2% **59.** 144 **61.** 512 **63.** C

6-2 Exercises

1–4. Possible answers given. **1.** 30 **2.** 40 **3.** 5 **4.** 15 **5.** Yes **6–17.** Possible answers given. **6.** 4 **7.** 8 **8.** 14 **9.** 7.7 **10.** $3.45 **11.** 26 **13.** 70 **15.** 216 **17.** 13 **19.** Betty's Boutique **21–35.** Possible answers given. **21.** 24 **23.** 12 **25.** 12 **27.** 24 **29.** 60 **31.** 3 **33.** 72 **35.** 15 **37.** about 26 oz **39.** about 16 hits **41.** Possible answer: 600 people **47.** 1.2 **49.** 0.375 **51.** C

6-3 Exercises

1. 24 **2.** 152 **3.** 20 **4.** 162 **5.** 8 **6.** 14 **7.** 423 **8.** 30 **9.** 171 students **11.** 11.2 **13.** 3,540 **15.** 0.04 **17.** 18 **19.** 13 **21.** 1.74 **23.** 39.6 **25.** 12.4 **27.** 6 **29.** 4.5 **31.** 11.75 **33.** 5,125 **35.** 80 **37.** 120 **39.** 0.61 **41.** 4.21 **43.** 2.25 grams **45.** 0.98 **51.** $x = 40$ **53.** $b = 1.1$ **55.** $\frac{\$1.75}{1 \text{ card}}$; $1.75 per card **57.** B

6-4 Exercises

1. 25% **2.** 40 **3.** 18% **4.** 200 **5.** 60 **6.** 80% **7.** 50 **8.** 90% **9.** 40% **11.** $33\frac{1}{3}$% **13.** 30 **15.** 100% **17.** 225 **19.** 300% **21.** 22 **23.** $235.00 **25.** 17.6 **27.** 63.2% **29.** 644.4 **31.** 62.5% **33.** 45 pieces **35.** about 700 words **39.** 25 **41.** $5\frac{5}{7}$ **43.** B

6-5 Exercises

1. 28% **2.** 133.3% **3.** 16.1% **4.** 242.9% **5.** $8.60; $34.39 **6.** $24.75 **7.** 37.5% **9.** 366.7% **11.** 22.2% **13.** $9.75, $55.25 **15.** $199.75 **17.** 100% **19.** 43.6% **21.** 30 **23. a.** $41,500 **b.** $17,845 **c.** 80.7% **25. a.** about 22,134 trillion Btu **b.** about 8,506 trillion Btu **27.** 64 pt **29.** 40 yd **31.** $\frac{1}{50}$ **33.** $\frac{12}{25}$

6-6 Exercises

1. $I = \$24$ **2.** $I = \$10$ **3.** $P = \$400$ **4.** $r = 3\%$ **5.** just over 4 years **7.** $I = \$3,240$ **9.** $P = \$2,200$ **11.** $r = 11\%$ **13.** almost 9 years **15.** $5,200 **17.** $212.75 **19.** 20 years **21.** $416.16 **25.** 5% **27.** 14% **29.** A

Chapter 6 Study Guide and Review

1. interest; simple interest; principal **2.** percent of increase **3.** percent of decrease **4.** percent **5.** 0.78 **6.** 0.4 **7.** 0.05 **8.** 1.19 **9.** 60% **10.** 16.7% **11.** 66.7% **12.** 8% **13–18.** Possible answers given. **13.** 8 **14.** 24 **15.** 40 **16.** 90 **17.** 32 **18.** 3 **19.** 68 **20.** 24 **21.** 4.41 **22.** 120 **23.** 27.3 **24.** 54 **25.** 125 **26.** 8% **27.** 12 **28.** 37.5% **29.** 8 **30.** 27.8% **31.** 50% **32.** 14.3% **33.** 30% **34.** 83.1% **35.** 23.1% **36.** 75% **37.** discount: $36.75; sale price: $208.25 **38.** $I = \$15$ **39.** $t = 3$ years **40.** $I = \$243$ **41.** $r = 3.9\%$ **42.** $P = \$2{,}300$ **43.** 7 years

Chapter 7

7-1 Exercises

1–5. Possible answers given: **1.** Q, R, S **2.** $\overleftrightarrow{QS}$, $\overleftrightarrow{RT}$ **3.** QRS **4.** $\overrightarrow{UQ}$, $\overrightarrow{UT}$, $\overrightarrow{US}$ **5.** $\overline{QU}$, $\overline{RU}$, $\overline{SU}$ **6.** $\overline{BA}$ and $\overline{BC}$, $\overline{AE}$ and $\overline{CE}$, $\overline{AD}$ and $\overline{CD}$ **7.** D, E, F **9.** DEF **11.** $\overline{DE}$, $\overline{EF}$, $\overline{DF}$ **21.** $\frac{7}{100}$ **23.** $\frac{31}{50}$ **25.** 4.5 **27.** C

7-2 Exercises

1. 60° **2.** 30° **3.** 120° **4.** 180° **5.** right angle **6.** acute angle **7.** straight angle **8.** $\angle MNL$ and $\angle ONP$, $\angle ONP$ and $\angle PNQ$ **9.** Possible answer: $\angle PNQ$ and $\angle MNP$, $\angle LNQ$ and $\angle PNQ$ **11.** 70° **13.** 120° **15.** right angle **17.** $\angle BAC$ and $\angle GAF$; $\angle EAF$ and $\angle GAF$ **19.** supplementary; 152° **21.** supplementary; 46° **23. a.** right angles **b.** about 39°N, 77°W **27.** 70% **29.** 145% **31.** 57.1% **33.** 46.7%

7-3 Exercises

1. parallel **2.** skew **3.** perpendicular **4.** 115° **5.** 115° **6.** 65° **7.** skew **9.** parallel **11.** 150° **13.** parallel **15.** perpendicular **17.** 45° **19. a.** parallel **b.** skew **c.** perpendicular **d.** none **23.** $33 **25.** $45

7-4 Exercises

1. $\overline{OQ}$, $\overline{OR}$, $\overline{OS}$, $\overline{OT}$ **2.** $\overline{RT}$ **3.** $\overline{RT}$, $\overline{RS}$, $\overline{ST}$, $\overline{TQ}$ **4.** 36° **5.** $\overline{CA}$, $\overline{CB}$, $\overline{CD}$, $\overline{CE}$, $\overline{CF}$ **7.** $\overline{GB}$, $\overline{BF}$, $\overline{DE}$, $\overline{FE}$, $\overline{AE}$ **9.** 10 cm **11.** 72° **13. a.** 133.2° **b.** 122.4° **17.** $w = 19.3$ **19.** $y = 1.96$ **21.** acute **23.** straight **25.** A

7-5 Exercises

1. no **2.** yes **3.** no **4.** octagon **5.** quadrilateral **6.** nonagon **7.** square; yes **8.** rectangle; no **9.** triangle; no **11.** no **13.** pentagon **15.** heptagon **17.** pentagon; no **21.** 16-gon **25.** 65° **27.** 24° **29.** 45° **31.** B

7-6 Exercises

1. isosceles right **2.** scalene obtuse **3.** isosceles acute **4.** 2 isosceles right; 1 isosceles acute; 2 scalene obtuse **5.** scalene right **7.** equilateral acute **9.** isosceles **11.** equilateral **13.** equilateral **15.** acute **17.** right **19.** right **21.** 8 in., isosceles **23.** scalene **29.** 384 **31.** 0.56 **33.** $h = 6$ **35.** $m = 10$

7-7 Exercises

1. parallelogram **2.** parallelogram, rectangle **3.** parallelogram, rhombus **4.** False **5.** True **7.** parallelogram **9.** parallelogram; rhombus **11.** parallelogram; rectangle **13.** False **15.** False **17.** parallelogram, rectangle, rhombus, square **19.** parallelogram, rhombus, rectangle, square **21.** 1 triangle, 1 pentagon, 2 trapezoids **23.** parallelogram, rhombus, square, rectangle, trapezoid, right triangle **27.** 16 **29.** 5 **31.** −3 **33.** −5 **35.** A

7-8 Exercises

1. 77° **2.** 110° **3.** 55° **4.** 720° **5.** 540° **6.** 360° **7.** 60° **9.** 32° **11.** 1,080° **13.** 23° **15.** 60° **17.** no **19. a.** 540° **b.** 108° **21.** 51° and 102° **27.** about 5 lb **29.** 7.88 **31.** 445.885

7-9 Exercises

1. The triangles on the gameboard are congruent; the holes on the gameboard are congruent. **2.** no congruent figures **3.** bowling pins are congruent **4.** yes **5.** no **6.** 90° **7.** 2.5 **9.** The triangles on the kite design are congruent. **11.** no **13.** 80°; 8 cm **15.** the lengths of all the sides **17.** the length of adjacent sides in each rectangle **21.** the squares and triangles **25.** 8.0 **27.** 1.2 **29.** 70% **31.** 50

7-10 Exercises

1. reflection **2.** translation

3.

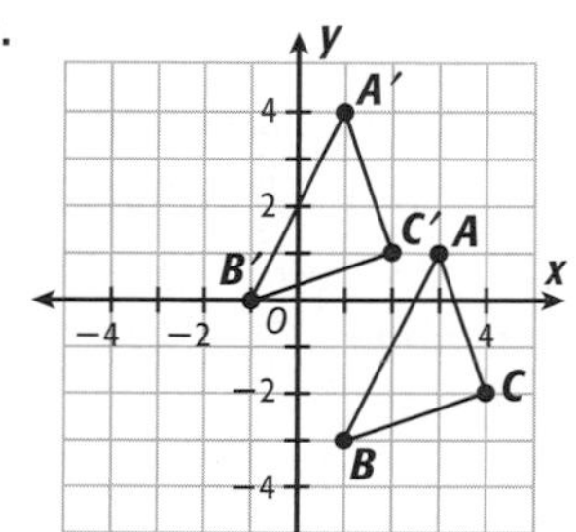

4.

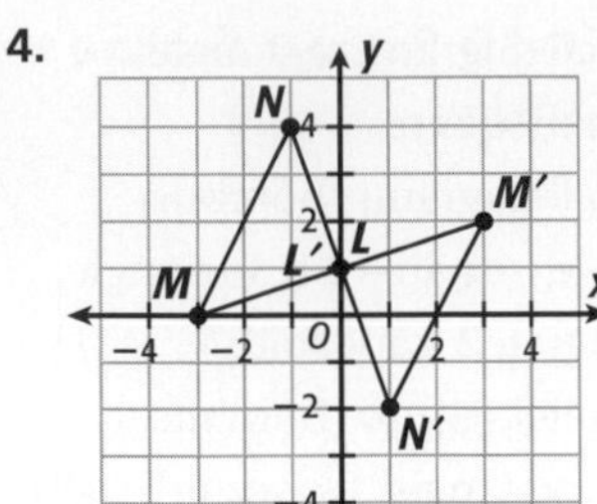

5.

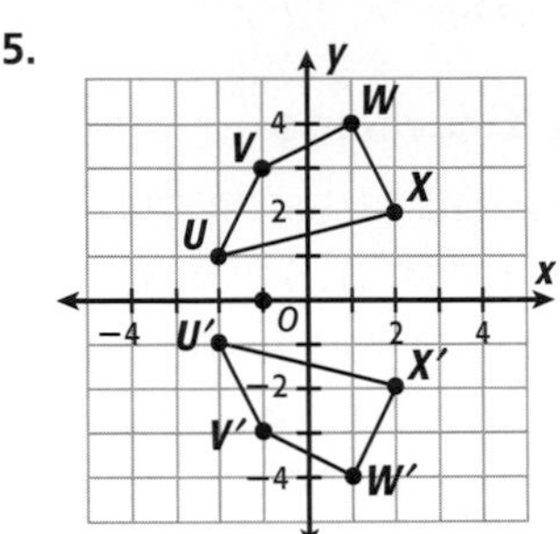

7. rotation

9.

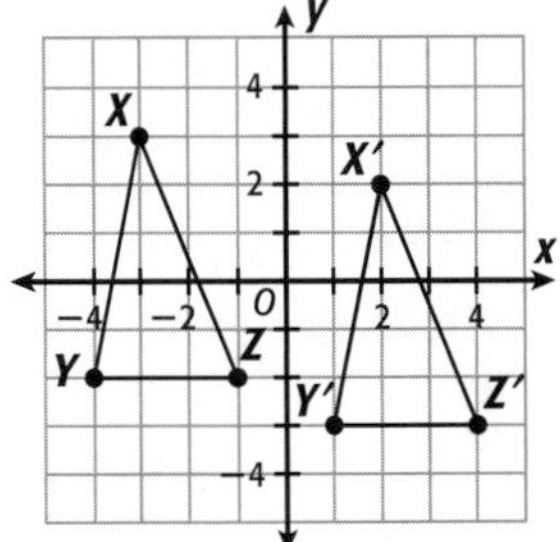

11. Rotation is shown, but not translation or reflection. **15.** none **17.** $y = 24$ **19.** A

7-11 Exercises

1. 5 lines **2.** 2 lines **3.** 4 lines **4.** horizontal line through center of flag **5.** none **6.** vertical line through center of flag **7.** 6 times **8.** 2 times **9.** 3 times **11.** 6 lines **13.** none **15.** horizontal and vertical lines through center of flag **17.** 8 times **21.** 1 line **23. a.** 4 times **b.** none **27.** $\frac{3}{4}$ **29.** 1 **31.** $x = 18$ **33.** $m = \frac{1}{9}$

Chapter 7 Study Guide and Review

1. acute isosceles **2.** parallel lines **3.** D, E, F **4.** $\overleftrightarrow{DF}$ **5.** plane DEF **6.** $\overrightarrow{ED}, \overrightarrow{FD}, \overrightarrow{DF}$ **7.** $\overline{DE}, \overline{DF}, \overline{EF}$ **8.** acute **9.** straight **10.** skew **11.** $\overline{HF}, \overline{FI}, \overline{FG}$ **12.** $\overline{GI}$ **13.** $\overline{HI}, \overline{GI}, \overline{GJ}, \overline{JI}$ **14.** a regular polygon (square) **15.** not a regular polygon **16.** equilateral acute **17.** scalene right **18.** parallelogram; rhombus **19.** parallelogram; rectangle **20.** 53° **21.** 133°

22.

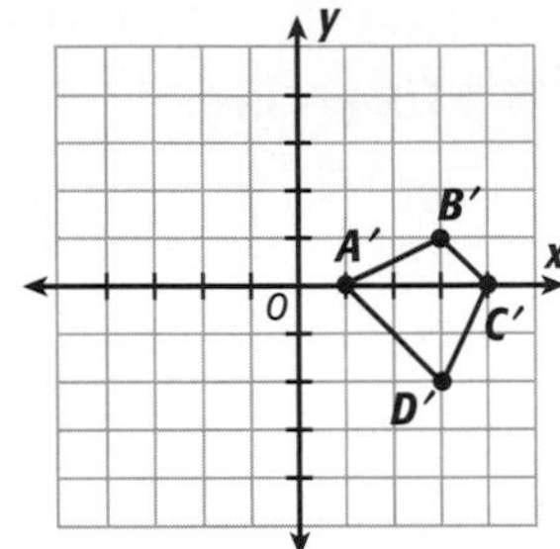

23. 1 line of symmetry

Chapter 8

8-1 Exercises

1. 576 hr **2.** 7.5 min **3.** 20 qt **4.** 10 hr **5.** 3.7 g **6.** 10,000 cm **7.** 42,240 ft **8.** 4.5 lb **9.** 6 gallon containers **11.** 2,700 s **13.** 498 min **15.** 1,000 lb **17.** 2.17 m **19.** 2.5 gal **21.** 25 ft **23.** 2,880 tributes **25.** 4 lb **27.** 3,520 yd **29.** 1,600 mm **31.** $>$ **33.** $=$ **35.** $>$ **37.** 2.4 mi **39.** 2,200 mL **41.** about 69,000 mi **47.** 1.82 **49.** 18.15 **51.** C

8-2 Exercises

1. 4 ft **2.** 21 mm **3.** $5\frac{1}{4}$ in. **4.** 4 **5.** 1 **6.** 2 **7.** 12 **8.** 4.2 **9.** 7 **10.** 180 **11.** 25 **12.** 6.3 **13.** 11 in. **15.** 14.2 km **17.** 2.8 m **19.** 1 **21.** 6 **23.** 3 **25.** 15.1 **27.** 1 **29.** 18.0 **31.** 1,800 **33.** 2.3 **35.** 300 **37.** foot **39.** milliliter **41.** 180,000 **43.** 280 **45.** 21.8 **47.** 2 **49.** 2,400 mg **53.** $\frac{31}{4}$ **55.** $\frac{52}{5}$ **57.** $m = 3.2$ **59.** $n = 24.21$ **61.** C

8-3 Exercises

1. 18 m **2.** 24 in. **3.** 32 ft **4.** 36 in. **5.** 20 m **6.** 12 ft **7.** 37.7 m **8.** 9.4 ft **9.** 50.2 in. **10.** 440 ft **11.** 48 cm **13.** 44 m **15.** 8 ft **17.** 25.1 cm **19.** 32.0 in. **21.** 2.8 m; 5.7 m **23.** 5.3 in.; 33.3 in. **25.** 1.98 m **27.** 141 mi **29.** 96 ft **33.** −12 **35.** 54 **37.** 24.5 **39.** 12

8-4 Exercises

1. 33.6 ft^2 **2.** 21 m^2 **3.** 147.6 cm^2 **4.** 48 in^2 **5.** $11\frac{1}{5}$ cm^2 **6.** 28.6 m^2 **7.** 40 tiles **9.** $131\frac{3}{4}$ in^2 **11.** 6 m^2 **13.** 31.98 cm^2 **15.** 72 yd^2 **17.** 192 cm^2 **19.** 14 $units^2$ **21.** 18 $units^2$ **23.** 6 in. **25. a.** 713 in^2 **b.** 108 in. **c.** 1,073 in^2 **31.** $d = 8$ **33.** $b = -4$ **35.** 43.2 **37.** 34.5

8-5 Exercises

1. 28 $units^2$ **2.** 12 $units^2$ **3.** 39.2 $units^2$ **4.** 6.5 cm^2 **5.** 64 m^2 **6.** 54 ft^2 **7.** 45 $units^2$ **9.** 72 $units^2$ **11.** 105 in^2 **13.** 4.5 cm **15.** 22 in. **17.** 15 $units^2$ **19.** 12 $units^2$ **21.** 3 m **23.** 108,000 mi^2 **25. a.** 6,800 mi^2 **b.** 181.7 **29.** prime **31.** prime **33.** 0.6 **35.** 1.5

8-6 Exercises

1. 78.5 in^2 **2.** 201.0 cm^2 **3.** 314 yd^2 **4.** 154 in^2 **5.** 17,662.5 mi^2 **7.** 803.8 ft^2 **9.** 63.6 ft^2 **11.** 3.8 m^2 **13.** 50.2 ft^2 **15.** 44.0 m; 153.9 m^2 **17.** 75.4 ft; 452.2 ft^2 **19.** 40.2 cm; 128.6 cm^2 **21.** $r = 6$ cm **23.** $r = 3$ in. **25. a.** 113 mi^2 **b.** 141 mi^2 **31.** 125 **33.** 243 **35.** −11 **37.** −18

8-7 Exercises

1. 16 **2.** 2.25 **3.** 81 **4.** 36 **5.** 4 **6.** 7 **7.** 9 **8.** 8 **9.** 11 mi **11.** 64 **13.** 20.25 **15.** 6 **17.** 10 **19.** 13 **21.** 15 **23.** 32 cm **25.** $\frac{9}{64}$ **27.** $\frac{1}{9}$ **29.** 20 **31.** 39 **33.** 12 **35.** 1

37. 192 **39.** 28 in. **41.** The window with the perimeter of 48 in. is larger. **43.** 12 ft **49.** 28.3 cm **51.** 31.4 m **53.** 108 m^2 **55.** A

8-8 Exercises

1. 20 m **2.** 8 ft **3.** about 8.7 ft **5.** 24 cm **7.** 16 in. **9.** 9.4 **11.** 12.4 **13.** yes **15. a.** 110 m **b.** 155.6 m **17.** 208.5 m **21.** 12 in. **23.** 22 ft **25.** 13 **27.** C

Chapter 8 Extension

1. 224 ft^2 **3.** 38 ft^2 **5.** 117 ft^2

Chapter 8 Study Guide and Review

1. hypotenuse **2.** circumference **3.** precision **4.** square root **5.** 13 **6.** 3.4 **7.** 288 **8.** 458,000 **9.** 3 significant digits **10.** 5 significant digits **11.** 1 significant digit **12.** 2 significant digits **13.** 4 significant digits **14.** 4 significant digits **15.** 83 m **16.** 81.4 cm **17.** 40.8 ft **18.** 49.0 in. **19.** 50.74 cm^2 **20.** 826.2 yd^2 **21.** 266 in^2 **22.** 108.75 cm^2 **23.** 50 yd^2 **24.** 2,163 in^2 **25.** 36.3 m^2 **26.** 226.9 ft^2 **27.** 5 **28.** 10 **29.** 10 **30.** 12 **31.** 16 ft **32.** 34 cm **33.** 60 ft **34.** 2 m **35.** 60 mm

Chapter 9

9-1 Exercises

1. 1 rectangular base; rectangular pyramid **2.** 2 octagonal bases; octagonal prism **3.** 2 triangular bases; triangular prism **4.** cylinder and cone **5.** cube and triangular prism **6.** hemispheres **7.** 1 triangular base; triangular pyramid **9.** 1 hexagonal base; hexagonal pyramid **11.** 2 triangular prisms **13.** cube **15.** triangular pyramid **17.** rectangular pyramid **19.** cylinder **21.** 1.06 **23.** 3.17 **25.** 1 **27.** 2 **29.** A

9-2 Exercises

1. 24 cubes; 24 cubic units **2.** 45 cubes; 45 cubic units **3.** 80 cubes; 80 cubic units **4.** 15 in^3 **5.** 118.1 cm^3 **6.** 88.2 m^3 **7.** 66.4 ft^3 **8.** 706.5 ft^3 **9.** 166.4 cm^3 **11.** 60 cubes; 60 cubic units **13.** 1,299.2 ft^3 **15.** 270 in^3 **17.** 243.0 cm^3 **19.** 2,198 m^3 **21.** 120 ft^3 **23.** 288 m^3 **25.** 41.5 m^3 **31.** 0.6 **33.** 0.3125 **35.** $640

9-3 Exercises

1. 10 ft^3 **2.** 35 cm^3 **3.** 32 m^3 **4.** 376.8 ft^3 **5.** 16.7 in^3 **6.** 235.5 m^3 **7.** 904.3 cm^3 **8.** 7,234.6 in^3 **9.** 3,367.6 m^3 **11.** 45 ft^3 **13.** 47.1 in^3 **15.** 3,768.0 m^3 **17.** 950.3 cm^3 **19.** 46.7 ft^3 **21.** 65.4 m^3 **23.** 60 ft^3 **27.** $a = 10$ **29.** $b = 7$ **31.** 115° **33.** B

9-4 Exercises

1. 286 ft^2 **2.** 856 cm^2 **3.** 244.9 m^2 **4.** 1,884.0 in^2 **5.** 50.2 yd^2 **6.** 1,017.4 ft^2 **7.** 928 in^2 **9.** 791.3 in^2 **11.** 314.0 ft^2 **13.** 38.4 ft^2 **15. b.** 158.0 cm^2 **c.** 85.4 cm^2 **21.** 5 and 6 **23.** 0 and −1 **25.** −7.06

9-5 Exercises

1. 93.6 cm^2 **2.** 0.5 m^2 **3.** 33,750 in^3 **4.** 486 lb **5.** 223.84 in^2 **7.** 8.2 cm^3 **9.** 58,750 ft^2; 937,500 ft^3 **11.** 1,130,400 cm^2; 113,040,000 cm^3 **13.** 259.8 cm or 2.598 m; 53.3 cm or 0.533 m **15.** 41,567,500 cm^2; 4,156.75 m^2 **19.** no **21.** no **23.** $\frac{3}{5}$ **25.** $\frac{9}{50}$

Chapter 9 Extension

1. 448 cm^2 **3.** 208 cm^2

Chapter 9 Study Guide and Review

1. cylinder **2.** surface area **3.** bases of a prism **4.** cone **5.** triangular prism **6.** rectangular pyramid **7.** 364 cm^3 **8.** 415.4 mm^3 **9.** 111.9 ft^3 **10.** 60 in^3 **11.** 471 cm^3 **12.** 50.9 mm^3 **13.** 677.7 in^3 **14.** 250 m^2 **15.** 262.3 cm^2 **16.** 254.3 ft^2 **17.** 143,799.4 m^2 **18.** 2,970 in^2 **19.** 4.1 ft^2

Chapter 10

10-1 Exercises

1. purple **2.** equally likely **3.** likely **4.** impossible **5.** Tim will likely be at the theater. **7.** black **9.** unlikely **11.** purple **13.** unlikely **17.** 50% **19.** B

10-2 Exercises

1. $\frac{3}{5}$ **2. a.** $\frac{13}{15}$ **b.** $\frac{2}{15}$ **3.** $\frac{13}{30}$ **5. a.** $\frac{9}{14}$ **b.** $\frac{5}{14}$ **7.** $\frac{2}{5}$ **9. a.** 9.1 in. **b.** 0 **c.** $\frac{1}{5}$ **11. a.** $\frac{3}{8}$ **b.** 0 **13.** 1 **15.** 5 **17.** B

10-3 Exercises

1. There are 4 possible outcomes in the sample space. **2.** There are 20 possible outcomes in the sample space. **3.** 24 possible outcomes **5.** There are 8 possible outcomes in the sample space. **7.** There are 12 possible outcomes in the sample space. **9.** 6 different ways **11. a.** 9 outcomes **b.** 6 outcomes **c.** 12 outcomes **17.** 75% **19.** 30% **21.** C

10-4 Exercises

1. $\frac{1}{6}$, 0.17, 17% **2.** $\frac{1}{4}$, 0.25, 25% **3. a.** $\frac{3}{7}$ **b.** $\frac{2}{7}$ **c.** $\frac{2}{7}$ **5.** $\frac{1}{4}$, 0.25, 25% **7. a.** $\frac{3}{7}$ **b.** $\frac{4}{7}$ **9. a.** 64% **b.** 25% **13.** $n = -21$ **15.** $s = -41$ **17.** B

10-5 Exercises

1. independent **2.** dependent **3.** $\frac{1}{6}$ **4.** $\frac{1}{60}$ **5.** $\frac{3}{20}$ **7.** independent **9.** $\frac{1}{4}$ **11.** dependent; $\frac{3}{14}$ **13.** $\frac{16}{25}$ **19.** 3 **21.** 15

10-6 Exercises

1. 6 **2.** 10 **3.** 10 **4.** 15 **5.** 10 **7.** 20 **9.** 21 **11.** 6 **13. a.** 3 **b.** $\frac{1}{3}$ **15. a.** 15 **b.** $\frac{1}{3}$ **17.** 5 **19.** 201.0 cm^2 **21.** 78.5 in^2 **23.** cone **25.** pentagonal prism

10-7 Exercises

1. a. 24 **b.** $\frac{1}{4}$ **2. a.** 120 **b.** $\frac{1}{120}$ **3.** 720 **4.** 5,040 **5. a.** 6 **b.** $\frac{1}{3}$ **7.** 362,880 **9.** combinations **11.** permutations **13.** 5,040 **15.** 40,320 different orders **17.** 3,628,800 **21.** 141.3 m^3 **23.** 58.5 cm^3 **25.** C

Chapter 10 Extension

1. 3:4 **3.** 1:6 **5.** 2:1 **7.** 1:2 **9.** 5:3 **11.** 1:7 **13.** 1:1 **15. a.** 5:21 **b.** 21:5 **17. a.** 2:3 **b.** 3:2

Chapter 10 Study Guide and Review

1. independent events **2.** combination **3.** sample space **4.** outcome **5.** likely **6.** unlikely **7.** impossible **8.** $\frac{2}{3}$ **9.** $\frac{1}{3}$ **10.** R1, R2, R3, R4, W1, W2, W3, W4, B1, B2, B3, B4 **11.** 12 possible outcomes **12.** $\frac{3}{7}$, 0.43, 43% **13.** $\frac{1}{8}$, 0.125, 12.5% **14.** $\frac{4}{195}$ **15.** $\frac{16}{121}$ **16.** 10 ways **17.** 35 committees **18.** 84 combinations **19.** 3,628,800 ways **20.** 720 ways **21.** 120 ways

Chapter 11

11-1 Exercises

1. $n = 7$ **2.** $m = -6$ **3.** $x = \frac{1}{3}$ **4.** $b = 24$ **5.** $y = 136$ **6.** $n = 72$ **7.** 12 refills **9.** $p = -12$ **11.** $d = \frac{1}{7}$ **13.** $y = 5$ **15.** $k = 85$ **17.** $m = -80$ **19.** $m = -112$ **21.** 6 more than a number divided by 3 equals 18; $m = 36$ **23.** 4 less than a number divided by 5 equals 2; $n = 30$ **25.** $x = 2$ **27.** $g = 20$ **29.** $w = -9$ **31.** $p = 2$ **33.** 120 minutes **35.** 2,200 calories **37.** 24 grams **39.** 8 **41.** B

11-2 Exercises

1. $n = 5$ **2.** $x = 9$ **3.** $p = 2$ **4.** $m = 42$ **5.** $y = 9$ **6.** $m = \frac{1}{2}$ **7.** 12 books **9.** $x = \frac{7}{8}$ **11.** $n = 6$ **13.** $x = -2$ **15.** $n = 9$ **17.** $r = 122$ **19.** $t = -3$ **21.** $x = 66$ **23.** $w = 8$ **25.** $a = 176$ **27.** $b = -7$ **29.** $x = 3$ **31.** $6.70 **33. a.** 1,063.$\overline{3}$°C **b.** −257.$\overline{8}$°C **39.** $y = \frac{1}{33}$ **41.** $n = -\frac{5}{6}$ **43.** 40,320 **45.** A

11-3 Exercises

1. $n = 32$ **2.** $-28 = 10x$ **3.** $12w = 32$ **4.** $y = 20$ **5.** $a = 2$ **6.** $d = 28$ **7.** 5 movies **9.** $-8 = 12p$ **11.** $-6 = 2c$ **13.** $6 = \frac{1}{10}a$ **15.** $b = -8$ **17.** $a = -0.8$ **19.** $c = -2$ **21.** $y = -7$ **23.** $r = 4$ **25.** $r = -2$ **27.** 67 **29.** $s = 11$ **31. a.** 194 mi per day **b.** 1,164 mi **35.** Possible answer: 30 **37.** Possible answer: 9 **39.** 288 in. **41.** 16 c **43.** C

11-4 Exercises

1. number of people ≤18
2. number of fish < 8
3. water level > 45
4. −5 −3 −1 1 3 5
5. −5 −3 −1 1 3 5
6. −5 −3 −1 1 3 5
7. −6 −4 −2 0 2 4
8. −4 −2 0 2 4 6
9. −4 −2 0 2 4 6
10. −4 −2 0 2 4 6
11. temperature < 40
13. number of tables ≤ 35
15. −4 −2 0 2 4 6
17. −4 −2 0 2 4 6
19. −8 −6 −4 −2 0 2
21. −4 −2 0 2 4 6
23. −4 −2 0 2 4 6
25. −4 −2 0 2 4 6

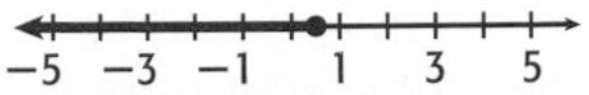

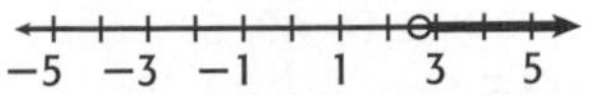

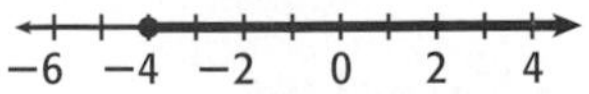

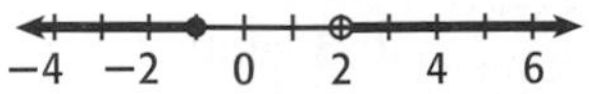

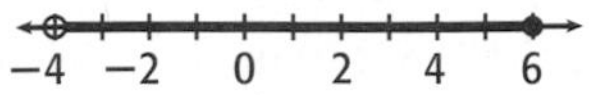

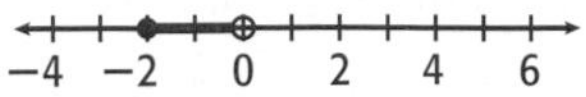

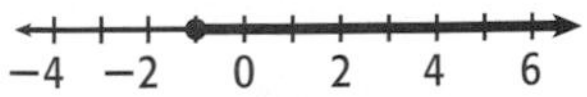

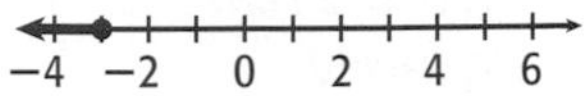

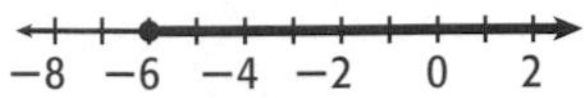

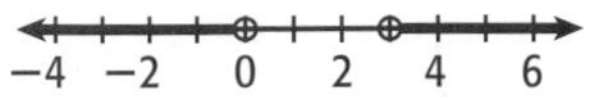

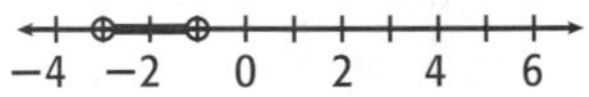

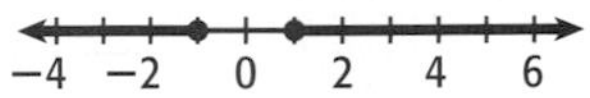

27. $-2 < c < 3$ **29.** $-3 < x < 1$ **31.** 0 ≤ depth ≤ −200 **33.** 0 ≥ *Manshu* depth measurement ≥ −32,190 ft; 0 ≥ *Challenger* depth measurement ≥ −35,640 ft; 0 ≥ *Horizon* depth measurement ≥ −34,884 ft; 0 ≥ *Vityaz* depth measurement ≥ −36,200 ft **37.** $\frac{1}{2}$, 0.5, 50% **39.** $\frac{1}{3}$, 0.33, 33% **41.** B

11-5 Exercises

1. $x < 27$ **2.** $y \geq 4$ **3.** $p \leq 7$ **4.** $n > 21$ **5.** $b \leq -24$ **6.** $k \geq -3$ **7.** at most 12°F warmer **9.** $m < 11$ **11.** $c \leq 11$ **13.** $x \geq 80$ **15.** $z > -12$ **17.** $f > -6$ **19.** $n \geq -4$ **21.** at most 24 birds **23.** $a > 0.3$ **25.** $m \leq -38$ **27.** $g < 6\frac{1}{3}$ **29.** $w \leq 15.7$ **31.** $t \geq -242$ **33.** $v \leq -0.6$ **35.** at least 0.5 meters **37.** at least 22 miles **39.** more than 12.4 km/h **43.** 144 in^2 **45.** 1,600 in^2 **47.** D

11-6 Exercises

1. $w < -32$ **2.** $z \leq -42$ **3.** $p < 48$ **4.** $m > -5$ **5.** $y > -\frac{11}{8}$ or $-1\frac{3}{8}$ **6.** $c \leq 8$ **7.** at least 27 candles **9.** $m > 52$ **11.** $c \geq -120$ **13.** $x \geq -1.1$ **15.** $z < \frac{3}{5}$ **17.** $f < -3$ **19.** $n \leq -10$ **21.** at least 46 wreaths **23.** $p > 4$ **25.** $y \geq 18$ **27.** $g > 0.63$ **29.** $w \leq \frac{3}{8}$ **31.** $t < \frac{5}{2}$ **33.** $v \geq -2.5$ **35. a.** more than 218,500 **b.** fewer than 655,500 **37.** 10 days **41.** $x = 11$ **43.** $r = 6$ **45.** B

11-7 Exercises

1. $x < 3$ **2.** $z \leq -294$ **3.** $y \geq 6$ **4.** $m > -4$ **5.** $p \leq -5$ **6.** $n < -24$ **7.** more than $26 each **9.** $b < 12$ **11.** $c \geq -3$ **13.** $x \leq -27$ **15.** $j \leq 2$ **17.** at most 6 bagels **19.** $y > -3$ **21.** $a \leq -54$ **23.** $n > -5$ **25.** $b < -1.6$ **27.** $k \geq -5$ **29.** $t > -3$ **31.** at least 155 prizes **33.** at least $5,000 **35.** at most 60% **39.** 24 **41.** $x = -15$ **43.** $n = 4$

Chapter 11 Study Guide and Review

1. inequality **2.** compound inequality **3.** solution set **4.** $y = 8$ **5.** $z = 30$ **6.** $w = 147$ **7.** $a = -7$ **8.** $b = 4$ **9.** $j = 9$ **10.** $y = 5$ **11.** $x = 12$ **12.** $b = \frac{1}{2}$ **13.** $c = 6$ **14.** $m = \frac{8}{3}$ or $2\frac{2}{3}$ **15.** $x = 20$ **16.** average ≥ 65 **17.** weight limit ≤ 9 tons **18.** age > 200 **19.** distance < 2 mi **20.** $r > 25$ **21.** $s \geq 14$ **22.** $x \leq -26$ **23.** $g < 8$ **24.** $t \leq \frac{1}{6}$ **25.** $9 > r$ **26.** $z < -30$ **27.** $u \geq -66$ **28.** $n < -55.2$ **29.** $x \leq 6$ **30.** $p \leq 6$ **31.** $k < -130$ **32.** $p < 5$ **33.** $v \geq 2.76$ **34.** $c > 33$ **35.** $y \leq 2.7$ **36.** $b < -2$ **37.** $d < -6$

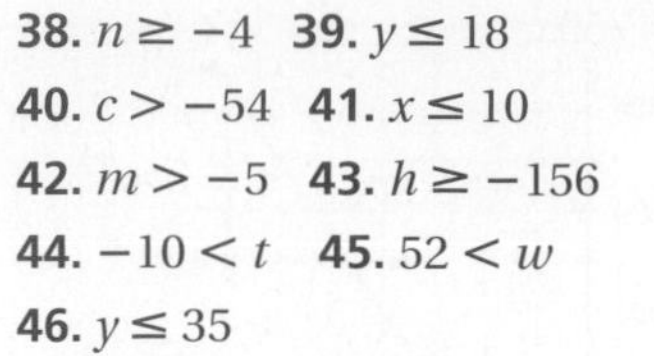

38. $n \geq -4$ **39.** $y \leq 18$ **40.** $c > -54$ **41.** $x \leq 10$ **42.** $m > -5$ **43.** $h \geq -156$ **44.** $-10 < t$ **45.** $52 < w$ **46.** $y \leq 35$

Chapter 12

12-1 Exercises

1. −5, 1, 3 **2.** 5, 3, 1 **3.** 50, 2, 18

4.

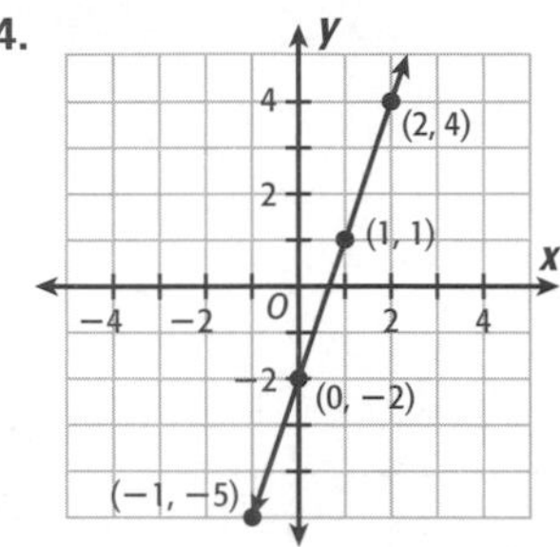

5.

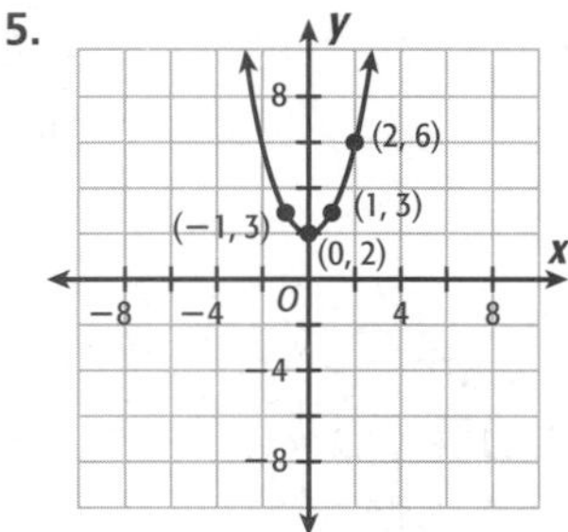

7. −7, −1, 8

9.

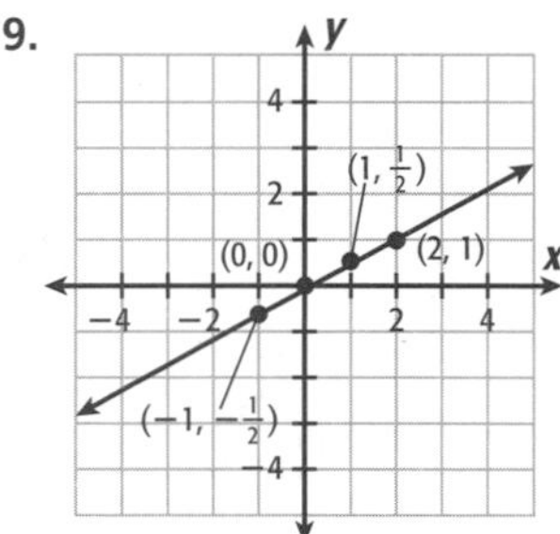

11.

Input	Rule	Output
C	$\frac{9}{5}C + 32$	F
−20°	$\frac{9}{5}(-20) + 32$	−4°
−5°	$\frac{9}{5}(-5) + 32$	23°
0°	$\frac{9}{5}(0) + 32$	32°
20°	$\frac{9}{5}(20) + 32$	68°
100°	$\frac{9}{5}(100) + 32$	212°

13.

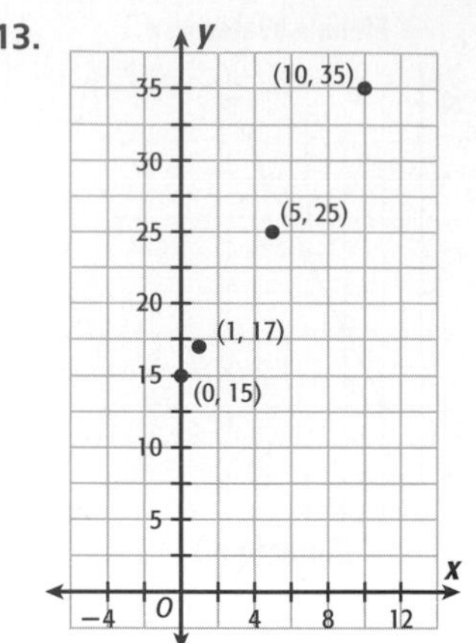

19. $\frac{1}{3}$ **21.** $x = 3$ **23.** $x = 5.5$

12-2 Exercises

1. $y = 3n$; 15, 18, 21 **2.** $y = n + 2$; 7, 8, 9 **3.** $y = n - 1$; 4, 5, 6 **4.** $y = 5n$; 25, 30, 35 **5.** 23 mi, 26 mi, 29 mi, 32 mi, 35 mi **7.** $y = \frac{n}{2}$; $\frac{5}{2}$, 3, $\frac{7}{2}$ **9.** $y = n - 3$; 2, 3, 4 **11.** $y = 2n + 1$; 11, 13, 15 **13.** $y = n^2$; 25, 36, 49 **15.** $52.50, $55.00, $57.50, $60.00, $62.50, $65.00 **17.** 4.7, 5.7, 6.7 **19.** 3, 4, 5 **21.** 26, 31, 36 **23.** $y = 2n - 2$; 18 **25.** $y = 5n - 2$; 48 **27.** $y = 4n + 2$; 42 **29.** $y = 2^n$ **33.** 8

12-3 Exercises

1. a

2.

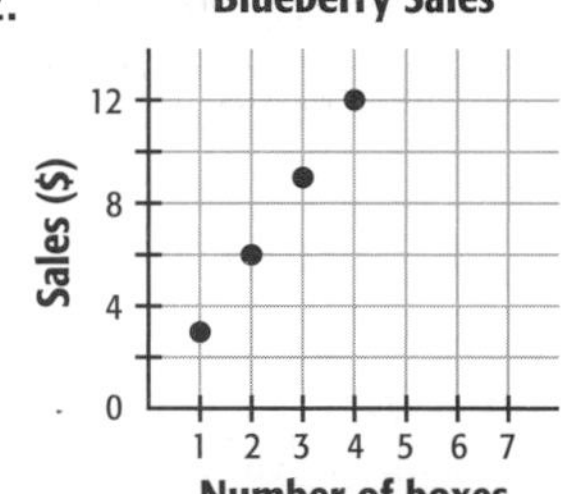

3. D: $-1 \leq x \leq 3$; R: $0 \leq y \leq 2$
4. D: $-4 \leq x \leq -1$; R: $-3 \leq y \leq -1$
5. D: $1 \leq x \leq 5$; R: $0 \leq y \leq 3$

7.

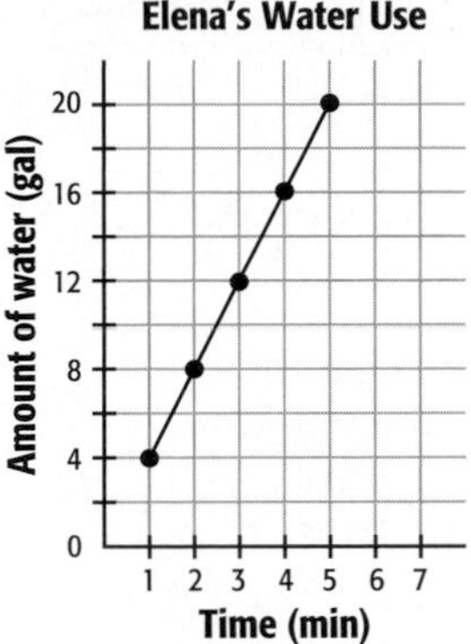

9. D: $-4 \le x \le 1$; R: $-1 \le y \le 1$

13.

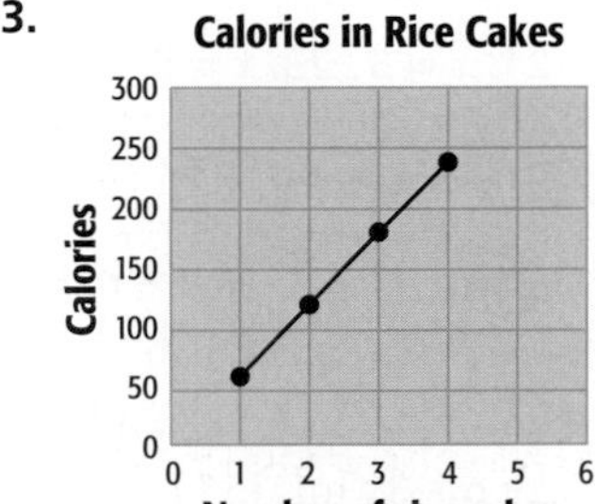

15. a. 1990, 1995, 2000, 2005, 2010 b. Possible answer: 15.5, 17, 18, 18.5, 19 19. 4 21. 20 23. five 25. A

12-4 Exercises

1.

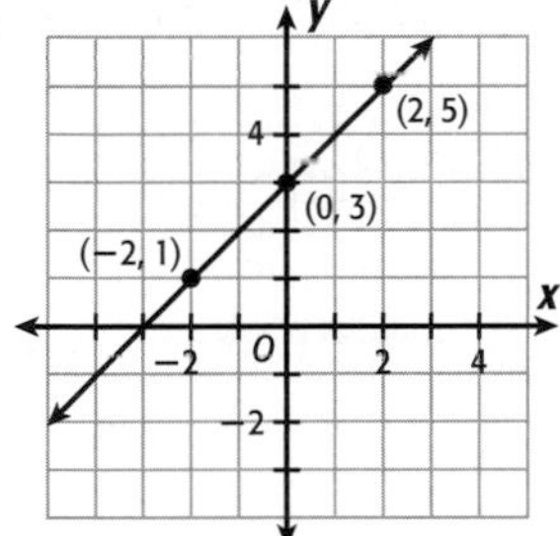

2.

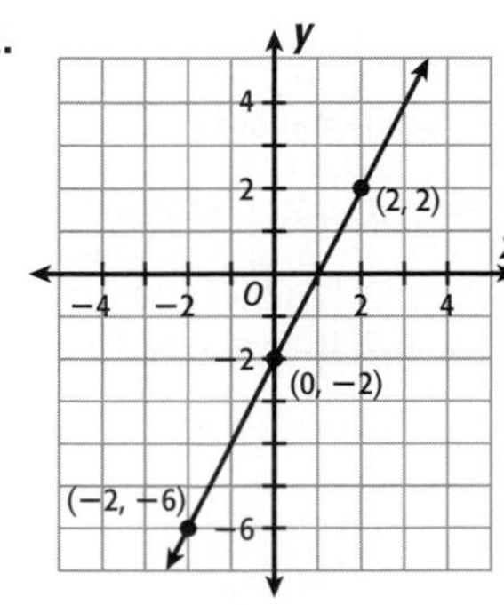

3. $y = 750x$;

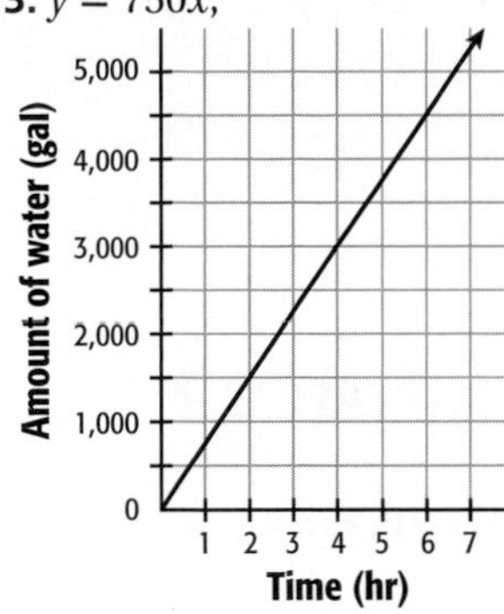

5.

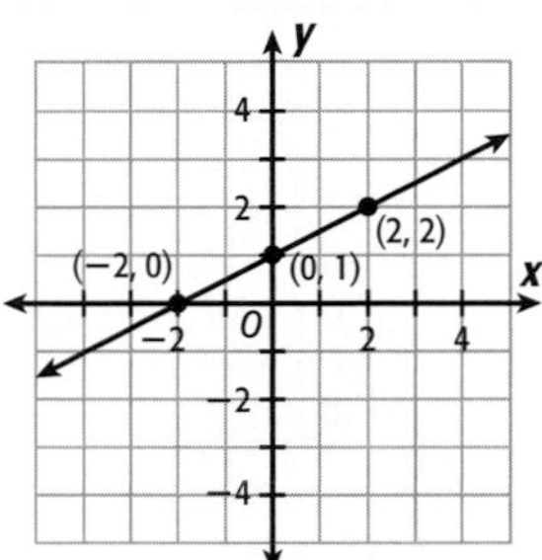

7.

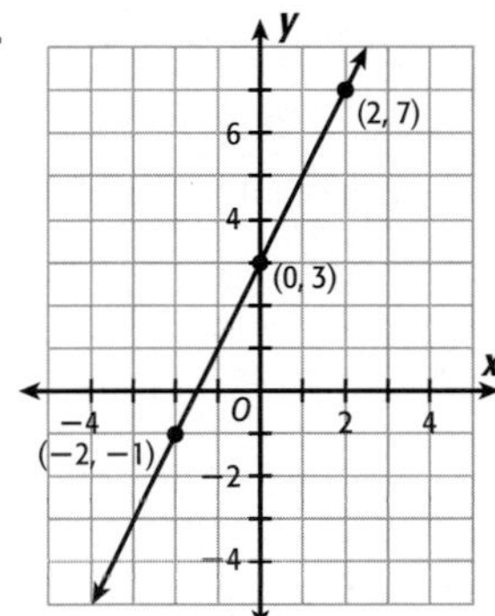

9. $y = x$;

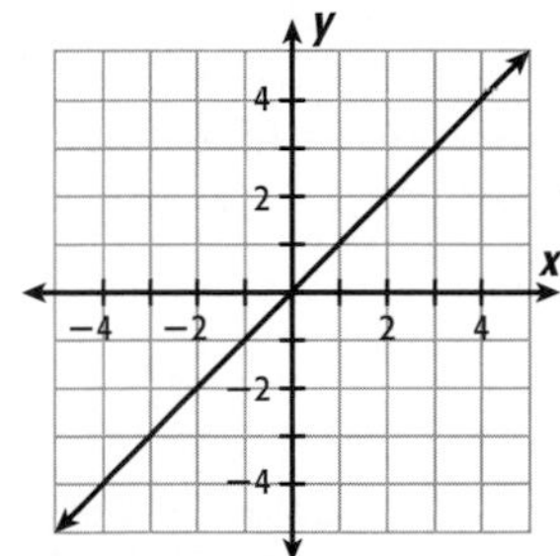

11. $y = -2$;

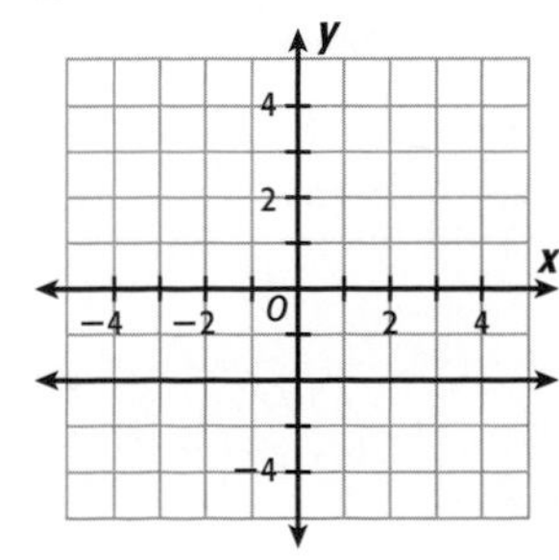

13. $y = 6$;

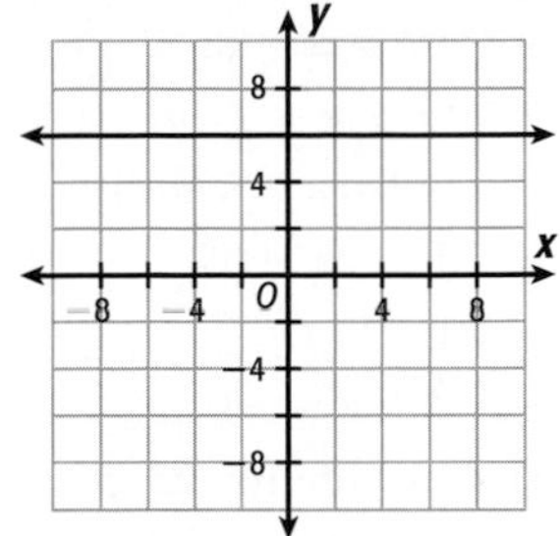

15. a. 5 ppm b. about 378 ppm
21. octagon 23. triangle 25. 96 in^2

12-5 Exercises

1. 1; 0 2. $-\frac{1}{2}$; 3 3. 3; −4
4. $\frac{1}{3}$; 0 5. −3; 8

6.

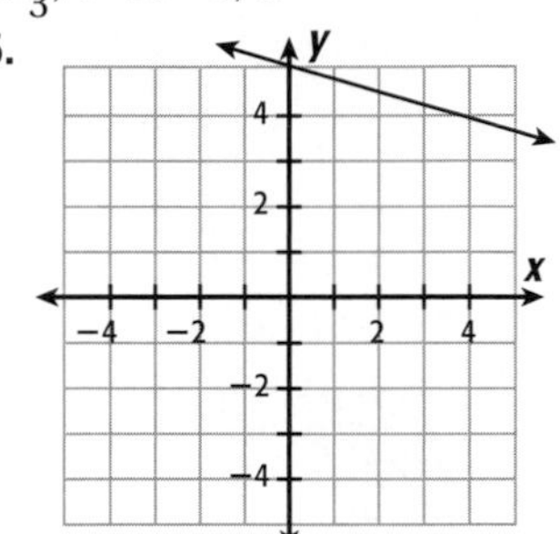

7.

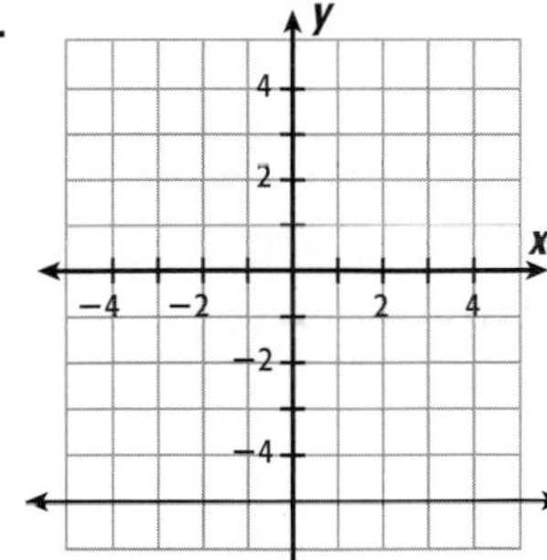

8.

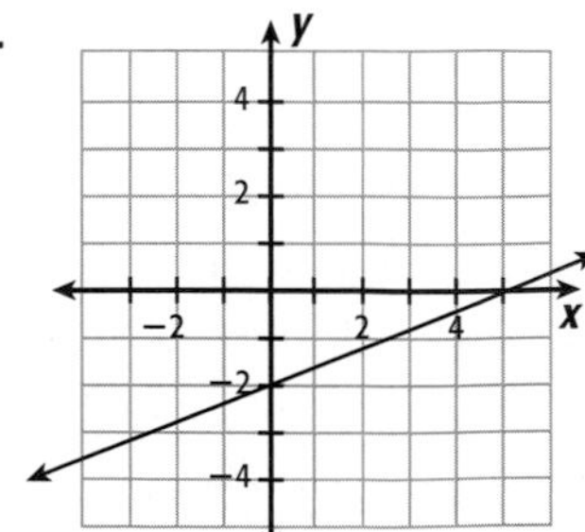

9. −3; −4 11. 1; −2 13. 4; 2

15.

17. $y = 3x + 8$; positive
19. $y = -5x + 1$; negative
21.

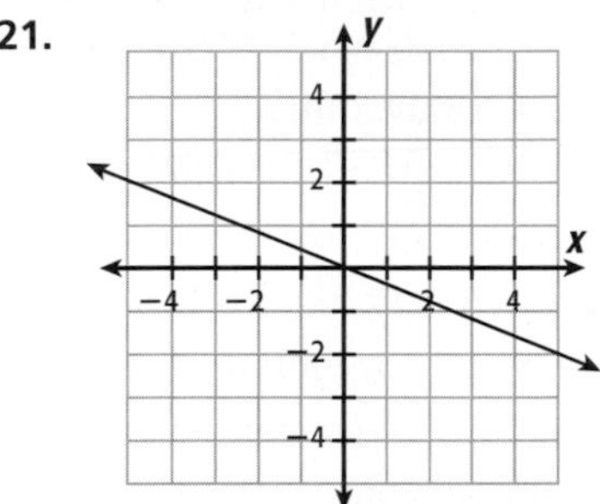

23. 1; −1 25. $\frac{1}{3}$; 4 27. 2; −2; $y = 2x - 2$ 29. $\frac{3}{2}$; 0; $y = \frac{3}{2}x$
31. 0; −3; $y = -3$ 33. b. −1,000
c. negative d. The older the car, the lower the value. 37. $b < 32$
39. $m > 0.3$

12-6 Exercises

1. nonlinear 2. linear
3. nonlinear 4. linear
5. nonlinear 6. nonlinear
7. nonlinear 9. nonlinear
11. nonlinear 13. b. linear
c. 98 m/s 17. $p = 4$ 19. $m = 4$
21. $a = 28$ 23. $m = 2$ 25. D

12-7 Exercises

1. from $x = -1$ to $x = 1$
2. from $x = -1$ to $x = 0$ and from $x = 0$ to $x = 1$ 3. from $x = -1$ to $x = 1$ 4. constant 5. variable
6. constant 7. constant; 15 mi/h
9. from $x = -3$ to $x = -1$ and from $x = -1$ to $x = 0$
11. constant 13. variable
14. constant; 1 in./h 17. variable
21. $x < 3$ 23. $-15 < x$

25.

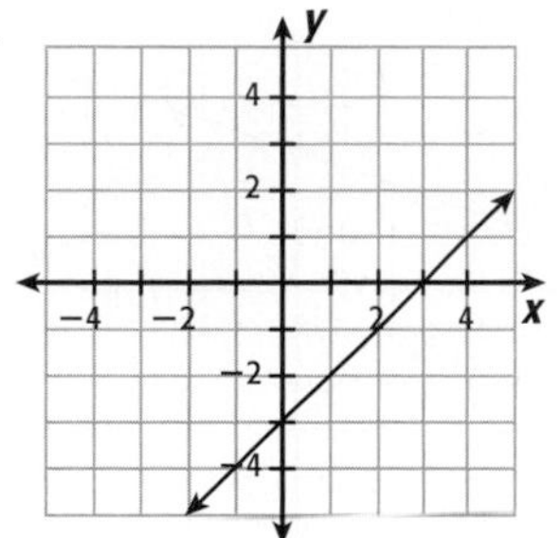

27.

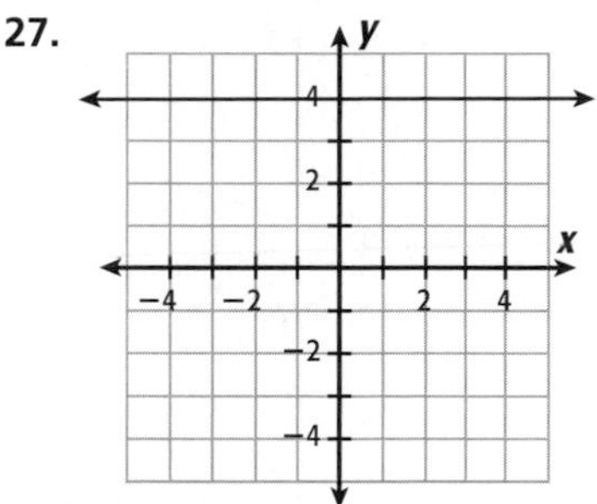

12-8 Exercises

1. {1, 3, 5, 7, 9, 11} 2. $\{x|x$ is an integer less than 0} 3. {−6, −5, −4} 4. $\{x|x > 10$ and x is even}
5. $A \cap B = \{15\}$; $A \cup B = \{5, 10, 11, 13, 15, 19, 20\}$; neither is a subset of the other. 6. $A \cap B = \{100, 200, 300\}$; $A \cup B = \{100, 200, 300, 400\}$; $A \subset B$ 7. $A \cap B = \emptyset$; $A \cup B = \{x|x$ is an integer}; neither is a subset of the other. 8. $A \cap B = \{0\}$; $A \cup B = \{x|x$ is a real number}; $B \subset A$
9. {3, 6, 9, 12, 15, 18} 11. $\{x|x$ is even} 13. $A \cap B = \{-2, 0\}$; $A \cup B = \{-6, -4, -2, 0, 4, 6\}$; neither is a subset of the other. 15. $A \cap B = \{1, 2, 3\}$; $A \cup B = \{1, 2, 3, 4, 5\}$; $B \subset A$ 17. $A \cap B$ = {Germany, France}; $A \cup B = \{x|x$ is a European country}; $B \subset A$
19. $B = \{4, 5\}$ 21. a. A = {New York, Chicago, Toronto}
b. B = {New York}
c. $A \cap B$ = {New York}
23. basketball players who are taller than 80 in. 29. 24
31. $x = 5$ 33. $n = 4$ 35. A

Chapter 12 Extension

1.

Math Club Members
Jason Juan
Keisha Elden
Harry
Doris
Tyrone Cara
Band Members

3. intersection: apple pie, cherry pie; union: pudding, cake, fruit, applesauce, ice cream, apple pie, cherry pie; subset of desserts: pies

Chapter 12 Study Guide and Review

1. sequence 2. function
3. linear function 4. 3, 8, 24
5. $y = 25n$; 125, 150, 175
6. $y = n - 4$; 1, 2, 3
7. $y = 3n - 7$; 8, 11, 14
8. $y = 2n + 2$; 12, 14, 16
9. D: $-4 \leq x \leq 5$; R: $-3 \leq y \leq 2$
10.

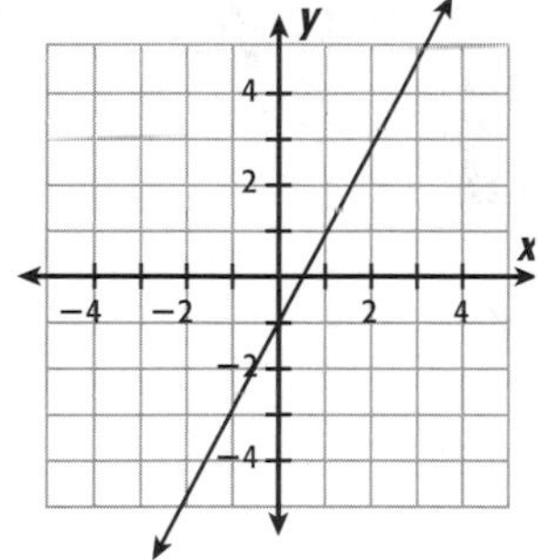

11.

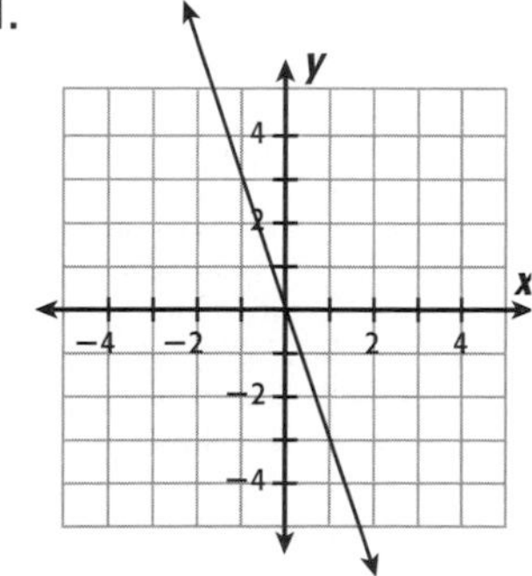

12.

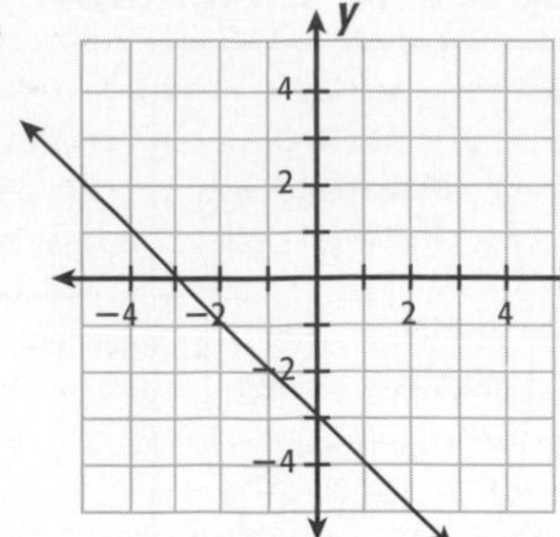

13.

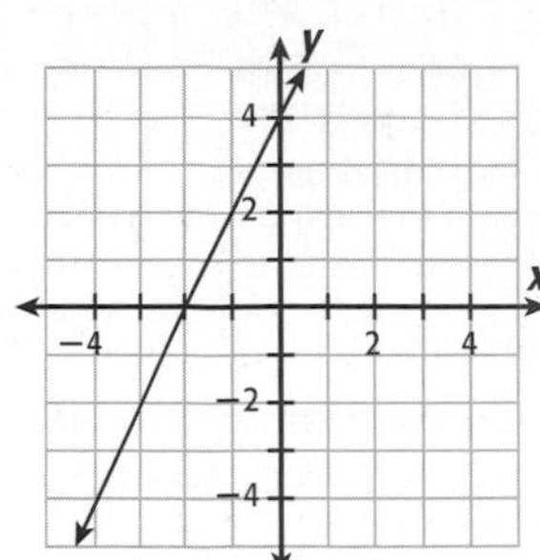

14.

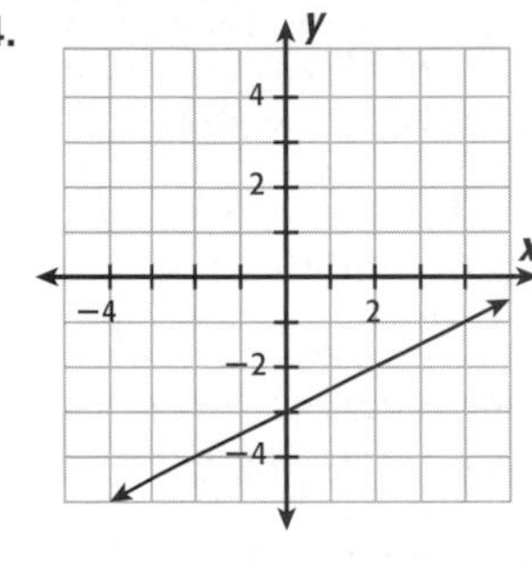

15.

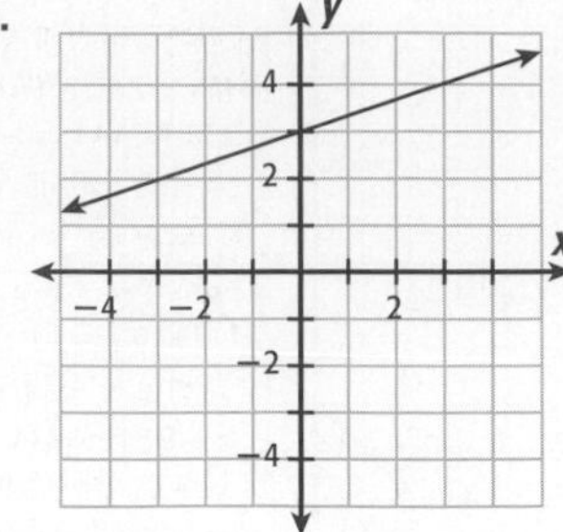

16.

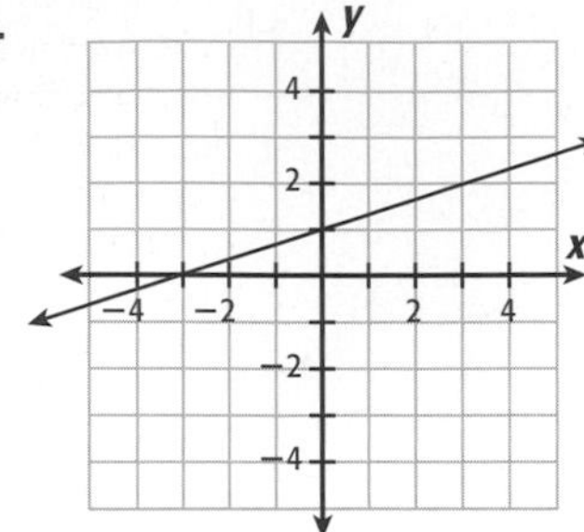

17.

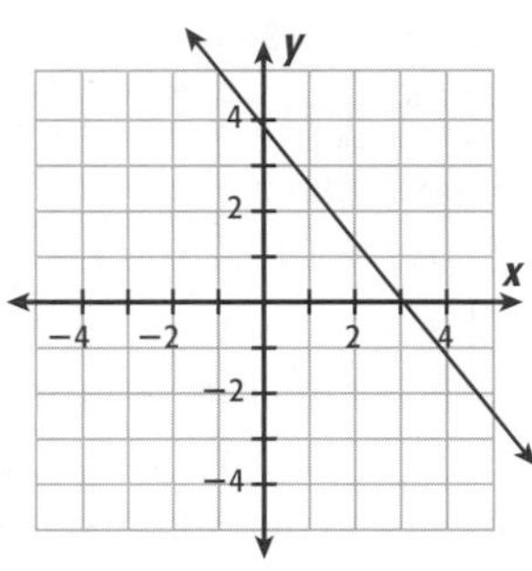

18.

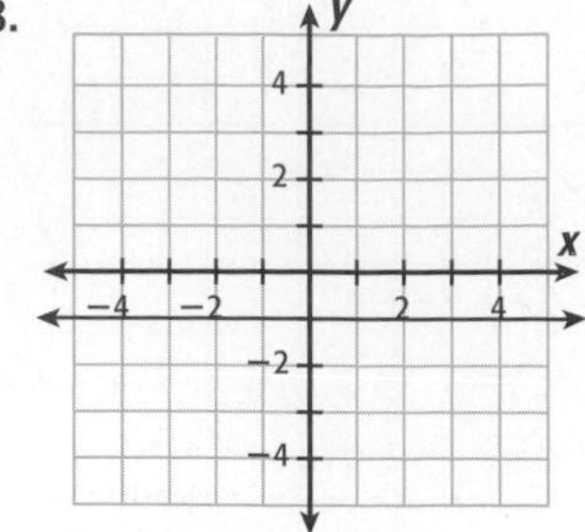

19. nonlinear **20.** linear
21. variable **22.** $A \cap B = \{3, 5, 7\}$; $A \cup B = \{1, 2, 3, 5, 7, 9\}$; neither is a subset of the other.
23. $A \cap B = \{x|x$ is divisible by $8\}$; $A \cup B = \{x|x$ is divisible by $2\}$, $B \subset A$ **24.** $A \cap B = \{x|x$ is a U.S. state beginning with the word *New*$\}$; $A \cup B = \{x|x$ is a state in the United States$\}$; $A \subset B$

Photo

Cover (All): Pronk & Associates.

Title page (All): Pronk & Associates.

Master icons — teens (All): Sam Dudgeon/HRW.

Author photos by Sam Dudgeon/HRW

Front Matter: S2 (border), Kim Steele/PhotoDisc/Getty Images; S2 (stamp), United States Postal Service; S2 (turtle), (c) David A. Northcott/CORBIS.

Problem Solving Handbook: xxxv Victoria Smith/HRW; xxxvi (t), © Stone, Lynn/Animals Animals/Earth Scenes; xxxvii (all), Victoria Smith/HRW/"Pride of the Plains" puzzle images used by permission of Sally J. Smith and FX Schmid, USA.; xxxviii (all), Peter Van Steen/HRW/cards created by Joyce Gonzalez; xxxix & xli Victoria Smith/HRW; xlii (all), © Richard Herrmann; xliii Victoria Smith/HRW. **Chapter 1:** 2–3 (bkgd), © Sam Fried/Photo Researchers, Inc.; 2 (br), Victoria Smith/HRW; 4 © Bill Banaszewski/Photolink; 5 Victoria Smith/HRW; 7 M*A*S*H © 1983 Twentieth Century Fox Television. All rights reserved./Courtesy Everett Collection; 10 © CORBIS; 11 (l), © James L. Amos/CORBIS/Collection of The Corning Museum of Glass, Corning, New York; 13 © Karl Weatherly/CORBIS; 14 Courtesy IMAX Corporation; 17 (tr), Klein/Hubert-BIOS/Peter Arnold, Inc.; 19 (b), © Stephen Frink/Index Stock Imagery/PictureQuest; 22 (all), PhotoDisc - Digital Image copyright © 2004 PhotoDisc; 23 (bl, cl & tr), © David J. & Janice L. Frent Collection/CORBIS; 23 (tl & cr), © CORBIS; 23 (br), © Bettmann/CORBIS; 24 (br), © Kathy deWet-Oleson/Lonely Planet Images; 24 (bl), © Jeffrey L. Rotman/CORBIS; 24 (tr), © Michael Aw/Lonely Planet Images; 25 (l), © Ron Sanford/Photo Researchers, Inc.; 26 (t), Sam Dudgeon/HRW; 28 © Bernd Mellmann/Alamy Photos; 31 © Reuters New Media, Inc./CORBIS; 35 (all), SuperStock; 38 SuperStock; 43 © Ecoscene/CORBIS; 45 (l), Copyright © 1998–2001 EyeWire, Inc. All rights reserved.; 45 (r), © James A. Sugar/CORBIS; 48 (bl), Courtesy of North Carolina Symphony/Michael Zirkle; 49 (tr), Bob Glesener; 49 (cr), Item from www.whitesquirrelshoppe.com/Photo by Sam Dudgeon/HRW; 49 (br), Item from www.whitesquirrelshoppe.com/Photo by Sam Dudgeon/HRW; 49 (tl), Digital Image copyright © 2004 PhotoDisc; 56 Victoria Smith/HRW. **Chapter 2:** 58–59 (bkgd), © AFP/CORBIS; 58 (br), David Gamble/Sygma; 61 © AFP/CORBIS; 63 (all), Bruce Iverson; 64 © Bill Frymire/Masterfile; 73 (br), Sam Dudgeon/HRW; 75 (bc & br), EyeWire - Digital Image copyright © (2004) EyeWire; 75 (bl), PhotoDisc - Digital Image copyright © (2004) PhotoDisc; 78 © Declan McCullagh; 82 Sam Dudgeon/HRW; 83 (all), Victoria Smith/HRW; 85 (l), Collection Walker Art Center, Minneapolis Gift of Fredrick R. Weisman in honor of his parents, William and Mary Weisman, 1988; 86 © 2001 Ron Kimball Studios; 89 (t), © D. Donne Bryant/DDB Stock Photo/All Rights Reserved; 89 (b), Erich Lessing/Art Resource, NY; 91 (b), Lisette LeBon/SuperStock; 92 (t), Everett Collection; 92 (b), Frederic De Lafosse/Sygma; 92 (c), Ulvis Alberts/Motion Picture & Television Photo Archive; 96 (b), Ken Taylor/Wildlife Images; 96 (tr), NASA; 97 (br), Ron Salmon/Simpson's Nature Photography; 97 (tr), Ron Salmon/Simpson's Nature Photography; 97 (cr), Ron Salmon/Simpson's Nature Photography; 99 (l), Photo Researchers, Inc.; 103 Courtesy of the National Grocers Association Best Bagger Contest; 104 Peter Van Steen/HRW; 105 (all), Sam Dudgeon/HRW/Courtesy Fast Forward Skate Shop, Austin, TX; 107 (br), James Urbach/SuperStock; 115 © Reuters NewMedia Inc./CORBIS; 118-119 (b), © Ali Kabas/Alamy Photos; 118 (cr), © 2001/ImageState Inc.; 118 (bl), © Joseph Sohm; ChromoSohm Inc./CORBIS; 120 (cl, cr & b), © Jenny Thomas/HRW; 126 Peter Van Steen/HRW. **Chapter 3:** 128–129 (bkgd), Tom Pantages Photography; 128 (br), © Jay Ireland & Georgienne E. Bradley; 130 (br), Chuck Nicklin/Al Giddings Images, Inc.; 130 (tl), Natalie Fobes/National Geographic Collection; 133 (l), © Neil Rabinowitz/CORBIS; 137 (l), © Stock Trek/PhotoDisc/Picture Quest; 140 © 2001 Jay Mallin; 143 (l), © Lee Foster/Words & Pictures/PictureQuest; 149 (t), © CORBIS; 151 © W. Faidley/WS Image; 153 © Ann Purcell; Carl Purcell/Words & Pictures/PictureQuest; 161 (b), Sam Dudgeon/HRW/Sheet music courtesy Martha Dudgeon.; 166 Victoria Smith/HRW; 169 (l), Michael Rosenfeld/Stone/Getty Images; 170 © Reuters New Media Inc./CORBIS; 173 (b), Classic PIO Partners; 173 (t), Image Copyright © Digital Vision; 173 (c), © Underwood & Underwood/CORBIS; 177 (l), © Buddy Mays/CORBIS; 180 (tr), Paul M. Hudy; 181 (cr), Jane Faircloth/Transparencies, Inc.; 181 (t), Chuck Eaton/Transparencies, Inc.; 182 (br), © Jenny Thomas/HRW; 188 (all), Victoria Smith/HRW. **Chapter 4:** 190–191 (bkgd), © Brian Leatart/FoodPix; 190 (b), Jenny Thomas/HRW; 192 Richard Nowitz/Photo Researchers, Inc.; 195 (l), © Paul Almasy/CORBIS; 196 © Lynn Stone/Index Stock Imagery/PictureQuest; 199 (l), AP Photo/The Fresno Bee, Richard Darby/Wide World Photos; 206 Darren Carroll/HRW; 213 © Galen Rowell/CORBIS; 214 Darren Carroll/HRW; 215 Victoria Smith/HRW/Courtesy Oshman's, Austin, TX; 217 (l), © Gail Mooney/CORBIS; 219 (b), Ken Karp/HRW; 222 © Glen Allison/Alamy Photos; 227 (l), Hulton Archive by Getty Images; 229 (r), Victoria Smith/HRW; 229 (tl), Richard Heinzen/SuperStock; 230 © Jeffrey L. Rotman/CORBIS; 239 (l), © Gallo Images/CORBIS; 239 (r), G.K. & Vikki Hart/Getty Images; 240 (l), Wolfgang Kaehler Photography; 243 (l), © Michael John Kielty/CORBIS; 244 Peter Van Steen/HRW/Courtesy Russell Korman Fine Jewelry, Austin, TX; 245 © Charles O'Rear/CORBIS; 248 (c), © Tony Freeman/PhotoEdit; 249 (t), Smithsonian Institution, Washington, DC; 249 (b), © Hulton-Deutsch Collection/CORBIS; 250 (b), © Jenny Thomas/HRW. **Chapter 5:** 258–259 (bkgd), Ship model by Jean K. Eckert/Photo Courtesy of © The Mariners' Museum, Newport News, Virginia; 258 (b), Gordon Chibroski/Press Herald/AP/Wide World Photos; 260 Darren Carrol/HRW; 263 (br), © Michael Lawrence/Lonely Planet Images; 263 (bc), © Chris Mellor/Lonely Planet Images; 263 (t), © Gavin Anderson/Lonely Planet Images; 263 (bl), © Stephen Frink/CORBIS; 263 (tr), © Stephen Frink/Alamy Photos; 264 Victoria Smith/HRW; 265 James L. Amos/SuperStock; 268 (all), Victoria Smith/HRW/Jenga ® is a registered trademark of Pokonobe Associates and is used with its permission. © 2002 Pokonobe Associates. All rights reserved. 269 (all), Victoria Smith/HRW/Jenga ® is a registered trademark of Pokonobe Associates and is used with its permission. © 2002 Pokonobe Associates. All rights reserved. 272 NASA; 275 (l), © Amos Nachoum/Corbis Stock Market; 275 (r), PhotoDisc - Digital Image © 2004; 277 (b), Sam Dudgeon/HRW; 280 Peter Van Steen/HRW; 284 © Francis E. Caldwell/Affordable Photo Stock; 288 (t), Sam Dudgeon/HRW/Courtesy Chuck and Nan Ellis; 288 (b), Victoria Smith/HRW; 289 Van Gogh Museum, Amsterdam/SuperStock; 291 (t & b), Library of Congress; 291 (c), Victoria Smith/HRW; 291 (t-frame) © 1999 Image Farm, Inc.; 292 (b), © Dennie Cody/Getty Images/Taxi; 293 (tr), roadsideamerica.com, Kirby, Smith & Wilkins; 293 (br), © Jim Hargan; 294 (b), Randall Hyman/HRW; 300 Sam Dudgeon/HRW. **Chapter 6:** 302–303 (bkgd), © Mark E. Gibson Photography; 302 (br), Victoria Smith/HRW; 308 Peter Van Steen/HRW; 315 (l), © Richard A. Cooke/CORBIS; 316 © Buddy Mays/CORBIS; 319 (l), © Ellen Senisi/The Image Works; 321 (b), © 1998 Joseph De Sciose; 324 Sam Dudgeon/HRW; 325 Peter Van Steen/HRW; 331 (l), © Archivo Iconografico, S.A./CORBIS; 332 (tr), AP Photo/The News & Observer, Chuck Liddy; 332 (bl), Burke/Triolo Productions/FoodPix; 333 (t), © Pat and Chuck Blackley; 333 (br), Ann & Rob Simpson/Simpson's Nature Photography; 334 (b), Randall Hyman/HRW; 340 © James Marshall/CORBIS. **Chapter 7:** 342–343 (bkgd), © 2002 Bruno Burklin/Aerial Aesthetics; 342 (br), © Stone/Getty Images/Stone; 344 The Art Archive/Private Collection /Harper Collins Publishers/© 2004 Artists Rights Society (ARS), New York/ADAGP, Paris; 346 Science Kit & Boreal Laboratories; 347 (t), © Burstein Collection/CORBIS/ © 2004 Mondrian/Holtzman Trust, c/o Beeldrecht/Artists Rights Society (ARS), New York; 347 (b), Copyright Tate Gallery, London, Great Britain/Art Resource, NY/© 2004 Artist Rights Society (ARS), New York/Pro Litteris, Zurich; 354 © Gisela Damm/eStock Photography/PictureQuest; 357 (l), John Burke/SuperStock; 362 (t & b), © Archivo Iconografico, S.A./CORBIS; 368 © Gianni Dagh Orti/CORBIS; 371 (t), John Warden/SuperStock; 371 (tc), © Roman Soumar/CORBIS; 371 (bc), © Jacqui Hurst/CORBIS; 371 (b), © Jacqui Hurst/CORBIS; 374 Sam Dudgeon/HRW; 377 (r), © Craig Aurness/CORBIS; 377 (l), © Mark E. Gibson; 378 © 1997 Jon Reis/Photolink; 381 (l), © Bob Krist/CORBIS; 385 (t), © CORBIS; 387 (b), © Craig Aurness/CORBIS; 388 (bl), Peter Van Steen/HRW; 388 (br), Sam Dudgeon/HRW; 390 (tl & tr), Peter Van Steen/HRW; 390 (tc & bc), Sam Dudgeon/HRW; 390 (br), © Mark Snyder, TetraLite Kites, Seattle, WA; 390 (bl), Peter Van Steen/Courtesy International Playthings, Inc. 392 (t), © Neal Preston/CORBIS; 394 (br) (twice) © 2001 Frank Siteman c/o MIRA; 395 (c & b), Werner Forman/Art Resource, NY; 395 (t), Smithsonian American Art Museum, Washington, DC/Art Resource, NY; 396 (t), Steve Vidler/SuperStock; 397 (tr), © Arthur Thévenart/CORBIS; 397 (c), © Karen Gowlett-Holmes; 397 (b), C. Zeiss/Bruce Coleman, Inc.; 399 (b), © William Panzer/Stock Connection/PictureQuest; 399 (t), © 2000 Joseph Scheer; 400 (t), Symmetry Drawing E121 by M.C. Escher © 1999 Cordon Art-Baarn-Holland. All rights reserved. 401 (t), Symmetry Drawing E25 by M.C. Escher © 1999 Cordon Art-Baarn-Holland. All rights reserved. 402 (bl), Photos Courtesy of NC Division of Tourism, Film, and Sports Development. 403 (r), Poster Art by Bob Rankin/Photo by Victoria Smith/HRW Photo; 403 (t), Tim Pflaum; 404 (b), Ken Karp/HRW; 410, © Jonathan Blair/CORBIS. **Chapter 8:** 412–413 (bkgd), © Mark E. Gibson c/o MIRA; 412 (br), Victoria Smith/HRW; 414 © Frank Lane Picture Agency/CORBIS; 417 (l), Associated Press, AP/Wide World Photos; 418 (t),

Michelle Bridwell/HRW; 418 (bl), © David A. Northcott/CORBIS; 418 (br), © Fogden, Michael/Animals Animals/Earth Scenes; 420 (t), Photo by Randall L. Ricklefs/ McDonald Observatory; 420 (b), Sam Dudgeon/HRW; 423 (t), Peter Van Steen/HRW; 423 (b), © John Kelly/Getty Images/Stone; 424 David Walberg/Sports Illustrated; 427 NOAA Costal Services Center; 429 (b), © Bill Truslow/Stock Connection/ PictureQuest; 430 (t), © Nik Wheeler/CORBIS; 433 © Christie's Images/CORBIS/ © 2004 Banco de México Diego Rivera & Frida Kahlo Museums Trust. Av. Cinco de Mayo No. 2, Col. Centro, Del. Cuauhtémoc 06059, México, D.F./Reproducción autorizada por el Instituto Nacional de Bellas Artes Y Literatura; 433 (frame), HRW; 439 (br), © Yann Arthus-Bertrand/CORBIS; 439 (tl), © Carl & Ann Purcell/CORBIS; 441 (t), Photo © Stefan Schott/Panoramic Images, Chicago 1998; 443 (b), © TONY FREEMAN/PhotoEdit; 445 © Johathan Blair/CORBIS; 447 Digital Image copyright © 2004 Karl Weatherly/PhotoDisc; 451 © K. H. Photo/International Stock/ ImageState; 453 (t), © Lary Lee Photography/CORBIS; 453 (bkgd), Corbis Images; 458 (tc), Kelly Culpepper/Transparencies, Inc.; 458 (bc), © David Muench/CORBIS; 459 (tr), John Elk/Bruce Coleman, Inc.; 459 (br), © John Elk III; 460 (b), Jenny Thomas/HRW; 466 (br), U.S. Postal Service. **Chapter 9:** 468–469 (bkgd), © Arvind Garg/CORBIS; 468 (br), Victoria Smith/HRW; 475 (tr), © Charles & Josette Lenars/CORBIS; 475 (bl), © Kevin Fleming/CORBIS; 475 (tl), Steve Vidler/SuperStock; 475 (br), R.M. Arakaki/Imagestate; 479 (r), Sam Dudgeon/HRW; 479 (l), © Natalie Fobes/CORBIS; 480 (t), Peter Van Steen/HRW; 483 (b), PhotoDisc - Digital Image copyright © 2002 PhotoDisc; 485 (b), Sam Dudgeon/HRW; 486 (t), Peter Van Steen/HRW; 487 (all), Victoria Smith/HRW; 488 (r), Mountain High Maps® Copyright © 1997 Digital Wisdom, Inc.; 490 (b), Sam Dudgeon/HRW; 491 (all), Peter Van Steen/HRW; 493 © Michael Newman/PhotoEdit; 495 (t), Hahn's Titanic Plans by Robert Hahn; 495 (c), Ulster Folk & Transport Museum; 495 (b), Bill Greenblatt/ UPI Photo Service/NewsCom; 500 (tr), © Gary W. Carter/CORBIS; 500 (bl), © R. Gilbert/Robertstock.com; 501 (tr), Courtesy of the Screen Gems Studios/Photo by Victoria Smith/HRW; 501 (b), Courtesy of the Screen Gems Studios; 501 (b), Courtesy of the Screen Gems Studios; 501 (b), Courtesy of the Screen Gems Studios; 501 (b), Courtesy of the Screen Gems Studios; 502 (b), Jenny Thomas/HRW. **Chapter 10:** 510–511 (bkgd), © Getty Images/Stone; 510 (br), Victoria Smith/ HRW; 516 © V.C.L. Tipp Howell/Getty Images/FPG International; 519 (b), REUTERS/ Gary Wiepert /NewsCom; 519 (t), Courtesy of National Weather Service, NOAA; 520 Peter Van Steen/HRW; 523 (l), © David Young-Wolff/PhotoEdit; 524 (all), Victoria Smith/HRW/SCRABBLE® is a trademark of Hasbro in the United States and Canada. © 2002 Hasbro, Inc. All Rights Reserved.; 528 (l), Peter Van Steen/HRW; 528 (r), Digital Image copyright © 2004 EyeWire; 529 (b), © Left Lane Productions/ CORBIS; 530 © Dennis Degnan/CORBIS; 532 (all), Sam Dudgeon/HRW; 536 Victoria Smith/HRW; 539 (tl), Pushkin Museum of Fine Arts, Moscow, Russia/SuperStock; 539 (tr), © Philadelphia Museum of Art/CORBIS; 540 Alden Pellett /AP/Wide World Photos; 543 (l), Photofest; 546 (tr), Jane Faircloth/Transparencies, Inc.; 546 (bl), © Pat and Chuck Blackley; 547 (t), © Harrison Shull/shullphoto.com; 547 (b), Ken Taylor/Wildlife Images; 548 (c), Sam Dudgeon/HRW; 548 (b), Ken Karp/HRW; 554 © Bettmann/CORBIS. **Chapter 11:** 556–557 (bkgd), NASA; 556 (b), NASA; 561 (r), © Barry Winiker/Index Stock Imagery, Inc.; 561 (l), Sam Dudgeon/HRW; 563 (b), Sam Dudgeon/HRW; 563 (t), CLOSE TO HOME © 1994 John McPherson. Reprinted with Permission of UNIVERSAL PRESS SYNDICATE. All rights reserved. 568 © L. Clarke/CORBIS; 571 The Holland Sentiel, Barbara Beal/AP/Wide World Photos; 573 (b), Victoria Smith/HRW; 573 (caricature) Sam Dudgeon/HRW/Caricature by Ryan Foerster, San Antonio; 577 (t), U.S. Geological Survey Western Region Costal and Marine Geology; 577 (b), James Wilson/Woodfin Camp & Associates; 578 Victoria Smith/HRW; 581 © Al Grotell 1990; 582, Spencer Tirey/AP/Wide World Photos; 583 Sam Dudgeon/HRW; 586 © 1995 Joseph De Sciose; 589 (tl), Dr. E. R. Degginger/Color-Pic, Inc.; 589 (tr), Sam Dudgeon/HRW; 589 (bl), Dr. E. R. Degginger/Color-Pic, Inc.; 589 (br), Pat Lanza/Bruce Coleman, Inc.; 590 © Michael S. Yamashita/CORBIS; 591 The Far Side ® by Gary Larson © 1985 FarWorks, Inc. All Rights Reserved. Used with permission.; 592 (bl), George H. Harrison/Bruce Coleman, Inc.; 592 (tr), Kenneth Fink/Bruce Coleman, Inc.; 593 (t), Ken Taylor/ Wildlife Images; 594 (b), Randall Hyman/HRW. **Chapter 12:** 602–603 (bkgd), © Kelly-Mooney Photography/CORBIS; 602 (br), Roberto Borea/AP/Wide World Photos; 604 United States Postal Service, © Rube Goldberg, Inc.; 607 (l), Ali Jarekji/REUTERS/TimePix; 608 (t), © Helen Norman/CORBIS; 608 (bl), © John Kaprielian/Photo Researchers, Inc.; 608 (bc), © Pointer, John/Animals Animals/Earth Scenes; 608 (r), © ChromaZone Images/Index Stock Imagery/PictureQuest; 611 (t), RO-MA Stock/Index Stock Imagery, Inc.; 611 (inset) RO-MA Stock/Index Stock Imagery, Inc.; 615 (all), John Langford/HRW; 616 Scott Vallance/VIP Photographic Associates; 617 NOAA Costal Services Center; 619 (l), © Ron Kimball Studios; 620 © Bill Horsman/Stock Boston; 623 © 1999 Ron Kimball Studios; 625 (b), © Michael T. Sedam/CORBIS; 628 Victoria Smith/HRW; 631 CALVIN AND HOBBES © 1987 Watterson. Reprinted with Permission of UNIVERSAL PRESS SYNDICATE. All rights reserved.; 635 Mary Ann Chastain/AP/Wide World Photos; 640 (all), Sam Dudgeon/HRW; 642 (r), Courtesy of Hugh Morton/USS Battleship North Carolina; 643 (tl), CBS-TV/The Kobal Collection; 643 (tr), © Pat and Chuck Blackley; 650 (t), Microworks/PhotoTake; 652 & 653 Sam Dudgeon/HRW

Glossary

internet connect
Multilingual* Glossary Online: *go.hrw.com*
Keyword: MS4 Glossary

*Languages: Cambodian, Chinese, Creole, Farsi, Hmong, Korean, Russian, Spanish, Tagalog, and Vietnamese

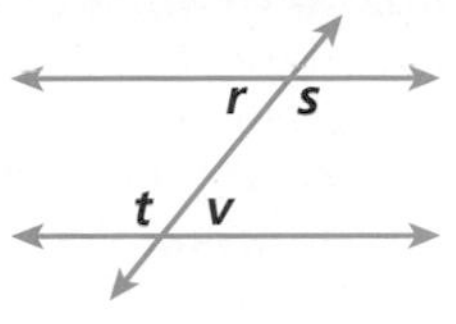

A

absolute value The distance of a number from zero on a number line; shown by | |. (p. 131)

Example: $|-5| = 5$

accuracy The closeness of a given measurement or value to the actual measurement or value. (p. 420)

acute angle An angle that measures less than 90°. (p. 349)

acute triangle A triangle with all angles measuring less than 90°. (p. 374)

addend A number added to one or more other numbers to form a sum.

Example: In the expression $4 + 6 + 7$, 4, 6, and 7 are addends.

Addition Property of Equality The property that states that if you add the same number to both sides of an equation, the new equation will have the same solution. (p. 110)

Addition Property of Opposites The property that states that the sum of a number and its opposite equals zero.

Example: $12 + (-12) = 0$

additive inverse The opposite of a number.

Example: The additive inverse of 6 is −6.

adjacent angles Angles in the same plane that have a common vertex and a common side; in the diagram, $\angle a$ and $\angle b$ are adjacent angles.

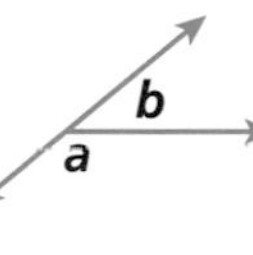

algebraic expression An expression that contains at least one variable. (p. 92)

Example: $x + 8$, $4(m - b)$

algebraic inequality An inequality that contains at least one variable. (p. 574)

Example: $x + 3 > 10$; $5a > b + 3$

alternate exterior angles A pair of angles formed by a transversal and two lines; in the diagram, the pairs of alternate exterior angles are $\angle a$ and $\angle d$ and $\angle b$ and $\angle c$. (p. 353)

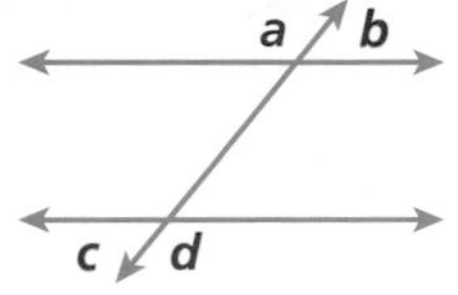

alternate interior angles A pair of angles formed by a transversal and two lines; in the diagram, the pairs of alternate interior angles are $\angle r$ and $\angle v$ and $\angle s$ and $\angle t$. (p. 353)

angle A figure formed by two rays with a common endpoint called the vertex. (p. 348)

arc A part of a circle named by its endpoints. (p. 362)

area The number of square units needed to cover a given surface. (p. 430)

Associative Property
Addition: The property that states that for all real numbers a, b, and c, the sum is always the same, regardless of their grouping: $a + b + c = (a + b) + c = a + (b + c)$. (p. 712)
Multiplication: The property that states that for all real numbers a, b, and c, their product is always the same, regardless of their grouping: $a \cdot b \cdot c = (a \cdot b) \cdot c = a \cdot (b \cdot c)$. (p. 712)

asymmetry Not identical on either side of a central line; not symmetrical. (p. 396)

axes The two perpendicular lines of a coordinate plane that intersect at the origin. (p. 134)

B

bar graph A graph that uses vertical or horizontal bars to display data. (p. 20)

base-10 system A number system in which all numbers are expressed using the digits 0–9.

base (in numeration) When a number is raised to a power, the number that is used as a factor is the base. (p. 60)

Example: $3^5 = 3 \cdot 3 \cdot 3 \cdot 3 \cdot 3$

base (of a polygon) A side of a polygon.

base (of a three-dimensional figure) A face of a three-dimensional figure by which the figure is measured or classified. (p. 472)

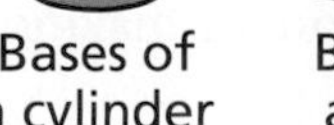
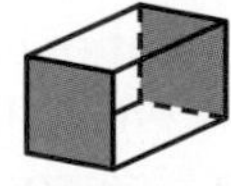

Bases of a cylinder | Bases of a prism | Base of a cone | Base of a pyramid

binary number system A number system in which all numbers are expressed using only two digits, 0 and 1. (p. 718)

bisect To divide into two congruent parts. (p. 358)

box-and-whisker plot A graph that displays the highest and lowest quarters of data as whiskers, the middle two quarters of the data as a box, and the median. (p. 28)

break (graph) A zigzag on a horizontal or vertical scale of a graph that indicates that some of the numbers on the scale have been omitted. (p. 44)

capacity The amount a container can hold when filled.

Celsius A metric scale for measuring temperature in which 0°C is the freezing point of water and 100°C is the boiling point of water; also called *centigrade*.

center (of a circle) The point inside a circle that is the same distance from all the points on the circle. (p. 362)

center (of rotation) The point about which a figure is rotated. (p. 397)

central angle An angle with its vertex at the center of a circle. (p. 363)

certain (probability) Sure to happen; having a probability of 1. (p. 512)

chord A line segment with endpoints on a circle. (p. 362)

circle The set of all points in a plane that are the same distance from a given point called the center. (p. 362)

circle graph A graph that uses sectors of a circle to compare parts to the whole and parts to other parts. (p. 24)

circumference The distance around a circle. (p. 425)

clockwise A circular movement to the right in the direction shown.

coefficient The number that is multiplied by the variable in an algebraic expression. (p. 100)

Example: 5 is the coefficient in $5b$.

combination An arrangement of items or events in which order does not matter. (p. 536)

commission A fee paid to a person for making a sale.

common denominator A denominator that is the same in two or more fractions.

Example: The common denominator of $\frac{5}{8}$ and $\frac{2}{8}$ is 8.

common factor A number that is a factor of two or more numbers.

Example: 8 is a common factor of 16 and 40.

common multiple A number that is a multiple of each of two or more numbers.

Example: 15 is a common multiple of 3 and 5.

Commutative Property

Addition: The property that states that two or more numbers can be added in any order without changing the sum. (p. 712)

Example: $8 + 20 = 20 + 8$; $a + b = b + a$

Multiplication: The property that states that two or more numbers can be multiplied in any order without changing the product. (p. 712)

Example: $6 \cdot 12 = 12 \cdot 6$; $a \cdot b = b \cdot a$

compatible numbers Numbers that are close to the given numbers that make estimation or mental calculation easier. (p. 192)

complementary angles Two angles whose measures add to 90°. (p. 349)

composite number A number greater than 1 that has more than two whole-number factors. (p. 78)

compound event An event made up of two or more simple events.

compound inequality A combination of more than one inequality. (p. 575)

Example: $-2 \le x < 10$. x is greater than or equal to -2 and less than 10.

cone A three-dimensional figure with one vertex and one circular base. (p. 473)

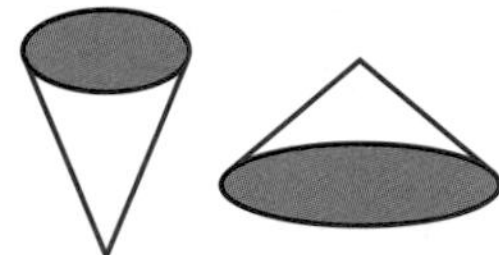

congruent Having the same size and shape. (p. 345)

congruent angles Angles that have the same measure. (p. 355)

congruent segments Segments that have the same length. (p. 345)

constant A value that does not change. (p. 92)

coordinate plane (coordinate grid) A plane formed by the intersection of a horizontal number line called the x-axis and a vertical number line called the y-axis. (p. 134)

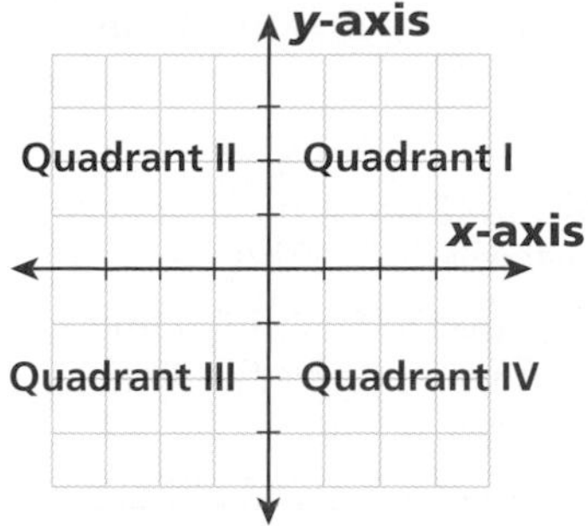

coordinate One of the numbers of an ordered pair that locate a point on a coordinate graph. (p. 134)

correlation The description of the relationship between two data sets. (p. 41)

correspondence The relationship between two or more objects that are matched.

corresponding angles (for lines) A pair of angles formed by a transversal and two lines; in the diagram, the pairs of corresponding angles are $\angle m$ and $\angle q$, $\angle n$ and $\angle r$, $\angle o$ and $\angle s$, and $\angle p$ and $\angle t$. (p. 353)

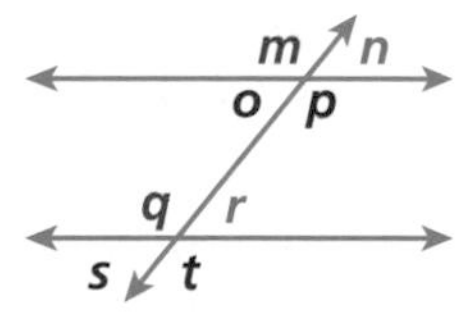

corresponding angles (in polygons) Matching angles of two or more polygons. (p. 280)

corresponding sides Matching sides of two or more polygons. (p. 280)

cosine (cos) In a right triangle, the ratio of the length of the side adjacent to an acute angle to the length of the hypotenuse. (p. 724)

counterclockwise A circular movement to the left in the direction shown.

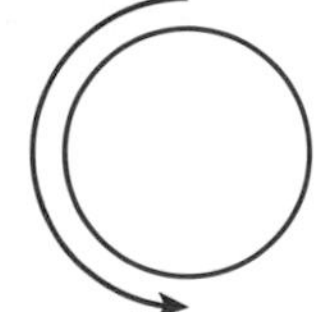

counterexample An example that shows that a statement is false. (p. 723)

cross product The product of numbers on the diagonal when comparing two ratios. (p. 268)

Example: 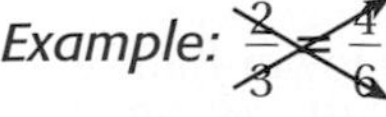 $2 \cdot 6 = 12$ $\quad 3 \cdot 4 = 12$

cube (geometric figure) A rectangular prism with six congruent square faces. (p. 472)

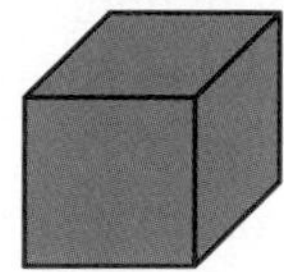

cube (in numeration) A number raised to the third power.

cumulative frequency The sum of successive data items. (p. 14)

customary system of measurement The measurement system often used in the United States. (p. 414)

Example: inches, feet, miles, ounces, pounds, tons, cups, quarts, gallons

cylinder A three-dimensional figure with two parallel, congruent circular bases connected by a curved lateral surface. (p. 472)

decagon A polygon with ten sides. (p. 369)

decimal system A base-10 place value system.

deductive reasoning Using logic to show that a statement is true. (p. 722)

degree The unit of measure for angles or temperature. (p. 348)

denominator The bottom number of a fraction that tells how many equal parts are in the whole.

Example: $\frac{3}{4}$ ← denominator

dependent events Events for which the outcome of one event affects the probability of the other. (p. 530)

diagonal A line segment that connects two non-adjacent vertices of a polygon. (p. 383)

diameter A line segment that passes through the center of a circle and has endpoints on the circle, or the length of that segment. (p. 362)

difference The result when one number is subtracted from another.

dimension The length, width, or height of a figure.

Distributive Property The property that states if you multiply a sum by a number, you will get the same result if you multiply each addend by that number and then add the products. (p. 712)

Example: $5(20 + 1) = (5 \cdot 20) + (5 \cdot 1)$

dividend The number to be divided in a division problem.

Example: In $8 \div 4 = 2$, 8 is the dividend.

divisible Can be divided by a number without leaving a remainder. (pp. 76, 715)

Division Property of Equality The property that states that if you divide both sides of an equation by the same nonzero number, the new equation will have the same solution. (p. 114)

divisor The number you are dividing by in a division problem.

Example: In $8 \div 4 = 2$, 4 is the divisor.

domain The set of all possible input values of a function. (p. 613)

double-bar graph A bar graph that compares two related sets of data. (p. 20)

double-line graph A line graph that shows how two related sets of data change over time. (p. 36)

edge The line segment along which two faces of a polyhedron intersect. (p. 472)

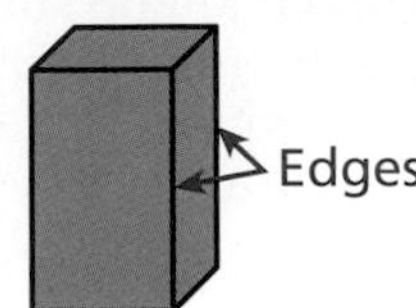

elements The words, numbers, or objects in a set. (p. 636)

empty set A set that has no elements. (p. 637)

endpoint A point at the end of a line segment or ray.

equally likely outcomes Outcomes that have the same probability. (p. 512)

equation A mathematical sentence that shows that two expressions are equivalent. (p. 104)

equilateral triangle A triangle with three congruent sides. (p. 374)

equivalent Having the same value.

equivalent fractions Fractions that name the same amount or part. (p. 166)

equivalent ratios Ratios that name the same comparison. (p. 264)

estimate (n) An answer that is close to the exact answer and is found by rounding, or other methods.

estimate (v) To find an answer close to the exact answer by rounding or other methods.

evaluate To find the value of a numerical or algebraic expression. (p. 92)

even number An integer that is divisible by two.

event An outcome or set of outcomes of an experiment or situation. (p. 512)

expanded form A number written as the sum of the values of its digits.

Example: 236,536 written in expanded form is $200{,}000 + 30{,}000 + 6{,}000 + 500 + 30 + 6$.

experiment In probability, any activity based on chance, such as tossing a coin. (p. 512)

experimental probability The ratio of the number of times an event occurs to the total number of trials, or times that the activity is performed. (p. 516)

exponent The number that indicates how many times the base is used as a factor. (p. 60)

Example: $2^3 = 2 \times 2 \times 2 = 8$; 3 is the exponent.

exponential decay Occurs in an exponential function when the output $f(x)$ gets smaller as the input x gets larger.

exponential form A number is in exponential form when it is written with a base and an exponent.

Example: 4^2 is the exponential form for $4 \cdot 4$.

exponential function A nonlinear function in which the variable is in the exponent.

exponential growth Occurs in an exponential function when the output $f(x)$ gets larger as the input x gets larger.

expression A mathematical phrase that contains operations, numbers, and/or variables.

face A flat surface of a polyhedron. (p. 472)

factor A number that is multiplied by another number to get a product. (p. 716)

factor tree A diagram showing how a whole number breaks down into its prime factors. (p. 78)

factorial The product of all whole numbers except zero that are less than or equal to a number. (p. 541)

Example: 4 factorial $= 4! = 4 \cdot 3 \cdot 2 \cdot 1$; $0!$ is defined to be 1.

Fahrenheit A temperature scale in which 32°F is the freezing point of water and 212°F is the boiling point of water.

fair When all outcomes of an experiment are equally likely, the experiment is said to be fair. (p. 524)

favorable outcome An outcome you are looking for when conducting an experiment. (p. 524)

Fibonacci sequence The infinite sequence of numbers (1, 1, 2, 3, 5, 8, 13, . . .); starting with the third term, each number is the sum of the two previous numbers; it is named after the thirteenth century mathematician Leonardo Fibonacci. (p. 608)

first quartile The median of the lower half of a set of data; also called *lower quartile.* (p. 28)

formula A rule showing relationships among quantities.

Example: $A = \ell w$ is the formula for the area of a rectangle.

fractal A structure with repeating patterns containing shapes that are like the whole but are of different sizes throughout. (p. 611)

fraction A number in the form $\frac{a}{b}$, where $b \neq 0$.

frequency table A table that lists items together according to the number of times, or frequency, that the items occur. (p. 14)

function An input-output relationship that has exactly one output for each input. (p. 604)

function table A table of ordered pairs that represent solutions of a function. (p. 604)

Fundamental Counting Principle If one event has m possible outcomes and a second event has n possible outcomes after the first event has occurred, then there are $m \cdot n$ total possible outcomes for the two events. (p. 520)

Golden Ratio A ratio approximately equal to 1.618. In a golden rectangle, the ratio of the length to the width is approximately 1.618. (p. 265)

graph of an equation A graph of the set of ordered pairs that are solutions of the equation. (p. 616)

greatest common factor (GCF) The largest common factor of two or more given numbers. (p. 82)

height In a pyramid or cone, the perpendicular distance from the base to the opposite vertex. (pp. 480, 481)

In a triangle or quadrilateral, the perpendicular distance from the base to the opposite vertex or side.

In a prism or cylinder, the perpendicular distance between the bases.

hemisphere A half of a sphere. (p. 473)

heptagon A seven-sided polygon. (p. 369)

hexagon A six-sided polygon. (p. 369)

histogram A bar graph that shows the frequency of data within equal intervals. (p. 21)

hypotenuse In a right triangle, the side opposite the right angle. (p. 450)

Identity Property of One The property that states that the product of 1 and any number is that number. (p. 712)

Identity Property of Zero The property that states the sum of zero and any number is that number. (p. 712)

image A figure resulting from a transformation. (p. 392)

impossible (probability) Can never happen; having a probability of 0. (p. 512)

improper fraction A fraction in which the numerator is greater than or equal to the denominator. (p. 166)

Example: $\frac{5}{5}, \frac{5}{3}$

independent events Events for which the outcome of one event does not affect the probability of the other. (p. 530)

indirect measurement The technique of using similar figures and proportions to find a measure. (p. 284)

inductive reasoning Using a pattern to make a conclusion. (p. 722)

inequality A mathematical sentence that shows the relationship between quantities that are not equivalent. (p. 574)

Example: $5 < 8$; $5x + 2 \geq 12$

input The value substituted into an expression or function. (p. 604)

integers The set of whole numbers and their opposites. (p. 130)

interest The amount of money charged for borrowing or using money, or the amount of money earned by saving money. (p. 328)

interior angles Angles on the inner sides of two lines cut by a transversal. In the diagram, $\angle c$, $\angle d$, $\angle e$, and $\angle f$ are interior angles. (p. 382)

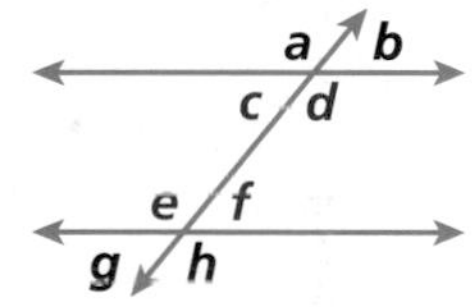

interquartile range The difference between the upper and lower quartiles in a box-and-whisker plot.

intersecting lines Lines that cross at exactly one point.

intersection (sets) The set of elements common to two or more sets. (p. 637)

interval The space between marked values on a number line or the scale of a graph.

inverse operations Operations that undo each other: addition and subtraction, or multiplication and division. (p. 110)

irrational number A number that cannot be expressed as a ratio of two integers or as a repeating or terminating decimal. (p. 178)

isolate the variable To get a variable alone on one side of an equation or inequality in order to solve the equation or inequality. (p. 110)

isosceles triangle A triangle with at least two congruent sides. (p. 374)

kite A quadrilateral with two pairs of adjacent, congruent sides. (p. 378)

lateral face In a prism or a pyramid, a face that is not a base.

lateral surface In a cylinder, the curved surface connecting the circular bases; in a cone, the curved surface that is not a base. (p. 473)

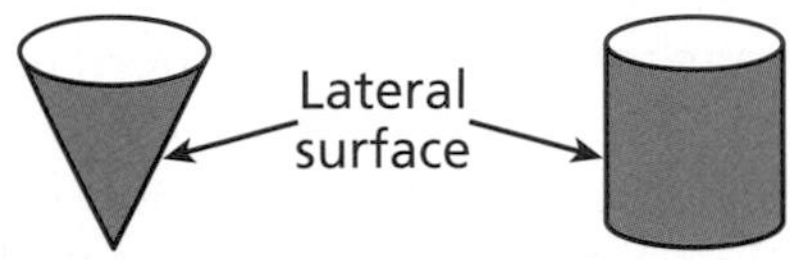

least common denominator (LCD) The least common multiple of two or more denominators.

least common multiple (LCM) The smallest number, other than zero, that is a multiple of two or more given numbers. (p. 86)

legs In a right triangle, the sides that include the right angle; in an isosceles triangle, the pair of congruent sides. (p. 450)

like terms Two or more terms that have the same variable raised to the same power. (p. 100)

Example: In the expression $3a + 5b + 12a$, $3a$ and $12a$ are like terms.

line A straight path that extends without end in opposite directions. (p. 344)

line graph A graph that uses line segments to show how data changes. (p. 35)

line of reflection A line that a figure is flipped across to create a mirror image of the original figure. (p. 392)

line of symmetry The imaginary "mirror" in line symmetry. (p. 396)

line plot A number line with marks or dots that show frequency.

line segment A part of a line between two endpoints. (p. 344)

line symmetry A figure has line symmetry if one-half is a mirror-image of the other half. (p. 396)

linear equation An equation whose solutions form a straight line on a coordinate plane. (p. 616)

linear function A function whose graph is a straight line. (p. 616)

lower extreme The least number in a set of data. (p. 28)

mean The sum of the items in a set of data divided by the number of items in the set; also called *average*. (p. 10)

measure of central tendency A measure used to describe the middle of a data set; the mean, median, and mode are measures of central tendency. (p. 10)

median The middle number, or the mean (average) of the two middle numbers, in an ordered set of data. (p. 10)

metric system of measurement A decimal system of weights and measures that is used universally in science and commonly throughout the world. (p. 414)

Example: centimeters, meters, kilometers, grams, kilograms, milliliters, liters

midpoint The point that divides a line segment into two congruent line segments.

mixed number A number made up of a whole number that is not zero and a fraction. (p. 167)

mode The number or numbers that occur most frequently in a set of data; when all numbers occur with the same frequency, we say there is no mode. (p. 10)

multiple The product of any number and a whole number is a multiple of that number. (p. 86)

Multiplication Property of Equality The property that states that if you multiply both sides of an equation by the same number, the new equation will have the same solution. (p. 114)

Multiplication Property of Zero The property that states that for all real numbers a, $a \times 0 = 0$ and $0 \times a = 0$. (p. 712)

mutually exclusive Two events are mutually exclusive if they cannot occur in the same trial of an experiment. (p. 721)

negative correlation Two data sets have a negative correlation, or relationship, if one set of data values increases while the other decreases. (p. 41)

negative integer An integer less than zero. (p. 130)

net An arrangement of two-dimensional figures that can be folded to form a polyhedron. (p. 486)

no correlation Two data sets have no correlation when there is no relationship between their data values. (p. 41)

nonlinear function A function whose graph is not a straight line. (p. 628)

nonterminating decimal A decimal that never ends. (p. 178)

numerator The top number of a fraction that tells how many parts of a whole are being considered.

Example: $\frac{4}{5}$ ⟵ numerator

numerical expression An expression that contains only numbers and operations. (p. 70)

obtuse angle An angle whose measure is greater than 90° but less than 180°. (p. 349)

obtuse triangle A triangle containing one obtuse angle. (p. 374)

octagon An eight-sided polygon. (p. 369)

odd number An integer that is not divisible by two.

odds A comparison of favorable outcomes and unfavorable outcomes. (p. 544)

opposites Two numbers that are an equal distance from zero on a number line; also called *additive inverse.* (p. 130)

order of operations A rule for evaluating expressions: first perform the operations in parentheses, then compute powers and roots, then perform all multiplication and division from left to right, and then perform all addition and subtraction from left to right. (p. 70)

ordered pair A pair of numbers that can be used to locate a point on a coordinate plane. (p. 134)

origin The point where the x-axis and y-axis intersect on the coordinate plane; (0, 0). (p. 134)

outcome A possible result of a probability experiment. (p. 512)

outlier A value much greater or much less than the others in a data set. (p. 11)

output The value that results from the substitution of a given input into an expression or function. (p. 604)

overestimate An estimate that is greater than the exact answer.

parallel lines Lines in a plane that do not intersect. (p. 354)

parallelogram A quadrilateral with two pairs of parallel sides. (p. 378)

Pascal's triangle A triangular arrangement of numbers in which each row starts and ends with 1 and each other number is the sum of the two numbers above it. (p. 534)

pentagon A five-sided polygon. (p. 369)

percent A ratio comparing a number to 100. (p. 304)

Example: $45\% = \frac{45}{100}$

percent of change The amount stated as a percent that a number increases or decreases. (p. 324)

percent of decrease A percent change describing a decrease in a quantity. (p. 324)

percent of increase A percent change describing an increase in a quantity. (p. 324)

perfect square A square of a whole number. (p. 444)

Example: $5 \cdot 5 = 25$, and $7^2 = 49$; 25 and 49 are perfect squares.

perimeter The distance around a polygon. (p. 424)

permutation An arrangement of items or events in which order is important. (p. 540)

perpendicular bisector A line that intersects a segment at its midpoint and is perpendicular to the segment. (p. 358)

perpendicular lines Lines that intersect to form right angles. (p. 354)

pi (π) The ratio of the circumference of a circle to the length of its diameter; $\pi \approx 3.14$ or $\frac{22}{7}$. (p. 425)

plane A flat surface that extends forever. (p. 344)

point An exact location in space. (p. 344)

polygon A closed plane figure formed by three or more line segments that intersect only at their endpoints (vertices). (p. 368)

polyhedron A three-dimensional figure in which all the surfaces or faces are polygons. (p. 472)

population The entire group of objects or individuals considered for a survey. (p. 4)

positive correlation Two data sets have a positive correlation, or relationship, when their data values increase or decrease together. (p. 41)

positive integer An integer greater than zero. (p. 130)

power A number produced by raising a base to an exponent. (p. 60)

Example: $2^3 = 8$, so 8 is the 3rd power of 2.

precision The level of detail of a measurement, determined by the unit of measure. (p. 420)

prime factorization A number written as the product of its prime factors. (p. 78)

Example: $10 = 2 \cdot 5$, $24 = 2^3 \cdot 3$

prime number A whole number greater than 1 that has exactly two factors, itself and 1. (p. 78)

principal The initial amount of money borrowed or saved. (p. 328)

prism A polyhedron that has two congruent polygon-shaped bases and other faces that are all parallelograms. (p. 472)

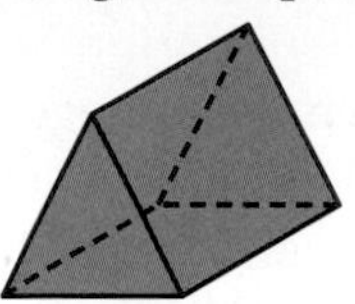

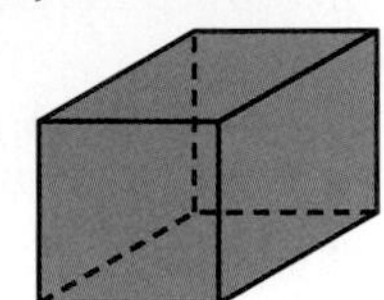

probability A number from 0 to 1 (or 0% to 100%) that describes how likely an event is to occur. (p. 512)

product The result when two or more numbers are multiplied.

proper fraction A fraction in which the numerator is less than the denominator.

Example: $\frac{3}{4}, \frac{1}{12}, \frac{7}{8}$

proportion An equation that states that two ratios are equivalent. (p. 264)

protractor A tool for measuring angles. (p. 348)

pyramid A polyhedron with a polygon base and triangular sides that all meet at a common vertex. (p. 472)

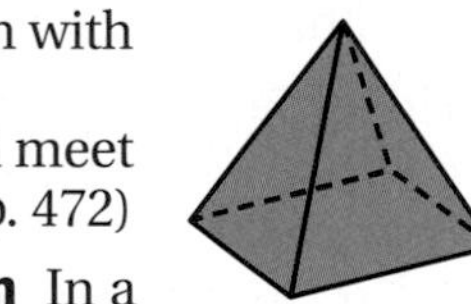

Pythagorean Theorem In a right triangle, the square of the length of the hypotenuse is equal to the sum of the squares of the lengths of the legs. (p. 450)

quadrant The x- and y-axes divide the coordinate plane into four regions. Each region is called a quadrant. (p. 134)

quadratic function A function of the form $y = ax^2 + bx + c$, where $a \neq 0$. (p. 729)

Example: $y = 2x^2 - 12x + 10$, $y = -3x^2$

quadrilateral A four-sided polygon. (p. 369)

quartile Three values, one of which is the median, that divide a data set into fourths. See also *first quartile, third quartile.* (p. 28)

quotient The result when one number is divided by another.

radical symbol The symbol $\sqrt{}$ used to represent the nonnegative square root of a number. (p. 444)

radius A line segment with one endpoint at the center of a circle and the other endpoint on the circle, or the length of that segment. (p. 362)

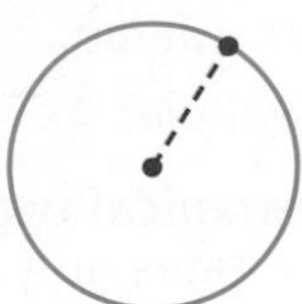

random numbers In a set of random numbers, each number has an equal chance of being selected. (p. 8)

random sample A sample in which each individual or object in the entire population has an equal chance of being selected. (p. 5)

range (in statistics) The difference between the greatest and least values in a data set. (p. 10)

range (in a function) The set of all possible output values of a function. (p. 613)

rate A ratio that compares two quantities measured in different units. (p. 260)

Example: The speed limit is 55 miles per hour, or 55 mi/h.

rate of interest The percent charged or earned on an amount of money; see *simple interest.* (p. 328)

ratio A comparison of two quantities by division. (p. 260)

Example: 12 to 25, 12:25, $\frac{12}{25}$

rational number Any number that can be expressed as a ratio of two integers. (p. 162)

Example: 6 can be expressed as $\frac{6}{1}$, and 0.5 as $\frac{1}{2}$.

ray A part of a line that starts at one endpoint and extends forever. (p. 344)

real number A rational or irrational number.

reciprocal One of two numbers whose product is 1; also called *multiplicative inverse.* (p. 226)

Example: The reciprocal of $\frac{2}{3}$ is $\frac{3}{2}$.

The reciprocal of 7 is $\frac{1}{7}$.

rectangle A parallelogram with four right angles. (p. 378)

rectangular prism A polyhedron whose bases are rectangles and whose other faces are parallelograms. (p. 472)

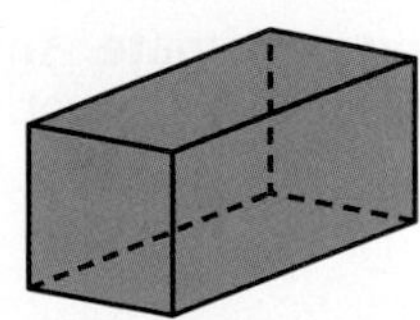

reflection A transformation of a figure that flips the figure across a line. (p. 392)

regular polygon A polygon with congruent sides and angles. (p. 369)

repeating decimal A decimal in which one or more digits repeat infinitely. (p. 171)

Example: $0.757575\ldots = 0.\overline{75}$

rhombus A parallelogram with all sides congruent. (p. 378)

right angle An angle that measures 90°. (p. 349)

right triangle A triangle containing a right angle. (p. 374)

rise The vertical change when the slope of a line is expressed as the ratio $\frac{\text{rise}}{\text{run}}$, or "rise over run." (p. 620)

rotation A transformation in which a figure is turned around a point. (p. 392)

rotational symmetry A figure has rotational symmetry if it can be rotated less than 360° around a central point and coincide with the original figure. (p. 397)

rounding Replacing a number with an estimate of that number to a given place value.

Example: 2,354 rounded to the nearest thousand is 2,000, and 2,354 rounded to the nearest 100 is 2,400.

run The horizontal change when the slope of a line is expressed as the ratio $\frac{\text{rise}}{\text{run}}$, or "rise over run." (p. 620)

sales tax A percent of the cost of an item, which is charged by governments to raise money.

sample A part of the population. (p. 4)

sample space All possible outcomes of an experiment. (p. 520)

scale The ratio between two sets of measurements. (p. 288)

scale drawing A drawing that uses a scale to make an object smaller than or larger than the real object. (p. 288)

scale factor The ratio used to enlarge or reduce similar figures. (p. 288)

scale model A proportional model of a three-dimensional object. (p. 288)

scalene triangle A triangle with no congruent sides. (p. 374)

scatter plot A graph with points plotted to show a possible relationship between two sets of data. (p. 40)

scientific notation A method of writing very large or very small numbers by using powers of 10. (p. 65)

second quartile The median of a set of data. (p. 28)

sector A region enclosed by two radii and the arc joining their endpoints. (p. 363)

sector (data) A section of a circle graph representing part of the data set. (p. 24)

segment A part of a line between two endpoints. (p. 344)

sequence An ordered list of numbers. (p. 608)

set A group of items. (p. 636)

side A line bounding a geometric figure; one of the faces forming the outside of an object. (p. 368)

Side-Side-Side (SSS) A rule stating that if three sides of one triangle are congruent to three sides of another triangle, then the triangles are congruent. (p. 388)

significant digits The digits used to express the precision of a measurement. (p. 420)

similar Figures with the same shape but not necessarily the same size are similar. (p. 280)

simple interest A fixed percent of the principal. It is found using the formula $I = Prt$, where P represents the principal, r the rate of interest, and t the time. (p. 328)

simplest form A fraction is in simplest form when the numerator and denominator have no common factors other than 1.

simplify To write a fraction or expression in simplest form.

sine (sin) In a right triangle, the ratio of the length of the side opposite an acute angle to the length of the hypotenuse. (p. 724)

skew lines Lines that lie in different planes that are neither parallel nor intersecting. (p. 354)

slope A measure of the steepness of a line on a graph; the rise divided by the run. (p. 620)

slope-intercept form A linear equation written in the form $y = mx + b$, where m represents slope and b represents the y-intercept. (p. 621)

solid figure A three-dimensional figure. (p. 472)

solution of an equation A value or values that make an equation true. (p. 104)

solution of an inequality A value or values that make an inequality true. (p. 574)

solution set The set of values that make a statement true. (p. 578)

solve To find an answer or a solution. (p. 110)

sphere A three-dimensional figure with all points the same distance from the center. (p. 473)

square (geometry) A rectangle with four congruent sides. (p. 378)

square (numeration) A number raised to the second power. (p. 444)

Example: In 5^2, the number 5 is squared.

square number The product of a number and itself. (p. 444)

Example: 25 is a square number. $5 \cdot 5 = 25$

square root One of the two equal factors of a number. (p. 444)

Example: $16 = 4 \cdot 4$, or $16 = -4 \cdot -4$, so 4 and -4 are square roots of 16.

standard form (in numeration) A way to write numbers by using digits. (p. 65)

Example: Five thousand, two hundred ten in standard form is 5,210.

stem-and-leaf plot A graph used to organize and display data so that the frequencies can be compared. (p. 15)

straight angle An angle that measures 180°. (p. 349)

subset A set contained within another set. (p. 636)

substitute To replace a variable with a number or another expression in an algebraic expression.

Subtraction Property of Equality The property that states that if you subtract the same number from both sides of an equation, the new equation will have the same solution. (p. 111)

sum The result when two or more numbers are added.

supplementary angles Two angles whose measures have a sum of 180°. (p. 349)

surface area The sum of the areas of the faces, or surfaces, of a three-dimensional figure. (p. 486)

tangent (tan) In a right triangle, the ratio of the length of the side opposite an acute angle to the length of the side adjacent to that acute angle. (p. 724)

term (in an expression) The parts of an expression that are added or subtracted. (p. 100)

Example: $5x^2$ is an expression with one term, -10 is an expression with one term, and $x + 1$ is an expression with two terms.

term (in a sequence) An element or number in a sequence. (p. 608)

terminating decimal A decimal number that ends or terminates. (p. 171)

Example: 6.75

tessellation A repeating pattern of plane figures that completely covers a plane with no gaps or overlaps. (p. 400)

tetrahedron A polyhedron with four faces. (p. 472)

theoretical probability The ratio of the number of equally likely outcomes in an event to the total number of possible outcomes. (p. 524)

third quartile The median of the upper half of a set of data; also called *upper quartile.* (p. 28)

transformation A change in the size or position of a figure. (p. 392)

translation A movement (slide) of a figure along a straight line. (p. 392)

transversal A line that intersects two or more lines. (p. 355)

trapezoid A quadrilateral with exactly one pair of parallel sides. (p. 378)

tree diagram A branching diagram that shows all possible combinations or outcomes of an event. (p. 521)

trial In probability, a single repetition or observation of an experiment. (p. 516)

triangle A three-sided polygon. (p. 369)

Triangle Sum Theorem The theorem that states that the measures of the angles in a triangle add to 180°.

triangular prism A polyhedron whose bases are triangles and whose other faces are parallelograms. (p. 472)

trigonometric ratios Ratios that compare the lengths of the sides of a right triangle; the common ratios are tangent, sine, and cosine. (p. 724)

underestimate An estimate that is less than the exact answer.

union The set of all elements that belong to two or more sets. (p. 637)

unit conversion The process of changing one unit of measure to another. (p. 272)

unit conversion factor A fraction used in unit conversion in which the numerator and denominator represent the same amount but are in different units. (p. 272)

Example: $\frac{60 \text{ min}}{1 \text{ h}}$ or $\frac{1 \text{ h}}{60 \text{ min}}$

unit price A unit rate used to compare prices.

unit rate A rate in which the second quantity in the comparison is one unit. (p. 260)

Example: 10 centimeters per minute

upper extreme The greatest number in a set of data. (p. 28)

V

variable A symbol used to represent a quantity that can change. (p. 92)

Venn diagram A diagram that is used to show relationships between sets. (p. 640)

verbal expression A word or phrase. (p. 96)

vertex On an angle or polygon, the point where two sides intersect (pp. 348, 368); on a polyhedron, the intersection of three or more faces (p. 472); on a cone or pyramid, the top point.

vertical angles A pair of opposite congruent angles formed by intersecting lines; in the diagram, $\angle a$ and $\angle c$ are congruent and $\angle b$ and $\angle d$ are congruent. (p. 355)

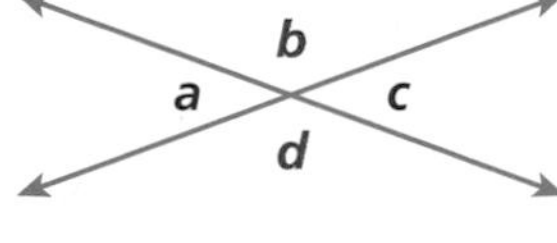

volume The number of cubic units needed to fill a given space. (p. 476)

x-axis The horizontal axis on a coordinate plane. (p. 134)

x-coordinate The first number in an ordered pair; it tells the distance to move right or left from the origin, (0, 0). (p. 134)

Example: 5 is the x-coordinate in (5, 3).

y-axis The vertical axis on a coordinate plane. (p. 134)

y-coordinate The second number in an ordered pair; it tells the distance to move up or down from the origin, (0, 0). (p. 134)

Example: 3 is the y-coordinate in (5, 3).

y-intercept The y-coordinate of the point where the graph of a line crosses the y-axis. (p. 620)

Index

Index

N

O

P

Index

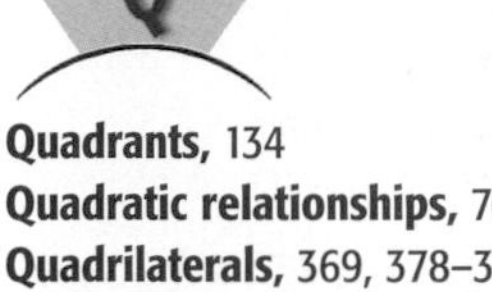

Index